PRINCIPLES *of* General, Organic, & Biological Chemistry

Second Edition

Janice Gorzynski Smith
University of Hawai'i at Mānoa

D1287226

Mc Graw Hill Education

PRINCIPLES OF GENERAL, ORGANIC, & BIOLOGICAL CHEMISTRY, SECOND EDITION

Published by McGraw-Hill Education, 2 Penn Plaza, New York, NY 10121. Copyright © 2015 by McGraw-Hill Education. All rights reserved. Printed in the United States of America. Previous edition © 2012. No part of this publication may be reproduced or distributed in any form or by any means, or stored in a database or retrieval system, without the prior written consent of McGraw-Hill Education, including, but not limited to, in any network or other electronic storage or transmission, or broadcast for distance learning.

Some ancillaries, including electronic and print components, may not be available to customers outside the United States.

This book is printed on acid-free paper.

1 2 3 4 5 6 7 8 9 0 DOW/DOW 1 0 9 8 7 6 5 4

ISBN 978–1–259–25227–3
MHID 1–259–25227–2

All credits appearing on page or at the end of the book are considered to be an extension of the copyright page.

The Internet addresses listed in the text were accurate at the time of publication. The inclusion of a website does not indicate an endorsement by the authors or McGraw-Hill Education, and McGraw-Hill Education does not guarantee the accuracy of the information presented at these sites.

www.mhhe.com

About the Author

Janice Gorzynski Smith was born in Schenectady, New York, and grew up following the Yankees, listening to the Beatles, and water skiing on Sacandaga Reservoir. She became interested in chemistry in high school and went on to major in chemistry at Cornell University where she received an A.B. degree *summa cum laude.* Jan earned a Ph.D. in Organic Chemistry from Harvard University under the direction of Nobel Laureate E. J. Corey, and she also spent a year as a National Science Foundation National Needs Postdoctoral Fellow at Harvard. During her tenure with the Corey group she completed the total synthesis of the plant growth hormone gibberellic acid.

Following her postdoctoral work, Jan joined the faculty of Mount Holyoke College, where she was employed for 21 years. During this time, she was active in teaching organic chemistry lecture and lab courses, conducting a research program in organic synthesis, and serving as department chair. Her organic chemistry class was named one of Mount Holyoke's "Don't-miss courses" in a survey by *Boston* magazine. After spending two sabbaticals amidst the natural beauty and diversity of Hawai'i in the 1990s, Jan and her family moved there permanently in 2000. She is a faculty member at the University of Hawai'i at Mānoa, where she has taught a one-semester organic and biological chemistry course for nursing students, as well as the two-semester organic chemistry lecture and lab courses. She has also served as the faculty advisor to the student affiliate chapter of the American Chemical Society. In 2003, she received the Chancellor's Citation for Meritorious Teaching.

Jan resides in Hawai'i with her husband Dan, an emergency medicine physician. She has four children and one grandchild. When not teaching, writing, or enjoying her family, Jan bikes, hikes, snorkels, and scuba dives in sunny Hawai'i, and time permitting, enjoys travel and Hawaiian quilting.

To my family

Contents in Brief

Contents

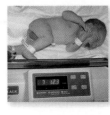

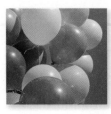

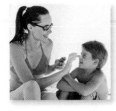

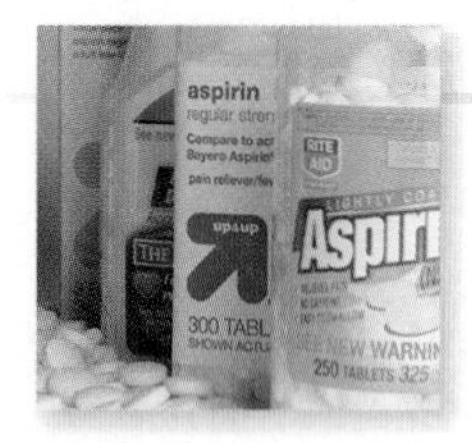
aspirin

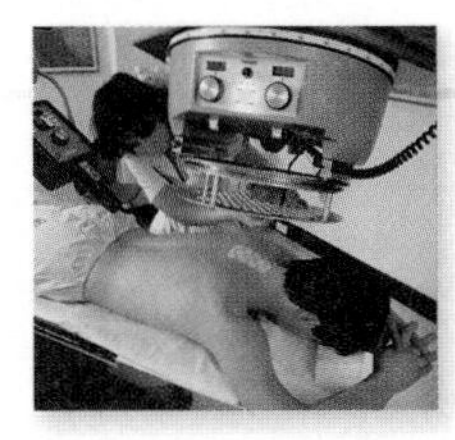

Preface

Students who are planning a career within the allied health field are required to gain exposure to the many ways in which chemistry is intrinsic to and influences life. This textbook is written for students who have an interest in nursing, nutrition, environmental science, food science, and a wide variety of other health-related professions. The content of this book is designed for an introductory chemistry course with no chemistry prerequisite and is suitable for either a one- or two-semester course. This text relates the principal concepts of general, organic, and biological chemistry to the world around us, and in this way illustrates how chemistry explains many aspects of daily life.

The learning style of today's students relies heavily on visual imagery. In this text, new concepts are introduced one at a time, keeping the basic themes in focus, and breaking down complex problems into manageable chunks of information. Relevant, interesting applications are provided for all basic chemical concepts. Diagrams and figures are annotated to help teach concepts and reinforce the major themes of chemistry, while molecular art illustrates and explains common everyday phenomena. Students learn step-by-step problem solving throughout the chapter within sample problems and *How To* boxes. Students are given enough detail to understand basic concepts, such as how oral contraceptives prevent pregnancy and how a catalytic converter removes pollutants from automobile exhaust.

Teaching chemistry for over 20 years at both a private liberal arts college and a large state university has given me a unique perspective with which to write this text. I have found that students arrive with vastly different levels of preparation and widely different expectations for their college experience. As an instructor and now an author I have tried to channel my love and knowledge of chemistry into a form that allows this spectrum of students to understand chemical science more clearly, and then see everyday phenomena in a new light. My interactions with thousands of students in my long teaching career have profoundly affected the way I teach and write about chemistry. My hope is that this text and its Learning System will help students better understand and appreciate the world of chemistry. Please feel free to email me with any comments or questions at jgsmith@hawaii.edu.

New to This Edition

Chapter-Specific Revisions

- A new section on determining the correct number of significant figures when using an electronic calculator has been added to Chapter 1. To help students understand density, a new sample problem and several problems with line art have been added as well.
- Atomic weights are now reported using four digits. This is reflected in the periodic tables in Chapter 2 and on the inside front cover, as well as in all problems that use atomic weights. Three new sample problems on isotopes, atomic size, and ionization energy have been added to Chapter 2 to further assist students in developing problem-solving skills.
- In response to reviewer feedback, the section on drawing Lewis structures for molecules that contain multiple bonds has now been expanded with Sample Problem 3.12.
- Chapter 4 now includes a new section on specific heat, material that was requested by several reviewers. Chapter 4 also contains a section that illustrates how to combine energy calculations, so that students can determine the total energy change when a substance undergoes both a temperature change and a change of state.

- Chapter 5 contains a new sample problem that illustrates how to balance equations with polyatomic ions. The important topics of equilibrium and Le Châtelier's principle have been added to Chapter 5, so that students can better understand reversible reactions.
- Sections 6.1 through 6.4 were revised to more clearly explain the relationship between the gas laws and the kinetic-molecular theory of gases. A new figure was added in Section 6.8 to clarify the discussion of Dalton's law.
- New material on colloids and suspensions was added to Chapter 7, a topic viewed as particularly useful for nursing students who sometimes give medications that must be shaken before they are administered. Also, Section 7.2 expands the discussion of electrolytes and now covers equivalents. It is hoped that this addition will be helpful to many nursing students who deal with equivalents in blood plasma and IV solutions.
- The topic of naming acids was added to Section 8.1 to aid students in identifying common acids. Equilibrium has now been added to the discussion of acidity, and Le Châtelier's principle is used to explain buffers in the blood in Section 8.9.
- Since nuclear chemistry is very different from the material students are exposed to in Chapters 1–8, two new sample problems on half-life and nuclear fission have been added to Chapter 9 to further assist students in developing problem-solving skills.
- A new sample problem was added to help students identify functional groups in Chapter 10. The discussion of physical properties in Section 10.9 was expanded and revised for further clarity.
- A discussion of alkyl halides was added to Chapter 12 to promote student understanding of both the beneficial and harmful effects of organic halides.
- In response to reviewer feedback, new material on the nomenclature of carboxylate salts was included in Chapter 13.
- Chapter 16 includes a new figure to help students visualize the four different levels of structure of proteins.
- A new figure was added to Chapter 17 to help students visualize the structure of the chromosome.

General Revisions

- **Problem sets.** More problems with molecular art and 3-D models have been added to the text and end-of-chapter. In response to reviewer feedback, problems have been added to the "Beyond the Classroom" section at the end of each chapter.
- **Design and layout.** An effort has been made with the revised second edition design and layout to move all photos, graphics, and tables closer to related material in the text.
- **Photos.** Roughly one-half of the chapter-opening photos have been replaced with photos emphasizing relevant material within the chapter. More marginal photos of applications have also been added.
- **Art.** The colors of subatomic particles in all nuclear art were revised for clarity and consistency (Chapters 2, 3, and 9).

The Construction of a Learning System

Writing a textbook and its supporting learning tools is a multifaceted endeavor. McGraw-Hill's 360° Development Process is an ongoing, market-oriented approach to building accurate and innovative Learning Systems. It is dedicated to continual large scale and incremental improvement, driven by multiple customer feedback loops and checkpoints. This is initiated during the early planning stages of new products and intensifies during the development and production stages, and then begins again upon publication, in anticipation of the next version of each print and digital product. This process is designed to provide a broad, comprehensive spectrum of feedback for refinement and innovation of learning tools for both student and instructor. The 360° Development Process includes market research, content reviews, faculty and student focus groups, course- and product-specific symposia, accuracy checks, and art reviews.

The Learning System Used in *Principles of General, Organic, & Biological Chemistry*, Second Edition

Writing Style

A succinct writing style weaves together key points of general, organic, and biological chemistry, along with attention-grabbing applications to consumer, environmental, and health-related fields. Concepts and topics are broken into small chunks of information that are more easily learned.

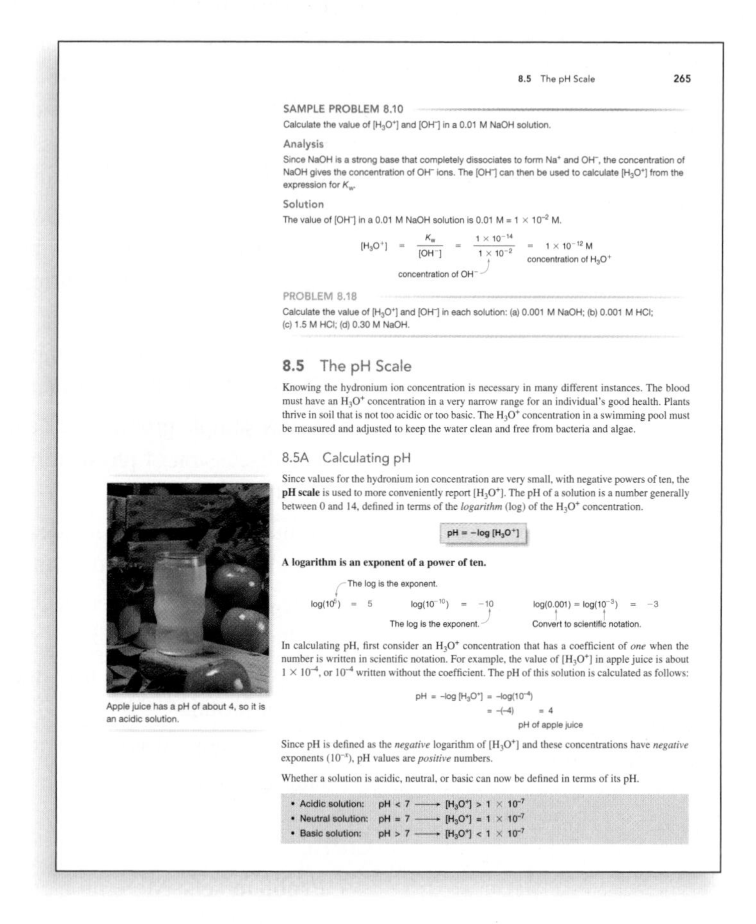

8.5 The pH Scale 265

SAMPLE PROBLEM 8.10

Calculate the value of $[H_3O^+]$ and $[OH^-]$ in a 0.01 M NaOH solution.

Analysis

Since NaOH is a strong base that completely dissociates to form Na^+ and OH^-, the concentration of NaOH gives the concentration of OH^- ions. The $[OH^-]$ can then be used to calculate $[H_3O^+]$ from the expression for K_w.

Solution

The value of $[OH^-]$ in a 0.01 M NaOH solution is 0.01 M = 1×10^{-2} M.

$$[H_3O^+] = \frac{K_w}{[OH^-]} = \frac{1 \times 10^{-14}}{1 \times 10^{-2}} = 1 \times 10^{-12}\ M$$

concentration of OH^- (pointing to 1×10^{-2}); concentration of H_3O^+ (under 1×10^{-12} M)

PROBLEM 8.18

Calculate the value of $[H_3O^+]$ and $[OH^-]$ in each solution: (a) 0.001 M NaOH; (b) 0.001 M HCl; (c) 1.5 M HCl; (d) 0.30 M NaOH.

8.5 The pH Scale

Knowing the hydronium ion concentration is necessary in many different instances. The blood must have an H_3O^+ concentration in a very narrow range for an individual's good health. Plants thrive in soil that is not too acidic or too basic. The H_3O^+ concentration in a swimming pool must be measured and adjusted to keep the water clean and free from bacteria and algae.

8.5A Calculating pH

Since values for the hydronium ion concentration are very small, with negative powers of ten, the **pH scale** is used to more conveniently report $[H_3O^+]$. The pH of a solution is a number generally between 0 and 14, defined in terms of the *logarithm* (log) of the H_3O^+ concentration.

$$pH = -\log [H_3O^+]$$

A logarithm is an exponent of a power of ten.

The log is the exponent.

$\log(10^5) = 5$ $\qquad \log(10^{-10}) = -10$ $\qquad \log(0.001) = \log(10^{-3}) = -3$

The log is the exponent. Convert to scientific notation.

In calculating pH, first consider an H_3O^+ concentration that has a coefficient of *one* when the number is written in scientific notation. For example, the value of $[H_3O^+]$ in apple juice is about 1×10^{-4}, or 10^{-4} written without the coefficient. The pH of this solution is calculated as follows:

$$pH = -\log [H_3O^+] = -\log(10^{-4}) = -(-4) = 4$$

pH of apple juice

Apple juice has a pH of about 4, so it is an acidic solution.

Since pH is defined as the *negative* logarithm of $[H_3O^+]$ and these concentrations have *negative* exponents (10^{-x}), pH values are *positive* numbers.

Whether a solution is acidic, neutral, or basic can now be defined in terms of its pH.

- Acidic solution: $pH < 7 \longrightarrow [H_3O^+] > 1 \times 10^{-7}$
- Neutral solution: $pH = 7 \longrightarrow [H_3O^+] = 1 \times 10^{-7}$
- Basic solution: $pH > 7 \longrightarrow [H_3O^+] < 1 \times 10^{-7}$

Chapter Goals, Tied to End-of-Chapter Key Concepts

Chapter Goals at the beginning of each chapter identify what students will learn, and are tied numerically to the end-of-chapter Key Concepts, which serve as bulleted summaries of the most important concepts for study.

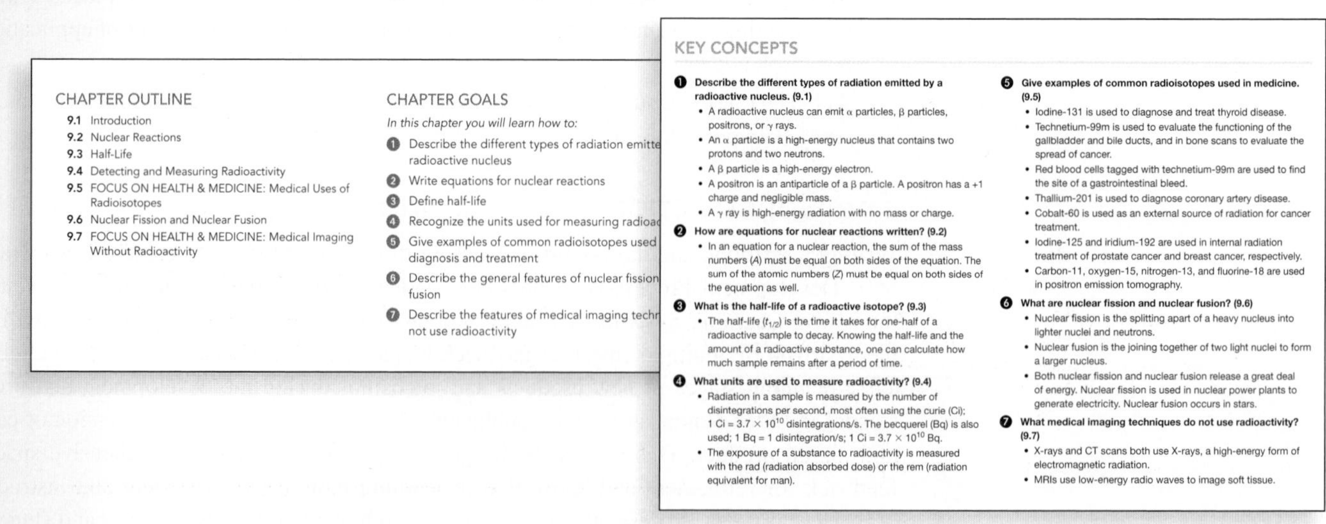

CHAPTER OUTLINE

9.1 Introduction
9.2 Nuclear Reactions
9.3 Half-Life
9.4 Detecting and Measuring Radioactivity
9.5 FOCUS ON HEALTH & MEDICINE: Medical Uses of Radioisotopes
9.6 Nuclear Fission and Nuclear Fusion
9.7 FOCUS ON HEALTH & MEDICINE: Medical Imaging Without Radioactivity

CHAPTER GOALS

In this chapter you will learn how to:

1. Describe the different types of radiation emitte[d by a] radioactive nucleus
2. Write equations for nuclear reactions
3. Define half-life
4. Recognize the units used for measuring radioa[ctivity]
5. Give examples of common radioisotopes used [in] diagnosis and treatment
6. Describe the general features of nuclear fission [and] fusion
7. Describe the features of medical imaging techn[iques that do] not use radioactivity

KEY CONCEPTS

1. **Describe the different types of radiation emitted by a radioactive nucleus. (9.1)**
   - A radioactive nucleus can emit α particles, β particles, positrons, or γ rays.
   - An α particle is a high-energy nucleus that contains two protons and two neutrons.
   - A β particle is a high-energy electron.
   - A positron is an antiparticle of a β particle. A positron has a +1 charge and negligible mass.
   - A γ ray is high-energy radiation with no mass or charge.
2. **How are equations for nuclear reactions written? (9.2)**
   - In an equation for a nuclear reaction, the sum of the mass numbers (*A*) must be equal on both sides of the equation. The sum of the atomic numbers (*Z*) must be equal on both sides of the equation as well.
3. **What is the half-life of a radioactive isotope? (9.3)**
   - The half-life ($t_{1/2}$) is the time it takes for one-half of a radioactive sample to decay. Knowing the half-life and the amount of a radioactive substance, one can calculate how much sample remains after a period of time.
4. **What units are used to measure radioactivity? (9.4)**
   - Radiation in a sample is measured by the number of disintegrations per second, most often using the curie (Ci); 1 Ci = 3.7×10^{10} disintegrations/s. The becquerel (Bq) is also used; 1 Bq = 1 disintegration/s; 1 Ci = 3.7×10^{10} Bq.
   - The exposure of a substance to radioactivity is measured with the rad (radiation absorbed dose) or the rem (radiation equivalent for man).
5. **Give examples of common radioisotopes used in medicine. (9.5)**
   - Iodine-131 is used to diagnose and treat thyroid disease.
   - Technetium-99m is used to evaluate the functioning of the gallbladder and bile ducts, and in bone scans to evaluate the spread of cancer.
   - Red blood cells tagged with technetium-99m are used to find the site of a gastrointestinal bleed.
   - Thallium-201 is used to diagnose coronary artery disease.
   - Cobalt-60 is used as an external source of radiation for cancer treatment.
   - Iodine-125 and iridium-192 are used in internal radiation treatment of prostate cancer and breast cancer, respectively.
   - Carbon-11, oxygen-15, nitrogen-13, and fluorine-18 are used in positron emission tomography.
6. **What are nuclear fission and nuclear fusion? (9.6)**
   - Nuclear fission is the splitting apart of a heavy nucleus into lighter nuclei and neutrons.
   - Nuclear fusion is the joining together of two light nuclei to form a larger nucleus.
   - Both nuclear fission and nuclear fusion release a great deal of energy. Nuclear fission is used in nuclear power plants to generate electricity. Nuclear fusion occurs in stars.
7. **What medical imaging techniques do not use radioactivity? (9.7)**
   - X-rays and CT scans both use X-rays, a high-energy form of electromagnetic radiation.
   - MRIs use low-energy radio waves to image soft tissue.

Macro-to-Micro Illustrations

Visualizing molecular-level representations of macroscopic phenomena is critical to the understanding of any chemistry course. Many illustrations in this text include photos or drawings of everyday objects, paired with their molecular representation, to help students visualize and understand the chemistry behind ordinary things. Many illustrations of the human body include magnifications for specific anatomic regions, as well as representations at the microscopic level, for today's visual learners.

6.7 The Ideal Gas Law 203

divided by the product of moles and Kelvin temperature is a constant, called the **universal gas constant** and symbolized by ***R***.

$$\frac{PV}{nT} = R \quad \text{universal gas constant}$$

More often the equation is rearranged and written in the following way:

$$PV = nRT$$

Ideal gas law

For atm: $R = 0.0821 \frac{\text{L} \cdot \text{atm}}{\text{mol} \cdot \text{K}}$

For mm Hg: $R = 62.4 \frac{\text{L} \cdot \text{mm Hg}}{\text{mol} \cdot \text{K}}$

The value of the universal gas constant *R* depends on its units. The two most common values of *R* are given using atmospheres or mm Hg for pressure, liters for volume, and kelvins for temperature. **Be careful to use the correct value of *R* for the pressure units in the problem you are solving.**

The ideal gas law can be used to find any value—*P*, *V*, *n*, or *T*—as long as three of the quantities are known. Solving a problem using the ideal gas law is shown in the stepwise *How To* procedure and in Sample Problem 6.8. Although the ideal gas law gives exact answers only for a perfectly "ideal" gas, it gives a good approximation for most real gases, such as the oxygen and carbon dioxide in breathing, as well (Figure 6.6).

Figure 6.6 Focus on the Human Body: The Lungs

- Humans have two lungs that contain a vast system of air passages, allowing gases to be exchanged between the atmosphere and with the bloodstream. The lungs contain about 1,500 miles of airways that have a total surface area about the size of a tennis court.
- Lungs are in a sense "overbuilt," in that their total air volume is large compared to the tidal volume, the amount of air taken in or expelled with each breath. This large reserve explains why people can smoke for years without noticing any significant change in normal breathing.
- In individuals with asthma, small airways are constricted and inflamed, making it difficult to breathe.

Applications

Relevant, interesting applications of chemistry to everyday life are included for all basic chemical concepts. These are interspersed in margin-placed Health Notes, Consumer Notes, and Environmental Notes, as well as sections entitled "Focus on Health & Medicine," "Focus on the Environment," and "Focus on the Human Body."

HEALTH NOTE

Lactic acid (Problem 8.14) accumulates in tissues during vigorous exercise, making muscles feel tired and sore. The formation of lactic acid is discussed in greater detail in Section 18.5.

298 Chapter 9 Nuclear Chemistry

Table 9.3 Units Used to Measure Radioactivity

1 Ci = 3.7×10^{10} disintegrations/s
1 Ci = 3.7×10^{10} Bq
1 Ci = 1,000 mCi
1 Ci = 1,000,000 μCi

The **becquerel** (Bq), an SI unit, is also used to measure radioactivity; 1 Bq = 1 disintegration/second. Since each nuclear decay corresponds to one becquerel, 1 Ci = 3.7×10^{10} Bq. Radioactivity units are summarized in Table 9.3.

Often a dose of radiation is measured in the number of millicuries that must be administered. For example, a diagnostic test for thyroid activity uses sodium iodide that contains iodine-131—that is, $Na^{131}I$. The radioisotope is purchased with a known amount of radioactivity per milliliter, such as 3.5 mCi/mL. By knowing the amount of radioactivity a patient must be given, as well as the concentration of radioactivity in the sample, one can calculate the volume of radioactive isotope that must be administered (Sample Problem 9.5).

SAMPLE PROBLEM 9.5

A patient must be given a 4.5-mCi dose of iodine-131, which is available as a solution that contains 3.5 mCi/mL. What volume of solution must be administered?

Analysis

Use the amount of radioactivity (mCi/mL) as a conversion factor to convert the dose of radioactivity from millicuries to a volume in milliliters.

Solution

The dose of radioactivity is known in millicuries, and the amount of radioactivity per unit volume (3.5 mCi/mL) is also known. Use 3.5 mCi/mL as a millicurie–milliliter conversion factor.

The curie is named for Polish chemist Marie Skłodowska Curie who discovered the radioactive elements polonium and radium, and received Nobel Prizes for both Chemistry and Physics in the early twentieth century.

mCi–mL conversion factor

$$4.5 \text{ mCi dose} \times \frac{1 \text{ mL}}{3.5 \text{ mCi}} = 1.3 \text{ mL dose}$$

Millicuries cancel.

Answer

PROBLEM 9.17

To treat a thyroid tumor, a patient must be given a 110-mCi dose of iodine-131, supplied in a vial containing 25 mCi/mL. What volume of solution must be administered?

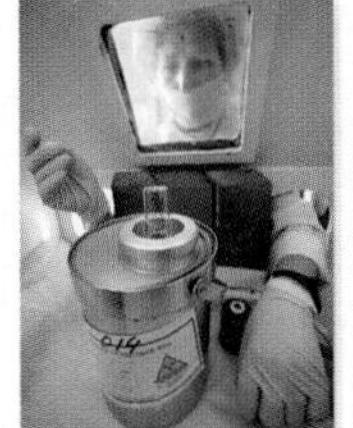

A lab worker must use protective equipment when working with radioactive substances.

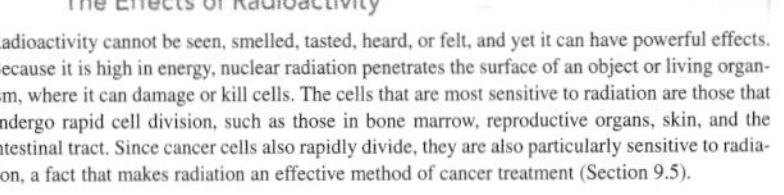

9.4B FOCUS ON HEALTH & MEDICINE
The Effects of Radioactivity

Radioactivity cannot be seen, smelled, tasted, heard, or felt, and yet it can have powerful effects. Because it is high in energy, nuclear radiation penetrates the surface of an object or living organism, where it can damage or kill cells. The cells that are most sensitive to radiation are those that undergo rapid cell division, such as those in bone marrow, reproductive organs, skin, and the intestinal tract. Since cancer cells also rapidly divide, they are also particularly sensitive to radiation, a fact that makes radiation an effective method of cancer treatment (Section 9.5).

Alpha (α) particles, β particles, and γ rays differ in the extent to which they can penetrate a surface. Alpha particles are the heaviest of the radioactive particles, and as a result they move the slowest and penetrate the least. Individuals who work with radioisotopes that emit α particles wear lab coats and gloves that provide a layer of sufficient protection. Beta particles move much faster since they have negligible mass, and they can penetrate into body tissue. Lab workers and health professionals must wear heavy lab coats and gloves when working with substances that give off β particles. Gamma rays travel the fastest and readily penetrate body tissue. Working with substances that emit γ rays is extremely hazardous, and a thick lead shield is required to halt their penetration.

Problem Solving

Stepwise practice problems lead students through the thought process tied to successful problem solving by employing *Analysis* and *Solution* steps. Sample Problems are categorized sequentially by topic to match chapter organization, and are often paired with practice problems to allow students to apply what they have just learned. Students can immediately verify their answers to the follow-up problems in the answers at the end of each chapter.

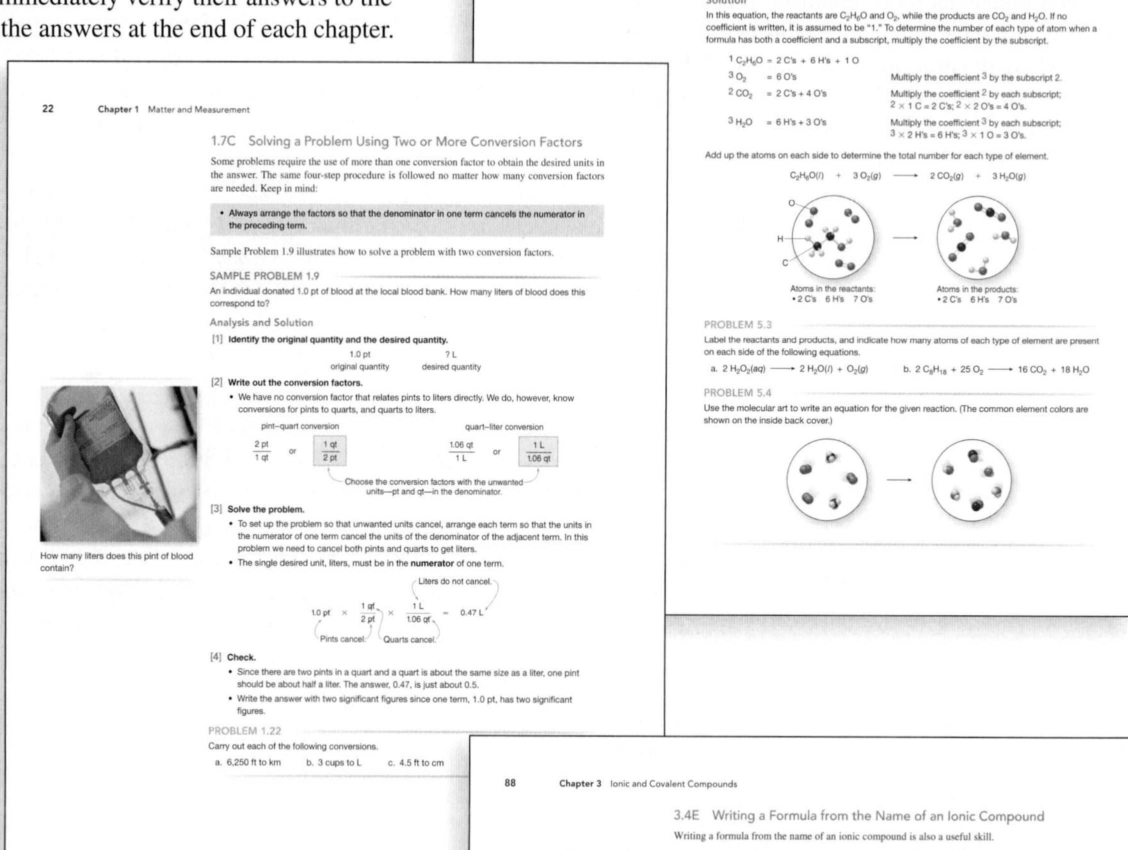

5.1 Introduction to Chemical Reactions 147

SAMPLE PROBLEM 5.2

Label the reactants and products, and indicate how many atoms of each type of element are present on each side of the equation.

$$C_2H_6O(l) + 3\,O_2(g) \longrightarrow 2\,CO_2(g) + 3\,H_2O(g)$$

Analysis

Reactants are on the left side of the arrow and products are on the right side in a chemical equation. When a formula contains a subscript, multiply its coefficient by the subscript to give the total number of atoms of a given type in the formula.

Solution

In this equation, the reactants are C_2H_6O and O_2, while the products are CO_2 and H_2O. If no coefficient is written, it is assumed to be "1." To determine the number of each type of atom when a formula has both a coefficient and a subscript, multiply the coefficient by the subscript.

$1\,C_2H_6O$ = 2 C's + 6 H's + 1 O

$3\,O_2$ = 6 O's — Multiply the coefficient 3 by the subscript 2.

$2\,CO_2$ = 2 C's + 4 O's — Multiply the coefficient 2 by each subscript; 2 × 1 C = 2 C's; 2 × 2 O's = 4 O's.

$3\,H_2O$ = 6 H's + 3 O's — Multiply the coefficient 3 by each subscript; 3 × 2 H's = 6 H's; 3 × 1 O = 3 O's.

Add up the atoms on each side to determine the total number for each type of element.

$$C_2H_6O(l) + 3\,O_2(g) \longrightarrow 2\,CO_2(g) + 3\,H_2O(g)$$

O
H
C

Atoms in the reactants:
• 2 C's 6 H's 7 O's

Atoms in the products:
• 2 C's 6 H's 7 O's

PROBLEM 5.3

Label the reactants and products, and indicate how many atoms of each type of element are present on each side of the following equations.

a. $2\,H_2O_2(aq) \longrightarrow 2\,H_2O(l) + O_2(g)$
b. $2\,C_8H_{18} + 25\,O_2 \longrightarrow 16\,CO_2 + 18\,H_2O$

PROBLEM 5.4

Use the molecular art to write an equation for the given reaction. (The common element colors are shown on the inside back cover.)

22 Chapter 1 Matter and Measurement

1.7C Solving a Problem Using Two or More Conversion Factors

Some problems require the use of more than one conversion factor to obtain the desired units in the answer. The same four-step procedure is followed no matter how many conversion factors are needed. Keep in mind:

- **Always arrange the factors so that the denominator in one term cancels the numerator in the preceding term.**

Sample Problem 1.9 illustrates how to solve a problem with two conversion factors.

SAMPLE PROBLEM 1.9

An individual donated 1.0 pt of blood at the local blood bank. How many liters of blood does this correspond to?

Analysis and Solution

[1] **Identify the original quantity and the desired quantity.**

1.0 pt (original quantity) ? L (desired quantity)

[2] **Write out the conversion factors.**

- We have no conversion factor that relates pints to liters directly. We do, however, know conversions for pints to quarts, and quarts to liters.

pint–quart conversion: $\frac{2\text{ pt}}{1\text{ qt}}$ or $\frac{1\text{ qt}}{2\text{ pt}}$

quart–liter conversion: $\frac{1.06\text{ qt}}{1\text{ L}}$ or $\frac{1\text{ L}}{1.06\text{ qt}}$

Choose the conversion factors with the unwanted units—pt and qt—in the denominator.

[3] **Solve the problem.**

- To set up the problem so that unwanted units cancel, arrange each term so that the units in the numerator of one term cancel the units of the denominator of the adjacent term. In this problem we need to cancel both pints and quarts to get liters.
- The single desired unit, liters, must be in the **numerator** of one term.

$$1.0\text{ pt} \times \frac{1\text{ qt}}{2\text{ pt}} \times \frac{1\text{ L}}{1.06\text{ qt}} = 0.47\text{ L}$$

Liters do not cancel. Pints cancel. Quarts cancel.

[4] **Check.**

- Since there are two pints in a quart and a quart is about the same size as a liter, one pint should be about half a liter. The answer, 0.47, is just about 0.5.
- Write the answer with two significant figures since one term, 1.0 pt, has two significant figures.

How many liters does this pint of blood contain?

PROBLEM 1.22

Carry out each of the following conversions.

a. 6,250 ft to km b. 3 cups to L c. 4.5 ft to cm

88 Chapter 3 Ionic and Covalent Compounds

3.4E Writing a Formula from the Name of an Ionic Compound

Writing a formula from the name of an ionic compound is also a useful skill.

How To Derive a Formula from the Name of an Ionic Compound

Example Write the formula for tin(IV) oxide.

Step [1] **Identify the cation and the anion and determine their charges.**

- The name of the cation appears first, followed by the anion.
- For metals with variable charge, the Roman numeral gives the charge on the cation.

In this example, tin is the cation. The Roman numeral tells us that its charge is +4, making the cation Sn^{4+}. Oxide is the name of the oxygen anion, O^{2-} (Table 3.3).

Step [2] **Balance charges.**

- Use the charge on the cation to determine the number of ions of the anion needed to balance charge.

Sn^{4+} (cation) O^{2-} (anion) — Two −2 anions are needed for each +4 cation.

Step [3] **Write the formula with the cation first, and use subscripts to show the number of each ion needed to have zero overall charge.**

Answer: SnO_2

PROBLEM 3.14

Write the formula for each ionic compound.

a. calcium bromide c. ferric bromide e. chromium(II) chloride
b. copper(I) iodide d. magnesium sulfide f. sodium oxide

3.5 Physical Properties of Ionic Compounds

Ionic compounds are crystalline solids composed of ions packed to maximize the interaction of the positive charge of the cations and negative charge of the anions. Ionic solids are held together by extremely strong interactions of the oppositely charged ions. How is this reflected in the melting point and boiling point of an ionic compound?

When a compound melts to form a liquid, energy is needed to overcome some of the attractive forces of the ordered solid, to form the less ordered liquid phase. Since an ionic compound is held together by very strong electrostatic interactions, it takes a great deal of energy to separate the ions from each other. As a result, **ionic compounds have very high melting points.** For example, the melting point of NaCl is 801 °C.

A great deal of energy is needed to overcome the attractive forces present in the liquid phase, too, to form ions that are far apart and very disorganized in the gas phase, so **ionic compounds have extremely high boiling points.** The boiling point of liquid NaCl is 1,413 °C.

A great many ionic compounds are soluble in water. When an ionic compound dissolves in water, the ions are separated, and each anion and cation is surrounded by water molecules, as shown in Figure 3.5. The interaction of the water solvent with the ions provides the energy needed to overcome the strong ion–ion attractions of the crystalline lattice.

How To's

Key processes are taught to students in a straightforward and easy-to-understand manner by using examples and multiple, detailed steps to solving problems.

Resources for Instructors and Students

McGraw-Hill Connect® Chemistry

connect plus+
chemistry
Featuring PerkinElmer® ChemDraw

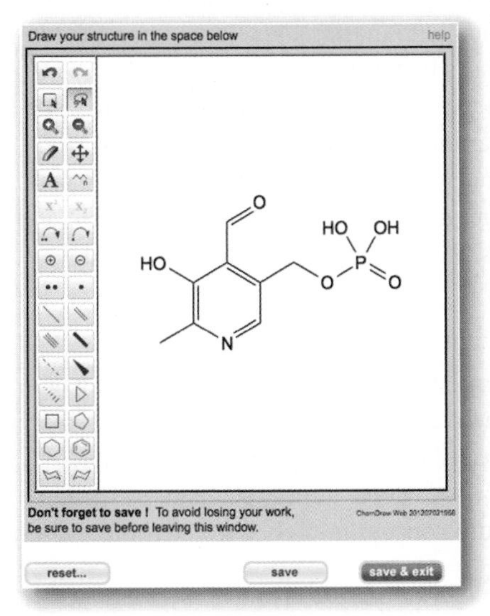

McGraw-Hill Connect® Chemistry provides online presentation, assignment, and assessment solutions. It connects your students with the tools and resources they'll need to achieve success. With Connect Chemistry, you can deliver assignments, quizzes, and tests online. A robust set of questions, problems, and interactives are presented and aligned with the textbook's learning goals. The integration of **ChemDraw by PerkinElmer,** the industry standard in chemical drawing software, allows students to create accurate chemical structures in their online homework assignments. As an instructor, you can edit existing questions and author entirely new problems. Track individual student performance—by question, assignment, or in relation to the class overall—with detailed grade reports. Integrate grade reports easily with Learning Management Systems (LMS), such as WebCT and Blackboard—and much more. **ConnectPlus Chemistry** provides students with all the advantages of Connect Chemistry, plus 24/7 online access to an eBook. This media-rich version of the book is available through the McGraw-Hill Connect platform and allows seamless integration of text, media, and assessments. To learn more, visit www.mcgrawhillconnect.com

McGraw-Hill LearnSmart™

McGraw-Hill LearnSmart™ is available as a stand-alone product as well as an integrated feature of McGraw-Hill Connect® Chemistry. It is an adaptive learning system designed to help students learn faster, study more efficiently, and retain more knowledge for greater success. LearnSmart assesses a student's knowledge of course content through a series of adaptive questions. It pinpoints concepts the student does not understand and maps out a personalized study plan for success. This innovative study tool also has features that allow instructors to see exactly what students have accomplished and a built-in assessment tool for graded assignments. Visit www.mhlearnsmart.com for a demonstration.

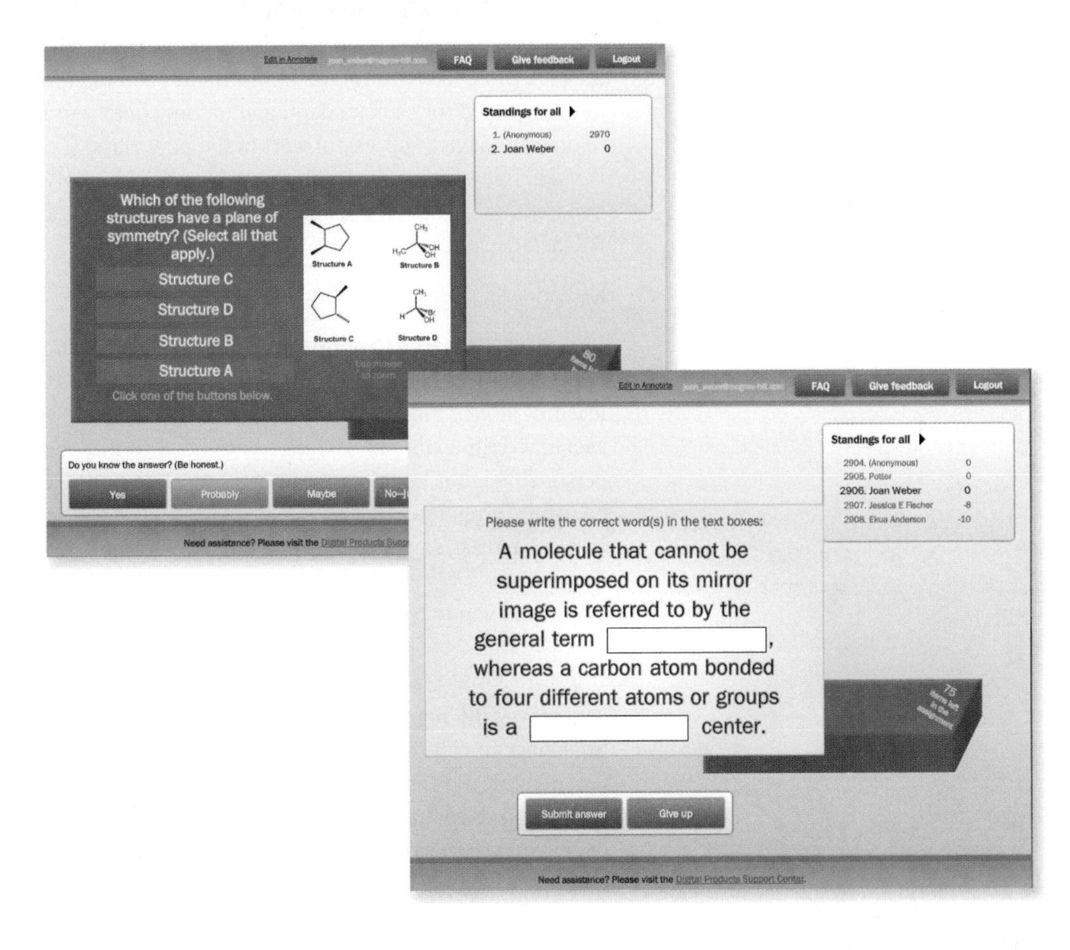

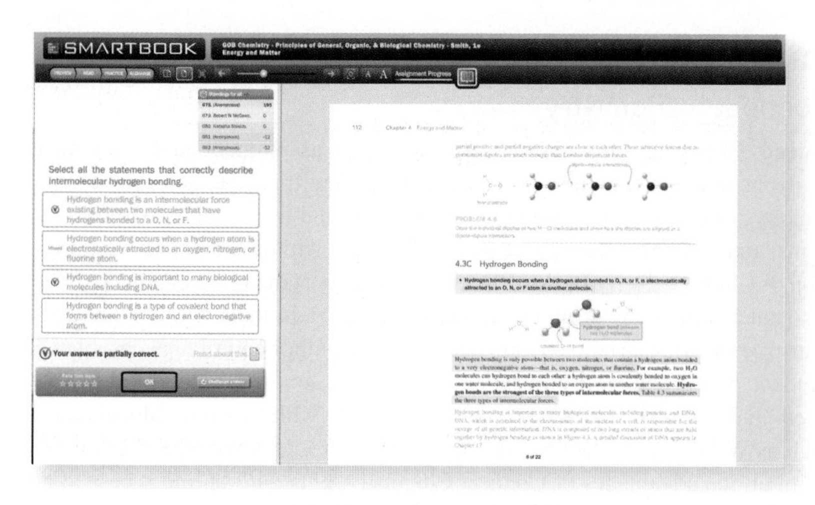

McGraw-Hill SmartBook™

Powered by the intelligent and adaptive LearnSmart engine, SmartBook is the first and only continuously adaptive reading experience available today. Distinguishing what students know from what they don't, and honing in on concepts they are most likely to forget, SmartBook personalizes content for each student. Reading is no longer a passive and linear experience but an engaging and dynamic one, where students are more likely to master and retain important concepts, coming to class better prepared.

SmartBook includes powerful reports that identify specific topics and learning objectives students need to study. These valuable reports also provide instructors insight into how students are progressing through textbook content and are useful for identifying class trends, focusing precious class time, providing personalized feedback to students, and tailoring assessment.

How does SmartBook work?
Each SmartBook contains four components: Preview, Read, Practice, and Recharge. Starting with an initial preview of each chapter and key learning objectives, students read the material and are guided to topics for which they need the most practice based on their responses to a continuously adapting diagnostic. Read and practice continue until SmartBook directs students to recharge important material they are most likely to forget to ensure concept mastery and retention.

McGraw-Hill LearnSmart Prep™

Fueled by LearnSmart—the most widely used and intelligent adaptive learning resource—LearnSmart Prep is designed to get students ready for a forthcoming course by quickly and effectively addressing prerequisite knowledge gaps that may cause problems down the road.

By distinguishing what students know from what they don't, and honing in on concepts they are most likely to forget, LearnSmart Prep maintains a continuously adapting learning path individualized for each student, and tailors content to focus on what the student needs to master in order to have a successful start in the new class.

This revolutionary technology is available only from McGraw-Hill Education and for hundreds of course areas, including general and organic chemistry, as part of the LearnSmart Advantage series.

McGraw-Hill Create™

With **McGraw-Hill Create™,** you can easily rearrange chapters, combine material from other content sources, and quickly upload content you have written, like your course syllabus or teaching notes. Find the content you need in Create by searching through thousands of leading McGraw-Hill textbooks. Arrange your book to fit your teaching style. Create even allows you to personalize your book's appearance by selecting the cover and adding your name, school, and course information. Order a Create book and you'll receive a complimentary print review copy in 3–5 business days or a complimentary electronic review copy (eComp) via e-mail in minutes. Go to www.mcgrawhillcreate.com today and register to experience how McGraw-Hill Create empowers you to teach your students *your* way.

My Lectures—Tegrity®

McGraw-Hill Tegrity® records and distributes your class lectures with just a click of a button. Students can view them anytime/anywhere via computer, iPod, or mobile device. It indexes as it records your PowerPoint® presentations and anything shown on your computer so students can use keywords to find exactly what they want to study. Tegrity is available as an integrated feature of McGraw-Hill Connect Chemistry and as a stand-alone product.

Presentation Tools

Within the Instructor's Resource Center, instructors have access to fully editable PowerPoint lecture outlines, which appear as ready-made presentations that combine art and lecture notes for each chapter of the text. For instructors who prefer to create their lectures from scratch, all illustrations, photos, and tables are pre-inserted by chapter into blank PowerPoint slides and are also available as downloadable jpeg files.

This online digital library contains photos, artwork, animations, and other media types from a wide variety of texts that can be used to create customized lectures, visually enhanced tests and quizzes, compelling course websites, or attractive printed support materials. All assets are copyrighted by McGraw-Hill Higher Education but can be used by instructors for classroom purposes. The visual resources in this collection include:

- **Art** Full-color digital files of all illustrations in the book can be readily incorporated into lecture presentations, exams, or custom-made classroom materials. In addition, all files are pre-inserted into PowerPoint slides for ease of lecture preparation.
- **Photos** The photo collection contains digital files of photographs from the text, which can be reproduced for multiple classroom uses.
- **Tables** Every table that appears in the text has been saved in electronic form for use in classroom presentations and/or quizzes.
- **Animations** Numerous full-color animations illustrating important processes are also provided. Harness the visual impact of concepts in motion by importing these files into classroom presentations or online course materials.

Instructor's Solutions Manual

This supplement contains complete, worked out solutions for all the end-of-chapter problems in the text. It can be found within the Instructor's Resources for this text from the Connect library tab.

Computerized Test Bank Online

A comprehensive bank of test questions prepared by Felix Ngassa of Grand Valley State University is provided within a computerized test bank, enabling professors to create paper and online tests or quizzes in an easy-to-use program anywhere, at any time. Instructors can create or edit questions, or drag-and-drop questions, to prepare tests quickly and easily. Tests may be published to their online course, or printed for paper-based assignments.

Additional Resources for the Student

Student Study Guide/Solutions Manual

The *Student Study Guide/Solutions Manual,* prepared by Erin Smith and Janice Gorzynski Smith, begins each chapter with a detailed chapter review that is organized around chapter goals and key concepts. The Problem-Solving section provides a number of examples for solving each type of problem essential to that chapter. The Self-Test section of each chapter quizzes on chapter highlights, with answers provided. Finally, each chapter ends with the solutions to all in-chapter problems, as well as the solutions to all odd-numbered end-of-chapter problems.

ConnectPlus® eBook

McGraw-Hill ConnectPlus eBook takes digital texts beyond a simple PDF. With the same content as the printed book, but optimized for the screen, ConnectPlus has embedded media, including animations and videos, which bring concepts to life and provide "just in time" learning for students. Additionally, Connect homework allows students to interact with the questions from the text and determine if they're gaining mastery of the content.

Acknowledgments

Publishing a modern chemistry textbook requires a team of knowledgeable and hard-working individuals who are able to translate an author's vision into a reality. Much thanks is due to Brand Manager Derek Elgin, who somehow handled the many responsibilities of his new position like an experienced editor. I was privileged to continue working with Senior Developmental Editor Mary Hurley and Senior Project Manager Jayne Klein, who both managed a very tight schedule with grace and professionalism. Designer Laurie Janssen has once again produced a stunning design that complements and emphasizes the many unique art features of the text. Thanks are also due to Photo Researcher Carrie Burger, Marketing Manager Heather Wagner, and Managing Director Thomas Timp, each of whom has ensured that this project provides students with a visually appealing, accurate, and well-thought-out text. I am especially grateful to freelance Developmental Editor John Murdzek, whose unique blend of humor, chemical knowledge, and attention to detail were key ingredients in revising the first edition. I have also greatly benefited from the many suggestions of Shirley Hino, a former Santa Rosa Junior College instructor and current member of the McGraw-Hill chemistry team, who read the entire first edition and worked every problem.

I would also like to acknowledge the following individuals for their masterful authoring of the ancillaries to accompany the second edition: Lauren McMills of Ohio University for the Instructor Solutions Manual; Jennifer Roberston–Honecker of West Virginia University for the PowerPoint Lecture Outlines; and Felix Ngassa of Grand Valley State for the Test Bank. David G. Jones of the University of North Carolina at Chapel Hill also served as the lead author on LearnSmart to accompany *Principles of General, Organic, & Biological Chemistry,* Second Edition, while Peter de Lijser of California State University—Fullerton, Adam I. Keller of Columbus State Community College, and Alexander J. Seed of Kent State University served as reviewers.

Finally, I thank my family for their support and patience during the long process of publishing a textbook. My husband Dan, an emergency medicine physician, took several photos that appear in the text, and served as a consultant for many medical applications. My daughter Erin co-authored the *Student Study Guide/Solutions Manual* with me.

The following individuals were instrumental in reading and providing feedback that helped to shape *Principles of General, Organic, & Biological Chemistry,* Second Edition:

Reviewers

Mamta Agarwal, *Chaffey College*
Kyle Backstrand, *Viterbo University*
Ling Chen, *Borough of Manhattan Community College/CUNY*
Eden Francis, *Clackamas Community College*
Ann Gaquere–Parker, *University of West Georgia*
Eric Goll, *Brookdale Community College*
Amy Grant, *El Camino College*
Shirley Hino, *Santa Rosa Junior College*
Mushtaq Khan, *Union County College*
Julie Lowe, *Bakersfield College*
Susanne McFadden, *Tarrant County Junior College Northwest*
John Muench, *Heartland Community College*
Lynda Nelson, *University of Arkansas–Fort Smith*
Odutayo Odunuga, *Stephen F. Austin State University*
David Peridian, *Broward College*
Doug Raynie, *South Dakota State University*
Nathan Tice, *Butler University*
John Vincent, *The University of Alabama*
Anne Marie Yunker, *Cuyahoga Community College*

List of *How To's*

How To boxes provide detailed instructions for key procedures that students need to master. Below is a list of each *How To* and where it is presented in the text.

List of Applications

Applications make any subject seem more relevant and interesting—for nonmajors and majors alike. The following is a list of the most important biological, medicinal, and environmental applications that have been integrated throughout *Principles of General, Organic, & Biological Chemistry,* Second Edition. Each chapter opener showcases an interesting and current application relating to the chapter's topic.

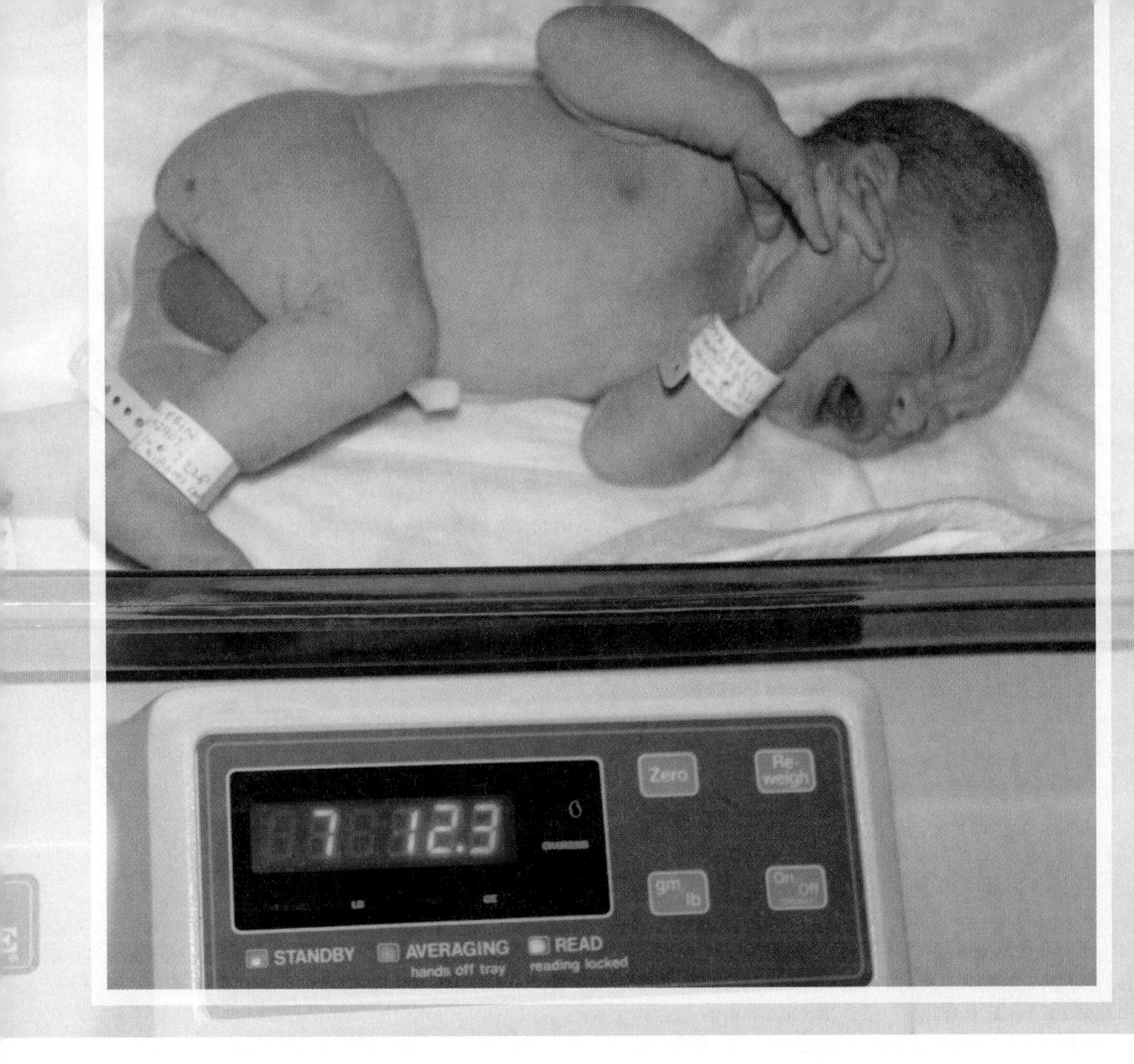

1

Determining the weight and length of a newborn are common measurements performed by healthcare professionals.

Matter and Measurement

CHAPTER OUTLINE

1.1 Chemistry—The Science of Everyday Experience
1.2 States of Matter
1.3 Classification of Matter
1.4 Measurement
1.5 Significant Figures
1.6 Scientific Notation
1.7 Problem Solving Using Conversion Factors
1.8 FOCUS ON HEALTH & MEDICINE: Problem Solving Using Clinical Conversion Factors
1.9 Temperature
1.10 Density and Specific Gravity

CHAPTER GOALS

In this chapter you will learn how to:

1. Describe the three states of matter
2. Classify matter as a pure substance, mixture, element, or compound
3. Report measurements using the metric units of length, mass, and volume
4. Use significant figures
5. Use scientific notation for very large and very small numbers
6. Use conversion factors to convert one unit to another
7. Convert temperature from one scale to another
8. Define density and specific gravity and use density to calculate the mass or volume of a substance

Everything you touch, feel, or taste is composed of chemicals—that is, **matter**—so an understanding of its composition and properties is crucial to our appreciation of the world around us. Some matter—lakes, trees, sand, and soil—is naturally occurring, while other examples of matter—aspirin, CDs, nylon fabric, plastic syringes, and vaccines—are made by humans. To understand the properties of matter, as well as how one form of matter is converted to another, we must also learn about measurements. Following a recipe, pumping gasoline, and figuring out drug dosages involve manipulating numbers. Thus, Chapter 1 begins our study of chemistry by examining the key concepts of matter and measurement.

1.1 Chemistry—The Science of Everyday Experience

What activities might occupy the day of a typical student? You may have done some or all of the following tasks: eaten some meals, drunk coffee or cola, taken notes in a class, checked email on a computer, watched some television, ridden a bike or car to a part-time job, taken an aspirin to relieve a headache, and spent some of the evening having snacks and refreshments with friends. Perhaps, without your awareness, your life was touched by chemistry in each of these activities. What, then, is this discipline we call **chemistry?**

- *Chemistry* is the study of matter—its composition, properties, and transformations.

What is **matter?**

- *Matter* is anything that has mass and takes up volume.

In other words, **chemistry studies anything that we touch, feel, see, smell, or taste,** from simple substances like water or salt, to complex substances like proteins and carbohydrates that combine to form the human body. Some matter—cotton, sand, an apple, and the cardiac drug digoxin—is **naturally occurring,** meaning it is isolated from natural sources. Other substances—nylon, Styrofoam, the plastic used in soft drink bottles, and the pain reliever ibuprofen—are **synthetic,** meaning they are produced by chemists in the laboratory (Figure 1.1).

Figure 1.1

Naturally Occurring and Synthetic Materials

a. Naturally occurring materials

b. Synthetic materials

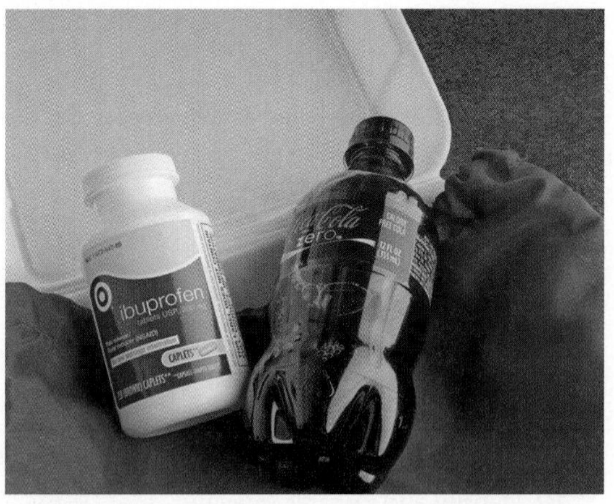

Matter occurs in nature or is synthesized in the lab. (a) Sand and apples are two examples of natural materials. Cotton fabric is woven from cotton fiber, obtained from the cotton plant. The drug digoxin (trade name Lanoxin), widely prescribed for decades for patients with congestive heart failure, is extracted from the leaves of the woolly foxglove plant. (b) Nylon was the first synthetic fiber made in the laboratory. It quickly replaced the natural fiber silk in parachutes and ladies' stockings. Styrofoam and PET (polyethylene terephthalate), the plastic used for soft drink bottles, are strong yet lightweight synthetic materials used for food storage. Over-the-counter pain relievers like ibuprofen are synthetic. The starting materials for all of these useful products are obtained from petroleum.

Figure 1.2

Transforming a Natural Material into a Useful Synthetic Product

a.

b.

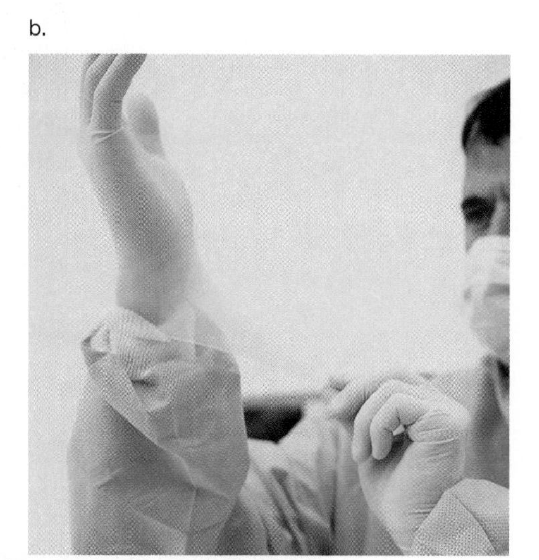

(a) Latex, the sticky liquid that oozes from a rubber tree when it is cut, is too soft for most applications. (b) Vulcanization converts latex to the stronger, elastic rubber used in tires and other products.

Sometimes a chemist studies what a substance is made of, while at other times he or she might be interested in its properties. Alternatively, the focus may be how to convert one material into a new material with unique and useful properties. As an example, naturally occurring rubber exists as the sticky liquid latex, which is too soft for most applications. The laboratory process of vulcanization converts it to the stronger, more elastic material used in tires and other products (Figure 1.2).

Chemistry is truly the science of everyday experience. Soaps and detergents, newspapers and CDs, condoms and oral contraceptives, Tylenol and penicillin—all of these items are products of chemistry. Without a doubt, advances in chemistry have transformed life in modern times.

PROBLEM 1.1

Imagine that your job as a healthcare professional is to take a blood sample from a patient and store it in a small container in a refrigerator until it is picked up for analysis in the hospital lab. You might have to put on gloves and a mask, use a plastic syringe with a metal needle, store the sample in a test tube or vial, and place it in a cold refrigerator. Pick five objects you might encounter during the process and decide if they are made of naturally occurring or synthetic materials.

1.2 States of Matter

Matter exists in three common states—solid, liquid, and gas.

- A *solid* has a definite volume, and maintains its shape regardless of the container in which it is placed. The particles of a solid lie close together, and are arranged in a regular three-dimensional array.
- A *liquid* has a definite volume, but takes on the shape of the container it occupies. The particles of a liquid are close together, but they can randomly move around, sliding past one another.
- A *gas* has no definite shape or volume. The particles of a gas move randomly and are separated by a distance much larger than their size. The particles of a gas expand to fill the volume and assume the shape of whatever container they are put in.

For example, water exists in its solid state as ice or snow, liquid state as liquid water, and gaseous state as steam or water vapor. Blow-up circles like those in Figure 1.3 will be used commonly in this text to indicate the composition and state of the particles that compose a substance. In this molecular art, different types of particles are shown in color-coded spheres, and the distance between the spheres signals its state—solid, liquid, or gas.

Matter is characterized by its **physical properties** and **chemical properties.**

- ***Physical properties*** **are those that can be observed or measured without changing the composition of the material.**

Figure 1.3 The Three States of Water—Solid, Liquid, and Gas

- The particles of a solid are close together and highly organized. (Photo: snow-capped Mauna Kea on the Big Island of Hawaii)
- The particles of a liquid are close together but more disorganized than the solid. (Photo: Akaka Falls on the Big Island of Hawaii)
- The particles of a gas are far apart and disorganized. (Photo: steam formed by a lava flow on the Big Island of Hawaii)

Each red sphere joined to two gray spheres represents a single water particle. In proceeding from left to right, from solid to liquid to gas, the molecular art shows that the level of organization of the water particles decreases. Color-coding and the identity of the spheres within the particles will be addressed in Chapter 2.

Common physical properties include melting point (mp), boiling point (bp), solubility, color, and odor. **A *physical change* alters a substance without changing its composition.** The most common physical changes are **changes in state.** Melting an ice cube to form liquid water, and boiling liquid water to form steam are two examples of physical changes. Water is the substance at the beginning and end of both physical changes. More details about physical changes are discussed in Chapter 4.

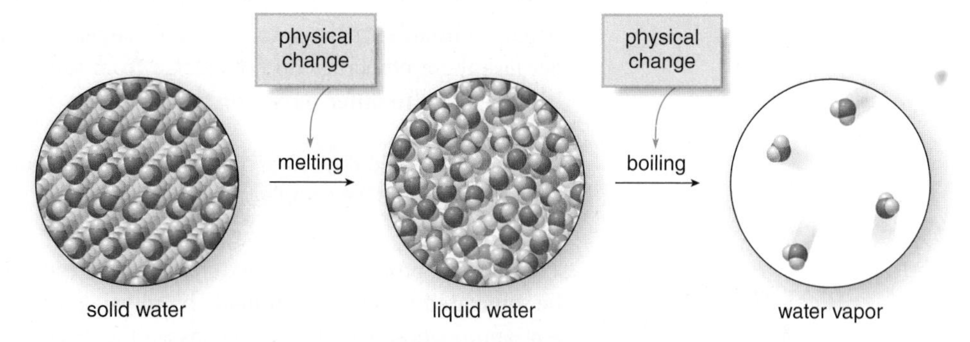

- ***Chemical properties*** **are those that determine how a substance can be converted to another substance.**

A *chemical change,* or a *chemical reaction,* converts one material to another. The conversion of hydrogen and oxygen to water is a chemical reaction because the composition of the material is different at the beginning and end of the process. Chemical reactions are discussed in Chapter 5.

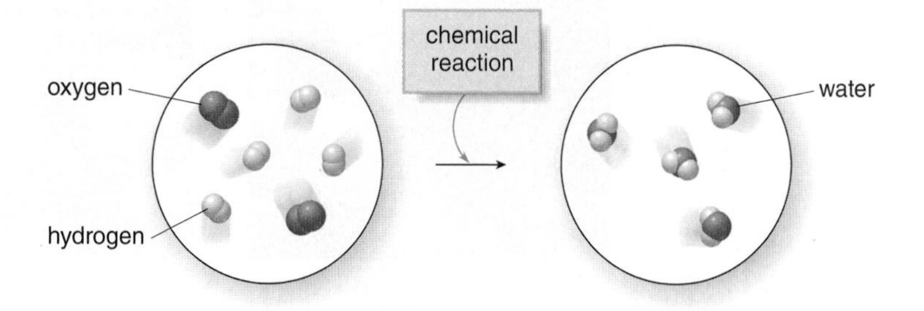

PROBLEM 1.2

Characterize each process as a physical change or a chemical change: (a) making ice cubes; (b) burning natural gas; (c) silver jewelry tarnishing; (d) a pile of snow melting; (e) fermenting grapes to produce wine.

PROBLEM 1.3

Does the molecular art represent a chemical change or a physical change? Explain your choice.

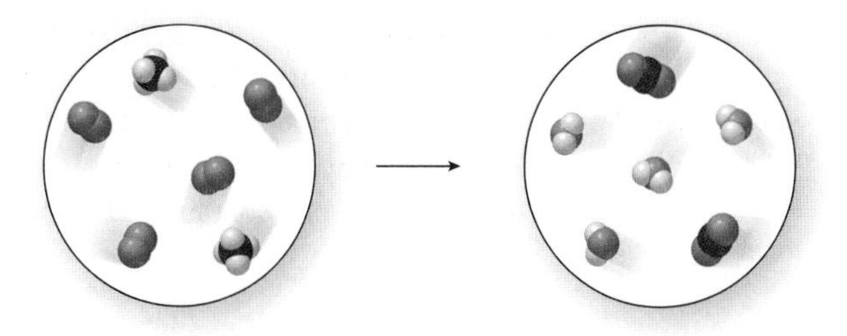

1.3 Classification of Matter

All matter can be classified as either a **pure substance** or a **mixture.**

- A *pure substance* is composed of a single component and has a constant composition, regardless of the sample size and the origin of the sample.

A pure substance, such as water or table sugar, can be characterized by its physical properties, because these properties do not change from sample to sample. **A pure substance cannot be broken down to other pure substances by any physical change.**

- A *mixture* is composed of more than one substance. The composition of a mixture can vary depending on the sample.

The physical properties of a mixture may also vary from one sample to another. **A mixture can be separated into its components by physical changes.** Dissolving table sugar in water forms a mixture, whose sweetness depends on the amount of sugar added. If the water is allowed to evaporate from the mixture, pure table sugar and pure water are obtained.

Mixtures can be formed from solids, liquids, and gases, as shown in Figure 1.4. The compressed air breathed by a scuba diver consists mainly of the gases oxygen and nitrogen. A saline solution used in an IV bag contains solid sodium chloride (table salt) dissolved in water.

Figure 1.4 Two Examples of Mixtures

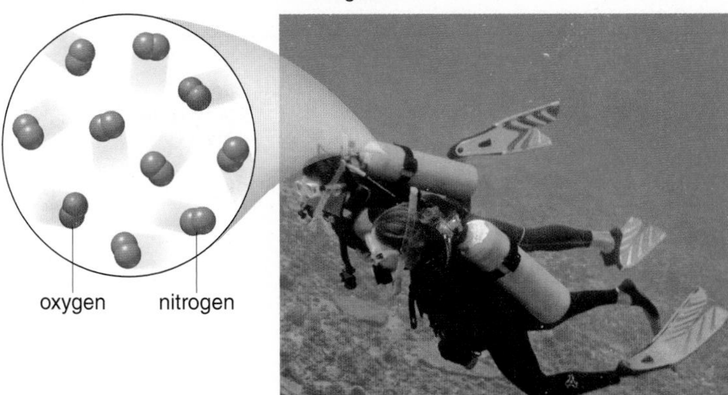

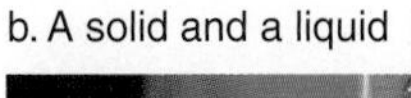

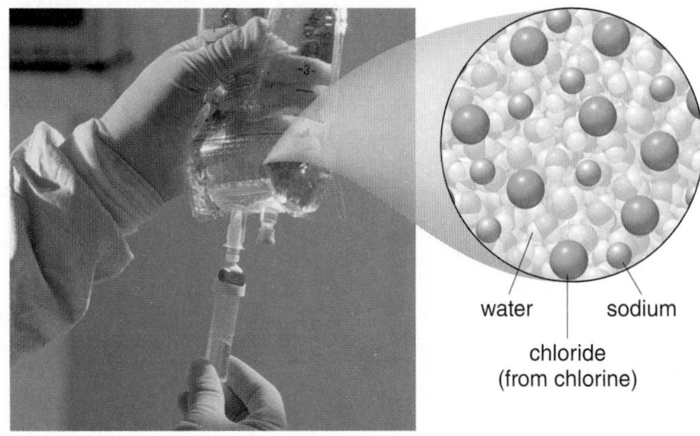

A pure substance is classified as either an **element** or a **compound.**

- **An *element* is a pure substance that cannot be broken down into simpler substances by a chemical reaction.**
- **A *compound* is a pure substance formed by chemically combining (joining together) two or more elements.**

Nitrogen gas, aluminum foil, and copper wire are all elements. Water is a compound because it is composed of the elements hydrogen and oxygen. Table salt, sodium chloride, is also a compound since it is formed from the elements sodium and chlorine (Figure 1.5). Although

Figure 1.5 Elements and Compounds

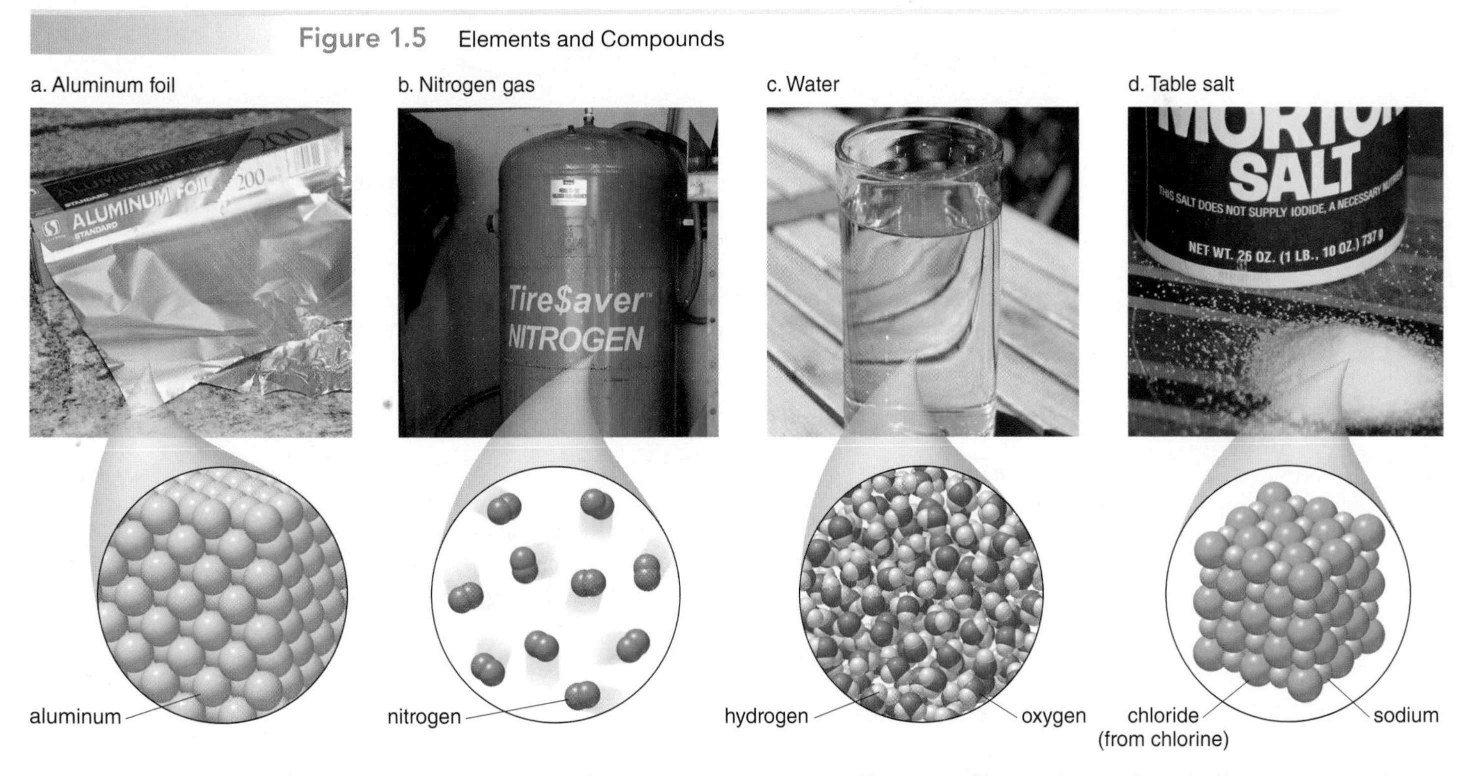

- Aluminum foil and nitrogen gas are elements. The molecular art used for an element shows spheres of one color only. Thus, aluminum is a solid shown with gray spheres, while nitrogen is a gas shown with blue spheres. Water and table salt are compounds. Color-coding of the spheres used in the molecular art indicates that water is composed of two elements—hydrogen shown as gray spheres and oxygen shown in red. Likewise, the gray (sodium) and green (chlorine) spheres illustrate that sodium chloride is formed from two elements as well.

Figure 1.6
Classification of Matter

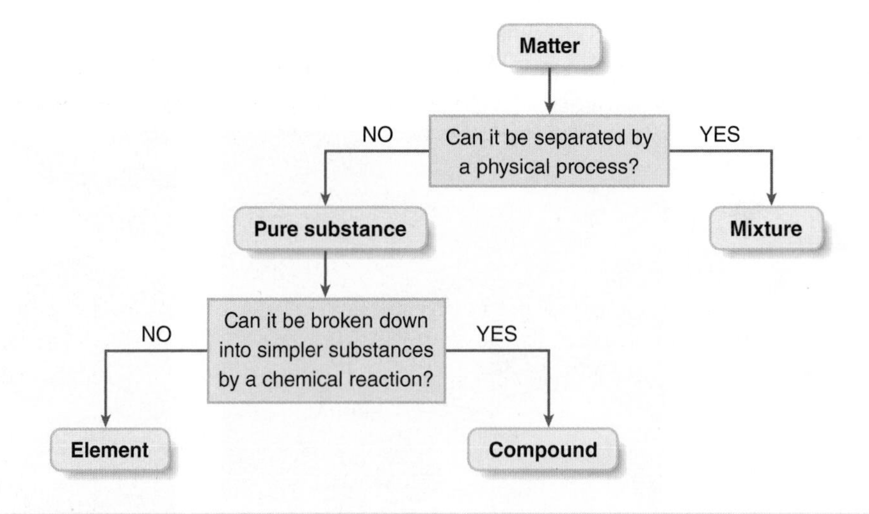

An alphabetical list of elements is located on the inside front cover of this text. The elements are commonly organized into a periodic table, also shown on the inside front cover, and discussed in much greater detail in Chapter 2.

only 118 elements are currently known, over 50 million compounds occur naturally or have been synthesized in the laboratory. We will learn much more about elements and compounds in Chapters 2 and 3. Figure 1.6 summarizes the categories into which matter is classified.

SAMPLE PROBLEM 1.1

Classify each example of molecular art as an element or a compound:

a.

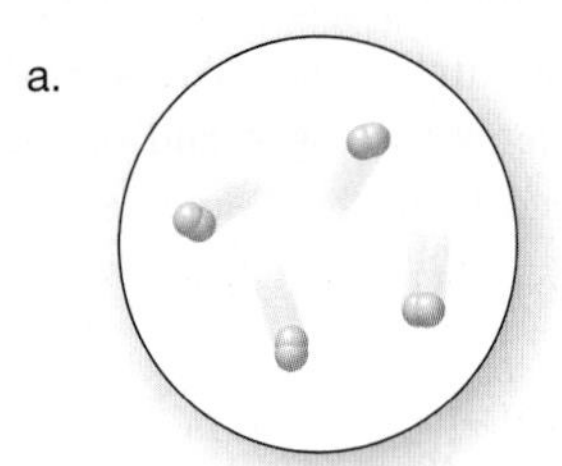

b.

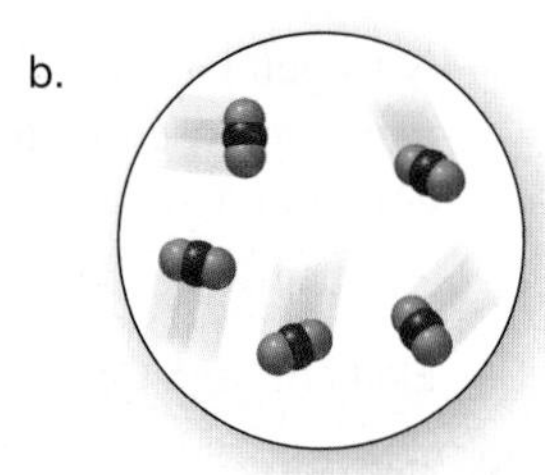

Analysis

In molecular art, an element is composed of spheres of the same color, while a compound is composed of spheres of different colors.

Solution

Representation (a) is an element since each particle contains only gray spheres. Representation (b) is a compound since each particle contains both red and black spheres.

PROBLEM 1.4

Classify each example of molecular art as a pure substance or a mixture:

a.

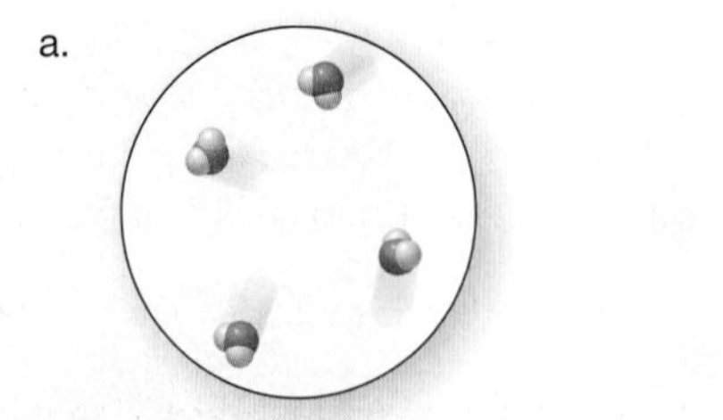

b.

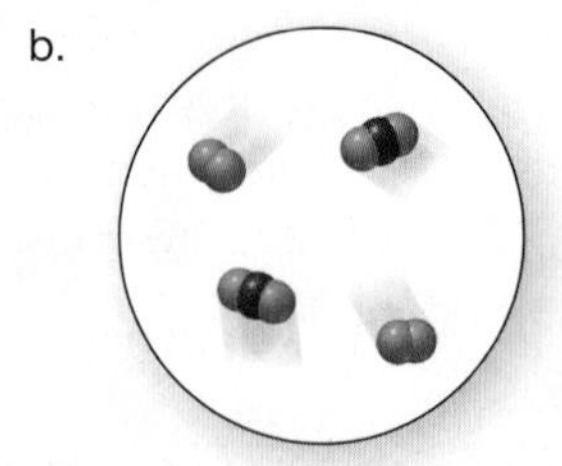

PROBLEM 1.5

Classify each item as a pure substance or a mixture: (a) blood; (b) ocean water; (c) a piece of wood; (d) a chunk of ice.

PROBLEM 1.6

Classify each item as an element or a compound: (a) the gas inside a helium balloon; (b) table sugar; (c) the rust on an iron nail; (d) aspirin. All elements are listed alphabetically on the inside front cover.

1.4 Measurement

Any time you check your weight on a scale, measure the ingredients of a recipe, or figure out how far it is from one location to another, you are measuring a quantity. Measurements are routine for healthcare professionals who use weight, blood pressure, pulse, and temperature to chart a patient's progress.

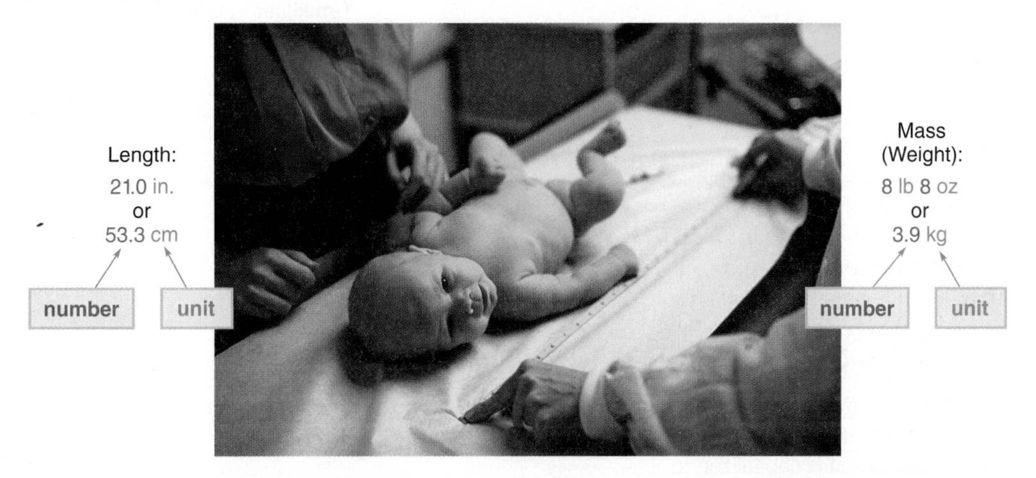

- **Every measurement is composed of a *number* and a *unit*.**

Reporting the value of a measurement is meaningless without its unit. For example, if you were told to give a patient an aspirin dosage of 325, does this mean 325 ounces, pounds, grams, milligrams, or tablets? Clearly there is a huge difference among these quantities.

In 1960, the **International System of Units** was formally adopted as the uniform system of units for the sciences. **SI units,** as they are called, are based on the metric system, but the system recommends the use of some metric units over others. SI stands for the French words, *Système Internationale.*

1.4A The Metric System

In the United States, most measurements are made with the **English system,** using units like miles (mi), gallons (gal), pounds (lb), and so forth. A disadvantage of this system is that the units are not systematically related to each other and require memorization. For example, 1 lb = 16 oz, 1 gal = 4 qt, and 1 mi = 5,280 ft.

The metric system is slowly gaining acceptance in the United States, as seen in the gallon jug of milk and the two-liter bottle of soda.

Scientists, health professionals, and people in most other countries use the **metric system,** with units like meter (m) for length, gram (g) for mass, and liter (L) for volume. The metric system is slowly gaining popularity in the United States. The weight of packaged foods is often given in both ounces and grams. Distances on many road signs are shown in miles and kilometers. Most measurements in this text will be reported using the metric system, but learning to convert English units to metric units is also a necessary skill that will be illustrated in Section 1.7.

The important features of the metric system are the following:

- **Each type of measurement has a base unit—the meter (m) for length; the gram (g) for mass; the liter (L) for volume; the second (s) for time.**
- **All other units are related to the base unit by powers of 10.**
- **The prefix of the unit name indicates if the unit is larger or smaller than the base unit.**

Table 1.1 Metric Units

Quantity	Base Unit	Symbol
Length	Meter	m
Mass	Gram	g
Volume	Liter	L
Time	Second	s

The base units of the metric system are summarized in Table 1.1, and the most common prefixes used to convert the base units to smaller or larger units are summarized in Table 1.2. **The same prefixes are used for all types of measurement.** For example, the prefix *kilo-* means 1,000 times as large. Thus,

1 **kilo**meter = **1,000** meters or 1 km = 1,000 m
1 **kilo**gram = **1,000** grams or 1 kg = 1,000 g
1 **kilo**liter = **1,000** liters or 1 kL = 1,000 L

The prefix *milli-* means one thousandth as large (1/1,000 or 0.001). Thus,

1 **milli**meter = **0.001** meters or 1 mm = 0.001 m
1 **milli**gram = **0.001** grams or 1 mg = 0.001 g
1 **milli**liter = **0.001** liters or 1 mL = 0.001 L

The metric symbols are all lower case except for the unit **liter** (L) and the prefixes **mega-** (M) and **giga-** (G). Liter is capitalized to distinguish it from the number *one.* Mega is capitalized to distinguish it from the symbol for the prefix *milli-* (m). Giga is capitalized to distinguish it from the abbreviation for gram (g).

Table 1.2 Common Prefixes Used for Metric Units

Prefix	Symbol	Meaning	Numerical Value[a]	Scientific Notation[b]
Giga-	G	Billion	1,000,000,000.	10^9
Mega-	M	Million	1,000,000.	10^6
Kilo-	k	Thousand	1,000.	10^3
Deci-	d	Tenth	0.1	10^{-1}
Centi-	c	Hundredth	0.01	10^{-2}
Milli-	m	Thousandth	0.001	10^{-3}
Micro-	μ[c]	Millionth	0.000 001	10^{-6}
Nano-	n	Billionth	0.000 000 001	10^{-9}

[a]Numbers that contain five or more digits to the right of the decimal point are written with a small space separating each group of three digits.
[b]How to express numbers in scientific notation is explained in Section 1.6.
[c]The symbol μ is the lower case Greek letter mu. The prefix *micro-* is sometimes abbreviated as **mc.**

PROBLEM 1.7

What term is used for each of the following units: (a) a million liters; (b) a thousandth of a second; (c) a hundredth of a gram; (d) a tenth of a liter?

1.4B Measuring Length

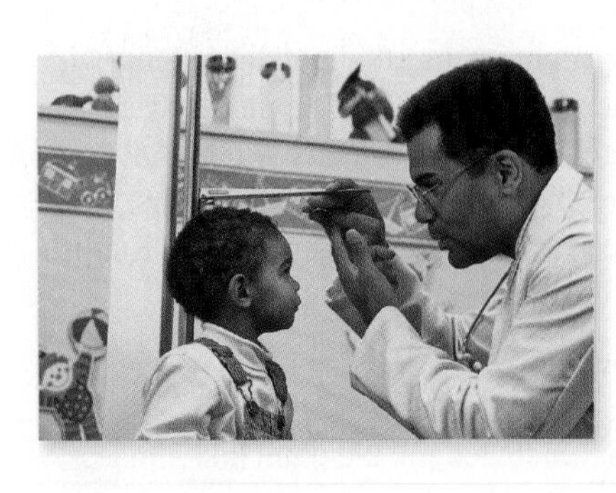

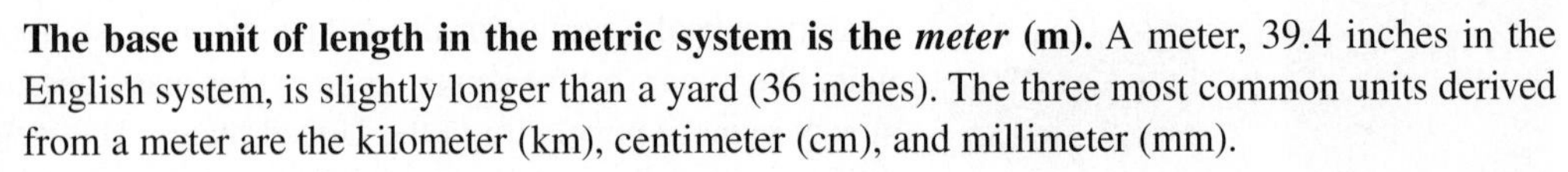

The base unit of length in the metric system is the *meter* (m). A meter, 39.4 inches in the English system, is slightly longer than a yard (36 inches). The three most common units derived from a meter are the kilometer (km), centimeter (cm), and millimeter (mm).

1,000 m = 1 km
1 m = 100 cm
1 m = 1,000 mm

Note how these values are related to those in Table 1.2. Since a centimeter is one *hundredth* of a meter (0.01 m), there are *100* centimeters in a meter.

PROBLEM 1.8

If a nanometer is one billionth of a meter (0.000 000 001 m), how many nanometers are there in one meter?

1.4C Measuring Mass

Although the terms mass and weight are often used interchangeably, they really have different meanings.

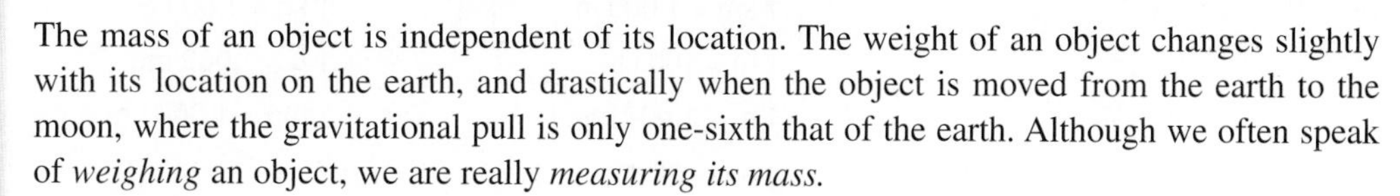

- *Mass* is a measure of the amount of matter in an object.
- *Weight* is the force that matter feels due to gravity.

The mass of an object is independent of its location. The weight of an object changes slightly with its location on the earth, and drastically when the object is moved from the earth to the moon, where the gravitational pull is only one-sixth that of the earth. Although we often speak of *weighing* an object, we are really *measuring its mass.*

The base unit of mass in the metric system is the *gram* (g), a small quantity compared to the English pound (1 lb = 454 g). The two most common units derived from a gram are the kilogram (kg) and milligram (mg).

$$1{,}000 \text{ g} = 1 \text{ kg}$$
$$1 \text{ g} = 1{,}000 \text{ mg}$$

PROBLEM 1.9

If a microgram is one millionth of a gram (0.000 001 g), how many micrograms are there in one gram?

1.4D Measuring Volume

The base unit of volume in the metric system is the *liter* (L), which is slightly larger than the English quart (1 L = 1.06 qt). One liter is defined as the volume of a cube 10 cm on an edge.

Note the difference between the units **cm** and **cm^3.** The centimeter (cm) is a unit of length. A cubic centimeter (cm^3 or cc) is a unit of volume.

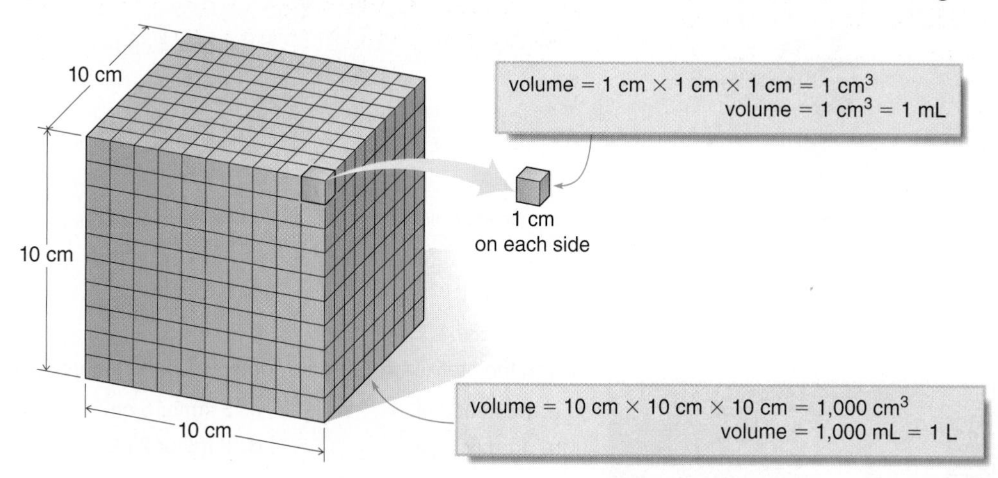

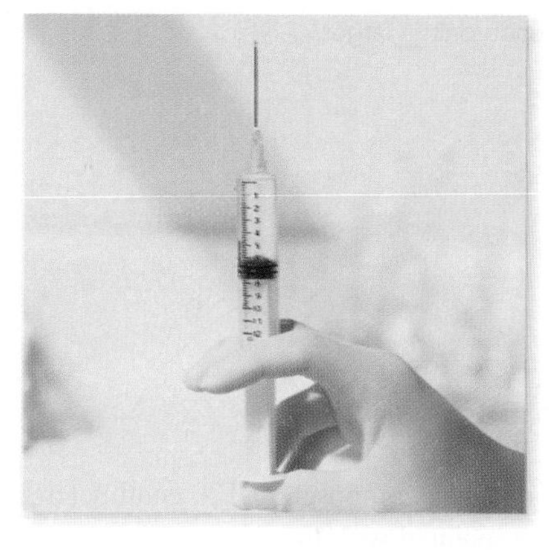

Three common units derived from a liter used in medicine and laboratory research are the deciliter (dL), milliliter (mL), and microliter (μL). **One milliliter is the same as one cubic centimeter (cm^3), which is abbreviated as cc.**

$$1 \text{ L} = 10 \text{ dL}$$
$$1 \text{ L} = 1{,}000 \text{ mL}$$
$$1 \text{ L} = 1{,}000{,}000 \text{ μL}$$
$$1 \text{ mL} = 1 \text{ cm}^3 = 1 \text{ cc}$$

Table 1.3 summarizes common metric units of length, mass, and volume. Table 1.4 lists English units of measurement, as well as their metric equivalents.

Table 1.3 Summary of the Common Metric Units of Length, Mass, and Volume

Length	Mass	Volume
1 km = 1,000 m	1 kg = 1,000 g	1 L = 10 dL
1 m = 100 cm	1 g = 1,000 mg	1 L = 1,000 mL
1 m = 1,000 mm	1 mg = 1,000 μg	1 L = 1,000,000 μL
1 cm = 10 mm		1 dL = 100 mL
		1 mL = 1 cm^3 = 1 cc

Table 1.4 English Units and Their Metric Equivalents

Quantity	English Unit	Metric–English Relationship
Length	1 ft = 12 in.	2.54 cm = 1 in.
	1 yd = 3 ft	1 m = 39.4 in.
	1 mi = 5,280 ft	1 km = 0.621 mi
Mass	1 lb = 16 oz	1 kg = 2.20 lb
	1 ton = 2,000 lb	454 g = 1 lb
		28.3 g = 1 oz
Volume	1 qt = 4 cups	946 mL = 1 qt
	1 qt = 2 pt	1 L = 1.06 qt
	1 qt = 32 fl oz	29.6 mL = 1 fl oz
	1 gal = 4 qt	

Common abbreviations for English units: inch (in.), foot (ft), yard (yd), mile (mi), pound (lb), ounce (oz), gallon (gal), quart (qt), pint (pt), and fluid ounce (fl oz).

PROBLEM 1.10

Using the prefixes in Table 1.2, determine which quantity in each pair is larger.

a. 3 mL or 3 cL
b. 1 ng or 1 μg
c. 5 km or 5 cm
d. 2 mL or 2 μL

A container of 71 macadamia nuts weighs 125 g. The number of nuts (71) is exact, while the mass of the nuts (125 g) is inexact.

1.5 Significant Figures

Numbers used in chemistry are either **exact** or **inexact.**

- **An *exact* number results from counting objects or is part of a definition.**

Our bodies have 10 fingers, 10 toes, and two kidneys. A meter is composed of 100 centimeters. These numbers are exact because there is no uncertainty associated with them.

- **An *inexact* number results from a measurement or observation and contains some uncertainty.**

Whenever we measure a quantity there is a degree of uncertainty associated with the result. The last number (farthest to the right) is an estimate, and it depends on the type of measuring device we use to obtain it. For example, the length of a fish caught on a recent outing could be reported as 53 cm or 53.5 cm depending on the tape measure used.

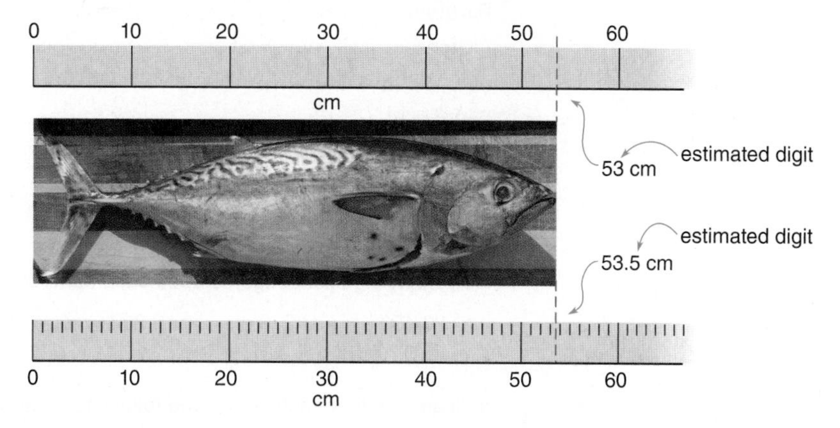

- ***Significant figures*** **are all the digits in a measured number including** ***one*** **estimated digit.**

Thus, the length 53 cm has two significant figures, and the length 53.5 cm has three significant figures.

1.5A Determining the Number of Significant Figures

How many significant figures are contained in a number?

- **All nonzero digits are always significant.**

65.2 g	three significant figures
1,265 m	four significant figures
25 μL	two significant figures
255.345 g	six significant figures

Whether a zero counts as a significant figure depends on its location in the number.

In reading a number with a decimal point from left to right, all digits starting with the first nonzero number are significant figures. The number 0.003 450 120 has seven significant figures, shown in red.

Rules to Determine When a Zero Is a Significant Figure

Rule [1] A zero *counts* as a significant figure when it occurs:

- Between two nonzero digits
  - 29.05 g—four significant figures
  - 1.0087 mL—five significant figures
- At the end of a number with a decimal point
  - 25.70 cm—four significant figures
  - 3.7500 g—five significant figures
  - 620. lb—three significant figures

Rule [2] A zero does *not* count as a significant figure when it occurs:

- At the beginning of a number
  - 0.0245 mg—three significant figures
  - 0.008 mL—one significant figure
- At the end of a number that does not have a decimal point
  - 2,570 m—three significant figures
  - 1,245,500 m—five significant figures

SAMPLE PROBLEM 1.2

How many significant figures does each number contain?

a. 34.08 b. 0.0054 c. 260.00 d. 260

Analysis

All nonzero digits are significant. A zero is significant only if it occurs between two nonzero digits, or at the end of a number with a decimal point.

Solution

Significant figures are shown in red.

a. 34.08 (four) b. 0.0054 (two) c. 260.00 (five) d. 260 (two)

PROBLEM 1.11

How many significant figures does each number contain?

a. 23.45 b. 230 c. 0.202 d. 0.003 60 e. 10,040 f. 1,004.00

PROBLEM 1.12

Indicate whether each zero in the following numbers is significant.

a. 0.003 04 b. 26,045 c. 1,000,034 d. 0.304 00

1.5B Using Significant Figures in Multiplication and Division

We often must perform calculations with numbers that contain a different number of significant figures. The number of significant figures in the answer of a problem depends on the type of mathematical calculation—multiplication (and division) or addition (and subtraction).

- **In multiplication and division, the answer has the same number of significant figures as the original number with the *fewest* significant figures.**

Let's say you drove a car 351.2 miles in 5.5 hours, and you wanted to calculate how many miles per hour you traveled. Entering these numbers on a calculator would give the following result:

four significant figures

$$\text{Miles per hour} = \frac{351.2 \text{ miles}}{5.5 \text{ hours}} = 63.854\ 545 \text{ miles per hour}$$

two significant figures

The answer must contain only **two** significant figures.

The answer to this problem can have only *two* significant figures, since one of the original numbers (5.5 hours) has only *two* significant figures. To write the answer in proper form, we must **round off the number** to give an answer with only two significant figures. Two rules are used in rounding off numbers.

- **If the first number that must be dropped is 4 or fewer, drop it and all remaining numbers.**
- **If the first number that must be dropped is 5 or greater, *round the number up* by adding one to the last digit that will be retained.**

In this problem:

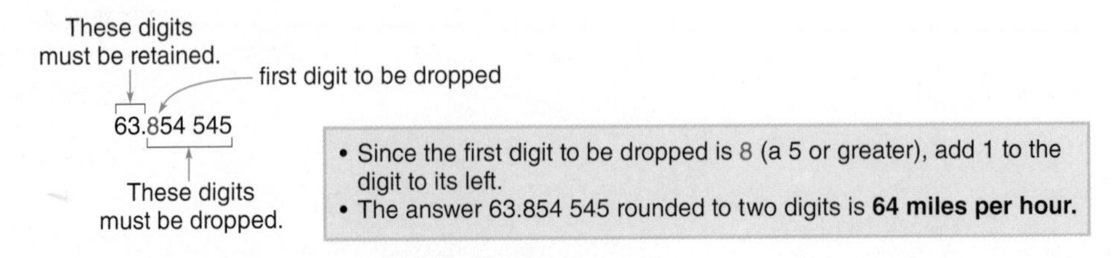
These digits must be retained.
first digit to be dropped
63.854 545
These digits must be dropped.
• Since the first digit to be dropped is 8 (a 5 or greater), add 1 to the digit to its left.
• The answer 63.854 545 rounded to two digits is 64 miles per hour.

Table 1.5 gives other examples of rounding off numbers.

When a calculator is used in calculations, sometimes the display shows more digits than are significant and sometimes it shows fewer digits. For example, in multiplying 23.2 by 1.1, the calculator displays the answer as 25.52. Since the quantity 1.1 has only two significant figures, the answer must contain only two significant figures and rounded to 26.

23.2	×	1.1	=	25.52	round off ------→	26
three significant figures		two significant figures		calculator display		two significant figures
						Answer

In contrast, dividing 25.0 by 0.50 displays the answer as 50, a quantity with only one significant figure. Since 0.50 has two significant figures, the answer must contain two significant figures, and this is achieved by adding a decimal point (50.).

25.0	×	0.50	=	50	add decimal --------→	50.
three significant figures		two significant figures		calculator display		two significant figures
						Answer

Table 1.5 Rounding Off Numbers

Original Number	Rounded to	Rounded Number
61.2537	Two places	61
61.2537	Three places	61.3
61.2537	Four places	61.25
61.2537	Five places	61.254

The first number to be dropped is indicated in red in each original number. When this number is 4 or fewer, it and all other digits to its right are dropped. When this number is 5 or greater, 1 is added to the digit to its left.

SAMPLE PROBLEM 1.3

Round off each number to three significant figures.

a. 1.2735 b. 0.002 536 22 c. 3,836.9

Analysis

If the answer is to have *three* significant figures, look at the *fourth* number from the left. If this number is 4 or fewer, drop it and all remaining numbers to the right. If the fourth number from the left is 5 or greater, round the number up by adding one to the third nonzero digit.

Solution

a. 1.27 b. 0.002 54 c. 3,840 (Omit the decimal point after the 0. The number 3,840. has four significant figures.)

PROBLEM 1.13

Round off each number in Sample Problem 1.3 to two significant figures.

SAMPLE PROBLEM 1.4

Carry out each calculation and give the answer using the proper number of significant figures.

a. 3.81×0.046 b. $120.085 \div 106$

Analysis

Since these calculations involve multiplication and division, the answer must have the same number of significant figures as the original number with the fewest number of significant figures.

Solution

a. $3.81 \times 0.046 = 0.1753$

- Since 0.046 has only two significant figures, round the answer to give it two significant figures.

0.1753 Since this number is 5 (5 or greater), round the 7 to its left up by one.

Answer: 0.18

b. $120.085 \div 106 = 1.132\ 877\ 36$

- Since 106 has three significant figures, round the answer to give it three significant figures.

1.132 877 36 Since this number is 2 (4 or fewer), drop it and all numbers to its right.

Answer: 1.13

PROBLEM 1.14

Carry out each calculation and give the answer using the proper number of significant figures.

a. 10.70 × 3.5   b. 0.206 ÷ 25,993   c. 1,300 ÷ 41.2   d. 120.5 × 26

1.5C Using Significant Figures in Addition and Subtraction

In determining significant figures in addition and subtraction, the decimal place of the last significant digit determines the number of significant figures in the answer.

- **In addition and subtraction, the answer has the same number of decimal places as the original number with the *fewest* decimal places.**

Suppose a baby weighed 3.6 kg at birth and 10.11 kg on his first birthday. To figure out how much weight the baby gained in his first year of life, we subtract these two numbers and report the answer using the proper number of significant figures.

weight at one year = 10.11 kg

weight at birth = 3.6 kg

10.11 kg (two digits after the decimal point)
−3.6 kg (one digit after the decimal point)
weight gain = 6.51 kg

last significant digit

- The answer can have only **one** digit after the decimal point.
- Round 6.51 to 6.5.
- The baby gained 6.5 kg during his first year of life.

Since 3.6 kg has only one significant figure after the decimal point, the answer can have only one significant figure after the decimal point as well.

SAMPLE PROBLEM 1.5

While on a diet, a woman lost 3.52 lb the first week, 2.2 lb the second week, and 0.59 lb the third week. How much weight did she lose in all?

Analysis

Add up the amount of weight loss each week to get the total weight loss. When adding, the answer has the same number of decimal places as the original number with the fewest decimal places.

Solution

3.52 lb
2.2 lb (one digit after the decimal point)
0.59 lb
6.31 lb → round off → 6.3 lb

last significant digit

- Since 2.2 lb has only one digit after the decimal point, the answer can have only one digit after the decimal point.
- Round 6.31 to 6.3.
- Total weight loss: 6.3 lb.

PROBLEM 1.15

Carry out each calculation and give the answer using the proper number of significant figures.

a. 27.8 cm + 0.246 cm
b. 102.66 mL + 0.857 mL + 24.0 mL
c. 54.6 mg – 25 mg
d. 2.35 s – 0.266 s

1.6 Scientific Notation

Hospital laboratory technicians determine thousands of laboratory results each day.

Healthcare professionals and scientists must often deal with very large and very small numbers. For example, the blood platelet count of a healthy adult might be 250,000 platelets per mL. At the other extreme, the level of the female sex hormone estriol during pregnancy might be 0.000 000 250 g per mL of blood plasma. Estriol is secreted by the placenta and its concentration is used as a measure of the health of the fetus.

To write numbers that contain many leading zeros (at the beginning) or trailing zeros (at the end), scientists use **scientific notation.**

- **In scientific notation, a number is written as $y \times 10^x$.**
- **The term *y*, called the coefficient, is a number between 1 and 10.**
- **The value *x* is an exponent, which can be any positive or negative whole number.**

First, let's recall what powers of 10 with *positive* exponents, such as 10^2 or 10^5, mean. These correspond to numbers greater than one, and the positive exponent tells how many zeros are to be written after the number one. Thus, $10^2 = 100$, a number with two zeros after the number one.

The product has two zeros.

$$10^2 = 10 \times 10 = 100$$

The exponent 2 means "multiply two 10s."

The product has five zeros.

$$10^5 = 10 \times 10 \times 10 \times 10 \times 10 = 100{,}000$$

The exponent 5 means "multiply five 10s."

Powers of 10 that contain *negative* exponents, such as 10^{-3}, correspond to numbers less than one. In this case the exponent tells how many places (*not* zeros) are located to the right of the decimal point.

The answer has three places to the right of the decimal point, including the number one.

$$10^{-3} = \frac{1}{10 \times 10 \times 10} = 0.001$$

The exponent −3 means "divide by three 10s."

To write a number in scientific notation, we follow a stepwise procedure.

How To Convert a Standard Number to Scientific Notation

Example Write each number in scientific notation: (a) 2,500; (b) 0.036.

Step [1] **Move the decimal point to give a number between 1 and 10.**

a. 2500.

Move the decimal point three places to the left to give the number 2.5.

b. 0.036

Move the decimal point two places to the right to give the number 3.6.

Step [2] **Multiply the result by 10^x, where *x* is the number of places the decimal point was moved.**

- If the decimal point is moved to the **left,** *x* is **positive.**
- If the decimal point is moved to the **right,** *x* is **negative.**

a. Since the decimal point was moved three places to the **left,** the exponent is +3, and the coefficient is multiplied by 10^3.

Answer: $2{,}500 = 2.5 \times 10^3$

b. Since the decimal point was moved two places to the **right,** the exponent is −2, and the coefficient is multiplied by 10^{-2}.

Answer: $0.036 = 3.6 \times 10^{-2}$

Figure 1.7

Numbers in Standard Form and Scientific Notation

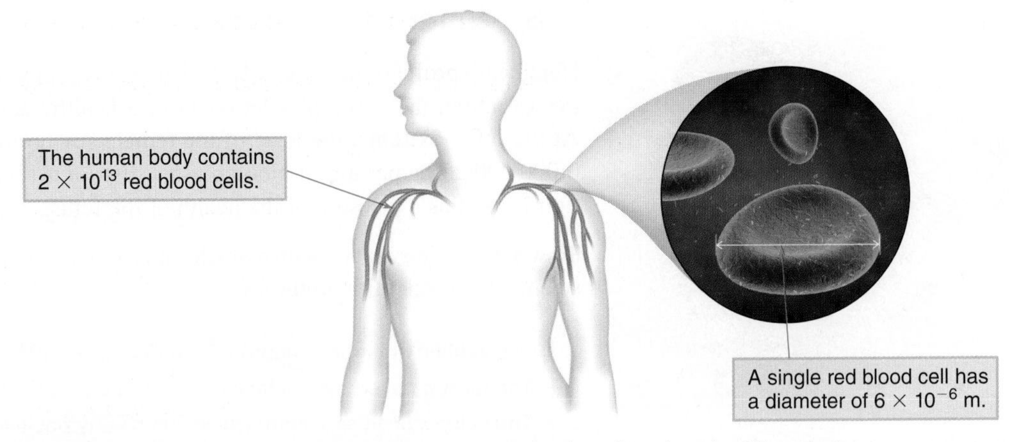
The human body contains
2 × 10^13 red blood cells.
A single red blood cell has
a diameter of 6 × 10^−6 m.

Very large and very small numbers are more conveniently written in scientific notation.

Quantity	**Number**	**Scientific Notation**
Number of red blood cells	20,000,000,000,000	2×10^{13}
Diameter of a red blood cell	0.000 006 m	6×10^{-6} m

Notice that the number of significant figures in the coefficient in scientific notation must equal the number of significant figures in the original number. Thus, the coefficients for both 2,500 and 0.036 need two significant figures and no more. Figure 1.7 shows two more examples of numbers written in standard form and scientific notation.

$2{,}500 = 2.5 \times 10^3$ *not* 2.50×10^3 (three significant figures)

two significant figures *not* 2.500×10^3 (four significant figures)

SAMPLE PROBLEM 1.6

Write the recommended daily dietary intake of each nutrient in scientific notation: (a) sodium, 2,400 mg; (b) vitamin B_{12}, 0.000 006 g.

Analysis

Move the decimal point to give a number between 1 and 10. Multiply the number by 10^x, where x is the number of places the decimal point was moved. The exponent x is (+) when the decimal point moves to the left and (–) when it moves to the right.

Solution

a.

$2400. = 2.4 \times 10^3$ ← the number of places the decimal point was moved to the left

Move the decimal point three places to the left.

- Write the coefficient as 2.4 (two significant figures), since 2,400 contains two significant figures.

b.

$0.000\,006 = 6 \times 10^{-6}$ ← the number of places the decimal point was moved to the right

Move the decimal point six places to the right.

- Write the coefficient as 6 (one significant figure), since 0.000 006 contains one significant figure.

PROBLEM 1.16

Lab results for a routine check-up showed an individual's iron level in the blood to be 0.000 098 g per deciliter, placing it in the normal range. Convert this number to scientific notation.

PROBLEM 1.17

Write each number in scientific notation.

a. 93,200 b. 0.000 725 c. 6,780,000 d. 0.000 030

To convert a number in scientific notation to a standard number, reverse the procedure, as shown in Sample Problem 1.7. It is often necessary to add leading or trailing zeros to write the number.

- **When the exponent x is positive, move the decimal point x places to the *right*.**

$$2.800 \times 10^2 \qquad 2.800 \dashrightarrow 280.0$$

Move the decimal point to the right two places.

- **When the exponent x is negative, move the decimal point x places to the *left*.**

$$2.80 \times 10^{-2} \qquad 002.80 \dashrightarrow 0.0280$$

Move the decimal point to the left two places.

SAMPLE PROBLEM 1.7

The element hydrogen is composed of two hydrogen atoms, separated by a distance of 7.4×10^{-11} m. Convert this value to a standard number.

Analysis

The exponent in 10^x tells how many places to move the decimal point in the coefficient to generate a standard number. The decimal point goes to the right when x is positive and to the left when x is negative.

Solution

$$7.4 \times 10^{-11} \qquad 000\ 000\ 000\ 07.4 \dashrightarrow 0.000\ 000\ 000\ 074 \text{ m}$$

Move the decimal point to the left 11 places. **Answer**

The answer, 0.000 000 000 074, has two significant figures, just like 7.4×10^{-11}.

PROBLEM 1.18

Convert each number to its standard form.

a. 6.5×10^3 b. 3.26×10^{-5} c. 3.780×10^{-2} d. 1.04×10^8

1.7 Problem Solving Using Conversion Factors

Often a measurement is recorded in one unit, and then it must be converted to another unit. For example, a patient may weigh 130 lb, but we may need to know her weight in kilograms to calculate a drug dosage. The recommended daily dietary intake of potassium is 3,500 mg, but we may need to know how many grams this corresponds to.

1.7A Conversion Factors

To convert one unit to another we use one or more **conversion factors.**

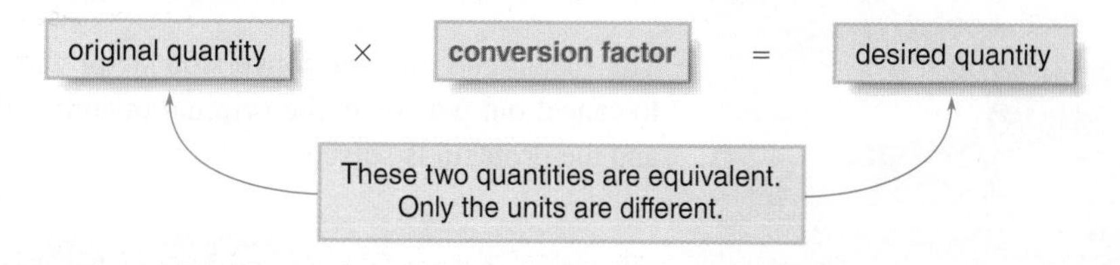

- **A *conversion factor* is a term that converts a quantity in one unit to a quantity in another unit.**

Refer to Tables 1.3 and 1.4 for metric and English units needed in problem solving. Common metric and English units are also listed on the inside back cover.

A conversion factor is formed by taking an equality, such as 2.20 lb = 1 kg, and writing it as a fraction. We can always write a conversion factor in two different ways.

$$\frac{2.20 \text{ lb}}{1 \text{ kg}} \quad \text{or} \quad \frac{1 \text{ kg}}{2.20 \text{ lb}}$$

numerator
denominator

conversion factors for pounds and kilograms

With pounds and kilograms, either of these values can be written above the division line of the fraction (the numerator) or below the division line (the denominator). The way the conversion factor is written will depend on the problem.

SAMPLE PROBLEM 1.8

Write two conversion factors for each pair of units: (a) kilograms and grams; (b) quarts and liters.

Analysis

Use the equalities in Tables 1.3 and 1.4 to write a fraction that shows the relationship between the two units.

Solution

a. Conversion factors for kilograms and grams:

$$\frac{1000 \text{ g}}{1 \text{ kg}} \quad \text{or} \quad \frac{1 \text{ kg}}{1000 \text{ g}}$$

b. Conversion factors for quarts and liters:

$$\frac{1.06 \text{ qt}}{1 \text{ L}} \quad \text{or} \quad \frac{1 \text{ L}}{1.06 \text{ qt}}$$

PROBLEM 1.19

Write two conversion factors for each pair of units: (a) miles and kilometers; (b) meters and millimeters.

1.7B Solving a Problem Using One Conversion Factor

When using conversion factors to solve a problem, a unit that appears in the numerator in one term and the denominator in another term will *cancel*. **The goal in setting up a problem is to make sure *all unwanted units cancel.***

Let's say we want to convert 130 lb to kilograms.

$$\underset{\text{original quantity}}{130 \text{ lb}} \times \boxed{\text{conversion factor}} = \underset{\text{desired quantity}}{? \text{ kg}}$$

Two possible conversion factors: $\frac{2.20 \text{ lb}}{1 \text{ kg}}$ or $\frac{1 \text{ kg}}{2.20 \text{ lb}}$

To solve this problem we must use a conversion factor that satisfies two criteria.

- **The conversion factor must relate the two quantities in question—pounds and kilograms.**
- **The conversion factor must cancel out the unwanted unit—pounds.**

This means choosing the conversion factor with the unwanted unit—pounds—*in the denominator* to cancel out pounds in the original quantity. This leaves kilograms as the only remaining unit, and the problem is solved.

conversion factor

$$130 \cancel{\text{lb}} \times \frac{1 \text{ kg}}{2.20 \cancel{\text{lb}}} = \boxed{59 \text{ kg}}$$ **answer in kilograms**

Pounds (lb) must be the denominator to cancel the unwanted unit (lb) in the original quantity.

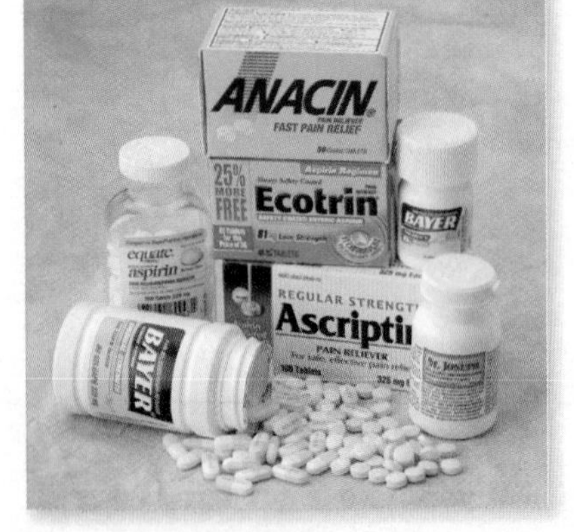

How many grams of aspirin are contained in a 325-mg tablet?

We must use the correct number of significant figures in reporting an answer to each problem. In this case, the value 1 kg is *defined* as 2.20 lb; in other words, 1 kg contains the exact number "1" with *no* uncertainty, so it does not limit the number of digits in the answer. Since 130 lb has two significant figures, the answer is rounded to two significant figures (59 kg).

As problems with units get more complicated, keep in mind the following general steps that are useful for solving any problem using conversion factors.

How To Solve a Problem Using Conversion Factors

Example How many grams of aspirin are contained in a 325-mg tablet?

Step [1] **Identify the original quantity and the desired quantity, including units.**

- In this problem the original quantity is reported in milligrams and the desired quantity is in grams.

325 mg	? g
original quantity	desired quantity

Step [2] **Write out the conversion factor(s) needed to solve the problem.**

- We need a conversion factor that relates milligrams and grams (Table 1.3). Since the unwanted unit is in milligrams, **choose the conversion factor that contains milligrams in the denominator so that the *units cancel.***

Two possible conversion factors: $\frac{1000\ \text{mg}}{1\ \text{g}}$ or $\frac{1\ \text{g}}{1000\ \text{mg}}$ ← Choose this factor to cancel the unwanted unit, mg.

- Sometimes one conversion factor is all that is needed in a problem. At other times (Section 1.7C) more than one conversion factor is needed.
- If the desired answer has a single unit (grams in this case), **the conversion factor must contain the desired unit in the numerator and the unwanted unit in the denominator.**

Step [3] **Set up and solve the problem.**

- Multiply the original quantity by the conversion factor to obtain the desired quantity.

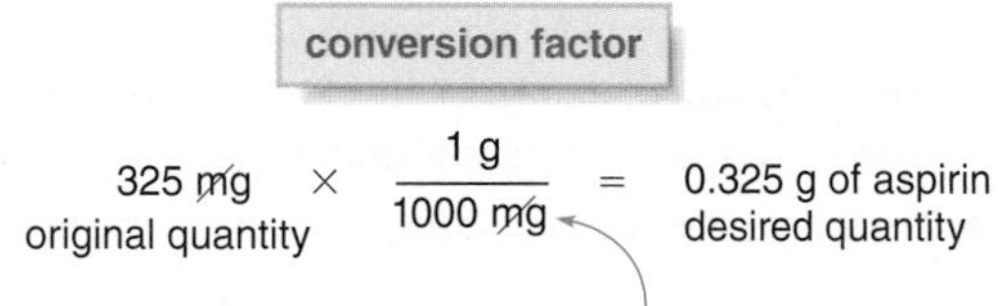

The number of mg (unwanted unit) cancels.

Step [4] **Write the answer using the correct number of significant figures and check it by estimation.**

- Use the number of significant figures in each inexact (measured) number to determine the number of significant figures in the answer. In this case the answer is limited to three significant figures by the original quantity (325 mg).
- Estimate the answer using a variety of methods. In this case we knew our answer had to be less than one, since it is obtained by dividing 325 by a number larger than itself.

PROBLEM 1.20

The distance between Honolulu, HI, and Los Angeles, CA, is 4,120 km. How many frequent flyer miles will you earn by traveling between the two cities?

PROBLEM 1.21

Carry out each of the following conversions.

a. 25 L to dL b. 40.0 oz to g c. 32 in. to cm d. 10 cm to mm

1.7C Solving a Problem Using Two or More Conversion Factors

Some problems require the use of more than one conversion factor to obtain the desired units in the answer. The same four-step procedure is followed no matter how many conversion factors are needed. Keep in mind:

- **Always arrange the factors so that the denominator in one term cancels the numerator in the preceding term.**

Sample Problem 1.9 illustrates how to solve a problem with two conversion factors.

SAMPLE PROBLEM 1.9

An individual donated 1.0 pt of blood at the local blood bank. How many liters of blood does this correspond to?

Analysis and Solution

[1] **Identify the original quantity and the desired quantity.**

1.0 pt	? L
original quantity	desired quantity

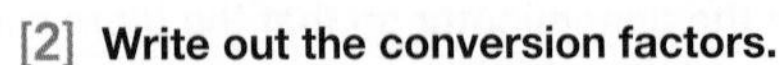

[2] **Write out the conversion factors.**

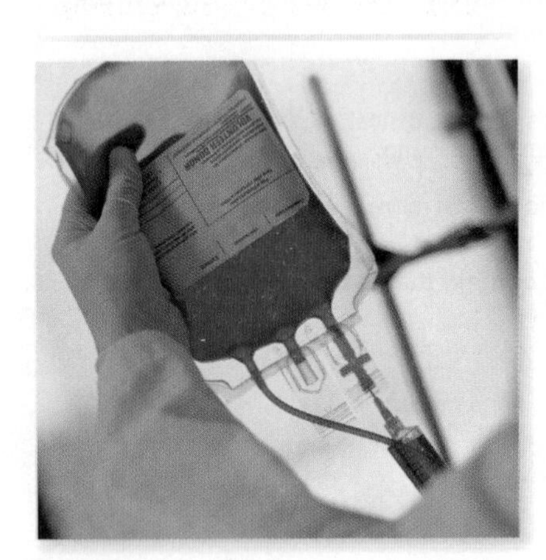

How many liters does this pint of blood contain?

- We have no conversion factor that relates pints to liters directly. We do, however, know conversions for pints to quarts, and quarts to liters.

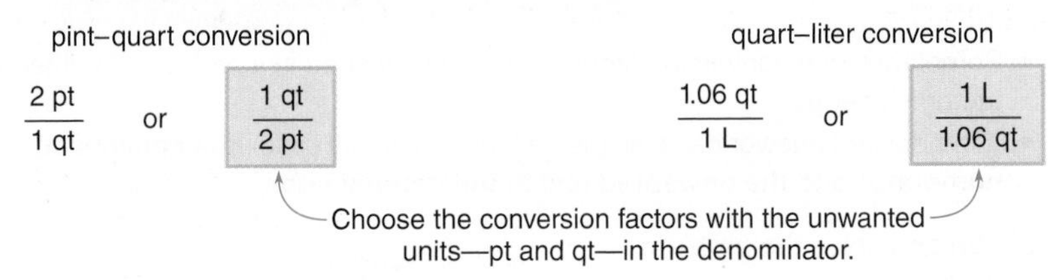

pint–quart conversion
2 pt / 1 qt or 1 qt / 2 pt
quart–liter conversion
1.06 qt / 1 L or 1 L / 1.06 qt
Choose the conversion factors with the unwanted units—pt and qt—in the denominator.

[3] **Solve the problem.**

- To set up the problem so that unwanted units cancel, arrange each term so that the units in the numerator of one term cancel the units of the denominator of the adjacent term. In this problem we need to cancel both pints and quarts to get liters.
- The single desired unit, liters, must be in the **numerator** of one term.

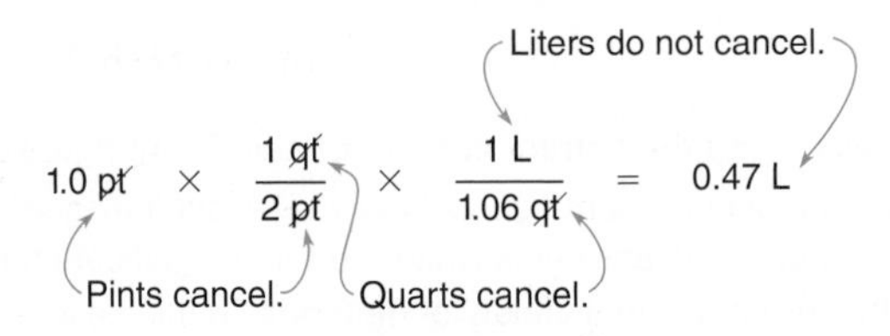

Liters do not cancel.
$1.0 \text{ pt} \times \frac{1 \text{ qt}}{2 \text{ pt}} \times \frac{1 \text{ L}}{1.06 \text{ qt}} = 0.47 \text{ L}$
Pints cancel.
Quarts cancel.

[4] **Check.**

- Since there are two pints in a quart and a quart is about the same size as a liter, one pint should be about half a liter. The answer, 0.47, is just about 0.5.
- Write the answer with two significant figures since one term, 1.0 pt, has two significant figures.

PROBLEM 1.22

Carry out each of the following conversions.

a. 6,250 ft to km b. 3 cups to L c. 4.5 ft to cm

1.8 FOCUS ON HEALTH & MEDICINE
Problem Solving Using Clinical Conversion Factors

Sometimes conversion factors don't have to be looked up in a table; they are stated in the problem. If a drug is sold as a 250-mg tablet, this fact becomes a conversion factor relating milligrams to tablets.

$$\frac{250 \text{ mg}}{1 \text{ tablet}} \quad \text{or} \quad \frac{1 \text{ tablet}}{250 \text{ mg}}$$

mg–tablet conversion factors

The active ingredient in Children's Tylenol is acetaminophen.

Alternatively, a drug could be sold as a liquid solution with a specific concentration. For example, Children's Tylenol contains 80 mg of the active ingredient acetaminophen in 2.5 mL. This fact becomes a conversion factor relating milligrams to milliliters.

$$\frac{80 \text{ mg}}{2.5 \text{ mL}} \quad \text{or} \quad \frac{2.5 \text{ mL}}{80 \text{ mg}}$$

mg of acetaminophen–mL conversion factors

Sample Problems 1.10 and 1.11 illustrate how these conversion factors are used in determining drug dosages.

SAMPLE PROBLEM 1.10

A patient is prescribed 1.25 g of amoxicillin, which is available in 250-mg tablets. How many tablets are needed?

Analysis and Solution

[1] **Identify the original quantity and the desired quantity.**

- We must convert the number of grams of amoxicillin needed to the number of tablets that must be administered.

1.25 g	? tablets
original quantity	desired quantity

[2] **Write out the conversion factors.**

- We have no conversion factor that relates grams to tablets directly. We do know, however, how to relate grams to milligrams, and milligrams to tablets.

g–mg conversion factors

$$\frac{1 \text{ g}}{1000 \text{ mg}} \quad \text{or} \quad \boxed{\frac{1000 \text{ mg}}{1 \text{ g}}}$$

mg–tablet conversion factors

$$\frac{250 \text{ mg}}{1 \text{ tablet}} \quad \text{or} \quad \boxed{\frac{1 \text{ tablet}}{250 \text{ mg}}}$$

Choose the conversion factors with the unwanted units–g and mg–in the denominator.

[3] **Solve the problem.**

- Arrange each term so that the units in the numerator of one term cancel the units in the denominator of the adjacent term. In this problem we need to cancel both grams and milligrams to get tablets.
- The single desired unit, tablets, must be located in the **numerator** of one term.

$$1.25 \cancel{\text{g}} \times \frac{1000 \cancel{\text{mg}}}{1 \cancel{\text{g}}} \times \frac{1 \text{ tablet}}{250 \cancel{\text{mg}}} = 5 \text{ tablets}$$

Tablets do not cancel.

Grams cancel.

Milligrams cancel.

[4] **Check.**

- The answer of 5 tablets of amoxicillin (not 0.5 or 50) is reasonable. Since the dose in a single tablet (250 mg) is a fraction of a gram, and the required dose is more than a gram, the answer must be greater than one.

SAMPLE PROBLEM 1.11

A dose of 240 mg of acetaminophen is prescribed for a 20-kg child. How many mL of Children's Tylenol (80. mg of acetaminophen per 2.5 mL) are needed?

Analysis and Solution

[1] Identify the original quantity and the desired quantity.

- We must convert the number of milligrams of acetaminophen needed to the number of mL that must be administered.

240 mg	? mL
original quantity	desired quantity

[2] Write out the conversion factors.

mg of acetaminophen–mL conversion factors

$$\frac{80.\ \text{mg}}{2.5\ \text{mL}} \quad \text{or} \quad \boxed{\frac{2.5\ \text{mL}}{80.\ \text{mg}}}$$

Choose the conversion factor to cancel mg.

[3] Solve the problem.

- Arrange the terms so that the units in the numerator of one term cancel the units of the denominator of the adjacent term. In this problem we need to cancel milligrams to obtain milliliters.
- In this problem we are given a fact we don't need to use—the child weighs 20 kg. We can ignore this quantity in carrying out the calculation.

$$240\ \cancel{\text{mg}} \times \frac{2.5\ \text{mL}}{80.\ \cancel{\text{mg}}} = 7.5\ \text{mL of Children's Tylenol}$$

Milligrams cancel.

[4] Check.

- The answer of 7.5 mL (not 0.75 or 75) is reasonable. Since the required dose is larger than the dose in 2.5 mL, the answer must be larger than 2.5 mL.

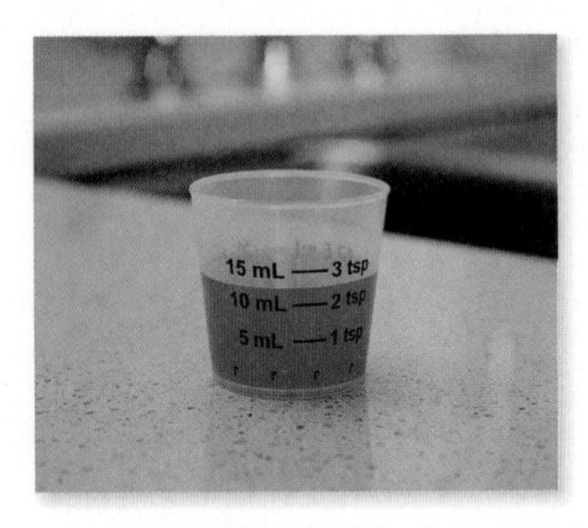

PROBLEM 1.23

(a) How many milliliters are contained in the dose of Children's Tylenol shown in the adjacent photo (1 teaspoon = 5 mL)? (b) If Children's Tylenol contains 80. mg of acetaminophen per 2.5 mL, how much acetaminophen (in mg) is contained in the dose?

PROBLEM 1.24

A patient is prescribed 0.100 mg of a drug that is available in 25-μg tablets. How many tablets are needed?

PROBLEM 1.25

How many milliliters of Children's Motrin (100 mg of ibuprofen per 5 mL) are needed to give a child a dose of 160 mg?

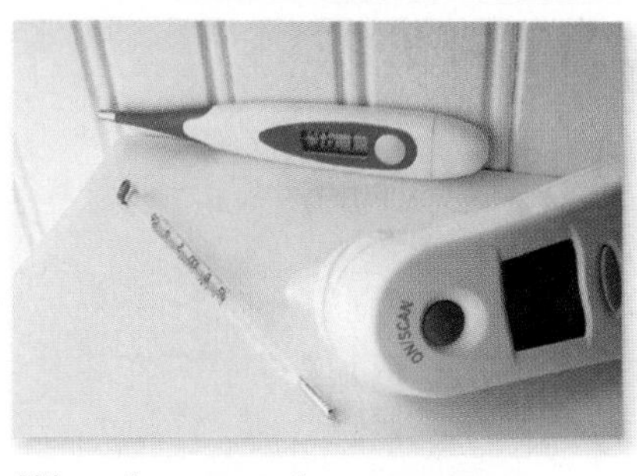

Although mercury thermometers were used in hospitals to measure temperature for many years, temperature is now more commonly recorded with a digital thermometer. Tympanic thermometers, which use an infrared sensing device placed in the ear, are also routinely used.

1.9 Temperature

Temperature is a measure of how hot or cold an object is. Three temperature scales are used: **Fahrenheit** (most common in the United States), **Celsius** (most commonly used by scientists and countries other than the United States), and **Kelvin** (Figure 1.8).

The Fahrenheit and Celsius scales are both divided into **degrees.** On the Fahrenheit scale, water freezes at 32 °F and boils at 212 °F. On the Celsius scale, water freezes at 0 °C and boils at 100 °C. To convert temperature values from one scale to another, we use two equations, where T_C is the Celsius temperature and T_F is the Fahrenheit temperature.

To convert from Celsius to Fahrenheit:

$$T_F = 1.8(T_C) + 32$$

To convert from Fahrenheit to Celsius:

$$T_C = \frac{T_F - 32}{1.8}$$

The Kelvin scale is divided into **kelvins** (K), not degrees. The only difference between the Kelvin scale and the Celsius scale is the zero point. A temperature of –273 °C corresponds to 0 K. The zero point on the Kelvin scale is called **absolute zero,** the lowest temperature possible. To convert temperature values from Celsius to Kelvin, or vice versa, use two equations.

To convert from Celsius to Kelvin:

$$T_K = T_C + 273$$

To convert from Kelvin to Celsius:

$$T_C = T_K - 273$$

Figure 1.8 Fahrenheit, Celsius, and Kelvin Temperature Scales Compared

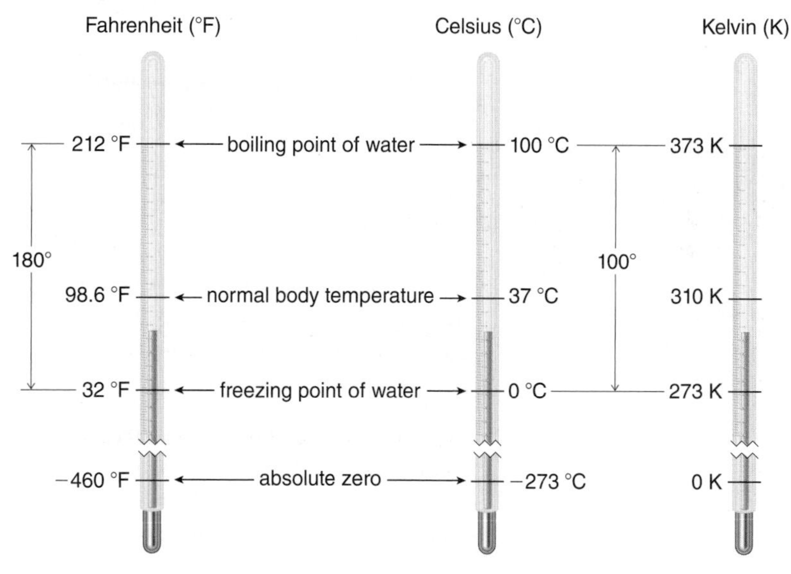

Since the freezing point and boiling point of water span 180° on the Fahrenheit scale, but only 100° on the Celsius scale, a Fahrenheit degree and a Celsius degree differ in size. The Kelvin scale is divided into kelvins (K), not degrees. Since the freezing point and boiling point of water span 100 kelvins, one kelvin is the same size as one Celsius degree.

SAMPLE PROBLEM 1.12

An infant had a temperature of 104 °F. Convert this temperature to both °C and K.

Analysis

First convert the Fahrenheit temperature to degrees Celsius using the equation $T_C = (T_F - 32)/1.8$. Then convert the Celsius temperature to kelvins by adding 273.

Solution

[1] **Convert T_F to T_C:**

$$T_C = \frac{T_F - 32}{1.8}$$

$$= \frac{104 - 32}{1.8} = 40.\ °C$$

[2] **Convert T_C to T_K:**

$$T_K = T_C + 273$$

$$= 40. + 273 = 313\ K$$

PROBLEM 1.26

When the human body is exposed to extreme cold, hypothermia can result and the body's temperature can drop to 28.5 °C. Convert this temperature to T_F and T_K.

PROBLEM 1.27

Convert each temperature to the requested temperature scale.

a. 20. °C to T_F b. 150 °F to T_C c. 75 °C to T_K

1.10 Density and Specific Gravity

Two additional quantities used to characterize substances are **density** and **specific gravity.**

1.10A Density

Density **is a physical property that relates the mass of a substance to its volume.** Density is reported in grams per milliliter (g/mL) or grams per cubic centimeter (g/cc).

$$\textbf{density} = \frac{\textbf{mass (g)}}{\textbf{volume (mL or cc)}}$$

The density of a substance depends on temperature. For most substances, the solid state is more dense than the liquid state, and as the temperature increases, the density decreases. This phenomenon occurs because the volume of a sample of a substance generally increases with temperature but the mass is always constant.

Water is an exception to this generalization. Solid water, ice, is *less* dense than liquid water, and from 0 °C to 4 °C, the density of water *increases*. Above 4 °C, water behaves like other liquids and its density decreases. Thus, water's maximum density of 1.00 g/mL occurs at 4 °C. Some representative densities are reported in Table 1.6.

Table 1.6 Representative Densities at 25 °C

Substance	Density [g/(mL or cc)]	Substance	Density [g/(mL or cc)]
Oxygen (0 °C)	0.001 43	Urine	1.003–1.030
Gasoline	0.66	Blood plasma	1.03
Ice (0 °C)	0.92	Table sugar	1.59
Water (4 °C)	1.00	Bone	1.80

Although a can of a diet soft drink floats in water because it is less dense, a can of a regular soft drink that contains sugar is more dense than water so it sinks.

The density (not the mass) of a substance determines whether it floats or sinks in a liquid.

- **A less dense substance floats on a more dense liquid.**

Ice floats on water because it is less dense. When petroleum leaks from an oil tanker or gasoline is spilled when fueling a boat, it floats on water because it is less dense. In contrast, a cannonball or torpedo sinks because it is more dense than water.

SAMPLE PROBLEM 1.13

The density of liquid **A** is twice the density of liquid **B.** (a) If you have an equal mass of **A** and **B,** which graduated cylinder ([1] or [2]) corresponds to **A** and which corresponds to **B?** (b) How do the masses of the liquids in graduated cylinders [2] and [3] compare?

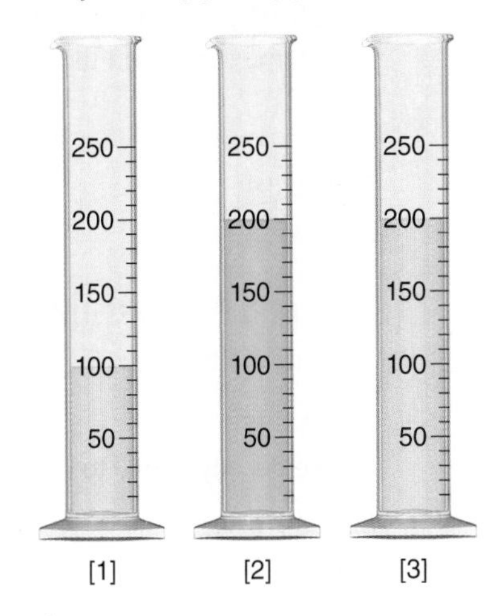

Analysis

Density is the number of grams per milliliter (g/mL) or grams per cubic centimeter (g/cc) of a substance.

Solution

a. If the density of **A** is twice the density of **B,** you need twice the volume of **B** to have the same mass as a sample of **A.** Thus, graduated cylinder [1] represents **A** (gold liquid) and graduated cylinder [2] represents **B** (green liquid).

b. Since graduated cylinders [2] and [3] have equal volumes of **A** and **B** but **A** is twice as dense as **B,** the mass of [3] (**A**) must be twice the mass of [2] (**B**).

PROBLEM 1.28

How does the mass of liquid **A** in cylinder [1] compare with the mass of liquid **B** in cylinder [2] in each case (greater than, less than, or equal to)? (a) The densities of **A** and **B** are the same. (b) The density of **A** is twice the density of **B.** (c) The density of **B** is twice the density of **A.**

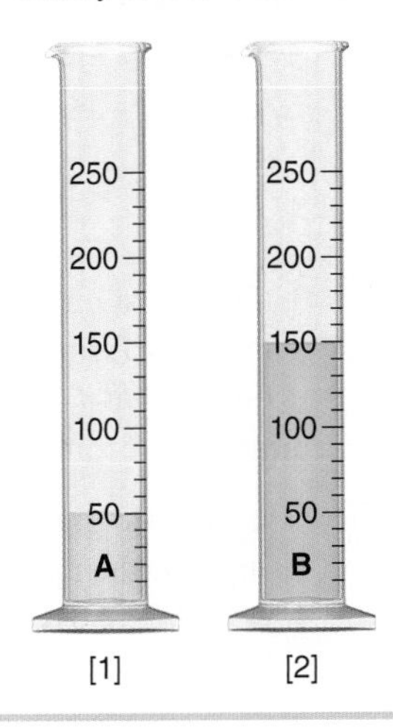

Knowing the density of a liquid allows us to convert the volume of a substance to its mass, or the mass of a substance to its volume.

To convert volume (mL) to mass (g):

$$\text{mL} \times \frac{\text{g}}{\text{mL}} = \text{g}$$

density — Milliliters cancel.

To convert mass (g) to volume (mL):

$$\text{g} \times \frac{\text{mL}}{\text{g}} = \text{mL}$$

inverse of the density — Grams cancel.

For example, one laboratory synthesis of aspirin uses the liquid acetic acid, which has a density of 1.05 g/mL. If we need 5.0 g for a synthesis, we could use density to convert this mass to a volume that could then be easily measured out using a syringe or pipette.

$$5.0 \text{ g acetic acid} \times \frac{1 \text{ mL}}{1.05 \text{ g}} = 4.8 \text{ mL of acetic acid}$$

Grams cancel.

SAMPLE PROBLEM 1.14

Calculate the mass in grams of 15.0 mL of a saline solution that has a density of 1.05 g/mL.

Analysis

Use density (g/mL) to interconvert the mass and volume of a liquid.

Solution

$$15.0 \text{ mL} \times \frac{1.05 \text{ g}}{1 \text{ mL}} = 15.8 \text{ g of saline solution}$$

density — Milliliters cancel.

The answer, 15.8 g, is rounded to three significant figures to match the number of significant figures in both factors in the problem.

PROBLEM 1.29

Calculate the mass in grams of 10.0 mL of diethyl ether, an anesthetic that has a density of 0.713 g/mL.

HEALTH NOTE

The specific gravity of a urine sample is measured to check if a patient has an imbalance in metabolism.

1.10B Specific Gravity

***Specific gravity* is a quantity that compares the density of a substance with the density of water at 4 °C.**

$$\textbf{specific gravity} = \frac{\textbf{density of a substance (g/mL)}}{\textbf{density of water (g/mL)}}$$

Unlike most other quantities, specific gravity is a quantity without units, since the units in the numerator (g/mL) cancel the units in the denominator (g/mL). Since the density of water is 1.00 g/mL at 4 °C, **the specific gravity of a substance equals its density, but it contains no units.** For example, if the density of a liquid is 1.5 g/mL at 4 °C, its specific gravity is 1.5.

The specific gravity of urine samples is often measured in a hospital lab. Normal urine has a density in the range of 1.003–1.030 g/mL (Table 1.6), so it has a specific gravity in the range of 1.003–1.030. Consistently high or low values can indicate an imbalance in metabolism. For example, the specific gravity of urine samples from patients with poorly controlled diabetes is abnormally high, because a large amount of glucose is excreted in the urine.

PROBLEM 1.30

(a) If the density of a liquid is 0.80 g/mL, what is its specific gravity? (b) If the specific gravity of a substance is 2.3, what is its density?

STUDY SKILLS PART I: CALCULATIONS IN CHEMISTRY

Many problems in Chapters 1–9 use mathematics in calculations. These problems often take one of two forms: using conversion factors to convert a value from one unit to another, or using a specific equation to find a missing value.

Most of the calculations in Chapters 1 and 5 utilize the problem-solving method described in the *How To* in Section 1.7. This method is used in any calculation with conversion factors to convert a quantity from one unit to another. To apply this method requires three steps: [1] identify the units of the original quantity and the desired quantity; [2] write out all needed conversion factors; and [3] arrange the conversion factors so that all unwanted units cancel to solve the problem. This strategy is the method needed to convert a measurement from one unit to another (Sample Problem 1.9), determine dosages for certain medications (Sample Problem 1.11), or convert grams to moles in Section 5.4. **The same three steps are used in every calculation,** making this procedure the single most valuable tool for solving a wide variety of calculations. Always determine what you are starting with and where you must end up; the conversion factors are the means to get there.

Other calculations require remembering a specific equation. Examples include converting temperature values from one scale to another (Section 1.9) and solving problems using the gas laws in Chapter 6. Always write down the equation, fill in the known quantities, and then solve the equation for the unknown. Keep in mind: **Only one quantity can be missing to solve the equation.** If there is more than one unknown, you are using the *wrong* equation.

A scientific calculator can be used in many of the calculations in Chapters 1–9. How to use a scientific calculator in a variety of mathematical operations is illustrated in the Appendix.

KEY TERMS

Celsius scale (1.9)
Chemical properties (1.2)
Chemistry (1.1)
Compound (1.3)
Conversion factor (1.7)
Cubic centimeter (1.4)
Density (1.10)
Element (1.3)
English system of measurement (1.4)
Exact number (1.5)
Fahrenheit scale (1.9)
Gas (1.2)
Gram (1.4)
Inexact number (1.5)
Kelvin scale (1.9)
Liquid (1.2)
Liter (1.4)
Mass (1.4)
Matter (1.1)
Meter (1.4)
Metric system (1.4)
Mixture (1.3)
Physical properties (1.2)
Pure substance (1.3)
Scientific notation (1.6)
SI units (1.4)
Significant figures (1.5)
Solid (1.2)
Specific gravity (1.10)
States of matter (1.2)
Temperature (1.9)
Weight (1.4)

KEY CONCEPTS

1 Describe the three states of matter. (1.1, 1.2)

- Matter is anything that has mass and takes up volume. Matter has three common states:
  - The solid state is composed of highly organized particles that lie close together. A solid has a definite shape and volume.
  - The liquid state is composed of particles that lie close together but are less organized than the solid state. A liquid has a definite volume but not a definite shape.
  - The gas state is composed of highly disorganized particles that lie far apart. A gas has no definite shape or volume.

2 How is matter classified? (1.3)

- Matter is classified in one of two categories:
  - A pure substance is composed of a single component with a constant composition. A pure substance is either an element, which cannot be broken down into simpler substances by a chemical reaction, or a compound, which is formed by combining two or more elements.
  - A mixture is composed of more than one substance and its composition can vary depending on the sample.

3 What are the key features of the metric system of measurement? (1.4)

- The metric system is a system of measurement in which each type of measurement has a base unit and all other units are related to the base unit by a prefix that indicates if the unit is larger or smaller than the base unit.
- The base units are meter (m) for length, gram (g) for mass, liter (L) for volume, and second (s) for time.

4 What are significant figures and how are they used in calculations? (1.5)

- Significant figures are all digits in a measured number, including one estimated digit. All nonzero digits are significant. A zero is significant only if it occurs between two nonzero digits, or at the end of a number with a decimal point. A trailing zero in a number without a decimal point is not considered significant.
- In multiplying and dividing with significant figures, the answer has the same number of significant figures as the original number with the fewest significant figures.
- In adding or subtracting with significant figures, the answer has the same number of decimal places as the original number with the fewest decimal places.

5 What is scientific notation? (1.6)

- Scientific notation is a method of writing a number as $y \times 10^x$, where y is a number between 1 and 10, and x is a positive or negative exponent.
- To convert a standard number to a number in scientific notation, move the decimal point to give a number between 1 and 10. Multiply the result by 10^x, where x is the number of places the decimal point was moved. When the decimal point is moved to the left, x is positive. When the decimal point is moved to the right, x is negative.

6 How are conversion factors used to convert one unit to another? (1.7, 1.8)

- A conversion factor is a term that converts a quantity in one unit to a quantity in another unit. To use conversion factors to solve a problem, set up the problem with any unwanted unit in the numerator of one term and the denominator of another term, so that unwanted units cancel.

7 What is temperature and how are the three temperature scales related? (1.9)

- Temperature is a measure of how hot or cold an object is. The Fahrenheit and Celsius temperature scales are divided into degrees. Both the size of the degree and the zero point of these scales differ. The Kelvin scale is divided into kelvins, and one kelvin is the same size as one degree Celsius.

8 What are density and specific gravity? (1.10)

- Density is a physical property reported in g/mL or g/cc that relates the mass of an object to its volume. A less dense substance floats on top of a more dense liquid.
- Specific gravity is a unitless quantity that relates the density of a substance to the density of water at 4 °C. Since the density of water is 1.00 g/mL at 4 °C, the specific gravity of a substance equals its density, but it contains no units.

UNDERSTANDING KEY CONCEPTS

Selected in-chapter and odd-numbered end-of-chapter problems have brief answers at the end of each chapter. The *Student Study Guide and Solutions Manual* contains detailed solutions to all in-chapter and odd-numbered end-of-chapter problems, as well as additional worked examples and a chapter self-test.

1.31 Classify each example of molecular art as a pure element, a pure compound, or a mixture.

a.

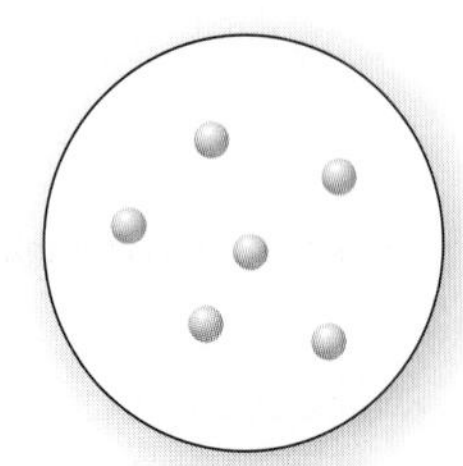

c.

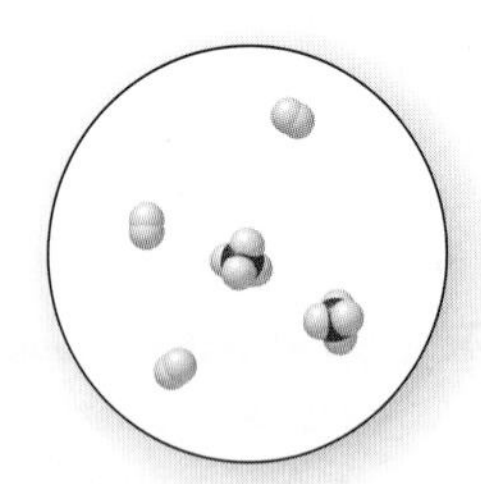

b.

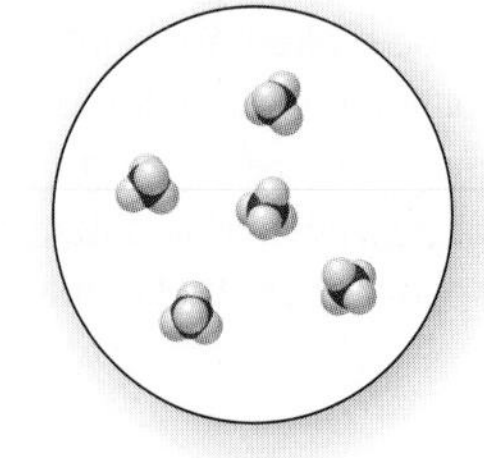

d.

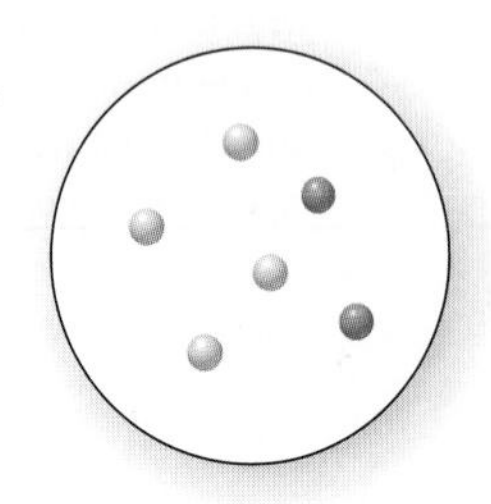

1.32 (a) Which representation(s) in Problem 1.31 illustrate a mixture of two elements? (b) Which representation(s) in Problem 1.31 illustrate a mixture of a compound and an element?

1.33 When a chunk of dry ice (solid carbon dioxide) is placed out in the air, the solid gradually disappears and a gas is formed above the solid. Does the molecular art drawn below indicate that a chemical or physical change has occurred? Explain your choice.

1.34 The inexpensive preparation of nitrogen-containing fertilizers begins with mixing together two elements, hydrogen and nitrogen, at high temperature and pressure in the presence of a metal. Does the molecular art depicted below indicate that a chemical or physical change occurs under these conditions? Explain your choice.

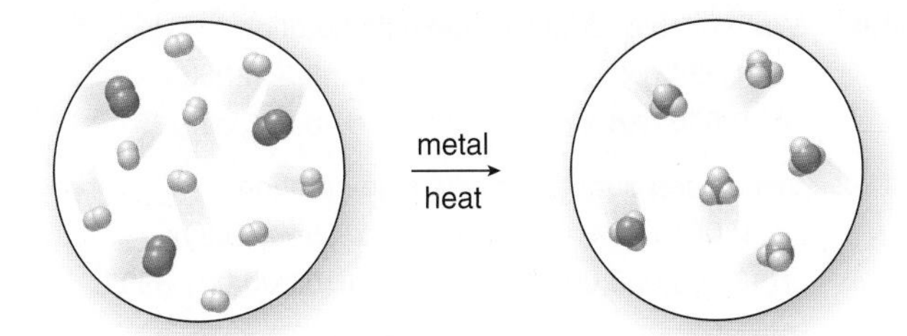

1.35

80

70

a. What is the temperature on the given Fahrenheit thermometer?

b. How many significant figures does your answer contain?

c. Convert this temperature into T_C.

1.36 (a) What is the length of the given crayon in centimeters? (b) How many significant figures does this value contain? (c) Convert this value to meters, and write the answer in scientific notation.

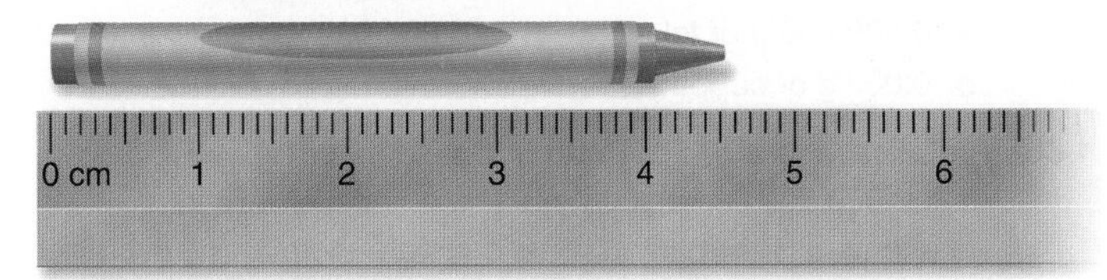

1.37 The given beaker contains 100 mL of water. Draw an illustration for what would be observed in each circumstance. (a) Hexane (50 mL, density = 0.65 g/mL) is added. (b) Dichloromethane (50 mL, density = 1.33 g/mL) is added.

1.38 (a) What can be said about the density of the liquid in beaker **A,** if the object in the beaker has a density of 2.0 g/cc? (b) What can be said about the density of the liquid in beaker **B,** if the object has a density of 0.90 g/cc?

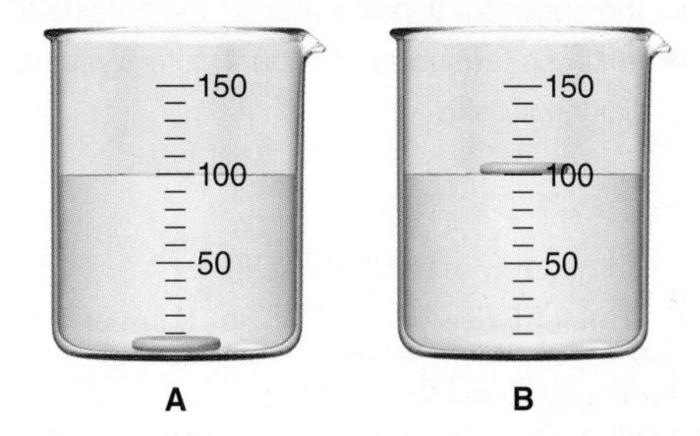

1.39 A blood vessel is 0.40 μm in diameter. (a) Convert this quantity to meters and write the answer in scientific notation. (b) Convert this quantity to inches and write the answer in scientific notation.

1.40 Red light has a wavelength of 683 nm. Convert this quantity to meters and write the answer in scientific notation.

1.41 What is the height in centimeters of a child who is 50. in. tall?

1.42 If the human body has 5.0 qt of blood that contains 2×10^{13} red blood cells, how many red blood cells are present in each μL of blood?

1.43 A woman was told to take a dose of 1.5 g of calcium daily. How many 500-mg tablets should she take?

1.44 The recommended daily calcium intake for a woman over 50 years of age is 1,200 mg. If one cup of milk has 306 mg of calcium, how many cups of milk provide this amount of calcium? (b) How many milliliters of milk does this correspond to?

ADDITIONAL PROBLEMS

Matter

1.45 Label each component in the molecular art as an element or a compound.

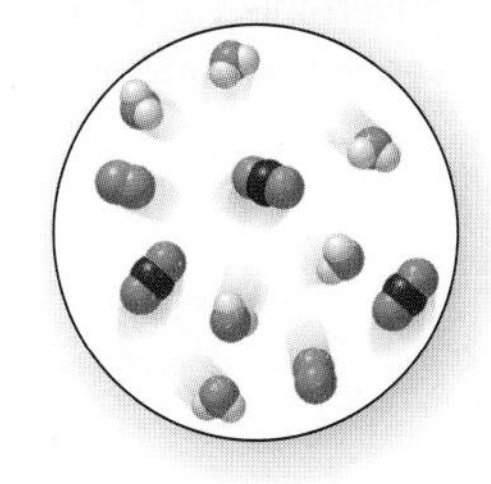

1.46 Label each component in the molecular art as an element or compound.

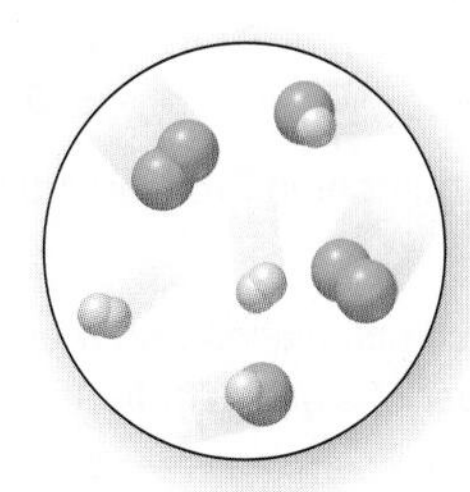

1.47 Describe solids, liquids, and gases in terms of (a) volume (how they fill a closed container); (b) shape; (c) level of organization of the particles that comprise them; (d) how close the particles that comprise them lie.

1.48 What is the difference between a compound and a mixture?

1.49 Classify each process as a chemical or physical change.
a. dissolving calcium chloride in water
b. burning gasoline to power a car
c. heating wax so that it melts

1.50 Classify each process as a chemical or physical change.
a. the condensation of water on the outside of a cold glass
b. mixing a teaspoon of instant coffee with hot water
c. baking a cake

Measurement

1.51 Which quantity in each pair is larger?
a. 5 mL or 5 dL
b. 10 mg or 10 μg
c. 5 cm or 5 mm
d. 10 Ms or 10 ms

1.52 Which quantity in each pair is larger?
a. 10 km or 10 m
b. 10 L or 10 mL
c. 10 g or 10 μg
d. 10 cm or 10 mm

1.53 Label each quantity as an exact or inexact number.
a. A recipe requires 10 cloves of garlic and two tablespoons of oil.
b. The four bicycles in the family have been ridden for a total of 250 mi.
c. A child fell and had a 4-cm laceration that required 12 stitches.

1.54 Rank the quantities in each group from smallest to largest.
a. 100 μL, 100 dL, and 100 mL
b. 10 g, 100 mg, and 0.1 kg
c. 1 km, 100 m, and 1,000 cm

Significant Figures

1.55 How many significant figures does each number contain?
a. 16.00
b. 160
c. 0.001 60
d. 1,600,000
e. 0.1600
f. 1.060×10^{10}

1.56 How many significant figures does each number contain?
a. 160.
b. 160.0
c. 0.000 16
d. 1,600.
e. 1.060
f. 1.600×10^{-10}

1.57 Round each number to three significant figures.
a. 25,401
b. 1,248,486
c. 0.001 265 982
d. 0.123 456

1.58 Round each number in Problem 1.57 to four significant figures.

1.59 Carry out each calculation and report the answer using the proper number of significant figures.
a. 53.6×0.41
b. $25.825 - 3.86$
c. $65.2 \div 12$
d. $41.0 + 9.135$

1.60 Carry out each calculation and report the answer using the proper number of significant figures.
a. $49{,}682 \times 0.80$
b. $66.815 + 2.82$
c. $1{,}000 \div 2.34$
d. $21 - 0.88$

Scientific Notation

1.61 Write each quantity in scientific notation.
a. 1,234 g
b. 0.000 016 2 m
c. 5,244,000 L
d. 0.005 62 g

1.62 Write each quantity in scientific notation.
a. 0.001 25 m
b. 8,100,000,000 lb
c. 54,235.6 m
d. 0.000 001 899 L

1.63 Convert each number to its standard form.
a. 3.4×10^{8}
b. 5.822×10^{-5}
c. 3×10^{2}
d. 6.86×10^{-8}

1.64 Convert each number to its standard form.
a. 4.02×10^{10}
b. 2.46×10^{-3}
c. 6.86×10^{9}
d. 1.00×10^{-7}

1.65 Which number in each pair is larger?
a. 4.44×10^{3} or 4.8×10^{2}
b. 5.6×10^{-6} or 5.6×10^{-5}
c. 1.3×10^{8} or 52,300,000
d. 9.8×10^{-4} or 0.000 089

1.66 Rank the numbers in each group from smallest to largest.
a. 5.06×10^{6}, 7×10^{4}, and 2.5×10^{8}
b. 6.3×10^{-2}, 2.5×10^{-4}, and 8.6×10^{-6}

1.67 Write the recommended daily intake of each nutrient in scientific notation.
a. 0.000 400 g of folate
b. 0.002 g of copper
c. 0.000 080 g of vitamin K
d. 3,400 mg of chloride

1.68 A picosecond is one trillionth of a second (0.000 000 000 001 s). (a) Write this number in scientific notation. (b) How many picoseconds are there in one second? Write this answer in scientific notation.

Problem Solving Using Conversion Factors

1.69 Carry out each of the following conversions.
a. 300 g to mg
b. 2 L to μL
c. 5.0 cm to m
d. 2 ft to m

1.70 Carry out each of the following conversions.
a. 25 μL to mL
b. 35 kg to g
c. 300 mL to qt
d. 3 cups to L

1.71

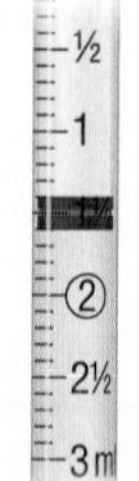

a. What is the volume of liquid contained in the given 3-mL syringe?
b. Convert this value to liters and write the answer in scientific notation.

1.72 a. What is the volume of liquid contained in the given 0.5-mL syringe?

b. Convert this value to microliters and write the answer in scientific notation.

1.73 Carry out each of the following conversions.

a. What is the mass in kilograms of an individual who weighs 234 lb?

b. A patient required 3.0 pt of blood during surgery. How many liters does this correspond to?

c. A patient had a body temperature of 37.7 °C. What is his body temperature in T_F?

1.74 Carry out each of the following conversions.

a. What is the mass in pounds of an individual who weighs 53.2 kg?

b. What is the height in inches of a child who is 90. cm tall?

c. A patient had a body temperature of 103.5 °F. What is his body temperature in T_C?

1.75 The average mass of a human liver is 1.5 kg. Convert this quantity to (a) grams; (b) pounds; (c) ounces.

1.76 The official distance of a marathon is 26 miles and 385 yards. Convert this value to kilometers.

Temperature

1.77 Carry out each of the following temperature conversions.

a. An over-the-counter pain reliever melts at 53 °C. Convert this temperature to T_F and T_K.

b. A cake is baked at 350. °F. Convert this temperature to T_C and T_K.

1.78 Which temperature in each pair is higher?

a. −10 °C or 10 °F   b. −50 °C or −50 °F

Density and Specific Gravity

1.79 If a urine sample has a mass of 122 g and a volume of 121 mL, what is its density in g/mL?

1.80 The density of sucrose, table sugar, is 1.56 g/cc. What volume (in cubic centimeters) does 20.0 g of sucrose occupy?

1.81 Isooctane is a high-octane component of gasoline. If the density of isooctane is 0.692 g/mL, what is the mass of 220 mL of gasoline?

1.82 A volume of saline solution has a mass of 25.6 g at 4 °C. An equal volume of water at the same temperature has a mass of 24.5 g. What is the density of the saline solution?

1.83 Which is the upper layer when each of the following liquids is added to water?

a. olive oil (density = 0.92 g/mL)

b. chloroform (density = 1.49 g/mL)

1.84 (a) What is the specific gravity of mercury, the liquid used in thermometers, if it has a density of 13.6 g/mL? (b) What is the density of ethanol if it has a specific gravity of 0.789?

Applications

1.85 A lab test showed an individual's cholesterol level to be 186 mg/dL. (a) Convert this quantity to g/dL. (b) Convert this quantity to mg/L.

1.86 Hemoglobin is a protein that transports oxygen from the lungs to the rest of the body. Lab results indicated a patient had a hemoglobin concentration in the blood of 15.5 g/dL, which is in the normal range. (a) Convert the number of grams to milligrams and write the answer in scientific notation. (b) Convert the number of grams to micrograms and write the answer in scientific notation.

1.87 Liposuction is a cosmetic procedure used to remove excess fat from the abdomen, thigh, or buttocks of a patient. (a) If 2.0 L of fat (density = 0.94 g/mL) is removed, what is the mass (in kg) of this fat? (b) How many pounds of fat have been removed?

1.88 A single 1-oz serving of tortilla chips contains 250 mg of sodium. If an individual ate the entire 13-oz bag, how many grams of sodium would he ingest? If the recommended daily intake of sodium is 2.4 g, does this provide more or less than the recommended daily value, and by how much?

1.89 A bottle of liquid medication contains 300 mL and costs $10.00. (a) If the usual dose is 20. mL, how much does each dose cost? (b) If the usual dose is two tablespoons (1 tablespoon = 15 mL), how much does each dose cost?

1.90 The average nicotine content of a Camel cigarette is 1.93 mg. (a) Convert this quantity to both grams and micrograms. (b) Nicotine patches, which are used to help quit smoking, release nicotine into the body by absorption through the skin. The patches come with different amounts of nicotine. A smoker begins with the amount of nicotine that matches his typical daily intake. The maximum amount of nicotine in one brand of patch supplies a smoker with 21 mg of nicotine per day. If an individual smoked one pack of 20 Camel cigarettes each day, would a smoker get more or less nicotine per day using this patch?

1.91 A chemist synthesized 0.510 kg of aspirin in the lab. If the normal dose of aspirin is two 325-mg tablets, how many doses did she prepare?

1.92 Maalox is the trade name for an antacid and antigas medication used for relief of heartburn, bloating, and acid indigestion. Each 5-mL portion of Maalox contains 400 mg of aluminum hydroxide, 400 mg of magnesium hydroxide, and 40 mg of simethicone. If the recommended dose is two teaspoons four times a day, how many grams of each substance would an individual take in a 24-hour period? (1 teaspoon = 5 mL)

CHALLENGE PROBLEMS

1.93 Children's Chewable Tylenol contains 80 mg of acetaminophen per tablet. If the recommended dosage is 10 mg/kg, how many tablets are needed for a 42-lb child?

1.94 Often the specific amount of a drug to be administered must be calculated from a given dose in mg per kilogram of body weight. This assures that individuals who have very different body mass get the proper dose. If the proper dosage of a drug is 2 mg/kg of body weight, how many milligrams would a 110-lb individual need?

1.95 Children's Liquid Motrin contains 100. mg of the pain reliever ibuprofen per 5 mL. If the dose for a 45-lb child is 1.5 teaspoons, how many grams of ibuprofen would the child receive? (1 teaspoon = 5 mL)

1.96 If a 180-lb patient is prescribed 20 mg of the cholesterol-lowering drug Lipitor daily, what dosage is the patient receiving in mg/kg of his body weight?

1.97 A soccer player weighed 70.7 kg before a match, drank 1.8 L of liquid (density 1.05 g/mL) during the match, and weighed 69.3 kg after the match. How many pounds of sweat did the soccer player lose?

1.98 A patient receives an intravenous (IV) solution that flows at the rate of 150 mL per hour. (a) How much fluid does the patient receive in 20. min? (b) How long does it take for the patient to receive 90. mL of fluid? (c) If the IV bag holds 600. mL of fluid, how many minutes does it take to empty the bag? (d) If the solution contains 90. mg of glucose per mL, how long will it take to give the patient 2.0 g of glucose?

BEYOND THE CLASSROOM

1.99 Examine the labels of several consumer products and list the amount of product each contains in both metric and English units. Examples might include a box of cereal, jar of peanut butter, carton of juice, can of tomatoes, etc. Is the product sold by volume or mass? What conversion factors are used to change one unit to another?

1.100 Research how specific gravity is used to monitor the fermentation of grain to alcohol during beer production. How does the specific gravity change as the fermentation forms alcohol?

1.101 A urinometer is a device for measuring the specific gravity of a urine sample. Values above or below the normal range (1.003–1.030) may result from a variety of conditions. Research how each of the following might affect the specific gravity of urine: (a) a high fever; (b) taking a diuretic (a drug that increases urine output); (c) diabetes; (d) dehydration after sustained exercise.

1.102 A typical soda contains 39 g of carbohydrates in a 12-oz can, while a diet soda with aspartame contains 180 mg of artificial sweetener. Estimate how many cans of regular or diet soft drinks your household consumes in a 30-day month. Determine how many grams of carbohydrates or milligrams of artificial sweetener this corresponds to. Compare your results with others in the class.

1.103 Despite the fact that humans vary widely in size and shape, many anatomical features of the human body are roughly constant from one individual to another. As an example, an adult's height is often about three times greater than the circumference of his or her head. Measure your height and head circumference in metric units, and compare your values with others in your class or household. How many significant figures do your measurements contain? Convert each metric unit to an English unit such as feet or inches. Are the measurements consistent with the height/circumference relationship mentioned earlier?

ANSWERS TO SELECTED PROBLEMS

1.1 gloves, mask, plastic syringe, stainless steel needle: synthetic; ice, blood: natural

1.3 This represents a chemical change because the "particles" on the left are different from the particles on the right. For example, on the left side there are particles consisting of only two red balls, while on the right there are none of these.

1.4 a. pure substance b. mixture

1.5 a,b,c: mixture d: pure substance

1.7 a. megaliter b. millisecond c. centigram d. deciliter

1.9 1,000,000

1.11 a. 4 b. 2 c. 3 d. 3 e. 4 f. 6

1.13 a. 1.3 b. 0.0025 c. 3,800

1.14 a. 37 b. 0.000 007 93 c. 32 d. 3,100

1.15 a. 28.0 cm b. 127.5 mL c. 30. mg d. 2.08 s

1.16 9.8×10^{-5} g/dL

1.17 a. 9.32×10^{4} c. 6.78×10^{6}
b. 7.25×10^{-4} d. 3.0×10^{-5}

1.18 a. 6,500 b. 0.000 032 6 c. 0.037 80 d. 104,000,000

1.19 a. 0.621 mi/1 km 1 km/0.621 mi
b. 1000 mm/1 m 1 m/1000 mm

1.21 a. 250 dL b. 1,140 g c. 81 cm d. 100 mm

1.22 a. 1.91 km b. 0.7 L c. 140 cm

1.23 a. 13 mL b. 420 mg

1.25 8 mL

1.26 83.3 °F or 302 K

1.27 a. 68 °F b. 66 °C c. 348 K

1.28 a. mass **B** > mass **A**
b. mass **B** > mass **A**
c. mass **B** > mass **A**

1.29 7.13 g

1.31 a. pure element b. pure compound c. mixture d. mixture

1.33 This is a physical change since the compound CO_2 is unchanged in this transition. The same "particles" exist at the beginning and end of the process.

1.35 a. 76.5 °F b. three c. 24.7 °C

1.37

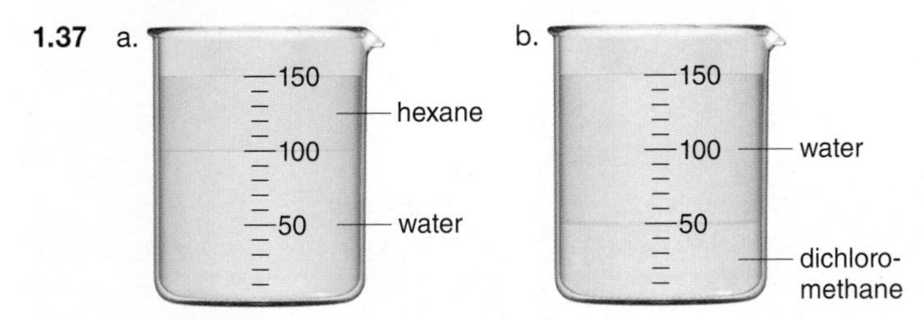

1.39 a. 4.0×10^{-7} m b. 1.6×10^{-5} in.

1.41 130 cm

1.43 three

1.45

compound
element
compound
element

1.47

Phase	a. Volume	b. Shape	c. Organization	d. Particle Proximity
Solid	Definite	Definite	Very organized	Very close
Liquid	Definite	Assumes shape of container	Less organized	Close
Gas	Not fixed, so assumes volume of container	Assumes shape of container	Disorganized	Far apart

1.49 a. physical b. chemical c. physical

1.51 a. 5 dL b. 10 mg c. 5 cm d. 10 Ms

1.53 a. 10 cloves: exact number; 2 tablespoons: inexact number
b. 4 bicycles: exact number; 250 mi: inexact number
c. 4 cm: inexact number; 12 stitches: exact number

1.55 a. 4 b. 2 c. 3 d. 2 e. 4 f. 4

1.57 a. 25,400 b. 1,250,000 c. 0.001 27 d. 0.123

1.59 a. 22 b. 21.97 c. 5.4 d. 50.1

1.61 a. 1.234×10^{3} g b. 1.62×10^{-5} m c. 5.244×10^{6} L d. 5.62×10^{-3} g

1.63 a. 340,000,000 b. 0.000 058 22 c. 300 d. 0.000 000 068 6

1.65 a. 4.44×10^{3} b. 5.6×10^{-5} c. 1.3×10^{8} d. 9.8×10^{-4}

1.67 a. 4.00×10^{-4} g b. 2×10^{-3} g c. 8.0×10^{-5} g d. 3.4×10^{3} mg

1.69 a. 300,000 mg b. 2,000,000 μL c. 0.050 m d. 0.6 m

1.71 a. 1.4 mL b. 1.4×10^{-3} L

1.73 a. 106 kg b. 1.4 L c. 99.9 °F

1.75 a. 1,500 g b. 3.3 lb c. 53 oz

1.77 a. 127 °F or 326 K b. 177 °C or 450 K

1.79 1.01 g/mL

1.81 152 g rounded to 150 g

1.83 a. olive oil b. water

1.85 a. 0.186 g/dL b. 1,860 mg/L

1.87 a. 1.9 kg b. 4.2 lb

1.89 a. $0.67 b. $1.00

1.91 784.6 doses, so 784 full doses

1.93 2 tablets (2.39)

1.95 0.150 g

1.97 7.3 lb

Because the element helium is lighter than air, balloons filled with helium must be secured with strings so they don't float away.

Atoms and the Periodic Table

CHAPTER OUTLINE

CHAPTER GOALS

In this chapter you will learn how to:

1. Identify an element by its symbol and classify it as a metal, nonmetal, or metalloid
2. Describe the basic parts of an atom
3. Distinguish isotopes
4. Describe the basic features of the periodic table
5. Understand the electronic structure of an atom
6. Write an electronic configuration for an element in periods 1–3
7. Relate the location of an element in the periodic table to its number of valence electrons
8. Draw an electron-dot symbol for an atom
9. Use the periodic table to predict the relative size and ionization energy of atoms

Examine the ingredients listed on a box of crackers. They may include flour, added vitamins, sugar for sweetness, a natural or synthetic coloring agent, baking soda, salt for flavor, and BHT as a preservative. No matter how simple or complex each of these substances is, it is composed of the basic building block, the **atom.** The word *atom* comes from the Greek word *atomos* meaning *unable to cut.* In Chapter 2, we examine the structure and properties of atoms, the building blocks that comprise all forms of matter.

2.1 Elements

Elements are named for people, places, and things. For example, *carbon* (C) comes from the Latin word *carbo,* meaning *coal* or *charcoal; neptunium* (Np) was named for the planet Neptune; *einsteinium* (Es) was named for scientist Albert Einstein; and *californium* (Cf) was named for the state of California.

You were first introduced to elements in Section 1.3.

- **An *element* is a pure substance that cannot be broken down into simpler substances by a chemical reaction.**

Of the 118 elements currently known, 90 are naturally occurring and the remaining 28 have been prepared by scientists in the laboratory. Some elements, like oxygen in the air we breathe and aluminum in a soft drink can, are familiar to you, while others, like samarium and seaborgium, are probably not. An alphabetical list of all elements appears on the inside front cover.

Each element is identified by a one- or two-letter symbol. The element carbon is symbolized by the single letter **C,** while the element chlorine is symbolized by **Cl.** When two letters are used in the element symbol, the first is upper case while the second is lower case. Thus, **Co** refers to the element cobalt, but CO is carbon monoxide, which is composed of the elements carbon (C) and oxygen (O). Table 2.1 lists common elements and their symbols.

While most element symbols are derived from the first one or two letters of the element name, 11 elements have symbols derived from Latin or German origins. Table 2.2 lists these elements and their symbols.

ENVIRONMENTAL NOTE

Carbon monoxide (CO), formed in small amounts during the combustion of fossil fuels like gasoline, is a toxic component of the smoggy air in many large cities. Carbon monoxide contains the elements carbon and oxygen. We will learn about carbon monoxide in Section 10.10.

PROBLEM 2.1

Give the symbol for each element.

a. calcium, a nutrient needed for strong teeth and bones
b. radon, a radioactive gas produced in the soil
c. nitrogen, the main component of the earth's atmosphere
d. gold, a precious metal used in coins and jewelry

Table 2.1 Common Elements and Their Symbols

Element	Symbol	Element	Symbol
Bromine	Br	Magnesium	Mg
Calcium	Ca	Manganese	Mn
Carbon	C	Molybdenum	Mo
Chlorine	Cl	Nitrogen	N
Chromium	Cr	Oxygen	O
Cobalt	Co	Phosphorus	P
Copper	Cu	Potassium	K
Fluorine	F	Sodium	Na
Hydrogen	H	Sulfur	S
Iodine	I	Zinc	Zn
Lead	Pb		

Table 2.2 Unusual Element Symbols

Element	Symbol
Antimony	Sb (stibium)
Copper	Cu (cuprum)
Gold	Au (aurum)
Iron	Fe (ferrum)
Lead	Pb (plumbum)
Mercury	Hg (hydrargyrum)
Potassium	K (kalium)
Silver	Ag (argentum)
Sodium	Na (natrium)
Tin	Sn (stannum)
Tungsten	W (wolfram)

PROBLEM 2.2

An alloy is a mixture of two or more elements that has metallic properties. Give the element symbol for the components of each alloy: (a) brass (copper and zinc); (b) bronze (copper and tin); (c) pewter (tin, antimony, and lead).

PROBLEM 2.3

Give the name corresponding to each element symbol: (a) Ne; (b) S; (c) I; (d) Si; (e) B; (f) Hg.

2.1A Elements and the Periodic Table

A periodic table appears on the inside front cover for easy reference.

Long ago it was realized that groups of elements have similar properties, and that these elements could be arranged in a schematic way called the **periodic table** (Figure 2.1). The position of an element in the periodic table tells us much about its chemical properties.

Figure 2.1 The Periodic Table of the Elements

	1A 1	2A 2	3B 3	4B 4	5B 5	6B 6	7B 7	← 8	8B 9	10 →	1B 11	2B 12	3A 13	4A 14	5A 15	6A 16	7A 17	8A 18	
1	1 **H** 1.008																	2 **He** 4.003	1
2	3 **Li** 6.941	4 **Be** 9.012											5 **B** 10.81	6 **C** 12.01	7 **N** 14.01	8 **O** 16.00	9 **F** 19.00	10 **Ne** 20.18	2
3	11 **Na** 22.99	12 **Mg** 24.31											13 **Al** 26.98	14 **Si** 28.09	15 **P** 30.97	16 **S** 32.07	17 **Cl** 35.45	18 **Ar** 39.95	3
4	19 **K** 39.10	20 **Ca** 40.08	21 **Sc** 44.96	22 **Ti** 47.88	23 **V** 50.94	24 **Cr** 52.00	25 **Mn** 54.94	26 **Fe** 55.85	27 **Co** 58.93	28 **Ni** 58.69	29 **Cu** 63.55	30 **Zn** 65.41	31 **Ga** 69.72	32 **Ge** 72.64	33 **As** 74.92	34 **Se** 78.96	35 **Br** 79.90	36 **Kr** 83.80	4
5	37 **Rb** 85.47	38 **Sr** 87.62	39 **Y** 88.91	40 **Zr** 91.22	41 **Nb** 92.91	42 **Mo** 95.94	43 **Tc** (98)	44 **Ru** 101.1	45 **Rh** 102.9	46 **Pd** 106.4	47 **Ag** 107.9	48 **Cd** 112.4	49 **In** 114.8	50 **Sn** 118.7	51 **Sb** 121.8	52 **Te** 127.6	53 **I** 126.9	54 **Xe** 131.3	5
6	55 **Cs** 132.9	56 **Ba** 137.3	57 **La** 138.9	72 **Hf** 178.5	73 **Ta** 180.9	74 **W** 183.8	75 **Re** 186.2	76 **Os** 190.2	77 **Ir** 192.2	78 **Pt** 195.1	79 **Au** 197.0	80 **Hg** 200.6	81 **Tl** 204.4	82 **Pb** 207.2	83 **Bi** 209.0	84 **Po** (209)	85 **At** (210)	86 **Rn** (222)	6
7	87 **Fr** (223)	88 **Ra** (226)	89 **Ac** (227)	104 **Rf** (267)	105 **Db** (268)	106 **Sg** (271)	107 **Bh** (272)	108 **Hs** (270)	109 **Mt** (276)	110 **Ds** (281)	111 **Rg** (280)	112 **Cn** (285)	113 **–** (284)	114 **Fl** (289)	115 **–** (289)	116 **Lv** (293)	117 **–** (294)	118 **–** (294)	7

6	58 **Ce** 140.1	59 **Pr** 140.9	60 **Nd** 144.2	61 **Pm** (145)	62 **Sm** 150.4	63 **Eu** 152.0	64 **Gd** 157.3	65 **Tb** 158.9	66 **Dy** 162.5	67 **Ho** 164.9	68 **Er** 167.3	69 **Tm** 168.9	70 **Yb** 173.0	71 **Lu** 175.0	6
7	90 **Th** 232.0	91 **Pa** 231.0	92 **U** 238.0	93 **Np** (237)	94 **Pu** (244)	95 **Am** (243)	96 **Cm** (247)	97 **Bk** (247)	98 **Cf** (251)	99 **Es** (252)	100 **Fm** (257)	101 **Md** (258)	102 **No** (259)	103 **Lr** (262)	7

metal metalloid nonmetal

- **Metals** are shiny substances that conduct heat and electricity. Metals are ductile, meaning they can be drawn into wires, and malleable, meaning they can be hammered into shapes.
- **Metalloids** have properties intermediate between metals and nonmetals.
- **Nonmetals** are poor conductors of heat and electricity.

The elements in the periodic table are divided into three categories—**metals, nonmetals,** and **metalloids.** The solid line that begins with boron (B) and angles in steps down to astatine (At) marks the three regions corresponding to these groups. All metals are located to the *left* of the line. All nonmetals except hydrogen are located to the *right.* Metalloids are located along the steps.

- *Metals* are shiny materials that are good conductors of heat and electricity. All metals are solids at room temperature except for mercury, which is a liquid.
- *Nonmetals* do not have a shiny appearance, and they are generally poor conductors of heat and electricity. Nonmetals like sulfur and carbon are solids at room temperature; bromine is a liquid; and nitrogen, oxygen, and nine other elements are gases.
- *Metalloids* have properties intermediate between metals and nonmetals. Only seven elements are categorized as metalloids: boron (B), silicon (Si), germanium (Ge), arsenic (As), antimony (Sb), tellurium (Te), and astatine (At).

PROBLEM 2.4

Locate each element in the periodic table and classify it as a metal, nonmetal, or metalloid.

a. titanium
b. chlorine
c. krypton
d. palladium
e. arsenic
f. cesium
g. selenium
h. osmium

2.1B FOCUS ON THE HUMAN BODY The Elements of Life

Because living organisms selectively take up elements from their surroundings, the abundance of elements in the human body is very different from the distribution of elements in the earth's crust. **Four nonmetals—oxygen, carbon, hydrogen, and nitrogen—comprise 96% of the mass of the human body, and are called the *building-block elements*** (Figure 2.2). Hydrogen and oxygen are the elements that form water, the most prevalent substance in the body. Carbon, hydrogen, and oxygen are found in the four main types of biological molecules—proteins, carbohydrates, lipids, and nucleic acids. Proteins and nucleic acids contain the element nitrogen as well. These biological molecules are discussed in Chapters 14–17.

Figure 2.2 The Elements of Life

Building-Block Elements

Oxygen (O)
Carbon (C)
Hydrogen (H)
Nitrogen (N)

These four elements compose almost 96% of the mass of the human body. Muscle tissue contains all four building-block elements.

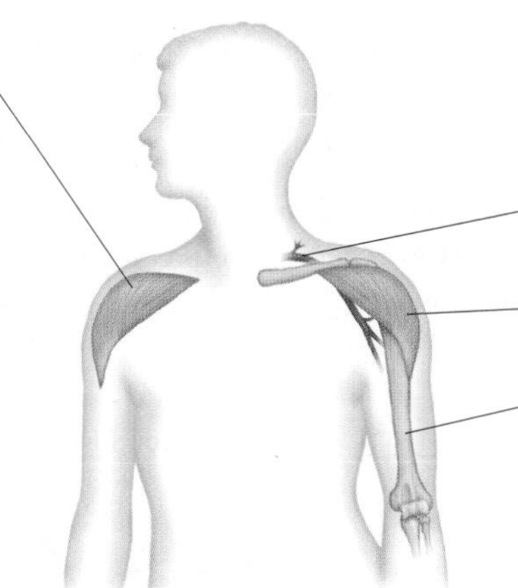

Major Minerals

Potassium (K), sodium (Na), and chlorine (Cl) are present in body fluids.

Magnesium (Mg) and sulfur (S) are present in the proteins found in muscle.

Calcium (Ca) and phosphorus (P) are present in teeth and bones.

Each major mineral is present in 0.1–2% by mass. At least 100 mg of each mineral is needed in the daily diet.

Trace Elements

Arsenic (As)	Fluorine (F)	Nickel (Ni)
Boron (B)	Iodine (I)	Selenium (Se)
Chromium (Cr)	Iron (Fe)	Silicon (Si)
Cobalt (Co)	Manganese (Mn)	Zinc (Zn)
Copper (Cu)	Molybdenum (Mo)	

Each trace element is present in less than 0.1% by mass. A small quantity (15 mg or less) of each element is needed in the daily diet.

Nutrition Facts

Serving Size 3/4 cup (30g)
Servings Per Container about 17

Amount Per Serving	Cereal	Cereal with 1/2 cup Fat Free Milk
Calories	120	160
Calories from Fat	15	15
	% Daily Value**	
Total Fat 1.5g*	**2%**	**2%**
Saturated Fat 0g	**0%**	**0%**
Trans Fat 0g		
Polyunsaturated Fat 0g		
Monounsaturated Fat 1g		
Cholesterol 0mg	**0%**	**0%**
Sodium 150mg	**6%**	**9%**
Potassium 60mg	**2%**	**7%**
Total Carbohydrate 25g	**8%**	**10%**
Dietary Fiber 2g	**8%**	**8%**
Sugars 6g		
Other Carbohydrate 17g		
Protein 2g		
Vitamin A	15%	20%
Vitamin C	0%	0%
Calcium	0%	15%
Iron	60%	60%
Vitamin D	10%	25%
Thiamin	25%	30%
Riboflavin	25%	35%
Niacin	25%	25%
Vitamin B6	25%	25%
Folic Acid	50%	50%
Vitamin B12	25%	35%
Phosphorus	4%	15%
Magnesium	4%	8%
Zinc	2%	6%
Copper	2%	2%

* Amount in Cereal. One half cup fat free milk contributes an additional 40 calories, 65mg sodium, 200mg potassium, 6g total carbohydrate (6g sugars), and 4g protein.

**Percent Daily Values are based on a 2,000 calorie diet. Your daily values may be higher or lower depending on your calorie needs:

	Calories:	2,000	2,500
Total Fat	Less than	65g	80g
Saturated Fat	Less than	20g	25g
Cholesterol	Less than	300mg	300mg
Sodium	Less than	2,400mg	2,400mg
Potassium		3,500mg	3,500mg
Total Carbohydrate		300g	375g
Dietary Fiber		25g	30g

INGREDIENTS: CORN, WHOLE GRAIN WHEAT, SUGAR, WHOLE GRAIN ROLLED OATS, BROWN SUGAR, HIGH OLEIC VEGETABLE OIL† (CANOLA OR SUNFLOWER OIL), RICE FLOUR, WHEAT FLOUR, MALTED BARLEY FLOUR, SALT, RICE, CORN SYRUP, WHEY (FROM MILK†), HONEY, MALTED CORN AND BARLEY SYRUP, CARAMEL COLOR, ARTIFICIAL FLAVOR, ANNATTO EXTRACT (COLOR). BHT ADDED TO PACKAGING MATERIAL TO PRESERVE PRODUCT FRESHNESS.

VITAMINS AND MINERALS: REDUCED IRON, NIACINAMIDE, VITAMIN B6, VITAMIN A PALMITATE, RIBOFLAVIN (VITAMIN B2), THIAMIN MONONITRATE (VITAMIN B1), ZINC OXIDE (SOURCE OF ZINC), FOLIC ACID, VITAMIN B12, VITAMIN D

Many breakfast cereals are fortified with iron to provide the consumer with this essential micronutrient.

Seven other elements, called the **major minerals** or **macronutrients,** are also present in the body in much smaller amounts (0.1–2% by mass). Sodium, potassium, and chlorine are present in body fluids. Magnesium and sulfur occur in proteins, and calcium and phosphorus are present in teeth and bones. Phosphorus is also contained in all nucleic acids, such as the DNA that transfers genetic information from one generation to another. At least 100 mg of each macronutrient is needed in the daily diet.

Many other elements occur in very small amounts in the body, but are essential to good health. These **trace elements** or **micronutrients** are required in the daily diet in small quantities, usually less than 15 mg. Each trace element has a specialized function that is important for proper cellular function. For example, iron is needed for hemoglobin, the protein that carries oxygen in red blood cells, and myoglobin, the protein that stores oxygen in muscle. Zinc is needed for the proper functioning of many enzymes in the liver and kidneys, and iodine is needed for proper thyroid function. Although most of the trace elements are metals, nonmetals like fluorine and selenium are micronutrients as well.

PROBLEM 2.5

Classify each micronutrient in Figure 2.2 as a metal, nonmetal, or metalloid.

2.1C Compounds

In Section 1.3 we learned that a ***compound*** **is a pure substance formed by chemically combining two or more elements together.** Element symbols are used to write chemical formulas for compounds.

- A *chemical formula* uses element symbols to show the identity of the elements forming a compound and subscripts to show the ratio of atoms (the building blocks of matter) contained in the compound.

For example, table salt is formed from sodium (Na) and chlorine (Cl) in a ratio of 1:1, so its formula is NaCl. Water, on the other hand, is formed from two hydrogens for each oxygen, so its formula is H_2O. The subscript "1" is understood when no subscript is written. Other examples of chemical formulas are shown below.

H_2O (2 H's for each O) — water

CO_2 (2 O's for each C) — carbon dioxide (dry ice)

C_3H_8 (3 C's, 8 H's) — propane

As we learned in Section 1.2, molecular art will often be used to illustrate the composition and state of elements and compounds. Color-coded spheres, shown in Figure 2.3, are used to identify the common elements that form compounds.

For example, a red sphere is used for the element oxygen and gray is used for the element hydrogen, so H_2O is represented as a red sphere joined to two gray spheres. Sometimes the spheres will be connected by "sticks" to generate a **ball-and-stick** representation for a compound. At

Figure 2.3 Common Element Colors Used in Molecular Art

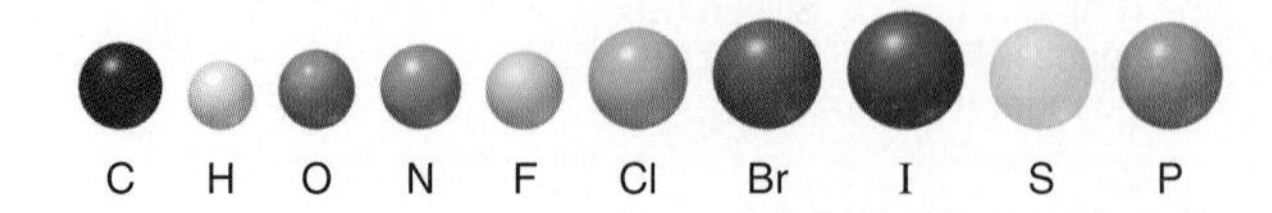

other times, the spheres will be drawn close together to form a **space-filling** representation. No matter how the spheres are depicted, H_2O always consists of one red sphere for the oxygen and two gray spheres for the two hydrogens.

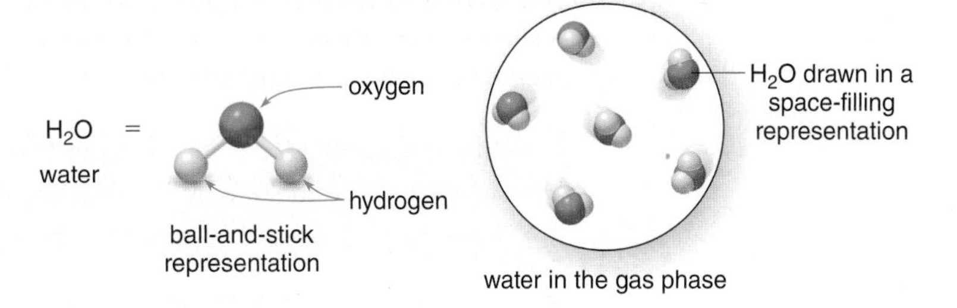

SAMPLE PROBLEM 2.1

Identify the elements used in each example of molecular art.

a. b.

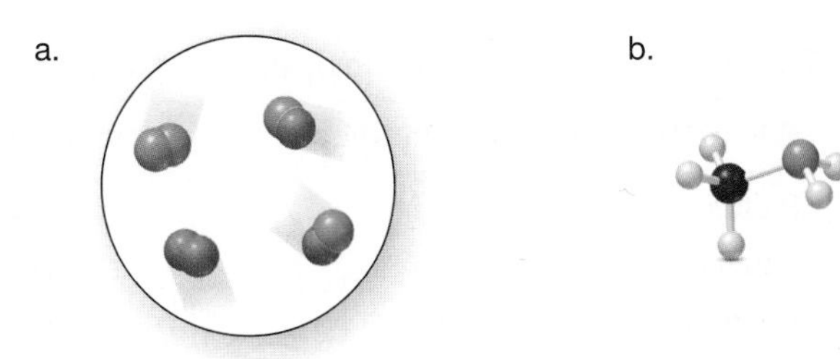

Analysis

Use Figure 2.3 to determine the identity of the color-coded spheres.

Solution

a. The blue spheres in this space-filling representation correspond to the element nitrogen and the red spheres correspond to the element oxygen. Thus, one "particle" contains two nitrogens, one contains two oxygens, and two "particles" contain one oxygen and one nitrogen.

b. This ball-and-stick representation contains the elements of carbon (black), nitrogen (blue), and hydrogen (gray).

PROBLEM 2.6

Identify the elements used in each example of molecular art.

PROBLEM 2.7

Identify the elements in each chemical formula, and give the number of atoms of each element.

a. NaCN (sodium cyanide)
b. H_2S (hydrogen sulfide)
c. C_2H_6 (ethane)
d. SnF_2 (stannous fluoride)
e. CO (carbon monoxide)
f. $C_3H_8O_3$ (glycerol)

PROBLEM 2.8

Halothane is an inhaled general anesthetic, commonly used since the 1950s. Identify the elements in the ball-and-stick representation of halothane.

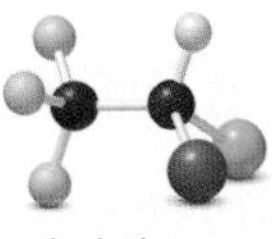

halothane

2.2 Structure of the Atom

All matter is composed of the same basic building blocks called *atoms*. An atom is much too small to be seen even by the most powerful light microscopes. The period at the end of this sentence holds about 1×10^8 atoms, and a human cheek cell contains about 1×10^{16} atoms. An atom is composed of three subatomic particles.

- **A proton, symbolized by p, has a positive (+) charge.**
- **An electron, symbolized by e^-, has a negative (–) charge.**
- **A neutron, symbolized by n, has no charge.**

Protons and neutrons have approximately the same, exceedingly small, mass as shown in Table 2.3. The mass of an electron is much less, 1/1,836 the mass of a proton. These subatomic particles are not evenly distributed in the volume of an atom. There are two main components of an atom.

Table 2.3 Summary: The Properties of the Three Subatomic Particles

Subatomic Particle	Charge	Mass (g)	Mass (amu)
Proton	+1	1.6726×10^{-24}	1
Neutron	0	1.6749×10^{-24}	1
Electron	–1	9.1093×10^{-28}	Negligible

- **The *nucleus* is a dense core that contains the protons and neutrons. Most of the mass of an atom resides in the nucleus.**
- **The *electron cloud* is composed of electrons that move rapidly in the almost empty space surrounding the nucleus. The electron cloud comprises most of the volume of an atom.**

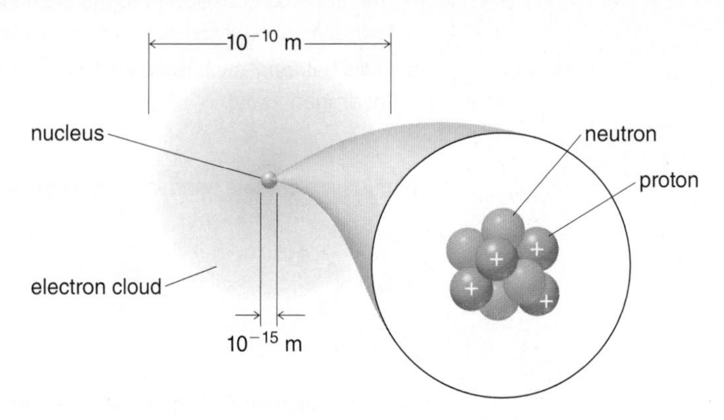

main components of an atom

While the diameter of an atom is about 10^{-10} m, the diameter of a nucleus is only about 10^{-15} m. For a macroscopic analogy, if the nucleus were the size of a baseball, an atom would be the size of Yankee Stadium!

The charged particles of an atom can either attract or repel each other.

- **Opposite charges attract while like charges repel each other.**

Thus, two electrons or two protons repel each other, while a proton and an electron attract each other.

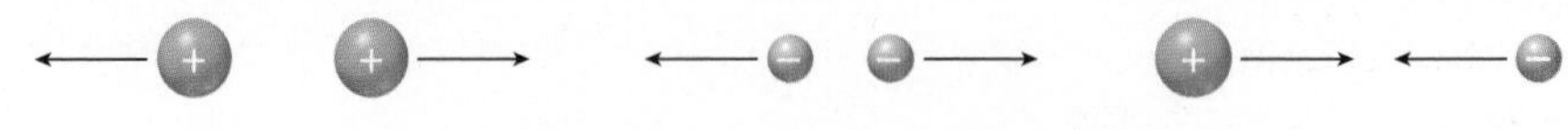

Positive charges repel. Negative charges repel. Opposite charges attract.

Since the mass of an individual atom is so small (on the order of 10^{-24} g), chemists use a standard mass unit, the **atomic mass unit,** which defines the mass of individual atoms relative to a standard mass.

- **One atomic mass unit (amu) equals one-twelfth the mass of a carbon atom that has six protons and six neutrons; 1 amu = 1.661×10^{-24} g.**

Using this scale, one proton has a mass of 1.0073 amu, a value typically rounded to 1 amu. One neutron has a mass of 1.0087 amu, a value also typically rounded to 1 amu. The mass of an electron is so small that it is ignored.

Every atom of a given type of element always has the *same* number of protons in the nucleus, a value called the *atomic number,* symbolized by ***Z.*** Conversely, two *different* elements have *different* atomic numbers.

- **The *atomic number* (*Z*) = the number of protons in the nucleus of an atom.**

Thus, the element hydrogen has one proton in its nucleus, so its atomic number is one. Lithium has three protons in its nucleus, so its atomic number is three. The periodic table is arranged in order of increasing atomic number beginning at the upper left-hand corner. The atomic number appears just above the element symbol for each entry in the table.

Since a neutral atom has no overall charge:

- ***Z* = the number of protons in the nucleus = the number of electrons.**

Thus, the atomic number tells us *both* the number of protons in the nucleus and the number of electrons in the electron cloud of a neutral atom.

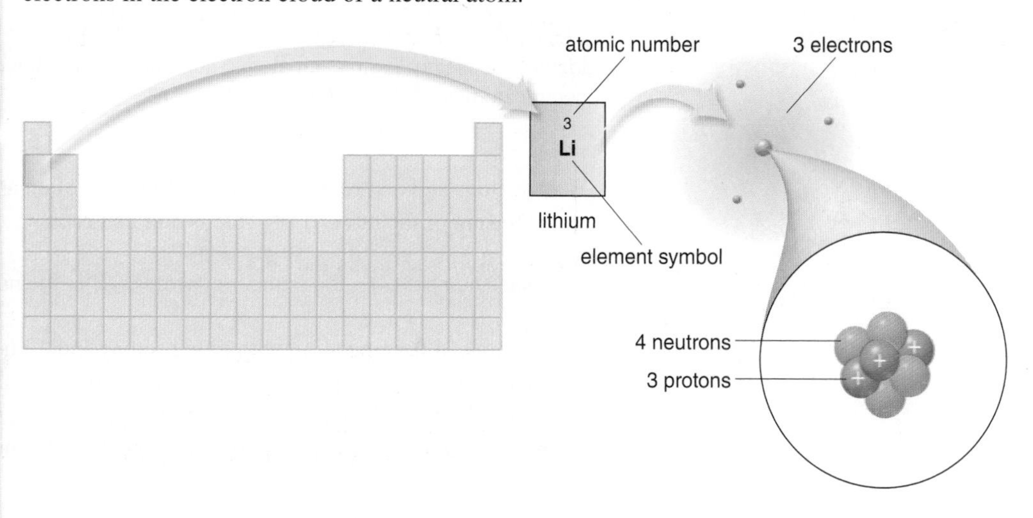

CONSUMER NOTE

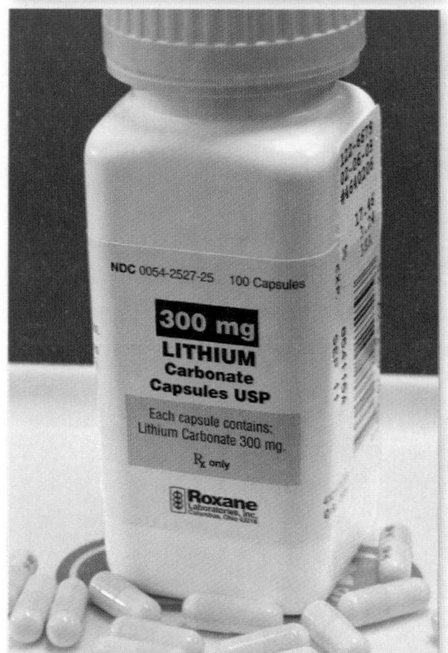

The element lithium is found in many consumer products, from long-lasting lithium batteries to the prescription medication lithium carbonate, used to treat individuals with bipolar disorder.

SAMPLE PROBLEM 2.2

For the given atom: (a) determine the number of protons, neutrons, and electrons in the neutral atom; (b) give the atomic number; (c) identify the element.

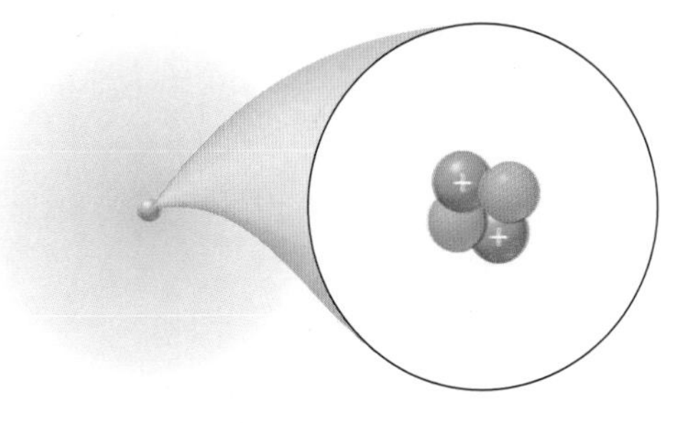

Analysis

- The number of protons = the number of positively charged particles = the atomic number.
- The number of neutrons = the number of uncharged particles.
- The number of protons = the number of electrons in a neutral atom.
- The atomic number determines the identity of an element.

Solution

a. The element contains two protons and two neutrons. Since a neutral atom has the same number of protons and electrons, the element has two electrons.

b. The two protons give the element an atomic number of two.

c. The element with two protons in the nucleus is **helium.**

PROBLEM 2.9

An element has nine protons and 10 neutrons in the neutral atom. (a) How many electrons are present in the neutral atom? (b) What is the atomic number of this element? (c) Identify the element.

SAMPLE PROBLEM 2.3

Identify the element that has an atomic number of 19, and give the number of protons and electrons in the neutral atom.

Analysis

The atomic number is unique to an element and tells the number of protons in the nucleus and the number of electrons in the electron cloud of a neutral atom.

Solution

According to the periodic table, the element potassium has atomic number 19. A neutral potassium atom has 19 protons and 19 electrons.

PROBLEM 2.10

Identify the element with each atomic number, and give the number of protons and electrons in the neutral atom: (a) 2; (b) 11; (c) 20; (d) 47; (e) 78.

Both protons and neutrons contribute to the mass of an atom. The **mass number,** symbolized by ***A*,** is the sum of the number of protons and neutrons.

- **Mass number (*A*) = the number of protons (*Z*) + the number of neutrons.**

For example, a fluorine atom with nine protons and 10 neutrons in the nucleus has a mass number of 19. Figure 2.4 lists the atomic number, mass number, and number of subatomic particles in the four building-block elements—hydrogen, carbon, nitrogen, and oxygen—found in a wide variety of compounds including caffeine (chemical formula $C_8H_{10}N_4O_2$), the bitter-tasting mild stimulant in coffee, tea, and cola beverages.

Figure 2.4

Atomic Composition of the Four Building-Block Elements

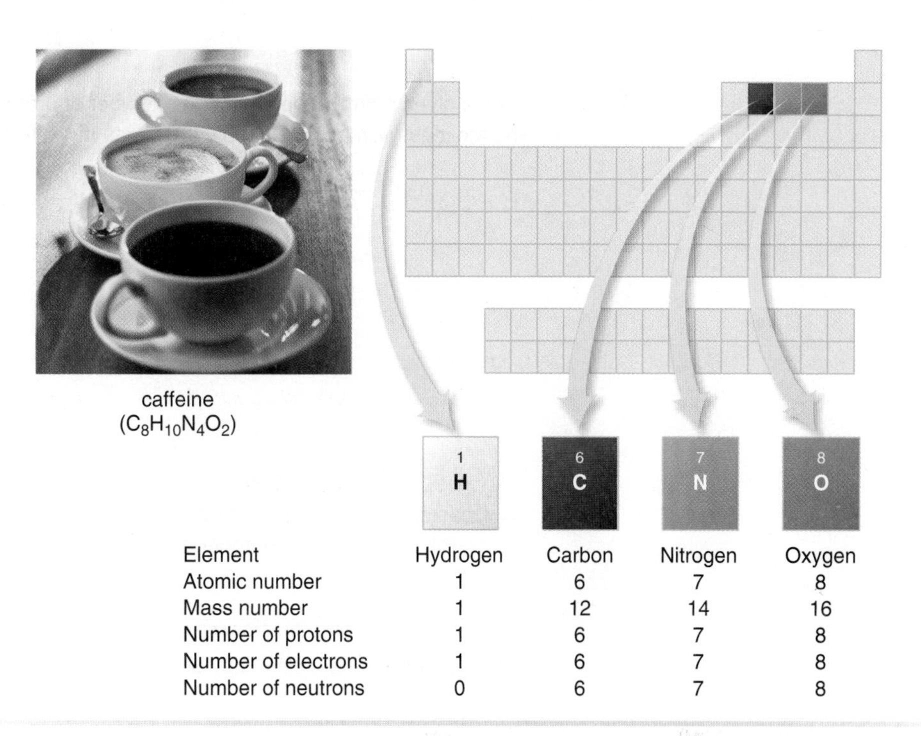

Element	Hydrogen	Carbon	Nitrogen	Oxygen
Atomic number	1	6	7	8
Mass number	1	12	14	16
Number of protons	1	6	7	8
Number of electrons	1	6	7	8
Number of neutrons	0	6	7	8

SAMPLE PROBLEM 2.4

For the given atom: (a) determine the number of protons and neutrons; (b) give the atomic number and the mass number; (c) identify the element.

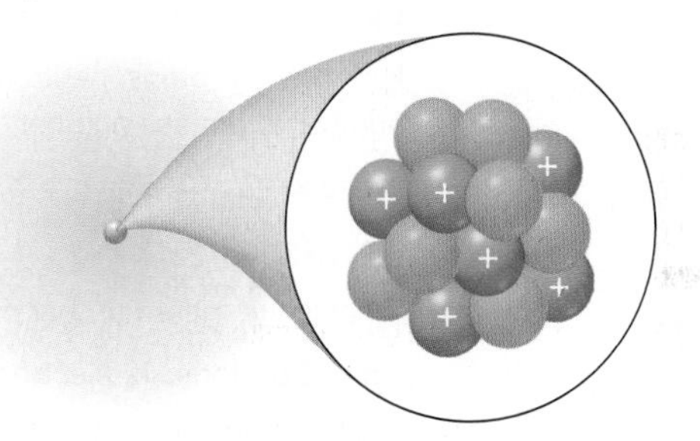

Analysis

- The number of protons = the number of positively charged particles = the atomic number. The atomic number determines the identity of an element.
- The number of neutrons = the number of uncharged particles.
- The mass number = the number of protons + the number of neutrons.

Solution

a. The element contains six protons and seven neutrons.

b. The six protons give the element an atomic number of six. The mass number = the number of protons + the number of neutrons = 6 + 7 = 13.

c. The element with six protons in the nucleus is **carbon.**

PROBLEM 2.11

For the given atom: (a) determine the number of protons and neutrons; (b) give the atomic number and the mass number; (c) identify the element.

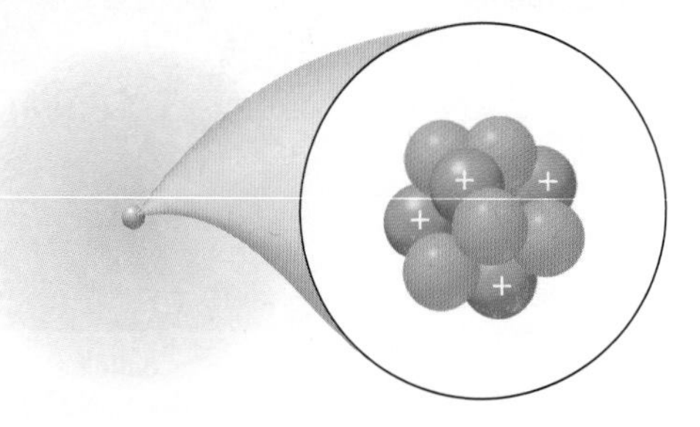

SAMPLE PROBLEM 2.5

How many protons, neutrons, and electrons are contained in an atom of argon, which has an atomic number of 18 and a mass number of 40?

Analysis

- In a neutral atom, the atomic number (Z) = the number of protons = the number of electrons.
- The mass number (A) = the number of protons + the number of neutrons.

Solution

The atomic number of 18 means that argon has 18 protons and 18 electrons. To find the number of neutrons, subtract the atomic number (Z) from the mass number (A).

$$\text{number of neutrons} = \text{mass number} - \text{atomic number}$$
$$= 40 - 18$$
$$= 22 \text{ neutrons}$$

PROBLEM 2.12

How many protons, neutrons, and electrons are contained in each atom with the given atomic number and mass number?

a. $Z = 17, A = 35$ b. $Z = 14, A = 28$ c. $Z = 92, A = 238$

PROBLEM 2.13

What is the mass number of an atom that contains

a. 42 protons, 42 electrons, and 53 neutrons? b. 24 protons, 24 electrons, and 28 neutrons?

2.3 Isotopes

Two atoms of the same element always have the same number of protons, but the number of neutrons can vary.

- ***Isotopes* are atoms of the same element having a different number of neutrons.**

2.3A Isotopes, Atomic Number, and Mass Number

Most elements in nature exist as a mixture of isotopes. For example, all atoms of the element chlorine contain 17 protons in the nucleus, but some of these atoms have 18 neutrons in the nucleus and some have 20 neutrons. Thus, chlorine has two isotopes with different mass numbers, 35 and 37. These isotopes are often referred to as chlorine-35 (or Cl-35) and chlorine-37 (or Cl-37).

An **isotope symbol** is also written using the element symbol with the atomic number as a subscript and the mass number as a superscript, both to the left.

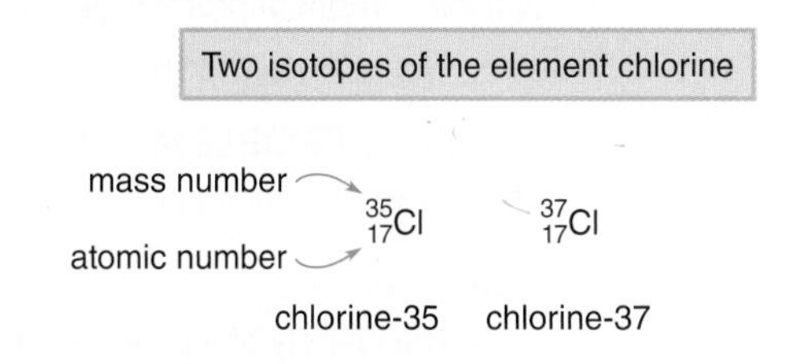

The element hydrogen has three isotopes. Most hydrogen atoms have one proton and no neutrons, giving them a mass number of one. About 1% of hydrogen atoms have one proton and one neutron, giving them a mass number of two. This isotope is called **deuterium,** and it is often symbolized as **D.** An even smaller number of hydrogen atoms contain one proton and two neutrons, giving them a mass number of three. This isotope is called **tritium,** often symbolized as **T.**

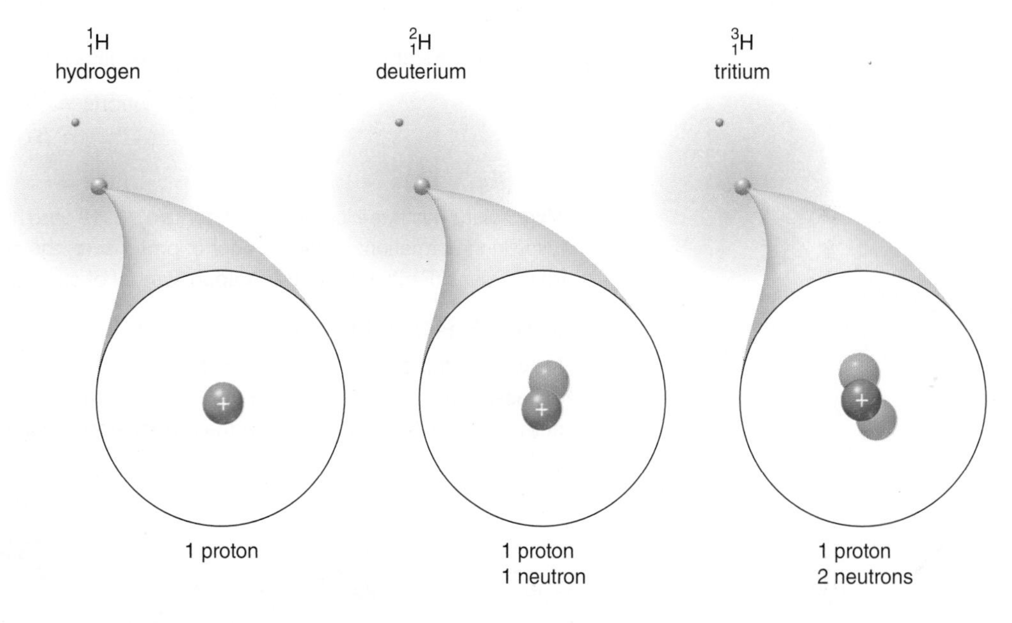

SAMPLE PROBLEM 2.6

For each atom give the following information: [1] the atomic number; [2] the mass number; [3] the number of protons; [4] the number of neutrons; [5] the number of electrons.

a. $^{118}_{50}Sn$ b. $^{195}_{78}Pt$

Analysis

- The superscript gives the mass number and the subscript gives the atomic number for each element.
- The atomic number = the number of protons = the number of electrons.
- The mass number = the number of protons + the number of neutrons.

Solution

	Atomic Number	Mass Number	Number of Protons	Number of Neutrons	Number of Electrons
a. $^{118}_{50}Sn$	50	118	50	118 – 50 = 68	50
b. $^{195}_{78}Pt$	78	195	78	195 – 78 = 117	78

PROBLEM 2.14

For each atom give the following information: [1] the atomic number; [2] the mass number; [3] the number of protons; [4] the number of neutrons; [5] the number of electrons.

a. $^{13}_{6}C$ b. $^{121}_{51}Sb$

SAMPLE PROBLEM 2.7

Determine the number of neutrons in each isotope: (a) carbon-14; (b) ^{81}Br.

Analysis

- The identity of the element tells us the atomic number.
- The number of neutrons = mass number (*A*) – atomic number (*Z*).

Solution

a. Carbon's atomic number (*Z*) is 6. Carbon-14 has a mass number (*A*) of 14.

$$\text{number of neutrons} = A - Z$$
$$= 14 - 6 = 8 \text{ neutrons}$$

b. Bromine's atomic number is 35 and the mass number of the given isotope is 81.

$$\text{number of neutrons} = A - Z$$
$$= 81 - 35 = 46 \text{ neutrons}$$

PROBLEM 2.15

Magnesium has three isotopes that contain 12, 13, and 14 neutrons. For each isotope give the following information: (a) the number of protons; (b) the number of electrons; (c) the atomic number; (d) the mass number. Write the isotope symbol of each isotope.

SAMPLE PROBLEM 2.8

Answer the following questions about the three elements whose nuclei are drawn below. (a) Which representation shows an isotope of [1]? (b) Which representation shows a different element from the other two? (c) Which element has the lowest atomic number?

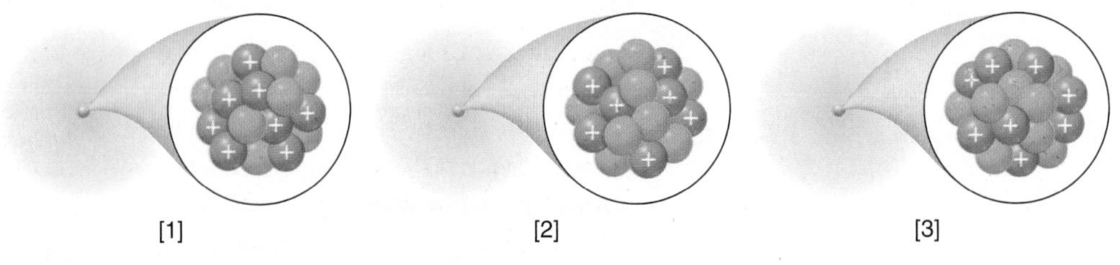

Analysis

- Two isotopes have the same number of protons but a different number of neutrons.
- Two different elements have different numbers of protons.
- The number of protons in the nucleus gives the atomic number.

Solution

a. Representation [1] has eight protons and eight neutrons in the nucleus. Representation [3] is an isotope because it has the same number of protons (eight), but it contains a different number of neutrons (nine).

b. Representation [2] is a different element because it contains only seven protons and representations [1] and [3] each have eight protons.

c. Since the number of protons determines the atomic number, representation [2] has the lowest atomic number (seven) because it has only seven protons in its nucleus.

PROBLEM 2.16

Use the representations ([1]–[3]) in Sample Problem 2.8 to answer the following questions.

a. Which species has the highest mass number?

b. How many electrons are present in the neutral atom of each element?

c. Give the isotope symbol for each species.

2.3B Atomic Weight

ENVIRONMENTAL NOTE

Although gasoline sold in the United States no longer contains lead, leaded gasoline is still used extensively in Asia, Africa, and Latin America. Gasoline exhaust containing lead pollutes the air and soil, and individuals exposed to high lead levels can suffer from circulatory, digestive, and nervous disorders. Lead (Pb) is a metal with atomic number 82 and atomic weight 207.2, as shown in the periodic table.

Some elements like fluorine occur naturally as a single isotope. More commonly, an element is a mixture of isotopes, and it is useful to know the average mass, called the **atomic weight** (or **atomic mass**), of the atoms in a sample.

- The *atomic weight* is the weighted average of the mass of the naturally occurring isotopes of a particular element reported in atomic mass units.

The atomic weight takes into account the amount of each isotope that occurs in nature. The alphabetical list of elements on the inside front cover gives the atomic weight for each element. The atomic weight is also given under the element symbol in the periodic table on the inside front cover.

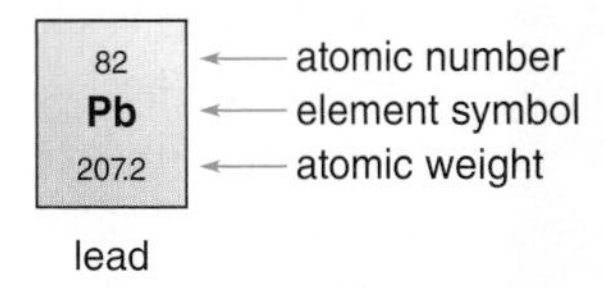

PROBLEM 2.17

The element cobalt is a micronutrient present in vitamin B_{12}. (a) What is the atomic number and atomic weight of cobalt? (b) How many protons and electrons does a neutral cobalt atom contain?

2.3C FOCUS ON HEALTH & MEDICINE Isotopes in Medicine

Generally the chemical properties of isotopes are identical. Sometimes, however, one isotope of an element is radioactive—that is, it emits particles or energy as some form of radiation. Radioactive isotopes have both diagnostic and therapeutic uses in medicine.

As an example, iodine-131 is used in at least two different ways for thyroid disease. Iodine is a micronutrient needed by the body to synthesize the thyroid hormone thyroxine, which contains four iodine atoms. To evaluate the thyroid gland, a patient can be given sodium iodide (NaI) that contains radioactive iodine-131. Iodine-131 is taken up in the thyroid gland and as it emits radiation, it produces an image in a thyroid scan, which is then used to determine the condition of the thyroid gland, as shown in Figure 2.5.

Other applications of radioactive isotopes in medicine are discussed in Chapter 9.

Higher doses of iodine-131 can also be used to treat thyroid disease. Since the radioactive isotope is taken up by the thyroid gland, the radiation it emits can kill overactive or cancerous cells in the thyroid.

Figure 2.5 Iodine-131 in Medicine

a.

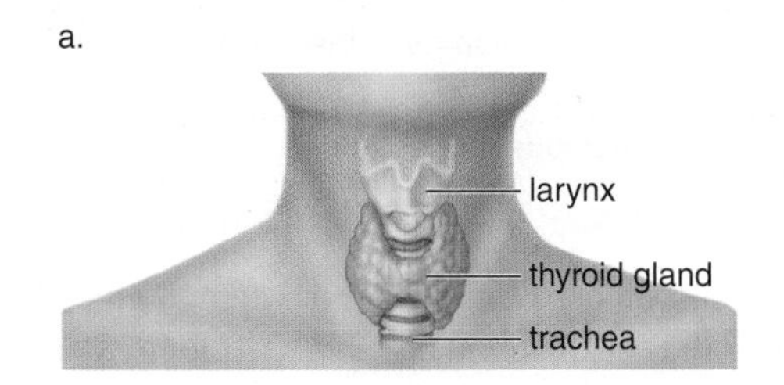

b.

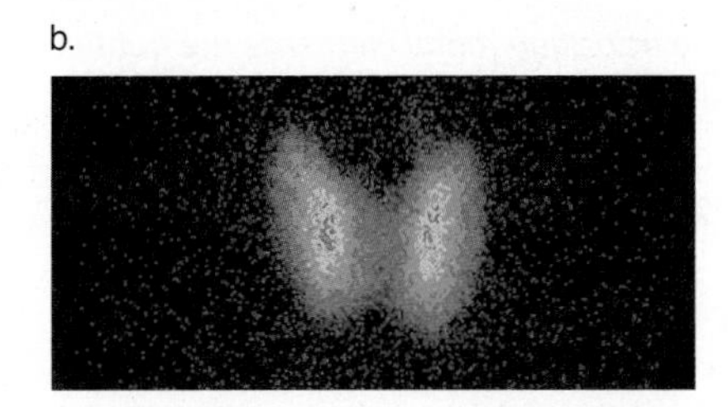

c.

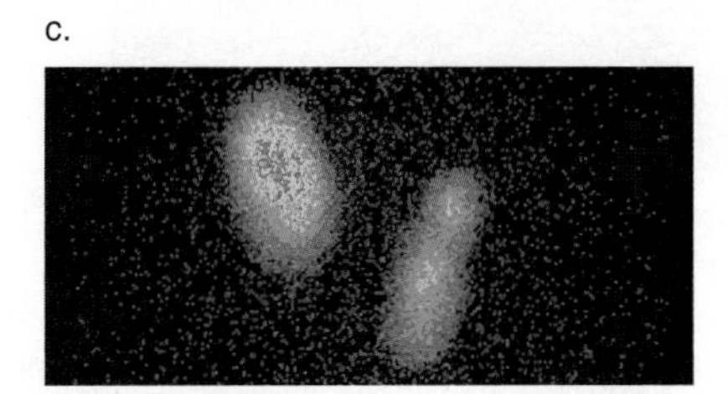

The thyroid gland is a butterfly-shaped gland in the neck, shown in (a). Uptake of radioactive iodine-131 can reveal the presence of a healthy thyroid as in (b), or an unsymmetrical thyroid gland with dense areas of iodine uptake as in (c), which may be indicative of cancer or other thyroid disease.

PROBLEM 2.18

Iodine-123 and iodine-131 are radioactive isotopes used for the diagnosis or treatment of thyroid disease. Complete the following table for both isotopes.

	Atomic Number	Mass Number	Number of Protons	Number of Neutrons	Isotope Symbol
Iodine-123					
Iodine-131					

HEALTH NOTE

Administering a zinc tablet dissolved in water to a child with diarrhea can save his life. Diarrhea kills more children worldwide than malaria or AIDS. Zinc is a metal located in group 2B (12) in the periodic table.

2.4 The Periodic Table

Every beginning chemistry text has a periodic table in a prominent location—often the inside front cover—because it is a valuable list of all known elements organized so that groups of elements with similar characteristics are arranged together. The periodic table evolved over many years, and it resulted from the careful observations and experiments of many brilliant scientists in the nineteenth century. Most prominent was Russian chemist Dmitri Mendeleev, whose arrangement in 1869 of the 60 known elements into groups having similar properties became the precursor of the modern periodic table (inside front cover and Figure 2.6).

2.4A Basic Features of the Periodic Table

The periodic table is arranged into seven horizontal rows and 18 vertical columns. The particular row and column tell us much about the properties of an element.

- **A row in the periodic table is called a *period*. Elements in the same row are similar in size.**
- **A column in the periodic table is called a *group*. Elements in the same group have similar electronic and chemical properties.**

The rows in the periodic table are numbered 1–7. The number of elements in each row varies. The first period has just two elements, hydrogen and helium. The second and third rows have eight elements each, and the fourth and fifth rows have 18 elements. Also note that two sets of fourteen elements appear at the bottom of the periodic table. The **lanthanides,** beginning with the element cerium ($Z = 58$), immediately follow the element lanthanum (La). The **actinides,** beginning with thorium ($Z = 90$), immediately follow the element actinium (Ac).

Each column in the periodic table is assigned a **group number.** Groups are numbered in two ways. In one system, the 18 columns of the periodic table are assigned the numbers 1–18, beginning with the column farthest to the left. An older but still widely used system numbers the groups 1–8, followed by the letter A or B.

- **The *main group elements* consist of the two columns on the far left and the six columns on the far right of the table. These groups are numbered 1A–8A.**
- **The *transition metal elements* are contained in the 10 short columns in the middle of the table, numbered 1B–8B.**
- **The *inner transition metal elements* consist of the lanthanides and actinides, and they are not assigned group numbers.**

HEALTH NOTE

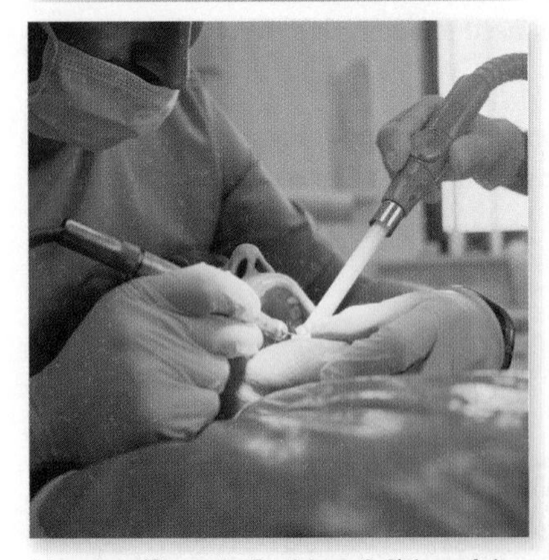

Mercury (Sample Problem 2.9) is safely used in dental amalgam to fill cavities in teeth. Mercury released into the environment, however, is converted to toxic methylmercury by microorganisms in water, so hazardous levels of this soluble mercury compound can accumulate in fish at the top of the food chain, such as sharks and swordfish.

The periodic table in Figure 2.6 has both systems of numbering groups. For example, the element carbon (C) is located in the second row (period 2) of the periodic table. Its group number is 4A (or 14).

PROBLEM 2.19

(a) How many elements are in the third period of the periodic table? (b) How many elements are in group 2A (or 2)?

Figure 2.6 Basic Features of the Periodic Table

Period — Group number

Period	1A 1	2A 2	3B 3	4B 4	5B 5	6B 6	7B 7	8B 8	8B 9	8B 10	1B 11	2B 12	3A 13	4A 14	5A 15	6A 16	7A 17	8A 18	Period
1	1 **H** 1.008																	2 **He** 4.003	1
2	3 **Li** 6.941	4 **Be** 9.012											5 **B** 10.81	6 **C** 12.01	7 **N** 14.01	8 **O** 16.00	9 **F** 19.00	10 **Ne** 20.18	2
3	11 **Na** 22.99	12 **Mg** 24.31											13 **Al** 26.98	14 **Si** 28.09	15 **P** 30.97	16 **S** 32.07	17 **Cl** 35.45	18 **Ar** 39.95	3
4	19 **K** 39.10	20 **Ca** 40.08	21 **Sc** 44.96	22 **Ti** 47.88	23 **V** 50.94	24 **Cr** 52.00	25 **Mn** 54.94	26 **Fe** 55.85	27 **Co** 58.93	28 **Ni** 58.69	29 **Cu** 63.55	30 **Zn** 65.41	31 **Ga** 69.72	32 **Ge** 72.64	33 **As** 74.92	34 **Se** 78.96	35 **Br** 79.90	36 **Kr** 83.80	4
5	37 **Rb** 85.47	38 **Sr** 87.62	39 **Y** 88.91	40 **Zr** 91.22	41 **Nb** 92.91	42 **Mo** 95.94	43 **Tc** (98)	44 **Ru** 101.1	45 **Rh** 102.9	46 **Pd** 106.4	47 **Ag** 107.9	48 **Cd** 112.4	49 **In** 114.8	50 **Sn** 118.7	51 **Sb** 121.8	52 **Te** 127.6	53 **I** 126.9	54 **Xe** 131.3	5
6	55 **Cs** 132.9	56 **Ba** 137.3	57 **La** 138.9	72 **Hf** 178.5	73 **Ta** 180.9	74 **W** 183.8	75 **Re** 186.2	76 **Os** 190.2	77 **Ir** 192.2	78 **Pt** 195.1	79 **Au** 197.0	80 **Hg** 200.6	81 **Tl** 204.4	82 **Pb** 207.2	83 **Bi** 209.0	84 **Po** (209)	85 **At** (210)	86 **Rn** (222)	6
7	87 **Fr** (223)	88 **Ra** (226)	89 **Ac** (227)	104 **Rf** (267)	105 **Db** (268)	106 **Sg** (271)	107 **Bh** (272)	108 **Hs** (270)	109 **Mt** (276)	110 **Ds** (281)	111 **Rg** (280)	112 **Cn** (285)	113 – (284)	114 **Fl** (289)	115 – (289)	116 **Lv** (293)	117 – (294)	118 – (294)	7

	Period															Period
Lanthanides	6	58 **Ce** 140.1	59 **Pr** 140.9	60 **Nd** 144.2	61 **Pm** (145)	62 **Sm** 150.4	63 **Eu** 152.0	64 **Gd** 157.3	65 **Tb** 158.9	66 **Dy** 162.5	67 **Ho** 164.9	68 **Er** 167.3	69 **Tm** 168.9	70 **Yb** 173.0	71 **Lu** 175.0	6
Actinides	7	90 **Th** 232.0	91 **Pa** 231.0	92 **U** 238.0	93 **Np** (237)	94 **Pu** (244)	95 **Am** (243)	96 **Cm** (247)	97 **Bk** (247)	98 **Cf** (251)	99 **Es** (252)	100 **Fm** (257)	101 **Md** (258)	102 **No** (259)	103 **Lr** (262)	7

Main group elements — Transition metal elements — Inner transition metal elements

- Each element of the periodic table is part of a horizontal row and a vertical column.
- The periodic table consists of seven rows, labeled periods 1–7, and 18 columns that are assigned a group number. Two different numbering systems are indicated.
- Elements are divided into three categories: main group elements (groups 1A–8A, shown in light blue), transition metals (groups 1B–8B, shown in tan), and inner transition metals (shown in light green).

SAMPLE PROBLEM 2.9

Give the period and group number for each element: (a) magnesium; (b) mercury.

Analysis

Use the element symbol to locate an element in the periodic table. Count down the rows of elements to determine the period. The group number is located at the top of each column.

Solution

a. Magnesium (Mg) is located in the third row (period 3), and has group number 2A (or 2).
b. Mercury (Hg) is located in the sixth row (period 6), and has group number 2B (or 12).

PROBLEM 2.20

Give the period and group number for each element: (a) oxygen; (b) calcium; (c) phosphorus; (d) platinum; (e) iodine.

2.4B Characteristics of Groups 1A, 2A, 7A, and 8A

Four columns of main group elements illustrate an important fact about the periodic table.

- **Elements that comprise a particular group have similar chemical properties.**

Alkali Metals (Group 1A) and Alkaline Earth Elements (Group 2A)

The alkali metals and the alkaline earth elements are located on the far left side of the periodic table.

Although hydrogen is also located in group 1A, it is *not* an alkali metal.

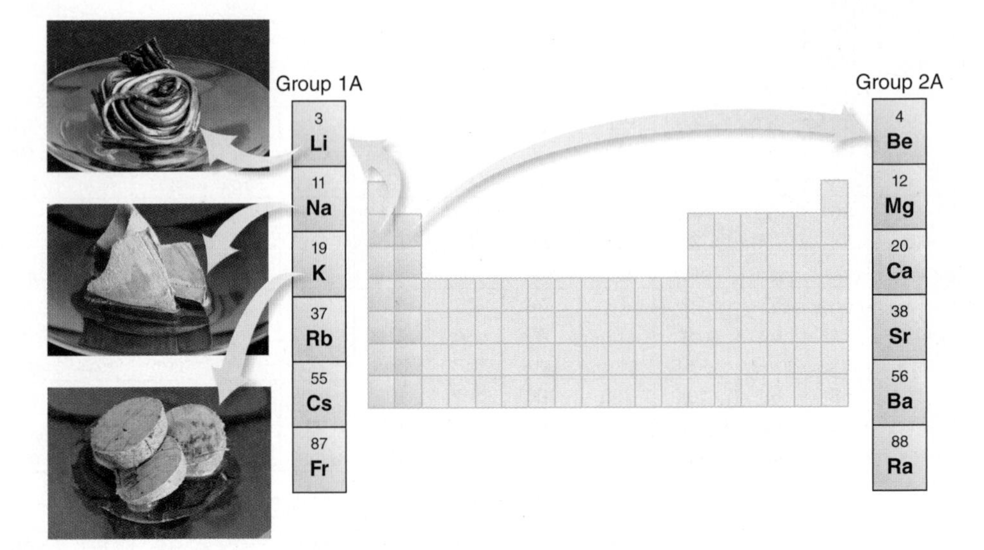

The **alkali metals,** located in group 1A (group 1), include lithium (Li), sodium (Na), potassium (K), rubidium (Rb), cesium (Cs), and francium (Fr). Alkali metals share the following characteristics:

- They are soft and shiny and have low melting points.
- They are good conductors of heat and electricity.
- They react readily with water to form basic solutions.

The **alkaline earth elements,** located in group 2A (group 2), include beryllium (Be), magnesium (Mg), calcium (Ca), strontium (Sr), barium (Ba), and radium (Ra). Alkaline earth metals are also shiny solids but less reactive than the alkali metals.

None of the metals in groups 1A or 2A exist in nature as pure elements; rather, they are always combined with other elements to form compounds. Examples of compounds from group 1A elements include sodium chloride (NaCl), table salt, and potassium iodide (KI), an essential nutrient added to make iodized salt. Examples of compounds from group 2A elements include magnesium sulfate ($MgSO_4$), an anticonvulsant used to prevent seizures in pregnant women; and barium sulfate ($BaSO_4$), which is used to improve the quality of X-ray images of the gastrointestinal tract.

ENVIRONMENTAL NOTE

CFCs are carbon compounds that contain the halogens fluorine and chlorine. CFCs such as $CFCl_3$ were once commonly used as aerosol propellants, but they have been shown to destroy ozone in the upper atmosphere. For this reason, the use of CFCs in spray cans was banned in the United States in 1978.

Halogens (Group 7A) and Noble Gases (Group 8A)

The halogens and noble gases are located on the far right side of the periodic table.

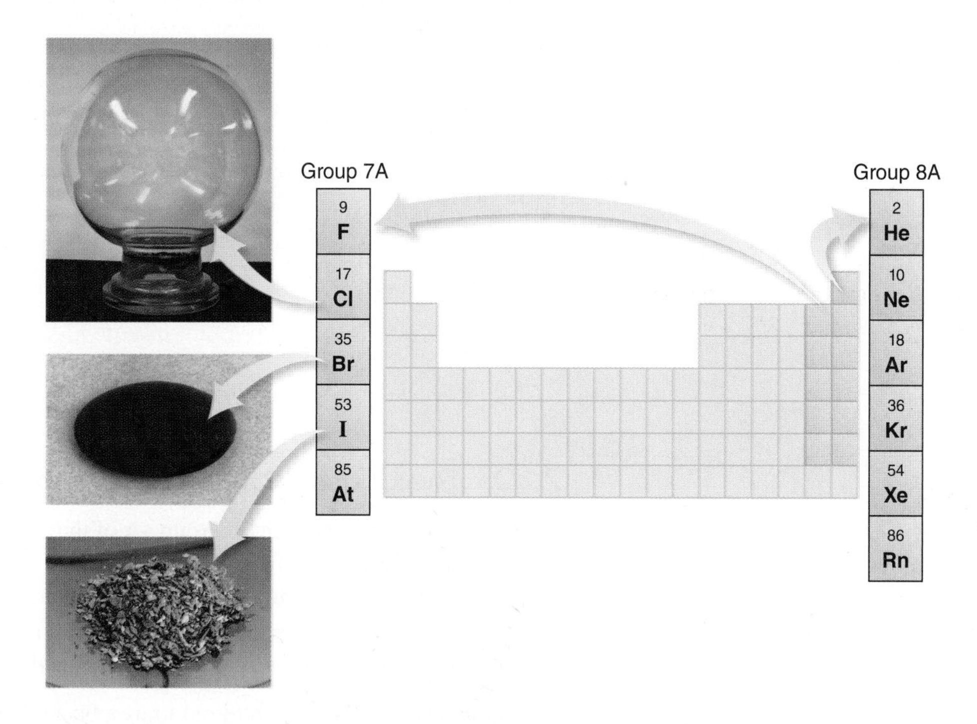

The **halogens,** located in group 7A (group 17), include fluorine (F), chlorine (Cl), bromine (Br), iodine (I), and the rare radioactive element astatine (At). In their elemental form, halogens contain two atoms joined together—F_2, Cl_2, Br_2, and I_2. Fluorine and chlorine are gases at room temperature, bromine is a liquid, and iodine is a solid. Halogens are very reactive and combine with many other elements to form compounds.

The **noble gases,** located in group 8A (group 18), include helium (He), neon (Ne), argon (Ar), krypton (Kr), xenon (Xe), and radon (Rn). Unlike other elements, the noble gases are especially stable as atoms, and so they rarely combine with other elements to form compounds.

The noble gas **radon** has received attention in recent years. Radon is a radioactive gas, and generally its concentration in the air is low and therefore its presence harmless. In some types of soil, however, radon levels can be high and radon detectors are recommended for the basement of homes to monitor radon levels. High radon levels are linked to an increased risk of lung cancer.

HEALTH NOTE

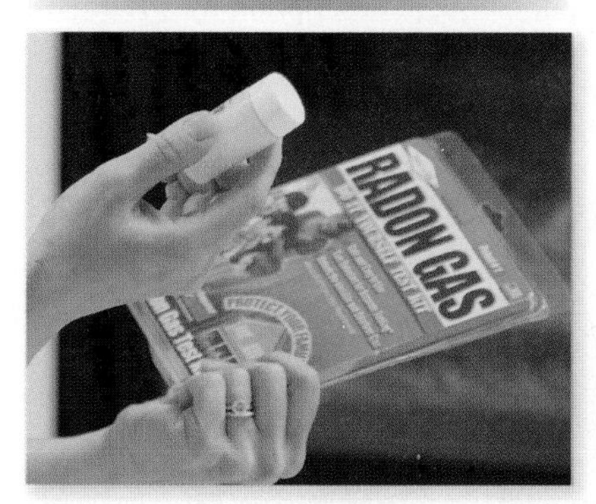

Radon detectors are used to measure high levels of radon, a radioactive noble gas linked to an increased incidence of lung cancer.

PROBLEM 2.21

Identify the element fitting each description.

a. an alkali metal in period 4
b. a second-row element in group 7A
c. a noble gas in the third period
d. a main group element in period 5 and group 2A
e. a transition metal in group 12, period 4
f. a transition metal in group 5, period 5

PROBLEM 2.22

Identify each highlighted element in the periodic table and give its [1] element name and symbol; [2] group number; [3] period; [4] classification (main group element, transition metal, or inner transition metal).

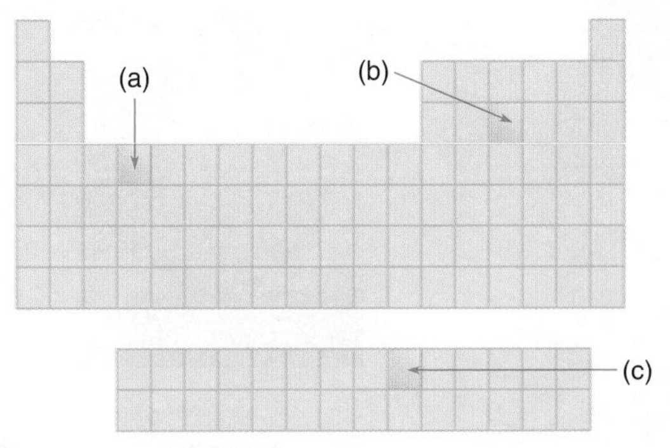

2.4C The Unusual Nature of Carbon

Carbon, a second-row element in group 4A of the periodic table, is different from most other elements in that it has three elemental forms (Figure 2.7). The two most common forms of carbon are diamond and graphite. **Diamond** is hard because it contains a dense three-dimensional network of carbon atoms in six-membered rings. **Graphite,** on the other hand, is a slippery black substance used as a lubricant. It contains parallel sheets of carbon atoms in flat six-membered rings.

Buckminsterfullerene, also referred to as a bucky ball, is a third form that contains 60 carbon atoms joined together in a sphere of 20 hexagons and 12 pentagons in a pattern that resembles a soccer ball. A component of soot, this form of carbon was not discovered until 1985. Its unusual name stems from its shape, which resembles the geodesic dome invented by R. Buckminster Fuller.

Carbon's ability to join with itself and other elements gives it versatility not seen with any other element in the periodic table. In the unscientific but eloquent description by writer Bill

Figure 2.7

Three Elemental Forms of Carbon

a. Diamond

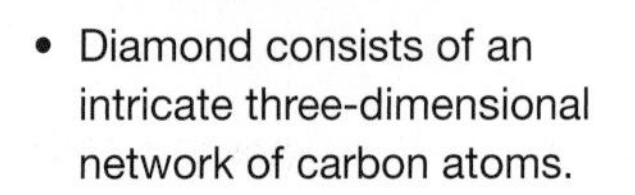

- Diamond consists of an intricate three-dimensional network of carbon atoms.

b. Graphite

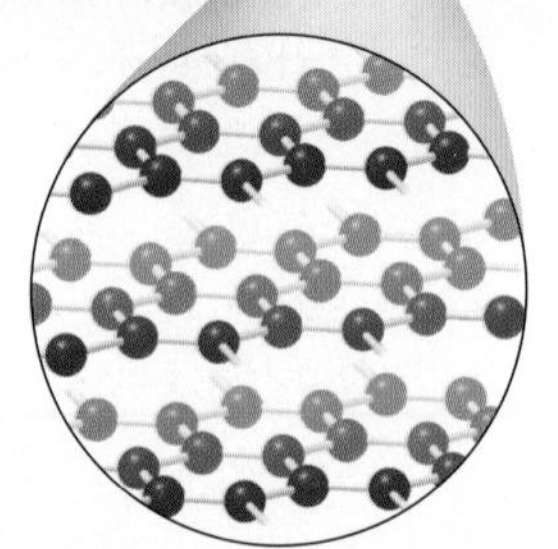

- Graphite contains parallel sheets of carbon atoms.

c. Buckminsterfullerene

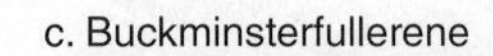

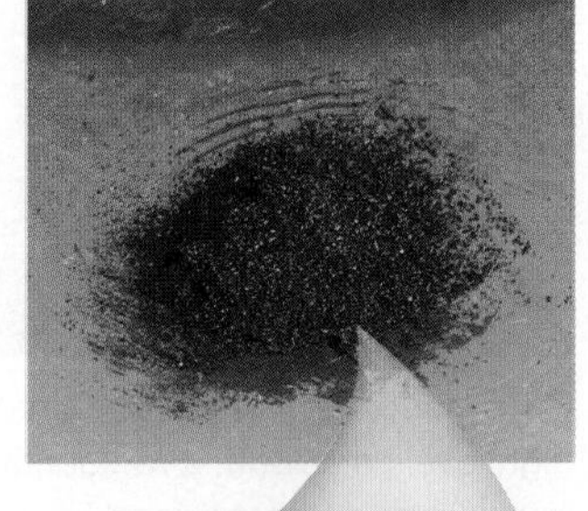

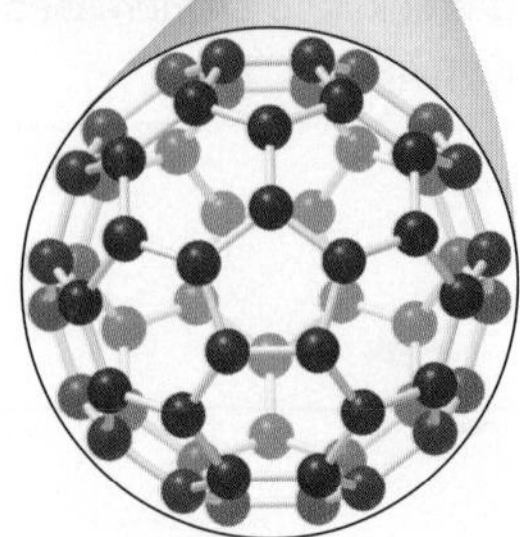

- Buckminsterfullerene contains a sphere with 60 carbon atoms.

Bryson in *A Short History of Nearly Everything,* carbon is described as "the party animal of the atomic world, latching on to many other atoms (including itself) and holding tight, forming molecular conga lines of hearty robustness—the very trick of nature necessary to build proteins and DNA." As a result, millions of compounds that contain the element carbon are known. The chemistry of these compounds is discussed at length in Chapters 10–18.

2.5 Electronic Structure

Why do elements in a group of the periodic table have similar chemical properties? **The chemical properties of an element are determined by the number of *electrons* in an atom.** To understand the properties of an element, therefore, we must learn more about the electrons that surround the nucleus.

The modern description of the electronic structure of an atom is based on the following principle.

- **Electrons do not move freely in space; rather, they occupy specific energy levels.**

The electrons that surround a nucleus are confined to regions called the **principal energy levels** or **shells.**

- **The shells are numbered, $n = 1, 2, 3, 4$, and so forth, beginning closest to the nucleus.**
- **Electrons closer to the nucleus are held more tightly and are lower in energy.**
- **Electrons farther from the nucleus are held less tightly and are higher in energy.**

The number of electrons that can occupy a given shell is determined by the value of n. **The farther an energy level is from the nucleus, the larger its volume becomes, and the more electrons it can hold.** Thus, the first energy level can hold only two electrons, the second holds eight, the third 18, and so forth. The maximum number of electrons is given by the formula $2n^2$, where n = the shell number.

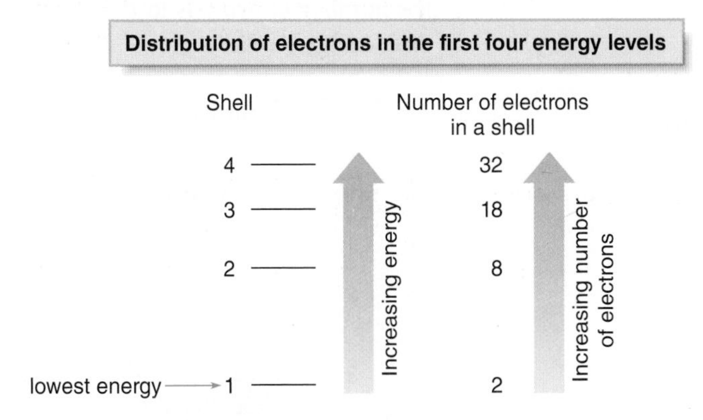

The energy levels consist of a set of **orbitals,** identified by the letters ***s, p, d,*** and ***f.***

- **An *orbital* is a region of space where the probability of finding an electron is high. Each orbital can hold *two* electrons.**

A particular energy level contains a specific number of a given type of orbital. There can be one *s* orbital, three *p* orbitals, five *d* orbitals, and seven *f* orbitals. The energy of orbitals shows the following trend:

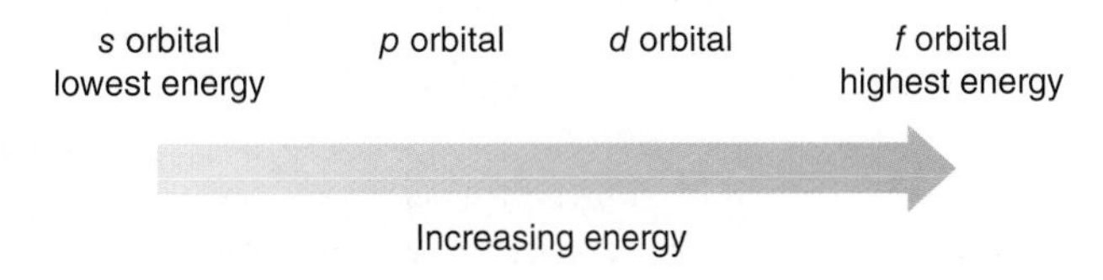

The first energy level of electrons around a nucleus (n = 1) has only one *s* orbital. This orbital is called the 1*s* orbital since it is the *s* orbital in the first shell. Since each orbital can hold two electrons and the first shell has only one orbital, the **first energy level can hold two electrons.**

shell number (principal energy level) → 1*s* = the *s* orbital in the first shell

The second energy level (n = 2) has two types of orbitals—one *s* and three *p* orbitals. These orbitals are called the 2*s* and 2*p* orbitals since they are located in the second energy level. Since each orbital can hold two electrons and there are four orbitals, the **second energy level can hold eight electrons.**

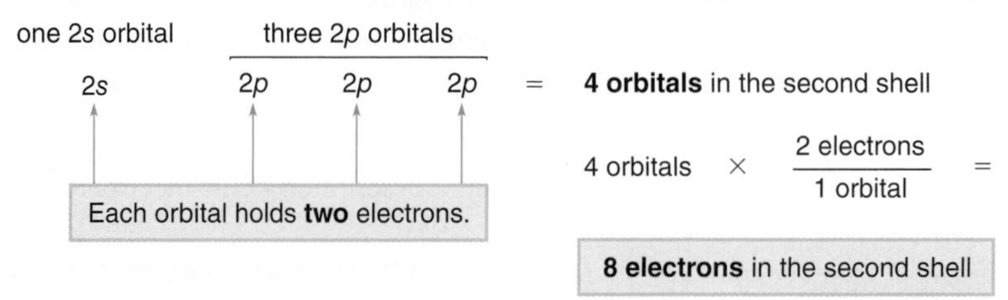

The third energy level (n = 3) has three types of orbitals—one *s,* three *p,* and five *d* orbitals. These orbitals are called the 3*s,* 3*p,* and 3*d* orbitals since they are located in the third energy level. Since each orbital can hold two electrons and the third shell has a total of nine orbitals, the **third energy level can hold 18 electrons.**

one 3*s* orbital — 3*s*; three 3*p* orbitals — 3*p* 3*p* 3*p*; five 3*d* orbitals — 3*d* 3*d* 3*d* 3*d* 3*d* = **9 orbitals** in the third shell

Each orbital holds **two** electrons.

9 orbitals × $\frac{\text{2 electrons}}{\text{1 orbital}}$ = **18 electrons** in the third shell

Thus, the maximum number of electrons that can occupy an energy level is determined by the number of orbitals in the shell. Table 2.4 summarizes the orbitals and electrons in the first three energy levels.

Table 2.4 Orbitals and Electrons Contained in the Principal Energy Levels (n = 1–3)

Shell	Orbitals	Number of Electrons in the Orbitals	Maximum Number of Electrons
1	1*s*	2	2
2	2*s* 2*p* 2*p* 2*p*	2 3 × 2 = 6	8
3	3*s* 3*p* 3*p* 3*p* 3*d* 3*d* 3*d* 3*d* 3*d*	2 3 × 2 = 6 5 × 2 = 10	18

Each type of orbital has a particular shape.

- An *s* orbital has a sphere of electron density. It is lower in energy than other orbitals in the same shell because electrons are kept closer to the positively charged nucleus.
- A *p* orbital has a dumbbell shape. A *p* orbital is higher in energy than an *s* orbital in the same shell because its electron density is farther from the nucleus.

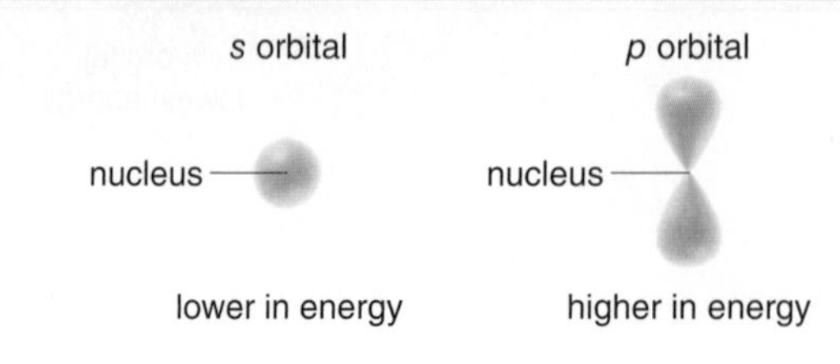

All *s* orbitals are spherical, but the orbital gets larger in size as the shell number increases. Thus, both a 1*s* orbital and a 2*s* orbital are spherical, but the 2*s* orbital is larger. The three *p* orbitals in a shell are perpendicular to each other along the *x*, *y*, and *z* axes.

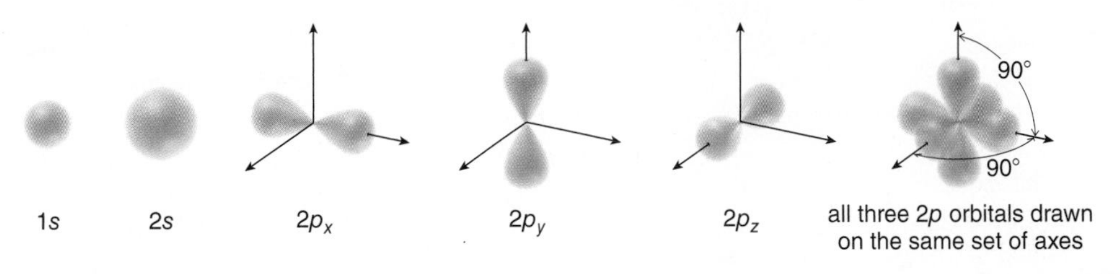

PROBLEM 2.23

What is the maximum number of electrons possible for each energy level or orbital?

a. a 2*p* orbital b. the second energy level c. a 3*s* orbital d. the third shell

2.6 Electronic Configurations

We can now examine the **electronic configuration** of an individual atom—that is, how the electrons are arranged in an atom's orbitals. **The lowest energy arrangement of electrons is called the *ground state*.** Two rules are followed.

Rules to Determine the Ground State Electronic Configuration of an Atom

Rule [1] Electrons are placed in the lowest energy orbitals beginning with the 1*s* orbital.

- In comparing similar types of orbitals from one shell to another (e.g., 1*s* and 2*s*), an orbital closer to the nucleus is lower in energy. Thus, the energy of a 1*s* orbital is lower than a 2*s* orbital.
- Within a shell, orbital energies increase in the following order: *s*, *p*, *d*, *f*.
- These guidelines result in the following order of energies in the first three periods: 1*s*, 2*s*, 2*p*, 3*s*, 3*p* (Figure 2.8).

Rule [2] Each orbital holds a maximum of two electrons.

Figure 2.8

Order of Orbital Filling

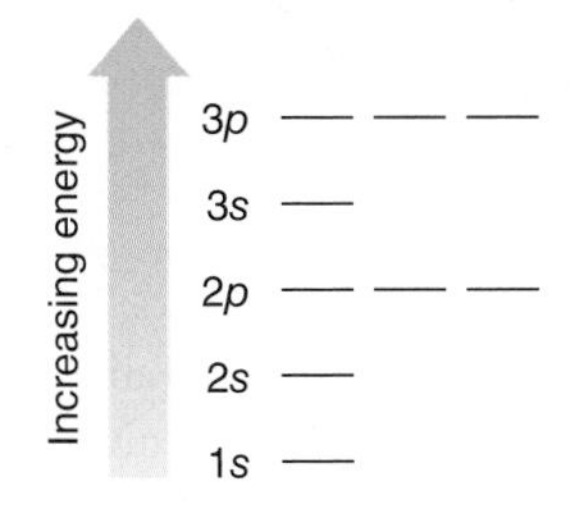

To illustrate how these rules are used, we can write the electronic configuration for several elements. **The electronic configuration shows what orbitals contain electrons and uses a superscript with each orbital to show how many electrons it contains.**

2.6A First-Row Elements (Period 1)

The first row of the periodic table contains only two elements—hydrogen and helium. Since the number of protons in the nucleus equals the number of electrons in a neutral atom, the **atomic number tells us how many electrons must be placed in orbitals.**

Hydrogen (H, Z = 1) has one electron. In the ground state, this electron is added to the lowest energy orbital, the 1*s* orbital. The electronic configuration of hydrogen is written as $1s^1$, meaning that the 1*s* orbital contains one electron.

H	$1s^1$ ← one electron in the 1*s* orbital
1 electron	electronic configuration

A blimp is an airship that contains helium, making it lighter than air. A blimp maneuvers with rudders and propellers and can hover for extended periods, thus making it a useful aircraft for aerial photography at sporting events.

Helium (He, $Z = 2$) has two electrons. In the ground state, both electrons are added to the 1s orbital. The electron configuration is written as $1s^2$, meaning the 1s orbital has two electrons. Helium has a filled first energy level of electrons.

He — $1s^2$ ← two electrons in the 1s orbital

2 electrons — electronic configuration

2.6B Second-Row Elements (Period 2)

To write electronic configurations for the second-row elements, we must now use the four orbitals in the second energy level—the 2s orbital and the three 2p orbitals. Since electrons are always added to the lowest energy orbitals first, all second-row elements have the 1s orbital filled with electrons, and then the remaining electrons are added to the orbitals in the second shell. Since the 2s orbital is lower in energy than the 2p orbitals, it is completely filled before adding electrons to the 2p orbitals.

Lithium (Li, $Z = 3$) has three electrons. In the ground state, two electrons are added to the 1s orbital and the remaining electron is added to the 2s orbital.

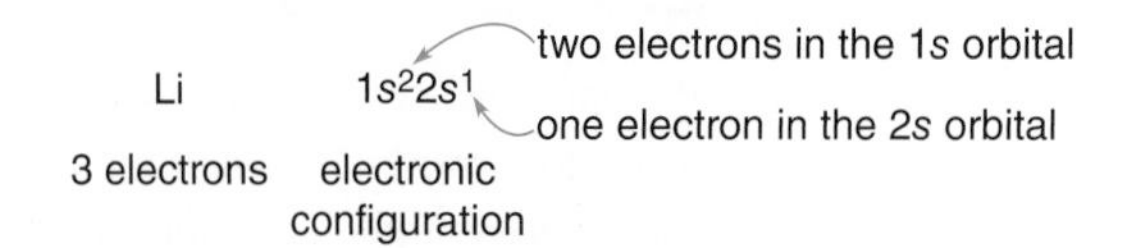

Carbon (C, $Z = 6$) has six electrons. In the ground state, two electrons are added to both the 1s and 2s orbitals. The two remaining electrons are added to the 2p orbitals, so the electronic configuration is written as $1s^22s^22p^2$.

C — $1s^22s^22p^2$ ← two electrons in the 2p orbitals

6 electrons — electronic configuration

Neon (Ne, $Z = 10$) has 10 electrons. In the ground state, two electrons are added to the 1s, 2s, and each of the three 2p orbitals, giving a total of six electrons in the 2p orbitals.

Ne — $1s^22s^22p^6$ ← six electrons in the 2p orbitals

10 electrons — electronic configuration

The electronic configurations of all the first- and second-row elements are listed in Table 2.5. The total number of electrons used for the electronic configuration of a neutral atom is always equal to the atomic number.

PROBLEM 2.24

What element(s) in the first and second period fit each description?

a. The element has one electron in the second energy level.

b. There are two electrons in the 2s orbital.

c. The electronic configuration is $1s^22s^22p^5$.

d. The element contains six electrons in the second energy level.

Table 2.5 Electronic Configurations of the First- and Second-Row Elements

Atomic Number	Element	Electronic Configuration	Total Number of Electrons
1	H	$1s^1$	1
2	He	$1s^2$	2
3	Li	$1s^22s^1$	3
4	Be	$1s^22s^2$	4
5	B	$1s^22s^22p^1$	5
6	C	$1s^22s^22p^2$	6
7	N	$1s^22s^22p^3$	7
8	O	$1s^22s^22p^4$	8
9	F	$1s^22s^22p^5$	9
10	Ne	$1s^22s^22p^6$	10

2.6C Other Elements

Electronic configurations can be written in a similar fashion for other elements in the periodic table. Sample Problems 2.10 and 2.11 illustrate two examples.

SAMPLE PROBLEM 2.10

Give the ground state electronic configuration of the element sulfur.

Analysis

- Use the atomic number to determine the number of electrons.
- Place electrons two at a time into the lowest energy orbitals, following the order of orbital filling in Figure 2.8.

Solution

The atomic number of sulfur is 16, so 16 electrons must be placed in orbitals. Twelve electrons are added in pairs to the 1*s*, 2*s*, three 2*p*, and 3*s* orbitals. The remaining four electrons are then added to the three 3*p* orbitals.

S

sulfur
16 electrons

Answer:
The electronic configuration is $1s^22s^22p^63s^23p^4$.

ENVIRONMENTAL NOTE

Coal that is high in **sulfur** content burns to form sulfur oxides, which in turn react with water to form sulfurous and sulfuric acids. Rain that contains these acids has destroyed acres of forests worldwide. Sulfur is a third-row element in the periodic table.

PROBLEM 2.25

Give the ground state electronic configuration for each element: (a) aluminum; (b) chlorine.

SAMPLE PROBLEM 2.11

What element has the ground state electronic configuration $1s^22s^22p^63s^2$?

Analysis

- Count the number of electrons in the electronic configuration.
- Since the number of electrons equals the atomic number in a neutral atom, identify the element from the atomic number in the periodic table.

Solution

The element has a total of 12 electrons (2 + 2 + 6 + 2). The element with an atomic number of 12 is magnesium.

PROBLEM 2.26

What element has each ground state electronic configuration?

a. $1s^2 2s^2 2p^6 3s^1$ b. $1s^2 2s^2 2p^6 3s^2 3p^2$

2.7 Valence Electrons

The chemical properties of an element depend on the most loosely held electrons—that is, those electrons in the outermost shell, called the **valence shell. The period number tells the number of the valence shell.**

- The electrons in the outermost shell are called the *valence electrons.*

2.7A Relating Valence Electrons to Group Number

To identify the electrons in the valence shell, always look for the shell with the *highest* number. Thus, beryllium has two valence electrons that occupy the 2*s* orbital. Chlorine has seven valence electrons since it has a total of seven electrons in the third shell, two in the 3*s* orbital and five in the 3*p* orbitals.

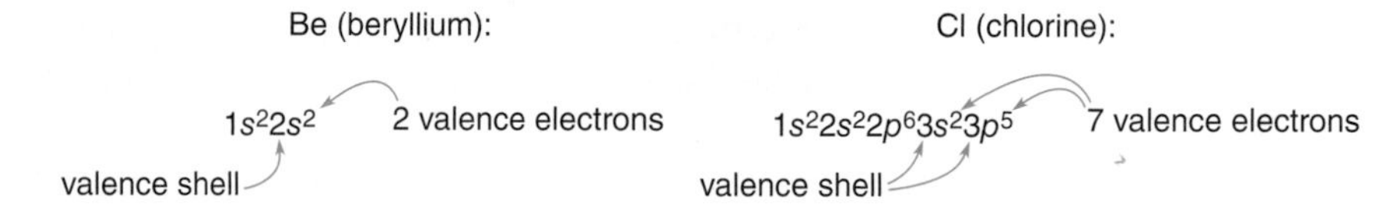

If we examine the electronic configuration of a group in the periodic table, two facts become apparent.

- Elements in the same group have the same number of valence electrons and similar electronic configurations.
- The group number (using the 1A–8A system) equals the number of valence electrons for main group elements (except helium).

Thus, the periodic table is organized into groups of elements with similar valence electronic configurations in the same column. The valence electronic configurations of the main group elements in the first three rows of the periodic table are given in Table 2.6. As an example, the alkali metals in group **1A** all have **one** valence electron that occupies an *s* orbital.

- The chemical properties of a group are similar because these elements contain the same electronic configuration of valence electrons.

Table 2.6 Valence Electronic Configurations for the Main Group Elements in Periods 1–3

Group Number	1A	2A	3A	4A	5A	6A	7A	8A
Period 1	H $1s^1$							He $1s^2$
Period 2	Li $2s^1$	Be $2s^2$	B $2s^2 2p^1$	C $2s^2 2p^2$	N $2s^2 2p^3$	O $2s^2 2p^4$	F $2s^2 2p^5$	Ne $2s^2 2p^6$
Period 3	Na $3s^1$	Mg $3s^2$	Al $3s^2 3p^1$	Si $3s^2 3p^2$	P $3s^2 3p^3$	S $3s^2 3p^4$	Cl $3s^2 3p^5$	Ar $3s^2 3p^6$

Take particular note of the electronic configuration of the noble gases in group 8A. **All of these elements have a completely filled outer shell of valence electrons.** Helium has a filled first shell ($1s^2$ configuration). The remaining elements have a completely filled valence shell of *s* and *p* orbitals (s^2p^6). This electronic arrangement is especially stable, and as a result, these elements exist in nature as single atoms. We will learn about the consequences of having a completely filled valence shell in Chapter 3.

SAMPLE PROBLEM 2.12

Identify the total number of electrons, the number of valence electrons, and the name of the element with each electronic configuration.

a. $1s^22s^2$ b. $1s^22s^22p^63s^23p^2$

Analysis

To obtain the total number of electrons, add up the superscripts. This gives the atomic number and identifies the element. To determine the number of valence electrons, add up the number of electrons in the shell with the highest number.

Solution

a.

valence shell

$1s^22s^2$

2 valence electrons

Total number of electrons = $2 + 2 = \mathbf{4}$

Answer: Beryllium (Be), 4 total electrons and 2 valence electrons

b.

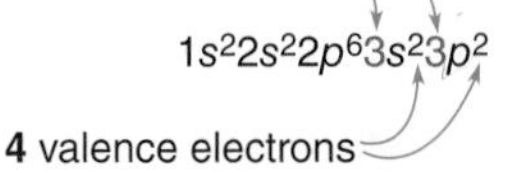

Total number of electrons = $2 + 2 + 6 + 2 + 2 = \mathbf{14}$

Answer: Silicon (Si), 14 total electrons and 4 valence electrons

PROBLEM 2.27

Identify the total number of electrons, the number of valence electrons, and the name of the element with each electronic configuration.

a. $1s^22s^1$ b. $1s^22s^22p^63s^2$ c. $1s^22s^22p^63s^23p^3$

SAMPLE PROBLEM 2.13

Determine the number of valence electrons of each element: (a) nitrogen; (b) potassium.

Analysis

The group number of a main group element = the number of valence electrons.

Solution

a. Nitrogen is located in group 5A, so it has five valence electrons.
b. Potassium is located in group 1A, so it has one valence electron.

PROBLEM 2.28

Determine the number of valence electrons of each element: (a) fluorine; (b) krypton; (c) magnesium; (d) germanium.

2.7B Electron-Dot Symbols

The number of valence electrons around an atom is often represented by an **electron-dot symbol.** Representative examples are shown.

	H	C	O	Cl
Number of valence electrons:	1	4	6	7
Electron-dot symbol:	H·	·Ċ· (4 dots)	·Ö· (6 dots)	·C̈l: (7 dots)

- **Each dot represents one valence electron.**
- **The dots are placed on the four sides of an element symbol.**
- **For one to four valence electrons, single dots are used. With more than four electrons, the dots are paired.**

The location of the dots around the symbol—side, top, or bottom—does not matter. Each of the following representations for the five valence electrons of nitrogen is equivalent.

SAMPLE PROBLEM 2.14

Write an electron-dot symbol for each element: (a) sodium; (b) phosphorus.

Analysis

Write the symbol for each element and use the group number to determine the number of valence electrons for a main group element. Represent each valence electron with a dot.

Solution

a. The symbol for sodium is Na. Na is in group 1A and has one valence electron. Electron-dot symbol:

Na·

b. The symbol for phosphorus is P. P is in group 5A and has five valence electrons. Electron-dot symbol:

·P̈·

PROBLEM 2.29

Give the electron-dot symbol for each element: (a) bromine; (b) lithium; (c) aluminum; (d) sulfur; (e) neon.

2.8 Periodic Trends

Many properties of atoms exhibit **periodic trends;** that is, they change in a regular way across a row or down a column of the periodic table. Two properties that illustrate this phenomenon are **atomic size** and **ionization energy.**

2.8A Atomic Size

The size of an atom is measured by its atomic radius—that is, the distance from the nucleus to the outer edge of the valence shell. Two periodic trends characterize the size of atoms.

- **The size of atoms increases down a column of the periodic table, as the valence electrons are farther from the nucleus.**

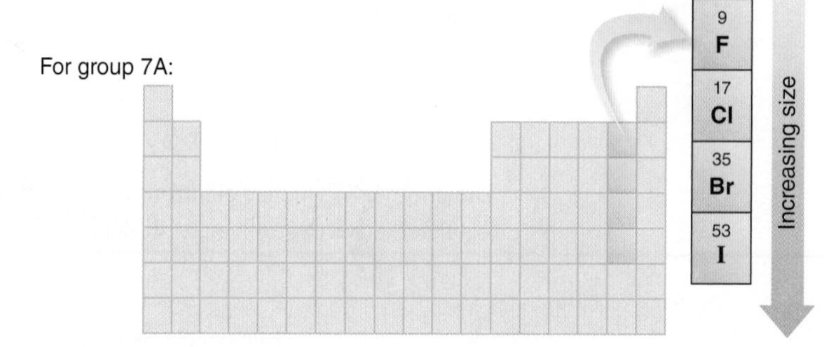

- The size of atoms decreases across a row of the periodic table as the number of protons in the nucleus increases. An increasing number of protons pulls the electrons closer to the nucleus, so the atom gets smaller.

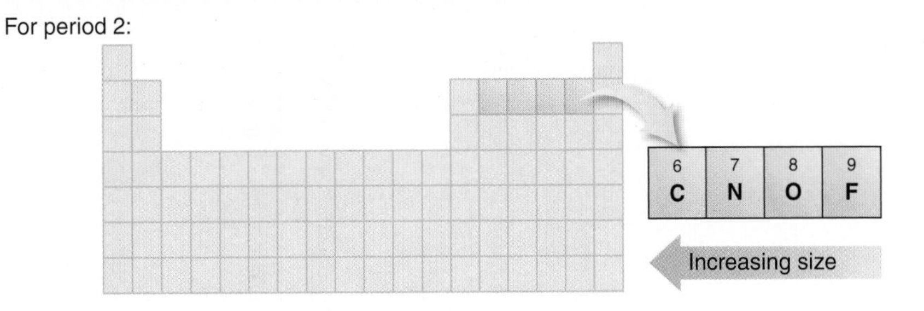

SAMPLE PROBLEM 2.15

(a) Which of the labeled atoms has the larger atomic radius? (b) Identify the elements.

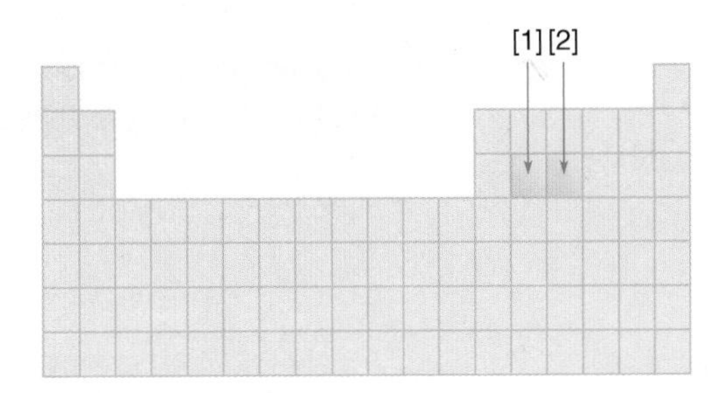

Analysis

Atomic radius (size) decreases across a row of the periodic table as the number of protons in the nucleus increases.

Solution

a. Element [1] is farther to the left in the third row, so it has the larger atomic radius.

b. Element [1] is silicon (period 3, group 4A) and element [2] is phosphorus (period 3, group 5A).

PROBLEM 2.30

Which element in each pair has the larger atomic radius?

a. nitrogen or arsenic b. silicon or sulfur c. fluorine or phosphorus

2.8B Ionization Energy

Since a negatively charged electron is attracted to a positively charged nucleus, energy is required to remove an electron from a neutral atom. The more tightly the electron is held, the greater the energy required to remove it. Removing an electron from a neutral atom forms a **cation.**

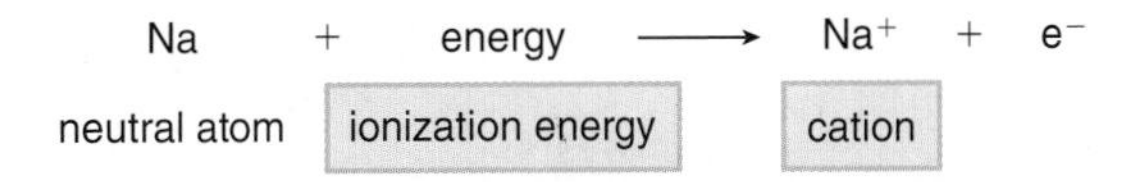

- The *ionization energy* is the energy needed to remove an electron from a neutral atom.
- A *cation* is positively charged, and has fewer electrons than the neutral atom.

Two periodic trends characterize ionization energy.

- **Ionization energies decrease down a column of the periodic table as the valence electrons get farther from the positively charged nucleus.**

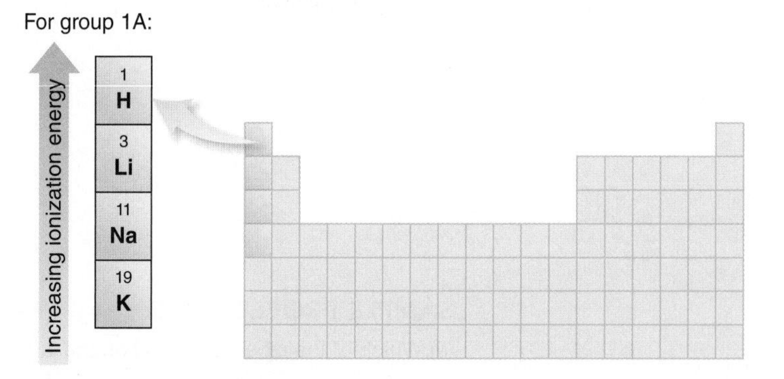

- **Ionization energies generally increase across a row of the periodic table as the number of protons in the nucleus increases.**

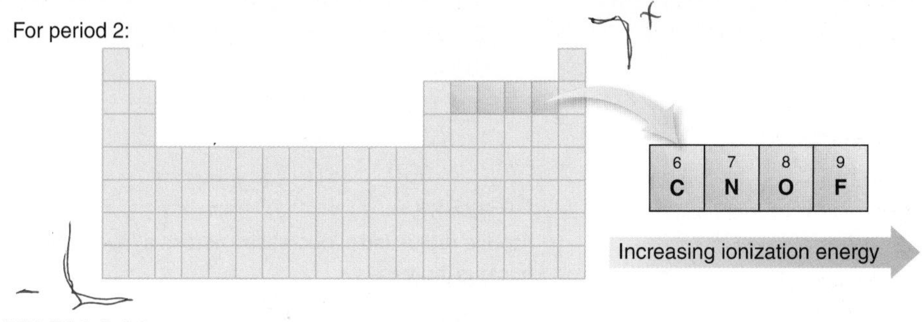

PROBLEM 2.31

Which element in each pair has the higher ionization energy?

a. silicon or sodium b. carbon or silicon c. sulfur or fluorine

SAMPLE PROBLEM 2.16

(a) Which of the indicated atoms has the larger atomic radius? (b) Which of the indicated atoms has the higher ionization energy?

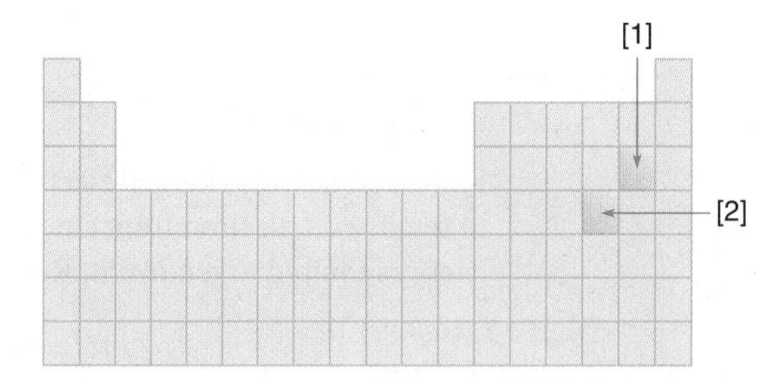

Analysis

- Atomic size decreases across a row and increases down a column of the periodic table.
- Ionization energy increases across a row and decreases down a column of the periodic table.

Solution

a. Element [2] is farther to the left and farther down the column, so it has the larger atomic radius.

b. Element [1] is farther to the right and closer to the top of the column (lower period number), so it has the higher ionization energy.

PROBLEM 2.32

(a) Which of the indicated atoms has the smaller atomic radius? (b) Which of the indicated atoms has the lower ionization energy? (c) Identify the elements.

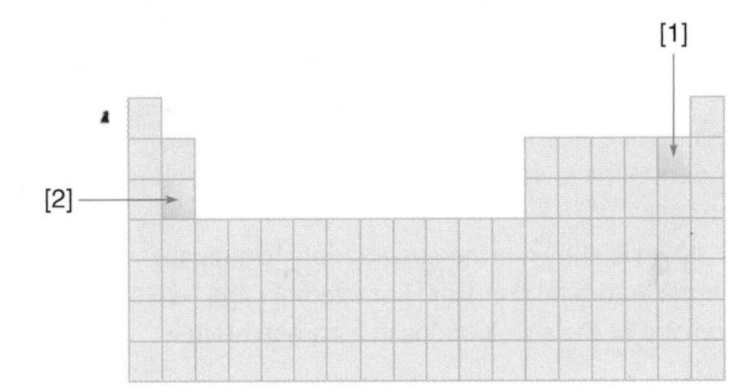

KEY TERMS

Actinide (2.4)
Alkali metal (2.4)
Alkaline earth element (2.4)
Atom (2.2)
Atomic mass unit (2.2)
Atomic number (2.2)
Atomic weight (2.3)
Building-block element (2.1)
Cation (2.8)
Chemical formula (2.1)
Compound (2.1)
Deuterium (2.3)
Electron (2.2)
Electron cloud (2.2)
Electron-dot symbol (2.7)
Electronic configuration (2.6)
Element (2.1)
Ground state (2.6)
Group (2.4)
Group number (2.4)
Halogen (2.4)
Inner transition metal element (2.4)
Ionization energy (2.8)
Isotope (2.3)
Lanthanide (2.4)
Main group element (2.4)
Major mineral (Macronutrient, 2.1)
Mass number (2.2)
Metal (2.1)
Metalloid (2.1)
Neutron (2.2)
Noble gas (2.4)
Nonmetal (2.1)
Nucleus (2.2)
Orbital (2.5)
Period (2.4)
Periodic table (2.1)
p Orbital (2.5)
Proton (2.2)
Shell (2.5)
s Orbital (2.5)
Trace element (Micronutrient, 2.1)
Transition metal element (2.4)
Tritium (2.3)
Valence electron (2.7)

KEY CONCEPTS

1 How is the name of an element abbreviated and how does the periodic table help to classify it as a metal, nonmetal, or metalloid? (2.1)

- An element is abbreviated by a one- or two-letter symbol. The periodic table contains a stepped line from boron to astatine. All metals are located to the left of the line. All nonmetals except hydrogen are located to the right of the line. The seven elements located along the line are metalloids.

2 What are the basic components of an atom? (2.2)

- An atom is composed of two parts: a dense nucleus containing positively charged protons and neutral neutrons, and an electron cloud containing negatively charged electrons. Most of the mass of an atom resides in the nucleus, while the electron cloud contains most of its volume.
- The atomic number (Z) of a neutral atom tells the number of protons and the number of electrons. The mass number (A) is the sum of the number of protons (Z) and the number of neutrons.

3 What are isotopes and how are they related to the atomic weight? (2.3)

- Isotopes are atoms that have the same number of protons but a different number of neutrons. The atomic weight is the weighted average of the mass of the naturally occurring isotopes of a particular element.

4 What are the basic features of the periodic table? (2.4)

- The periodic table is a schematic of all known elements, arranged in rows (periods) and columns (groups), organized so that elements with similar properties are grouped together.
- The vertical columns are assigned group numbers using two different numbering schemes: 1–8 plus the letters A or B; or 1–18.
- The periodic table is divided into the main group elements (groups 1A–8A), the transition metals (groups 1B–8B), and the inner transition metals located in the two rows below the main table.

5 How are electrons arranged around an atom? (2.5)

- Electrons occupy discrete energy levels (numbered 1, 2, 3, and so on) that contain orbitals (*s, p, d,* and *f*).
- Each orbital can hold two electrons.

6 What rules determine the electronic configuration of an atom? (2.6)

- To write the ground state electronic configuration of an atom, electrons are added to the lowest energy orbitals, giving each orbital two electrons.
- Electron configuration is shown using superscripts to indicate how many electrons an orbital contains. For example, the electron configuration of the six electrons in a carbon atom is $1s^2 2s^2 2p^2$.

7 How is the location of an element in the periodic table related to its number of valence electrons? (2.7)

- Elements in the same group have the same number of valence electrons.

8 What is an electron-dot symbol? (2.7)

- An electron-dot symbol uses a dot to represent each valence electron around the symbol for an element.

9 How are atomic size and ionization energy related to location in the periodic table? (2.8)

- The size of an atom decreases across a row and increases down a column.
- Ionization energy—the energy needed to remove an electron from an atom—increases across a row and decreases down a column.

UNDERSTANDING KEY CONCEPTS

Selected in-chapter and odd-numbered end-of-chapter problems have brief answers at the end of each chapter. The *Student Study Guide and Solutions Manual* contains detailed solutions to all in-chapter and odd-numbered end-of-chapter problems, as well as additional worked examples and a chapter self-test.

2.33 Identify the elements used in each example of molecular art.

a. b.

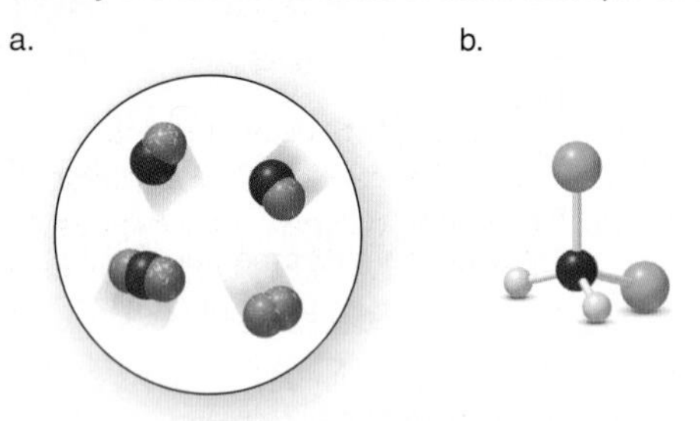

2.34 Write a chemical formula for each example of molecular art.

a. b. c.

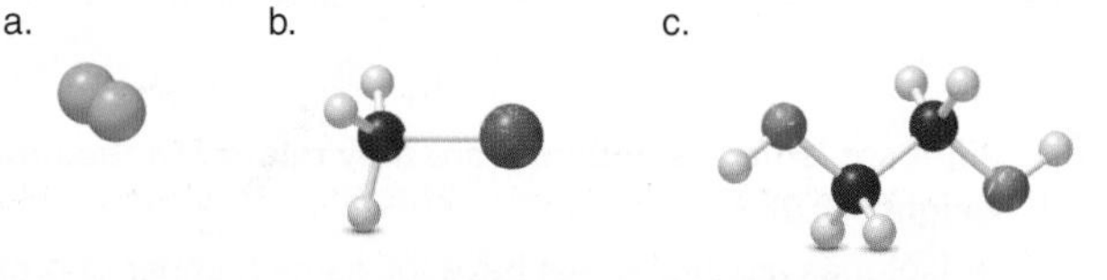

2.35 Give the following information about the atom shown: (a) the number of protons and neutrons in the nucleus; (b) the atomic number; (c) the mass number; (d) the number of electrons in the neutral atom; and (e) the element symbol.

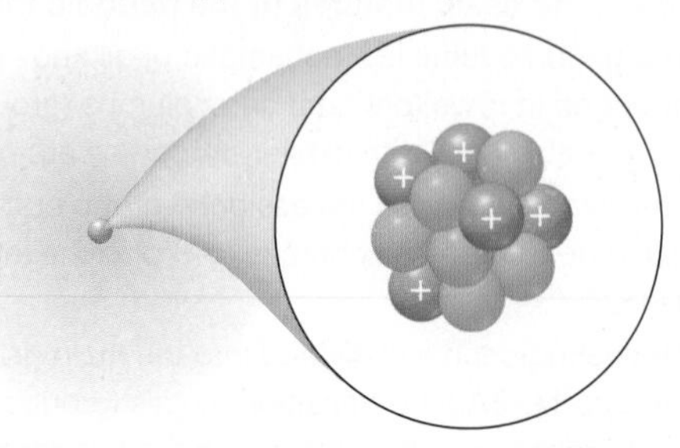

2.36 Give the following information about the atom shown: (a) the number of protons and neutrons in the nucleus; (b) the atomic number; (c) the mass number; (d) the number of electrons in the neutral atom; and (e) the element symbol.

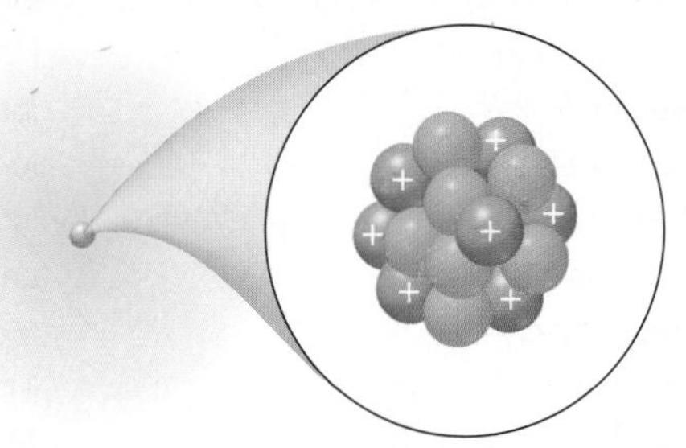

2.37 Selenium is a micronutrient necessary for certain enzymes that block unwanted oxidation reactions. Selenium is also needed for proper functioning of the thyroid gland. Answer the following questions about the element selenium.

a. What is its element symbol?
b. To what group number and period does selenium belong?
c. Is selenium a main group element, transition metal, or inner transition metal?
d. How many valence electrons does selenium contain?
e. Draw an isotope symbol for a selenium atom that contains 46 neutrons in the nucleus.

2.38 Answer the following questions about the element silicon, a micronutrient needed for healthy bones, nails, skin, and hair.

a. What is its element symbol?
b. To what group number and period does silicon belong?
c. Is silicon a main group element, transition metal, or inner transition metal?
d. Draw an isotope symbol for a silicon atom that contains 14 neutrons in the nucleus.

2.39 (a) What element has the ground state electronic configuration $1s^22s^22p^63s^23p^1$? (b) How many valence electrons does this element contain? (c) Give the group number and the period number for the element.

2.40 (a) Write the ground state electronic configuration for the element silicon. (b) How many valence electrons does silicon contain? (c) Give an electron-dot symbol for silicon.

2.41 Which element in each pair is larger?

a. bromine or iodine
b. carbon or nitrogen

2.42 Which element in each pair has its valence electrons farther from the nucleus?

a. sodium or magnesium
b. neon or krypton

2.43 Consider the following two spheres, which represent atoms in the same group in the periodic table. (a) Which atom has the higher atomic number? (b) Which atom has the larger ionization energy? (c) Which atom gives up an electron more easily?

2.44 Consider the following two spheres, which represent atoms in the same period in the periodic table. (a) Which atom has the higher atomic number? (b) Which atom has the larger ionization energy? (c) Which atom gives up an electron more easily?

ADDITIONAL PROBLEMS

Elements

2.45 Give the name of the elements in each group of three element symbols.

a. Au, At, Ag
b. N, Na, Ni
c. S, Si, Sn
d. Ca, Cr, Cl

2.46 What element(s) are designated by each symbol or group of symbols?

a. CU and Cu
b. Os and OS
c. Ni and NI
d. BIN, BiN, and BIn

2.47 Does each chemical formula represent an element or a compound?

a. H_2 b. H_2O_2 c. S_8 d. Na_2CO_3 e. C_{60}

2.48 Identify the elements in each chemical formula and tell how many atoms of each are present.

a. $K_2Cr_2O_7$
b. $C_5H_8NNaO_4$ (MSG, flavor enhancer)
c. $C_{10}H_{16}N_2O_3S$ (vitamin B_7)

2.49 Identify the element that fits each description.

a. an alkali metal in period 6
b. a transition metal in period 5, group 8
c. a main group element in period 3, group 7A
d. a halogen in period 2

2.50 Identify the element that fits each description.

a. an alkaline earth element in period 3
b. a noble gas in period 6
c. a transition metal in period 4, group 11
d. a transition metal in period 6, group 10

2.51 Give all of the terms that apply to each element: [1] metal; [2] nonmetal; [3] metalloid; [4] alkali metal; [5] alkaline earth element; [6] halogen; [7] noble gas; [8] main group element; [9] transition metal; [10] inner transition metal.

a. sodium
b. silver
c. xenon
d. platinum

2.52 Give all of the terms that apply to each element: [1] metal; [2] nonmetal; [3] metalloid; [4] alkali metal; [5] alkaline earth element; [6] halogen; [7] noble gas; [8] main group element; [9] transition metal; [10] inner transition metal.

a. bromine
b. calcium
c. cesium
d. gold

Atomic Structure

2.53 Complete the following table for neutral elements.

Element Symbol	Atomic Number	Mass Number	Number of Protons	Number of Neutrons	Number of Electrons
a. C		12			
b.		31			15
c.				35	30
d. Mg		24			

2.54 For the given atomic number (*Z*) and mass number (*A*): [1] identify the element; [2] give the element symbol; [3] give the number of protons, neutrons, and electrons.

a. $Z = 10, A = 20$
b. $Z = 13, A = 27$
c. $Z = 38, A = 88$
d. $Z = 55, A = 133$

Periodic Table

2.55 Label each region on the periodic table.

a. noble gases
b. period 3
c. group 4A
d. alkaline earth elements
e. transition metals
f. group 10

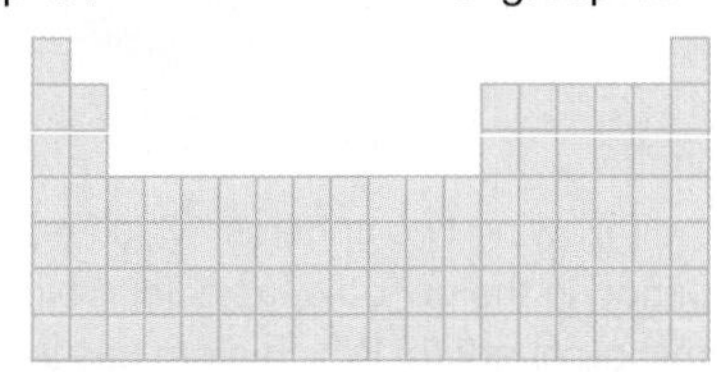

2.56 Identify each highlighted element in the periodic table and give its [1] element name and symbol; [2] group number; [3] period; [4] classification (i.e., main group element, transition metal, or inner transition metal).

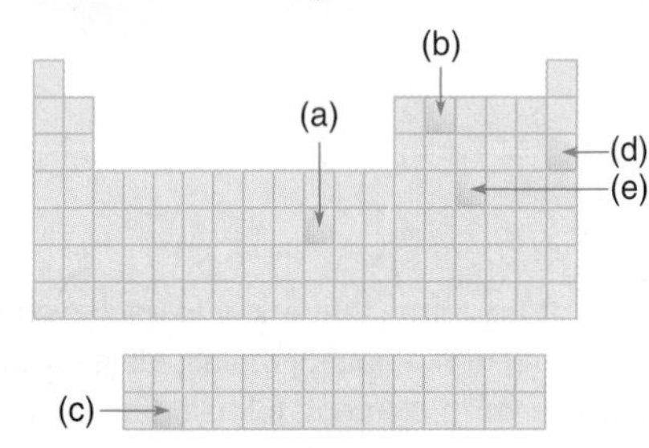

2.57 Classify each element in the fourth row of the periodic table as a metal, nonmetal, or metalloid.

2.58 Which group(s) in the periodic table contain only nonmetals?

Isotopes and Atomic Weight

2.59 The most common isotope of oxygen has a mass number of 16, but two other isotopes having mass numbers of 17 and 18 are also known. For each isotope, give the following information: (a) the number of protons; (b) the number of neutrons; (c) the number of electrons in the neutral atom; (d) the group number; (e) the element symbols using superscripts and subscripts.

2.60 The three most common isotopes of tin have mass numbers 116, 118, and 120. For each isotope, give the following information: (a) the number of protons; (b) the number of neutrons; (c) the number of electrons in the neutral atom; (d) the group number; (e) the element symbols using superscripts and subscripts.

2.61 How many protons, neutrons, and electrons are contained in each element?

a. $^{27}_{13}Al$ b. $^{35}_{17}Cl$ c. $^{34}_{16}S$

2.62 Give the number of protons, neutrons, and electrons in each element: (a) silver-115; (b) Au-197; (c) Rn-222; (d) osmium-192.

2.63 Write the element symbol that fits each description, using a superscript for the mass number and a subscript for the atomic number.

a. an element that contains 53 protons and 74 neutrons
b. an element with 35 electrons and a mass number of 79

2.64 Write the element symbol that fits each description. Use a superscript for the mass number and a subscript for the atomic number.

a. an element that contains 10 protons and 12 neutrons
b. an element with atomic number 24 and mass number 52

2.65 Can the neutral atoms of two different elements have the same number of electrons? Explain.

2.66 Can the neutral atoms of two different elements have the same number of neutrons? Explain.

Electronic Configuration

2.67 What element(s) in the first three periods fit each description?

a. The element contains five electrons in the 3*p* orbitals.
b. There is one valence electron.
c. The element contains four electrons in the second energy level.

2.68 What element(s) in the first three periods fit each description?

a. The element contains three electrons in the 3*p* orbitals.
b. There are two valence electrons.
c. The element contains five electrons in the second shell.

2.69 Write out the electronic configuration for each element: (a) B; (b) Mg.

2.70 Write out the electronic configuration for each element: (a) nitrogen; (b) argon.

2.71 Give the total number of electrons, the number of valence electrons, and the identity of the element with each electronic configuration.

a. $1s^22s^22p^63s^23p^4$ b. $1s^22s^22p^4$

2.72 Give the total number of electrons, the number of valence electrons, and the identity of the element with each electronic configuration.

a. $1s^22s^22p^63s^23p^5$ b. $1s^22s^22p^3$

2.73 For each element, give the following information: [1] total number of electrons; [2] group number; [3] number of valence electrons; [4] period.

a. carbon b. calcium c. krypton

2.74 For each element, give the following information: [1] total number of electrons; [2] group number; [3] number of valence electrons; [4] period.

a. oxygen b. sodium c. phosphorus

2.75 How many valence electrons does an element in each group contain: (a) 2A; (b) 4A; (c) 7A?

2.76 In what shell do the valence electrons reside for an element in period: (a) 2; (b) 3; (c) 4?

2.77 Give the number of valence electrons in each element.

a. sulfur b. chlorine c. barium

2.78 Give the number of valence electrons in each element.

a. neon b. rubidium c. aluminum

2.79 Write an electron-dot symbol for each element: (a) beryllium; (b) iodine; (c) magnesium; (d) argon.

2.80 Write an electron-dot symbol for each element: (a) K; (b) B; (c) F; (d) Ca.

Periodic Trends

2.81 Which element in each pair has its valence electrons farther from the nucleus?

a. carbon or fluorine b. argon or bromine

2.82 For each pair of elements in Problem 2.81, label the element from which it is easier to remove an electron.

2.83 Rank the atoms in each group in order of increasing size.

a. boron, carbon, neon
b. calcium, magnesium, beryllium
c. silicon, sulfur, magnesium
d. krypton, neon, xenon

2.84 Arrange the elements in each group in order of increasing ionization energy.

a. phosphorus, silicon, sulfur
b. magnesium, calcium, beryllium
c. carbon, fluorine, beryllium
d. neon, krypton, argon

2.85 Rank the following elements in order of increasing size: sulfur, silicon, oxygen, magnesium, and fluorine.

2.86 Rank the following elements in order of increasing ionization energy: nitrogen, fluorine, magnesium, sodium, and phosphorus.

Applications

2.87 Answer the following questions about macronutrients [1], [2], and [3] highlighted in the periodic table.

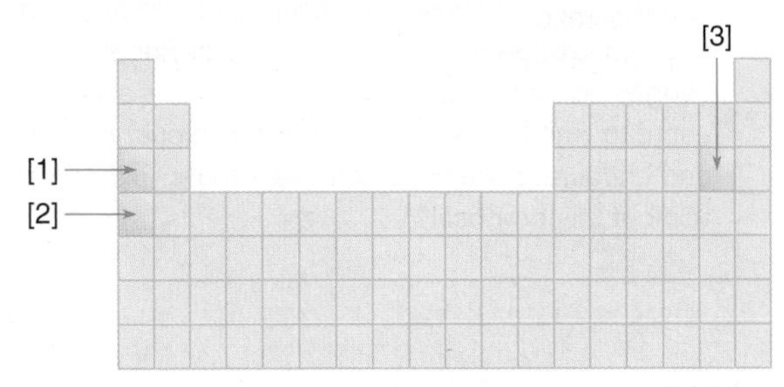

a. What elements correspond to macronutrients [1]–[3]?
b. Is each element classified as a metal, nonmetal, or metalloid?
c. Which element has the smallest atomic radius?
d. Which element has the largest atomic radius?
e. Which element has the largest ionization energy?
f. Which element has the smallest ionization energy?
g. How many valence electrons does each element possess?

2.88 Platinum is a precious metal used in a wide variety of products. Besides fine jewelry, platinum is also the catalyst found in the catalytic converters of automobile exhaust systems, and platinum-containing drugs like cisplatin are used to treat some lung and ovarian cancers. Answer the following questions about the element platinum.

a. What is its element symbol?
b. What group number and period are assigned to platinum?
c. What is its atomic number?
d. Is platinum classified as a main group element, transition metal, or inner transition metal?

2.89 (a) What is the chemical formula for bupropion (trade name Zyban), an antidepressant also used to reduce nicotine cravings? (b) Which element in bupropion has valence electrons that are farthest from the nucleus?

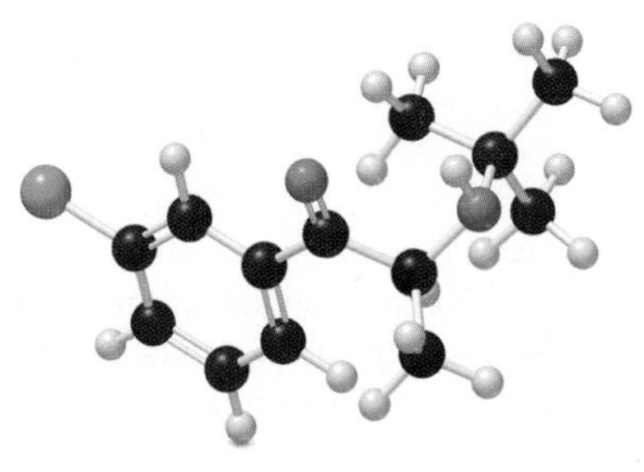

bupropion

2.90 (a) What is the chemical formula for brompheniramine, an antihistamine used to relieve the runny nose and sneezing associated with allergies? (b) Which element in brompheniramine has valence electrons that are farthest from the nucleus?

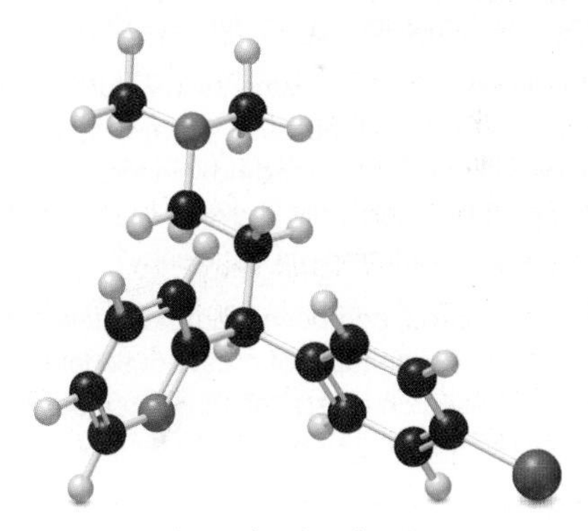

brompheniramine

CHALLENGE PROBLEMS

2.91 Strontium-90 is a radioactive element formed in nuclear reactors. When an unusually high level of strontium is released into the air, such as occurred during the Chernobyl nuclear disaster in 1986, the strontium can be incorporated into the bones of exposed individuals. High levels of strontium can cause bone cancer and leukemia. Why does Sr-90 cause this particular health problem? (Hint: What macronutrient has similar chemical properties to strontium?)

2.92 Sesame seeds, sunflower seeds, and peanuts are good dietary sources of the trace element copper. Copper is needed for the synthesis of neurotransmitters, compounds that transmit nerve signals from one nerve cell to another. Copper is also needed for the synthesis of collagen, a protein found in bone, tendons, teeth, and blood vessels.

a. Give the element symbol, group number, and period number for copper.
b. Classify copper as a main group element, transition metal, or inner transition metal.
c. If a 60.-kg individual contains 60. mg of copper in his body, how many grams of copper are present in each gram of body mass? Write the number in scientific notation.

BEYOND THE CLASSROOM

2.93 Research why long-term exposure to high levels of lead causes harmful effects on the human body. Besides leaded gasoline, where might an individual be exposed to lead? How is lead removed from the body when a person is diagnosed with lead poisoning?

2.94 Calculate how much lead your car would emit into the atmosphere in a year if one gallon of gasoline contained about 2 g of lead. Make needed assumptions—such as the number of miles you drive in a year and the number of miles per gallon of gas your car gets—based on your family's driving habits, and include them in your calculation. Compare results with other class members with different driving habits.

2.95 Pick one of the trace elements in Figure 2.2 and research why it is needed in the body. From what dietary sources do we obtain the nutrient and how much do we need? What is the recommended daily intake of the trace element, and what symptoms result from its deficiency?

2.96 Soft drink cans, composed almost entirely of the element aluminum, are a staple of most households in the United States. Research where the aluminum comes from and in what form it is found in the earth. Why is aluminum used instead of other substances like copper, tin, or stainless steel? A soda can contains about 15 g of aluminum. Estimate how many aluminum cans are used in your household or class each month, and how many pounds of aluminum this amounts to. If your state or town has a recycling program, see if you can find out how many pounds of aluminum cans are recycled each month.

2.97 While the element helium usually brings to mind party balloons, helium is also used in scuba diving, medicine, and the aerospace industry. Much of the helium comes from natural gas reserves present below the central plains of the United States, and these reserves are rapidly being depleted. Research what applications helium is crucial for, and what is being done to conserve present supplies and find other sources. Nobel-Prize-winning physicist Robert Richardson has suggested that helium-filled party balloons should cost $100 each, to limit the use of helium in "frivolous" circumstances, and encourage major helium users to recycle it. What do you think of this proposal?

ANSWERS TO SELECTED PROBLEMS

2.1 a. Ca b. Rn c. N d. Au

2.3 a. neon b. sulfur c. iodine d. silicon e. boron f. mercury

2.5 As, B, Si: metalloids
Cr, Co, Cu, Fe, Mn, Mo, Ni, Zn: metals
F, I, Se: nonmetals

2.6 a. 4 hydrogens, 1 carbon
b. 3 hydrogens, 1 nitrogen
c. 6 hydrogens, 2 carbons, 1 oxygen

2.7 a. 1 sodium, 1 carbon, 1 nitrogen
b. 2 hydrogens, 1 sulfur
c. 2 carbons, 6 hydrogens
d. 1 tin, 2 fluorines
e. 1 carbon, 1 oxygen
f. 3 carbons, 8 hydrogens, 3 oxygens

2.9 a. 9 b. 9 c. fluorine

2.10

	Atomic Number	Element	Protons	Electrons
a.	2	Helium	2	2
b.	11	Sodium	11	11
c.	20	Calcium	20	20
d.	47	Silver	47	47
e.	78	Platinum	78	78

2.11 a. 4 protons and 5 neutrons
b. atomic number 4, mass number 9
c. beryllium

2.12

	Protons	Neutrons	Electrons
a.	17	18	17
b.	14	14	14
c.	92	146	92

2.13 95, 52

2.14

	Atomic Number	Mass Number	Protons	Neutrons	Electrons
a.	6	13	6	7	6
b.	51	121	51	70	51

2.15

	Protons	Electrons	Atomic Number	Mass Number
$^{24}_{12}Mg$	12	12	12	24
$^{25}_{12}Mg$	12	12	12	25
$^{26}_{12}Mg$	12	12	12	26

2.16 a. [3]
b. [1] 8; [2] 7; [3] 8
c. [1] $^{16}_{8}O$; [2] $^{16}_{7}N$; [3] $^{17}_{8}O$

2.17 a. atomic number 27 and atomic weight 58.93 amu
b. 27 protons and 27 electrons

2.19 a. 8 b. 6

2.20

Element	Period	Group
a. Oxygen	2	6A (or 16)
b. Calcium	4	2A (or 2)
c. Phosphorus	3	5A (or 15)
d. Platinum	6	8B (or 10)
e. Iodine	5	7A (or 17)

2.21 a. K b. F c. Ar d. Sr e. Zn f. Nb

2.23 a. 2 b. 8 c. 2 d. 18

2.25 a. $1s^22s^22p^63s^23p^1$ b. $1s^22s^22p^63s^23p^5$

2.26 a. sodium b. silicon

2.27 a. 3 electrons, 1 valence electron, lithium
b. 12 electrons, 2 valence electrons, magnesium
c. 15 electrons, 5 valence electrons, phosphorus

2.28 a. 7 b. 8 c. 2 d. 4

2.29 a. :B̤̈r: b. Li· c. ·Al· d. ·S̤̈· e. :N̤̈e:

2.30 a. arsenic b. silicon c. phosphorus

2.31 a. silicon b. carbon c. fluorine

2.32 a. [1] b. [2] c. [1] fluorine; [2] magnesium

2.33 a. carbon (black), oxygen (red)
b. carbon (black), hydrogen (gray), chlorine (green)

2.35 a. 5 protons and 6 neutrons
b. 5
c. 11
d. 5
e. B

2.37 a. Se
b. group number 6A (16) and period 4
c. main group element
d. 6
e. $^{80}_{34}Se$

2.39 a. aluminum
b. 3
c. group number 3A (13) and period 3

2.41 a. iodine b. carbon

2.43 a. [2] b. [1] c. [2]

2.45 a. gold, astatine, silver
b. nitrogen, sodium, nickel
c. sulfur, silicon, tin
d. calcium, chromium, chlorine

2.47 a,c,e: element b,d: compound

2.49 a. cesium b. ruthenium c. chlorine d. fluorine

2.51 a. sodium: metal, alkali metal, main group element
b. silver: metal, transition metal
c. xenon: nonmetal, noble gas, main group element
d. platinum: metal, transition metal

2.53

	Element Symbol	Atomic Number	Mass Number	Number of Protons	Number of Neutrons	Number of Electrons
a.	C	6	12	6	6	6
b.	P	15	31	15	16	15
c.	Zn	30	65	30	35	30
d.	Mg	12	24	12	12	12

2.55

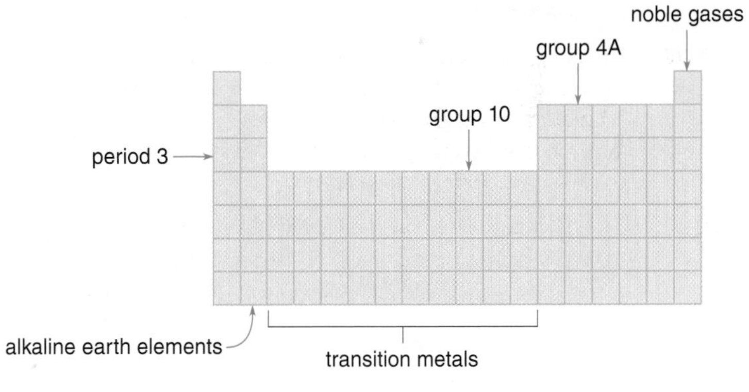

2.57 K, Ca, Sc, Ti, V, Cr, Mn, Fe, Co, Ni, Cu, Zn, Ga: metals
Ge, As: metalloids
Se, Br, Kr: nonmetals

2.59

Mass Number	Protons	Neutrons	Electrons	Group	Symbol
16	8	8	8	6A (16)	$^{16}_{8}O$
17	8	9	8	6A (16)	$^{17}_{8}O$
18	8	10	8	6A (16)	$^{18}_{8}O$

2.61

	Protons	Neutrons	Electrons
a.	13	14	13
b.	17	18	17
c.	16	18	16

2.63 a. $^{127}_{53}I$ b. $^{79}_{35}Br$

2.65 No, two different elements must have a different number of protons and so, in the neutral atom, they must have a different number of electrons.

2.67 a. chlorine b. hydrogen, lithium, sodium c. carbon

2.69 a. $1s^22s^22p^1$ b. $1s^22s^22p^63s^2$

2.71 a. 16 electrons, 6 valence electrons, sulfur
b. 8 electrons, 6 valence electrons, oxygen

2.73

	Group	Valence		
	Electrons	**Number**	**Electrons**	**Period**
a. Carbon	6	4A	4	2
b. Calcium	20	2A	2	4
c. Krypton	36	8A	8	4

2.75 a. 2 b. 4 c. 7

2.77 a. 6 b. 7 c. 2

2.79 a. Be· b. ·Ï: c. ·Mg· d. :Är:

2.81 a. carbon b. bromine

2.83 a. neon < carbon < boron
b. beryllium < magnesium < calcium
c. sulfur < silicon < magnesium
d. neon < krypton < xenon

2.85 fluorine, oxygen, sulfur, silicon, magnesium

2.87

a. Element	b. Type	c,d: Radius	e,f: Ionization Energy	g. Valence Electrons
Sodium	Metal			1
Potassium	Metal	Largest	Lowest	1
Chlorine	Nonmetal	Smallest	Highest	7

2.89 a. $C_{13}H_{18}ClNO$ b. chlorine

2.91 Strontium is in the same group as calcium, so it has similar chemical properties.

3

Zinc oxide is an ionic compound widely used in sunblocks to protect the skin from harmful ultraviolet radiation.

Ionic and Covalent Compounds

CHAPTER OUTLINE

CHAPTER GOALS

In this chapter you will learn how to:

1. Describe the basic features of ionic and covalent bonds
2. Use the periodic table to determine the charge of an ion using the group number
3. Describe the octet rule
4. Write the formula for an ionic compound
5. Name ionic compounds
6. Describe the properties of ionic compounds
7. Recognize the structures of common polyatomic ions and name compounds that contain them
8. Draw Lewis structures for covalent compounds
9. Name covalent compounds that contain two types of elements
10. Predict the shape around an atom in a molecule
11. Use electronegativity to determine whether a bond is polar or nonpolar
12. Determine whether a molecule is polar or nonpolar

Although much of the discussion in Chapter 2 focused on atoms, individual atoms are rarely encountered in nature. Instead, atoms are far more commonly joined together to form compounds. There are two types of chemical compounds, **ionic** and **covalent. Ionic compounds** are composed of positively and negatively charged ions held together by strong **electrostatic forces**—the electrical attraction between oppositely charged ions. Examples of ionic compounds include the sodium chloride (NaCl) in table salt and the calcium carbonate ($CaCO_3$) in snail shells. **Covalent compounds** are composed of individual molecules, discrete groups of atoms that share electrons. Covalent compounds include water (H_2O) and methane (CH_4), the main component of natural gas. Chapter 3 focuses on the structure and properties of ionic and covalent compounds.

3.1 Introduction to Bonding

You were first introduced to elements and compounds in Section 1.3.

It is rare in nature to encounter individual atoms. Instead, anywhere from two to hundreds or thousands of atoms tend to join together. The oxygen we breathe, for instance, consists of two oxygen atoms joined together, whereas the hemoglobin that transports it to our tissues consists of thousands of carbon, hydrogen, oxygen, nitrogen, and sulfur atoms joined together. We say **two atoms are *bonded* together.**

- ***Bonding* is the joining of two atoms in a stable arrangement.**

Only the noble gases in group 8A of the periodic table are particularly stable as individual atoms; that is, the **noble gases do *not* readily react to form bonds,** because the electronic configuration of the noble gases is especially stable to begin with. As a result, one overriding principle explains the process of bonding.

- **In bonding, elements gain, lose, or share electrons to attain the electronic configuration of the noble gas closest to them in the periodic table.**

Bonding involves only the valence electrons of an atom. There are two different kinds of bonding: **ionic** and **covalent.**

- ***Ionic bonds* result from the transfer of electrons from one element to another.**
- ***Covalent bonds* result from the sharing of electrons between two atoms.**

The position of an element in the periodic table determines the type of bonds it makes. **Ionic bonds form between a metal on the left side of the periodic table and a nonmetal on the right side.** As shown in Figure 3.1, when the metal sodium (Na) bonds to the nonmetal chlorine (Cl_2), the ionic compound sodium chloride (NaCl) forms. Ionic compounds are composed of ***ions*—charged species in which the number of protons and electrons in an atom is *not* equal.**

Covalent bonds are formed when two nonmetals combine, or when a metalloid bonds to a nonmetal. **A *molecule* is a compound or element containing two or more atoms joined together with covalent bonds.** For example, when two hydrogen atoms bond they form the molecule H_2, and two electrons are shared.

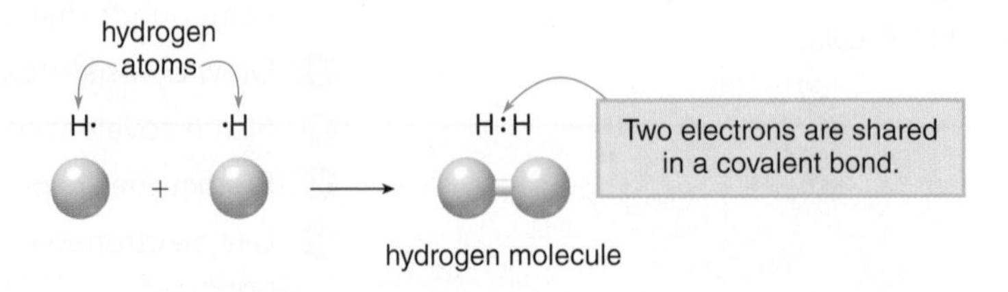

Figure 3.1

Sodium Chloride, an Ionic Compound

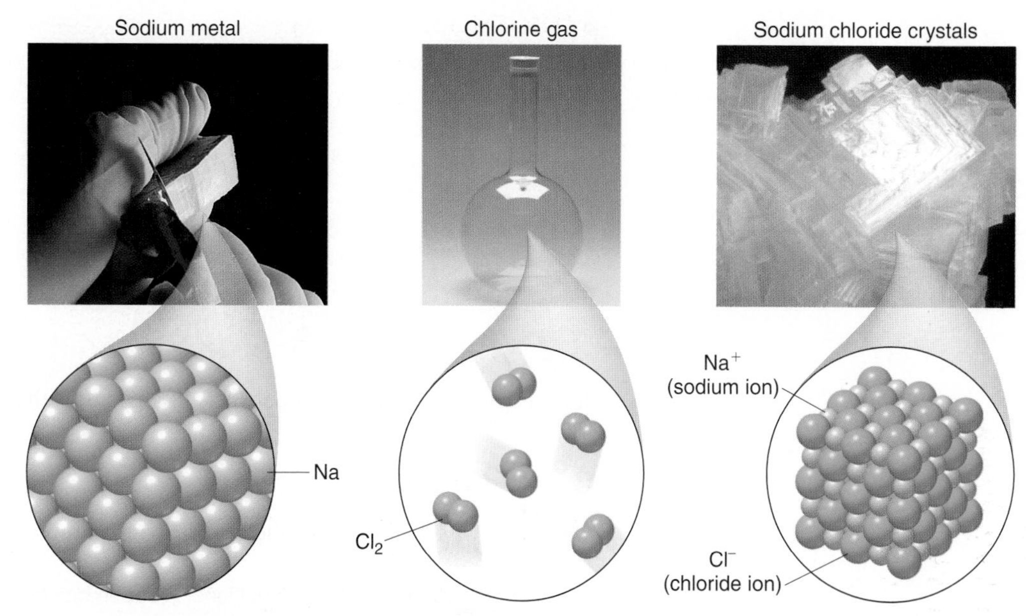

Sodium metal and chlorine gas are both elements. Sodium chloride is an ionic compound composed of sodium ions and chloride ions.

HEALTH NOTE

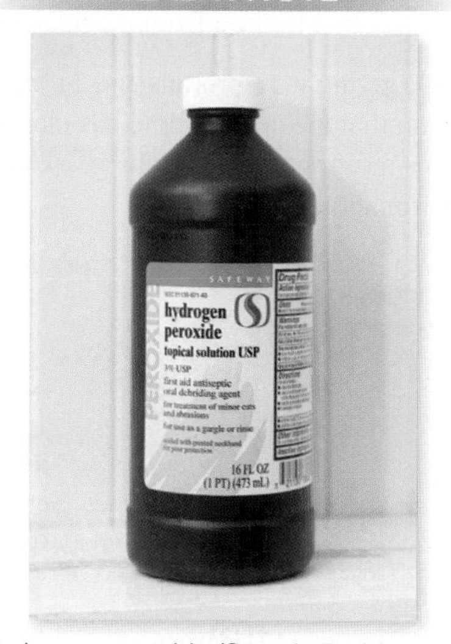

Hydrogen peroxide (Sample Problem 3.1) is used to disinfect wounds.

SAMPLE PROBLEM 3.1

Predict whether the bonds in the following compounds are ionic or covalent: (a) NaI (sodium iodide); (b) H_2O_2 (hydrogen peroxide).

Analysis

When a metal and nonmetal combine, the bond is ionic. When two nonmetals combine, or a metalloid bonds to a nonmetal, the bond is covalent.

Solution

a. Since Na is a metal on the left side and I is a nonmetal on the right side of the periodic table, the bonds in NaI are ionic.

b. Since H_2O_2 contains only the nonmetals hydrogen and oxygen, the bonds must be covalent.

PROBLEM 3.1

Predict whether the bonds in the following species are ionic or covalent.

a. CO b. CaF_2 c. MgO d. Cl_2 e. HF f. NaF

SAMPLE PROBLEM 3.2

(a) Classify each example of molecular art as a compound or an element. (b) Which of the species are molecules?

Analysis

- In molecular art, an element is composed of spheres of the same color, while a compound is composed of spheres of different colors. The common element colors are shown in Figure 2.3 (and on the inside back cover).
- A molecule is a compound or element that contains covalent bonds.

Solution

a. **A** is an element composed of two fluorines, while **B** is a compound composed of hydrogen and fluorine.

b. Both **A** and **B** are molecules because they are composed solely of covalent bonds between nonmetals.

PROBLEM 3.2

(a) Classify each example of molecular art as a compound or element. (b) Which of the species are molecules?

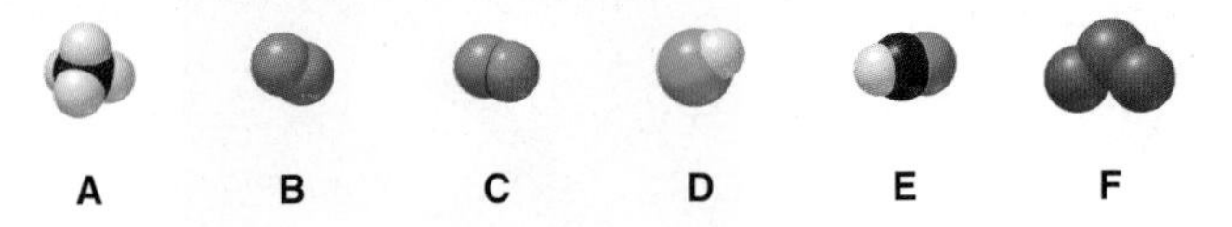

3.2 Ions

Ionic compounds consist of oppositely charged **ions** that have a strong attraction for each other.

3.2A Cations and Anions

There are two types of ions called **cations** and **anions.**

- ***Cations* are positively charged ions. A cation has fewer electrons than protons.**
- ***Anions* are negatively charged ions. An anion has more electrons than protons.**

The charge on an ion depends on the position of an element in the periodic table. In forming an ion, an atom of a main group element loses or gains electrons to obtain the electronic configuration of the noble gas closest to it in the periodic table. This gives the ion an especially stable electronic arrangement in which the electrons completely fill the shell farthest from the nucleus.

For example, sodium (group 1A) has an atomic number of 11, giving it 11 protons and 11 electrons in the neutral atom. This gives sodium one *more* electron than neon, the noble gas closest to it in the periodic table. In losing one electron, sodium forms a cation with a +1 charge, which still has 11 protons, but now has only 10 electrons in its electron cloud.

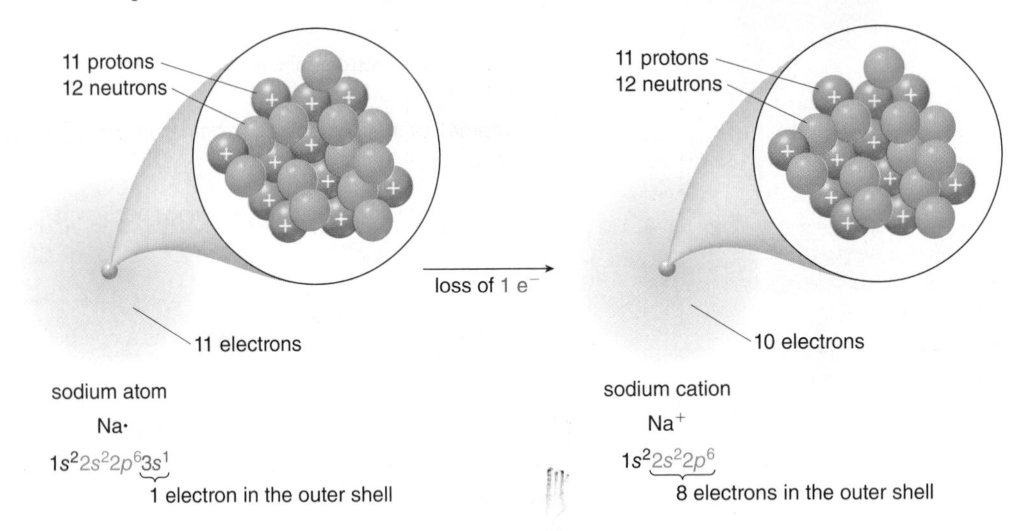

What does this mean in terms of valence electrons? A neutral sodium atom, with an electronic configuration of $1s^22s^22p^63s^1$, has a single valence electron. Loss of this valence electron forms a **sodium cation,** symbolized as **Na^+,** which has the especially stable electronic configuration of the noble gas neon, $1s^22s^22p^6$. The sodium cation now has **eight electrons** that fill the 2*s* and three 2*p* orbitals.

Sodium is an example of a metal.

Some metals—notably tin and lead—can lose *four* electrons to form cations.

- **Metals form *cations.* By losing one, two, or three electrons, an atom forms a cation with a completely filled outer shell of electrons.**

A neutral chlorine atom (group 7A), on the other hand, has 17 protons and 17 electrons. This gives it one *fewer* electron than argon, the noble gas closest to it in the periodic table. By gaining one electron, chlorine forms an anion with a –1 charge because it still has 17 protons, but now has 18 electrons in its electron cloud.

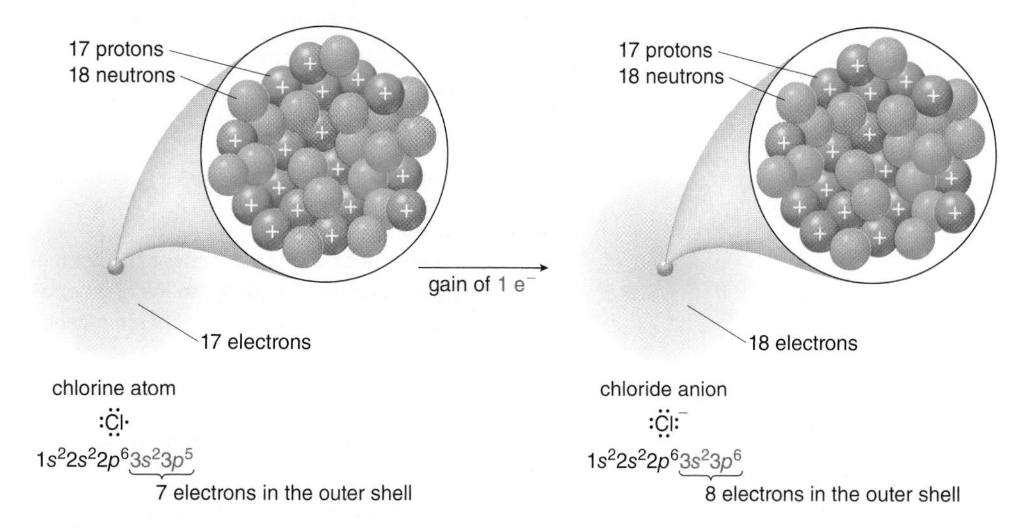

In terms of valence electrons, a neutral chlorine atom, with an electronic configuration of $1s^22s^22p^63s^23p^5$, has seven valence electrons. Gain of one electron forms a **chloride anion,** symbolized as **Cl^-,** which has the especially stable electronic configuration of the noble gas argon, $1s^22s^22p^63s^23p^6$. The chloride anion now has **eight valence electrons** that fill the 3*s* and three 3*p* orbitals.

Chlorine is an example of a nonmetal.

Ions are written with the element symbol followed by a superscript to indicate the charge. The number "1" is omitted in ions that have a +1 or –1 charge, as in Na^+ or Cl^-. When the charge is "2" or greater, it is written as 2+ or 2–, as in Mg^{2+} or O^{2-}.

- **Nonmetals form *anions.* By gaining one, two, or sometimes three electrons, an atom forms an anion with a completely filled outer shell of electrons.**

Each of these ions formed from a main group element has **eight valence electrons.** This illustrates the **octet rule.**

- **A main group element is especially stable when it possesses an *octet* of electrons in its outer shell.**

SAMPLE PROBLEM 3.3

Write the ion symbol for an atom with: (a) nine protons and 10 electrons; (b) three protons and two electrons.

Analysis

Since the number of protons equals the atomic number (Section 2.2), this quantity identifies the element. The charge is determined by comparing the number of protons and electrons. If the number of electrons is greater than the number of protons, the charge is negative (an anion). If the number of protons is greater than the number of electrons, the charge is positive (a cation).

Solution

a. An element with nine protons has an atomic number of nine, identifying it as fluorine (F). Since there is one more electron than proton (10 vs. 9), the charge is –1. **Answer:** F^-

b. An element with three protons has an atomic number of three, identifying it as lithium (Li). Since there is one more proton than electron (3 vs. 2), the charge is +1. **Answer:** Li^+

PROBLEM 3.3

Write the ion symbol for an atom with the given number of protons and electrons.

a. 19 protons and 18 electrons
b. seven protons and 10 electrons
c. 35 protons and 36 electrons
d. 23 protons and 21 electrons

SAMPLE PROBLEM 3.4

How many protons and electrons are present in each ion: (a) Ca^{2+}; (b) O^{2-}?

Analysis

Use the identity of the element to determine the number of protons. The charge tells how many more or fewer electrons there are compared to the number of protons. A positive charge means more protons than electrons, while a negative charge means more electrons than protons.

Solution

a. Ca^{2+}: The element calcium (Ca) has an atomic number of 20, so it has 20 protons. Since the charge is +2, there are two more protons than electrons, giving the ion 18 electrons.

b. O^{2-}: The element oxygen (O) has an atomic number of eight, so it has eight protons. Since the charge is –2, there are two more electrons than protons, giving the ion 10 electrons.

PROBLEM 3.4

How many protons and electrons are present in each ion?

a. Ni^{2+} b. Se^{2-} c. Fe^{3+}

3.2B Relating Group Number to Ionic Charge for Main Group Elements

Since elements with similar electronic configurations are grouped together in the periodic table, **elements in the same group form ions of similar charge.** The group number of a main group element can be used to determine the charge on an ion derived from that element.

- **Metals form cations. For metals in groups 1A, 2A, and 3A, the group number = the charge on the cation.**

Group **1**A elements (Li, Na, K, Rb, and Cs) have **one** valence electron. Loss of this electron forms a cation with a **+1** charge. Group **2**A elements (Be, Mg, Ca, Sr, and Ba) have **two** valence electrons. Loss of both electrons forms a cation with a **+2** charge. Group **3**A elements (Al, Ga, In, and Tl) form cations, too, but only aluminum is commonly found in ionic compounds. It has **three** valence electrons, so loss of three electrons from aluminum forms a cation with a **+3** charge.

- **Nonmetals form anions. For nonmetals in groups 5A, 6A, and 7A, the anion charge = 8 – (the group number).**

Group **5**A elements have **five** valence electrons. A gain of **three** electrons forms an anion with a **–3** charge (anion charge = 8 – 5). Group **6**A elements have **six** valence electrons. A gain of **two** electrons forms an anion with a **–2** charge (anion charge = 8 – 6). Group **7**A elements have **seven** valence electrons. A gain of one electron forms an anion with a **–1** charge (anion charge = 8 – 7).

The periodic table in Figure 3.2 gives the common ions formed by the main group elements.

Figure 3.2

Common Ions Formed by Main Group Elements

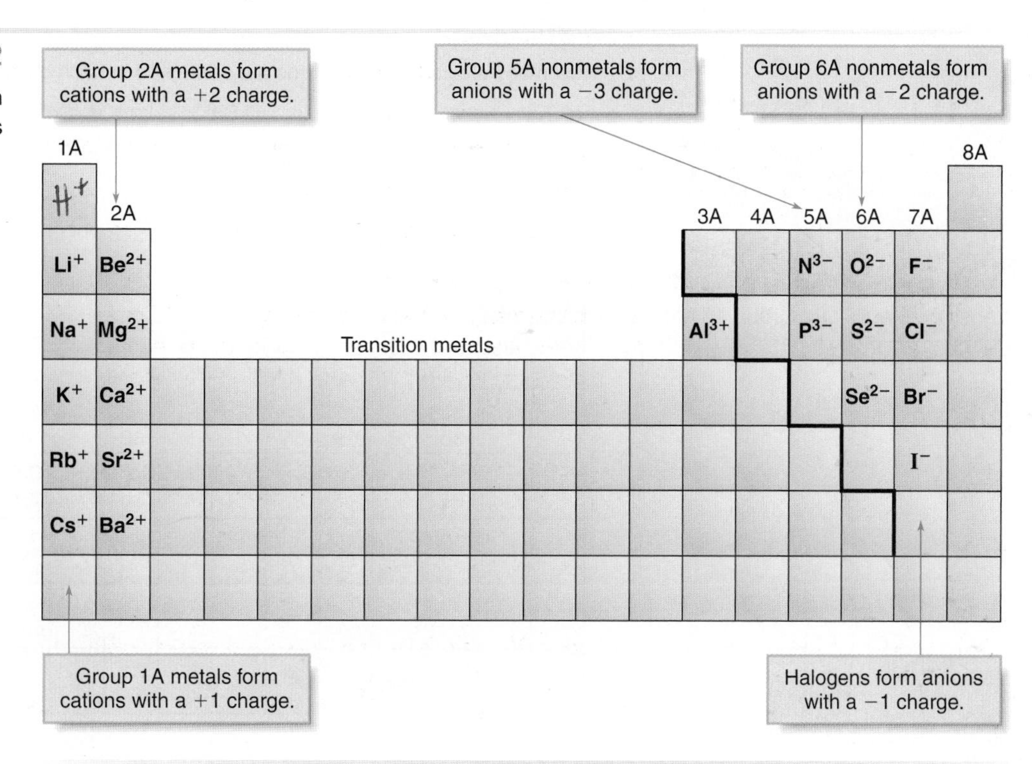

SAMPLE PROBLEM 3.5

Use the group number to determine the charge on an ion derived from each element: (a) barium; (b) sulfur.

Analysis

Locate the element in the periodic table. A metal in groups 1A, 2A, or 3A forms a cation equal in charge to the group number. A nonmetal in groups 5A, 6A, and 7A forms an anion whose charge equals 8 – (the group number).

Solution

a. Barium (Ba) is located in group 2A, so it forms a cation with a +2 charge; Ba^{2+}.
b. Sulfur (S) is located in group 6A, so it forms an anion with a negative charge of 8 – 6 = 2; S^{2-}.

PROBLEM 3.5

Use the group number to determine the charge on an ion derived from each element.

a. magnesium b. iodine c. selenium d. rubidium

3.2C Metals with Variable Charge

The transition metals form cations like other metals, but the magnitude of the charge on the cation is harder to predict. Some transition metals, and a few main group metals as well, form more than one type of cation. For example, iron forms two different cations, Fe^{2+} and Fe^{3+}. Figure 3.3 illustrates the common cations formed from transition metals, as well as some main group elements that form more than one cation.

PROBLEM 3.6

How many electrons and protons are contained in each cation?

a. Au^+ b. Au^{3+} c. Sn^{2+} d. Sn^{4+}

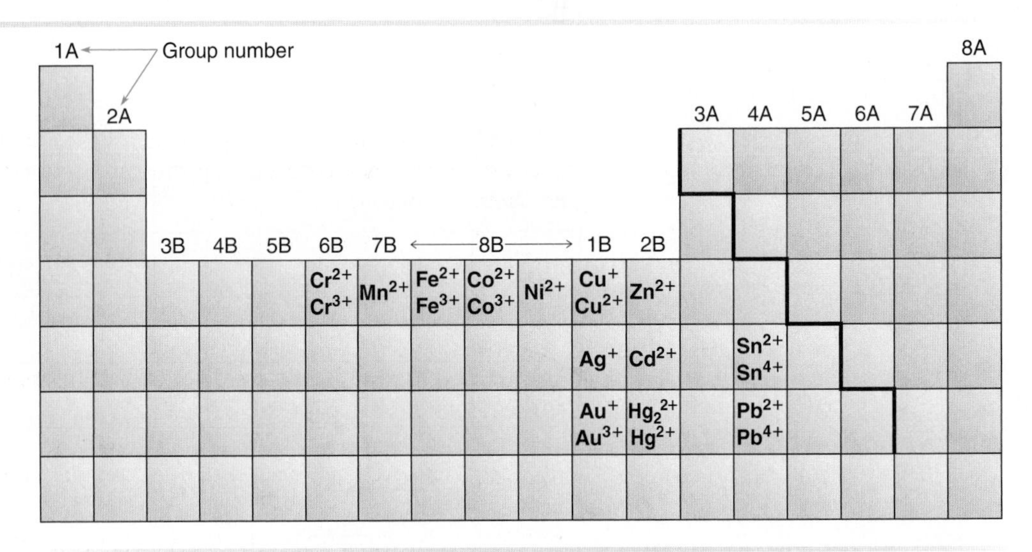

Figure 3.3 Common Cations Derived from Transition Metals and Group 4A Metals

3.2D FOCUS ON THE HUMAN BODY
Important Ions in the Body

Many different ions are required for proper cellular and organ function. The major cations in the body are Na^+, K^+, Ca^{2+}, and Mg^{2+}. K^+ and Mg^{2+} are present in high concentrations inside cells, while Na^+ and Ca^{2+} are present in a higher concentration outside of cells, in the extracellular fluids. Na^+ is the major cation present in blood and extracellular bodily fluids and its concentration is carefully regulated to maintain blood volume and blood pressure within acceptable ranges that permit organ function. Ca^{2+} is found mainly in solid body parts such as teeth and bones, but it is also needed for proper nerve conduction and muscle contraction, as is Mg^{2+}.

In addition to these four cations, Fe^{2+} and Cl^- are also important ions. Fe^{2+} is essential for oxygen transport by red blood cells. Cl^- is present in red blood cells, gastric juices, and other body fluids. Along with Na^+, it plays a major role in regulating the fluid balance in the body.

Although Na^+ is an essential mineral needed in the daily diet, the average American consumes three to five times the recommended daily allowance (RDA) of 2,400 mg. Excess sodium intake is linked to high blood pressure and heart disease. Dietary Na^+ comes from salt, NaCl, added during cooking or at the table. Na^+ is also added during the preparation of processed foods and canned products. For example, one 3.5-oz serving of fresh asparagus has only 1 mg of Na^+, but the same serving size of canned asparagus contains 236 mg of Na^+. Potato chips, snack foods, ketchup, processed meats, and many cheeses are particularly high in Na^+. Table 3.1 lists the Na^+ content of some common foods.

HEALTH NOTE

All of these foods are high in sodium.

Table 3.1 Na^+ Content in Common Foods

Foods High in Na^+		Foods Low in Na^+	
Food	**Na^+ (mg)**	**Food**	**Na^+ (mg)**
Potato chips (30)	276	Banana (1)	1
Hot dog (1)	504	Orange juice (1 cup)	2
Ham, smoked (3 oz)	908	Oatmeal, cooked (1 cup)	2
Chicken soup, canned (1 cup)	1,106	Cereal, shredded wheat (3.5 oz)	3
Tomato sauce, canned (1 cup)	1,402	Raisins, dried (3.5 oz)	27
Parmesan cheese (1 cup)	1,861	Salmon (3 oz)	55

PROBLEM 3.7

Mn^{2+} is an essential nutrient needed for blood clotting and the formation of the protein collagen. (a) How many protons and electrons are found in a neutral manganese atom? (b) How many protons and electrons are found in the cation Mn^{2+}?

3.3 Ionic Compounds

When a metal (on the left side of the periodic table) transfers one or more electrons to a nonmetal (on the right side), **ionic bonds** are formed.

- Ionic compounds are composed of cations and anions.

The ions in an ionic compound are arranged to maximize the attractive force between the oppositely charged species. For example, sodium chloride, NaCl, is composed of sodium cations (Na^+) and chloride anions (Cl^-), packed together in a regular arrangement in a crystal lattice. Each Na^+ cation is surrounded by six Cl^- anions, and each Cl^- anion is surrounded by six Na^+ cations.

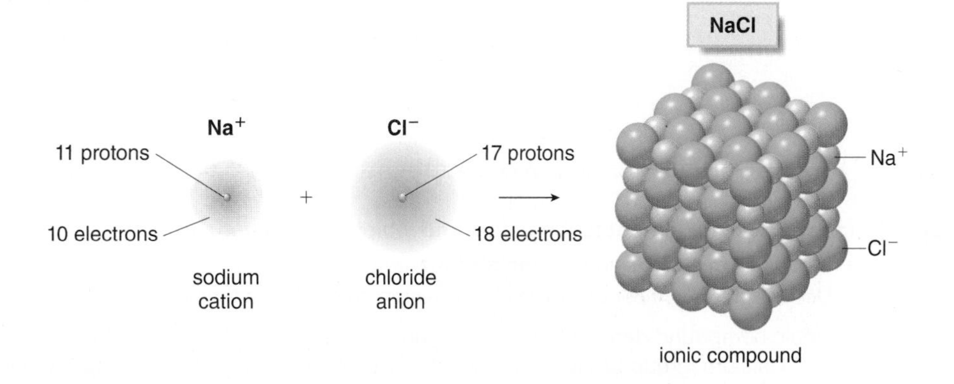

- The sum of the charges in an ionic compound must always be zero overall.

The formula for an ionic compound shows the ratio of ions that combine to give zero charge. Since the sodium cation has a +1 charge and the chloride anion has a –1 charge, there must be one Na^+ cation for each Cl^- anion; thus, the formula is **NaCl.**

When cations and anions having charges of different magnitude combine, the number of cations per anion is not equal. Consider an ionic compound formed from calcium (Ca) and fluorine (F). Since calcium is located in group 2A, it loses two valence electrons to form Ca^{2+}. Since fluorine is located in group 7A, it gains one electron to form F^- like other halogens. When Ca^{2+} combines with the fluorine anion F^-, **there must be two F^- anions for each Ca^{2+} cation to have an overall charge of zero.**

In writing a formula for an ionic compound, we use subscripts when the number of ions needed to achieve zero charge is greater than one. Since two F^- anions are needed for each calcium cation, the formula is **CaF_2.**

Figure 3.4 illustrates additional examples of how cations and anions combine to give zero overall charge.

Figure 3.4
Examples of Ionic Compounds with Zero Overall Charge

$NaCl$	Li_2O	BaI_2	Al_2O_3
Na^+ Cl^-	Li^+ O^{2-}	Ba^{2+} I^-	Al^{3+} O^{2-}
+1 −1	+2 −2	+2 −2	+6 −6

The ratio of oppositely charged ions that combine to form an ionic compound depends on the charge of the ions.

- $NaCl$: One Na^+ cation (+1 charge) combines with one Cl^- anion (–1 charge).
- Li_2O: Two Li^+ cations (+2 charge total) combine with one O^{2-} anion (–2 charge).
- BaI_2: One Ba^{2+} cation (+2 charge) combines with two I^- anions (–2 charge total).
- Al_2O_3: Two Al^{3+} cations (+6 charge total) combine with three O^{2-} anions (–6 charge total).

3.3A Formulas for Ionic Compounds

Writing a formula for an ionic compound from two elements is a useful skill that can be practiced by following a series of steps.

How To Write a Formula for an Ionic Compound

Step [1] Identify which element is the cation and which is the anion.

- **Metals form *cations* and nonmetals form *anions.***
- Use the group number of a main group element to determine the charge.

An ionic compound derived from calcium and oxygen has the metal calcium as the cation and the nonmetal oxygen as the anion. Calcium (group 2A) loses two electrons to form Ca^{2+}. Oxygen (group 6A) gains two electrons to form O^{2-}.

Step [2] Determine how many of each ion type are needed for an overall charge of zero.

- When the cation and anion have the *same* charge only *one* of each is needed.

The charges are equal in magnitude, +2 and −2.

$$Ca^{2+} + O^{2-} \longrightarrow \boxed{CaO}$$

One of each ion is needed to balance charge.

How To, continued . . .

- When the cation and anion have *different* charges, as is the case with the Ca^{2+} cation and Cl^{-} anion, use the ion charges to determine the number of ions of each needed. **The charges on the ions tell us how many of the *oppositely* charged ions are needed to balance charge.**
- Write a subscript for the cation that is equal in magnitude to the charge on the anion. Write a subscript for the anion that is equal in magnitude to the charge on the cation.

The charges are not equal in magnitude, +2 and −1.

Ca^{2+} Cl^{1-} The "1" is written for emphasis.

Ca_1Cl_2 = **$CaCl_2$** 2 Cl^{-} anions for each Ca^{2+}

Step [3] To write the formula, place the cation first and then the anion, and omit charges.

- Use subscripts to show the number of each ion needed to have zero overall charge. When no subscript is written it is assumed to be "1."

As shown in step [2], the formula for the ionic compound formed from one calcium cation (Ca^{2+}) and one oxygen anion (O^{2-}) is CaO. The formula for the ionic compound formed from one calcium cation (Ca^{2+}) and two chlorine anions (Cl^{-}) is $CaCl_2$.

The tarnish on sterling silver is composed of an ionic compound formed from silver and sulfur (Sample Problem 3.6).

SAMPLE PROBLEM 3.6

When sterling silver tarnishes it forms an ionic compound derived from silver and sulfur. Write the formula for this ionic compound.

Analysis

- Identify the cation and the anion, and use the periodic table to determine the charges.
- When ions of equal charge combine, one of each ion is needed. When ions of unequal charge combine, use the ionic charges to determine the relative number of each ion.
- Write the formula with the cation first and then the anion, omitting charges, and using subscripts to indicate the number of each ion.

Solution

Silver is a metal, so it forms the cation. Sulfur is a nonmetal, so it forms the anion. The charge on silver is +1 (Ag^{+}), as shown in Figure 3.3. Sulfur (group 6A) is a main group element with a −2 charge (S^{2-}). Since the charges are unequal, use their magnitudes to determine the relative number of each ion to give an overall charge of zero.

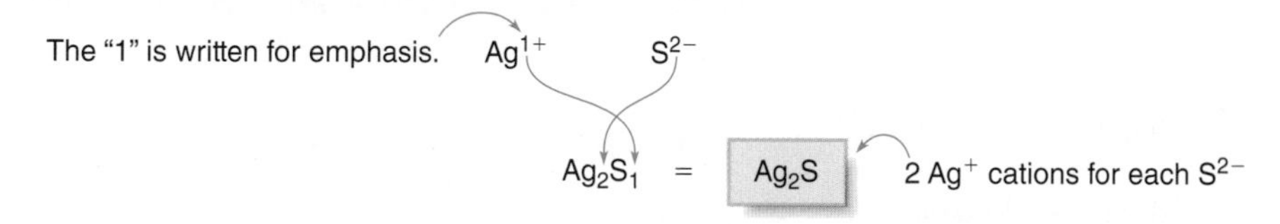

Answer: Since two Ag^{+} cations are needed for each S^{2-} anion, the formula is **Ag_2S**.

PROBLEM 3.8

Write the formula for the ionic compound formed from each pair of elements.

a. sodium and bromine
b. barium and oxygen
c. magnesium and iodine
d. lithium and oxygen

HEALTH NOTE

Potassium is a critical cation for normal heart and skeletal muscle function and nerve impulse conduction. Drinking electrolyte replacement beverages like Gatorade or Powerade can replenish K^+ lost in sweat.

3.3B FOCUS ON HEALTH & MEDICINE Ionic Compounds in Consumer Products

Simple ionic compounds are added to food or consumer products to prevent disease or maintain good health. For example, **potassium iodide** (KI) is an essential nutrient added to table salt. Iodine is needed to synthesize thyroid hormones. A deficiency of iodine in the diet can lead to insufficient thyroid hormone production. In an attempt to compensate, the thyroid gland may become enlarged, producing a swollen thyroid referred to as a goiter. **Sodium fluoride** (NaF) is added to toothpaste to strengthen tooth enamel and help prevent tooth decay.

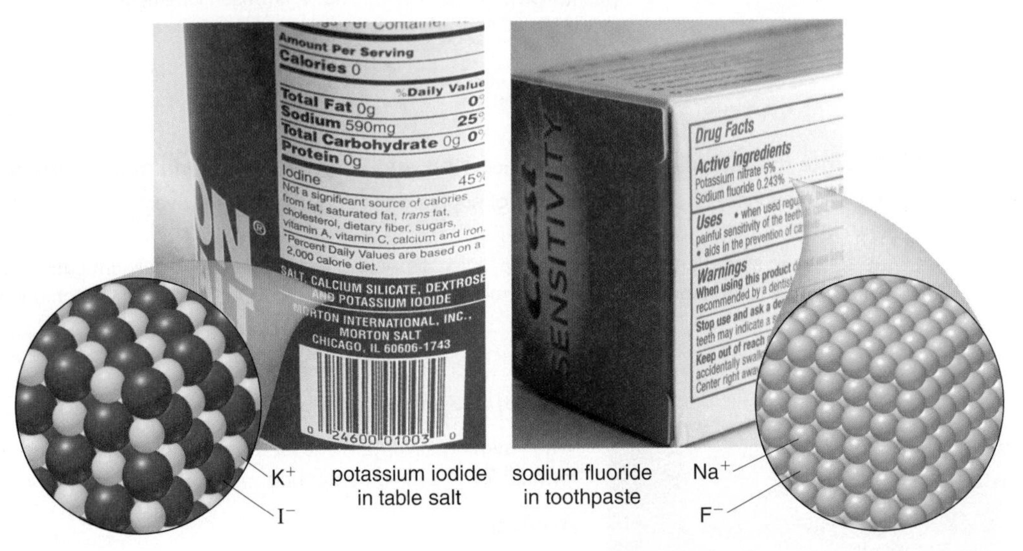

Potassium chloride (KCl), sold under trade names such as K–Dur, Klor–Con, and Micro–K, is an ionic compound used for patients whose potassium levels are low. Potassium chloride can be given as tablets, an oral suspension, or intravenously. Adequate potassium levels are needed for proper fluid balance and organ function. Although potassium is readily obtained from many different food sources (e.g., potatoes, beans, melon, bananas, and spinach), levels can become low when too much potassium is lost in sweat and urine or through the use of certain medications.

PROBLEM 3.9

Zinc oxide, an ionic compound formed from zinc and oxygen, is a common component of sunblocks, as mentioned in the chapter opener. The zinc oxide crystals reflect sunlight away from the skin, and in this way, protect it from sun exposure. What is the ionic formula for zinc oxide?

3.4 Naming Ionic Compounds

Now that we have learned how to write the formulas of some simple ionic compounds, we must learn how to name them. Assigning an unambiguous name to each compound is called chemical **nomenclature.**

3.4A Naming Cations

Cations of main group metals are given the name of the element from which they are formed.

Na^+	K^+	Ca^{2+}	Mg^{2+}
sodium	potassium	calcium	magnesium

It is common to add the word "ion" after the name of the metal cation to distinguish it from the neutral metal itself. For example, when the concentration of sodium in a blood sample is determined, what is really measured is the concentration of sodium *ions* (Na^+).

When a metal is able to form two different cations, a method is needed to distinguish these cations. Two systems are used, the systematic method and the common method. The systematic method (Method [1]) will largely be followed in this text. Since many ions are still identified by older names, however, the common method (Method [2]) is also given.

- **Method [1]: Follow the name of the cation by a Roman numeral in parentheses to indicate its charge.**
- **Method [2]: Use the suffix *-ous* for the cation with the lesser charge, and the suffix *-ic* for the cation with the higher charge. These suffixes are often added to the Latin names of the elements.**

For example, the element iron (Fe) forms two cations, Fe^{2+} and Fe^{3+}, which are named in the following way:

	Systematic Name	Common Name
Fe^{2+}	iron(II)	ferr***ous***
Fe^{3+}	iron(III)	ferr***ic***

Table 3.2 lists the systematic and common names for several cations.

Table 3.2 Systematic and Common Names for Some Metal Ions

Element	Ion Symbol	Systematic Name	Common Name
Copper	Cu^{+}	Copper(I)	Cuprous
	Cu^{2+}	Copper(II)	Cupric
Chromium	Cr^{2+}	Chromium(II)	Chromous
	Cr^{3+}	Chromium(III)	Chromic
Iron	Fe^{2+}	Iron(II)	Ferrous
	Fe^{3+}	Iron(III)	Ferric
Mercury	Hg_2^{2+}	Mercury(I)[a]	Mercurous
	Hg^{2+}	Mercury(II)	Mercuric
Tin	Sn^{2+}	Tin(II)	Stannous
	Sn^{4+}	Tin(IV)	Stannic

[a]Mercury(I) exists as Hg_2^{2+}, containing two atoms of mercury, each with a +1 charge.

3.4B Naming Anions

Anions are named by replacing the ending of the element name by the suffix *-ide*. For example:

Cl chlor*ine*	---→	Cl^{-} chlor***ide***	[Change *-ine* to ***-ide.***]
O o*xygen*	---→	O^{2-} ox***ide***	[Change *-ygen* to ***-ide.***]

Table 3.3 lists the names of common anions derived from nonmetal elements.

Table 3.3 Names of Common Anions

Element	Ion Symbol	Name
Bromine	Br^{-}	Bromide
Chlorine	Cl^{-}	Chloride
Fluorine	F^{-}	Fluoride
Iodine	I^{-}	Iodide
Nitrogen	N^{3-}	Nitride
Oxygen	O^{2-}	Oxide
Phosphorus	P^{3-}	Phosphide
Sulfur	S^{2-}	Sulfide

PROBLEM 3.10

Give the name of each ion.

a. S^{2-} b. Cu^{+} c. Cs^{+} d. Al^{3+} e. Sn^{4+}

PROBLEM 3.11

Give the symbol for each ion.

a. stannous b. iodide c. manganese ion d. lead(II)

3.4C Naming Ionic Compounds with Cations from Main Group Metals

To name an ionic compound with a main group metal cation whose charge never varies, **name the cation and then the anion.** Do *not* specify the charge on the cation. Do *not* specify how many ions of each type are needed to balance charge.

Na^+	F^-	---→	NaF
sodium	fluoride		sodium fluoride
Mg^{2+}	Cl^-	---→	$MgCl_2$
magnesium	chloride		magnesium chloride

SAMPLE PROBLEM 3.7

Name each ionic compound: (a) Na_2S; (b) $AlBr_3$.

Analysis

Name the cation and then the anion.

Solution

a. Na_2S: The cation is sodium and the anion is sulfide (derived from sulfur); thus, the name is sodium sulfide.
b. $AlBr_3$: The cation is aluminum and the anion is bromide (derived from bromine); thus, the name is aluminum bromide.

PROBLEM 3.12

Name each ionic compound.

a. NaF b. MgO c. $SrBr_2$ d. Li_2O e. TiO_2

3.4D Naming Ionic Compounds Containing Metals with Variable Charge

To name an ionic compound that contains a metal with variable charge, we must specify the charge on the cation. The formula of the ionic compound—that is, how many cations there are per anion—allows us to determine the charge on the cation.

How To Name an Ionic Compound That Contains a Metal with Variable Charge

Example Give the name for $CuCl_2$.

Step [1] Determine the charge on the cation.

- Since there are two Cl^- anions, each of which has a –1 charge, the copper cation must have a +2 charge to make the overall charge zero.

$CuCl_2$ 2 Cl^- anions ---→ The total negative charge is –2.

Cu must have a +2 charge to balance the –2 charge of the anions.

→ Cu^{2+}

How To, continued . . .

Step [2] Name the cation and anion.

- Name the cation using its element name followed by a Roman numeral to indicate its charge. In the common system, use the suffix *-ous* or *-ic* to indicate charge.
- Name the anion by changing the ending of the element name to the suffix *-ide.*

Cu^{2+} ---→ copper(II) or cupric

Cl^- ---→ chloride

Step [3] Write the name of the cation first, then the anion.

- **Answer:** Copper(II) chloride or cupric chloride.

Sample Problem 3.8 illustrates the difference in naming ionic compounds derived from metals that have fixed or variable charge.

HEALTH NOTE

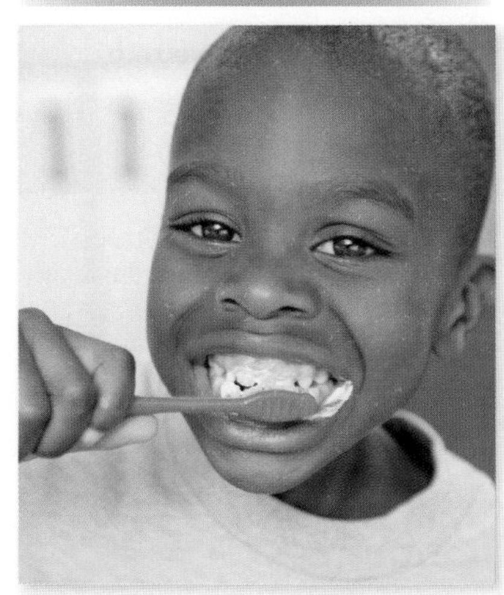

Some toothpastes contain the ionic compounds SnF_2 as a source of fluoride and Al_2O_3 as an abrasive.

SAMPLE PROBLEM 3.8

SnF_2 and Al_2O_3 are both ingredients in commercial toothpastes. SnF_2 contains fluoride, which strengthens tooth enamel. Al_2O_3 is an abrasive that helps to scrub the teeth clean when they are brushed. Give names for (a) SnF_2; (b) Al_2O_3.

Analysis

First determine if the cation has a fixed or variable charge. To name an ionic compound that contains a cation that always has the same charge, name the cation and then the anion (using the suffix *-ide*). When the metal has a variable charge, use the overall anion charge to determine the charge on the cation. Then name the cation (using a Roman numeral or the suffix *-ous* or *-ic*), followed by the anion.

Solution

a. SnF_2: Sn cations have variable charge so the overall anion charge determines the cation charge.

SnF_2 → 2 F^- anions ---→ The total negative charge is −2.

Sn must have a +2 charge to balance the −2 charge of the anions.

→ tin(II) or stannous → fluoride

Answer: tin(II) fluoride or stannous fluoride

b. Al_2O_3: Al has a fixed charge of +3. To name the compound, name the cation as the element (aluminum), and the anion by changing the ending of the element name to the suffix *-ide* (oxygen → oxide).

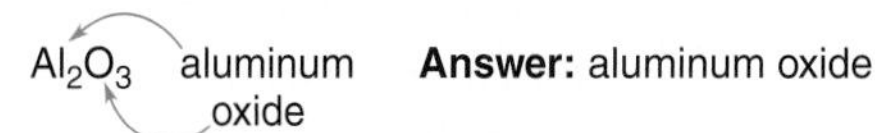

Answer: aluminum oxide

PROBLEM 3.13

Name each ionic compound.

a. $CrCl_3$ b. PbS c. SnF_4 d. PbO_2 e. $FeBr_2$

3.4E Writing a Formula from the Name of an Ionic Compound

Writing a formula from the name of an ionic compound is also a useful skill.

How To Derive a Formula from the Name of an Ionic Compound

Example Write the formula for tin(IV) oxide.

Step [1] Identify the cation and the anion and determine their charges.

- The name of the cation appears first, followed by the anion.
- For metals with variable charge, the Roman numeral gives the charge on the cation.

In this example, tin is the cation. The Roman numeral tells us that its charge is +4, making the cation Sn^{4+}. Oxide is the name of the oxygen anion, O^{2-} (Table 3.3).

Step [2] Balance charges.

- Use the charge on the cation to determine the number of ions of the anion needed to balance charge.

Sn^{4+} O^{2-} Two −2 anions are needed for each +4 cation.

cation anion

Step [3] Write the formula with the cation first, and use subscripts to show the number of each ion needed to have zero overall charge.

Answer: SnO_2

PROBLEM 3.14

Write the formula for each ionic compound.

a. calcium bromide
b. copper(I) iodide
c. ferric bromide
d. magnesium sulfide
e. chromium(II) chloride
f. sodium oxide

3.5 Physical Properties of Ionic Compounds

Ionic compounds are crystalline solids composed of ions packed to maximize the interaction of the positive charge of the cations and negative charge of the anions. Ionic solids are held together by extremely strong interactions of the oppositely charged ions. How is this reflected in the melting point and boiling point of an ionic compound?

When a compound melts to form a liquid, energy is needed to overcome some of the attractive forces of the ordered solid, to form the less ordered liquid phase. Since an ionic compound is held together by very strong electrostatic interactions, it takes a great deal of energy to separate the ions from each other. As a result, **ionic compounds have very high melting points.** For example, the melting point of NaCl is 801 °C.

A great deal of energy is needed to overcome the attractive forces present in the liquid phase, too, to form ions that are far apart and very disorganized in the gas phase, so **ionic compounds have extremely high boiling points.** The boiling point of liquid NaCl is 1,413 °C.

A great many ionic compounds are soluble in water. When an ionic compound dissolves in water, the ions are separated, and each anion and cation is surrounded by water molecules, as shown in Figure 3.5. The interaction of the water solvent with the ions provides the energy needed to overcome the strong ion–ion attractions of the crystalline lattice.

Figure 3.5

Dissolving NaCl in Water

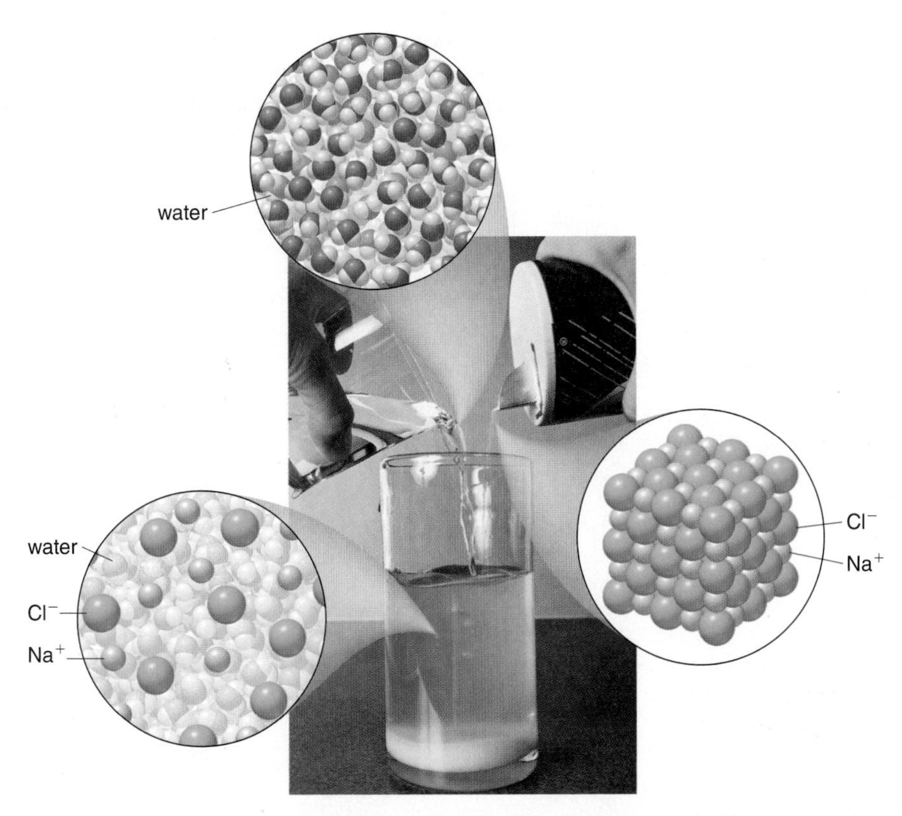

When NaCl dissolves in water, each Na^+ ion and each Cl^- ion are surrounded by water molecules. The interactions of these ions with water molecules provide the energy needed to break apart the ions of the crystal lattice.

An **aqueous solution** contains a substance dissolved in liquid water.

When an ionic compound dissolves in water, the resulting aqueous solution conducts an electric current. This distinguishes ionic compounds from other compounds that dissolve in water but do not form ions and therefore do not conduct electricity.

PROBLEM 3.15

List four physical properties of ionic compounds.

3.6 Polyatomic Ions

Sometimes ions are composed of more than one element. The ion bears a charge because the total number of electrons it contains is different from the total number of protons in the nuclei of all of the atoms.

- A *polyatomic ion* is a cation or anion that contains more than one atom.

The atoms in the polyatomic ion are held together by covalent bonds, but since the ion bears a charge, it bonds to other ions by ionic bonding. For example, calcium sulfate, $CaSO_4$, is composed of a calcium cation, Ca^{2+}, and the polyatomic anion sulfate, SO_4^{2-}. $CaSO_4$ is used to make plaster casts for broken bones.

We will encounter only two polyatomic cations: **H_3O^+, the hydronium ion,** which will play a key role in the acid–base chemistry discussed in Chapter 8, and **NH_4^+, the ammonium ion.**

In contrast, there are several common polyatomic anions, most of which contain a nonmetal like carbon, sulfur, or phosphorus, usually bonded to one or more oxygen atoms. Common examples include **carbonate (CO_3^{2-}), sulfate (SO_4^{2-}),** and **phosphate (PO_4^{3-}).** Table 3.4 lists the most common polyatomic anions.

Table 3.4 Names of Common Polyatomic Anions

Nonmetal	Formula	Name
Carbon	CO_3^{2-}	**Carbonate**
	HCO_3^-	Hydrogen carbonate or bicarbonate
	$CH_3CO_2^-$	Acetate
	CN^-	Cyanide
Nitrogen	NO_3^-	**Nitrate**
	NO_2^-	Nitrite
Oxygen	OH^-	**Hydroxide**
Phosphorus	PO_4^{3-}	**Phosphate**
	HPO_4^{2-}	Hydrogen phosphate
	$H_2PO_4^-$	Dihydrogen phosphate
Sulfur	SO_4^{2-}	**Sulfate**
	HSO_4^-	Hydrogen sulfate or bisulfate
	SO_3^{2-}	Sulfite
	HSO_3^-	Hydrogen sulfite or bisulfite

The names of most polyatomic anions end in the suffix ***-ate.*** Exceptions to this generalization include hydroxide (OH^-) and cyanide (CN^-). Two other aspects of nomenclature are worthy of note.

- **The suffix *-ite* is used for an anion that has one fewer oxygen atom than a similar anion named with the *-ate* ending. Thus, SO_4^{2-} is sulf*ate,* but SO_3^{2-} is sulf*ite.***
- **When two anions differ in the presence of a hydrogen, the word *hydrogen* or the prefix *bi-* is added to the name of the anion. Thus, SO_4^{2-} is sulfate, but HSO_4^- is *hydrogen* sulfate or *bi*sulfate.**

HEALTH NOTE

SPAM
Classic
Hormel

Spam, a canned meat widely consumed in Alaska, Hawaii, and other parts of the United States, contains the preservative sodium nitrite, $NaNO_2$. Sodium nitrite inhibits the growth of *Clostridium botulinum,* a bacterium responsible for a lethal form of food poisoning.

3.6A Writing Formulas for Ionic Compounds with Polyatomic Ions

Writing the formula for an ionic compound with a polyatomic ion is no different than writing a formula for an ion with a single charged atom, so we follow the procedure outlined in Section 3.3A. When the cation and anion have the *same* charge, only *one* of each ion is needed for an overall charge of zero.

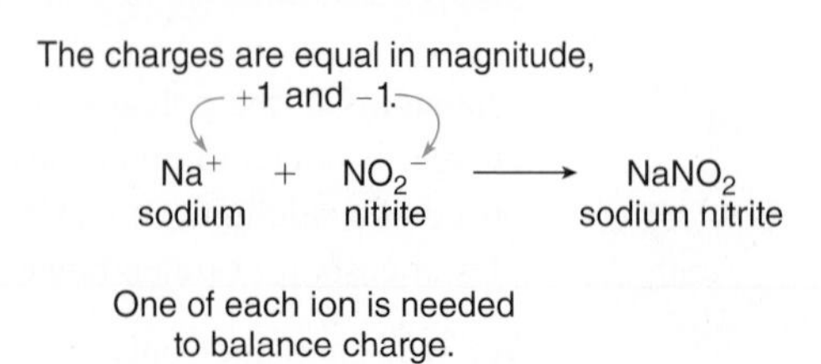
The charges are equal in magnitude, +1 and −1.
Na^+ + NO_2^- ⟶ $NaNO_2$
sodium
nitrite
sodium nitrite
One of each ion is needed to balance charge.

In a compound formed from ions of unequal charge, such as magnesium (Mg^{2+}) and hydroxide (OH^-), **the charges on the ions tell us how many of the *oppositely* charged ions are needed to balance the charge.**

The charges are not equal in magnitude, +2 and −1.

$$Mg^{2+} + OH^- \longrightarrow Mg(OH)_2$$

Two OH^- anions are needed to balance charge.

Use parentheses around all atoms of the ion.

Use a subscript outside the parentheses.

Parentheses are used around the polyatomic ion, and a subscript indicates how many of each are needed to balance charge. The formula is written as $Mg(OH)_2$ *not* MgO_2H_2.

PROBLEM 3.16

Write the formula for the compound formed when K^+ combines with each anion.

a. OH^- b. NO_2^- c. SO_4^{2-} d. HSO_3^- e. PO_4^{3-}

PROBLEM 3.17

Write the formula of the ionic compound formed from each pair of cations and anions.

a. sodium and bicarbonate
b. potassium and nitrate
c. ammonium and sulfate
d. magnesium and phosphate
e. calcium and bisulfate
f. barium and hydroxide

3.6B Naming Ionic Compounds with Polyatomic Ions

Naming ionic compounds derived from polyatomic anions follows the same procedures outlined in Sections 3.4C and 3.4D. There is no easy trick for remembering the names and structures of the anions listed in Table 3.4. The names of the anions in boldface type are especially common and should be committed to memory.

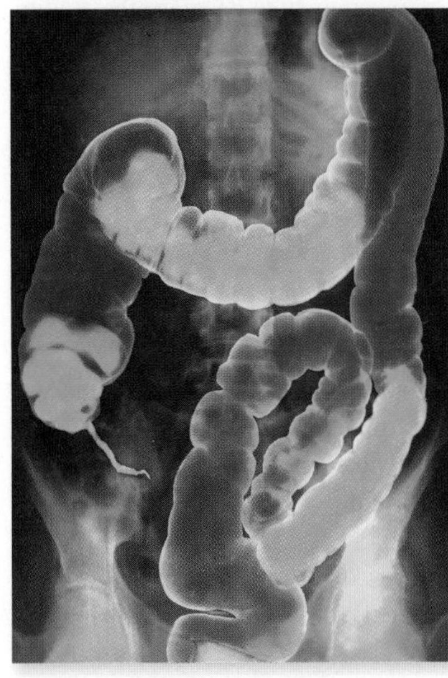

Barium sulfate is used to visualize the digestive system during an X-ray procedure.

SAMPLE PROBLEM 3.9

Name each ionic compound: (a) $NaHCO_3$, the active ingredient in baking soda; (b) $BaSO_4$, a compound used in X-ray imaging.

Analysis

First determine if the cation has a fixed or variable charge. To name an ionic compound that contains a cation that always has the same charge, name the cation and then the anion. When the metal has a variable charge, use the overall anion charge to determine the charge on the cation. Then name the cation (using a Roman numeral or the suffix *-ous* or *-ic*), followed by the anion.

Solution

a. $NaHCO_3$: Sodium cations have a fixed charge of +1. The anion HCO_3^- is called bicarbonate or hydrogen carbonate.

Answer: sodium bicarbonate or sodium hydrogen carbonate

b. $BaSO_4$: Barium cations have a fixed charge of +2. The anion SO_4^{2-} is called sulfate.

Answer: barium sulfate

PROBLEM 3.18

Name each compound.

a. Na_2CO_3
b. $Ca(OH)_2$
c. $Mg(NO_3)_2$
d. $Mn(CH_3CO_2)_2$
e. $Fe(HSO_3)_3$
f. $Mg_3(PO_4)_2$

The shells of oysters and other mollusks are composed largely of calcium carbonate, $CaCO_3$.

3.6C FOCUS ON HEALTH & MEDICINE Useful Ionic Compounds

Ionic compounds are the active ingredients in several over-the-counter drugs. Examples include **calcium carbonate ($CaCO_3$),** the antacid in Tums; **magnesium hydroxide [$Mg(OH)_2$],** one of the active components in the antacids Maalox and milk of magnesia; and **iron(II) sulfate ($FeSO_4$),** an iron supplement used to treat anemia.

HEALTH NOTE

Some calcium supplements contain calcium phosphate (Problem 3.19), which provides the consumer with calcium and phosphorus, both needed for healthy teeth and bones.

Some ionic compounds are given as intravenous drugs. Bicarbonate (HCO_3^-) is an important polyatomic anion that controls the acid–base balance in the blood. When the blood becomes too acidic, sodium bicarbonate ($NaHCO_3$) is administered intravenously to decrease the acidity. Magnesium sulfate ($MgSO_4$), an over-the-counter laxative, is also given intravenously to prevent seizures caused by extremely high blood pressure associated with some pregnancies.

PROBLEM 3.19

Write the formula for the dietary supplement formed from calcium and phosphate.

3.7 Covalent Bonding

In Section 3.1 we learned that **covalent bonds result from the *sharing* of electrons between two atoms.** For example, when two hydrogen atoms with one electron each (H·) combine, they form the hydrogen molecule, H_2, with a covalent bond that contains two electrons. We use a **solid line between two element symbols to represent a two-electron bond.**

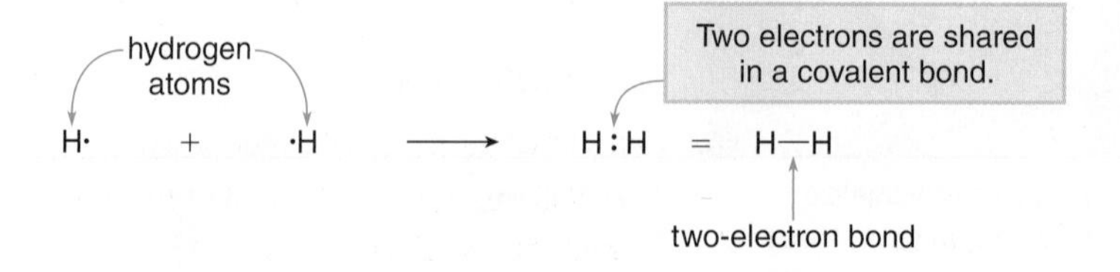

- A covalent bond is a two-electron bond in which the bonding atoms share the electrons.

Hydrogen is called a **diatomic molecule** because it contains just two atoms. In addition to hydrogen, six other elements exist as diatomic molecules: nitrogen (N_2), oxygen (O_2), fluorine (F_2), chlorine (Cl_2), bromine (Br_2), and iodine (I_2).

Hydrogen fluoride, HF, is an example of a diatomic molecule formed between two different atoms, hydrogen and fluorine. Hydrogen has one valence electron and fluorine has seven. H and F each donate one electron to form a single two-electron bond.

H· + ·F: ⟶ H:F: or H—F: **three lone pairs**

two electrons around H **eight electrons** around F

Nonbonded electron pair = lone pair.

The resulting molecule gives both H and F a filled valence shell: H is surrounded by two electrons, giving it the noble gas configuration of helium, and F is surrounded by eight electrons, giving it the noble gas configuration of neon. The F atom shares two electrons in one covalent bond, and it also contains three pairs of electrons that it does not share with hydrogen. These unshared electron pairs are called **nonbonded electron pairs** or **lone pairs.**

- **In covalent bonding, atoms share electrons to attain the electronic configuration of the noble gas closest to them in the periodic table.**
- **As a result, hydrogen shares two electrons.**
- **Other main group elements are especially stable when they possess an *octet* of electrons in their outer shell.**

PROBLEM 3.20

Use electron-dot symbols to show how a hydrogen atom and a chlorine atom form the diatomic molecule HCl.

Methane (CH_4) and water (H_2O) are two examples of covalent molecules in which each main group element is surrounded by eight electrons. Methane, the main component of natural gas, contains four covalent carbon–hydrogen bonds, each having two electrons. The oxygen atom in H_2O is also surrounded by an octet since it has two bonds and two lone pairs.

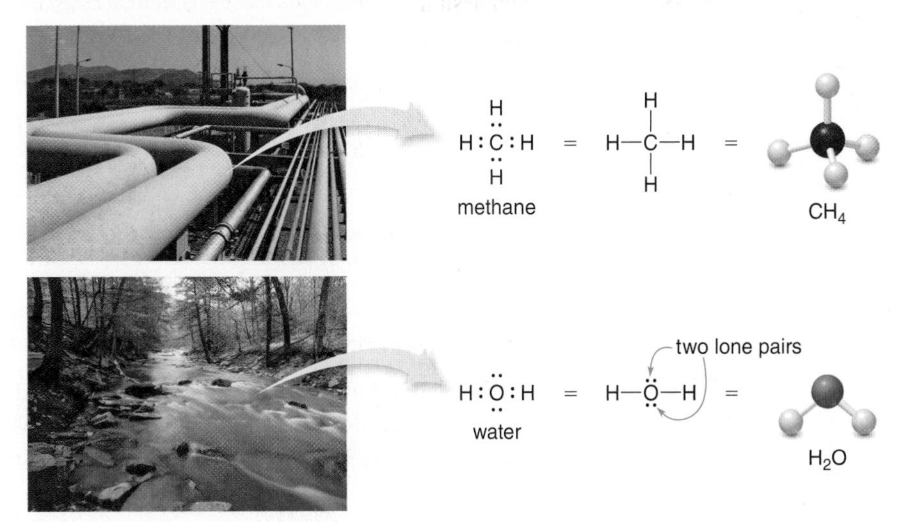

These electron-dot structures for molecules are called Lewis structures. Lewis structures show the location of all valence electrons in a molecule, both the shared electrons in bonds, and the nonbonded electron pairs.

How many covalent bonds will a particular atom typically form? In the first row, hydrogen forms one covalent bond with its one valence electron. Other main group elements generally have no more than eight electrons around them. For neutral molecules, two consequences result.

- **Atoms with one, two, or three valence electrons generally form one, two, or three bonds, respectively.**
- **Atoms with four or more valence electrons form enough bonds to give an octet. Thus, for atoms with four or more valence electrons:**

$$\textbf{Predicted number of bonds} = \textbf{8} - \textbf{number of valence electrons}$$

These guidelines are used in Figure 3.6 to summarize the usual number of covalent bonds formed by some common atoms. Except for hydrogen, **the number of bonds plus the number of lone pairs equals four** for common atoms.

Figure 3.6

Bonding Patterns for Common Main Group Elements

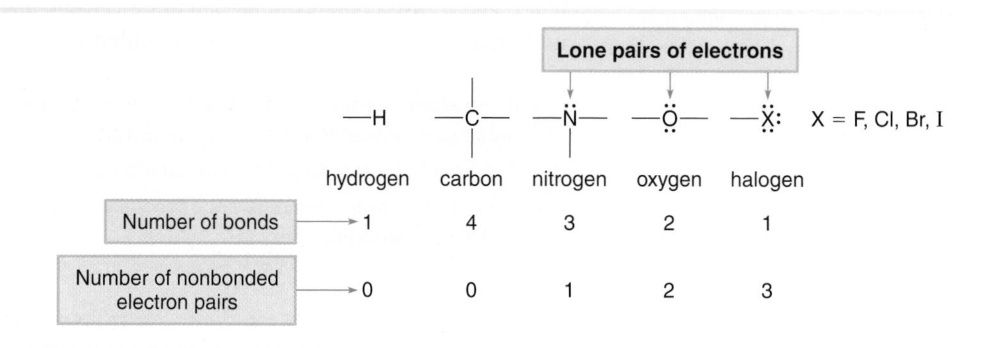

SAMPLE PROBLEM 3.10

Without referring to Figure 3.6, how many covalent bonds are predicted for each atom: (a) B; (b) N?

Analysis

Atoms with one, two, or three valence electrons form one, two, or three bonds, respectively. Atoms with four or more valence electrons form enough bonds to give an octet.

Solution

a. B has three valence electrons. Thus, it is expected to form three bonds.

b. N has five valence electrons. Since it contains more than four valence electrons, it is expected to form $8 - 5 = 3$ bonds.

PROBLEM 3.21

How many covalent bonds are predicted for each atom: (a) F; (b) Si; (c) Br; (d) O; (e) P; (f) S?

3.8 Lewis Structures

A **molecular formula** shows the number and identity of all of the atoms in a compound, but it does not tell us what atoms are bonded to each other. A **Lewis structure,** in contrast, shows the connectivity between the atoms, as well as where all the bonding and nonbonding valence electrons reside.

3.8A Drawing Lewis Structures

There are three general rules for drawing Lewis structures.

1. Draw only the *valence electrons.*
2. Give every main group element (except hydrogen) an *octet* of electrons.
3. Give each hydrogen two electrons.

Sample Problem 3.11 illustrates how to draw a Lewis structure in a molecule that contains only single bonds. In any Lewis structure, **always place hydrogens and halogens on the periphery, since these atoms form only one bond.**

The covalent molecule CH_3Cl is one of many gases released into the air from an erupting volcano.

SAMPLE PROBLEM 3.11

Draw a Lewis structure for chloromethane, CH_3Cl, a compound produced by giant kelp and a component of volcanic emissions.

Analysis

To draw a Lewis structure:

- Arrange the atoms, placing hydrogens and halogens on the periphery.
- Count the valence electrons from all atoms.
- Add the bonds, and use the remaining electrons to fill octets with lone pairs.

Solution

[1] **Arrange the atoms.**

```
   H
H  C  Cl
   H
```

- Place C in the center and 3 H's and 1 Cl on the periphery.
- In this arrangement, C is surrounded by four atoms, its usual number.

[2] **Count the electrons.**

$$1\ \text{C} \times 4\ e^- = 4\ e^-$$
$$3\ \text{H} \times 1\ e^- = 3\ e^-$$
$$1\ \text{Cl} \times 7\ e^- = 7\ e^-$$

14 e^- total

[3] **Add the bonds and lone pairs.**

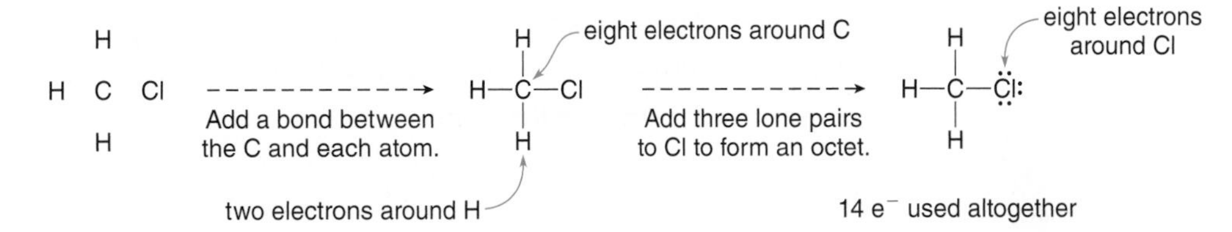

First add four single bonds, three C—H bonds and one C—Cl bond. This uses eight valence electrons, and gives carbon an octet (four two-electron bonds) and each hydrogen two electrons. Next, give Cl an octet by adding three lone pairs. This uses all 14 valence electrons, and gives a valid Lewis structure for CH_3Cl.

PROBLEM 3.22

Draw a Lewis structure for each covalent molecule.

a. HBr b. CH_3F c. H_2O_2 d. N_2H_4 e. C_2H_6 f. CH_2Cl_2

3.8B Multiple Bonds

Sometimes it is not possible to give every main group element (except hydrogen) an octet of electrons by placing only single bonds in a molecule. In this case, the Lewis structures must contain one or more *multiple* bonds.

For example, in drawing a Lewis structure for N_2, each N has five valence electrons, so there are 10 electrons to place. If there is only one N—N bond, adding lone pairs gives one or both N's fewer than eight electrons.

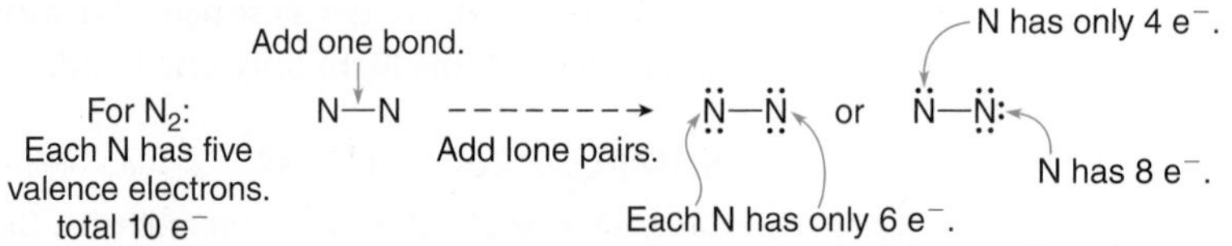

For N_2:
Each N has five valence electrons.
total 10 e^-

Add one bond. N—N → Add lone pairs. → :N̈—N̈: (Each N has only 6 e^-.) or N̈—N̈: (N has only 4 e^-. N has 8 e^-.)

In this case, we must convert a lone pair to a bonding pair of electrons to form a multiple bond. Since we have four fewer electrons than needed, **we must convert *two* lone pairs to *two* bonding pairs of electrons and form a *triple* bond.**

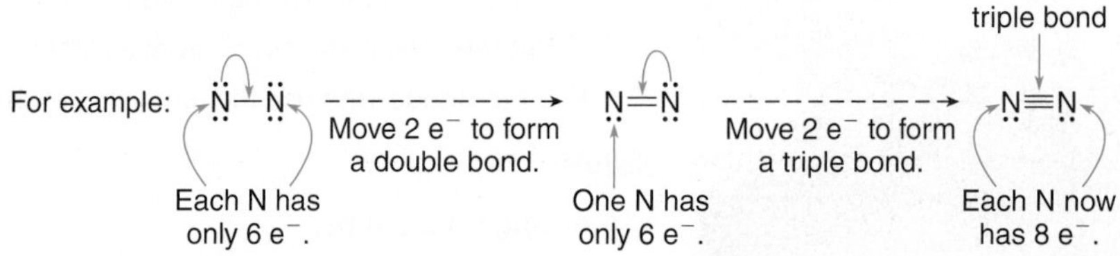

For example: N̈—N̈ (Each N has only 6 e^-.) → Move 2 e^- to form a double bond. → N=N̈ (One N has only 6 e^-.) → Move 2 e^- to form a triple bond. → :N≡N: (triple bond; Each N now has 8 e^-.)

- **A triple bond contains six electrons in three two-electron bonds.**

Sample Problem 3.12 illustrates another example of a Lewis structure that contains a double bond.

- **A double bond contains four electrons in two two-electron bonds.**

SAMPLE PROBLEM 3.12

Draw a Lewis structure for ethylene, a compound of molecular formula C_2H_4 in which each carbon is bonded to two hydrogens.

Analysis and Solution

Follow steps [1]–[3] to draw a Lewis structure.

[1] **Arrange the atoms.**

H C C H
H H

- Each C gets 2 H's.

[2] **Count the electrons.**

$$2\text{ C} \times 4\ e^- = 8\ e^-$$
$$4\text{ H} \times 1\ e^- = 4\ e^-$$

12 e^- total

[3] **Add the bonds and lone pairs.**

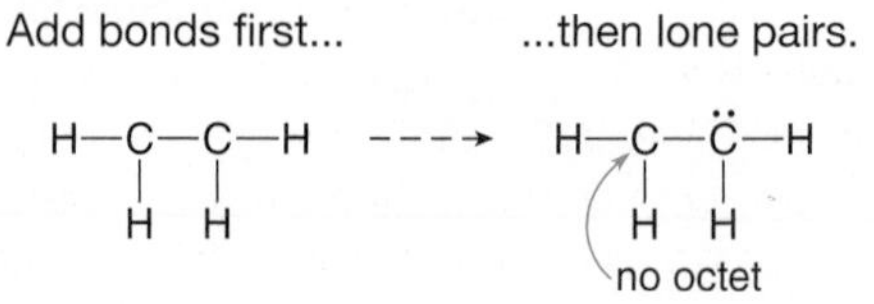

Add bonds first... H—C—C—H (each C bonded to H below) → ...then lone pairs. H—C—C̈—H (no octet)

After placing five bonds between the atoms and adding the two remaining electrons as a lone pair, one C still has no octet.

[4] **To give both C's an octet, change *one* lone pair into *one* bonding pair of electrons between the two C atoms, forming a *double* bond.**

Move a lone pair. ethylene — double bond

- Each C now has four bonds.
- Each C is now surrounded by eight electrons.

This uses all 12 electrons, each C has an octet, and each H has two electrons. The Lewis structure is valid. **Ethylene contains a carbon–carbon double bond.**

- **After placing all electrons in bonds and lone pairs, use a lone pair to form a multiple bond if an atom does not have an octet.**

Formic acid (CH_2O_2, Problem 3.24) is responsible for the sting of some types of ants.

PROBLEM 3.23

Draw a valid Lewis structure for each compound, using the given arrangement of atoms.

a. HCN (hydrogen cyanide) H C N

b. CH_2O (formaldehyde) H C O, with H below C

PROBLEM 3.24

Draw a Lewis structure for formic acid with the given arrangement of atoms.

H C O H, with O above C

3.9 Naming Covalent Compounds

Although some covalent compounds are always referred to by their common names—H_2O (water) and NH_3 (ammonia)—these names tell us nothing about the atoms that the molecule contains. Other covalent compounds with two elements are named to indicate the identity and number of elements they contain.

How To Name a Covalent Molecule with Two Elements

Example Name each covalent molecule: (a) NO_2; (b) N_2O_4.

Step [1] Name the first nonmetal by its element name and the second using the suffix *-ide*.

- In both compounds the first nonmetal is nitrogen.
- To name the second element, change the name oxygen to **ox*ide*.**

Step [2] Add prefixes to show the number of atoms of each element.

- Use a prefix from Table 3.5 for each element.
- The prefix *mono-* is omitted when only one atom of the first element is present, but retained for the second element.
- When the prefix and element name would place two vowels next to each other, omit the first vowel. For example, mono- + oxide = monoxide (*not* monooxide).

a. NO_2 contains one N atom, so the prefix *mono-* is understood. Since NO_2 contains two O atoms, use the prefix *di-* → *di*oxide. Thus, NO_2 is **nitrogen dioxide.**

b. N_2O_4 contains two N atoms, so use the prefix *di-* → *di*nitrogen. Since N_2O_4 contains four O atoms, use the prefix *tetra-* and omit the *a* → *tetr*oxide (*not* tetraoxide). Thus, N_2O_4 is **dinitrogen tetroxide.**

Table 3.5 Common Prefixes in Nomenclature

Number of Atoms	Prefix
1	Mono
2	Di
3	Tri
4	Tetra
5	Penta
6	Hexa
7	Hepta
8	Octa
9	Nona
10	Deca

PROBLEM 3.25

Name each compound: (a) CS_2; (b) SO_2; (c) PCl_5; (d) BF_3.

To write a formula from a name, write the element symbols in the order of the elements in the name. Then use the prefixes to determine the subscripts of the formula, as shown in Sample Problem 3.13.

SAMPLE PROBLEM 3.13

Give the formula for each compound: (a) silicon tetrafluoride; (b) diphosphorus pentoxide.

Analysis

- Determine the symbols for the elements in the order given in the name.
- Use the prefixes to write the subscripts.

Solution

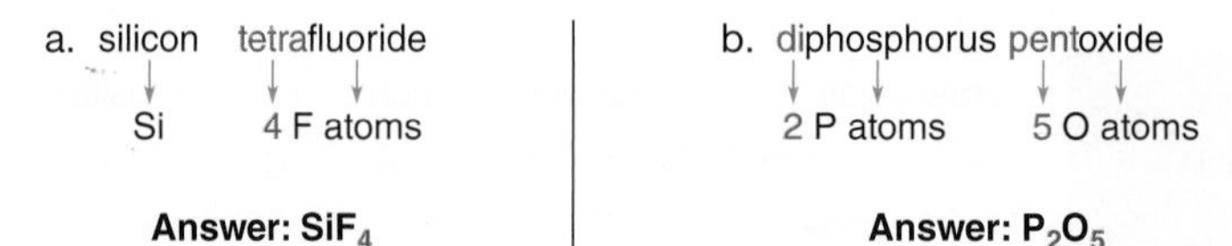

PROBLEM 3.26

Give the formula for each compound: (a) silicon dioxide; (b) phosphorus trichloride; (c) sulfur trioxide; (d) dinitrogen trioxide.

3.10 Molecular Shape

We can now use Lewis structures to determine the shape around a particular atom in a molecule. Consider the H_2O molecule. The Lewis structure tells us only which atoms are connected to each other, but it implies nothing about the geometry. Is H_2O a bent or linear molecule?

What is the bond angle?

H—Ö—H

To determine the shape around a given atom, we must first determine how many groups surround the atom. **A group is either an atom or a lone pair of electrons.** Then we use the **valence shell electron pair repulsion (VSEPR) theory** to determine the shape. VSEPR is based on the fact that electron pairs repel each other; thus:

- **The most stable arrangement keeps these groups as far away from each other as possible.**

In general, an atom has three possible arrangements of the groups that surround it.

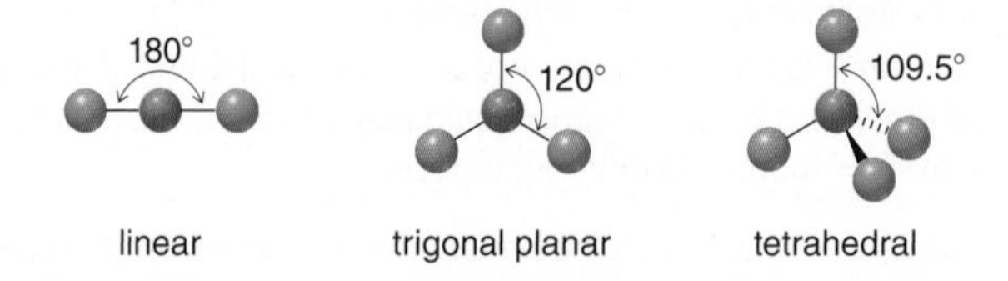

- **An atom surrounded by two groups is linear and has a bond angle of 180°.**
- **An atom surrounded by three groups is trigonal planar and has bond angles of 120°.**
- **An atom surrounded by four groups is tetrahedral and has bond angles of 109.5°.**

HEALTH NOTE

Cassava is a widely grown root crop, first introduced to Africa by Portuguese traders from Brazil in the sixteenth century. The root must be boiled or roasted to remove the compound linamarin before ingestion. Linamarin is not toxic itself, but it forms HCN in the presence of water and some enzymes. Eating the root without processing affords high levels of HCN, a cellular poison with a characteristic almond odor.

ENVIRONMENTAL NOTE

Over time, some adhesives and insulation made from formaldehyde can decompose back to formaldehyde, a reactive and potentially hazardous substance. Spider plants act as natural air purifiers by removing formaldehyde (H_2CO) from the air.

Trigonal = three-sided.

3.10A Two Groups Around an Atom

Any atom surrounded by only two groups is linear and has a bond angle of 180°. Two examples illustrating this geometry are **CO_2** (carbon dioxide) and **HCN** (hydrogen cyanide).

The Lewis structure for CO_2 contains a central carbon atom surrounded by two oxygen atoms. To give every atom an octet and the usual number of bonds requires two carbon–oxygen double bonds. Since the carbon atom is surrounded by two groups, the molecule is linear and the O—C—O bond angle is 180°.

four bonds
:Ö=C=Ö:
two bonds two bonds

180°
:Ö=C=Ö: =
two atoms around C
two groups

linear molecule

Carbon dioxide illustrates another important feature of VSEPR theory: ***ignore multiple bonds in predicting geometry.*** **Count only atoms and lone pairs.**

Similarly, the Lewis structure for HCN contains a central carbon atom surrounded by one hydrogen and one nitrogen. To give carbon and nitrogen an octet and the usual number of bonds requires a carbon–nitrogen triple bond. The carbon atom is surrounded by two groups, making the molecule linear and the H—C—N bond angle 180°.

four bonds
H—C≡N:
three bonds

180°
H—C≡N: =
two atoms around C
two groups

linear molecule

3.10B Three Groups Around an Atom

Any atom surrounded by three groups is trigonal planar and has bond angles of 120°. Formaldehyde (**$H_2C{=}O$**) illustrates this geometry.

H
120° C=Ö: =
H

three atoms around C
three groups

trigonal planar molecule

The carbon atom in $H_2C{=}O$ is surrounded by three atoms (two H's and one O) and no lone pairs—that is, three groups. To keep the three groups as far from each other as possible, they are arranged in a trigonal planar fashion, with bond angles of 120°.

3.10C Four Groups Around an Atom

Any atom surrounded by four groups is tetrahedral and has bond angles of (approximately) 109.5°. For example, the simple organic compound methane, **CH_4,** has a central carbon atom with four bonds to hydrogen, each pointing to the corners of a tetrahedron.

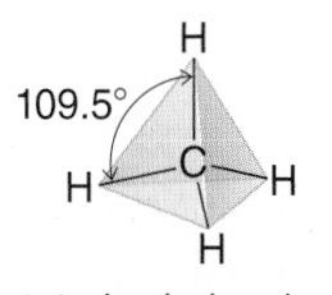

tetrahedral carbon

How can we represent the three-dimensional geometry of a tetrahedron on a two-dimensional piece of paper? **Place two of the bonds in the plane of the paper, one bond in front, and one bond behind,** using the following conventions:

- A solid line is used for bonds in the plane.
- A wedge is used for a bond in front of the plane.
- A dashed line is used for a bond behind the plane.

bonds **in the plane** — bond **behind** — bond **in front** = tetrahedral molecule

Up to now, each of the groups around the central atom has been another atom. **A group can also be a lone pair of electrons.** NH_3 and H_2O represent two examples of molecules with atoms surrounded by four groups, some of which are lone pairs.

The Lewis structure for ammonia, $\mathbf{NH_3}$, has an N atom surrounded by three hydrogen atoms and one lone pair of electrons—four groups. To keep four groups as far apart as possible, the three H atoms and the one lone pair around N point to the corners of a tetrahedron. The H—N—H bond angle of 107° is close to the tetrahedral bond angle of 109.5°. This shape is referred to as a **trigonal pyramid,** since one of the groups around the N is a nonbonded electron pair, not another atom.

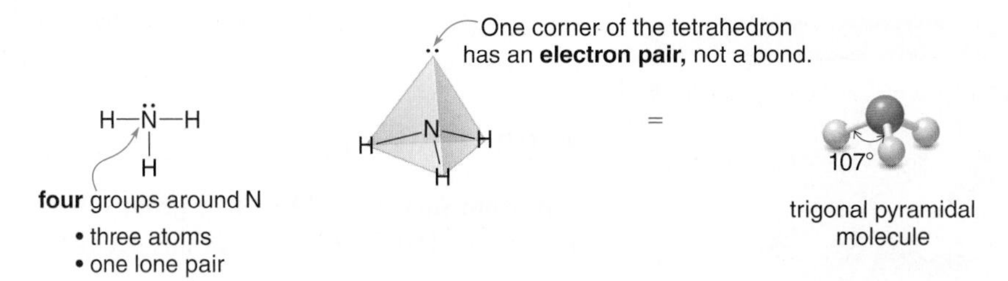

The Lewis structure for water, $\mathbf{H_2O}$, has an O atom surrounded by two hydrogen atoms and two lone pairs of electrons—four groups. In $\mathbf{H_2O}$, the two H atoms and the two lone pairs around O point to the corners of a tetrahedron. The H—O—H bond angle of 105° is close to the tetrahedral bond angle of 109.5°. Water has a ***bent*** shape, because two of the groups around oxygen are lone pairs of electrons.

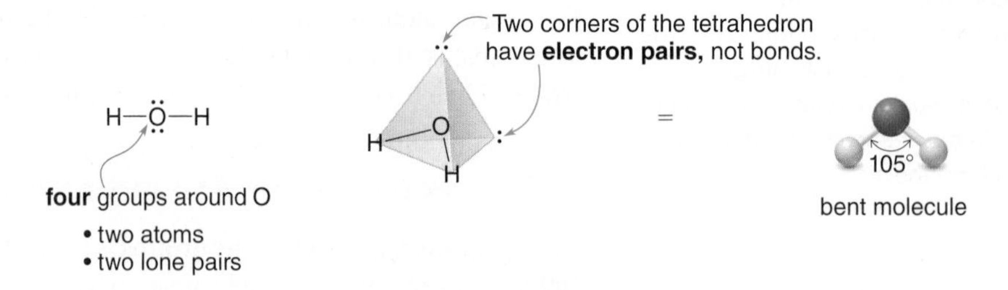

Common molecular shapes are summarized in Table 3.6.

Table 3.6 Common Molecular Shapes Around Atoms

Total Number of Groups	Number of Atoms	Number of Lone Pairs	Shape Around an Atom (A)	Approximate Bond Angle (°)	Example
2	2	0	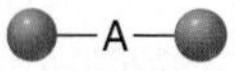 linear	180	CO_2, HC≡N
3	3	0	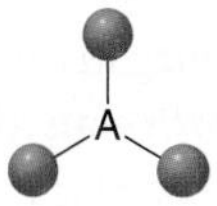 trigonal planar	120	$H_2C{=}O$
4	4	0	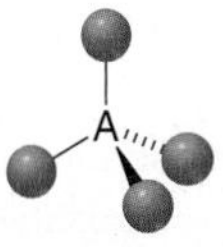 tetrahedral	109.5	CH_4
4	3	1	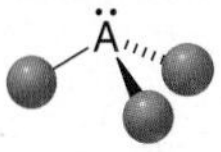 trigonal pyramidal	~109.5[a]	NH_3
4	2	2	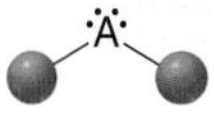 bent	~109.5[a]	H_2O

[a]The symbol "~" means approximately.

SAMPLE PROBLEM 3.14

Using the given Lewis structure, determine the shape around the second-row elements in each compound.

a. H—C≡C—H (acetylene)

b. 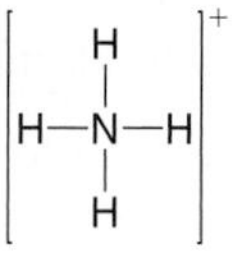 $[NH_4]^+$ (ammonium ion)

Analysis

To predict the shape around an atom, we count groups around the atom in the Lewis structure and use the information in Table 3.6.

Solution

a. Each C in H—C≡C—H is surrounded by two atoms (one C and one H) and no lone pairs—that is, two groups. An atom surrounded by two groups is linear with a 180° bond angle.

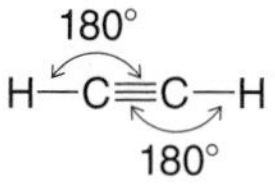

180° H—C≡C—H 180°

b. The N atom in NH_4^+ is surrounded by four H atoms—that is, four groups. An atom surrounded by four groups is tetrahedral, with 109.5° bond angles.

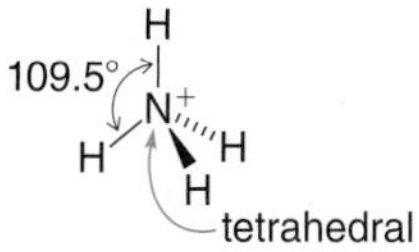

109.5° tetrahedral

PROBLEM 3.27

What is the shape around the indicated atom in each molecule?

a. $H_2\ddot{S}$:

b. CH_2Cl_2

c. :NCl_3

d. 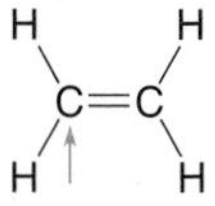 $H_2C{=}CH_2$

3.11 Electronegativity and Bond Polarity

When two atoms share electrons in a covalent bond, are the electrons in the bond attracted to both nuclei to the same extent? That depends on the **electronegativity** of the atoms in the bond.

- ***Electronegativity*** **is a measure of an atom's attraction for electrons in a bond. Electronegativity tells us how much a particular atom "*wants*" electrons.**

The electronegativity of an atom is assigned a value from 0 to 4; **the *higher* the value, the *more electronegative* an atom is, and the *more it is attracted* to the electrons in a bond.** The electronegativity values for main group elements are shown in Figure 3.7. The noble gases are not assigned values, since they do not typically form bonds.

Electronegativity values exhibit periodic trends.

- **Electronegativity *increases* across a row of the periodic table as the nuclear charge increases (excluding the noble gases).**
- **Electronegativity *decreases* down a column of the periodic table as the atomic radius increases, pushing the valence electrons farther from the nucleus.**

PROBLEM 3.28

Using the trends in the periodic table, rank the following atoms in order of increasing electronegativity.

a. Li, Na, H   b. O, C, Be   c. Cl, I, F   d. B, O, N

Electronegativity values are used as a guideline to indicate whether the electrons in a bond are *equally* shared or *unequally* shared between two atoms. For example, whenever two *identical* atoms are bonded together, each atom attracts the electrons in the bond to the same extent. The electrons are equally shared, and the bond is said to be **nonpolar.** Thus, a **carbon–carbon bond is nonpolar,** as is the fluorine–fluorine bond in F_2. The same is true whenever two different atoms having *similar* electronegativities are bonded together. **C—H bonds are considered to be nonpolar,** because the electronegativity difference between C (2.5) and H (2.1) is small.

2.5 —C—C— 2.5
nonpolar bond

2.5 —C—H 2.1
nonpolar bond
The small electronegativity difference between C and H is ignored.

Figure 3.7
Electronegativity Values for Main Group Elements

Increasing electronegativity (→ across a row); Increasing electronegativity (↑ up a column)

1A	2A	3A	4A	5A	6A	7A	8A
H 2.1							
Li 1.0	Be 1.5	B 2.0	C 2.5	N 3.0	O 3.5	F 4.0	
Na 0.9	Mg 1.2	Al 1.5	Si 1.8	P 2.1	S 2.5	Cl 3.0	
K 0.8	Ca 1.0	Ga 1.6	Ge 1.8	As 2.0	Se 2.4	Br 2.8	
Rb 0.8	Sr 1.0	In 1.7	Sn 1.8	Sb 1.9	Te 2.1	I 2.5	

In contrast, bonding between atoms of *different* electronegativity results in the *unequal* sharing of electrons. For example, in a C—O bond, the electrons are pulled away from C (2.5) towards the element of higher electronegativity, O (3.5). **The bond is *polar*, or *polar covalent*.** The bond is said to have a **dipole**—that is, **a partial separation of charge.**

$$\overset{\delta^+}{—C}—\overset{\delta^-}{O}—$$

a dipole

A C–O bond is a *polar* bond.

The direction of polarity in a bond is often indicated by an arrow, with the head of the arrow pointing towards the more electronegative element. The tail of the arrow, with a perpendicular line drawn through it, is drawn at the less electronegative element. Alternatively, the lower case Greek letter delta (δ) with a positive or negative charge is used, resulting in the symbols δ^+ and δ^- to indicate this unequal sharing of electron density.

- **The symbol δ^+ is given to the less electronegative atom.**
- **The symbol δ^- is given to the more electronegative atom.**

Students often wonder how large an electronegativity difference must be to consider a bond polar. That's hard to say. **Usually, a polar bond will be one in which the electronegativity difference between two atoms is 0.5 units or greater.**

As the electronegativity difference between the two atoms in a bond increases, the shared electrons are pulled more and more towards the more electronegative element. When the electronegativity difference is larger than 1.9 units, the electrons are essentially transferred from the less electronegative element to the more electronegative element and the bond is considered ionic. Table 3.7 summarizes the relationship between the electronegativity difference of the atoms in a bond and the type of bond formed.

Table 3.7 Electronegativity Difference and Bond Type

Electronegativity Difference	Bond Type	Electron Sharing
Less than 0.5 units	Nonpolar	Electrons are equally shared.
0.5–1.9 units	Polar covalent	Electrons are unequally shared; they are pulled towards the more electronegative element.
Greater than 1.9 units	Ionic	Electrons are transferred from the less electronegative element to the more electronegative element.

SAMPLE PROBLEM 3.15

Use electronegativity values to classify each bond as nonpolar, polar covalent, or ionic: (a) Cl_2; (b) HCl; (c) NaCl.

Analysis

Calculate the electronegativity difference between the two atoms and use the following rules: less than 0.5 (nonpolar); 0.5–1.9 (polar covalent); and greater than 1.9 (ionic).

Solution

	Electronegativity Difference	Bond Type
a. Cl_2	3.0 (Cl) – 3.0 (Cl) = 0	Nonpolar
b. HCl	3.0 (Cl) – 2.1 (H) = 0.9	Polar covalent
c. NaCl	3.0 (Cl) – 0.9 (Na) = 2.1	Ionic

PROBLEM 3.29

Use electronegativity values to classify the bond(s) in each compound as nonpolar, polar covalent, or ionic.

a. HF b. MgO c. F_2 d. ClF e. H_2O

PROBLEM 3.30

Show the direction of the dipole in each bond. Label the atoms with δ^+ and δ^-.

a. H—F b. —B—C— c. —C—Li d. —C—Cl

3.12 Polarity of Molecules

Thus far, we have been concerned with the polarity of a single bond. Is an entire covalent molecule polar or nonpolar? That depends on two factors: the polarity of the individual bonds and the overall shape. When a molecule contains zero or one polar bond, the following can be said:

- **A molecule with no polar bonds is a nonpolar molecule.**
- **A molecule with one polar bond is a polar molecule.**

Thus, CH_4 is a nonpolar molecule because all of the C—H bonds are nonpolar. In contrast, CH_3Cl contains only one polar bond, so it is a polar molecule. The dipole is in the same direction as the dipole of the only polar bond.

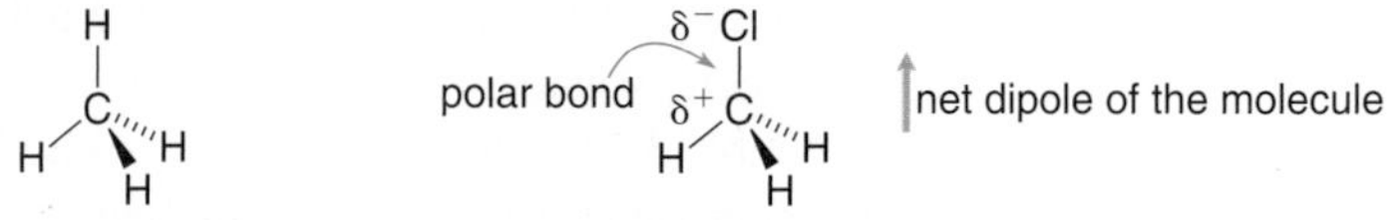

CH_4
no polar bonds
nonpolar molecule

polar bond
net dipole of the molecule

CH_3Cl
one polar bond
polar molecule

With covalent compounds that have more than one polar bond, the shape of the molecule determines the overall polarity.

- **If the individual bond dipoles do not cancel, the molecule is polar.**
- **If the individual bond dipoles cancel, the molecule is nonpolar.**

To determine the polarity of a molecule that has two or more polar bonds:

1. **Identify all polar bonds based on electronegativity differences.**
2. **Determine the shape around individual atoms by counting groups.**
3. **Decide if individual dipoles cancel or reinforce.**

Figure 3.8 illustrates several examples of polar and nonpolar molecules that contain polar bonds. The net dipole is the sum of all the bond dipoles in a molecule.

Figure 3.8
Examples of Polar and Nonpolar Molecules

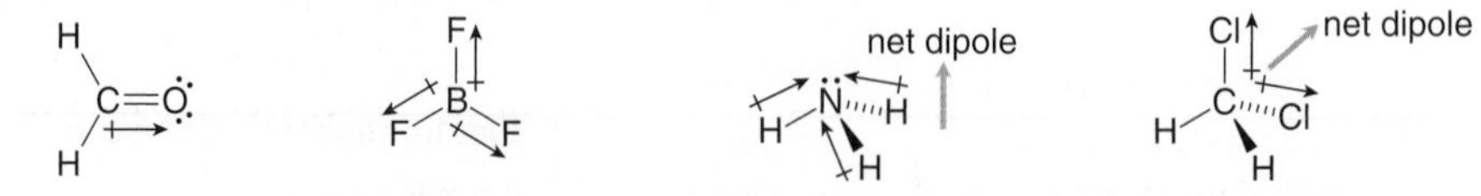

one polar bond — **polar** molecule

three polar bonds. All dipoles cancel. NO net dipole — **nonpolar** molecule

net dipole

three polar bonds. All dipoles reinforce. — **polar** molecule

net dipole
two polar bonds. Two dipoles reinforce. — **polar** molecule

SAMPLE PROBLEM 3.16

Determine whether each molecule is polar or nonpolar: (a) H_2O; (b) CO_2.

Analysis

To determine the overall polarity of a molecule: identify the polar bonds; determine the shape around individual atoms; decide if the individual bond dipoles cancel or reinforce.

Solution

a. **H_2O:** Each O—H bond is polar because the electronegativity difference between O (3.5) and H (2.1) is 1.4. Since the O atom of H_2O has two atoms and two lone pairs around it, H_2O is a bent molecule around the O atom. The two dipoles reinforce (both point *up*), so **H_2O has a net dipole;** that is, **H_2O is a polar molecule.**

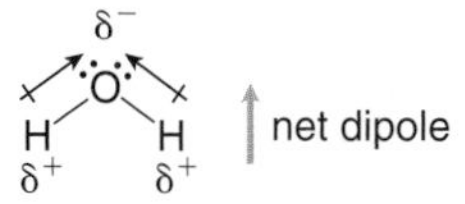

The dipoles reinforce.

b. **CO_2:** Each C—O bond is polar because the electronegativity difference between O (3.5) and C (2.5) is 1.0. The Lewis structure of CO_2 (Section 3.10A) shows that the C atom is surrounded by two groups (two O atoms), making it linear. In this case, the two dipoles are equal and opposite in direction so they cancel. Thus, CO_2 is a **nonpolar molecule** with **no net dipole.**

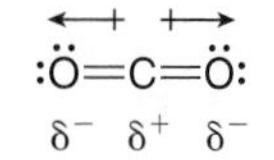

NO net dipole

PROBLEM 3.31

Label the polar bonds in each molecule, and then decide if the molecule is polar or nonpolar.

a. HCl b. CH_2F_2 c. HCN d. CCl_4

The principles learned in Sections 3.10–3.12 apply to all molecules regardless of size. Count groups around each atom individually to determine its shape. Look for electronegativity differences between the two atoms in a bond to determine polarity.

For example, ethanol, the "alcohol" in alcoholic beverages and the world's most widely abused drug, has molecular formula C_2H_6O. The Lewis structure of ethanol shows that each carbon atom is surrounded by four atoms (four groups), making each carbon tetrahedral. The oxygen atom is surrounded by two atoms and two lone pairs, giving it a bent shape. Ethanol contains two polar bonds, the C—O and the O—H bonds, because of the large electronegativity difference between the atoms in each bond (C and O or O and H).

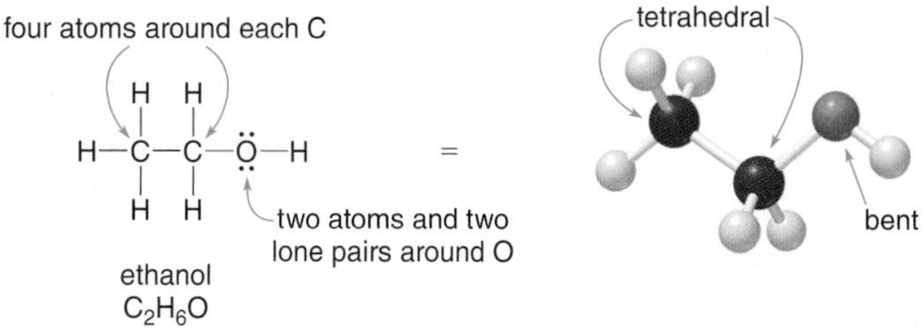

Ethanol contains two polar bonds, drawn in red.

HEALTH NOTE

The ethanol in wine is formed by the fermentation of carbohydrates in grapes. The chronic and excessive consumption of alcoholic beverages has become a major health and social crisis in the United States.

PROBLEM 3.32

Ethanol is metabolized in the body to acetaldehyde, a toxic compound that produces some of the ill effects of ingesting too much ethanol. Determine the geometry around each carbon in acetaldehyde. Label each bond as polar or nonpolar.

```
   H  :O:
   |  ||
H—C—C—H
   |
   H
```

acetaldehyde

KEY TERMS

Ammonium ion (3.6)
Anion (3.2)
Bonding (3.1)
Carbonate (3.6)
Cation (3.2)
Covalent bond (3.7)
Diatomic molecule (3.7)
Dipole (3.11)
Double bond (3.8)
Electronegativity (3.11)
Hydronium ion (3.6)
Hydroxide (3.6)
Ion (3.1)
Ionic bond (3.1)
Lewis structure (3.7)
Lone pair (3.7)
Molecular formula (3.8)
Molecule (3.1)
Nomenclature (3.4)
Nonbonded electron pair (3.7)
Nonpolar bond (3.11)
Octet rule (3.2)
Phosphate (3.6)
Polar bond (3.11)
Polyatomic ion (3.6)
Sulfate (3.6)
Triple bond (3.8)
Valence shell electron pair repulsion (VSEPR) theory (3.10)

KEY CONCEPTS

1 What are the basic features of ionic and covalent bonds? (3.1)

- Both ionic and covalent bonding follows one general rule: Elements gain, lose, or share electrons to attain the electronic configuration of the noble gas closest to them in the periodic table.
- Ionic bonds result from the transfer of electrons from one element to another. Ionic compounds consist of oppositely charged ions that feel a strong electrostatic attraction for each other.
- Covalent bonds result from the sharing of electrons between two atoms. Covalent bonding forms discrete molecules.

2 How can the periodic table be used to determine the charge of an ion? (3.2)

- Metals form cations and nonmetals form anions.
- The charge on main group ions can be predicted from the position in the periodic table. For metals in groups 1A, 2A, and 3A, the group number = the charge on the cation. For nonmetals in groups 5A, 6A, and 7A, the anion charge = 8 – (the group number).

3 What is the octet rule? (3.2)

- Main group elements are especially stable when they possess an octet of electrons. Main group elements gain or lose one, two, or three electrons to form ions with eight outer shell electrons.

4 What determines the formula of an ionic compound? (3.3)

- Cations and anions always form ionic compounds that have zero overall charge.
- Ionic compounds are written with the cation first, and then the anion, with subscripts to show how many of each are needed to have zero net charge.

5 How are ionic compounds named? (3.4)

- With cations having a fixed charge, the cation has the same name as its neutral element. The name of the anion usually ends in the suffix *-ide* if it is derived from a single atom or *-ate* (or *-ite*) if it is polyatomic.
- When the metal has a variable charge, use the overall anion charge to determine the charge on the cation. Then name the cation using a Roman numeral or the suffix *-ous* (for the ion with the lesser charge) or *-ic* (for the ion with the larger charge).

6 Describe the properties of ionic compounds. (3.5)

- Ionic compounds are crystalline solids.
- Ionic compounds have high melting points and boiling points.
- Most ionic compounds are soluble in water and their aqueous solutions conduct an electric current.

7 What are polyatomic ions and how are they named? (3.6)

- Polyatomic ions are charged species that are composed of more than one element.
- The names for polyatomic cations end in the suffix *-onium*.
- Many polyatomic anions have names that end in the suffix *-ate*.

8 What are Lewis structures and how are they drawn? (3.7, 3.8)

- Lewis structures are electron-dot representations of molecules. Two-electron bonds are drawn with a solid line and nonbonded electrons are drawn with dots (:).
- Lewis structures contain only valence electrons. Each H gets two electrons and main group elements generally get eight.

9 How are covalent compounds with two elements named? (3.9)

- Name the first nonmetal by its element name and the second using the suffix *-ide*. Add prefixes to indicate the number of atoms of each element.

10 How is the molecular shape around an atom determined? (3.10)

- To determine the shape around an atom, count groups—atoms and lone pairs—and keep the groups as far away from each other as possible.
- Two groups = linear, 180° bond angle; three groups = trigonal planar, 120° bond angle; four groups = tetrahedral, 109.5° bond angle.

11 How does electronegativity determine bond polarity? (3.11)

- Electronegativity is a measure of an atom's attraction for electrons in a bond.
- When two atoms have the same electronegativity value, or the difference is less than 0.5 units, the electrons are equally shared and the bond is nonpolar.
- When two atoms have very different electronegativity values—a difference of 0.5–1.9 units—the electrons are unequally shared and the bond is polar.

12 When is a molecule polar or nonpolar? (3.12)

- A polar molecule has either one polar bond, or two or more bond dipoles that do not cancel.
- A nonpolar molecule has either all nonpolar bonds, or two or more bond dipoles that cancel.

UNDERSTANDING KEY CONCEPTS

Selected in-chapter and odd-numbered end-of-chapter problems have brief answers at the end of each chapter. The *Student Study Guide and Solutions Manual* contains detailed solutions to all in-chapter and odd-numbered end-of-chapter problems, as well as additional worked examples and a chapter self-test.

3.33 Which formulas represent ionic compounds and which represent covalent compounds?

a. CO_2 b. H_2SO_4 c. KF d. CH_5N

3.34 Which pairs of elements are likely to form ionic bonds and which pairs are likely to form covalent bonds?

a. potassium and oxygen
b. sulfur and carbon
c. two bromine atoms
d. carbon and oxygen

3.35 Complete the following table by filling in the formula of the ionic compound derived from the cations on the left and each of the anions across the top.

	Br^-	OH^-	HCO_3^-	SO_3^{2-}	PO_4^{3-}
Na^+					
Co^{2+}					
Al^{3+}					

3.36 Complete the following table by filling in the formula of the ionic compound derived from the cations on the left and each of the anions across the top.

	I^-	CN^-	NO_3^-	SO_4^{2-}	HPO_4^{2-}
K^+					
Mg^{2+}					
Cr^{3+}					

3.37 Use the element colors in Figure 2.3 (also shown on the inside back cover) and the information in Table 3.4 to identify each polyatomic ion. Give the name and proper formula, including charge.

a.  b. 

3.38 Use the element colors in Figure 2.3 (also shown on the inside back cover) and the information in Table 3.4 to identify each polyatomic ion. Give the name and proper formula, including charge.

a.  b. 

3.39 Write the formula for silver nitrate, an antiseptic and germ killing agent.

3.40 Ammonium carbonate is the active ingredient in smelling salts. Write its formula.

3.41 $CaSO_3$ is used to preserve cider and fruit juices. Name this ionic compound.

3.42 Ammonium nitrate is the most common source of the element nitrogen in fertilizers. When it is mixed with water, the solution gets cold, so it is used in instant cold packs. When mixed with diesel fuel it forms an explosive mixture that can be used as a bomb. Write the formula for ammonium nitrate.

3.43 Draw a valid Lewis structure for each molecule.

a. HI b. CH_2F_2

3.44 Draw a valid Lewis structure for each molecule.

a. CH_3Br b. PH_3

3.45 Answer the following questions about the molecule Cl_2O.
a. Draw a valid Lewis structure.
b. Label all polar bonds and show bond dipoles.
c. What is the shape around the O atom?
d. Is Cl_2O a polar molecule?

3.46 Explain why $CHCl_3$ is a polar molecule but CCl_4 is not.

3.47 (a) Translate each ball-and-stick model to a Lewis structure and include all nonbonded electron pairs. (b) Label all polar bonds. (c) Is the molecule polar or nonpolar?

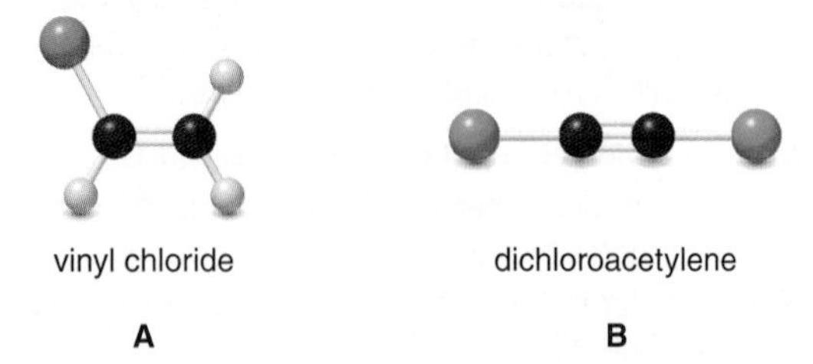

3.48 (a) Translate each ball-and-stick model to a Lewis structure and include all nonbonded electron pairs. (b) Label all polar bonds. (c) Is the molecule polar or nonpolar?

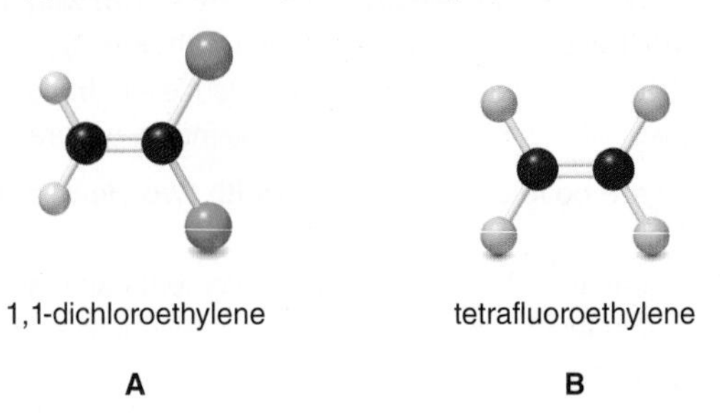

ADDITIONAL PROBLEMS

Ions

3.49 Write the ion symbol for an atom with the given number of protons and electrons.
a. four protons and two electrons
b. 22 protons and 20 electrons
c. 16 protons and 18 electrons
d. 13 protons and 10 electrons

3.50 How many protons and electrons are present in each ion?
a. K^+ b. Mn^{2+} c. Fe^{2+} d. Cs^+

3.51 What species fits each description?
a. a period 2 element that forms a +2 cation
b. an ion from group 7A with 18 electrons
c. a cation from group 1A with 36 electrons

3.52 What species fits each description?
a. a period 3 element that forms an ion with a –1 charge
b. an ion from group 2A with 36 electrons
c. an ion from group 6A with 18 electrons

3.53 Give the ion symbol for each ion.
a. sodium ion c. gold(III)
b. manganese ion d. stannic

3.54 Give the ion symbol for each ion.
a. barium ion c. oxide
b. iron(II) d. ferrous

3.55 Label each of the following regions in the periodic table.

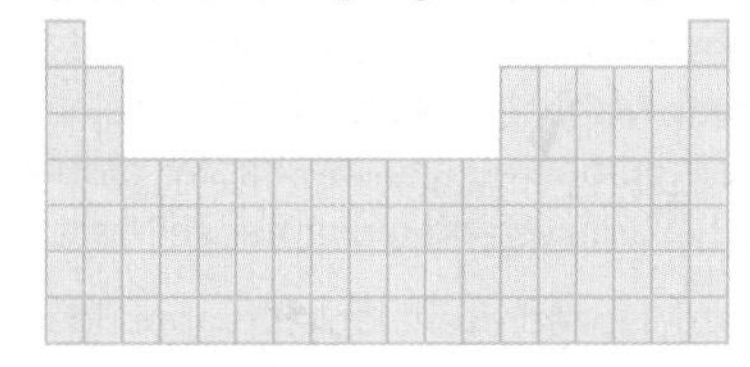

a. a group that forms cations with a +2 charge
b. a group that forms anions with a –2 charge
c. a group that forms cations with a +1 charge
d. a group that forms anions with a –1 charge

3.56 For each of the general electron-dot formulas for elements, give the following information: [1] the number of valence electrons; [2] the group number of the element; [3] how many electrons must be gained or lost to achieve a completely filled outer shell of electrons; [4] the charge on the resulting ion.
a. X· b. ·Z̈· (with a pair of dots above and a pair below)

3.57 Use the element colors in Figure 2.3 (also shown on the inside back cover) and the information in Table 3.4 to identify each of the following anions. Give the name and proper formula, including the charge.

a.  b. c.

3.58 Use the element colors in Figure 2.3 (also shown on the inside back cover) and the information in Table 3.4 to identify each of the following anions. Give the name and proper formula, including the charge.

a.  b.  c. 

Ionic Compounds

3.59 What is the charge on the cation M in each of the following ionic compounds?

a. MCl_2 b. MO c. M_2O_3 d. M_3N

3.60 What is the charge on the cation M in each of the following ionic compounds?

a. M_2S b. M_3P_2 c. MBr_3 d. MO_2

3.61 Write the formula for the ionic compound formed from each pair of elements.

a. calcium and sulfur
b. aluminum and bromine
c. lithium and iodine
d. nickel and chlorine

3.62 Write the formula for the ionic compound formed from each pair of elements.

a. barium and bromine
b. aluminum and sulfur
c. zinc and sulfur
d. magnesium and fluorine

3.63 Write the formula for the ionic compound formed from each cation and anion.

a. lithium and nitrite
b. calcium and acetate
c. sodium and bisulfite

3.64 Write the formula for the ionic compound formed from each cation and anion.

a. potassium and bicarbonate
b. magnesium and nitrate
c. lithium and carbonate

3.65 Write the formula for the ionic compound formed from the bisulfate anion (HSO_4^-) and each cation: (a) K^+; (b) Ba^{2+}; (c) Al^{3+}; (d) Zn^{2+}.

3.66 Write the formula for the ionic compound formed from the sulfite anion (SO_3^{2-}) and each cation: (a) K^+; (b) Ba^{2+}; (c) Al^{3+}; (d) Zn^{2+}.

3.67 Write the formula for the ionic compound formed from the barium cation (Ba^{2+}) and each anion: (a) CN^-; (b) PO_4^{3-}; (c) HPO_4^{2-}; (d) $H_2PO_4^-$.

3.68 Write the formula for the ionic compound formed from the iron(III) cation (Fe^{3+}) and each anion: (a) CN^-; (b) PO_4^{3-}; (c) HPO_4^{2-}; (d) $H_2PO_4^-$.

Naming Ionic Compounds

3.69 Name each ionic compound.

a. Na_2O
b. BaS
c. PbS_2
d. AgCl
e. $CoBr_2$
f. RbBr

3.70 Name each ionic compound.

a. KF
b. $ZnCl_2$
c. Cu_2S
d. SnO
e. $AuBr_3$
f. Li_2S

3.71 Write formulas to illustrate the difference between each pair of compounds.

a. sodium sulfide and sodium sulfate
b. magnesium oxide and magnesium hydroxide
c. magnesium sulfate and magnesium bisulfate

3.72 Write formulas to illustrate the difference between each pair of compounds.

a. lithium sulfite and lithium sulfide
b. sodium carbonate and sodium hydrogen carbonate
c. calcium phosphate and calcium dihydrogen phosphate

3.73 Name each ionic compound.

a. NH_4Cl
b. $PbSO_4$
c. $Cu(NO_3)_2$
d. $Ca(HCO_3)_2$

3.74 Name each ionic compound.

a. $(NH_4)_2SO_4$
b. NaH_2PO_4
c. $Cr(CH_3CO_2)_3$
d. $Sn(HPO_4)_2$

3.75 Write a formula from each name.

a. magnesium carbonate
b. nickel sulfate
c. copper(II) hydroxide
d. potassium hydrogen phosphate
e. gold(III) nitrate

3.76 Write a formula from each name.

a. copper(I) sulfite
b. aluminum nitrate
c. tin(II) acetate
d. lead(IV) carbonate
e. zinc hydrogen phosphate

Properties of Ionic Compounds

3.77 Label each statement as "true" or "false." Correct any false statement to make it true.

a. Ionic compounds have high melting points.
b. Ionic compounds can be solid, liquid, or gas at room temperature.
c. Most ionic compounds are insoluble in water.
d. An ionic solid like sodium chloride consists of discrete pairs of sodium cations and chloride anions.

3.78 Label each statement as "true" or "false." Correct any false statement to make it true.

a. Ionic compounds have high boiling points.
b. The ions in a crystal lattice are arranged randomly and the overall charge is zero.
c. When an ionic compound dissolves in water, the solution conducts electricity.
d. In an ionic crystal, ions having like charges are arranged close to each other.

Covalent Bonding and Lewis Structures

3.79 For each pair of compounds, classify the bonding as ionic or covalent and explain your choice.

a. LiCl and HCl
b. KBr and HBr

3.80 For each pair of compounds, classify the bonding as ionic or covalent and explain your choice.

a. BeH_2 and $BeCl_2$
b. Na_3N and NH_3

3.81 Draw a valid Lewis structure for each molecule: (a) H_2Se; (b) C_2Cl_6; (c) C_2H_2 (Assume each C is bonded to one H).

3.82 Draw a valid Lewis structure for each molecule: (a) HCl; (b) SiF_4; (c) CCl_2O (Assume C is bonded to both Cl's and the O atom).

3.83 Convert the 3-D model of oxalic acid into a Lewis structure and include all nonbonded electron pairs on atoms that contain them. Oxalic acid occurs naturally in spinach and rhubarb. Although oxalic acid is toxic, you would have to eat about nine pounds of spinach at one time to ingest a fatal dose.

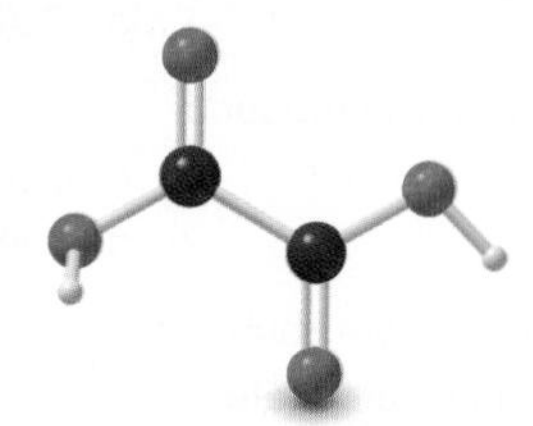

oxalic acid

3.84 Convert the 3-D model of the general anesthetic methoxyflurane into a Lewis structure and include all nonbonded electron pairs on atoms that contain them.

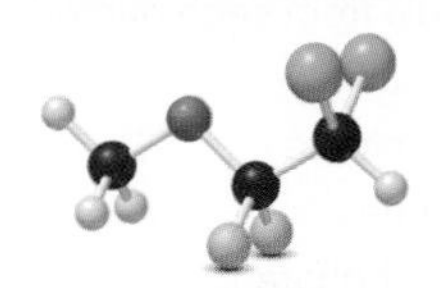

methoxyflurane

Naming Covalent Compounds

3.85 Name each covalent compound.
a. PBr_3 b. SO_3 c. NCl_3 d. P_2S_5

3.86 Name each covalent compound.
a. SF_6 b. CBr_4 c. N_2O d. P_4O_{10}

3.87 Write a formula that corresponds to each name.
a. selenium dioxide
b. carbon tetrachloride
c. dinitrogen pentoxide

3.88 Write a formula that corresponds to each name.
a. silicon tetrafluoride
b. nitrogen monoxide
c. phosphorus triiodide

Molecular Shape

3.89 Determine the shape around each indicated atom.

a. H—C(H)(H)—Ö—H
b. :NF_3

3.90 Determine the shape around each indicated atom.

a. H—C(H)(H)—S̈—H
b. H—C≡C—C(H)(H)—C̈l:

3.91 Add lone pairs to the N and O atoms to give octets and then determine the shape around each indicated atom.

a. H—N(H)—O—H
b. H—C(H)(H)—C(=O)—OH

3.92 Add lone pairs to the N and O atoms to give octets and then determine the shape around each indicated atom.

a. H—N(H)—N(H)—H
b. H—C(H)(H)—O—O—H

3.93 Predict the bond angles around the indicated atoms in each compound.

a. H—C(H)(H)—F̈:
b. H—C(H)=C=Ö:

3.94 Predict the bond angles around the indicated atoms in each compound.

a. H—C(H)(H)—C(H)(H)—C̈l:
b. H—C≡C—C(H)(H)—N̈(H)—H

Electronegativity and Polarity

3.95 Rank the atoms in each group in order of increasing electronegativity.
a. Se, O, S
b. P, Na, Cl
c. Cl, S, F
d. O, P, N

3.96 Rank the atoms in each group in order of increasing electronegativity.
a. Si, P, S
b. Be, Mg, Ca
c. Se, Cl, Br
d. Li, Be, Na

3.97 Label the bond formed between carbon and each of the following elements as nonpolar, polar, or ionic.
a. carbon
b. oxygen
c. lithium
d. chlorine
e. hydrogen

3.98 Label the bond formed between fluorine and each of the following elements as nonpolar, polar, or ionic.
a. hydrogen
b. fluorine
c. carbon
d. lithium
e. sulfur

3.99 Which bond in each pair is more polar—that is, has the larger electronegativity difference between atoms?
a. C—O or C—N
b. C—F or C—Cl

3.100 Which bond in each pair is more polar—that is, has the larger electronegativity difference between atoms?
a. Si—O or Si—S
b. H—F or H—Br

Applications

3.101 Zinc is an essential nutrient needed by many enzymes to maintain proper cellular function. Zinc is obtained in many dietary sources, including oysters, beans, nuts, whole grains, and sunflower seeds. (a) How many protons and electrons are found in a neutral zinc atom? (b) How many electrons and protons are found in the Zn^{2+} cation?

3.102 Wilson's disease is an inherited defect in copper metabolism in which copper accumulates in tissues, causing neurological problems and liver disease. The disease can be treated with compounds that bind to copper and thus remove it from the tissues. (a) How many protons and electrons are found in a neutral copper atom? (b) How many electrons and protons are found in the Cu^{+} cation? (c) How many electrons and protons are found in the Cu^{2+} cation? (d) Zinc acetate inhibits copper absorption and so it is used to treat Wilson's disease. What is the chemical formula of zinc acetate?

3.103 Isobutyl cyanoacrylate is used in medical glues to close wounds, thus avoiding the need for sutures.

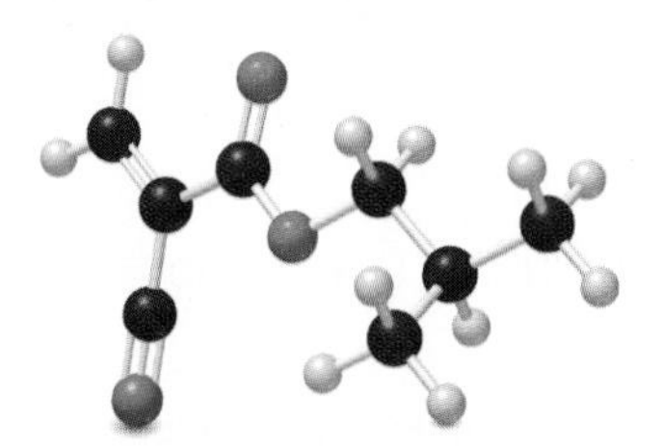

isobutyl cyanoacrylate

a. Convert the ball-and-stick model to a Lewis structure with all bonds and lone pairs drawn in.
b. Determine the shape around each carbon atom.
c. Label all polar bonds.

3.104 3-Methyl-2-hexenoic acid is one of the compounds that give human sweat its characteristic odor.

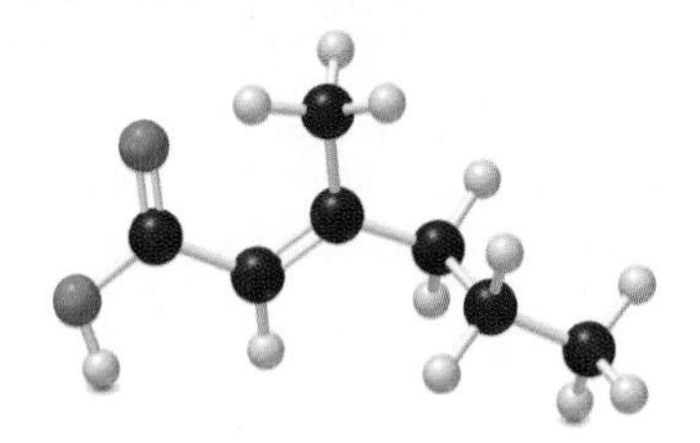

3-methyl-2-hexenoic acid

a. Convert the ball-and-stick model to a Lewis structure with all bonds and lone pairs drawn in.
b. Determine the shape around each carbon atom.
c. Label all polar bonds.

CHALLENGE PROBLEMS

3.105 Answer the following questions about the molecule OCS.
a. How many valence electrons does OCS contain?
b. Draw a valid Lewis structure. Assume that OCS contains a carbon–oxygen double bond, as well as a carbon–sulfur double bond.
c. Label all polar bonds.
d. What is the shape around the C atom?
e. Is OCS a polar molecule? Explain.

3.106 Glycine is a building block used to make proteins, such as those in heart muscle.

```
        H  O
        |  ||
  H—N—C—C—O—H
     |  |
     H  H
```

glycine

a. Add lone pairs where needed to give the second-row elements octets, and then count the total number of valence electrons in glycine.
b. Determine the shape around the four indicated atoms.
c. Label all of the polar bonds.
d. Is glycine a polar or nonpolar molecule? Explain.

BEYOND THE CLASSROOM

3.107 Energy bars contain protein, carbohydrates, and added nutrients for athletes and others looking for an on-the-go meal. Look at an ingredients label for some common energy bars, concentrating on the vitamins and minerals added in small amounts, itemized at the end of the list. Examples might include calcium phosphate, ferrous fumarate, copper gluconate, riboflavin, and folic acid. Try to determine the chemical formulas of these ingredients and why they are added. Is there any significant difference among popular energy bars? Can you determine which compounds are ionic and which are covalent?

3.108 Although mercury is safely used in dental fillings and thermometers, some forms of mercury, such as methylmercury and mercuric chloride, are highly toxic, and breathing mercury vapor is also a health hazard. In some waters, mercury-contaminated bluefin tuna, albacore, and swordfish have high levels of methylmercury and are considered unsafe to eat. Research the uses of mercury and how it enters the environment. Discuss the different forms of mercury encountered in the environment and why some are much more harmful than others. What laws have been enacted in the United States and other countries to limit an individual's exposure to toxic mercury compounds?

3.109 Compare the ingredients in several commercial toothpastes. Research the chemical formula for each component and what function it serves. Does the compound act as an abrasive, antiseptic, source of fluoride, coloring agent, flavoring agent, or binder (a compound that holds all other components together)? From the chemical formula, is it likely that an ingredient contains ionic or covalent bonds? What are the major differences between toothpaste brands?

3.110 In Section 3.2D, we learned that the body contains several ions—such as Na^+, K^+, Ca^{2+}, Mg^{2+}, Fe^{2+}, and Cl^-—that are crucial for cellular and organ function. Pick one or more of these common ions and research the following information: Where does the ion occur in the body? What function does the ion serve? From what dietary sources do we obtain the ion? What happens when we have too little or too much of the ion in our diet?

3.111 Ozone, O_3, is a covalent molecule formed in the upper atmosphere by the reaction of oxygen molecules with oxygen atoms. Research the role that ozone plays in the upper atmosphere and the negative consequences of a decrease in its concentration at high altitudes. Why is ozone at ground level considered a pollutant? What are some sources of ozone closer to the earth's surface?

ANSWERS TO SELECTED PROBLEMS

3.1 a,d,e: covalent b,c,f: ionic

3.2 a. compound: **A, C, D, E** element: **B, F**
b. All are molecules.

3.3 a. K^+ b. N^{3-} c. Br^- d. V^{2+}

3.4 a. 28 protons, 26 electrons c. 26 protons, 23 electrons
b. 34 protons, 36 electrons

3.5 a. +2 b. −1 c. −2 d. +1

3.7 Mn, 25 protons, 25 electrons
Mn^{2+}, 25 protons, 23 electrons

3.8 a. NaBr b. BaO c. MgI_2 d. Li_2O

3.9 ZnO

3.11 a. Sn^{2+} b. I^- c. Mn^{2+} d. Pb^{2+}

3.12 a. sodium fluoride d. lithium oxide
b. magnesium oxide e. titanium oxide
c. strontium bromide

3.13 a. chromium(III) chloride, chromic chloride
b. lead(II) sulfide
c. tin(IV) fluoride, stannic fluoride
d. lead(IV) oxide
e. iron(II) bromide, ferrous bromide

3.14 a. $CaBr_2$ c. $FeBr_3$ e. $CrCl_2$
b. CuI d. MgS f. Na_2O

3.15 Ionic compounds have high melting points and high boiling points. They usually dissolve in water, and their solutions conduct electricity.

3.17 a. $NaHCO_3$ c. $(NH_4)_2SO_4$ e. $Ca(HSO_4)_2$
b. KNO_3 d. $Mg_3(PO_4)_2$ f. $Ba(OH)_2$

3.18 a. sodium carbonate d. manganese acetate
b. calcium hydroxide e. iron(III) hydrogen sulfite, ferric bisulfite
c. magnesium nitrate f. magnesium phosphate

3.19 $Ca_3(PO_4)_2$

3.21 a. 1 b. 4 c. 1 d. 2 e. 3 f. 2

3.22 a. H—Br: c. H—O—O—H e. H—C(H)(H)—C(H)(H)—H
b. H—C(H)(H)—F: d. H—N(H)—N(H)—H f. H—C(H)(:Cl:)—Cl:

3.23 a. H—C≡N: b. H—C(H)=O:

3.25 a. carbon disulfide c. phosphorus pentachloride
b. sulfur dioxide d. boron trifluoride

3.26 a. SiO_2 b. PCl_3 c. SO_3 d. N_2O_3

3.27 a. bent c. trigonal pyramidal
b. tetrahedral d. trigonal planar

3.29 a. polar c. nonpolar e. polar
b. ionic d. polar

3.31 a. $\overset{\delta+}{H}$—$\overset{\delta-}{Cl}$ (⟷) polar c. $\overset{\delta+}{H}$—C≡$\overset{\delta-}{N}$ (⟷) polar
b. CH_2F_2: H(δ+)—C(H)(Fδ−)(Fδ−) polar d. CCl_4: C with four Cl(δ−) nonpolar

3.32 H—C(H)(H)—C(=O:)(δ−)—H (δ+); tetrahedral (CH₃ carbon), trigonal planar (C=O carbon)
All C–H and C–C bonds are nonpolar.

3.33 a,b,d: covalent c: ionic

3.35

	Br^-	OH^-	HCO_3^-	SO_3^{2-}	PO_4^{3-}
Na^+	NaBr	NaOH	$NaHCO_3$	Na_2SO_3	Na_3PO_4
Co^{2+}	$CoBr_2$	$Co(OH)_2$	$Co(HCO_3)_2$	$CoSO_3$	$Co_3(PO_4)_2$
Al^{3+}	$AlBr_3$	$Al(OH)_3$	$Al(HCO_3)_3$	$Al_2(SO_3)_3$	$AlPO_4$

3.37 a. OH^-, hydroxide b. NH_4^+, ammonium

3.39 $AgNO_3$

3.41 calcium sulfite

3.43 a. H—I: b. H—C(H)(:F:)—F:

3.45 a,b: 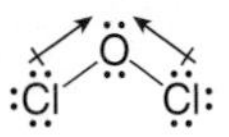

c. bent

d. The compound is polar since the two bond dipoles do not cancel.

3.47 **A** a. 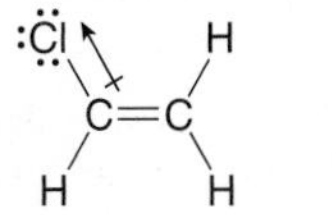

b. one polar bond

c. polar molecule

B a. 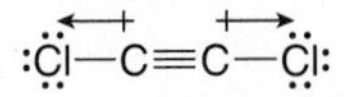

b. two polar bonds

c. nonpolar molecule

3.49 a. Be^{2+} b. Ti^{2+} c. S^{2-} d. Al^{3+}

3.51 a. Be b. Cl^{-} c. Rb^{+}

3.53 a. Na^{+} b. Mn^{2+} c. Au^{3+} d. Sn^{4+}

3.55 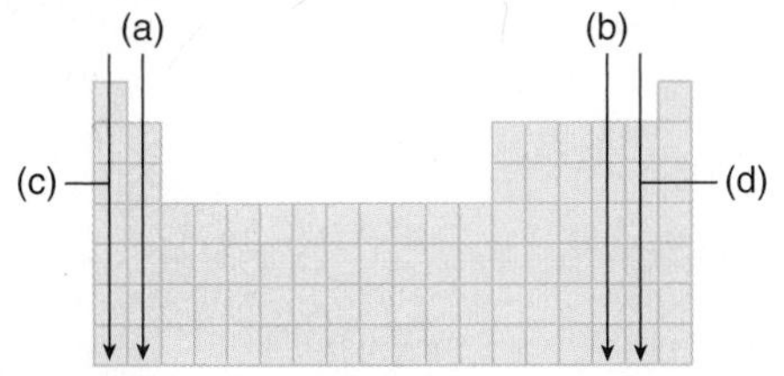

3.57 a. SO_4^{2-} sulfate b. NO_2^{-} nitrite c. S^{2-} sulfide

3.59 a. +2 b. +2 c. +3 d. +1

3.61 a. CaS b. $AlBr_3$ c. LiI d. $NiCl_2$

3.63 a. $LiNO_2$ c. $NaHSO_3$
b. $Ca(CH_3COO)_2$

3.65 a. $KHSO_4$ c. $Al(HSO_4)_3$
b. $Ba(HSO_4)_2$ d. $Zn(HSO_4)_2$

3.67 a. $Ba(CN)_2$ c. $BaHPO_4$
b. $Ba_3(PO_4)_2$ d. $Ba(H_2PO_4)_2$

3.69 a. sodium oxide d. silver chloride
b. barium sulfide e. cobalt bromide
c. lead(IV) sulfide f. rubidium bromide

3.71 a. Na_2S Na_2SO_4
b. MgO $Mg(OH)_2$
c. $MgSO_4$ $Mg(HSO_4)_2$

3.73 a. ammonium chloride
b. lead(II) sulfate
c. copper(II) nitrate, cupric nitrate
d. calcium bicarbonate, calcium hydrogen carbonate

3.75 a. $MgCO_3$ d. K_2HPO_4
b. $NiSO_4$ e. $Au(NO_3)_3$
c. $Cu(OH)_2$

3.77 a. True.
b. False—ionic compounds are solids at room temperature.
c. False—most ionic compounds are soluble in water.
d. False—ionic solids exist as crystalline lattices with the ions arranged to maximize the electrostatic interactions of anions and cations.

3.79 a. LiCl: ionic; the metal Li donates electrons to chlorine. HCl: covalent; H and Cl share electrons since both are nonmetals and the electronegativity difference is not large enough for electron transfer to occur.
b. KBr: ionic; the metal K donates electrons to bromine. HBr: covalent; H and Br share electrons since the electronegativity difference is not large enough for electron transfer to occur.

3.81 a. 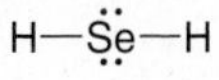 b. 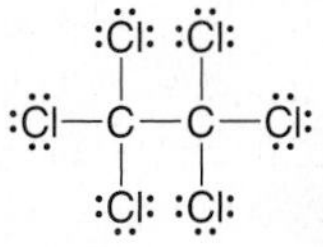 c. H—C≡C—H

3.83 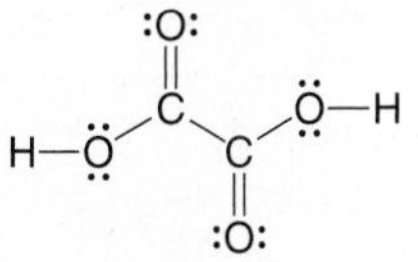

3.85 a. phosphorus tribromide c. nitrogen trichloride
b. sulfur trioxide d. diphosphorus pentasulfide

3.87 a. SeO_2 b. CCl_4 c. N_2O_5

3.89 a. 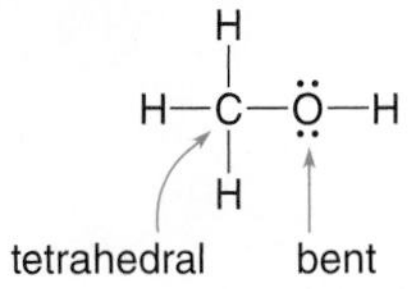

b. 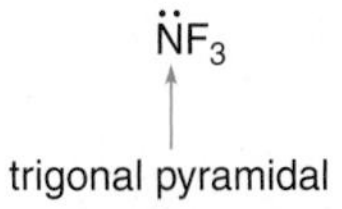

3.91 a. 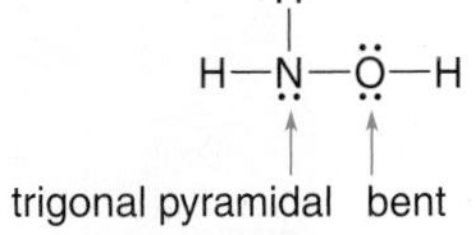

b. 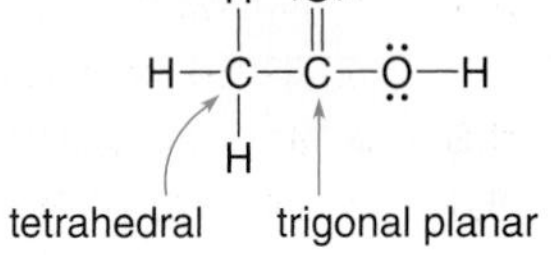

3.93 a. 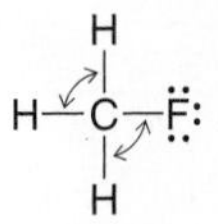

both 109.5°

b. 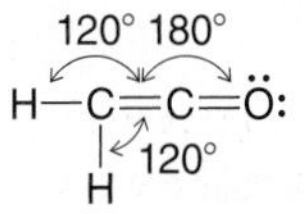

3.95 a. Se < S < O c. S < Cl < F
b. Na < P < Cl d. P < N < O

3.97 a. nonpolar b. polar c. polar d. polar e. nonpolar

3.99 a. C—O b. C—F

3.101 a. 30 protons, 30 electrons
b. 30 protons, 28 electrons

3.103 a. 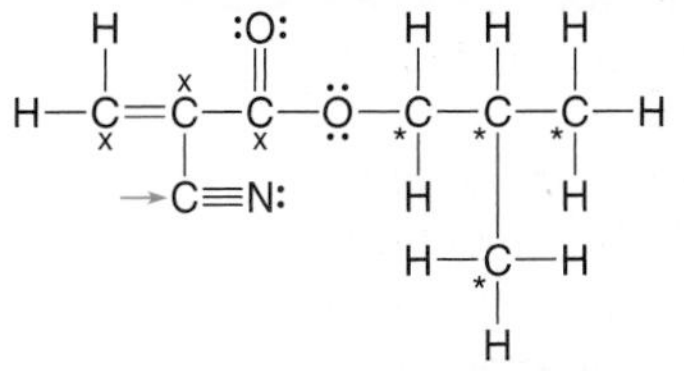

b. C's with x — trigonal planar
C's with * — tetrahedral
C with arrow — linear

c. All C—N and C—O bonds are polar.

3.105 a. 16 valence electrons [1 O atom (6 valence electrons) + 1 C atom (4 valence electrons) + 1 S atom (6 valence electrons)]

b,c: 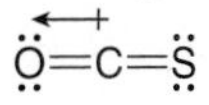

d. OCS has a linear shape since C is surrounded by two atoms and no lone pairs.

e. Yes, the compound is polar since there is only one bond dipole. The bond between carbon and sulfur is nonpolar since the electronegativity values of both C and S are 2.5.

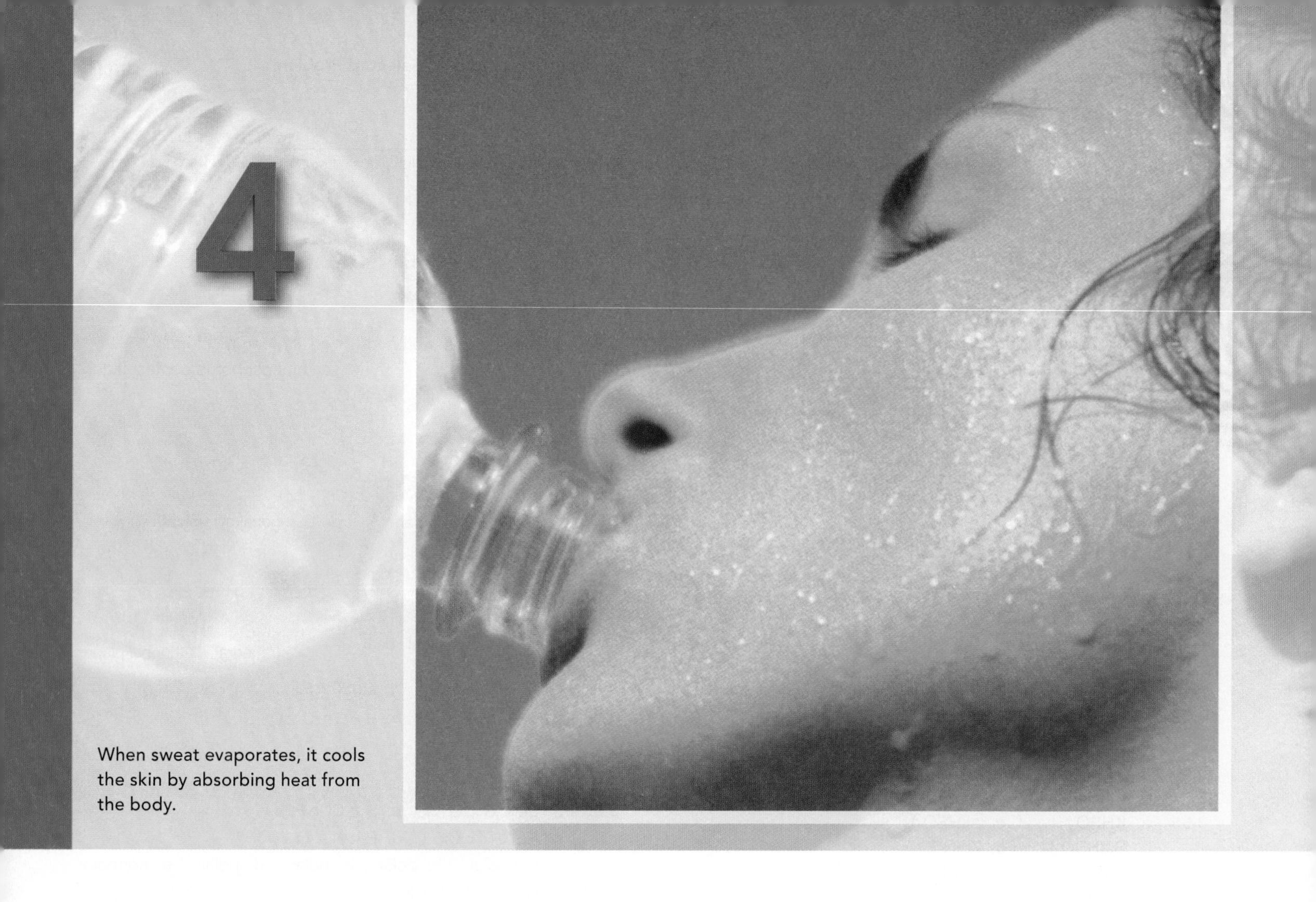

When sweat evaporates, it cools the skin by absorbing heat from the body.

Energy and Matter

CHAPTER OUTLINE

CHAPTER GOALS

In this chapter you will learn how to:

1. Define energy and become familiar with the units of energy
2. Identify the characteristics of the three states of matter
3. Determine what types of intermolecular forces a compound possesses
4. Relate the strength of intermolecular forces to a compound's boiling point and melting point
5. Define specific heat and use specific heat to determine the amount of heat gained or lost by a substance
6. Describe the energy changes that accompany changes of state
7. Interpret the changes depicted in heating and cooling curves

The water at the top of a waterfall has potential energy because of its position. This potential energy becomes kinetic energy as the water falls.

Why does water (H_2O) have a much higher boiling point than methane (CH_4), the main component of natural gas? Why does chloroethane, a local anesthetic, numb an injury when it is sprayed on a wound? To answer these questions, we turn our attention in Chapter 4 to energy and how energy changes are related to the forces of attraction that exist between molecules.

4.1 Energy

***Energy* is the capacity to do work.** Whenever you throw a ball, ride a bike, or read a newspaper, you use energy to do work. There are two types of energy.

- **Potential energy is stored energy.**
- **Kinetic energy is the energy of motion.**

A ball at the top of a hill or the water in a reservoir behind a dam are examples of potential energy. When the ball rolls down the hill or the water flows over the dam, the stored potential energy is converted to the kinetic energy of motion. Although energy can be converted from one form to another, one rule, the **law of conservation of energy,** governs the process.

- **The total energy in a system does not change. Energy cannot be created or destroyed.**

Throwing a ball into the air illustrates the interplay of potential and kinetic energy (Figure 4.1).

Figure 4.1

Potential and Kinetic Energy

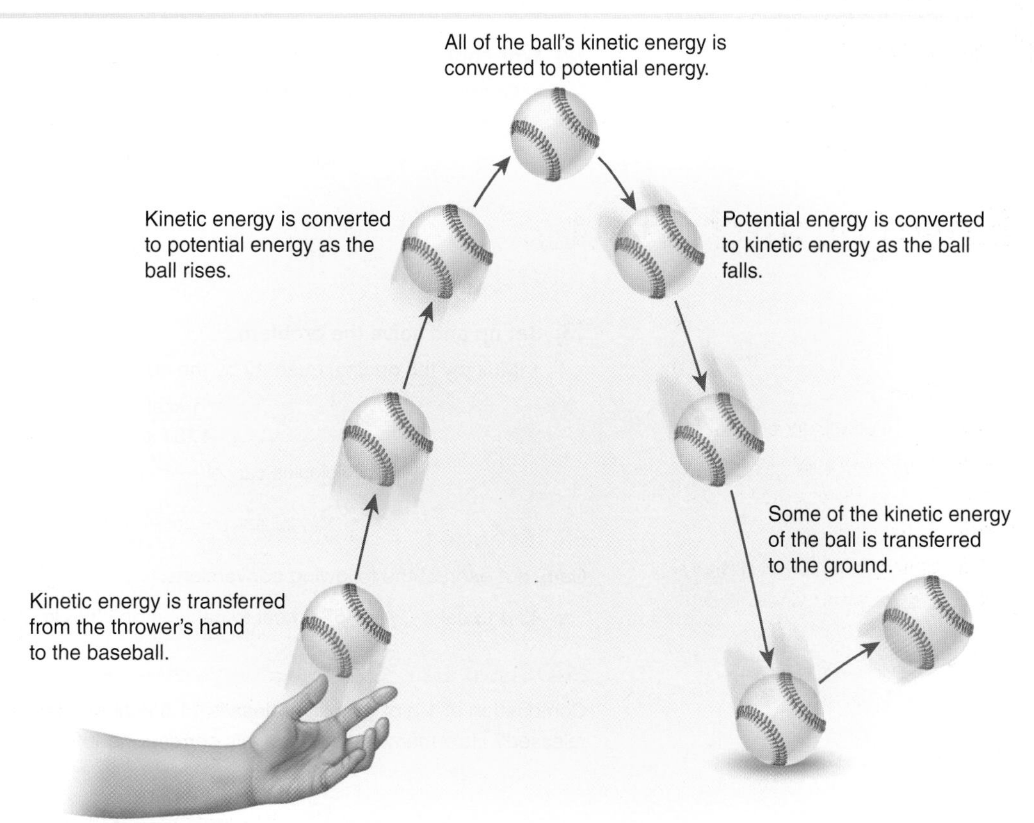

All of the ball's kinetic energy is converted to potential energy.
Kinetic energy is converted to potential energy as the ball rises.
Potential energy is converted to kinetic energy as the ball falls.
Some of the kinetic energy of the ball is transferred to the ground.
Kinetic energy is transferred from the thrower's hand to the baseball.

When a ball is thrown into the air it possesses kinetic energy, which is converted to potential energy as it reaches its maximum height. When the ball descends, the potential energy is converted back to kinetic energy, until the ball rests motionless on the ground.

4.1A The Units of Energy

The joule, named after the nineteenth-century English physicist James Prescott Joule, is pronounced *jewel*.

Energy can be measured using two different units, **calories (cal)** and **joules (J).** A **calorie** is the amount of energy needed to raise the temperature of 1 g of water 1 °C. Joules and calories are related in the following way.

$$1 \text{ cal} = 4.184 \text{ J}$$

Since both the calorie and the joule are small units of measurement, more often energies in reactions are reported with kilocalories (kcal) and kilojoules (kJ). Recall from Table 1.2 that the prefix *kilo* means 1,000.

$$1 \text{ kcal} = 1{,}000 \text{ cal}$$
$$1 \text{ kJ} = 1{,}000 \text{ J}$$
$$1 \text{ kcal} = 4.184 \text{ kJ}$$

To convert a quantity from one unit of measurement to another, set up conversion factors and use the method first shown in Section 1.7B and illustrated in Sample Problem 4.1.

SAMPLE PROBLEM 4.1

A reaction releases 421 kJ of energy. How many kilocalories does this correspond to?

Analysis and Solution

[1] **Identify the original quantity and the desired quantity.**

421 kJ	? kcal
original quantity	desired quantity

[2] **Write out the conversion factors.**

- Choose the conversion factor that places the unwanted unit, kilojoules, in the denominator so that the units cancel.

kJ–kcal conversion factors

$$\frac{4.184 \text{ kJ}}{1 \text{ kcal}} \quad \text{or} \quad \boxed{\frac{1 \text{ kcal}}{4.184 \text{ kJ}}}$$

Choose this conversion factor to cancel kJ.

[3] **Set up and solve the problem.**

- Multiply the original quantity by the conversion factor to obtain the desired quantity.

$$421 \cancel{\text{kJ}} \times \frac{1 \text{ kcal}}{4.184 \cancel{\text{kJ}}} = 100.6 \text{ kcal, rounded to } 101 \text{ kcal}$$

Kilojoules cancel. **Answer**

PROBLEM 4.1

Carry out each of the following conversions.

a. 42 J to cal b. 55.6 kcal to cal c. 326 kcal to kJ d. 25.6 kcal to J

PROBLEM 4.2

Combustion of 1 g of gasoline releases 11.5 kcal of energy. How many kilojoules of energy is released? How many joules does this correspond to?

4.1B FOCUS ON THE HUMAN BODY Energy and Nutrition

When we eat food, the protein, carbohydrates, and fat (lipid) in the food are metabolized to form small molecules that in turn are used to prepare new molecules that cells need for maintenance

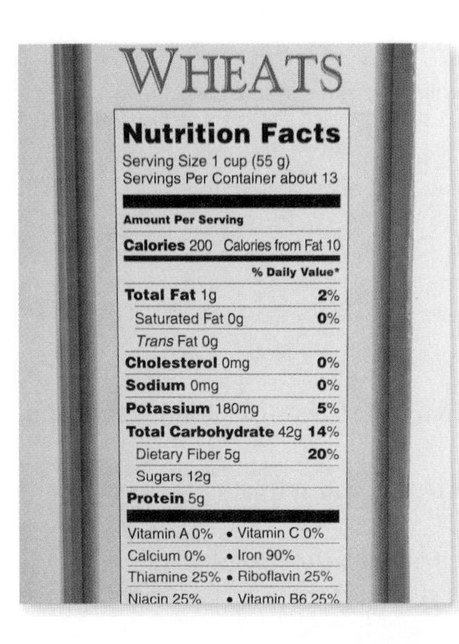

Table 4.1 Caloric Value for Three Classes of Compounds

	Cal/g	cal/g
Protein	4	4,000
Carbohydrate	4	4,000
Fat	9	9,000

One nutritional Calorie (1 Cal) = 1,000 cal = 1 kcal.

and growth. This process also generates the energy needed for the organs to function, allowing the heart to beat, the lungs to breathe, and the brain to think.

The amount of stored energy in food is measured using nutritional Calories (upper case C), where 1 Cal = 1,000 cal. Since 1,000 cal = 1 kcal, the following relationships exist.

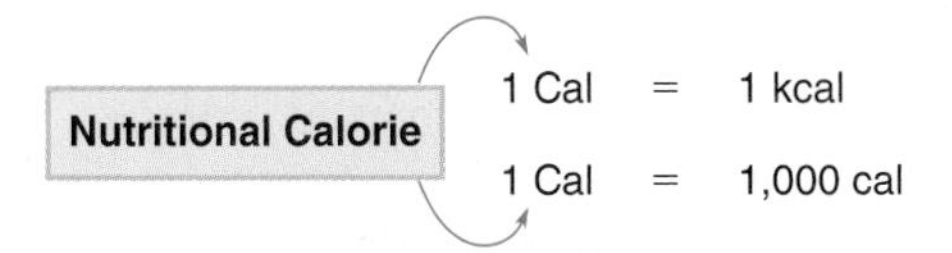

Upon metabolism, proteins, carbohydrates, and fat each release a predictable amount of energy, the **caloric value** of the substance. For example, one gram of protein or one gram of carbohydrate typically releases about 4 Cal/g, while fat releases 9 Cal/g (Table 4.1). If we know the amount of each of these substances contained in a food product, we can make a first approximation of the number of Calories it contains by using caloric values as conversion factors, as illustrated in Sample Problem 4.2.

When an individual eats more Calories than are needed for normal bodily maintenance, the body stores the excess as fat. The average body fat content for men and women is about 20% and 25%, respectively. This stored fat can fill the body's energy needs for two or three months. Frequent ingestion of a large excess of Calories results in a great deal of stored fat, causing an individual to be overweight.

SAMPLE PROBLEM 4.2

If a baked potato contains 3 g of protein, a trace of fat, and 23 g of carbohydrates, estimate its number of Calories.

Analysis

Use the caloric value (Cal/g) of each class of molecule to form a conversion factor to convert the number of grams to Calories and add up the results.

Solution

[1] Identify the original quantity and the desired quantity.

3 g protein 23 g carbohydrates	? Cal
original quantities	desired quantity

[2] Write out the conversion factors.

- Write out conversion factors that relate the number of grams to the number of Calories for each substance. Each conversion factor must place the unwanted unit, grams, in the denominator so that the units cancel.

Cal–g conversion factor for protein: $\frac{4\text{ Cal}}{1\text{ g protein}}$

Cal–g conversion factor for carbohydrates: $\frac{4\text{ Cal}}{1\text{ g carbohydrate}}$

[3] Set up and solve the problem.

- Multiply the original quantity by the conversion factor for both protein and carbohydrates and add up the results to obtain the desired quantity.

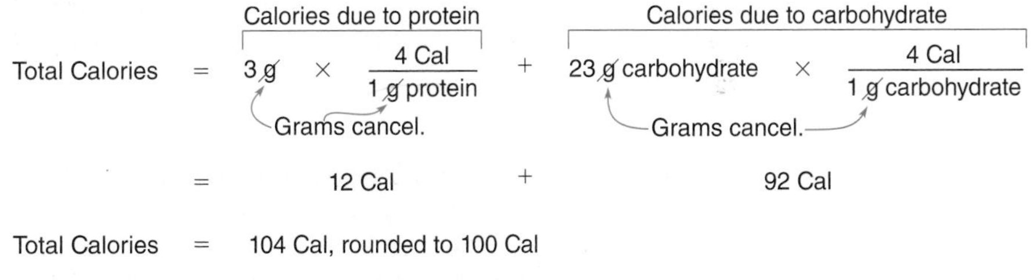

CONSUMER NOTE

Knowing the number of grams of protein, carbohydrates, and fat allows you to estimate how many Calories a food product contains. A quarter-pound burger with cheese with 29 g of protein, 40 g of carbohydrates, and 26 g of fat contains 510 Calories.

PROBLEM 4.3

How many Calories are contained in one tablespoon of olive oil, which has 14 g of fat?

PROBLEM 4.4

One serving (36 crackers) of wheat crackers contains 6 g of fat, 20 g of carbohydrates, and 2 g of protein. Estimate the number of calories.

4.2 The Three States of Matter

As we first learned in Section 1.2, matter exists in three common states—**gas, liquid,** and **solid.**

- A gas consists of particles that are far apart and move rapidly and independently from each other.
- A liquid consists of particles that are much closer together but are still somewhat disorganized since they can move about. The particles in a liquid are close enough that they exert a force of attraction on each other.
- A solid consists of particles—atoms, molecules, or ions—that are close to each other and are often highly organized. The particles in a solid have little freedom of motion and are held together by attractive forces.

As shown in Figure 4.2, air is composed largely of N_2 and O_2 molecules, along with small amounts of argon (Ar), carbon dioxide (CO_2), and water molecules that move about rapidly. Liquid water is composed of H_2O molecules that have no particular organization. Sand is a solid composed of SiO_2, which contains a network of covalent silicon–oxygen bonds.

Figure 4.2

The Three States of Matter—Solid, Liquid, and Gas

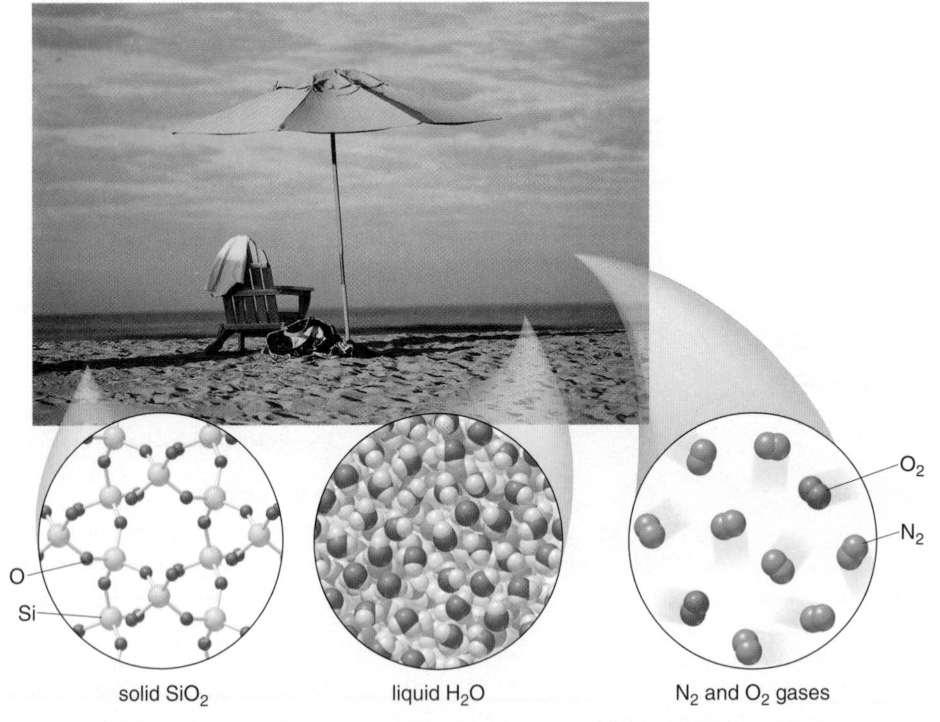

Most sand is composed of silicon dioxide (SiO_2), which forms a three-dimensional network of covalent bonds. Liquid water is composed of H_2O molecules, which can move past each other but are held close together by a force of attraction (Section 4.3). Air contains primarily N_2 and O_2 molecules that move rapidly with no force of attraction for each other.

Whether a substance exists as a gas, liquid, or solid depends on the balance between the kinetic energy of its particles and the strength of the interactions between the particles. In a gas, the kinetic energy of motion is high and the particles are far apart from each other. As a result, the attractive forces between the molecules are negligible and gas molecules move freely. In a liquid, attractive forces hold the molecules much more closely together, so the distance between molecules and the kinetic energy is much less than the gas. In a solid, the attractive forces between molecules are even stronger, so the distance between individual particles is small and there is little freedom of motion. The properties of gases, liquids, and solids are summarized in Table 4.2.

PROBLEM 4.5

How do gaseous, liquid, and solid methanol (CH_4O) compare in each of the following features: (a) density; (b) the space between the molecules; (c) the attractive force between the molecules?

PROBLEM 4.6

Why is a gas much more easily compressed into a smaller volume than a liquid or solid?

Table 4.2 Properties of Gases, Liquids, and Solids

Property	Gas	Liquid	Solid
Shape and volume	Expands to fill its container	A fixed volume that takes the shape of the container it occupies	A definite shape and volume
Arrangement of particles	Randomly arranged, disorganized, and far apart	Randomly arranged but close	Fixed arrangement of very close particles
Density	Low (< 0.01 g/mL)	High (~1 g/mL)[a]	High (1–10 g/mL)
Particle movement	Very fast	Moderate	Slow
Interaction between particles	None	Strong	Very strong

[a]The symbol "~" means approximately.

4.3 Intermolecular Forces

To understand many of the properties of solids, liquids, and gases, we must learn about the forces of attraction that exist between particles—atoms, molecules, or ions.

Ionic compounds are composed of extensive arrays of oppositely charged ions that are held together by strong electrostatic interactions. **These ionic interactions are much stronger than the forces between covalent molecules,** so it takes a great deal of energy to separate ions from each other (Section 3.5).

In covalent compounds, the nature and strength of the attraction between individual molecules depend on the identity of the atoms.

- Intermolecular forces are the attractive forces that exist *between* molecules.

One atom has to do with anothe atom of the same type.

There are three different types of intermolecular forces in covalent molecules, presented in order of ***increasing strength:***

- London dispersion forces
- Dipole–dipole interactions
- Hydrogen bonding

Thus, a compound that exhibits hydrogen bonding has stronger intermolecular forces than a compound of similar size that has dipole–dipole interactions. Likewise, a compound that has dipole–dipole interactions has stronger intermolecular forces than a compound of similar size that has only London dispersion forces. The strength of the intermolecular forces determines whether a compound has a high or low melting point and boiling point, and thus if the compound is a solid, liquid, or gas at a given temperature.

4.3A London Dispersion Forces

London dispersion forces can also be called **van der Waals forces.**

- **London dispersion forces are very weak interactions due to the momentary changes in electron density in a molecule.**

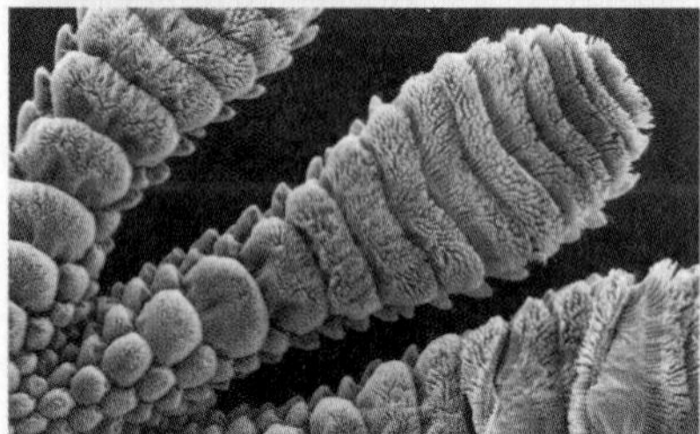

Although any single interaction is weak, a large number of London dispersion forces creates a strong force. For example, geckos stick to walls and ceilings by London dispersion forces between the surfaces and the 500,000 tiny hairs on each foot.

For example, although a nonpolar methane molecule (CH_4) has no net dipole, at any one instant its electron density may not be completely symmetrical. If more electron density is present in one region of the molecule, less electron density must be present some place else, and this creates a *temporary* dipole. A temporary dipole in one CH_4 molecule induces a temporary dipole in another CH_4 molecule, with the partial positive and negative charges arranged close to each other. **The weak interaction between these temporary dipoles constitutes London dispersion forces.**

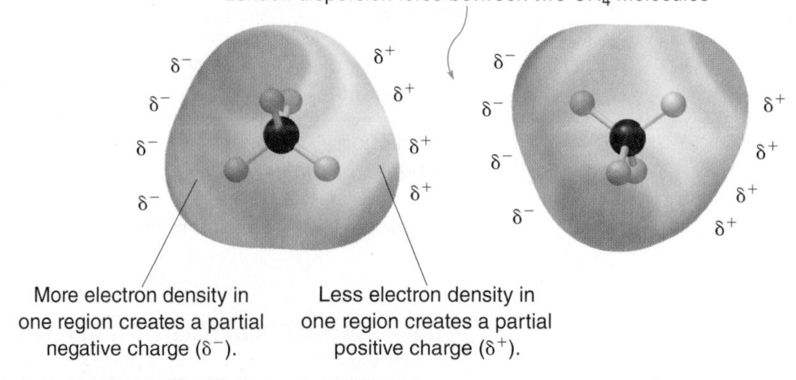
London dispersion force between two CH_4 molecules
δ^-
δ^+
More electron density in one region creates a partial negative charge (δ^-).
Less electron density in one region creates a partial positive charge (δ^+).

All **covalent compounds exhibit London dispersion forces.** These intermolecular forces are the only intermolecular forces present in nonpolar compounds. The strength of these forces is related to the size of the molecule.

- **The larger the molecule, the larger the attractive force between two molecules, and the stronger the intermolecular forces.**

Atoms like the noble gases helium and argon exhibit London dispersion forces, too. The attractive forces between argon atoms are stronger than the attractive forces between helium atoms because argon atoms are considerably larger in size.

PROBLEM 4.7

Which of the following compounds exhibit London dispersion forces: (a) NH_3; (b) H_2O; (c) HCl; (d) ethane (C_2H_6)?

PROBLEM 4.8

Which species in each pair exhibits the stronger London dispersion forces?

a. CH_4 (H–C–H with H above and below) or $CH_3CH_2CH_3$ (H–C–C–C–H with H above and below each C)

b. He atoms or Ne atoms

4.3B Dipole–Dipole Interactions

How to determine whether a molecule is polar is shown in Section 3.12.

- **Dipole–dipole interactions are the attractive forces between the permanent dipoles of two polar molecules.**

For example, the carbon–oxygen bond in formaldehyde, $H_2C{=}O$, is polar because oxygen is more electronegative than carbon. This polar bond gives formaldehyde a permanent dipole, making it a polar molecule. The dipoles in adjacent formaldehyde molecules can align so that the partial positive and partial negative charges are close to each other. These attractive forces due to permanent dipoles are much stronger than London dispersion forces.

dipole–dipole interactions

formaldehyde = δ^+ δ^- δ^+ δ^- δ^+ δ^-

PROBLEM 4.9

Draw the individual dipoles of two H—Cl molecules and show how the dipoles are aligned in a dipole–dipole interaction.

4.3C Hydrogen Bonding

- **Hydrogen bonding occurs when a hydrogen atom bonded to O, N, or F, is electrostatically attracted to an O, N, or F atom in another molecule.**

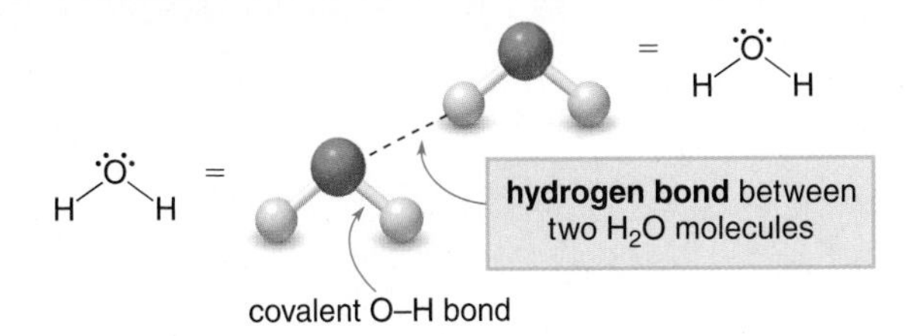

Hydrogen bonding can occur only when molecules contain a hydrogen atom bonded to a very electronegative atom—that is, oxygen, nitrogen, or fluorine. For example, two H_2O molecules can hydrogen bond to each other: a hydrogen atom is covalently bonded to oxygen in one water molecule, and hydrogen bonded to an oxygen atom in another water molecule. **Hydrogen bonds are the strongest of the three types of intermolecular forces.** Table 4.3 summarizes the three types of intermolecular forces.

Table 4.3 Summary of the Types of Intermolecular Forces

Type of Force	Relative Strength	Exhibited by	Example
London dispersion	Weak	All molecules	CH_4, H_2CO, H_2O
Dipole–dipole	Moderate	Molecules with a net dipole	H_2CO, H_2O
Hydrogen bonding	Strong	Molecules with an O—H, N—H, or H—F bond	H_2O

Hydrogen bonding is important in many biological molecules, including proteins and DNA. DNA, which is contained in the chromosomes of the nucleus of a cell, is responsible for the storage of all genetic information. DNA is composed of two long strands of atoms that are held together by hydrogen bonding as shown in Figure 4.3. A detailed discussion of DNA appears in Chapter 17.

Figure 4.3 Hydrogen Bonding and DNA

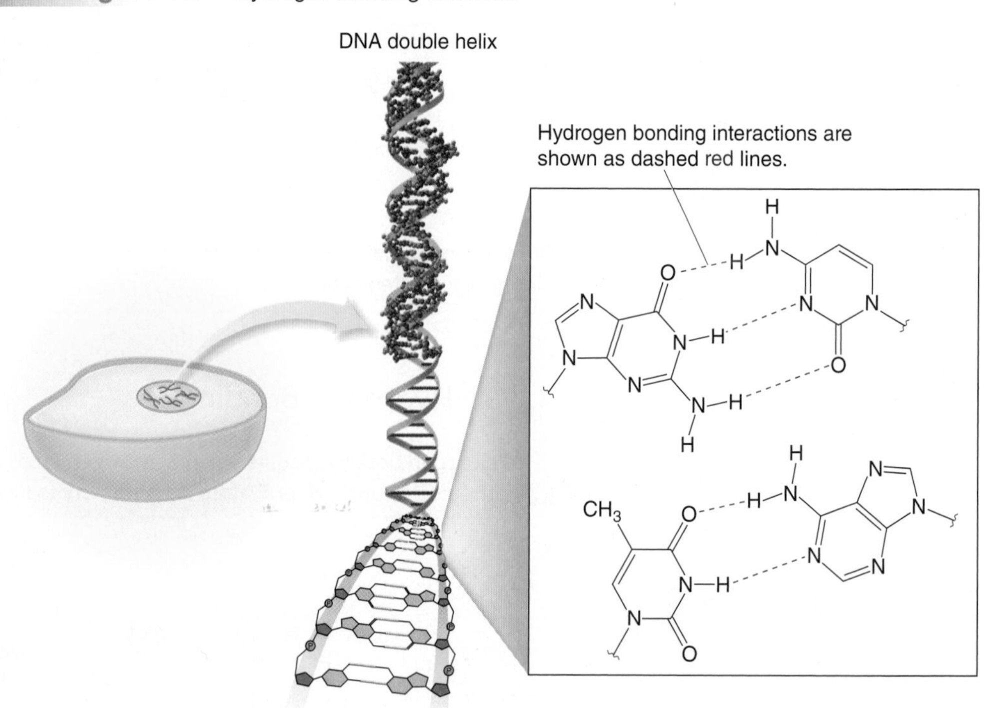

DNA is composed of two long strands of atoms that wind around each other in an arrangement called a double helix. The two strands are held together by an extensive network of hydrogen bonds. In each hydrogen bond, a hydrogen atom of an N—H bond on one chain is intermolecularly hydrogen bonded to an oxygen or nitrogen atom on an adjacent chain. Five hydrogen bonds are indicated.

SAMPLE PROBLEM 4.3

What types of intermolecular forces are present in each compound: (a) HCl; (b) C_2H_6 (ethane); (c) NH_3?

Analysis

- London dispersion forces are present in all covalent compounds.
- Dipole–dipole interactions are present only in polar compounds with a permanent dipole.
- Hydrogen bonding occurs only in compounds that contain an O—H, N—H, or H—F bond.

Solution

a. polar bond
δ^+ H—Cl δ^-

- HCl has London forces like all covalent compounds.
- HCl has a polar bond, so it exhibits dipole–dipole interactions.
- HCl has no H atom on an O, N, or F, so it has no intermolecular hydrogen bonding.

b. H H
H—C—C—H
H H
nonpolar molecule

- C_2H_6 is a nonpolar molecule since it has only nonpolar C—C and C—H bonds. Thus, it exhibits only London forces.

c. δ^- N, H δ^+, H δ^+, H δ^+
net dipole

- NH_3 has London forces like all covalent compounds.
- NH_3 has a net dipole from its three polar bonds (Section 3.12), so it exhibits dipole–dipole interactions.
- NH_3 has a H atom bonded to N, so it exhibits intermolecular hydrogen bonding.

PROBLEM 4.10

What types of intermolecular forces are present in each molecule?

a. Cl_2 b. HCN c. HF d. CH_3Cl e. H_2

PROBLEM 4.11

Which of the compounds in each pair has stronger intermolecular forces?

a. CO_2 or H_2O b. CO_2 or HBr c. HBr or H_2O d. CH_4 or C_2H_6

4.4 Boiling Point and Melting Point

The **boiling point (bp)** of a compound is the temperature at which a liquid is converted to the gas phase, while the **melting point (mp)** is the temperature at which a solid is converted to the liquid phase. The strength of the intermolecular forces determines the boiling point and melting point of compounds.

- **The *stronger* the intermolecular forces, the *higher* the boiling point and melting point.**

In boiling, energy must be supplied to overcome the attractive forces of the liquid state and separate the molecules to the gas phase. Similarly, in melting, energy must be supplied to overcome the highly ordered solid state and convert it to the less ordered liquid phase. A stronger force of attraction between molecules means that more energy must be supplied to overcome those intermolecular forces, increasing the boiling point and melting point.

In comparing compounds of similar size, the following trend is observed:

Compounds with London dispersion forces only → Compounds with dipole–dipole interactions → Compounds that can hydrogen bond

Increasing strength of intermolecular forces
Increasing boiling point
Increasing melting point

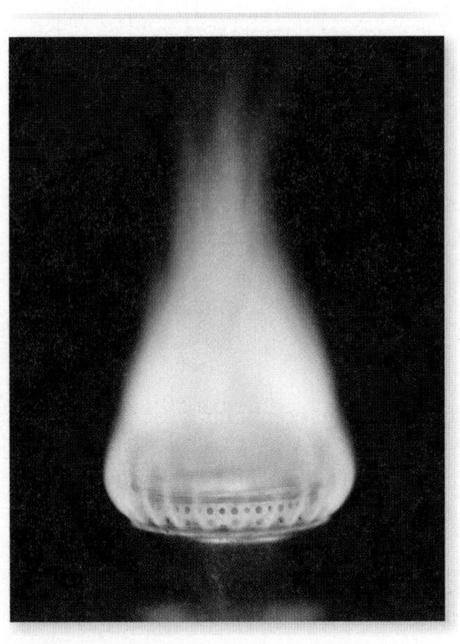
Methane, the main component of natural gas, is a gas at room temperature because the CH_4 molecules have weak forces of attraction for each other.

Methane (CH_4) and water (H_2O) are both small molecules with hydrogen atoms bonded to a second-row element, so you might expect them to have similar melting points and boiling points. Methane, however, is a nonpolar molecule that exhibits only London dispersion forces, whereas water is a polar molecule that can form intermolecular hydrogen bonds. As a result, the melting point and boiling point of water are *much higher* than those of methane. In fact, the hydrogen bonds in water are so strong that it is a liquid at room temperature, whereas methane is a gas.

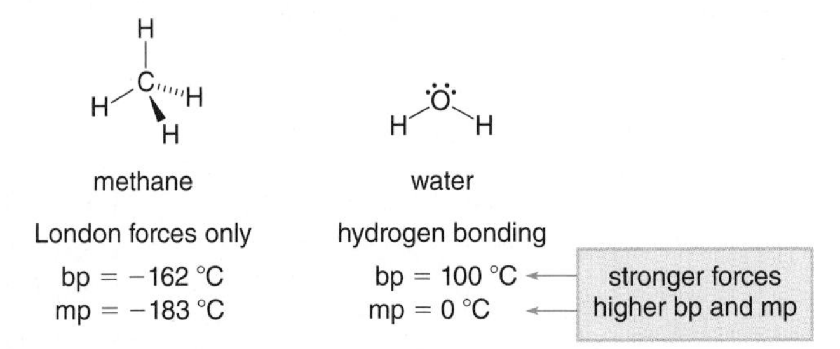

In comparing two compounds with similar types of intermolecular forces, the larger compound generally has more surface area and therefore a larger force of attraction, giving it the higher boiling point and melting point. Thus, propane (C_3H_8) and butane (C_4H_{10}) have only nonpolar bonds and London forces, but butane is larger and therefore has the higher boiling point and melting point.

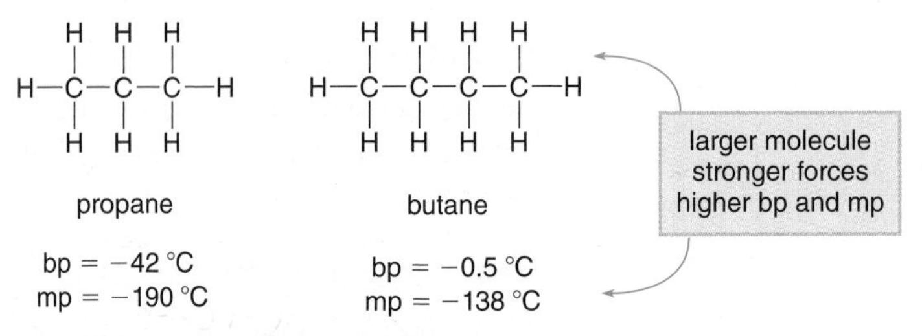

SAMPLE PROBLEM 4.4

(a) Which compound, **A** or **B,** has the higher boiling point? (b) Which compound, **C** or **D,** has the higher melting point?

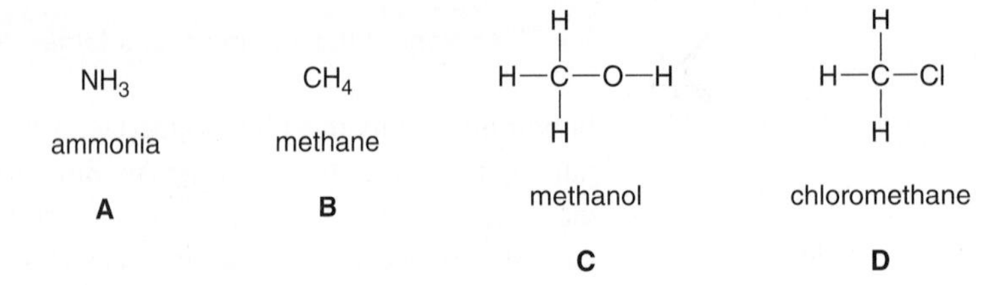

Analysis

Determine the types of intermolecular forces in each compound. The compound with the stronger forces has the higher boiling point or melting point.

Solution

a. NH_3 (**A**) has an N—H bond, so it exhibits intermolecular hydrogen bonding. CH_4 (**B**) has only London forces since it has only nonpolar C—H bonds. NH_3 has stronger forces and the higher boiling point.

b. Methanol (**C**) has an O—H bond, so it can intermolecularly hydrogen bond. Chloromethane (**D**) has a polar C—Cl bond, so it has dipole–dipole interactions, but it cannot hydrogen bond. **C** has stronger forces, so **C** has the higher melting point.

PROBLEM 4.12

Which compound in each pair has the higher boiling point? Which compound in each pair has the higher melting point?

a. CH_4 or C_2H_6 b. C_2H_6 or CH_3OH c. HBr or HCl d. C_2H_6 or CH_3Br

PROBLEM 4.13

Explain why CO_2 is a gas at room temperature but H_2O is a liquid.

4.5 Specific Heat

In addition to boiling point and melting point, there are many other physical properties that characterize a substance. For example, **specific heat (*SH*)** is a physical property that reflects the ability of a substance to absorb heat. Specific heat relates energy, mass, and temperature change (ΔT).

- **The specific heat is the amount of heat energy (in calories or joules) needed to raise the temperature of 1 g of a substance by 1 °C.**

$$\text{specific heat} = \frac{\text{heat}}{\text{mass} \times \Delta T} = \frac{\text{cal (or J)}}{\text{g} \cdot {}^\circ\text{C}}$$

The specific heat of water is 1.00 cal/(g · °C), meaning that 1.00 cal of heat must be added to increase the temperature of 1.00 g of water by 1.00 °C. The amount of heat that must be added for a particular temperature increase depends on the amount of sample present. To increase the temperature of 2.00 g of water by 1 °C requires 2.00 cal of heat. Table 4.4 lists the specific heat values for a variety of substances.

- **The *larger* the specific heat of a substance, the *less* its temperature will change when it absorbs a particular amount of heat energy.**

Table 4.4 Specific Heats of Some Substances

Substance	cal/(g · °C)	J/(g · °C)	Substance	cal/(g · °C)	J/(g · °C)
Aluminum	0.214	0.895	2-Propanol	0.612	2.56
Carbon (graphite)	0.169	0.707	Rock	0.200	0.837
Copper	0.0900	0.377	Sand	0.200	0.837
Ethanol	0.583	2.44	Silver	0.0560	0.234
Gold	0.0310	0.130	Water(*l*)	1.00	4.18
Iron	0.107	0.448	Water(*g*)	0.481	2.01
Mercury	0.0335	0.140	Water(*s*)	0.486	2.03

The specific heat of graphite is about twice that of copper. Adding 1.00 cal of heat will raise the temperature of 1.00 g of graphite by 5.9 °C, but raise the temperature of 1.00 g of copper by 11.1 °C. Because metals like copper have low specific heats, they absorb and transfer heat readily. We use copper, iron, or aluminum cookware because the metals readily transfer heat from the stove to the food in the pan.

The specific heat of water is very high compared to other liquids. As a result, **water absorbs a large amount of heat with only a small change in temperature.** Water has a high specific heat because of its strong intermolecular forces. Since water molecules are held together with an extensive array of hydrogen bonds, it takes a great deal of heat energy to break these bonds and increase the disorder of the water molecules.

Moreover, since the amount of heat absorbed for a given temperature increase equals the amount of heat released upon cooling, water releases a great deal of energy when its temperature drops even a few degrees. This explains why the climate in a coastal city is often more moderate than that of a city located 100 miles inland. A large body of water acts as a reservoir to absorb or release heat as temperature increases or decreases, so it moderates the climate of the land nearby.

SAMPLE PROBLEM 4.5

Consider the elements aluminum, copper, gold, and iron in Table 4.4. (a) If 10 kcal of heat is added to 10 g of each element, which element will have the highest temperature? (b) Which element would require the largest amount of heat to raise the temperature of a 5-g sample by 5 °C?

Analysis

- The larger the specific heat, the less the temperature of a substance will change when it absorbs heat energy.
- The larger the specific heat, the more heat that must be added to increase the temperature of a substance a given number of degrees.

Solution

The specific heats of the metals in Table 4.4 increase in the following order: gold $<$ copper $<$ iron $<$ aluminum. In part (a), gold has the lowest specific heat, so its temperature will be the highest if the same amount of heat is added to the same mass of all four elements. In part (b), aluminum has the largest specific heat, so it will require the largest amount of heat to raise its temperature the same number of degrees as the same mass of the other elements.

PROBLEM 4.14

A student has two containers—one with 10 g of sand and one with 10 g of ethanol. (a) Which substance has the higher temperature after 10.0 cal of heat is added to each container? (b) Which substance requires the larger amount of heat to raise its temperature by 10 °C?

PROBLEM 4.15

The human body is composed of about 70% water. How does this help the body to maintain a steady internal temperature?

We can use the specific heat as a conversion factor to calculate how much heat is absorbed or lost from a substance as long as its mass and change in temperature are known, using the equation:

heat absorbed or released

$$\text{heat} = \text{mass} \times \underset{\Delta T}{\text{temperature change}} \times \text{specific heat}$$

$$\text{cal} = \text{g} \times {}^\circ\text{C} \times \frac{\text{cal}}{\text{g} \cdot {}^\circ\text{C}}$$

Sample Problem 4.6 shows how to use specific heat to calculate the amount of heat absorbed by a substance when the mass and temperature change are known.

SAMPLE PROBLEM 4.6

How many calories are needed to heat a pot of 1,600 g of water from 25 °C to 100. °C?

Analysis

Use specific heat as a conversion factor to calculate the amount of heat absorbed given the known mass and temperature change.

Solution

[1] **Identify the known quantities and the desired quantity.**

known quantities	desired quantity
mass = 1,600 g T_1 = 25 °C T_2 = 100. °C	? calories

- Subtract the initial temperature (T_1) from the final temperature (T_2) to determine the temperature change: $T_2 - T_1 = \Delta T = 100. - 25 = 75$ °C.
- The specific heat of water is 1.00 cal/(g · °C).

[2] **Write the equation.**

- The specific heat is a conversion factor that relates the heat absorbed to the temperature change (ΔT) and mass.

$$\text{heat} = \text{mass} \times \Delta T \times \text{specific heat}$$

$$\text{cal} = \text{g} \times {}^\circ\text{C} \times \frac{\text{cal}}{\text{g} \cdot {}^\circ\text{C}}$$

[3] **Solve the equation.**

- Substitute the known quantities into the equation and solve for heat in calories.

$$\text{cal} = 1600\ \cancel{\text{g}} \times 75\ \cancel{{}^\circ\text{C}} \times \frac{1.00\ \text{cal}}{1\ \cancel{\text{g}} \cdot 1\ \cancel{{}^\circ\text{C}}} = 1.2 \times 10^5\ \text{cal}$$

Answer

PROBLEM 4.16

How much energy is required to heat 28.0 g of iron from 19 °C to 150. °C? Report your answer in calories and joules.

PROBLEM 4.17

How much energy is released when 200. g of water is cooled from 55 °C to 12 °C? Report your answer in calories and kilocalories.

Building on what you have learned in Sample Problem 4.6, Sample Problem 4.7 shows how to determine the temperature change of a given mass of a substance when the amount of heat absorbed is known.

SAMPLE PROBLEM 4.7

If 400. cal of heat is added to 25.0 g of 2-propanol at 21 °C, what is the final temperature?

Analysis

Use specific heat as a conversion factor to determine the temperature change (ΔT) given the known mass of the substance and the amount of heat absorbed. Add the temperature change to the initial temperature (T_1) to obtain the final temperature (T_2).

Solution

[1] Identify the known quantities and the desired quantity.

mass = 25.0 g
heat added = 400. cal
T_1 = 21 °C — T_2 = ?

known quantities — desired quantity

- According to Table 4.4, the specific heat of 2-propanol is 0.612 cal/(g · °C).

[2] Write the equation and rearrange it to isolate ΔT on one side.

- Divide both sides of the equation by mass (in g) and specific heat [in cal/(g · °C)] to place ΔT (in °C) on one side.

$$\text{heat} = \text{mass} \times \Delta T \times \text{specific heat}$$

$$\frac{\text{heat}}{\text{mass} \cdot \text{specific heat}} = \Delta T$$

$$\frac{\text{cal}}{\text{g} \cdot \text{cal/(g} \cdot {}^\circ\text{C)}} = \Delta T$$

[3] Solve the equation to determine the change in temperature.

$$\Delta T = \frac{\text{cal}}{\text{g} \cdot \text{cal/(g} \cdot {}^\circ\text{C)}} = \frac{400.\ \cancel{\text{cal}}}{25.0\ \cancel{\text{g}} \cdot 0.612\ \cancel{\text{cal}}/(\cancel{\text{g}} \cdot {}^\circ\text{C})} = 26.1\ {}^\circ\text{C}$$

temperature change

[4] Add the change in temperature (ΔT) to T_1 to obtain the final temperature T_2.

$$T_2 = 21 + 26.1 = 47.1\ {}^\circ\text{C rounded to } 47\ {}^\circ\text{C}$$

Answer

PROBLEM 4.18

If 20. cal of heat is added to 10.0 g each of copper and mercury at 15 °C, what is the final temperature of each element?

PROBLEM 4.19

If the initial temperature of 120. g of ethanol is 20. °C, what will be the final temperature after 950. cal of heat is added?

4.6 Energy and Phase Changes

In Section 4.4 we learned how the strength of intermolecular forces in a liquid and solid affect a compound's boiling point and melting point. Let's now look in more detail at the energy changes that occur during phase changes.

- **When energy is absorbed, a process is said to be *endothermic.***
- **When energy is released, a process is said to be *exothermic.***

In a phase change, the physical state of a substance is altered without changing its composition.

4.6A Converting a Solid to a Liquid

Converting a solid to a liquid is called *melting*. Melting is a phase change because the highly organized water molecules in the solid phase become more disorganized in the liquid phase, but the chemical bonds do not change. Each water molecule is composed of two O—H bonds in both the solid and the liquid phases.

Melting is an *endothermic* process. Energy must be absorbed to overcome some of the attractive intermolecular forces that hold the organized solid molecules together to form the more random liquid phase. The amount of energy needed to melt 1 g of a substance is called its **heat of fusion.**

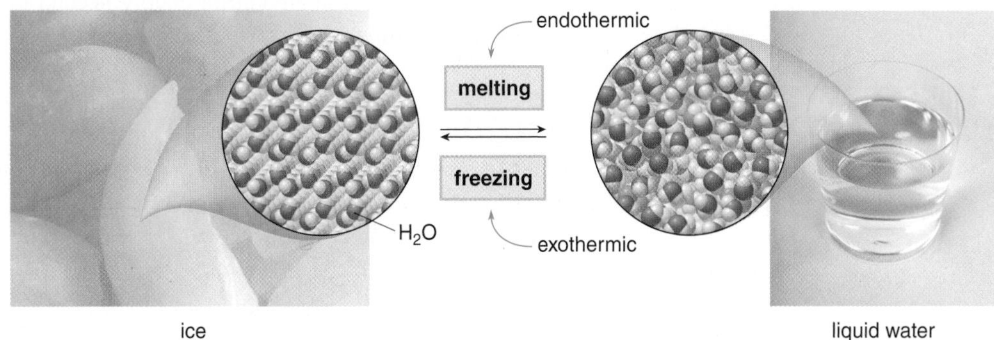

Freezing is the opposite of melting; that is, ***freezing* converts a liquid to a solid.** Freezing is an *exothermic* process because energy is released as the faster moving liquid molecules form an organized solid in which particles have little freedom of motion. For a given mass of a particular substance, the amount of energy released in freezing is the same as the amount of energy absorbed during melting.

Heats of fusion are reported in calories per gram (cal/g). A heat of fusion can be used as a conversion factor to determine how much energy is absorbed when a particular amount of a substance melts, as shown in Sample Problem 4.8.

When an ice cube is added to a liquid at room temperature, the ice cube melts. The energy needed for melting is "pulled" from the warmer liquid molecules and the liquid cools down.

SAMPLE PROBLEM 4.8

How much energy in calories is absorbed when 50.0 g of ice cubes melt? The heat of fusion of H_2O is 79.7 cal/g.

Analysis

Use the heat of fusion as a conversion factor to determine the amount of energy absorbed in melting.

Solution

[1] Identify the original quantity and the desired quantity.

50.0 g	? calories
original quantity	desired quantity

[2] Write out the conversion factors.

- Use the heat of fusion as a conversion factor to convert grams to calories.

g–cal conversion factors

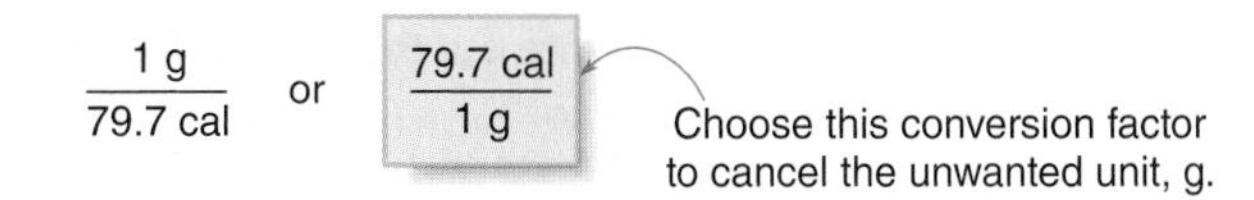

$$\frac{1\ \text{g}}{79.7\ \text{cal}} \quad \text{or} \quad \frac{79.7\ \text{cal}}{1\ \text{g}}$$

[3] Solve the problem.

$$50.0\ \cancel{\text{g}} \times \frac{79.7\ \text{cal}}{1\ \cancel{\text{g}}} = 3{,}985\ \text{cal rounded to } 3{,}990\ \text{cal}$$

Grams cancel.

Answer

PROBLEM 4.20

Use the heat of fusion of water from Sample Problem 4.8 to answer each question.

a. How much energy in calories is released when 50.0 g of water freezes?
b. How much energy in calories is absorbed when 35.0 g of water melts?
c. How much energy in kilocalories is absorbed when 35.0 g of water melts?

4.6B Converting a Liquid to a Gas

Converting a liquid to a gas is called *vaporization*. Vaporization is an *endothermic* process. Energy must be absorbed to overcome the attractive intermolecular forces of the liquid phase to form gas molecules. The amount of energy needed to vaporize 1 g of a substance is called its **heat of vaporization.**

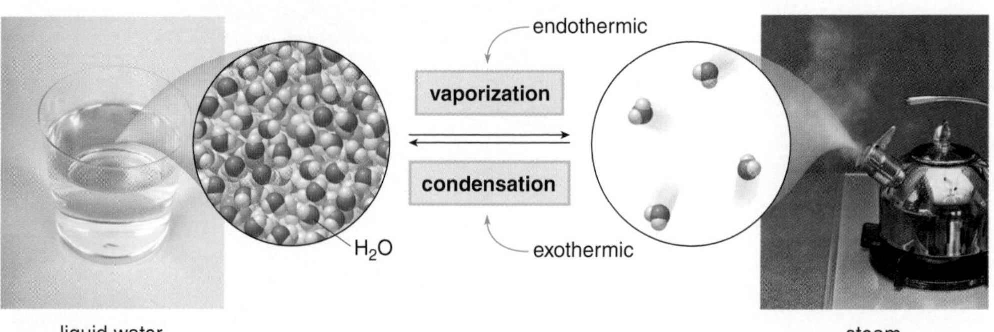

Condensation is the opposite of vaporization; that is, ***condensation* converts a gas to a liquid.** Condensation is an *exothermic* process because energy is released as the faster moving gas molecules form the more organized liquid phase. For a given mass of a particular substance, the amount of energy released in condensation equals the amount of energy absorbed during vaporization.

Heats of vaporization are reported in calories per gram (cal/g). A high heat of vaporization means that a substance absorbs a great deal of energy as it is converted from a liquid to a gas. **Water has a high heat of vaporization.** As a result, the evaporation of sweat from the skin is a very effective cooling mechanism for the body. The heat of vaporization can be used as a conversion factor to determine how much energy is absorbed when a particular amount of a substance vaporizes, as shown in Sample Problem 4.9.

HEALTH NOTE

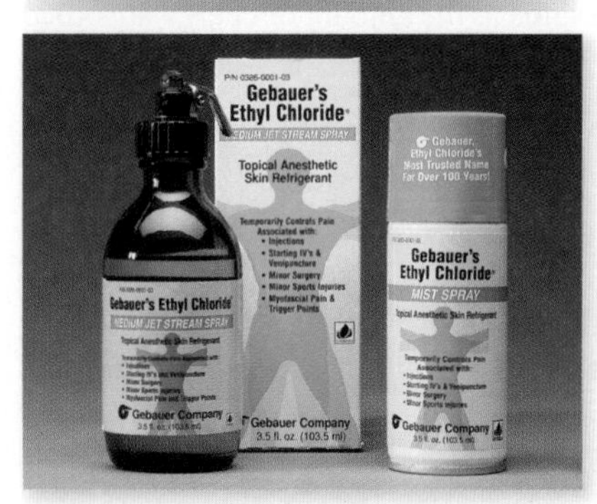

Chloroethane (CH_3CH_2Cl), commonly called ethyl chloride, is a local anesthetic. When chloroethane is sprayed on a wound it quickly evaporates, causing a cooling sensation that numbs the site of an injury.

SAMPLE PROBLEM 4.9

How much heat in kilocalories is absorbed when 22.0 g of 2-propanol, rubbing alcohol, evaporates after being rubbed on the skin? The heat of vaporization of 2-propanol is 159 cal/g.

Analysis

Use the heat of vaporization to convert grams to an energy unit, calories. Calories must also be converted to kilocalories using a cal–kcal conversion factor.

Solution

[1] **Identify the original quantity and the desired quantity.**

22.0 g	? kilocalories
original quantity	desired quantity

[2] **Write out the conversion factors.**

- We have no conversion factor that directly relates grams and kilocalories. We do know, however, how to relate grams to calories using the heat of vaporization, and calories to kilocalories.

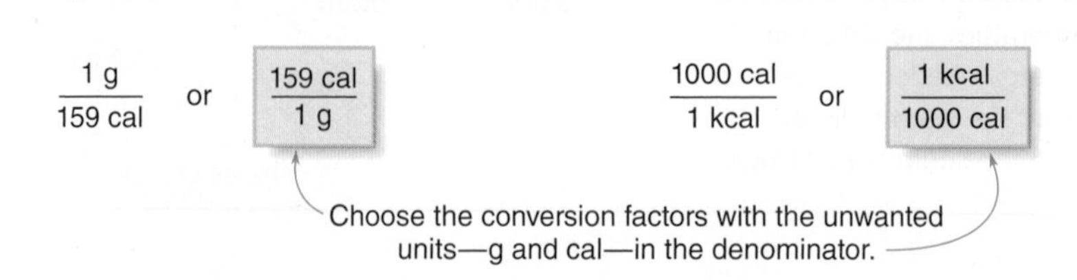

g–cal conversion factors
$\frac{1\text{ g}}{159\text{ cal}}$ or $\frac{159\text{ cal}}{1\text{ g}}$
cal–kcal conversion factors
$\frac{1000\text{ cal}}{1\text{ kcal}}$ or $\frac{1\text{ kcal}}{1000\text{ cal}}$
Choose the conversion factors with the unwanted units—g and cal—in the denominator.

[3] **Solve the problem.**

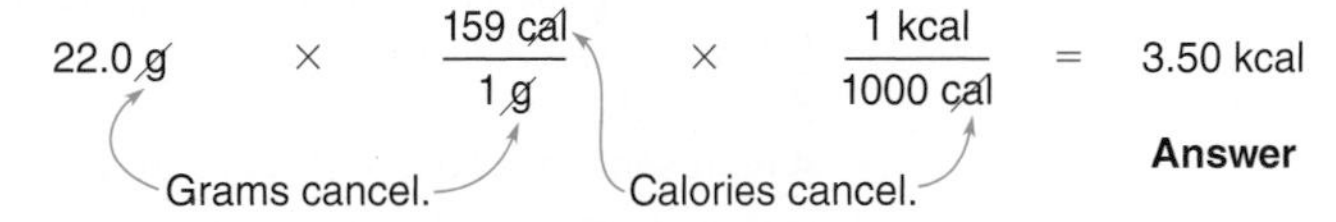

22.0 g × 159 cal / 1 g × 1 kcal / 1000 cal = 3.50 kcal
Answer
Grams cancel.
Calories cancel.

PROBLEM 4.21

Answer the following questions about water, which has a heat of vaporization of 540 cal/g.

a. How much energy in calories is absorbed when 42 g of water is vaporized?
b. How much energy in calories is released when 42 g of water is condensed?

4.6C Converting a Solid to a Gas

CONSUMER NOTE

Freeze-drying removes water from foods by the process of sublimation. These products can be stored almost indefinitely, since bacteria cannot grow in them without water.

Occasionally a solid phase forms a gas phase without passing through the liquid state. This process is called **sublimation.** The reverse process, conversion of a gas directly to a solid, is called **deposition.** Carbon dioxide is called *dry ice* because solid carbon dioxide (CO_2) sublimes to gaseous CO_2 without forming liquid CO_2.

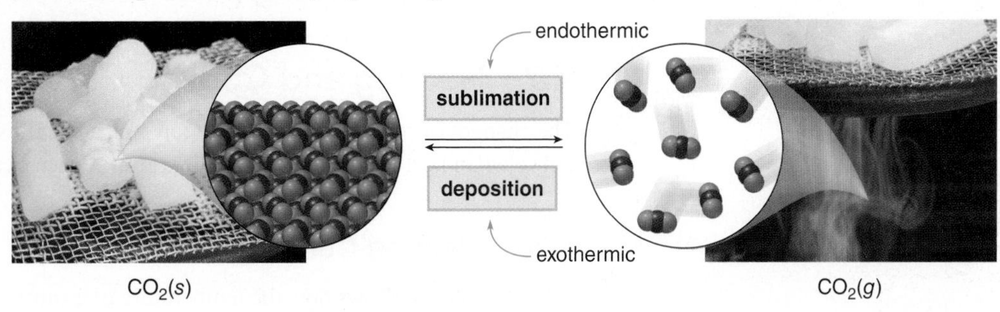

Carbon dioxide is a good example of a solid that undergoes this process at atmospheric pressure. At reduced pressure other substances sublime. For example, freeze-dried foods are prepared by subliming water from a food product at low pressure.

PROBLEM 4.22

Label each process as endothermic or exothermic and explain your reasoning: (a) sublimation; (b) deposition.

Sample Problem 4.10 illustrates how molecular art can be used to depict and identify phase changes.

SAMPLE PROBLEM 4.10

What phase change is shown in the accompanying molecular art? Is the process endothermic or exothermic?

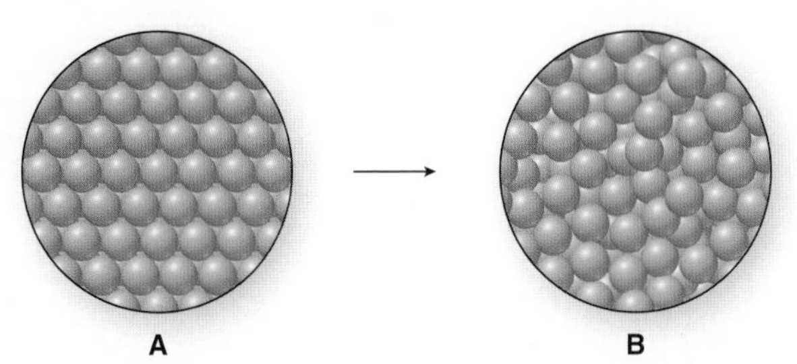

Analysis

Identify the phase by the distance between the spheres and their level of organization. A solid has closely packed spheres that are well organized; a liquid has closely packed but randomly arranged spheres;

a gas has randomly arranged spheres that are far apart. Then, classify the transformation as melting, freezing, vaporization, condensation, sublimation, or deposition, depending on the phases depicted.

Solution

A represents a solid and **B** represents a liquid, so the molecular art represents melting. Melting is an endothermic process because energy must be absorbed to convert the more ordered solid state to the less ordered liquid state.

PROBLEM 4.23

What phase change is shown in the accompanying molecular art? Is the process endothermic or exothermic?

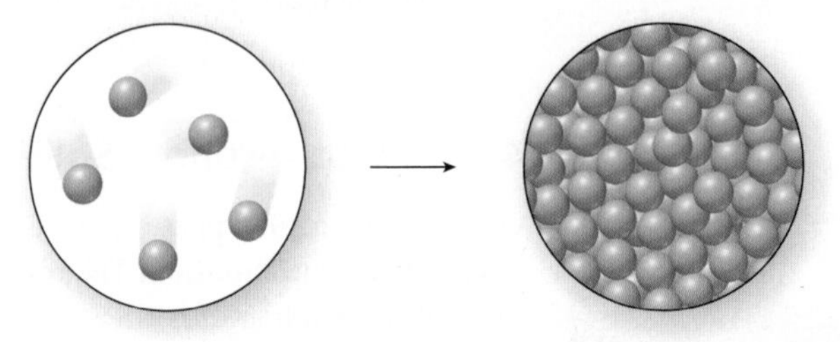

4.7 Heating and Cooling Curves

The changes of state described in Section 4.6 can be illustrated on a single graph called a **heating curve** when heat is added and a **cooling curve** when heat is removed.

4.7A Heating Curves

A **heating curve** shows how the temperature of a substance (plotted on the vertical axis) changes as heat is *added.* A general heating curve is shown in Figure 4.4.

Figure 4.4 Heating Curve

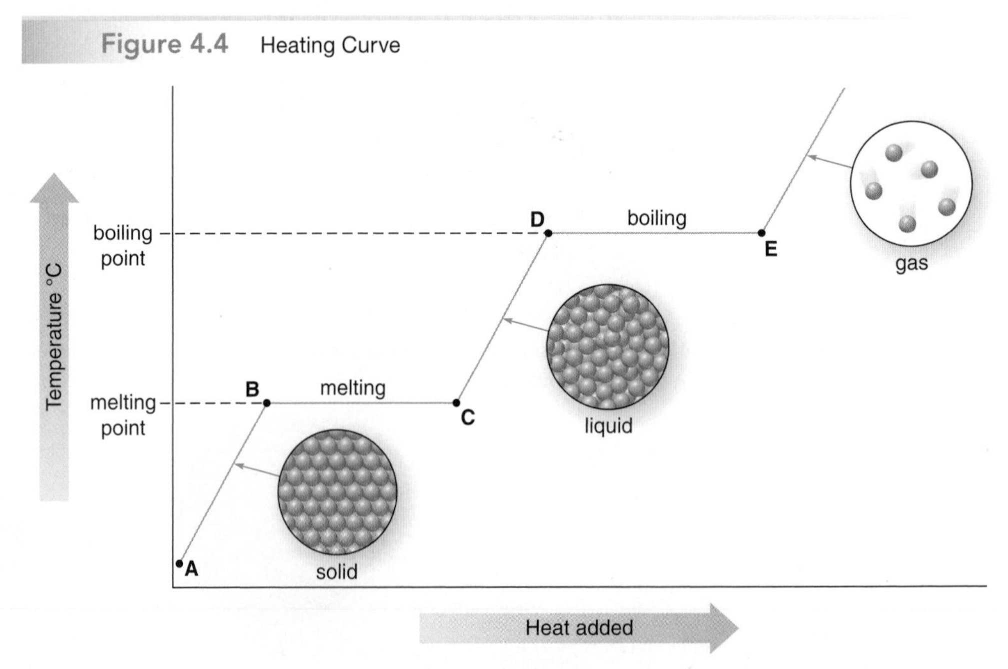

A heating curve shows how the temperature of a substance changes as heat is added. The plateau **B** ⟶ **C** occurs at the melting point, while the plateau **D** ⟶ **E** occurs at the boiling point.

A solid is present at point **A.** As the solid is heated it increases in temperature until its melting point is reached at **B.** More heat causes the solid to melt to a liquid, without increasing its temperature (the plateau from **B** ⟶ **C**). Added heat increases the temperature of the liquid until its boiling point is reached at **D.** More heat causes the liquid to boil to form a gas, without increasing its temperature (the plateau from **D** ⟶ **E**). Additional heat then increases the temperature of the gas. Each diagonal line corresponds to the presence of a single phase—solid, liquid, or gas—while horizontal lines correspond to phase changes—solid to liquid or liquid to gas.

PROBLEM 4.24

Answer the following questions about the graph.

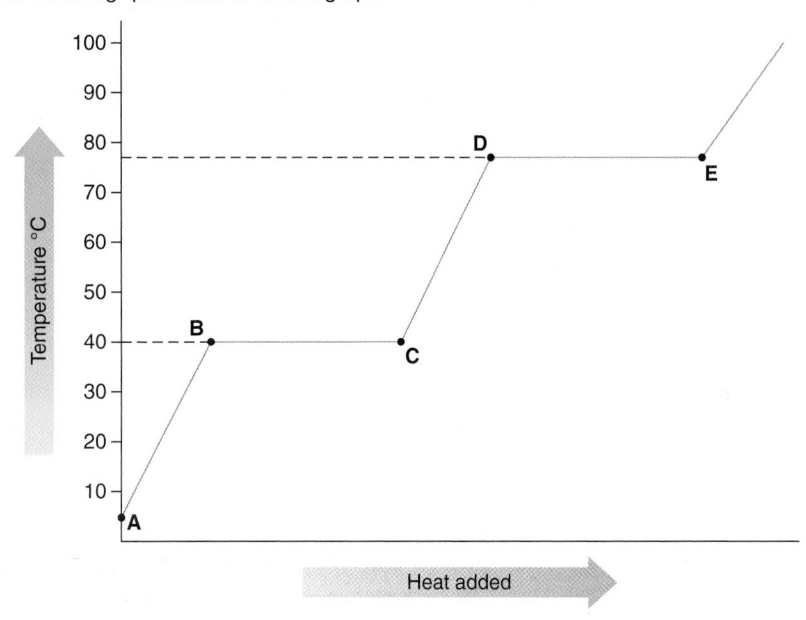

a. What is the melting point of the substance?
b. What is the boiling point of the substance?
c. What phase(s) are present at plateau **B ⟶ C?**
d. What phase(s) are present along the diagonal **C ⟶ D?**

PROBLEM 4.25

If the substance shown in the heating curve in Figure 4.4 has a melting point of 50 °C and a boiling point of 75 °C, what state or states of matter are present at each temperature?

a. 85 °C b. 50 °C c. 65 °C d. 10 °C e. 75 °C

4.7B Cooling Curves

A **cooling curve** illustrates how the temperature of a substance (plotted on the vertical axis) changes as heat is *removed.* A cooling curve for water is shown in Figure 4.5.

Figure 4.5 Cooling Curve for Water

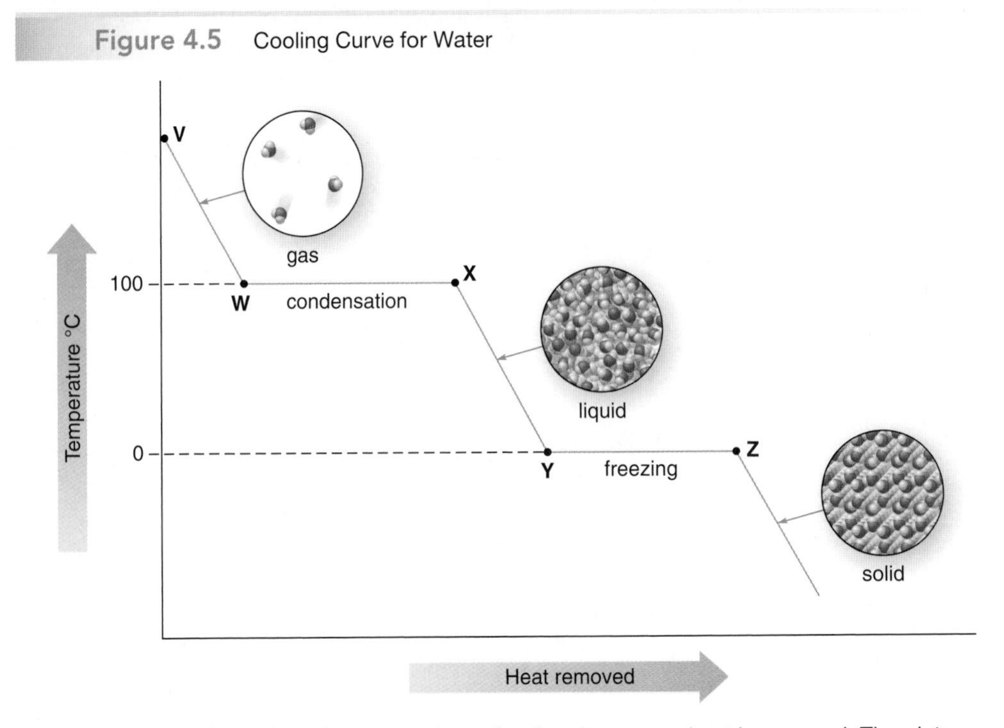

The cooling curve shows how the temperature of water changes as heat is removed. The plateau **W** ⟶ **X** occurs at the boiling point, while the plateau **Y** ⟶ **Z** occurs at the freezing point.

Gaseous water is present at point **V.** As the gas is cooled it decreases in temperature until its boiling point is reached at **W.** Condensation at 100 °C forms liquid water, represented by the plateau from **W** ⟶ **X.** Further cooling of the liquid water takes place until its freezing point (melting point) is reached at **Y.** Freezing water forms ice at 0 °C, represented by the plateau from **Y** ⟶ **Z.** Cooling the ice further decreases its temperature below its freezing point.

PROBLEM 4.26

If the cooling curve in Figure 4.5 represented a substance with a melting point of 40 °C and a boiling point of 85 °C, what state or states of matter would be present at each temperature?

a. 75 °C b. 50 °C c. 65 °C d. 10 °C e. 85 °C

4.7C Combining Energy Calculations

We have now performed two different types of energy calculations. In Section 4.5 we used specific heat to calculate the amount of energy needed to raise the temperature of a substance in a single phase—such as heating a liquid from a lower to a higher temperature. In Section 4.6 we learned how to calculate energy changes during changes of state using heats of fusion or vaporization. Sometimes these calculations must be combined together to determine the total energy change when both a temperature change and a change of state occur.

Sample Problem 4.11 illustrates how to calculate the total energy needed to convert a given mass of liquid water to steam at its boiling point.

SAMPLE PROBLEM 4.11

How much energy is required to heat 25.0 g of water from 25 °C to a gas at its boiling point of 100. °C? The specific heat of water is 1.00 cal/(g · °C), and the heat of vaporization of water is 540 cal/g.

Analysis

Use specific heat to calculate how much energy is required to heat the given mass of water to its boiling point. Then use the heat of vaporization to calculate how much energy is required to convert liquid water to a gas.

Solution

[1] Identify the original quantities and the desired quantity.

known quantities	desired quantity
mass = 25.0 g	
T_1 = 25 °C	
T_2 = 100. °C	? calories

- Subtract the initial temperature (T_1) from the final temperature (T_2) to determine the temperature change: $T_2 - T_1 = \Delta T = 100. - 25 = 75$ °C.

[2] Write out the conversion factors.

- Conversion factors are needed for both the specific heat and the heat of vaporization.

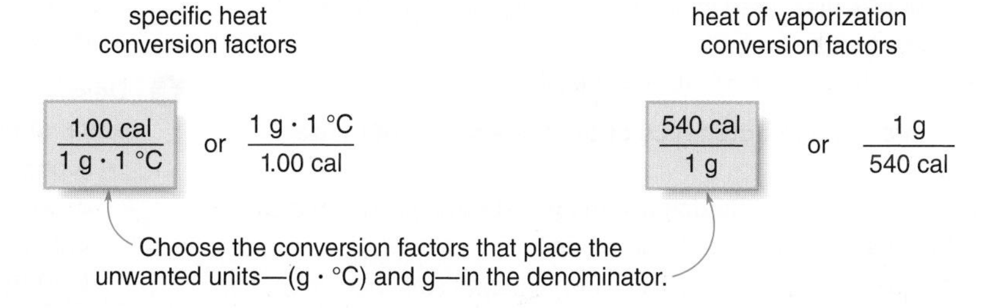

specific heat
conversion factors
heat of vaporization
conversion factors
1.00 cal / 1 g · 1 °C
or
1 g · 1 °C / 1.00 cal
540 cal / 1 g
or
1 g / 540 cal
Choose the conversion factors that place the
unwanted units—(g · °C) and g—in the denominator.

[3] Solve the problem.

- Calculate the heat needed to change the temperature of water 75 °C using specific heat.

$$\text{heat} = \text{mass} \times \Delta T \times \text{specific heat}$$

$$\text{cal} = 25.0\,\cancel{\text{g}} \times 75.0\,\cancel{°\text{C}} \times \frac{1.00\text{ cal}}{1\,\cancel{\text{g}} \cdot 1\,\cancel{°\text{C}}} = 1{,}875\text{ cal rounded to }1{,}900\text{ cal}$$

- Calculate the heat needed for the phase change (liquid water ⟶ gaseous water) using the heat of vaporization.

$$\text{cal} = 25.0\,\cancel{\text{g}} \times \frac{540\text{ cal}}{1\,\cancel{\text{g}}} = 13{,}500\text{ cal rounded to }14{,}000\text{ cal}$$

- Add the two values together to obtain the total energy required.

$$\text{Total energy} = 1{,}900\text{ cal} + 14{,}000\text{ cal} = 15{,}900\text{ cal rounded to }16{,}000\text{ cal}$$

Answer

PROBLEM 4.27

How much energy (in calories) is released when 50.0 g of water is cooled from 25 °C to solid ice at 0.0 °C? The specific heat of water is 1.00 cal/(g · °C), and the heat of fusion of water is 79.7 cal/g.

PROBLEM 4.28

How much energy (in calories) is required to melt 25.0 g of ice to water at 0.0 °C, heat the liquid water to 100. °C, and vaporize the water to steam at 100. °C? The specific heat of water is 1.00 cal/(g · °C), the heat of fusion of water is 79.7 cal/g, and the heat of vaporization of water is 540 cal/g.

KEY TERMS

Boiling point (bp, 4.4)
Calorie (4.1)
Condensation (4.6)
Cooling curve (4.7)
Deposition (4.6)
Dipole–dipole interactions (4.3)
Endothermic (4.6)
Energy (4.1)
Exothermic (4.6)
Freezing (4.6)
Heating curve (4.7)
Heat of fusion (4.6)
Heat of vaporization (4.6)
Hydrogen bonding (4.3)
Intermolecular forces (4.3)
Joule (4.1)
Kinetic energy (4.1)
Law of conservation of energy (4.1)
London dispersion forces (4.3)
Melting (4.6)
Melting point (mp, 4.4)
Potential energy (4.1)
Specific heat (*SH*, 4.5)
Sublimation (4.6)
Vaporization (4.6)

KEY CONCEPTS

1 What is energy and what units are used to measure energy? (4.1)

- Energy is the capacity to do work. Kinetic energy is the energy of motion, whereas potential energy is stored energy.
- Energy is measured in calories (cal) or joules (J), where 1 cal = 4.184 J.
- One nutritional calorie (Cal) = 1 kcal = 1,000 cal.

2 What are the characteristics of the three states of matter? (4.2)

- A gas consists of randomly arranged, disorganized particles that are far apart and move very fast.
- A liquid consists of randomly arranged particles that are much closer and held together by attractive interactions.
- A solid consists of highly organized, very close particles held together by strong attractive forces.

3 What types of intermolecular forces exist? (4.3)

- Intermolecular forces are the forces of attraction between molecules. Three types of intermolecular forces exist in covalent compounds. London dispersion forces are due to momentary changes in electron density in a molecule. Dipole–dipole interactions are due to permanent dipoles. Hydrogen bonding, the strongest intermolecular force, results when a H atom bonded to an O, N, or F, is attracted to an O, N, or F atom in another molecule.

4 How are intermolecular forces related to a compound's boiling point and melting point? (4.4)

- The stronger the intermolecular forces, the higher the boiling point and melting point of a compound.

5 What is specific heat? (4.5)

- Specific heat is the amount of energy needed to raise the temperature of 1 g of a substance by 1 °C.
- Specific heat is used as a conversion factor to calculate how much heat a known mass of a substance absorbs or how much its temperature changes.

6 Describe the energy changes that accompany changes of state. (4.6)

- A phase change converts one state to another. Energy is absorbed when a more organized state is converted to a less organized state. Thus, energy is absorbed when a solid melts to form a liquid, or when a liquid vaporizes to form a gas.
- Energy is released when a less organized state is converted to a more organized state. Thus, energy is released when a gas condenses to form a liquid, or a liquid freezes to form a solid.
- The heat of fusion is the energy needed to melt 1 g of a substance, while the heat of vaporization is the energy needed to vaporize 1 g of a substance.

7 What changes are depicted on heating and cooling curves? (4.7)

- A heating curve shows how the temperature of a substance changes as heat is added. Diagonal lines show the temperature increase of a single phase. Horizontal lines correspond to phase changes—solid to liquid or liquid to gas.
- A cooling curve shows how the temperature of a substance changes as heat is removed. Diagonal lines show the temperature decrease of a single phase. Horizontal lines correspond to phase changes—gas to liquid or liquid to solid.

UNDERSTANDING KEY CONCEPTS

Selected in-chapter and odd-numbered end-of-chapter problems have brief answers at the end of each chapter. The *Student Study Guide and Solutions Manual* contains detailed solutions to all in-chapter and odd-numbered end-of-chapter problems, as well as additional worked examples and a chapter self-test.

4.29 What phase change is shown in the accompanying molecular art? Is energy absorbed or released during the process?

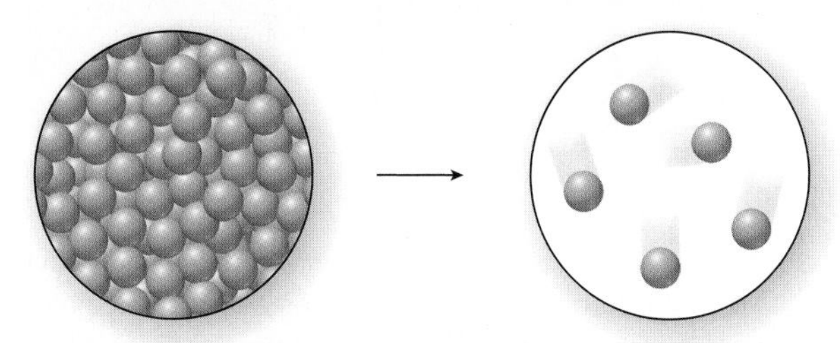

4.30 What phase change is shown in the accompanying molecular art? Is energy absorbed or released during the process?

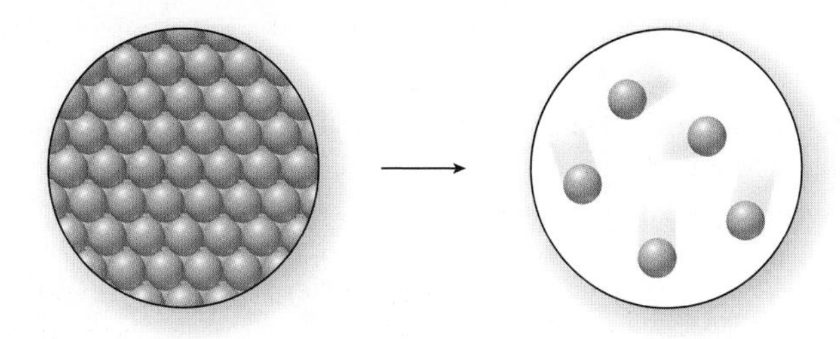

4.31 Consider the cooling curve drawn below.

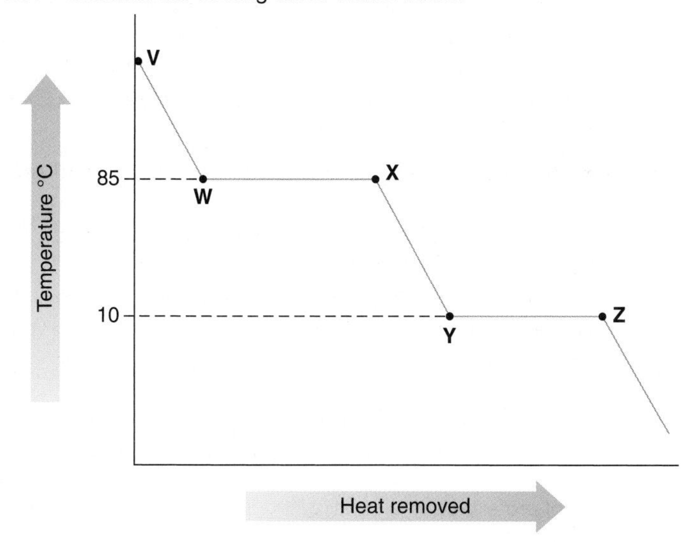

a. Which line segment corresponds to the following changes of state?

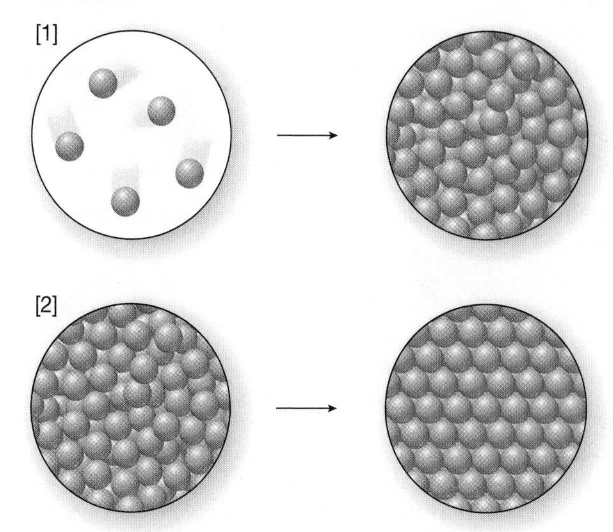

b. What is the melting point of the substance?

c. What is the boiling point of the substance?

4.32 Which line segments on the cooling curve in Problem 4.31 correspond to each of the following physical states?

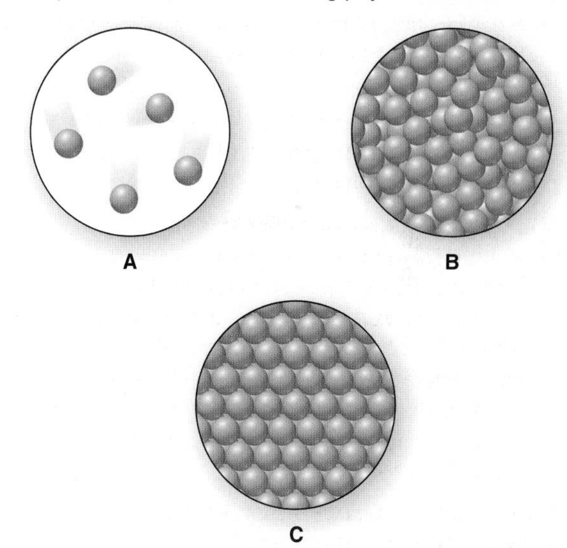

4.33 Riding a bicycle at 12–13 miles per hour uses 563 Calories in an hour. Convert this value to (a) calories; (b) kilocalories; (c) joules; (d) kilojoules.

4.34 Estimate the number of Calories in two tablespoons of peanut butter, which contain 16 g of protein, 7 g of carbohydrates, and 16 g of fat.

4.35 What types of intermolecular forces are exhibited by each compound? Acetaldehyde is formed when ethanol, the alcohol in alcoholic beverages, is metabolized, and acetic acid gives vinegar its biting odor and taste.

a. 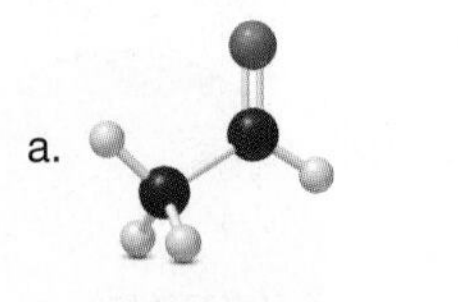

b. 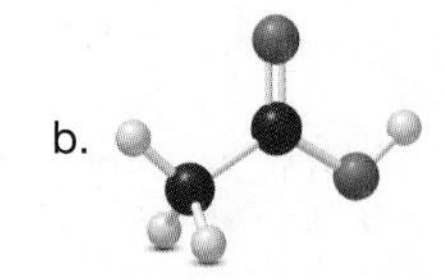

4.36 Ethanol and dimethyl ether have the same molecular formula.

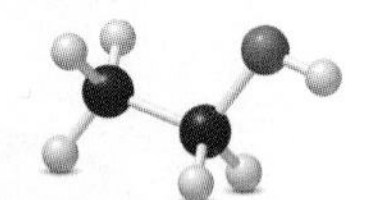

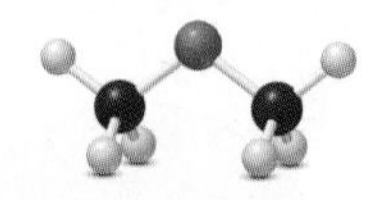

a. What types of intermolecular forces are present in each compound?
b. Which compound has the higher boiling point?

4.37 Consider the two beakers containing the same mass of **X** and **Y** that were at the same initial temperature. Which compound, **X** or **Y**, has the higher specific heat if the same amount of heat was added to both substances?

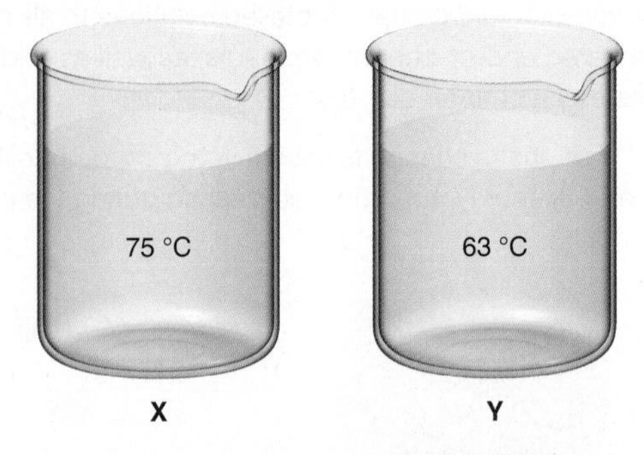

4.38 Consider the following three containers (**A–C**) drawn below. (a) If the same amount of heat was added to all three samples, which sample has the highest temperature? (b) Which sample has the lowest temperature?

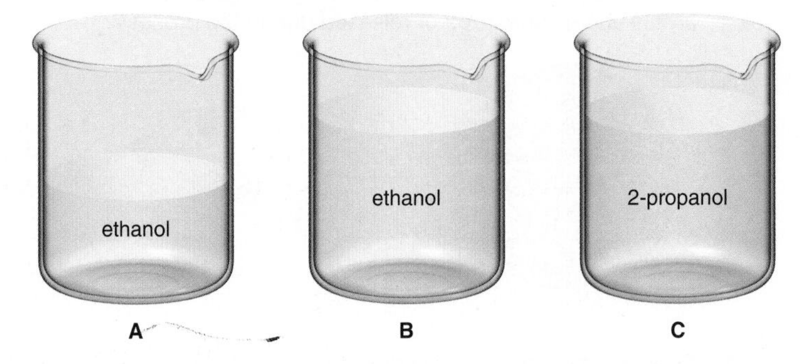

ADDITIONAL PROBLEMS

Energy

4.39 Carry out each of the following conversions.

a. 50 cal to kcal
b. 56 cal to kJ
c. 0.96 kJ to cal
d. 4,230 kJ to cal

4.40 Carry out each of the following conversions.

a. 5 kcal to cal
b. 2,560 cal to kJ
c. 1.22 kJ to cal
d. 4,230 J to kcal

4.41 Running at a rate of 6 mi/h uses 704 Calories in an hour. Convert this value to (a) calories; (b) kilocalories; (c) joules; (d) kilojoules.

4.42 Estimate the number of Calories in a serving of oatmeal that has 4 g of protein, 19 g of carbohydrates, and 2 g of fat.

4.43 A can of soda contains 120 Calories, and no protein or fat. How many grams of carbohydrates are present in each can?

4.44 Alcohol releases 29.7 kJ/g when it burns. Convert this value to the number of Calories per gram.

4.45 Which food has more Calories: 3 oz of salmon, which contains 17 g of protein and 5 g of fat, or 3 oz of chicken, which contains 20 g of protein and 3 g of fat?

4.46 Which food has more Calories: one egg, which contains 6 g of protein and 6 g of fat, or 1 cup of nonfat milk, which contains 9 g of protein and 12 g of carbohydrates?

Intermolecular Forces, Boiling Point, and Melting Point

4.47 Why is H_2O a liquid at room temperature, but H_2S, which has a higher molecular weight and a larger surface area, is a gas at room temperature?

4.48 Why is Cl_2 a gas, Br_2 a liquid, and I_2 a solid at room temperature?

4.49 What types of intermolecular forces are present in each compound?

a. 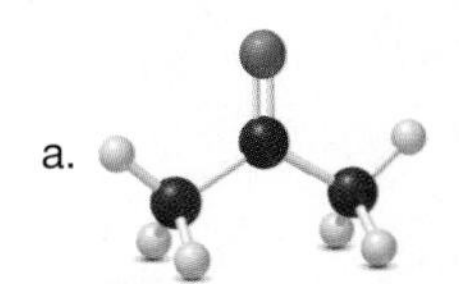

b. 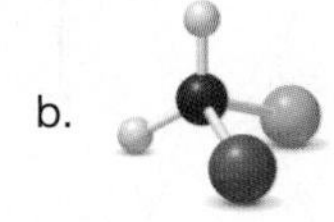

4.50 What types of intermolecular forces are present in each compound?

a. 

b. 

4.51 What types of intermolecular forces are exhibited by each compound? Chloroethane is a local anesthetic and cyclopropane is a general anesthetic.

a. 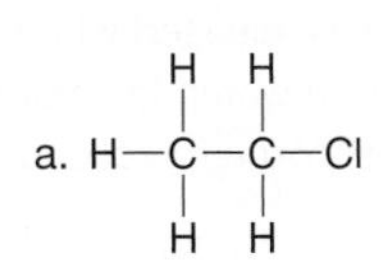

chloroethane

b. 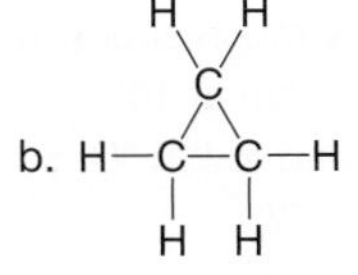

cyclopropane

4.52 Consider two compounds, ethylene and methanol.

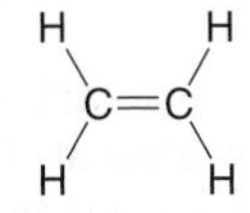

ethylene

methanol: H—C(H)(H)—O—H

a. What types of intermolecular forces are present in each compound?

b. Which compound has the higher boiling point?

4.53 Which molecules are capable of intermolecular hydrogen bonding?

a. H—C≡C—H b. CO_2 c. Br_2 d. H—C(H)(H)—N(H)—H

4.54 Which molecules are capable of intermolecular hydrogen bonding?

a. N_2 b. H—C(H)(H)—F c. HI d. H—C(H)(H)—O—H

4.55 Can two molecules of formaldehyde ($H_2C{=}O$) intermolecularly hydrogen bond to each other? Explain why or why not.

4.56 Why is the melting point of NaCl (801 °C) much higher than the melting point of water (0 °C)?

4.57 Which compound, undecane or pentane, has the higher melting point? Explain.

undecane

pentane

4.58 Explain why the boiling point of **A** is higher than the boiling point of **B** despite the fact that **A** and **B** have the same chemical formula (C_3H_9N).

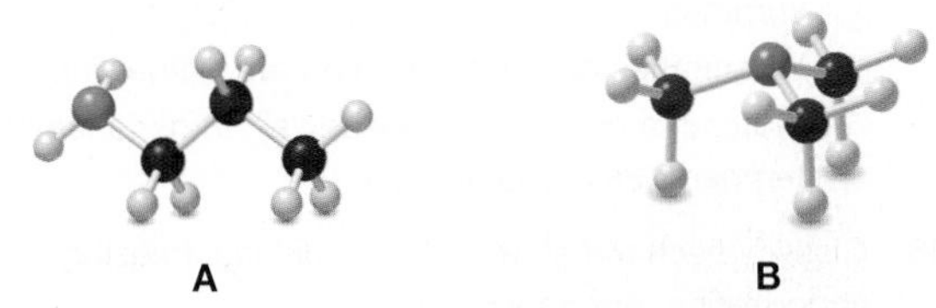

4.59 Consider two compounds, formaldehyde ($H_2C{=}O$) and ethylene ($H_2C{=}CH_2$).

a. Which compound exhibits the stronger intermolecular forces?

b. Which compound has the higher boiling point?

c. Which compound has the higher melting point?

4.60 Consider two compounds, formaldehyde ($H_2C{=}O$) and methanol (CH_3OH).

a. Which compound exhibits the weaker intermolecular forces?

b. Which compound has the lower boiling point?

c. Which compound has the lower melting point?

Specific Heat

4.61 How much energy is absorbed or lost in each of the following? Calculate your answer in both calories and joules.
a. the energy needed to heat 50. g of water from 15 °C to 50. °C
b. the energy lost when 250 g of aluminum is cooled from 125 °C to 50. °C

4.62 How many calories of heat are needed to increase the temperature of 55 g of ethanol from 18 °C to 48 °C?

4.63 Which of the following samples has the higher temperature?
a. 100. g of liquid water at 16.0 °C that absorbs 200. cal of heat
b. 50.0 g of liquid water at 16.0 °C that absorbs 350. J of heat

4.64 Which has the higher final temperature, 10.0 g of aluminum at 18 °C that absorbs 25.0 cal of heat or 12.0 g of iron at 22 °C that absorbs 65.0 J of heat?

4.65 If it takes 37.0 cal of heat to raise the temperature of 12.0 g of a substance by 8.5 °C, what is its specific heat?

4.66 Why does it take weeks for a lake to freeze in the winter even if the outdoor temperature is consistently below 0 °C?

Energy and Phase Changes

4.67 Classify each transformation as melting, freezing, vaporization, or condensation.
a. Beads of water form on the glass of a cool drink in the summer.
b. Wet clothes dry when hung on the clothesline in the sun.
c. Water in a puddle on the sidewalk turns to ice when the temperature drops overnight.

4.68 Classify each transformation as melting, freezing, vaporization, or condensation.
a. Fog forms on the mirror of the bathroom when a hot shower is taken.
b. A puddle of water slowly disappears.
c. A dish of ice cream becomes a bowl of liquid when left on the kitchen counter on a hot day.

4.69 Indicate whether heat is absorbed or released in each process.
a. melting 100 g of ice
b. freezing 25 g of water
c. condensing 20 g of steam
d. vaporizing 30 g of water

4.70 What is the difference between the heat of fusion and the heat of vaporization?

4.71 Which process requires more energy, melting 250 g of ice or vaporizing 50.0 g of water? The heat of fusion of water is 79.7 cal/g and the heat of vaporization is 540 cal/g.

4.72 How much energy in kilocalories is needed to vaporize 255 g of water? The heat of vaporization of water is 540 cal/g.

Heating and Cooling Curves

4.73 Draw the heating curve that is observed when octane is warmed from –70 °C to 130 °C. Octane, a component of gasoline, has a melting point of –57 °C and a boiling point of 126 °C.

4.74 Draw the heating curve that is observed when ice is warmed from –20 °C to 120 °C. Which sections of the curve correspond to the molecular art in **A, B,** and **C?**

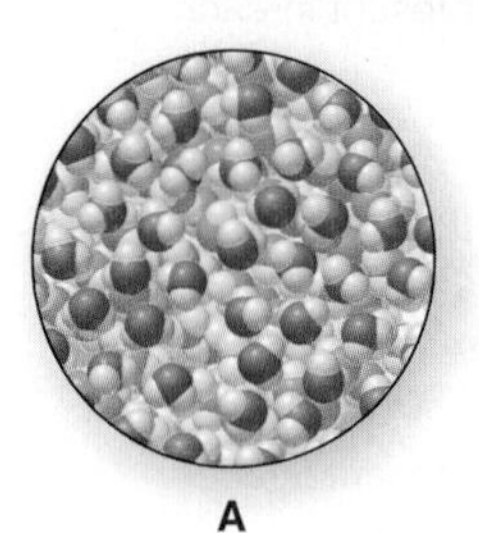
A

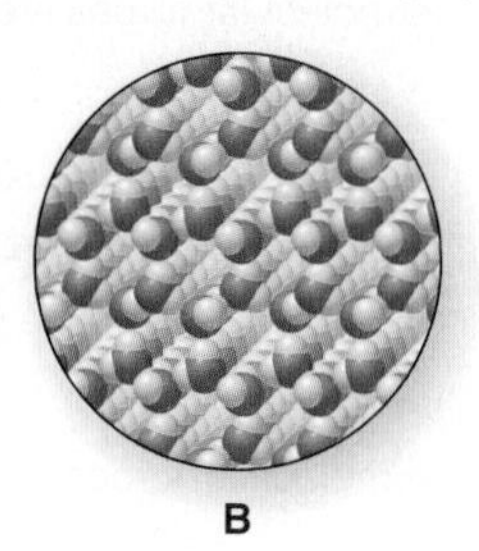
B

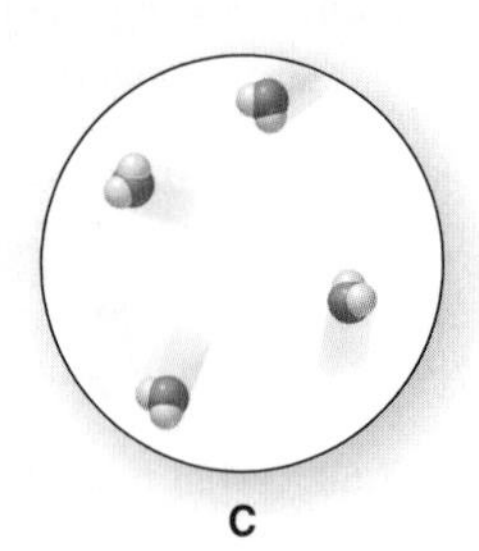
C

Combined Energy Calculations

4.75 Use the following values to answer each part. The specific heat of water is 1.00 cal/(g · °C); the heat of fusion of water is 79.7 cal/g; and the heat of vaporization of water is 540 cal/g.
a. How much energy (in calories) is needed to melt 45 g of ice at 0.0 °C and warm it to 55 °C?
b. How much energy (in calories) is released when 45 g of water at 55 °C is cooled to 0.0 °C, and frozen to solid ice at 0.0 °C?
c. How much energy (in kilocalories) is released when 35 g of steam at 100. °C is condensed to water, the water is cooled to 0.0 °C, and the water is frozen to solid ice at 0.0 °C?

4.76 Use the values in Problem 4.75 to solve each part.
a. How much energy (in calories) is needed to heat 150 g of water from 35 °C to 100. °C and vaporize the water to steam at 100. °C?
b. How much energy (in kilocalories) is released when 42 g of steam is condensed to water at 100. °C, the water is cooled to 0.0 °C, and the water is frozen to solid ice at 0.0 °C?

Applications

4.77 Explain why you feel cool when you get out of a swimming pool, even when the air temperature is quite warm. Then explain why the water feels warmer when you get back into the swimming pool.

4.78 To keep oranges from freezing when the outdoor temperature drops near 32 °F, an orchard is sprayed with water. Explain why this strategy is used.

4.79 A patient receives 2,000 mL of a glucose solution that contains 5 g of glucose in 100 mL. How many Calories does the glucose, a simple carbohydrate, contain?

4.80 Why does steam form when hot lava falls into the ocean?

4.81 Walking at a brisk pace burns off about 280 Cal/h. How long would you have to walk to burn off the Calories obtained from eating a cheeseburger that contained 32 g of protein, 29 g of fat, and 34 g of carbohydrates?

4.82 How many kilocalories does a runner expend when he runs for 4.5 h and uses 710 Cal/h? How many pieces of pizza that each contain 12 g of protein, 11 g of fat, and 30 g of carbohydrates could be eaten after the race to replenish these Calories?

CHALLENGE PROBLEMS

4.83 Burning gasoline releases 11.5 kcal of energy per gram. How many joules of energy are released when 1.0 gal of gasoline is burned? Write the answer in scientific notation. Assume the density of gasoline is 0.74 g/mL.

4.84 An energy bar contains 4 g of fat, 12 g of protein, and 24 g of carbohydrates. How many kilojoules of energy are obtained from eating two bars per day for a month? Write the answer in scientific notation.

4.85 How much heat (in kcal) must be added to raise the temperature of the water in a 400.-gal hot tub from 60. °F to 110. °F? (Recall that the density of water is 1.00 g/mL.)

BEYOND THE CLASSROOM

4.86 Some studies suggest that recycling one aluminum beverage can saves the energy equivalent of 0.5 gallons of gasoline. Estimate how many aluminum cans your household uses per week, and calculate how much gasoline would be saved by recycling these cans. If burning one gallon of gasoline releases about 3.1×10^4 kcal of energy, calculate how much energy is saved each week in recycling.

4.87 Obtain Calorie data from a fast-food restaurant and calculate how many Calories you ingest in a typical meal. How long would you have to walk or run to burn off those Calories? Assume that walking at a moderate pace expends 280 Cal/h and that running at a vigorous pace expends 590 Cal/h. Compare results for meals at different restaurants.

4.88 The strength of intermolecular forces can be used to explain many characteristics of liquids. For example, surface tension is a measure of the resistance of a liquid to spread out. Research how intermolecular forces are related to surface tension and why water has a high surface tension. Use this information to explain why insects such as water striders can walk across the surface of water. Is it possible to "float" a paper clip on water?

4.89 Identify two cities that are geographically close to each other, with one located directly on the coast and one located several miles inland. Using information from your local newspaper or the web, record the high and low temperatures for both cities each day for a period of time. How do the temperature ranges compare? Share your data with other members of your class who picked different cities. Explain any trends you observe. What other factors—such as terrain and population density—might also affect your data?

ANSWERS TO SELECTED PROBLEMS

4.1 a. 10. cal b. 55,600 cal c. 1,360 kJ d. 107,000 J

4.3 126 Cal, rounded to 100 Cal

4.5

	a. Density	b. Intermolecular Spacing	c. Intermolecular Attraction
Gas	Lowest	Greatest	Lowest
Liquid	Higher	Smaller	Higher
Solid	Highest	Smallest	Highest

4.7 all: a–d

4.9 $\overset{\delta^+}{\text{H}}-\overset{\delta^-}{\text{Cl}}$ $\overset{\delta^+}{\text{H}}-\overset{\delta^-}{\text{Cl}}$

4.10

	London Dispersion	Dipole–Dipole	Hydrogen Bonding
a. Cl_2	+		
b. HCN	+	+	
c. HF	+	+	+
d. CH_3Cl	+	+	
e. H_2	+		

4.11 a. H_2O b. HBr c. H_2O d. C_2H_6

4.12 a. C_2H_6 b. CH_3OH c. HBr d. CH_3Br

4.13 Water has stronger intermolecular forces since it can hydrogen bond. This explains why it is a liquid at room temperature, whereas CO_2 is a gas.

4.14 a. sand b. ethanol

4.15 Water has a high specific heat, so it can absorb or release a great deal of energy with only a small temperature change.

4.16 392 cal, 1,640 J

4.17 8,600 cal, 8.6 kcal

4.18 Cu 37 °C, Hg 75 °C

4.19 34 °C

4.20 a. 3,990 cal b. 2,790 cal c. 2.79 kcal

4.21 a. 23,000 cal b. 23,000 cal

4.23 condensation; exothermic

4.25 a. gas b. solid and liquid c. liquid d. solid e. liquid and gas

4.27 5,300 cal

4.29 vaporization; energy absorbed

4.31 a. [1] **W** $\longrightarrow$ **X**; [2] **Y** $\longrightarrow$ **Z** b. 10 °C c. 85 °C

4.33 a. 563,000 cal/h b. 563 kcal/h c. 2.36×10^6 J/h d. 2,360 kJ/h

4.35 a. London forces and dipole–dipole
b. London forces, dipole–dipole, hydrogen bonding

4.37 **Y**

4.39 a. 0.05 kcal b. 0.23 kJ c. 230 cal d. 1.01×10^6 cal

4.41 a. 704,000 cal b. 704 kcal c. 2.95×10^6 J d. 2,950 kJ

4.43 30 g

4.45 salmon, 113 Calories vs. chicken, 107 Calories

4.47 Water is capable of hydrogen bonding and these strong intermolecular attractive forces give it a higher boiling point than H_2S.

4.49 a,b: London dispersion forces, dipole–dipole interactions

4.51 a. London forces, dipole–dipole
b. London forces only

4.53 d.

4.55 No; $H_2C{=}O$ has no H on the O atom.

4.57 Undecane has the higher melting point because its size is larger, so its London forces are stronger.

4.59 a. formaldehyde b. formaldehyde c. formaldehyde

4.61 a. 1.8×10^3 cal, 7.5×10^3 J
b. 4.0×10^3 cal, 1.7×10^4 J

4.63 a.

4.65 0.36 cal/(g · °C)

4.67 a. condensation b. vaporization c. freezing

4.69 a,d: absorbed b,c: released

4.71 Vaporizing 50.0 g of water takes more energy; 27,000 cal vs. 20,000 cal.

4.73

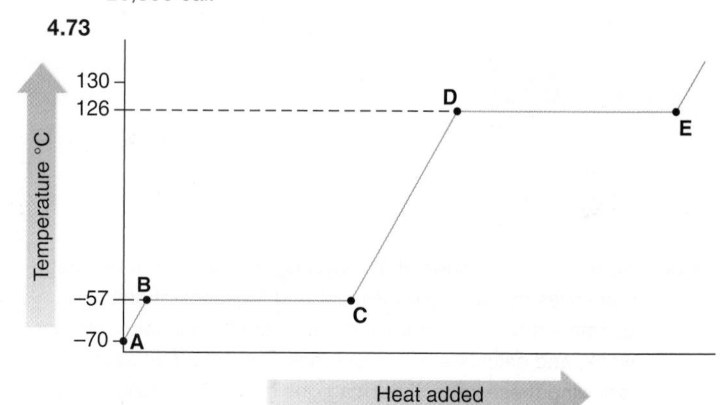

4.75 25 kcal

4.77 When you get out of a pool, the water on your body evaporates and this cools your skin. When you re-enter the water, the water feels warmer because the skin is cooler.

4.79 400 Calories

4.81 1.9 h, rounded to 2 h

4.83 1.3×10^8 J

4.85 4.2×10^4 kcal

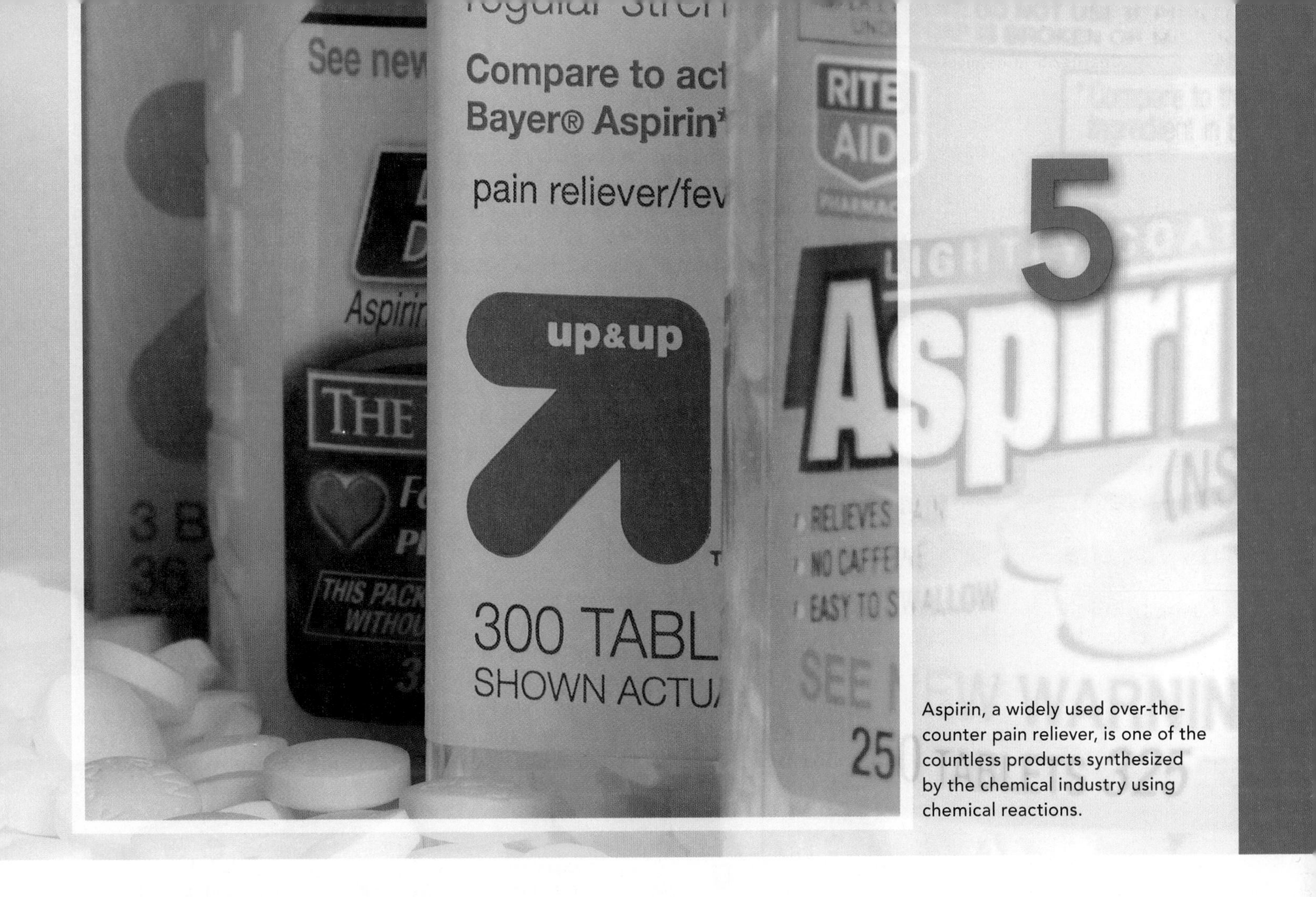

Aspirin, a widely used over-the-counter pain reliever, is one of the countless products synthesized by the chemical industry using chemical reactions.

5

Chemical Reactions

CHAPTER OUTLINE

5.1 Introduction to Chemical Reactions
5.2 Balancing Chemical Equations
5.3 The Mole and Avogadro's Number
5.4 Mass to Mole Conversions
5.5 Mole Calculations in Chemical Equations
5.6 Mass Calculations in Chemical Equations
5.7 Oxidation and Reduction
5.8 Energy Changes in Reactions
5.9 Reaction Rates
5.10 Equilibrium
5.11 FOCUS ON THE HUMAN BODY: Body Temperature

CHAPTER GOALS

In this chapter you will learn how to:

1. Write and balance chemical equations
2. Define a mole and use Avogadro's number in calculations
3. Calculate molar mass
4. Relate the mass of a substance to its number of moles
5. Carry out mole and mass calculations in chemical equations
6. Define oxidation and reduction and recognize the components of a redox reaction
7. Describe energy changes in reactions and classify a reaction as endothermic or exothermic
8. Understand the factors that affect the rate of a reaction
9. Describe the basic features of chemical equilibrium, and use Le Châtelier's principle to predict what happens when equilibrium is disturbed
10. Understand how temperature is regulated in the body

Having learned about atoms, ionic compounds, and covalent molecules in Chapters 2 and 3, we now turn our attention to chemical reactions. Reactions are at the heart of chemistry. An understanding of chemical processes has made possible the conversion of natural substances into new compounds with different and sometimes superior properties. Aspirin, ibuprofen, and nylon are all products of chemical reactions utilizing substances derived from petroleum. Chemical reactions are not limited to industrial processes. The metabolism of food involves a series of reactions that both forms new compounds and also provides energy for the body's maintenance and growth. Burning gasoline, baking a cake, and photosynthesis involve chemical reactions. In Chapter 5 we learn the basic principles about chemical reactions.

5.1 Introduction to Chemical Reactions

Now that we have learned about compounds and the atoms that compose them, we can better understand the chemical changes first discussed in Section 1.2.

5.1A General Features

- A chemical change—chemical reaction—converts one substance into another.

Chemical reactions involve breaking bonds in the starting materials, called *reactants,* and forming new bonds in the *products.* The combustion of methane (CH_4), the main constituent of natural gas, in the presence of oxygen (O_2) to form carbon dioxide (CO_2) and water (H_2O) is an example of a chemical reaction. The carbon–hydrogen bonds in methane and the oxygen–oxygen bond in elemental oxygen are broken, and new carbon–oxygen and hydrogen–oxygen bonds are formed in the products.

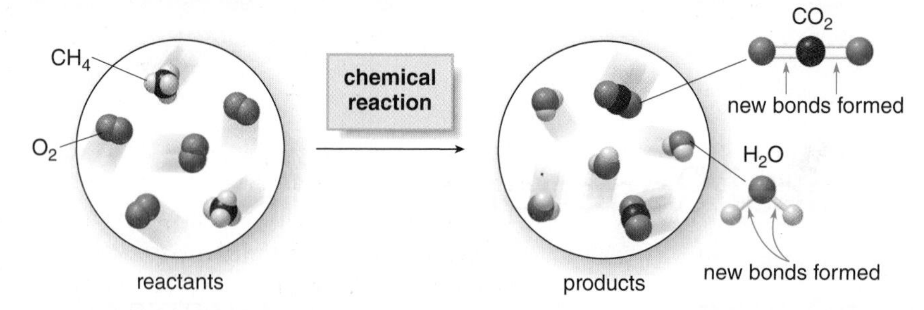

A chemical reaction is thus fundamentally different from a physical change such as melting or boiling discussed in Section 4.6. When water melts, for example, the highly organized water molecules in the solid phase become more disorganized in the liquid phase, but the bonds do *not* change.

SAMPLE PROBLEM 5.1

Identify each process as a chemical reaction or a physical change.

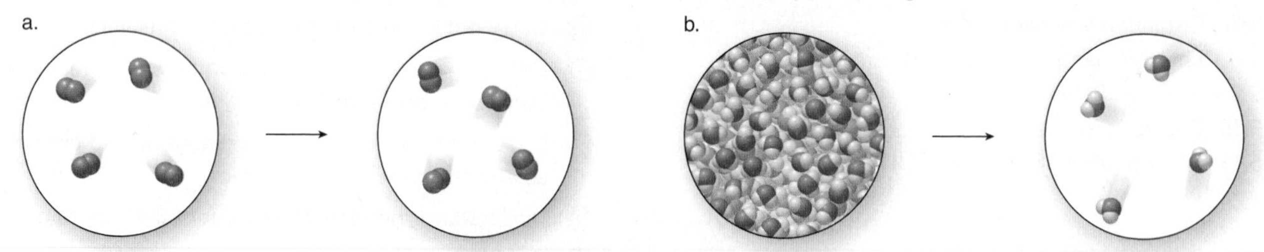

Analysis

A chemical reaction occurs when the bonds in the reactants are broken and new bonds are formed in the products. A physical change occurs when the bonds in the reactants are the same as the bonds in the products.

Solution

Part (a) represents a chemical reaction—the reactants contain two N_2 molecules (with blue spheres joined) and two O_2 molecules (two red spheres joined), while the product contains four NO molecules (a red sphere joined to a blue sphere). Part (b) represents a physical change—boiling—since liquid H_2O molecules are converted to gaseous H_2O molecules and the bonds do not change.

PROBLEM 5.1

Use the molecular art to identify the process as a chemical reaction or a physical change, and explain your choice.

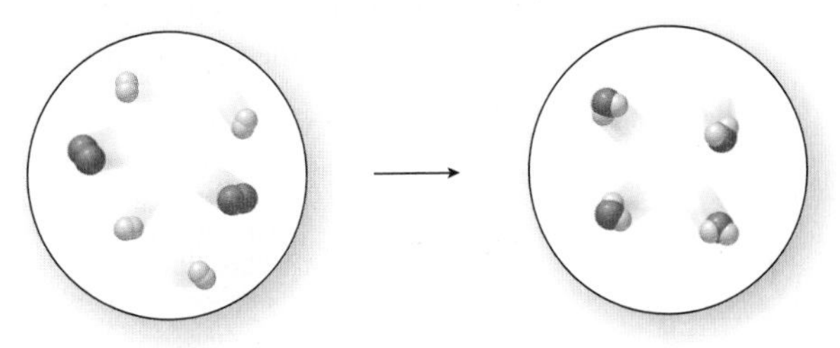

PROBLEM 5.2

Use the molecular art to identify the process as a chemical reaction or a physical change, and explain your choice.

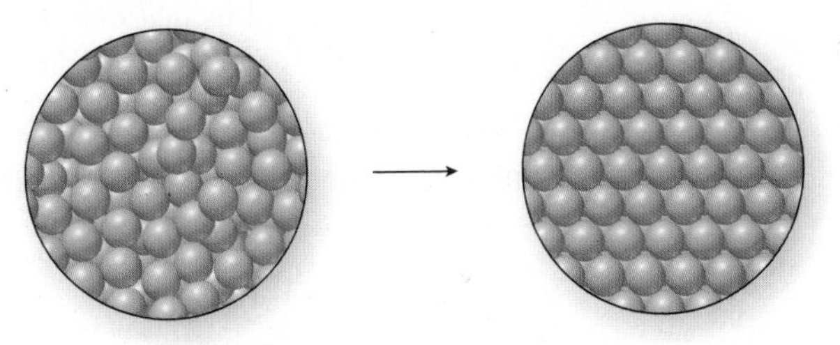

A chemical reaction may be accompanied by a visible change: two colorless reactants can form a colored product; a gas may be given off; two liquid reactants may yield a solid product. Sometimes heat is produced so that a reaction flask feels hot. A reaction having a characteristic visible change occurs when hydrogen peroxide (H_2O_2) is used to clean a bloody wound. An enzyme in the blood called catalase converts the H_2O_2 to water (H_2O) and oxygen (O_2), and bubbles of oxygen appear as a foam, as shown in Figure 5.1.

Figure 5.1 Treating Wounds with Hydrogen Peroxide—A Visible Chemical Reaction

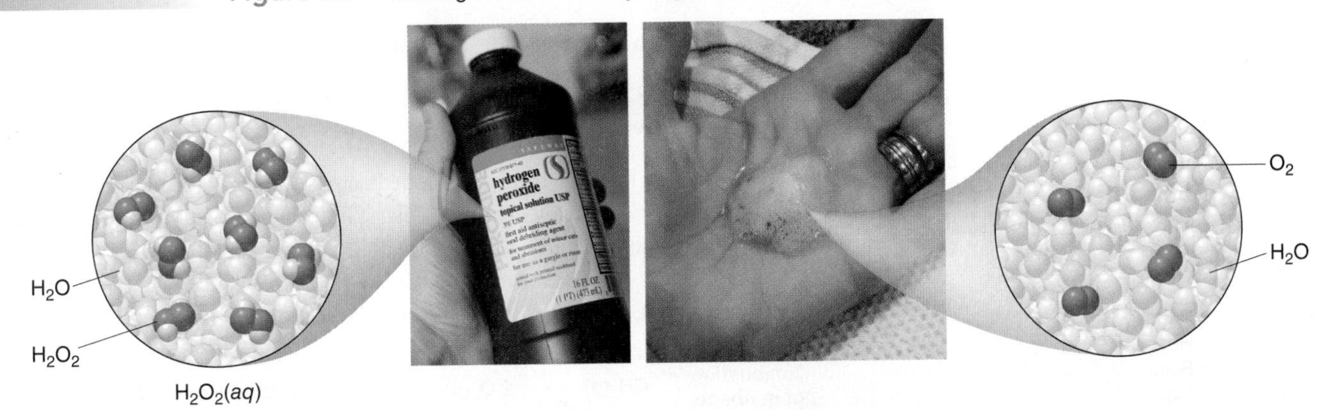

The enzyme catalase in red blood converts hydrogen peroxide (H_2O_2) to water and oxygen gas, which appears as a visible white foam on the bloody surface. Hydrogen peroxide does not foam when it comes in contact with skin because skin cells do not contain the catalase needed for the reaction to occur.

5.1B Writing Chemical Equations

- A *chemical equation* is an expression that uses chemical formulas and other symbols to illustrate what reactants constitute the starting materials in a reaction and what products are formed.

Chemical equations are written with the **reactants on the left** and the **products on the right,** separated by a horizontal arrow—a **reaction arrow**—that points from the reactants to the products. In the combustion of methane, methane (CH_4) and oxygen (O_2) are the reactants on the left side of the arrow, and carbon dioxide (CO_2) and water (H_2O) are the products on the right side.

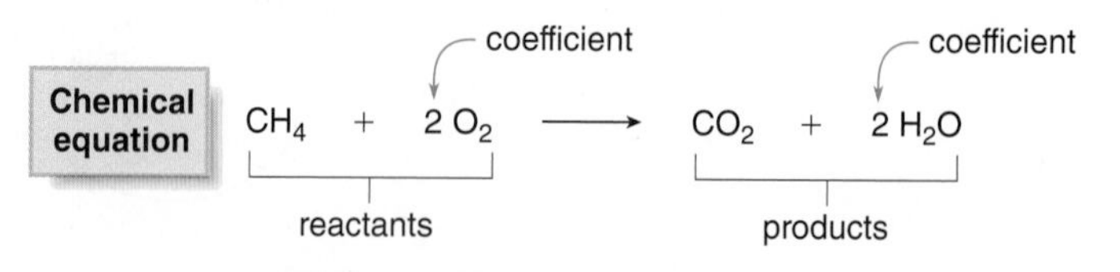

The numbers written in front of any formula are called **coefficients. Coefficients show the number of molecules of a given element or compound that react or are formed.** When no number precedes a formula, the coefficient is assumed to be "1." In the combustion of methane, the coefficients tell us that one molecule of CH_4 reacts with two molecules of O_2 to form one molecule of CO_2 and two molecules of H_2O.

When a formula contains a subscript, **multiply its coefficient by the subscript** to give the total number of atoms of a given type in that formula.

$$2\ O_2 = 4\ \text{O atoms}$$
$$2\ H_2O = 4\ \text{H atoms} + 2\ \text{O atoms}$$

Coefficients are used because all chemical reactions follow a fundamental principle of nature, the **law of conservation of matter,** which states:

- Atoms cannot be created or destroyed in a chemical reaction.

Although bonds are broken and formed in reactions, the number of atoms of each element in the reactants must be the same as the number of atoms of each type in the products. **Coefficients are used to *balance* an equation,** making the number of atoms of each element the same on both sides of the equation.

$$CH_4 + 2\ O_2 \longrightarrow CO_2 + 2\ H_2O$$

Atoms in the reactants:
- 1 C atom
- 4 H atoms
- 4 O atoms

Atoms in the products:
- 1 C atom
- 4 H atoms
- 4 O atoms

Two other features are worthy of note. If heat is needed for a reaction to occur, the Greek letter delta (Δ) may be written over the arrow. The physical states of the reactants and products are sometimes indicated next to each formula—solid (*s*), liquid (*l*), or gas (*g*). If an aqueous solution is used—that is, if a reactant is dissolved in water—the symbol (*aq*) is used next to the reactant. When these features are added, the equation for the combustion of methane becomes:

Combustion of methane

$$CH_4(g) + 2\ O_2(g) \xrightarrow{\Delta} CO_2(g) + 2\ H_2O(g)$$

The symbols used for chemical equations are summarized in Table 5.1.

Table 5.1 Symbols Used in Chemical Equations

Symbol	Meaning
⟶	Reaction arrow
Δ	Heat
(*s*)	Solid
(*l*)	Liquid
(*g*)	Gas
(*aq*)	Aqueous solution

SAMPLE PROBLEM 5.2

Label the reactants and products, and indicate how many atoms of each type of element are present on each side of the equation.

$$C_2H_6O(l) + 3\ O_2(g) \longrightarrow 2\ CO_2(g) + 3\ H_2O(g)$$

Analysis

Reactants are on the left side of the arrow and products are on the right side in a chemical equation. When a formula contains a subscript, multiply its coefficient by the subscript to give the total number of atoms of a given type in the formula.

Solution

In this equation, the reactants are C_2H_6O and O_2, while the products are CO_2 and H_2O. If no coefficient is written, it is assumed to be "1." To determine the number of each type of atom when a formula has both a coefficient and a subscript, multiply the coefficient by the subscript.

$1\ C_2H_6O$	= 2 C's + 6 H's + 1 O	
$3\ O_2$	= 6 O's	Multiply the coefficient 3 by the subscript 2.
$2\ CO_2$	= 2 C's + 4 O's	Multiply the coefficient 2 by each subscript; 2 × 1 C = 2 C's; 2 × 2 O's = 4 O's.
$3\ H_2O$	= 6 H's + 3 O's	Multiply the coefficient 3 by each subscript; 3 × 2 H's = 6 H's; 3 × 1 O = 3 O's.

Add up the atoms on each side to determine the total number for each type of element.

$$C_2H_6O(l) \quad + \quad 3\ O_2(g) \quad \longrightarrow \quad 2\ CO_2(g) \quad + \quad 3\ H_2O(g)$$

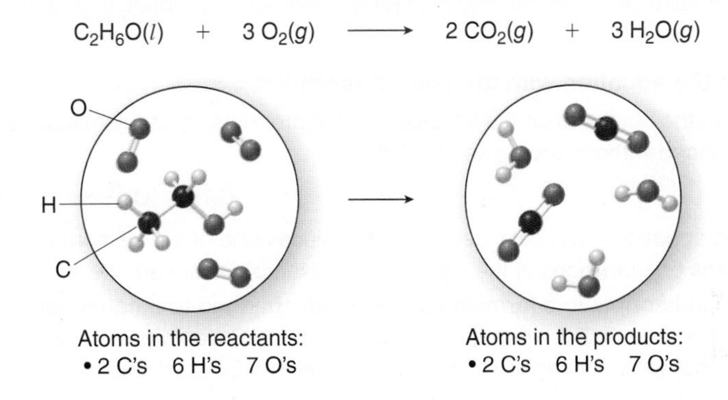

PROBLEM 5.3

Label the reactants and products, and indicate how many atoms of each type of element are present on each side of the following equations.

a. $2\ H_2O_2(aq) \longrightarrow 2\ H_2O(l) + O_2(g)$

b. $2\ C_8H_{18} + 25\ O_2 \longrightarrow 16\ CO_2 + 18\ H_2O$

PROBLEM 5.4

Use the molecular art to write an equation for the given reaction. (The common element colors are shown on the inside back cover.)

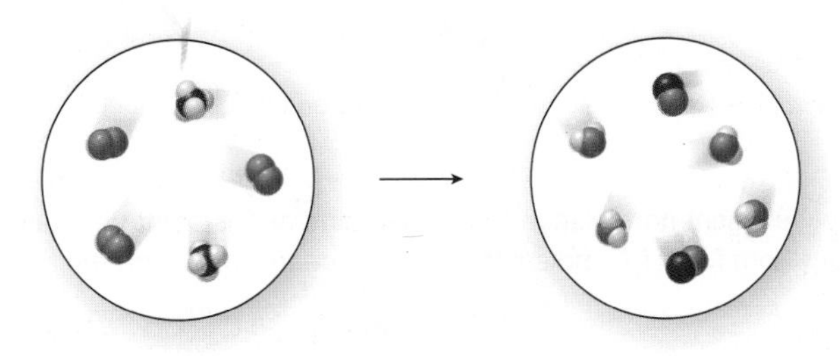

ENVIRONMENTAL NOTE

The reaction of propane with oxygen forms carbon dioxide, water, and a great deal of energy that can be used for cooking, heating homes, drying clothes, and powering generators and vehicles. The combustion of propane and other fossil fuels adds a tremendous amount of CO_2 to the atmosphere each year, with clear environmental consequences.

PROBLEM 5.5

Write a chemical equation from the following description of a reaction: One molecule of gaseous methane (CH_4) is heated with four molecules of gaseous chlorine (Cl_2), forming one molecule of liquid carbon tetrachloride (CCl_4) and four molecules of gaseous hydrogen chloride (HCl).

5.2 Balancing Chemical Equations

Sometimes a chemical equation is balanced as written and the coefficient of each formula is "1." For example, when charcoal is burned, the carbon (C) it contains reacts with oxygen (O_2) to form carbon dioxide (CO_2). One carbon atom reacts with one oxygen molecule to form one molecule of carbon dioxide.

$$C(s) + O_2(g) \longrightarrow CO_2(g)$$

More often, however, an equation must be balanced by adding coefficients in front of some formulas so that the **number of atoms of each element is equal on both sides of the equation.**

How To Balance a Chemical Equation

Example Write a balanced chemical equation for the reaction of propane (C_3H_8) with oxygen (O_2) to form carbon dioxide (CO_2) and water (H_2O).

Step [1] Write the equation with the correct formulas.

- Write the reactants on the left side and the products on the right side of the reaction arrow, and check if the equation is balanced without adding any coefficients.

$$C_3H_8 + O_2 \longrightarrow CO_2 + H_2O$$

- This equation is not balanced as written since none of the elements—carbon, hydrogen, and oxygen—has the same number of atoms on both sides of the equation. For example, there are 3 C's on the left and only 1 C on the right.
- **The subscripts in a formula can *never* be changed to balance an equation.** Changing a subscript changes the identity of the compound. For example, changing CO_2 to CO would balance oxygen (there would be 2 O's on both sides of the equation), but that would change CO_2 (carbon dioxide) into CO (carbon monoxide).

Step [2] Balance the equation with coefficients one element at a time.

- Begin with the most complex formula, and **start with an element that appears in only one formula on both sides of the equation.** Save the element found in multiple reactants or products for last. In this example, begin with either the C's or H's in C_3H_8. Since there are 3 C's on the left, place the coefficient 3 before CO_2 on the right.

$$C_3H_8 + O_2 \longrightarrow 3\,CO_2 + H_2O$$

3 C's on the left — Place a 3 to balance C's.

- To balance the 8 H's in C_3H_8, place the coefficient 4 before H_2O on the right.

$$C_3H_8 + O_2 \longrightarrow 3\,CO_2 + 4\,H_2O$$

8 H's on the left — Place a 4 to balance H's. (4×2 H's in H_2O = 8 H's)

- The only element not balanced is oxygen, and at this point there are a total of 10 O's on the right—six from three CO_2 molecules and four from four H_2O molecules. To balance the 10 O's on the right, place the coefficient 5 before O_2 on the left.

$$C_3H_8 + 5\,O_2 \longrightarrow 3\,CO_2 + 4\,H_2O$$

Place a 5 to balance O's. — 10 O's on the right

How To, continued . . .

Step [3] Check to make sure that the smallest set of whole numbers is used.

$$C_3H_8 + 5\ O_2 \longrightarrow 3\ CO_2 + 4\ H_2O$$

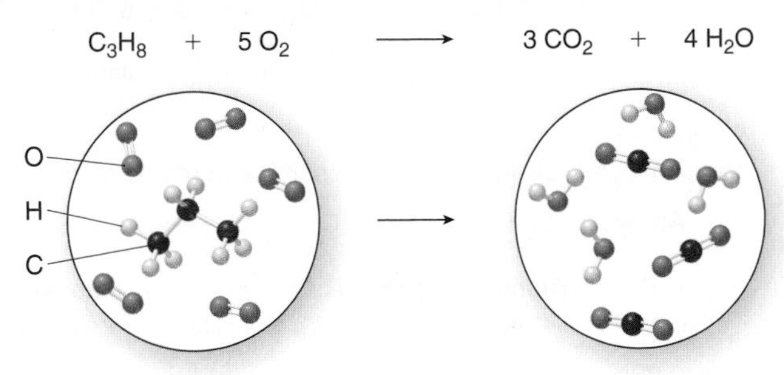

Atoms in the reactants:
- 3 C's 8 H's 10 O's (5 × 2 O's)

Atoms in the products:
- 3 C's 8 H's 10 O's [(3 × 2 O's) + (4 × 1 O)]

- This equation is balanced because the same number of C's, O's, and H's is present on both sides of the equation.
- Sometimes an equation is balanced but the lowest set of whole numbers is not used as coefficients. Say, for example, that balancing yielded the following equation:

$$2\ C_3H_8 + 10\ O_2 \longrightarrow 6\ CO_2 + 8\ H_2O$$

- This equation has the same number of C's, O's, and H's on both sides, but *each coefficient must be divided by two* to give the lowest set of whole numbers for the balanced equation, as drawn in the first equation in step [3].

Sample Problems 5.3–5.5 illustrate additional examples of balancing chemical equations. Sample Problem 5.4 shows how to balance an equation when there is an odd–even relationship of atoms in the reactants and products. Sample Problem 5.5 illustrates how to balance an equation that contains polyatomic ions.

Bagels, pasta, bread, and rice are high in starch, which is hydrolyzed to the simple carbohydrate glucose after ingestion. The metabolism of glucose forms CO_2 and H_2O and provides energy for bodily functions.

SAMPLE PROBLEM 5.3

Write a balanced equation for the reaction of glucose ($C_6H_{12}O_6$) with oxygen (O_2) to form carbon dioxide (CO_2) and water (H_2O).

Analysis

Balance an equation with coefficients, one element at a time, beginning with the most complex formula and starting with an element that appears in only one formula on both sides of the equation. Continue placing coefficients until the **number of atoms of each element is equal on both sides of the equation.**

Solution

[1] Write the equation with correct formulas.

$$\underset{\text{glucose}}{C_6H_{12}O_6} + O_2 \longrightarrow CO_2 + H_2O$$

- None of the elements is balanced in this equation. As an example, there are 6 C's on the left side, but only 1 C on the right side.

[2] Balance the equation with coefficients one element at a time.

- Begin with glucose, since its formula is most complex. Balance the 6 C's of glucose by placing the coefficient 6 before CO_2. Balance the 12 H's of glucose by placing the coefficient 6 before H_2O.

Place a 6 to balance C's.

$$C_6H_{12}O_6 + O_2 \longrightarrow 6\ CO_2 + 6\ H_2O$$

Place a 6 to balance H's.

- The right side of the equation now has 18 O's. Since glucose already has 6 O's on the left side, 12 additional O's are needed on the left side. The equation will be balanced if the coefficient 6 is placed before O_2.

$$C_6H_{12}O_6 + 6\,O_2 \longrightarrow 6\,CO_2 + 6\,H_2O$$

Place a 6 to balance O's.

[3] Check.

- The equation is balanced since the number of atoms of each element is the same on both sides.

Answer: $C_6H_{12}O_6 + 6\,O_2 \longrightarrow 6\,CO_2 + 6\,H_2O$

Atoms in the reactants:	Atoms in the products:
• 6 C's	• 6 C's (6 × 1 C)
• 12 H's	• 12 H's (6 × 2 H's)
• 18 O's (1 × 6 O's) + (6 × 2 O's)	• 18 O's (6 × 2 O's) + (6 × 1 O)

PROBLEM 5.6

Write a balanced equation for each reaction.

a. $H_2 + O_2 \longrightarrow H_2O$

b. $NO + O_2 \longrightarrow NO_2$

c. $Fe + O_2 \longrightarrow Fe_2O_3$

d. $CH_4 + Cl_2 \longrightarrow CH_2Cl_2 + HCl$

PROBLEM 5.7

Write a balanced equation for the following reaction, shown with molecular art.

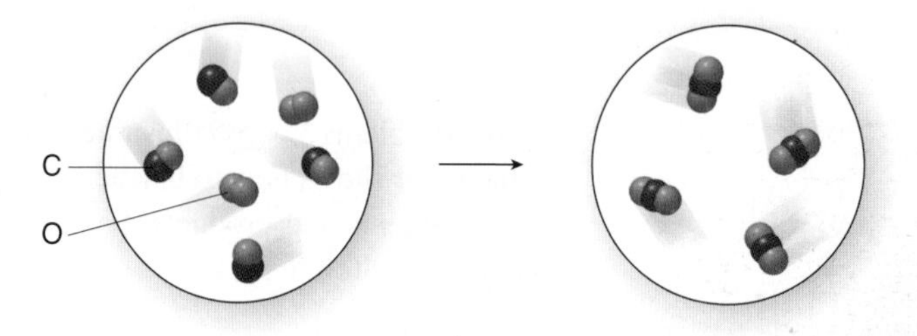

SAMPLE PROBLEM 5.4

The airbag in an automobile inflates when ionic sodium azide (NaN_3), which is composed of Na^+ cations and the polyatomic anion, N_3^- (azide), rapidly decomposes to sodium (Na) and gaseous N_2 (Figure 5.2). Write a balanced equation for this reaction.

Figure 5.2 Chemistry of an Automobile Airbag

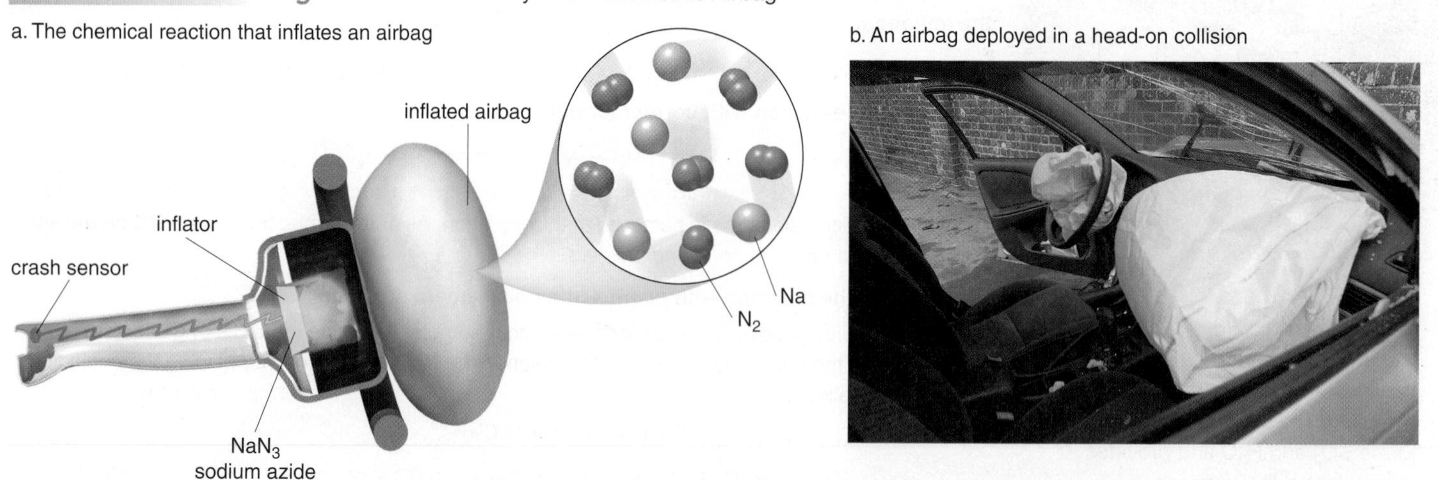

A severe car crash triggers an airbag to deploy when an electric sensor causes sodium azide (NaN_3) to ignite, converting it to sodium (Na) and nitrogen gas (N_2). The nitrogen gas causes the bag to inflate fully in 40 milliseconds, helping to protect passengers from serious injury. The sodium atoms formed in this first reaction are hazardous and subsequently converted to a safe sodium salt. It took 30 years to develop a reliable airbag system for automobiles.

Analysis

Balance an equation with coefficients, one element at a time, beginning with the most complex formula and starting with an element that appears in only one formula on both sides of the equation. Continue placing coefficients until the **number of atoms of each element is equal on both sides of the equation.**

Solution

[1] Write the equation with correct formulas.

$$NaN_3 \longrightarrow Na + N_2$$

sodium azide

- The N atoms are not balanced since there are 3 N's on the left side and only 2 N's on the right.

[2] Balance the equation with coefficients.

- To balance the odd number of N atoms (3 N's) on the reactant side with the even number of N atoms on the product side (2 N's) requires the placement of two coefficients. Place the coefficient 2 on the left side (for a total of 6 N's in the reactants). Then place the coefficient 3 before N_2 (for a total of 6 N's in the product). Placing two coefficients is necessary whenever there is an odd–even relationship of atoms in the reactants and products (for any odd number other than one).

Place a 2 to give 6 N's on the left.

$$2\ NaN_3 \longrightarrow Na + 3\ N_2$$

Place a 3 to give 6 N's on the right.

- Balance the 2 Na atoms on the left side by placing a 2 before the Na atoms on the right.

$$2\ NaN_3 \longrightarrow 2\ Na + 3\ N_2$$

Place a 2 to balance Na's.

[3] Check.

- The equation is balanced since the number of atoms of each element is the same on both sides.

$$2\ NaN_3 \longrightarrow 2\ Na + 3\ N_2$$

Atoms in the reactants:	**Atoms in the products:**
• 2 Na's	• 2 Na's
• 6 N's (2 × 3 N's)	• 6 N's (3 × 2 N's)

PROBLEM 5.8

Write a balanced equation for the reaction of ethane (C_2H_6) with O_2 to form CO_2 and H_2O.

PROBLEM 5.9

The Haber process is an important industrial reaction that converts N_2 and H_2 to ammonia (NH_3), an agricultural fertilizer and starting material for the synthesis of nitrate fertilizers. Write a balanced equation for the Haber process.

SAMPLE PROBLEM 5.5

Balance the following equation.

$$Ca_3(PO_4)_2 + H_2SO_4 \longrightarrow CaSO_4 + H_3PO_4$$

calcium phosphate, sulfuric acid, calcium sulfate, phosphoric acid

Analysis

Balance an equation with coefficients, one element at a time, beginning with the most complex formula and starting with an element that appears in only one formula on both sides of the equation. Continue placing coefficients until the **number of atoms of each element is equal on both sides of the equation.**

Solution

[1] **Write the equation with correct formulas.**

- The correct formula for each compound is given in the problem statement. When the reactants and products contain polyatomic ions, PO_4^{3-} and SO_4^{2-} in this case, **balance each ion as a *unit*,** rather than balancing the individual atoms. Thus, phosphate is not balanced in the equation as written, because the left side has two PO_4^{3-} anions while the right side has only one.

[2] **Balance the equation with coefficients.**

- Begin with $Ca_3(PO_4)_2$. Balance the 3 Ca's by placing the coefficient 3 before $CaSO_4$. Balance the 2 PO_4^{3-} anions by placing the coefficient 2 before H_3PO_4.

Place a 3 to balance Ca's.

$$Ca_3(PO_4)_2 + H_2SO_4 \longrightarrow 3\ CaSO_4 + 2\ H_3PO_4$$

Place a 2 to balance PO_4^{3-}.

- Two components are still not balanced—H atoms and sulfate anions (SO_4^{2-}). Both can be balanced by placing the coefficient 3 before H_2SO_4 on the left.

$$Ca_3(PO_4)_2 + \underbrace{3\ H_2SO_4}_{6\ H's} \longrightarrow 3\ CaSO_4 + \underbrace{2\ H_3PO_4}_{6\ H's}$$

3 SO_4^{2-} in both

Place a 3 to balance H and SO_4^{2-}.

[3] **Check.**

- The equation is balanced since the number of atoms and polyatomic anions is the same on both sides.

Answer: $$Ca_3(PO_4)_2 + 3\ H_2SO_4 \longrightarrow 3\ CaSO_4 + 2\ H_3PO_4$$

Atoms or ions in the reactants:		**Atoms or ions in the products:**	
• 3 Ca's	• 6 H's	• 3 Ca's	• 6 H's
• 2 PO_4^{3-}	• 3 SO_4^{2-}	• 2 PO_4^{3-}	• 3 SO_4^{2-}

PROBLEM 5.10

Balance each chemical equation. Balance each polyatomic ion as a unit, rather than balancing the individual atoms.

a. $Al + H_2SO_4 \longrightarrow Al_2(SO_4)_3 + H_2$

b. $Na_2SO_3 + H_3PO_4 \longrightarrow H_2SO_3 + Na_3PO_4$

5.3 The Mole and Avogadro's Number

Although the chemical equations in Section 5.2 were discussed in terms of individual atoms and molecules, atoms are exceedingly small. It is more convenient to talk about larger quantities of atoms, and for this reason, scientists use the **mole.** A mole defines a quantity, much like a dozen items means 12, and a case of soda means 24 cans. The only difference is that a mole is much larger.

- **A *mole* is a quantity that contains 6.02×10^{23} items—usually atoms, molecules, or ions.**

The definition of a mole is based on the number of atoms contained in exactly 12 g of the carbon-12 isotope. This number is called **Avogadro's number,** after the Italian scientist Amadeo Avogadro, who first proposed the concept of a mole in the nineteenth century. One mole, abbreviated as **mol,** always contains an Avogadro's number of particles.

$$\text{1 mole of C atoms} = 6.02 \times 10^{23} \text{ C atoms}$$
$$\text{1 mole of } H_2O \text{ molecules} = 6.02 \times 10^{23}\ H_2O \text{ molecules}$$
$$\text{1 mole of vitamin C molecules} = 6.02 \times 10^{23} \text{ vitamin C molecules}$$

PROBLEM 5.11

How many items are contained in one mole of (a) baseballs; (b) bicycles; (c) Cheerios; (d) CH_4 molecules?

Each sample contains one mole of the substance—water (H_2O molecules), salt (NaCl, one mole of Na^+ and one mole of Cl^-), and aspirin ($C_9H_8O_4$ molecules). Pictured is a mole of aspirin *molecules,* not a mole of aspirin *tablets,* which is a quantity too large to easily represent. If a mole of aspirin tablets were arranged next to one another to cover a football field and then stacked on top of each other, they would occupy a volume 100 yards long, 53.3 yards wide, and over 20,000,000,000 miles high!

For a number written in scientific notation as $y \times 10^x$, y is the coefficient and x is the exponent in the power of 10 (Section 1.6).

We can use Avogadro's number as a conversion factor to relate the number of moles of a substance to the number of atoms or molecules it contains.

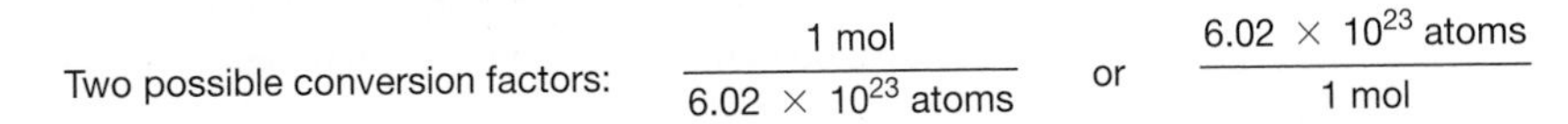

These conversion factors allow us to determine how many atoms or molecules are contained in a given number of moles. To carry out calculations that contain numbers written in scientific notation, we must first learn how to multiply and divide numbers written in this form.

- **To multiply two numbers in scientific notation, multiply the coefficients together and add the exponents in the powers of 10.**

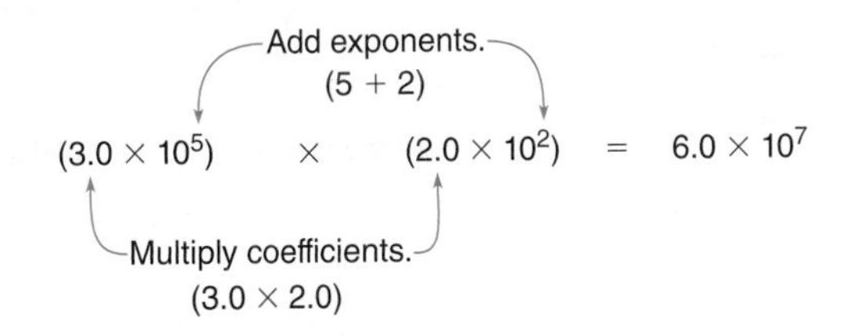

Add exponents.
(5 + 2)
(3.0 × 10^5) × (2.0 × 10^2) = 6.0 × 10^7
Multiply coefficients.
(3.0 × 2.0)

- **To divide two numbers in scientific notation, divide the coefficients and subtract the exponents in the powers of 10.**

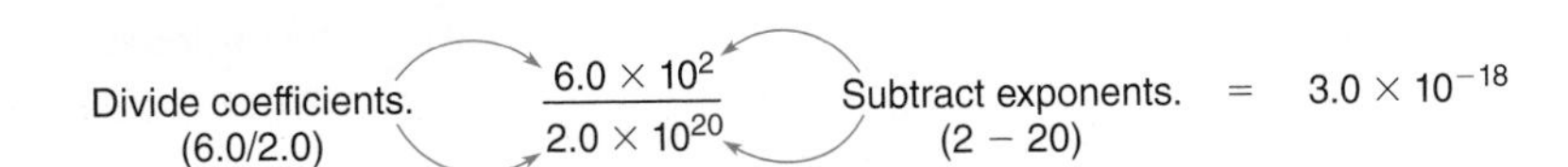

Sample Problems 5.6 and 5.7 illustrate how to interconvert moles and molecules.

SAMPLE PROBLEM 5.6

How many molecules are contained in 5.0 moles of carbon dioxide (CO_2)?

Analysis and Solution

[1] Identify the original quantity and the desired quantity.

5.0 mol of CO_2	? number of molecules of CO_2
original quantity	desired quantity

[2] Write out the conversion factors.

- Choose the conversion factor that places the unwanted unit, mol, in the denominator, so that the units cancel.

$$\frac{1 \text{ mol}}{6.02 \times 10^{23} \text{ molecules}} \quad \text{or} \quad \boxed{\frac{6.02 \times 10^{23} \text{ molecules}}{1 \text{ mol}}}$$

Choose this conversion factor to cancel mol.

[3] Set up and solve the problem.

- Multiply the original quantity by the conversion factor to obtain the desired quantity.

$$5.0 \text{ mol} \times \frac{6.02 \times 10^{23} \text{ molecules}}{1 \text{ mol}} = 30. \times 10^{23} \text{ molecules}$$

Moles cancel.

Convert to a number between 1 and 10.

$$= 3.0 \times 10^{24} \text{ molecules of } CO_2$$

Answer

- Multiplication first gives an answer that is not written in scientific notation since the coefficient (30.) is greater than 10. Moving the decimal point one place to the *left* and *increasing* the exponent by one gives the answer written in the proper form.

PROBLEM 5.12

How many carbon atoms are contained in each of the following number of moles: (a) 2.00 mol; (b) 6.00 mol; (c) 0.500 mol; (d) 25.0 mol?

PROBLEM 5.13

How many molecules are contained in each of the following number of moles?

a. 2.5 mol of penicillin molecules
b. 0.25 mol of NH_3 molecules
c. 0.40 mol of sugar molecules
d. 55.3 mol of acetaminophen molecules

SAMPLE PROBLEM 5.7

How many moles of aspirin contain 8.62×10^{25} molecules?

Analysis and Solution

[1] Identify the original quantity and the desired quantity.

8.62×10^{25} molecules of aspirin	? mole of aspirin
original quantity	desired quantity

[2] Write out the conversion factors.

- Choose the conversion factor that places the unwanted unit, number of molecules, in the denominator so that the units cancel.

$$\frac{6.02 \times 10^{23} \text{ molecules}}{1 \text{ mol}} \quad \text{or} \quad \frac{1 \text{ mol}}{6.02 \times 10^{23} \text{ molecules}}$$

Choose this conversion factor to cancel molecules.

[3] Set up and solve the problem.

- Multiply the original quantity by the conversion factor to obtain the desired quantity.
- To divide numbers using scientific notation, divide the coefficients (8.62/6.02) and subtract the exponents (25 – 23).

$$8.62 \times 10^{25} \text{ molecules} \times \frac{1 \text{ mol}}{6.02 \times 10^{23} \text{ molecules}} = 1.43 \times 10^{2} \text{ mol}$$

Molecules cancel.

$$= 143 \text{ mol of aspirin}$$

Answer

PROBLEM 5.14

How many moles of water contain each of the following number of molecules?

a. 6.02×10^{25} molecules
b. 3.01×10^{22} molecules
c. 9.0×10^{24} molecules

5.4 Mass to Mole Conversions

In Section 2.3, we learned that the ***atomic weight* is the average mass of an element,** reported in atomic mass units (amu). Thus, carbon has an atomic weight of 12.01 amu. We use atomic weights to calculate the mass of a compound.

- **The *formula weight* is the sum of the atomic weights of all the atoms in a compound, reported in atomic mass units (amu).**

The term "formula weight" is used for both ionic and covalent compounds. Often the term **"molecular weight"** is used in place of formula weight for covalent compounds, since they are composed of molecules, not ions. The formula weight of ionic sodium chloride (NaCl) is 58.44 amu, which is determined by adding up the atomic weights of Na (22.99 amu) and Cl (35.45 amu).

Formula weight of NaCl:

Atomic weight of 1 Na	=	22.99 amu
Atomic weight of 1 Cl	=	35.45 amu
Formula weight of NaCl	=	58.44 amu

PROBLEM 5.15

Calculate the formula weight of each ionic compound.

a. $CaCO_3$, a common calcium supplement
b. KI, the essential nutrient added to NaCl to make iodized salt

5.4A Molar Mass

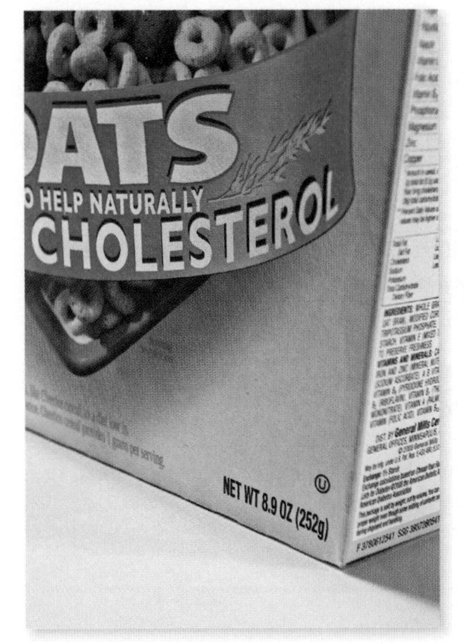

When a consumer product contains a great many lightweight small objects—for example, Cheerios—it is typically sold by weight, not by the number of objects. We buy Cheerios in an 8.9-oz box, not a box that contains 2,554 Cheerios.

When reactions are carried out in the laboratory, single atoms and molecules are much too small to measure out. Instead, substances are weighed on a balance and amounts are typically reported in grams, not atomic mass units. To determine how many atoms or molecules are contained in a given mass, we use its **molar mass.**

- **The *molar mass* is the mass of one mole of any substance, reported in grams per mole.**

The value of the molar mass of an element in the periodic table (in grams per mole) is the same as the value of its atomic weight (in amu). Thus, the molar mass of carbon is 12.01 g/mol, since its atomic weight is 12.01 amu; that is, one mole of carbon atoms weighs 12.01 g.

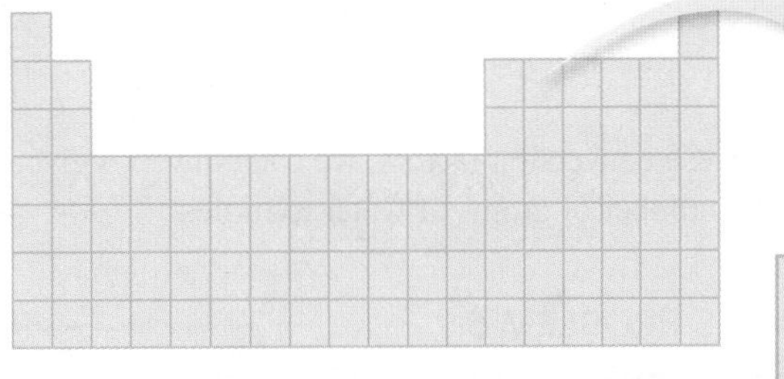
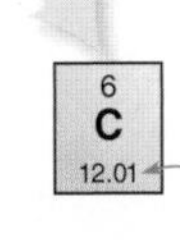

- Carbon's **atomic weight** is 12.01 amu.
- Carbon's **molar mass** is 12.01 g/mol.
- **One mole of carbon atoms weighs 12.01 g.**

- **The value of the molar mass of a compound in grams equals the value of its formula weight in amu.**

Since the formula weight of NaCl is 58.44 amu, its molar mass is 58.44 g/mol. One mole of NaCl weighs 58.44 g. We use a compound's formula weight to calculate its molar mass, as shown in Sample Problem 5.8.

SAMPLE PROBLEM 5.8

What is the molar mass of nicotine ($C_{10}H_{14}N_2$), the toxic and addictive stimulant in tobacco?

Analysis

Determine the number of atoms of each element from the subscripts in the chemical formula, multiply the number of atoms of each element by the atomic weight, and add up the results.

Solution

10 C atoms × 12.01 amu	=	120.1 amu
14 H atoms × 1.008 amu	=	14.11 amu
2 N atoms × 14.01 amu	=	28.02 amu
Formula weight of nicotine	=	162.23 amu rounded to 162.2 amu

Answer: Since the formula weight of nicotine is 162.2 amu, the molar mass of nicotine is 162.2 g/mol.

PROBLEM 5.16

What is the molar mass of each compound?

a. Li_2CO_3 (lithium carbonate), a drug used to treat bipolar disorder
b. C_2H_5Cl (ethyl chloride), a local anesthetic
c. $C_{13}H_{21}NO_3$ (albuterol), a drug used to treat asthma

PROBLEM 5.17

The unmistakable odor of a freshly cut cucumber is due to cucumber aldehyde. (a) What is the chemical formula of cucumber aldehyde? (b) Calculate its molar mass.

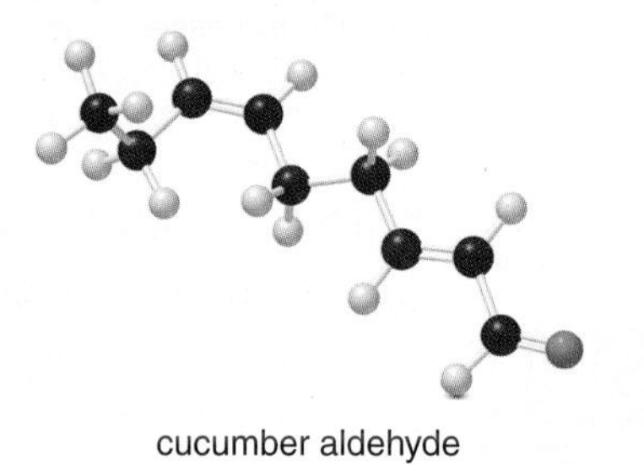
cucumber aldehyde

Cucumber aldehyde is a powerful odorant that can be detected in very low concentrations when a cucumber is cut.

5.4B Relating Grams to Moles

The molar mass is a very useful quantity because it relates the number of *moles* to the number of *grams* of a substance. In this way, the molar mass can be used as a conversion factor. For example, since the molar mass of H_2O is 18.02 g/mol, two conversion factors can be written.

$$\frac{18.02 \text{ g } H_2O}{1 \text{ mol}} \quad \text{or} \quad \frac{1 \text{ mol}}{18.02 \text{ g } H_2O}$$

Using these conversion factors, we can convert a given number of moles of water to grams, or a specific number of grams of water to moles.

SAMPLE PROBLEM 5.9

Converting moles to mass: What is the mass of 0.25 moles of water?

Analysis and Solution

[1] **Identify the original quantity and the desired quantity.**

0.25 mol of H_2O	? g of H_2O
original quantity	desired quantity

[2] **Write out the conversion factors.**

- Choose the conversion factor that places the unwanted unit, moles, in the denominator, so that the units cancel.

$$\frac{1\text{ mol}}{18.02\text{ g }H_2O} \quad \text{or} \quad \boxed{\frac{18.02\text{ g }H_2O}{1\text{ mol}}}$$

Choose this conversion factor to cancel mol.

[3] **Set up and solve the problem.**

- Multiply the original quantity by the conversion factor to obtain the desired quantity.

$$0.25\text{ mol} \times \frac{18.02\text{ g }H_2O}{1\text{ mol}} = 4.5\text{ g of }H_2O$$

Moles cancel.

Answer

PROBLEM 5.18

Calculate the number of grams contained in each of the following number of moles.

a. 0.500 mol of NaCl

b. 2.00 mol of KI

c. 3.60 mol of C_2H_4 (ethylene)

d. 0.820 mol of CH_4O (methanol)

SAMPLE PROBLEM 5.10

Converting mass to moles: How many moles are present in 100. g of aspirin ($C_9H_8O_4$, molar mass 180.2 g/mol)?

Analysis and Solution

[1] **Identify the original quantity and the desired quantity.**

100. g of aspirin	? mol of aspirin
original quantity	desired quantity

[2] **Write out the conversion factors.**

- Choose the conversion factor that places the unwanted unit, grams, in the denominator, so that the units cancel.

$$\frac{180.2\text{ g aspirin}}{1\text{ mol}} \quad \text{or} \quad \boxed{\frac{1\text{ mol}}{180.2\text{ g aspirin}}}$$

Choose this conversion factor to cancel g.

[3] **Set up and solve the problem.**

- Multiply the original quantity by the conversion factor to obtain the desired quantity.

$$100.\text{ g} \times \frac{1\text{ mol}}{180.2\text{ g aspirin}} = 0.555\text{ mol of aspirin}$$

Grams cancel.

Answer

PROBLEM 5.19

How many moles are contained in each of the following?

a. 100. g of NaCl

b. 25.5 g of CH_4

c. 0.250 g of aspirin ($C_9H_8O_4$)

d. 25.0 g of H_2O

5.5 Mole Calculations in Chemical Equations

Having learned about moles and molar mass, we can now return to balanced chemical equations. As we learned in Section 5.2, the coefficients in a balanced chemical equation tell us the number of *molecules* of each compound that react or are formed in a given reaction.

ENVIRONMENTAL NOTE

NO, nitrogen monoxide, is formed from N_2 and O_2 at very high temperature in automobile engines and coal-burning furnaces. NO is a reactive air pollutant that goes on to form other air pollutants, such as ozone (O_3) and nitric acid (HNO_3). HNO_3 is one component of acid rain that can devastate forests and acidify streams, making them unfit for fish and other wildlife.

- **A balanced chemical equation also tells us the number of *moles* of each reactant that combine and the number of *moles* of each product formed.**

$$1\ N_2(g) \quad + \quad 1\ O_2(g) \xrightarrow{\Delta} 2\ NO(g)$$

$1\ N_2(g)$	$1\ O_2(g)$	$2\ NO(g)$
one molecule of N_2	one molecule of O_2	two molecules of NO
one mole of N_2	one mole of O_2	two moles of NO

[The coefficient "1" has been written for emphasis.]

For example, the balanced chemical equation for the high temperature reaction of N_2 and O_2 to form nitrogen monoxide, NO, shows that one *molecule* of N_2 combines with one *molecule* of O_2 to form two *molecules* of NO. It also shows that one *mole* of N_2 combines with one *mole* of O_2 to form two *moles* of NO.

Coefficients are used to form mole ratios, which can serve as conversion factors. These ratios tell us the relative number of moles of reactants that combine in a reaction, as well as the relative number of moles of product formed from a given reactant, as shown in Sample Problem 5.11.

Mole ratios:

$$\frac{1\ \text{mol}\ N_2}{1\ \text{mol}\ O_2} \qquad \frac{1\ \text{mol}\ N_2}{2\ \text{mol NO}} \qquad \frac{1\ \text{mol}\ O_2}{2\ \text{mol NO}}$$

two reactants	reactant–product	reactant–product
N_2 and O_2	N_2 and NO	O_2 and NO

- **Use the mole ratio from the coefficients in the balanced equation to convert the number of moles of one compound (A) into the number of moles of another compound (B).**

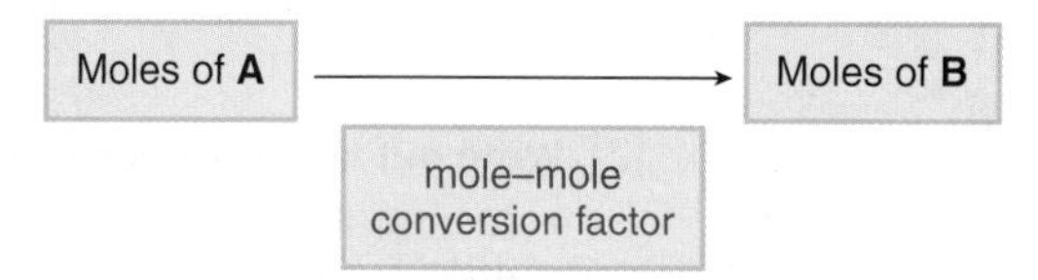

HEALTH NOTE

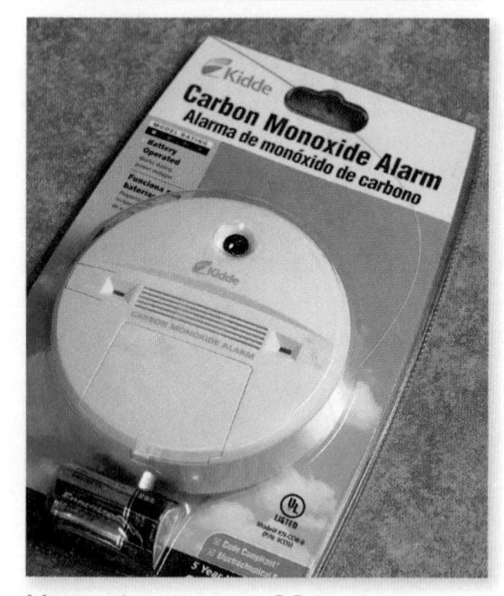

Meters that measure CO levels in homes are sold commercially. CO, a colorless, odorless gas, is a minor product formed whenever fossil fuels and wood are burned. In poorly ventilated rooms, such as those found in modern, well-insulated homes, CO levels can reach unhealthy levels.

SAMPLE PROBLEM 5.11

Carbon monoxide (CO) is a poisonous gas that combines with hemoglobin in the blood, thus reducing the amount of oxygen that can be delivered to tissues. Under certain conditions, CO is formed when ethane (C_2H_6) in natural gas is burned in the presence of oxygen. Using the balanced equation, how many moles of CO are produced from 3.5 mol of C_2H_6?

$$2\ C_2H_6(g) \quad + \quad 5\ O_2(g) \xrightarrow{\Delta} 4\ CO(g) \quad + \quad 6\ H_2O(g)$$

Analysis and Solution

[1] Identify the original quantity and the desired quantity.

3.5 mol of C_2H_6	? mol of CO
original quantity	desired quantity

[2] Write out the conversion factors.

- Use the coefficients in the balanced equation to write mole–mole conversion factors for the two compounds, C_2H_6 and CO. Choose the conversion factor that places the unwanted unit, moles of C_2H_6, in the denominator, so that the units cancel.

$$\frac{2\ \text{mol}\ C_2H_6}{4\ \text{mol CO}} \quad \text{or} \quad \boxed{\frac{4\ \text{mol CO}}{2\ \text{mol}\ C_2H_6}}$$

Choose this conversion factor to cancel mol C_2H_6.

[3] Set up and solve the problem.

- Multiply the original quantity by the conversion factor to obtain the desired quantity.

$$3.5 \text{ mol } C_2H_6 \times \frac{4 \text{ mol CO}}{2 \text{ mol } C_2H_6} = 7.0 \text{ mol CO}$$

Moles C_2H_6 cancel.

Answer

PROBLEM 5.20

Use the balanced equation for the reaction of N_2 and O_2 to form NO at the beginning of Section 5.5 to answer each question.

a. How many moles of NO are formed from 3.3 moles of N_2?
b. How many moles of NO are formed from 0.50 moles of O_2?
c. How many moles of O_2 are needed to completely react with 1.2 moles of N_2?

PROBLEM 5.21

Use the balanced equation in Sample Problem 5.11 to answer each question.

a. How many moles of O_2 are needed to react completely with 3.0 moles of C_2H_6?
b. How many moles of H_2O are formed from 0.50 moles of C_2H_6?
c. How many moles of C_2H_6 are needed to form 3.0 moles of CO?

5.6 Mass Calculations in Chemical Equations

Since a mole represents an enormously large number of very small molecules, there is no way to directly count the number of moles or molecules used in a chemical reaction. Instead, we utilize a balance to measure the number of grams of a compound used and the number of grams of product formed. The number of grams of a substance and the number of moles it contains are related by the molar mass (Section 5.4).

ENVIRONMENTAL NOTE

Lightning produces O_3 from O_2 during an electrical storm. O_3 at the ground level is an unwanted pollutant. In the stratosphere, however, it protects us from harmful radiation from the sun.

5.6A Converting Moles of Reactant to Grams of Product

To determine how many grams of product are expected from a given number of moles of reactant, two operations are necessary. First, we must determine how many moles of product to expect using the coefficients of the balanced chemical equation (Section 5.5). Then, we convert the number of moles of product to the number of grams using the molar mass (Section 5.4). Each step needs a conversion factor. The stepwise procedure is outlined in the accompanying *How To*, and then illustrated with an example in Sample Problem 5.12.

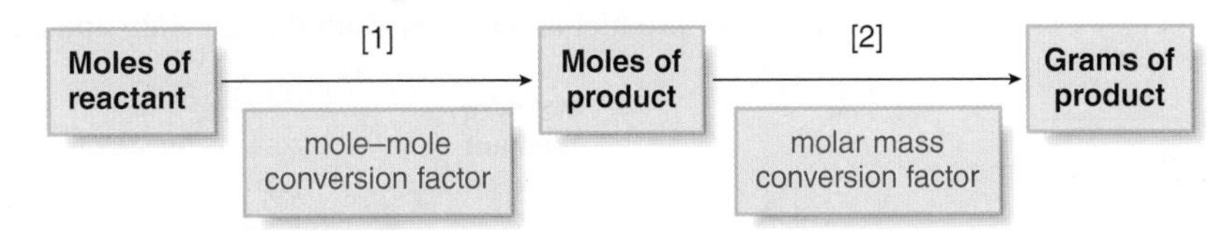

How To Convert Moles of Reactant to Grams of Product

Example In the upper atmosphere, high-energy radiation from the sun converts oxygen (O_2) to ozone (O_3). Using the balanced equation, how many grams of O_3 are formed from 9.0 mol of O_2?

$$3\ O_2(g) \xrightarrow{\text{sunlight}} 2\ O_3(g)$$

—Continued

How To, continued . . .

Step [1] Convert the number of moles of reactant to the number of moles of product using a mole–mole conversion factor.

- Use the coefficients in the balanced chemical equation to write mole–mole conversion factors.

$$\frac{3 \text{ mol } O_2}{2 \text{ mol } O_3} \quad \text{or} \quad \boxed{\frac{2 \text{ mol } O_3}{3 \text{ mol } O_2}}$$

Choose this conversion factor to cancel mol O_2.

- Multiply the number of moles of starting material (9.0 mol) by the conversion factor to give the number of moles of product. In this example, 6.0 mol of O_3 are formed.

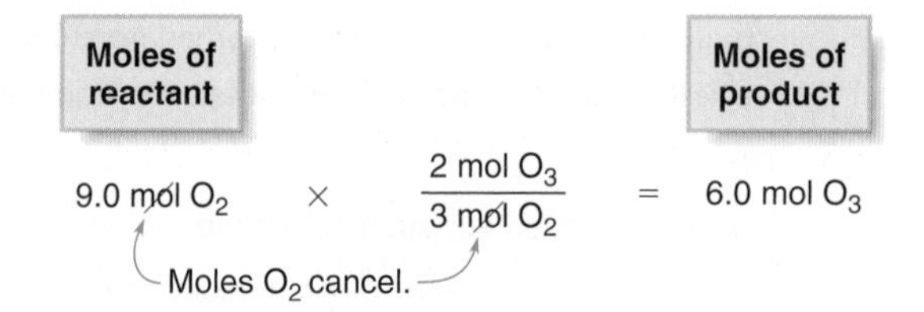

Moles of reactant
Moles of product
9.0 mol O2 × 2 mol O3 / 3 mol O2 = 6.0 mol O3
Moles O2 cancel.

Step [2] Convert the number of moles of product to the number of grams of product using the product's molar mass.

- Use the molar mass of the product (O_3) to write a conversion factor. The molar mass of O_3 is 48.00 g/mol (3 O atoms × 16.00 g/mol for each O atom = 48.00 g/mol).

$$\frac{1 \text{ mol } O_3}{48.00 \text{ g } O_3} \quad \text{or} \quad \boxed{\frac{48.00 \text{ g } O_3}{1 \text{ mol } O_3}}$$

Choose this conversion factor to cancel mol.

- Multiply the number of moles of product (from step [1]) by the conversion factor to give the number of grams of product.

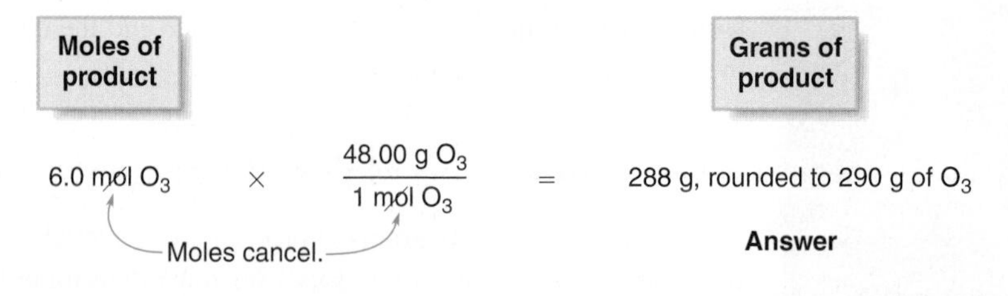

Moles of product
Grams of product
6.0 mol O3 × 48.00 g O3 / 1 mol O3 = 288 g, rounded to 290 g of O3
Moles cancel.
Answer

It is also possible to combine the multiplication operations from steps [1] and [2] into a single operation using both conversion factors. This converts the moles of starting material to grams of product all at once. Both the one-step and stepwise approaches give the same overall result.

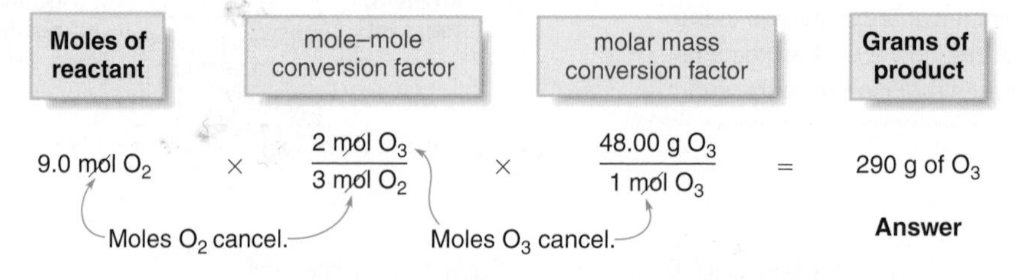

Moles of reactant
mole–mole conversion factor
molar mass conversion factor
Grams of product
9.0 mol O2 × 2 mol O3 / 3 mol O2 × 48.00 g O3 / 1 mol O3 = 290 g of O3
Moles O2 cancel.
Moles O3 cancel.
Answer

HEALTH NOTE

Ethanol (C_2H_6O) is the alcohol in red wine, formed by the fermentation of grapes. Ethanol depresses the central nervous system, increases the production of stomach acid, and dilates blood vessels. Excessive alcohol consumption is a major health problem in the United States.

ENVIRONMENTAL NOTE

Ethanol is used as a gasoline additive. Although some of the ethanol used for this purpose comes from corn and other grains, much of it is still produced by the reaction of ethylene with water (see *How To,* p. 163). Ethanol produced from grains is a renewable resource, whereas ethanol produced from ethylene is not, because ethylene is made from crude oil. Thus, running your car on gasohol (gasoline mixed with ethanol) reduces our reliance on fossil fuels only if the ethanol is produced from renewable sources such as grains or sugarcane.

SAMPLE PROBLEM 5.12

Wine is produced by the fermentation of grapes. In fermentation, the carbohydrate glucose ($C_6H_{12}O_6$) is converted to ethanol and carbon dioxide according to the given balanced equation. How many grams of ethanol (C_2H_6O, molar mass 46.07 g/mol) are produced from 5.00 mol of glucose?

$$\underset{\text{glucose}}{C_6H_{12}O_6(aq)} \longrightarrow \underset{\text{ethanol}}{2\ C_2H_6O(aq)} + 2\ CO_2(g)$$

Analysis and Solution

[1] **Convert the number of moles of reactant to the number of moles of product using a mole–mole conversion factor.**

- Use the coefficients in the balanced chemical equation to write mole–mole conversion factors for the two compounds—one mole of glucose ($C_6H_{12}O_6$) forms two moles of ethanol (C_2H_6O).
- Multiply the number of moles of reactant (glucose) by the conversion factor to give the number of moles of product (ethanol).

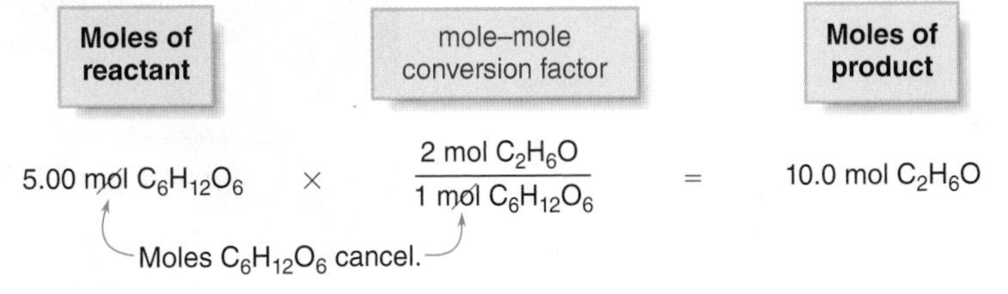

[2] **Convert the number of moles of product to the number of grams of product using the product's molar mass.**

- Use the molar mass of the product (C_2H_6O, molar mass 46.07 g/mol) to write a conversion factor.
- Multiply the number of moles of product (from step [1]) by the conversion factor to give the number of grams of product.

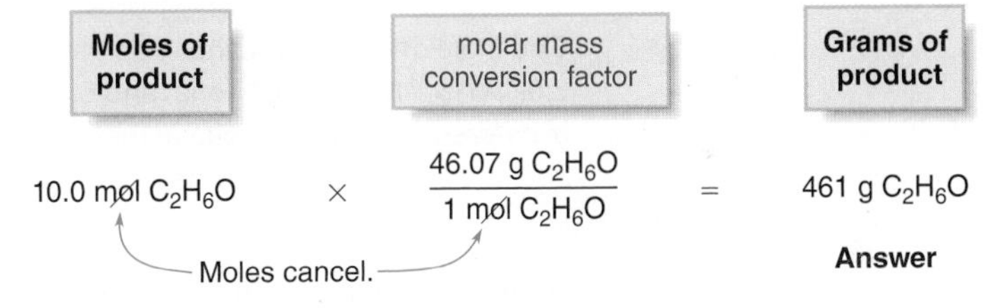

PROBLEM 5.22

Using the balanced equation for fermentation written in Sample Problem 5.12, answer the following questions.

a. How many grams of ethanol are formed from 0.55 mol of glucose?
b. How many grams of CO_2 are formed from 0.25 mol of glucose?
c. How many grams of glucose are needed to form 1.0 mol of ethanol?

PROBLEM 5.23

Using the balanced equation for the combustion of ethanol, answer the following questions.

$$\underset{\text{ethanol}}{C_2H_6O(l)} + 3\ O_2(g) \longrightarrow 2\ CO_2(g) + 3\ H_2O(g)$$

a. How many grams of CO_2 are formed from 0.50 mol of ethanol?
b. How many grams of H_2O are formed from 2.4 mol of ethanol?
c. How many grams of O_2 are needed to react with 0.25 mol of ethanol?

5.6B Converting Grams of Reactant to Grams of Product

The **coefficients in chemical equations tell us the ratio of the number of *molecules* or *moles* that are involved in a chemical reaction.** The coefficients do *not,* however, tell us directly about the number of grams. That's because the molar mass—the number of grams in one mole—of a substance depends on the identity of the elements that compose it. One mole of H_2O molecules weighs 18.02 g, one mole of NaCl weighs 58.44 g, and one mole of sugar molecules weighs 342.3 g (Figure 5.3).

Figure 5.3 One Mole of Water, Table Salt, and Table Sugar

One mole of each substance has the same number of units—6.02×10^{23} H_2O molecules, 6.02×10^{23} Na^+ and Cl^- ions, and 6.02×10^{23} sugar molecules. The molar mass of each substance is *different,* however, because they are each composed of *different* elements.

In the laboratory, we measure out the number of grams of a reactant on a balance. This does not tell us directly the number of grams of a particular product that will form, because in all likelihood, the molar masses of the reactant and product are different. To carry out this type of calculation—grams of one compound (reactant) to grams of another compound (product)—three operations are necessary.

First, we must determine how many moles of reactant are contained in the given number of grams using the molar mass. Then, we can determine the number of moles of product expected using the coefficients of the balanced chemical equation. Finally, we convert the number of moles of product to the number of grams of product using its molar mass. Now there are three steps and three conversion factors. The stepwise procedure is outlined in the accompanying *How To,* and then illustrated with an example in Sample Problem 5.13.

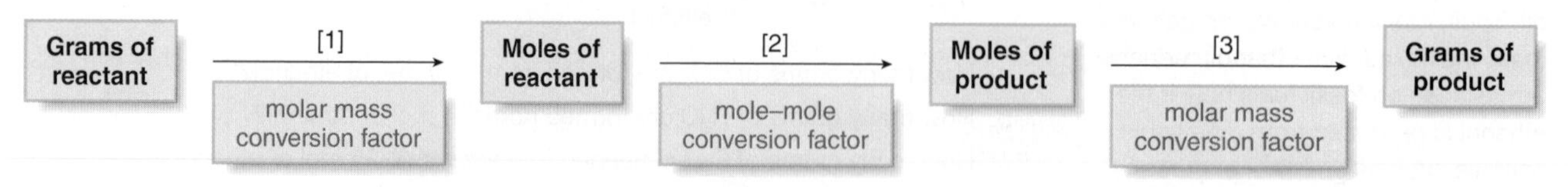

How To Convert Grams of Reactant to Grams of Product

Example Ethanol (C_2H_6O, molar mass 46.07 g/mol) is synthesized by reacting ethylene (C_2H_4, molar mass 28.05 g/mol) with water. How many grams of ethanol are formed from 14 g of ethylene?

$$H_2C{=}CH_2 + H_2O \longrightarrow H{-}CH_2{-}CH_2{-}O{-}H$$

ethylene ethanol

Step [1] Convert the number of grams of reactant to the number of moles of reactant using the reactant's molar mass.

- Use the molar mass of the reactant (C_2H_4) to write a conversion factor.

$$\frac{28.05 \text{ g } C_2H_4}{1 \text{ mol } C_2H_4} \quad \text{or} \quad \boxed{\frac{1 \text{ mol } C_2H_4}{28.05 \text{ g } C_2H_4}}$$

Choose this conversion factor to cancel g.

- Multiply the number of grams of reactant by the conversion factor to give the number of moles of reactant.

Grams of reactant → **Moles of reactant**

$$14 \text{ g } C_2H_4 \times \frac{1 \text{ mol } C_2H_4}{28.05 \text{ g } C_2H_4} = 0.50 \text{ mol } C_2H_4$$

Grams cancel.

Step [2] Convert the number of moles of reactant to the number of moles of product using a mole–mole conversion factor.

- Use the coefficients in the balanced chemical equation to write mole–mole conversion factors.

$$\frac{1 \text{ mol } C_2H_4}{1 \text{ mol } C_2H_6O} \quad \text{or} \quad \boxed{\frac{1 \text{ mol } C_2H_6O}{1 \text{ mol } C_2H_4}}$$

Choose this conversion factor to cancel mol C_2H_4.

- Multiply the number of moles of reactant by the conversion factor to give the number of moles of product. In this example, 0.50 mol of C_2H_6O is formed.

Moles of reactant → **Moles of product**

$$0.50 \text{ mol } C_2H_4 \times \frac{1 \text{ mol } C_2H_6O}{1 \text{ mol } C_2H_4} = 0.50 \text{ mol } C_2H_6O$$

Moles C_2H_4 cancel.

Step [3] Convert the number of moles of product to the number of grams of product using the product's molar mass.

- Use the molar mass of the product (C_2H_6O) to write a conversion factor.

$$\frac{1 \text{ mol } C_2H_6O}{46.07 \text{ g } C_2H_6O} \quad \text{or} \quad \boxed{\frac{46.07 \text{ g } C_2H_6O}{1 \text{ mol } C_2H_6O}}$$

Choose this conversion factor to cancel mol C_2H_6O.

- Multiply the number of moles of product (from step [2]) by the conversion factor to give the number of grams of product.

Moles of product → **Grams of product**

$$0.50 \text{ mol } C_2H_6O \times \frac{46.07 \text{ g } C_2H_6O}{1 \text{ mol } C_2H_6O} = 23 \text{ g } C_2H_6O$$

Moles C_2H_6O cancel.

Answer

It is also possible to combine the multiplication operations from steps [1], [2], and [3] into a single operation using all three conversion factors. This converts grams of starting material to grams of product all at once. Both the one-step and stepwise approaches give the same overall result.

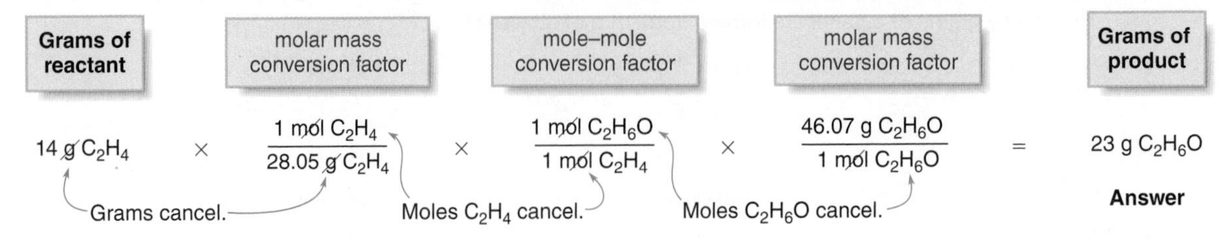

SAMPLE PROBLEM 5.13

How many grams of aspirin are formed from 10.0 g of salicylic acid using the given balanced equation?

$$\underset{\text{salicylic acid}}{C_7H_6O_3(s)} + \underset{\text{acetic acid}}{C_2H_4O_2(l)} \longrightarrow \underset{\text{aspirin}}{C_9H_8O_4(s)} + H_2O(l)$$

Analysis and Solution

[1] **Convert the number of grams of reactant to the number of moles of reactant using the reactant's molar mass.**

- Use the molar mass of the reactant ($C_7H_6O_3$, molar mass 138.1 g/mol) to write a conversion factor. Multiply the number of grams of reactant by the conversion factor to give the number of moles of reactant.

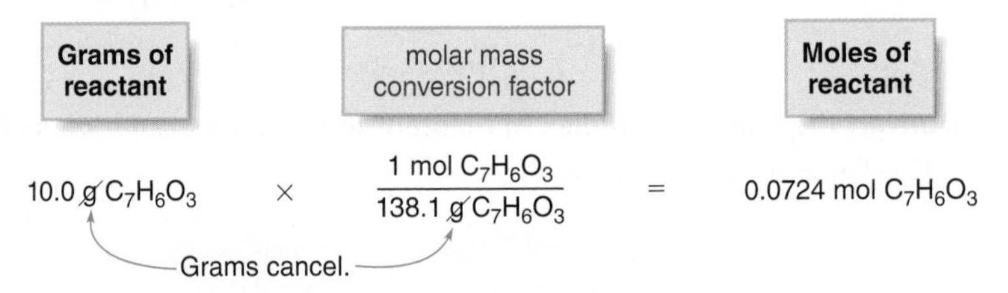

[2] **Convert the number of moles of reactant to the number of moles of product using a mole–mole conversion factor.**

- Use the coefficients in the balanced chemical equation to write mole–mole conversion factors for the two compounds—one mole of salicylic acid ($C_7H_6O_3$) forms one mole of aspirin ($C_9H_8O_4$).
- Multiply the number of moles of reactant (salicylic acid) by the conversion factor to give the number of moles of product (aspirin).

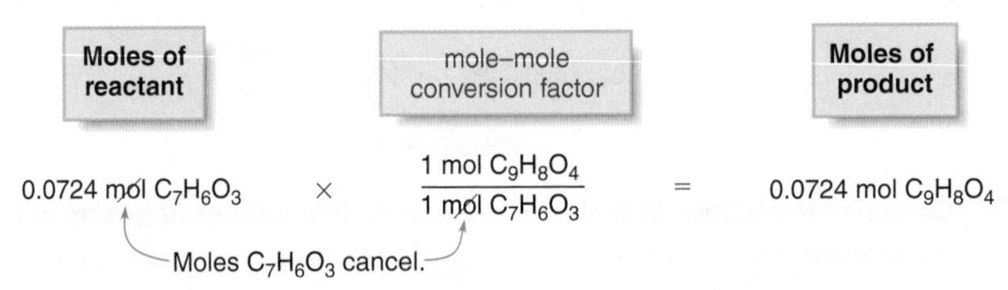

[3] **Convert the number of moles of product to the number of grams of product using the product's molar mass.**

- Use the molar mass of the product ($C_9H_8O_4$, molar mass 180.2 g/mol) to write a conversion factor. Multiply the number of moles of product (from step [2]) by the conversion factor to give the number of grams of product.

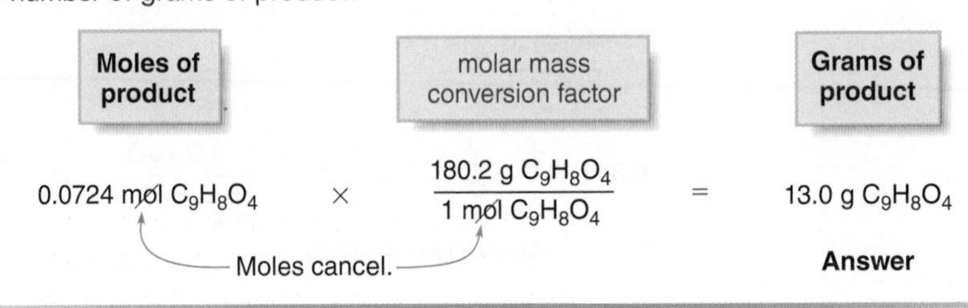

PROBLEM 5.24

Use the balanced equation in Sample Problem 5.13 for the conversion of salicylic acid and acetic acid to aspirin to answer the following questions.

a. How many grams of aspirin are formed from 55.5 g of salicylic acid?
b. How many grams of acetic acid are needed to react with 55.5 g of salicylic acid?
c. How many grams of water are formed from 55.5 g of salicylic acid?

PROBLEM 5.25

Use the balanced equation, $N_2 + O_2 \longrightarrow 2\ NO$, to answer the following questions.

a. How many grams of NO are formed from 10.0 g of N_2?
b. How many grams of NO are formed from 10.0 g of O_2?
c. How many grams of O_2 are needed to react completely with 10.0 g of N_2?

5.7 Oxidation and Reduction

Another group of reactions—acid–base reactions—is discussed in Chapter 8.

Thus far we have examined features that are common to all types of chemical reactions. We now examine one class of reactions that involves electron transfer—oxidation–reduction reactions.

5.7A General Features of Oxidation–Reduction Reactions

A common type of chemical reaction involves the transfer of electrons from one element to another. When iron rusts, methane and wood burn, and a battery generates electricity, one element gains electrons and another loses them. These reactions involve **oxidation** and **reduction.**

- Oxidation is the loss of electrons from an atom.
- Reduction is the gain of electrons by an atom.

Oxidation and reduction are opposite processes, and both occur together in a single reaction called an **oxidation–reduction** or **redox reaction.** A redox reaction always has two components—one that is oxidized and one that is reduced.

- A redox reaction involves the transfer of electrons from one element to another.

An example of an oxidation–reduction reaction occurs when Zn metal reacts with Cu^{2+} cations, as shown in Figure 5.4.

Figure 5.4

A Redox Reaction—The Transfer of Electrons from Zn to Cu^{2+}

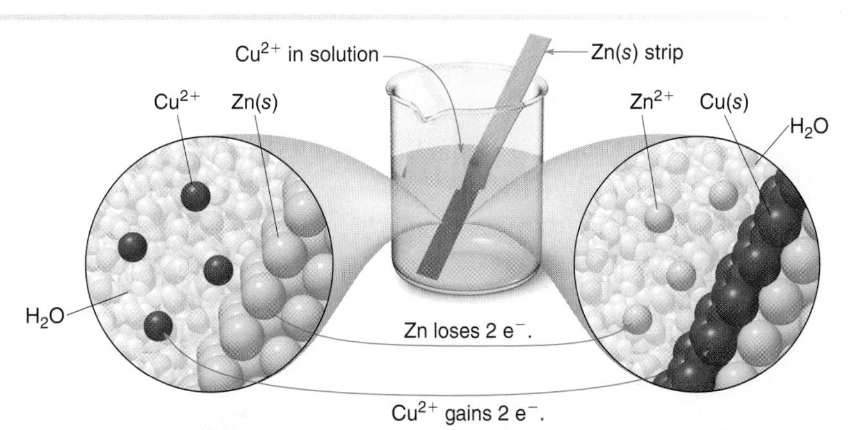

A redox reaction occurs when a strip of Zn metal is placed in a solution of Cu^{2+} ions. In this reaction, Zn loses two electrons to form Zn^{2+}, which goes into solution. Cu^{2+} gains two electrons to form Cu metal, which precipitates out of solution, forming a coating on the zinc strip.

CONSUMER NOTE

Benzoyl peroxide ($C_{14}H_{10}O_4$) is the active ingredient in several acne medications. Benzoyl peroxide kills bacteria by oxidation reactions.

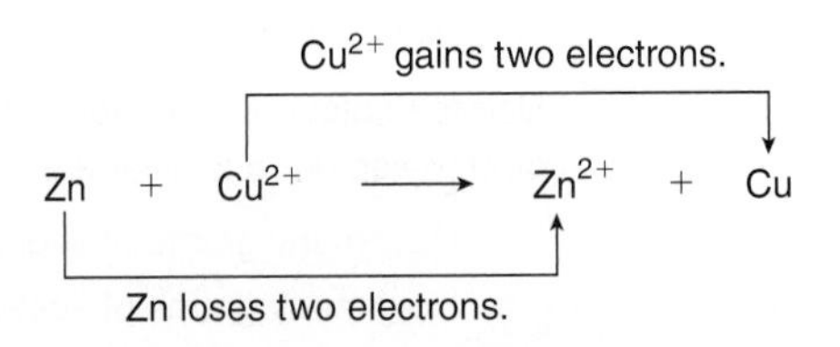

- **Zn loses two electrons to form Zn^{2+}, so Zn is oxidized.**
- **Cu^{2+} gains two electrons to form Cu metal, so Cu^{2+} is reduced.**

Each of these processes can be written as individual reactions, called **half reactions,** to emphasize which electrons are gained and lost.

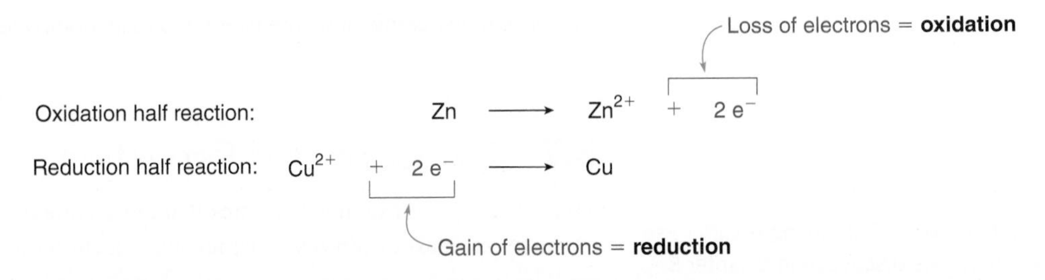

- **A compound that gains electrons (is reduced) while causing another compound to be oxidized is called an *oxidizing agent.***
- **A compound that loses electrons (is oxidized) while causing another compound to be reduced is called a *reducing agent.***

In this example, Zn loses electrons to Cu^{2+}. We can think of Zn as a **reducing agent** since it causes Cu^{2+} to gain electrons and become reduced. We can think of Cu^{2+} as an **oxidizing agent** since it causes Zn to lose electrons and become oxidized.

To draw the products of an oxidation–reduction reaction, we must decide which element or ion gains electrons and which element or ion loses electrons. Use the following guidelines.

- **When considering neutral atoms, metals lose electrons and nonmetals gain electrons.**
- **When considering ions, cations tend to gain electrons and anions tend to lose electrons.**

Thus, the metals sodium (Na) and magnesium (Mg) readily lose electrons to form the cations Na^+ and Mg^{2+}, respectively; that is, they are oxidized. The nonmetals O_2 and Cl_2 readily gain electrons to form 2 O^{2-} and 2 Cl^-, respectively; that is, they are reduced. A positively charged ion like Cu^{2+} is reduced to Cu by gaining two electrons, while two negatively charged Cl^- anions are oxidized to Cl_2 by losing two electrons. These reactions and additional examples are shown in Figure 5.5.

Figure 5.5

Examples of Oxidation and Reduction Reactions

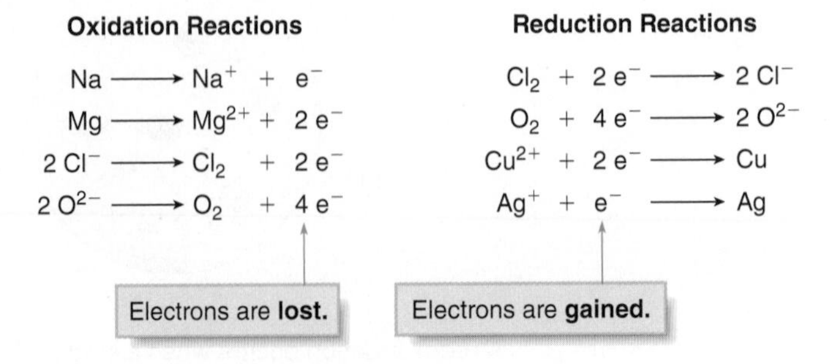

SAMPLE PROBLEM 5.14

Identify the species that is oxidized and the species that is reduced in the following reaction. Write out half reactions to show how many electrons are gained or lost by each species.

$$Mg(s) + 2\ H^+(aq) \longrightarrow Mg^{2+}(aq) + H_2(g)$$

Analysis

Metals and anions tend to lose electrons and thus undergo oxidation. Nonmetals and cations tend to gain electrons and thus undergo reduction.

Solution

The metal Mg is oxidized to Mg^{2+}, thus losing two electrons. Two H^+ cations gain a total of two electrons, and so are reduced to the nonmetal H_2.

$$Mg(s) \longrightarrow Mg^{2+}(aq) + 2\ e^-$$

Mg is oxidized.

$$2\ H^+(aq) + 2\ e^- \longrightarrow H_2(g)$$

H^+ is reduced. Two electrons are needed to balance charge.

We need enough electrons so that **the total charge is the same on both sides of the equation.** Since 2 H^+ cations have a +2 overall charge, this means that 2 e^- must be gained so that the total charge on both sides of the equation is zero.

PROBLEM 5.26

Identify the species that is oxidized and the species that is reduced in each reaction. Write out half reactions to show how many electrons are gained or lost by each species.

a. $Zn(s) + 2\ H^+(aq) \longrightarrow Zn^{2+}(aq) + H_2(g)$

b. $Fe^{3+}(aq) + Al(s) \longrightarrow Al^{3+}(aq) + Fe(s)$

c. $2\ I^- + Br_2 \longrightarrow I_2 + 2\ Br^-$

d. $2\ AgBr \longrightarrow 2\ Ag + Br_2$

5.7B Examples of Oxidation–Reduction Reactions

Many common processes involve oxidation and reduction. For example, common antiseptics like iodine (I_2) and hydrogen peroxide (H_2O_2) are oxidizing agents that clean wounds by oxidizing, thereby killing bacteria that might cause infection.

When iron (Fe) rusts, it is oxidized by the oxygen in air to form iron(III) oxide, Fe_2O_3. In this redox reaction, neutral iron atoms are oxidized to Fe^{3+} cations, and elemental O_2 is reduced to O^{2-} anions.

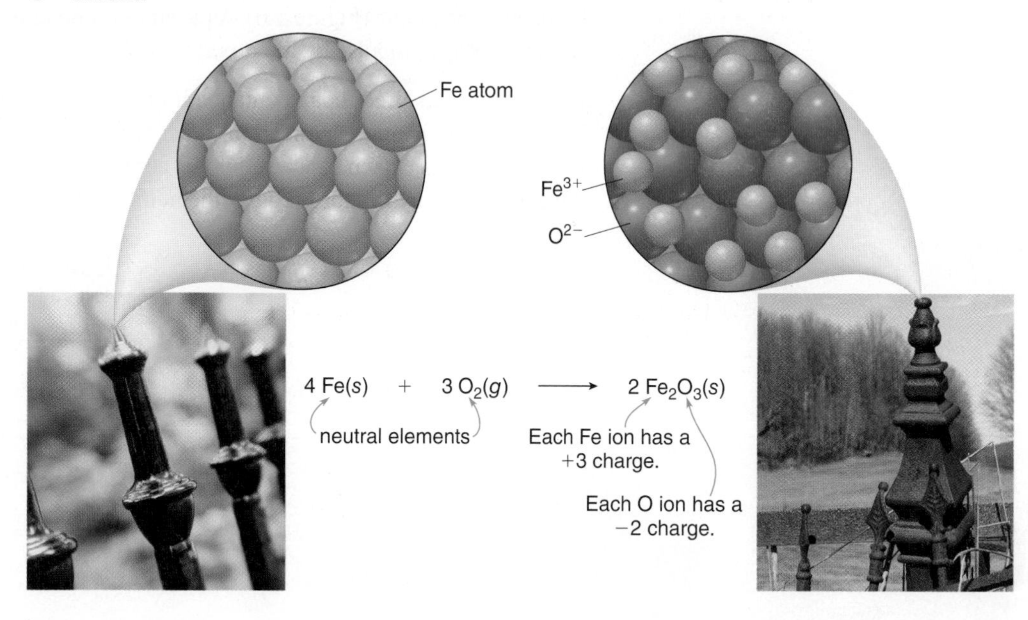

In some reactions it is much less apparent which reactant is oxidized and which is reduced. For example, in the combustion of methane (CH_4) with oxygen to form CO_2 and H_2O, there are no metals or cations that obviously lose or gain electrons, yet this is a redox reaction. In these instances, it is often best to count oxygen and hydrogen atoms.

- **Oxidation results in the gain of oxygen atoms or the loss of hydrogen atoms.**
- **Reduction results in the loss of oxygen atoms or the gain of hydrogen atoms.**

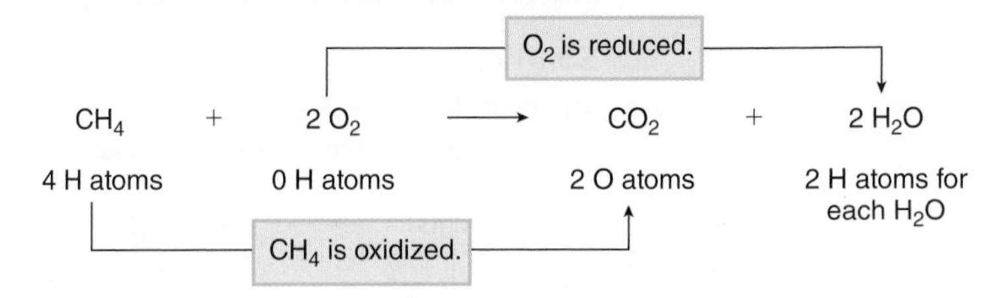

CH_4 is oxidized since it gains two oxygen atoms to form CO_2. O_2 is reduced since it gains two hydrogen atoms to form H_2O.

PROBLEM 5.27

The following redox reaction occurs in mercury batteries for watches. Identify the species that is oxidized and the species that is reduced, and write out two half reactions to show how many electrons are gained or lost. In this reaction, oxide (O^{2-}) is unchanged; that is, it is neither oxidized nor reduced.

$$Zn + HgO \longrightarrow ZnO + Hg$$

PROBLEM 5.28

Identify the species that is oxidized and the species that is reduced in the following redox reaction. Explain your choices.

$$C_2H_4O_2 + 2\ H_2 \longrightarrow C_2H_6O + H_2O$$

5.7C FOCUS ON HEALTH & MEDICINE Pacemakers

A pacemaker is a small electrical device implanted in an individual's chest and used to maintain an adequate heart rate (Figure 5.6). A pacemaker contains a small, long-lasting battery that generates an electrical impulse by a redox reaction.

Most pacemakers used today contain a lithium–iodine battery. Each neutral lithium atom is oxidized to Li^+ by losing one electron. Each I_2 molecule is reduced by gaining two electrons and forming $2\ I^-$. Since the balanced equation contains two Li atoms for each I_2 molecule, the number of electrons lost by Li atoms equals the number of electrons gained by I_2.

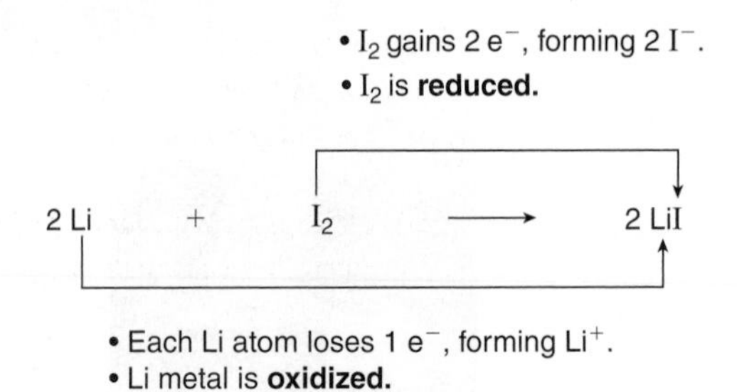

Figure 5.6

The Lithium–Iodine Battery in a Pacemaker

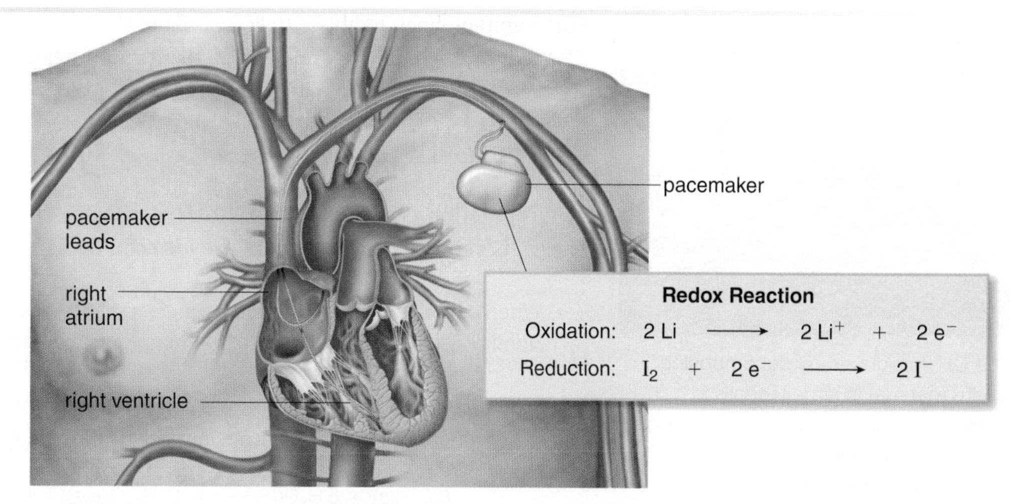

A pacemaker generates a small electrical impulse that triggers the heart to beat. Today's pacemakers sense when the heart beats normally and provide an electrical signal only when the heart rate slows. Such devices are called "demand" pacemakers, and they quickly replaced earlier "fixed" rate models that continuously produced impulses to set the heart rate at a fixed value.

PROBLEM 5.29

Early pacemakers generated an electrical impulse by the following reaction. What species is the oxidizing agent and what species is the reducing agent in this reaction?

$$Zn + Hg^{2+} \longrightarrow Zn^{2+} + Hg$$

5.8 Energy Changes in Reactions

When molecules come together and react, bonds are broken in the reactants and new bonds are formed in the products. **Breaking a bond requires energy, while forming a bond releases energy.**

5.8A Heat of Reaction

The energy absorbed or released in any reaction is called the **heat of reaction** or the **enthalpy change,** symbolized by ΔH. The heat of reaction is given a positive (+) or negative (–) sign depending on whether energy is absorbed or released.

- When energy is absorbed, the reaction is said to be *endothermic* and ΔH is positive (+).
- When energy is released, the reaction is said to be *exothermic* and ΔH is negative (–).

The heat of reaction measures the difference between the energy needed to break bonds in the reactants and the energy released from the bonds formed in the products. In other words, **ΔH indicates the relative strength of bonds broken and formed in a reaction.**

- When ΔH is negative, more energy is released in forming bonds than is needed to break bonds. The products are lower in energy than the reactants.

ENVIRONMENTAL NOTE

The CH_4 produced by decomposing waste material in large landfills is burned to produce energy for heating and generating electricity.

For example, when methane (CH_4) burns in the presence of oxygen (O_2) to form CO_2 and H_2O, 213 kcal/mol of energy is released in the form of heat.

$$CH_4(g) + 2\,O_2(g) \longrightarrow CO_2(g) + 2\,H_2O(l) \quad \Delta H = -213 \text{ kcal/mol}$$

Heat is released.

In this reaction energy is released, ΔH is negative (–), and the reaction is exothermic. **Since energy is released, the products are *lower* in energy than the reactants.**

- When ΔH is positive, more energy is needed to break bonds than is released in forming bonds. The reactants are lower in energy than the products.

For example, in the process of photosynthesis, green plants use chlorophyll to convert CO_2 and H_2O to glucose ($C_6H_{12}O_6$, a simple carbohydrate) and O_2 and 678 kcal of energy is absorbed.

$$6\,CO_2(g) + 6\,H_2O(l) \longrightarrow C_6H_{12}O_6(aq) + 6\,O_2(g) \quad \Delta H = +678 \text{ kcal/mol}$$

In this reaction energy is absorbed, ΔH is positive (+), and the reaction is endothermic. **Since energy is absorbed, the products are *higher* in energy than the reactants.**

Photosynthesis is an endothermic reaction. Energy from sunlight is absorbed in the reaction and stored in the bonds of the products.

Table 5.2 summarizes the characteristics of energy changes in reactions.

Table 5.2 Endothermic and Exothermic Reactions

Endothermic Reaction	Exothermic Reaction
• Heat is absorbed.	• Heat is released.
• ΔH is positive.	• ΔH is negative.
• The products are higher in energy than the reactants.	• The products are lower in energy than the reactants.

PROBLEM 5.30

Answer the following questions using the given equation and ΔH. (a) Is heat absorbed or released? (b) Are the reactants or products lower in energy? (c) Is the reaction endothermic or exothermic?

$$2\,NH_3(g) \longrightarrow 3\,H_2(g) + N_2(g) \quad \Delta H = +22.0 \text{ kcal/mol}$$

PROBLEM 5.31

The ΔH for the combustion of propane (C_3H_8) with O_2 according to the given balanced chemical equation is –531 kcal/mol. (a) Is heat absorbed or released? (b) Are the reactants or products lower in energy? (c) Is the reaction endothermic or exothermic?

$$\underset{\text{propane}}{C_3H_8(g)} + 5\,O_2(g) \longrightarrow 3\,CO_2(g) + 4\,H_2O(l) \quad \Delta H = -531 \text{ kcal/mol}$$

5.8B Energy Diagrams

On a molecular level, what happens when a reaction occurs? In order for two molecules to react, they must collide, and in the collision, the kinetic energy they possess is used to break bonds.

The energy changes in a reaction are often illustrated on an **energy diagram,** which plots energy on the vertical axis, and the progress of the reaction—the **reaction coordinate**—on the horizontal axis. The reactants are written on the left side and the products on the right side, and a smooth curve that illustrates how energy changes with time connects them. Consider a general reaction between two starting materials, **A—B** and **C,** in which the **A—B** bond is broken and a new **B—C** bond is formed.

Let's assume that the products, **A** and **B—C,** are lower in energy than the reactants, **A—B** and **C.**

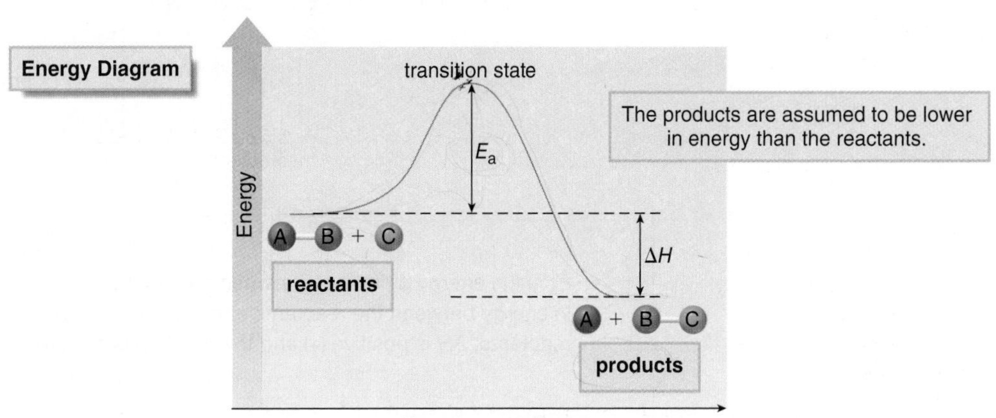

When the reactants **A—B** and **C** approach each other, their electron clouds feel some repulsion, causing an increase in energy until a maximum value is reached. This point is called the **transition state.** In the transition state, the bond between **A** and **B** is partially broken and the bond between **B** and **C** is partially formed.

At the transition state, the bond between **A** and **B** can re-form to regenerate reactants, or the bond between **B** and **C** can form to generate products. As the bond forms between **B** and **C,** the energy decreases until some stable energy minimum is reached. The products are drawn lower in energy than the reactants to reflect the initial assumption about their relative energies.

- The difference in energy between the reactants and the transition state is called the *energy of activation,* symbolized by E_a.

The energy of activation is the minimum amount of energy needed for a reaction to occur. The energy of activation is often called the **energy barrier** that must be crossed. The height of the energy barrier—the magnitude of the energy of activation—determines the **reaction rate, how fast the reaction occurs.**

- When the energy of activation is *high,* few molecules have enough energy to cross the energy barrier and the reaction is *slow.*
- When the energy of activation is *low,* many molecules have enough energy to cross the energy barrier and the reaction is *fast.*

The difference in energy between the reactants and products is the ΔH, which is also labeled on the energy diagram. When the products are lower in energy than the reactants, as is the case here, ΔH is negative (–) and the reaction is exothermic.

Energy diagrams can be drawn for any reaction. In the endothermic reaction shown in Figure 5.7, the products are higher in energy than the reactants.

Figure 5.7 Energy Diagram for an Endothermic Reaction

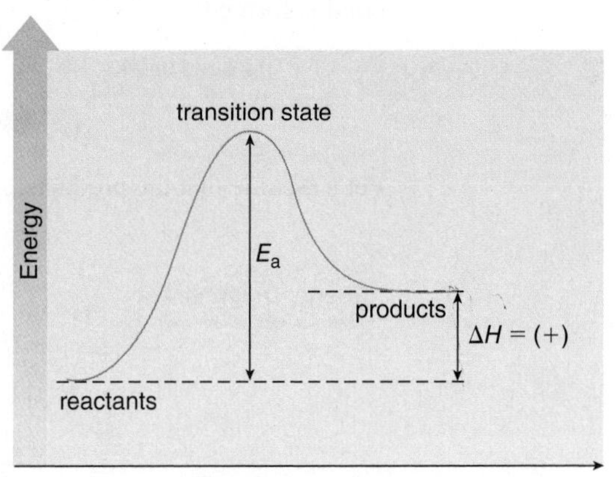

E_a is the energy difference between the reactants and the transition state. ΔH is the difference in energy between the reactants and products. Since the products are higher in energy than the reactants, ΔH is positive (+) and the reaction is endothermic.

SAMPLE PROBLEM 5.15

Draw an energy diagram for a reaction with a low energy of activation and a ΔH of −10 kcal/mol. Label the axes, reactants, products, transition state, E_a, and ΔH.

Analysis

A low energy of activation means a low energy barrier and a small hill that separates reactants and products. When ΔH is (–), the products are lower in energy than the reactants.

Solution

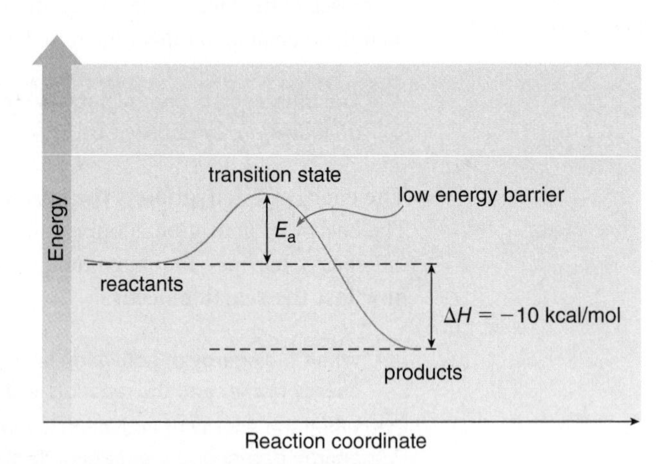

PROBLEM 5.32

Draw an energy diagram for a reaction with a high E_a and a ΔH = +20 kcal/mol.

PROBLEM 5.33

Draw an energy diagram for the following reaction: $H_2O + HCl \longrightarrow H_3O^+ + Cl^-$. Assume the energy of activation is low and the products are lower in energy than the reactants. Clearly label the reactants and products on the energy diagram.

5.9 Reaction Rates

CONSUMER NOTE

We store food in a cold refrigerator to slow the reactions that cause food to spoil.

Even though we may not realize it, the rate of chemical processes affects many facets of our lives. Aspirin is an effective pain reliever because it rapidly blocks the synthesis of pain-causing molecules. DDT is a persistent environmental pollutant because it does not react appreciably with water, oxygen, or any other chemical with which it comes into contact. These processes occur at different rates, resulting in beneficial or harmful effects.

The energy of activation, the minimum amount of energy needed for a reaction to occur, is a fundamental characteristic of a reaction. Some reactions are fast because they have low energies of activation. Other reactions are slow because the energy of activation is high.

5.9A How Concentration and Temperature Affect Reaction Rate

As we learned in Section 5.8, chemical reactions occur when molecules collide. How do changes in concentration and temperature affect the reaction rate?

- **Increasing the concentration of the reactants increases the number of collisions, so the reaction rate increases.**
- **Increasing the temperature increases the reaction rate.**

Increasing the temperature increases the kinetic energy, which increases the number of collisions. Increasing the temperature also increases the *average* kinetic energy of the reactants. Because the kinetic energy of colliding molecules is used for bond cleavage, more molecules have sufficient energy to cause bond breaking, and the reaction rate increases.

PROBLEM 5.34

Consider the reaction of ozone (O_3) with nitrogen monoxide (NO), which occurs in smog. What effect would each of the following changes have on the rate of this reaction?

$$O_3(g) + NO(g) \longrightarrow O_2(g) + NO_2(g)$$

a. increasing the concentration of O_3
b. decreasing the concentration of NO
c. increasing the temperature
d. decreasing the temperature

5.9B Catalysts

Some reactions do not occur in a reasonable period of time unless a **catalyst** is added.

- **A *catalyst* is a substance that speeds up the rate of a reaction. A catalyst is recovered unchanged in a reaction, and it does not appear in the product.**

Catalysts accelerate a reaction by lowering the energy of activation (Figure 5.8). They have no effect on the energies of the reactants and products. Thus, the addition of a catalyst lowers E_a but does not affect ΔH.

Figure 5.8 The Effect of a Catalyst on a Reaction

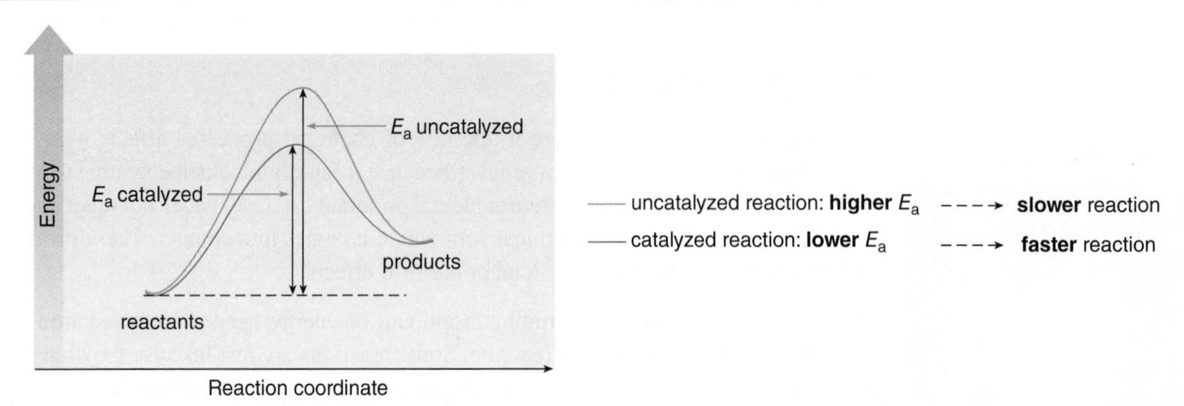

A catalyst lowers the energy of activation, thus increasing the rate of the catalyzed reaction. The energy of the reactants and products is the same in both the uncatalyzed and catalyzed reactions.

PROBLEM 5.35

The reaction of acetic acid ($C_2H_4O_2$) and ethanol (C_2H_6O) to form ethyl acetate ($C_4H_8O_2$) and water is catalyzed by sulfuric acid (H_2SO_4). What effect does H_2SO_4 have on the relative energies of the reactants and products? What effect does H_2SO_4 have on the energy of activation?

HEALTH NOTE

There is a direct link between the poor air quality in large metropolitan areas like Los Angeles and an increase in respiratory diseases.

5.9C FOCUS ON THE ENVIRONMENT Catalytic Converters

The combustion of gasoline with oxygen provides a great deal of energy, and this energy is used to power vehicles. As the number of automobiles increased in the twentieth century, the air pollution they were responsible for became a major problem, especially in congested urban areas. In addition to CO_2 and H_2O formed during combustion, auto exhaust also contained unreacted gasoline molecules (general formula C_xH_y), the toxic gas carbon monoxide (CO, Section 5.5), and nitrogen monoxide (NO, Section 5.5, a contributing component of acid rain). **Catalytic converters** were devised to clean up these polluting automobile emissions.

The newest catalytic converters use a metal as a surface to catalyze three reactions, as shown in Figure 5.9. Both the unreacted gasoline molecules and carbon monoxide (CO) are oxidized to CO_2 and H_2O. Nitrogen monoxide is converted to oxygen and nitrogen. As a result, the only materials in the engine exhaust are CO_2, H_2O, N_2, and O_2.

PROBLEM 5.36

Nitrogen dioxide, NO_2, also an undesired product formed during combustion, is converted to N_2 and O_2 in a catalytic converter. Write a balanced equation for this reaction.

Figure 5.9

How a Catalytic Converter Works

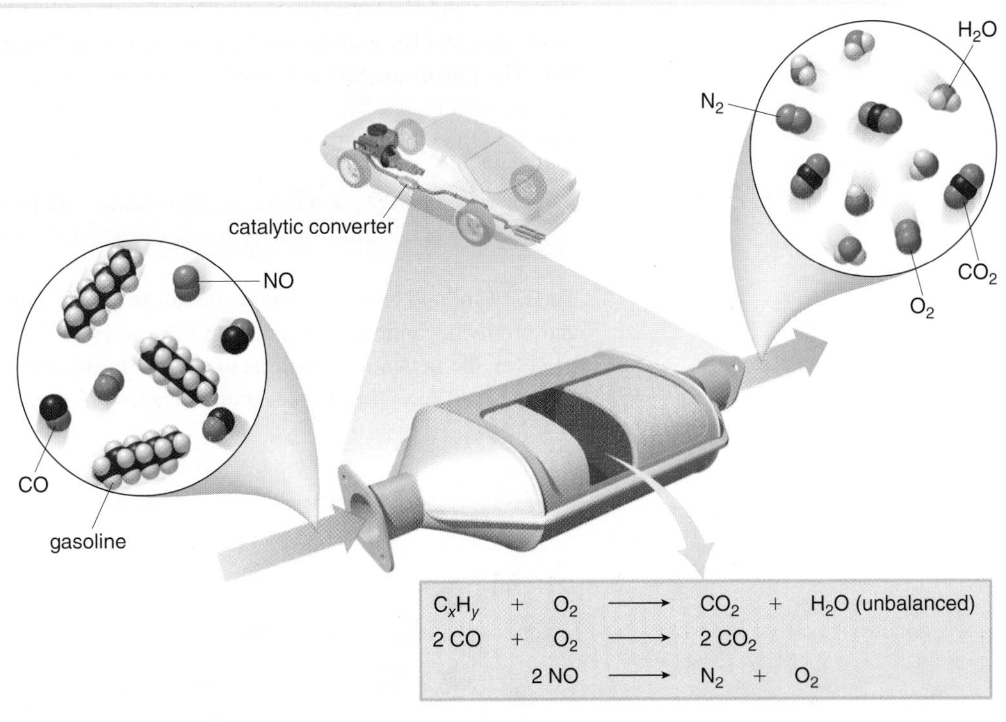

A catalytic converter uses a metal catalyst—rhodium, platinum, or palladium—to catalyze three reactions that clean up the exhaust from an auto engine.

5.10 Equilibrium

Thus far in discussing reactions we have assumed that the reactants are completely converted to products. A reaction of this sort is said to **go to completion.** Sometimes, however, a reaction is **reversible;** that is, reactants can come together and form products, and products can come together to re-form reactants.

- A *reversible reaction* can occur in either direction, from reactants to products or from products to reactants.

5.10A Reversible Reactions

Consider the reversible reaction of carbon monoxide (CO) with water to form carbon dioxide (CO_2) and hydrogen. Two full-headed arrows ($\rightleftarrows$) are used to show that the reaction can proceed from left to right and right to left as written.

The **forward** reaction proceeds to the ***right.***

$$CO(g) + H_2O(g) \rightleftarrows CO_2(g) + H_2(g)$$

The **reverse** reaction proceeds to the ***left.***

- The *forward* reaction proceeds from left to right as drawn.
- The *reverse* reaction proceeds from right to left as drawn.

When CO and H_2O are mixed together they react to form CO_2 and H_2 by the forward reaction. Once CO_2 and H_2 are formed, they can react together to form CO and H_2O by the reverse reaction. The rate of the forward reaction is rapid at first, and then decreases as the concentration of reactants decreases. The rate of the reverse reaction is slow at first, but speeds up as the concentration of the products increases.

- **When the rate of the forward reaction equals the rate of the reverse reaction, the net concentrations of all species do not change and the system is at *equilibrium.***

The forward and reverse reactions do not stop once equilibrium has been reached. The reactants and products continue to react. Since the rates of the forward and reverse reactions are equal, however, the **net concentrations of all reactants and products do *not* change.** Molecular art in Figure 5.10 shows how the reversible reaction **A** $\rightleftharpoons$ **B** reaches equilibrium.

Figure 5.10 Attaining Equilibrium in a Reaction, **A** $\rightleftharpoons$ **B**

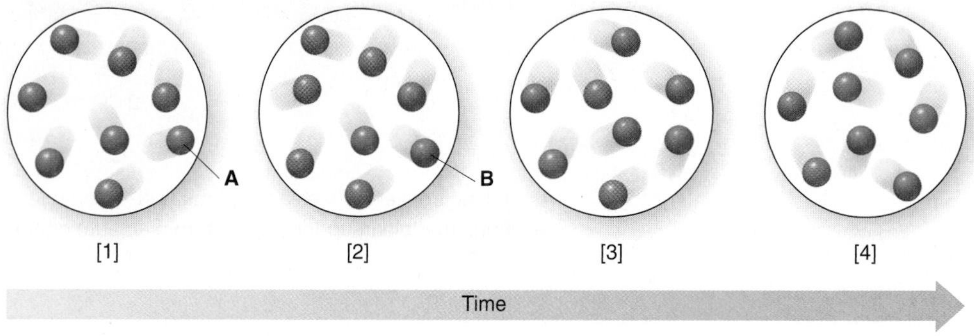

The conversion of **A** (shown in blue spheres) to **B** (shown in red spheres) is a reversible reaction. Initially, the reaction contains only **A** (diagram [1]). As **A** reacts to form **B,** the amount of **B** increases (diagram [2]). As **B** is formed, it can be re-converted to **A.** At equilibrium (diagrams [3] and [4]), the amounts of **A** and **B** do not change, but the conversion of **A** to **B** and the conversion of **B** to **A** still occur.

PROBLEM 5.37

Identify the forward and reverse reactions in each of the following reversible reactions.

a. $2\ SO_2(g) + O_2(g) \rightleftharpoons 2\ SO_3(g)$
b. $N_2(g) + O_2(g) \rightleftharpoons 2\ NO(g)$
c. $C_2H_4O_2 + CH_4O \rightleftharpoons C_3H_6O_2 + H_2O$

5.10B Le Châtelier's Principle

What happens when a reaction is at equilibrium and something changes? For example, what happens when some additional reactant is added? **Le Châtelier's principle** is a general rule used to explain the effect of a change in reaction conditions on equilibrium. Le Châtelier's principle states:

- **If a chemical system at equilibrium is disturbed or stressed, the system will react in the direction that counteracts the disturbance or relieves the stress.**

Consider the reaction of carbon monoxide (CO) with oxygen (O_2) to form carbon dioxide (CO_2).

$$2\ CO(g) + O_2(g) \rightleftharpoons 2\ CO_2(g)$$

If the reactants and products are at equilibrium, what happens if the concentration of CO is increased? Now the equilibrium is disturbed and, as a result, the rate of the forward reaction increases to produce more CO_2. We can think of **added reactant as driving the equilibrium to the *right*.**

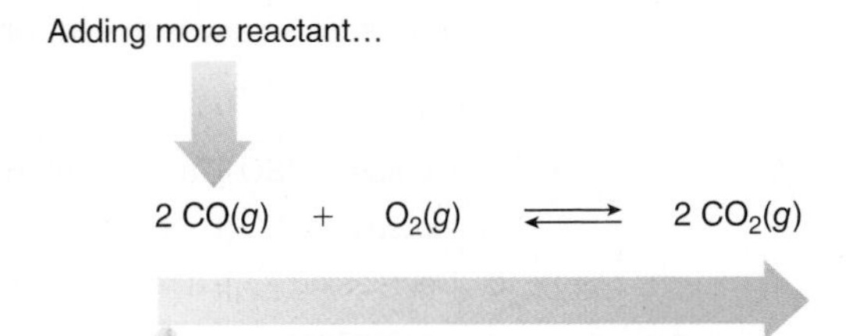
Adding more reactant...
2 CO(g) + O2(g) ⇌ 2 CO2(g)
...drives the reaction to the right.

What happens, instead, if the concentration of CO_2 is increased? Now the equilibrium is disturbed but there is more product than there should be. As a result, the rate of the reverse reaction increases to produce more of both reactants, CO and O_2. We can think of **added product as driving the equilibrium to the *left*.**

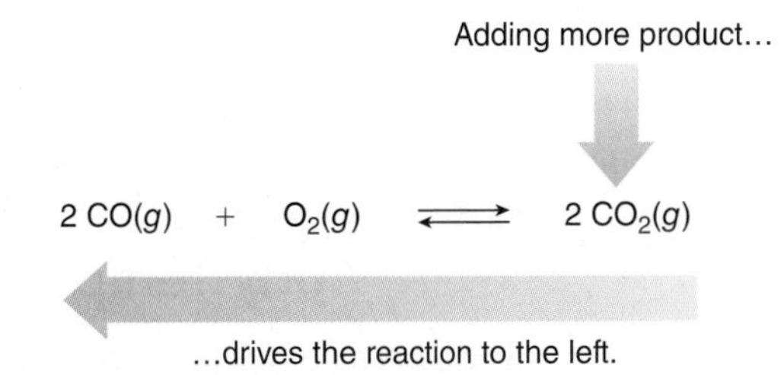
Adding more product...
2 CO(g) + O2(g) ⇌ 2 CO2(g)
...drives the reaction to the left.

Similar arguments can be made about the effect of *decreasing* the concentration of a reactant or product. For example, ethanol (C_2H_6O) can be converted to ethylene ($CH_2{=}CH_2$) and water in the presence of a small amount of acid, but equilibrium does not favor the products.

$$C_2H_6O \rightleftharpoons H_2C{=}CH_2 + H_2O$$

ethanol

Removing a product...

...drives the reaction to the right.

In this case, water can be removed from the reaction as it is formed. A decrease in concentration of one product results in more of the forward reaction to form more product. **This process drives the equilibrium to the right.** If water is continuously removed, essentially all of the starting material can be converted to product. Table 5.3 summarizes the effects of concentration changes on the direction of equilibrium.

Table 5.3 The Effect of Concentration Changes on Equilibrium

Change	Effect on Equilibrium
Adding reactant	Equilibrium favors the products.
Removing reactant	Equilibrium favors the reactants.
Adding product	Equilibrium favors the reactants.
Removing product	Equilibrium favors the products.

SAMPLE PROBLEM 5.16

In which direction is the equilibrium shifted with each of the following concentration changes for the given reaction: (a) increase $[SO_2]$; (b) increase $[SO_3]$; (c) decrease $[O_2]$; (d) decrease $[SO_3]$?

$$2\ SO_2(g) + O_2(g) \rightleftharpoons 2\ SO_3(g)$$

Analysis

Use Le Châtelier's principle to predict the effect of a change in concentration on equilibrium. Adding more reactant or removing product drives the equilibrium to the right. Adding more product or removing reactant drives the equilibrium to the left.

Solution

a. Increasing $[SO_2]$, a reactant, drives the equilibrium to the right to form more product.
b. Increasing $[SO_3]$, a product, drives the equilibrium to the left to form more reactants.
c. Decreasing $[O_2]$, a reactant, drives the equilibrium to the left to form more reactants.
d. Decreasing $[SO_3]$, a product, drives the equilibrium to the right to form more product.

PROBLEM 5.38

In which direction is the equilibrium shifted with each of the following concentration changes for the given reaction: (a) increase $[H_2]$; (b) increase [HCl]; (c) decrease $[Cl_2]$; (d) decrease [HCl]?

$$H_2(g) + Cl_2(g) \rightleftharpoons 2\ HCl(g)$$

5.11 FOCUS ON THE HUMAN BODY
Body Temperature

The human body is an enormously complex organism that illustrates important features of energy and reaction rates. At any moment, millions of reactions occur in the body, when nutrients are metabolized and new cell materials are synthesized.

Normal body temperature, 37 °C, reflects a delicate balance between the amount of heat absorbed and released in all of the reactions and other processes. Since reaction rate increases with increasing temperature, it is crucial to maintain the right temperature for proper body function. When temperature increases, reactions proceed at a faster rate. An individual must breathe more rapidly and the heart must pump harder to supply oxygen for the faster metabolic processes. When temperature decreases, reactions slow down, less heat is generated in exothermic reactions, and it becomes harder and harder to maintain an adequate body temperature.

Thermoregulation—regulating temperature—is a complex mechanism that involves the brain, the circulatory system, and the skin (Figure 5.11). Temperature sensors in the skin and body core signal when there is a temperature change. The hypothalamus region of the brain, in turn, responds to changes in its environment.

When the temperature increases, the body must somehow rid itself of excess heat. Blood vessels near the surface of the skin are dilated to release more heat. Sweat glands are stimulated, so the body can be cooled by the evaporation of water from the skin's surface.

When the temperature decreases, the body must generate more heat as well as slow down the loss of heat to the surroundings. Blood vessels constrict to reduce heat loss from the skin and muscles shiver to generate more heat.

An infection in the body is often accompanied by a fever; that is, the temperature in the body increases. A fever is part of the body's response to increase the rates of defensive reactions that kill bacteria. The respiratory rate and heart rate increase to supply more oxygen needed for faster reactions.

Figure 5.11

Temperature Regulation in the Body

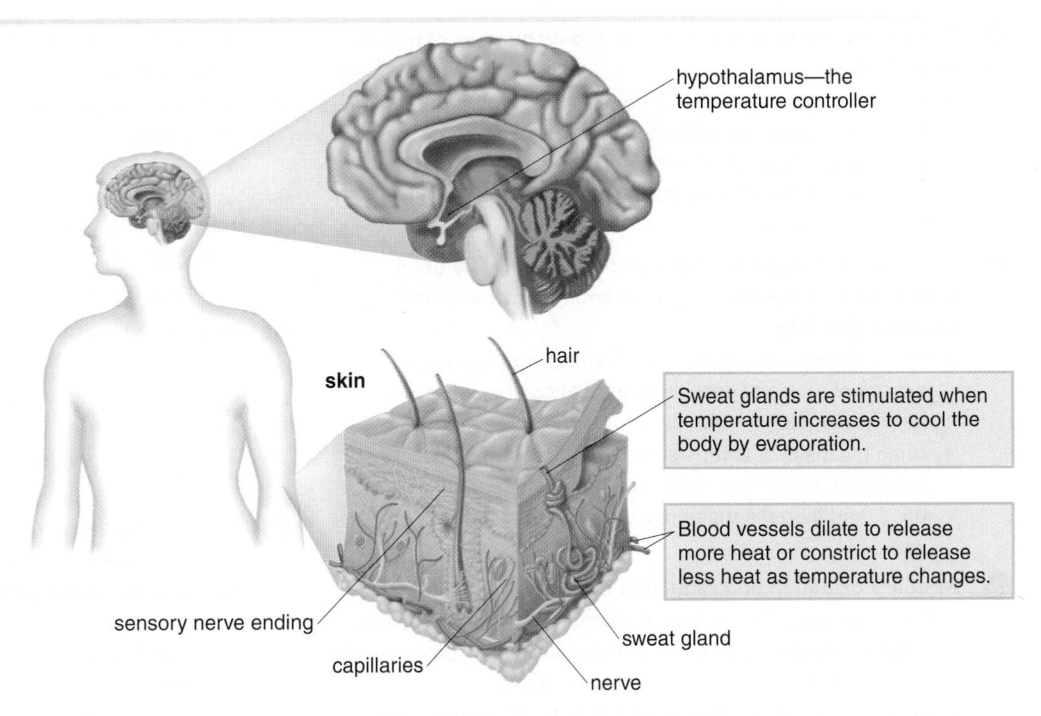
hypothalamus—the temperature controller
hair
skin
Sweat glands are stimulated when temperature increases to cool the body by evaporation.
Blood vessels dilate to release more heat or constrict to release less heat as temperature changes.
sensory nerve ending
sweat gland
capillaries
nerve

When the temperature in the environment around the body changes, the body works to counteract the change. The hypothalamus acts as a thermostat, which signals the body to respond to temperature changes. When the temperature increases, the body must dissipate excess heat by dilating blood vessels and sweating. When the temperature decreases, blood vessels constrict and the body shivers.

KEY TERMS

Avogadro's number (5.3)
Balanced chemical equation (5.2)
Catalyst (5.9)
Chemical equation (5.1)
Endothermic reaction (5.8)
Energy diagram (5.8)
Energy of activation (5.8)
Enthalpy change (5.8)
Equilibrium (5.10)
Exothermic reaction (5.8)
Formula weight (5.4)
Forward reaction (5.10)
Half reaction (5.7)
Heat of reaction (5.8)
Law of conservation of matter (5.1)
Le Châtelier's principle (5.10)
Molar mass (5.4)
Mole (5.3)
Molecular weight (5.4)
Oxidation (5.7)
Oxidizing agent (5.7)
Product (5.1)
Reactant (5.1)
Reaction rate (5.9)
Redox reaction (5.7)
Reducing agent (5.7)
Reduction (5.7)
Reverse reaction (5.10)
Reversible reaction (5.10)
Transition state (5.8)

KEY CONCEPTS

1 What do the terms in a chemical equation mean and how is an equation balanced? (5.1, 5.2)

- A chemical equation contains the reactants on the left side of an arrow and the products on the right. The coefficients tell how many molecules or moles of a substance react or are formed.
- A chemical equation is balanced by placing coefficients in front of chemical formulas one at a time, beginning with the most complex formula, so that the number of atoms of each element is the same on both sides.

2 Define the terms mole and Avogadro's number. (5.3)

- A mole is a quantity that contains 6.02×10^{23} atoms, molecules, or ions.
- Avogadro's number is the number of particles in a mole—6.02×10^{23}.

3 How is molar mass calculated? (5.4)

- The molar mass is the mass of one mole of a substance, reported in grams. The molar mass is numerically equal to the formula weight but the units are different (g/mol not amu).

4 How are the mass of a substance and its number of moles related? (5.4)

- The molar mass is used as a conversion factor to determine how many grams are contained in a given number of moles of a substance. Similarly, the molar mass is used to determine how many moles of a substance are contained in a given number of grams.

5 How can a balanced equation and molar mass be used to calculate the number of moles and mass of a reaction product? (5.5, 5.6)

- The coefficients in a balanced chemical equation tell us the number of moles of each reactant that combine and the number of moles of each product formed.
- When the mass of a substance in a reaction must be calculated, first its number of moles is determined using mole ratios, and then the molar mass is used to convert moles to grams.

6 What are oxidation and reduction reactions? (5.7)

- Oxidation results in the loss of electrons. Metals and anions tend to undergo oxidation. In some reactions, oxidation results in the gain of O atoms or the loss of H atoms.
- Reduction results in the gain of electrons. Nonmetals and cations tend to undergo reduction. In some reactions, reduction results in the loss of O atoms or the gain of H atoms.

7 What is the heat of reaction and what is the difference between an endothermic and an exothermic reaction? (5.8)

- The heat of reaction, also called the enthalpy change and symbolized by ΔH, is the energy absorbed or released in a reaction.
- In an endothermic reaction, energy is absorbed, ΔH is positive (+), and the products are higher in energy than the reactants.
- In an exothermic reaction, energy is released, ΔH is negative (–), and the reactants are higher in energy than the products.
- An energy diagram illustrates the energy changes that occur during the course of a reaction. Energy is plotted on the vertical axis and reaction coordinate is plotted on the horizontal axis. The transition state is located at the top of the energy barrier that separates the reactants and products.

8 How do temperature, concentration, and catalysts affect the rate of a reaction? (5.9)

- Increasing the temperature and concentration increases the reaction rate.
- A catalyst speeds up the rate of a reaction without affecting the energies of the reactants and products.

9 What are the basic features of equilibrium and how does Le Châtelier's principle explain what happens when equilibrium is disturbed? (5.10)

- At equilibrium, the rates of the forward and reverse reactions are equal and the net concentrations of all substances do not change.
- Le Châtelier's principle states that a system at equilibrium reacts to counteract any disturbance to the equilibrium. How changes in concentration affect equilibrium are summarized in Table 5.3.

10 How is temperature regulated in the body? (5.11)

- Increasing temperature increases the rates of reactions in the body. When temperature increases, reactions proceed at a faster rate, an individual breathes faster, and the heart pumps harder. When temperature decreases, reactions slow down and less heat is generated in exothermic reactions.

UNDERSTANDING KEY CONCEPTS

Selected in-chapter and odd-numbered end-of-chapter problems have brief answers at the end of each chapter. The *Student Study Guide and Solutions Manual* contains detailed solutions to all in-chapter and odd-numbered end-of-chapter problems, as well as additional worked examples and a chapter self-test.

5.39 Use the molecular art to identify the process as a chemical reaction or a physical change. If a chemical reaction occurs, write the equation from the molecular art. If a physical change occurs, identify the phase change.

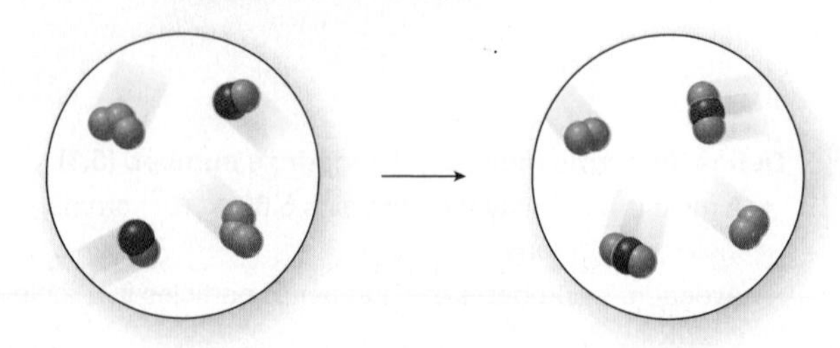

5.40 Diagram [1] represents the compound **AB,** which contains two elements, **A** shown with red spheres, and **B** shown with blue spheres. Label each transformation as a physical change or a chemical reaction: (a) [1] ⟶ [2]; (b) [1] ⟶ [3].

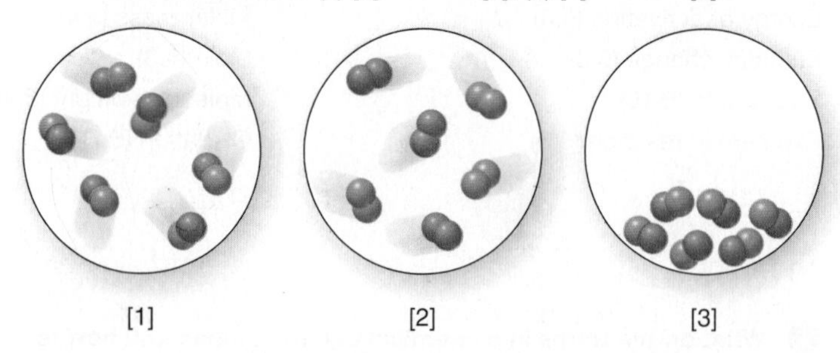

5.41 Use the molecular art to write a balanced equation for the given reaction.

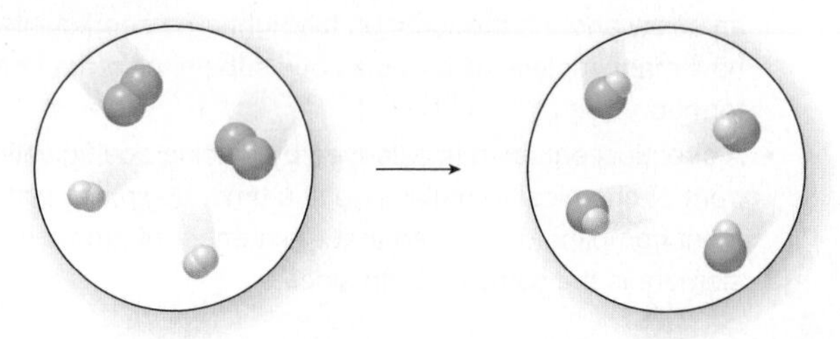

5.42 Use the molecular art to write a balanced equation for the given reaction.

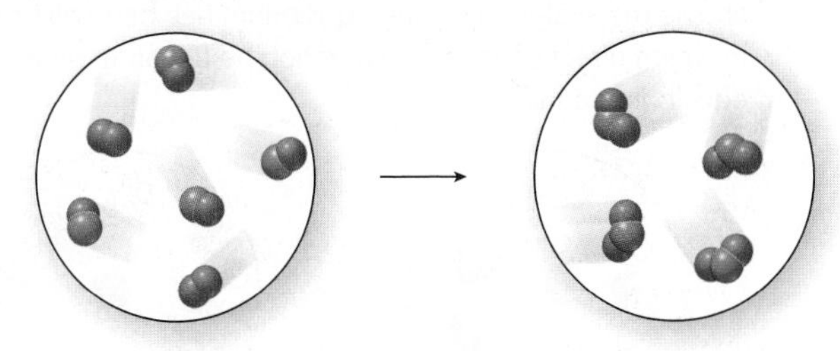

5.43 Some coal is high in sulfur (S) content, and when it burns, it forms sulfuric acid (H_2SO_4), a major component of acid rain, by a series of reactions. Balance the equation for the overall conversion drawn below.

$$S(s) + O_2(g) + H_2O(l) \longrightarrow H_2SO_4(l)$$

5.44 Balance the equation for the formation of magnesium hydroxide [$Mg(OH)_2$], one of the active ingredients in milk of magnesia.

$$MgCl_2 + NaOH \longrightarrow Mg(OH)_2 + NaCl$$

5.45 What is the mass in grams of each quantity of lactic acid ($C_3H_6O_3$, molar mass 90.08 g/mol), the compound responsible for the aching feeling of tired muscles during vigorous exercise?

a. 3.60 mol b. 7.3×10^{24} molecules

5.46 Spinach, cabbage, and broccoli are excellent sources of vitamin K (molar mass 450.7 g/mol), which is needed in adequate amounts for blood to clot. The recommended daily intake of vitamin K is 120 μg. How many molecules of vitamin K does this correspond to?

5.47 Fats, such as butter, and oils, such as corn oil, are formed from compounds called fatty acids, one of which is linolenic acid ($C_{18}H_{30}O_2$). Linolenic acid undergoes reactions with hydrogen and oxygen to form the products shown in each equation.

[1] $C_{18}H_{30}O_2 + H_2 \longrightarrow C_{18}H_{36}O_2$
linolenic acid

[2] $C_{18}H_{30}O_2 + O_2 \longrightarrow CO_2 + H_2O$
linolenic acid

a. Calculate the molar mass of linolenic acid.
b. Balance Equation [1], which shows the reaction with hydrogen.
c. Balance Equation [2], which shows the reaction with oxygen.
d. How many grams of product are formed from 10.0 g of linolenic acid in Equation [1]?

5.48 Iron, like most metals, does not occur naturally as the pure metal. Rather, it must be produced from iron ore, which contains iron(III) oxide, according to the given balanced equation.

$$Fe_2O_3(s) + 3\,CO(g) \longrightarrow 2\,Fe(s) + 3\,CO_2(g)$$

a. How many grams of Fe are formed from 10.0 g of Fe_2O_3?
b. How many grams of Fe are formed from 25.0 g of Fe_2O_3?

5.49 Zinc–silver oxide batteries are used in cameras and hearing aids. Identify the species that is oxidized and the species that is reduced in the following redox reaction. Identify the oxidizing agent and the reducing agent. Oxide (O^{2-}) is unchanged in this reaction.

$$Zn + Ag_2O \longrightarrow ZnO + 2\,Ag$$

5.50 Rechargeable nickel–cadmium batteries are used in appliances and power tools. Identify the species that is oxidized and the species that is reduced in the following redox reaction. Identify the oxidizing agent and the reducing agent.

$$Cd + Ni^{4+} \longrightarrow Cd^{2+} + Ni^{2+}$$

5.51 Consider the energy diagram drawn below.

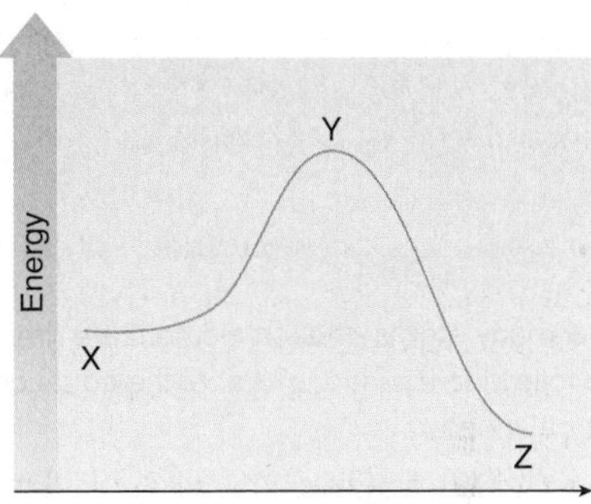

a. Which point on the graph corresponds to reactants?
b. Which point on the graph corresponds to products?
c. Which point on the graph corresponds to the transition state?
d. The difference in energy between which two points equals the energy of activation?
e. The difference in energy between which two points equals the ΔH?
f. Which point is highest in energy?
g. Which point is lowest in energy?

5.52 Compound **A** can be converted to either **B** or **C.** The energy diagrams for both processes are drawn on the graph below.

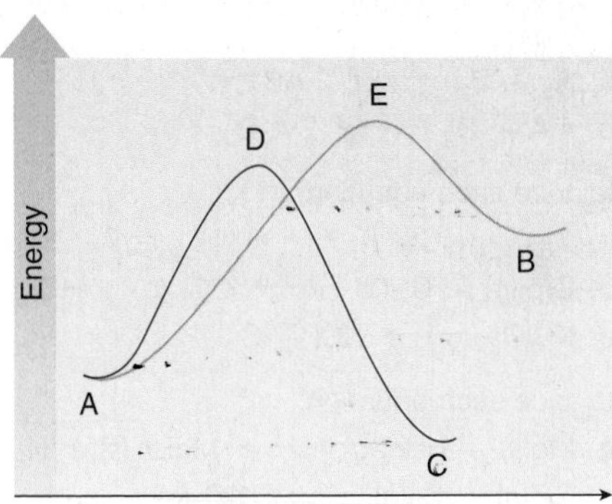

a. Label each reaction as endothermic or exothermic.
b. Which reaction is faster?
c. Which reaction generates the product lower in energy?
d. Which points on the graphs correspond to transition states?
e. Label the energy of activation for each reaction.
f. Label the ΔH for each reaction.

5.53 Consider the reversible reaction $\mathbf{A} + \mathbf{B} \rightleftharpoons \mathbf{AB}$, shown in representation [1] at equilibrium. If representation [2] shows the system after the equilibrium has been disturbed, in which direction must the reaction be driven to achieve equilibrium again?

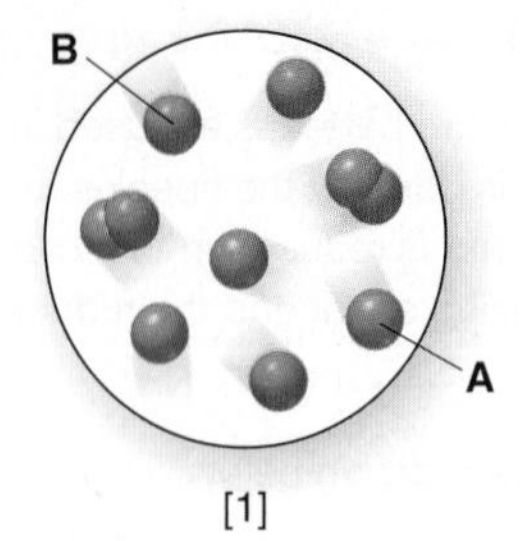

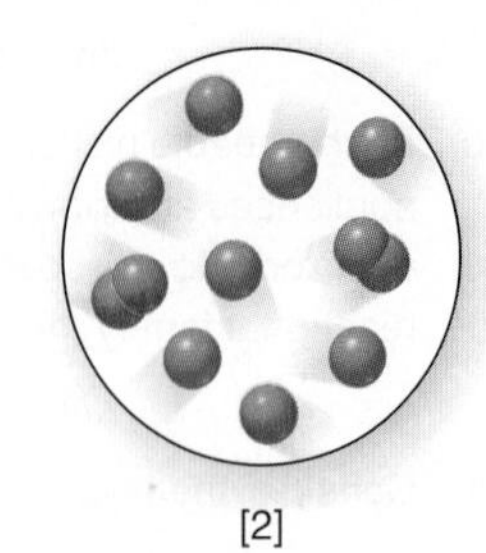

5.54 Consider the reversible reaction $2\,\mathbf{A} + \mathbf{B}_2 \rightleftharpoons 2\,\mathbf{AB}$, shown in representation [3] at equilibrium. If representation [4] shows the system after the equilibrium has been disturbed, in which direction must the reaction be driven to achieve equilibrium again?

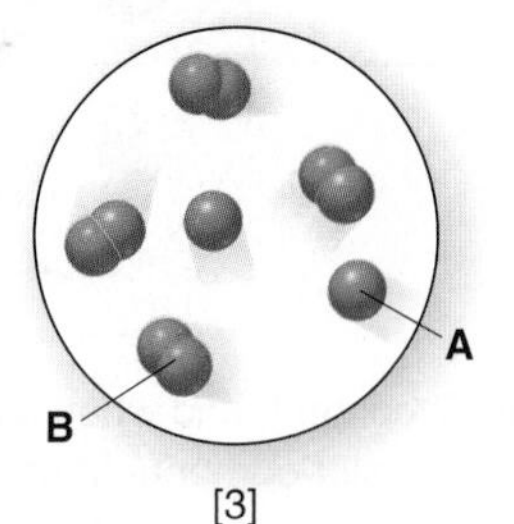

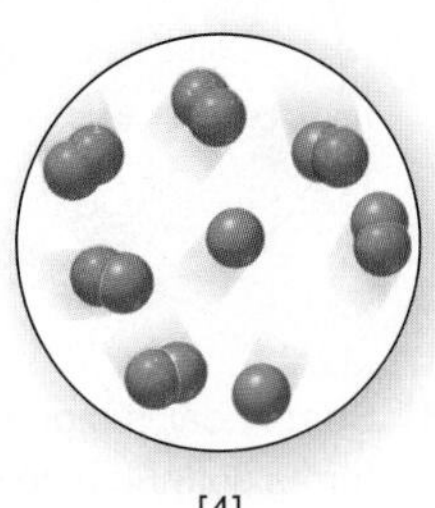

ADDITIONAL PROBLEMS

Chemical Equations

5.55 How many atoms of each element are drawn on each side of the following equations? Label the equations as balanced or not balanced.

a. $2\,HCl(aq) + Ca(s) \longrightarrow CaCl_2(aq) + H_2(g)$
b. $Al(OH)_3 + H_3PO_4 \longrightarrow AlPO_4 + 3\,H_2O$

5.56 How many atoms of each element are drawn on each side of the following equations? Label the equations as balanced or not balanced.

a. $3\,NO_2 + H_2O \longrightarrow HNO_3 + 2\,NO$
b. $Ca(OH)_2 + 2\,HNO_3 \longrightarrow 2\,H_2O + Ca(NO_3)_2$

5.57 Draw representations using molecular art for each equation. Use a red sphere for each **A** atom and a blue sphere for each **B** atom.

a. $2\,\mathbf{A} + 2\,\mathbf{B} \longrightarrow 2\,\mathbf{AB}$
b. $\mathbf{A}_2 + 4\,\mathbf{B} \longrightarrow 2\,\mathbf{AB}_2$

5.58 Draw representations using molecular art for each equation. Use a red sphere for each **A** atom and a blue sphere for each **B** atom.

a. $\mathbf{A}_2 + \mathbf{B}_2 \longrightarrow 2\,\mathbf{AB}$
b. $4\,\mathbf{A} + \mathbf{B}_2 \longrightarrow 2\,\mathbf{A}_2\mathbf{B}$

5.59 Balance each equation.

a. $Ni(s) + HCl(aq) \longrightarrow NiCl_2(aq) + H_2(g)$
b. $CH_4(g) + Cl_2(g) \longrightarrow CCl_4(g) + HCl(g)$
c. $KClO_3 \longrightarrow KCl + O_2$

5.60 Balance each equation.

a. $Mg(s) + HBr(aq) \longrightarrow MgBr_2(s) + H_2(g)$
b. $CO(g) + O_2(g) \longrightarrow CO_2(g)$
c. $H_2SO_4 + NaOH \longrightarrow Na_2SO_4 + H_2O$

5.61 Hydrocarbons are compounds that contain only C and H atoms. When a hydrocarbon reacts with O_2, CO_2 and H_2O are formed. Write a balanced equation for the combustion of each of the following hydrocarbons, both of which are high-octane components of gasoline.

a. C_6H_6 (benzene)
b. C_7H_8 (toluene)

5.62 MTBE ($C_5H_{12}O$) is a high-octane gasoline additive with a sweet, nauseating odor. Because small amounts of MTBE have contaminated the drinking water in some towns, it is now banned as a fuel additive in some states. MTBE reacts with O_2 to form CO_2 and H_2O. Write a balanced equation for the combustion of MTBE.

5.63 Consider the reaction, $O_3 + CO \longrightarrow O_2 + CO_2$. Molecular art is used to show the starting materials for this reaction. Fill in the molecules of the products using the balanced equation and following the law of conservation of matter.

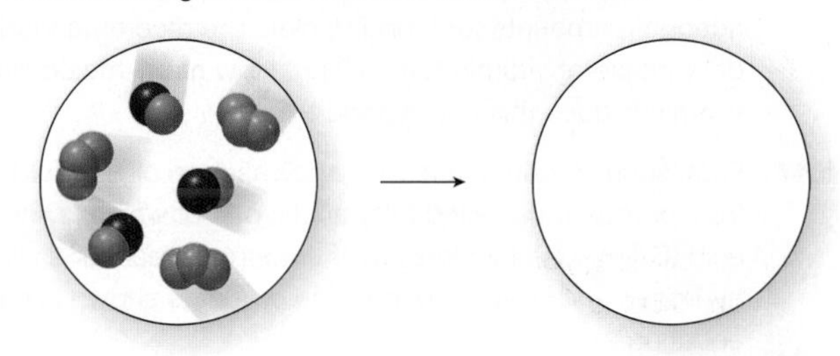

5.64 Consider the reaction, $2\,NO + 2\,CO \longrightarrow N_2 + 2\,CO_2$. Molecular art is used to show the starting materials for this reaction. Fill in the molecules of the products using the balanced equation and following the law of conservation of matter.

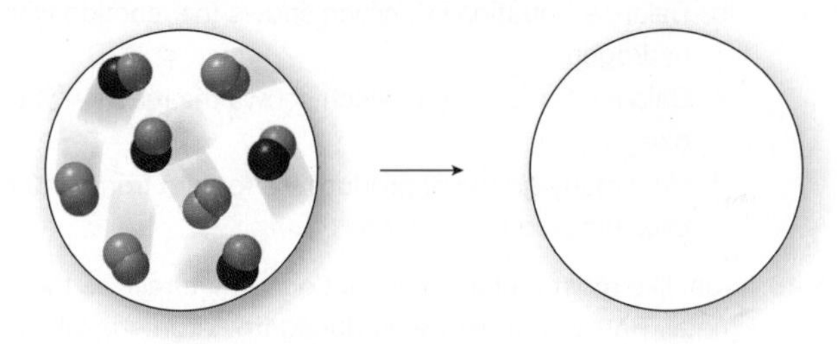

Formula Weight and Molar Mass

5.65 Calculate the formula weight and molar mass of each compound.

a. $NaNO_2$ (sodium nitrite), a preservative in hot dogs, ham, and other cured meats

b. C_2H_4 (ethylene), the industrial starting material for the plastic polyethylene

c. $C_9H_{13}NO_2$ (phenylephrine), a decongestant in Sudafed PE

5.66 L-Dopa is a drug used to treat Parkinson's disease.

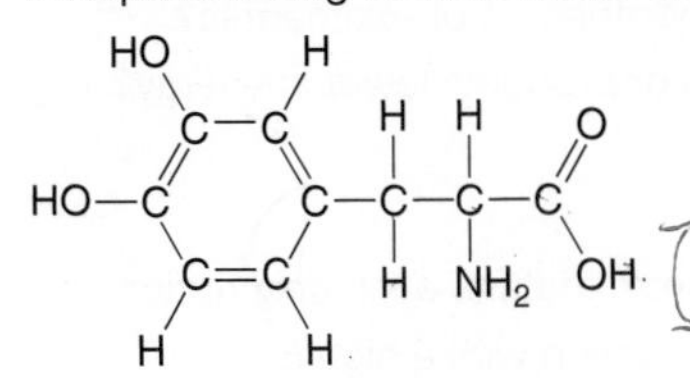

L-dopa

a. What is the molecular formula of L-dopa?

b. What is the formula weight of L-dopa?

c. What is the molar mass of L-dopa?

Moles, Mass, and Avogadro's Number

5.67 How many grams are contained in 5.00 mol of each compound?

a. HCl b. Na_2SO_4 c. C_2H_2

5.68 How many grams are contained in 0.50 mol of each compound?

a. NaOH b. $CaSO_4$ c. C_3H_6

5.69 How many moles are contained in each number of grams of table sugar ($C_{12}H_{22}O_{11}$, molar mass 342.3 g/mol)?

a. 0.500 g b. 5.00 g c. 25.0 g

5.70 How many moles are contained in each number of grams of fructose ($C_6H_{12}O_6$, molar mass 180.2 g/mol), a carbohydrate that is about twice as sweet as table sugar? "Lite" food products use half as much fructose as table sugar to achieve the same sweet taste, but with fewer calories.

a. 0.500 g b. 5.00 g c. 25.0 g

5.71 Mescaline is a hallucinogen in peyote, a cactus native to the southwestern United States and Mexico. (a) What is the chemical formula of mescaline? (b) Calculate its molar mass. (b) How many moles are contained in 7.50 g of mescaline?

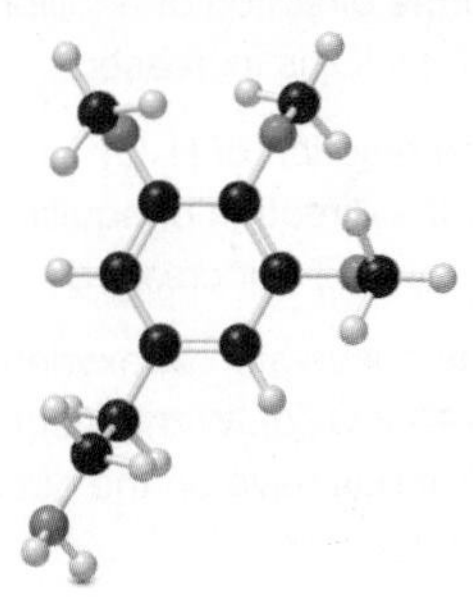

mescaline

5.72 (a) What is the chemical formula of **A?** (b) Calculate its molar mass. (b) How many moles are contained in 10.50 g of **A?**

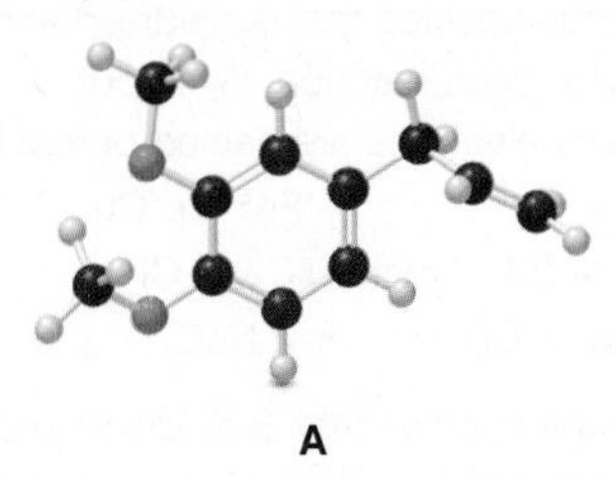

A

5.73 Which has the greater mass: 0.050 mol of aspirin or 10.0 g of aspirin ($C_9H_8O_4$)?

5.74 What is the mass in grams of 2.02×10^{20} molecules of the pain reliever ibuprofen ($C_{13}H_{18}O_2$, molar mass 206.3 g/mol)?

Mass and Mole Calculations in Chemical Equations

5.75 Using the balanced equation for the combustion of acetylene, answer the following questions.

$$2\ H{-}C{\equiv}C{-}H + 5\ O_2 \longrightarrow 4\ CO_2 + 2\ H_2O$$

acetylene

a. How many moles of O_2 are needed to react completely with 5.00 mol of C_2H_2?

b. How many moles of CO_2 are formed from 6.0 mol of C_2H_2?

c. How many moles of H_2O are formed from 0.50 mol of C_2H_2?

5.76 Sodium metal (Na) reacts violently when added to water according to the following balanced equation.

$$2\ Na(s) + 2\ H_2O(l) \longrightarrow 2\ NaOH(aq) + H_2(g)$$

a. How many moles of H_2O are needed to react completely with 3.0 mol of Na?

b. How many moles of H_2 are formed from 0.38 mol of Na?

c. How many moles of H_2 are formed from 3.64 mol of H_2O?

5.77 Using the balanced equation for the combustion of acetylene in Problem 5.75, answer the following questions.

a. How many grams of CO_2 are formed from 2.5 mol of C_2H_2?

b. How many grams of CO_2 are formed from 0.50 mol of C_2H_2?

c. How many grams of H_2O are formed from 0.25 mol of C_2H_2?

5.78 Using the balanced equation for the reaction of Na with H_2O in Problem 5.76, answer the following questions.

a. How many grams of NaOH are formed from 3.0 mol of Na?

b. How many grams of H_2 are formed from 0.30 mol of Na?

c. How many grams of H_2O are needed to react completely with 0.20 mol of Na?

Oxidation–Reduction Reactions

5.79 Identify the species that is oxidized and the species that is reduced in each reaction. Write out two half reactions to show how many electrons are gained or lost by each species.

a. $Fe + Cu^{2+} \longrightarrow Fe^{2+} + Cu$
b. $Cl_2 + 2\,I^- \longrightarrow I_2 + 2\,Cl^-$
c. $2\,Na + Cl_2 \longrightarrow 2\,NaCl$

5.80 Identify the species that is oxidized and the species that is reduced in each reaction. Write out two half reactions to show how many electrons are gained or lost by each species.

a. $Mg + Fe^{2+} \longrightarrow Mg^{2+} + Fe$
b. $Cu^{2+} + Sn \longrightarrow Sn^{2+} + Cu$
c. $4\,Na + O_2 \longrightarrow 2\,Na_2O$

5.81 The reaction of magnesium metal (Mg) with oxygen (O_2) forms MgO. Write a balanced equation for this redox reaction. Write two half reactions to show how many electrons are gained or lost by each species.

5.82 When Cl_2 is used to disinfect drinking water, Cl^- is formed. Is Cl_2 oxidized or reduced in this process?

Energy Changes in Reactions

5.83 Do each of the following statements describe an endothermic or exothermic reaction?

a. ΔH is a negative value.
b. The energy of the reactants is lower than the energy of the products.
c. Energy is absorbed in the reaction.

5.84 Do each of the following statements describe an endothermic or exothermic reaction?

a. ΔH is a positive value.
b. The energy of the products is lower than the energy of the reactants.
c. Energy is released in the reaction.

5.85 The combustion of coal with oxygen forms CO_2 according to the given equation.

$$C(s) + O_2(g) \longrightarrow CO_2(g) \qquad \Delta H = -94 \text{ kcal/mol}$$

a. Is heat absorbed or released?
b. Are the reactants or products lower in energy?
c. Is the reaction endothermic or exothermic?

5.86 Ammonia (NH_3) is formed from hydrogen and nitrogen according to the given equation.

$$3\,H_2(g) + N_2(g) \longrightarrow 2\,NH_3(g) \qquad \Delta H = -22.0 \text{ kcal/mol}$$

a. Is heat absorbed or released?
b. Are the reactants or products lower in energy?
c. Is the reaction endothermic or exothermic?

5.87 The metabolism of glucose with oxygen forms CO_2 and H_2O and releases 678 kcal/mol of energy.

$$\underset{\text{glucose}}{C_6H_{12}O_6(aq)} + 6\,O_2(g) \longrightarrow 6\,CO_2(g) + 6\,H_2O(l)$$

a. Is ΔH for this reaction positive or negative?
b. Is the reaction endothermic or exothermic?
c. Are the reactants or products lower in energy?

5.88 Ethanol (C_2H_6O), a gasoline additive, is formed by the reaction of ethylene ($CH_2{=}CH_2$) with water. This reaction releases 9.0 kcal/mol of energy.

$$\underset{\text{ethylene}}{H_2C{=}CH_2} + H_2O \longrightarrow \underset{\text{ethanol}}{C_2H_6O}$$

a. Is ΔH for this reaction positive or negative?
b. Is the reaction endothermic or exothermic?
c. Are the reactants or products lower in energy?

Energy Diagrams

5.89 Draw an energy diagram that fits each description.

a. an endothermic reaction with a high E_a
b. a reaction that has a low E_a and ΔH is negative

5.90 Draw an energy diagram for the following reaction in which $\Delta H = -12$ kcal/mol and $E_a = 5$ kcal: $A_2 + B_2 \longrightarrow 2\,AB$. Label the axes, reactants, products, transition state, E_a, and ΔH. Is the reaction endothermic or exothermic?

Reaction Rates

5.91 Which value (if any) in each pair corresponds to a faster reaction? Explain your choice.

a. $E_a = 10$ kcal or $E_a = 1$ kcal
b. $\Delta H = -2$ kcal/mol or $\Delta H = +2$ kcal/mol

5.92 Which of the following affect the rate of a reaction: (a) concentration; (b) ΔH; (c) energy difference between the reactants and the transition state?

5.93 Which of the following affect the rate of a reaction: (a) catalyst; (b) E_a; (c) temperature?

5.94 How does a catalyst affect each of the following: (a) reaction rate; (b) ΔH; (c) E_a; (d) relative energy of the reactants and products?

Equilibrium and Le Châtelier's Principle

5.95 Consider the reaction of $N_2(g) + O_2(g) \rightleftharpoons 2\,NO(g)$. What happens to the direction of equilibrium when (a) $[O_2]$ is increased; (b) [NO] is increased?

5.96 Consider the reaction of $H_2(g) + F_2(g) \rightleftharpoons 2\,HF(g)$. What happens to the direction of equilibrium when (a) $[H_2]$ is decreased; (b) [HF] is increased?

5.97 Consider the conversion of oxygen to ozone: $3\,O_2(g) \rightleftharpoons 2\,O_3(g)$. What effect does each of the following changes have on the direction of equilibrium?

a. decrease $[O_3]$
b. decrease $[O_2]$
c. increase $[O_3]$

5.98 Consider the reaction: $H_2(g) + I_2(g) \rightleftharpoons 2\,HI(g)$. What effect does each of the following changes have on the direction of equilibrium?

a. decrease [HI]
b. increase $[H_2]$
c. decrease $[I_2]$

General Questions and Applications

5.99 Answer the following questions about the conversion of the sucrose ($C_{12}H_{22}O_{11}$) in sugarcane to ethanol (C_2H_6O) and CO_2 according to the following unbalanced equation. In this way sugarcane is used as a renewable source of ethanol, which is used as a fuel additive in gasoline.

$$\underset{\text{sucrose}}{C_{12}H_{22}O_{11}(s)} + H_2O(l) \longrightarrow \underset{\text{ethanol}}{C_2H_6O(l)} + CO_2(g)$$

a. What is the molar mass of sucrose?
b. Balance the given equation.
c. How many moles of ethanol are formed from 2 mol of sucrose?
d. How many grams of ethanol are formed from 34.2 g of sucrose?

5.100 Answer the following questions about diethyl ether ($C_4H_{10}O$), the first widely used general anesthetic. Diethyl ether can be prepared from ethanol according to the following unbalanced equation.

$$\underset{\text{ethanol}}{C_2H_6O(l)} \longrightarrow \underset{\text{diethyl ether}}{C_4H_{10}O(l)} + H_2O(l)$$

a. What is the molar mass of diethyl ether?
b. Balance the given equation.
c. How many moles of diethyl ether are formed from 2 mol of ethanol?
d. How many grams of diethyl ether are formed from 4.60 g of ethanol?

5.101 One dose of Maalox contains 500. mg each of $Mg(OH)_2$ and $Al(OH)_3$. How many moles of each compound are contained in a single dose?

5.102 The average nicotine ($C_{10}H_{14}N_2$, molar mass 162.3 g/mol) content of a Camel cigarette is 1.93 mg. Suppose an individual smokes one pack of 20 cigarettes a day. How many moles of nicotine are smoked in a day?

CHALLENGE PROBLEMS

5.103 TCDD, also called dioxin ($C_{12}H_4Cl_4O_2$, molar mass 322.0 g/mol), is a potent poison. The average lethal dose in humans is estimated to be 3.0×10^{-2} mg per kg of body weight. (a) How many grams constitute a lethal dose for a 70.-kg individual? (b) How many molecules of TCDD does this correspond to?

5.104 The amount of energy released when a fuel burns is called its heat content. The heat content of fuels is often reported in kcal/g not kcal/mol so that fuels with different molar masses can be compared on a mass basis. The heat content of propane (C_3H_8), used as the fuel in gas grills, is 531 kcal/mol, while the heat content of butane (C_4H_{10}), used in lighters, is 688 kcal/mol. Show that the heat content of these two fuels is similar when converted to kcal/g.

BEYOND THE CLASSROOM

5.105 Choose a product that consists of only one compound. Determine the molar mass of the compound and calculate the number of moles contained in a package sold at the market. Possible products include table salt, sugar, baking soda, or bottled water.

5.106 Choose an over-the-counter pain reliever such as aspirin, acetaminophen (the active ingredient in Tylenol), or naproxen (the active ingredient in Aleve). Calculate the number of molecules of the drug contained in a typical dose.

5.107 The combustion of gasoline is a redox reaction in which O_2 oxidizes the C—C and C—H bonds in the gasoline to form carbon dioxide (CO_2) and water (H_2O). One estimate suggests that driving an automobile 10,000 miles per year at 25 miles per gallon releases 10,000 lb of CO_2 into the atmosphere. Using these values as conversion factors, calculate how many pounds of CO_2 your family releases into the atmosphere given your car's gas mileage and the distance it is driven each year. Convert this value to the number of moles of CO_2, as well as the number of molecules of CO_2. Compare your result with others in the class whose driving habits are different from yours.

5.108 A 12-oz can of diet soda contains 180 mg of the artificial sweetener aspartame (trade names NutraSweet and Equal). When aspartame is ingested, it is metabolized to phenylalanine, aspartic acid, and methanol, according to the balanced equation:

$$\underset{\text{aspartame}}{C_{14}H_{18}N_2O_5} + 2\,H_2O \longrightarrow \underset{\text{phenylalanine}}{C_9H_{11}NO_2} + \underset{\text{aspartic acid}}{C_4H_7NO_4} + \underset{\text{methanol}}{CH_3OH}$$

a. Using this equation, determine how much methanol is formed on metabolism of the aspartame in each 12-oz can. The molar mass of aspartame is 294.2 g/mol, and the molar mass of methanol is 32.04 g/mol.
b. Methanol is a toxic compound, with the average lethal dose in humans estimated at 428 mg/kg of body weight. How many cans of soda would you have to ingest to obtain a lethal dose of methanol? Compare your results with others in your class or family.
c. Estimate how many cans of diet soda you drink in a 30-day month, and determine how much methanol you consume during that interval.

5.109 Pick a prescription medication such as atorvastatin (trade name Lipitor), fluoxetine (trade name Prozac), levothyroxine (trade name Synthroid), or digoxin (trade name Lanoxin). Determine the chemical formula and molar mass. Research how much drug is present in a typical dose and calculate how many grams would be taken in a 30-day month. How many moles does this correspond to? Compare the values with others in your class who chose different medications. Are any of the drugs taken in a particularly large or small dose?

ANSWERS TO SELECTED PROBLEMS

5.1 The process is a chemical reaction because the reactants contain two gray spheres joined (indicating H_2) and two red spheres joined (indicating O_2), while the product (H_2O) contains a red sphere joined to two gray spheres (indicating O—H bonds).

5.3 reactants products
a. $2\ H_2O_2(aq) \longrightarrow 2\ H_2O(l) + O_2(g)$ (4 H, 4 O)
b. $2\ C_8H_{18} + 25\ O_2 \longrightarrow 16\ CO_2 + 18\ H_2O$ (16 C, 50 O, 36 H)

5.5 $CH_4(g) + 4\ Cl_2(g) \xrightarrow{\Delta} CCl_4(l) + 4\ HCl(g)$

5.6 a. $2\ H_2 + O_2 \longrightarrow 2\ H_2O$
b. $2\ NO + O_2 \longrightarrow 2\ NO_2$
c. $4\ Fe + 3\ O_2 \longrightarrow 2\ Fe_2O_3$
d. $CH_4 + 2\ Cl_2 \longrightarrow CH_2Cl_2 + 2\ HCl$

5.7 $2\ CO + O_2 \longrightarrow 2\ CO_2$

5.8 $2\ C_2H_6 + 7\ O_2 \longrightarrow 4\ CO_2 + 6\ H_2O$

5.9 $N_2 + 3\ H_2 \longrightarrow 2\ NH_3$

5.10 a. $2\ Al + 3\ H_2SO_4 \longrightarrow Al_2(SO_4)_3 + 3\ H_2$
b. $3\ Na_2SO_3 + 2\ H_3PO_4 \longrightarrow 3\ H_2SO_3 + 2\ Na_3PO_4$

5.11 a,b,c,d: 6.02×10^{23}

5.12 a. 1.20×10^{24} atoms c. 3.01×10^{23} atoms
b. 3.61×10^{24} atoms d. 1.51×10^{25} atoms

5.13 a. 1.5×10^{24} molecules c. 2.4×10^{23} molecules
b. 1.5×10^{23} molecules d. 3.33×10^{25} molecules

5.14 a. 100. mol b. 0.0500 mol c. 15 mol

5.15 a. 100.09 amu b. 166.0 amu

5.16 a. 73.89 g/mol b. 64.51 g/mol c. 239.3 g/mol

5.17 a. $C_9H_{14}O$ b. 138.2 g/mol

5.18 a. 29.2 g b. 332 g c. 101 g d. 26.3 g

5.19 a. 1.71 mol c. 1.39×10^{-3} mol
b. 1.59 mol d. 1.39 mol

5.20 a. 6.6 mol b. 1.0 mol c. 1.2 mol

5.21 a. 7.5 mol b. 1.5 mol c. 1.5 mol

5.22 a. 51 g b. 22 g c. 90. g

5.23 a. 44 g b. 130 g c. 24 g

5.24 a. 72.4 g b. 24.1 g c. 7.24 g

5.25 a. 21.4 g b. 18.8 g c. 11.4 g

5.26 (oxidized) (reduced)
a. $Zn(s) + 2\ H^+(aq) \longrightarrow Zn^{2+}(aq) + H_2(g)$
$Zn \longrightarrow Zn^{2+} + 2\ e^-$
$2\ H^+ + 2\ e^- \longrightarrow H_2$
(reduced) (oxidized)
b. $Fe^{3+}(aq) + Al(s) \longrightarrow Al^{3+}(aq) + Fe(s)$
$Al \longrightarrow Al^{3+} + 3\ e^-$
$Fe^{3+} + 3\ e^- \longrightarrow Fe$
(oxidized) (reduced)
c. $2\ I^- + Br_2 \longrightarrow I_2 + 2\ Br^-$
$2\ I^- \longrightarrow I_2 + 2\ e^-$
$Br_2 + 2\ e^- \longrightarrow 2\ Br^-$
d. $2\ AgBr \longrightarrow 2\ Ag + Br_2$
$2\ Br^-$ (oxidized) $2\ Ag^+$ (reduced)
$2\ Br^- \longrightarrow Br_2 + 2\ e^-$
$2\ Ag^+ + 2\ e^- \longrightarrow 2\ Ag$

5.27 Zn is oxidized, and Hg^{2+} is reduced.
$Zn \longrightarrow Zn^{2+} + 2\ e^-$
$Hg^{2+} + 2\ e^- \longrightarrow Hg$

5.29 Zn is the reducing agent and Hg^{2+} is the oxidizing agent.

5.31 a. released b. products c. exothermic

5.32

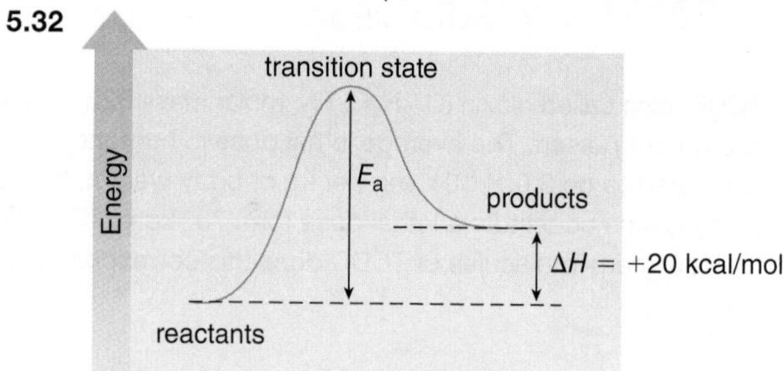

5.33

Energy
E_a
$H_2O + HCl$
ΔH is negative.
$H_3O^+ + Cl^-$
Reaction coordinate

5.35 The catalyst does not affect the relative energies of reactants and products. The catalyst lowers the energy of activation.

5.37 a. $2\ SO_2(g) + O_2(g) \longrightarrow 2\ SO_3(g)$ forward reaction
$2\ SO_3(g) \longrightarrow 2\ SO_2(g) + O_2(g)$ reverse reaction
b. $N_2(g) + O_2(g) \longrightarrow 2\ NO(g)$ forward reaction
$2\ NO(g) \longrightarrow N_2(g) + O_2(g)$ reverse reaction
c. $C_2H_4O_2 + CH_4O \longrightarrow C_3H_6O_2 + H_2O$ forward reaction
$C_3H_6O_2 + H_2O \longrightarrow C_2H_4O_2 + CH_4O$ reverse reaction

5.38 a,d: right b,c: left

5.39 $2\ CO + 2\ O_3 \longrightarrow 2\ CO_2 + 2\ O_2$ (not balanced)

5.41 $H_2 + Cl_2 \longrightarrow 2\ HCl$

5.43 $2\ S(s) + 3\ O_2(g) + 2\ H_2O(l) \longrightarrow 2\ H_2SO_4(l)$

5.45 a. 324 g b. 1.1×10^3 g

5.47 a. 278.4 g/mol
b. $C_{18}H_{30}O_2 + 3\ H_2 \longrightarrow C_{18}H_{36}O_2$
c. $2\ C_{18}H_{30}O_2 + 49\ O_2 \longrightarrow 36\ CO_2 + 30\ H_2O$
d. 10.2 g

5.49

Zn	Ag^+
oxidized	reduced
reducing agent	oxidizing agent

5.51 a. X b. Z c. Y d. X,Y e. X,Z f. Y g. Z

5.53 right

5.55 a. 2 H, 2 Cl, 1 Ca on both sides; therefore balanced
b. 1 Al, 1 P, 7 O, 6 H on both sides; therefore balanced

5.57

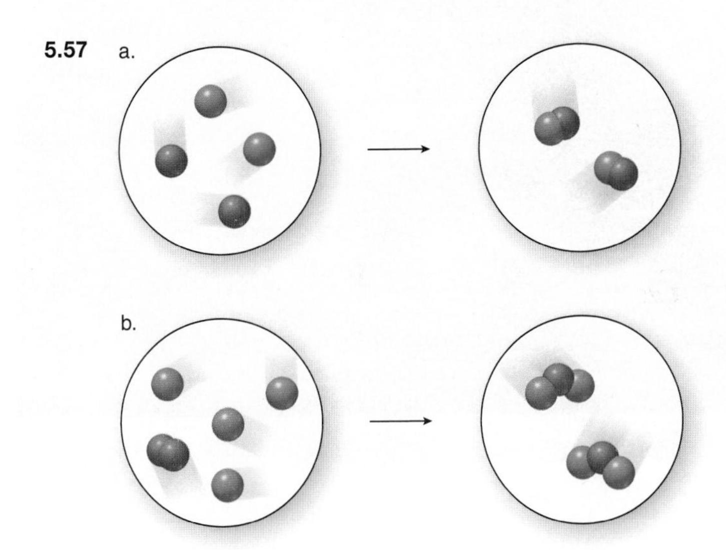

5.59 a. $Ni(s) + 2\ HCl(aq) \longrightarrow NiCl_2(aq) + H_2(g)$
b. $CH_4(g) + 4\ Cl_2(g) \longrightarrow CCl_4(g) + 4\ HCl(g)$
c. $2\ KClO_3 \longrightarrow 2\ KCl + 3\ O_2$

5.61 a. $2\ C_6H_6 + 15\ O_2 \longrightarrow 12\ CO_2 + 6\ H_2O$
b. $C_7H_8 + 9\ O_2 \longrightarrow 7\ CO_2 + 4\ H_2O$

5.63

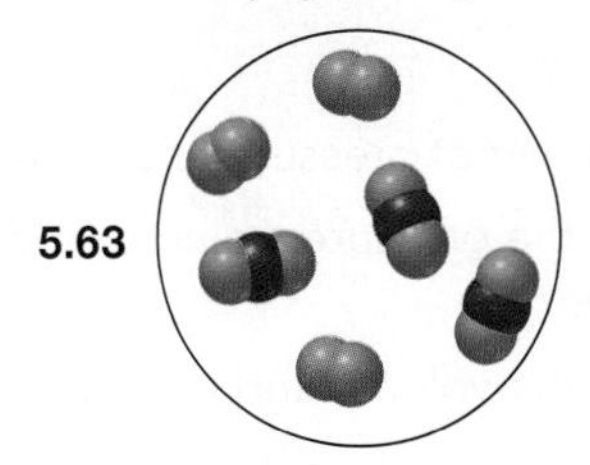

5.65 a. 69.00 amu, 69.00 g/mol
b. 28.05 amu, 28.05 g/mol
c. 167.2 amu, 167.2 g/mol

5.67 a. 182 g b. 710. g c. 130. g

5.69 a. 1.46×10^{-3} mol c. 0.0730 mol
b. 0.0146 mol

5.71 a. $C_{11}H_{17}NO_3$
b. 211.3 g/mol
c. 0.0355 mol

5.73 10.0 g of aspirin

5.75 a. 12.5 mol b. 12 mol c. 0.50 mol

5.77 a. 220 g b. 44 g c. 4.5 g

5.79 a. Fe (oxidized) Cu^{2+} (reduced)
$Fe \longrightarrow Fe^{2+} + 2\ e^-$
$Cu^{2+} + 2\ e^- \longrightarrow Cu$
b. Cl_2 (reduced) $2\ I^-$ (oxidized)
$2\ I^- \longrightarrow I_2 + 2\ e^-$
$Cl_2 + 2\ e^- \longrightarrow 2\ Cl^-$
c. 2 Na (oxidized) Cl_2 (reduced)
$2\ Na \longrightarrow 2\ Na^+ + 2\ e^-$
$Cl_2 + 2\ e^- \longrightarrow 2\ Cl^-$

5.81 $2\ Mg + O_2 \longrightarrow 2\ MgO$
$2\ Mg \longrightarrow 2\ Mg^{2+} + 4\ e^-$ $O_2 + 4\ e^- \longrightarrow 2\ O^{2-}$

5.83 a. exothermic b,c: endothermic

5.85 a. released b. products c. exothermic

5.87 a. negative b. exothermic c. products

5.89 a.

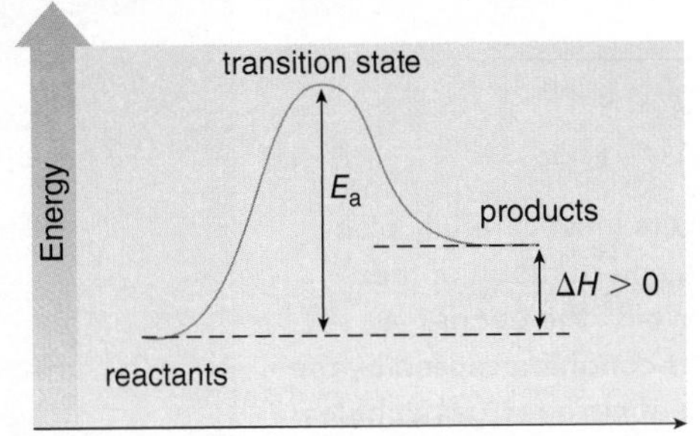

b.

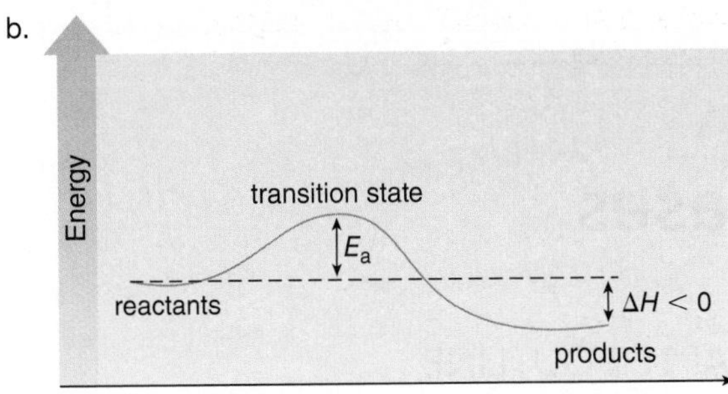

5.91 a. The reaction with $E_a = 1$ kcal will proceed faster because the energy of activation is lower.
b. One cannot predict which reaction will proceed faster from the ΔH.

5.93 a, b, and c

5.95 a. proceeds to right
b. proceeds to left

5.97 a. proceeds to right
b,c: proceeds to left

5.99 a. 342.3 g/mol
b. $C_{12}H_{22}O_{11}(s) + H_2O(l) \longrightarrow 4\ C_2H_6O(l) + 4\ CO_2(g)$
c. 8 mol
d. 18.4 g

5.101 8.57×10^{-3} mol $Mg(OH)_2$ and 6.41×10^{-3} mol $Al(OH)_3$

5.103 a. 2.1×10^{-3} g b. 3.9×10^{18} molecules

Scuba divers must carefully plan the depth and duration of their dives to avoid "the bends," a dangerous condition caused by the formation of nitrogen gas bubbles in the bloodstream.

Gases

CHAPTER OUTLINE

6.1 Gases and Pressure
6.2 Boyle's Law Relating Gas Pressure and Volume
6.3 Charles's Law Relating Gas Volume and Temperature
6.4 Gay–Lussac's Law Relating Gas Pressure and Temperature
6.5 The Combined Gas Law
6.6 Avogadro's Law Relating Gas Volume and Moles
6.7 The Ideal Gas Law
6.8 Dalton's Law and Partial Pressures
6.9 FOCUS ON THE ENVIRONMENT: Ozone and Carbon Dioxide in the Atmosphere

CHAPTER GOALS

In this chapter you will learn how to:

1. Measure pressure and convert one unit of pressure to another
2. Describe the relationship between the pressure, volume, and temperature of a gas using gas laws
3. Describe the relationship between the volume and number of moles of a gas
4. Write the equation for the ideal gas law and use it in calculations
5. Use Dalton's law to determine the partial pressure and total pressure of a gas mixture
6. Understand the importance of two minor components of the earth's atmosphere, ozone and carbon dioxide

In Chapter 6 we study the properties of gases. Why is air pulled into the lungs when we expand our rib cage and diaphragm? Why does a lid pop off a container of food when it is heated in the microwave? Why is a hyperbaric chamber used to treat a scuba diver suffering from the bends, a painful condition that may result from surfacing too quickly? To answer questions of this sort, we must understand the particulate properties of the gas state, as well as the laws that govern the behavior of all gases.

6.1 Gases and Pressure

Anyone who has ridden a bike against the wind knows that even though we can't see the gas molecules of the air, we can feel them as we move through them. Air is a mixture of 78% nitrogen (N_2), 21% oxygen (O_2), and 1% other gases, including carbon dioxide (CO_2), argon (Ar), water (H_2O), and ozone (O_3) (Figure 6.1).

Simple gases in the atmosphere—oxygen (O_2), carbon dioxide (CO_2), and ozone (O_3)—are vital to life. Oxygen, which constitutes 21% of the earth's atmosphere, is needed for metabolic processes that convert carbohydrates to energy. Green plants use carbon dioxide, a minor component of the atmosphere, to store the energy of the sun in the bonds of carbohydrate molecules during photosynthesis. Ozone forms a protective shield in the upper atmosphere to filter out harmful radiation from the sun, thus keeping it from the surface of the earth (Section 6.9).

Figure 6.1

Composition of the Atmosphere

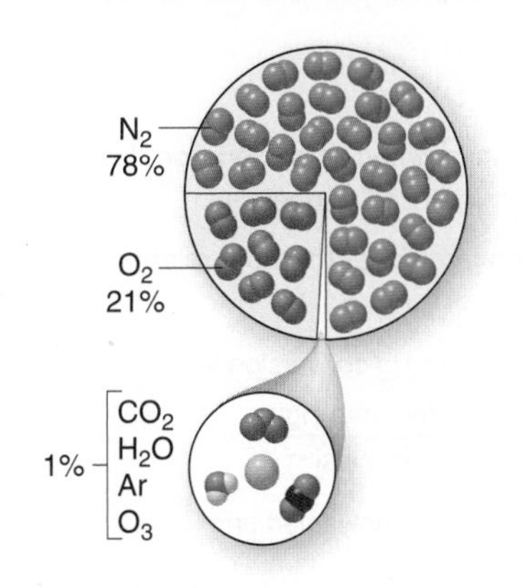

6.1A Properties of Gases

Helium, a noble gas composed of He atoms, and oxygen, a gas composed of diatomic O_2 molecules, behave differently in chemical reactions. Many of their physical properties, however, and the properties of all gases, can be explained by the **kinetic-molecular theory of gases,** a set of principles based on the following assumptions:

- A gas consists of particles—atoms or molecules—that move randomly and rapidly.
- The size of gas particles is negligible compared to the space between the particles.
- Because the space between gas particles is large, gas particles exert no attractive forces on each other.
- The kinetic energy of gas particles increases with increasing temperature.
- When gas particles collide with each other, they rebound and travel in new directions. When gas particles collide with the walls of a container, they exert a pressure.

Because gas particles move rapidly, two gases mix together quickly. Moreover, when a gas is added to a container, the particles rapidly move to fill the entire container.

Figure 6.2

A Barometer—A Device for Measuring Atmospheric Pressure

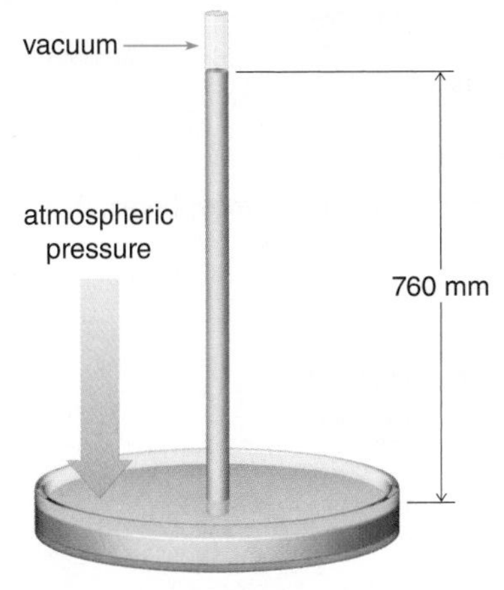

A barometer measures atmospheric pressure. Air pressure on the Hg in the dish pushes Hg up a sealed glass tube to a height that equals the atmospheric pressure.

6.1B Gas Pressure

When many gas molecules strike a surface, they exert a measurable pressure. **Pressure (*P*) is the force (*F*) exerted per unit area (*A*) when gas particles hit a surface.**

$$\text{Pressure} = \frac{\text{Force}}{\text{Area}} = \frac{F}{A}$$

Gas pressure depends on the number of collisions of gas particles. Factors that increase the number of collisions increase pressure. All of the gases in the atmosphere collectively exert **atmospheric pressure** on the surface of the earth. The value of the atmospheric pressure varies with location, decreasing with increasing altitude. Atmospheric pressure also varies slightly from day to day, depending on the weather.

Atmospheric pressure is measured with a **barometer** (Figure 6.2). A barometer consists of a column of mercury (Hg) sealed at one end and inverted in a dish of mercury. The downward pressure exerted by the mercury in the column equals the atmospheric pressure on the mercury in the dish.

Thus, the height of the mercury in the column measures the atmospheric pressure. Atmospheric pressure at sea level corresponds to a column of mercury 760. mm in height.

Many different units are used for pressure. The two most common units are the **atmosphere (atm),** and **millimeters of mercury (mm Hg),** where **1 atm = 760. mm Hg.** One millimeter of mercury is also called one **torr.** In the United States, the common pressure unit is **pounds per square inch (psi),** where **1 atm = 14.7 psi.** Pressure can also be measured in pascals (Pa), where 1 mm Hg = 133.32 Pa.

$$\begin{aligned} 1 \text{ atm} &= 760. \text{ mm Hg} \\ &= 760. \text{ torr} \\ &= 14.7 \text{ psi} \\ &= 101{,}325 \text{ Pa} \end{aligned}$$

To convert a value from one pressure unit to another, set up conversion factors and use the method in Sample Problem 6.1.

SAMPLE PROBLEM 6.1

A scuba diver typically begins a dive with a compressed air tank at 3,000. psi. Convert this value to (a) atmospheres; (b) mm Hg.

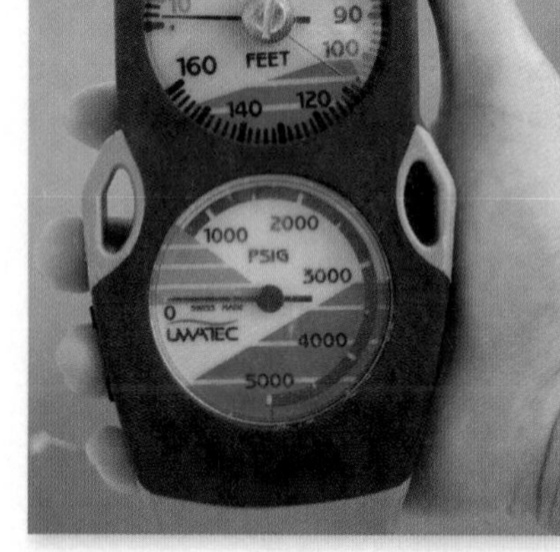

A scuba diver's pressure gauge shows the amount of air (usually measured in psi) in his tank before and during a dive.

Analysis

To solve each part, set up conversion factors that relate the two units under consideration. Use conversion factors that place the unwanted unit, psi, in the denominator to cancel.

In part (a), the conversion factor must relate psi and atm:

psi–atm conversion factor

$$\frac{1 \text{ atm}}{14.7 \text{ psi}}$$

(psi = unwanted unit)

In part (b), the conversion factor must relate psi and mm Hg:

psi–mm Hg conversion factor

$$\frac{760. \text{ mm Hg}}{14.7 \text{ psi}}$$

(psi = unwanted unit)

Solution

a. Convert the original unit (3,000. psi) to the desired unit (atm) using the conversion factor:

$$3000. \text{ psi} \times \frac{1 \text{ atm}}{14.7 \text{ psi}} = 204 \text{ atm}$$

Psi cancels. **Answer**

b. Convert the original unit (3,000. psi) to the desired unit (mm Hg) using the conversion factor:

$$3000. \text{ psi} \times \frac{760. \text{ mm Hg}}{14.7 \text{ psi}} = 155{,}000 \text{ mm Hg}$$

Psi cancels. **Answer**

PROBLEM 6.1

Typical atmospheric pressure in Denver is 630 mm Hg. Convert this value to (a) atmospheres; (b) psi.

PROBLEM 6.2

Convert each pressure unit to the indicated unit.

a. 3.0 atm to mm Hg b. 720 mm Hg to psi c. 424 mm Hg to atm

6.1C FOCUS ON HEALTH & MEDICINE Blood Pressure

Taking a patient's blood pressure is an important part of most physical examinations. Blood pressure measures the pressure in an artery of the upper arm using a device called a **sphygmomanometer.** A blood pressure reading consists of two numbers such as 120/80, where both

Figure 6.3 Measuring Blood Pressure

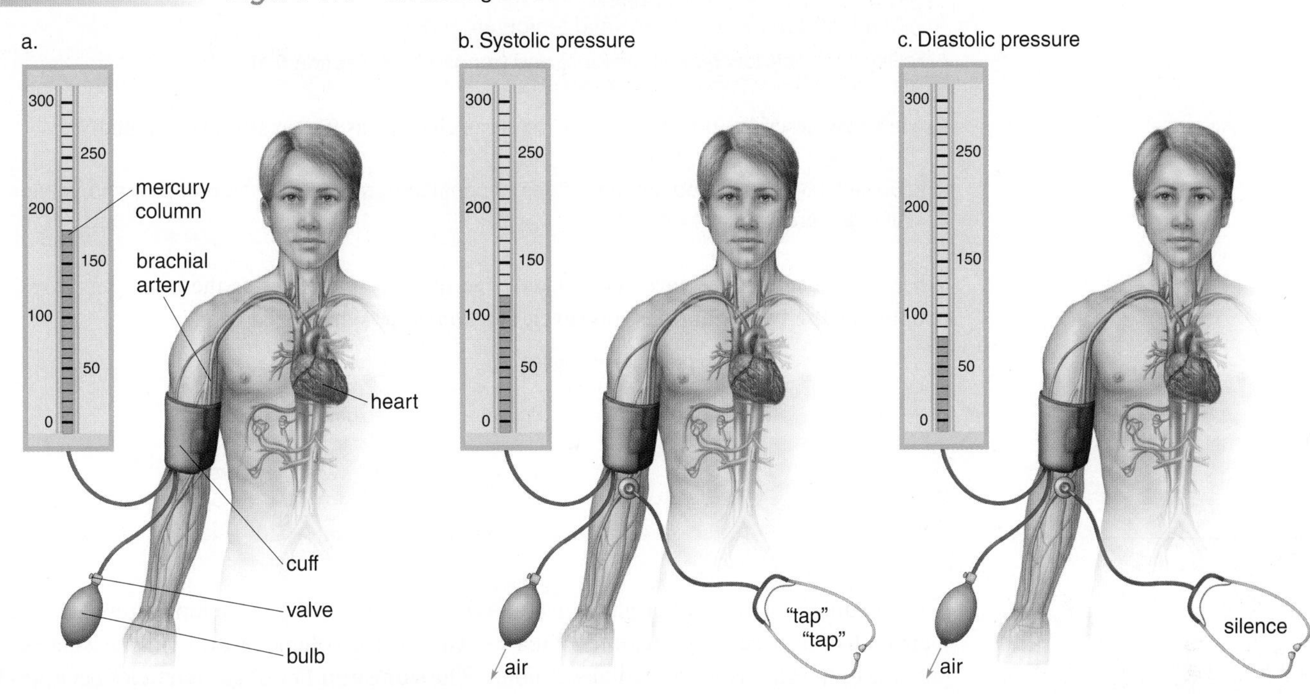

(a) To measure blood pressure, a cuff is inflated around the upper arm and a stethoscope is used to listen to the sound of blood flowing through the brachial artery. When the pressure in the cuff is high, it constricts the artery, so that no blood can flow to the lower arm. (b) Slowly the pressure in the cuff is decreased, and when it gets to the point that blood begins to spurt into the artery, a tapping sound is heard in the stethoscope. This value corresponds to the systolic blood pressure. (c) When the pressure in the cuff is further decreased, so that blood once again flows freely in the artery, the tapping sound disappears and the diastolic pressure is recorded.

values represent pressures in mm Hg. The higher number is the systolic pressure and refers to the maximum pressure in the artery right after the heart contracts. The lower number is the diastolic pressure and represents the minimum pressure when the heart muscle relaxes. A desirable systolic pressure is in the range of 100–120 mm Hg. A desirable diastolic pressure is in the range of 60–80 mm Hg. Figure 6.3 illustrates how a sphygmomanometer records pressure in a blood vessel.

When a patient's systolic pressure is routinely 140 mm Hg or greater or diastolic pressure is 90 mm Hg or greater, an individual is said to have **hypertension**—that is, high blood pressure. Consistently high blood pressure leads to increased risk of stroke and heart attacks.

PROBLEM 6.3

Suppose blood pressure readings were reported in cm Hg rather than mm Hg. If this were the case, how would the pressure 140/90 be reported?

6.2 Boyle's Law Relating Gas Pressure and Volume

Four variables are important in discussing the behavior of gases—pressure (*P*), volume (*V*), temperature (*T*), and number of moles (*n*). The relationship of these variables is described by equations called **gas laws** that explain and predict the behavior of all gases as conditions change. Three gas laws illustrate the interrelationship of pressure, volume, and temperature.

- **Boyle's law relates pressure and volume (Section 6.2).**
- **Charles's law relates volume and temperature (Section 6.3).**
- **Gay–Lussac's law relates pressure and temperature (Section 6.4).**

Boyle's law describes how the volume of a gas changes as the pressure is changed.

- **Boyle's law: For a fixed amount of gas at constant temperature, the pressure and volume of a gas are inversely related.**

When two quantities are *inversely* related, one quantity *increases* as the other *decreases*. **The product of the two quantities, however, is a *constant,*** symbolized by *k*.

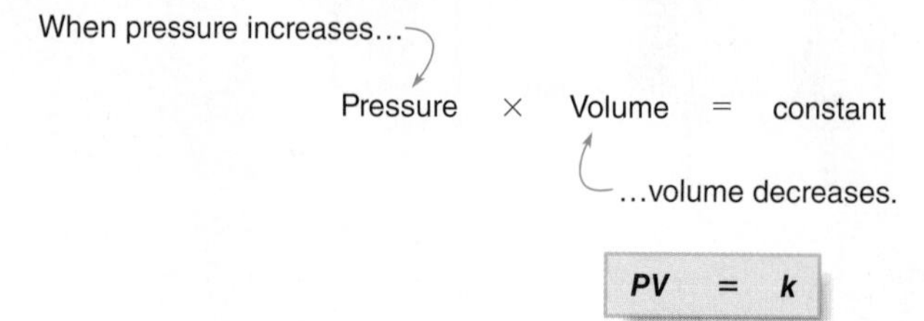

If the volume of a cylinder of gas is decreased, the particles of gas are more crowded together, more collisions occur, and pressure increases. Thus, if the volume of the cylinder is halved, the pressure of the gas inside the cylinder doubles. **The same number of gas particles occupies half the volume and exerts two times the pressure.**

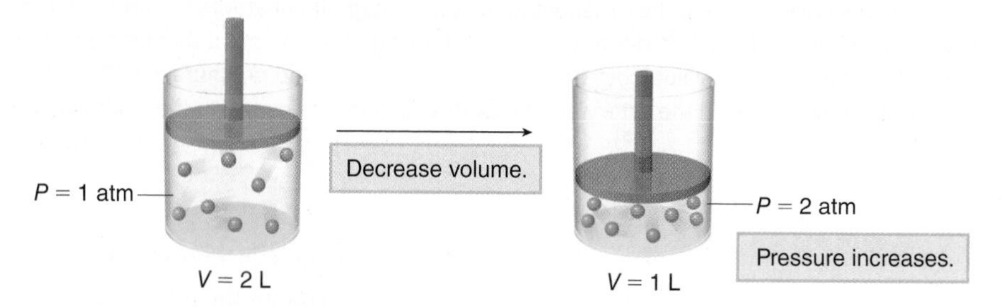

If we know the pressure and volume under an initial set of conditions (P_1 and V_1), we can calculate the pressure or volume under a different set of conditions (P_2 and V_2), since the product of pressure and volume is a constant.

$$P_1V_1 = P_2V_2$$

initial conditions — new conditions

How To Use Boyle's Law to Calculate a New Gas Volume or Pressure

Example If a 4.0-L container of helium gas has a pressure of 10.0 atm, what pressure does the gas exert if the volume is increased to 6.0 L?

Step [1] Identify the known quantities and the desired quantity.

- To solve an equation using Boyle's law, we must know three quantities and solve for one quantity. In this case P_1, V_1, and V_2 are known and the final pressure, P_2, must be determined.

P_1 = 10.0 atm
V_1 = 4.0 L V_2 = 6.0 L
known quantities

P_2 = ?
desired quantity

—*Continued*

How To, continued . . .

Step [2] Write the equation and rearrange it to isolate the desired quantity on one side.

- Rearrange the equation for Boyle's law so that the unknown quantity, P_2, is present alone on one side.

$$P_1V_1 = P_2V_2 \quad \text{Solve for } P_2 \text{ by dividing both sides by } V_2.$$

$$\frac{P_1V_1}{V_2} = P_2$$

Step [3] Solve the problem.

- Substitute the known quantities into the equation and solve for P_2. Identical units must be used for two similar quantities (liters in this case) so that the units cancel.

$$P_2 = \frac{P_1V_1}{V_2} = \frac{(10.0 \text{ atm})(4.0 \text{ L})}{6.0 \text{ L}} = 6.7 \text{ atm}$$

Liters cancel. **Answer**

- In this example, the volume increased so the pressure decreased.

SAMPLE PROBLEM 6.2

A tank of compressed air for scuba diving contains 8.5 L of gas at 204 atm pressure. What volume of air does this gas occupy at 1.0 atm?

Analysis

Boyle's law can be used to solve this problem since an initial pressure and volume (P_1 and V_1) and a final pressure (P_2) are known, and a final volume (V_2) must be determined.

Solution

[1] **Identify the known quantities and the desired quantity.**

$P_1 = 204$ atm, $V_1 = 8.5$ L, $P_2 = 1.0$ atm — known quantities

$V_2 = ?$ — desired quantity

[2] **Write the equation and rearrange it to isolate the desired quantity, V_2, on one side.**

$$P_1V_1 = P_2V_2 \quad \text{Solve for } V_2 \text{ by dividing both sides by } P_2.$$

$$\frac{P_1V_1}{P_2} = V_2$$

[3] **Solve the problem.**

- Substitute the three known quantities into the equation and solve for V_2.

$$V_2 = \frac{P_1V_1}{P_2} = \frac{(204 \text{ atm})(8.5 \text{ L})}{1.0 \text{ atm}} = 1{,}734 \text{ rounded to } 1{,}700 \text{ L}$$

Atm cancels. **Answer**

- Thus, the volume increased because the pressure decreased.

PROBLEM 6.4

A sample of helium gas has a volume of 2.0 L at a pressure of 4.0 atm. What is the volume of gas at each of the following pressures?

a. 5.0 atm b. 2.5 atm c. 10.0 atm d. 380 mm Hg

Figure 6.4

Focus on the Human Body: Boyle's Law and Breathing

a.

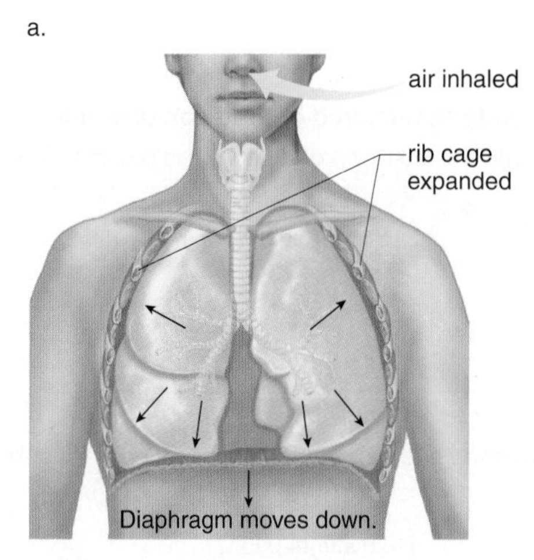

(a) When an individual inhales, the rib cage expands and the diaphragm is lowered, thus increasing the volume of the lungs. According to Boyle's law, increasing the volume of the lungs decreases the pressure inside the lungs. The decrease in pressure draws air into the lungs.

b.

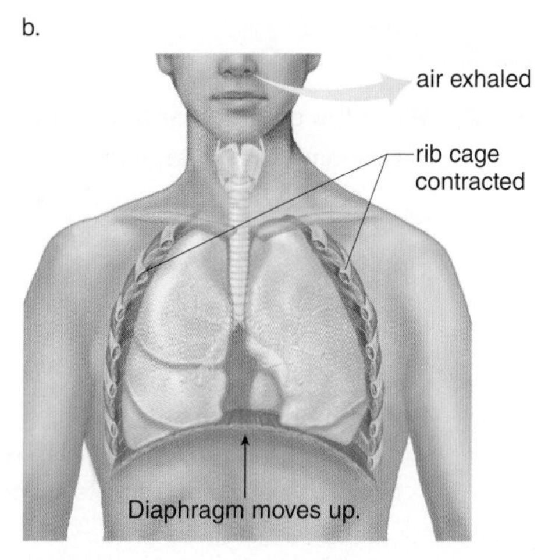

(b) When an individual exhales, the rib cage contracts and the diaphragm is raised, thus decreasing the volume of the lungs. Since the volume is now decreased, the pressure inside the lungs increases, causing air to be expelled into the surroundings.

PROBLEM 6.5

A sample of nitrogen gas has a volume of 15.0 mL at a pressure of 0.50 atm. What is the pressure exerted by the gas if the volume is changed to each of the following values?

a. 30.0 mL b. 5.0 mL c. 100. mL d. 1.0 L

Boyle's law explains how air is brought into or expelled from the lungs as the rib cage and diaphragm expand and contract when we breathe (Figure 6.4).

6.3 Charles's Law Relating Gas Volume and Temperature

All gases expand when they are heated and contract when they are cooled. Charles's law describes how the volume of a gas changes as the Kelvin temperature is changed.

- **Charles's law: For a fixed amount of gas at constant pressure, the volume of a gas is proportional to its Kelvin temperature.**

Volume and temperature are ***proportional;*** that is, as one quantity *increases,* the other *increases* as well. Thus, **dividing volume by temperature is a constant (*k*).**

$$\frac{V}{T} = k$$

A hot air balloon illustrates Charles's law. Heating the air inside the balloon causes it to expand and fill the balloon. When the air inside the balloon becomes less dense than the surrounding air, the balloon rises.

Increasing the temperature increases the kinetic energy of the gas particles, and they move faster and spread out, thus occupying a larger volume. Note that **Kelvin temperature** must be used in calculations involving gas laws. Any temperature reported in °C or °F must be converted to kelvins (K) prior to carrying out the calculation.

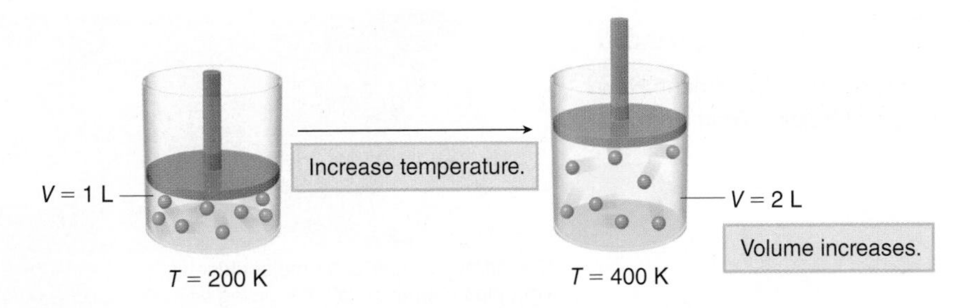

Since dividing the volume of a gas by the temperature gives a constant, knowing the volume and temperature under an initial set of conditions (V_1 and T_1) means we can calculate the volume or temperature under another set of conditions (V_2 and T_2) when either volume or temperature is changed.

Equations for converting one temperature unit to another are given in Section 1.9.

$$\underset{\text{initial conditions}}{\frac{V_1}{T_1}} = \underset{\text{new conditions}}{\frac{V_2}{T_2}}$$

To solve a problem of this sort, we follow the same three steps listed in the *How To* outlined in Section 6.2, except we use the equation for Charles's law in step [2] in place of the equation for Boyle's law. This procedure is illustrated in Sample Problem 6.3.

SAMPLE PROBLEM 6.3

A balloon that contains 0.50 L of air at 25 °C is cooled to –196 °C. What volume does the balloon now occupy?

Analysis

Since this question deals with volume and temperature, Charles's law is used to determine a final volume because three quantities are known—the initial volume and temperature (V_1 and T_1), and the final temperature (T_2).

Solution

[1] Identify the known quantities and the desired quantity.

V_1 = 0.50 L
T_1 = 25 °C $\quad$ T_2 = –196 °C
known quantities

V_2 = ?
desired quantity

- Both temperatures must be converted to Kelvin temperatures using the equation $T_K = T_C + 273$.
- T_1 = 25 °C + 273 = 298 K
- T_2 = –196 °C + 273 = 77 K

[2] Write the equation and rearrange it to isolate the desired quantity, V_2, on one side.

- Use Charles's law.

$$\frac{V_1}{T_1} = \frac{V_2}{T_2}$$

Solve for V_2 by multiplying both sides by T_2.

$$\frac{V_1 T_2}{T_1} = V_2$$

[3] Solve the problem.

- Substitute the three known quantities into the equation and solve for V_2.

$$V_2 = \frac{V_1 T_2}{T_1} = \frac{(0.50\ \text{L})(77\ \text{K})}{298\ \text{K}} = 0.13\ \text{L}$$

Kelvins cancel.

Answer

- Since the temperature has decreased, the volume of gas must decrease as well.

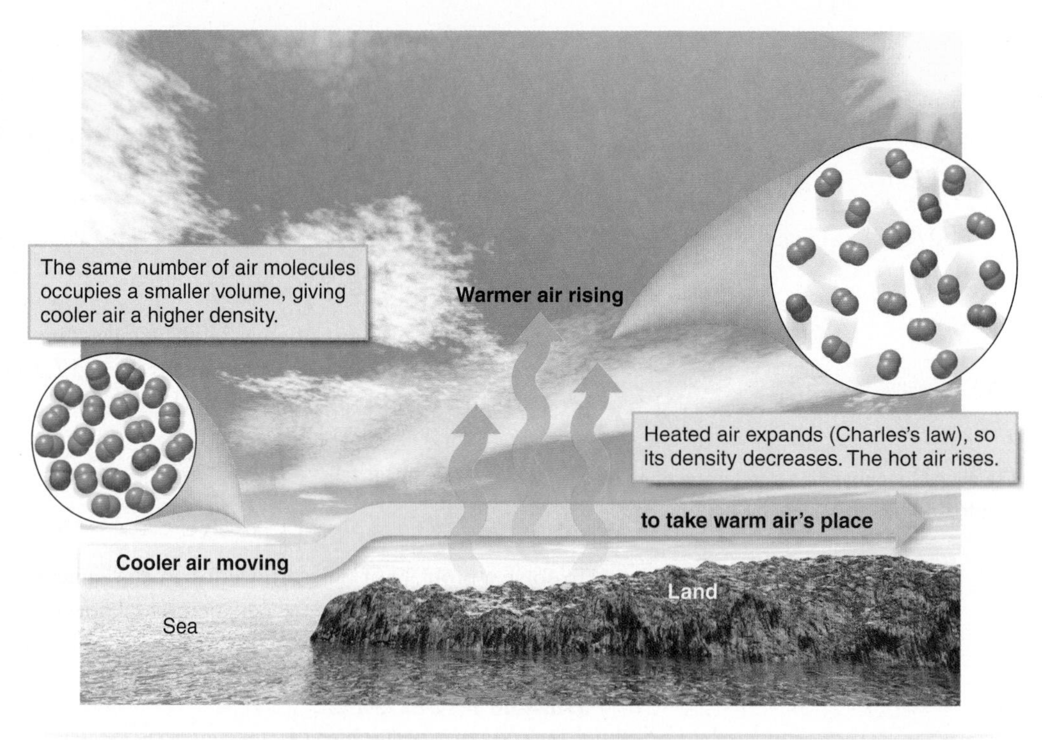
The same number of air molecules occupies a smaller volume, giving cooler air a higher density.
Warmer air rising
Heated air expands (Charles's law), so its density decreases. The hot air rises.
to take warm air's place
Cooler air moving
Land
Sea

Figure 6.5

Focus on the Environment: How Charles's Law Explains Wind Currents

PROBLEM 6.6

A volume of 0.50 L of air at 37 °C is expelled from the lungs into cold surroundings at 0.0 °C. What volume does the expelled air occupy at this temperature?

PROBLEM 6.7

(a) A volume (25.0 L) of gas at 45 K is heated to 450 K. What volume does the gas now occupy?
(b) A volume (50.0 mL) of gas at 400. °C is cooled to 50. °C. What volume does the gas now occupy?

PROBLEM 6.8

Calculate the Kelvin temperature to which 10.0 L of a gas at 27 °C would have to be heated to change the volume to 12.0 L.

Charles's law can be used to explain how wind currents form at the beach (Figure 6.5). The air above land heats up faster than the air above water. As the temperature of the air above the land increases, the volume that it occupies increases; that is, the air expands, and as a result, its density decreases. This warmer, less dense air then rises, and the cooler denser air above the water moves toward the land as wind, filling the space left vacant by the warm, rising air.

6.4 Gay–Lussac's Law Relating Gas Pressure and Temperature

Gay–Lussac's law describes how the pressure of a gas changes as the Kelvin temperature is changed.

- **Gay–Lussac's law: For a fixed amount of gas at constant volume, the pressure of a gas is proportional to its Kelvin temperature.**

Pressure and temperature are *proportional;* that is, as one quantity *increases,* the other *increases*. Thus, **dividing the pressure by the temperature is a constant (k).**

$$\frac{P}{T} = k$$

Increasing the temperature increases the kinetic energy of the gas particles, thus increasing both the number of collisions and the force they exert. If the volume is kept constant, the pressure exerted by the particles increases. Since dividing the pressure of a gas by the temperature gives a constant, knowing the pressure and Kelvin temperature under an initial set of conditions (P_1 and T_1) means we can calculate the pressure or temperature under another set of conditions (P_2 and T_2) when either pressure or temperature is changed.

$$\underset{\text{initial conditions}}{\frac{P_1}{T_1}} = \underset{\text{new conditions}}{\frac{P_2}{T_2}}$$

We solve this type of problem by following the same three steps in the *How To* in Section 6.2, using the equation for Gay–Lussac's law in step [2].

SAMPLE PROBLEM 6.4

The tire on a bicycle stored in a cool garage at 18 °C had a pressure of 80. psi. What is the pressure inside the tire after riding the bike at 43 °C?

Analysis

Since this question deals with pressure and temperature, Gay–Lussac's law is used to determine a final pressure because three quantities are known—the initial pressure and temperature (P_1 and T_1), and the final temperature (T_2).

Solution

[1] Identify the known quantities and the desired quantity.

$P_1 = 80.$ psi
$T_1 = 18$ °C $\quad T_2 = 43$ °C $\quad P_2 = ?$
known quantities $\quad$ desired quantity

- Both temperatures must be converted to Kelvin temperatures.
- $T_1 = T_C + 273 = 18\ °C + 273 = 291\ K$
- $T_2 = T_C + 273 = 43\ °C + 273 = 316\ K$

[2] Write the equation and rearrange it to isolate the desired quantity, P_2, on one side.

- Use Gay–Lussac's law. Since the initial pressure is reported in psi, the final pressure will be calculated in psi.

$$\frac{P_1}{T_1} = \frac{P_2}{T_2}$$ Solve for P_2 by multiplying both sides by T_2.

$$\frac{P_1 T_2}{T_1} = P_2$$

[3] Solve the problem.

- Substitute the three known quantities into the equation and solve for P_2.

$$P_2 = \frac{P_1 T_2}{T_1} = \frac{(80.\ \text{psi})(316\ \text{K})}{291\ \text{K}} = 87\ \text{psi}$$

Kelvins cancel. **Answer**

- Since the temperature has increased, the pressure of the gas must increase as well.

CONSUMER NOTE

Food cooks faster in a pressure cooker because the reactions involved in cooking occur at a faster rate at a higher temperature.

PROBLEM 6.9

A pressure cooker is used to cook food in a closed pot. By heating the contents of a pressure cooker at constant volume, the pressure increases. If the steam inside the pressure cooker is initially at 100. °C and 1.00 atm, what is the final temperature of the steam if the pressure is increased to 1.05 atm?

PROBLEM 6.10

The temperature of a 0.50-L gas sample at 25 °C and 1.00 atm is changed to each of the following temperatures. What is the final pressure of the system?

a. 310. K b. 150. K c. 50. °C d. 200. °C

PROBLEM 6.11

Use Gay–Lussac's law to answer the question posed at the beginning of the chapter: Why does a lid pop off a container of food when it is heated in a microwave?

6.5 The Combined Gas Law

All three gas laws—Boyle's, Charles's, and Gay–Lussac's laws—can be combined in a single equation, the **combined gas law,** that relates pressure, volume, and temperature.

$$\underset{\text{initial conditions}}{\frac{P_1V_1}{T_1}} = \underset{\text{new conditions}}{\frac{P_2V_2}{T_2}}$$

The combined gas law contains six terms that relate the pressure, volume, and temperature of an initial and final state of a gas. It can be used to calculate one quantity when the other five are known, as long as the amount of gas is constant. The combined gas law is used for determining the effect of changing two factors—such as pressure and temperature—on the third factor, volume.

We solve this type of problem by following the same three steps in the *How To* in Section 6.2, using the equation for the combined gas law in step [2]. Sample Problem 6.5 shows how this is done. Table 6.1 summarizes the equations for the gas laws presented in Sections 6.2–6.5.

Table 6.1 Summary of the Gas Laws That Relate Pressure, Volume, and Temperature

Law	Equation	Relationship
Boyle's law	$P_1V_1 = P_2V_2$	As P increases, V decreases for constant T and n.
Charles's law	$\frac{V_1}{T_1} = \frac{V_2}{T_2}$	As T increases, V increases for constant P and n.
Gay–Lussac's law	$\frac{P_1}{T_1} = \frac{P_2}{T_2}$	As T increases, P increases for constant V and n.
Combined gas law	$\frac{P_1V_1}{T_1} = \frac{P_2V_2}{T_2}$	The combined gas law shows the relationship of P, V, and T when two quantities are changed.

SAMPLE PROBLEM 6.5

A weather balloon contains 222 L of helium at 20. °C and 760 mm Hg. What is the volume of the balloon when it ascends to an altitude where the temperature is –40. °C and 540 mm Hg?

Analysis

Since this question deals with pressure, volume, and temperature, the combined gas law is used to determine a final volume (V_2) because five quantities are known—the initial pressure, volume, and temperature (P_1, V_1, and T_1), and the final pressure and temperature (P_2 and T_2).

Solution

[1] **Identify the known quantities and the desired quantity.**

$P_1 = 760$ mm Hg $\quad P_2 = 540$ mm Hg
$T_1 = 20.$ °C $\quad T_2 = -40.$ °C
$V_1 = 222$ L $\quad V_2 = ?$

known quantities — desired quantity

- Both temperatures must be converted to Kelvin temperatures.
- $T_1 = T_C + 273 = 20.\ °C + 273 = 293\ K$
- $T_2 = T_C + 273 = -40.\ °C + 273 = 233\ K$

[2] **Write the equation and rearrange it to isolate the desired quantity, V_2, on one side.**

- Use the combined gas law.

$$\frac{P_1V_1}{T_1} = \frac{P_2V_2}{T_2}$$

Solve for V_2 by multiplying both sides by $\frac{T_2}{P_2}$.

$$\frac{P_1V_1T_2}{T_1P_2} = V_2$$

[3] **Solve the problem.**

- Substitute the five known quantities into the equation and solve for V_2.

$$V_2 = \frac{P_1V_1T_2}{T_1P_2} = \frac{(760\ \text{mm Hg})(222\ \text{L})(233\ \text{K})}{(293\ \text{K})(540\ \text{mm Hg})} = 248.5\ \text{L rounded to 250 L}$$

Kelvins and mm Hg cancel.

Answer

PROBLEM 6.12

The pressure inside a 1.0-L balloon at 25 °C was 750 mm Hg. What is the pressure inside the balloon when it is cooled to –40. °C and expands to 2.0 L in volume?

6.6 Avogadro's Law Relating Gas Volume and Moles

Each equation in Table 6.1 was written for a constant amount of gas; that is, the number of moles (n) did not change. **Avogadro's law** describes the relationship between the number of moles of a gas and its volume.

- **Avogadro's law: When the pressure and temperature are held constant, the volume of a gas is proportional to the number of moles present.**

As the number of moles of a gas *increases*, its volume *increases* as well. Thus, dividing the volume by the number of moles is a constant (k). **The value of k is the same regardless of the identity of the gas.**

$$\frac{V}{n} = k$$

A balloon inflates when someone blows into it because the volume increases with an increased number of moles of air.

Thus, if the pressure and temperature of a system are held constant, **increasing the number of moles increases the volume of a gas.**

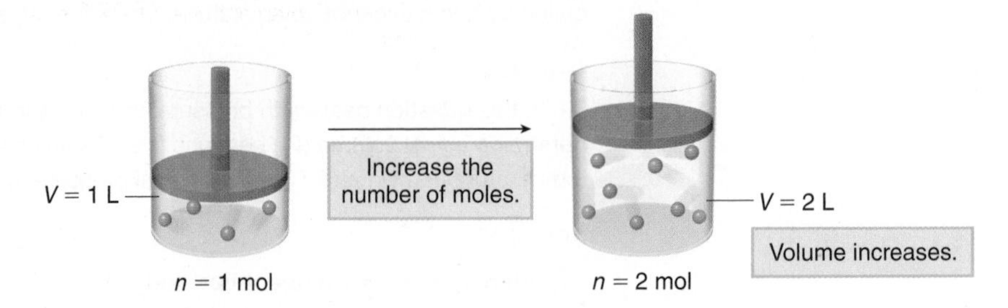

Since dividing the volume of a gas by the number of moles is a constant, knowing the volume and number of moles initially (V_1 and n_1) means we can calculate a new volume or number of moles (V_2 and n_2) when one of these quantities is changed.

$$\underset{\text{initial conditions}}{\frac{V_1}{n_1}} = \underset{\text{new conditions}}{\frac{V_2}{n_2}}$$

To solve a problem of this sort, we follow the same three steps listed in the *How To* outlined in Section 6.2, using Avogadro's law in step [2].

SAMPLE PROBLEM 6.6

The lungs of an average male hold 0.25 mol of air in a volume of 5.8 L. How many moles of air do the lungs of an average female hold if the volume is 4.6 L?

Analysis

This question deals with volume and number of moles, so Avogadro's law is used to determine a final number of moles when three quantities are known—the initial volume and number of moles (V_1 and n_1), and the final volume (V_2).

Solution

[1] **Identify the known quantities and the desired quantity.**

$V_1 = 5.8$ L $V_2 = 4.6$ L
$n_1 = 0.25$ mol
known quantities

$n_2 = ?$
desired quantity

[2] **Write the equation and rearrange it to isolate the desired quantity, n_2, on one side.**

- Use Avogadro's law. To solve for n_2, we must invert the numerator and denominator on *both* sides of the equation, and then multiply by V_2.

$$\frac{V_1}{n_1} = \frac{V_2}{n_2} \xrightarrow{\text{Switch } V \text{ and } n \text{ on both sides.}} \frac{n_1}{V_1} = \frac{n_2}{V_2}$$

Solve for n_2 by multiplying both sides by V_2.

$$\frac{n_1 V_2}{V_1} = n_2$$

[3] **Solve the problem.**

- Substitute the three known quantities into the equation and solve for n_2.

$$n_2 = \frac{n_1 V_2}{V_1} = \frac{(0.25 \text{ mol})(4.6 \cancel{\text{L}})}{(5.8 \cancel{\text{L}})} = 0.20 \text{ mol}$$

Liters cancel. **Answer**

PROBLEM 6.13

A balloon that contains 0.30 mol of helium in a volume of 6.4 L develops a leak so that its volume decreases to 3.85 L at constant temperature and pressure. How many moles of helium does the balloon now contain?

PROBLEM 6.14

A sample of nitrogen gas contains 5.0 mol in a volume of 3.5 L. Calculate the new volume of the container if the pressure and temperature are kept constant but the number of moles of nitrogen is changed to each of the following values: (a) 2.5 mol; (b) 3.65 mol; (c) 21.5 mol.

Avogadro's law allows us to compare the amounts of any two gases by comparing their volumes. Often amounts of gas are compared at a set of **standard conditions of temperature and pressure,** abbreviated as **STP.**

- **STP conditions are: 1 atm (760 mm Hg) for pressure**
  273 K (0 °C) for temperature
- **At STP, one mole of any gas has the same volume, 22.4 L, called the *standard molar volume.***

Under STP conditions, one mole of nitrogen gas and one mole of helium gas each contain 6.02×10^{23} particles of gas and occupy a volume of 22.4 L at 0 °C and 1 atm pressure. Since the molar masses of nitrogen and helium are different (28.02 g for N_2 compared to 4.003 g for He), one mole of each substance has a *different* mass.

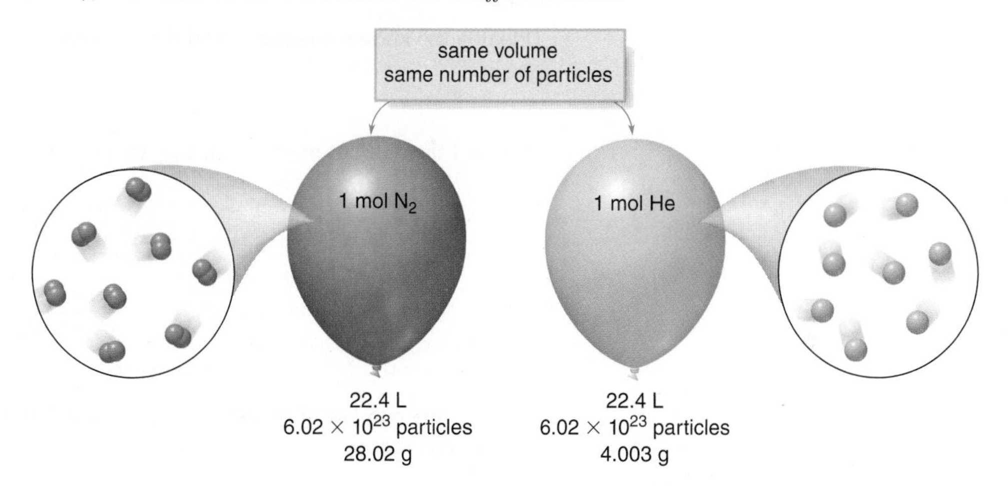

The standard molar volume can be used to set up conversion factors that relate the volume and number of moles of a gas at STP, as shown in the following stepwise procedure.

How To Convert Moles of Gas to Volume at STP

Example How many moles are contained in 2.0 L of N_2 at standard temperature and pressure?

Step [1] Identify the known quantities and the desired quantity.

2.0 L of N_2	? moles of N_2
original quantity	desired quantity

Step [2] Write out the conversion factors.

- Set up conversion factors that relate the number of moles of a gas to volume at STP. Choose the conversion factor that places the unwanted unit, liters, in the denominator so that the units cancel.

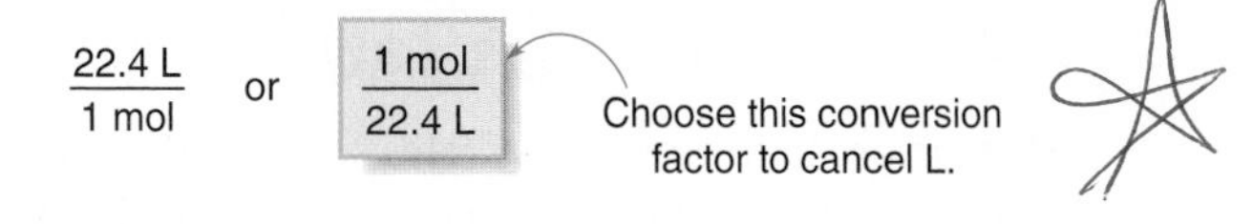

$\frac{22.4\ L}{1\ mol}$ or $\frac{1\ mol}{22.4\ L}$
Choose this conversion factor to cancel L.

—Continued

How To, continued . . .

Step [3] Solve the problem.

- Multiply the original quantity by the conversion factor to obtain the desired quantity.

$$2.0 \text{ L} \times \frac{1 \text{ mol}}{22.4 \text{ L}} = 0.089 \text{ mol of } N_2$$

Liters cancel.

Answer

By using the molar mass of a gas, we can determine the volume of a gas from a given number of grams, as shown in Sample Problem 6.7.

SAMPLE PROBLEM 6.7

Burning one mole of propane in a gas grill adds 132.0 g of carbon dioxide (CO_2) to the atmosphere. What volume of CO_2 does this correspond to at STP?

Analysis

To solve this problem, we must convert the number of grams of CO_2 to moles using the molar mass. The number of moles of CO_2 can then be converted to its volume using a mole–volume conversion factor (1 mol/22.4 L).

Solution

[1] **Identify the known quantities and the desired quantity.**

132.0 g CO_2	? L CO_2
known quantity	desired quantity

[2] **Convert the number of grams of CO_2 to the number of moles of CO_2 using the molar mass.**

molar mass conversion factor

$$132.0 \text{ g } CO_2 \times \frac{1 \text{ mol } CO_2}{44.01 \text{ g } CO_2} = 3.00 \text{ mol } CO_2$$

Grams cancel.

[3] **Convert the number of moles of CO_2 to the volume of CO_2 using a mole–volume conversion factor.**

mole–volume conversion factor

$$3.00 \text{ mol } CO_2 \times \frac{22.4 \text{ L}}{1 \text{ mol}} = 67.2 \text{ L } CO_2$$

Moles cancel.

Answer

PROBLEM 6.15

How many liters does each of the following quantities of O_2 occupy at STP: (a) 4.5 mol; (b) 0.35 mol; (c) 18.0 g?

PROBLEM 6.16

How many moles are contained in the following volumes of air at STP: (a) 1.5 L; (b) 8.5 L; (c) 25 mL?

6.7 The Ideal Gas Law

All four properties of gases—pressure, volume, temperature, and number of moles—can be combined into a single equation called the **ideal gas law.** The product of pressure and volume divided by the product of moles and Kelvin temperature is a constant, called the **universal gas constant** and symbolized by ***R***.

$$\frac{PV}{nT} = R \quad \text{universal gas constant}$$

More often the equation is rearranged and written in the following way:

$$PV = nRT$$

Ideal gas law

For atm: $R = 0.0821 \frac{\text{L} \cdot \text{atm}}{\text{mol} \cdot \text{K}}$

For mm Hg: $R = 62.4 \frac{\text{L} \cdot \text{mm Hg}}{\text{mol} \cdot \text{K}}$

The value of the universal gas constant *R* depends on its units. The two most common values of *R* are given using atmospheres or mm Hg for pressure, liters for volume, and kelvins for temperature. **Be careful to use the correct value of *R* for the pressure units in the problem you are solving.**

The ideal gas law can be used to find any value—*P*, *V*, *n*, or *T*—as long as three of the quantities are known. Solving a problem using the ideal gas law is shown in the stepwise *How To* procedure and in Sample Problem 6.8. Although the ideal gas law gives exact answers only for a perfectly "ideal" gas, it gives a good approximation for most real gases, such as the oxygen and carbon dioxide in breathing, as well (Figure 6.6).

Figure 6.6 Focus on the Human Body: The Lungs

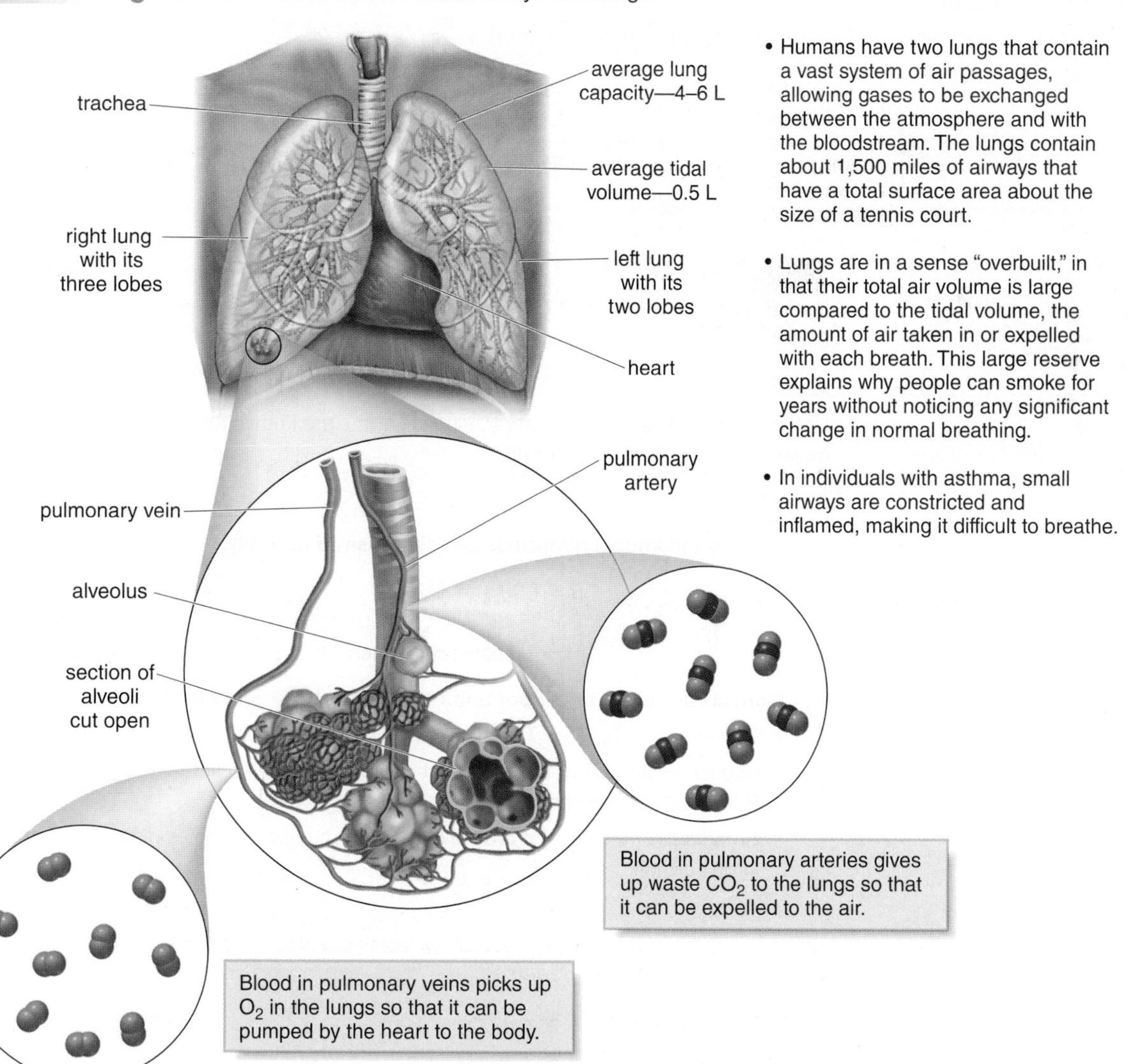

- Humans have two lungs that contain a vast system of air passages, allowing gases to be exchanged between the atmosphere and with the bloodstream. The lungs contain about 1,500 miles of airways that have a total surface area about the size of a tennis court.
- Lungs are in a sense "overbuilt," in that their total air volume is large compared to the tidal volume, the amount of air taken in or expelled with each breath. This large reserve explains why people can smoke for years without noticing any significant change in normal breathing.
- In individuals with asthma, small airways are constricted and inflamed, making it difficult to breathe.

How To Carry Out Calculations with the Ideal Gas Law

Example How many moles of gas are contained in a typical human breath that takes in 0.50 L of air at 1.0 atm pressure and 37 °C?

Step [1] Identify the known quantities and the desired quantity.

$$P = 1.0 \text{ atm}$$
$$V = 0.50 \text{ L}$$
$$T = 37 \text{ °C} \qquad n = ? \text{ mol}$$

known quantities — desired quantity

Step [2] Convert all values to proper units and choose the value of R that contains these units.

- Convert T_C to T_K. $T_K = T_C + 273 = 37\ °C + 273 = 310.\ K$
- Use the value of R in atm since the pressure is given in atm; that is, $R = 0.0821\ L \cdot atm/(mol \cdot K)$.

Step [3] Write the equation and rearrange it to isolate the desired quantity on one side.

- Use the ideal gas law and solve for n by dividing both sides by RT.

$$PV = nRT$$ Solve for n by dividing both sides by RT.

$$\frac{PV}{RT} = n$$

Step [4] Solve the problem.

- Substitute the known quantities into the equation and solve for n.

$$n = \frac{PV}{RT} = \frac{(1.0 \text{ atm})(0.50 \text{ L})}{\left(0.0821 \frac{\text{L} \cdot \text{atm}}{\text{mol} \cdot \text{K}}\right)(310.\ \text{K})} = 0.0196 \text{ rounded to } 0.020 \text{ mol}$$

Answer

SAMPLE PROBLEM 6.8

If a person exhales 25.0 g of CO_2 in an hour, what volume does this amount occupy at 1.00 atm and 37 °C?

Analysis

Use the ideal gas law to calculate V, since P and T are known and n can be determined by using the molar mass of CO_2 (44.01 g/mol).

Solution

[1] Identify the known quantities and the desired quantity.

$$P = 1.00 \text{ atm}$$
$$T = 37 \text{ °C} \qquad 25.0 \text{ g } CO_2 \qquad V = ? \text{ L}$$

known quantities — desired quantity

[2] Convert all values to proper units and choose the value of R that contains these units.

- Convert T_C to T_K. $T_K = T_C + 273 = 37\ °C + 273 = 310.\ K$
- Use the value of R with atm since the pressure is given in atm; that is, $R = 0.0821\ L \cdot atm/(mol \cdot K)$.
- Convert the number of grams of CO_2 to the number of moles of CO_2 using the molar mass (44.01 g/mol).

molar mass conversion factor

$$25.0 \text{ g } CO_2 \times \frac{1 \text{ mol } CO_2}{44.01 \text{ g } CO_2} = 0.568 \text{ mol } CO_2$$

Grams cancel.

[3] Write the equation and rearrange it to isolate the desired quantity, *V*, on one side.

- Use the ideal gas law and solve for *V* by dividing both sides by *P*.

$$PV = nRT \quad \text{Solve for } V \text{ by dividing both sides by } P.$$

$$V = \frac{nRT}{P}$$

[4] Solve the problem.

- Substitute the three known quantities into the equation and solve for *V*.

$$V = \frac{nRT}{P} = \frac{(0.568\ \text{mol})\left(0.0821\ \frac{\text{L}\cdot\text{atm}}{\text{mol}\cdot\text{K}}\right)(310.\ \text{K})}{1.0\ \text{atm}} = 14.5\ \text{L}$$

Answer

PROBLEM 6.17

How many moles of oxygen (O_2) are contained in a 5.0-L cylinder that has a pressure of 175 atm and a temperature of 20. °C?

PROBLEM 6.18

Determine the pressure of N_2 (in atm) under each of the following conditions.

a. 0.45 mol at 25 °C in 10.0 L

b. 10.0 g at 20. °C in 5.0 L

PROBLEM 6.19

Determine the volume (in L) of 8.50 g of He gas at 25 °C and 750 mm Hg.

6.8 Dalton's Law and Partial Pressures

Since the partial pressure of O_2 is low at very high altitudes, most mountain climbers use supplemental O_2 tanks above about 24,000 ft.

Since gas particles are very far apart compared to the size of an individual particle, gas particles behave independently. As a result, the identity of the components of a gas mixture does not matter, and **a mixture of gases behaves like a pure gas.** Each component of a gas mixture is said to exert a pressure called its **partial pressure. Dalton's law** describes the relationship between the partial pressures of the components and the total pressure of a gas mixture.

- **Dalton's law: The *total pressure* (P_{total}) of a gas mixture is the sum of the partial pressures of its component gases.**

Thus, if a mixture has two gases (**A** and **B**) with partial pressures P_A and P_B, respectively, the total pressure of the system (P_{total}) is the sum of the two partial pressures. The partial pressure of a component of a mixture is the same pressure that the gas would exert if it were a pure gas.

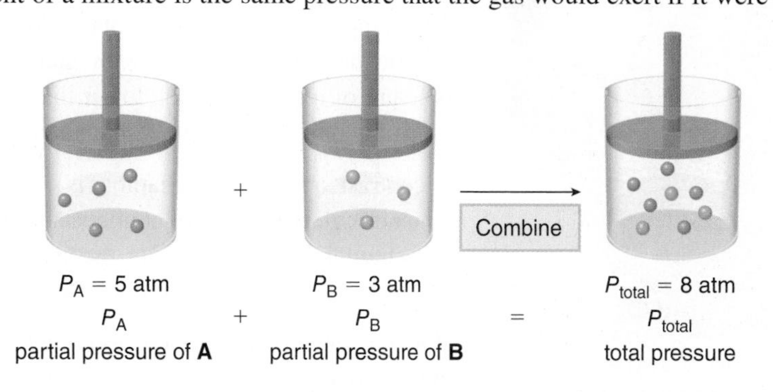

$$P_A + P_B = P_{total}$$

partial pressure of **A** + partial pressure of **B** = total pressure

SAMPLE PROBLEM 6.9

A sample of exhaled air from the lungs contains four gases with the following partial pressures: N_2 (563 mm Hg), O_2 (118 mm Hg), CO_2 (30. mm Hg), and H_2O (50. mm Hg). What is the total pressure of the sample?

Analysis

Using Dalton's law, the total pressure is the sum of the partial pressures.

Solution

Adding up the four partial pressures gives the total:

$$563 + 118 + 30. + 50. = 761 \text{ mm Hg (total pressure)}$$

PROBLEM 6.20

CO_2 was added to a cylinder containing 2.5 atm of O_2 to give a total pressure of 4.0 atm of gas. What is the partial pressure of O_2 and CO_2 in the final mixture?

We can also calculate the partial pressure of each gas in a mixture if two quantities are known—[1] the total pressure and [2] the percent of each component—as shown in Sample Problem 6.10.

SAMPLE PROBLEM 6.10

Air is a mixture of 21% O_2, 78% N_2, and 1% argon by volume. What is the partial pressure of each gas at sea level, where the total pressure is 760 mm Hg?

Analysis

Convert each percent to a decimal by moving the decimal point two places to the left. Multiply each decimal by the total pressure to obtain the partial pressure for each component.

Solution

			Partial pressure
Fraction O_2:	21% = 0.21	0.21 × 760 mm Hg =	160 mm Hg (O_2)
Fraction N_2:	78% = 0.78	0.78 × 760 mm Hg =	590 mm Hg (N_2)
Fraction Ar:	1% = 0.01	0.01 × 760 mm Hg =	8 mm Hg (Ar)
			758 rounded to 760 mm Hg

PROBLEM 6.21

A sample of natural gas at 750 mm Hg contains 85% methane, 10.% ethane, and 5.0% propane. What are the partial pressures of each gas in this mixture?

HEALTH NOTE

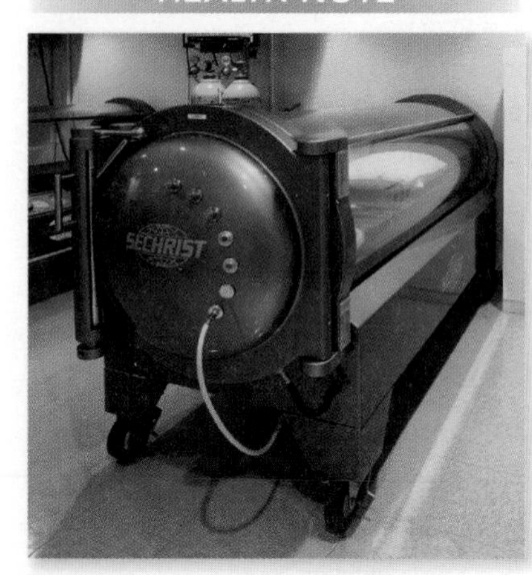

The high pressures of a hyperbaric chamber can be used to treat patients fighting infections and scuba divers suffering from the bends.

The composition of the atmosphere does not change with location, even though the total atmospheric pressure decreases with increasing altitude. At high altitudes, therefore, the partial pressure of oxygen is much lower than it is at sea level, making breathing difficult. This is why mountain climbers use supplemental oxygen at altitudes above 8,000 meters.

In contrast, a hyperbaric chamber is a device that maintains air pressure two to three times higher than normal. Hyperbaric chambers have many uses. At this higher pressure the partial pressure of O_2 is higher. For burn patients, the higher pressure of O_2 increases the amount of O_2 in the blood, where it can be used by the body for reactions that fight infections.

When a scuba diver surfaces too quickly, the N_2 dissolved in the blood can form microscopic bubbles that cause pain in joints and can occlude small blood vessels, causing organ injury. This condition, called the bends, is treated by placing a diver in a hyperbaric chamber, where the elevated pressure decreases the size of the N_2 bubbles, which are then eliminated as N_2 gas from the lungs as the pressure is slowly decreased.

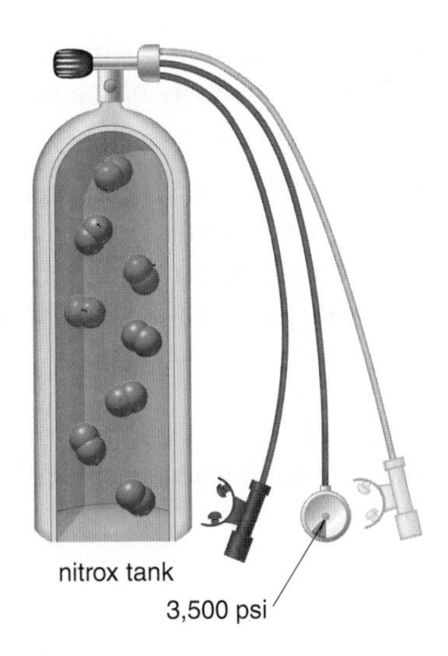

PROBLEM 6.22

Air contains 21% O_2 and 78% N_2. What are the partial pressures of N_2 and O_2 (in atm) in a hyperbaric chamber that contains air at 2.5 atm?

PROBLEM 6.23

At the summit of Kilimanjaro, the highest mountain in Africa with an elevation of 5,895 m, the partial pressure of oxygen is 78.5 mm Hg. What is the atmospheric pressure at the top of Mount Kilimanjaro?

PROBLEM 6.24

Nitrox is a gas mixture used in scuba diving that contains a higher-than-normal level of oxygen and a lower-than-normal level of nitrogen. A lower amount of nitrogen reduces the risk of "the bends."

a. What percent of N_2 and O_2 is present in the given nitrox tank?
b. What is the partial pressure of each gas (in psi) in the tank?

6.9 FOCUS ON THE ENVIRONMENT Ozone and Carbon Dioxide in the Atmosphere

Although both ozone and carbon dioxide are present in only minor amounts in the earth's atmosphere, each plays an important role in the dynamics of the environment on the surface of the earth.

6.9A The Ozone Layer

Ozone (O_3) is a gas formed in the upper atmosphere (the stratosphere) by the reaction of oxygen molecules (O_2) with oxygen atoms (O). Stratospheric ozone acts as a shield that protects the earth by absorbing destructive ultraviolet radiation before it reaches the earth's surface (Figure 6.7). A decrease in ozone concentration in this protective layer would have some immediate negative consequences, including an increase in the incidence of skin cancer and eye cataracts.

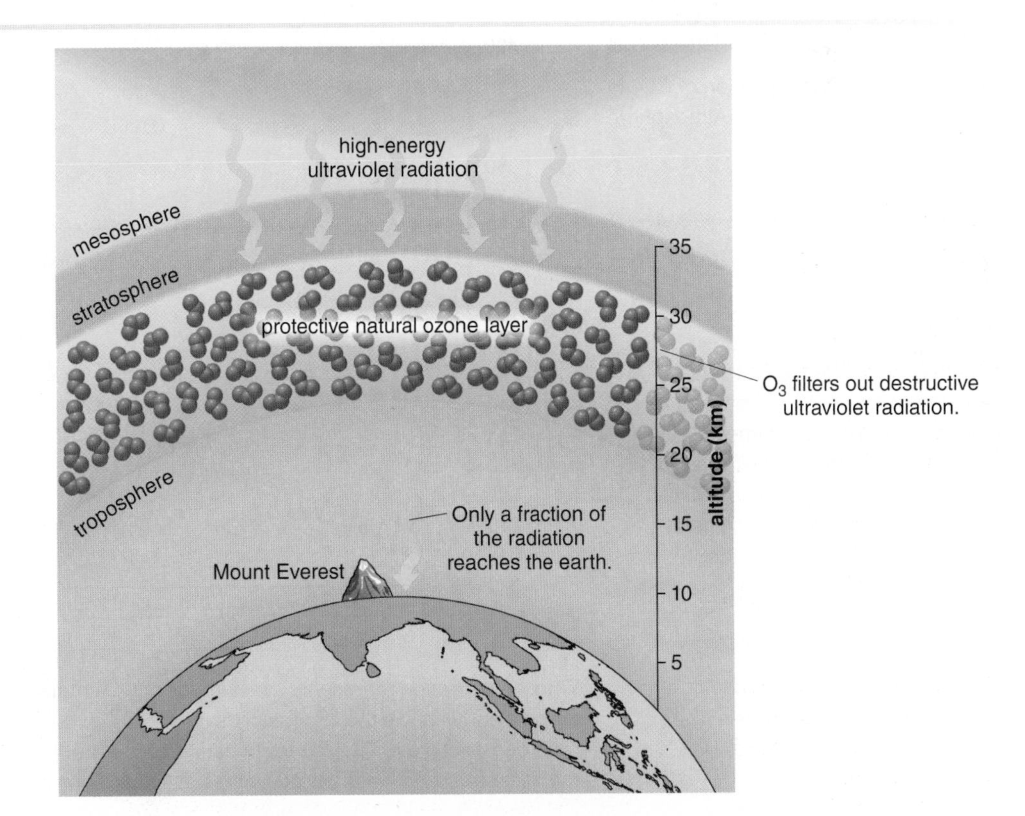

Figure 6.7 Ozone in the Upper Atmosphere

ENVIRONMENTAL NOTE

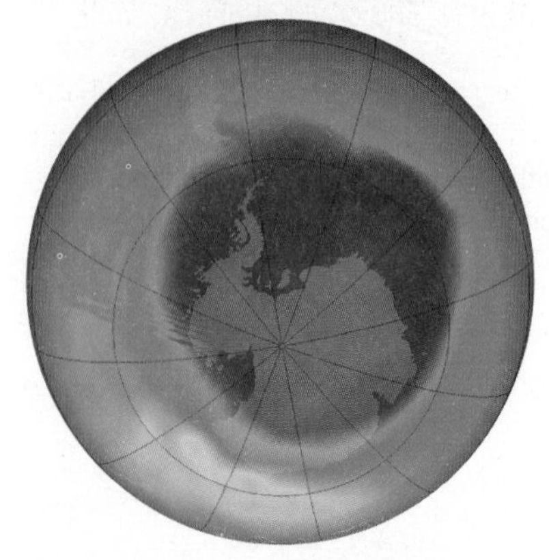

O_3 destruction is most severe in the region of the South Pole, where a large ozone hole (shown in purple) is visible with satellite imaging.

Other possible long-term effects include a reduced immune response, interference with photosynthesis in plants, and harmful effects on the growth of plankton, the mainstay of the ocean food chain.

Research over the last 40 years has shown that **chlorofluorocarbons (CFCs),** simple compounds that contain the elements of carbon, chlorine, and fluorine, destroy ozone in the upper atmosphere. CFCs are synthetic compounds that were once widely used as refrigerants and aerosol propellants, but these findings led to a ban on the use of CFCs in aerosol propellants in the United States in 1978.

PROBLEM 6.25

CF_2Cl_2 is a chlorofluorocarbon once used as an aerosol propellant. Draw a Lewis structure for CF_2Cl_2.

6.9B Carbon Dioxide and Global Warming

Carbon dioxide (CO_2) is a very minor component of the earth's atmosphere; there are only approximately 396 CO_2 molecules found in every one million molecules of air. Carbon dioxide is produced when heating oil, natural gas, and gasoline are burned in the presence of oxygen, as we learned in Section 5.7. This reaction releases energy for heating homes, powering vehicles, and cooking food.

$$\underset{\text{methane (natural gas)}}{CH_4} + 2\,O_2 \xrightarrow[\text{flame}]{} CO_2 + 2\,H_2O + \text{energy}$$

The combustion of these fossil fuels adds a tremendous amount of CO_2 to the atmosphere each year. Quantitatively, data show over a 20% increase in the atmospheric concentration of CO_2 in the last 49 years (Figure 6.8). Although the composition of the atmosphere has changed over the lifetime of the earth, this is likely the first time that the actions of mankind have altered that composition significantly and so quickly.

Figure 6.8

The Changing Concentration of CO_2 in the Atmosphere

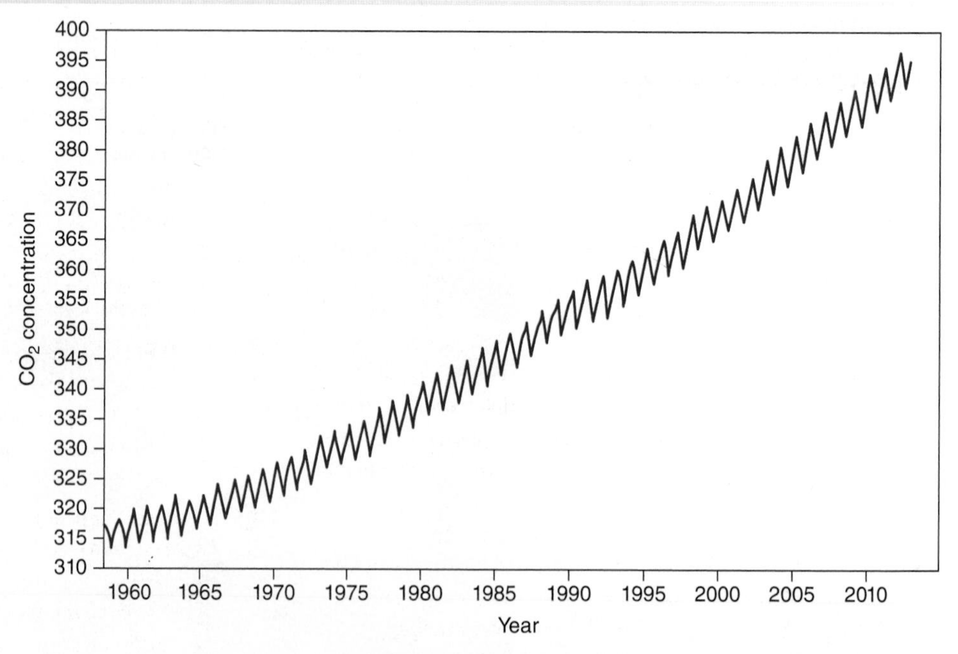

The concentration represents the number of CO_2 molecules in 1,000,000 molecules of air. The graph clearly shows the increasing level of CO_2 in the atmosphere (1958–2012). Two data points are recorded each year. The sawtooth nature of the graph is due to seasonal variation of CO_2 level with the seasonal variation in photosynthesis. (Data recorded at Mauna Loa, Hawaii.)

The 2007 Nobel Peace Prize was awarded to former Vice President Al Gore and the Intergovernmental Panel on Climate Change for their roles in focusing attention on the potentially disastrous effects of rapid climate change caused by human activity.

An increased CO_2 concentration in the atmosphere may have long-range and far-reaching effects. CO_2 is a **greenhouse gas** because it absorbs thermal energy that normally radiates from the earth's surface, and redirects it back to the surface. Higher levels of CO_2 may therefore contribute to an increase in the average temperature of the earth's atmosphere. The global climate change resulting from these effects may lead to the melting of polar ice caps, the rise in sea level, and many more unforeseen consequences.

PROBLEM 6.26

Write a balanced equation for each combustion reaction. Propane and butane are minor components of natural gas. Balancing equations was described in Section 5.2.

a. $CH_3CH_2CH_3 + O_2 \xrightarrow{\text{flame}}$ (propane)

b. $CH_3CH_2CH_2CH_3 + O_2 \xrightarrow{\text{flame}}$ (butane)

KEY TERMS

Atmosphere (6.1)
Avogadro's law (6.6)
Barometer (6.1)
Boyle's law (6.2)
Charles's law (6.3)
Chlorofluorocarbons (CFCs, 6.9)
Combined gas law (6.5)
Dalton's law (6.8)
Gas laws (6.2)
Gay–Lussac's law (6.4)
Ideal gas law (6.7)
Kinetic-molecular theory (6.1)
Millimeters mercury (6.1)
Partial pressure (6.8)
Pressure (6.1)
Standard molar volume (6.6)
STP (6.6)
Universal gas constant (6.7)

KEY CONCEPTS

❶ What is pressure and what units are used to measure it? (6.1)

- Pressure is the force per unit area. The pressure of a gas is the force exerted when gas particles strike a surface. Pressure is measured by a barometer and recorded in atmospheres (atm), millimeters of mercury (mm Hg), or pounds per square inch (psi).
- 1 atm = 760 mm Hg = 14.7 psi.

❷ What are gas laws and how are they used to describe the relationship between the pressure, volume, and temperature of a gas? (6.2–6.5)

- Because gas particles are far apart and behave independently, a set of gas laws describes the behavior of all gases regardless of their identity. Three gas laws—Boyle's law, Charles's law, and Gay–Lussac's law—describe the relationship between the pressure, volume, and temperature of a gas. These gas laws are summarized in "Key Equations—The Gas Laws" (p. 210).
- For a constant amount of gas, the following relationships exist.
  - The pressure and volume of a gas are inversely related, so increasing the pressure decreases the volume at constant temperature.
  - The volume of a gas is proportional to its Kelvin temperature, so increasing the temperature increases the volume at constant pressure.
  - The pressure of a gas is proportional to its Kelvin temperature, so increasing the temperature increases the pressure at constant volume.

❸ Describe the relationship between the volume and number of moles of a gas. (6.6)

- Avogadro's law states that when temperature and pressure are held constant, the volume of a gas is proportional to its number of moles.
- One mole of any gas has the same volume, the standard molar volume of 22.4 L, at 1 atm and 273 K (STP).

❹ What is the ideal gas law? (6.7)

- The ideal gas law is an equation that relates the pressure (P), volume (V), temperature (T), and number of moles (n) of a gas; $PV = nRT$, where R is the universal gas constant. The ideal gas law can be used to calculate any one of the four variables, as long as the other three variables are known.

❺ What is Dalton's law and how is it used to relate partial pressures and the total pressure of a gas mixture? (6.8)

- Dalton's law states that the total pressure of a gas mixture is the sum of the partial pressures of its component gases. The partial pressure is the pressure exerted by each component of a mixture.

❻ Discuss the importance of two minor components of the atmosphere—ozone and carbon dioxide. (6.9)

- Ozone in the stratosphere shields the earth's surface by absorbing ultraviolet radiation.
- Carbon dioxide is a greenhouse gas that absorbs thermal energy and redirects it back to the earth's surface. An increase in CO_2 concentration due to the combustion of fossil fuels may increase the average temperature of the earth's atmosphere, causing global climate change.

KEY EQUATIONS—THE GAS LAWS

Name	Equation	Variables Related	Constant Terms
Boyle's law	$P_1V_1 = P_2V_2$	P, V	T, n
Charles's law	$\frac{V_1}{T_1} = \frac{V_2}{T_2}$	V, T	P, n
Gay–Lussac's law	$\frac{P_1}{T_1} = \frac{P_2}{T_2}$	P, T	V, n
Combined gas law	$\frac{P_1V_1}{T_1} = \frac{P_2V_2}{T_2}$	P, V, T	n
Avogadro's law	$\frac{V_1}{n_1} = \frac{V_2}{n_2}$	V, n	P, T
Ideal gas law	$PV = nRT$	P, V, T, n	R

UNDERSTANDING KEY CONCEPTS

Selected in-chapter and odd-numbered end-of-chapter problems have brief answers at the end of each chapter. The *Student Study Guide and Solutions Manual* contains detailed solutions to all in-chapter and odd-numbered end-of-chapter problems, as well as additional worked examples and a chapter self-test.

6.27 (a) What is the pressure in psi (pounds per square inch) on the given pressure gauge used for scuba diving? (b) How many mm Hg does this correspond to?

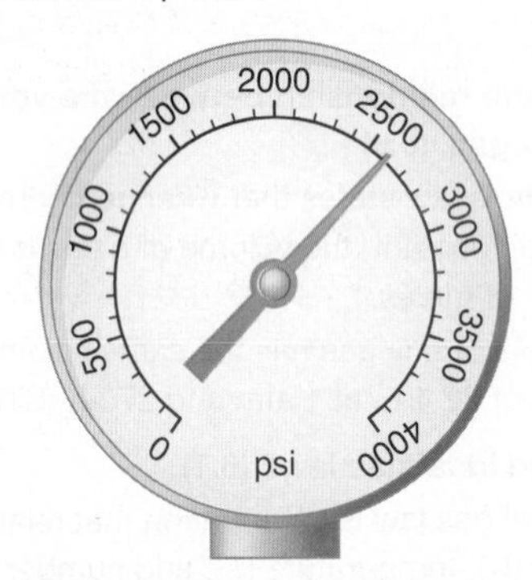

6.28 (a) What is the pressure in psi (pounds per square inch) on the given pressure gauge used for scuba diving? (b) How many pascals (Pa) does this correspond to?

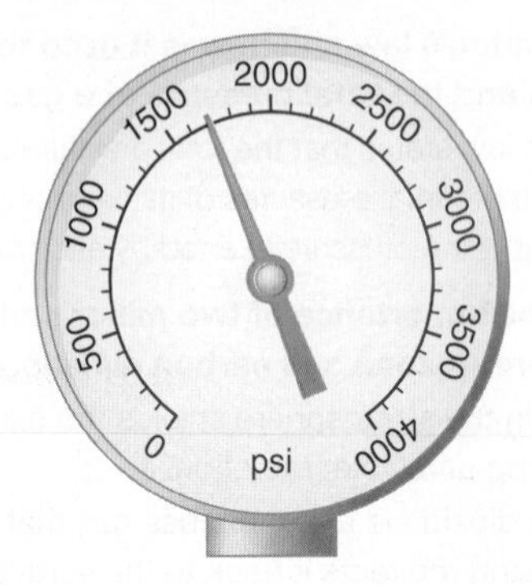

6.29 **X** consists of a flexible container with eight particles of a gas as shown. What happens to the pressure of the system when **X** is converted to the representations in parts (a), (b), and (c)?

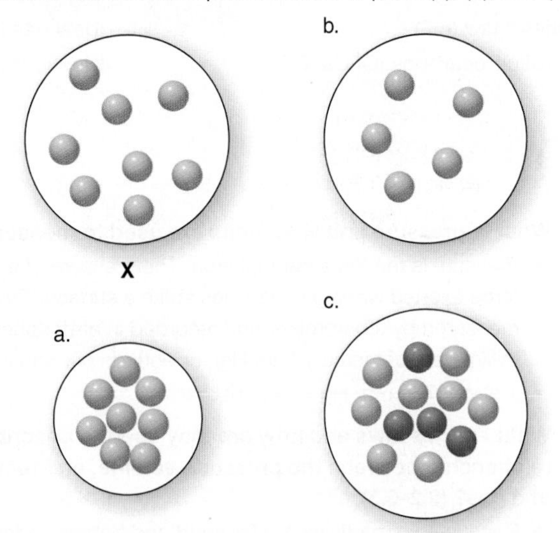

6.30 Suppose **A** represents a balloon that can expand or contract. **A** contains 10 particles of a gas as shown. Draw a diagram that shows the volume of **A** when each change occurs.

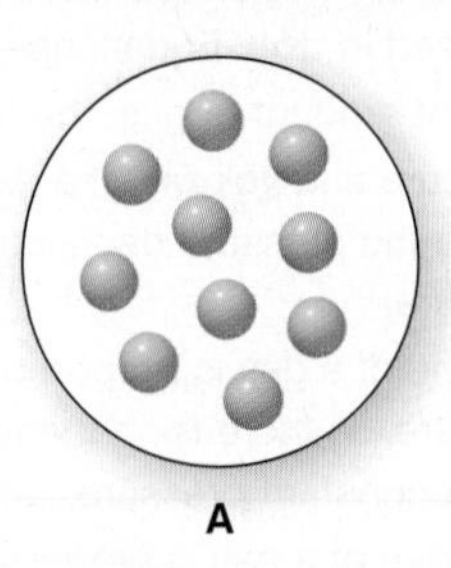

a. The volume is halved but the temperature and number of particles remain the same.
b. The pressure is doubled but the temperature and number of particles remain the same.
c. The temperature is increased but the pressure and number of particles remain the same.
d. The number of particles is doubled but the pressure and temperature remain the same.

6.31 A gas syringe contains 50. mL of CO_2 at 1.0 atm pressure. What is the pressure inside the syringe when the plunger is depressed to 25 mL?

6.32 A cylinder contains 3.4 L of helium at 200. atm. How many 1.0-L balloons can be filled at 1 atm, assuming the temperature is kept constant?

6.33 Draw a picture that represents the given balloon when each of the following changes occurs.

a. The balloon is inflated outside on a cold winter day and then taken inside a building at 75 °F.
b. The balloon is taken to the top of Mauna Kea, Hawaii (elevation 13,796 ft). Assume the temperature is constant.
c. The balloon is taken inside an airplane pressurized at 0.8 atm.

6.34 Which representation ([1], [2], or [3]) shows what balloon **Z** resembles after each change occurs?

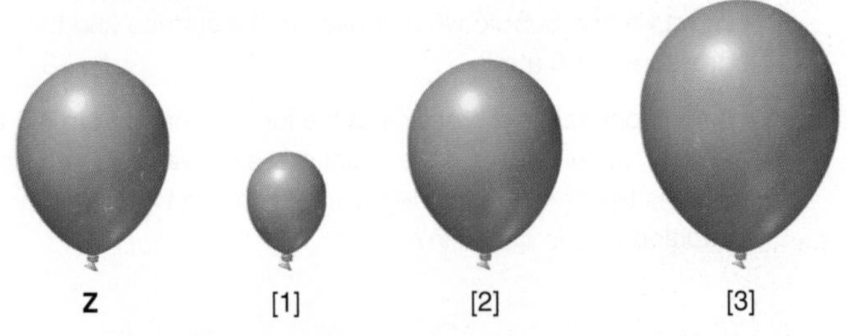

a. The balloon is cooled to a lower temperature.
b. Some gas leaks out.
c. The balloon is allowed to rise to a higher altitude.

6.35 If a compressed air cylinder for scuba diving contains 6.0 L of gas at 18 °C and 200. atm pressure, what volume does the gas occupy (in L) at 1.0 atm and 25 °C?

6.36 What happens to the pressure of a sample with each of the following changes?
a. Double the volume and halve the Kelvin temperature.
b. Double the volume and double the Kelvin temperature.
c. Halve the volume and double the Kelvin temperature.

6.37 How many moles of helium are contained in each volume at STP: (a) 5.0 L; (b) 11.2 L; (c) 50.0 mL?

6.38 How many moles of argon are contained in each volume at STP: (a) 4.0 L; (b) 31.2 L; (c) 120 mL?

6.39 The partial pressure of N_2 in the air is 593 mm Hg at 1 atm. What is the partial pressure of N_2 in a bubble of air a scuba diver breathes when he is 66 ft below the surface of the water where the pressure is 3.00 atm?

6.40 If N_2 is added to a balloon that contains O_2 (partial pressure 450 mm Hg) and CO_2 (partial pressure 150 mm Hg) to give a total pressure of 850 mm Hg, what is the partial pressure of each gas in the final mixture?

ADDITIONAL PROBLEMS

Pressure

6.41 The highest atmospheric pressure ever measured is 814.3 mm Hg, recorded in Mongolia in December, 2001. Convert this value to atmospheres.

6.42 The lowest atmospheric pressure ever measured is 652.5 mm Hg, recorded during Typhoon Tip on October 12, 1979. Convert this value to atmospheres.

6.43 Convert each quantity to the indicated unit.
a. 2.8 atm to psi
b. 520 mm Hg to atm
c. 20.0 atm to torr
d. 100. mm Hg to Pa

6.44 The compressed air tank of a scuba diver reads 3,200 psi at the beginning of a dive and 825 psi at the end of a dive. Convert each of these values to atm and mm Hg.

Boyle's Law

6.45 Assuming a fixed amount of gas at constant temperature, complete the following table.

	P_1	V_1	P_2	V_2
a.	2.0 atm	3.0 L	8.0 atm	?
b.	55 mm Hg	0.35 L	18 mm Hg	?
c.	705 mm Hg	215 mL	?	1.52 L

6.46 Assuming a fixed amount of gas at constant temperature, complete the following table.

	P_1	V_1	P_2	V_2
a.	2.5 atm	1.5 L	3.8 atm	?
b.	2.0 atm	350 mL	750 mm Hg	?
c.	75 mm Hg	9.1 mL	?	890 mL

6.47 If a scuba diver releases a 10.-mL air bubble below the surface where the pressure is 3.5 atm, what is the volume (in mL) of the bubble when it rises to the surface and the pressure is 1.0 atm?

6.48 If someone takes a breath and the lungs expand from 4.5 L to 5.6 L in volume, and the initial pressure was 756 mm Hg, what is the pressure inside the lungs (in mm Hg) before any additional air is pulled in?

Charles's Law

6.49 Assuming a fixed amount of gas at constant pressure, complete the following table.

	V_1	T_1	V_2	T_2
a.	5.0 L	310 K	?	250 K
b.	150 mL	45 K	?	45 °C
c.	60.0 L	0.0 °C	180 L	?

6.50 Assuming a fixed amount of gas at constant pressure, complete the following table.

	V_1	T_1	V_2	T_2
a.	10.0 mL	210 K	?	450 K
b.	255 mL	55 °C	?	150 K
c.	13 L	−150 °C	52 L	?

6.51 If a balloon containing 2.2 L of gas at 25 °C is cooled to −78 °C, what is its new volume (in L)?

6.52 How hot must the air in a balloon be heated if initially it has a volume of 750. L at 20 °C and the final volume must be 1,000. L?

Gay–Lussac's Law

6.53 Assuming a fixed amount of gas at constant volume, complete the following table.

	P_1	T_1	P_2	T_2
a.	3.25 atm	298 K	?	398 K
b.	550 mm Hg	273 K	?	−100. °C
c.	0.50 atm	250 °C	955 mm Hg	?

6.54 Assuming a fixed amount of gas at constant volume, complete the following table.

	P_1	T_1	P_2	T_2
a.	1.74 atm	120 °C	?	20. °C
b.	220 mm Hg	150 °C	?	300. K
c.	0.75 atm	198 °C	220 mm Hg	?

6.55 An autoclave is a pressurized container used to sterilize medical equipment by heating it to a high temperature under pressure. If an autoclave containing steam at 100. °C and 1.0 atm pressure is then heated to 150. °C, what is the pressure inside it (in atm)?

6.56 If a plastic container at 1.0 °C and 750. mm Hg is heated in a microwave oven to 80. °C, what is the pressure inside the container (in mm Hg)?

Combined Gas Law

6.57 Assuming a fixed amount of gas, complete the following table.

	P_1	V_1	T_1	P_2	V_2	T_2
a.	0.90 atm	4.0 L	265 K	?	3.0 L	310 K
b.	1.2 atm	75 L	5.0 °C	700. mm Hg	?	50 °C
c.	200. mm Hg	125 mL	298 K	100. mm Hg	0.62 L	?

6.58 Assuming a fixed amount of gas, complete the following table.

	P_1	V_1	T_1	P_2	V_2	T_2
a.	0.55 atm	1.1 L	340 K	?	3.0 L	298 K
b.	735 mm Hg	1.2 L	298 K	1.1 atm	?	0.0 °C
c.	7.5 atm	230 mL	−120 °C	15 atm	0.45 L	?

Avogadro's Law

6.59 What is the difference between STP and standard molar volume?

6.60 Given the same number of moles of two gases at STP conditions, how do the volumes of two gases compare? How do the masses of the two gas samples compare?

6.61 Calculate the volume of each substance at STP.
a. 4.2 mol Ar b. 3.5 g CO_2 c. 2.1 g N_2

6.62 Calculate the volume of each substance at STP.
a. 4.2 mol N_2 b. 6.5 g He c. 22.0 g CH_4

6.63 What volume does 3.01×10^{21} molecules of N_2 occupy at STP?

6.64 What volume does 1.50×10^{24} molecules of CO_2 occupy at STP?

Ideal Gas Law

6.65 How many moles of gas are contained in a human breath that occupies 0.45 L and has a pressure of 747 mm Hg at 37 °C?

6.66 How many moles of gas are contained in a compressed air tank for scuba diving that has a volume of 7.0 L and a pressure of 210 atm at 25 °C?

6.67 How many moles of air are present in the lungs if they occupy a volume of 5.0 L at 37 °C and 740 mm Hg? How many molecules of air does this correspond to?

6.68 If a cylinder contains 10.0 g of CO_2 in 10.0 L at 325 K, what is the pressure?

6.69 Which sample contains more moles: 2.0 L of O_2 at 273 K and 500 mm Hg, or 1.5 L of N_2 at 298 K and 650 mm Hg? Which sample has the greater mass?

6.70 If 2.00 g of solid carbon dioxide (dry ice, CO_2) is placed in a 1.5-L closed container at 25 °C, what is the pressure of the CO_2 after it has all sublimed to the gas phase?

6.71 An unknown amount of gas occupies 30.0 L at 2.1 atm and 298 K. How many moles does the sample contain? What is the mass if the gas is helium? What is the mass if the gas is argon?

6.72 If a cylinder contains 12.0 lb of liquid propane (C_3H_8), what volume does the propane occupy when it has completely vaporized at STP?

Dalton's Law and Partial Pressure

6.73 The molecular art shows a closed container with two gases, **A** (red spheres) and **B** (blue spheres) at a pressure of 630 mm Hg. What are the partial pressures of **A** and **B?**

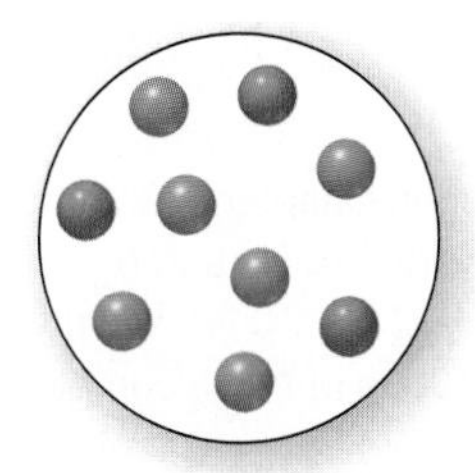

6.74 If the pressure due to the red spheres in the molecular art is 480 mm Hg, what is the pressure due to the blue spheres? What is the total pressure of the system?

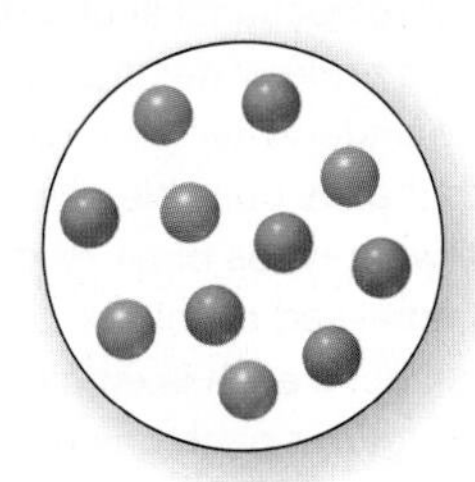

6.75 Air pressure on the top of Mauna Loa, a 13,000-ft mountain in Hawaii, is 460 mm Hg. What are the partial pressures of O_2 and N_2, which compose 21% and 78% of the atmosphere, respectively?

6.76 Nitrox is a gas mixture used in scuba diving that contains a higher-than-normal level of oxygen and a lower-than-normal level of nitrogen.

a. What percent of N_2 and O_2 is present in the given nitrox tank?
b. What is the partial pressure of each gas (in psi) in the tank?

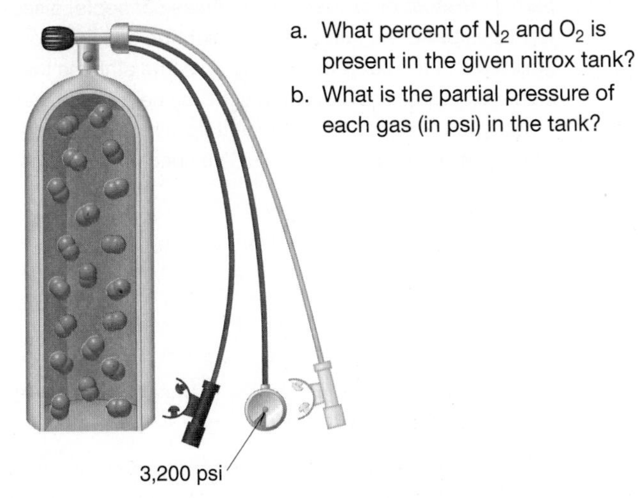

General Problems

6.77 Assume that each of the following samples is at the same temperature. (a) Which diagram represents the sample at the highest pressure? (b) Which diagram represents the sample at the lowest pressure?

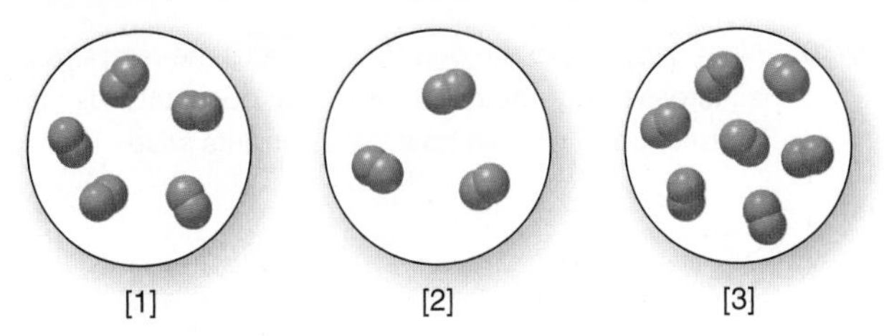

6.78 Use the diagrams in Problem 6.77 to answer the following questions.

a. What happens to the pressure when sample [1] is heated and the volume is kept constant?
b. What happens to the pressure of sample [2] when the volume is halved and the temperature is kept constant?
c. What happens to the pressure of sample [3] when its volume is halved and its Kelvin temperature is doubled?

6.79 A balloon is filled with helium at sea level. What happens to the volume of the balloon in each instance? Explain each answer.

a. The balloon floats to a higher altitude.
b. The balloon is placed in a bath of liquid nitrogen at –196 °C.
c. The balloon is placed inside a hyperbaric chamber at a pressure of 2.5 atm.
d. The balloon is heated inside a microwave.

6.80 Suppose you have a fixed amount of gas in a container with a movable piston, as drawn. Re-draw the container and piston to illustrate what it looks like after each of the following changes takes place.

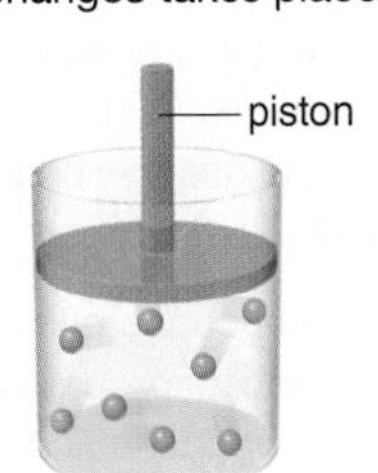

a. The temperature is held constant and the pressure is doubled.
b. The pressure is held constant and the Kelvin temperature is doubled.
c. The pressure is halved and the Kelvin temperature is halved.

Applications

6.81 What is the difference between the systolic and diastolic blood pressure?

6.82 What is hypertension and what are some of its complications?

6.83 If you pack a bag of potato chips for a snack on a plane ride, the bag appears to have inflated when you take it out to open. Explain why this occurs. If the initial volume of air in the bag was 250 mL at 760 mm Hg, and the plane is pressurized at 650 mm Hg, what is the final volume of the bag?

6.84 Why does a bubble at the bottom of a glass of a soft drink get larger as it rises to the surface?

6.85 If a scuba diver inhales 0.50 L of air at a depth of 100. ft and 4.0 atm pressure, what volume does this air occupy at the surface of the water, assuming air pressure is 1.0 atm? When a scuba diver must make a rapid ascent to the surface, he is told to exhale slowly as he ascends. How does your result support this recommendation?

6.86 What happens to the density of a gas if the temperature is increased but the pressure is held constant? Use this information to explain how wind currents arise.

6.87 A common laboratory test for a patient is to measure blood gases—that is, the partial pressures of O_2 and CO_2 in oxygenated blood. Normal values are 100 mm Hg for O_2 and 40 mm Hg for CO_2. A high or low level of one or both readings has some underlying cause. Explain the following. If a patient comes in agitated and hyperventilating—breathing very rapidly—the partial pressure of O_2 is normal but the partial pressure of CO_2 is 22 mm Hg.

6.88 Use the information in Problem 6.87 to explain the following. A patient with chronic lung disease has a partial pressure of O_2 of 60 mm Hg and a partial pressure of CO_2 of 60 mm Hg.

CHALLENGE PROBLEMS

6.89 A gas (4.0 g) occupies 11.2 L at 2 atm and 273 K. What is the molar mass of the gas? What is the identity of the gas?

6.90 As we learned in Chapter 5, an automobile airbag inflates when NaN_3 is converted to Na and N_2 according to the equation, $2\ NaN_3 \longrightarrow 2\ Na + 3\ N_2$. What volume of N_2 would be produced if 100. g of NaN_3 completely reacted at STP?

BEYOND THE CLASSROOM

6.91 Research what is known about the air quality in your city or region. What factors contribute to making the air unhealthy? Factors might include automobile exhaust, industrial emissions, or natural sources (e.g., volcanic gases or forest fires). What are the chemical formulas (or chemical structures) of the major pollutants? What problems are peculiar to your location? Are efforts in place to improve unhealthy air quality?

6.92 Several compounds besides carbon dioxide are classified as greenhouse gases, compounds that absorb thermal energy and redirect it back to the earth's surface. Research what other gases in the atmosphere fall into this category, and give their chemical structures. Is there a natural source of each gas, or is the gas produced solely by human activity? Is it known whether the concentration of the gas has changed significantly in the last 50–100 years?

6.93 As we learned in Chapter 6, CFCs, compounds once extensively used as refrigerants and aerosol propellants, deplete the ozone layer, and so their production and use have been eliminated in some countries. Research what compounds are currently used in place of CFCs in refrigeration, air conditioners, and propellants. Draw the structures of these compounds and explain why they are not expected to harm the ozone layer.

6.94 Although most recreational scuba divers use compressed air for breathing underwater, trained technical divers must sometimes be underwater for prolonged periods or at a greater depth than is possible using a tank of compressed air. Research what types of air mixtures are used by divers, including compressed air, nitrox, trimix, and heliox, and under what circumstances each is employed. How does the partial pressure of oxygen in each gas mixture compare to the partial pressure of oxygen in the atmosphere? How does the partial pressure of oxygen in the various mixtures compare? Why can't a diver use a tank filled with 100% oxygen?

6.95 When a major sporting event such as the World Cup or the Olympics occurs at a location that is considerably above sea level, the effects of higher altitude are often mentioned. Discuss some consequences of holding an athletic event at a venue with significantly lower atmospheric pressure and lower partial pressure of oxygen. Using these concepts, discuss what positive or negative effects might be observed on the athletes and the sport itself. Why do some athletes train at higher altitudes for an event that will be held at sea level? What types of sports might benefit from higher altitude and what types might be hindered? Give specific examples.

ANSWERS TO SELECTED PROBLEMS

6.1 a. 0.83 atm b. 12 psi

6.3 14/9

6.4 a. 1.6 L b. 3.2 L c. 0.80 L d. 16 L

6.5 a. 0.25 atm b. 1.5 atm c. 0.075 atm d. 0.0075 atm

6.6 0.44 L

6.7 a. 250 L b. 24.0 mL

6.9 392 K or 119 °C

6.11 As a sealed container is heated, the gases inside expand and increase the container's internal pressure, eventually popping the lid off.

6.12 290 mm Hg

6.13 0.18 mol

6.15 a. 1.0×10^2 L b. 7.8 L c. 12.6 L

6.17 36 mol

6.19 53 L

6.20 O_2: 2.5 atm CO_2: 1.5 atm

6.21 methane: 640 mm Hg ethane: 75 mm Hg
propane: 38 mm Hg

6.23 370 mm Hg

6.25

```
       :F:
        |
  :Cl—C—Cl:
        |
       :F:
```

6.27 a. 2,600 psi b. 1.3×10^5 mm Hg

6.29 a. increases b. decreases c. increases

6.31 2.0 atm

6.33 a. Volume increases.
b. Volume increases.
c. Volume increases.

6.35 1,200 L

6.37 a. 0.22 mol b. 0.500 mol c. 0.002 23 mol

6.39 1,780 mm Hg

6.41 1.07 atm

6.43 a. 41 psi b. 0.68 atm c. 15,200 torr d. 13,300 Pa

6.45 a. 0.75 L b. 1.1 L c. 99.7 mm Hg

6.47 35 mL

6.49 a. 4.0 L b. 1.1 L c. 820 K

6.51 1.4 L

6.53 a. 4.34 atm b. 350 mm Hg c. 1,300 K

6.55 1.1 atm

6.57 a. 1.4 atm b. 110 L c. 740 K

6.59 STP is "standard temperature and pressure," or 0 °C at 760 mm Hg. The standard molar volume is the volume that one mole of a gas occupies at STP, or 22.4 liters.

6.61 a. 94 L b. 1.8 L c. 1.7 L

6.63 0.112 L or 112 mL

6.65 0.017 mol

6.67 0.19 mol or 1.1×10^{23} molecules

6.69 O_2 has more moles and also has the greater mass.

6.71 2.6 mol 10. g He 1.0×10^2 g Ar

6.73 **A:** 210 mm Hg **B:** 420 mm Hg

6.75 97 mm Hg for O_2 360 mm Hg for N_2

6.77 a. [3] b. [2]

6.79 a. Volume increases as outside atmospheric pressure decreases.
b. Volume decreases at the lower temperature.
c. Volume decreases as external pressure increases.
d. Volume increases as temperature increases.

6.81 The systolic pressure is the maximum pressure generated with each heartbeat. The diastolic pressure is the lowest pressure recorded between heartbeats.

6.83 290 mL
The gases inside the bag had a volume of 250 mL at 760 mm Hg and take up a greater volume at the reduced pressure of 650 mm Hg.

6.85 The volume of air will increase from 0.50 L to 2.0 L as the scuba diver ascends to the surface. Therefore, it is necessary for the scuba diver to exhale as he rises to the surface of the water in order to eliminate the excess volume of gas.

6.87 As a person breathes faster, he eliminates more CO_2 from the lungs; therefore, the measured value of CO_2 is lower than the normal value of 40 mm Hg.

6.89 The molar mass is 4.0 g/mol and the gas is helium.

A sports drink is a solution of dissolved ions and carbohydrates, used to provide energy and hydration during strenuous exercise.

7

Solutions

CHAPTER OUTLINE

CHAPTER GOALS

In this chapter you will learn how to:

1. Describe the fundamental properties of a solution and determine whether a mixture is a solution, colloid, or suspension
2. Classify a substance as an electrolyte or nonelectrolyte
3. Predict whether a substance is soluble in water or a nonpolar solvent
4. Predict the effect of temperature and pressure on solubility
5. Calculate the concentration of a solution
6. Prepare a dilute solution from a more concentrated solution
7. Describe the process of osmosis and how it relates to biological membranes and dialysis

In Chapter 7 we study solutions—homogeneous mixtures of two or more substances. Why are table salt (NaCl) and sugar (sucrose) soluble in water but vegetable oil and gasoline are not? How does a healthcare professional take a drug as supplied by the manufacturer and prepare a dilute solution to administer a proper dose to a patient? An understanding of solubility and concentration is needed to explain each of these phenomena.

7.1 Mixtures

A pepperoni pizza is a heterogeneous mixture, while a soft drink is a homogeneous solution.

Thus far we have concentrated primarily on **pure substances**—elements, covalent compounds, and ionic compounds. Most matter with which we come into contact, however, is a **mixture** composed of two or more pure substances. The air we breathe is composed of nitrogen and oxygen, together with small amounts of argon, water vapor, carbon dioxide, and other gases. Seawater is composed largely of sodium chloride and water. A mixture may be **heterogeneous** or **homogeneous.**

- A *heterogeneous* mixture does not have a uniform composition throughout a sample.
- A *homogeneous* mixture has a uniform composition throughout a sample.

7.1A Solutions

The most common type of a homogeneous mixture is a solution.

- A *solution* is a homogeneous mixture that contains small particles. Liquid solutions are transparent.

A cup of hot coffee, vinegar, and gasoline are solutions. Any phase of matter can form a solution (Figure 7.1). Air is a solution of gases. An intravenous saline solution contains solid sodium chloride (NaCl) in liquid water. A dental filling contains liquid mercury (Hg) in solid silver.

Figure 7.1

Three Different Types of Solutions

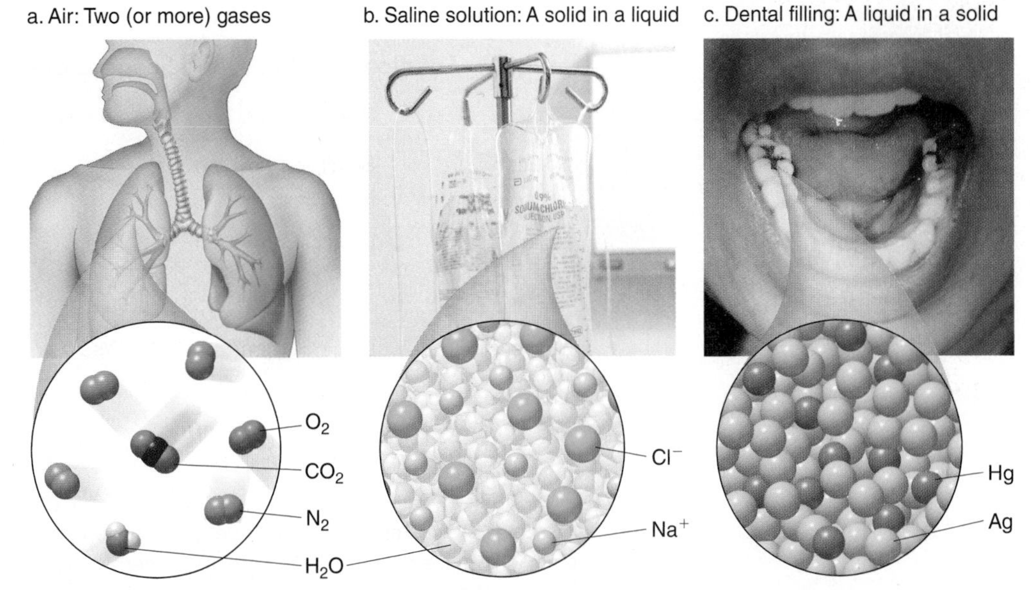

a. Air is a solution of gases, primarily N_2 and O_2. The lungs contain significant amounts of H_2O and CO_2 as well.
b. An IV saline solution contains solid sodium chloride (NaCl) dissolved in liquid water.
c. A dental filling contains a liquid, mercury (Hg), dissolved in solid silver (Ag).

When two substances form a solution, the substance present in the lesser amount is called the **solute,** and the substance present in the larger amount is the **solvent.** A solution with water as the solvent is called an **aqueous solution.**

Although a solution can be separated into its pure components, one component of a solution cannot be filtered away from the other component, because the particles of the solute are similar in size to the particles of the solvent. For a particular solute and solvent, solutions having different compositions are possible. For example, 1.0 g of NaCl can be mixed with 50.0 g of water or 10.0 g of NaCl can be mixed with 50.0 g of water.

7.1B Colloids and Suspensions

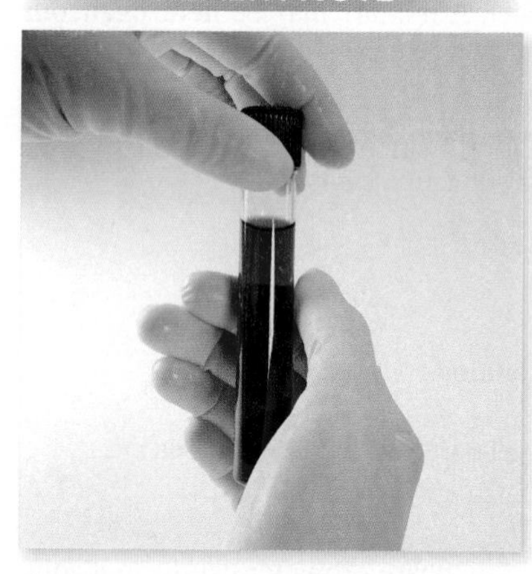

Blood is a suspension containing blood cells that can be separated from liquid blood plasma by centrifugation.

Colloids and suspensions are mixtures that contain larger particles than the particles in a solution.

- A *colloid* is a homogeneous mixture with larger particles, often having an opaque appearance.
- A *suspension* is a heterogeneous mixture that contains large particles suspended in a liquid.

Like solutions, the particles in a colloid generally cannot be filtered from its other components, and they do not settle out on standing. By definition, **a colloid contains particles that are 1 nm–1 μm in diameter.** Milk is a colloid. Homogenized milk is an opaque homogeneous mixture that contains large protein and fat molecules that do not dissolve in water. Homogenized milk does not have to be shaken before drinking because the particles of fat and protein do not settle out.

A suspension is a heterogeneous mixture that contains particles greater than 1 μm in diameter. The particles are so large that they do not dissolve in a liquid, and they can either be filtered away from the liquid, or separated using a centrifuge. The particles in a suspension settle out on standing and the suspension must be shaken to disperse the particles. Milk of magnesia, a common antacid, is a suspension that must be shaken before using because the active ingredients settle out.

Table 7.1 summarizes the properties of solutions, colloids, and suspensions.

Table 7.1 Solutions, Colloids, and Suspensions

Mixture	Particle Size	Settling	Separation
Solution	Small, < 1 nm	No	The components cannot be separated by filtration.
Colloid	Larger, 1 nm–1 μm	No	The components cannot be separated by common filtration techniques.
Suspension	Large, > 1 μm	Yes	The components can be separated by filtration or centrifugation.

PROBLEM 7.1

Classify each substance as a heterogeneous mixture, solution, or colloid: (a) Cherry Garcia ice cream (cherry ice cream + chocolate bits + cherries); (b) mayonnaise; (c) seltzer water; (d) nail polish remover; (e) brass (an alloy of Cu and Zn).

PROBLEM 7.2

Classify each product as a solution, colloid, or suspension: (a) hand cream; (b) Gatorade sports drink; (c) Kaopectate, a medication used to treat diarrhea; (d) shaving cream; (e) apple juice.

7.2 Electrolytes and Nonelectrolytes

A solute that dissolves in water to form ions conducts an electric current, whereas a solute that contains only neutral molecules does not.

- **A substance that conducts an electric current in water is called an *electrolyte.***
- **A substance that does not conduct an electric current in water is called a *nonelectrolyte.***

7.2A Classification

Consider the difference between an aqueous solution of sodium chloride, NaCl, and an aqueous solution of hydrogen peroxide, H_2O_2. A solution of sodium chloride contains Na^+ cations and Cl^- anions and conducts electricity. A solution of hydrogen peroxide contains only uncharged H_2O_2 molecules in H_2O, so it does not conduct electricity.

Electrolytes can be classified as strong or weak, depending on the extent that the compound dissociates (forms ions).

A **strong electrolyte** dissociates completely to form ions when it dissolves in water. NaCl is a strong electrolyte because all of the species present in solution are Na^+ cations and Cl^- anions.

A **weak electrolyte** dissolves in water to yield largely uncharged molecules, but a small fraction of the molecules form ions. When ammonia (NH_3) dissolves in water, the predominant species is uncharged NH_3 molecules, but some NH_3 molecules react with water to form NH_4^+ cations and

OH^- anions. Since a weak electrolyte contains some ions, it conducts an electric current, but not as well as a strong electrolyte. Common weak electrolytes are the weak acids and bases that we will learn about in Chapter 8.

NH3 is a weak electrolyte because it exists in H2O largely as uncharged NH3 molecules, together with a small number of ions (NH4+ and OH−).
NH3 dissolved in H2O
NH4+
OH−
NH3

Table 7.2 summarizes the characteristics of strong electrolytes, weak electrolytes, and nonelectrolytes, and lists examples of each.

Table 7.2 Strong Electrolytes, Weak Electrolytes, and Nonelectrolytes

Solute	Species in Solution	Conductivity	Examples
Strong electrolyte	Ions	Conducts an electric current	NaCl, KOH, HCl, KBr
Weak electrolyte	Molecules with a few ions	Conducts an electric current	NH_3, CH_3CO_2H, HF
Nonelectrolyte	Molecules	Does not conduct an electric current	CH_3CH_2OH, H_2O_2

PROBLEM 7.3

Consider the following diagrams for an aqueous solution of a compound **AB** (with **A** represented by red spheres and **B** represented by blue spheres). Label each diagram as a strong electrolyte, weak electrolyte, or nonelectrolyte.

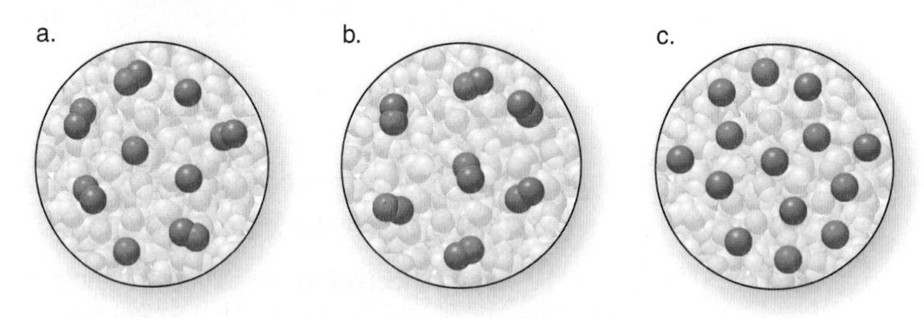
a.
b.
c.

PROBLEM 7.4

Classify each solution as an electrolyte or nonelectrolyte: (a) KCl in H_2O; (b) sucrose ($C_{12}H_{22}O_{11}$) in H_2O; (c) KI in H_2O.

Table 7.3 Moles and Equivalents

Ion	Equivalents
1 mol Na^+	1 Eq Na^+
1 mol Ca^{2+}	2 Eq Ca^{2+}
1 mol HCO_3^-	1 Eq HCO_3^-
1 mol PO_4^{3-}	3 Eq PO_4^{3-}

7.2B Equivalents

Blood plasma contains electrolytes with dissolved ions such as Na^+, K^+, Cl^-, and HCO_3^-. An **equivalent (Eq)** is a unit used to express the amount of each ion in solution. An equivalent relates the amount of an ion to its charge.

- An *equivalent* is the number of moles of charge that a mole of ions contributes to a solution.

Thus, one mole of K^+ equals one equivalent of K^+ ions because potassium bears a +1 charge. One mole of Ca^{2+} equals two equivalents of Ca^{2+} because calcium bears a +2 charge. **The number of equivalents per mole of an ion equals the charge on the ion,** as shown in Table 7.3.

SAMPLE PROBLEM 7.1

How many equivalents of sulfate (SO_4^{2-}) are present in a solution that contains 3.2 moles of SO_4^{2-}?

Analysis

Use a conversion factor that relates the number of moles of the ion to its number of equivalents.

Solution

Since sulfate has a –2 charge, there are 2 Eq of SO_4^{2-} ions per mole of SO_4^{2-}.

$$3.2 \cancel{\text{mol}}\ SO_4^{2-} \times \frac{2\ \text{Eq}\ SO_4^{2-}}{1\ \cancel{\text{mol}}\ SO_4^{2-}} = 6.4\ \text{Eq}\ SO_4^{2-}$$

Answer

PROBLEM 7.5

Using the given number of moles, determine how many equivalents of each ion are present.

a. 1 mol Na^+ b. 1 mol Mg^{2+} c. 0.5 mol K^+ d. 0.5 mol PO_4^{3-}

The amount of an electrolyte in blood plasma and IV solutions is often reported in the number of milliequivalents per liter of solution (mEq/L), where 1,000 mEq = 1 Eq. In solutions that contain ions, **there must be a balance between the total positive charge and total negative charge of all the ions present.** Thus, if a solution of KCl contains 40 mEq/L of K^+ ions, it must also contain 40 mEq/L of Cl^- ions.

SAMPLE PROBLEM 7.2

If an intravenous aqueous NaCl solution contains 154 mEq/L of Na^+ ions, how many milliequivalents of Na^+ is a patient given in 800. mL of solution?

Analysis and Solution

[1] **Identify the known quantities and the desired quantity.**

154 mEq/L Na^+ in solution
800. mL of solution
known quantities

? mEq Na^+
desired quantity

[2] **Write out the conversion factors.**

- Set up a conversion factor that relates milliequivalents to volume. Since the volume of the solution is given in mL, a mL–L conversion factor is needed as well.

mEq–L conversion factors

$$\frac{154\ \text{mEq}\ Na^+}{1\ \text{L}} \quad \text{or} \quad \frac{1\ \text{L}}{154\ \text{mEq}\ Na^+}$$

L–mL conversion factors

$$\frac{1\ \text{L}}{1000\ \text{mL}} \quad \text{or} \quad \frac{1000\ \text{mL}}{1\ \text{L}}$$

Use the conversion factors with the unwanted units—L and mL—in the denominator.

[3] **Solve the problem.**

- Multiply the original quantity by the conversion factors to obtain the desired quantity

$$800.\,\cancel{mL} \times \frac{1\,\cancel{L}}{1000\,\cancel{mL}} \times \frac{154\text{ mEq Na}^+}{1\,\cancel{L}} = 123\text{ mEq Na}^+$$

Answer

PROBLEM 7.6

If a KCl solution given intravenously to a patient that has a low potassium level contains 40. mEq of K^+ per liter, how many milliequivalents of K^+ are present in 550 mL of the solution?

PROBLEM 7.7

A solution contains the following ions: Na^+ (15 mEq/L), K^+ (10. mEq/L), and Ca^{2+} (4 mEq/L). If Cl^- is the only anion present in the solution, what is its concentration in mEq/L?

PROBLEM 7.8

If a solution contains 125 mEq of Na^+ per liter, how many milliliters of solution contain 25 mEq of Na^+?

7.3 Solubility—General Features

Solubility is the amount of solute that dissolves in a given amount of solvent, usually reported in grams of solute per 100 mL of solution (g/100 mL). A solution that has less than the maximum number of grams of solute is said to be **unsaturated.** A solution that has the maximum number of grams of solute that can dissolve is said to be **saturated.** If we added more solute to a saturated solution, the additional solute would remain undissolved in the flask.

What determines if a compound dissolves in a particular solvent? Whether a compound is soluble in a given solvent depends on the strength of the interactions between the compound and the solvent. As a result, compounds are soluble in solvents to which they are strongly attracted—that is, compounds are soluble in solvents with similar intermolecular forces. Solubility is often summed up in three words: **"Like dissolves like."**

- **Most ionic and polar covalent compounds are soluble in water, a polar solvent.**
- **Nonpolar compounds are soluble in nonpolar solvents.**

Water-soluble compounds are ionic or are small polar molecules that can hydrogen bond with the water solvent. For example, solid sodium chloride (NaCl) is held together by very strong electrostatic interactions of the oppositely charged ions. When it is mixed with water, the Na^+ and Cl^- ions are separated from each other and surrounded by polar water molecules (Figure 7.2). Each Na^+ is surrounded by water molecules arranged with their O atoms (which bear a partial negative charge) in close proximity to the positive charge of the cation. Each Cl^- is surrounded by water molecules arranged with their H atoms (which bear a partial positive charge) in close proximity to the negative charge of the anion.

- **The attraction of an ion with a dipole in a molecule is called an ion–dipole interaction.**

Figure 7.2 Dissolving Sodium Chloride in Water

When ionic NaCl dissolves in water, the Na^+ and Cl^- interactions of the crystal are replaced by new interactions of Na^+ and Cl^- ions with the solvent. Each ion is surrounded by a loose shell of water molecules arranged so that oppositely charged species are close to each other.

The ion–dipole interactions between Na^+, Cl^-, and water provide the energy needed to break apart the ions from the crystal lattice. The water molecules form a loose shell of solvent around each ion. The process of surrounding particles of a solute with solvent molecules is called **solvation.**

Small uncharged molecules that can hydrogen bond with water are also soluble. Thus, ethanol (C_2H_5OH), which is present in alcoholic beverages, dissolves in water because hydrogen bonding occurs between the OH group in ethanol and the OH group of water.

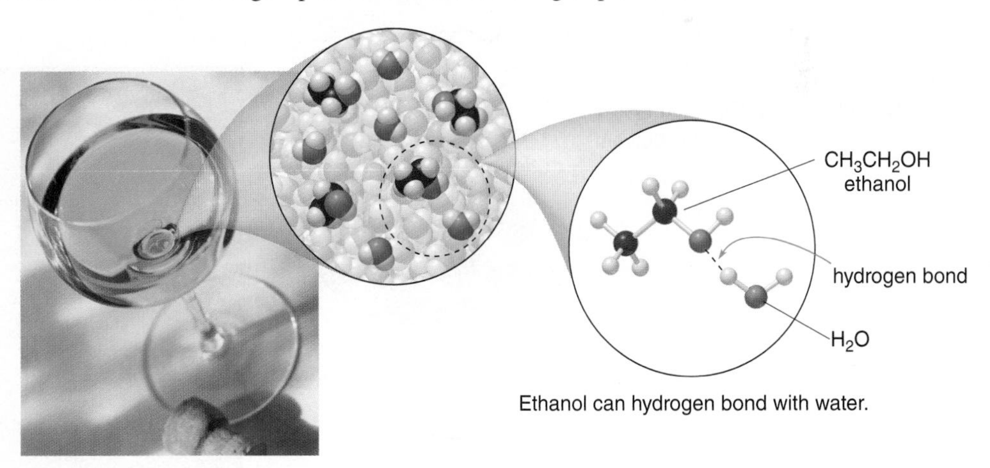

Ethanol can hydrogen bond with water.

Water solubility for uncharged molecules occurs only with small polar molecules or those with many O or N atoms that can hydrogen bond to water. Thus, stearic acid ($C_{18}H_{36}O_2$), a component of animal fats, is *insoluble* in water because its nonpolar part (C—C and C—H bonds) is large compared to its polar part (C—O and O—H bonds). On the other hand, glucose ($C_6H_{12}O_6$), a simple carbohydrate, is *soluble* in water because it has many OH groups and thus many opportunities for hydrogen bonding with water.

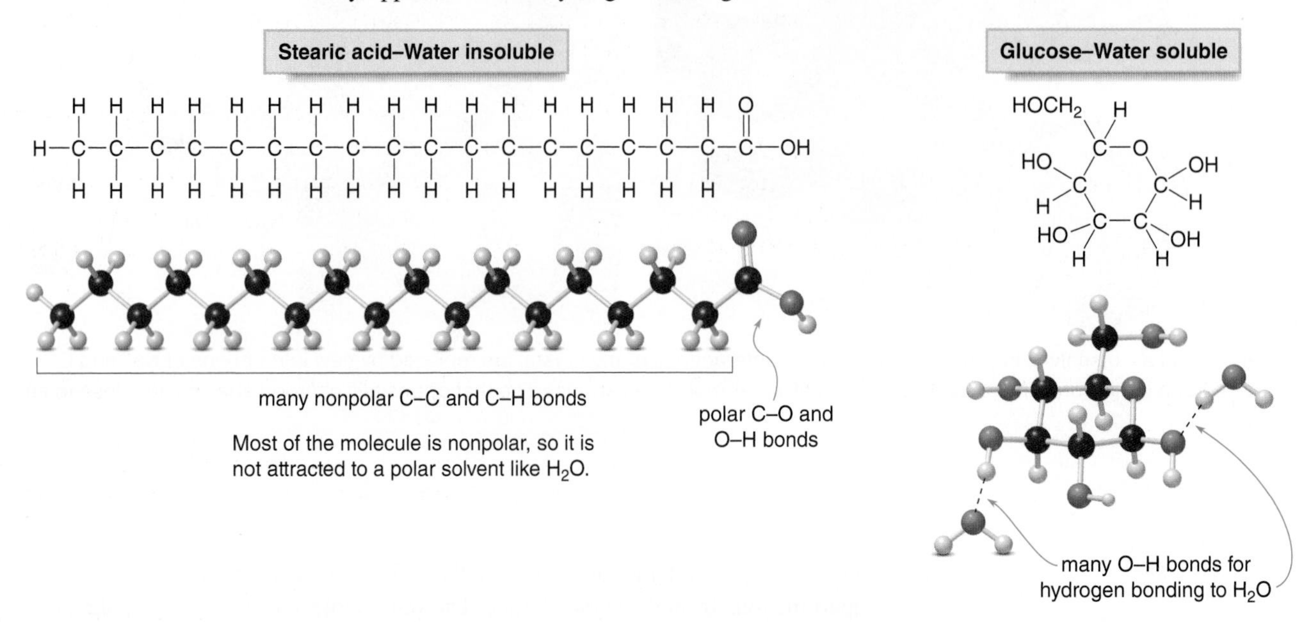

Nonpolar compounds are soluble in nonpolar solvents. As a result, octane (C_8H_{18}), a component of gasoline, dissolves in the nonpolar solvent carbon tetrachloride (CCl_4), as shown in Figure 7.3. Animal fat and vegetable oils, which are composed largely of nonpolar C—C and C—H bonds, are soluble in CCl_4, but are insoluble in a polar solvent like water. These solubility properties explain why "oil and water don't mix."

Figure 7.3

Solubility—A Nonpolar Compound in a Nonpolar Solvent

Octane (C_8H_{18}) dissolves in CCl_4 because both are nonpolar liquids that exhibit only London dispersion forces.

Dissolving a solute in a solvent is a physical process that is accompanied by an energy change. Breaking up the particles of the solute requires energy, and forming new attractive forces between the solute and the solvent releases energy.

- **When solvation releases more energy than that required to separate particles, the overall process is exothermic (heat is released).**
- **When the separation of particles requires more energy than is released during solvation, the process is endothermic (heat is absorbed).**

These energy changes are used to an advantage in commercially available hot packs and cold packs. A hot pack, sometimes used for pain relief of sore muscles, contains calcium chloride ($CaCl_2$) or magnesium sulfate ($MgSO_4$) and water. Breaking the seal that separates them allows the salt to dissolve in the water, releasing heat, and the pouch gets warm. In contrast, ammonium nitrate (NH_4NO_3) absorbs heat on mixing with water, so this salt is found in instant cold packs used to reduce swelling.

SAMPLE PROBLEM 7.3

Predict the water solubility of each compound: (a) KCl; (b) methanol (CH_3OH); (c) hexane (C_6H_{14}).

Analysis

Use the general solubility rule—"like dissolves like." Generally, ionic and small polar compounds that can hydrogen bond are soluble in water. Nonpolar compounds are soluble in nonpolar solvents.

Solution

a. KCl is an ionic compound, so it dissolves in water, a polar solvent.
b. CH_3OH is a small polar molecule that contains an OH group. As a result, it can hydrogen bond to water, making it soluble.
c. Hexane (C_6H_{14}) has only nonpolar C—C and C—H bonds, making it a nonpolar molecule that is therefore water insoluble.

PROBLEM 7.9

Which compounds are water soluble?

a. $NaNO_3$
b. CH_4
c. $\begin{array}{c} \text{H}\quad\text{H} \\ |\quad\;\,| \\ \text{HO—C—C—OH} \\ |\quad\;\,| \\ \text{H}\quad\text{H} \end{array}$
d. KBr
e. NH_2OH

PROBLEM 7.10

Which pairs of compounds will form a solution?

a. benzene (C_6H_6) and hexane (C_6H_{14})
b. Na_2SO_4 and H_2O
c. NaCl and hexane (C_6H_{14})
d. H_2O and CCl_4

7.4 Solubility—Effects of Temperature and Pressure

Both temperature and pressure can affect solubility.

7.4A Temperature Effects

For most ionic and molecular solids, solubility generally increases as temperature increases. Thus, sugar is much more soluble in a cup of hot coffee than in a glass of iced tea. If a solid is dissolved in a solvent at high temperature and then the solution is slowly cooled, the solubility of the solute decreases and it precipitates from the solution. Sometimes, however, if cooling is very slow, the solution becomes **supersaturated** with solute; that is, the solution contains more than the predicted maximum amount of solute at a given temperature. Such a solution is unstable, and when it is disturbed, the solute precipitates rapidly.

In contrast, **the solubility of gases *decreases* with increasing temperature.** Because increasing temperature increases the kinetic energy, more molecules escape into the gas phase and fewer remain in solution. Increasing temperature decreases the solubility of oxygen in lakes and streams. In cases where industrial plants operating near lakes or streams have raised water temperature, marine life dies from lack of sufficient oxygen in solution.

PROBLEM 7.11

Why does a soft drink become "flat" faster when it is left open at room temperature compared to when it is left open in the refrigerator?

7.4B Pressure Effects

Scuba divers often hold onto a rope so that they ascend to the surface slowly after a dive, in order to avoid forming bubbles of nitrogen gas in joints and blood vessels.

Pressure changes do not affect the solubility of liquids and solids, but pressure affects the solubility of gases a great deal. **Henry's law** describes the effect of pressure on gas solubility.

- Henry's law: The solubility of a gas in a liquid is proportional to the partial pressure of the gas above the liquid.

Thus, the **higher the pressure, the higher the solubility of a gas in a solvent.** A practical demonstration of Henry's law occurs whenever we open a carbonated soft drink. Soft drinks containing dissolved CO_2 are sealed under greater than 1 atm pressure. When a can is opened, the pressure above the liquid decreases to 1 atm, so the solubility of the CO_2 in the soda decreases as well and some of the dissolved CO_2 fizzes out of solution (Figure 7.4).

As we learned in Section 6.6, increasing gas solubility affects scuba divers because more N_2 is dissolved in the blood under the higher pressures experienced under water. Divers must ascend slowly to avoid forming bubbles of N_2 in joints and small blood vessels. If a diver ascends slowly, the external pressure around the diver slowly decreases and by Henry's law, the solubility of the gas in the diver's blood slowly decreases as well.

PROBLEM 7.12

Predict the effect each change has on the solubility of [1] $Na_2CO_3(s)$; [2] $N_2(g)$.

a. increasing the temperature
b. decreasing the temperature
c. increasing the pressure
d. decreasing the pressure

Figure 7.4

Henry's Law and Carbonated Beverages

The gas pressure in a closed can of soda is approximately 2 atm. When the can is opened, the pressure above the liquid in the can decreases to 1 atm, so the CO_2 concentration in the soda decreases as well, and the gas fizzes from the soda.

HEALTH NOTE

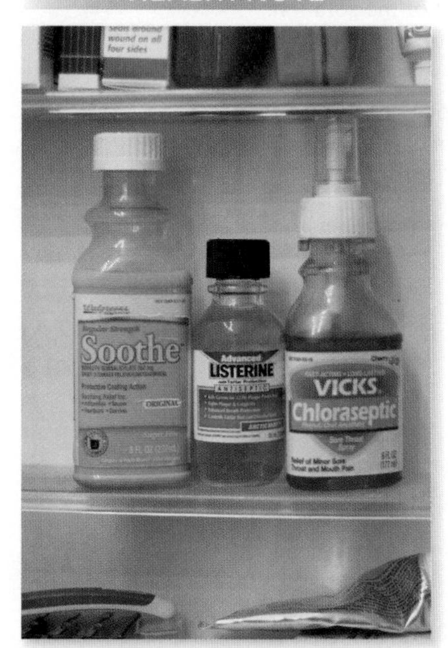

Mouthwash, sore throat spray, and many other over-the-counter medications contain ingredients whose concentrations are reported in (w/v)%.

7.5 Concentration Units—Percent Concentration

In using a solution in the laboratory or in administering the proper dose of a liquid medication, we must know its ***concentration*—how much solute is dissolved in a given amount of solution.** Concentration can be measured in several different ways that use mass, volume, or moles. Two useful measures of concentration are reported as percentages—that is, the number of grams or milliliters of solute per 100 mL of solution.

7.5A Weight/Volume Percent

One of the most common measures of concentration is **weight/volume percent concentration, (w/v)%—that is, the number of grams of solute dissolved in 100 mL of solution.** Mathematically, weight/volume percent is calculated by dividing the number of grams of solute in a given number of milliliters of solution, and multiplying by 100%.

Weight/volume percent concentration

$$(\text{w/v})\% = \frac{\text{mass of solute (g)}}{\text{volume of solution (mL)}} \times 100\%$$

For example, vinegar contains 5 g of acetic acid dissolved in 100 mL of solution, so the acetic acid concentration is 5% (w/v).

$$(\text{w/v})\% = \frac{\text{5 g acetic acid}}{\text{100 mL vinegar solution}} \times 100\% = 5\%\ (\text{w/v})\ \text{acetic acid}$$

Note that the volume used to calculate concentration is the *final* volume of the solution, not the volume of solvent added to make the solution. A special flask called a **volumetric flask** is used to make a solution of a given concentration (Figure 7.5). The solute is placed in the flask and then

Figure 7.5 Making a Solution with a Particular Concentration

a. Add the solute.

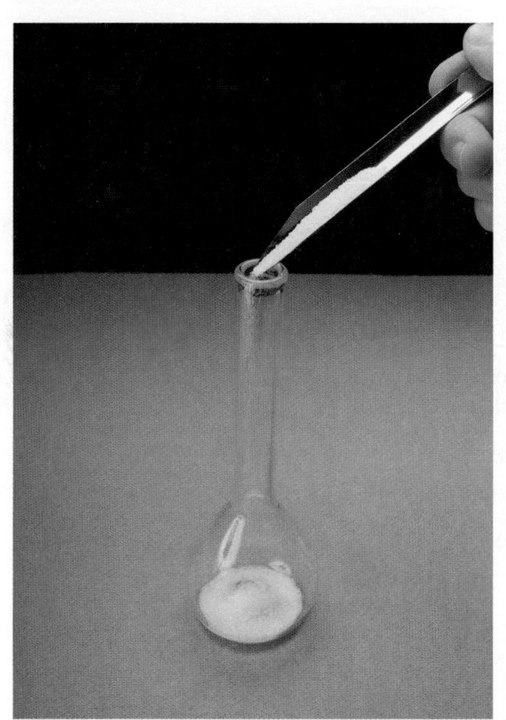

b. Add the solvent.

To make a solution of a given concentration: (a) add a measured number of grams of solute to a volumetric flask; (b) then add solvent to dissolve the solid, bringing the level of the solvent to the calibrated mark on the neck of the flask.

enough solvent is added to dissolve the solute by mixing. Next, additional solvent is added until it reaches a calibrated line that measures the final volume of the solution.

SAMPLE PROBLEM 7.4

Chloraseptic sore throat spray contains 0.35 g of the antiseptic phenol dissolved in 25 mL of solution. What is the weight/volume percent concentration of phenol?

Analysis

Use the formula (w/v)% = (grams of solute)/(mL of solution) × 100%.

Solution

$$(\text{w/v})\% = \frac{0.35 \text{ g phenol}}{25 \text{ mL solution}} \times 100\% = 1.4\% \text{ (w/v) phenol}$$

Answer

PROBLEM 7.13

Pepto-Bismol, an over-the-counter medication used for upset stomach and diarrhea, contains 525 mg of bismuth subsalicylate in each 15-mL tablespoon. What is the weight/volume percent concentration of bismuth subsalicylate?

PROBLEM 7.14

A commercial mouthwash contains 4.3 g of ethanol and 0.021 g of antiseptic in each 30.-mL portion. Calculate the weight/volume percent concentration of each component.

7.5B Volume/Volume Percent

When the solute in a solution is a liquid, its concentration is often reported using **volume/volume percent concentration, (v/v)%—that is, the number of milliliters of solute dissolved in 100 mL of solution.** Mathematically, volume/volume percent is calculated by dividing the number of milliliters of solute in a given number of milliliters of solution, and multiplying by 100%.

Volume/volume percent concentration

$$(\text{v/v})\% = \frac{\text{volume of solute (mL)}}{\text{volume of solution (mL)}} \times 100\%$$

For example, a bottle of rubbing alcohol that contains 70 mL of 2-propanol in 100 mL of solution has a 70% (v/v) concentration of 2-propanol.

$$(\text{v/v})\% = \frac{70 \text{ mL 2-propanol}}{100 \text{ mL rubbing alcohol}} \times 100\% = 70\% \text{ (v/v) 2-propanol}$$

The alcohol (ethanol) content of wine, beer, and other alcoholic beverages is reported using volume/volume percent concentration. Wines typically contain 10–13% (v/v) ethanol, whereas beer usually contains 3–5%.

SAMPLE PROBLEM 7.5

A 750-mL bottle of wine contains 101 mL of ethanol. What is the volume/volume percent concentration of ethanol?

Analysis

Use the formula (v/v)% = (mL of solute)/(mL of solution) × 100%.

Solution

$$(\text{v/v})\% = \frac{101 \text{ mL ethanol}}{750 \text{ mL wine}} \times 100\% = 14\% \text{ (v/v) ethanol}$$

Answer

PROBLEM 7.15

A 250-mL bottle of mouthwash contains 21 mL of ethanol. What is the volume/volume percent concentration of ethanol?

HEALTH NOTE

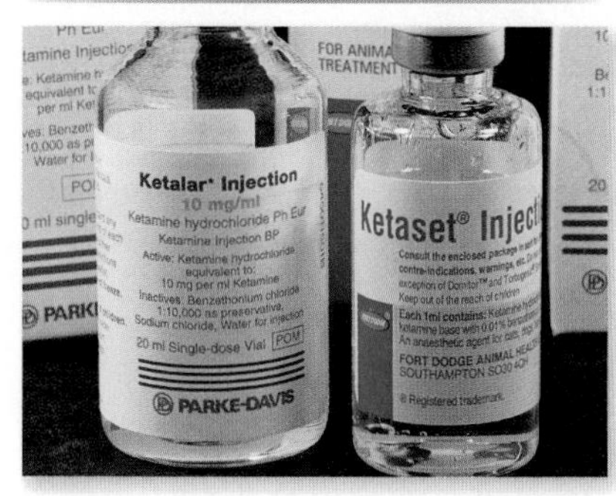

Ketamine is a widely used anesthetic in both human and veterinary medicine. It has been illegally used as a recreational drug because it can produce hallucinations.

7.5C Using a Percent Concentration as a Conversion Factor

Percent concentration can be used as a conversion factor to relate the amount of solute (either grams or milliliters) to the amount of solution. For example, ketamine, an anesthetic especially useful for children, is supplied as a 5.0% (w/v) solution, meaning that 5.0 g of ketamine are present in 100 mL of solution. Two conversion factors derived from the percent concentration can be written.

$$\underset{\text{weight/volume percent concentration}}{\text{5.0\% (w/v) ketamine}} \qquad \frac{\text{5.0 g ketamine}}{\text{100 mL solution}} \quad \text{or} \quad \frac{\text{100 mL solution}}{\text{5.0 g ketamine}}$$

We can use these conversion factors to determine the amount of solute contained in a given volume of solution (Sample Problem 7.6), or to determine how much solution contains a given number of grams of solute (Sample Problem 7.7).

SAMPLE PROBLEM 7.6

A saline solution used in intravenous drips for patients who cannot take oral fluids contains 0.92% (w/v) NaCl in water. How many grams of NaCl are contained in 250 mL of this solution?

Analysis and Solution

[1] Identify the known quantities and the desired quantity.

0.92% (w/v) NaCl solution
250 mL — known quantities

? g NaCl — desired quantity

[2] Write out the conversion factors.

- Set up conversion factors that relate grams of NaCl to the volume of the solution using the weight/volume percent concentration.

$$\frac{\text{100 mL solution}}{\text{0.92 g NaCl}} \quad \text{or} \quad \frac{\text{0.92 g NaCl}}{\text{100 mL solution}}$$

Choose this conversion factor to cancel mL.

[3] Solve the problem.

- Multiply the original quantity by the conversion factor to obtain the desired quantity.

$$250\ \cancel{\text{mL}} \times \frac{\text{0.92 g NaCl}}{100\ \cancel{\text{mL}}\text{ solution}} = \text{2.3 g NaCl}$$

Answer

SAMPLE PROBLEM 7.7

What volume of a 5.0% (w/v) solution of ketamine contains 75 mg?

Analysis and Solution

[1] Identify the known quantities and the desired quantity.

5.0% (w/v) ketamine solution
75 mg — known quantities

? mL ketamine — desired quantity

[2] Write out the conversion factors.

- Use the weight/volume percent concentration to set up conversion factors that relate grams of ketamine to mL of solution. Since percent concentration is expressed in grams, a mg–g conversion factor is needed as well.

mg–g conversion factors

$$\frac{\text{1000 mg}}{\text{1 g}} \quad \text{or} \quad \frac{\text{1 g}}{\text{1000 mg}}$$

g–mL solution conversion factors

$$\frac{\text{5.0 g ketamine}}{\text{100 mL solution}} \quad \text{or} \quad \frac{\text{100 mL solution}}{\text{5.0 g ketamine}}$$

Choose the conversion factors with the unwanted units—mg and g—in the denominator.

CONSUMER NOTE

Alcohol-based hand sanitizers clean hands by killing many types of disease-causing bacteria and viruses.

[3] **Solve the problem.**

- Multiply the original quantity by the conversion factors to obtain the desired quantity.

$$75\ \cancel{\text{mg}}\ \text{ketamine} \times \frac{1\ \cancel{\text{g}}}{1000\ \cancel{\text{mg}}} \times \frac{100\ \text{mL solution}}{5.0\ \cancel{\text{g}}\ \text{ketamine}} = 1.5\ \text{mL solution}$$

Answer

PROBLEM 7.16

A drink sold in a health food store contains 0.50% (w/v) of vitamin C. What volume would you have to ingest to obtain 1,000. mg of vitamin C?

PROBLEM 7.17

A cough medicine contains 0.20% (w/v) dextromethorphan, a cough suppressant, and 2.0% (w/v) guaifenisin, an expectorant. How many milligrams of each drug would you obtain from 3.0 tsp of cough syrup? (1 tsp = 5 mL)

PROBLEM 7.18

How many mL of ethanol are contained in an 8.0-oz bottle of hand sanitizer that has a concentration of 62% (v/v) of ethanol? (1 fl oz = 29.6 mL)

7.5D Parts Per Million

When a solution contains a very small concentration of solute, concentration is often expressed in **parts per million (ppm).** Whereas percent concentration is the number of "parts"—grams or milliliters—in 100 parts (100 mL) of solution, parts per million is the number of "parts" in 1,000,000 parts of solution. The "parts" may be expressed in either mass or volume units as long as the *same* unit is used for both the numerator and denominator.

Parts per million

$$\text{ppm} = \frac{\text{mass of solute (g)}}{\text{mass of solution (g)}} \times 10^6$$

or

$$\text{ppm} = \frac{\text{volume of solute (mL)}}{\text{volume of solution (mL)}} \times 10^6$$

A sample of seawater that contains 1.3 g of magnesium ions in 10^6 g of solution contains 1.3 ppm of magnesium.

$$\text{ppm} = \frac{1.3\ \text{g magnesium}}{10^6\ \text{g seawater}} \times 10^6 = 1.3\ \text{ppm magnesium}$$

Parts per million is used as a concentration unit for very dilute solutions. When water is the solvent, the density of the solution is close to the density of pure water, which is 1.0 g/mL at room temperature. In this case, the numerical value of the denominator is the same no matter if the unit is grams or milliliters. Thus, an aqueous solution that contains 2 ppm of MTBE, a gasoline additive and environmental pollutant, can be written in the following ways:

$$\frac{2\ \text{g MTBE}}{10^6\ \text{g solution}} \times 10^6 = \frac{2\ \text{g MTBE}}{10^6\ \text{mL solution}} \times 10^6 = 2\ \text{ppm MTBE}$$

10^6 mL has a mass of 10^6 g.

ENVIRONMENTAL NOTE

Seabirds such as osprey that feed on fish contaminated with the pesticide DDT accumulate an average of 25 parts per million of DDT in their fatty tissues. When DDT concentration is high, mother osprey produce eggs with very thin shells that are easily crushed, so fewer osprey chicks hatch.

SAMPLE PROBLEM 7.8

What is the concentration in parts per million of DDT in the tissues of a seabird that contains 50. mg of DDT in 1,900 g of tissue? DDT, a nonbiodegradable pesticide that is a persistent environmental pollutant, has been banned from use in the United States since 1973.

Analysis

Use the formula ppm = (g of solute)/(g of solution) $\times 10^6$.

Solution

[1] Convert milligrams of DDT to grams of DDT so that both the solute and solution have the same unit.

$$50.\ \cancel{mg}\ \text{DDT} \times \frac{1\ \text{g}}{1000\ \cancel{mg}} = 0.050\ \text{g DDT}$$

[2] Use the formula to calculate parts per million.

$$\frac{0.050\ \text{g DDT}}{1900\ \text{g tissue}} \times 10^6 = 26\ \text{ppm DDT}$$

Answer

PROBLEM 7.19

What is the concentration in parts per million of DDT in each of the following?

a. 0.042 mg in 1,400 g plankton

b. 5×10^{-4} g in 1.0 kg minnow tissue

c. 2.0 mg in 1.0 kg needlefish tissue

d. 225 μg in 1.0 kg breast milk

7.6 Concentration Units—Molarity

The most common measure of concentration in the laboratory is ***molarity*****—the number of moles of solute per liter of solution,** abbreviated as M.

$$\textbf{Molarity} = \textbf{M} = \frac{\textbf{moles of solute (mol)}}{\textbf{liter of solution (L)}}$$

A solution that is formed from 1.00 mol (58.44 g) of NaCl in enough water to give 1.00 L of solution has a molarity of 1.00 M. A solution that is formed from 2.50 mol (146 g) of NaCl in enough water to give 2.50 L of solution is also a 1.00 M solution. Both solutions contain the *same number of moles per unit volume.*

$$\text{M} = \frac{\text{moles of solute (mol)}}{V\ \text{(L)}} = \frac{1.00\ \text{mol NaCl}}{1.00\ \text{L solution}} = 1.00\ \text{M}$$

$$\text{M} = \frac{\text{moles of solute (mol)}}{V\ \text{(L)}} = \frac{2.50\ \text{mol NaCl}}{2.50\ \text{L solution}} = 1.00\ \text{M}$$

same concentration
same number of moles per unit volume (V)

Since quantities in the laboratory are weighed on a balance, we must learn how to determine molarity beginning with a particular number of grams of a substance, as shown in the accompanying stepwise procedure.

How To Calculate Molarity from a Given Number of Grams of Solute

Example Calculate the molarity of a solution made from 20.0 g of NaOH in 250 mL of solution.

Step [1] Identify the known quantities and the desired quantity.

known quantities	desired quantity
20.0 g NaOH	
250 mL solution	? M (mol/L)

How To, continued . . .

Step [2] Convert the number of grams of solute to the number of moles. Convert the volume of the solution to liters, if necessary.

- Use the molar mass to convert grams of NaOH to moles of NaOH (molar mass 40.00 g/mol).

molar mass conversion factor

$$20.0 \text{ g NaOH} \times \frac{1 \text{ mol}}{40.00 \text{ g NaOH}} = 0.500 \text{ mol NaOH}$$

Grams cancel.

- Convert milliliters of solution to liters of solution using a mL–L conversion factor.

mL–L conversion factor

$$250 \text{ mL solution} \times \frac{1 \text{ L}}{1000 \text{ mL}} = 0.25 \text{ L solution}$$

Milliliters cancel.

Step [3] Divide the number of moles of solute by the number of liters of solution to obtain the molarity.

$$\underset{\text{molarity}}{\text{M}} = \frac{\text{moles of solute (mol)}}{V \text{ (L)}} = \frac{0.500 \text{ mol NaOH}}{0.25 \text{ L solution}} = \underset{\textbf{Answer}}{2.0 \text{ M}}$$

SAMPLE PROBLEM 7.9

What is the molarity of an intravenous glucose solution prepared from 108 g of glucose in 2.0 L of solution?

Analysis and Solution

[1] Identify the known quantities and the desired quantity.

known quantities	desired quantity
108 g glucose 2.0 L solution	? M (mol/L)

[2] Convert the number of grams of glucose to the number of moles using the molar mass (180.2 g/mol).

$$108 \text{ g glucose} \times \frac{1 \text{ mol}}{180.2 \text{ g}} = 0.599 \text{ mol glucose}$$

Grams cancel.

- Since the volume of the solution is given in liters, no conversion is necessary for volume.

[3] Divide the number of moles of solute by the number of liters of solution to obtain the molarity.

$$\underset{\text{molarity}}{\text{M}} = \frac{\text{moles of solute (mol)}}{V \text{ (L)}} = \frac{0.599 \text{ mol glucose}}{2.0 \text{ L solution}} = \underset{\textbf{Answer}}{0.30 \text{ M}}$$

PROBLEM 7.20

Calculate the molarity of each aqueous solution with the given amount of NaCl (molar mass 58.44 g/mol) and final volume.

a. 1.0 mol in 0.50 L
b. 2.0 mol in 250 mL
c. 0.050 mol in 5.0 mL
d. 12.0 g in 2.0 L

Molarity is a conversion factor that relates the number of moles of solute to the volume of solution it occupies. Thus, if we know the molarity and volume of a solution, we can calculate the number of moles it contains. If we know the molarity and number of moles, we can calculate the volume in liters.

To calculate the moles of solute...

$$\frac{\text{moles of solute (mol)}}{V\text{ (L)}} = \text{M}$$

...rearrange the equation for molarity (M):

$$\text{moles of solute (mol)} = \text{M} \times V\text{ (L)}$$

To calculate the volume of solution...

$$\frac{\text{moles of solute (mol)}}{V\text{ (L)}} = \text{M}$$

...rearrange the equation for molarity (M):

$$V\text{ (L)} = \frac{\text{moles of solute (mol)}}{\text{M}}$$

SAMPLE PROBLEM 7.10

What volume in milliliters of a 0.30 M solution of glucose contains 0.025 mol of glucose?

Analysis

Use the equation, V = (moles of solute)/M, to find the volume in liters, and then convert the liters to milliliters.

Solution

[1] **Identify the known quantities and the desired quantity.**

known quantities	desired quantity
0.30 M 0.025 mol glucose	? V (L) solution

[2] **Divide the number of moles by molarity to obtain the volume in liters.**

$$V\text{ (L)} = \frac{\text{moles of solute (mol)}}{\text{M}}$$

$$= \frac{0.025\ \cancel{\text{mol}}\text{ glucose}}{0.30\ \cancel{\text{mol}}/\text{L}} = 0.083\text{ L solution}$$

[3] **Use a mL–L conversion factor to convert liters to milliliters.**

mL–L conversion factor

$$0.083\ \cancel{\text{L}}\text{ solution} \times \frac{1000\text{ mL}}{1\ \cancel{\text{L}}} = 83\text{ mL glucose solution}$$

Liters cancel.

Answer

PROBLEM 7.21

How many milliliters of a 1.5 M glucose solution contain each of the following number of moles?

a. 0.15 mol
b. 0.020 mol
c. 0.0030 mol
d. 3.0 mol

PROBLEM 7.22

How many moles of NaCl are contained in each volume of aqueous NaCl solution?

a. 2.0 L of a 2.0 M solution
b. 2.5 L of a 0.25 M solution
c. 25 mL of a 2.0 M solution
d. 250 mL of a 0.25 M solution

Since the number of grams and moles of a substance is related by the molar mass, we can convert a given volume of solution to the number of grams of solute it contains by carrying out the stepwise calculation shown in Sample Problem 7.11.

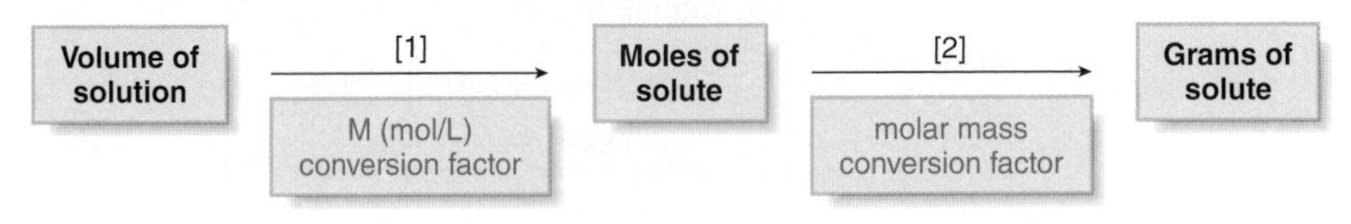

SAMPLE PROBLEM 7.11

How many grams of aspirin are contained in 50.0 mL of a 0.050 M solution?

Analysis

Use the molarity to convert the volume of the solution to moles of solute. Then use the molar mass to convert moles to grams.

Solution

[1] **Identify the known quantities and the desired quantity.**

known quantities	desired quantity
0.050 M	? g aspirin
50.0 mL solution	

[2] **Determine the number of moles of aspirin using the molarity.**

volume — molarity — mL–L conversion factor

$$50.0 \text{ mL solution} \times \frac{0.050 \text{ mol aspirin}}{1 \text{ L}} \times \frac{1 \text{ L}}{1000 \text{ mL}} = 0.0025 \text{ mol aspirin}$$

[3] **Convert the number of moles of aspirin to grams using the molar mass (180.2 g/mol).**

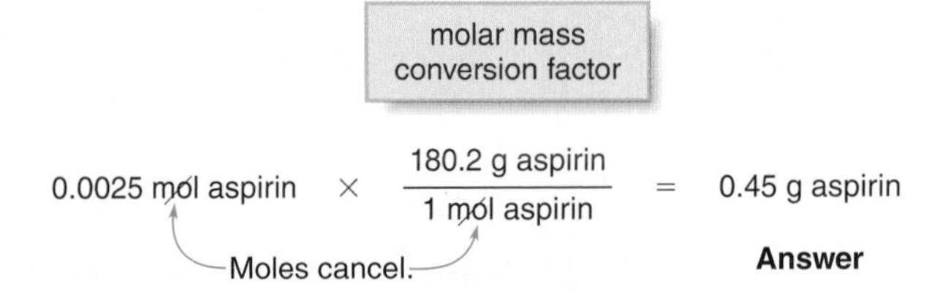

molar mass conversion factor
0.0025 mol aspirin × 180.2 g aspirin / 1 mol aspirin = 0.45 g aspirin
Moles cancel.
Answer

PROBLEM 7.23

How many grams of NaCl are contained in each of the following volumes of a 1.25 M solution?

a. 0.10 L b. 2.0 L c. 0.55 L d. 50. mL

PROBLEM 7.24

How many milliliters of a 0.25 M sucrose solution contain each of the following number of grams? The molar mass of sucrose ($C_{12}H_{22}O_{11}$) is 342.3 g/mol.

a. 0.500 g b. 2.0 g c. 1.25 g d. 50.0 mg

CONSUMER NOTE

Some cleaning products are sold as concentrated solutions, which are then diluted prior to use.

7.7 Dilution

Sometimes a solution has a higher concentration than is needed. ***Dilution* is the addition of solvent to decrease the concentration of solute.** For example, a stock solution of a drug is often supplied in a concentrated form to take up less space on a pharmacy shelf, and then it is diluted so that it can be administered in a reasonable volume and lower concentration that allows for more accurate dosing.

A key fact to keep in mind is that the **amount of solute is *constant*.** Only the volume of the solution is changed by adding solvent.

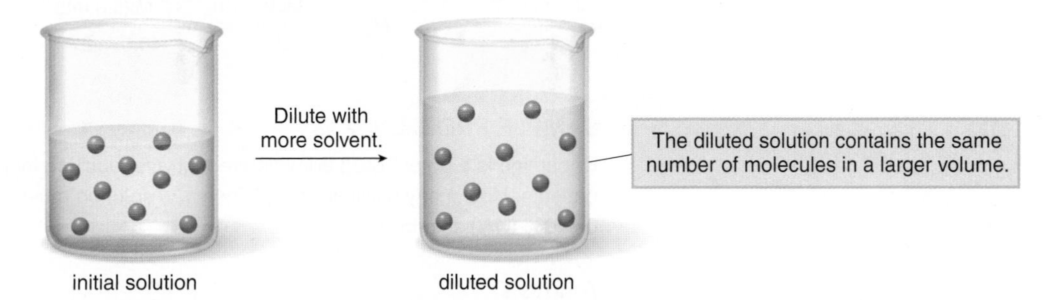
Dilute with
more solvent.
The diluted solution contains the same
number of molecules in a larger volume.
initial solution
diluted solution

In using molarity as a measure of concentration in Section 7.6, we learned that the number of moles of solute can be calculated from the molarity and volume of a solution.

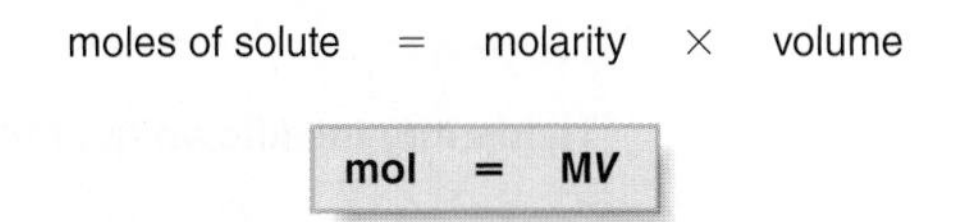
moles of solute = molarity × volume
mol = MV

Thus, if we have initial values for the molarity and volume (M_1 and V_1), we can calculate a new value for the molarity or volume (M_2 or V_2), since the product of the molarity and volume equals the number of moles, a constant.

$$\underset{\text{initial values}}{M_1V_1} = \underset{\text{final values}}{M_2V_2}$$

Although molarity is the most common concentration measure in the laboratory, the same facts hold in diluting solutions reported in other concentration units—percent concentration and parts per million—as well. In general, therefore, if we have initial values for the concentration and volume (C_1 and V_1), we can calculate a new value for the concentration or volume (C_2 or V_2), since the product of the concentration and volume is a constant.

$$\underset{\text{initial values}}{C_1V_1} = \underset{\text{final values}}{C_2V_2}$$

SAMPLE PROBLEM 7.12

What is the concentration of a solution formed by diluting 5.0 mL of a 3.2 M glucose solution to 40.0 mL?

Analysis

Since we know an initial molarity and volume (M_1 and V_1) and a final volume (V_2), we can calculate a new molarity (M_2) using the equation $M_1V_1 = M_2V_2$.

Solution

[1] **Identify the known quantities and the desired quantity.**

$M_1 = 3.2$ M
$V_1 = 5.0$ mL $V_2 = 40.0$ mL $M_2 = ?$
known quantities   desired quantity

[2] **Write the equation and rearrange it to isolate the desired quantity, M_2, on one side.**

$$M_1V_1 = M_2V_2$$ Solve for M_2 by dividing both sides by V_2.

$$\frac{M_1V_1}{V_2} = M_2$$

[3] Solve the problem.

- Substitute the three known quantities into the equation and solve for M_2.

$$M_2 = \frac{M_1V_1}{V_2} = \frac{(3.2\ M)(5.0\ \cancel{mL})}{(40.0\ \cancel{mL})} = 0.40\ M \text{ glucose solution}$$

Answer

SAMPLE PROBLEM 7.13

Dopamine is a potent drug administered intravenously to increase blood pressure in seriously ill patients. How many milliliters of a 4.0% (w/v) solution must be used to prepare 250 mL of a 0.080% (w/v) solution?

Analysis

Since we know an initial concentration (C_1), a final concentration (C_2), and a final volume (V_2), we can calculate the volume (V_1) of the initial solution that must be used with the equation, $C_1V_1 = C_2V_2$.

Solution

[1] Identify the known quantities and the desired quantity.

$C_1 = 4.0\%$ (w/v) $C_2 = 0.080\%$ (w/v)
$V_2 = 250$ mL
known quantities

$V_1 = ?$
desired quantity

[2] Write the equation and rearrange it to isolate the desired quantity, V_1, on one side.

$$C_1V_1 = C_2V_2$$

Solve for V_1 by dividing both sides by C_1.

$$V_1 = \frac{C_2V_2}{C_1}$$

[3] Solve the problem.

- Substitute the three known quantities into the equation and solve for V_1.

$$V_1 = \frac{(0.080\%)(250\ mL)}{4.0\%} = 5.0\ mL \text{ dopamine solution}$$

Answer

PROBLEM 7.25

What is the concentration of a solution formed by diluting 25.0 mL of a 3.8 M glucose solution to 275 mL?

PROBLEM 7.26

How many milliliters of a 6.0 M NaOH solution would be needed to prepare each solution?

a. 525 mL of a 2.5 M solution
b. 750 mL of a 4.0 M solution
c. 450 mL of a 0.10 M solution
d. 25 mL of a 3.5 M solution

PROBLEM 7.27

Ketamine (Sample Problem 7.7) is supplied in a solution of 100. mg/mL. If 2.0 mL of this solution is diluted to a volume of 10.0 mL, how much of the diluted solution should be administered to supply a dose of 75 mg?

7.8 Osmosis and Dialysis

The membrane that surrounds living cells is an example of a **semipermeable membrane**—a membrane that allows water and small molecules to pass across, but ions and large molecules cannot.

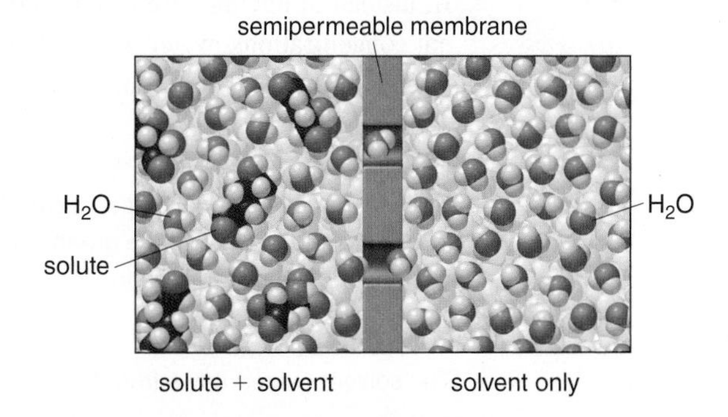

- *Osmosis* is the passage of solvent (usually water) across a semipermeable membrane from a solution of low solute concentration to a solution of higher solute concentration.

7.8A Osmotic Pressure

What happens when water and an aqueous glucose solution are separated by a semipermeable membrane? Water flows back and forth across the membrane, but more water flows from the side that has pure solvent towards the side that has dissolved glucose. This decreases the volume of pure solvent on one side of the membrane and increases the volume of the glucose solution on the other side.

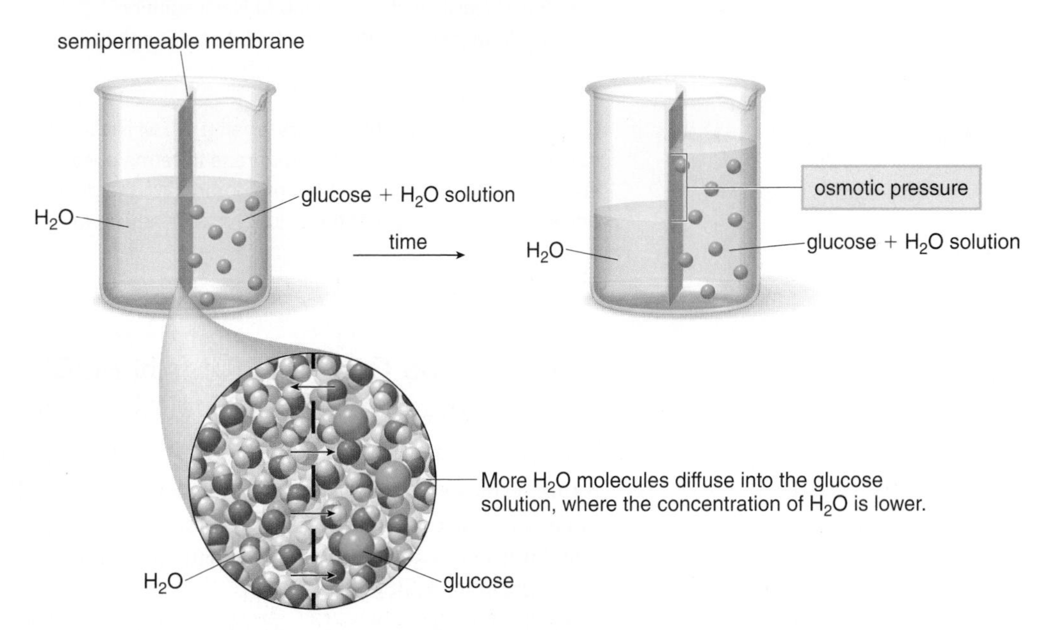

The increased weight of the glucose solution creates increased pressure on one side of the membrane. When the increased pressure gets to a certain point, it prevents more water movement to further dilute the glucose solution. Water continues to diffuse back and forth across the membrane, but the level of the two liquids does not change any further.

- *Osmotic pressure* is the pressure that prevents the flow of additional solvent into a solution on one side of a semipermeable membrane.

Osmotic pressure depends only on the number of particles in a solution. **The greater the number of dissolved particles, the greater the osmotic pressure.** A 0.1 M NaCl solution has twice the osmotic pressure as a 0.1 M glucose solution, since each NaCl is composed of two particles, Na^+ cations and Cl^- anions.

If, instead of having pure water on one side of the membrane, there were two solutions of different concentrations, water would flow from the side of the *less* concentrated solution to dilute the *more* concentrated solution.

SAMPLE PROBLEM 7.14

A 0.1 M glucose solution is separated from a 0.2 M glucose solution by a semipermeable membrane. (a) Which solution exerts the greater osmotic pressure? (b) In which direction will water flow between the two solutions? (c) Describe the level of the two solutions when equilibrium is reached.

Analysis

The solvent (water) flows from the less concentrated solution to the more concentrated solution.

Solution

a. The greater the number of dissolved particles, the higher the osmotic pressure, so the 0.2 M glucose solution exerts the greater pressure.
b. Water will flow from the less concentrated solution (0.1 M) to the more concentrated solution (0.2 M).
c. Since water flows into the 0.2 M solution, its height will increase, and the height of the 0.1 M glucose solution will decrease.

PROBLEM 7.28

Which solution in each pair exerts the greater osmotic pressure?

a. 1.0% sugar solution or 5.0% sugar solution
b. 3.0 M NaCl solution or a 4.0 M NaCl solution
c. 1.0 M glucose solution or a 0.75 M NaCl solution

PROBLEM 7.29

Describe the process that occurs when a 1.0 M NaCl solution is separated from a 1.5 M NaCl solution by a semipermeable membrane in terms of each of the following: (a) the identity of the substances that flow across the membrane; (b) the direction of flow before and after equilibrium is achieved; (c) the height of the solutions after equilibrium is achieved.

7.8B FOCUS ON THE HUMAN BODY Osmosis and Biological Membranes

Since cell membranes are semipermeable and biological fluids contain dissolved ions and molecules, osmosis is an ongoing phenomenon in living cells. Fluids on both sides of a cell membrane must have the same osmotic pressure to avoid pressure buildup inside or outside the cell. Any intravenous solution given to a patient, therefore, must have the same osmotic pressure as the fluids in the body.

- **Two solutions with the same osmotic pressure are said to be *isotonic.***

Isotonic solutions used in hospitals include 0.92% (w/v) NaCl solution (or 0.15 M NaCl solution) and 5.0% (w/v) glucose solution. Although these solutions do not contain exactly the same ions or molecules present in body fluids, they exert the same osmotic pressure.

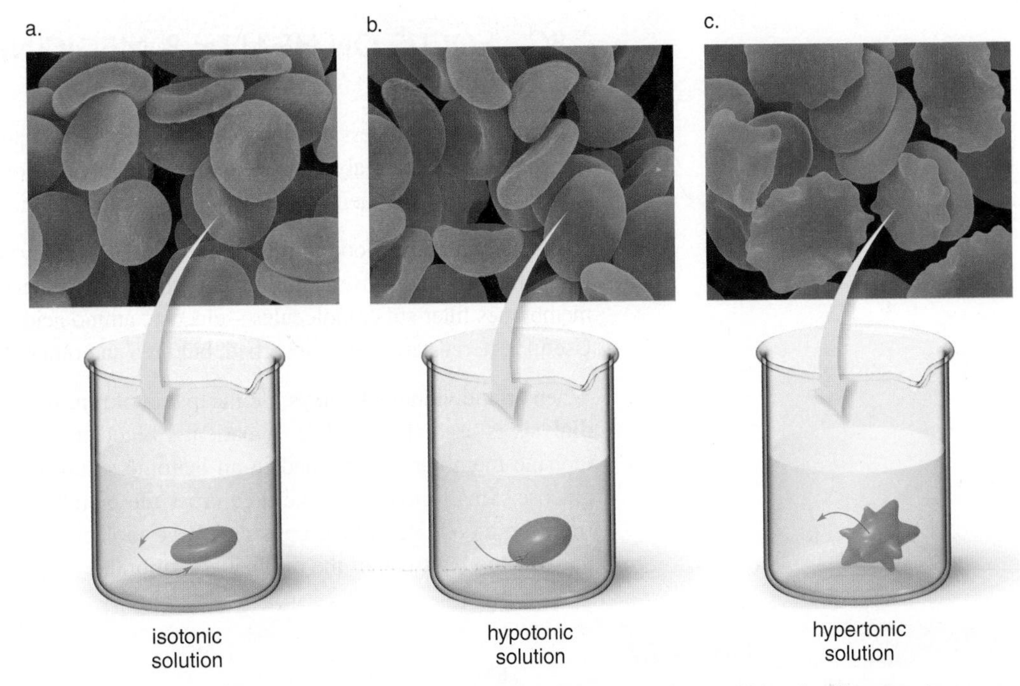

Figure 7.6

The Effect of Osmotic Pressure Differences on Red Blood Cells

(a) In an isotonic solution, the movement of water into and out of the red blood cell occurs to an equal extent and the red blood cell keeps its normal volume. (b) In a hypotonic solution, more water moves into the cell than diffuses out, so the cell swells and eventually it can rupture (hemolysis). (c) In a hypertonic solution, more water moves out of the cell than diffuses in, so the cell shrivels (crenation).

If a red blood cell is placed in an isotonic NaCl solution, called physiological saline solution, the red blood cells retain their same size and shape because the osmotic pressure inside and outside the cell is the same (Figure 7.6a). What happens if a red blood cell is placed in a solution having a different osmotic pressure?

- **A *hypotonic* solution has a lower osmotic pressure than body fluids.**
- **A *hypertonic* solution has a higher osmotic pressure than body fluids.**

In a hypotonic solution, the concentration of particles outside the cell is lower than the concentration of particles inside the cell. In other words, the concentration of water outside the cell is *higher* than the concentration of water inside the cell, so water diffuses inside (Figure 7.6b). As a result, the cell swells and eventually bursts. This swelling and rupture of red blood cells is called **hemolysis.**

In a hypertonic solution, the concentration of particles outside the cell is higher than the concentration of particles inside the cell. In other words, the concentration of water inside the cell is *higher* than the concentration of water outside the cell, so water diffuses out of the cell (Figure 7.6c). As a result, the cell shrinks. This process is called **crenation.**

PROBLEM 7.30

What happens to a red blood cell when it is placed in each of the following solutions: (a) 3% (w/v) glucose solution; (b) 0.15 M KCl solution?

7.8C FOCUS ON HEALTH & MEDICINE Dialysis

Dialysis is also a process that involves the selective passage of substances across a semipermeable membrane, called a dialyzing membrane. In dialysis, however, water, small molecules, and ions can travel across the membrane; only large biological molecules like proteins and starch cannot.

In the human body, blood is filtered through the kidneys by the process of dialysis (Figure 7.7). Each kidney contains over a million nephrons, tubelike structures with filtration membranes. These membranes filter small molecules—glucose, amino acids, urea, ions, and water—from the blood. Useful materials are then reabsorbed, but urea and other waste products are eliminated in urine.

When an individual's kidneys are incapable of removing waste products from the blood, **hemodialysis** is used (Figure 7.8). A patient's blood flows through a long tube connected to a cellophane membrane suspended in an isotonic solution that contains NaCl, KCl, $NaHCO_3$, and glucose. Small molecules like urea cross the membrane into the solution, thus removing them from the blood. Red blood cells and large molecules are not removed from the blood because they are too big to cross the dialyzing membrane.

Figure 7.7

Dialysis of Body Fluids by the Kidneys

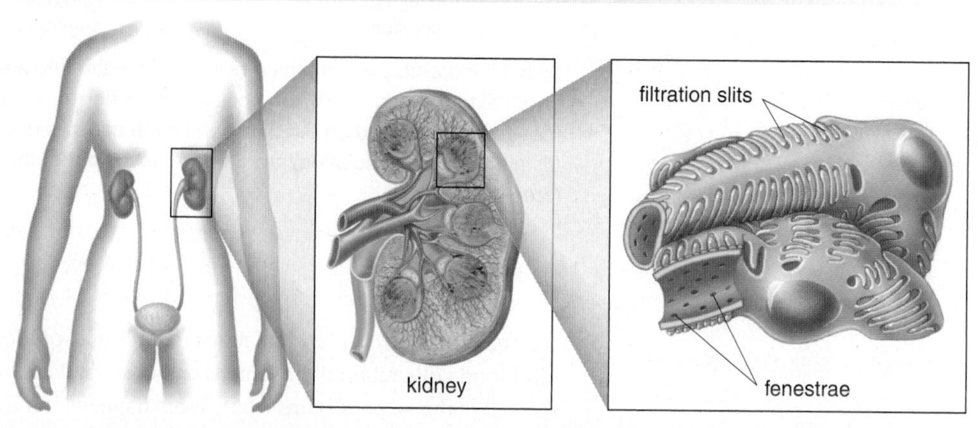

Body fluids are dialyzed by passage through the kidneys, which contain more than a million nephrons that filter out small molecules and ions from the blood. Useful materials are then reabsorbed while urea and other waste products are eliminated in urine.

Figure 7.8

Hemodialysis

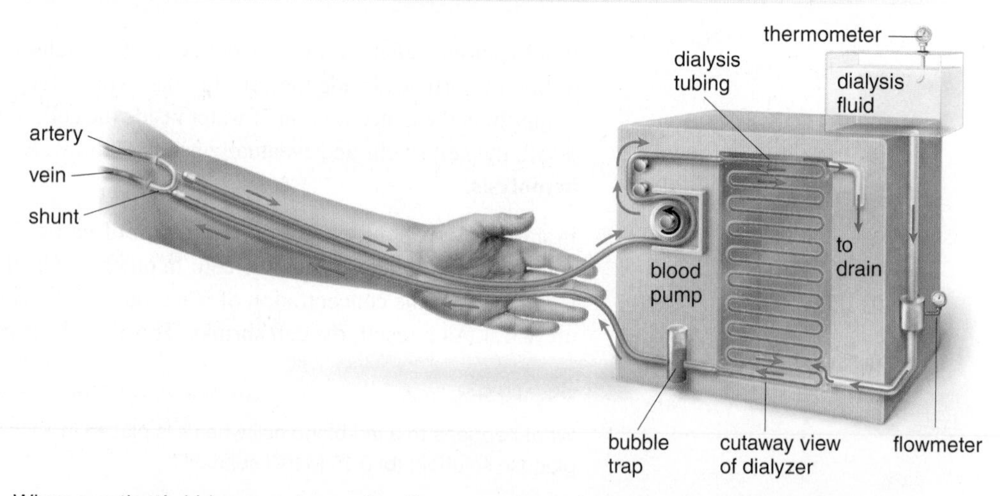

When a patient's kidneys no longer function properly, periodic dialysis treatments are used to remove waste products from the blood. Blood is passed through a dialyzer, which contains a membrane that allows small molecules to pass through, thus acting as an artificial kidney. Each treatment takes several hours. Patients usually require two to three treatments per week.

KEY TERMS

Aqueous solution (7.1)
Colloid (7.1)
Concentration (7.5)
Dialysis (7.8)
Dilution (7.7)
Electrolyte (7.2)
Equivalent (7.2)
Henry's law (7.4)
Heterogeneous mixture (7.1)
Homogeneous mixture (7.1)
Hypertonic solution (7.8)
Hypotonic solution (7.8)
Ion–dipole interaction (7.3)
Isotonic solution (7.8)
Molarity (7.6)
Nonelectrolyte (7.2)
Osmosis (7.8)
Osmotic pressure (7.8)
Parts per million (7.5)
Saturated solution (7.3)
Semipermeable membrane (7.8)
Solubility (7.3)
Solute (7.1)
Solution (7.1)
Solvation (7.3)
Solvent (7.1)
Supersaturated solution (7.4)
Suspension (7.1)
Unsaturated solution (7.3)
Volume/volume percent concentration (7.5)
Weight/volume percent concentration (7.5)

KEY CONCEPTS

1 What are the characteristics of solutions, colloids, and suspensions? (7.1)

- A solution is a homogeneous mixture that contains small dissolved particles. The substance present in the lesser amount is called the solute, and the substance present in the larger amount is the solvent.
- A colloid is a homogeneous mixture with larger particles (1 nm–1 μm), often having an opaque appearance.
- A suspension is a heterogeneous mixture that contains larger particles (> 1 μm) suspended in a liquid.

2 What is the difference between an electrolyte and a nonelectrolyte? (7.2)

- An electrolyte dissolves in water to form ions, so it conducts an electric current. A strong electrolyte dissociates completely to form ions, whereas a weak electrolyte contains mostly uncharged molecules in water, together with a small number of ions.
- A nonelectrolyte dissolves in water to form uncharged molecules that do not conduct an electric current.

3 What determines whether a substance is soluble in water or a nonpolar solvent? (7.3)

- One rule summarizes solubility: "Like dissolves like."
- Most ionic compounds are soluble in water.
- Small polar compounds that can hydrogen bond are soluble in water.
- Nonpolar compounds are soluble in nonpolar solvents. Compounds with many nonpolar C—C and C—H bonds are soluble in nonpolar solvents.

4 What effect do temperature and pressure have on solubility? (7.4)

- The solubility of solids in a liquid solvent generally increases with increasing temperature. The solubility of gases decreases with increasing temperature.
- Increasing pressure increases the solubility of a gas in a solvent. Pressure changes do not affect the solubility of liquids and solids.

5 How is the concentration of a solution expressed? (7.5, 7.6)

- Concentration is a measure of how much solute is dissolved in a given amount of solution, and can be measured using mass, volume, or moles.
- Weight/volume (w/v) percent concentration is the number of grams of solute dissolved in 100 mL of solution.
- Volume/volume (v/v) percent concentration is the number of milliliters of solute dissolved in 100 mL of solution.
- Parts per million (ppm) is the number of parts of solute in 1,000,000 parts of solution, where the units for both the solute and the solution are the same.
- Molarity (M) is the number of moles of solute per liter of solution.

6 How are dilutions performed? (7.7)

- Dilution is the addition of solvent to decrease the concentration of a solute. Since the number of moles of solute is constant in carrying out a dilution, a new molarity or volume (M_2 and V_2) can be calculated from a given molarity and volume (M_1 and V_1) using the equation $M_1V_1 = M_2V_2$, as long as three of the four quantities are known.

7 What is osmosis? (7.8)

- Osmosis is the passage of solvent, usually water, across a semipermeable membrane. Solvent always moves from the less concentrated solution to the more concentrated solution, until the osmotic pressure prevents additional flow of solvent.

KEY EQUATIONS—CONCENTRATION

Weight/volume percent concentration

$$(w/v)\% = \frac{\text{mass of solute (g)}}{\text{volume of solution (mL)}} \times 100\%$$

Volume/volume percent concentration

$$(v/v)\% = \frac{\text{volume of solute (mL)}}{\text{volume of solution (mL)}} \times 100\%$$

Parts per million

$$\text{ppm} = \frac{\text{parts of solute (g or mL)}}{\text{parts of solution (g or mL)}} \times 10^6$$

Molarity

$$\text{M} = \frac{\text{moles of solute (mol)}}{\text{liter of solution (L)}}$$

UNDERSTANDING KEY CONCEPTS

Selected in-chapter and odd-numbered end-of-chapter problems have brief answers at the end of each chapter. The *Student Study Guide and Solutions Manual* contains detailed solutions to all in-chapter and odd-numbered end-of-chapter problems, as well as additional worked examples and a chapter self-test.

7.31 Which representation of molecular art better shows a solution of KI dissolved in water? Explain your choice. Will this solution conduct an electric current?

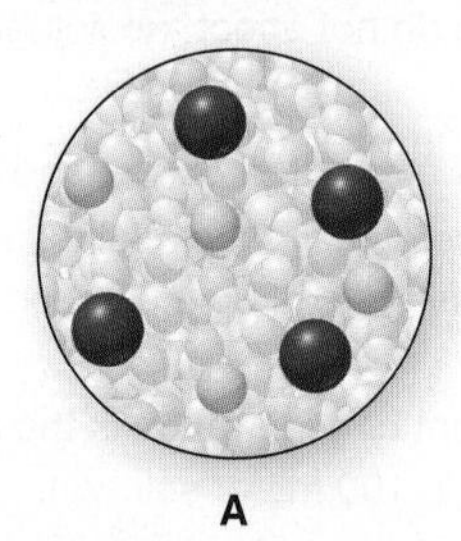

A

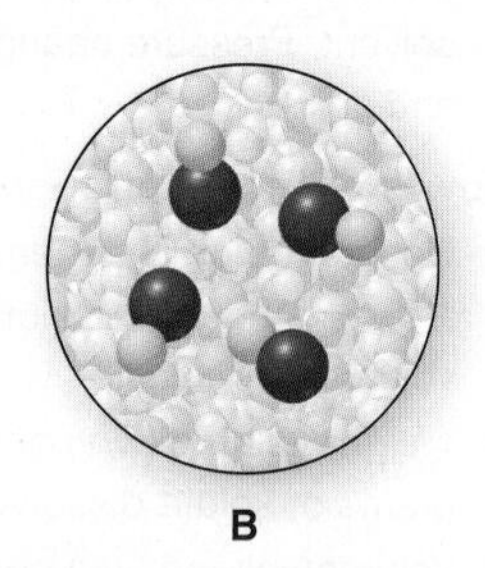

B

7.32 Which representation of molecular art better shows a solution of NaF dissolved in water? Explain your choice. Will this solution conduct an electric current?

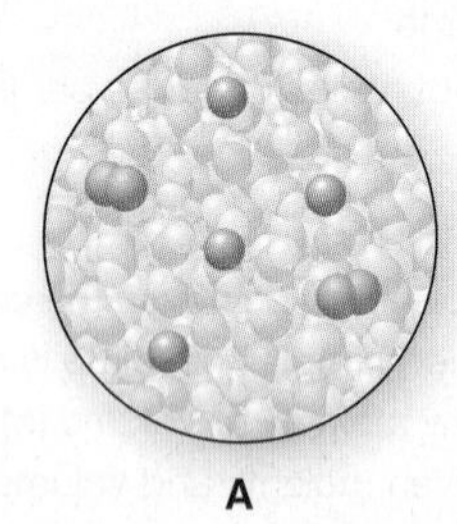

A

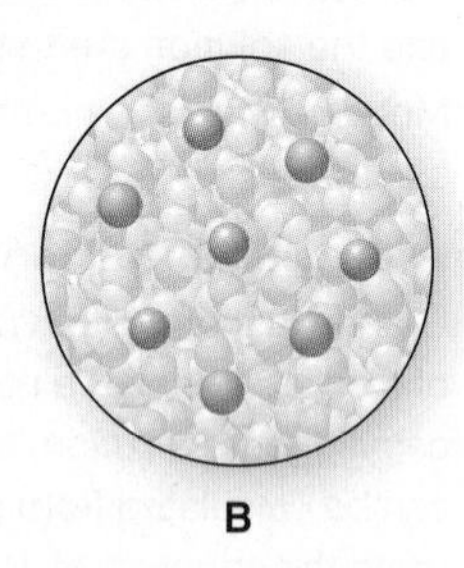

B

7.33 Using the ball-and-stick model for methanol (CH_3OH), draw a representation for the solution that results when methanol is dissolved in water.

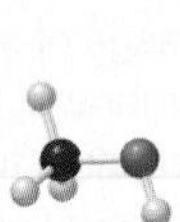

7.34 Using the ball-and-stick model for dimethyl ether [$(CH_3)_2O$], draw a representation for the solution that results when dimethyl ether is dissolved in water.

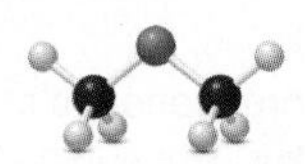

7.35 The molecular art represents O_2 molecules dissolved in water at 25 °C. Draw a representation for the solution that results (a) after heating to 50 °C; (b) after cooling to 10 °C.

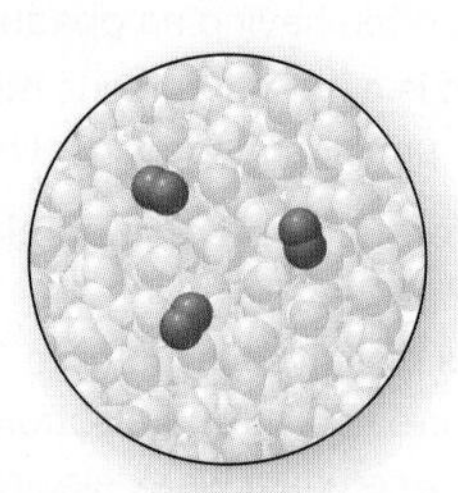

7.36 The molecular art represents NaCl dissolved in water at 25 °C. Assume that this solution is saturated in NaCl and that solid NaCl remains at the bottom of the flask in which the solution is stored. Draw a representation for the solution that results (a) after heating to 50 °C; (b) after cooling to 10 °C.

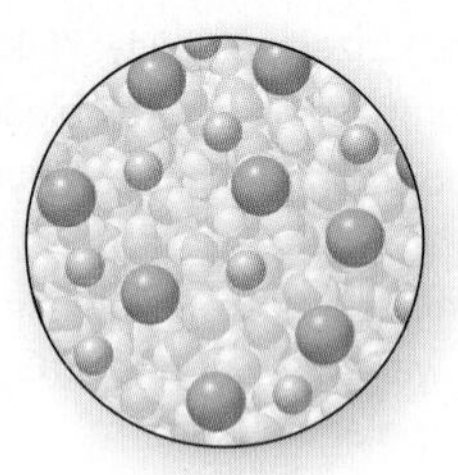

7.37 What is the weight/volume percent concentration of a 30.0% (w/v) solution of vitamin C after each of the following dilutions?

a. 100. mL diluted to 200. mL

b. 250. mL diluted to 1.5 L

7.38 One gram (1.00 g) of vitamin B_3 (niacin) is dissolved in water to give 10.0 mL of solution. (a) What is the weight/volume percent concentration of this solution? (b) What is the concentration of a solution formed by diluting 1.0 mL of this solution to each of the following volumes: [1] 10.0 mL; [2] 2.5 mL?

7.39 Cefdinir is an antibiotic used to treat infections of the middle ear, tonsils, and lungs. It is sold as a suspension in water.

a. What is the weight/volume percent concentration in a suspension of cefdinir that contains 250 mg/5 mL?

b. If the typical pediatric dose is 7.0 mg/kg every 12 hours, what volume of drug is given for a 21-kg child?

7.40 Amoxil is an antibiotic used to treat infections of the ear, nose, and throat. It is sold as a suspension in water.

a. What is the weight/volume percent concentration in a suspension of amoxil that contains 50 mg/mL?

b. If the typical pediatric dose is 25 mg/kg per day, what volume of drug is given for a 15-kg child?

7.41 A flask contains two compartments (**A** and **B**) with equal volumes of solution separated by a semipermeable membrane. Describe the final level of the liquids when **A** and **B** contain each of the following solutions.

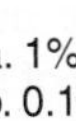

A	B
a. 1% (w/v) glucose solution	pure water
b. 0.10 M glucose solution	0.20 M glucose solution
c. 0.10 M NaCl solution	0.10 M NaI solution
d. 0.10 M $CaCl_2$ solution	0.10 M NaCl solution
e. 0.20 M glucose solution	0.10 M NaCl solution

7.42 A flask contains two compartments (**A** and **B**) with equal volumes of solution separated by a semipermeable membrane. Which diagram represents the final level of the liquids when **A** and **B** contain each of the following solutions?

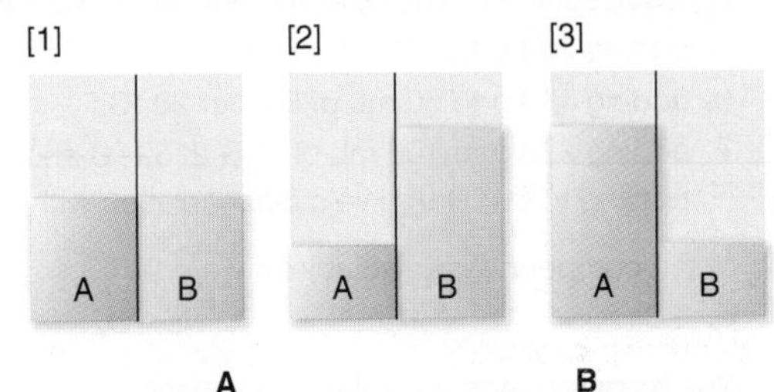

A	B
a. 10% (w/v) glucose	20% (w/v) glucose
b. 0.20 M NaCl	0.30 M glucose
c. pure water	5% (w/v) glucose
d. 2.0 M NaCl	pure water
e. 3% (w/v) sucrose	1% (w/v) sucrose

ADDITIONAL PROBLEMS

Solutions, Colloids, and Suspensions

7.43 Classify each of the following as a solution, a colloid, or a suspension.

a. bronze (an alloy of Sn and Cu)

b. orange juice with pulp

c. gasoline

d. fog

7.44 Classify each of the following as a solution, a colloid, or a suspension.

a. soft drink

b. cream

c. muddy water

d. bleach

Electrolytes and Equivalents

7.45 Label each diagram as a strong electrolyte, weak electrolyte, or nonelectrolyte.

a.

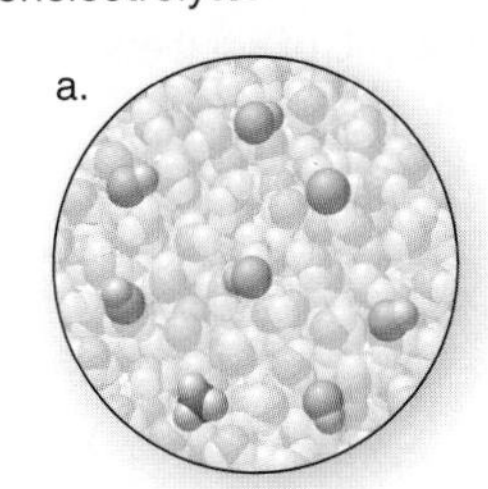

b.

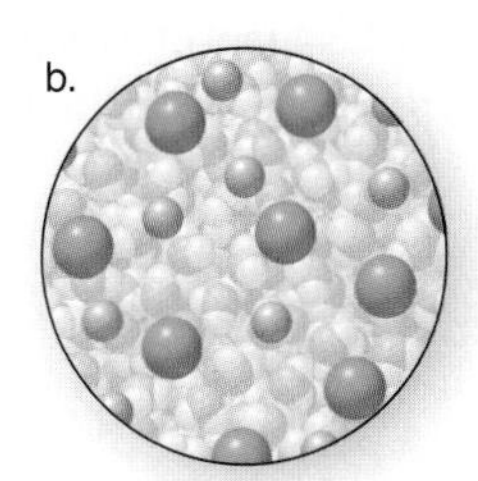

7.46 Label each diagram as a strong electrolyte, weak electrolyte, or nonelectrolyte.

a.

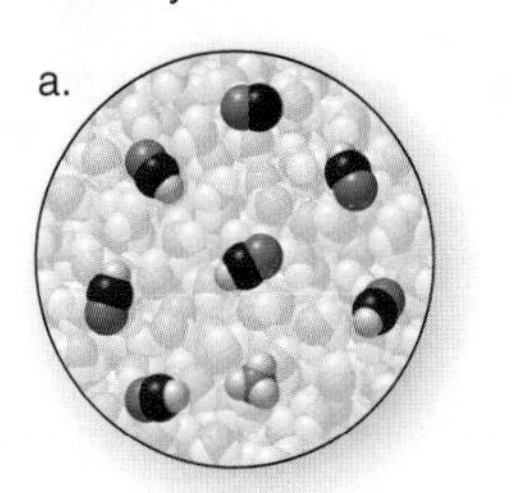

b.

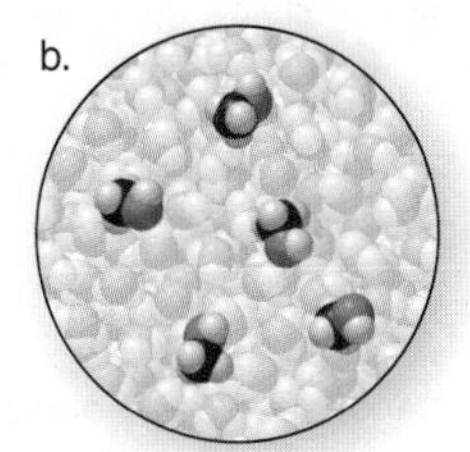

7.47 A solution contains 2.5 mol KCl and 2.5 mol $CaSO_4$. How many equivalents of each ion are present?

7.48 A dextrose solution contains 40. mEq/L Na^+ and 15 mEq/L of HPO_4^{2-}, in addition to other ions. How many milliequivalents of each ion are present in 320 mL of solution?

7.49 If a solution contains 132 mEq/L of Na^+, how many milliequivalents of Na^+ are present in each of the following volumes of solution: (a) 1.0 L; (b) 2.0 L; (c) 440 mL?

7.50 Ringer's solution is an intravenous solution used for the replacement of electrolytes when an individual is dehydrated. Ringer's solution contains the following cations: Na^+ (147 mEq/L), K^+ (4 mEq/L), and Ca^{2+} (4 mEq/L). (a) How many milliequivalents of Na^+ are present in 125 mL of Ringer's solution? (b) If the only anion in Ringer's solution is Cl^-, what is the concentration of Cl^- in mEq/L?

Solubility

7.51 If the solubility of KCl in 100 mL of H_2O is 34 g at 20 °C and 43 g at 50 °C, label each of the following solutions as unsaturated, saturated, or supersaturated. If more solid is added than can dissolve in the solvent, assume that undissolved solid remains at the bottom of the flask.

a. adding 30 g to 100 mL of H_2O at 20 °C
b. adding 65 g to 100 mL of H_2O at 50 °C
c. adding 42 g to 100 mL of H_2O at 50 °C and slowly cooling to 20 °C to give a clear solution with no precipitate

7.52 If the solubility of sucrose in 100 mL of H_2O is 204 g at 20 °C and 260 g at 50 °C, label each of the following solutions as unsaturated, saturated, or supersaturated. If more solid is added than can dissolve in the solvent, assume that undissolved solid remains at the bottom of the flask.

a. adding 200 g to 100 mL of H_2O at 20 °C
b. adding 110 g to 50 mL of H_2O at 20 °C
c. adding 220 g to 100 mL of H_2O at 50 °C and slowly cooling to 20 °C to give a clear solution with no precipitate

7.53 Which compounds are soluble in water?

a. LiCl b. C_7H_8 c. Na_3PO_4

7.54 Which compounds are soluble in water?

a. C_5H_{12} b. $CaCl_2$ c. CH_3Br

7.55 Label each compound as water soluble or water insoluble.

a.
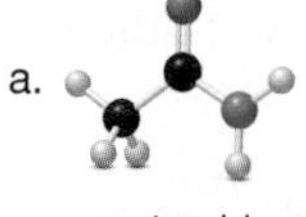
acetamide

b.
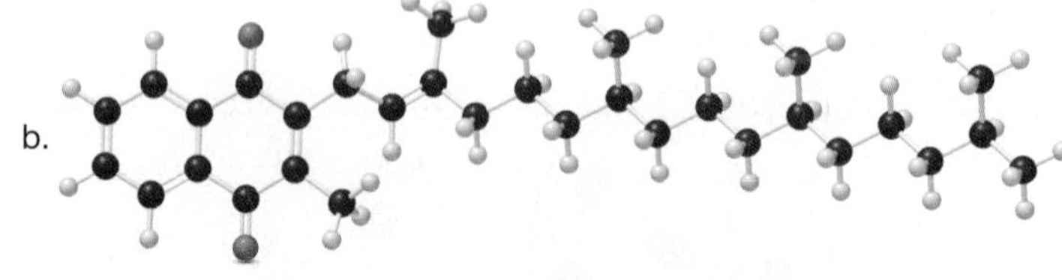
vitamin K

7.56 Label each compound as water soluble or water insoluble.

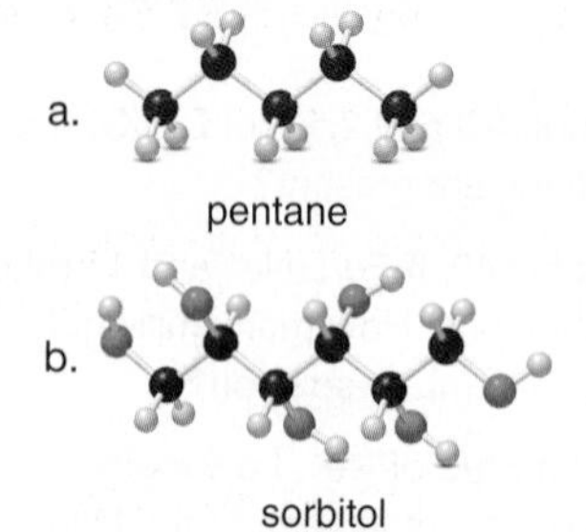
a. pentane

b. sorbitol

7.57 Explain the statement, "Oil and water don't mix."

7.58 Explain why a bottle of salad dressing that contains oil and vinegar has two layers.

7.59 Explain why cholesterol, a compound with molecular formula $C_{27}H_{46}O$ and one OH group, is soluble in CCl_4 but insoluble in water.

7.60 Which of the following pairs of compounds form a solution?

a. KCl and CCl_4
b. 1-propanol (C_3H_8O) and H_2O
c. pentane (C_5H_{12}) and hexane (C_6H_{14})

7.61 How is the solubility of solid NaCl in water affected by each of the following changes?

a. increasing the temperature from 25 °C to 50 °C
b. decreasing the temperature from 25 °C to 0 °C
c. increasing the pressure from 1 atm to 2 atm
d. decreasing the pressure from 5 atm to 1 atm

7.62 How is the solubility of helium gas in water affected by each of the following changes?

a. increasing the temperature from 25 °C to 50 °C
b. decreasing the temperature from 25 °C to 0 °C
c. increasing the pressure from 1 atm to 2 atm
d. decreasing the pressure from 5 atm to 1 atm

Concentration

7.63 What is the weight/volume percent concentration using the given amount of solute and total volume of solution?

a. 10.0 g of LiCl in 750 mL of solution
b. 25 g of $NaNO_3$ in 150 mL of solution
c. 40.0 g of NaOH in 500. mL of solution

7.64 What is the weight/volume percent concentration using the given amount of solute and total volume of solution?

a. 5.5 g of LiCl in 550 mL of solution
b. 12.5 g of $NaNO_3$ in 250 mL of solution
c. 20.0 g of NaOH in 400. mL of solution

7.65 What is the volume/volume percent concentration of a solution prepared from 25 mL of ethyl acetate in 150 mL of solution?

7.66 What is the volume/volume percent concentration of a solution prepared from 75 mL of acetone in 250 mL of solution?

7.67 What is the molarity of a solution prepared using the given amount of solute and total volume of solution?

a. 3.5 mol of KCl in 1.50 L of solution
b. 0.44 mol of $NaNO_3$ in 855 mL of solution
c. 25.0 g of NaCl in 650 mL of solution

7.68 What is the molarity of a solution prepared using the given amount of solute and total volume of solution?
a. 2.4 mol of NaOH in 1.50 L of solution
b. 0.48 mol of KNO_3 in 750 mL of solution
c. 25.0 g of KCl in 650 mL of solution

7.69 How would you use a 250-mL volumetric flask to prepare each of the following solutions?
a. 4.8% (w/v) acetic acid in water
b. 22% (v/v) ethyl acetate in water
c. 2.5 M NaCl solution

7.70 How would you use a 250-mL volumetric flask to prepare each of the following solutions?
a. 2.0% (w/v) KCl in water
b. 34% (v/v) ethanol in water
c. 4.0 M NaCl solution

7.71 How many moles of solute are contained in each solution?
a. 150 mL of a 0.25 M $NaNO_3$ solution
b. 45 mL of a 2.0 M HNO_3 solution
c. 2.5 L of a 1.5 M HCl solution

7.72 How many moles of solute are contained in each solution?
a. 250 mL of a 0.55 M $NaNO_3$ solution
b. 145 mL of a 4.0 M HNO_3 solution
c. 6.5 L of a 2.5 M HCl solution

7.73 How many grams of solute are contained in each solution in Problem 7.71?

7.74 How many grams of solute are contained in each solution in Problem 7.72?

7.75 How many mL of ethanol are contained in a 750-mL bottle of wine that contains 11.0% (v/v) of ethanol?

7.76 What is the molarity of a 20.0% (v/v) aqueous ethanol solution? The density of ethanol (C_2H_6O, molar mass 46.07 g/mol) is 0.790 g/mL.

7.77 A 1.89-L bottle of vinegar contains 5.0% (w/v) of acetic acid ($C_2H_4O_2$, molar mass 60.05 g/mol) in water.
a. How many grams of acetic acid are present in the container?
b. How many moles of acetic acid are present in the container?
c. Convert the weight/volume percent concentration to molarity.

7.78 What is the molarity of a 15% (w/v) glucose solution?

7.79 The maximum safe level of each compound in drinking water is given below. Convert each value to parts per million.
a. chloroform ($CHCl_3$, a solvent), 80 µg/kg
b. glyphosate (a pesticide), 700 µg/kg

7.80 The maximum safe level of each metal in drinking water is given below. Convert each value to parts per million.
a. copper, 1,300 µg/kg
b. arsenic, 10 µg/kg
c. chromium, 100 µg/kg

Dilution

7.81 If the solution in **X** is diluted, which representation (**A**–**C**) represents the final solution?

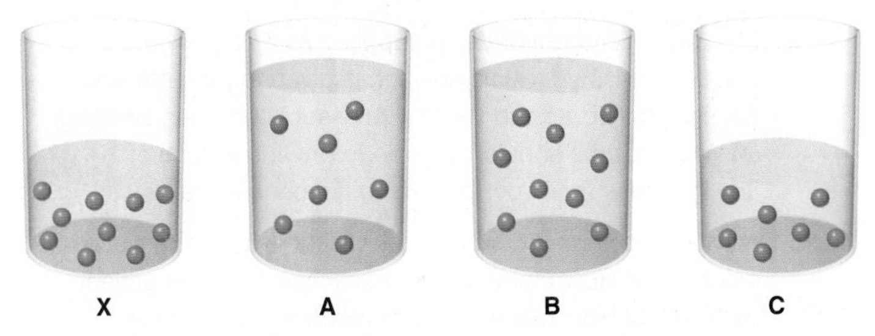

7.82 Consider solutions **A**, **B**, and **C** in Problem 7.81. If the volumes of **A** and **B** are twice the volume of **C**, which solution is most concentrated? Which solution is least concentrated?

7.83 What is the concentration of a solution formed by diluting 125 mL of 12.0 M HCl solution to 850 mL?

7.84 What is the concentration of a solution formed by diluting 250 mL of 6.0 M NaOH solution to 0.45 L?

7.85 How many milliliters of a 2.5 M NaCl solution would be needed to prepare each solution?
a. 25 mL of a 1.0 M solution
b. 1.5 L of a 0.75 M solution
c. 15 mL of a 0.25 M solution

7.86 How many milliliters of a 5.0 M sucrose solution would be needed to prepare each solution?
a. 45 mL of a 4.0 M solution
b. 150 mL of a 0.5 M solution
c. 1.2 L of a 0.025 M solution

Osmosis

7.87 What is the difference between osmosis and osmotic pressure?

7.88 What is the difference between a hypotonic solution and an isotonic solution?

Applications

7.89 Explain why opening a warm can of soda causes a louder "whoosh" and more fizzing than opening a cold can of soda.

7.90 Explain why more sugar dissolves in a cup of hot coffee than a glass of iced coffee.

7.91 If the concentration of glucose in the blood is 90 mg/100 mL, what is the weight/volume percent concentration of glucose? What is the molarity of glucose (molar mass 180.2 g/mol) in the blood?

7.92 If the human body contains 5.0 L of blood, how many grams of glucose are present in the blood if the concentration is 90. mg/100. mL?

7.93 A sports drink contains 15 g of soluble complex carbohydrates in 8.0 oz (1 fl oz = 29.6 mL). What weight/volume percent concentration does this represent?

7.94 A sports drink contains 25 mg of magnesium in an 8.0-oz portion (1 fl oz = 29.6 mL). How many parts per million does this represent? Assume that the mass of 1.0 mL of the solution is 1.0 g.

7.95 Mannitol, a carbohydrate, is supplied as a 25% (w/v) solution. This hypertonic solution is given to patients who have sustained a head injury with associated brain swelling. (a) What volume should be given to provide a dose of 70. g? (b) How does the hypertonic mannitol benefit brain swelling?

7.96 A patient receives 750 mL of a 10.% (w/v) aqueous glucose solution. (a) How many grams of glucose does the patient receive? (b) How many moles of glucose (molar mass 180.2 g/mol) does the patient receive?

7.97 Explain why a cucumber placed in a concentrated salt solution shrivels.

7.98 Explain why a raisin placed in water swells.

7.99 Each day, the stomach produces 2.0 L of gastric juice that contains 0.10 M HCl. How many grams of HCl does this correspond to?

7.100 Describe what happens when a red blood cell is placed in pure water.

7.101 An individual is legally intoxicated with a blood alcohol level of 0.08% (w/v) of ethanol. How many milligrams of ethanol are contained in 5.0 L of blood with this level?

7.102 A bottle of vodka labeled "80 proof" contains 40.% (v/v) ethanol in water. How many mL of ethanol are contained in 250 mL of vodka?

CHALLENGE PROBLEMS

7.103 The therapeutic concentration—the concentration needed to be effective—of acetaminophen ($C_8H_9NO_2$, molar mass 151.2 g/mol) is 10–20 μg/mL. Assume that the density of blood is 1.0 g/mL.

a. If the concentration of acetaminophen in the blood was measured at 15 ppm, is this concentration in the therapeutic range?
b. How many moles of acetaminophen are present at this concentration in 5.0 L of blood?

7.104 Very dilute solutions can be measured in parts per billion—that is, the number of parts in 1,000,000,000 parts of solution. To be effective, the concentration of digoxin, a drug used to treat congestive heart failure, must be 0.5–2.0 ng/mL. Convert both values to parts per billion (ppb).

BEYOND THE CLASSROOM

7.105 Pick two over-the-counter cough medicines that list ingredients as a percent concentration and are available in a local drug store or supermarket. Determine how many milligrams of each active ingredient are contained in a single dose. Determine how many milligrams (or grams) of each active ingredient are contained in the entire package. If you have chosen two cough medicines with the same active ingredients, is there a relationship between the cost and the quantity of the active ingredients?

7.106 Carry out a practical experiment to demonstrate osmosis. Place a few small pieces of fresh fruit (such as grapes, blueberries, cherries, or cranberries) in a glass of water. Place other pieces of fresh fruit in a glass of saturated salt solution. Then, add some dried fruit (such as raisins, Craisins, or dried blueberries) to a glass of water, as well as some dried fruit to a glass of saturated salt solution. What trends do you observe? Can you explain the results based on what you learned about osmosis in Section 7.8?

7.107 Many sports drinks contain carbohydrates, sodium, and potassium, in addition to other dissolved minerals and vitamins. Pick a sports drink and calculate the number of grams or milligrams of carbohydrates, sodium, and potassium in 100 mL. Compare the values with sports drinks chosen by other members of the class. Is there any significant difference in commercial products? Is any difference reflected in price?

7.108 Body fluids fall into three categories: blood plasma (the liquid part of the blood), interstitial fluid (the fluid between cells), and intracellular fluid (the fluid inside cells). Research the types of ions present in each type of body fluid. Which ions dominate in each environment? How is ion composition related to fluid balance and the conduction of electrical signals? How is electrolyte composition affected by heavy perspiration, severe diarrhea, or vomiting? When electrolytes are extremely out of balance, intravenous fluids can be given to restore proper electrolyte levels. What types of IV solutions are commonly used, and for what conditions?

ANSWERS TO SELECTED PROBLEMS

7.1 a. heterogeneous mixture b. colloid c,d,e: solution

7.3 a. weak electrolyte
b. nonelectrolyte
c. strong electrolyte

7.5 a. 1 Eq Na^+ c. 0.5 Eq K^+
b. 2 Eq Mg^{2+} d. 1.5 Eq PO_4^{3-}

7.6 22 mEq

7.7 29 mEq/L

7.9 a,c,d,e: water soluble

7.11 CO_2 comes out of solution faster at room temperature because the solubility of gases in liquids decreases as temperature increases.

7.13 3.5% (w/v)

7.15 8.4% (v/v)

7.16 2.0×10^2 mL

7.17 30. mg dextromethorphan 3.0×10^2 mg guaifenisin

7.19 a. 0.030 ppm b. 0.5 ppm c. 2.0 ppm d. 0.23 ppm

7.20 a. 2.0 M c. 10. M
b. 8.0 M d. 0.10 M

7.21 a. 1.0×10^2 mL c. 2.0 mL
b. 13 mL d. 2.0×10^3 mL

7.23 a. 7.3 g b. 150 g c. 40. g d. 3.7 g

7.25 0.35 M

7.27 3.8 mL

7.28 a. 5.0% sugar solution b. 4.0 M NaCl c. 0.75 M NaCl

7.29 a. Water flows across the membrane.
b. More water initially flows from the 1.0 M side to the more concentrated 1.5 M side. When equilibrium is reached there is equal flow of water in both directions.
c. The height of the 1.5 M side will be higher and the height of the 1.0 M NaCl will be lower.

7.31 **A.** The K^+ and I^- ions are separated when KI is dissolved in water. The solution contains ions so it conducts an electric current.

7.33

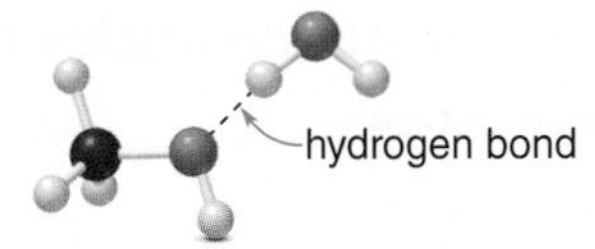

7.35 a. b.

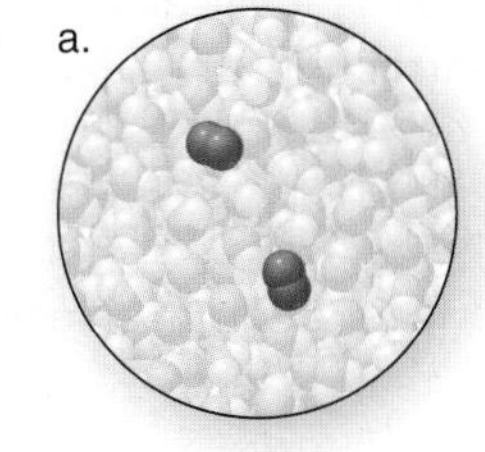

7.37 a. 15.0% (w/v) b. 5.0% (w/v)

7.39 a. 5% (w/v) b. 3 mL

7.41 a. **A** > **B** c. no change e. no change
b. **B** > **A** d. **A** > **B**

7.43 a,c: solution b. suspension d. colloid

7.45 a. weak electrolyte b. strong electrolyte

7.47 2.5 Eq K^+ 5.0 Eq Ca^{2+}
2.5 Eq Cl^- 5.0 Eq SO_4^{2-}

7.49 a. 130 mEq b. 260 mEq c. 58 mEq

7.51 a. unsaturated b. saturated c. supersaturated

7.53 a and c

7.55 a. water soluble b. water insoluble

7.57 Water-soluble compounds are ionic or are small polar molecules that can hydrogen bond with the water solvent, but nonpolar compounds, such as oil, are soluble in nonpolar solvents.

7.59 Cholesterol is not water soluble because it is a large nonpolar molecule with a single OH group.

7.61 a. increased b. decreased c,d: no change

7.63 a. 1.3% (w/v) b. 17% (w/v) c. 8.00% (w/v)

7.65 17% (v/v)

7.67 a. 2.3 M b. 0.51 M c. 0.66 M

7.69 a. Add 12 g of acetic acid to the flask and then water to bring the volume to 250 mL.
b. Add 55 mL of ethyl acetate to the flask and then water to bring the volume to 250 mL.
c. Add 37 g of NaCl to the flask and then water to bring the volume to 250 mL.

7.71 a. 0.038 mol b. 0.090 mol c. 3.8 mol

7.73 a. 3.2 g b. 5.7 g c. 140 g

7.75 83 mL

7.77 a. 95 g b. 1.6 mol c. 0.85 M

7.79 a. 0.08 ppm b. 0.7 ppm

7.81 **B**

7.83 1.8 M

7.85 a. 10. mL b. 450 mL c. 1.5 mL

7.87 Osmosis is the selective diffusion of solvent, usually water, across a semipermeable membrane. Osmotic pressure is the pressure that prevents the flow of additional solvent into a solution on one side of a semipermeable membrane.

7.89 At warmer temperatures, CO_2 is less soluble in water and more is in the gas phase and escapes as the can is opened and pressure is reduced.

7.91 0.09% (w/v) 0.005 M

7.93 6.3% (w/v)

7.95 a. 280 mL would have to be given.
b. The hypertonic mannitol draws water out of swollen brain cells and thus reduces the pressure on the brain.

7.97 Water moves out of the cells of the cucumber to the hypertonic salt solution, so the cucumber shrinks and loses its crispness.

7.99 7.3 g

7.101 4,000 mg

7.103 a. yes b. 0.000 50 mol

Citrus fruits contain citric acid, one of the many acids we routinely encounter in foods and consumer products.

Acids and Bases

CHAPTER OUTLINE

CHAPTER GOALS

In this chapter you will learn how to:

1. Identify acids and bases and describe their characteristics
2. Write equations for acid–base reactions
3. Classify an acid or base as strong or weak
4. Define the ion–product of water and use it to calculate hydronium or hydroxide ion concentration
5. Calculate pH
6. Draw the products of common acid–base reactions
7. Use a titration to determine the concentration of an acid or a base
8. Describe the fundamental features of a buffer
9. Understand the importance of buffers in maintaining pH in the body

Chemical terms such as *anion* and *cation* may be unfamiliar to most nonscientists, but **acid** has found a place in everyday language. Commercials advertise the latest remedy for the heartburn caused by excess stomach *acid.* The nightly news may report the latest environmental impact of *acid* rain. Wine lovers often know that wine sours because its alcohol has turned to *acid. Acid* comes from the Latin word *acidus,* meaning sour, because when tasting compounds was a routine method of identification, these compounds were found to be sour. Acids commonly react with **bases,** and many products, including antacid tablets, glass cleaners, and drain cleaners, all contain bases. In Chapter 8 we learn about the characteristics of acids and bases and the reactions they undergo.

8.1 Introduction to Acids and Bases

The earliest definition of acids and bases was suggested by Swedish chemist Svante Arrhenius in the late nineteenth century. According to Arrhenius,

- An *acid* contains a hydrogen atom and dissolves in water to form a hydrogen ion, H^+.
- A *base* contains hydroxide and dissolves in water to form OH^-.

By the Arrhenius definition, HCl is an acid because it forms aqueous H^+ and Cl^- when it dissolves in water. Sodium hydroxide (NaOH) is a base because it contains OH^- and forms solvated Na^+ and OH^- ions when it dissolves in water.

H^+ is formed from HCl.

$$\underset{\text{acid}}{HCl(g)} \longrightarrow H^+(aq) + Cl^-(aq)$$

$$\underset{\text{base}}{NaOH(s)} \longrightarrow Na^+(aq) + OH^-(aq)$$

OH^- is formed from NaOH.

The names of common acids are derived from the anions formed when they dissolve in water.

- For acids derived from anions whose names end in the suffix *-ide,* add the prefix *hydro* and change the *-ide* ending to *-ic acid.*
- For acids derived from polyatomic anions whose names end in the suffix *-ate,* change the *-ate* ending to *-ic acid.*
- For acids derived from polyatomic anions whose names end in the suffix *-ite,* change the *-ite* ending to *-ous acid.*

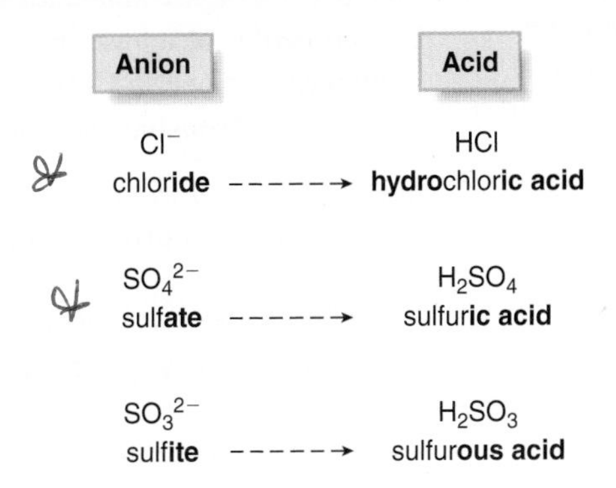

Additional examples are listed in Table 8.1.

Table 8.1 Names of Common Acids

Anion	Name of Anion	Acid	Name of Acid
Br^-	Bromide	HBr	Hydrobromic acid
PO_4^{3-}	Phosphate	H_3PO_4	Phosphoric acid
CO_3^{2-}	Carbonate	H_2CO_3	Carbonic acid
$CH_3CO_2^-$	Acetate	CH_3CO_2H	Acetic acid
NO_2^-	Nitrite	HNO_2	Nitrous acid

HNO3 Nitric acid

PROBLEM 8.1

Name each acid: (a) HF; (b) HNO_3; (c) HCN.

PROBLEM 8.2

If the polyatomic anion ClO_2^- is called chlorite, what is the name for the acid $HClO_2$?

Common Arrhenius bases are formed from cations of Group 1A or Group 2A metals (such as Na^+, K^+, and Ca^{2+}) and the hydroxide anion (OH^-).

8.1A Brønsted–Lowry Definition

While the Arrhenius definition correctly predicts the behavior of many acids and bases, this definition is limited and sometimes inaccurate. We now know, for example, that the hydrogen ion, H^+, does *not* exist in water. H^+ is a naked proton with no electrons, and this concentrated positive charge reacts rapidly with a molecule of H_2O to form the **hydronium ion, H_3O^+.** Although $H^+(aq)$ will sometimes be written in an equation for emphasis, $H_3O^+(aq)$ is actually the reacting species.

$$\underset{\text{hydrogen ion}}{H^+(aq)} + H_2O(l) \longrightarrow \underset{\text{hydronium ion}}{H_3O^+(aq)}$$

$H^+(aq)$: does not really exist in aqueous solution

$H_3O^+(aq)$: actually present in aqueous solution

$H^+(aq)$ and $H_3O^+(aq)$ are sometimes used interchangeably by chemists. Keep in mind, however, that $H^+(aq)$ does not really exist in aqueous solution.

Moreover, several compounds contain no hydroxide anions, yet they still exhibit the characteristic properties of a base. Examples include the neutral molecule ammonia (NH_3) and the salt sodium carbonate (Na_2CO_3). As a result, a more general definition of acids and bases, proposed by Johannes **Brønsted** and Thomas **Lowry** in the early twentieth century, is widely used today.

In the Brønsted–Lowry definition, acids and bases are classified according to whether they can donate or accept a **proton**—a positively charged hydrogen ion, H^+.

- A Brønsted–Lowry acid is a *proton donor.*
- A Brønsted–Lowry base is a *proton acceptor.*

Consider what happens when HCl is dissolved in water.

$$\underset{\text{Brønsted–Lowry acid}}{HCl(g)} + \underset{\text{Brønsted–Lowry base}}{H_2O(l)} \longrightarrow H_3O^+(aq) + Cl^-(aq)$$

HCl: This proton is donated.

H_2O: H_2O accepts a proton.

- **HCl is a Brønsted–Lowry *acid* because it *donates* a proton to the solvent water.**
- **H_2O is a Brønsted–Lowry *base* because it *accepts* a proton from HCl.**

Before we learn more about the details of this process, we must first learn about the characteristics of Brønsted–Lowry acids and bases.

8.1B Brønsted–Lowry Acids

A Brønsted–Lowry acid must contain a hydrogen atom. HCl is a Brønsted–Lowry acid because it *donates* a proton (H^+) to water when it dissolves, forming the hydronium ion (H_3O^+) and chloride (Cl^-).

This proton is donated to H_2O.

$$\underset{\textbf{Brønsted–Lowry acid}}{HCl(g)} + H_2O(l) \longrightarrow H_3O^+(aq) + Cl^-(aq)$$

Although hydrogen chloride, HCl, is a covalent molecule and a gas at room temperature, when it dissolves in water it reacts to form two ions, H_3O^+ and Cl^-. **Hydrochloric acid** is an aqueous solution of hydrogen chloride.

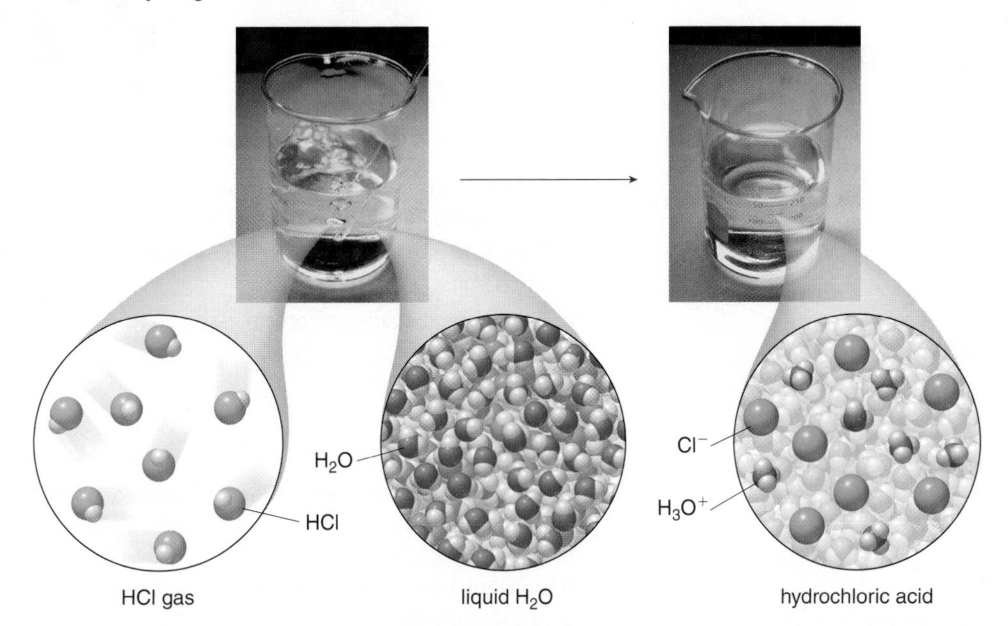

Because a Brønsted–Lowry acid contains a hydrogen atom, a general Brønsted–Lowry acid is often written as **HA. A** can be a single atom such as Cl or Br. Thus, HCl and HBr are Brønsted–Lowry acids. **A** can also be a polyatomic ion. Sulfuric acid (H_2SO_4) and nitric acid (HNO_3) are Brønsted–Lowry acids, as well. **Carboxylic acids** are a group of Brønsted–Lowry acids that contain the atoms COOH arranged so that the carbon atom is doubly bonded to one O atom and singly bonded to another. Acetic acid, CH_3COOH, is a simple carboxylic acid. Although carboxylic acids may contain several hydrogen atoms, the **H atom of the OH group is the acidic proton that is donated.**

Common Brønsted–Lowry Acids

HCl hydrochloric acid	H_2SO_4 sulfuric acid	$CH_3C(=O)O\text{–}H$ (acidic H atom)
HBr hydrobromic acid	HNO_3 nitric acid	acetic acid a carboxylic acid

Although a Brønsted–Lowry acid must contain a hydrogen atom, **it may be a neutral molecule or contain a net positive or negative charge.** Thus, H_3O^+, HCl, and HSO_4^- are all Brønsted–Lowry acids even though their net charges are +1, 0, and –1, respectively. Vinegar, citrus fruits, and carbonated soft drinks all contain Brønsted–Lowry acids, as shown in Figure 8.1.

Figure 8.1

Examples of Brønsted–Lowry Acids in Food Products

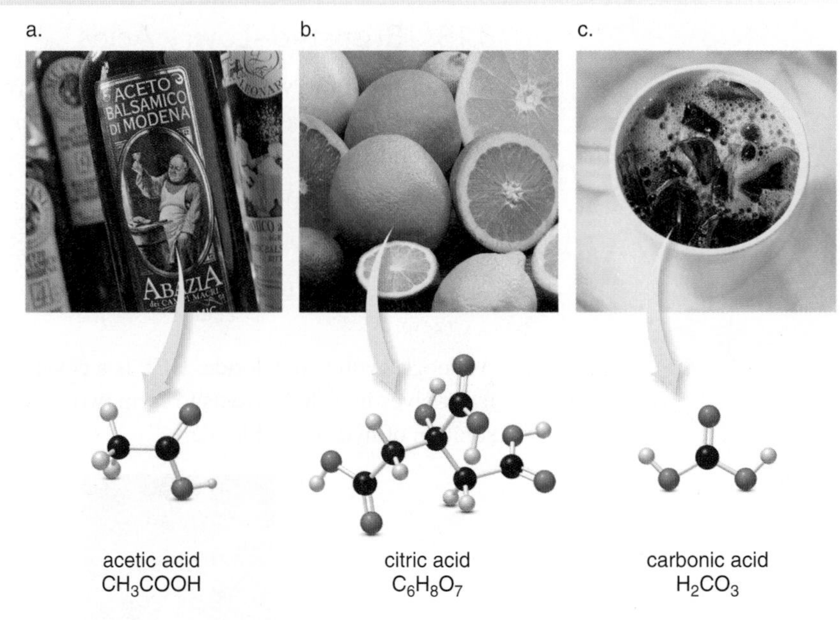

a. Acetic acid is the sour-tasting component of vinegar. The air oxidation of ethanol to acetic acid is the process that makes "bad" wine taste sour.

b. Citric acid (mentioned in the chapter opener) imparts a sour taste to oranges, lemons, and other citrus fruits.

c. Carbonated beverages contain carbonic acid, H_2CO_3.

SAMPLE PROBLEM 8.1

Which of the following species can be Brønsted–Lowry acids: (a) HF; (b) HSO_3^-; (c) Cl_2?

Analysis

A Brønsted–Lowry acid must contain a hydrogen atom, but it may be neutral or contain a net positive or negative charge.

Solution

a. HF is a Brønsted–Lowry acid since it contains a H.
b. HSO_3^- is a Brønsted–Lowry acid since it contains a H.
c. Cl_2 is not a Brønsted–Lowry acid because it does not contain a H.

PROBLEM 8.3

Which of the following species can be Brønsted–Lowry acids: (a) HI; (b) SO_4^{2-}; (c) $H_2PO_4^-$; (d) Cl^-?

8.1C Brønsted–Lowry Bases

A Brønsted–Lowry base is a proton acceptor and as such, it must be able to form a bond to a proton. Because a proton has no electrons, **a base must contain a lone pair of electrons** that can be donated to form a new bond. Thus, ammonia (NH_3) is a Brønsted–Lowry base because it contains a nitrogen atom with a lone pair of electrons. When NH_3 is dissolved in water, its N atom accepts a proton from H_2O, forming an ammonium cation (NH_4^+) and hydroxide (OH^-).

This electron pair forms a new bond to a H from H_2O.

$$\underset{\textbf{Brønsted–Lowry base}}{H{-}\ddot{N}H{-}H} \;+\; H_2O(l) \longrightarrow [NH_4]^+ \;+\; OH^-(aq)$$

A general Brønsted–Lowry base is often written as **B:** to emphasize that the base must contain a lone pair of electrons to bond to a proton. A base may be neutral or, more commonly, have a net negative charge. Hydroxide (OH^-) is the most common Brønsted–Lowry base. The source of hydroxide anions can be a variety of metal salts, including NaOH, KOH, $Mg(OH)_2$, and $Ca(OH)_2$. Ammonia (NH_3) and water (H_2O) are both Brønsted–Lowry bases because each contains an atom with a lone pair of electrons.

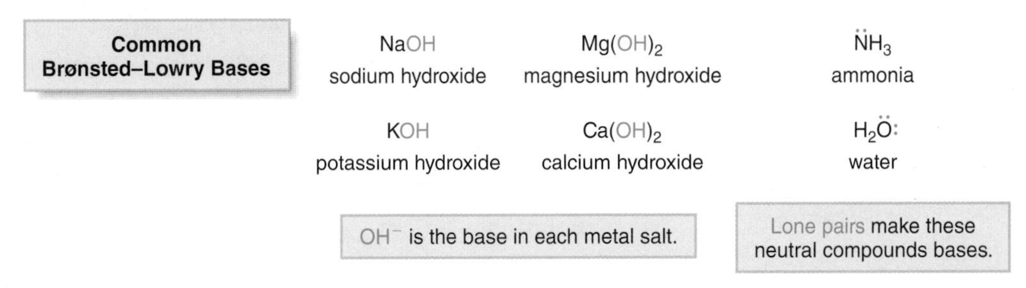

Many consumer products contain Brønsted–Lowry bases, as shown in Figure 8.2.

Figure 8.2 Examples of Brønsted–Lowry Bases in Consumer Products

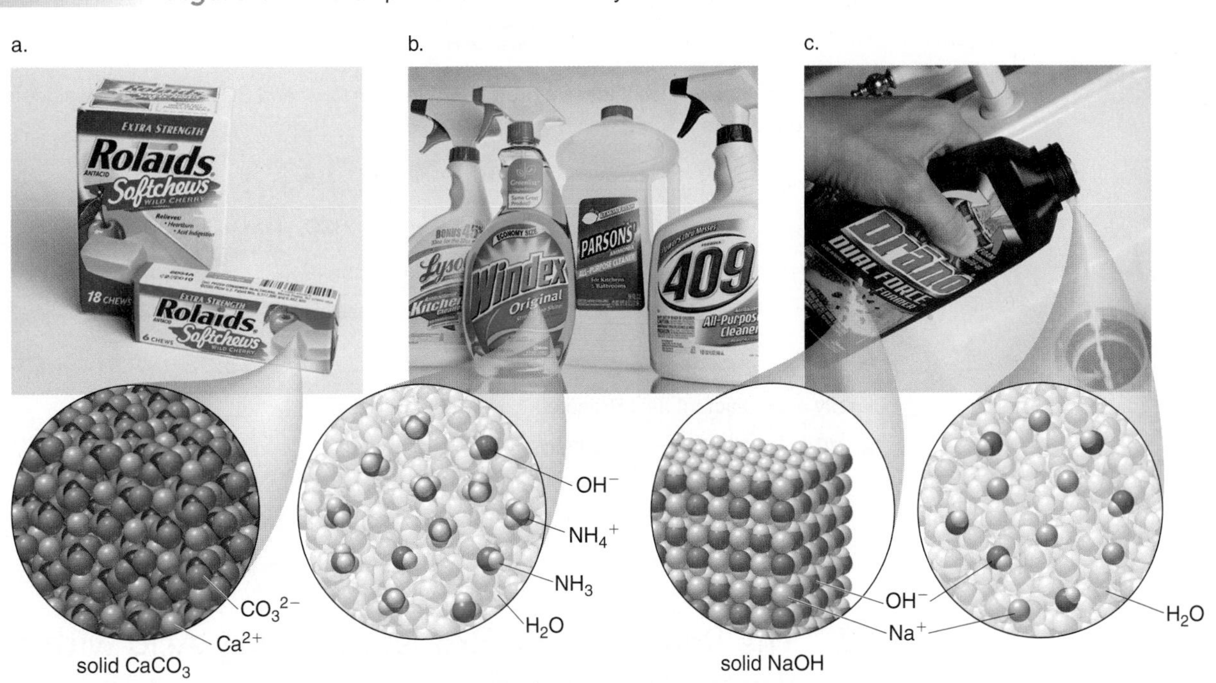

a. Calcium carbonate ($CaCO_3$), a base, is the active ingredient in the antacid Rolaids.

b. Windex and other household cleaners contain ammonia (NH_3) dissolved in water, forming NH_4^+ cations and OH^- anions.

c. Drain cleaners contain pellets of solid sodium hydroxide (NaOH), which form Na^+ cations and OH^- anions when mixed with water.

SAMPLE PROBLEM 8.2

Which of the following species can be Brønsted–Lowry bases: (a) LiOH; (b) Cl^-; (c) CH_4?

Analysis

A Brønsted–Lowry base must contain a lone pair of electrons, but it may be neutral or have a net negative charge.

Solution

a. LiOH is a base since it contains hydroxide, OH^-, which has three lone pairs on its O atom.
b. Cl^- is a base since it has four lone pairs.
c. CH_4 is not a base since it has no lone pairs.

PROBLEM 8.4

Which of the following species can be Brønsted–Lowry bases: (a) $Al(OH)_3$; (b) Br^-; (c) NH_4^+; (d) CN^-?

SAMPLE PROBLEM 8.3

Classify each reactant as a Brønsted–Lowry acid or base.

a. $HF(g) + H_2O(l) \longrightarrow F^-(aq) + H_3O^+(aq)$
b. $SO_4^{2-}(aq) + H_2O(l) \longrightarrow HSO_4^-(aq) + OH^-(aq)$

Analysis

In each equation, the Brønsted–Lowry acid is the species that loses a proton and the Brønsted–Lowry base is the species that gains a proton.

Solution

a. HF is the acid since it loses a proton (H^+) to form F^-, and H_2O is the base since it gains a proton to form H_3O^+.

gain of H^+

$$\underset{\text{acid}}{HF(g)} + \underset{\text{base}}{H_2O(l)} \longrightarrow F^-(aq) + H_3O^+(aq)$$

loss of H^+

b. H_2O is the acid since it loses a proton (H^+) to form OH^-, and SO_4^{2-} is the base since it gains a proton to form HSO_4^-.

gain of H^+

$$\underset{\text{base}}{SO_4^{2-}(aq)} + \underset{\text{acid}}{H_2O(l)} \longrightarrow HSO_4^-(aq) + OH^-(aq)$$

loss of H^+

PROBLEM 8.5

Classify each reactant as a Brønsted–Lowry acid or base.

a. $HCl(g) + NH_3(g) \longrightarrow Cl^-(aq) + NH_4^+(aq)$
b. $CH_3COOH(l) + H_2O(l) \longrightarrow CH_3COO^-(aq) + H_3O^+(aq)$
c. $OH^-(aq) + HSO_4^-(aq) \longrightarrow H_2O(l) + SO_4^{2-}(aq)$

8.2 The Reaction of a Brønsted–Lowry Acid with a Brønsted–Lowry Base

When a Brønsted–Lowry acid reacts with a Brønsted–Lowry base, a proton is *transferred* from the acid to the base. **The Brønsted–Lowry acid donates a proton to the Brønsted–Lowry base, which accepts it.**

Consider, for example, the reaction of the general acid H—A with the general base B:. **In an acid–base reaction, one bond is broken and one bond is formed.** The electron pair of the base B: forms a new bond to the proton of the acid, forming $H—B^+$. The acid H—A loses a proton, leaving the electron pair in the H—A bond on A, forming $A:^-$.

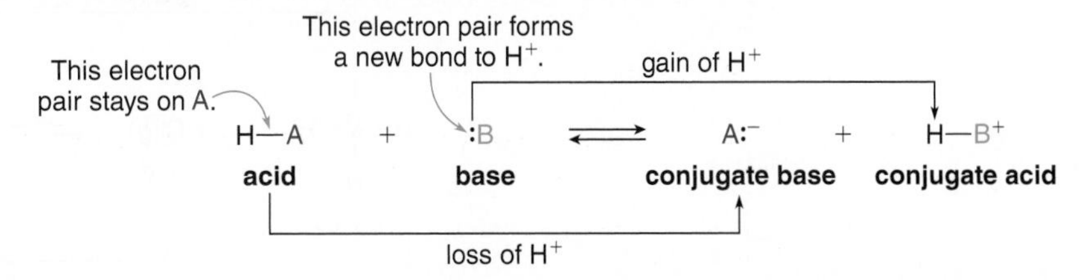

- **The product formed by loss of a proton from an acid is called its *conjugate base.***
- **The product formed by gain of a proton by a base is called its *conjugate acid.***

Thus, the conjugate base of the acid HA is $A:^-$. The conjugate acid of the base B: is HB^+.

- **Two species that differ by the presence of a proton are called a conjugate acid–base pair.**

Thus, in an acid–base reaction, the acid and the base on the left side of the equation (HA and B:) form two products that are also an acid and a base (HB^+ and $A:^-$). Equilibrium arrows ($\rightleftarrows$) are often used to separate reactants and products, because the reaction can proceed in either the forward or reverse directions. In some reactions, the products are greatly favored, as discussed in Section 8.3.

When HBr is dissolved in water, for example, the acid HBr loses a proton to form its conjugate base Br^-, and the base H_2O gains a proton to form H_3O^+.

gain of H^+

$$\underset{\text{acid}}{H—Br} + \underset{\text{base}}{H_2O} \rightleftarrows \underset{\text{conjugate base}}{Br^-} + \underset{\text{conjugate acid}}{H_3O^+}$$

loss of H^+

Thus, HBr and Br^- are a conjugate acid–base pair since these two species differ by the presence of a proton (H^+). H_2O and H_3O^+ are also a conjugate acid–base pair because these two species differ by the presence of a proton as well.

The net charge must be the same on both sides of the equation. In this example, the two reactants are neutral (zero net charge), and the sum of the –1 and +1 charges in the products is also zero.

Take particular note of what happens to the charges in each conjugate acid–base pair. **When a species gains a proton (H^+), it gains a +1 charge.** Thus, if a reactant is neutral to begin with, it ends up with a +1 charge. **When a species loses a proton (H^+), it effectively gains a –1 charge** since the product has one fewer proton (+1 charge) than it started with. Thus, if a reactant is neutral to begin with, it ends up with a –1 charge.

0 charge → $\underset{\text{base}}{H_2O} \xrightarrow{\text{add } H^+} H_3O^+$ ← +1 charge

0 charge → $\underset{\text{acid}}{H—Br} \xrightarrow{\text{lose } H^+} Br^-$ ← –1 charge

Take away +1 charge.

The reaction of ammonia (NH_3) with HCl is also a Brønsted–Lowry acid–base reaction. In this example, NH_3 is the base since it gains a proton to form its conjugate acid, NH_4^+. HCl is the acid since it donates a proton, forming its conjugate base, Cl^-.

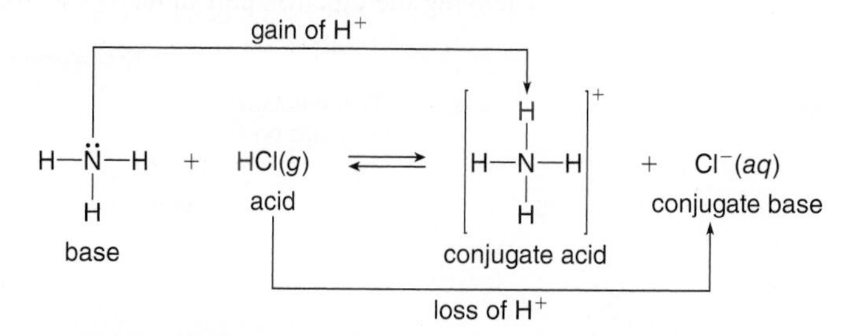

- **A Brønsted–Lowry acid–base reaction is a *proton transfer reaction* since it always results in the transfer of a proton from an acid to a base.**

The ability to identify and draw a conjugate acid or base from a given starting material is a necessary skill, illustrated in Sample Problems 8.4 and 8.5.

SAMPLE PROBLEM 8.4

Draw the conjugate acid of each base: (a) F^-; (b) NO_3^-.

Analysis

Conjugate acid–base pairs differ by the presence of a proton. To draw a conjugate acid from a base, *add* a proton, H^+. This adds +1 to the charge of the base to give the charge on the conjugate acid.

Solution

a. F^- + H^+ gives HF as the conjugate acid.

b. NO_3^- + H^+ gives HNO_3 (nitric acid) as the conjugate acid.

PROBLEM 8.6

Draw the conjugate acid of each species: (a) H_2O; (b) I^-; (c) HCO_3^-.

SAMPLE PROBLEM 8.5

Draw the conjugate base of each acid: (a) H_2O; (b) HCO_3^-.

Analysis

Conjugate acid–base pairs differ by the presence of a proton. To draw a conjugate base from an acid, *remove* a proton, H^+. This adds −1 to the charge of the acid to give the charge on the conjugate base.

Solution

a. Remove H^+ from H_2O to form OH^-, the conjugate base.

b. Remove H^+ from HCO_3^- to form CO_3^{2-}, the conjugate base. CO_3^{2-} has a −2 charge since −1 is added to an anion that had a −1 charge to begin with.

PROBLEM 8.7

Draw the conjugate base of each species: (a) H_2S; (b) HCN; (c) HSO_4^-.

PROBLEM 8.8

Draw the structure of the conjugate base of each acid.

a. 

b. 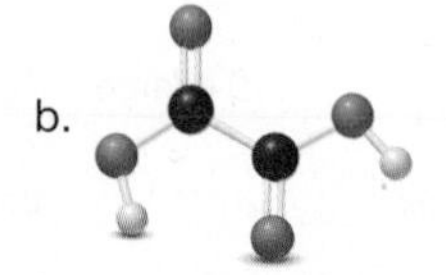

A compound that contains both a hydrogen atom and a lone pair of electrons can be either an acid or a base, depending on the particular reaction. Such a compound is said to be **amphoteric.** For example, when H_2O acts as a base it gains a proton, forming H_3O^+. Thus, H_2O and H_3O^+ are a conjugate acid–base pair. When H_2O acts as an acid it loses a proton, forming OH^-. H_2O and OH^- are also a conjugate acid–base pair.

H_2O as a base: H—Ö—H (base) —add H^+→ $[H_3O]^+$ (conjugate acid)

H_2O as an acid: H—Ö—H (acid) —remove H^+→ $[H—\ddot{O}:]^-$ (conjugate base)

vitamin C

SAMPLE PROBLEM 8.6

Label the acid and the base and the conjugate acid and the conjugate base in the following reaction.

$$NH_4^+(aq) + OH^-(aq) \longrightarrow NH_3(g) + H_2O(l)$$

Analysis

The Brønsted–Lowry acid loses a proton to form its conjugate base. The Brønsted–Lowry base gains a proton to form its conjugate acid.

Solution

NH_4^+ is the acid since it loses a proton to form NH_3, its conjugate base. OH^- is the base since it gains a proton to form its conjugate acid, H_2O.

$$\underset{\text{acid}}{NH_4^+(aq)} + \underset{\text{base}}{OH^-(aq)} \rightleftharpoons \underset{\text{conjugate base}}{NH_3(g)} + \underset{\text{conjugate acid}}{H_2O(l)}$$

gain of H^+ (OH⁻ → H_2O); loss of H^+ (NH_4^+ → NH_3)

PROBLEM 8.9

Label the acid and the base and the conjugate acid and the conjugate base in each reaction.

a. $H_2O(l) + HI(g) \rightleftharpoons I^-(aq) + H_3O^+(aq)$
b. $CH_3COOH(l) + NH_3(g) \rightleftharpoons CH_3COO^-(aq) + NH_4^+(aq)$
c. $Br^-(aq) + HNO_3(aq) \rightleftharpoons HBr(aq) + NO_3^-(aq)$

PROBLEM 8.10

Ammonia, NH_3, is amphoteric. (a) Draw the conjugate acid of NH_3. (b) Draw the conjugate base of NH_3.

PROBLEM 8.11

When ascorbic acid (vitamin C, molecular formula $C_6H_8O_6$) is dissolved in water, the following acid–base reaction occurs. Label the conjugate acid–base pairs in the given equation.

$$\underset{\text{vitamin C}}{C_6H_8O_6(aq)} + H_2O(l) \rightleftharpoons C_6H_7O_6^-(aq) + H_3O^+(aq)$$

HEALTH NOTE

Citrus fruits (oranges, grapefruit, and lemons) are well known sources of vitamin C (Problem 8.11), but guava, kiwifruit, and rose hips are excellent sources, too. Vitamin C is needed for the formation of collagen, a common protein in connective tissues in muscles and blood vessels.

8.3 Acid and Base Strength

Although all Brønsted–Lowry acids contain protons, some acids readily donate protons while others do not. Similarly, some Brønsted–Lowry bases accept a proton much more readily than others. How readily proton transfer occurs is determined by the strength of the acid and base.

When a covalent acid dissolves in water, proton transfer forms H_3O^+ and an anion. The splitting apart of a covalent molecule (or an ionic compound) into individual ions is called **dissociation.** Acids differ in their tendency to donate a proton; that is, acids differ in the extent to which they *dissociate* in water.

- **A strong acid readily donates a proton. When a strong acid dissolves in water, essentially 100% of the acid dissociates into ions.**
- **A weak acid less readily donates a proton. When a weak acid dissolves in water, only a small fraction of the acid dissociates into ions.**

Common strong acids include **HI, HBr, HCl, H_2SO_4,** and **HNO_3** (Table 8.2). When each acid is dissolved in water, 100% of the acid dissociates, forming H_3O^+ and the conjugate base, as shown for HCl and H_2SO_4.

- Use a single reaction arrow.
- The product is greatly favored at equilibrium.

$$\underset{\text{strong acid}}{HCl(g)} + H_2O(l) \longrightarrow H_3O^+(aq) + \underset{\text{conjugate base}}{Cl^-(aq)}$$

$$\underset{\text{strong acid}}{H_2SO_4(l)} + H_2O(l) \longrightarrow H_3O^+(aq) + \underset{\text{conjugate base}}{HSO_4^-(aq)}$$

HCl, hydrochloric acid, is secreted by the stomach to digest food (Figure 8.3), and **H_2SO_4,** sulfuric acid, is an important industrial starting material in the synthesis of phosphate fertilizers. A single reaction arrow ($\longrightarrow$) is drawn to show that essentially all of the reactants are converted to products.

Acetic acid, **CH_3COOH,** is a weak acid. When acetic acid dissolves in water, only a small fraction of acetic acid molecules donate a proton to water to form H_3O^+ and the conjugate base,

Table 8.2 Relative Strength of Acids and Their Conjugate Bases

Acid		Conjugate Base	
Strong Acids			
Hydroiodic acid	HI	I^-	Iodide ion
Hydrobromic acid	HBr	Br^-	Bromide ion
Hydrochloric acid	HCl	Cl^-	Chloride ion
Sulfuric acid	H_2SO_4	HSO_4^-	Hydrogen sulfate ion
Nitric acid	HNO_3	NO_3^-	Nitrate ion
Hydronium ion	**H_3O^+**	**H_2O**	**Water**
Weak Acids			
Phosphoric acid	H_3PO_4	$H_2PO_4^-$	Dihydrogen phosphate ion
Hydrofluoric acid	HF	F^-	Fluoride ion
Acetic acid	CH_3COOH	CH_3COO^-	Acetate ion
Carbonic acid	H_2CO_3	HCO_3^-	Bicarbonate ion
Ammonium ion	NH_4^+	NH_3	Ammonia
Hydrocyanic acid	HCN	CN^-	Cyanide ion
Water	H_2O	OH^-	Hydroxide ion

Increasing acid strength (upward arrow); Increasing base strength (downward arrow)

Figure 8.3

Focus on the Human Body: Hydrochloric Acid in the Stomach

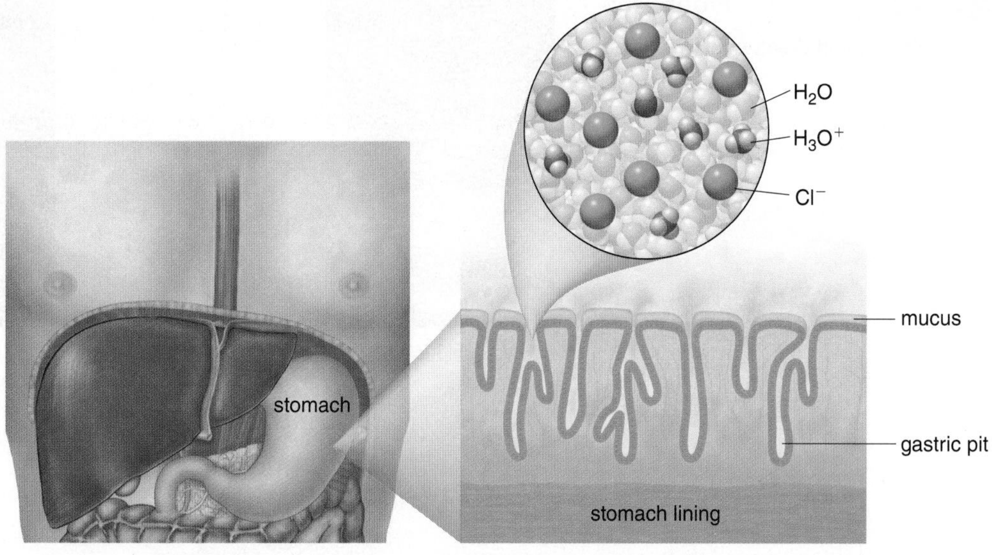

The thick mucous layer protects the stomach lining.

Although HCl is a corrosive acid secreted in the stomach, a thick layer of mucus covering the stomach wall protects it from damage by the strong acid. The strong acid HCl is completely dissociated to H_3O^+ and Cl^-.

CH_3COO^-. The major species in solution is the undissociated acid, CH_3COOH. Two arrows that are unequal in length ($\leftleftharpoons$) are used between the reactants and products to show that both are present in solution. The longer arrow points towards the reactants, since few molecules of acetic acid dissociate. Other weak acids and their conjugate bases are listed in Table 8.2.

- Use unequal reaction arrows.
- The reactants are favored at equilibrium.

$$\underset{\text{weak acid}}{CH_3COOH(l)} + H_2O(l) \rightleftharpoons H_3O^+(aq) + \underset{\text{conjugate base}}{CH_3COO^-(aq)}$$

Figure 8.4 illustrates the difference between an aqueous solution of a strong acid that is completely dissociated and a weak acid that contains much undissociated acid. Sample Problem 8.7 illustrates how the extent of dissociation and acid strength are related.

SAMPLE PROBLEM 8.7

Diagrams **A–C** represent three acids (HA) dissolved in water. Which representation shows the strongest acid? Which representation shows the weakest acid?

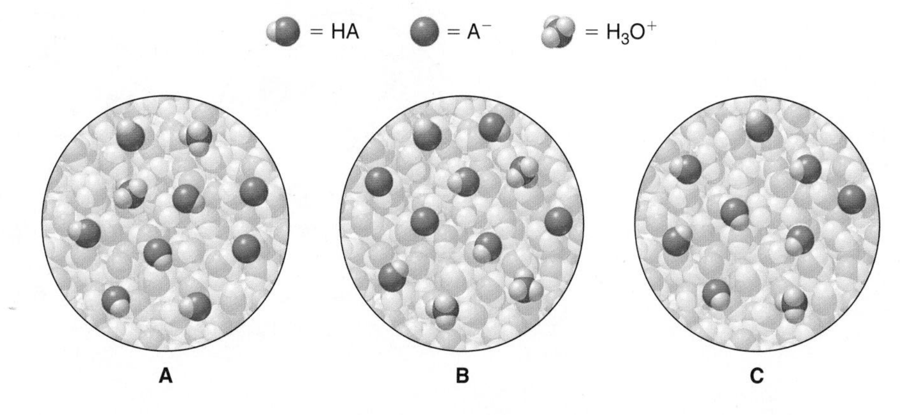

Figure 8.4

A Strong and Weak Acid Dissolved in Water

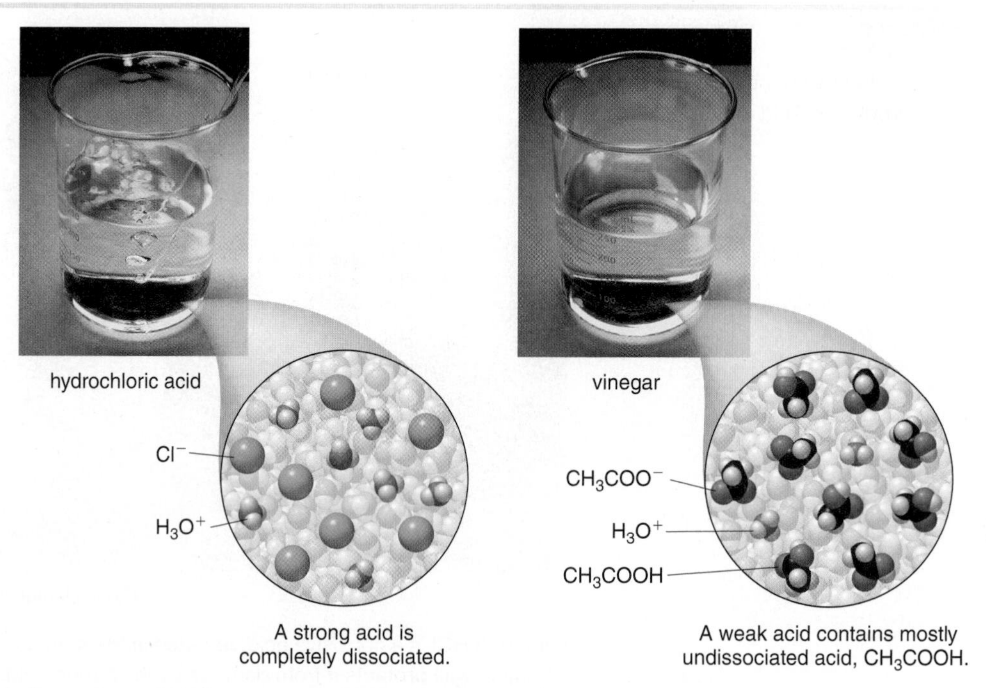

- The strong acid HCl completely dissociates into H_3O^+ and Cl^- in water.
- Vinegar contains CH_3COOH dissolved in H_2O. The weak acid CH_3COOH is only slightly dissociated into H_3O^+ and CH_3COO^-, so mostly CH_3COOH is present in solution.

Analysis

The stronger the acid, the more readily it dissociates to form its conjugate base A^- and H_3O^+. In molecular art, the strongest acid has the most A^- and H_3O^+ ions, and the fewest molecules of undissociated HA.

Solution

B is the strongest acid since it contains the largest number of A^- and H_3O^+ ions (three each), and the smallest number (five) of undissociated HA molecules. **C** is the weakest acid since it contains the smallest number of A^- and H_3O^+ ions (one each), and the largest number (seven) of undissociated HA molecules.

PROBLEM 8.12

Diagrams **D–F** represent three acids (HA) dissolved in water. Rank the three acids (**D–F**) in order of increasing acidity.

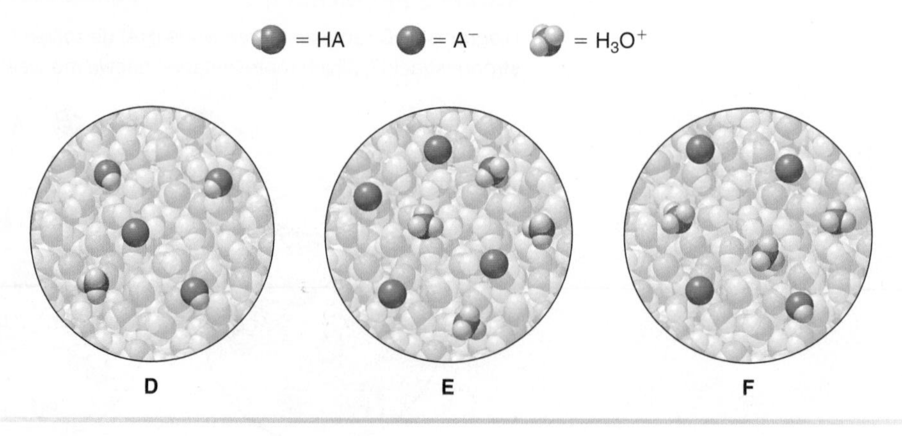

Bases also differ in their ability to accept a proton.

- **A strong base readily accepts a proton. When a strong base dissolves in water, 100% of the base dissociates into ions.**
- **A weak base less readily accepts a proton. When a weak base dissolves in water, only a small fraction of the base forms ions.**

The most common strong base is hydroxide, **OH^-,** used as a variety of metal salts, including NaOH and KOH. Solid NaOH dissolves in water to form solvated Na^+ cations and OH^- anions. In contrast, when **NH_3,** a weak base, dissolves in water, only a small fraction of NH_3 molecules react to form NH_4^+ and OH^-. The major species in solution is the undissociated molecule, NH_3. Figure 8.5 illustrates the difference between aqueous solutions of strong and weak bases. Table 8.2 lists common bases.

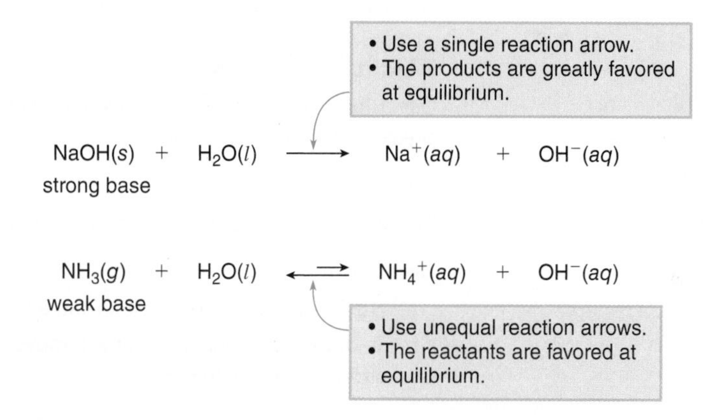
• Use a single reaction arrow.
• The products are greatly favored at equilibrium.
$NaOH(s) + H_2O(l) \longrightarrow Na^+(aq) + OH^-(aq)$
strong base
$NH_3(g) + H_2O(l) \rightleftharpoons NH_4^+(aq) + OH^-(aq)$
weak base
• Use unequal reaction arrows.
• The reactants are favored at equilibrium.

Figure 8.5
A Strong and Weak Base Dissolved in Water

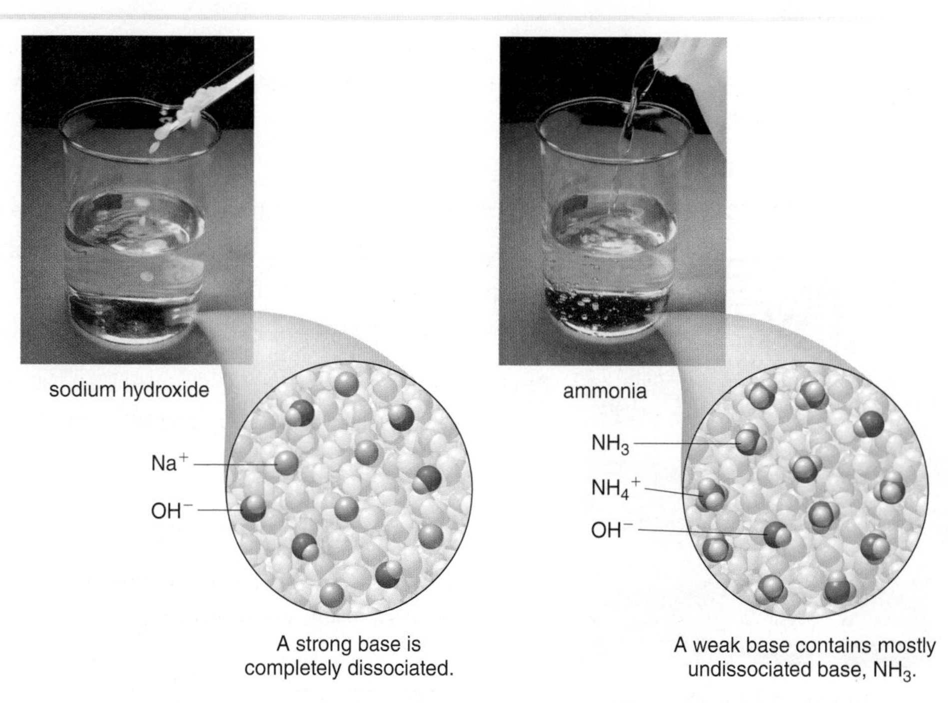
sodium hydroxide
Na+
OH−
ammonia
NH3
NH4+
OH−

A strong base is completely dissociated.

A weak base contains mostly undissociated base, NH_3.

- The strong base NaOH completely dissociates into Na^+ and OH^- in water.
- The weak base NH_3 is only slightly dissociated into NH_4^+ and OH^-, so mostly NH_3 is present in solution.

An inverse relationship exists between acid and base strength.

- A strong acid readily donates a proton, forming a *weak* conjugate base.
- A strong base readily accepts a proton, forming a *weak* conjugate acid.

Why does this inverse relationship exist? Since a strong acid readily donates a proton, it forms a conjugate base that has little ability to accept a proton. Since a strong base readily accepts a proton, it forms a conjugate acid that tightly holds onto its proton, making it a weak acid.

Thus, a *strong* acid like HCl forms a *weak* conjugate base (Cl^-), and a *strong* base like OH^- forms a *weak* conjugate acid (H_2O). The entries in Table 8.2 are arranged in order of *decreasing* acid strength. This means that Table 8.2 is also arranged in order of *increasing* strength of the resulting conjugate bases. Knowing the relative strength of two acids makes it possible to predict the relative strength of their conjugate bases.

SAMPLE PROBLEM 8.8

Using Table 8.2: (a) Is H_3PO_4 or HF the stronger acid? (b) Draw the conjugate base of each acid and predict which base is stronger.

Analysis

The stronger the acid, the weaker the conjugate base.

Solution

a. H_3PO_4 is located above HF in Table 8.2, making it the stronger acid.
b. To draw each conjugate base, remove a proton (H^+). Since HF is the weaker acid, F^- is the stronger conjugate base.

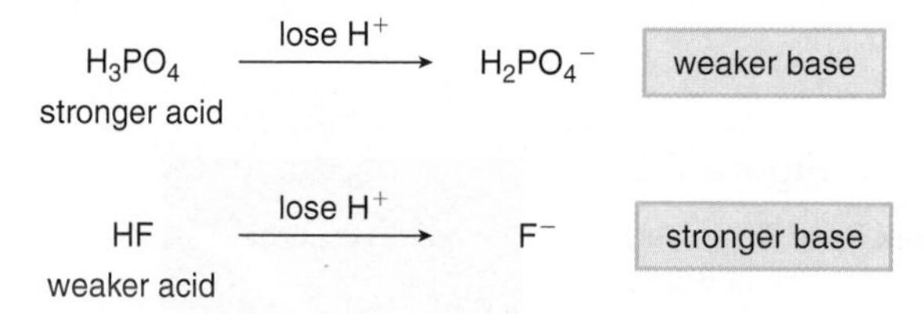

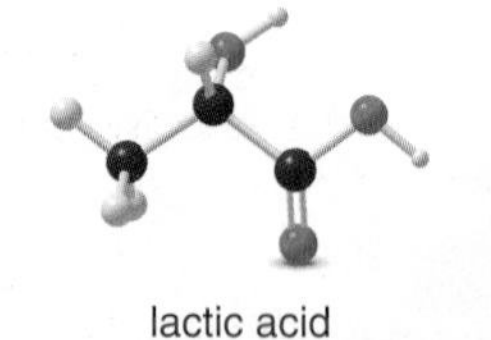
lactic acid

PROBLEM 8.13

Label the stronger acid in each pair. Which acid has the stronger conjugate base?

a. H_2SO_4 or H_3PO_4 b. HF or HCl c. H_2CO_3 or NH_4^+ d. HCN or HF

PROBLEM 8.14

If lactic acid ($C_3H_6O_3$) is a stronger acid than HCN, which compound forms the stronger conjugate base?

PROBLEM 8.15

(a) Draw the conjugate acids of NO_2^- and NO_3^-. (b) If NO_2^- is the stronger base, which acid is stronger?

HEALTH NOTE

Lactic acid (Problem 8.14) accumulates in tissues during vigorous exercise, making muscles feel tired and sore. The formation of lactic acid is discussed in greater detail in Section 18.5.

8.4 Dissociation of Water

In Section 8.2 we learned that water can behave as *both* a Brønsted–Lowry acid and a Brønsted–Lowry base. As a result, two molecules of water can react together in an acid–base reaction.

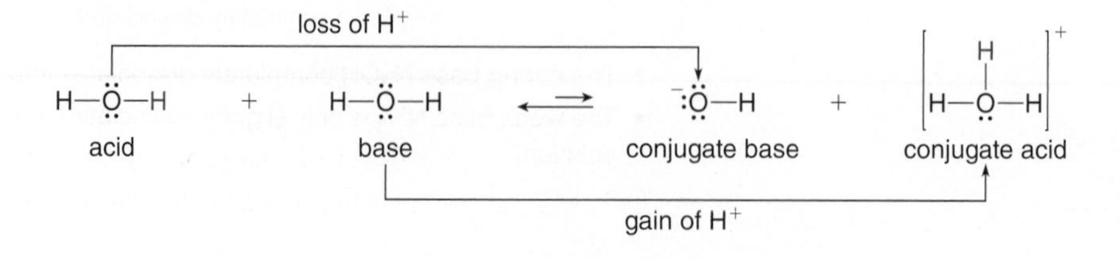

- One molecule of H_2O donates a proton (H^+), forming its conjugate base OH^-.
- One molecule of H_2O accepts a proton, forming its conjugate acid H_3O^+.

Since water is a very weak acid (Table 8.2), pure water contains an exceedingly low concentration of ions, H_3O^+ and OH^-. Since one H_3O^+ ion and one OH^- ion are formed in each reaction, the concentrations of H_3O^+ and OH^- are *equal* in pure water. Experimentally it can be shown that the $[H_3O^+] = [OH^-] = 1.0 \times 10^{-7}$ M at 25 °C.

- In pure water, $[H_3O^+] = [OH^-] = 1.0 \times 10^{-7}$ M.

Multiplying these concentrations together gives the **ion–product constant** for water, symbolized by $\boldsymbol{K_w}$.

$$K_w = [H_3O^+][OH^-]$$

ion–product constant

Substituting the concentrations for H_3O^+ and OH^- into the expression for K_w gives the following result.

$$K_w = [H_3O^+][OH^-]$$
$$K_w = (1.0 \times 10^{-7}) \times (1.0 \times 10^{-7})$$
$$K_w = 1.0 \times 10^{-14}$$

- The product, $[H_3O^+][OH^-]$, is a constant, 1.0×10^{-14}, for all aqueous solutions at 25 °C.

Thus, **the value of $\boldsymbol{K_w}$ applies to any aqueous solution,** not just pure water. If we know the concentration of one ion, H_3O^+ or OH^-, we can find the concentration of the other by rearranging the expression for K_w.

To calculate $[OH^-]$ when $[H_3O^+]$ is known:

$$K_w = [H_3O^+][OH^-]$$
$$[OH^-] = \frac{K_w}{[H_3O^+]}$$
$$[OH^-] = \frac{1.0 \times 10^{-14}}{[H_3O^+]}$$

To calculate $[H_3O^+]$ when $[OH^-]$ is known:

$$K_w = [H_3O^+][OH^-]$$
$$[H_3O^+] = \frac{K_w}{[OH^-]}$$
$$[H_3O^+] = \frac{1.0 \times 10^{-14}}{[OH^-]}$$

How to write numbers in scientific notation was presented in Section 1.6. Multiplying and dividing numbers written in scientific notation was described in Section 5.3.

Thus, if the concentration of H_3O^+ in a cup of coffee is 1.0×10^{-5} M, we can use this value to calculate $[OH^-]$.

$$[OH^-] = \frac{K_w}{[H_3O^+]} = \frac{1.0 \times 10^{-14}}{1.0 \times 10^{-5}}$$
$$[OH^-] = 1.0 \times 10^{-9}\ \text{M}$$

hydroxide ion concentration in a cup of coffee

In a cup of coffee, therefore, the concentration of H_3O^+ ions is greater than the concentration of OH^- ions, but the product of these concentrations, 1.0×10^{-14}, is a constant, K_w.

Coffee is an *acidic* solution since the concentration of H_3O^+ is *greater* than the concentration of OH^-.

Pure water and any solution that has an equal concentration of H_3O^+ and OH^- ions (1.0×10^{-7}) is said to be *neutral*. Other solutions are classified as **acidic** or **basic,** depending on which ion is present in a higher concentration.

- In an *acidic* solution, $[H_3O^+] > [OH^-]$; thus, $[H_3O^+] > 10^{-7}$ M.
- In a *basic* solution, $[OH^-] > [H_3O^+]$; thus, $[OH^-] > 10^{-7}$ M.

In an acidic solution, the concentration of the acid H_3O^+ is greater than the concentration of the base OH^-. In a basic solution, the concentration of the base OH^- is greater than the concentration of the acid H_3O^+. Table 8.3 summarizes information about neutral, acidic, and basic solutions.

Table 8.3 Neutral, Acidic, and Basic Solutions

Type	$[H_3O^+]$ and $[OH^-]$	$[H_3O^+]$	$[OH^-]$
Neutral	$[H_3O^+] = [OH^-]$	10^{-7} M	10^{-7} M
Acidic	$[H_3O^+] > [OH^-]$	$> 10^{-7}$ M	$< 10^{-7}$ M
Basic	$[H_3O^+] < [OH^-]$	$< 10^{-7}$ M	$> 10^{-7}$ M

SAMPLE PROBLEM 8.9

If $[H_3O^+]$ in blood is 4.0×10^{-8} M, what is the value of $[OH^-]$? Is blood acidic, basic, or neutral?

Analysis

Use the equation $[OH^-] = K_w/[H_3O^+]$ to calculate the hydroxide ion concentration.

Solution

Substitute the given value of $[H_3O^+]$ in the equation to find $[OH^-]$.

$$[OH^-] = \frac{K_w}{[H_3O^+]} = \frac{1.0 \times 10^{-14}}{4.0 \times 10^{-8}} = 2.5 \times 10^{-7}\ M$$

hydroxide ion concentration in the blood

Since $[OH^-] > [H_3O^+]$, blood is a basic solution.

PROBLEM 8.16

Calculate the value of $[OH^-]$ from the given $[H_3O^+]$ in each solution and label the solution as acidic or basic: (a) $[H_3O^+] = 10^{-3}$ M; (b) $[H_3O^+] = 10^{-11}$ M; (c) $[H_3O^+] = 2.8 \times 10^{-10}$ M; (d) $[H_3O^+] = 5.6 \times 10^{-4}$ M.

PROBLEM 8.17

Calculate the value of $[H_3O^+]$ from the given $[OH^-]$ in each solution and label the solution as acidic or basic: (a) $[OH^-] = 10^{-6}$ M; (b) $[OH^-] = 10^{-9}$ M; (c) $[OH^-] = 5.2 \times 10^{-11}$ M; (d) $[OH^-] = 7.3 \times 10^{-4}$ M.

Since a strong acid like HCl is completely dissociated in aqueous solution, the concentration of the acid tells us the concentration of hydronium ions present. Thus, a 0.1 M HCl solution completely dissociates, so the concentration of H_3O^+ is 0.1 M. This value can then be used to calculate the hydroxide ion concentration. Similarly, a strong base like NaOH completely dissociates, so the concentration of the base gives the concentration of hydroxide ions present. Thus, the concentration of OH^- in a 0.1 M NaOH solution is 0.1 M.

In 0.1 M HCl solution: strong acid — $[H_3O^+] = 0.1\ M = 1 \times 10^{-1}\ M$

In 0.1 M NaOH solution: strong base — $[OH^-] = 0.1\ M = 1 \times 10^{-1}\ M$

SAMPLE PROBLEM 8.10

Calculate the value of $[H_3O^+]$ and $[OH^-]$ in a 0.01 M NaOH solution.

Analysis

Since NaOH is a strong base that completely dissociates to form Na^+ and OH^-, the concentration of NaOH gives the concentration of OH^- ions. The $[OH^-]$ can then be used to calculate $[H_3O^+]$ from the expression for K_w.

Solution

The value of $[OH^-]$ in a 0.01 M NaOH solution is 0.01 M = 1×10^{-2} M.

$$[H_3O^+] = \frac{K_w}{[OH^-]} = \frac{1 \times 10^{-14}}{1 \times 10^{-2}} = 1 \times 10^{-12}\text{ M}$$

concentration of OH^- (the denominator 1×10^{-2}); 1×10^{-12} M = concentration of H_3O^+

PROBLEM 8.18

Calculate the value of $[H_3O^+]$ and $[OH^-]$ in each solution: (a) 0.001 M NaOH; (b) 0.001 M HCl; (c) 1.5 M HCl; (d) 0.30 M NaOH.

8.5 The pH Scale

Knowing the hydronium ion concentration is necessary in many different instances. The blood must have an H_3O^+ concentration in a very narrow range for an individual's good health. Plants thrive in soil that is not too acidic or too basic. The H_3O^+ concentration in a swimming pool must be measured and adjusted to keep the water clean and free from bacteria and algae.

8.5A Calculating pH

Apple juice has a pH of about 4, so it is an acidic solution.

Since values for the hydronium ion concentration are very small, with negative powers of ten, the **pH scale** is used to more conveniently report $[H_3O^+]$. The pH of a solution is a number generally between 0 and 14, defined in terms of the *logarithm* (log) of the H_3O^+ concentration.

$$\mathbf{pH = -log\,[H_3O^+]}$$

A logarithm is an exponent of a power of ten.

$$\log(10^5) = 5 \qquad \log(10^{-10}) = -10 \qquad \log(0.001) = \log(10^{-3}) = -3$$

The log is the exponent. The log is the exponent. Convert to scientific notation.

In calculating pH, first consider an H_3O^+ concentration that has a coefficient of *one* when the number is written in scientific notation. For example, the value of $[H_3O^+]$ in apple juice is about 1×10^{-4}, or 10^{-4} written without the coefficient. The pH of this solution is calculated as follows:

$$\text{pH} = -\log\,[H_3O^+] = -\log(10^{-4}) = -(-4) = 4$$

pH of apple juice

Since pH is defined as the *negative* logarithm of $[H_3O^+]$ and these concentrations have *negative* exponents (10^{-x}), pH values are *positive* numbers.

Whether a solution is acidic, neutral, or basic can now be defined in terms of its pH.

- Acidic solution: pH < 7 ⟶ $[H_3O^+] > 1 \times 10^{-7}$
- Neutral solution: pH = 7 ⟶ $[H_3O^+] = 1 \times 10^{-7}$
- Basic solution: pH > 7 ⟶ $[H_3O^+] < 1 \times 10^{-7}$

Note the relationship between $[H_3O^+]$ and pH.

- The *lower* the pH, the *higher* the concentration of H_3O^+.

The pH of a solution can be measured using a pH meter as shown in Figure 8.6. Approximate pH values are determined using pH paper or indicators that turn different colors depending on the pH of the solution. The pH of various substances is shown in Figure 8.7.

Converting a given H_3O^+ concentration to a pH value is shown in Sample Problem 8.11. The reverse process, converting a pH value to an H_3O^+ concentration, is shown in Sample Problem 8.12.

Figure 8.6 Measuring pH

a.

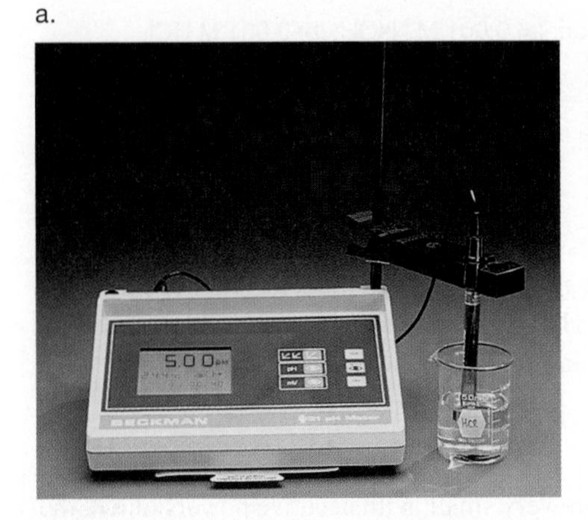

b.

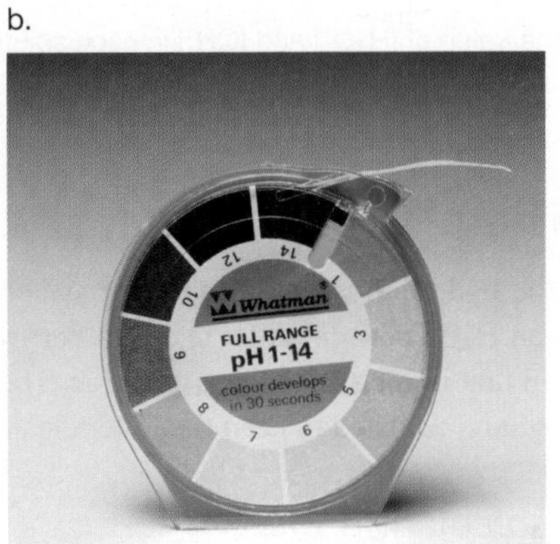

c.

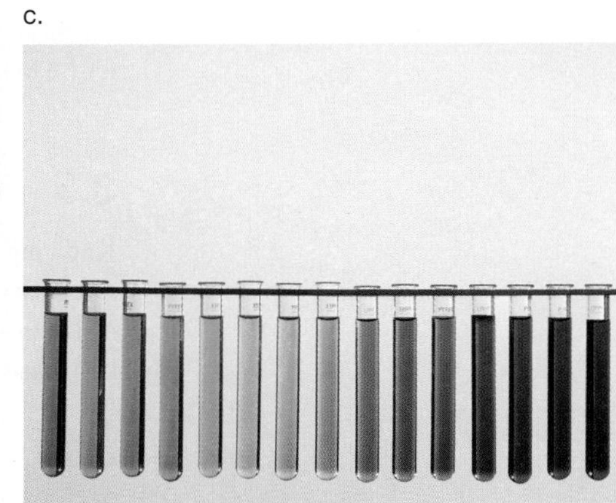

a. A pH meter is a small electronic device that measures pH when an electrode is dipped into a solution.

b. Paper strips called pH paper change color corresponding to a particular pH, when a drop of an aqueous solution is applied to them.

c. An acid–base indicator can be used to give an approximate pH. The indicator is a dye that changes color depending on the pH of the solution.

Figure 8.7 The pH of Some Common Substances

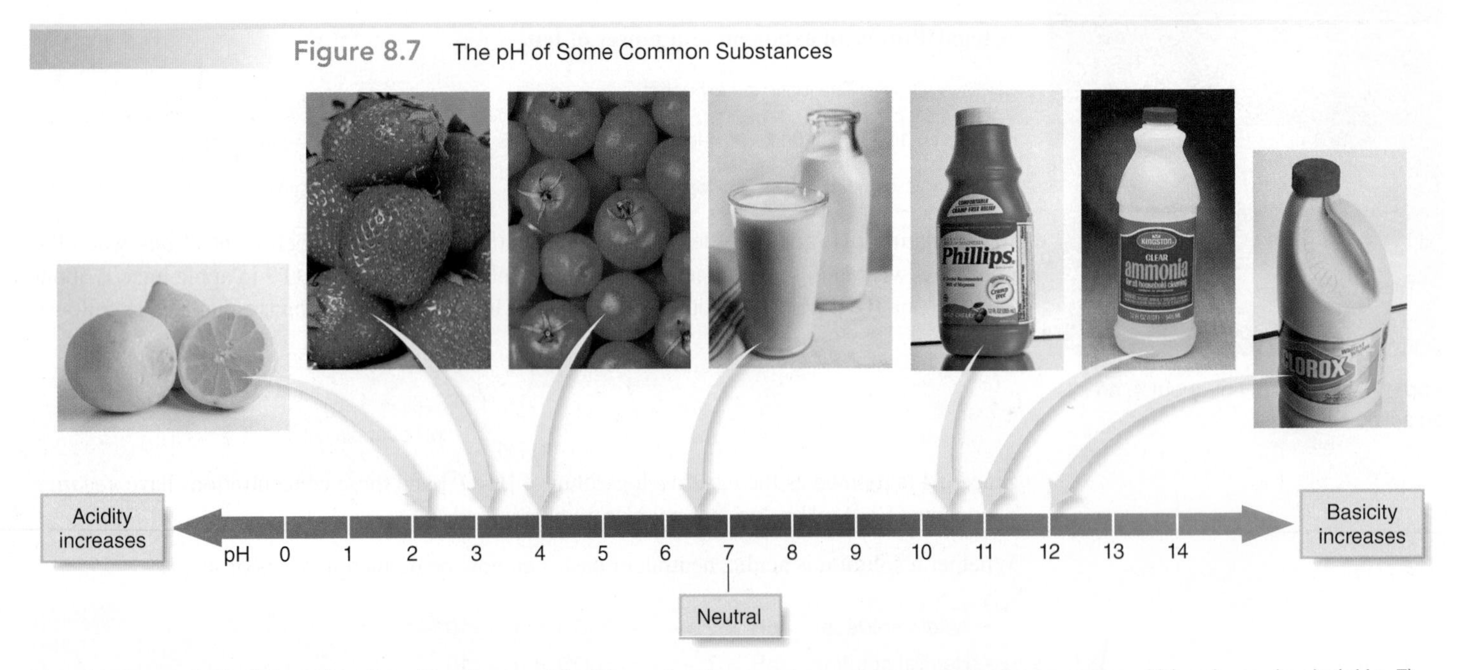

The pH of many fruits is less than 7, making them acidic. Many cleaning agents, such as household ammonia and bleach, are basic (pH > 7).

SAMPLE PROBLEM 8.11

What is the pH of a urine sample that has an H_3O^+ concentration of 1×10^{-5} M? Classify the solution as acidic, basic, or neutral.

Analysis

Use the formula pH = $-\log [H_3O^+]$. When the coefficient of a number written in scientific notation is one, the pH equals the value *x* in 10^{-x}.

Solution

$$\text{pH} = -\log [H_3O^+] = -\log(10^{-5})$$
$$= -(-5) = 5 \quad \text{pH of urine sample}$$

Answer

The urine sample is acidic since the pH < 7.

PROBLEM 8.19

Convert each H_3O^+ concentration to a pH value.

a. 1×10^{-6} M b. 1×10^{-12} M c. 0.000 01 M d. 0.000 000 000 01 M

SAMPLE PROBLEM 8.12

What is the H_3O^+ concentration in lemon juice that has a pH of about 2? Classify the solution as acidic, basic, or neutral.

Analysis

To find $[H_3O^+]$ from a pH, which is logarithm, we must determine what number corresponds to the given logarithm. When the pH is a whole number *x,* the value of *x* becomes the exponent in the expression $1 \times 10^{-x} = [H_3O^+]$.

Solution

If the pH of lemon juice is 2, $[H_3O^+] = 1 \times 10^{-2}$ M. Since the pH is less than 7, the lemon juice is acidic.

PROBLEM 8.20

What H_3O^+ concentration corresponds to each pH value: (a) 13; (b) 7; (c) 3? Label the solution as acidic, basic, or neutral.

8.5B Calculating pH Using a Calculator

Determining logarithms and antilogarithms using an electronic calculator is shown in the Appendix.

To calculate the pH of a solution in which the hydronium ion concentration has a coefficient in scientific notation that is *not* equal to one—as in 2.0×10^{-3}—you need a calculator that has a log function. How the keys are labeled and the order of the steps depends on your particular calculator.

Enter this number on your calculator.

$$\text{pH} = -\log [H_3O^+] = -\log(2.0 \times 10^{-3})$$
$$= -(-2.70) = 2.70$$

Because seawater contains dissolved salts, its pH is 8.50, making it slightly basic, not neutral like pure water.

Similarly, when a reported pH is *not* a whole number—as in the pH = 8.50 for a sample of seawater—you need a calculator to calculate an *antilogarithm*—that is, the number that has a logarithm of 8.50. To make sure your calculation is correct, note that since the pH of seawater is between 8 and 9, the H_3O^+ concentration must be between 10^{-8} and 10^{-9}.

$$[H_3O^+] = \text{antilog}(-\text{pH}) = \text{antilog}(-8.50)$$
$$= 3.2 \times 10^{-9} \text{ M}$$

Care must be taken in keeping track of significant figures when using logarithms.

- **A logarithm has the same number of digits to the right of the decimal point as are contained in the coefficient of the original number.**

$$[H_3O^+] = 3.2 \times 10^{-9}\ M \qquad pH = 8.50$$

two significant figures — two digits after the decimal point

SAMPLE PROBLEM 8.13

What is the H_3O^+ concentration in sweat that has a pH of 5.8?

Analysis

Use a calculator to determine the antilogarithm of the negative of the pH; $[H_3O^+]$ = antilog(–pH).

Solution

The order of the steps in using an electronic calculator, as well as the labels on the calculator buttons, vary. In some cases it is possible to calculate $[H_3O^+]$ by the following steps: enter the pH value; press the change sign key; and press the *2nd* + *log* buttons. Since the pH has only one number to the right of the decimal point, the H_3O^+ concentration must have only one significant figure in its coefficient.

one digit to the right of the decimal point

$$[H_3O^+] = \text{antilog}(-\text{pH}) = \text{antilog}(-5.8)$$

$$[H_3O^+] = 2 \times 10^{-6}\ M$$

one significant figure

PROBLEM 8.21

What H_3O^+ concentration corresponds to each pH value: (a) 10.2; (b) 7.8; (c) 4.3?

SAMPLE PROBLEM 8.14

What is the pH of wine that has an H_3O^+ concentration of 3.2×10^{-4} M?

Analysis

Use a calculator to determine the logarithm of a number that contains a coefficient in scientific notation that is not a whole number; pH = –log $[H_3O^+]$.

Solution

The order of the steps in using an electronic calculator, as well as the labels on the calculator buttons, vary. In some cases it is possible to calculate the pH by following three steps: enter the number (H_3O^+ concentration); press the *log* button; and press the change sign key. Consult your calculator manual if these steps do not give the desired value. Because the coefficient in the original number had two significant figures, the pH must have two digits to the right of the decimal point.

two significant figures

$$pH = -\log [H_3O^+] = -\log(3.2 \times 10^{-4})$$

$$= -(-3.49) = 3.49$$

two digits to the right of the decimal point

PROBLEM 8.22

Convert each H_3O^+ concentration to a pH value.

a. 1.8×10^{-6} M b. 9.21×10^{-12} M c. 0.000 088 M d. 0.000 000 000 076 2 M

8.5C FOCUS ON THE HUMAN BODY
The pH of Body Fluids

The human body contains fluids that vary in pH as shown in Figure 8.8. While saliva is slightly acidic, the gastric juice in the stomach has the lowest pH found in the body. The strongly acidic environment of the stomach aids in the digestion of food. It also kills many types of bacteria that might be inadvertently consumed along with food and drink. When food leaves the stomach, it passes to the basic environment of the small intestines. Bases in the small intestines react with acid from the stomach.

The pH of some body fluids must occupy a very narrow range. For example, a healthy individual has a blood pH in the range of 7.35–7.45. Maintaining this pH is accomplished by a complex mechanism described in Section 8.9. The pH of other fluids can be more variable. Urine has a pH anywhere from 4.6–8.0, depending on an individual's recent diet and exercise.

Figure 8.8

Variation in pH Values in the Human Body

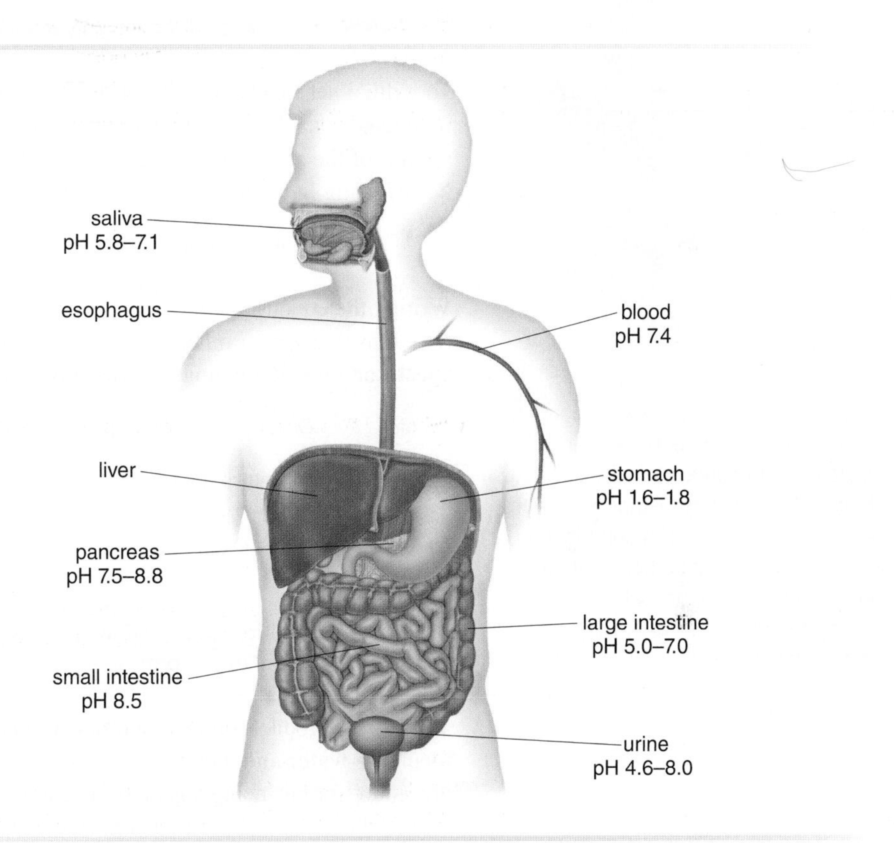

8.6 Common Acid–Base Reactions

Although we have already seen a variety of acid–base reactions in Sections 8.2–8.3, two common reactions deserve additional attention—reaction of acids with hydroxide bases (OH^-), and reaction of acids with bicarbonate (HCO_3^-) or carbonate (CO_3^{2-}).

8.6A Reaction of Acids with Hydroxide Bases

The reaction of a Brønsted–Lowry acid (HA) with the metal salt of a hydroxide base (MOH) is an example of a ***neutralization*** **reaction—an acid–base reaction that produces a salt and water as products.**

$$\underset{\text{acid}}{HA(aq)} + \underset{\text{base}}{MOH(aq)} \longrightarrow \underset{\text{water}}{H{-}OH(l)} + \underset{\text{salt}}{MA(aq)}$$

- **The acid HA donates a proton (H^+) to the OH^- base to form H_2O.**
- **The anion A^- from the acid combines with the cation M^+ from the base to form the salt MA.**

For example, hydrochloric acid, HCl, reacts with sodium hydroxide, NaOH, to form water and sodium chloride, NaCl.

$$\underset{\text{acid}}{HCl(aq)} + \underset{\text{base}}{NaOH(aq)} \longrightarrow \underset{\text{water}}{H{-}OH(l)} + \underset{\text{salt}}{NaCl(aq)}$$

HEALTH NOTE

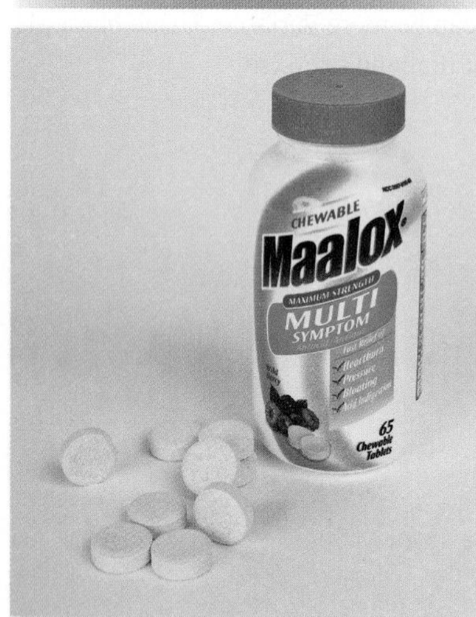

The antacid products Maalox and Mylanta both contain two bases—$Mg(OH)_2$ and $Al(OH)_3$—that react with excess stomach acid. A combination of bases is used so that the constipating effect of the aluminum salt is counteracted by the laxative effect of the magnesium salt.

The important reacting species in this reaction are H^+ from the acid HCl and OH^- from the base NaOH. To more clearly see the acid–base reaction, we can write an equation that contains only the species that are actually involved in the reaction. Such an equation is called a **net ionic equation.**

- A *net ionic equation* contains only the species involved in a reaction.

To write a net ionic equation for an acid–base reaction, we first write the acid, base, and salt as individual ions in solution. This process is simplified if we use H^+ (not H_3O^+) as the reacting species of the acid, since it is the H^+ ion that is transferred to the base. The reaction of HCl with NaOH using individual ions is then drawn as:

$$H^+(aq) + Cl^-(aq) + Na^+(aq) + OH^-(aq) \longrightarrow H{-}OH(l) + Na^+(aq) + Cl^-(aq)$$

Writing the equation in this manner shows that the Na^+ and Cl^- ions are unchanged in the reaction. Ions that appear on both sides of an equation but undergo no change in a reaction are called **spectator ions.** Removing the spectator ions from the equation gives the net ionic equation.

$$H^+(aq) + \cancel{Cl^-}(aq) + \cancel{Na^+}(aq) + OH^-(aq) \longrightarrow H{-}OH(l) + \cancel{Na^+}(aq) + \cancel{Cl^-}(aq)$$

Omit the spectator ions. Omit the spectator ions.

Net ionic equation $H^+(aq) + OH^-(aq) \longrightarrow H{-}OH(l)$

- **Whenever a strong acid and strong base react, the net ionic equation is always the same—H^+ reacts with OH^- to form H_2O.**

To draw the products of these neutralization reactions, keep in mind that **two products are always formed—water and a metal salt.** Balancing an acid–base equation can be done with the stepwise procedure for balancing a general reaction outlined in Section 5.2. The coefficients in a balanced chemical equation illustrate that one H^+ ion is always needed to react with each OH^- anion.

How To Draw a Balanced Equation for a Neutralization Reaction Between HA and MOH

Example Write a balanced equation for the reaction of $Mg(OH)_2$, an active ingredient in the antacid product Maalox, with the hydrochloric acid (HCl) in the stomach.

Step [1] Identify the acid and base in the reactants and draw H_2O as one product.

- HCl is the acid and $Mg(OH)_2$ is the base. H^+ from the acid reacts with OH^- from the base to form H_2O.

$$\underset{\text{acid}}{HCl(aq)} + \underset{\text{base}}{Mg(OH)_2(aq)} \longrightarrow H_2O(l) + \text{salt}$$

Step [2] Determine the structure of the salt formed as product.

- The salt is formed from the elements of the acid and base that are *not* used to form H_2O. The anion of the salt comes from the acid and the cation of the salt comes from the base.
- In this case, Cl^- (from HCl) and Mg^{2+} [from $Mg(OH)_2$] combine to form the salt $MgCl_2$.

How To, continued . . .

Step [3] Balance the equation.

- Follow the procedure in Section 5.2 to balance an equation. The balanced equation shows that *two* moles of HCl are needed for *each* mole of $Mg(OH)_2$, since each mole of $Mg(OH)_2$ contains two moles of OH^-.

$$2\,HCl(aq) + Mg(OH)_2(aq) \longrightarrow 2\,H_2O(l) + MgCl_2$$

Place a 2 to balance Cl. Place a 2 to balance H and O.

PROBLEM 8.23

Write a balanced equation for each acid–base reaction.

a. $HNO_3(aq) + NaOH(aq) \longrightarrow$

b. $H_2SO_4(aq) + KOH(aq) \longrightarrow$

PROBLEM 8.24

Write the net ionic equation for each reaction in Problem 8.23.

8.6B Reaction of Acids with Bicarbonate and Carbonate

Acids react with the bases bicarbonate (HCO_3^-) and carbonate (CO_3^{2-}). A bicarbonate base reacts with *one* proton to form carbonic acid, H_2CO_3. A carbonate base reacts with *two* protons. The carbonic acid formed in these reactions is unstable and decomposes to form CO_2 and H_2O. Thus, when an acid reacts with either base, bubbles of CO_2 gas are given off.

$$H^+(aq) + HCO_3^-(aq) \longrightarrow [H_2CO_3(aq)] \longrightarrow H_2O(l) + CO_2(g)$$

1 H^+ needed; bicarbonate

$$2\,H^+(aq) + CO_3^{2-}(aq) \longrightarrow [H_2CO_3(aq)] \longrightarrow H_2O(l) + CO_2(g)$$

2 H^+ needed; carbonate; bubbles of CO_2

Sodium bicarbonate ($NaHCO_3$), an ingredient in the over-the-counter antacid Alka-Seltzer, is the metal salt of a bicarbonate base that reacts with excess stomach acid, releasing CO_2. Like the neutralization reactions in Section 8.6A, a salt, NaCl, is formed in which the cation (Na^+) comes from the base and the anion (Cl^-) comes from the acid.

$$\underset{acid}{HCl(aq)} + \underset{base}{NaHCO_3(aq)} \longrightarrow \underset{salt}{NaCl(aq)} + H_2CO_3(aq)$$

$$H_2CO_3(aq) \longrightarrow H_2O(l) + CO_2(g)$$

Calcium carbonate ($CaCO_3$), a calcium supplement and antacid in Tums, also reacts with excess stomach acid with release of CO_2. Since each carbonate ion reacts with two protons, the balanced equation shows a 2:1 ratio of HCl to $CaCO_3$.

$$\underset{acid}{2\,HCl(aq)} + \underset{base}{CaCO_3(aq)} \longrightarrow \underset{salt}{CaCl_2(aq)} + H_2CO_3(aq)$$

$$H_2CO_3(aq) \longrightarrow H_2O(l) + CO_2(g)$$

HEALTH NOTE

Like taking other over-the-counter medications, care must be exercised when using antacids. Ingestion of large amounts of $CaCO_3$ can increase the incidence of kidney stones.

SAMPLE PROBLEM 8.15

Write a balanced equation for the reaction of H_2SO_4 with $NaHCO_3$.

Analysis

The acid and base react to form a salt and carbonic acid (H_2CO_3), which decomposes to CO_2 and H_2O.

Solution

H_2SO_4 is the acid and $NaHCO_3$ is the base. H^+ from the acid reacts with HCO_3^- from the base to give H_2CO_3, which decomposes to H_2O and CO_2. A salt (Na_2SO_4) is also formed from the cation of the base (Na^+) and the anion of the acid (SO_4^{2-}).

Unbalanced equation: $$\underset{\text{acid}}{H_2SO_4(aq)} + \underset{\text{base}}{NaHCO_3(aq)} \longrightarrow \underset{\text{salt}}{Na_2SO_4(aq)} + \underbrace{H_2O(l) + CO_2(g)}_{\text{from } H_2CO_3}$$

To balance the equation, place coefficients so the number of atoms on both sides of the arrow is the same.

Place a 2 to balance Na...

$$\underset{\text{acid}}{H_2SO_4(aq)} + \underset{\text{base}}{2\ NaHCO_3(aq)} \longrightarrow \underset{\text{salt}}{Na_2SO_4(aq)} + 2\ H_2O(l) + 2\ CO_2(g)$$

...then place 2's to balance C, H, and O.

PROBLEM 8.25

The acid in acid rain is generally sulfuric acid (H_2SO_4). When this rainwater falls on statues composed of marble ($CaCO_3$), the H_2SO_4 slowly dissolves the $CaCO_3$. Write a balanced equation for this acid–base reaction.

PROBLEM 8.26

Write a balanced equation for the reaction of nitric acid (HNO_3) with each base: (a) $NaHCO_3$; (b) $MgCO_3$.

8.7 Titration

Sometimes it is necessary to know the exact concentration of acid or base in a solution. To determine the molarity of a solution, we carry out a **titration.** A titration uses a *buret,* a calibrated tube with a stopcock at the bottom that allows a solution of known molarity to be added in small quantities to a solution of unknown molarity. The procedure for determining the total acid concentration of a solution of HCl is illustrated in Figure 8.9.

How does a titration tell us the concentration of an HCl solution? A titration is based on the acid–base reaction that occurs between the acid in the flask (HCl) and the base that is added (NaOH). **When the number of moles of base added equals the number of moles of acid in the flask, the acid is *neutralized,* forming a salt and water.**

At the end point, the number of moles of the acid (H^+) and the base (OH^-) are equal.

$$HCl(aq) + NaOH(aq) \longrightarrow NaCl(aq) + H_2O(aq)$$

To determine an unknown molarity from titration data requires three operations.

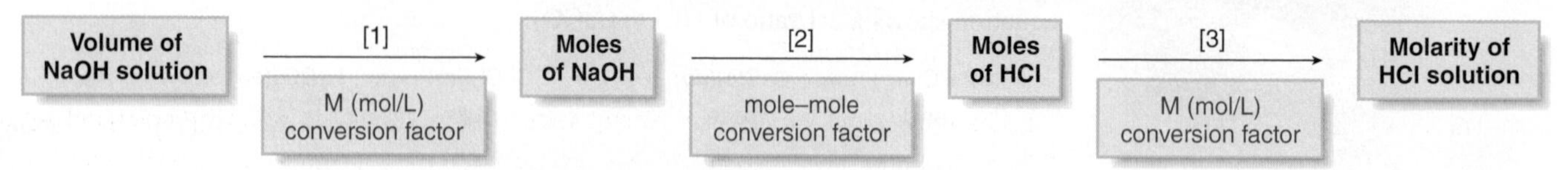

First, we determine the number of moles of base added using its known molarity and volume. Then we use coefficients in the balanced acid–base equation to tell us the number of moles of acid that react with the base. Finally, we determine the molarity of the acid from the calculated number of moles and the known volume of the acid.

Figure 8.9 Titration of an Acid with a Base of Known Concentration

a.

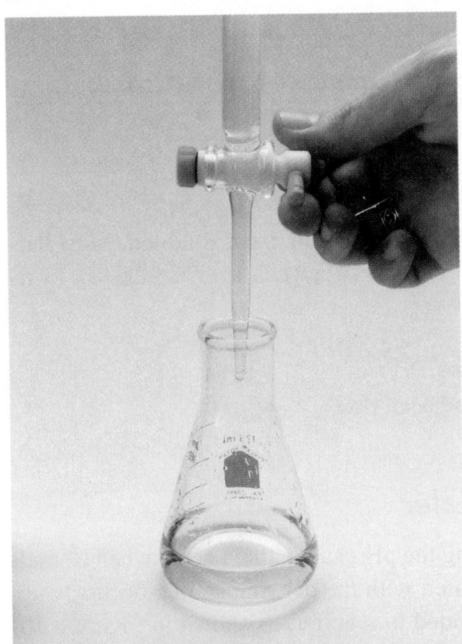

b.

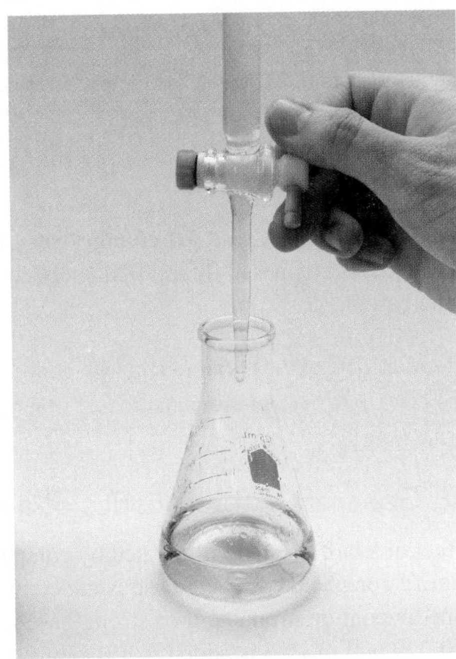

c.

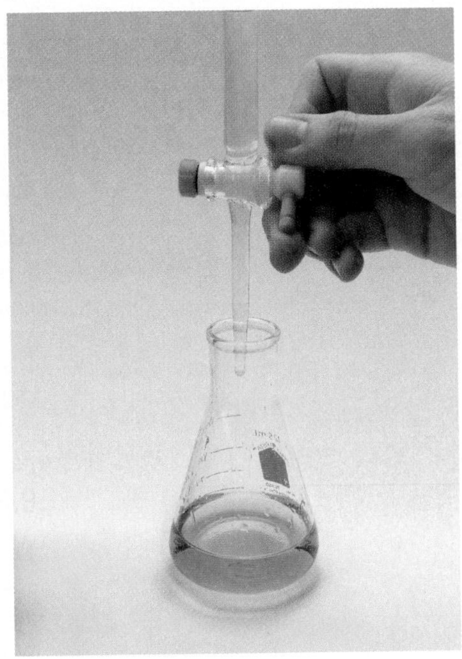

Steps in determining the molarity of a solution of HCl:

a. Add a measured volume of HCl solution to a flask. Add an acid–base indicator, often phenolphthalein, which is colorless in acid but turns bright pink in base.

b. Fill a buret with an NaOH solution of known molarity and slowly add it to the HCl solution.

c. Add NaOH solution until the *end point* is reached, the point at which the indicator changes color. At the end point, the **number of moles of NaOH added equals the number of moles of HCl** in the flask. In other words, all of the HCl has reacted with NaOH and the solution is no longer acidic. Read the volume of NaOH solution added from the buret. Using the known volume and molarity of the NaOH solution and the known volume of HCl solution, the molarity of the HCl solution can be calculated.

SAMPLE PROBLEM 8.16

What is the molarity of an HCl solution if 22.5 mL of a 0.100 M NaOH solution are needed to titrate 25.0 mL of the sample? The balanced equation for this acid–base reaction is given.

$$\text{HCl}(aq) + \text{NaOH}(aq) \longrightarrow \text{NaCl}(aq) + \text{H}_2\text{O}(l)$$

Analysis and Solution

[1] Determine the number of moles of base used to neutralize the acid.

- Use the molarity (M) and volume (V) of the base to calculate the number of moles (mol = MV).

$$22.5\ \cancel{\text{mL}}\ \text{NaOH} \times \frac{1\ \cancel{\text{L}}}{1000\ \cancel{\text{mL}}} \times \frac{0.100\ \text{mol NaOH}}{1\ \cancel{\text{L}}} = 0.002\ 25\ \text{mol NaOH}$$

[2] Determine the number of moles of acid that react from the balanced chemical equation.

- One mole of HCl reacts with one mole of NaOH, so the number of moles of NaOH equals the number of moles of HCl at the end point.

$$0.002\ 25\ \text{mol NaOH} \times \frac{1\ \text{mol HCl}}{1\ \text{mol NaOH}} = 0.002\ 25\ \text{mol HCl}$$

[3] Determine the molarity of the acid from the number of moles and known volume.

$$\underset{\text{molarity}}{\text{M}} = \frac{\text{mol}}{\text{L}} = \frac{0.002\ 25\ \text{mol HCl}}{25.0\ \cancel{\text{mL}}\ \text{solution}} \times \frac{1000\ \cancel{\text{mL}}}{1\ \text{L}} = \underset{\textbf{Answer}}{0.0900\ \text{M HCl}}$$

PROBLEM 8.27

What is the molarity of an HCl solution if 25.5 mL of a 0.24 M NaOH solution are needed to neutralize 15.0 mL of the sample?

PROBLEM 8.28

How many milliliters of 2.0 M NaOH are needed to neutralize 5.0 mL of a 6.0 M H_2SO_4 solution?

8.8 Buffers

A *buffer* is a solution whose pH changes very little when acid or base is added. Most buffers are solutions composed of approximately equal amounts of a weak acid and the salt of its conjugate base.

- The weak acid of the buffer reacts with added base, OH^-.
- The conjugate base of the buffer reacts with added acid, H_3O^+.

8.8A General Characteristics of a Buffer

The effect of a buffer can be illustrated by comparing the pH change that occurs when a small amount of strong acid or strong base is added to water, with the pH change that occurs when the same amount of strong acid or strong base is added to a buffer, as shown in Figure 8.10. When 0.020 mol of HCl is added to 1.0 L of water, the pH changes from 7 to 1.7, and when 0.020 mol of NaOH is added to 1.0 L of water, the pH changes from 7 to 12.3. In this example, addition of a small quantity of a strong acid or strong base to neutral water changes the pH by over 5 pH units.

In contrast, a buffer prepared from 0.50 M acetic acid (CH_3COOH) and 0.50 M sodium acetate ($NaCH_3COO$) has a pH of 4.74. Addition of the same quantity of acid, 0.020 mol HCl, changes the pH to 4.70, and addition of the same quantity of base, 0.020 mol of NaOH, changes the pH to 4.77. In this example, the change of pH in the presence of the buffer is no more than 0.04 pH units!

Why is a buffer able to absorb acid or base with very little pH change? Let's use as an example a buffer that contains equal concentrations of acetic acid (CH_3COOH), and the sodium salt of its conjugate base, sodium acetate ($NaCH_3COO$). CH_3COOH is a weak acid, so when it dissolves in water, only a small fraction dissociates to form its conjugate base CH_3COO^-. In the buffer solution, however, the sodium acetate provides an equal amount of the conjugate base.

$$CH_3COOH(aq) + H_2O(l) \rightleftharpoons H_3O^+(aq) + CH_3COO^-(aq)$$

approximately equal amounts

Suppose a small amount of strong acid is added to the buffer. Added H_3O^+ reacts with CH_3COO^- to form CH_3COOH, so that $[CH_3COO^-]$ decreases slightly and $[CH_3COOH]$ increases slightly, but the $[H_3O^+]$ and therefore the pH change only slightly.

Adding H_3O^+...

This increases slightly.

$$CH_3COOH(aq) + H_2O(l) \rightleftharpoons H_3O^+(aq) + CH_3COO^-(aq)$$

This decreases slightly.

...drives the reaction to the left.

Figure 8.10
The Effect of a Buffer on pH Changes

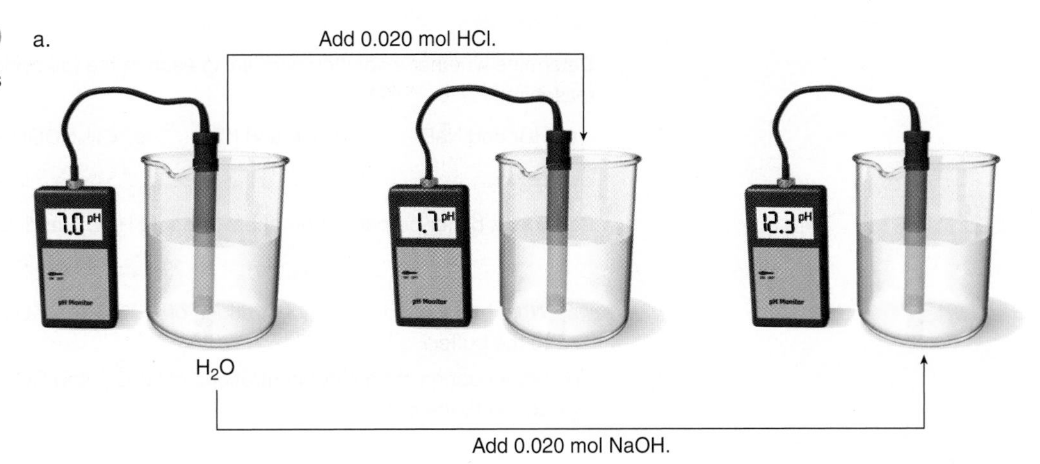

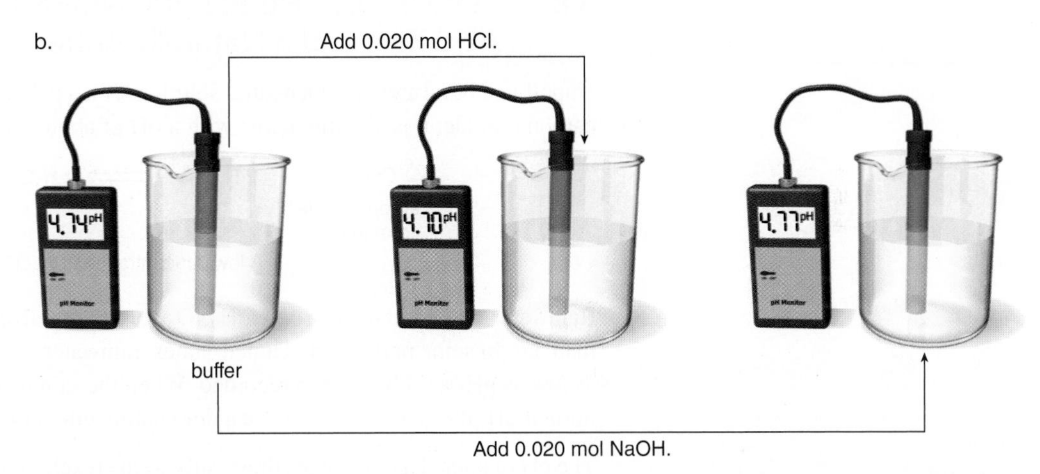

a. The pH of pure water changes drastically when a small amount of strong acid or strong base is added.

b. The pH of a buffer changes very little when the same amount of strong acid or strong base is added.

On the other hand, if a small amount of strong base is added to the buffer, OH^- reacts with CH_3COOH to form CH_3COO^-, so that $[CH_3COOH]$ decreases slightly and $[CH_3COO^-]$ increases slightly but the $[H_3O^+]$ and therefore the pH change only slightly.

Adding OH^-...

This decreases slightly.

$$CH_3COOH(aq) + OH^-(aq) \rightleftharpoons H_2O(l) + CH_3COO^-(aq)$$

This increases slightly.

...drives the reaction to the right.

For a buffer to be effective, the amount of added acid or base must be small compared to the amount of buffer present. When a large amount of acid or base is added to a buffer, the H_3O^+ concentration and therefore the pH change a great deal.

PROBLEM 8.29

Determine whether a solution containing each of the following substances is a buffer. Explain your reasoning.

a. HBr and NaBr b. HF and KF c. CH_3COOH alone

PROBLEM 8.30

Consider a buffer prepared from the weak acid HCO_3^- and its conjugate base CO_3^{2-}.

$$HCO_3^-(aq) + H_2O(l) \rightleftharpoons CO_3^{2-}(aq) + H_3O^+(aq)$$

a. What happens to the concentrations of HCO_3^- and CO_3^{2-} when a small amount of acid is added to the buffer?

b. What happens to the concentrations of HCO_3^- and CO_3^{2-} when a small amount of base is added to the buffer?

8.8B FOCUS ON THE ENVIRONMENT Acid Rain and a Naturally Buffered Lake

Unpolluted rainwater is not a neutral solution with a pH of 7; rather, because it contains dissolved carbon dioxide, it is slightly acidic with a pH of about 5.6.

$$\underset{\text{carbon dioxide from the air}}{CO_2(g)} + 2\,H_2O(l) \rightleftharpoons H_3O^+(aq) + HCO_3^-(aq)$$

A low concentration of H_3O^+ gives rainwater a pH < 7.

Rainwater that contains dissolved H_2SO_4 (or HNO_3) from burning fossil fuels has a pH lower than 5.6. In some parts of the United States, rainwater often has a pH range of 4–5, and readings as low as pH = 1.8 have been recorded. When the rain in a region consistently has a lower-than-normal pH, this acid rain can have a devastating effect on plant and animal life.

The pH of some lakes changes drastically as the result of acid rain, whereas the pH of other lakes does not. In fact, the ability of some lakes to absorb acid rain without much pH change is entirely due to buffers (Figure 8.11). Lakes that are surrounded by limestone-rich soil are in contact with

Figure 8.11 Acid Rain and a Naturally Buffered Lake

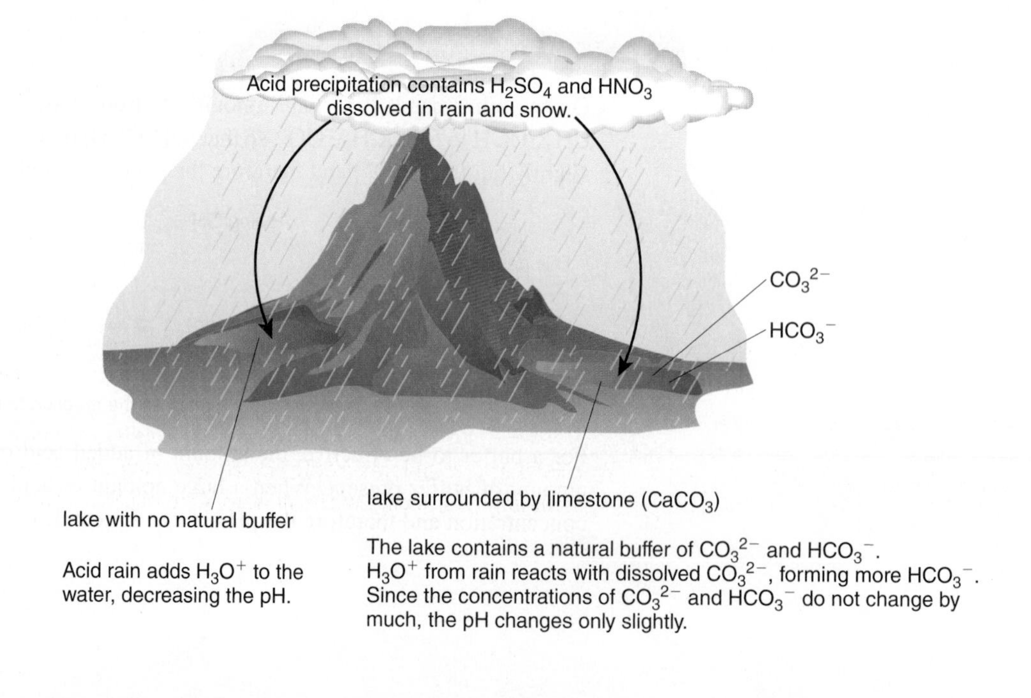
Acid precipitation contains H2SO4 and HNO3 dissolved in rain and snow.
CO3 2−
HCO3 −
lake with no natural buffer
Acid rain adds H3O+ to the water, decreasing the pH.
lake surrounded by limestone (CaCO3)
The lake contains a natural buffer of CO3 2− and HCO3 −. H3O+ from rain reacts with dissolved CO3 2−, forming more HCO3 −. Since the concentrations of CO3 2− and HCO3 − do not change by much, the pH changes only slightly.

solid calcium carbonate, $CaCO_3$. As a result, the lake contains a natural carbonate/bicarbonate buffer. When acid precipitation falls on the lake, the dissolved carbonate (CO_3^{2-}) reacts with the acid to form bicarbonate (HCO_3^-).

The buffer reacts with added acid from rain.

$$CO_3^{2-}(aq) + H_3O^+(aq) \rightleftharpoons HCO_3^-(aq) + H_2O(l)$$

A lake surrounded by limestone contains a natural CO_3^{2-}/HCO_3^- buffer.

The carbonate/bicarbonate buffer thus allows the lake to resist large pH changes when acid is added. In some areas acidic lakes have been treated with limestone, thus adding calcium carbonate to neutralize the acid and restore the natural pH. This procedure is expensive and temporary because with time and more acid rain, the pH of the lakes decreases again.

HEALTH NOTE

Individuals with cystic fibrosis, the most common genetic disease in Caucasians, produce thick mucus in the lungs, resulting in a higher-than-normal level of CO_2 and respiratory acidosis.

HEALTH NOTE

During strenuous exercise, the lungs expel more CO_2 than usual and the pH of the blood increases.

8.9 FOCUS ON THE HUMAN BODY Buffers in the Blood

The normal blood pH of a healthy individual is in the range of 7.35 to 7.45. A pH above or below this range is generally indicative of an imbalance in respiratory or metabolic processes. The body is able to maintain a very stable pH because the blood and other tissues are buffered. The principal buffer in the blood is carbonic acid/bicarbonate (H_2CO_3/HCO_3^-).

In examining the carbonic acid/bicarbonate buffer system in the blood, two reactions are important. First of all, carbonic acid (H_2CO_3) is formed from CO_2 dissolved in the bloodstream (Section 8.6). Second, since carbonic acid is a weak acid, it is also dissociated in water to form its conjugate base, bicarbonate (HCO_3^-). Bicarbonate is also generated in the kidneys.

$$CO_2(g) + H_2O(l) \rightleftharpoons \underset{\text{carbonic acid}}{H_2CO_3(aq)} \overset{H_2O}{\rightleftharpoons} H_3O^+(aq) + \underset{\text{bicarbonate}}{HCO_3^-(aq)}$$

principal buffer in the blood

CO_2 is constantly produced by metabolic processes in the body and then transported to the lungs to be eliminated. Le Châtelier's principle explains the effect of increasing or decreasing the level of dissolved CO_2 on the pH of the blood. A higher-than-normal CO_2 concentration shifts the equilibrium to the right, increasing the H_3O^+ concentration and lowering the pH. **Respiratory acidosis** results when the body fails to eliminate adequate amounts of CO_2 through the lungs. This may occur in patients with advanced lung disease or respiratory failure.

A lower respiratory rate increases $[CO_2]$.

$$CO_2(g) + 2\,H_2O(g) \rightleftharpoons H_3O^+(aq) + HCO_3^-(aq)$$

This drives the reaction to the right, increasing $[H_3O^+]$.
Blood has a higher $[H_3O^+]$ ⟶ lower pH

A lower-than-normal CO_2 concentration shifts the equilibrium to the left, decreasing the H_3O^+ concentration and raising the pH. **Respiratory alkalosis** is caused by hyperventilation, very rapid breathing that occurs when an individual experiences excitement or panic.

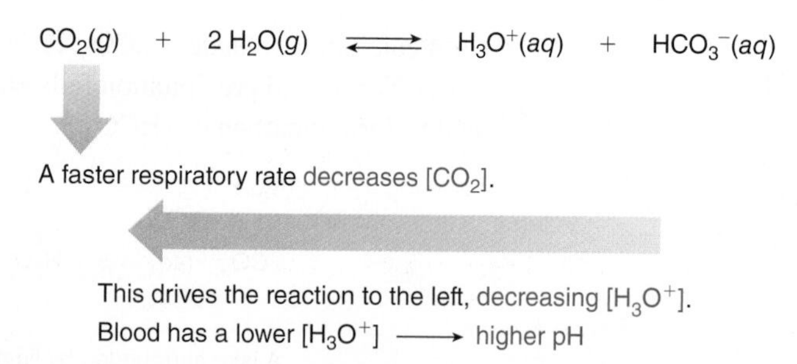

$CO_2(g)$ + $2 H_2O(g)$ ⇌ $H_3O^+(aq)$ + $HCO_3^-(aq)$
A faster respiratory rate decreases $[CO_2]$.
This drives the reaction to the left, decreasing $[H_3O^+]$.
Blood has a lower $[H_3O^+]$ ⟶ higher pH

The pH of the blood may also be altered when the metabolic processes of the body are not in balance. **Metabolic acidosis** results when excessive amounts of acid are produced and the blood pH falls. This may be observed in patients with severe infections (sepsis). It may also occur in poorly controlled diabetes. **Metabolic alkalosis** may occur when recurrent vomiting decreases the amount of acid in the stomach, thus causing a rise in pH.

KEY TERMS

Acid (8.1)
Acidic solution (8.4)
Amphoteric (8.2)
Base (8.1)
Basic solution (8.4)
Brønsted–Lowry acid (8.1)
Brønsted–Lowry base (8.1)
Buffer (8.8)
Conjugate acid (8.2)
Conjugate acid–base pair (8.2)
Conjugate base (8.2)
Dissociation (8.3)
Ion–product constant (8.4)
Net ionic equation (8.6)
Neutral solution (8.4)
Neutralization reaction (8.6)
pH scale (8.5)
Proton transfer reaction (8.2)
Spectator ion (8.6)
Titration (8.7)

KEY CONCEPTS

1 Describe the principal features of acids and bases. (8.1)

- A Brønsted–Lowry acid is a proton donor, often symbolized by HA. A Brønsted–Lowry acid must contain one or more hydrogen atoms.
- A Brønsted–Lowry base is a proton acceptor, often symbolized by B:. To form a bond to a proton, a Brønsted–Lowry base must contain a lone pair of electrons.

2 What are the principal features of an acid–base reaction? (8.2)

- In a Brønsted–Lowry acid–base reaction, a proton is transferred from the acid (HA) to the base (B:). In this reaction, the acid loses a proton to form its conjugate base ($A:^-$) and the base gains a proton to form its conjugate acid (HB^+).

3 What are the characteristics of a strong acid and a strong base? (8.3)

- A strong acid readily donates a proton, and when dissolved in water, 100% of the acid dissociates into ions, forming H_3O^+. A strong base readily accepts a proton, and when dissolved in water, 100% of the base dissociates into ions, forming OH^-.
- An inverse relationship exists between acid and base strength. A strong acid forms a weak conjugate base, whereas a weak acid forms a strong conjugate base.

4 What is the ion–product of water and how is it used to calculate hydronium or hydroxide ion concentration? (8.4)

- The ion–product of water, K_w, is a constant for all aqueous solutions; $K_w = [H_3O^+][OH^-] = 1.0 \times 10^{-14}$ at 25 °C. If either $[H_3O^+]$ or $[OH^-]$ is known, the other value can be calculated from K_w.

5 What is pH? (8.5)

- The pH of a solution measures the concentration of H_3O^+; $pH = -\log [H_3O^+]$.
- A pH = 7 means $[H_3O^+] = [OH^-]$ and the solution is neutral.
- A pH < 7 means $[H_3O^+] > [OH^-]$ and the solution is acidic.
- A pH > 7 means $[H_3O^+] < [OH^-]$ and the solution is basic.

6 Draw the products of some common acid–base reactions. (8.6)

- In a Brønsted–Lowry acid–base reaction with hydroxide bases (MOH), the acid HA donates a proton to OH^- to form H_2O. The anion from the acid HA combines with the cation M^+ of the base to form the salt MA. This reaction is called a neutralization reaction.
- In acid–base reactions with bicarbonate (HCO_3^-) or carbonate (CO_3^{2-}) bases, carbonic acid (H_2CO_3) is formed, which decomposes to form H_2O and CO_2.

7 How is a titration used to determine the concentration of an acid or base? (8.7)

- A titration is a procedure that uses a base (or acid) of known volume and molarity to react with a known volume of acid (or base) of unknown molarity. The volume and molarity of the base are used to calculate the number of moles of base that react, and from this value, the molarity of the acid can be determined.

8 What is a buffer? (8.8)

- A buffer is a solution whose pH changes very little when acid or base is added. Most buffers are composed of approximately equal amounts of a weak acid and the salt of its conjugate base.

9 What is the principal buffer present in the blood? (8.9)

- The principal buffer in the blood is carbonic acid/bicarbonate. Since carbonic acid (H_2CO_3) is formed from dissolved CO_2, the amount of CO_2 in the blood affects its pH, which is normally maintained in the range of 7.35–7.45. When the CO_2 concentration in the blood is higher than normal, more H_3O^+ is formed and the pH decreases. When the CO_2 concentration in the blood is lower than normal, $[H_3O^+]$ decreases and the pH increases.

UNDERSTANDING KEY CONCEPTS

Selected in-chapter and odd-numbered end-of-chapter problems have brief answers at the end of each chapter. The *Student Study Guide and Solutions Manual* contains detailed solutions to all in-chapter and odd-numbered end-of-chapter problems, as well as additional worked examples and a chapter self-test.

8.31 Draw the structure of the conjugate base of each acid.

a. 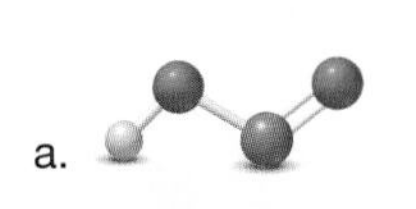

b. 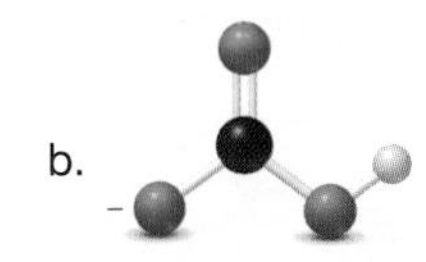

8.32 Draw the structure of the conjugate base of each acid.

a. 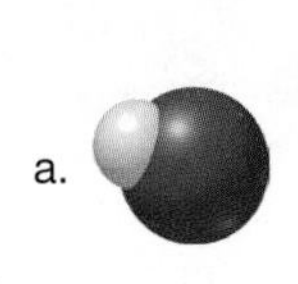

b. 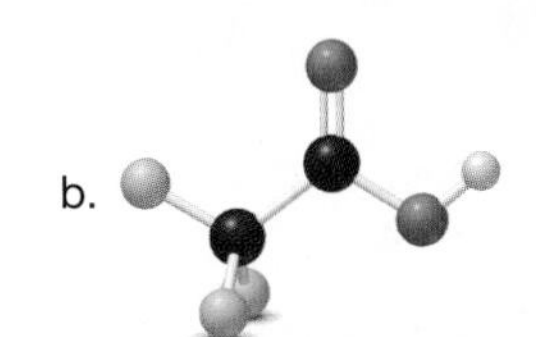

8.33 (a) Which of the following represents a strong acid HZ dissolved in water? (b) Which represents a weak acid HZ dissolved in water?

= HZ   = Z^-

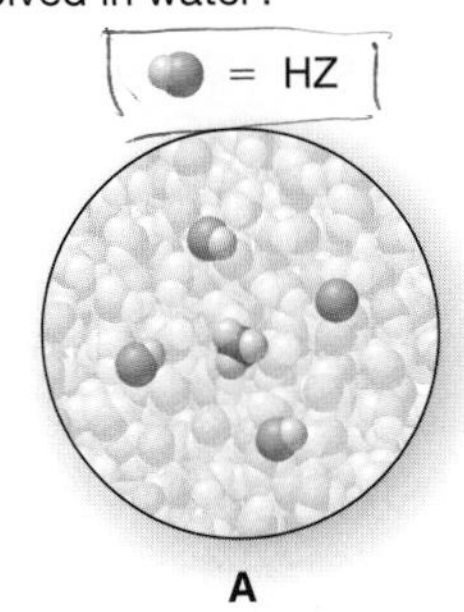

A

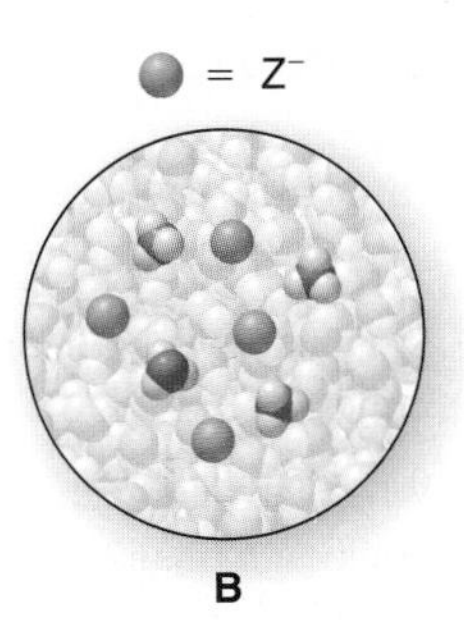

B

8.34 (a) Using molecular art, draw a diagram that represents an aqueous solution of a strong acid H_2Z. (b) Using molecular art, draw a diagram that represents an aqueous solution of a weak acid H_2Z.

= H_2Z = HZ^-

8.35 Identify the acid, base, conjugate acid, and conjugate base in each diagram. Gray spheres correspond to H atoms.

a.   

b. 

8.36 Use the given representations for H_2Z, HZ^-, and Z^{2-}, as well as the space-filling structures of H_2O, H_3O^+, and OH^- that appear in Chapter 8, to depict each equation. Label the acid, base, conjugate acid, and conjugate base in each equation.

= H_2Z = HZ^- = Z^{2-}

a. $H_2Z + H_2O \longrightarrow HZ^- + H_3O^+$
b. $HZ^- + OH^- \longrightarrow Z^{2-} + H_2O$

8.37 Why is NH_3 a Brønsted–Lowry base but CH_4 is not?

8.38 Why is HCl a Brønsted–Lowry acid but NaCl is not?

8.39 Label the acid in the reactants and the conjugate acid in the products in each reaction.

a. $H_3PO_4(aq) + CN^-(aq) \longrightarrow H_2PO_4^-(aq) + HCN(aq)$
b. $Br^-(aq) + HSO_4^-(aq) \longrightarrow SO_4^{2-}(aq) + HBr(aq)$

8.40 Label the acid in the reactants and the conjugate acid in the products in each reaction.

a. $HF(g) + NH_3(g) \longrightarrow NH_4^+(aq) + F^-(aq)$
b. $Br^-(aq) + H_2O(l) \longrightarrow HBr(aq) + OH^-(aq)$

8.41 If a urine sample has a pH of 5.90, calculate the concentrations of H_3O^+ and OH^- in the sample.

8.42 If pancreatic fluids have a pH of 8.2, calculate the concentrations of H_3O^+ and OH^- in the pancreas.

8.43 Marble statues, which are composed of calcium carbonate ($CaCO_3$), are slowly eaten away by the nitric acid (HNO_3) in acid rain. Write a balanced equation for the reaction of $CaCO_3$ with HNO_3.

8.44 Some liquid antacids contain suspensions of aluminum hydroxide [$Al(OH)_3$]. Write a balanced equation for the reaction of $Al(OH)_3$ with the HCl in stomach acid.

8.45 Consider a buffer prepared from the weak acid HNO_2 and its conjugate base NO_2^-. What happens to the concentrations of HNO_2 and NO_2^- when a small amount of acid is added to the buffer?

$$HNO_2(aq) + H_2O(l) \rightleftharpoons NO_2^-(aq) + H_3O^+(aq)$$

8.46 Referring to the equation in Problem 8.45, state what happens to the concentrations of HNO_2 and NO_2^- when a small amount of base is added to the buffer.

ADDITIONAL PROBLEMS

Acids and Bases

8.47 Which of the following species can be Brønsted–Lowry acids?

a. HBr c. $AlCl_3$ e. NO_2^-
b. Br_2 d. HCOOH f. HNO_2

8.48 Which of the following species can be Brønsted–Lowry acids?

a. H_2O c. HOCl e. CH_3CH_2COOH
b. I^- d. $FeBr_3$ f. CO_2

8.49 Which of the following species can be Brønsted–Lowry bases?

a. OH^- c. C_2H_6 e. OCl^-
b. Ca^{2+} d. PO_4^{3-} f. $MgCO_3$

8.50 Which of the following species can be Brønsted–Lowry bases?

a. Cl^- c. H_2O e. $Ca(OH)_2$
b. BH_3 d. Na^+ f. $HCOO^-$

8.51 Draw the conjugate acid of each base.

a. HS^- b. CO_3^{2-} c. NO_2^-

8.52 Draw the conjugate acid of each base.

a. Br^- b. HPO_4^{2-} c. CH_3COO^-

8.53 Draw the conjugate base of each acid.

a. HNO_2 b. NH_4^+ c. H_2O_2

8.54 Draw the conjugate base of each acid.

a. H_3O^+ b. H_2Se c. HSO_4^-

8.55 Label the conjugate acid–base pairs in each equation.

a. $HI(g) + NH_3(g) \longrightarrow NH_4^+(aq) + I^-(aq)$
b. $HCOOH(l) + H_2O(l) \longrightarrow H_3O^+(aq) + HCOO^-(aq)$
c. $HSO_4^-(aq) + H_2O(l) \longrightarrow H_2SO_4(aq) + OH^-(aq)$

8.56 Label the conjugate acid–base pairs in each equation.

a. $Cl^-(aq) + HSO_4^-(aq) \longrightarrow HCl(aq) + SO_4^{2-}(aq)$
b. $HPO_4^{2-}(aq) + OH^-(aq) \longrightarrow PO_4^{3-}(aq) + H_2O(l)$
c. $NH_3(g) + HF(g) \longrightarrow NH_4^+(aq) + F^-(aq)$

8.57 Like H_2O, HCO_3^- is amphoteric. (a) Draw the conjugate acid of HCO_3^-. (b) Draw the conjugate base of HCO_3^-.

8.58 Like H_2O, $H_2PO_4^-$ is amphoteric. (a) Draw the conjugate acid of $H_2PO_4^-$. (b) Draw the conjugate base of $H_2PO_4^-$.

Acid and Base Strength

8.59 Which diagram represents an aqueous solution of HF and which represents HCl? Explain your choice.

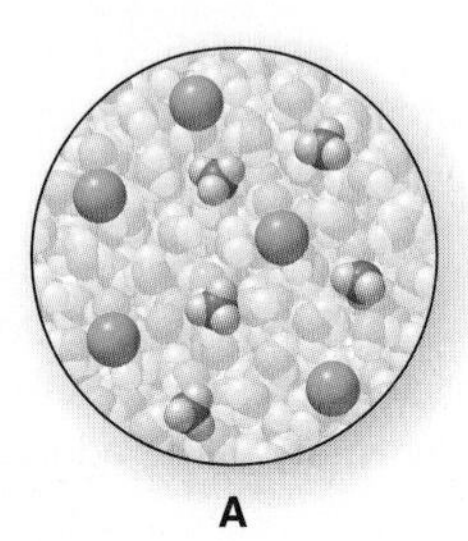
A

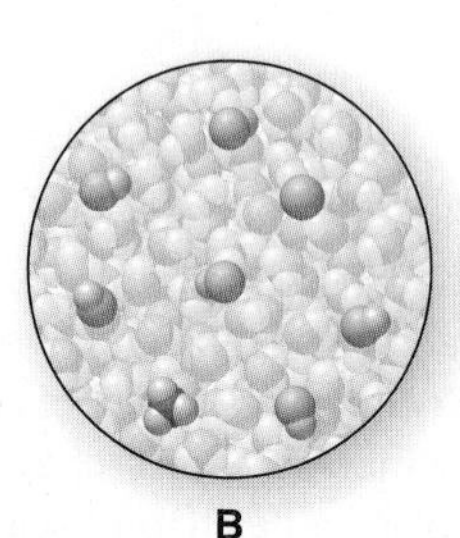
B

8.60 Which diagram represents what happens when HCN dissolves in water? Explain your choice.

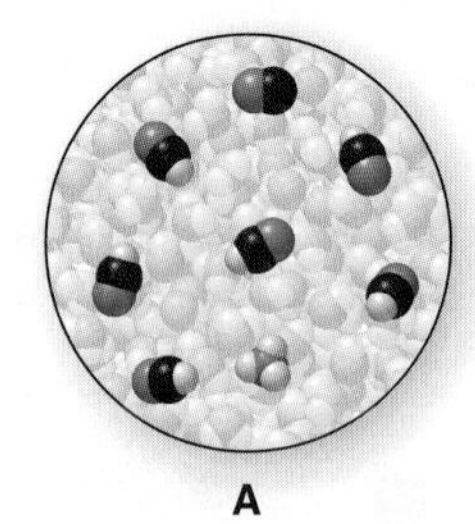
A

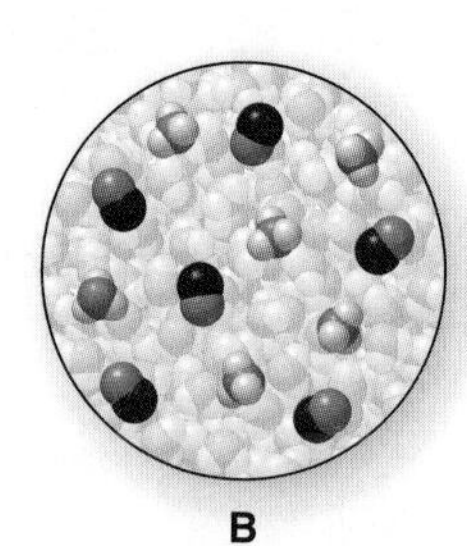
B

8.61 Use the data in Table 8.2 to label the stronger acid in each pair.

a. H_2O or CH_3COOH
b. H_3PO_4 or H_2SO_4

8.62 Use the data in Table 8.2 to label the stronger acid in each pair.

a. H_2SO_4 or NH_4^+
b. H_2O or HF

8.63 Which acid in each pair in Problem 8.61 has the stronger conjugate base?

8.64 Which acid in each pair in Problem 8.62 has the stronger conjugate base?

8.65 Which diagram illustrates the stronger acid?

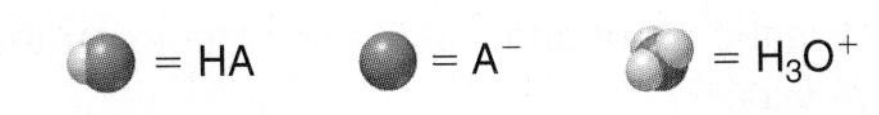

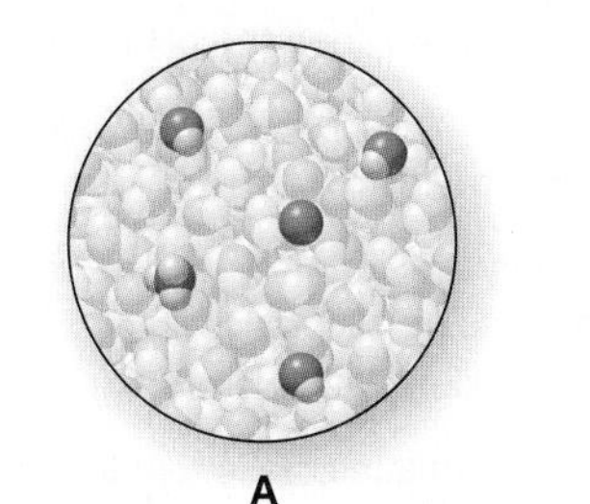

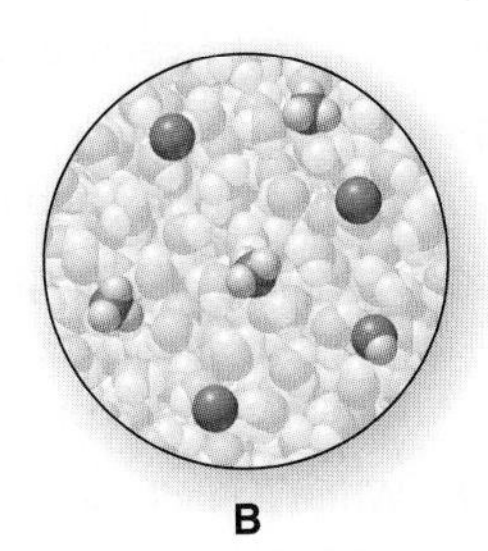

8.66 Which diagram in Problem 8.65 contains the stronger conjugate base? Explain your choice.

8.67 Which acid, **A** or **B,** is stronger in each part?
a. **A** dissociates to a greater extent in water.
b. The conjugate base of **A** is stronger than the conjugate base of **B.**

8.68 Which acid, **A** or **B,** is stronger in each part?
a. **B** dissociates to a greater extent in water.
b. The conjugate base of **B** is stronger than the conjugate base of **A.**

Water and the pH Scale

8.69 Calculate the value of $[OH^-]$ from the given $[H_3O^+]$ and label the solution as acidic or basic.
a. 10^{-8} M
b. 10^{-10} M
c. 3.0×10^{-4} M
d. 2.5×10^{-11} M

8.70 Calculate the value of $[OH^-]$ from the given $[H_3O^+]$ and label the solution as acidic or basic.
a. 10^{-1} M
b. 10^{-13} M
c. 2.6×10^{-7} M
d. 1.2×10^{-12} M

8.71 Calculate the value of $[H_3O^+]$ from the given $[OH^-]$ and label the solution as acidic or basic.
a. 10^{-2} M
b. 4.0×10^{-8} M
c. 6.2×10^{-7} M
d. 8.5×10^{-13} M

8.72 Calculate the value of $[H_3O^+]$ from the given $[OH^-]$ and label the solution as acidic or basic.
a. 10^{-12} M
b. 5.0×10^{-10} M
c. 6.0×10^{-4} M
d. 8.9×10^{-11} M

8.73 Calculate the pH from each H_3O^+ concentration determined in Problem 8.71.

8.74 Calculate the pH from each H_3O^+ concentration determined in Problem 8.72.

8.75 Calculate the H_3O^+ concentration from each pH: (a) 12; (b) 1; (c) 1.80; (d) 8.90.

8.76 Calculate the H_3O^+ concentration from each pH: (a) 4; (b) 8; (c) 2.60; (d) 11.30.

8.77 What are the concentrations of H_3O^+ and OH^- in tomatoes that have a pH of 4.10?

8.78 What are the concentrations of H_3O^+ and OH^- in a cola beverage that has a pH of 3.15?

8.79 Calculate the pH of each aqueous solution: (a) 0.0025 M HCl; (b) 0.015 M KOH.

8.80 Calculate the pH of each aqueous solution: (a) 0.015 M HNO_3; (b) 0.0025 M NaOH.

Acid–Base Reactions

8.81 Write a balanced equation for each reaction.
a. $HBr(aq) + KOH(aq) \longrightarrow$
b. $HNO_3(aq) + Ca(OH)_2(aq) \longrightarrow$
c. $HCl(aq) + NaHCO_3(aq) \longrightarrow$
d. $H_2SO_4(aq) + Mg(OH)_2(aq) \longrightarrow$

8.82 Write a balanced equation for each reaction.
a. $HNO_3(aq) + LiOH(aq) \longrightarrow$
b. $H_2SO_4(aq) + NaOH(aq) \longrightarrow$
c. $K_2CO_3(aq) + HCl(aq) \longrightarrow$
d. $HI(aq) + NaHCO_3(aq) \longrightarrow$

Titration

8.83 What is the molarity of an HCl solution if 35.5 mL of 0.10 M NaOH are needed to neutralize 25.0 mL of the sample?

8.84 What is the molarity of an HCl solution if 17.2 mL of 0.15 M NaOH are needed to neutralize 5.00 mL of the sample?

8.85 What is the molarity of an acetic acid (CH_3COOH) solution if 15.5 mL of 0.20 M NaOH are needed to neutralize 25.0 mL of the sample?

8.86 What is the molarity of an H_2SO_4 solution if 18.5 mL of 0.18 M NaOH are needed to neutralize 25.0 mL of the sample?

8.87 How many milliliters of 1.0 M NaOH solution are needed to neutralize 10.0 mL of 2.5 M CH_3COOH solution?

8.88 How many milliliters of 2.0 M NaOH solution are needed to neutralize 8.0 mL of 3.5 M H_2SO_4 solution?

Buffers

8.89 Consider a weak acid H_2A and its conjugate base HA^-. Which diagram represents a buffer? Explain your choice.

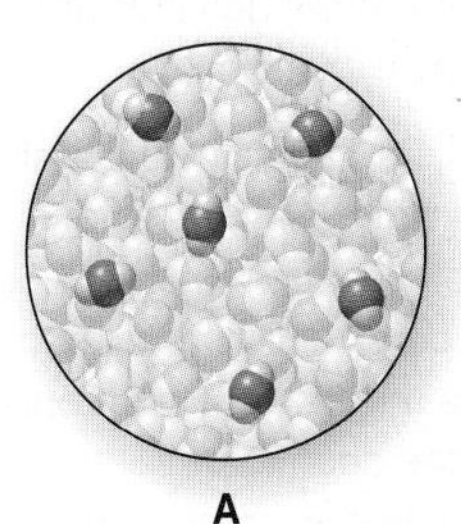

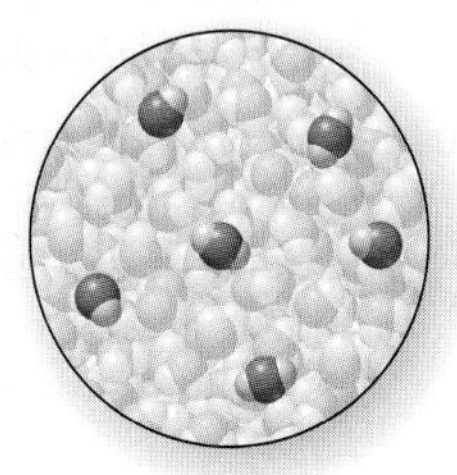

8.90 Consider a weak acid H_2A and its conjugate base HA^-. Which diagram represents a buffer? Explain your choice.

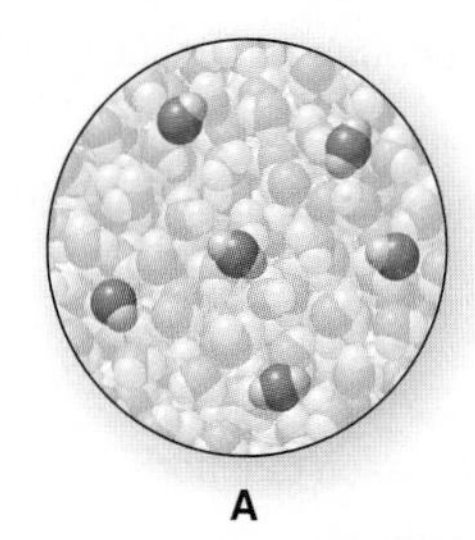
A

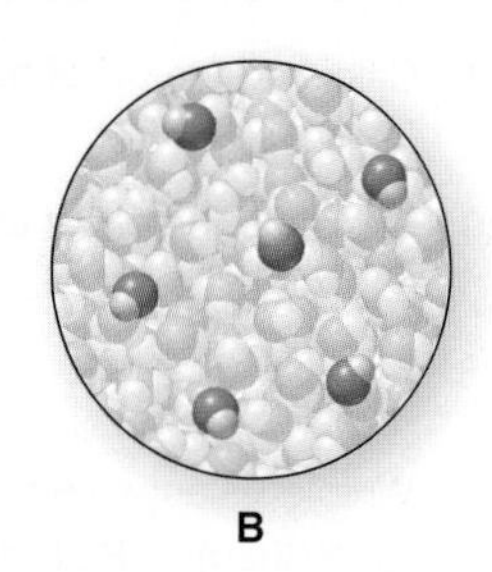
B

8.91 Consider a buffer prepared from the weak acid HF and its conjugate base F^-.

$$HF(aq) + H_2O(l) \rightleftharpoons F^-(aq) + H_3O^+(aq)$$

a. What happens to the concentrations of HF and F^- when a small amount of acid is added to the buffer?

b. What happens to the concentrations of HF and F^- when a small amount of base is added to the buffer?

8.92 Explain why both HF and F^- are needed to prepare the buffer in Problem 8.91.

Applications

8.93 Why is the pH of unpolluted rainwater lower than the pH of pure water?

8.94 Why is the pH of acid rain lower than the pH of rainwater?

8.95 The optimum pH of a swimming pool is 7.50. Calculate the value of $[H_3O^+]$ and $[OH^-]$ at this pH.

8.96 A sample of rainwater has a pH of 4.18. (a) Calculate the H_3O^+ concentration in the sample. (b) Suggest a reason why this pH differs from the pH of unpolluted rainwater (5.6).

8.97 When an individual hyperventilates, he is told to blow into a paper bag held over his mouth. What effect should this process have on the CO_2 concentration and pH of the blood?

8.98 What is the difference between respiratory acidosis and respiratory alkalosis?

8.99 How is CO_2 concentration related to the pH of the blood?

8.100 Explain why a lake on a bed of limestone is naturally buffered against the effects of acid rain.

CHALLENGE PROBLEMS

8.101 Calcium hypochlorite $[Ca(OCl)_2]$ is used to chlorinate swimming pools. $Ca(OCl)_2$ acts as a source of the weak acid hypochlorous acid, HOCl, a disinfectant that kills bacteria. Write the acid–base reaction that occurs when OCl^- dissolves in water and explain why this reaction makes a swimming pool more basic.

8.102 Most buffer solutions are prepared using a weak acid and a salt of its conjugate base. Explain how the following combination can also form a buffer solution: 0.20 M H_3PO_4 and 0.10 M NaOH.

BEYOND THE CLASSROOM

8.103 Acid rain, rainwater that has a lower-than-normal pH, can have a severe negative impact on the plant and animal life in a region. What are the main components of acid rain? Research the sources of these components and what steps are currently in place to improve the pH of rainwater. Explain why acid rain is considered a regional problem, whereas ozone depletion is considered a global problem. What negative effects of acid rain are observed in your town or region?

8.104 Heartburn, acid reflux, and stomach ulcers are all medical conditions that can result from the strong acid in an individual's stomach. Cimetidine (trade name Tagamet), famotidine (trade name Pepcid), and nizatidine (trade name Axid) are three medications marketed to treat one or more of these conditions. Draw the chemical structure of one of these drugs and research why it is effective in treating excess stomach acid. Why are these drugs fundamentally different in tackling this problem than antacids such as Maalox, Mylanta, and Rolaids?

8.105 Ascorbic acid (vitamin C) and pantothenic acid (vitamin B_5) are two acids needed by humans in the diet for normal cellular function. Pick one of these vitamins and draw its chemical structure. How much is required in the daily diet? What are some dietary sources of the vitamin? Why is the vitamin needed in the body and what symptoms result from its deficiency?

8.106 Many conditions that cause an imbalance of respiratory or metabolic processes affect the pH of the blood. Research how the normal pH of the blood is altered by some or all of the following conditions: pneumonia, excessive alcohol consumption, emphysema, starvation, holding one's breath, hyperventilation, and the excessive use of antacids. Using the principles presented in Section 8.9, explain why the imbalance results. What other conditions affect the pH of the blood?

ANSWERS TO SELECTED PROBLEMS

8.1 a. hydrofluoric acid
b. nitric acid
c. hydrocyanic acid

8.3 a,c

8.4 a,b,d

8.5 a. $HCl(g)$ (acid) + $NH_3(g)$ (base)   c. $OH^-(aq)$ (base) + $HSO_4^-(aq)$ (acid)
b. $CH_3COOH(l)$ (acid) + $H_2O(l)$ (base)

8.6 a. H_3O^+ b. HI c. H_2CO_3

8.7 a. HS^- b. CN^- c. SO_4^{2-}

8.9 a. $H_2O(l)$ (base) + $HI(g)$ (acid) $\rightleftharpoons$ $I^-(aq)$ (conjugate base) + $H_3O^+(aq)$ (conjugate acid)
b. $CH_3COOH(l)$ (acid) + $NH_3(g)$ (base) $\rightleftharpoons$ $CH_3COO^-(aq)$ (conjugate base) + $NH_4^+(aq)$ (conjugate acid)
c. $Br^-(aq)$ (base) + $HNO_3(aq)$ (acid) $\rightleftharpoons$ $HBr(aq)$ (conjugate acid) + $NO_3^-(aq)$ (conjugate base)

8.11 $C_6H_8O_6(aq)$ (acid) + $H_2O(l)$ (base) $\rightleftharpoons$ $C_6H_7O_6^-(aq)$ (conjugate base) + $H_3O^+(aq)$ (conjugate acid)

8.12 **D < F < E**

8.13 a. H_2SO_4 is the stronger acid; H_3PO_4 has the stronger conjugate base.
b. HCl is the stronger acid; HF has the stronger conjugate base.
c. H_2CO_3 is the stronger acid; NH_4^+ has the stronger conjugate base.
d. HF is the stronger acid; HCN has the stronger conjugate base.

8.15 a. HNO_2 and HNO_3 b. HNO_3 is the stronger acid.

8.16 a. 10^{-11} M acidic   c. 3.6×10^{-5} M basic
b. 10^{-3} M basic   d. 1.8×10^{-11} M acidic

8.17 a. 10^{-8} M basic   c. 1.9×10^{-4} M acidic
b. 10^{-5} M acidic   d. 1.4×10^{-11} M basic

8.18

	$[H_3O^+]$	$[OH^-]$
a.	10^{-11} M	10^{-3} M
b.	10^{-3} M	10^{-11} M
c.	1.5 M	6.7×10^{-15} M
d.	3.3×10^{-14} M	3.0×10^{-1} M

8.19 a. 6 b. 12 c. 5 d. 11

8.20 a. 1×10^{-13} M basic   c. 1×10^{-3} M acidic
b. 1×10^{-7} M neutral

8.21 a. 6×10^{-11} M b. 2×10^{-8} M c. 5×10^{-5} M

8.22 a. 5.74 b. 11.036 c. 4.06 d. 10.118

8.23 a. $HNO_3(aq) + NaOH(aq) \longrightarrow H_2O(l) + NaNO_3(aq)$
b. $H_2SO_4(aq) + 2\ KOH(aq) \longrightarrow 2\ H_2O(l) + K_2SO_4(aq)$

8.25 $H_2SO_4(aq) + CaCO_3(s) \longrightarrow CaSO_4(aq) + \underbrace{H_2O(l) + CO_2(g)}_{\text{from } H_2CO_3}$

8.27 0.41 M

8.29 a. Not a buffer since it contains a strong acid, HBr.
b. A buffer since HF is a weak acid and F^- is its conjugate base.
c. Not a buffer since it contains a weak acid only.

8.31 a. NO_2^- b. CO_3^{2-}

8.33 a. **B** b. **A**

8.35 a. acid + base ⟶ conjugate base + conjugate acid
b. base + acid ⟶ conjugate acid + conjugate base

8.37 NH_3 can accept a proton because the N atom has a lone pair, but CH_4 does not have a lone pair, so it cannot.

8.39 a. $H_3PO_4(aq)$ (acid) + $CN^-(aq) \longrightarrow H_2PO_4^-(aq) + HCN(aq)$ (conjugate acid)
b. $Br^-(aq) + HSO_4^-(aq)$ (acid) $\longrightarrow SO_4^{2-}(aq) + HBr(g)$ (conjugate acid)

8.41 $[H_3O^+] = 1.3 \times 10^{-6}$ M $[OH^-] = 7.7 \times 10^{-9}$ M

8.43 $2\ HNO_3(aq) + CaCO_3(s) \longrightarrow Ca(NO_3)_2(aq) + CO_2(g) + H_2O(l)$

8.45 The concentration of HNO_2 increases and that of NO_2^- decreases when a small amount of acid is added to the buffer.

8.47 a,d,f

8.49 a,d,e,f

8.51 a. H_2S b. HCO_3^- c. HNO_2

8.53 a. NO_2^- b. NH_3 c. HO_2^-

8.55 a. $HI(g)$ acid $I^-(aq)$ conjugate base
$NH_3(g)$ base $NH_4^+(aq)$ conjugate acid
b. $HCOOH(l)$ acid $HCOO^-(aq)$ conjugate base
$H_2O(l)$ base $H_3O^+(aq)$ conjugate acid
c. $HSO_4^-(aq)$ base $H_2SO_4(aq)$ conjugate acid
$H_2O(l)$ acid $OH^-(aq)$ conjugate base

8.57 a. H_2CO_3 b. CO_3^{2-}

8.59 **A** represents HCl because it shows a fully dissociated acid. **B** represents HF because it is only partially dissociated.

8.61 a. CH_3COOH b. H_2SO_4

8.63 a. H_2O b. H_3PO_4

8.65 **B**

8.67 a. **A** b. **B**

8.69 a. 10^{-6} M basic   c. 3.3×10^{-11} M acidic
b. 10^{-4} M basic   d. 4.0×10^{-4} M basic

8.71 a. 10^{-12} M basic   c. 1.6×10^{-8} M basic
b. 2.5×10^{-7} M acidic   d. 1.2×10^{-2} M acidic

8.73 a. 12 b. 6.60 c. 7.80 d. 1.92

8.75 a. 1×10^{-12} M c. 1.6×10^{-2} M
b. 1×10^{-1} M d. 1.3×10^{-9} M

8.77 $[H_3O^+] = 7.9 \times 10^{-5}$ M $[OH^-] = 1.3 \times 10^{-10}$ M

8.79 a. 2.60 b. 12.17

8.81 a. $HBr(aq) + KOH(aq) \longrightarrow KBr(aq) + H_2O(l)$
b. $2\ HNO_3(aq) + Ca(OH)_2(aq) \longrightarrow 2\ H_2O(l) + Ca(NO_3)_2(aq)$
c. $HCl(aq) + NaHCO_3(aq) \longrightarrow NaCl(aq) + H_2O(l) + CO_2(g)$
d. $H_2SO_4(aq) + Mg(OH)_2(aq) \longrightarrow 2\ H_2O(l) + MgSO_4(aq)$

8.83 0.14 M

8.84 0.12 M

8.87 25 mL

8.89 **B** represents a buffer because it has equal amounts of H_2A and HA^-.

8.91 a. When a small amount of acid is added to the buffer, the concentration of HF increases and the concentration of F^- decreases.
b. The concentration of HF decreases and the concentration of F^- increases when a small amount of base is added to the buffer.

8.93 CO_2 combines with rainwater to form H_2CO_3, which is acidic, lowering the pH.

8.95 $[H_3O^+] = 3.2 \times 10^{-8}$ M $[OH^-] = 3.1 \times 10^{-7}$ M

8.97 By breathing into a bag, the individual breathes in air with a higher CO_2 concentration. Thus, the CO_2 concentration in the lungs and the blood increases, thereby lowering the pH.

8.99 A rise in CO_2 concentration leads to an increase in $H^+(aq)$ concentration by the following reactions:
$CO_2(g) + H_2O(l) \rightleftharpoons H_2CO_3(aq) \rightleftharpoons HCO_3^-(aq) + H^+(aq)$

8.101 $OCl^-(aq) + H_2O(l) \longrightarrow HOCl(aq) + OH^-(aq)$
OH^- is formed by this reaction, so the water is basic.

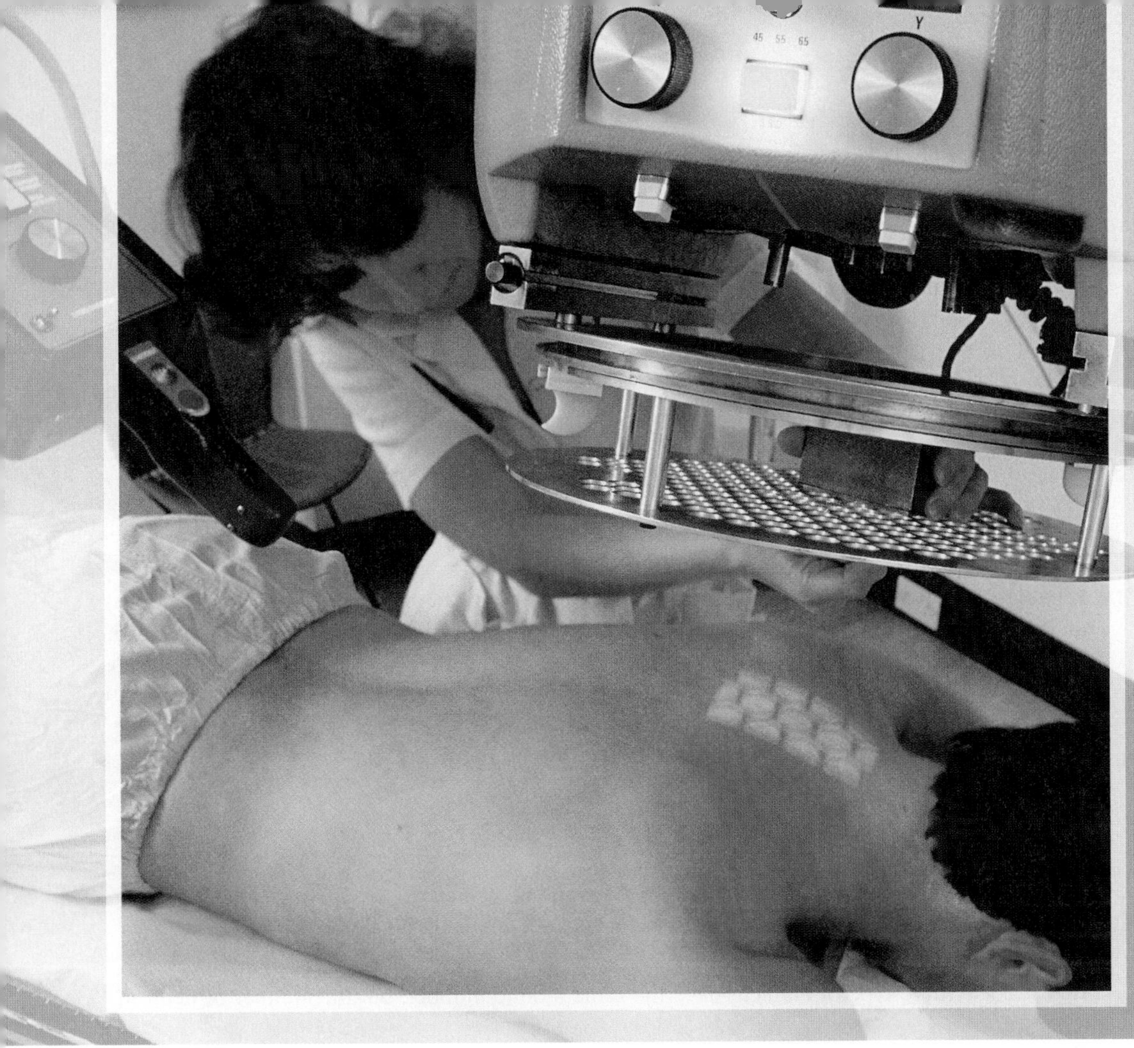

9

Radiation produced by cobalt-60 and other radioactive isotopes is used to treat many different forms of cancer.

Nuclear Chemistry

CHAPTER OUTLINE

- **9.1** Introduction
- **9.2** Nuclear Reactions
- **9.3** Half-Life
- **9.4** Detecting and Measuring Radioactivity
- **9.5** FOCUS ON HEALTH & MEDICINE: Medical Uses of Radioisotopes
- **9.6** Nuclear Fission and Nuclear Fusion
- **9.7** FOCUS ON HEALTH & MEDICINE: Medical Imaging Without Radioactivity

CHAPTER GOALS

In this chapter you will learn how to:

1. Describe the different types of radiation emitted by a radioactive nucleus
2. Write equations for nuclear reactions
3. Define half-life
4. Recognize the units used for measuring radioactivity
5. Give examples of common radioisotopes used in medical diagnosis and treatment
6. Describe the general features of nuclear fission and nuclear fusion
7. Describe the features of medical imaging techniques that do not use radioactivity

Thus far our study of reactions has concentrated on processes that involve the valence electrons of atoms. In these reactions, bonds that join atoms are broken and new bonds between atoms are formed, but the identity of the atoms does not change. In Chapter 9, we turn our attention to **nuclear reactions,** processes that involve changes in the nucleus of an atom. While certainly much less common than chemical reactions that occur with electrons, nuclear reactions form a useful group of processes with a wide range of applications. Nuclear medicine labs in hospitals use radioactive isotopes to diagnose disease, visualize organs, and treat tumors. Generating energy in nuclear power plants, dating archaeological objects using the isotope carbon-14, and designing a simple and reliable smoke detector all utilize the concepts of nuclear chemistry discussed here in Chapter 9.

9.1 Introduction

Although most reactions involve valence electrons, a small but significant group of reactions, **nuclear reactions,** involves the subatomic particles of the nucleus. To understand nuclear reactions we must first review facts presented in Chapter 2 regarding isotopes and the characteristics of the nucleus.

9.1A Isotopes

The nucleus of an atom is composed of protons and neutrons.

- **The atomic number (*Z*) = the number of protons in the nucleus.**
- **The mass number (*A*) = the number of protons and neutrons in the nucleus.**

Atoms of the same type of element have the same atomic number, but the number of neutrons may vary.

- **Isotopes are atoms of the same element having a different number of neutrons.**

As a result, isotopes have the same atomic number (*Z*) but different mass numbers (*A*). Carbon, for example, has three naturally occurring isotopes. Each isotope has six protons in the nucleus (i.e., $Z = 6$), but the number of neutrons may be six, seven, or eight. Thus, the mass numbers (*A*) of these isotopes are 12, 13, and 14, respectively. As we learned in Chapter 2, we can refer to these isotopes as carbon-12, carbon-13, and carbon-14. Isotopes are also written with the mass number to the upper left of the element symbol and the atomic number to the lower left.

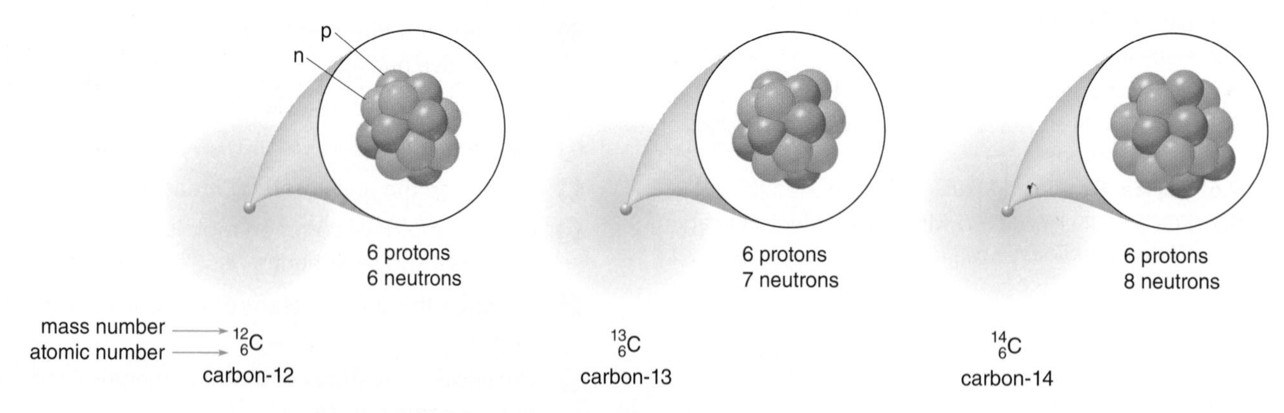

Many isotopes are stable, but a larger number are not.

- **A *radioactive isotope,* called a *radioisotope,* is unstable and spontaneously emits energy to form a more stable nucleus.**

Radioactivity **is the nuclear radiation emitted by a radioactive isotope.** Of the known isotopes of all the elements, 264 are stable and 300 are naturally occurring but unstable. An even larger number of radioactive isotopes, called **artificial isotopes,** have been produced in the laboratory. Both carbon-12 and carbon-13 are stable isotopes and occur in higher natural abundance than carbon-14, a radioactive isotope.

SAMPLE PROBLEM 9.1

Complete the following table for two isotopes of cobalt. Cobalt-60 is commonly used in cancer therapy.

	Atomic Number	Mass Number	Number of Protons	Number of Neutrons	Isotope Symbol
Cobalt-59					
Cobalt-60					

Analysis

- The atomic number = the number of protons.
- The mass number = the number of protons + the number of neutrons.
- Isotopes are written with the mass number to the upper left of the element symbol and the atomic number to the lower left.

Solution

	Atomic Number	Mass Number	Number of Protons	Number of Neutrons	Isotope Symbol
Cobalt-59	27	59	27	59 – 27 = 32	${}^{59}_{27}Co$
Cobalt-60	27	60	27	60 – 27 = 33	${}^{60}_{27}Co$

PROBLEM 9.1

Each of the following radioisotopes is used in medicine. For each isotope give its: [1] atomic number; [2] mass number; [3] number of protons; [4] number of neutrons.

a. ${}^{85}_{38}Sr$ used in bone scans

b. ${}^{67}_{31}Ga$ used in abdominal scans

c. selenium-75 used in pancreas scans

9.1B Types of Radiation

Different forms of radiation are emitted when a radioactive nucleus is converted to a more stable nucleus, including **alpha particles, beta particles, positrons,** and **gamma radiation.**

- An *alpha particle* is a high-energy particle that contains two protons and two neutrons.

alpha particle: α or ${}^{4}_{2}He$

An alpha particle, symbolized by the Greek letter **alpha (α)** or the element symbol for helium, has a +2 charge and a mass number of 4.

- A *beta particle* is a high-energy electron.

beta particle: β or ${}^{0}_{-1}e$

An electron has a –1 charge and a negligible mass compared to a proton. A beta particle, symbolized by the **Greek letter beta (β),** is also drawn with the symbol for an electron, **e,** with a mass number of 0 in the upper left corner and a charge of –1 in the lower left corner. A β particle is formed when a neutron (n) is converted to a proton (p) and an electron.

$$\underset{\text{neutron}}{{}^{1}_{0}\text{n}} \longrightarrow \underset{\text{proton}}{{}^{1}_{1}\text{p}} + \underset{\beta\ \text{particle}}{{}^{0}_{-1}\text{e}}$$

- A *positron* is called an *antiparticle* of a β particle, since their charges are different but their masses are the same.

Thus, a **positron** has a negligible mass like a β particle, but is opposite in charge, +1. A positron, symbolized as β^+, is also drawn with the symbol for an electron, **e,** with a mass number of 0 in the upper left corner and a charge of +1 in the lower left corner. A positron, which can be thought of as a "positive electron," is formed when a proton is converted to a neutron.

Symbol: $\underset{\text{positron}}{{}^{0}_{+1}\text{e} \text{ or } \beta^+}$ **Formation:** $\underset{\text{proton}}{{}^{1}_{1}\text{p}} \longrightarrow \underset{\text{neutron}}{{}^{1}_{0}\text{n}} + \underset{\text{positron}}{{}^{0}_{+1}\text{e}}$

- *Gamma rays* are high-energy radiation released from a radioactive nucleus.

Gamma rays, symbolized by the **Greek letter gamma (γ),** are a form of energy and thus they have no mass or charge. Table 9.1 summarizes the properties of some of the different types of radiation.

gamma ray: γ

Table 9.1 Types of Radiation

Type of Radiation	Symbol	Charge	Mass Number
Alpha particle	α or ${}^{4}_{2}\text{He}$	+2	4
Beta particle	β or ${}^{0}_{-1}\text{e}$	–1	0
Positron	β^+ or ${}^{0}_{+1}\text{e}$	+1	0
Gamma ray	γ	0	0

PROBLEM 9.2

What is the difference between an α particle and a helium atom?

PROBLEM 9.3

What is the difference between an electron and a positron?

PROBLEM 9.4

Identify Q in each of the following symbols.

a. ${}^{0}_{-1}\text{Q}$ b. ${}^{4}_{2}\text{Q}$ c. ${}^{0}_{+1}\text{Q}$

9.2 Nuclear Reactions

Radioactive decay **is the process by which an unstable radioactive nucleus emits radiation, forming a nucleus of new composition.** A nuclear equation can be written for this process, which contains the original nucleus, the new nucleus, and the radiation emitted. Unlike a chemical equation that balances atoms, in a nuclear equation the mass numbers and the atomic numbers of the nuclei must be balanced.

- **The sum of the mass numbers (*A*) must be equal on both sides of a nuclear equation.**
- **The sum of the atomic numbers (*Z*) must be equal on both sides of a nuclear equation.**

9.2A Alpha Emission

***Alpha emission* is the decay of a nucleus by emitting an α particle.** For example, uranium-238 decays to form thorium-234 by loss of an α particle.

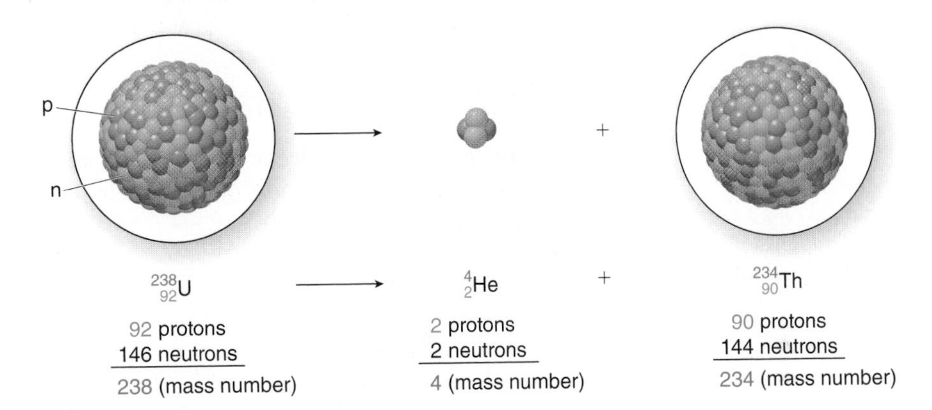

HEALTH NOTE

Americium-241 is a radioactive element contained in smoke detectors. The decay of α particles creates an electric current that is interrupted when smoke enters the detector, sounding an alarm.

Since an α particle has two protons, **the new nucleus has *two fewer protons* than the original nucleus.** Because it has a *different* number of protons, the new nucleus represents a *different* element. Uranium-238 has 92 protons, so loss of two forms the element thorium with 90 protons. The thorium nucleus has a mass number that is four fewer than the original—234—because it has been formed by loss of an α particle with a mass number of four.

As a result, the sum of the mass numbers is equal on both sides of the equation—238 = 4 + 234. The sum of the atomic numbers is also equal on both sides of the equation—92 = 2 + 90.

How To Balance an Equation for a Nuclear Reaction

Example Write a balanced nuclear equation showing how americium-241, a radioactive element used in smoke detectors, decays to form an α particle.

Step [1] Write an incomplete equation with the original nucleus on the left and the particle emitted on the right.

- Include the mass number and atomic number (from the periodic table) in the equation.

$$^{241}_{95}Am \longrightarrow {}^{4}_{2}He + ?$$

Step [2] Calculate the mass number and atomic number of the newly formed nucleus on the right.

- Mass number: The sum of the mass numbers must be equal on both sides of the equation. Subtract the mass of an α particle (4) to obtain the mass of the new nucleus; 241 – 4 = 237.
- Atomic number: The sum of the atomic numbers must be equal on both sides of the equation. Subtract the two protons of an α particle to obtain the atomic number of the new nucleus; 95 – 2 = 93.

Step [3] Use the atomic number to identify the new nucleus and complete the equation.

- From the periodic table, the element with an atomic number of 93 is neptunium, Np.
- Write the mass number and the atomic number with the element symbol to complete the equation.

241 = 4 + 237
95 = 2 + 93

$$^{241}_{95}Am \longrightarrow {}^{4}_{2}He + {}^{237}_{93}Np$$

PROBLEM 9.5

Radon, a radioactive gas formed in the soil, can cause lung cancers when inhaled in high concentrations for a long period of time. Write a balanced nuclear equation for the decay of radon-222, which emits an α particle.

PROBLEM 9.6

Radon (Problem 9.5) is formed in the soil as a product of radioactive decay that produces an α particle. Write a balanced nuclear equation for the formation of radon-222 and an α particle.

PROBLEM 9.7

Write a balanced equation showing how each nucleus decays to form an α particle:
(a) polonium-218; (b) thorium-230; (c) Es-252.

9.2B Beta Emission

***Beta emission* is the decay of a nucleus by emitting a β particle.** For example, carbon-14 decays to nitrogen-14 by loss of a β particle. The decay of carbon-14 is used to date archaeological specimens (Section 9.3).

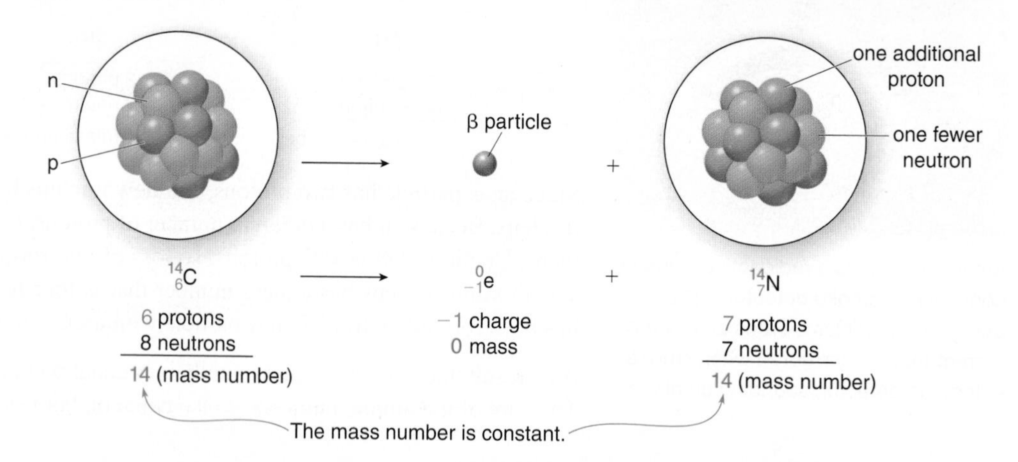

In β emission, one neutron of the original nucleus decays to a β particle and a proton. As a result, the **new nucleus has *one more proton* and *one fewer neutron* than the original nucleus.** In this example, a carbon atom with six protons decays to a nitrogen atom with seven protons. Since the total number of particles in the nucleus does not change, the **mass number is constant.**

The subscripts that represent the atomic numbers are balanced because the β particle has a charge of –1. Seven protons on the right side plus a –1 charge for the β particle gives a total "charge" of +6, the atomic number of carbon on the left. The mass numbers are also balanced since a β particle has zero mass, and both the original nucleus and the new nucleus contain 14 subatomic particles (protons + neutrons).

Radioactive elements that emit β radiation are widely used in medicine. Iodine-131, a radioactive element that emits β radiation, is used to treat hyperthyroidism, a condition resulting from an overactive thyroid gland (Figure 9.1). Moreover, since β radiation is composed of high-energy electrons that penetrate tissue in a small, localized region, radioactive elements situated in close contact with tumor cells kill them. Although both healthy and diseased cells are destroyed by this internal radiation therapy, rapidly dividing tumor cells are more sensitive to its effects and therefore their growth and replication are affected the most.

Figure 9.1

The Use of Iodine-131 to Treat Hyperthyroidism

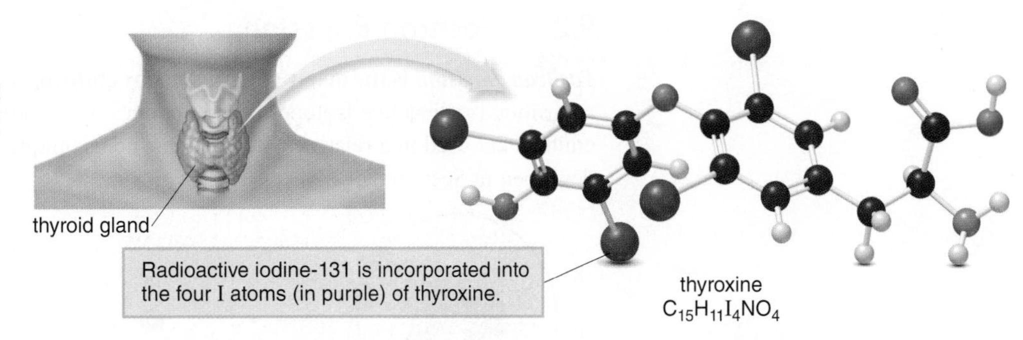
thyroid gland
Radioactive iodine-131 is incorporated into the four I atoms (in purple) of thyroxine.
thyroxine
$C_{15}H_{11}I_4NO_4$

Iodine-131 is incorporated into the thyroid hormone thyroxine. Beta radiation emitted by the radioactive isotope destroys nearby thyroid cells, thus decreasing the activity of the thyroid gland and bringing the disease under control.

SAMPLE PROBLEM 9.2

Write a balanced nuclear equation for the β emission of phosphorus-32, a radioisotope used to treat leukemia and other blood disorders.

Analysis

Balance the atomic numbers and mass numbers on both sides of a nuclear equation. With β emission, treat the β particle as an electron with zero mass in balancing mass numbers, and a −1 charge when balancing the atomic numbers.

Solution

[1] **Write an incomplete equation with the original nucleus on the left and the particle emitted on the right.**

- Use the identity of the element to determine the atomic number; phosphorus has an atomic number of 15.

$$^{32}_{15}\mathrm{P} \longrightarrow {}^{0}_{-1}\mathrm{e} + ?$$

[2] **Calculate the mass number and the atomic number of the newly formed nucleus on the right.**

- Mass number: Since a β particle has no mass, the masses of the new particle and the original particle are the same, 32.
- Atomic number: Since β emission converts a neutron into a proton, the new nucleus has one more proton than the original nucleus; 15 = −1 + ?. Thus the new nucleus has an atomic number of 16.

[3] **Use the atomic number to identify the new nucleus and complete the equation.**

- From the periodic table, the element with an atomic number of 16 is sulfur, S.
- Write the mass number and the atomic number with the element symbol to complete the equation.

$$^{32}_{15}\mathrm{P} \longrightarrow {}^{0}_{-1}\mathrm{e} + {}^{32}_{16}\mathrm{S}$$

PROBLEM 9.8

Write a balanced nuclear equation for the β emission of iodine-131.

PROBLEM 9.9

Write a balanced nuclear equation for the β emission of each of the following isotopes.

a. $^{20}_{9}\mathrm{F}$ b. $^{92}_{38}\mathrm{Sr}$ c. chromium-55

9.2C Positron Emission

Positron emission **is the decay of a nucleus by emitting a positron (β^+).** For example, carbon-11, an artificial radioactive isotope of carbon, decays to boron-11 by loss of a β^+ particle. Positron emitters are used in a relatively new diagnostic technique, positron emission tomography (PET), described in Section 9.5.

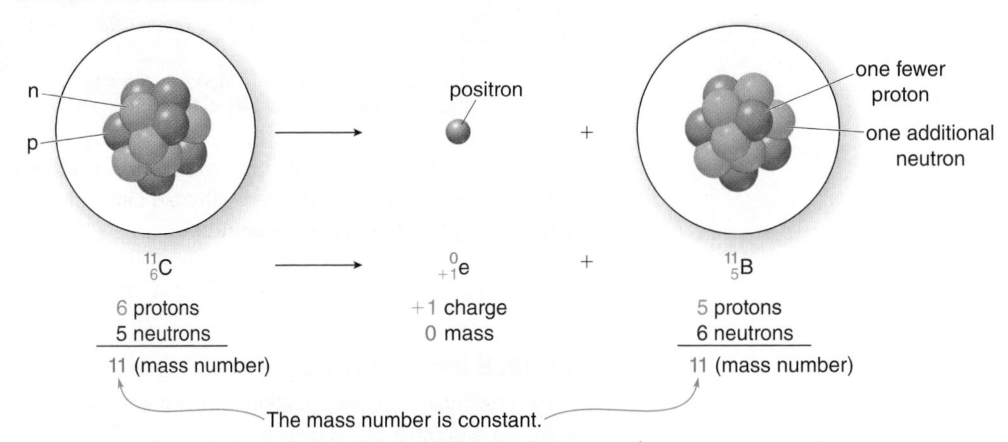

In positron emission, one proton of the original nucleus decays to a β^+ particle and a neutron. As a result, the **new nucleus has *one fewer proton* and *one more neutron* than the original nucleus.** In this example, a carbon atom with six protons decays to a boron atom with five protons. Since the total number of particles in the nucleus does not change, the **mass number is constant.**

SAMPLE PROBLEM 9.3

Write a balanced nuclear equation for the positron emission of fluorine-18, a radioisotope used for imaging in PET scans.

Analysis

Balance the atomic numbers and mass numbers on both sides of a nuclear equation. With β^+ emission, treat the positron as a particle with zero mass when balancing mass numbers, and a +1 charge when balancing the atomic numbers.

Solution

[1] **Write an incomplete equation with the original nucleus on the left and the particle emitted on the right.**

- Use the identity of the element to determine the atomic number; fluorine has an atomic number of 9.

$$^{18}_{9}F \longrightarrow \,^{0}_{+1}e + ?$$

[2] **Calculate the mass number and the atomic number of the newly formed nucleus on the right.**

- Mass number: Since a β^+ particle has no mass, the masses of the new particle and the original particle are the same, 18.
- Atomic number: Since β^+ emission converts a proton into a neutron, the new nucleus has one fewer proton than the original nucleus; 9 – 1 = 8. Thus, the new nucleus has an atomic number of 8.

[3] **Use the atomic number to identify the new nucleus and complete the equation.**

- From the periodic table, the element with an atomic number of 8 is oxygen, O.
- Write the mass number and the atomic number with the element symbol to complete the equation.

$$^{18}_{9}F \longrightarrow \,^{0}_{+1}e + \,^{18}_{8}O$$

PROBLEM 9.10

Write a balanced nuclear equation for the positron emission of each of the following nuclei: (a) arsenic-74; (b) oxygen-15.

9.2D Gamma Emission

***Gamma emission* is the decay of a nucleus by emitting γ radiation.** Since γ rays are simply a form of energy, their emission causes **no change in the atomic number or mass number** of a radioactive nucleus. Gamma emission sometimes occurs alone. For example, one form of technetium-99, written as technetium-99m, is an energetic form of the technetium nucleus that decays with emission of γ rays to technetium-99, a more stable but still radioactive element.

The *m* in technetium-99m stands for *metastable.* This designation is meant to indicate that the isotope decays to a more stable form of the same isotope.

$$^{99m}_{43}\text{Tc} \longrightarrow {}^{99}_{43}\text{Tc} + \gamma$$

The mass number and atomic number are the same.

Technetium-99m is a widely used radioisotope in medical imaging. Because it emits high-energy γ rays but decays in a short period of time, it is used to image the brain, thyroid, lungs, liver, skeleton, and many other organs. It has also been used to detect ulcers in the gastrointestinal system, and combined with other compounds, it is used to map the circulatory system and gauge damage after a heart attack.

More commonly, γ emission accompanies α or β emission. For example, cobalt-60 decays with both β and γ emission. Because a β particle is formed, decay generates an element with the *same* mass but a *different* number of protons, and thus a new element, nickel-60.

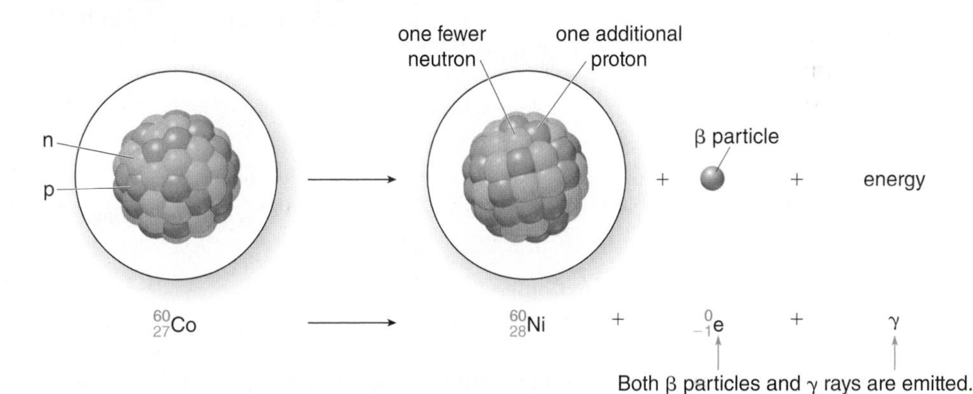

$$^{60}_{27}\text{Co} \longrightarrow {}^{60}_{28}\text{Ni} + {}^{0}_{-1}\text{e} + \gamma$$

Both β particles and γ rays are emitted.

Cobalt-60 is used in external radiation treatment for cancer. Radiation generated by cobalt-60 decay is focused on a specific site in the body that contains cancerous cells (Figure 9.2). By directing the radiation on the tumor, damage to surrounding healthy tissues is minimized.

Figure 9.2 Focus on Health & Medicine: External Radiation Treatment for Tumors

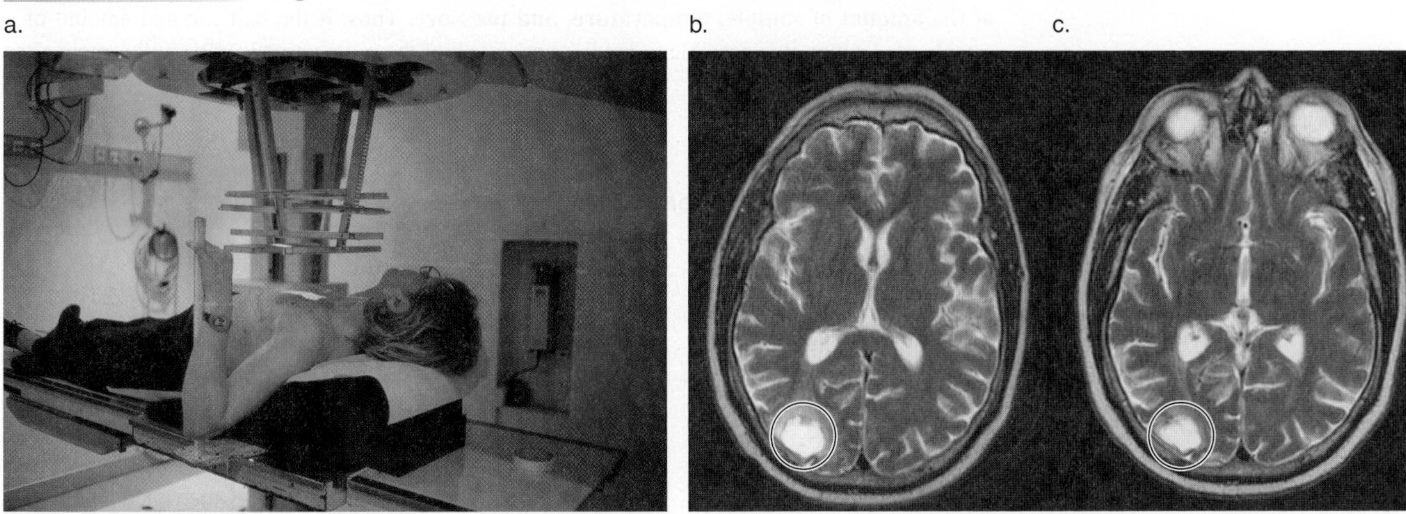

a. Gamma radiation from the decay of cobalt-60 is used to treat a variety of tumors, especially those that cannot be surgically removed.
b. A tumor (bright area in circle) before radiation treatment
c. A tumor (bright area in circle) that has decreased in size after six months of radiation treatment

PROBLEM 9.11

Write a nuclear equation for the decay of iridium-192 with β and γ emission. Iridium implants have been used to treat breast cancer. After the correct dose is administered, the iridium source is removed.

9.3 Half-Life

How fast do radioactive isotopes decay? It depends on the isotope.

- **The *half-life* ($t_{1/2}$) of a radioactive isotope is the time it takes for one-half of the sample to decay.**

9.3A General Features

Suppose we have a sample that contains 16 g of phosphorus-32, a radioactive isotope that decays to sulfur-32 by β emission (Sample Problem 9.2). Phosphorus-32 has a half-life of approximately 14 days. Thus, after 14 days, the sample contains only half the amount of P-32—8.0 g. After another 14 days (a total of two half-lives), the 8.0 g of P-32 is again halved to 4.0 g. After another 14 days (a total of three half-lives), the 4.0 g of P-32 is halved to 2.0 g, and so on. Every 14 days, half of the P-32 decays.

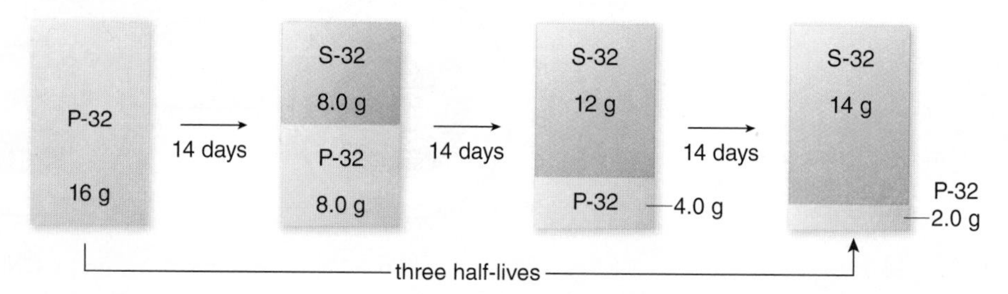

Many naturally occurring isotopes have long half-lives. Examples include carbon-14 (5,730 years) and uranium-235 (7.0×10^8 years). Radioisotopes that are used for diagnosis and imaging in medicine have short half-lives so they do not linger in the body. Examples include technetium-99m (6.0 hours) and iodine-131 (8.0 days). The half-lives of several elements are given in Table 9.2.

The half-life of a radioactive isotope is a property of a given isotope and is independent of the amount of sample, temperature, and pressure. Thus, if the half-life and amount of a sample are known, it is possible to predict how much of the radioactive isotope will remain after a period of time.

Table 9.2 Half-Lives of Some Common Radioisotopes

Radioisotope	Symbol	Half-Life	Use
Carbon-14	${}^{14}_{6}C$	5,730 years	Archaeological dating
Cobalt-60	${}^{60}_{27}Co$	5.3 years	Cancer therapy
Iodine-131	${}^{131}_{53}I$	8.0 days	Thyroid therapy
Potassium-40	${}^{40}_{19}K$	1.3×10^9 years	Geological dating
Phosphorus-32	${}^{32}_{15}P$	14.3 days	Leukemia treatment
Technetium-99m	${}^{99m}_{43}Tc$	6.0 hours	Organ imaging
Uranium-235	${}^{235}_{92}U$	7.0×10^8 years	Nuclear reactors

How To Use a Half-Life to Determine the Amount of Radioisotope Present

Example If the half-life of iodine-131 is 8.0 days, how much of a 100. mg sample of iodine-131 remains after 32 days? Iodine-131 forms xenon-131 by β emission.

Step [1] Determine how many half-lives occur in the given amount of time.

- Use the half-life of iodine-131 as a conversion factor to convert the number of days to the number of half-lives.

$$32 \text{ days} \times \frac{1 \text{ half-life}}{8.0 \text{ days}} = 4.0 \text{ half-lives}$$

Step [2] For each half-life, multiply the initial mass by one-half to obtain the final mass.

- Since 32 days corresponds to *four* half-lives, multiply the initial mass by ½ *four* times to obtain the final mass. After four half-lives, 6.25 mg of iodine-131 remains.

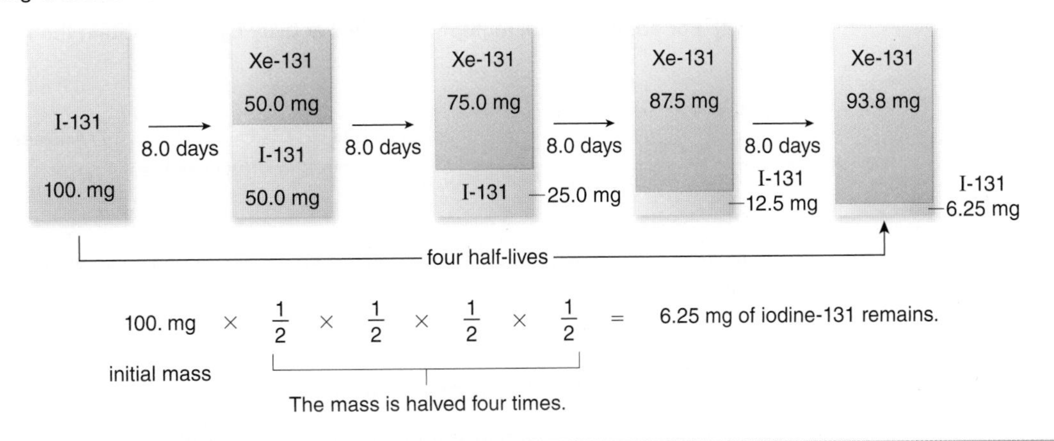

$$100. \text{ mg} \times \frac{1}{2} \times \frac{1}{2} \times \frac{1}{2} \times \frac{1}{2} = 6.25 \text{ mg of iodine-131 remains.}$$

initial mass

The mass is halved four times.

PROBLEM 9.12

A radioactive element (shown with blue spheres) forms a decay product (shown with green spheres) with a half-life of 22 min. (a) How many half-lives have elapsed to form the sample in diagram **A?** (b) How many minutes have elapsed in the conversion of the initial sample to **A?**

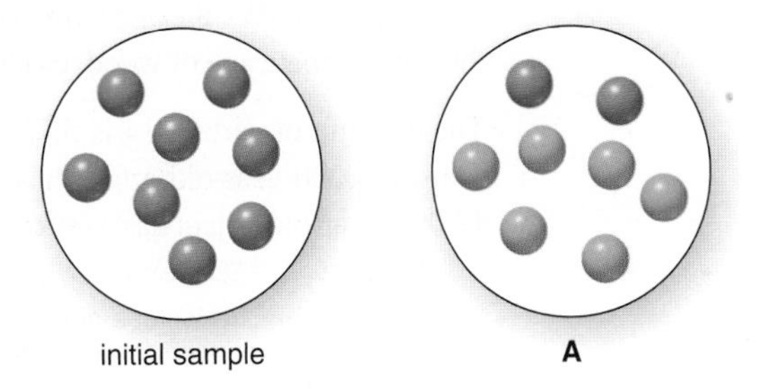

PROBLEM 9.13

How much phosphorus-32 remains from a 1.00 g sample after each of the following number of half-lives: (a) 2; (b) 4; (c) 8?

PROBLEM 9.14

If a 160. mg sample of technetium-99m is used for a diagnostic procedure, how much Tc-99m remains after each interval: (a) 6.0 h; (b) 18.0 h; (c) 24.0 h; (d) 2 days?

Half-life can also be used to determine the amount of radioactivity remaining in a sample after a known interval of time.

SAMPLE PROBLEM 9.4

A patient is injected with a sample of technetium-99m ($t_{1/2}$ = 6.0 h), which has an activity of 32 mCi. What activity is observed after 12 h?

Analysis

First determine how many half-lives have occurred in the time interval. Then, multiply the initial activity by the number of half-lives to determine the remaining activity.

Solution

[1] Use the half-life of technetium-99m as a conversion factor to convert the number of hours to the number of half-lives.

$$12\ \cancel{h} \times \frac{1\ \text{half-life}}{6.0\ \cancel{h}} = 2.0\ \text{half-lives}$$

[2] For each half-life, multiply the initial activity by 1/2 to obtain the final activity.

$$\underset{\text{initial activity}}{32\ \text{mCi}} \times \underbrace{\frac{1}{2} \times \frac{1}{2}}_{12\ \text{h} = \text{two half-lives}} = \underset{\textbf{Answer}}{8.0\ \text{mCi}}$$

PROBLEM 9.15

A sample of iodine-131 ($t_{1/2}$ = 8.0 days) has an activity of 240 mCi. What is the activity of the sample after 24 days?

9.3B Archaeological Dating

Archaeologists use the half-life of carbon-14 to determine the age of carbon-containing material derived from plants or animals. The technique, **radiocarbon dating,** is based on the fact that the ratio of radioactive carbon-14 to stable carbon-12 is a constant value in a living organism that is constantly taking in CO_2 and other carbon-containing nutrients from its surroundings. Once the organism dies, however, the radioactive isotope (C-14) decays (Section 9.2B) without being replenished, thus decreasing its concentration, while the stable isotope of carbon (C-12) remains at a constant value. By comparing the ratio of C-14 to C-12 in an artifact to the ratio of C-14 to C-12 in organisms today, the age of the artifact can be determined. Radiocarbon dating can be used to give the approximate age of wood, cloth, bone, charcoal, and many other substances that contain carbon.

The half-life of carbon-14 is 5,730 years, so half of the C-14 has decayed after about 6,000 years. Thus, a 6,000-year-old object has a ratio of C-14 to C-12 that has decreased by a factor of two, a 12,000-year-old object has a ratio of C-14 to C-12 that has decreased by a factor of four, and so forth.

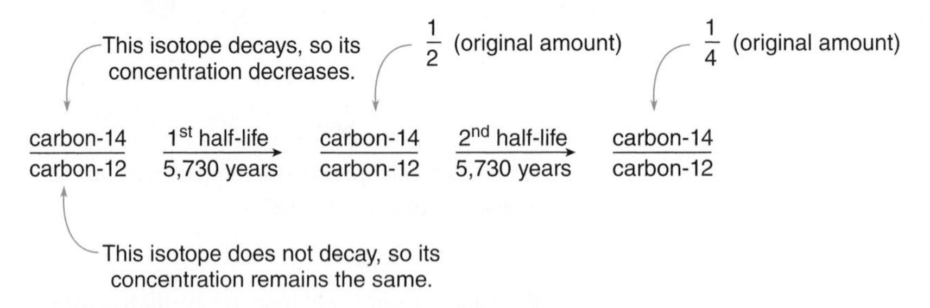

Using this technique, archaeologists have determined the age of the paintings on cave walls in Algeria to be about 8,000 years old (Figure 9.3). Because the amount of carbon-14 decreases with time, artifacts older than about 20,000 years have too little carbon-14 to accurately estimate their age.

PROBLEM 9.16

Estimate the age of an artifact that has 1/8 of the amount of C-14 (relative to C-12) compared to living organisms.

Figure 9.3

Radiocarbon Dating

Radiocarbon dating has been used to estimate the age of this Algerian cave painting at about 8,000 years.

9.4 Detecting and Measuring Radioactivity

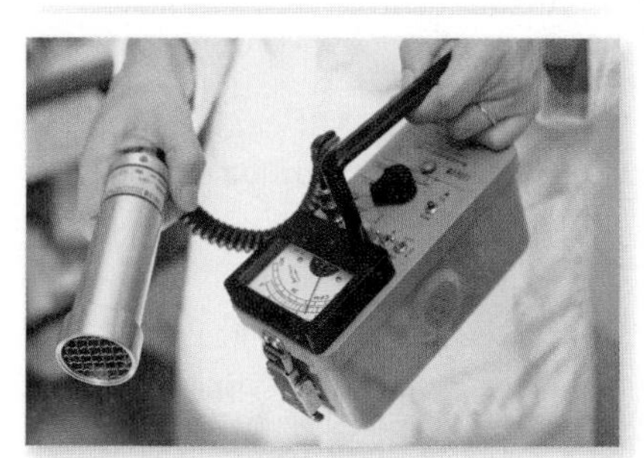

A Geiger counter is a device used to detect radiation.

Individuals who work with radioactivity wear badges to monitor radiation levels.

We all receive a miniscule daily dose of radiation from cosmic rays and radioactive substances in the soil. Additional radiation exposure comes from television sets, dental X-rays, and other man-made sources. Moreover, we are still exposed to nuclear fallout, residual radiation resulting from the testing of nuclear weapons in the atmosphere decades ago.

Although this background radiation is unavoidable and minute, higher levels can be harmful and life-threatening, because radiation is composed of high-energy particles and waves that damage cells and disrupt key biological processes, often causing cell death. How can radiation be detected and measured when it can't be directly observed by any of the senses?

A **Geiger counter** is a small portable device used for measuring radioactivity. It consists of a tube filled with argon gas that is ionized when it comes into contact with nuclear radiation. This in turn generates an electric current that produces a clicking sound or registers on a meter. Geiger counters are used to locate a radiation source or a site that has become contaminated by radioactivity.

Individuals who work with radioactivity wear radiation badges. A radiation badge contains photographic film that fogs when it comes into contact with radioactivity. These badges are regularly monitored to assure that these individuals are not exposed to unhealthy levels of harmful radiation.

9.4A Measuring the Radioactivity in a Sample

The amount of radioactivity in a sample is measured by the number of nuclei that decay per unit time—disintegrations per second. The most common unit is the **curie** (Ci), and smaller units derived from it, the **millicurie** (mCi) and the **microcurie** (μCi). One curie equals 3.7×10^{10} disintegrations/second, which corresponds to the decay rate of 1 g of the element radium.

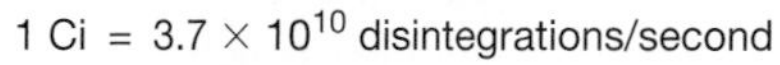

$$1\ \text{Ci} = 3.7 \times 10^{10}\ \text{disintegrations/second}$$
$$1\ \text{Ci} = 1{,}000\ \text{mCi}$$
$$1\ \text{Ci} = 1{,}000{,}000\ \mu\text{Ci}$$

Table 9.3 Units Used to Measure Radioactivity

1 Ci = 3.7×10^{10} disintegrations/s
1 Ci = 3.7×10^{10} Bq
1 Ci = 1,000 mCi
1 Ci = 1,000,000 μCi

The **becquerel** (Bq), an SI unit, is also used to measure radioactivity; 1 Bq = 1 disintegration/second. Since each nuclear decay corresponds to one becquerel, 1 Ci = 3.7×10^{10} Bq. Radioactivity units are summarized in Table 9.3.

Often a dose of radiation is measured in the number of millicuries that must be administered. For example, a diagnostic test for thyroid activity uses sodium iodide that contains iodine-131—that is, $Na^{131}I$. The radioisotope is purchased with a known amount of radioactivity per milliliter, such as 3.5 mCi/mL. By knowing the amount of radioactivity a patient must be given, as well as the concentration of radioactivity in the sample, one can calculate the volume of radioactive isotope that must be administered (Sample Problem 9.5).

SAMPLE PROBLEM 9.5

A patient must be given a 4.5-mCi dose of iodine-131, which is available as a solution that contains 3.5 mCi/mL. What volume of solution must be administered?

Analysis

Use the amount of radioactivity (mCi/mL) as a conversion factor to convert the dose of radioactivity from millicuries to a volume in milliliters.

Solution

The dose of radioactivity is known in millicuries, and the amount of radioactivity per unit volume (3.5 mCi/mL) is also known. Use 3.5 mCi/mL as a millicurie–milliliter conversion factor.

The curie is named for Polish chemist Marie Skłodowska Curie who discovered the radioactive elements polonium and radium, and received Nobel Prizes for both Chemistry and Physics in the early twentieth century.

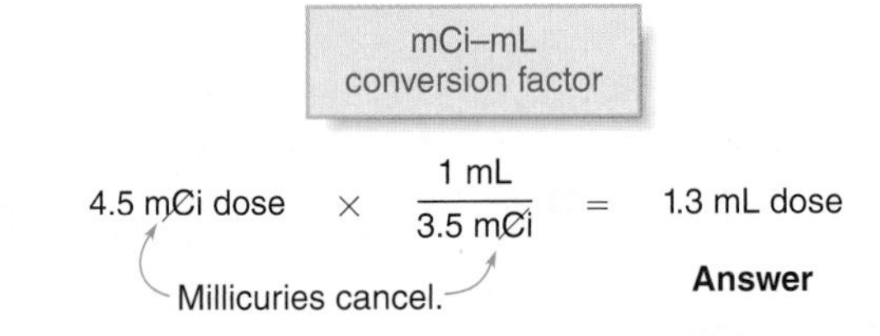

PROBLEM 9.17

To treat a thyroid tumor, a patient must be given a 110-mCi dose of iodine-131, supplied in a vial containing 25 mCi/mL. What volume of solution must be administered?

9.4B FOCUS ON HEALTH & MEDICINE The Effects of Radioactivity

Radioactivity cannot be seen, smelled, tasted, heard, or felt, and yet it can have powerful effects. Because it is high in energy, nuclear radiation penetrates the surface of an object or living organism, where it can damage or kill cells. The cells that are most sensitive to radiation are those that undergo rapid cell division, such as those in bone marrow, reproductive organs, skin, and the intestinal tract. Since cancer cells also rapidly divide, they are also particularly sensitive to radiation, a fact that makes radiation an effective method of cancer treatment (Section 9.5).

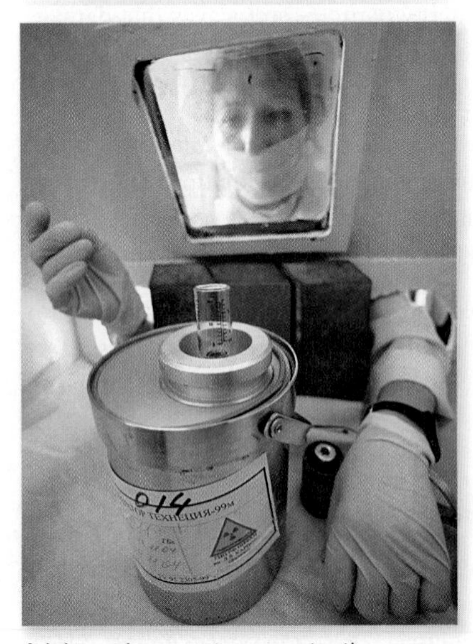

A lab worker must use protective equipment when working with radioactive substances.

Alpha (α) particles, β particles, and γ rays differ in the extent to which they can penetrate a surface. Alpha particles are the heaviest of the radioactive particles, and as a result they move the slowest and penetrate the least. Individuals who work with radioisotopes that emit α particles wear lab coats and gloves that provide a layer of sufficient protection. Beta particles move much faster since they have negligible mass, and they can penetrate into body tissue. Lab workers and health professionals must wear heavy lab coats and gloves when working with substances that give off β particles. Gamma rays travel the fastest and readily penetrate body tissue. Working with substances that emit γ rays is extremely hazardous, and a thick lead shield is required to halt their penetration.

CONSUMER NOTE

Strawberries that have been irradiated (on left) show no mold growth after two weeks, compared to strawberries that have not been irradiated (on right), which are moldy.

That γ rays kill cells is used to an advantage in the food industry. To decrease the incidence of harmful bacteria in foods, certain fruits and vegetables are irradiated with γ rays that kill any bacteria contained in them. Foods do not come into contact with radioisotopes and the food is not radioactive after radiation. Gamma rays merely penetrate the food and destroy any live organism, and often as a result, the food product has a considerably longer shelf life.

9.4C Measuring Human Exposure to Radioactivity

Several units are used to measure the amount of radiation *absorbed* by an organism.

- **The rad—radiation absorbed dose—is the amount of radiation absorbed by one gram of a substance. The amount of energy absorbed varies with both the nature of the substance and the type of radiation.**
- **The rem—radiation equivalent for man—is the amount of radiation that also factors in its energy and potential to damage tissue. Using rem as a measure of radiation, one rem of any type of radiation produces the same amount of tissue damage.**

Other units to measure absorbed radiation include the **gray** (1 Gy = 100 rad) and the **sievert** (1 Sv = 100 rem).

Although background radiation varies with location, the average radiation dose per year for an individual is estimated at 0.27 rem. Generally, no detectable biological effects are noticed when the dose of radiation is less than 25 rem. A single dose of 25–100 rem causes a temporary decrease in white blood cell count. The symptoms of radiation sickness—nausea, vomiting, fatigue, and prolonged decrease in white blood cell count—are visible at a dose of more than 100 rem.

Death results at still higher doses of radiation. The **LD_{50}—the lethal dose that kills 50% of a population**—is 500 rem in humans, and exposure to 600 rem of radiation is fatal for an entire population.

PROBLEM 9.18

The unit millirem (1 rem = 1,000 mrem) is often used to measure the amount of radiation absorbed. (a) The average yearly dose of radiation from radon gas is 200 mrem. How many rem does this correspond to? (b) If a thyroid scan exposes a patient to 0.014 rem of radiation, how many mrem does this correspond to? (c) Which represents the larger dose?

9.5 FOCUS ON HEALTH & MEDICINE Medical Uses of Radioisotopes

Radioactive isotopes are used for both diagnostic and therapeutic procedures in medicine. In a diagnostic test to measure the function of an organ or to locate a tumor, low doses of radioactivity are generally given. When the purpose of using radiation is therapeutic, such as to kill diseased cells or cancerous tissue, a much higher dose of radiation is required.

9.5A Radioisotopes Used in Diagnosis

Radioisotopes are routinely used to determine if an organ is functioning properly or to detect the presence of a tumor. The isotope is ingested or injected and the radiation it emits can be used to produce a scan. Sometimes the isotope is an atom or ion that is not part of a larger molecule. Examples include iodine-131, which is administered as the salt sodium iodide ($Na^{131}I$), and xenon-133, which is a gas containing radioactive xenon atoms. At other times the radioactive atom is bonded to a larger molecule that targets a specific organ. An organ that has increased or decreased uptake of the radioactive element can indicate disease, the presence of a tumor, or other conditions.

A HIDA scan (hepatobiliary iminodiacetic acid scan) uses a technetium-99m-labeled molecule to evaluate the functioning of the gallbladder and bile ducts (Figure 9.4). After injection, the technetium-99m travels through the bloodstream and into the liver, gallbladder, and bile ducts, where, in a healthy individual, the organs are all clearly visible on a scan. When the gallbladder is inflamed or the bile ducts are obstructed by gallstones, uptake of the radioisotope does not occur and these organs are not visualized because they do not contain the radioisotope.

Red blood cells tagged with technetium-99m are used to identify the site of internal bleeding in an individual. Bone scans performed with technetium-99m can show the location of metastatic cancer, so that specific sites can be targeted for radiation therapy (Figure 9.5).

Figure 9.4 HIDA Scan Using Technetium-99m

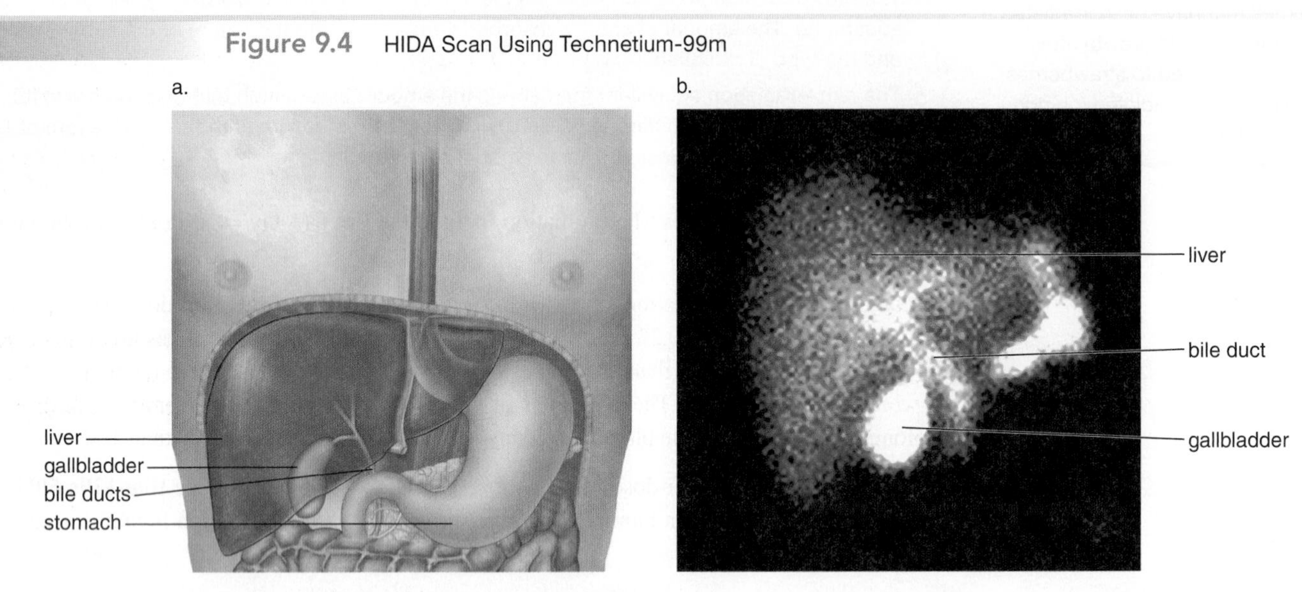

a. Schematic showing the location of the liver, gallbladder, and bile ducts

b. A scan using technetium-99m showing bright areas for the liver, gallbladder, and bile ducts, indicating normal function

Figure 9.5 Bone Scan Using Technetium-99m

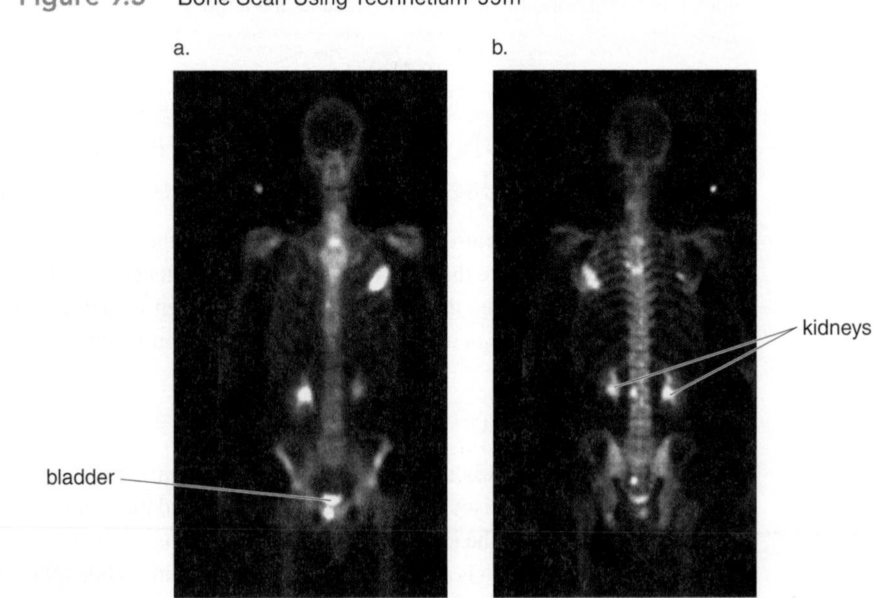

The bone scan of a patient whose lung cancer has spread to other organs. The anterior view [from the front in (a)] shows the spread of disease to the ribs, while the posterior view [from the back in (b)] shows spread of disease to the ribs and spine. The bright areas in the mid-torso and lower pelvis are due to a collection of radioisotope in the kidneys and bladder, before it is eliminated in the urine.

Thallium-201 is used in stress tests to diagnose coronary artery disease. Thallium injected into a vein crosses cell membranes into normal heart muscle. Little radioactive thallium is found in areas of the heart that have a poor blood supply. This technique is used to identify individuals who may need bypass surgery or other interventions because of blocked coronary arteries.

PROBLEM 9.19

The half-life of thallium-201 is three days. What fraction of thallium-201 is still present in an individual after nine days?

9.5B Radioisotopes Used in Treatment

The high-energy radiation emitted by radioisotopes can be used to kill rapidly dividing tumor cells. Two techniques are used. Sometimes the radiation source is external to the body. For example, a beam of radiation produced by decaying cobalt-60 can be focused at a tumor. Such a radiation source must have a much longer half-life—5.3 years in this case—than radioisotopes that are ingested for diagnostic purposes. With this method some destruction of healthy tissue often occurs, and a patient may experience some signs of radiation sickness, including vomiting, fatigue, and hair loss.

A more selective approach to cancer treatment involves using a radioactive isotope internally at the site of the tumor within the body. Using iodine-131 to treat hyperthyroidism has already been discussed (Section 9.1). Other examples include using radioactive "seeds" or wire that can be implanted close to a tumor. Iodine-125 seeds are used to treat prostate cancer and iridium-192 wire is used to treat some cancers of the breast.

Figure 9.6 illustrates radioisotopes that are used for diagnosis or treatment.

Figure 9.6 Common Radioisotopes Used in Medicine

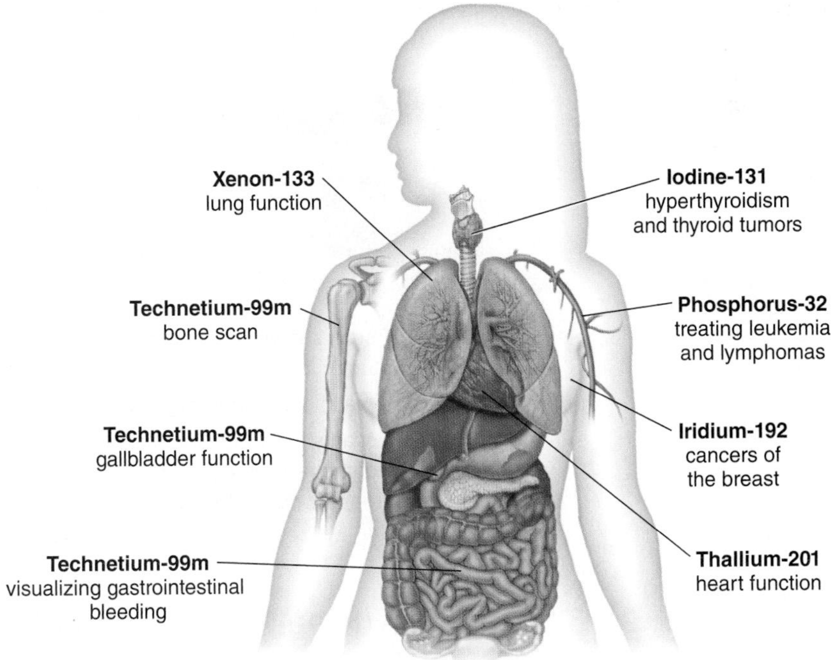

Xenon-133
lung function
Iodine-131
hyperthyroidism and thyroid tumors
Technetium-99m
bone scan
Phosphorus-32
treating leukemia and lymphomas
Technetium-99m
gallbladder function
Iridium-192
cancers of the breast
Technetium-99m
visualizing gastrointestinal bleeding
Thallium-201
heart function

9.5C Positron Emission Tomography—PET Scans

Positron emission tomography (PET) scans use radioisotopes that emit positrons when the nucleus decays. Once formed, a positron combines with an electron to form two γ rays, which create a scan of an organ.

$$ {}^{0}_{+1}\text{e} + {}^{0}_{-1}\text{e} \longrightarrow 2\,\gamma $$

positron electron gamma rays

Carbon-11, oxygen-15, nitrogen-13, and fluorine-18 are common radioactive isotopes used in PET scans. For example, a carbon-11 or fluorine-18 isotope can be incorporated in a glucose molecule. When this radioactive molecule is taken internally, its concentration becomes highest in areas in the body that continually use glucose. A healthy brain shows a high level of radioactivity from labeled glucose. When an individual suffers a stroke or has Alzheimer's disease, brain activity is significantly decreased and radioactivity levels are decreased.

PET scans are also used to detect tumors and coronary artery disease, and determine whether cancer has spread to other organs of the body. A PET scan is also a noninvasive method of monitoring whether cancer treatment has been successful (Figure 9.7).

PROBLEM 9.20

Write a nuclear equation for the emission of a positron from nitrogen-13.

Figure 9.7 PET Scans

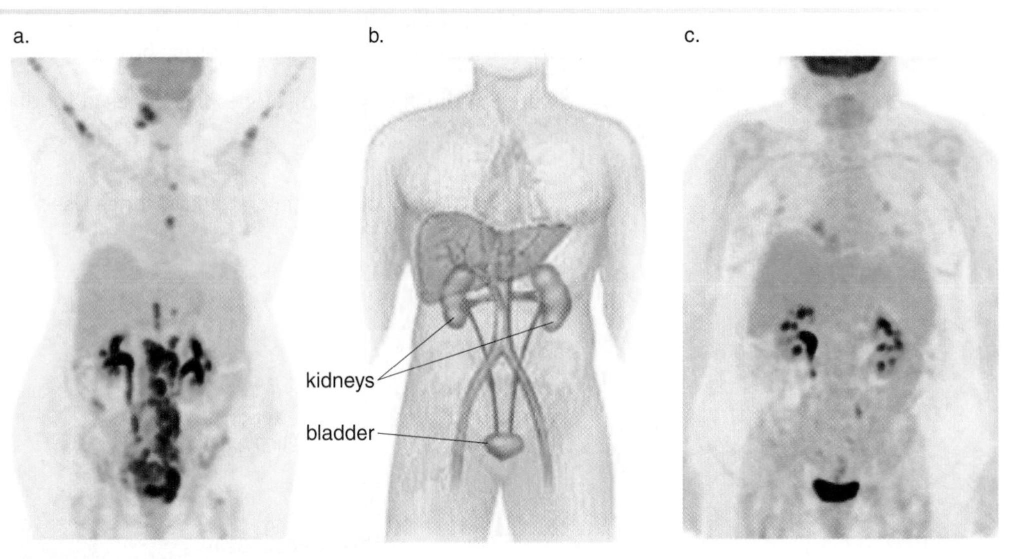

a. The PET scan shows cancer of the lymph nodes in the neck and abdomen, as well as scattered areas of tumor in the bone marrow of the arms and spine before treatment.
b. The schematic of selected organs in the torso and pelvis
c. The PET scan shows significant clearing of disease after chemotherapy by the decrease in intensity of the radioisotope. The dark regions in the kidneys (in the torso) and bladder (in the lower pelvis) are due to the concentration of the radioisotope before elimination in the urine.

9.6 Nuclear Fission and Nuclear Fusion

The nuclear reactions used in nuclear power plants occur by a process called *nuclear fission,* whereas the nuclear reactions that take place in the sun occur by a process called *nuclear fusion.*

- *Nuclear fission* is the splitting apart of a heavy nucleus into lighter nuclei and neutrons.
- *Nuclear fusion* is the joining together of two light nuclei to form a larger nucleus.

9.6A Nuclear Fission

When uranium-235 is bombarded by a neutron, it undergoes **nuclear fission** and splits apart into two lighter nuclei. Several different fission products have been identified. One common nuclear reaction is the fission of uranium-235 into krypton-91 and barium-142.

$$^{235}_{92}U + ^{1}_{0}n \longrightarrow ^{91}_{36}Kr + ^{142}_{56}Ba + 3\,^{1}_{0}n$$

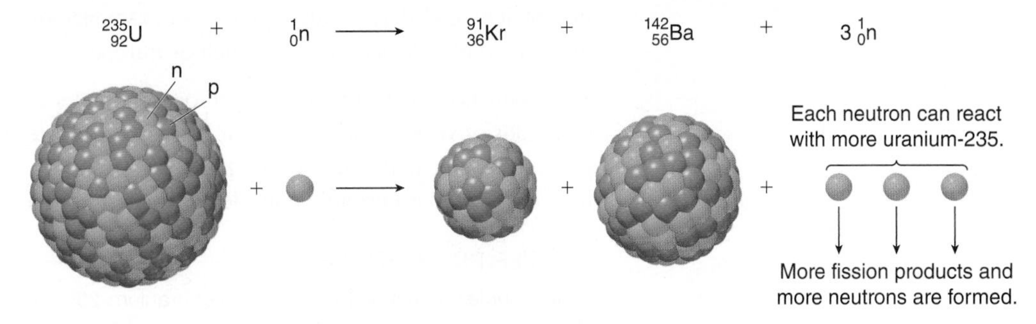

Three high-energy neutrons are also produced in the reaction as well as a great deal of energy. Whereas burning 1 g of methane in natural gas releases 13 kcal of energy, fission of 1 g of uranium-235 releases 3.4×10^8 kcal. Each neutron produced during fission can go on to bombard three other uranium-235 nuclei to produce more nuclei and more neutrons. Such a process is called a **chain reaction.**

In order to sustain a chain reaction there must be a sufficient amount of uranium-235. When that amount—the **critical mass**—is present, the chain reaction occurs over and over again and an atomic explosion occurs. When less than the critical mass of uranium-235 is present, there is a more controlled production of energy, as is the case in a nuclear power plant.

A nuclear power plant utilizes the tremendous amount of energy produced by fission of the uranium-235 nucleus to heat water to steam, which powers a generator to produce electricity (Figure 9.8). While nuclear energy accounts for a small but significant fraction of the electricity needs in the United States, most of the electricity generated in some European countries comes from nuclear power.

Figure 9.8 A Nuclear Power Plant

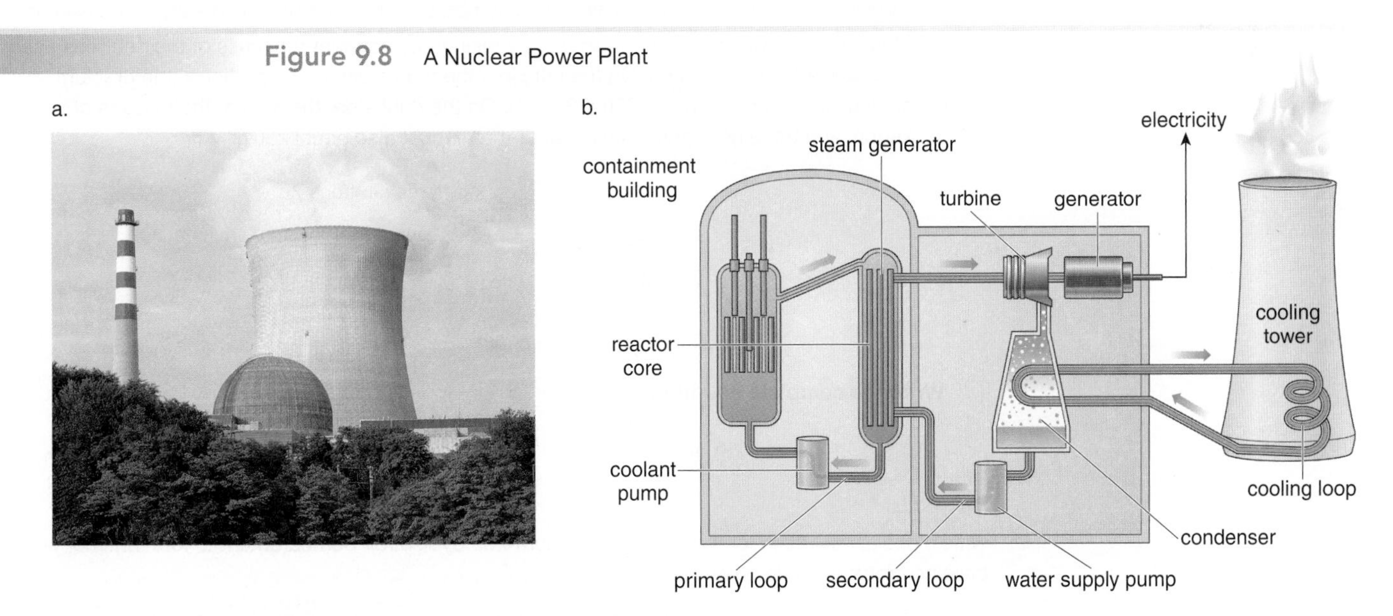

a. Nuclear power plant with steam rising from a cooling tower

b. Fission occurs in a nuclear reactor core that is housed in a containment facility. Water surrounding the reactor is heated by the energy released during fission, and this energy drives a turbine, which produces electricity. Once the steam has been used to drive the turbine, it is cooled and re-circulated around the core of the reactor. To prevent the loss of any radioactive material to the environment, the water that surrounds the reactor core never leaves the containment building.

Two problems that surround nuclear power generation are the possibility of radiation leaks and the disposal of nuclear waste. Plants are designed and monitored to contain the radioactive materials within the nuclear reactor. The reactor core itself is located in a containment facility with thick walls, so that should a leak occur, the radiation should in principle be kept within the building. The nuclear reactor in Chernobyl, Russia, was built without a containment facility and in 1986 it exploded, releasing high levels of radioactivity to the immediate environment and sending a cloud of radioactivity over much of Europe.

The products of nuclear fission are radioactive nuclei with long half-lives, often hundreds or even thousands of years. As a result, nuclear fission generates radioactive waste that must be stored in a secure facility so that it does not pose a hazard to the immediate surroundings. Burying waste far underground is currently considered the best option, but this issue is still unresolved.

SAMPLE PROBLEM 9.6

Write a nuclear equation for the fission of uranium-235 by neutron bombardment to form strontium-90, an isotope of xenon, and three neutrons.

Analysis

Balance the atomic numbers and mass numbers on both sides of the nuclear equation. In fission reactions, include the neutron used for bombardment on the left and the high-energy neutrons produced on the right. Each neutron has a mass of one and zero charge.

Solution

[1] **Write an incomplete equation with the original nucleus and neutron used for bombardment on the left, and the particles formed on the right.**

- Use the identity of each element to determine its atomic number. Uranium has an atomic number of 92, and xenon has an atomic number of 54.
- Include one neutron on the left side. In this reaction, three high-energy neutrons are formed, so include three neutrons on the right side.

$$^{235}_{92}\text{U} + ^{1}_{0}\text{n} \longrightarrow ^{90}_{38}\text{Sr} + ^{?}_{54}\text{Xe} + 3\ ^{1}_{0}\text{n}$$

[2] **Calculate the mass numbers and atomic numbers of all newly formed nuclei on the right.**

- Atomic number: In this problem all atomic numbers are known from the identity of the elements.
- Mass number: Balance mass numbers by taking into account the mass of the neutrons used or produced in the reaction. On the left side, the total mass of the particles (the uranium nucleus and one neutron) is 236 (235 + 1). On the right side, the sum of the masses of Sr-90, xenon, and three neutrons (total mass of three) must also equal 236.

$$236 = 90 + ? + 3(1)$$
$$236 = 93 + ?$$
$$143 = ?$$

three neutrons

one mass unit from each neutron

The mass number of xenon is 143.

[3] **Write the complete equation.**

$$^{235}_{92}\text{U} + ^{1}_{0}\text{n} \longrightarrow ^{90}_{38}\text{Sr} + ^{143}_{54}\text{Xe} + 3\ ^{1}_{0}\text{n}$$

PROBLEM 9.21

Write a nuclear equation for the fission of uranium-235 by neutron bombardment to form antimony-133, three neutrons, and one other isotope.

9.6B Nuclear Fusion

Nuclear fusion occurs when two light nuclei join together to form a larger nucleus. For example, fusion of a deuterium nucleus with a tritium nucleus forms helium and a neutron. Recall from Section 2.3 that deuterium is an isotope of hydrogen that contains one proton and one neutron in its nucleus, while tritium is an isotope of hydrogen that contains one proton and two neutrons in its nucleus.

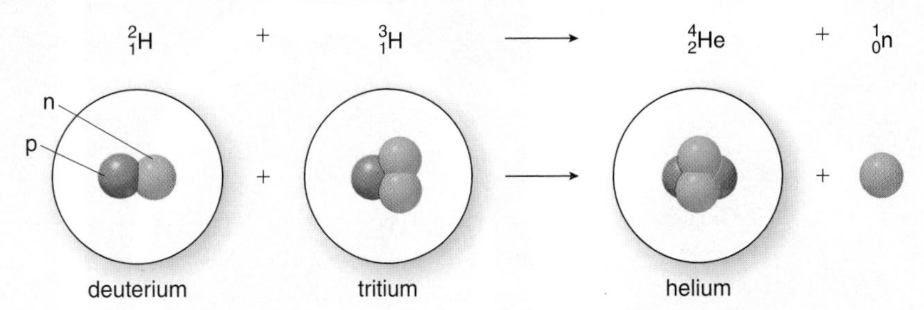

Like fission, fusion also releases a great deal of energy—namely, 5.3×10^{8} kcal/mol of helium produced. The light and heat of the sun and other stars result from nuclear fusion.

One limitation of using fusion to provide energy for mankind is the extreme experimental conditions needed to produce it. Because it takes a considerable amount of energy to overcome the repulsive forces of the like charges of two nuclei, fusion can only be accomplished at high temperatures (greater than 100,000,000 °C) and pressures (greater than 100,000 atm). Since these conditions are not easily achieved, using controlled nuclear fusion as an energy source has yet to become a reality.

Controlled nuclear fusion has the potential of providing cheap and clean power. It is not plagued by the nuclear waste issues of fission reactors, and the needed reactants are readily available.

PROBLEM 9.22

Nuclear fusion in the stars occurs by a series of reactions. Identify **X, Y,** and **Z** in the following nuclear reactions that ultimately convert hydrogen into helium.

a. $^{1}_{1}H + \mathbf{X} \longrightarrow ^{2}_{1}H + ^{0}_{+1}e$

b. $^{1}_{1}H + ^{2}_{1}H \longrightarrow \mathbf{Y}$

c. $^{1}_{1}H + ^{3}_{2}He \longrightarrow ^{4}_{2}He + \mathbf{Z}$

9.7 FOCUS ON HEALTH & MEDICINE Medical Imaging Without Radioactivity

X-rays, CT scans, and **MRIs** are also techniques that provide an image of an organ or extremity that is used for diagnosis of a medical condition. Unlike PET scans and other procedures discussed thus far, however, **these procedures are *not* based on nuclear reactions and they do *not* utilize radioactivity.** In each technique, an energy source is directed towards a specific region in the body, and a scan is produced that is analyzed by a trained medical professional.

X-rays are a high-energy type of radiation called electromagnetic radiation. Tissues of different density interact differently with an X-ray beam, and so a map of bone and internal organs is created on an X-ray film. Dense bone is clearly visible in an X-ray, making it a good diagnostic technique for finding fractures (Figure 9.9a). Although X-rays are a form of high-energy radiation, they are lower in energy than the γ rays produced in nuclear reactions. Nonetheless, X-rays still cause adverse biological effects on the cells with which they come in contact, and the exposure of both the patient and X-ray technician must be limited.

Figure 9.9 Imaging the Human Body

a.

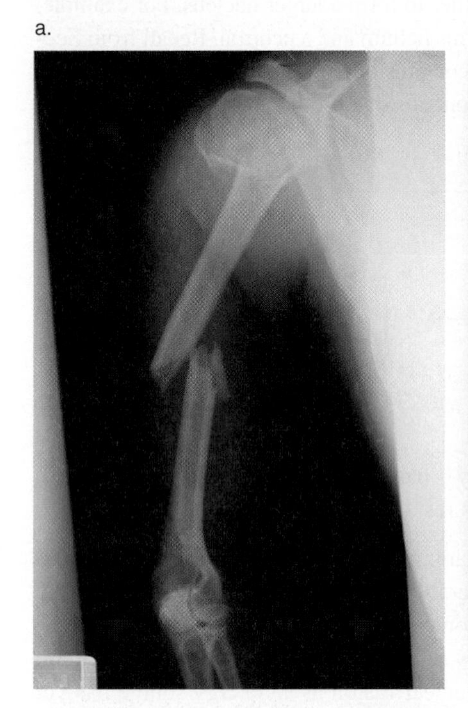

b.

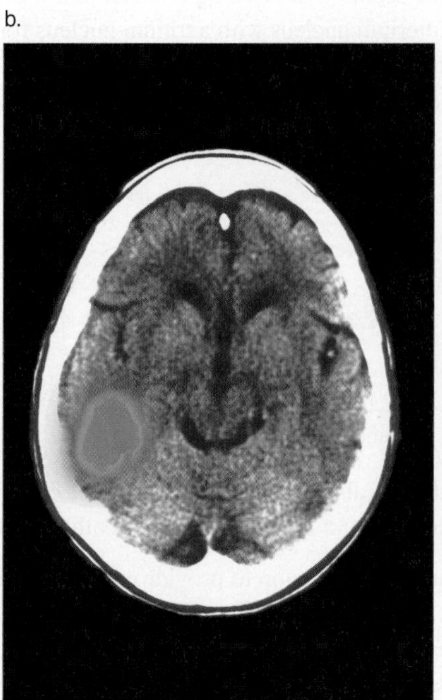

c.

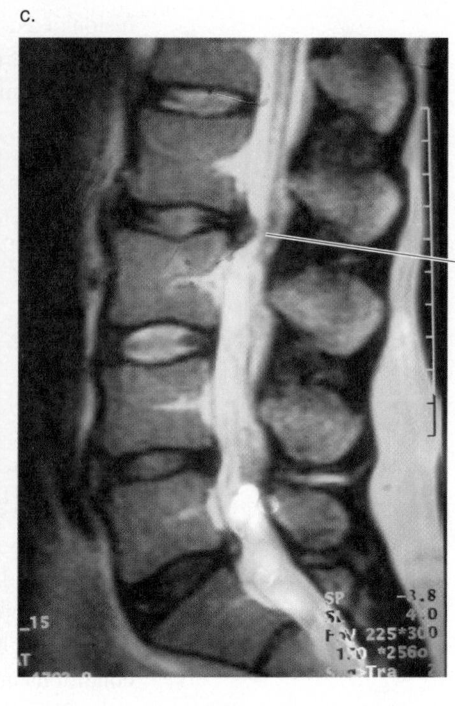

a. X-ray of a broken humerus in a patient's arm

b. A color-enhanced CT scan of the head showing the site of a stroke

c. MRI of the spinal cord showing spinal compression from a herniated disc

CT (computed tomography) scans, which also use X-rays, provide high resolution images of "slices" of the body. Historically, CT images have shown a slice of tissue perpendicular to the long axis of the body. Modern CT scanners can now provide a three-dimensional view of the body's organs. CT scans of the head are used to diagnose bleeding and tumors in the brain (Figure 9.9b).

MRI (magnetic resonance imaging) uses low-energy radio waves to visualize internal organs. Unlike methods that use high-energy radiation, MRIs do not damage cells. An MRI is a good diagnostic method for visualizing soft tissue (Figure 9.9c), and thus it complements X-ray techniques.

KEY TERMS

Alpha (α) particle (9.1)
Becquerel (9.4)
Beta (β) particle (9.1)
Chain reaction (9.6)
Critical mass (9.6)
Curie (9.4)
Gamma (γ) ray (9.1)
Geiger counter (9.4)
Gray (9.4)
Half-life (9.3)
LD_{50} (9.4)
Nuclear fission (9.6)
Nuclear fusion (9.6)
Nuclear reaction (9.2)
Positron (9.1)
Rad (9.4)
Radioactive decay (9.2)
Radioactive isotope (9.1)
Radioactivity (9.1)
Radiocarbon dating (9.3)
Rem (9.4)
Sievert (9.4)
X-ray (9.7)

KEY CONCEPTS

1 Describe the different types of radiation emitted by a radioactive nucleus. (9.1)

- A radioactive nucleus can emit α particles, β particles, positrons, or γ rays.
- An α particle is a high-energy nucleus that contains two protons and two neutrons.
- A β particle is a high-energy electron.
- A positron is an antiparticle of a β particle. A positron has a +1 charge and negligible mass.
- A γ ray is high-energy radiation with no mass or charge.

2 How are equations for nuclear reactions written? (9.2)

- In an equation for a nuclear reaction, the sum of the mass numbers (*A*) must be equal on both sides of the equation. The sum of the atomic numbers (*Z*) must be equal on both sides of the equation as well.

3 What is the half-life of a radioactive isotope? (9.3)

- The half-life ($t_{1/2}$) is the time it takes for one-half of a radioactive sample to decay. Knowing the half-life and the amount of a radioactive substance, one can calculate how much sample remains after a period of time.

4 What units are used to measure radioactivity? (9.4)

- Radiation in a sample is measured by the number of disintegrations per second, most often using the curie (Ci); 1 Ci = 3.7×10^{10} disintegrations/s. The becquerel (Bq) is also used; 1 Bq = 1 disintegration/s; 1 Ci = 3.7×10^{10} Bq.
- The exposure of a substance to radioactivity is measured with the rad (radiation absorbed dose) or the rem (radiation equivalent for man).

5 Give examples of common radioisotopes used in medicine. (9.5)

- Iodine-131 is used to diagnose and treat thyroid disease.
- Technetium-99m is used to evaluate the functioning of the gallbladder and bile ducts, and in bone scans to evaluate the spread of cancer.
- Red blood cells tagged with technetium-99m are used to find the site of a gastrointestinal bleed.
- Thallium-201 is used to diagnose coronary artery disease.
- Cobalt-60 is used as an external source of radiation for cancer treatment.
- Iodine-125 and iridium-192 are used in internal radiation treatment of prostate cancer and breast cancer, respectively.
- Carbon-11, oxygen-15, nitrogen-13, and fluorine-18 are used in positron emission tomography.

6 What are nuclear fission and nuclear fusion? (9.6)

- Nuclear fission is the splitting apart of a heavy nucleus into lighter nuclei and neutrons.
- Nuclear fusion is the joining together of two light nuclei to form a larger nucleus.
- Both nuclear fission and nuclear fusion release a great deal of energy. Nuclear fission is used in nuclear power plants to generate electricity. Nuclear fusion occurs in stars.

7 What medical imaging techniques do not use radioactivity? (9.7)

- X-rays and CT scans both use X-rays, a high-energy form of electromagnetic radiation.
- MRIs use low-energy radio waves to image soft tissue.

UNDERSTANDING KEY CONCEPTS

Selected in-chapter and odd-numbered end-of-chapter problems have brief answers at the end of each chapter. The *Student Study Guide and Solutions Manual* contains detailed solutions to all in-chapter and odd-numbered end-of-chapter problems, as well as additional worked examples and a chapter self-test.

9.23 Compare fluorine-18 and fluorine-19 with regard to each of the following: (a) atomic number; (b) number of protons; (c) number of neutrons; (d) mass number. Give the isotope symbol for each isotope. F-19 is a stable nucleus and F-18 is used in PET scans.

9.24 Compare nitrogen-13 and nitrogen-14 with regard to each of the following: (a) atomic number; (b) number of protons; (c) number of neutrons; (d) mass number. Give the isotope symbol for each isotope. N-14 is a stable nucleus and N-13 is used in PET scans.

9.25 Complete the nuclear equation by drawing the nucleus of the missing atom. Give the symbol for each atom and type of radiation. (The blue spheres represent protons and the orange spheres represent neutrons.)

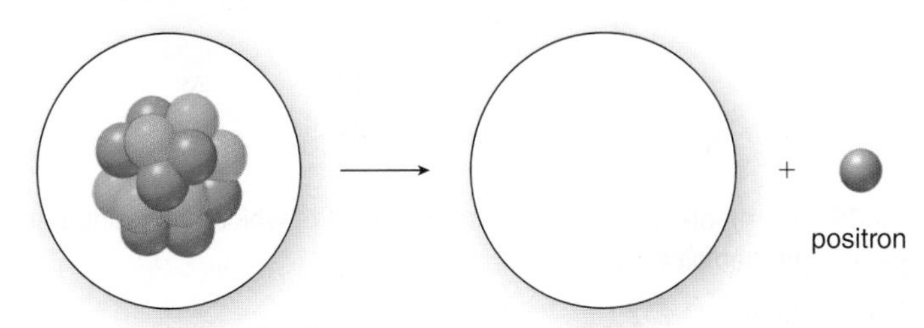

9.26 Complete the nuclear equation by drawing the nucleus of the missing atom. Give the symbol for each atom and type of radiation. (The blue spheres represent protons and the orange spheres represent neutrons.)

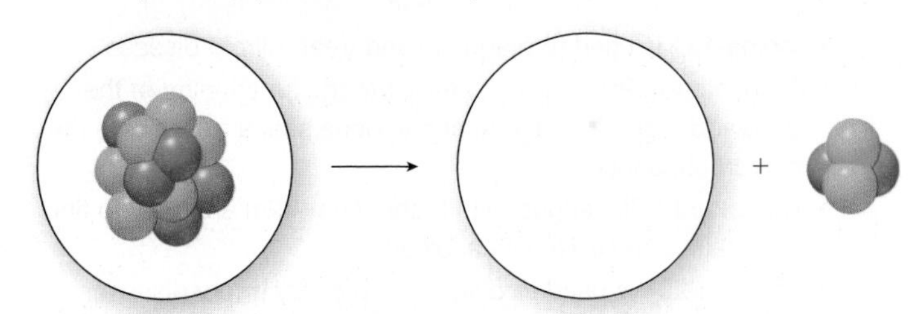

9.27 A radioactive element (shown with blue spheres) forms a decay product (shown with green spheres) with a half-life of three days. How many half-lives have elapsed in diagram **A?**

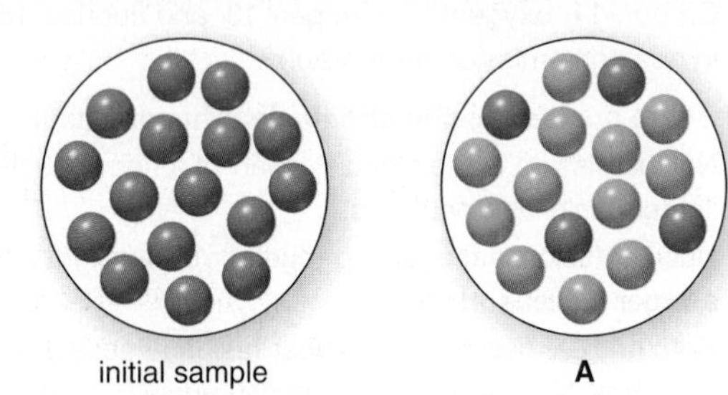

9.28 A radioactive element (shown with blue spheres) forms a decay product (shown with green spheres) with a half-life of 10 min. How many minutes have elapsed to form the sample in diagram **B?**

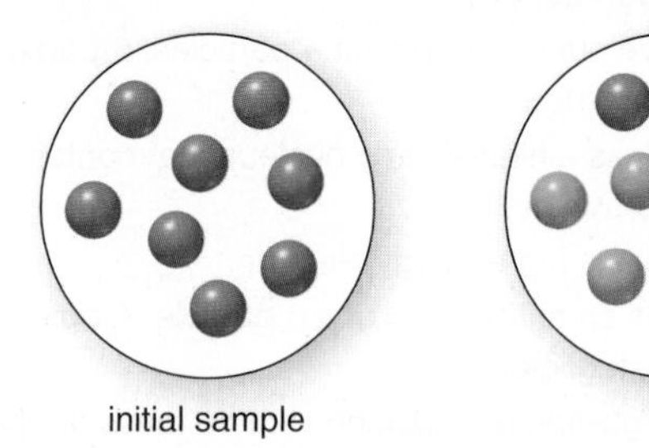

9.29 Arsenic-74 is a radioisotope used for locating brain tumors.

a. Write a balanced nuclear equation for the positron emission of arsenic-74.

b. If $t_{1/2}$ for As-74 is 18 days, how much of a 120-mg sample remains after 90. days?

c. If the radioactivity of a 2.0-mL vial of arsenic-74 is 10.0 mCi, what volume must be administered to give a 7.5-mCi dose?

9.30 Sodium-24 is a radioisotope used for examining circulation.

a. Write a balanced nuclear equation for the β decay of sodium-24.

b. If $t_{1/2}$ for Na-24 is 15 h, how much of an 84-mg sample remains after 2.5 days?

c. If the radioactivity of a 5.0-mL vial of sodium-24 is 10.0 mCi, what volume must be administered to give a 6.5-mCi dose?

ADDITIONAL PROBLEMS

Isotopes and Radiation

9.31 Complete the table of isotopes, each of which has found use in medicine.

	Atomic Number	Mass Number	Number of Protons	Number of Neutrons	Isotope Symbol
a. Chromium-51					
b.	46	103			
c.			19	23	
d.		133	54		

9.32 Complete the table of isotopes, each of which has found use in medicine.

	Atomic Number	Mass Number	Number of Protons	Number of Neutrons	Isotope Symbol
a. Sodium-24					
b.		89		51	
c.		59	26		
d. Samarium-153					

9.33 How much does the mass and charge of a nucleus change when each type of radiation is emitted: (a) α particle; (b) β particle; (c) γ ray; (d) positron?

9.34 Compare α particles, β particles, and γ rays with regard to each of the following: (a) speed the radiation travels; (b) penetrating power; (c) protective equipment that must be worn when handling.

9.35 What is the mass and charge of radiation that has each of the following symbols: (a) α; (b) n; (c) γ; (d) β?

9.36 What is the mass and charge of radiation that has each of the following symbols?

a. ${}^{0}_{-1}e$ b. ${}^{0}_{+1}e$ c. ${}^{4}_{2}He$ d. β^+

Nuclear Reactions

9.37 Complete each nuclear equation.

a. ${}^{59}_{26}Fe \longrightarrow ? + {}^{0}_{-1}e$

b. ${}^{190}_{78}Pt \longrightarrow ? + {}^{4}_{2}He$

c. ${}^{178}_{80}Hg \longrightarrow ? + {}^{0}_{+1}e$

9.38 Complete each nuclear equation.

a. ${}^{77}_{37}Rb \longrightarrow ? + {}^{0}_{+1}e$

b. ${}^{251}_{102}No \longrightarrow ? + {}^{4}_{2}He$

c. ${}^{66}_{29}Cu \longrightarrow ? + {}^{0}_{-1}e$

9.39 Complete each nuclear equation.

a. ${}^{90}_{39}Y \longrightarrow {}^{90}_{40}Zr + ?$

b. $? \longrightarrow {}^{135}_{59}Pr + {}^{0}_{+1}e$

c. ${}^{210}_{83}Bi \longrightarrow ? + {}^{4}_{2}He$

9.40 Complete each nuclear equation.

a. $? \longrightarrow {}^{90}_{39}Y + {}^{0}_{-1}e$

b. ${}^{29}_{15}P \longrightarrow {}^{29}_{14}Si + ?$

c. ${}^{214}_{84}Po \longrightarrow ? + {}^{4}_{2}He$

9.41 Bismuth-214 can decay to form either polonium-214 or thallium-210, depending on what type of radiation is emitted. Write a balanced nuclear equation for each process.

9.42 Lead-210 can be formed by the decay of either thallium-210 or polonium-214, depending on what type of radiation is emitted. Write a balanced nuclear equation for each process.

9.43 Write a balanced nuclear equation for each reaction.

a. decay of thorium-232 by α emission
b. decay of sodium-25 by β emission
c. decay of xenon-118 by positron emission
d. decay of curium-243 by α emission

9.44 Write a balanced nuclear equation for each reaction.

a. decay of sulfur-35 by β emission
b. decay of thorium-225 by α emission
c. decay of rhodium-93 by positron emission
d. decay of silver-114 by β emission

Half-Life

9.45 If the amount of a radioactive element decreases from 2.4 g to 0.30 g in 12 days, what is its half-life?

9.46 If the amount of a radioactive element decreases from 0.36 g to 90. mg in 22 min, what is its half-life?

9.47 Radioactive iodine-131 ($t_{1/2}$ = 8.0 days) decays to form xenon-131 by emission of a β particle. How much of each isotope is present after each time interval if 64 mg of iodine-131 was present initially: (a) 8.0 days; (b) 16 days; (c) 24 days; (d) 32 days?

9.48 Radioactive phosphorus-32 decays to form sulfur-32 by emission of a β particle. Estimating the half-life to be 14 days, how much of each isotope is present after each time interval if 124 mg of phosphorus-32 was present initially: (a) 14 days; (b) 28 days; (c) 42 days; (d) 56 days?

9.49 If the half-life of an isotope is 24 hours, has all the isotope decayed in 48 hours?

9.50 Explain how the half-life of carbon-14 is used to date objects.

9.51 Why can't radiocarbon dating be used to determine the age of an artifact that is over 50,000 years old?

9.52 Why can't radiocarbon dating be used to estimate the age of rocks?

9.53 A patient is injected with a sample of technetium-99m ($t_{1/2}$ = 6.0 h), which has an activity of 20 mCi. What activity is observed after each interval: (a) 6 h; (b) 12 h; (c) 24 h?

9.54 A sample of iodine-131 ($t_{1/2}$ = 8.0 days) has an activity of 200. mCi. What activity is observed after each interval: (a) 8.0 days; (b) 24 days; (c) 48 days?

Measuring Radioactivity

9.55 A patient must be administered a 28-mCi dose of technetium-99m, which is supplied in a vial containing a solution with an activity of 12 mCi/mL. What volume of solution must be given?

9.56 A radioactive isotope used for imaging is supplied in an 8.0-mL vial containing a solution with an activity of 108 mCi. What volume must be given to a patient who needs a 12-mCi dose?

9.57 Radioactive sodium-24, administered as $^{24}NaCl$, is given to treat leukemia. If a patient must receive 190 μCi/kg and the isotope is supplied as a solution that contains 5.0 mCi/mL, what volume is needed for a 68-kg patient?

9.58 Radioactive phosphorus-32, administered as sodium phosphate ($Na_3{}^{32}PO_4$), is used to treat chronic leukemia. The activity of an intravenous solution is 670 μCi/mL. What volume of solution must be used to supply a dose of 15 mCi?

9.59 The initial responders to the Chernobyl nuclear disaster were exposed to 20 Sv of radiation. Convert this value to rem. Did these individuals receive a fatal dose of radiation?

9.60 Many individuals who fought fires at the Chernobyl nuclear disaster site were exposed to 0.25 Sv of radiation. Convert this value to rem. Did these individuals receive a fatal dose of radiation? Would you expect any of these individuals to have shown ill health effects?

Nuclear Fission and Nuclear Fusion

9.61 What is the difference between nuclear fission and nuclear fusion?

9.62 What is the difference between the nuclear fission process that takes place in a nuclear reactor and the nuclear fission that occurs in an atomic bomb?

9.63 For which process does each statement apply—nuclear fission, nuclear fusion, both fission and fusion?

a. The reaction occurs in the sun.
b. A neutron is used to bombard a nucleus.
c. A large amount of energy is released.
d. Very high temperatures are required.

9.64 For which process does each statement apply—nuclear fission, nuclear fusion, both fission and fusion?

a. The reaction splits a nucleus into lighter nuclei.
b. The reaction joins two lighter nuclei into a heavier nucleus.
c. The reaction is used to generate energy in a nuclear power plant.
d. The reaction generates radioactive waste with a long half-life.

9.65 Complete the nuclear fusion equation by drawing in the nucleus of the missing atom. Blue spheres represent protons and orange spheres represent neutrons.

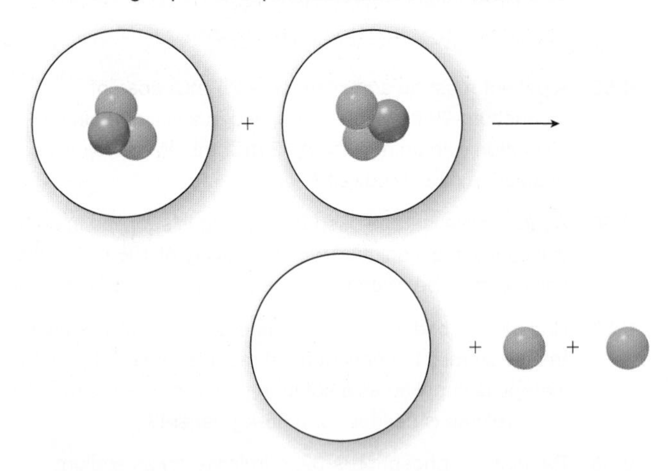

9.66 Complete the nuclear fusion equation by drawing in the nucleus of the missing atom. Blue spheres represent protons and orange spheres represent neutrons.

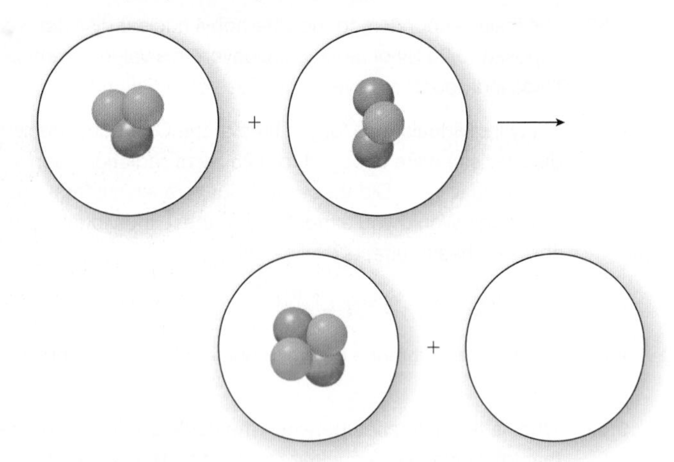

9.67 Complete each nuclear fission equation.

a. ${}^{235}_{92}U + {}^{1}_{0}n \longrightarrow ? + {}^{97}_{42}Mo + 2\,{}^{1}_{0}n$

b. ${}^{235}_{92}U + {}^{1}_{0}n \longrightarrow ? + {}^{140}_{56}Ba + 3\,{}^{1}_{0}n$

9.68 Complete each nuclear fission equation.

a. ${}^{235}_{92}U + {}^{1}_{0}n \longrightarrow ? + {}^{139}_{57}La + 2\,{}^{1}_{0}n$

b. ${}^{235}_{92}U + {}^{1}_{0}n \longrightarrow ? + {}^{140}_{58}Ce + 2\,{}^{1}_{0}n + 6\,{}^{0}_{-1}e$

9.69 The fusion of two deuterium nuclei (hydrogen-2) forms a hydrogen nucleus (hydrogen-1) as one product. What other product is formed?

9.70 Fill in the missing product in the following nuclear fusion reaction.

$${}^{3}_{2}He + {}^{3}_{2}He \longrightarrow ? + 2\,{}^{1}_{1}H$$

9.71 Discuss two problems that surround the generation of electricity from a nuclear power plant.

9.72 Why are there as yet no nuclear power plants that use nuclear fusion to generate electricity?

General Questions

9.73 Answer the following questions about radioactive iridium-192.

a. Write a balanced nuclear equation for the decay of iridium-192, which emits both a β particle and a γ ray.

b. If $t_{1/2}$ for Ir-192 is 74 days, estimate how much of a 120-mg sample remains after five 30-day months.

c. If a sample of Ir-192 had an initial activity of 36 Ci, estimate how much activity remained in the sample after ten 30-day months.

9.74 Answer the following questions about radioactive samarium-153.

a. Write a balanced nuclear equation for the decay of samarium-153, which emits both a β particle and a γ ray.

b. If $t_{1/2}$ for Sm-153 is 46 h, estimate how much of a 160-mg sample remains after four days.

c. If a sample of Sm-153 had an initial activity of 48 Ci, estimate how much activity remained in the sample after six days.

9.75 All nuclei with atomic numbers around 100 or larger do not exist naturally; rather, they have been synthesized by fusing two lighter-weight nuclei together. Using the principles for balancing nuclear reactions in Section 9.2, complete the nuclear equation by drawing the nucleus of the missing atom **X.** Give the element symbol, atomic number, and mass number of **X.**

9.76 Using the principles for balancing nuclear reactions in Section 9.2, complete the nuclear equation by drawing the nucleus of the missing atom **Y.** Give the element symbol, atomic number, and mass number of **Y.**

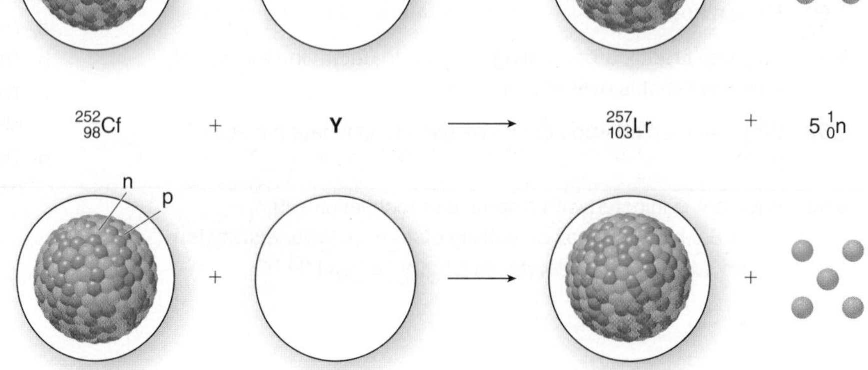

9.77 Complete the following nuclear equation by giving the name, atomic number, and mass number of the element made by this reaction.

$$^{209}_{83}Bi + ^{58}_{26}Fe \longrightarrow ? + ^{1}_{0}n$$

9.78 Complete the following nuclear equation, and give the name, atomic number, and mass number of the element made by this reaction.

$$^{235}_{92}U + ^{14}_{7}N \longrightarrow ? + 5\ ^{1}_{0}n$$

Applications

9.79 Explain how each isotope is used in medicine.

a. iodine-131 b. iridium-192 c. thallium-201

9.80 Explain how each isotope is used in medicine.

a. iodine-125 b. technetium-99m c. cobalt-60

9.81 How does the half-life of each of the following isotopes of iodine affect the manner in which it is administered to a patient: (a) iodine-125, $t_{1/2}$ = 60 days; (b) iodine-131, $t_{1/2}$ = 8 days?

9.82 Explain why food is irradiated with γ rays.

9.83 A mammogram is an X-ray of the breast. Why does an X-ray technician leave the room or go behind a shield when a mammogram is performed on a patient?

9.84 Why is a lead apron placed over a patient's body when dental X-rays are taken?

9.85 One of the radioactive isotopes that contaminated the area around Chernobyl after the nuclear accident in 1986 was iodine-131. Suggest a reason why individuals in the affected region were given doses of NaI that contained the stable iodine-127 isotope.

9.86 The element strontium has similar properties to calcium. Suggest a reason why exposure to strontium-90, a product of nuclear testing in the atmosphere, is especially hazardous for children.

CHALLENGE PROBLEMS

9.87 An article states that the fission of 1.0 g of uranium-235 releases 3.4×10^8 kcal, the same amount of energy as burning one ton (2,000. lb) of coal. If this report is accurate, how much energy is released when 1.0 g of coal is burned?

9.88 Radioactive isotopes with high atomic numbers often decay to form isotopes that are themselves radioactive, and once formed, decay to form new isotopes. Sometimes a series of such decays occurs over many steps until a stable nucleus is formed. The following series of decays occurs: Polonium-218 decays with emission of an α particle to form **X,** which emits a β particle to form **Y,** which emits an α particle to form **Z.** Identify **X, Y,** and **Z.**

BEYOND THE CLASSROOM

9.89 The incident at the Three Mile Island Nuclear Generating Station in Pennsylvania in 1979 is considered the most serious accident in the history of the nuclear power industry in the United States. What factors contributed to the accident? How much radiation was released into the atmosphere and what radioactive isotopes were present? What improvements were made in employee training and reactor design as a result of this incident? Were any short- or long-term effects observed on the general health or cancer rate of individuals living near the plant?

9.90 The use of radioisotopes in medicine is a rapidly expanding field. Pick an isotope not discussed extensively in Chapter 9. Possibilities might include the radioactive isotopes of bismuth, lutetium, chromium, phosphorus, or samarium. Report the mass number of the radioisotope as well as the mass numbers of stable isotopes of the element. What is its half-life? How is the isotope used in medicine? What type of radiation does it emit? What advantage (if any) does the chosen isotope have over other radioisotopes?

9.91 Although the irradiation of certain food products provides an effective method for destroying harmful bacteria and disease-carrying pathogens, some consumer groups question the safety of irradiated food. Discuss the advantages and disadvantages of food irradiation. How might large-scale irradiation affect product availability? What products are typically irradiated, and how large of a radiation dose is used? Does your local market carry food that has been irradiated?

9.92 Polonium-210 is a naturally occurring radioactive element that was used to poison Alexander Litvinenko, an ex-KGB agent who became a critic of the Russian government, in 2006. What is known about polonium-210? By what path does it decay? What is it used for? What is known about the source of polonium-210 used in the poisoning? How long is its half-life and what are the biological hazards of coming into contact with polonium-210? What is the current status of the investigation into Litvinenko's death?

ANSWERS TO SELECTED PROBLEMS

9.1

	Atomic Number	Mass Number	Number of Protons	Number of Neutrons
a. $^{85}_{38}Sr$	38	85	38	47
b. $^{67}_{31}Ga$	31	67	31	36
c. Selenium-75	34	75	34	41

9.3 An electron has a negative charge of –1 and a positron has a positive charge of +1.

9.5 $^{222}_{86}Rn \longrightarrow {}^{4}_{2}He + {}^{218}_{84}Po$

9.7 a. $^{218}_{84}Po \longrightarrow {}^{4}_{2}He + {}^{214}_{82}Pb$

b. $^{230}_{90}Th \longrightarrow {}^{4}_{2}He + {}^{226}_{88}Ra$

c. $^{252}_{99}Es \longrightarrow {}^{4}_{2}He + {}^{248}_{97}Bk$

9.8 $^{131}_{53}I \longrightarrow {}^{0}_{-1}e + {}^{131}_{54}Xe$

9.9 a. $^{20}_{9}F \longrightarrow {}^{0}_{-1}e + {}^{20}_{10}Ne$

b. $^{92}_{38}Sr \longrightarrow {}^{0}_{-1}e + {}^{92}_{39}Y$

c. $^{55}_{24}Cr \longrightarrow {}^{0}_{-1}e + {}^{55}_{25}Mn$

9.10 a. $^{74}_{33}As \longrightarrow {}^{0}_{+1}e + {}^{74}_{32}Ge$

b. $^{15}_{8}O \longrightarrow {}^{0}_{+1}e + {}^{15}_{7}N$

9.11 $^{192}_{77}Ir \longrightarrow {}^{192}_{78}Pt + {}^{0}_{-1}e + \gamma$

9.13 a. 0.250 g b. 0.0625 g c. 0.003 91 g

9.15 30. mCi

9.17 4.4 mL

9.19 1/8

9.21 $^{235}_{92}U + {}^{1}_{0}n \longrightarrow {}^{133}_{51}Sb + {}^{100}_{41}Nb + 3\,{}^{1}_{0}n$

9.23

	a. Atomic Number	b. Number of Protons	c. Number of Neutrons	d. Mass Number	Isotope Symbol
Fluorine-18	9	9	9	18	$^{18}_{9}F$
Fluorine-19	9	9	10	19	$^{19}_{9}F$

9.25

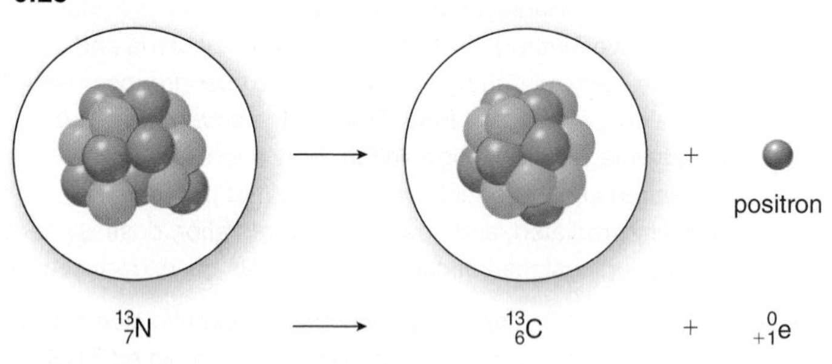

9.27 two half-lives

9.29 a. $^{74}_{33}As \longrightarrow {}^{0}_{+1}e + {}^{74}_{32}Ge$

b. 3.8 mg

c. 1.5 mL

9.31

	Atomic Number	Mass Number	Number of Protons	Number of Neutrons	Isotope Symbol
a. Chromium-51	24	51	24	27	$^{51}_{24}Cr$
b. Palladium-103	46	103	46	57	$^{103}_{46}Pd$
c. Potassium-42	19	42	19	23	$^{42}_{19}K$
d. Xenon-133	54	133	54	79	$^{133}_{54}Xe$

9.33

	Change in Mass	Change in Charge
a. α particle	–4	–2
b. β particle	0	+1
c. γ ray	0	0
d. positron	0	–1

9.35

	Mass	Charge
a. α	4	+2
b. n	1	0
c. γ	0	0
d. β	0	–1

9.37 a. $^{59}_{26}Fe \longrightarrow {}^{59}_{27}Co + {}^{0}_{-1}e$

b. $^{190}_{78}Pt \longrightarrow {}^{186}_{76}Os + {}^{4}_{2}He$

c. $^{178}_{80}Hg \longrightarrow {}^{178}_{79}Au + {}^{0}_{+1}e$

9.39 a. $^{90}_{39}Y \longrightarrow {}^{90}_{40}Zr + {}^{0}_{-1}e$

b. $^{135}_{60}Nd \longrightarrow {}^{135}_{59}Pr + {}^{0}_{+1}e$

c. $^{210}_{83}Bi \longrightarrow {}^{206}_{81}Tl + {}^{4}_{2}He$

9.41 $^{214}_{83}Bi \longrightarrow {}^{214}_{84}Po + {}^{0}_{-1}e$

$^{214}_{83}Bi \longrightarrow {}^{210}_{81}Tl + {}^{4}_{2}He$

9.43 a. $^{232}_{90}Th \longrightarrow {}^{4}_{2}He + {}^{228}_{88}Ra$

b. $^{25}_{11}Na \longrightarrow {}^{25}_{12}Mg + {}^{0}_{-1}e$

c. $^{118}_{54}Xe \longrightarrow {}^{118}_{53}I + {}^{0}_{+1}e$

d. $^{243}_{96}Cm \longrightarrow {}^{4}_{2}He + {}^{239}_{94}Pu$

9.45 4.0 days

9.47

	Iodine-131	Xenon-131
a.	32 mg	32 mg
b.	16 mg	48 mg
c.	8.0 mg	56 mg
d.	4.0 mg	60. mg

9.49 No, 25% remains.

9.51 In artifacts over 50,000 years old, the percentage of carbon-14 is too small to accurately measure.

9.53 a. 10 mCi b. 5 mCi c. 1 mCi

9.55 2.3 mL

9.57 2.6 mL

9.59 20 Sv = 2,000 rem
This represents a fatal dose because 600 rem is uniformly fatal.

9.61 Nuclear fission refers to the splitting of nuclei and fusion refers to the joining of small nuclei to form larger ones.

9.63 a. fusion
b. fission
c. both
d. fusion

9.65

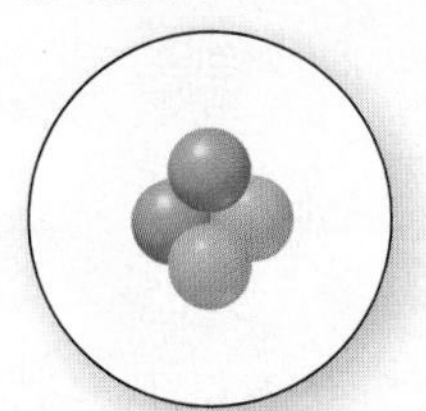

9.67 a. $^{235}_{92}U + ^{1}_{0}n \longrightarrow ^{137}_{50}Sn + ^{97}_{42}Mo + 2\ ^{1}_{0}n$
b. $^{235}_{92}U + ^{1}_{0}n \longrightarrow ^{93}_{36}Kr + ^{140}_{56}Ba + 3\ ^{1}_{0}n$

9.69 tritium; $^{3}_{1}H$

9.71 The containment of radiation leaks and disposal of radioactive waste are two problems that are associated with nuclear power production.

9.73 a. $^{192}_{77}Ir \longrightarrow ^{192}_{78}Pt + ^{0}_{-1}e + \gamma$
b. 30. mg
c. 2.3 Ci

9.75 $^{12}_{6}C$

9.77 $^{209}_{83}Bi + ^{58}_{26}Fe \longrightarrow ^{266}_{109}Mt + ^{1}_{0}n$
meitnerium-266

9.79 a. used for the treatment and diagnosis of thyroid diseases
b. used for the treatment of breast cancer
c. used for the diagnosis of heart disease

9.81 a. Iodine-125, with its longer half-life, is used for the treatment of prostate cancers using implanted radioactive seeds.
b. Iodine-131, with its shorter half-life, is used for the diagnostic and therapeutic treatment of thyroid diseases and tumors. A patient is administered radioactive iodine-131, which is then incorporated into the thyroid hormone, thyroxine. Since its half-life is short, the radioactive iodine isotope decays, so that little remains after a month or so.

9.83 Radiology technicians must be shielded to avoid exposure to excessive and dangerous doses of radiation.

9.85 High doses of stable iodine will prevent the absorption and uptake of the radioactive iodine-131.

9.87 370 kcal

Many new varieties of lawn furniture are built from recycled polyethylene, a synthetic alkane routinely used in milk jugs and other plastic containers.

Introduction to Organic Molecules

CHAPTER OUTLINE

CHAPTER GOALS

In this chapter you will learn how to:

1. Distinguish organic compounds from ionic inorganic compounds
2. Recognize the characteristic features of organic compounds
3. Use shorthand methods to draw organic molecules
4. Recognize the common functional groups and understand their importance
5. Identify and draw acyclic alkanes and cycloalkanes
6. Identify constitutional isomers
7. Name alkanes using the IUPAC system of nomenclature
8. Predict the physical properties of alkanes
9. Determine the products of complete and incomplete combustion of alkanes

Consider for a moment the activities that occupied your past 24 hours. You likely showered with soap, drank a caffeinated beverage, ate at least one form of starch, took some medication, read a newspaper, listened to a CD, and traveled in a vehicle that had rubber tires and was powered by fossil fuels. If you did any *one* of these, your life was touched by organic chemistry. In Chapter 10, we first examine the characteristic features of all organic molecules. Then, we study the alkanes, the major components of petroleum and the simplest organic molecules.

10.1 Introduction to Organic Chemistry

Some compounds that contain the element carbon are *not* organic compounds. Examples include carbon dioxide (CO_2), sodium carbonate (Na_2CO_3), and sodium bicarbonate ($NaHCO_3$).

What is organic chemistry?

- **Organic chemistry is the study of compounds that contain the element carbon.**

Clothes, foods, medicines, gasoline, refrigerants, and soaps are composed almost solely of organic compounds. By studying the principles and concepts of organic chemistry, you can learn more about compounds present in these substances and how they affect the world around you. Figure 10.1 illustrates some common products of organic chemistry used in medicine.

Because organic compounds are composed of covalent bonds, their properties differ a great deal from those of ionic inorganic compounds.

The intermolecular forces in covalent compounds were discussed in Section 4.3.

- **Organic compounds exist as discrete molecules with much weaker intermolecular forces—the forces that exist *between* molecules—than the very strong interactions of oppositely charged ions seen in ionic compounds.**

As a result, organic compounds resemble other covalent compounds in that they have much lower melting points and boiling points than ionic compounds. While ionic compounds are generally solids at room temperature, many organic compounds are liquids and some are even gases. Table 10.1 compares these and other properties of a typical organic compound (butane, $CH_3CH_2CH_2CH_3$) and a typical ionic inorganic compound (sodium chloride, NaCl).

Figure 10.1 Some Common Products of Organic Chemistry Used in Medicine

a. Oral contraceptives

b. Plastic syringes

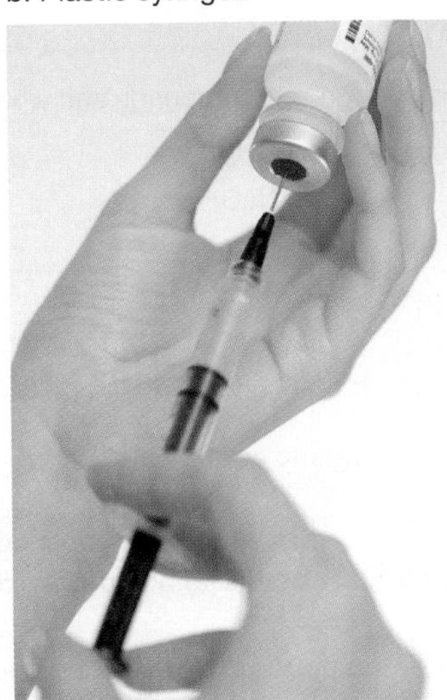

c. Antibiotics

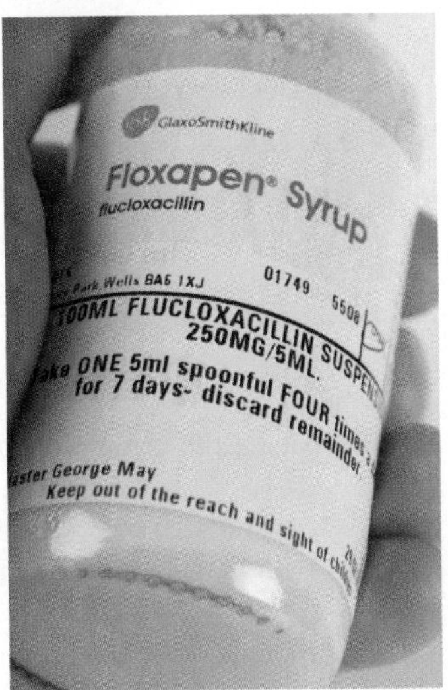

d. Synthetic heart valves

- Organic chemistry has given us contraceptives, plastics, antibiotics, synthetic heart valves, and a myriad of other materials. Our lives would be vastly different today without these products of organic chemistry.

Table 10.1 Comparing the Properties of an Organic Compound ($CH_3CH_2CH_2CH_3$) and an Ionic Inorganic Compound (NaCl)

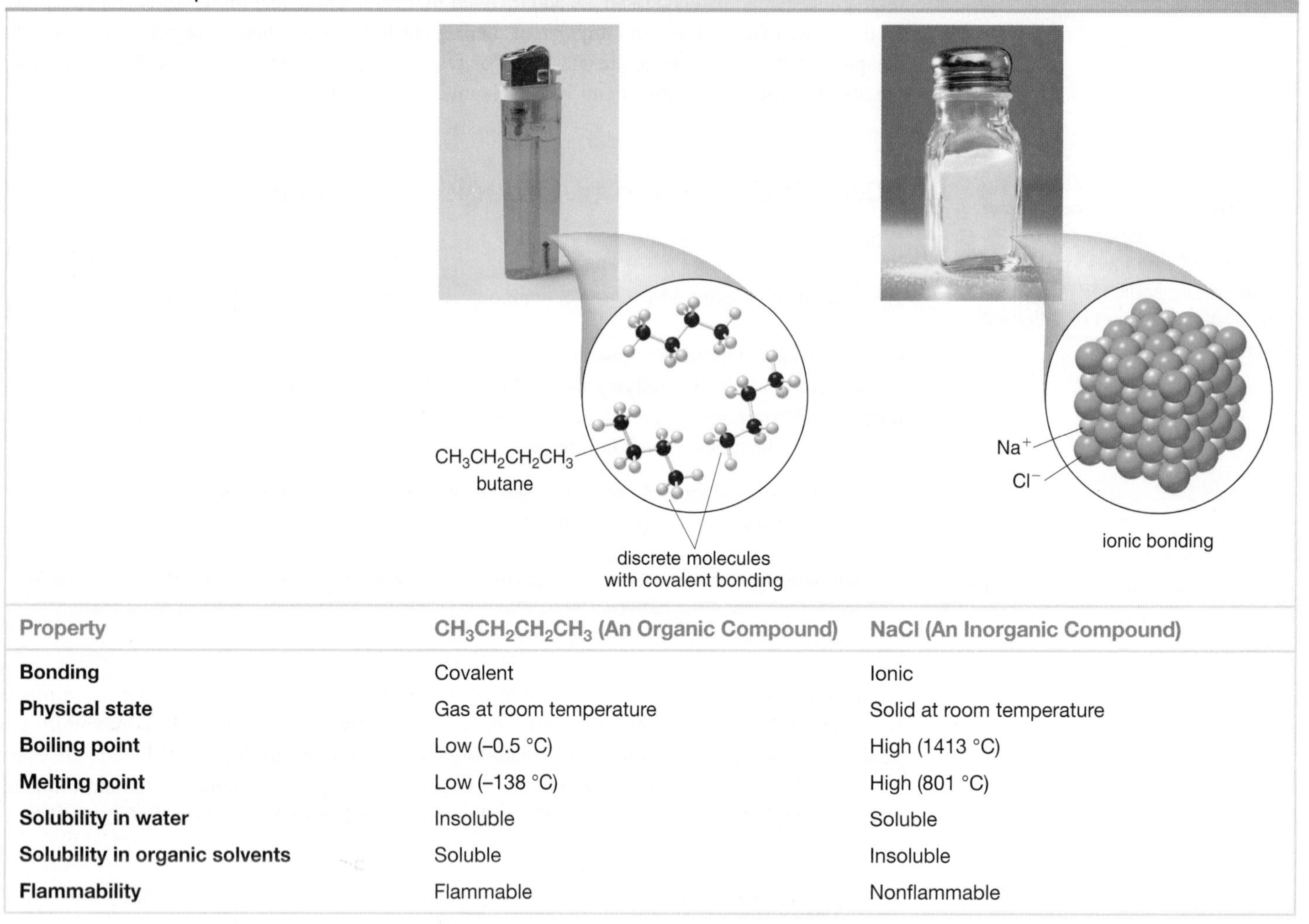

Property	$CH_3CH_2CH_2CH_3$ (An Organic Compound)	NaCl (An Inorganic Compound)
Bonding	Covalent	Ionic
Physical state	Gas at room temperature	Solid at room temperature
Boiling point	Low (–0.5 °C)	High (1413 °C)
Melting point	Low (–138 °C)	High (801 °C)
Solubility in water	Insoluble	Soluble
Solubility in organic solvents	Soluble	Insoluble
Flammability	Flammable	Nonflammable

PROBLEM 10.1

Which chemical formulas represent organic compounds and which represent inorganic compounds?

a. C_6H_{12}
b. H_2O
c. KI
d. $MgSO_4$
e. CH_4O
f. NaOH

10.2 Characteristic Features of Organic Compounds

What are the common features of organic compounds?

[1] All organic compounds contain carbon atoms and most contain hydrogen atoms. Carbon always forms four covalent bonds, and hydrogen forms one covalent bond.

Carbon is located in group 4A of the periodic table, so a carbon atom has four valence electrons available for bonding (Section 3.7). Since hydrogen has a single valence electron, methane (CH_4)

Methane, the main component of natural gas, burns in the presence of oxygen. The natural gas we use today was formed by the decomposition of organic material millions of years ago.

Ethylene is an important starting material in the preparation of the plastic polyethylene.

Because **acetylene** produces a very hot flame on burning, it is often used in welding torches.

consists of four single bonds, each formed from one electron from a hydrogen atom and one electron from carbon.

H : C : H (with H above and below) = H—C—H (with H above and below)

methane

two-electron bond

[2] Carbon forms single, double, and triple bonds to other carbon atoms.

When a compound contains two or more carbon atoms, the type of bonding is determined by the number of atoms around carbon. Consider the three compounds drawn below:

H : C : C : H (each C with H above and below)	H : C : : C : H (each C with H below)	H : C : : : C : H
Each C forms four single bonds.	**A double bond contains four electrons.**	**A triple bond contains six electrons.**
H—C—C—H (each C with H above and below)	$H_2C{=}CH_2$	H—C≡C—H
ethane	ethylene	acetylene

- **A C atom surrounded by four atoms forms four single bonds.** In ethane (C_2H_6), each carbon atom is bonded to three hydrogen atoms and one carbon atom. All bonds are single bonds.
- **A C atom surrounded by three atoms forms one double bond.** In ethylene (C_2H_4), each carbon atom is surrounded by three atoms (two hydrogens and one carbon); thus, each C forms a single bond to each hydrogen atom and a double bond to carbon.
- **A C atom surrounded by two atoms generally forms one triple bond.** In acetylene (C_2H_2), each carbon atom is surrounded by two atoms (one hydrogen and one carbon); thus, each C forms a single bond to hydrogen and a triple bond to carbon.

[3] Some compounds have chains of atoms and some compounds have rings.

For example, three carbon atoms can bond in a row to form propane, or form a ring called cyclopropane. Propane is the fuel burned in gas grills, and cyclopropane is an anesthetic.

H—C—C—C—H (each C with H above and below)	three-membered ring of C atoms, each C bonded to two H
propane C_3H_8	cyclopropane C_3H_6

[4] Organic compounds may also contain elements other than carbon and hydrogen. Any atom that is not carbon or hydrogen is called a *heteroatom.*

The most common heteroatoms are nitrogen, oxygen, and the halogens (F, Cl, Br, and I).

- **Each heteroatom forms a characteristic number of bonds, determined by its location in the periodic table.**
- **The common heteroatoms also have nonbonding, lone pairs of electrons, so that each atom is surrounded by eight electrons.**

The number of bonds formed by common elements was first discussed in Section 3.7.

Thus, nitrogen forms three bonds and has one lone pair of electrons, while oxygen forms two bonds and has two additional lone pairs. The halogens form one bond and have three additional lone pairs. Common bonding patterns for atoms in organic compounds are summarized in Table 10.2. Except for hydrogen, these common elements in organic compounds follow one rule in bonding:

Oxygen and nitrogen form both single and multiple bonds to carbon. **The most common multiple bond between carbon and a heteroatom is a carbon–oxygen double bond (C=O).** The bonding patterns remain the same even when an atom is part of a multiple bond, as shown with methanol (CH_3OH) and formaldehyde ($H_2C{=}O$, a preservative).

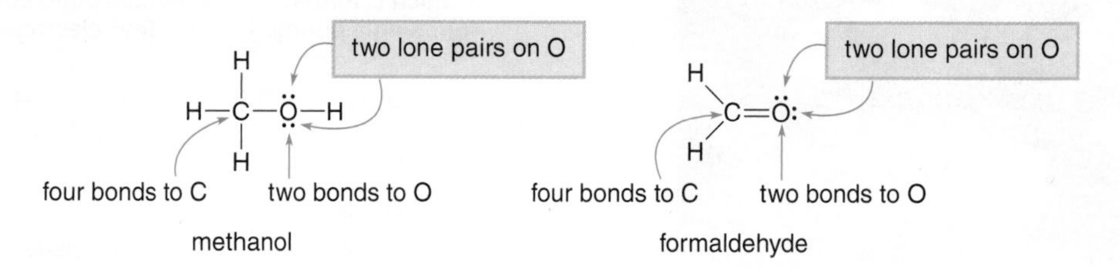

Table 10.2 Common Bonding Patterns for Atoms in Organic Compounds

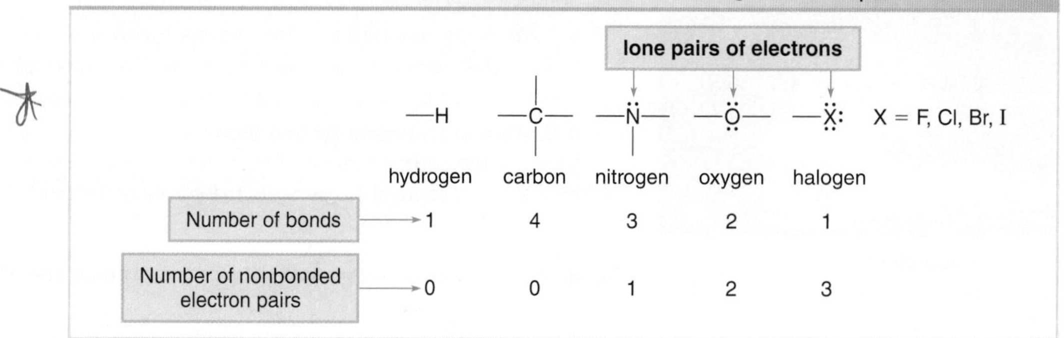

	hydrogen	carbon	nitrogen	oxygen	halogen
Number of bonds	1	4	3	2	1
Number of nonbonded electron pairs	0	0	1	2	3

SAMPLE PROBLEM 10.1

Draw in all H's and lone pairs in each compound.

a. C—C—Cl b. C—C=O c. C—C≡N

Analysis

Each C and heteroatom must be surrounded by eight electrons. Use the common bonding patterns in Table 10.2 to fill in the needed H's and lone pairs. C needs four bonds; Cl needs one bond and three lone pairs; O needs two bonds and two lone pairs; N needs three bonds and one lone pair.

Solution

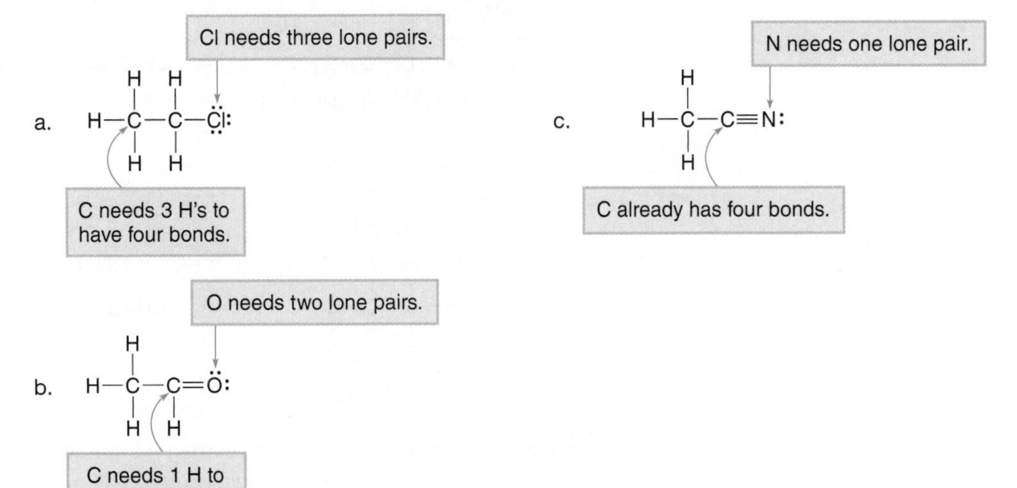

PROBLEM 10.2

Fill in all H's and lone pairs in each compound.

a. C—C=C—C b. C—C(—O)—C c. C—C(=O)—C d. C—C(=O)—N—C e. (three-membered ring of C, C, O)

10.3 Drawing Organic Molecules

Because organic molecules often contain many atoms, we need shorthand methods to simplify their structures. The two main types of shorthand representations used for organic compounds are **condensed structures** and **skeletal structures.**

10.3A Condensed Structures

Condensed structures are most often used for a compound having a chain of atoms bonded together, rather than a ring. The following conventions are used.

- **All of the atoms are drawn in, but the two-electron bond lines are generally omitted.**
- **Lone pairs on heteroatoms are omitted.**

To interpret a condensed formula, it is usually best to start at the *left side* of the molecule and remember that the ***carbon atoms must have four bonds.***

- **A carbon bonded to 3 H's becomes CH_3.**
- **A carbon bonded to 2 H's becomes CH_2.**
- **A carbon bonded to 1 H becomes CH.**

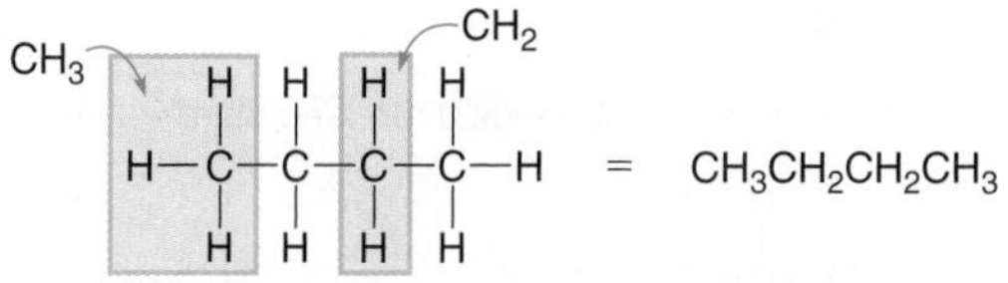

CH3
CH2
= CH3CH2CH2CH3

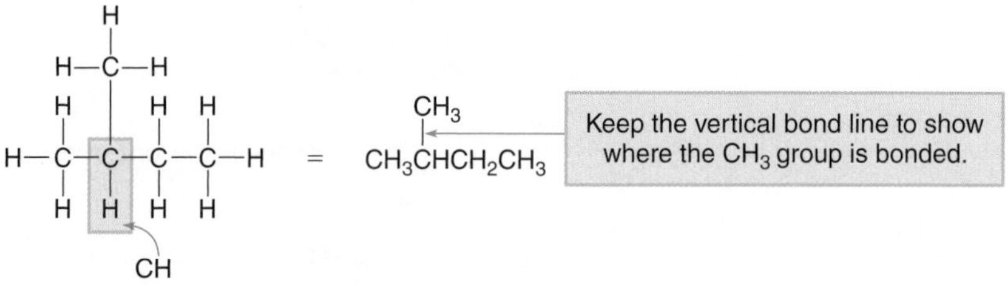

CH
CH3
= CH3CHCH2CH3
Keep the vertical bond line to show where the CH3 group is bonded.

Sometimes these structures are further simplified by using parentheses around like groups. Two CH_2 groups bonded together become $(CH_2)_2$. Two CH_3 groups bonded to the same carbon become $(CH_3)_2C$.

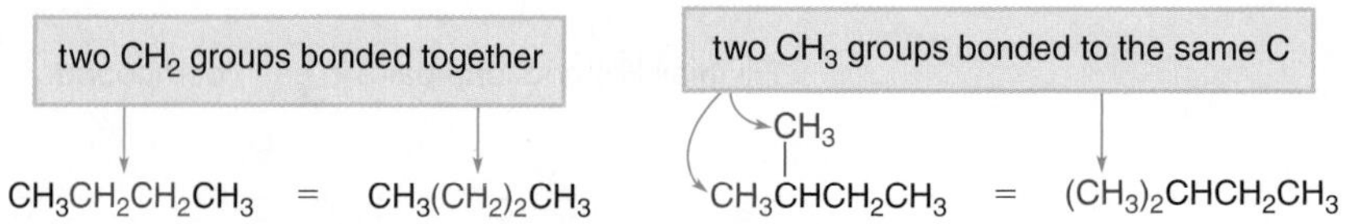

two CH2 groups bonded together
CH3CH2CH2CH3 = CH3(CH2)2CH3
two CH3 groups bonded to the same C
CH3
CH3CHCH2CH3 = (CH3)2CHCH2CH3

SAMPLE PROBLEM 10.2

Convert each compound into a condensed structure.

a. 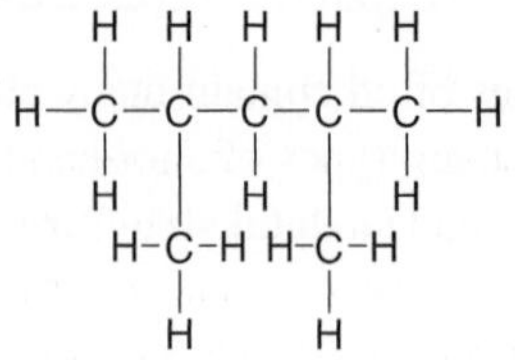

b. 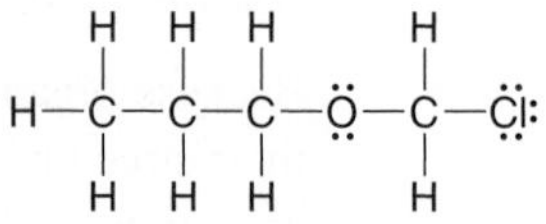

Analysis

Start at the left and proceed to the right, making sure that each carbon has four bonds. Omit lone pairs on the heteroatoms O and Cl. When like groups are bonded together or bonded to the same atom, use parentheses to further simplify the structure.

Solution

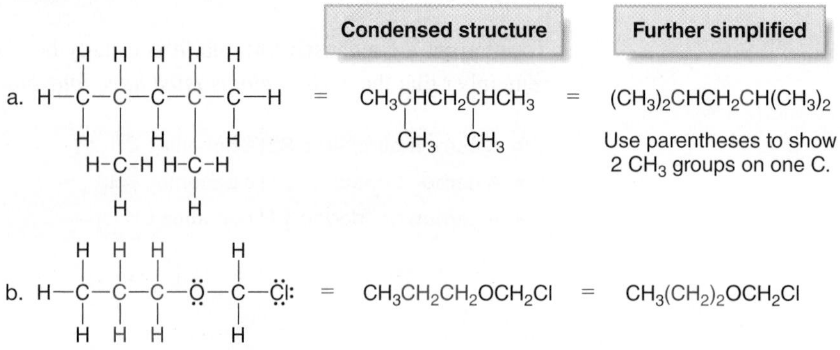

Condensed structure
Further simplified
a.
= CH3CHCH2CHCH3 = (CH3)2CHCH2CH(CH3)2
CH3 CH3
Use parentheses to show 2 CH3 groups on one C.
b.
= CH3CH2CH2OCH2Cl = CH3(CH2)2OCH2Cl

PROBLEM 10.3

Convert each compound to a condensed formula.

a. 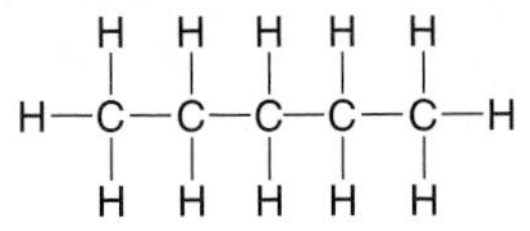 b. 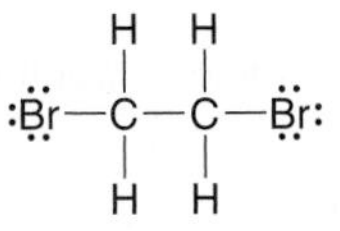 c. d.

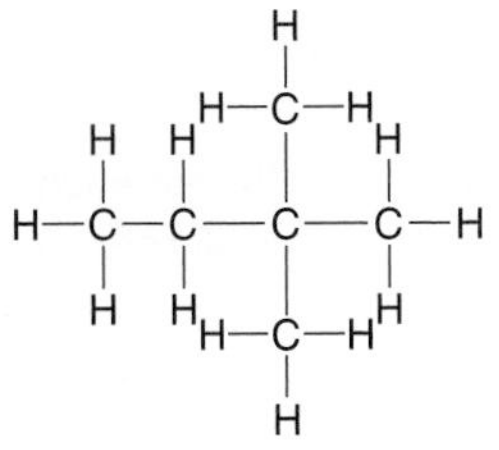

PROBLEM 10.4

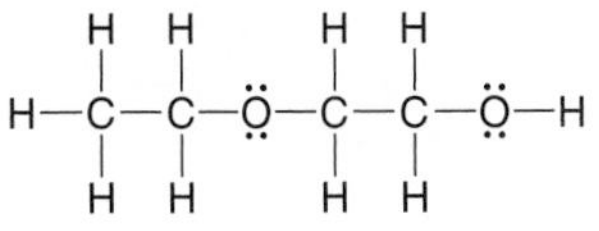

Convert each condensed formula to a complete structure with lone pairs on heteroatoms.

a. $CH_3(CH_2)_8CH_3$
b. $CH_3(CH_2)_4OH$
c. CH_3CCl_3
d. $CH_3(CH_2)_4CH(CH_3)_2$
e. $(CH_3)_2CHCH_2NH_2$

10.3B Skeletal Structures

Skeletal structures are used for organic compounds containing both rings and chains of atoms. Three important rules are used in drawing them.

- **Assume there is a carbon atom at the junction of any two lines or at the end of any line.**
- **Assume there are enough hydrogens around each carbon to give it four bonds.**
- **Draw in all heteroatoms and the hydrogens directly bonded to them.**

Rings are drawn as polygons with a carbon atom "understood" at each vertex, as shown for cyclohexane and cyclopentanol. All carbons and hydrogens in these molecules are understood, except for H's bonded to heteroatoms.

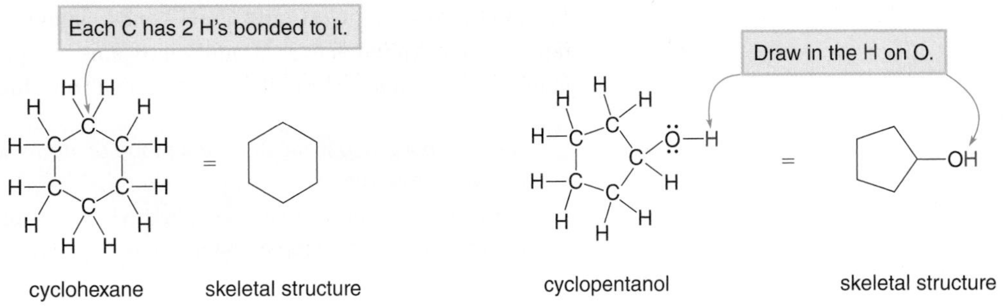

Each C has 2 H's bonded to it.

cyclohexane = skeletal structure

Draw in the H on O.

cyclopentanol = skeletal structure

SAMPLE PROBLEM 10.3

Convert each skeletal structure to a complete structure with all C's and H's drawn in.

a. 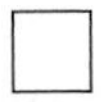 b. 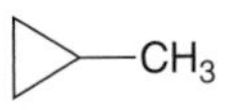 CH_3

Analysis

To draw each complete structure, place a C atom at the corner of each polygon and add H's to give carbon four bonds.

Solution

a. 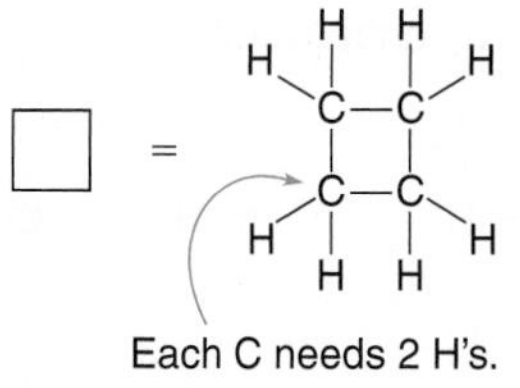 [square] = cyclobutane with each C bonded to 2 H's. Each C needs 2 H's.

b. 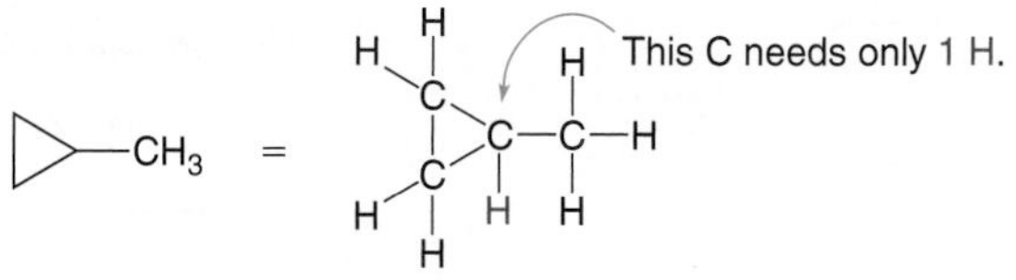 [cyclopropane]—CH_3 = complete structure with H's drawn in. This C needs only 1 H.

PROBLEM 10.5

Convert each skeletal structure to a complete structure with all C's and H's drawn in. Add lone pairs on all heteroatoms.

a. 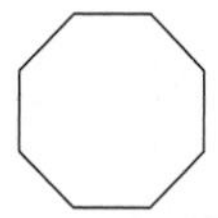 cyclooctane

b. 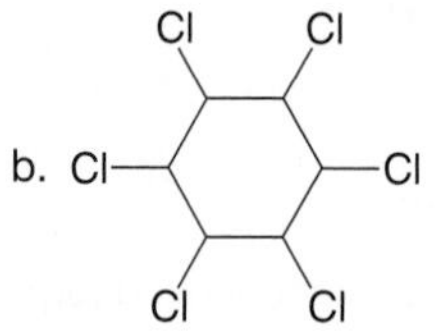 lindane (an insecticide) — six-membered ring with six Cl atoms

c. 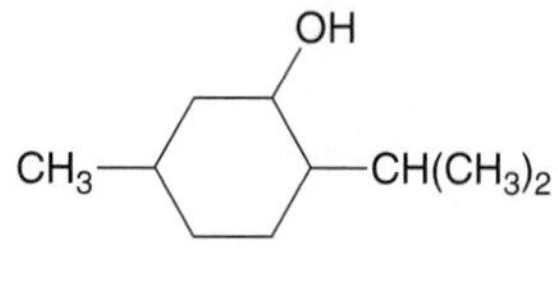 menthol (isolated from peppermint) — CH_3, OH, $CH(CH_3)_2$ on a six-membered ring

PROBLEM 10.6

How many H's are bonded to each indicated carbon (C1–C5) in the following drugs?

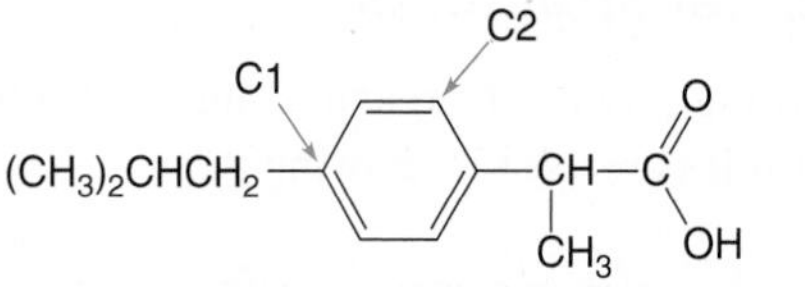

ibuprofen (anti-inflammatory agent): $(CH_3)_2CHCH_2$—[benzene ring]—$CH(CH_3)$—$COOH$, with C1 and C2 indicated

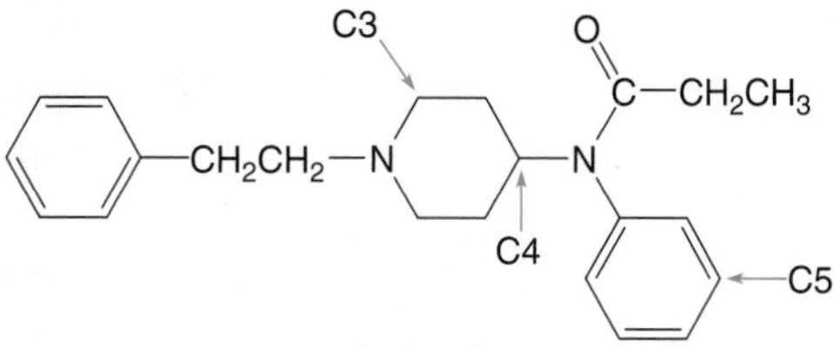

fentanyl (narcotic pain reliever): [benzene ring]—CH_2CH_2—N[piperidine ring]—N(phenyl)—$C(=O)CH_2CH_3$, with C3, C4 and C5 indicated

10.4 Functional Groups

In addition to strong C—C and C—H bonds, organic molecules may have other structural features as well. Although over 50 million organic compounds are currently known, only a limited number of common structural features, called **functional groups,** are found in these molecules.

- A *functional group* is an atom or a group of atoms with characteristic chemical and physical properties.
- A functional group contains a heteroatom, a multiple bond, or sometimes both a heteroatom *and* a multiple bond.

A functional group determines a molecule's shape, properties, and the type of reactions it undergoes. A functional group behaves the same whether it is bonded to a carbon backbone having as few as two or as many as 20 carbons. For this reason, we often abbreviate the carbon and hydrogen portion of the molecule by a capital letter **R,** and draw the **R** bonded to a particular functional group.

R—Functional group

Carbon backbone… bonded to… a particular functional group.

Ethanol (CH_3CH_2OH), for example, has two carbons and five hydrogens in its carbon backbone, as well as an OH group, a functional group called a **hydroxyl group.** The hydroxyl group deter-

Ethanol, the alcohol present in wine and other alcoholic beverages, is formed by the fermentation of sugar. Ethanol can also be made in the lab by a totally different process. **Ethanol produced in the lab is identical to the ethanol produced by fermentation.**

mines the physical properties of ethanol as well as the type of reactions it undergoes. Moreover, any organic molecule containing a hydroxyl group has properties similar to ethanol. Compounds that contain a hydroxyl group are called **alcohols.**

carbon backbone → CH_3CH_2— ; hydroxyl group → —$\ddot{O}$—H

ethanol

The most common functional groups can be subdivided into three types.

- Hydrocarbons
- Compounds containing a single bond to a heteroatom
- Compounds containing a C=O group

10.4A Hydrocarbons

Hydrocarbons **are compounds that contain only the elements of carbon and hydrogen,** as shown in Table 10.3.

- ***Alkanes*** have only C—C single bonds and no functional group. Ethane, $\mathbf{CH_3CH_3}$, is a simple alkane.
- ***Alkenes*** have a C—C double bond as their functional group. Ethylene, $\mathbf{CH_2{=}CH_2}$, is a simple alkene.
- ***Alkynes*** have a C—C triple bond as their functional group. Acetylene, **HC≡CH,** is a simple alkyne.
- ***Aromatic hydrocarbons*** contain a benzene ring, a six-membered ring with three double bonds.

Table 10.3 Hydrocarbons

Type of Compound	General Structure	Example	3-D Structure	Functional Group
Alkane	R—H	CH_3CH_3		—
Alkene	C=C	$H_2C{=}CH_2$		Carbon–carbon double bond
Alkyne	—C≡C—	H—C≡C—H		Carbon–carbon triple bond
Aromatic compound	(benzene ring)	(benzene)		Benzene ring

CONSUMER NOTE

Polyethylene was first produced in the 1930s and initially used as insulating material for radar during World War II. It is now a plastic in milk containers, sandwich bags, and plastic wrapping. Over 100 billion pounds of polyethylene are manufactured each year.

All hydrocarbons other than alkanes contain multiple bonds. Alkanes, which have no functional groups and therefore no reactive sites, are notoriously unreactive except under very drastic conditions. For example, **polyethylene** is a synthetic plastic and high molecular weight alkane, consisting of long chains of $—CH_2—$ groups bonded together, hundreds or even thousands of atoms long. Because it has no reactive sites, it is a very stable compound that does not readily degrade and thus persists for years in landfills.

polyethylene

The chain continues in both directions.

10.4B Compounds Containing a Single Bond to a Heteroatom

Several types of functional groups contain a carbon atom singly bonded to a heteroatom. Common examples include alkyl halides, alcohols, ethers, amines, and thiols, as shown in Table 10.4.

Molecules containing these functional groups may be simple or very complex. It doesn't matter what else is present in other parts of the molecule. **Always dissect it into small pieces to identify the functional groups.** For example, diethyl ether, the first general anesthetic, is an ether because it has an O atom bonded to two C's. Tetrahydrocannabinol (THC), the active component in marijuana, is also an ether because it contains an O atom bonded to two carbon atoms. In this case, the O atom is also part of a ring.

ether

$CH_3CH_2—\ddot{O}—CH_2CH_3$

diethyl ether

CH_3 OH CH_3 CH_3 $(CH_2)_4CH_3$ ether

tetrahydrocannabinol (THC)

HEALTH NOTE

Tetrahydrocannabinol (THC) is the primary active constituent in marijuana. Although the recreational use of cannabis is illegal in the United States, THC can be used legally in some states for medical purposes.

Table 10.4 Compounds Containing a Carbon–Heteroatom Single Bond

Type of Compound	General Structure	Example	3-D Structure	Functional Group
Alkyl halide	$R—\ddot{X}:$ (X = F, Cl, Br, I)	$CH_3—\ddot{Br}:$		$—X$ halo group
Alcohol	$R—\ddot{O}H$	$CH_3—\ddot{O}H$		$—OH$ hydroxyl group
Ether	$R—\ddot{O}—R$	$CH_3—\ddot{O}—CH_3$		$—OR$ alkoxy group
Amine	$R—\ddot{N}H_2$ or $R_2\ddot{N}H$ or $R_3\ddot{N}$	$CH_3—\ddot{N}H_2$		$—NH_2$ amino group
Thiol	$R—\ddot{S}H$	$CH_3—\ddot{S}H$		$—SH$ sulfhydryl group

PROBLEM 10.7

Identify the functional groups in each compound. Some compounds contain more than one functional group.

a. 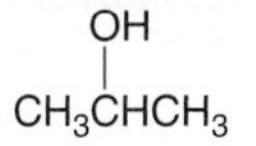 $CH_3CH(OH)CH_3$ — 2-propanol (disinfectant in rubbing alcohol)

b. 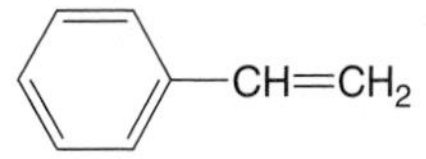 C_6H_5–CH=CH_2 — styrene (starting material used to synthesize Styrofoam)

c. $H_2NCH_2CH_2CH_2CH_2NH_2$ — putrescine (putrid odor of rotting fish)

10.4C Compounds Containing a C=O Group

Many different kinds of compounds contain a carbon–oxygen double bond **(C=O, carbonyl group),** as shown in Table 10.5. Carbonyl compounds include aldehydes, ketones, carboxylic acids, esters, and amides. The type of atom bonded to the carbonyl carbon—hydrogen, carbon, or a heteroatom—determines the specific class of carbonyl compound.

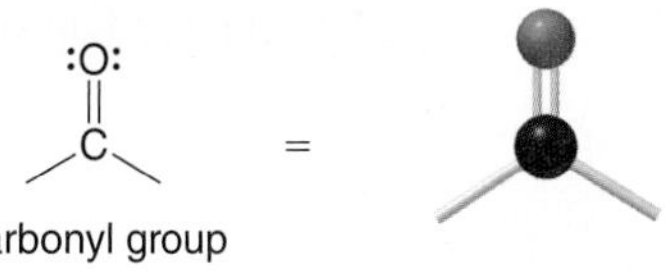

carbonyl group

Table 10.5 Compounds Containing a C=O Group

Type of Compound	General Structure	Example	3-D Structure	Functional Group
Aldehyde	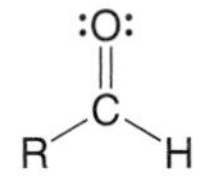 RCHO	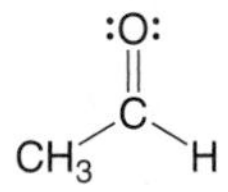 CH_3CHO	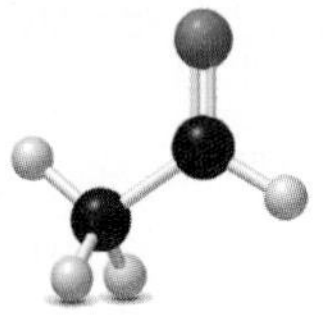	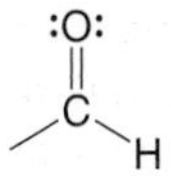 –CHO
Ketone	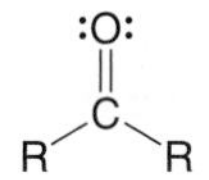 RCOR	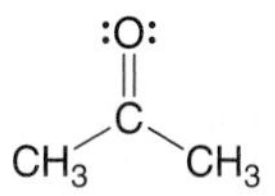 CH_3COCH_3		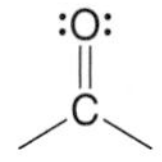 C=O
Carboxylic acid	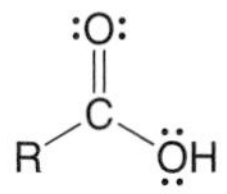 RCOOH	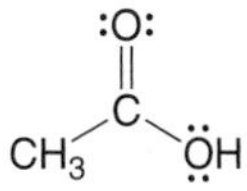 CH_3COOH	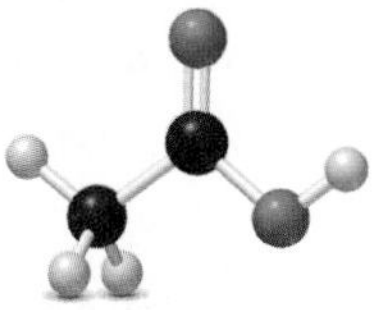	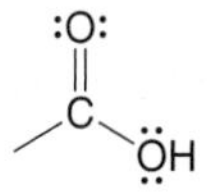 –COOH carboxyl group
Ester	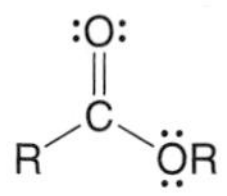 RCOOR	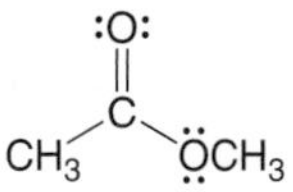 CH_3COOCH_3	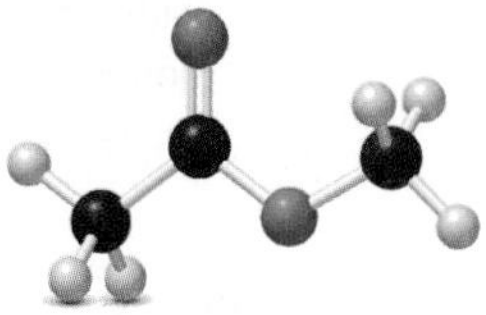	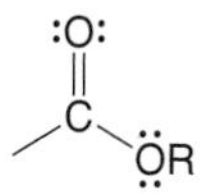 –COOR
Amide	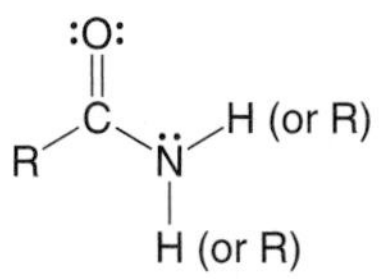 RCON(H or R)(H or R)	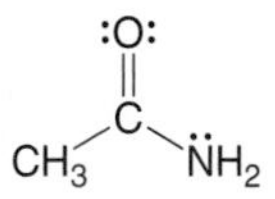 CH_3CONH_2	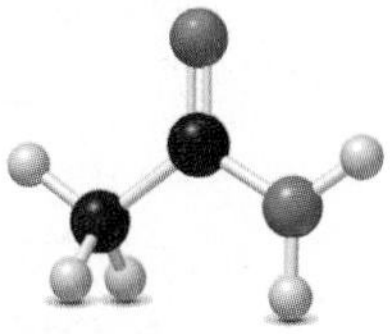	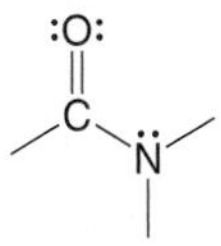 –CON<

CONSUMER NOTE

The characteristic odor of many fruits is due to low molecular weight esters.

$CH_3COOCH_2CH_2CH(CH_3)_2$

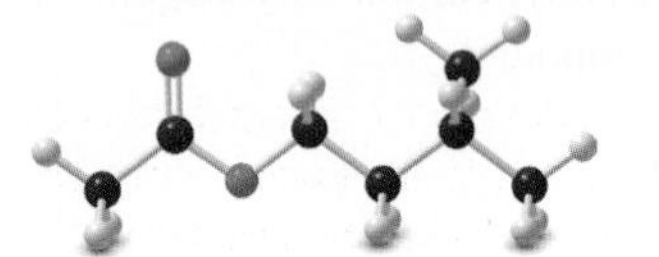

isoamyl acetate
odor of banana

Take special note of the condensed structures used to draw aldehydes, carboxylic acids, and esters.

- **An aldehyde has a hydrogen atom bonded directly to the carbonyl carbon.**

acetaldehyde = CH_3CHO (condensed structure) — C is bonded to both H and O.

- **A carboxylic acid contains an OH group bonded directly to the carbonyl carbon.**

acetic acid = CH_3COOH or CH_3CO_2H (condensed structures) — C is bonded to both O atoms.

- **An ester contains an OR group bonded directly to the carbonyl carbon.**

ethyl acetate = $CH_3COOCH_2CH_3$ or $CH_3CO_2CH_2CH_3$ (condensed structures) — C is bonded to both O atoms.

SAMPLE PROBLEM 10.4

Identify the functional group in each compound.

a. A b. B c. C

Analysis

Concentrate on the multiple bonds and heteroatoms and refer to Tables 10.3, 10.4, and 10.5.

Solution

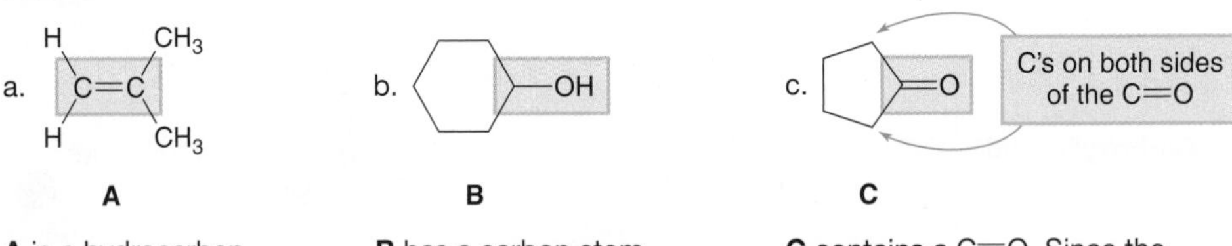

A is a hydrocarbon with a carbon–carbon double bond, making it an alkene.

B has a carbon atom bonded to a hydroxyl group (OH), making it an alcohol.

C contains a C=O. Since the carbonyl carbon is bonded to two other carbons in the ring, **C** is a ketone.

PROBLEM 10.8

For each compound: [1] Identify the functional group; [2] draw out the complete compound, including lone pairs on heteroatoms.

a. (cyclohexane)–CHO

b. $CH_3CH_2CH_2CO_2H$

c. (cyclohexanone)=O

d. $CH_3CH_2CO_2CH_2CH_3$

e. (cyclopentyl)–C(=O)–$NHCH_3$

Many useful organic compounds contain complex structures with two or more functional groups. Sample Problem 10.5 illustrates an example of identifying several functional groups in a single molecule. Sample Problem 10.6 shows examples of identifying functional groups in simple and complex molecules beginning with ball-and-stick models.

SAMPLE PROBLEM 10.5

Tamiflu is an antiviral drug effective against avian influenza. Identify all of the functional groups in Tamiflu.

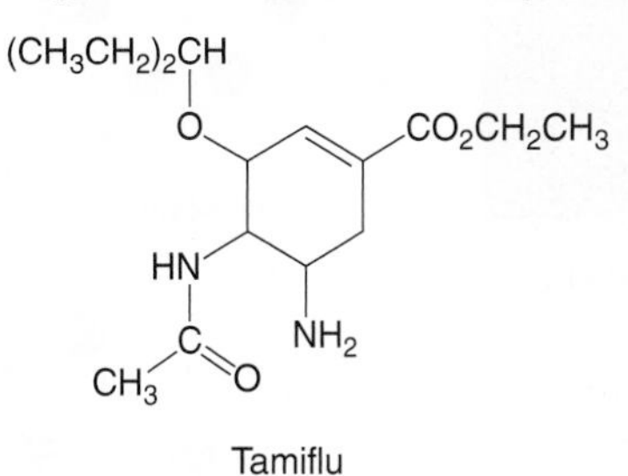

Tamiflu

HEALTH NOTE

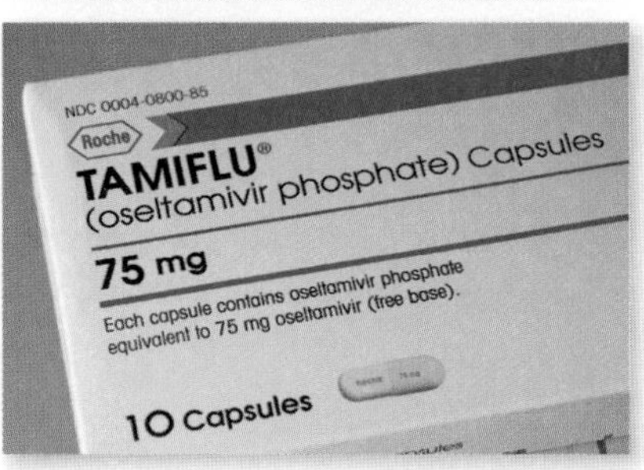

Tamiflu is the trade name for oseltamivir, an antiviral drug used to treat influenza.

Analysis

To identify functional groups, look for multiple bonds and heteroatoms. With functional groups that contain O atoms, look at what is bonded to the O's to decide if the group is an alcohol, ether, or other group. With carbonyl-containing groups, look at what is bonded to the carbonyl carbon.

Solution

Re-draw Tamiflu to further clarify the functional groups:

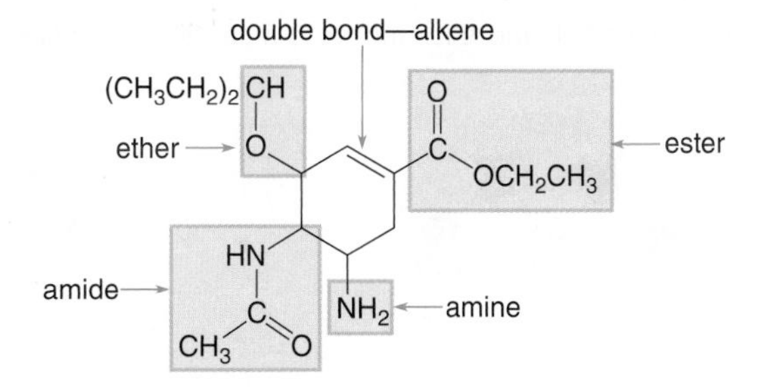

- Tamiflu contains a total of five functional groups: a carbon–carbon double bond (an alkene), ether, ester, amide, and amine.
- Note the difference between an amine and an amide. **An amine contains a N atom but no C═O. An amide contains a N atom bonded directly to a C═O.**

PROBLEM 10.9

Identify all of the functional groups in each drug. Atenolol is an antihypertensive agent; that is, it is used to treat high blood pressure. Mestranol is a synthetic estrogen used in oral contraceptives.

a. H_2N–C(=O)–CH_2–C_6H_4–O–$CH_2CHCH_2NHCH(CH_3)_2$ (OH on CH)

atenolol

b. CH_3O, CH_3, OH, C≡CH

mestranol

SAMPLE PROBLEM 10.6

Convert each ball-and-stick model to a condensed or skeletal structure and identify the functional groups.

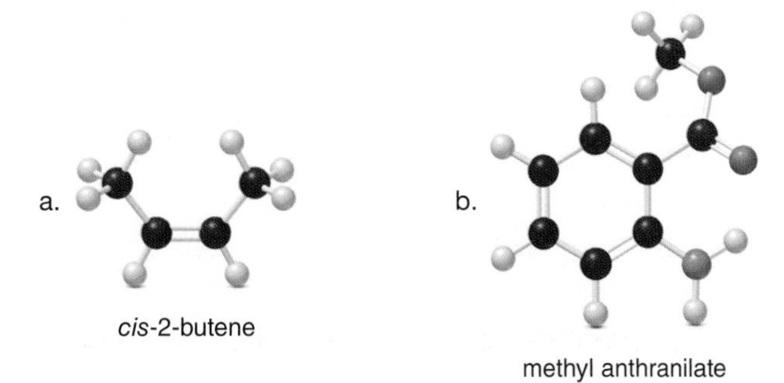

a. *cis*-2-butene

b. methyl anthranilate

Analysis

Use the common element colors shown on the inside back cover to convert the ball-and-stick model to a shorthand representation. Look for multiple bonds and heteroatoms to locate the functional groups, and refer to Tables 10.3, 10.4, and 10.5.

Solution

a.

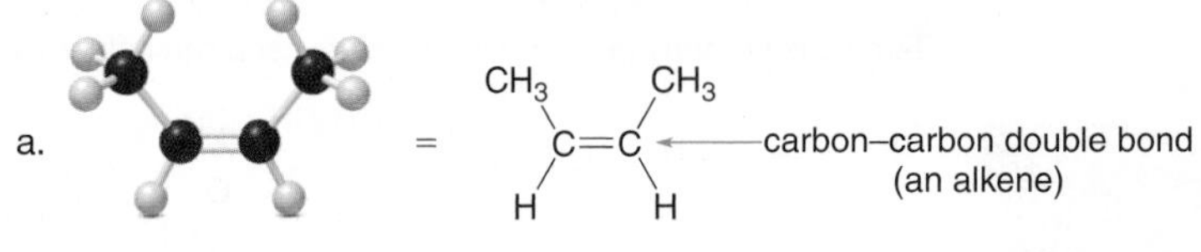

cis-2-butene

b.

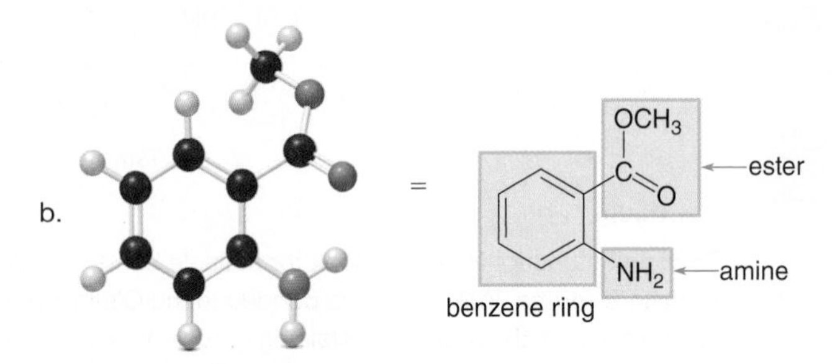

methyl anthranilate

PROBLEM 10.10

Convert each ball-and-stick model to a condensed or skeletal structure and identify the functional groups.

PROBLEM 10.11

Convert the ball-and-stick model of the local anesthetic lidocaine to a shorthand structure and identify the functional groups.

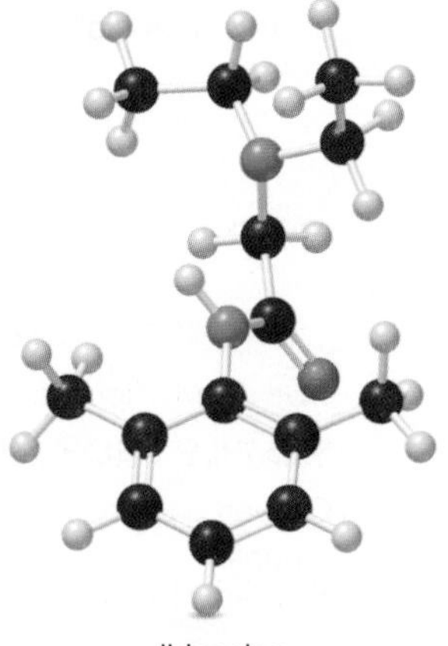

lidocaine

CONSUMER NOTE

Methyl anthranilate (Sample Problem 10.6) is a component of neroli oil, a fragrant oil obtained from the flowers of the bitter orange tree and used in perfumery. Commercial products that contain methyl anthranilate are sprayed on fruit trees to keep birds away.

HEALTH NOTE

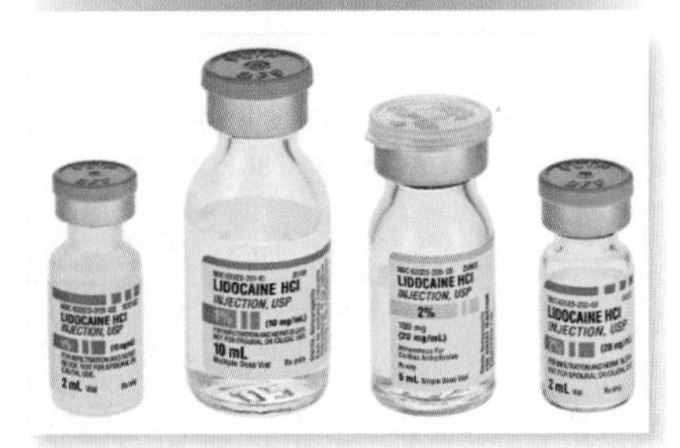

Lidocaine (trade name Xylocaine) is a local anesthetic used in dentistry and minor surgery.

10.5 Alkanes

10.5A Introduction

Now that we have learned some general features of organic compounds, we can examine the alkanes, the simplest organic molecules, in more detail. ***Alkanes* are hydrocarbons having only C—C and C—H single bonds.** The carbons of an alkane can be joined together to form chains or rings of atoms.

The prefix *a-* means *not,* so an **a**cyclic alkane is *not* cyclic.

- **Alkanes that contain chains of carbon atoms but no rings are called *acyclic alkanes.*** An acyclic alkane has the molecular formula C_nH_{2n+2}, where *n* is the number of carbons it contains. Acyclic alkanes are also called **saturated hydrocarbons** because they have the maximum number of hydrogen atoms per carbon.
- ***Cycloalkanes* contain carbons joined in one or more rings.** Since a cycloalkane has two fewer H's than an acyclic alkane with the same number of carbons, its general formula is C_nH_{2n}.

Undecane and cyclohexane are examples of two naturally occurring alkanes. Undecane is an acyclic alkane with molecular formula $C_{11}H_{24}$. Undecane is a ***pheromone,* a chemical substance used for communication** in a specific animal species, most commonly an insect population. Secretion of undecane by a cockroach causes other members of the species to aggregate. Cyclohexane, a cycloalkane with molecular formula C_6H_{12}, is one component of the mango, the most widely consumed fruit in the world.

$CH_3CH_2CH_2CH_2CH_2CH_2CH_2CH_2CH_2CH_2CH_3$

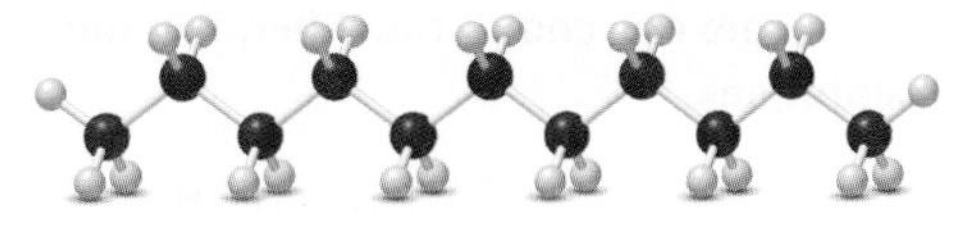

undecane

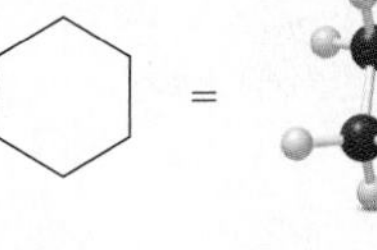

cyclohexane

PROBLEM 10.12

How many hydrogen atoms are present in each compound?

a. an acyclic alkane with three carbons
b. a cycloalkane with four carbons
c. a cycloalkane with nine carbons
d. an acyclic alkane with seven carbons

PROBLEM 10.13

Which formulas represent acyclic alkanes and which represent cycloalkanes?

a. C_5H_{12} b. C_4H_8 c. $C_{12}H_{24}$ d. $C_{10}H_{22}$

10.5B Acyclic Alkanes Having Fewer Than Five Carbons

The structures for the two simplest acyclic alkanes were given in Section 10.2.

- **Methane, CH_4,** has a single carbon atom surrounded by four hydrogens to give it four bonds.
- **Ethane, CH_3CH_3,** has two carbon atoms joined together by a single bond. Each carbon is also bonded to three hydrogens to give it four bonds total.

Recall how to draw a tetrahedron from Section 3.10: Place two bonds in the plane of the page (on solid lines), one bond in front (on a wedge), and one bond behind the plane (on a dashed line).

The shape around atoms in organic molecules is determined by counting groups using the principles of VSEPR theory (Section 3.10). Since each carbon in an alkane is surrounded by four atoms, each carbon is **tetrahedral,** and all bond angles are 109.5°.

```
                  H
                  |
CH4       =   H—C—H
methane           |
                  H
```

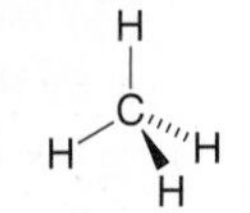

3-D representation

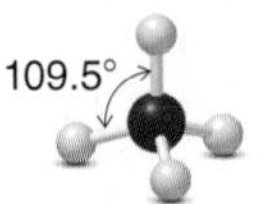

ball-and-stick model

```
                  H   H
                  |   |
CH3CH3    =   H—C—C—H
ethane            |   |
                  H   H
```

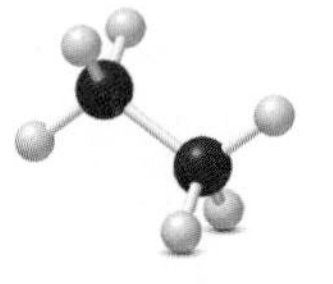

To draw a three-carbon alkane, draw three carbons joined together with single bonds and add enough hydrogens to give each carbon four bonds. This forms **propane, $CH_3CH_2CH_3$.**

```
                  H   H   H
                  |   |   |
CH3CH2CH3   =  H—C—C—C—H
propane           |   |   |
                  H   H   H
```

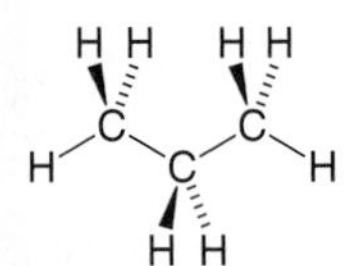

3-D representation

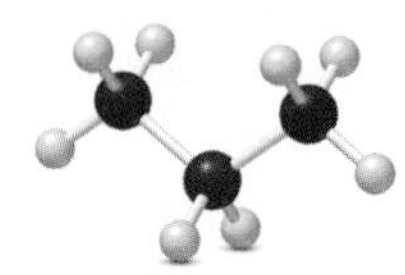

ball-and-stick model

CONSUMER NOTE

Propane is the principal alkane in LPG (liquefied petroleum gas), a fuel used for vehicles and cooking. LPG has also been used as an aerosol propellant, replacing the ozone-depleting chlorofluorocarbons (Section 6.9).

The carbon skeleton in propane and other alkanes can be drawn in a variety of different ways and still represent the same molecule. For example, the three carbons of propane can be drawn in a horizontal row or with a bend. *These representations are equivalent.* **If you follow the carbon chain from one end to the other, you move across the *same* three carbon atoms in both representations.**

```
   H   H   H               H     H
   |   |   |               |     |
H—C—C—C—H      =      H—C—— C—H
   |   |   |               |     |
   H   H   H               H  H—C—H
                                 |
                                 H

  3 C's in a row        3 C's with a bend
```

- The bends in a carbon chain don't matter when it comes to identifying different compounds.

There are two different ways to arrange four carbons, giving two compounds with molecular formula C_4H_{10}.

- **Butane, $CH_3CH_2CH_2CH_3$,** has four carbon atoms in a row. Butane is a **straight-chain alkane,** an alkane that has all of its carbons in one continuous chain.
- **Isobutane, $(CH_3)_3CH$,** has three carbon atoms in a row and one carbon bonded to the middle carbon. Isobutane is a **branched-chain alkane,** an alkane that contains one or more carbon branches bonded to a carbon chain.

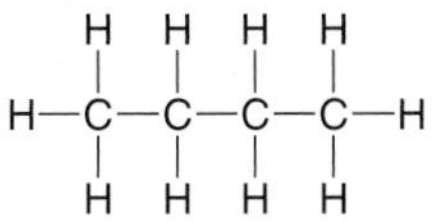

$CH_3CH_2CH_2CH_3$ = 4 C's in a row

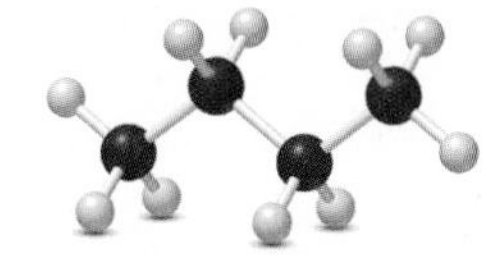

butane — **straight-chain alkane**

$CH_3-CH(CH_3)-CH_3$ = 3 C's with a one-carbon branch

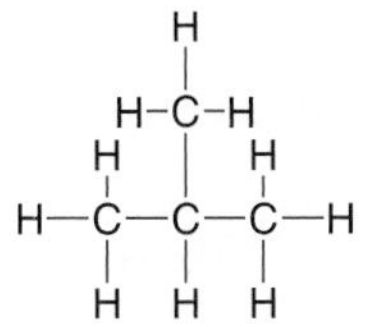

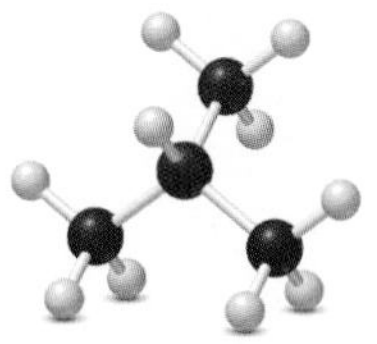

isobutane (2-methylpropane) — **branched-chain alkane**

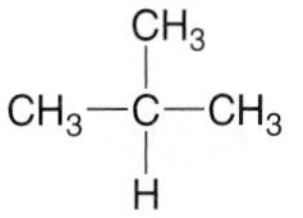

Butane and isobutane are ***isomers,*** **two different compounds with the same molecular formula.** They belong to one of the two major classes of isomers called **constitutional isomers.**

Constitutional isomers are also called structural isomers.

- *Constitutional isomers* differ in the way the atoms are connected to each other.

Constitutional isomers like butane and isobutane belong to the same family of compounds: they are both **alkanes.** This is not always the case. For example, there are two different arrangements of atoms for a compound of molecular formula C_2H_6O.

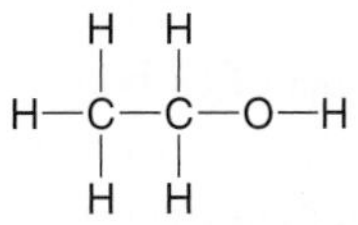

CH_3CH_2OH — ethanol

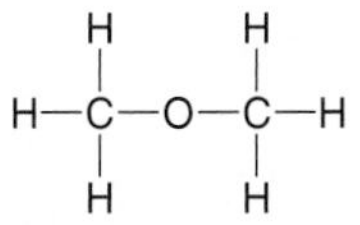

CH_3OCH_3 — dimethyl ether

Ethanol (CH_3CH_2OH) and dimethyl ether (CH_3OCH_3) are constitutional isomers with different functional groups: CH_3CH_2OH is an **alcohol** and CH_3OCH_3 is an **ether.**

SAMPLE PROBLEM 10.7

Are the compounds in each pair constitutional isomers or are they not constitutional isomers of each other?

a. $CH_3CH_2CH_3$ and $H_2C=CHCH_3$

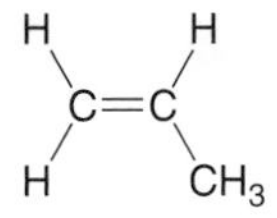

b. $CH_3CH_2CH_2NH_2$ and $CH_3-N(CH_3)-CH_3$

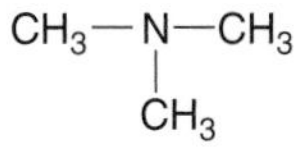

Analysis

First compare molecular formulas; two compounds are isomers only if they have the same molecular formula. Then, check how the atoms are connected to each other. Constitutional isomers have atoms bonded to different atoms.

Solution

a. $CH_3CH_2CH_3$ and (H)(H)C=C(H)(CH_3)

molecular formula C_3H_8 — molecular formula C_3H_6

The two compounds have the same number of C's but a different number of H's, so they have different molecular formulas. Thus, they are *not* isomers of each other.

b. $CH_3CH_2CH_2NH_2$ and $CH_3—N(CH_3)—CH_3$

Both compounds have molecular formula C_3H_9N. Since one compound has C–C bonds and the other does not, the atoms are connected differently. These compounds are *constitutional isomers.*

PROBLEM 10.14

Are the compounds in each pair constitutional isomers or are they not constitutional isomers of each other?

a. $CH_3CH_2CH_2CH_3$ and $CH_3CH_2CH_3$

b. $CH_3CH_2CH_2OH$ and $CH_3OCH_2CH_3$

c. (cyclohexane) and (cyclopentane)$—CH_3$

PROBLEM 10.15

Label each representation as butane or isobutane (2-methylpropane).

a. H—C(H)(H)—C(H)(H)—C(H)(CH_3)—H

b. $HC(CH_3)_3$

c. H—C(H)(H)—C(H)—H, with H—C(H)(H)—C(H)—H bonded below

d. H—C(CH_3)(H)—C(CH_3)(H)—H

10.5C Acyclic Alkanes Having Five or More Carbons

As the number of carbon atoms in an alkane increases, so does the number of isomers. There are three constitutional isomers for the five-carbon alkane, each having molecular formula C_5H_{12}: **pentane, isopentane** (or 2-methylbutane), and **neopentane** (or 2,2-dimethylpropane).

$CH_3CH_2CH_2CH_2CH_3$ — pentane

$CH_3—C(CH_3)(H)—CH_2CH_3$ — isopentane (2-methylbutane)

$CH_3—C(CH_3)(CH_3)—CH_3$ — neopentane (2,2-dimethylpropane)

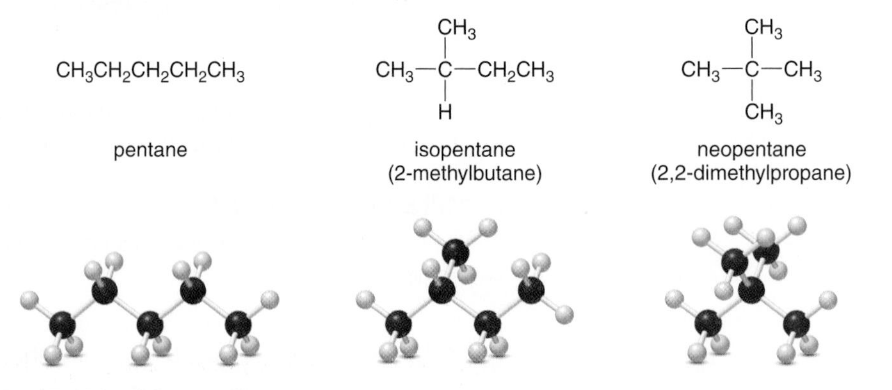

With alkanes having five or more carbons, the names of the straight-chain isomers are derived from Greek roots: ***pent*ane** for **five** carbons, ***hex*ane** for **six,** and so on. Table 10.6 lists the names and structures for the straight-chain alkanes having up to 10 carbons. The suffix ***-ane*** identifies a molecule as an alk***ane.*** The remainder of the name—meth-, eth-, prop-, and so forth—indicates the number of carbons in the long chain.

The suffix ***-ane*** identifies a molecule as an alk***ane.***

Table 10.6 Straight-Chain Alkanes

Number of C's	Molecular Formula	Structure	Name
1	CH_4	CH_4	*meth*ane
2	C_2H_6	CH_3CH_3	*eth*ane
3	C_3H_8	$CH_3CH_2CH_3$	*prop*ane
4	C_4H_{10}	$CH_3CH_2CH_2CH_3$	*but*ane
5	C_5H_{12}	$CH_3CH_2CH_2CH_2CH_3$	*pent*ane
6	C_6H_{14}	$CH_3CH_2CH_2CH_2CH_2CH_3$	*hex*ane
7	C_7H_{16}	$CH_3CH_2CH_2CH_2CH_2CH_2CH_3$	*hept*ane
8	C_8H_{18}	$CH_3CH_2CH_2CH_2CH_2CH_2CH_2CH_3$	*oct*ane
9	C_9H_{20}	$CH_3CH_2CH_2CH_2CH_2CH_2CH_2CH_2CH_3$	*non*ane
10	$C_{10}H_{22}$	$CH_3CH_2CH_2CH_2CH_2CH_2CH_2CH_2CH_2CH_3$	*dec*ane

SAMPLE PROBLEM 10.8

Draw two isomers with molecular formula C_6H_{14} that have five carbon atoms in the longest chain and a one-carbon branch coming off the chain.

Analysis

Since isomers are different compounds with the same molecular formula, we must add a one-carbon branch to two different carbons to form two different products. Then add enough H's to give each C four bonds.

Solution

Add a C here. or Add a C here.

C—C—C—C—C

→ C—C(C)—C—C—C --Add H's--> $CH_3CH(CH_3)CH_2CH_2CH_3$ **A**

isomers

→ C—C—C(C)—C—C --Add H's--> $CH_3CH_2CH(CH_3)CH_2CH_3$ **B**

Compounds **A** and **B** are isomers because the CH_3 group is bonded to different atoms in the five-carbon chain. Note, too, that we cannot add the one-carbon branch to an *end* carbon because that creates a continuous six-carbon chain. Remember that **bends in the chain don't matter.**

Adding a C to an end C makes a six-carbon chain.

C—C—C—C—C ----> C—C—C—C—C(C) = C—C—C—C—C—C

equivalent structures

PROBLEM 10.16

Draw two isomers with molecular formula C_6H_{14} that have four carbon atoms in th[illegible] and two one-carbon branches coming off the chain.

10.6 Alkane Nomenclature

10.6A The IUPAC System of Nomenclature

$CH_2{=}CHCH_2S(=O)SCH_2CH{=}CH_2$

allicin
(odor of garlic)

How are organic compounds named? Long ago, the name of a compound was often based on the plant or animal source from which it was obtained. For example, the name allicin, the principal component of the odor of garlic, is derived from the botanical name for garlic, *Allium sativum.*

With the isolation and preparation of thousands of new organic compounds it became obvious that each organic compound must have an unambiguous name. A systematic method of naming compounds (a system of **nomenclature**) was developed by the ***I***nternational ***U***nion of ***P***ure and ***A***pplied ***C***hemistry. It is referred to as the **IUPAC** system of nomenclature.

Garlic has been used in Chinese herbal medicine for over 4,000 years. Today it is sold as a dietary supplement because of its reported health benefits. **Allicin,** the molecule responsible for garlic's odor, is not stored in the garlic bulb, but rather is produced by the action of enzymes when the bulb is crushed or bruised.

10.6B The Basic Features of Alkane Nomenclature

Although the names of the straight-chain alkanes having 10 carbons or fewer were already given in Table 10.6, we must also learn how to name alkanes that have carbon branches, called **substituents,** bonded to a long chain. The names of these organic molecules have three parts.

- The **parent name** indicates the number of carbons in the longest continuous carbon chain in the molecule.
- The **suffix** indicates what functional group is present.
- The **prefix** tells us the identity, location, and number of substituents attached to the carbon chain.

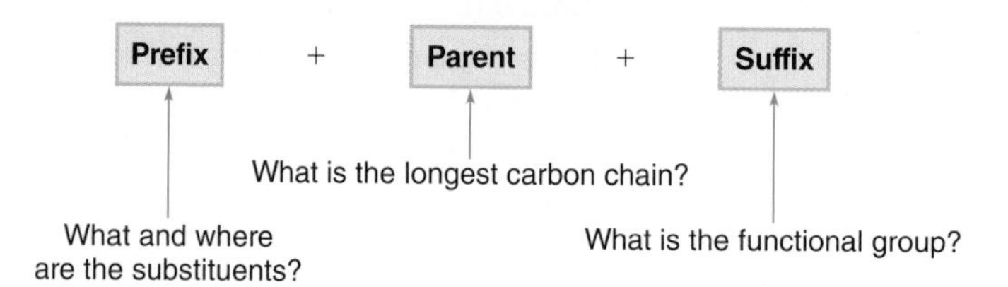

The names of the straight-chain alkanes in Table 10.6 consist of two parts. The suffix *-ane* indicates that the compounds are alkanes. The remainder of the name is the parent name, which indicates the number of carbon atoms in the longest carbon chain. The parent name for **one carbon is *meth-*,** for **two carbons is *eth-*,** and so on. Thus, we are already familiar with two parts of the name of an organic compound.

To determine the third part of a name, the prefix, we must learn how to name the substituents that are bonded to the longest carbon chain.

10.6C Naming Substituents

Carbon substituents bonded to a long carbon chain are called *alkyl groups.*

- An *alkyl group* is formed by removing one hydrogen from an alkane.

An alkyl group is a part of a molecule that is now able to bond to another atom or a functional group. **To name an alkyl group, change the *-ane* ending of the parent alkane to *-yl.*** Thus, methane (CH_4) becomes methyl (CH_3–) and ethane (CH_3CH_3) becomes ethyl (CH_3CH_2–).

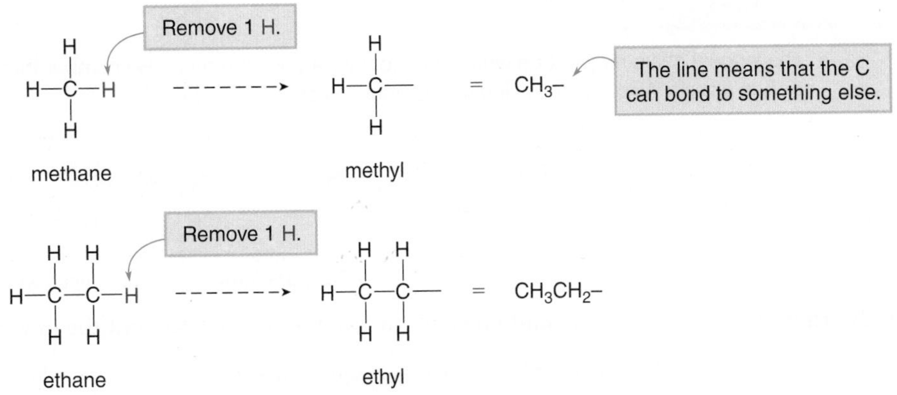

methane --(Remove 1 H.)--> methyl = CH_3- (The line means that the C can bond to something else.)

ethane --(Remove 1 H.)--> ethyl = CH_3CH_2-

Removing one hydrogen from an *end* carbon in any straight-chain alkane forms other alkyl groups named in a similar fashion. Thus, propane ($CH_3CH_2CH_3$) becomes propyl ($CH_3CH_2CH_2-$) and butane ($CH_3CH_2CH_2CH_3$) becomes butyl ($CH_3CH_2CH_2CH_2-$). The names of alkyl groups having six carbons or fewer are summarized in Table 10.7.

Table 10.7 Some Common Alkyl Groups

Number of C's	Structure	Name
1	CH_3-	methyl
2	CH_3CH_2-	ethyl
3	$CH_3CH_2CH_2-$	propyl
4	$CH_3CH_2CH_2CH_2-$	butyl
5	$CH_3CH_2CH_2CH_2CH_2-$	pentyl
6	$CH_3CH_2CH_2CH_2CH_2CH_2-$	hexyl

10.6D Naming an Acyclic Alkane

Four steps are needed to name an alkane.

How To Name an Alkane Using the IUPAC System

Step [1] Find the parent carbon chain and add the suffix.

- **Find the longest continuous carbon chain,** and name the molecule using the parent name for that number of carbons, given in Table 10.6. To the name of the parent, add the suffix ***-ane*** for an alkane. Each functional group has its own suffix.

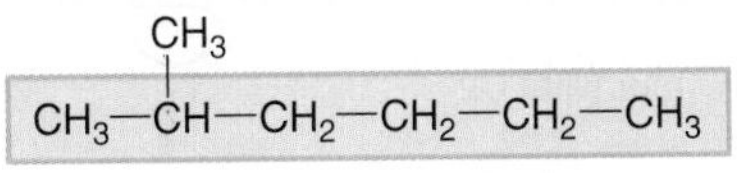

$CH_3-CH(CH_3)-CH_2-CH_2-CH_2-CH_3$

6 C's in the longest chain

6 C's ---> **hexane**

—Continued

How To, continued . . .

- The longest chain may not be written horizontally across the page. Remember that **it does not matter if the chain is *straight* or has *bends.*** All of the following representations are equivalent.

$CH_3-CH(CH_3)-CH_2-CH_2-CH_2-CH_3$ = $CH_3-CH(CH_3)-CH_2-CH_2-CH_2-CH_3$ = $CH_3-CH_2-CH_2-CH_2-C(CH_3)(CH_3)-H$

6 C's in the longest chain of each structure

Step [2] Number the atoms in the carbon chain to give the first substituent the lower number.

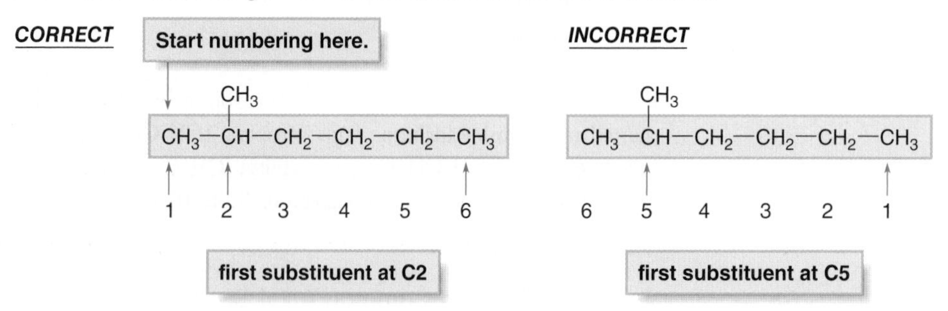

Step [3] Name and number the substituents.

- Name the substituents as alkyl groups, and use the numbers from step [2] to designate their location.

methyl at C2

$CH_3-CH(CH_3)-CH_2-CH_2-CH_2-CH_3$ ← longest chain (6 C's) **hexane**

1 2

- Every carbon belongs to *either* the longest chain or a substituent, but *not both.*
- Each substituent needs its *own* number.
- If two or more identical substituents are bonded to the longest chain, use **prefixes** to indicate how many: **di-** for two groups, **tri-** for three groups, **tetra-** for four groups, and so forth. The following compound has two methyl groups so its name contains the prefix di- before methyl → *di*methyl.

methyl at C2

$CH_3-CH(CH_3)-CH(CH_3)-CH_2-CH_2-CH_3$

1 2 3

methyl at C3

Use the prefix di-.

Step [4] Combine substituent names and numbers + parent + suffix.

- Precede the name of the parent by the names of the substituents. **Alphabetize** the names of the substituents, ignoring any prefixes like *di-*. For example, triethyl precedes dimethyl because the *e* of **e**thyl comes before the *m* of **m**ethyl in the alphabet.
- Precede the name of each substituent by the number that indicates its location. There must be ***one number for each substituent.***
- Separate numbers by commas and separate numbers from letters by dashes.

To help identify which carbons belong to the longest chain and which are substituents, always box in the atoms of the long chain. Every other carbon atom then becomes a substituent that needs its own name as an alkyl group.

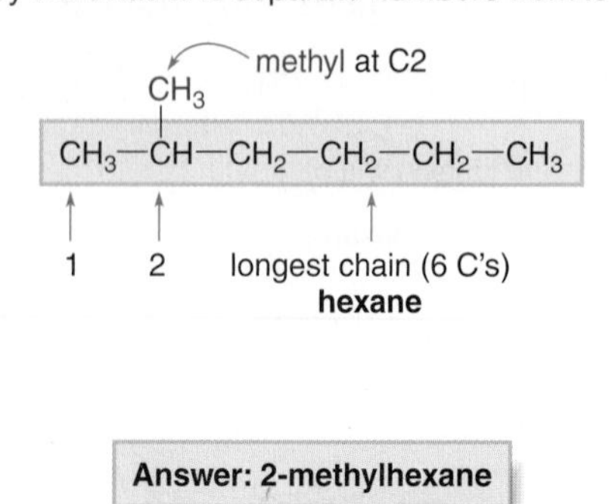

Answer: 2-methylhexane

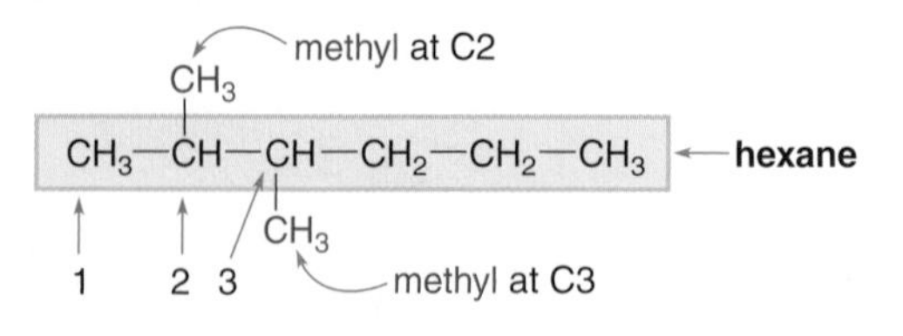

- Two numbers are needed, one for each methyl group.

Answer: 2,3-dimethylhexane

SAMPLE PROBLEM 10.9

Give the IUPAC name for the following compound.

```
        CH3         CH3CH2 H
        |                |  |
  CH3—C—CH2CH2—C—C—CH2CH3
        |                |  |
        H                H  CH3
```

Analysis and Solution

[1] **Name the parent and use the suffix *-ane* since the molecule is an alkane.**

```
        CH3         CH3CH2 H
        |                |  |
  CH3—C—CH2CH2—C—C—CH2CH3
        |                |  |
        H                H  CH3
```

8 C's in the longest chain ----→ **octane**

- Box in the atoms of the longest chain to clearly show which carbons are part of the longest chain and which carbons are substituents.

[2] **Number the chain to give the first substituent the lower number.**

```
        CH3         CH3CH2 H
        |                |  |
  CH3—C—CH2CH2—C—C—CH2CH3
        |                |  |
        H                H  CH3
  1       2            5  6
```

- Numbering from left to right puts the first substituent at C2.

[3] **Name and number the substituents.**

```
methyl at C2                      ethyl at C5
        CH3         CH3CH2 H
        |                |  |
  CH3—C—CH2CH2—C—C—CH2CH3
        |                |  |
        H                H  CH3   methyl at C6
          2            5  6
```

- This compound has three substituents: two methyl groups at C2 and C6 and an ethyl group at C5.

[4] **Combine the parts.**

- Write the name as one word and use the prefix di- before methyl since there are two methyl groups.
- Alphabetize the *e* of **e**thyl before the *m* of **m**ethyl. The prefix di- is ignored when alphabetizing.

```
methyl at C2                      ethyl at C5
        CH3         CH3CH2 H
        |                |  |
  CH3—C—CH2CH2—C—C—CH2CH3   ←—— octane
        |                |  |
        H                H  CH3   methyl at C6
          2            5  6
```

Answer: 5-ethyl-2,6-dimethyloctane

PROBLEM 10.17

Give the IUPAC name for each compound.

```
a. CH3CH2CHCH2CH3
            |
            CH3

          CH2CH3
          |
b.  H—C—CH2—CHCH3
          |          |
          CH3        CH3
```

PROBLEM 10.18

Give the IUPAC name for each compound.

a. $(CH_3)_2CHCH(CH_3)_2$

b.
$$\begin{array}{lcccl} & CH_3 & & H & \\ & | & & | & \\ CH_3CH_2CH_2CH_2- & C & - & C & -CH_2CH_3 \\ & | & & | & \\ & CH_2 & & CH_3 & \\ & | & & & \\ & CH_3 & & & \end{array}$$

You must also know how to derive a structure from a given name. Sample Problem 10.10 demonstrates a stepwise method.

SAMPLE PROBLEM 10.10

Give the structure with the following IUPAC name: 2,3,3-trimethylpentane.

Analysis

To derive a structure from a name, first look at the end of the name to find the parent name and suffix. From the parent we know the number of C's in the longest chain, and the suffix tells us the functional group; the suffix *-ane* = an alkane. Then, number the carbon chain from either end and add the substituents. Finally, add enough H's to give each C four bonds.

Solution

2,3,3-Trimethyl**pentane** has **pentane** (5 C's) as the longest chain and three methyl groups at carbons 2, 3, and 3.

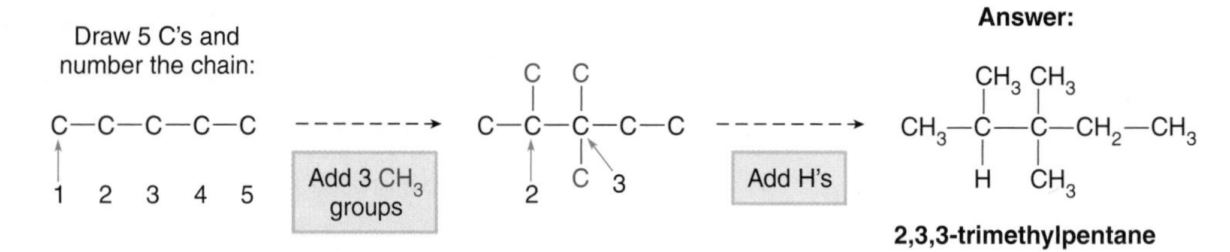

PROBLEM 10.19

Give the structure corresponding to each IUPAC name.

a. 3-methylhexane
b. 3,3-dimethylpentane
c. 3,5,5-trimethyloctane
d. 3-ethyl-4-methylhexane

PROBLEM 10.20

Give the structure corresponding to each IUPAC name.

a. 2,2-dimethylbutane
b. 6-butyl-3-methyldecane
c. 4,4,5,5-tetramethylnonane
d. 3-ethyl-5-propylnonane

10.6E FOCUS ON HEALTH & MEDICINE Naming New Drugs

Naming organic compounds has become big business for drug companies. The IUPAC name of an organic compound can be long and complex. As a result, most drugs have three names:

- **Systematic:** The systematic name follows the accepted rules of nomenclature; this is the IUPAC name.
- **Generic:** The generic name is the official, internationally approved name for the drug.
- **Trade:** The trade name for a drug is assigned by the company that manufactures it. Trade names are often "catchy" and easy to remember. Companies hope that the public will continue to purchase a drug with an easily recalled trade name long after a cheaper generic version becomes available.

Consider the world of over-the-counter anti-inflammatory agents. The compound a chemist calls 2-[4-(2-methylpropyl)phenyl]propanoic acid has the generic name ibuprofen. It is marketed under a variety of trade names, including Motrin and Advil.

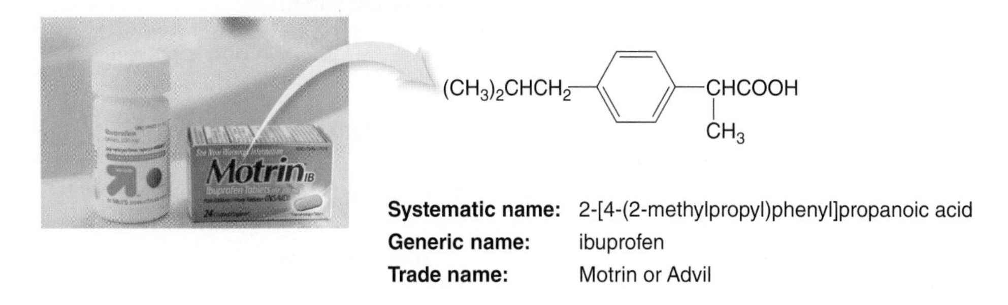

Systematic name: 2-[4-(2-methylpropyl)phenyl]propanoic acid
Generic name: ibuprofen
Trade name: Motrin or Advil

10.7 Cycloalkanes

Cycloalkanes contain carbon atoms arranged in a ring. Think of a cycloalkane as being formed by removing two H's from the end carbons of a chain, and then bonding the two carbons together.

10.7A Simple Cycloalkanes

Simple cycloalkanes are named by adding the prefix *cyclo-* to the name of the acyclic alkane having the same number of carbons. Cycloalkanes having three to six carbon atoms are shown in the accompanying figure. They are drawn using polygons in skeletal representations (Section 10.3). Each corner of the polygon has a carbon atom with two hydrogen atoms to give it four bonds.

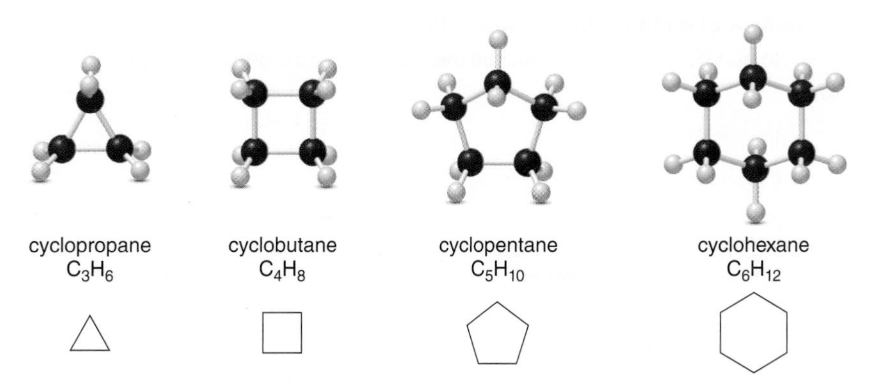

Although we draw cycloalkanes as flat polygons, in reality cycloalkanes with more than three carbons are not planar molecules. Cyclohexane, for example, adopts a puckered arrangement called the **chair** form, in which all bond angles are 109.5°.

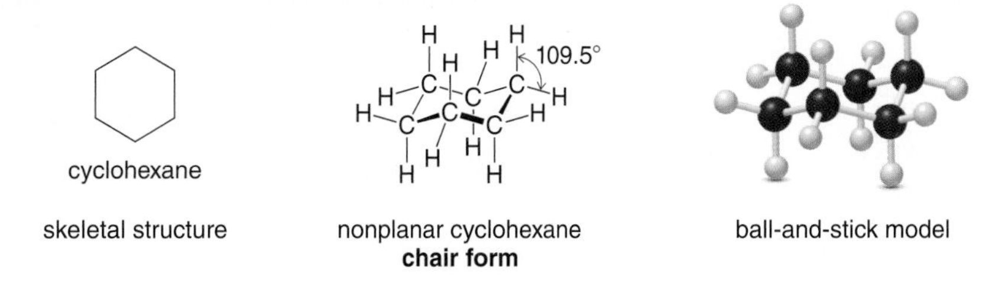

skeletal structure

nonplanar cyclohexane
chair form

ball-and-stick model

10.7B Naming Cycloalkanes

Cycloalkanes are named using the rules in Section 10.6, but the prefix *cyclo-* immediately precedes the name of the parent.

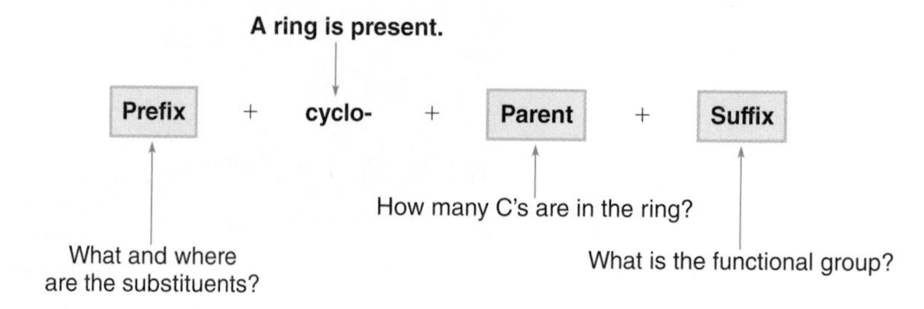

How To Name a Cycloalkane Using the IUPAC System

Step [1] Find the parent cycloalkane.

- Count the number of carbon atoms in the ring and use the parent name for that number of carbons. Add the prefix *cyclo-* and the suffix *-ane* to the parent name.

6 C's in the ring → (ring with CH_3 and CH_2CH_3)

cyclohexane

Step [2] Name and number the substituents.

- No number is needed to indicate the location of a **single** substituent.

methylcyclohexane (CH_3 on cyclohexane) butylcyclopentane ($CH_2CH_2CH_2CH_3$ on cyclopentane)

- For rings with more than one substituent, begin numbering at one substituent, and then give the **second substituent** the lower number. With two **different** substituents, number the ring to assign the lower number to the substituents **alphabetically.**

• Place CH_3 groups at C1 and C3.

1,3-dimethylcyclohexane
(*not* 1,5-dimethylcyclohexane)

Earlier letter ---→ lower number

• ethyl group at **C1**
• methyl group at **C3**

1-ethyl-3-methylcyclohexane
(*not* 3-ethyl-1-methylcyclohexane)

SAMPLE PROBLEM 10.11

Give the IUPAC name for the following compound.

CH_2CH_3

$CH_2CH_2CH_3$

Analysis and Solution

[1] **Name the ring.** The ring has 5 C's so the molecule is named as a cyclopentane.

[2] **Name and number the substituents.**

- There are two substituents: $CH_3CH_2–$ is an ethyl group and $CH_3CH_2CH_2–$ is a propyl group.
- Number to put the two groups at C1 and C2 (not C1 and C5).
- Place the ethyl group at C1 since the *e* of **e**thyl comes before the *p* of **p**ropyl in the alphabet.

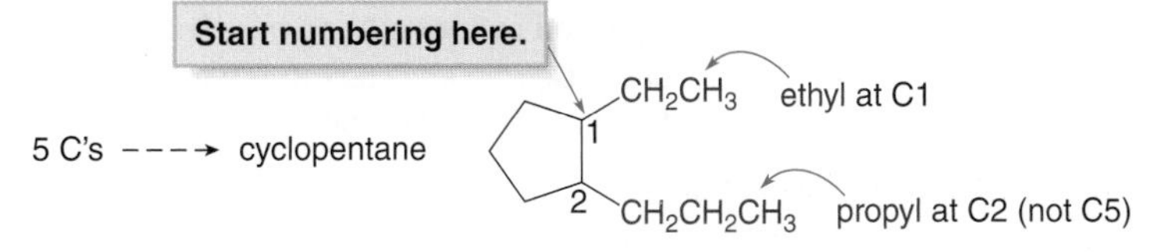

Answer: 1-ethyl-2-propylcyclopentane

PROBLEM 10.21

Give the IUPAC name for each compound.

a. CH_3 b. CH_3 CH_3 c. CH_3CH_2 $CH_2CH_2CH_3$ d. CH_2CH_3 CH_3

PROBLEM 10.22

Give the structure corresponding to each IUPAC name.

a. propylcyclopentane
b. 1,2-dimethylcyclobutane
c. 1,1,2-trimethylcyclopropane
d. 4-ethyl-1,2-dimethylcyclohexane

CONSUMER NOTE

Natural gas is odorless. The smell observed in a gas leak is due to minute amounts of a sulfur additive such as methanethiol, CH_3SH, which provides an odor for easy detection and safety.

ENVIRONMENTAL NOTE

Methane is formed and used in a variety of ways. The CH_4 released from decaying vegetable matter in New York City's main landfill is used for heating homes. CH_4 generators in China convert cow manure into energy in rural farming towns.

10.8 FOCUS ON THE ENVIRONMENT Fossil Fuels

Many alkanes occur in nature, primarily in natural gas and petroleum. Both of these fossil fuels serve as energy sources, formed long ago by the degradation of organic material.

Natural gas is composed largely of **methane** (60–80% depending on its source), with lesser amounts of ethane, propane, and butane. These organic compounds burn in the presence of oxygen, releasing energy for cooking and heating (Section 10.10).

Figure 10.2

Refining Crude Petroleum into Usable Fuel and Other Petroleum Products

a.

(a) **An oil refinery.** At an oil refinery, crude petroleum is separated into fractions of similar boiling point.

b.

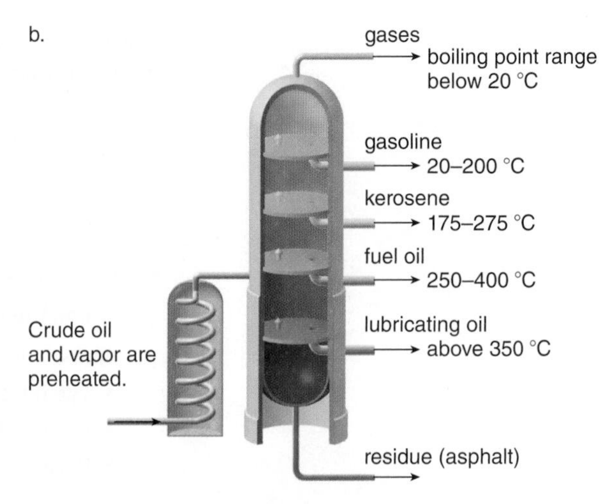

(b) **A refinery tower.** As crude petroleum is heated, the lower-boiling components come off at the top of the tower, followed by fractions of higher boiling point.

Petroleum is a complex mixture of compounds, most of which are hydrocarbons containing 1–40 carbon atoms. Distilling crude petroleum, a process called **refining,** separates it into usable fractions that differ in boiling point (Figure 10.2). Most products of petroleum refining provide fuel for home heating, automobiles, diesel engines, and airplanes. Each fuel type has a different composition of hydrocarbons: gasoline (C_5H_{12}–$C_{12}H_{26}$), kerosene ($C_{12}H_{26}$–$C_{16}H_{34}$), and diesel fuel ($C_{15}H_{32}$–$C_{18}H_{38}$).

Petroleum provides more than fuel. About 3% of crude oil is used to make plastics and other synthetic compounds, including drugs, fabrics, dyes, and pesticides. These products are responsible for many of the comforts we now take for granted in industrialized countries. Imagine what life would be like without air-conditioning, refrigeration, anesthetics, and pain relievers, all products of the petroleum industry.

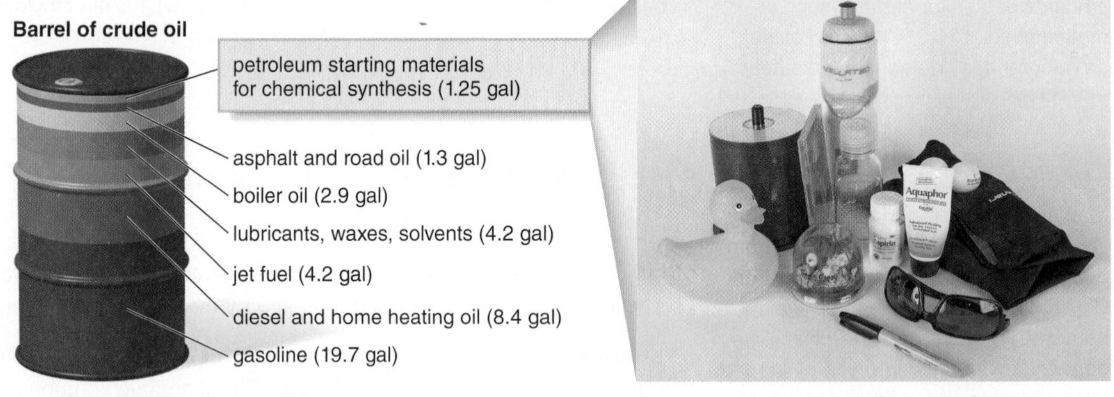

products made from petroleum

Energy from petroleum is *nonrenewable,* and the remaining known oil reserves are limited. Given our dependence on petroleum, not only for fuel, but also for the many necessities of modern society, it becomes obvious that we must both conserve what we have and find alternate energy sources.

10.9 Physical Properties

Organic molecules resemble other covalent compounds in that three types of intermolecular forces are possible, depending on the functional group. Intermolecular forces were first discussed in Section 4.3.

- **London dispersion forces are weak intermolecular forces present in all covalent compounds, and the only forces present in nonpolar molecules.**
- **Dipole–dipole interactions are stronger intermolecular forces due to the attraction of permanent dipoles in polar molecules.**
- **Hydrogen bonding is the strongest intermolecular force that occurs when a hydrogen atom bonded to O, N, or F is electrostatically attracted to an O, N, or F atom in another molecule.**

ENVIRONMENTAL NOTE

Crude oil that leaks into the sea forms an insoluble layer on the surface.

Since **alkanes contain only nonpolar C—C and C—H bonds,** they cannot exhibit dipole–dipole interactions or hydrogen bonding. **Alkanes are nonpolar molecules that exhibit only weak London dispersion forces.** Because the strength of the intermolecular forces determines the boiling point and melting point of compounds, **alkanes have low boiling points and low melting points.** Low molecular weight alkanes are gases at room temperature, and alkanes used in gasoline are all liquids.

The melting points and boiling points of alkanes increase as the number of carbons increases. Increased surface area increases the force of attraction between molecules, thus raising the boiling point and melting point. This is seen in comparing the boiling points of three straight-chain alkanes.

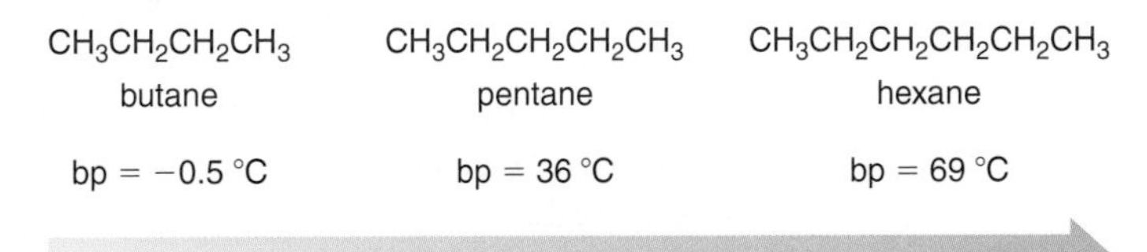

Increasing surface area
Increasing boiling point

The insolubility of nonpolar oil and very polar water leads to the expression, "Oil and water don't mix."

Because nonpolar alkanes are water insoluble and less dense than water, crude petroleum spilled into the sea from a ruptured oil tanker creates an insoluble oil slick on the surface. The insoluble hydrocarbon oil poses a special threat to birds whose feathers are coated with natural nonpolar oils for insulation. Because these oils dissolve in the crude petroleum, birds lose their layer of natural protection and many die.

HEALTH NOTE

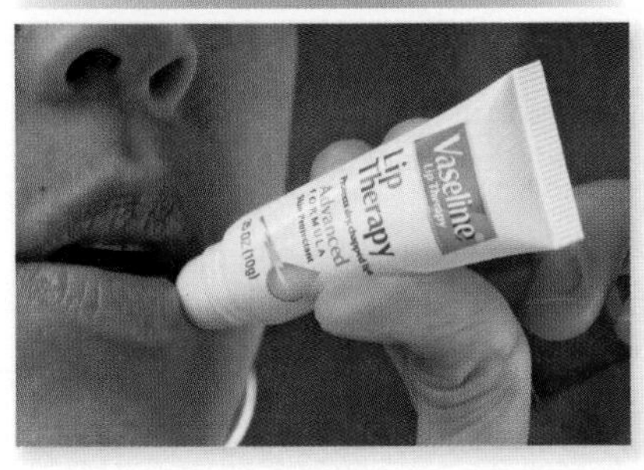

Vaseline petroleum jelly (Problem 10.24), a skin lubricant often used for chapped lips and diaper rash, is a mixture of high molecular weight alkanes.

PROBLEM 10.23

Answer the following questions about pentane (C_5H_{12}), heptane (C_7H_{16}), and decane ($C_{10}H_{22}$).

a. Which compound has the highest boiling point?
b. Which compound has the lowest boiling point?
c. Which compound has the highest melting point?
d. Which compound has the lowest melting point?

PROBLEM 10.24

Explain why Vaseline, a complex mixture of hydrocarbons used as a skin lubricant and ointment, is insoluble in water and soluble in dichloromethane (CH_2Cl_2).

PROBLEM 10.25

The coating of apple skins contains an alkane with 27 carbon atoms. (a) How many hydrogen atoms does this alkane contain? (b) What purpose does this alkane serve?

10.10 FOCUS ON THE ENVIRONMENT Combustion

Combustion was first discussed in Sections 5.7B and 6.9B.

Alkanes are the only family of organic molecules that has no functional group, so **alkanes undergo few reactions.** In this chapter, we consider only one reaction of alkanes—**combustion**—that is, **alkanes burn in the presence of oxygen to form carbon dioxide (CO_2) and water.** This is a practical example of oxidation. Every C—H and C—C bond in the starting material is converted to a C—O bond in the product.

$$2\ (CH_3)_3CCH_2CH(CH_3)_2 + 25\ O_2 \xrightarrow{\text{flame}} 16\ CO_2 + 18\ H_2O + \text{energy}$$

isooctane (high-octane component of gasoline)

CONSUMER NOTE

A spark or a flame is needed to initiate combustion. Gasoline, which is composed largely of alkanes, can be safely handled and stored in the air, but the presence of a spark or open flame causes immediate and violent combustion.

Note that the products, **CO_2 + H_2O,** are the same regardless of the identity of the starting material. Combustion of alkanes in the form of natural gas, gasoline, or heating oil releases energy for heating homes, powering vehicles, and cooking food.

When there is not enough oxygen available to completely burn a hydrocarbon, **incomplete combustion** may occur and **carbon monoxide** (CO) is formed instead of carbon dioxide (CO_2).

Incomplete combustion

$$2\ CH_4 + 3\ O_2 \xrightarrow{\text{flame}} 2\ CO + 4\ H_2O + \text{energy}$$

methane; carbon monoxide

Carbon monoxide is a poisonous gas that binds to hemoglobin in the blood, thus reducing the amount of oxygen that can be transported through the bloodstream to cells. CO can be formed whenever hydrocarbons burn. When an automobile engine burns gasoline, unwanted carbon monoxide can be produced. Yearly car inspections measure CO and other pollutant levels and are designed to prevent cars from emitting potentially hazardous substances into ambient air. Carbon monoxide is also formed when cigarettes burn, so heavy smokers have an unhealthy concentration of CO in their bloodstream.

HEALTH NOTE

Meters that measure CO levels can be purchased, as described in Section 5.5. When wood is burned in a poorly ventilated fireplace situated in a well-insulated room, the CO concentration can reach an unhealthy level.

PROBLEM 10.26

(a) What products are formed when the ethane (CH_3CH_3) in natural gas undergoes combustion in the presence of a flame? (b) Write a balanced equation for the incomplete combustion of ethane (CH_3CH_3) to form carbon monoxide as one product.

STUDY SKILLS PART II: ORGANIC CHEMISTRY

The organic chemistry in Chapters 10–13 is very different from the general chemistry in Chapters 1–9. One clear difference is that most problems in organic chemistry do *not* require math.

While many breathe a sigh of relief about the absence of math, understand that *working problems with a pencil in hand is absolutely necessary to master the concepts.* Do not merely think through an answer; **write out an answer.** You must become comfortable with drawing organic molecules and recognizing the functional groups they possess. No matter how many carbon atoms a molecule contains, its nomenclature, physical properties, and reactions are determined by the few atoms that comprise its functional groups.

Several types of problems will appear again and again. These include naming compounds, drawing structures from a name, and drawing the products of a chemical reaction. Each type of problem can be mastered by repeated practice. Some students learn reactions by making flash cards with reactants on one side of an index card and products on the other. In studying, look at the starting materials, *write out* the structure of the product, and then check the answer on the back of the index card.

In addition, predicting the shapes of molecules and the type of intermolecular forces they possess, two topics first examined in Chapter 3, will allow us to understand the physical properties of organic compounds. Re-read Sections 3.10–3.12 to refresh your memory on these useful concepts.

Finally, while the general chemistry presented in Chapters 1–9 is firmly grounded in high school chemistry courses, few beginning students of college chemistry have learned much background in organic chemistry. Don't let this fact intimidate you! Organic chemistry is based on a set of fundamental themes. Moreover, because many interesting chemical phenomena involve organic molecules, you will learn about relevant examples of organic chemistry in your daily lives.

KEY TERMS

Acyclic alkane (10.5)
Alcohol (10.4)
Aldehyde (10.4)
Alkane (10.4)
Alkene (10.4)
Alkyl group (10.6)
Alkyl halide (10.4)
Alkyne (10.4)
Amide (10.4)
Amine (10.4)
Amino group (10.4)
Aromatic compound (10.4)
Branched-chain alkane (10.5)
Carbonyl group (10.4)
Carboxyl group (10.4)
Carboxylic acid (10.4)
Combustion (10.10)
Condensed structure (10.3)
Constitutional isomer (10.5)
Cycloalkane (10.5)
Ester (10.4)
Ether (10.4)
Functional group (10.4)
Heteroatom (10.2)
Hydrocarbon (10.4)
Hydroxyl group (10.4)
Incomplete combustion (10.10)
Isomer (10.5)
IUPAC nomenclature (10.6)
Ketone (10.4)
Organic chemistry (10.1)
Parent name (10.6)
Pheromone (10.5)
Saturated hydrocarbon (10.5)
Skeletal structure (10.3)
Straight-chain alkane (10.5)
Thiol (10.4)

KEY CONCEPTS

1 How do organic compounds differ from ionic inorganic compounds? (10.1)

- Organic compounds are composed of discrete molecules with covalent bonds. Ionic inorganic compounds are composed of cations and anions, held together by the strong attraction of oppositely charged ions. Other properties that are consequences of these bonding differences are summarized in Table 10.1.

2 What are the characteristic features of organic compounds? (10.2)

- Organic compounds contain carbon atoms and most contain hydrogen atoms. Carbon forms four bonds.
- Carbon forms single, double, and triple bonds to itself and other atoms.
- Carbon atoms can bond to form chains or rings.
- Organic compounds often contain heteroatoms, commonly N, O, and the halogens.

3 What shorthand methods are used to draw organic molecules? (10.3)

- In condensed structures, atoms are drawn in but the two-electron bonds are generally omitted. Lone pairs are omitted as well. Parentheses are used around like groups bonded together or to the same atom.
- Three assumptions are used in drawing skeletal structures: [1] There is a carbon at the intersection of two lines or at the end of any line. [2] Each carbon has enough hydrogens to give it four bonds. [3] Heteroatoms and the hydrogens bonded to them are drawn in.

4 What is a functional group and why are functional groups important? (10.4)

- A functional group is an atom or a group of atoms with characteristic chemical and physical properties.
- A functional group determines all of the properties of a molecule—its shape, physical properties, and the type of reactions it undergoes.

5 What are the characteristics of an alkane? (10.5)

- Alkanes are hydrocarbons having only nonpolar C—C and C—H single bonds.
- There are two types of alkanes: Acyclic alkanes (C_nH_{2n+2}) have no rings. Cycloalkanes (C_nH_{2n}) have one or more rings.

6 What are constitutional isomers? (10.5)

- Isomers are different compounds with the same molecular formula.
- Constitutional isomers differ in the way the atoms are connected to each other. $CH_3CH_2CH_2CH_3$ and $HC(CH_3)_3$ are constitutional isomers because they have molecular formula C_4H_{10}, but one compound has a chain of four carbons in a row and the other does not.

7 How are alkanes named? (10.6, 10.7)

- Alkanes are named using the IUPAC system of nomenclature. A name has three parts: the parent indicates the number of carbons in the longest chain or the ring; the suffix indicates the functional group (*-ane* = alkane); the prefix tells the number and location of substituents coming off the chain or ring.
- Alkyl groups are formed by removing one hydrogen from an alkane. Alkyl groups are named by changing the *-ane* ending of the parent alkane to the suffix *-yl*.

8 Characterize the physical properties of alkanes. (10.9)

- Alkanes are nonpolar, so they have weak intermolecular forces, low melting points, and low boiling points.
- The melting points and boiling points of alkanes increase as the number of carbons increases due to increased surface area.
- Alkanes are insoluble in water.

9 What are the products of the combustion and incomplete combustion of an alkane? (10.10)

- Alkanes burn in the presence of air. Combustion forms CO_2 and H_2O as products. Incomplete combustion forms CO and H_2O.

UNDERSTANDING KEY CONCEPTS

Selected in-chapter and odd-numbered end-of-chapter problems have brief answers at the end of each chapter. The *Student Study Guide and Solutions Manual* contains detailed solutions to all in-chapter and odd-numbered end-of-chapter problems, as well as additional worked examples and a chapter self-test.

10.27 Identify the functional group in each ball-and-stick model.

a. 

b. 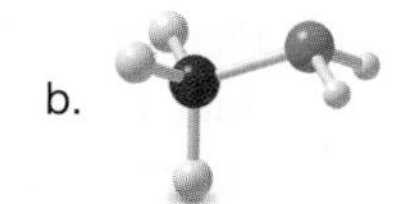

10.28 GHB is an addictive, illegal recreational drug that depresses the central nervous system and results in intoxication. Identify the functional groups in GHB.

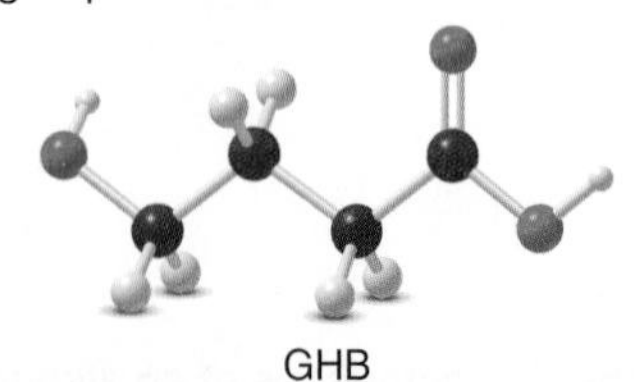

GHB

10.29 Convert each shorthand structure to a complete structure with all atoms and lone pairs drawn in.

a. $(CH_3)_2CH(CH_2)_6CH_3$ b. $(CH_3)_3COH$

10.30 Convert each shorthand structure to a complete structure with all atoms and lone pairs drawn in.

a. $(CH_3)_2CHO(CH_2)_4CH_3$ b. $(CH_3)_3C(CH_2)_3CBr_3$

10.31 The waxy coating that covers tobacco leaves contains a straight-chain alkane having 31 carbons. How many hydrogens does this alkane contain?

10.32 The largest known cycloalkane with a single ring has 288 carbons. What is its molecular formula?

10.33 Draw three constitutional isomers having molecular formula C_7H_{14} that contain a five-membered ring and two CH_3 groups bonded to the ring. Give the IUPAC name for each isomer.

10.34 Draw four constitutional isomers having molecular formula C_6H_{12} that contain a four-membered ring. Give the IUPAC name for each isomer.

10.35 Answer the following questions about the alkane drawn below.

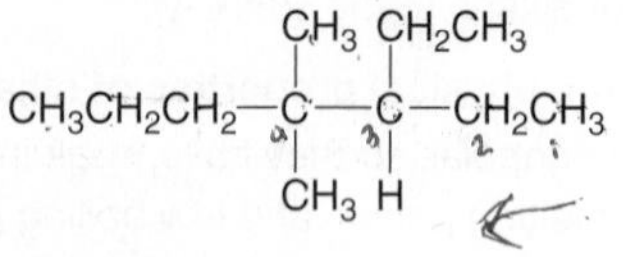

CH_3 CH_2CH_3
$CH_3CH_2CH_2$—C—C—CH_2CH_3
CH_3 H

a. Give the IUPAC name.
b. Draw one constitutional isomer.
c. Predict the solubility in water.
d. Predict the solubility in an organic solvent.
e. Write a balanced equation for complete combustion.

10.36 Answer the questions in Problem 10.35 for the alkane drawn below.

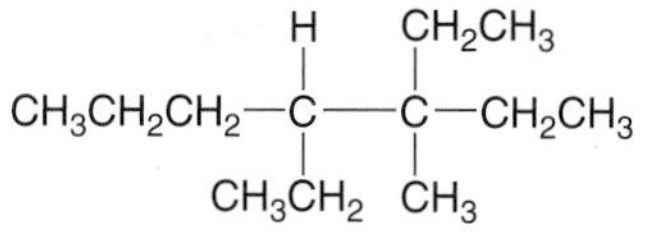

H CH_2CH_3
$CH_3CH_2CH_2$—C—C—CH_2CH_3
CH_3CH_2 CH_3

10.37 Methyl salicylate is responsible for the characteristic odor of the oil wintergreen.

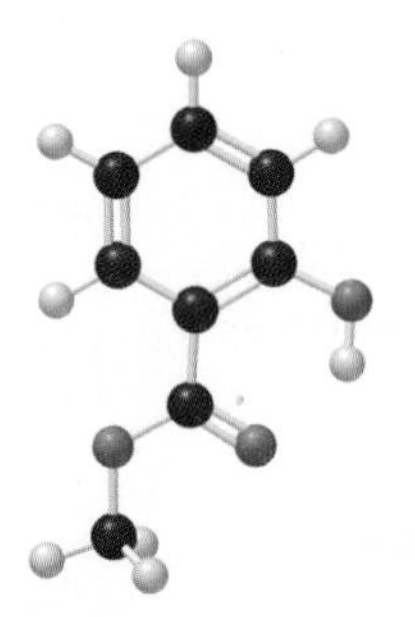

methyl salicylate

a. Give the molecular formula for methyl salicylate.
b. Draw in all lone pairs on heteroatoms using a skeletal structure.
c. Predict the water solubility of methyl salicylate.

10.38 Procaine (trade name Novocain) is a local anesthetic once commonly used in medicine and dentistry.

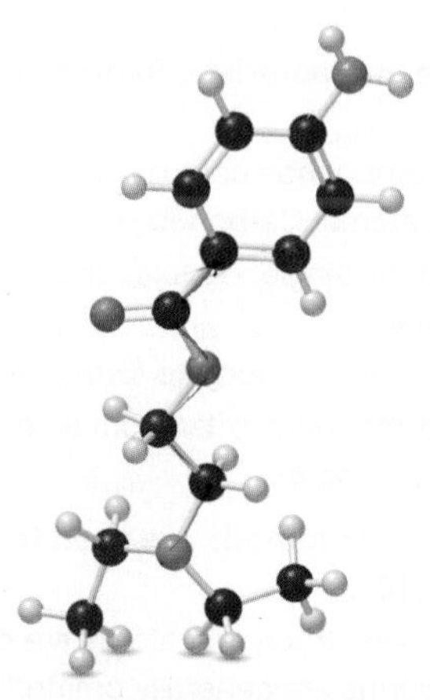

procaine

a. What is the molecular formula of procaine?
b. Draw in all the lone pairs on heteroatoms using a skeletal structure.
c. Identify the functional groups.

ADDITIONAL PROBLEMS

General Characteristics of Organic Molecules

10.39 Which molecular formulas represent organic compounds and which represent inorganic compounds: (a) H_2SO_4; (b) Br_2; (c) C_5H_{12}?

10.40 Which chemical formulas represent organic compounds and which represent inorganic compounds: (a) LiBr; (b) HCl; (c) CH_5N?

10.41 Complete each structure by filling in all H's and lone pairs.

a. C—C=C—C≡C

b. C=C—C—O

c. C≡C—C—C (Cl on the third C)

d. C=C—C—C(=O)—N

10.42 Complete each structure by filling in all H's and lone pairs.

a. C(—O)—C(—O)—C=O

glyceraldehyde
(simple carbohydrate)

b. C—C(—O)—C(—O)=O

lactic acid
(product of carbohydrate metabolism)

c. N—C(—C)—C(=O)—O

alanine
(amino acid in proteins)

d. O—C—C(=O)—C—O

dihydroxyacetone
(ingredient in artificial tanning agents)

Properties of Organic Compounds

10.43 You are given two unlabeled bottles of solids, one containing sodium chloride (NaCl) and one containing cholesterol ($C_{27}H_{46}O$). You are also given two labeled bottles of liquids, one containing water and one containing dichloromethane (CH_2Cl_2). How can you determine which solid is sodium chloride and which solid is cholesterol?

10.44 State how potassium iodide (KI) and pentane ($CH_3CH_2CH_2CH_2CH_3$) differ with regards to each of the following properties: (a) type of bonding; (b) solubility in water; (c) solubility in an organic solvent; (d) melting point; (e) boiling point.

10.45 Spermaceti wax [$CH_3(CH_2)_{14}CO_2(CH_2)_{15}CH_3$], a compound isolated from sperm whales, was once a common ingredient in cosmetics. Its use is now banned to help protect whales. (a) Identify the functional group in spermaceti wax. (b) Predict its solubility properties in water and organic solvents and explain your reasoning.

10.46 Acetic acid (CH_3CO_2H) and palmitic acid [$CH_3(CH_2)_{14}CO_2H$] are both carboxylic acids. Acetic acid is the water-soluble carboxylic acid in vinegar, while palmitic acid is a water-insoluble fatty acid. Give a reason for this solubility difference.

Drawing Organic Molecules

10.47 Convert each compound to a condensed structure.

a. H—C—C—C—C—C—H (fully substituted with H; two CH₃ groups on the fourth C)

b. H—C—C—C—Ö—C—C—C—H (CH₃ branch on the second C)

c. H—C—C—C—C(=:O:)—H

10.48 Convert each compound to a condensed structure.

a. H—C—C—C—C—C—C—Ö—H

b. H—C—C—C—C—N̈—C—H

c. H—C—C(—CH₃)—C(=:O:)—Ö—C—C—H

10.49 Convert each compound to a skeletal structure.

a. (six-membered ring containing N—H, with —Ö—CH₃ substituent)

b. (six-membered ring with two :C̈l: substituents)

10.50 Convert each compound to a skeletal structure.

a. (benzene ring with CH₃, —B̈r:, and :Ö—H substituents)

b. (four-membered ring with —C(=:O:)—Ö—H substituent)

10.51 Convert each shorthand structure to a complete structure with all atoms and lone pairs drawn in.

a. $CH_3CO_2(CH_2)_3CH_3$

b. HO–(cyclohexane with Cl, Cl)–$OCH(CH_3)_2$

10.52 Convert each shorthand structure to a complete structure with all atoms and lone pairs drawn in.

a. $(CH_3)_2CHCONH_2$

b. $(CH_3)_2CH$–(benzene ring)–CHO

10.53 Albuterol (trade names: Proventil and Ventolin) is a bronchodilator, a drug that widens airways, thus making it an effective treatment for individuals suffering from asthma. Draw a complete structure for albuterol with all atoms and lone pairs.

HO, $HOCH_2$, OH, $CHCH_2NHC(CH_3)_3$

albuterol

10.54 Naproxen is the anti-inflammatory agent in Aleve and Naprosyn. Draw a complete structure for naproxen including all atoms and lone pairs.

CH_3, $CHCO_2H$, CH_3O

naproxen

Functional Groups

10.55 Identify the functional groups in each molecule.

a. isoprene (emitted by plants)

b. $CH_3COOCH_2CH_2CH_3$ propyl acetate (from pears)

10.56 Identify the functional groups in each molecule.

a. HO OH, C–CHCH–C, HO, OH, O, O

tartaric acid (from grapes)

b. CH_3–C(=O)–C(=O)–CH_3

butanedione (component of butter flavor)

10.57 Identify the functional groups in each molecule.

a. $CH_2{=}CHCH(OH)CH_2{-}C{\equiv}C{-}C{\equiv}C{-}CH_2CH{=}CH(CH_2)_5CH_3$

carotatoxin (neurotoxin from carrots)

b. CH_3O, HO, C(=O)H

vanillin (isolated from vanilla beans)

10.58 Identify the functional groups in each drug.

a. OH CH_3, C–C–N–CH_3, H H H

pseudoephedrine (nasal decongestant)

b. HO, H, N, C–CH_3, O

acetaminophen (analgesic in Tylenol)

10.59 Convert the ball-and-stick model of neral, a compound isolated from lemons and limes, to a condensed or skeletal structure and identify the functional groups.

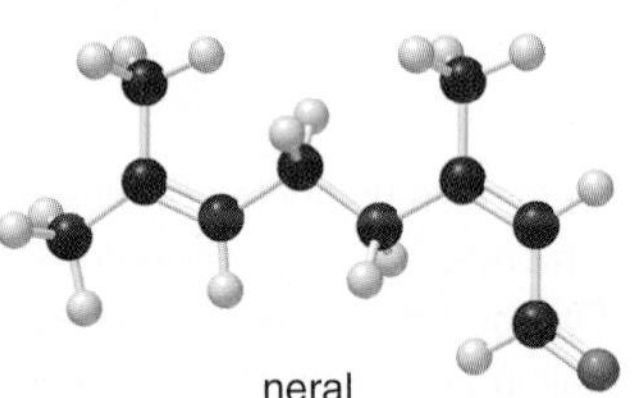

neral

10.60 Convert the ball-and-stick model of estragole, a compound isolated from tarragon, to a skeletal structure and identify the functional groups.

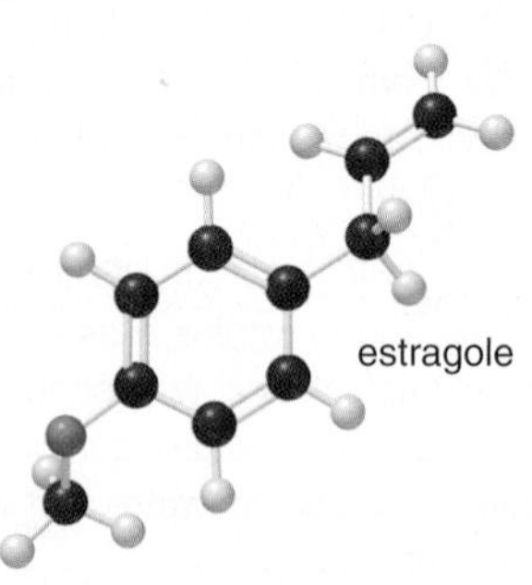

estragole

10.61 Draw an organic compound that fits each of the following criteria.

a. a hydrocarbon having molecular formula C_3H_4 that contains a triple bond

b. an alcohol containing three carbons

c. an aldehyde containing three carbons

d. a ketone having molecular formula C_4H_8O

10.62 Draw an organic compound that fits each of the following criteria.

a. an amine containing two CH_3 groups bonded to the N atom
b. an alkene that has the double bond in a ring
c. an ether having two different R groups bonded to the ether oxygen
d. an amide that has molecular formula C_3H_7NO

10.63 Draw the structure of an alkane, an alkene, and an alkyne, each having five carbon atoms. What is the molecular formula for each compound?

10.64 An alkane has 20 hydrogen atoms. How many carbon atoms would it contain if it were (a) a straight-chain alkane; (b) a branched-chain alkane; (c) a cycloalkane?

10.65 Label each pair of compounds as constitutional isomers or identical molecules.

a. $CH_3CH(CH_2CH_3)CH(CH_2CH_3)CH_3$ (drawn as $CH_3CHCHCH_3$ with CH_2CH_3 above and CH_2CH_3 below) and $CH_3CH_2CH(CH_3)CH_2CH(CH_3)_2$

b. $CH_3CH_2CH(CH_3)CH(CH_3)CH_2CH_3$ (drawn as $CH_3CH_2CHCHCH_2CH_3$ with CH_3 above and CH_3 below) and $CH_3CH(CH_2CH_3)CH(CH_2CH_3)CH_3$ (drawn as $CH_3CHCHCH_3$ with CH_2CH_3 above and CH_2CH_3 below)

c. cyclopropane with two CH_3 groups on one carbon and cyclopropane—CH_2CH_3

10.66 Label each pair of compounds as constitutional isomers, identical molecules, or not isomers of each other.

a. $CH_3CH_2CH_2CH_3$ and $CH_3CH_2CH_2CH_2CH_3$

b. CH_3CH_2—CH_2—CH_2—CH_2CH_3 (drawn as $CH_2—CH_2$ above with CH_3CH_2 and CH_2CH_3 below) and $CH_3(CH_2)_4CH_3$

c. $CH_3—C(=O)—OCH_3$ and $CH_3CH_2—C(=O)—OH$

10.67 Draw structures that fit the following descriptions:

a. two cycloalkanes that are constitutional isomers with molecular formula C_7H_{14}
b. an ether and an alcohol that are constitutional isomers with molecular formula $C_5H_{12}O$
c. two constitutional isomers of molecular formula C_3H_7Cl

10.68 Draw the five constitutional isomers having molecular formula C_6H_{14}.

10.69 Draw three constitutional isomers with molecular formula C_3H_6O. Draw the structure of one alcohol, one ketone, and one cyclic ether.

10.70 Draw one constitutional isomer of each compound.

a. $CH_3CH_2—C(=O)—CH_3$

b. cyclohexane—OH

Alkane Nomenclature

10.71 Give the IUPAC name for each compound.

a. $CH_3CH_2CH(CH_3)CH_2CH_2CH_2CH_2CH_2CH_2CH_3$

b. $CH_3CH_2CH(CH_3)CH_2CH(CH_3)CH_2CH_2CH_3$

c. $CH_3CH_2CH_2C(CH_2CH_3)_3$

d. $CH_3CH(CH_2CH_3)CH_2CH(CH_2CH(CH_3)CH_3)CH_2CH_3$ (drawn as $CH_3CHCH_2CHCH_2CH_3$ with CH_2CH_3 below the second carbon and CH_2CHCH_3 bearing CH_3 above the fourth carbon)

10.72 Give the IUPAC name for each compound.

a. $CH_3—C(H)(CH_2CH_3)—CH(CH_3)CH_3$ (drawn as $CH_3—C—CHCH_3$ with H above C, CH_3 above CH, and CH_2CH_3 below C)

b. $CH_3CH_2—C(CH_3)_2—CH_2CH_2CH_3$

c. $CH_3CH_2CH(CH_2CH_2CH_2CH_3)CH_2CH_2CH_2CH_3$

d. $(CH_3CH_2)_2CHCH(CH_2CH_3)_2$

10.73 Give the IUPAC name for each compound.

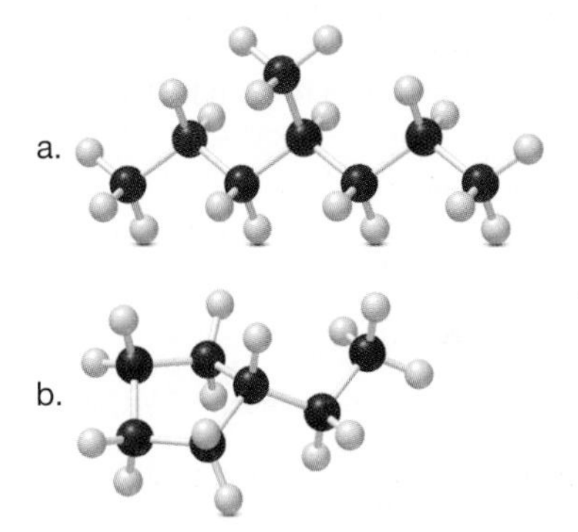

10.74 Give the IUPAC name for each compound.

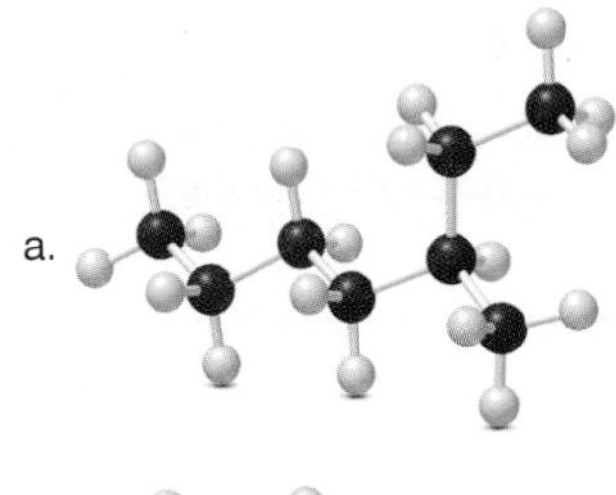

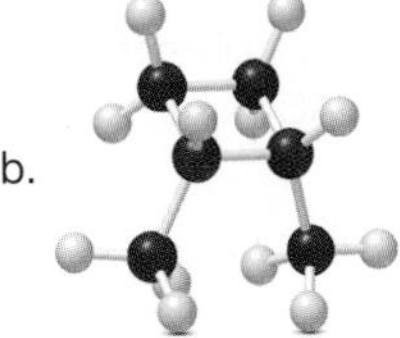

10.75 Give the IUPAC name for each cycloalkane.

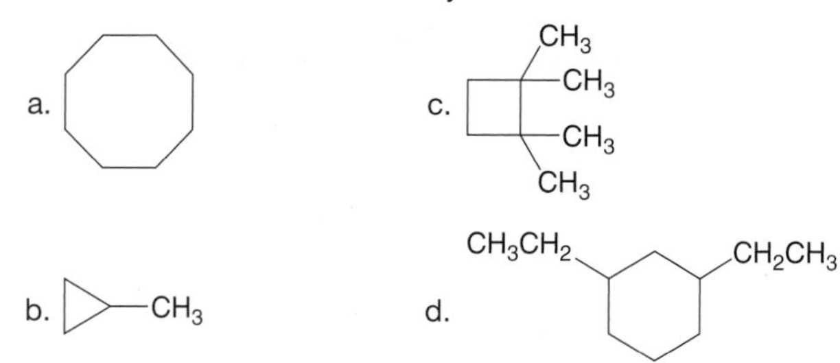

b. —CH_3

d. CH_3CH_2 CH_2CH_3

10.76 Give the IUPAC name for each cycloalkane.

a. CH_2CH_3

b. $CH_2CH_2CH_3$ $CH_2CH_2CH_2CH_3$

c. $CH_2CH_2CH_3$ CH_2CH_3

d. CH_3— —$CH_2CH_2CH_2CH_3$ CH_3CH_2

10.77 Give the structure corresponding to each IUPAC name.

a. 3-ethylhexane
b. 3-ethyl-3-methyloctane
c. 2,3,4,5-tetramethyldecane
d. cyclononane
e. 1,1,3-trimethylcyclohexane

10.78 Give the structure corresponding to each IUPAC name.

a. 3-ethyl-3-methylhexane
b. 2,2,3,4-tetramethylhexane
c. 4-ethyl-2,2-dimethyloctane
d. 1,3,5-triethylcycloheptane
e. 3-ethyl-3,4-dimethylnonane

10.79 Each of the following IUPAC names is incorrect. Explain why it is incorrect and give the correct IUPAC name.

a. 3-methylbutane
b. 1-methylcyclopentane
c. 1,3-dimethylbutane
d. 5-ethyl-2-methylhexane

10.80 Each of the following IUPAC names is incorrect. Explain why it is incorrect and give the correct IUPAC name.

a. 4-methylpentane
b. 2,3,3-trimethylbutane
c. 1,3-dimethylpentane
d. 3-methyl-5-ethylhexane

Physical Properties

10.81 Which compound in each pair has the higher melting point?

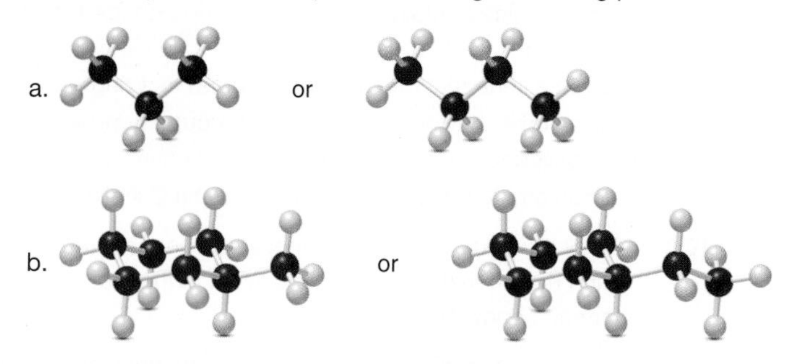

10.82 Which compound in each pair has the higher boiling point?

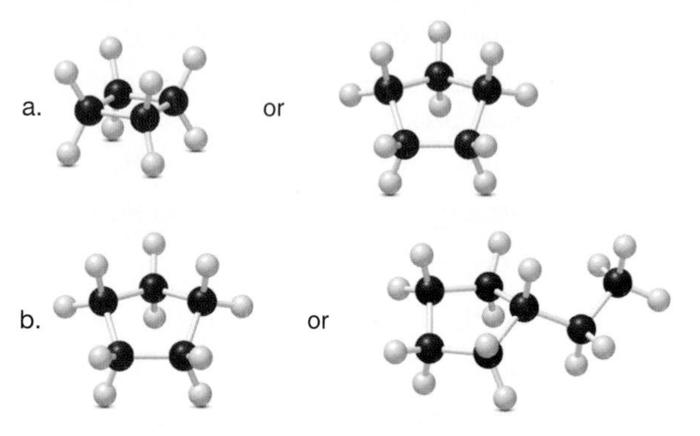

10.83 Explain why hexane is more soluble in dichloromethane (CH_2Cl_2) than in water.

10.84 Mineral oil and Vaseline are both mixtures of alkanes, but mineral oil is a liquid at room temperature and Vaseline is a solid. Which product is composed of alkanes that contain a larger number of carbon atoms? Explain your choice.

Reactions

10.85 Write a balanced equation for the combustion of each alkane: (a) CH_3CH_3; (b) $(CH_3)_2CHCH_2CH_3$.

10.86 Write a balanced equation for the combustion of each cycloalkane.

10.87 Write a balanced equation for the incomplete combustion of each alkane: (a) $CH_3CH_2CH_3$; (b) $CH_3CH_2CH_2CH_3$.

10.88 Benzene (C_6H_6) is a fuel additive sometimes used to make gasoline burn more efficiently. Write a balanced equation for the incomplete combustion of benzene.

Applications and General Problems

10.89 The gasoline industry seasonally changes the composition of gasoline in locations where it gets very hot in the summer and very cold in the winter. Gasoline is refined to contain a larger fraction of higher molecular weight alkanes in warmer weather. In colder weather, it is refined to contain a larger fraction of lower molecular weight alkanes. What is the purpose of producing different types of gasoline for different temperatures?

10.90 Polyethylene (Section 10.4) is a high molecular weight alkane that contains hundreds or even thousands of carbon atoms, bonded together in long carbon chains. When a new home is built, the concrete foundation is often wrapped with polyethylene (sold under the trade name of Tyvek). What purpose does the polyethylene serve?

10.91 Cabbage leaves are coated with a hydrocarbon of molecular formula $C_{29}H_{60}$. What purpose might this hydrocarbon coating serve?

10.92 Skin moisturizers come in two types. (a) One type of moisturizer is composed mainly of hydrocarbon material. Suggest a reason as to how a moisturizer of this sort helps to keep the skin from drying out. (b) A second type of moisturizer is composed mainly of propylene glycol [$CH_3CH(OH)CH_2OH$]. Suggest a reason as to how a moisturizer of this sort helps to keep the skin from drying out.

10.93 Answer the following questions for the cycloalkane depicted in the given ball-and-stick model.

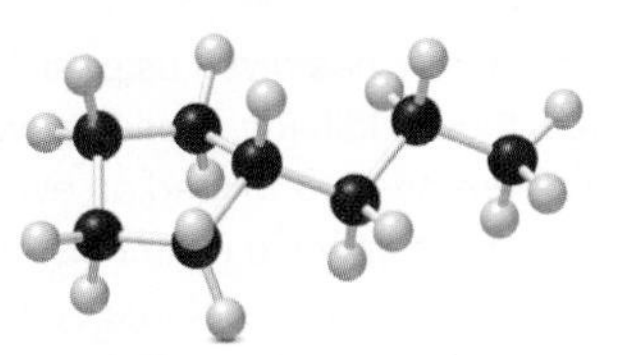

a. Give the IUPAC name.
b. Draw one constitutional isomer.
c. Predict the solubility in water.
d. Predict the solubility in an organic solvent.
e. Draw a skeletal structure.

10.94 Answer the questions in Problem 10.93 for the cycloalkane depicted in the given ball-and-stick model.

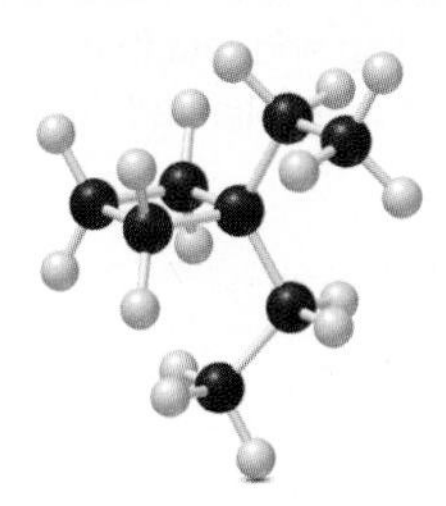

CHALLENGE PROBLEMS

10.95 THC is the active component in marijuana (Section 10.4), and ethanol (CH_3CH_2OH) is the alcohol in alcoholic beverages. Which compound is more water soluble? Explain why drug screenings are able to detect the presence of THC but not ethanol weeks after these substances have been introduced into the body.

10.96 Cocaine is a widely abused, addicting drug. Cocaine is usually obtained as its hydrochloride salt (cocaine hydrochloride, an ionic salt). This salt can be converted to crack (a neutral molecule) by treatment with base.

cocaine (crack)
neutral organic molecule

cocaine hydrochloride
a salt

a. Identify the functional groups in cocaine.
b. Given what you have learned about ionic and covalent bonding, which of the two compounds—crack or cocaine hydrochloride—has a higher boiling point?
c. Which compound is more soluble in water?
d. Can you use the relative solubility to explain why crack is usually smoked but cocaine hydrochloride is injected directly into the bloodstream?

BEYOND THE CLASSROOM

10.97 Organic chemistry has given us pesticides, fertilizers, herbicides, food additives, preservatives, and refrigerants, all of which are needed to feed an ever-growing population. Give one or more specific examples of synthetic compounds in one of these categories. Name each compound, give its chemical formula, and draw its structure. How does the chosen molecule, and others like it, help to feed the world's population? Can you make an argument that one of these types of synthetic compounds is more crucial than the others in producing and preserving the world's food supply?

10.98 What is meant by the octane rating of gasoline? How is octane rating calculated? How is octane rating related to engine efficiency? Give an example of an alkane that has a low octane rating and one that has a high octane rating. Give an example of a gasoline additive that boosts octane ratings.

10.99 Research one of the following drugs: propoxyphene, meperidine, salmeterol, sertraline, donepezil, or fexofenadine (or choose another drug). Give its trade name, chemical formula, and chemical structure. Identify all of the functional groups. What is the drug used for? Is its mechanism of action known?

10.100 Several localities now have waste-to-energy facilities—that is, plants that convert garbage to energy. What are the pros and cons of a garbage-burning facility? Research the nearest waste-burning plant near your home and determine how much garbage is combusted annually. If 27,000 lb of garbage provide the energy equivalent of 21 barrels of crude oil, how much oil does this translate into? What happens to the garbage that is not combustible as well as the ash that is produced after combustion? How much impact does a waste-burning facility have on the local landfill?

ANSWERS TO SELECTED PROBLEMS

10.1 organic: (a), (e)
inorganic: (b), (c), (d), (f)

10.2
a. H—C—C=C—C—H
b. H—C—C—C—H
c. H—C—C—C—H
d. H—C—C—N—C—H
e.

10.3
a. $CH_3(CH_2)_3CH_3$
b. $Br(CH_2)_2Br$
c. $CH_3CH_2O(CH_2)_2OH$
d. $CH_3CH_2C(CH_3)_3$

10.5
a.
b.
c.

10.7
a. hydroxyl
b. alkene; aromatic ring
c. two amines

10.8
a. aldehyde
b. carboxylic acid H—C—C—C—C—O—H
c. ketone
d. ester H—C—C—C—O—C—C—H
e. amide

10.9
a. amide, aromatic ring, ether, hydroxyl, amine
b. ether, aromatic ring, hydroxyl, alkyne

10.10 a. 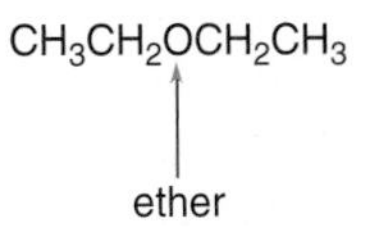 $CH_3CH_2OCH_2CH_3$ (ether) b. 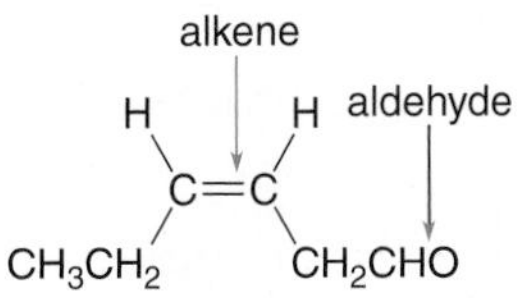 $CH_3CH_2(H)C=C(H)CH_2CHO$ (alkene, aldehyde)

10.11 aromatic ring (CH_3, CH_3) —N(H)—$\overset{O}{\overset{\|}{C}}CH_2N(CH_2CH_3)_2$ (amide, amine)

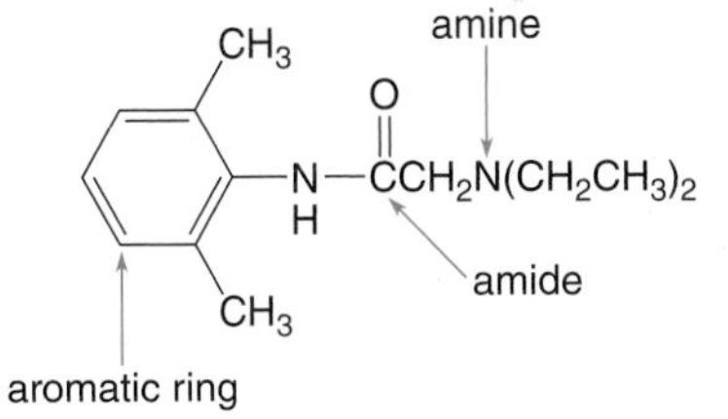

10.13 a. acyclic b. cyclic c. cyclic d. acyclic

10.14 a. not isomers b. isomers c. isomers

10.15 a. butane b. isobutane c. butane d. butane

10.16 $CH_3CH(CH_3)CH(CH_3)CH_3$ $CH_3C(CH_3)_2CH_2CH_3$

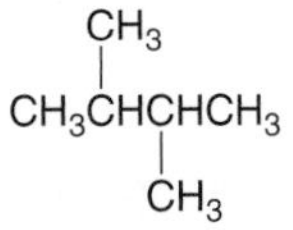

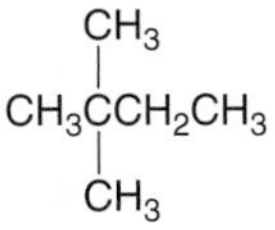

10.17 a. 3-methylpentane
b. 2,4-dimethylhexane

10.19 a. 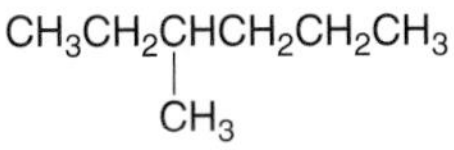 $CH_3CH_2CH(CH_3)CH_2CH_2CH_3$
b. 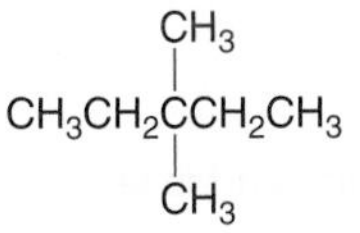 $CH_3CH_2C(CH_3)_2CH_2CH_3$
c. $CH_3CH_2CH(CH_3)CH_2C(CH_3)_2CH_2CH_2CH_3$
d. 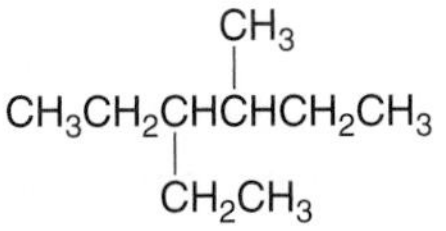 $CH_3CH_2CH(CH_2CH_3)CH(CH_3)CH_2CH_3$

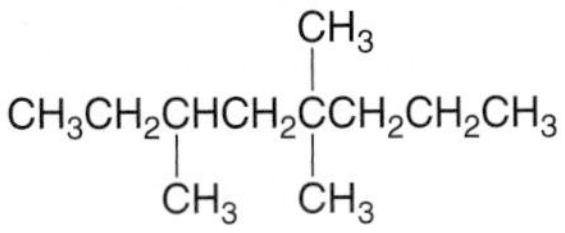

10.21 a. methylcyclobutane
b. 1,1-dimethylcyclohexane
c. 1-ethyl-3-propylcyclopentane
d. 1-ethyl-4-methylcyclohexane

10.23 a. decane b. pentane c. decane d. pentane

10.25 a. $C_{31}H_{64}$
b. The nonpolar alkane keeps the moisture (a polar medium) inside.

10.27 a. aldehyde b. amine

10.29 a. 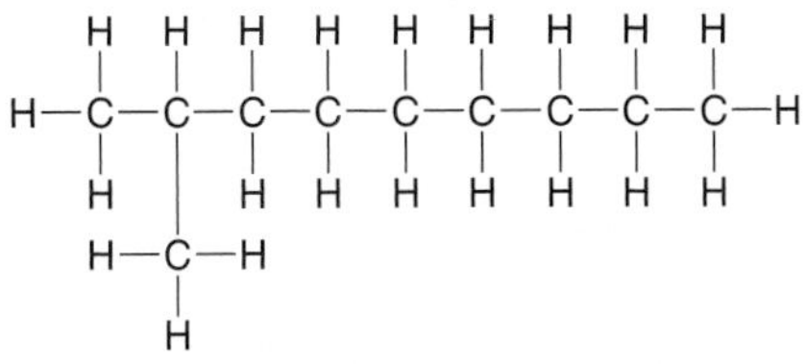 H—C(H)(H)—C(H)(CH$_3$)—C(H)(H)—C(H)(H)—C(H)(H)—C(H)(H)—C(H)(H)—C(H)(H)—C(H)(H)—H
b. 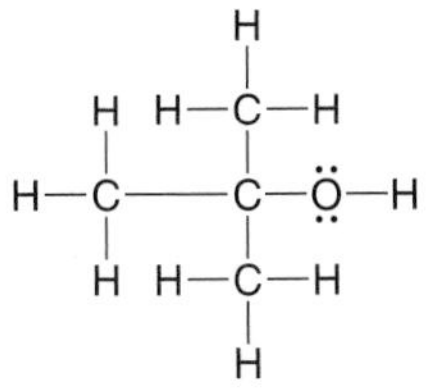 H—C(H)(H)—C(CH$_3$)(CH$_3$)—$\ddot{O}$—H

10.31 64 H's

10.33 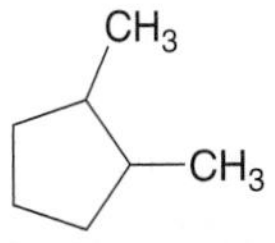 1,2-dimethylcyclopentane 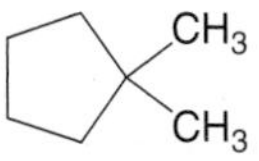 1,1-dimethylcyclopentane 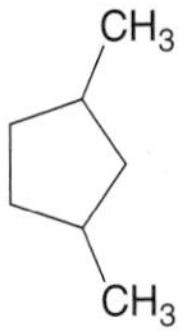 1,3-dimethylcyclopentane

10.35 a. 3-ethyl-4,4-dimethylheptane
b. 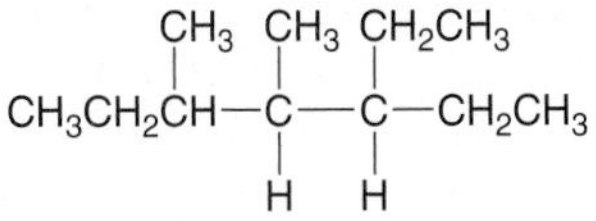 $CH_3CH_2CH(CH_3)—C(CH_3)(H)—C(CH_2CH_3)(H)—CH_2CH_3$
c. not water soluble
d. soluble in organic solvents
e. $C_{11}H_{24} + 17\ O_2 \longrightarrow 11\ CO_2 + 12\ H_2O$

10.37 a. $C_8H_8O_3$
b. 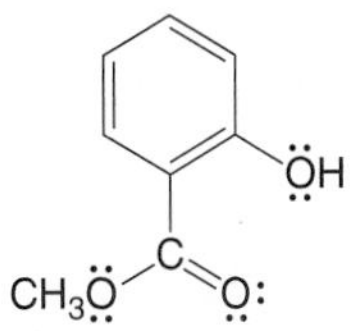 benzene ring with $\ddot{O}H$ and $C(=\ddot{O}:)\ddot{O}CH_3$
c. water soluble

10.39 inorganic: (a), (b)
organic: (c)

10.41 a. 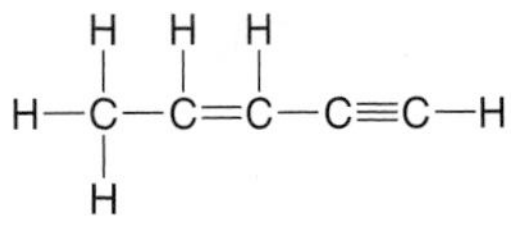 H—C(H)(H)—C(H)=C(H)—C≡C—H
b. 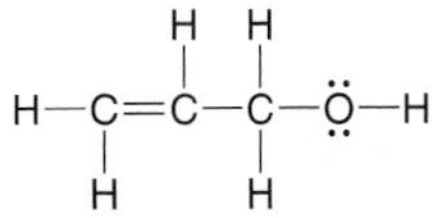 H—C(H)=C(H)—C(H)(H)—$\ddot{O}$—H
c. H—C≡C—C(H)(:$\ddot{Cl}$:)—C(H)(H)—H
d. 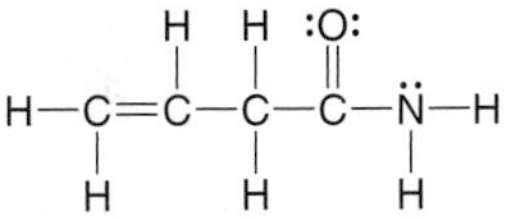 H—C(H)=C(H)—C(H)(H)—C(=:O:)—$\ddot{N}$(H)—H

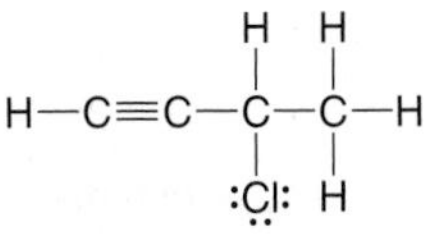

10.43 NaCl dissolves in water but not in dichloromethane, and cholesterol dissolves in dichloromethane but not in water.

10.45 a. Spermaceti wax is an ester.
b. The very long hydrocarbon chains would make it insoluble in water and soluble in organic solvents.

10.47 a. $CH_3(CH_2)_2C(CH_3)_3$ b. $(CH_3)_2CHCH_2O(CH_2)_2CH_3$ c. $CH_3(CH_2)_2CHO$

10.49 a. 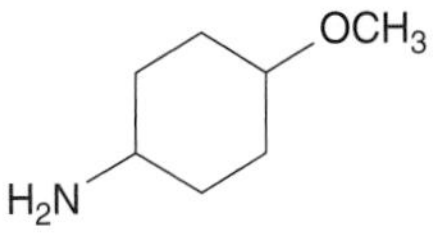 cyclohexane with OCH_3 and H_2N b. 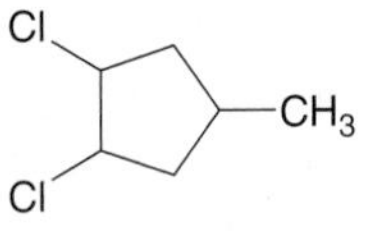 cyclopentane with Cl, Cl, and CH_3

10.51 a. 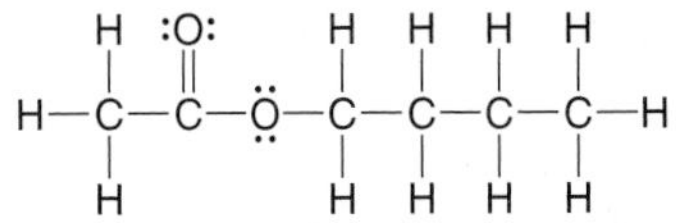 H—C(H)(H)—C(=:O:)—$\ddot{O}$—C(H)(H)—C(H)(H)—C(H)(H)—C(H)(H)—H
b. 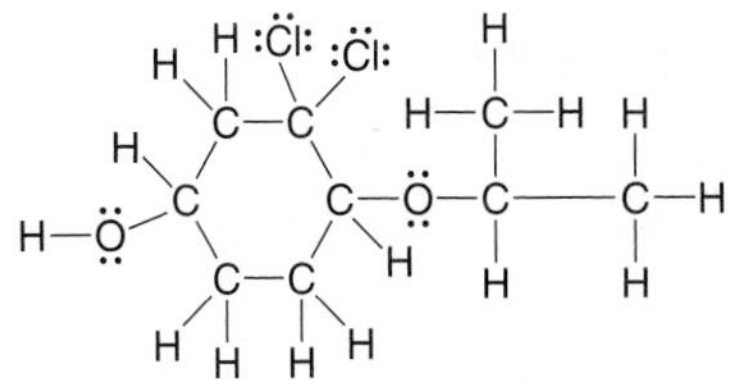 ring with H—$\ddot{O}$—, :$\ddot{Cl}$:, :$\ddot{Cl}$:, and —$\ddot{O}$—C(H)(CH$_3$)—C(H)(H)—H

10.53 (structural formula with the printed labels: H, H, H, H, $C=C$, :Ö:, H, H, $H-\ddot{O}-C$, C, $C-C-\ddot{N}-C-C-H$, $C-C$, H, H, H, H, $H-C-H$, $H-\ddot{O}-C-H$, H, H, $H-C-H$, H)

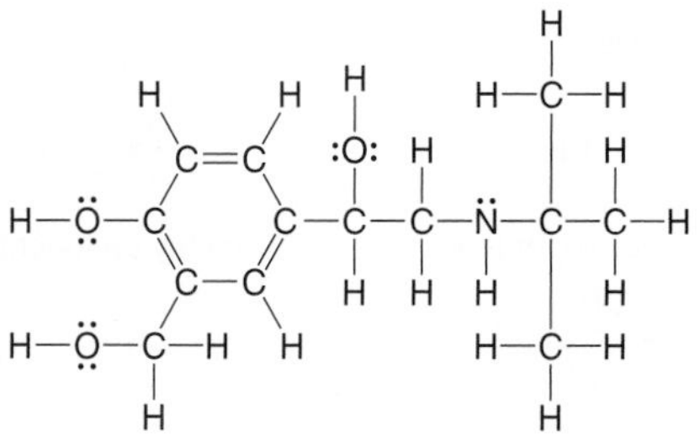

10.55 a. two alkenes
b. ester

10.57 a. two alkenes, two alkynes, one hydroxyl
b. hydroxyl, ether, aldehyde, aromatic ring

10.59 $(CH_3)_2C{=}CH(CH_2)_2C(CH_3){=}CHCHO$ — alkenes (the two C=C), aldehyde (CHO)

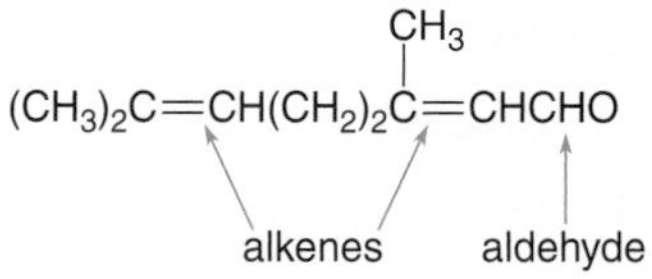

10.61 a. $HC{\equiv}C{-}CH_3$
b. $CH_3CH_2CH_2OH$
c. CH_3CH_2CHO
d. $CH_3CH_2CCH_3$ (with C=O on the third carbon)

10.63 alkane $CH_3CH_2CH_2CH_2CH_3$ C_5H_{12}

alkene $CH_3{-}CH_2{-}CH{=}CH{-}CH_3$ (drawn with H atoms) C_5H_{10}

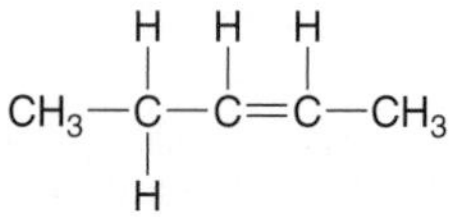

alkyne $CH_3{-}CH_2{-}C{\equiv}C{-}CH_3$ (drawn with H atoms) C_5H_8

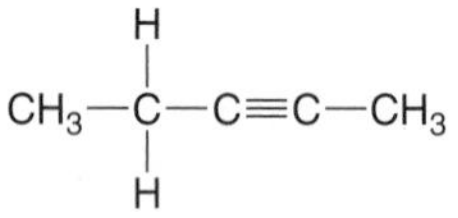

10.65 a,c: constitutional isomers
b. identical

10.67 a. 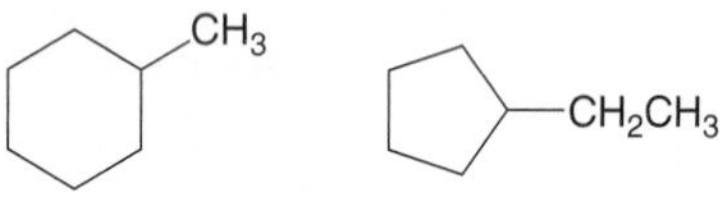 [cyclohexane ring with CH_3] [cyclopentane ring]$-CH_2CH_3$
b. $CH_3CH_2CH_2CH_2CH_2OH$ $CH_3OCH_2CH_2CH_2CH_3$
c. $CH_3CH_2CH_2Cl$ CH_3CHCH_3 (Cl on the middle carbon)

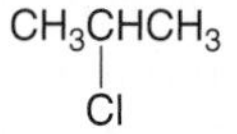

10.69 [cyclopropane ring]$-OH$ CH_3CCH_3 (with C=O on the middle carbon) [epoxide ring with O]$-CH_3$

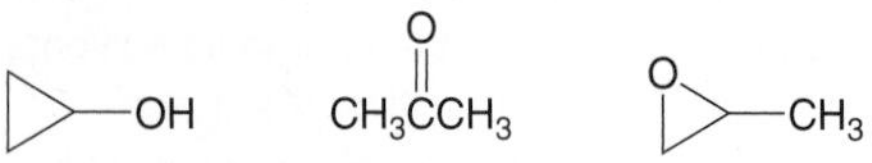

10.71 a. 3-methyldecane
b. 3,5-dimethyloctane
c. 3,3-diethylhexane
d. 4-ethyl-2,6-dimethyloctane

10.73 a. 4-methylheptane
b. ethylcyclopentane

10.75 a. cyclooctane
b. methylcyclopropane
c. 1,1,2,2-tetramethylcyclobutane
d. 1,3-diethylcyclohexane

10.77 a. 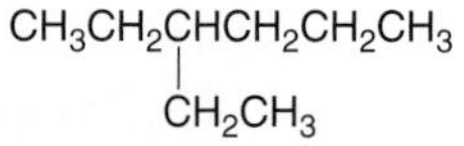 $CH_3CH_2CHCH_2CH_2CH_3$ with CH_2CH_3 on the third carbon
b. 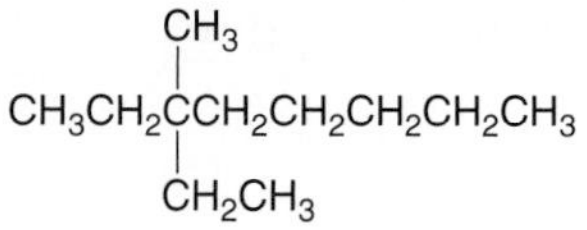 $CH_3CH_2CCH_2CH_2CH_2CH_2CH_3$ with CH_3 and CH_2CH_3 on the third carbon
c. 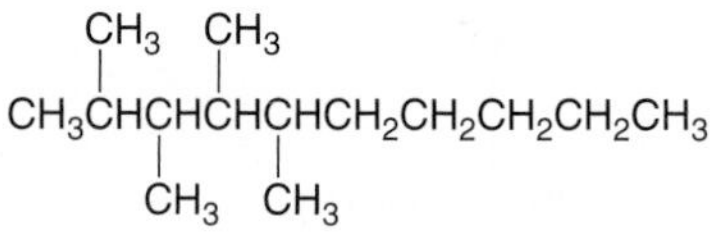 $CH_3CHCHCHCHCH_2CH_2CH_2CH_2CH_3$ with CH_3 groups on carbons 2, 3, 4 and 5
d. 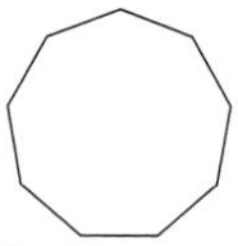 [cyclononane ring]
e. 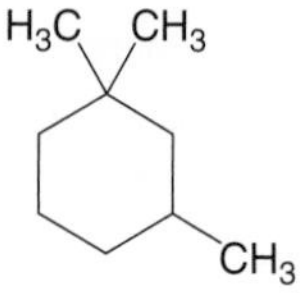 [cyclohexane ring with H_3C and CH_3 on one carbon and CH_3 on another]

10.79 a. 2-methylbutane: Number to give CH_3 the lower number, 2 not 3.
b. methylcyclopentane: no number assigned if only one substituent
c. 2-methylpentane: five-carbon chain
d. 2,5-dimethylheptane: longest chain not chosen

10.81 a. $CH_3CH_2CH_2CH_3$
b. 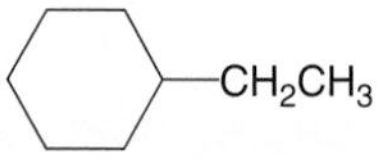 [cyclohexane ring]$-CH_2CH_3$

10.83 Hexane is a nonpolar hydrocarbon and is soluble in organic solvents but not in water.

10.85 a. $2\ CH_3CH_3 + 7\ O_2 \longrightarrow 4\ CO_2 + 6\ H_2O$
b. $(CH_3)_2CHCH_2CH_3 + 8\ O_2 \longrightarrow 5\ CO_2 + 6\ H_2O$

10.87 a. $2\ CH_3CH_2CH_3 + 7\ O_2 \longrightarrow 6\ CO + 8\ H_2O$
b. $2\ CH_3CH_2CH_2CH_3 + 9\ O_2 \longrightarrow 8\ CO + 10\ H_2O$

10.89 Higher molecular weight alkanes in warmer weather means less evaporation. Lower molecular weight alkanes in colder weather means the gasoline won't freeze.

10.91 The waxy coating will prevent loss of water from leaves and will keep leaves crisp.

10.93 a,e: [cyclopentane ring with a propyl chain]

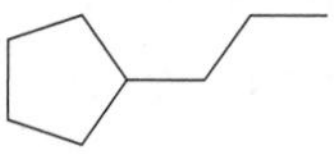

propylcyclopentane

b. 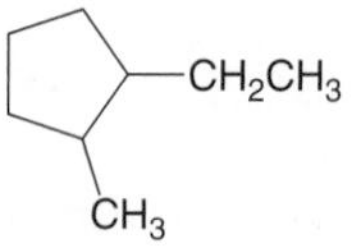 [cyclopentane ring]$-CH_2CH_3$ (one possibility), with CH_3 on the adjacent ring carbon
c. insoluble
d. soluble

10.95 Ethanol is more water soluble. THC has many nonpolar C—C and C—H bonds that make it soluble in fatty tissues; therefore, it will persist in tissues for an extended period of time. Since ethanol is water soluble, it will be quickly excreted in the urine, which is mostly water.

11

Candlenuts, known as kukui nuts in Hawaii, are rich in oils derived from unsaturated fatty acids, such as linoleic acid and linolenic acid.

Unsaturated Hydrocarbons

CHAPTER OUTLINE

CHAPTER GOALS

In this chapter you will learn how to:

1. Identify the three major types of unsaturated hydrocarbons—alkenes, alkynes, and aromatic compounds
2. Name alkenes, alkynes, and substituted benzenes
3. Recognize the difference between constitutional isomers and stereoisomers, as well as identify cis and trans isomers
4. Identify saturated and unsaturated fatty acids and predict their relative melting points
5. Draw the products of addition reactions of alkenes
6. Explain what products are formed when a vegetable oil is partially hydrogenated
7. Draw the structure of polymers formed from alkene monomers

In Chapter 11 we continue our study of hydrocarbons by examining three families of compounds that contain carbon–carbon multiple bonds. **Alkenes** contain a double bond and **alkynes** contain a triple bond. **Aromatic hydrocarbons** contain a benzene ring, a six-membered ring with three double bonds. These compounds differ from the alkanes of Chapter 10 because they each have a functional group, making them much more reactive. Thousands of biologically active molecules contain these functional groups, and many useful synthetic products result from their reactions.

11.1 Alkenes and Alkynes

Alkenes and alkynes are two families of organic molecules that contain multiple bonds.

- **Alkenes are compounds that contain a carbon–carbon double bond.**

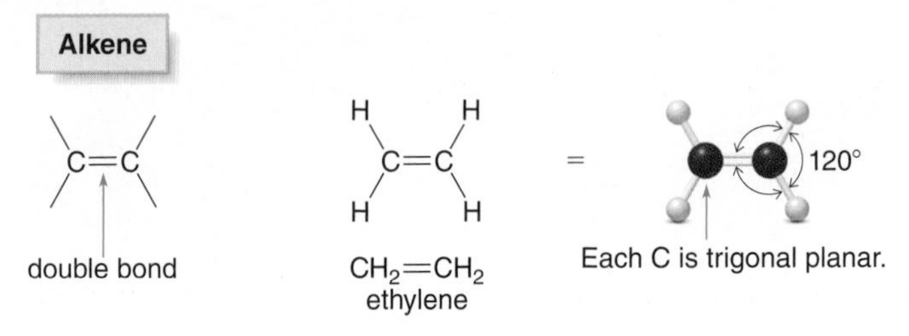

The general molecular formula of an alkene is $\mathbf{C_nH_{2n}}$, so an alkene has **two** fewer hydrogens than an acyclic alkane, which has a general molecular formula of C_nH_{2n+2}. Ethylene (C_2H_4) is the simplest alkene. Since each carbon of ethylene is surrounded by three atoms, each carbon is **trigonal planar.** All six atoms of ethylene lie in the same plane, and all bond angles are **120°.**

CONSUMER NOTE

A ripe banana speeds up the ripening of green tomatoes because the banana gives off ethylene, a plant growth hormone. Fruit grown in faraway countries can be picked green, and then sprayed with ethylene when ripening is desired upon arrival at its destination.

- **Alkynes are compounds that contain a carbon–carbon triple bond.**

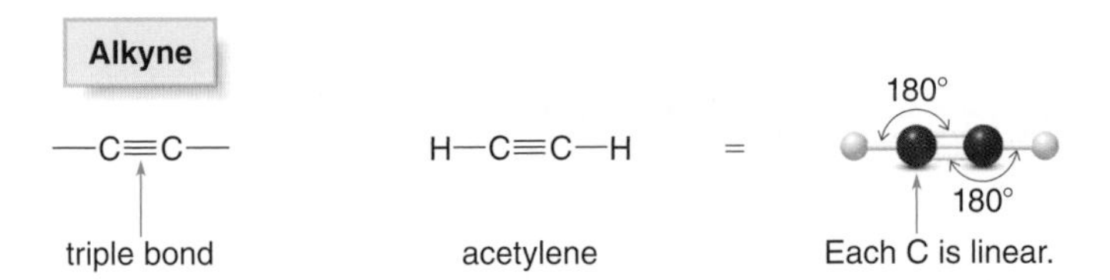

The general molecular formula for an alkyne is $\mathbf{C_nH_{2n-2}}$, so an alkyne has **four** fewer hydrogens than an acyclic alkane. Acetylene (C_2H_2) is the simplest alkyne. Each carbon of acetylene is surrounded by two atoms, making each carbon **linear** with bond angles of **180°.**

Because alkenes and alkynes are composed of nonpolar carbon–carbon and carbon–hydrogen bonds, their physical properties are similar to other hydrocarbons. Like alkanes:

- **Alkenes and alkynes have low melting points and boiling points and are insoluble in water.**

Recall from Chapter 10 that acyclic alkanes are called saturated hydrocarbons, because they contain the maximum number of hydrogen atoms per carbon. In contrast, **alkenes and alkynes are called *unsaturated* hydrocarbons.**

- **Unsaturated hydrocarbons are compounds that contain fewer than the maximum number of hydrogen atoms per carbon.**

The multiple bond of an alkene or alkyne is always drawn in a condensed structure. To translate a condensed structure to a complete structure with all bond lines drawn in, make sure that each carbon of a double bond has three atoms around it, and each carbon of a triple bond has two atoms around it, as shown in Sample Problem 11.1.

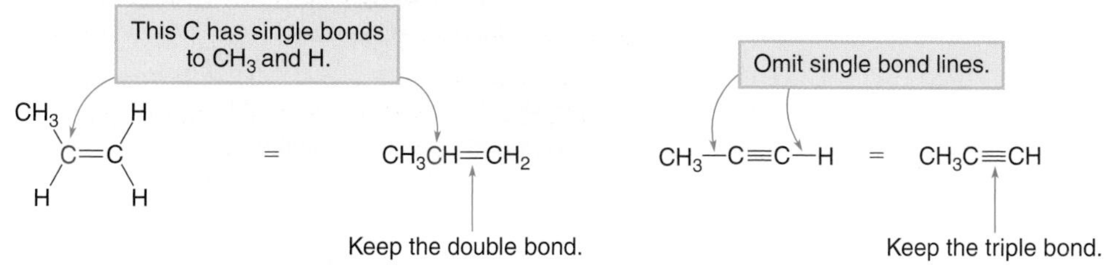

This C has single bonds to CH_3 and H.

$CH_3(H)C{=}CH_2$ = $CH_3CH{=}CH_2$

Keep the double bond.

Omit single bond lines.

$CH_3{-}C{\equiv}C{-}H$ = $CH_3C{\equiv}CH$

Keep the triple bond.

SAMPLE PROBLEM 11.1

Draw a complete structure for each alkene or alkyne.

a. $CH_2{=}CHCH_2CH_3$ b. $CH_3C{\equiv}CCH_2CH_3$

Analysis

First, draw the multiple bond in each structure. Draw an alkene so that each C of the double bond has three atoms around it. Draw an alkyne so that each C of the triple bond has two atoms around it. All other C's have four single bonds.

Solution

a. The C labeled in red, has single bonds to 2 H's. The C labeled in blue has single bonds to 1 H and a CH_2CH_3 group.

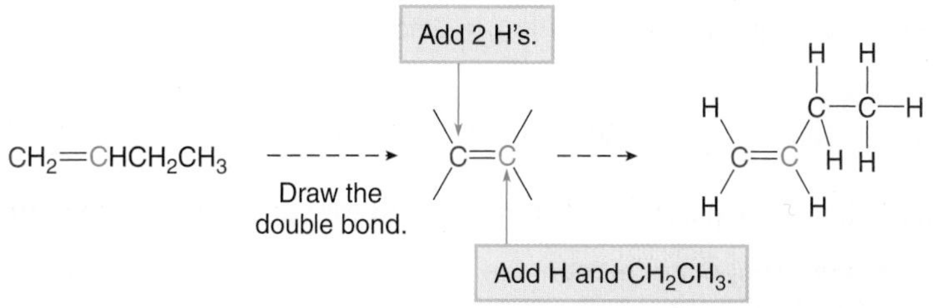

$CH_2{=}CHCH_2CH_3$ ------→ C=C ---→ complete structure

Draw the double bond.

Add 2 H's.

Add H and CH_2CH_3.

b. The C labeled in red has a single bond to a CH_3 group. The C labeled in blue has a single bond to a CH_2CH_3 group.

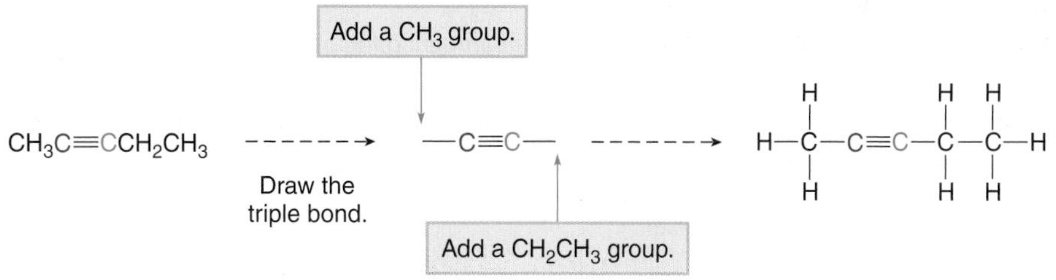

$CH_3C{\equiv}CCH_2CH_3$ -------→ —C≡C— -------→ H—C(H)(H)—C≡C—C(H)(H)—C(H)(H)—H

Draw the triple bond.

Add a CH_3 group.

Add a CH_2CH_3 group.

PROBLEM 11.1

Convert each condensed structure to a complete structure with all atoms, bond lines, and lone pairs drawn in.

a. $CH_2{=}CHCH_2OH$ b. $(CH_3)_2C{=}CH(CH_2)_2CH_3$ c. $(CH_3)_2CHC{\equiv}CCH_2C(CH_3)_3$

PROBLEM 11.2

Determine whether each molecular formula corresponds to a saturated hydrocarbon, an alkene, or an alkyne.

a. C_3H_6 b. C_5H_{12} c. C_8H_{14} d. C_6H_{12}

PROBLEM 11.3

Give the molecular formula for each of the following compounds.

a. an alkene that has four carbons
b. a saturated hydrocarbon that has six carbons
c. an alkyne that has seven carbons
d. an alkene that has five carbons

11.2 Nomenclature of Alkenes and Alkynes

Whenever we encounter a new functional group, we must learn how to use the IUPAC system to name it. In the IUPAC system:

- **An alkene is identified by the suffix *-ene.***
- **An alkyne is identified by the suffix *-yne.***

How To Name an Alkene or an Alkyne

Example Give the IUPAC name of each alkene and alkyne.

a. $CH_2{=}CHCH(CH_3)CH_3$ b. $CH_3CH_2CH(CH_2CH_3)C{\equiv}CCH_3$

Step [1] Find the longest chain that contains *both* carbon atoms of the double or triple bond.

a. $CH_2{=}CHCH(CH_3)CH_3$

4 C's in the longest chain

but**ane** ---→ but**ene**

- Since the compound is an alkene, change the ***-ane*** ending of the parent alkane to ***-ene.***

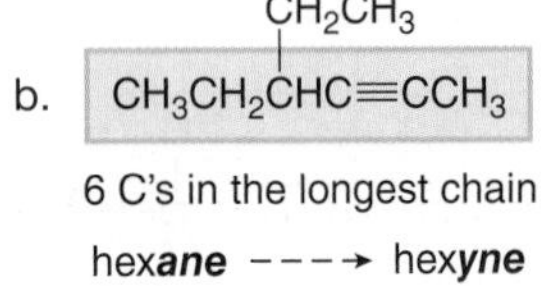

- Since the compound is an alkyne, change the ***-ane*** ending of the parent alkane to ***-yne.***

Step [2] Number the carbon chain from the end that gives the multiple bond the lower number.

For each compound, number the chain and name the compound using the *first* number assigned to the multiple bond.

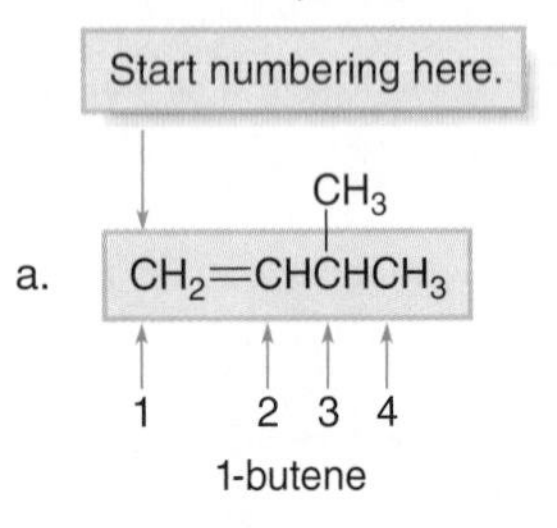

- Numbering the chain from left to right puts the double bond at C1 (not C3). The alkene is named using the *first* number assigned to the double bond, making it 1-butene.

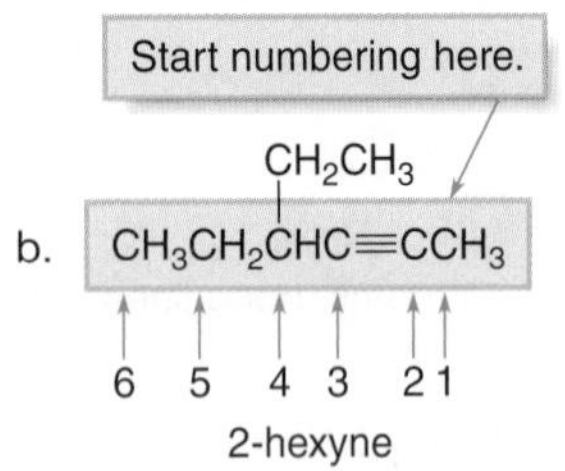

- Numbering the chain from right to left puts the triple bond at C2 (not C4). The alkyne is named using the *first* number assigned to the triple bond, making it 2-hexyne.

Step [3] Number and name the substituents, and write the name.

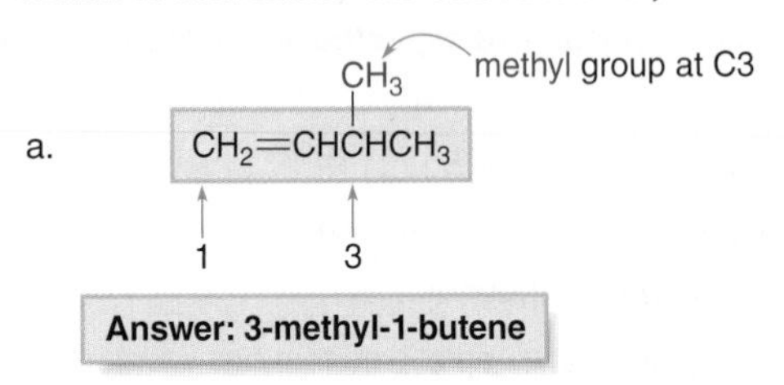

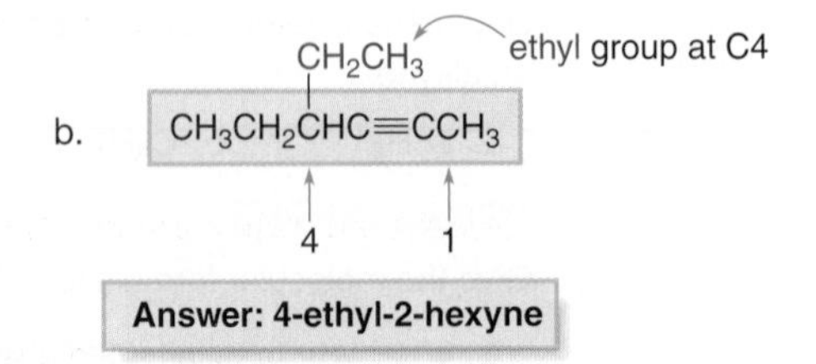

A few simple alkenes and alkynes have names that do not follow the IUPAC system. The simplest alkene, $CH_2{=}CH_2$, is called *ethene* in the IUPAC system, but it is commonly called **ethylene.** The simplest alkyne, $HC{\equiv}CH$, is called *ethyne* in the IUPAC system, but it is commonly named **acetylene.** We will use these common names since they are more widely used than their systematic IUPAC names.

SAMPLE PROBLEM 11.2

Give the IUPAC name for the following compound.

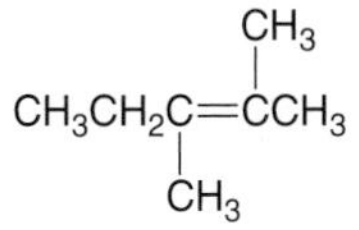

$$CH_3CH_2C(CH_3){=}C(CH_3)CH_3$$

(one CH_3 shown above the right C of the double bond, one CH_3 shown below the left C)

Analysis and Solution

[1] Find the longest chain containing both carbon atoms of the multiple bond.

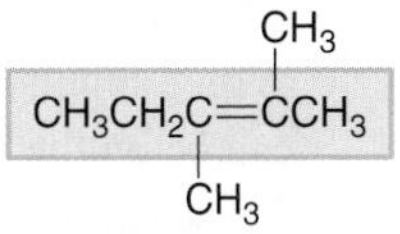

$$CH_3CH_2C(CH_3){=}C(CH_3)CH_3$$

5 C's in the longest chain ---→ **pentene**

[2] Number the chain to give the double bond the lower number.

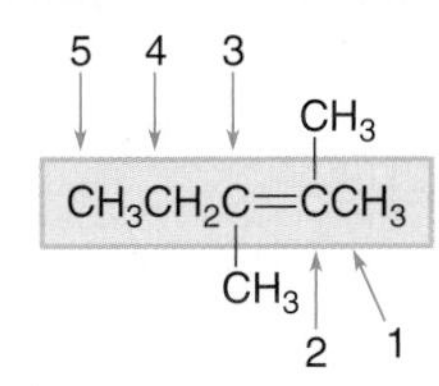

- Numbering from right to left is preferred since the double bond begins at C2 (not C3). The molecule is named as a **2-pentene.**

[3] Name and number the substituents and write the complete name.

- The alkene has two methyl groups located at C2 and C3. Use the prefix di- before methyl → 2,3-**di**methyl.

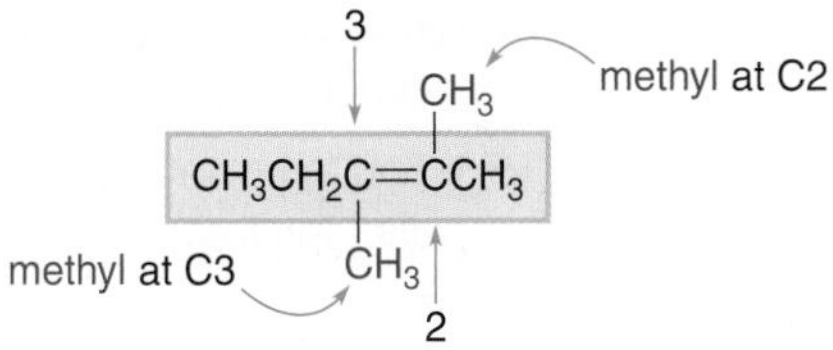

$$CH_3CH_2C(CH_3){=}C(CH_3)CH_3$$

3; methyl at C2; methyl at C3; 2

Answer: 2,3-dimethyl-2-pentene

PROBLEM 11.4

Give the IUPAC name for each alkene.

a. 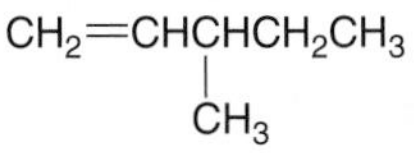 $CH_2{=}CHCH(CH_3)CH_2CH_3$

b. $(CH_3CH_2)_2C{=}CHCH_2CH_2CH_3$

PROBLEM 11.5

Give the IUPAC name for each alkyne.

a. $CH_3CH_2CH_2CH_2CH_2C{\equiv}CCH(CH_3)_2$

b. 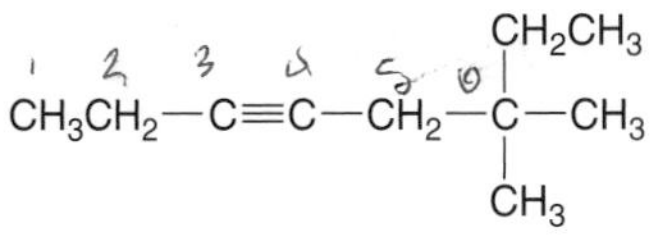 $CH_3CH_2{-}C{\equiv}C{-}CH_2{-}C(CH_2CH_3)(CH_3){-}CH_3$

PROBLEM 11.6

Give the structure corresponding to each name.

a. 4-methyl-1-hexene

b. 5-ethyl-2-methyl-2-heptene

c. 2,5-dimethyl-3-hexyne

d. 4-ethyl-1-decyne

11.3 Cis–Trans Isomers

As we learned in Section 10.5 on alkanes, constitutional isomers are possible for alkenes of a given molecular formula. For example, there are three constitutional isomers for an alkene of molecular formula C_4H_8—1-butene, 2-butene, and 2-methylpropene.

$CH_2{=}CHCH_2CH_3$	$CH_3CH{=}CHCH_3$	$CH_2{=}C(CH_3)CH_3$
1-butene	2-butene	2-methylpropene

11.3A Stereoisomers—A New Class of Isomer

2-Butene illustrates another important aspect about alkenes. There is **restricted rotation** around the carbon atoms of a double bond. As a result, the groups on one side of the double bond *cannot* rotate to the other side.

With 2-butene, there are two ways to arrange the atoms on the double bond. The two CH_3 groups can be on the *same side* of the double bond or they can be on *opposite sides* of the double bond. These molecules are *different* compounds with the same molecular formula; that is, they are **isomers.**

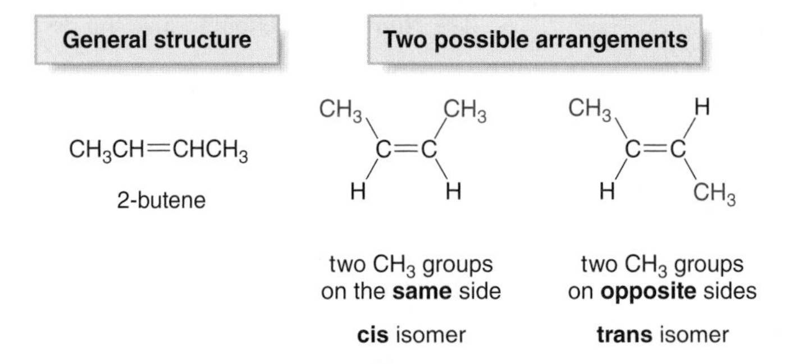

- When the two CH_3 groups are on the *same side* of the double bond, the compound is called the *cis* isomer.
- When the two CH_3 groups are on *opposite sides* of the double bond, the compound is called the *trans* isomer.

Thus, one isomer of 2-butene is called ***cis*-2-butene,** and the other isomer is called ***trans*-2-butene.**

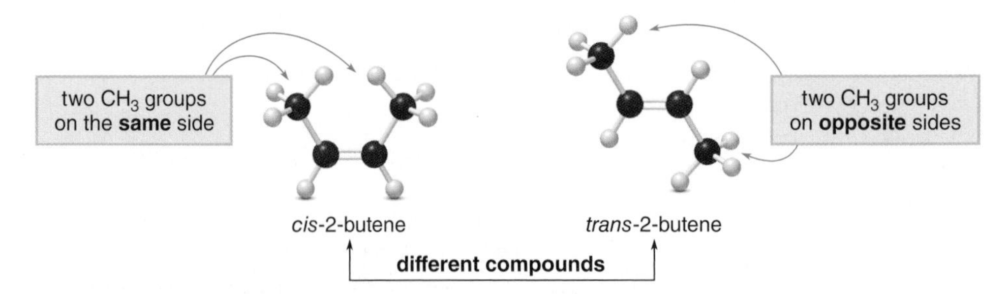

The cis and trans isomers of 2-butene are a specific example of a general class of isomer that occurs at carbon–carbon double bonds. **Whenever the two groups on *each* end of a C=C are *different from each other,* two isomers are possible.**

These two groups must be **different from each other...** (A and B) — $AB\,C{=}C\,CD$ — ***...and*** these two groups must also be **different from each other.** (C and D)

When the two groups on one end of the double bond are identical, there is still restricted rotation, but no cis and trans isomers are possible. With 1-butene, $CH_2{=}CHCH_2CH_3$, one end of the double bond has two hydrogens, so the ethyl group (CH_2CH_3) is always cis to a hydrogen, no matter how the molecule is drawn.

two identical groups — 1-butene: H, H / C=C / H, CH_2CH_3 ---→ identical: H, CH_2CH_3 / C=C / H, H

No cis and trans isomers are possible.

SAMPLE PROBLEM 11.3

Draw *cis*- and *trans*-3-hexene.

Analysis

First, use the parent name to draw the carbon skeleton, and place the double bond at the correct carbon; 3-hexene indicates a 6 C chain with the double bond beginning at C3. Then use the definitions of cis and trans to draw the isomers.

Solution

Each C of the double bond is bonded to a CH_3CH_2 group and a hydrogen. A cis isomer has the CH_3CH_2 groups bonded to the same side of the double bond. A trans isomer has the two CH_3CH_2 groups bonded to the opposite sides of the double bond.

$CH_3CH_2CH{=}CHCH_2CH_3$ (1 2 3 4 5 6) — 3-hexene ---→

cis-3-hexene: CH_3CH_2, CH_2CH_3 / C=C / H, H

trans-3-hexene: CH_3CH_2, H / C=C / H, CH_2CH_3

PROBLEM 11.7

Draw the structure of each compound: (a) *cis*-2-octene; (b) *trans*-3-heptene; (c) *trans*-4-methyl-2-pentene.

PROBLEM 11.8

Bombykol is secreted by the female silkworm moth (*Bombyx mori*) to attract mates. Bombykol contains two double bonds, and each double bond must have a particular three-dimensional arrangement of groups around it to be biologically active. Label the double bonds of bombykol as cis or trans.

bombykol: $HOCH_2(CH_2)_7CH_2$–CH=CH–CH=CH–$CH_2CH_2CH_3$ (H, H / C=C; H / C=C / H)

ENVIRONMENTAL NOTE

The female silkworm moth, *Bombyx mori,* secretes the sex pheromone bombykol (Problem 11.8). Pheromones like bombykol have been used to control insect populations. In one method, the pheromone is placed in a trap containing a poison or sticky substance, and the male is lured to the trap by the pheromone.

Cis and trans compounds are isomers, but they are *not* constitutional isomers. Each carbon atom of *cis*- and *trans*-2-butene is bonded to the same atoms. The only difference is the three-dimensional arrangement of the groups around the double bond. Isomers of this sort are called **stereoisomers.**

- *Stereoisomers* are isomers that differ *only* in the 3-D arrangement of atoms.

Cis–trans isomers are also called **geometric isomers.**

Thus, *cis*-2-butene and *trans*-2-butene are a specific type of stereoisomer, but each of these compounds is a constitutional isomer of 1-butene, as shown in Figure 11.1. We will learn about another type of stereoisomer in Section 12.10.

Figure 11.1

Comparing Three Isomers: 1-Butene, *cis*-2-Butene, and *trans*-2-Butene

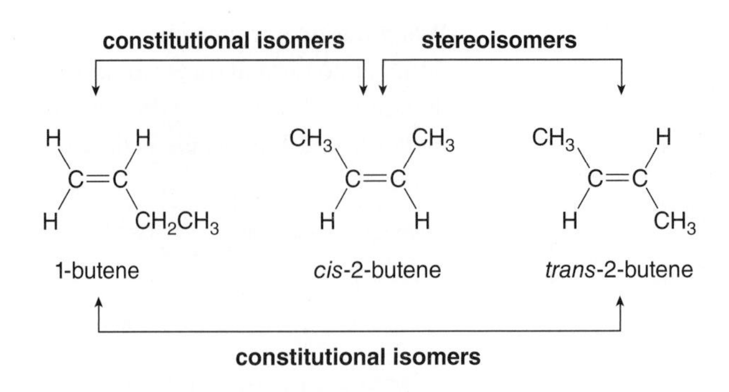

We have now learned the two major classes of isomers.

- Constitutional isomers differ in the way the atoms are bonded to each other.
- Stereoisomers differ only in the three-dimensional arrangement of atoms.

PROBLEM 11.9

Label each pair of alkenes as constitutional isomers or stereoisomers.

a. $CH_3CH{=}CHCH_2CH_3$ and $CH_2{=}CHCH_2CH_2CH_3$

b. (CH_3CH_2)(H)C=C(H)(CH_3) [CH_3CH_2 and CH_3 on the same side] and (CH_3CH_2)(H)C=C(CH_3)(H) [CH_3CH_2 and CH_3 on opposite sides]

PROBLEM 11.10

Double bonds in rings can also be characterized as cis or trans. Label each double bond in the following compounds as cis or trans.

a. [cyclohexene] b. [cyclodecadiene]

HEALTH NOTE

Linoleic and linolenic acids are **essential fatty acids,** meaning they cannot be synthesized in the human body and must therefore be obtained in the diet. A common source of these essential fatty acids is whole milk. Babies fed a diet of nonfat milk in their early months do not thrive because they do not obtain enough of these essential fatty acids.

11.3B FOCUS ON HEALTH & MEDICINE
Saturated and Unsaturated Fatty Acids

Naturally occurring animal fats and vegetable oils are formed from fatty acids. **Fatty acids are carboxylic acids (RCOOH) with long carbon chains of 12–20 carbon atoms.** Because a fatty acid has many nonpolar C—C and C—H bonds and few polar bonds, fatty acids are insoluble in water. There are two types of fatty acids.

- *Saturated* fatty acids have no double bonds in their long hydrocarbon chains.
- *Unsaturated* fatty acids have one or more double bonds in their long hydrocarbon chains.

Table 11.1 lists the structure and melting point of four fatty acids containing 18 carbon atoms. Stearic acid is one of the two most common saturated fatty acids, while oleic and linoleic acids are the most common unsaturated ones. One structural feature of unsaturated fatty acids is especially noteworthy.

- Generally, double bonds in naturally occurring fatty acids are cis.

The presence of cis double bonds affects the melting point of these fatty acids greatly.

- As the number of double bonds in the fatty acid *increases,* the melting point *decreases.*

Table 11.1 Common Saturated and Unsaturated Fatty Acids

Name	Structure	Mp (°C)
Stearic acid (0 C=C)	$CH_3CH_2CH_2CH_2CH_2CH_2CH_2CH_2CH_2CH_2CH_2CH_2CH_2CH_2CH_2CH_2CH_2COOH$	71
Oleic acid (1 C=C)	$CH_3CH_2CH_2CH_2CH_2CH_2CH_2CH_2$–C(H)=C(H)–$CH_2CH_2CH_2CH_2CH_2CH_2CH_2COOH$ (cis)	16
Linoleic acid (2 C=C)	$CH_3CH_2CH_2CH_2CH_2$–C(H)=C(H)–CH_2–C(H)=C(H)–$CH_2CH_2CH_2CH_2CH_2CH_2CH_2COOH$ (cis)	-5
Linolenic acid (3 C=C)	CH_3CH_2–C(H)=C(H)–CH_2–C(H)=C(H)–CH_2–C(H)=C(H)–$CH_2CH_2CH_2CH_2CH_2CH_2CH_2COOH$ (cis)	−11

The cis double bonds introduce kinks in the long hydrocarbon chain, as shown in Figure 11.2. This makes it difficult for the molecules to pack closely together in a solid. **The larger the number of cis double bonds, the more kinks in the hydrocarbon chain, and the lower the melting point.**

Fats and oils are organic molecules synthesized in plant and animal cells from fatty acids. Fats and oils have different physical properties.

- **Fats are solids at room temperature. Fats are generally formed from fatty acids having few double bonds.**
- **Oils are liquids at room temperature. Oils are generally formed from fatty acids having a larger number of double bonds.**

Figure 11.2 The Three-Dimensional Structure of Four Fatty Acids

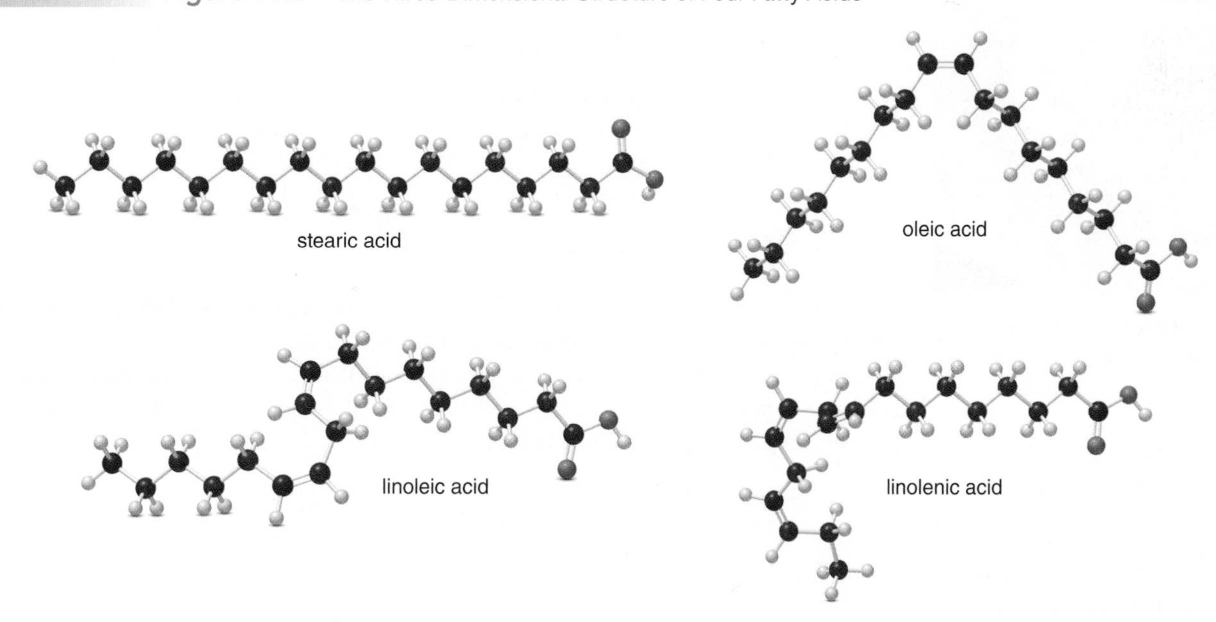

Saturated fats are typically obtained from animal sources, while unsaturated oils are common in vegetable sources. Thus, butter and lard are formed from saturated fatty acids, while olive oil and safflower oil are formed from unsaturated fatty acids. An exception to this generalization is coconut oil, which is composed largely of saturated fatty acids.

Considerable evidence suggests that an elevated cholesterol level is linked to increased risk of heart disease. Saturated fats stimulate cholesterol synthesis in the liver, resulting in an increase in cholesterol concentration in the blood. We will learn more about fats and oils in Section 11.6 and Chapter 15.

PROBLEM 11.11

You have two fatty acids, one with a melting point of 63 °C, and one with a melting point of 1 °C. Which structure corresponds to each melting point?

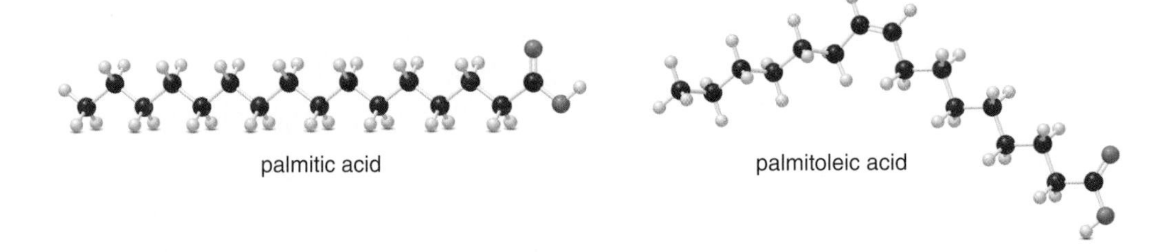

11.4 FOCUS ON HEALTH & MEDICINE Oral Contraceptives

The development of synthetic oral contraceptives in the 1960s revolutionized the ability to control fertility. Prior to that time, women ingested all sorts of substances—iron rust, gunpowder, tree bark, sheep's urine, elephant dung, and others—in the hope of preventing pregnancy.

Synthetic birth control pills are similar in structure to the female sex hormones **estradiol** and **progesterone,** but they also contain a carbon–carbon triple bond. Most oral contraceptives contain two synthetic hormones that are more potent than these natural hormones, so they can be administered in lower doses.

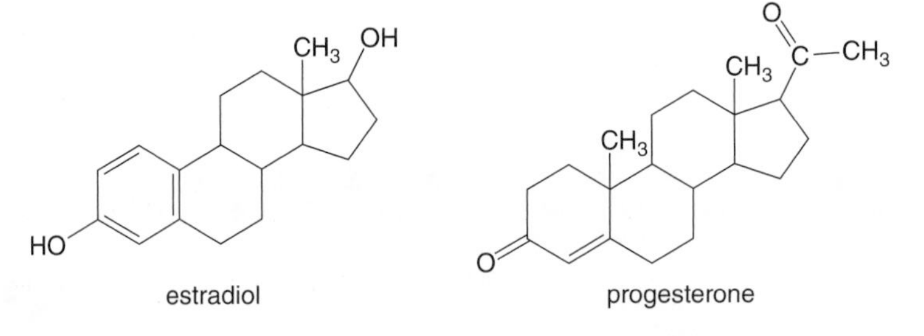

HEALTH NOTE

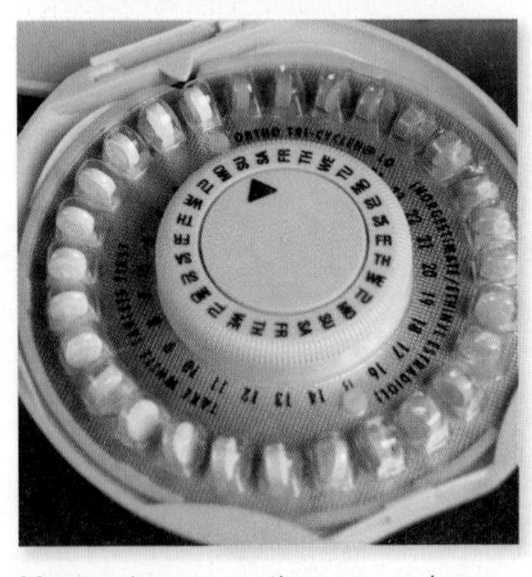

Most oral contraceptives are made up of two synthetic female sex hormones, each containing a carbon–carbon triple bond.

Two common components of birth control pills are **ethynylestradiol** and **norethindrone.** Ethynylestradiol is a synthetic estrogen that resembles the structure and biological activity of estradiol. Norethindrone is a synthetic progesterone that is similar to the natural hormone progesterone. These compounds act by artificially elevating hormone levels in a woman, and this prevents pregnancy, as illustrated in Figure 11.3.

Figure 11.3 How Oral Contraceptives Work

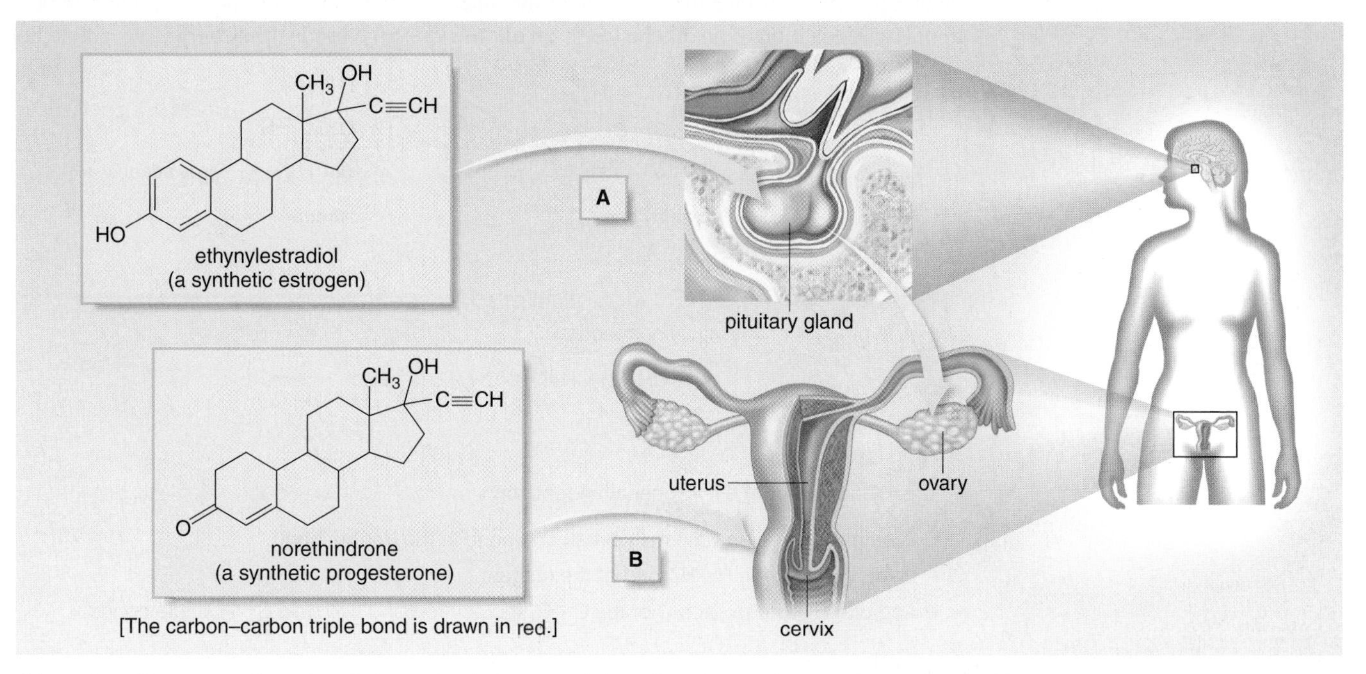

Monthly cycles of hormones from the pituitary gland cause ovulation, the release of an egg from an ovary. To prevent pregnancy, the two synthetic hormones in many oral contraceptives have different effects on the female reproductive system. **A:** The elevated level of **ethynylestradiol,** a synthetic estrogen, "fools" the pituitary gland into thinking a woman is pregnant, so ovulation does not occur. **B:** The elevated level of **norethindrone,** a synthetic progesterone, stimulates the formation of a thick layer of mucus in the cervix, making it difficult for sperm to reach the uterus.

11.5 Reactions of Alkenes

Most families of organic compounds undergo a characteristic type of reaction. **Alkenes undergo addition reactions.** In an addition reaction, new groups X and Y are added to a starting material. One bond of the double bond is broken and two new single bonds are formed.

- **Addition is a reaction in which elements are added to a compound.**

Addition

$$>C=C< \; + \; X-Y \longrightarrow -\underset{X}{\overset{|}{\underset{|}{C}}}-\underset{Y}{\overset{|}{\underset{|}{C}}}-$$

One bond is broken. Two single bonds are formed.

Why does addition occur? A double bond is composed of one strong bond and one weak bond. In an addition reaction, the weak bond is broken and two new strong single bonds are formed. For example, alkenes react with hydrogen (H_2, Section 11.5A) and water (H_2O, Section 11.5B).

11.5A Addition of Hydrogen—Hydrogenation

Hydrogenation is the addition of hydrogen (H_2) to an alkene. Two bonds are broken—one bond of the carbon–carbon double bond and the H—H bond—and two new C—H bonds are formed.

Hydrogenation

$$\underset{\text{alkene}}{>C=C<} \; + \; H-H \xrightarrow[\text{catalyst}]{\text{metal}} \underset{\text{alkane}}{-\underset{H}{\overset{|}{\underset{|}{C}}}-\underset{H}{\overset{|}{\underset{|}{C}}}-}$$

H_2 is added.

The addition of H_2 occurs only in the presence of a **metal catalyst** such as palladium (Pd). The metal provides a surface that binds both the alkene and H_2, and this speeds up the rate of reaction. Hydrogenation of an alkene forms an **alkane** since the product has only C—C single bonds.

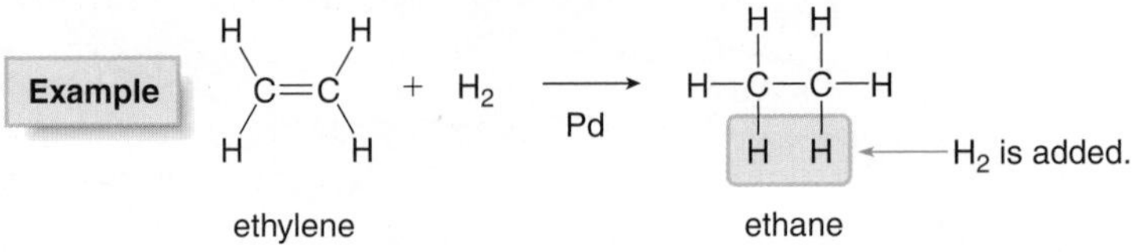

Example

$$H_2C{=}CH_2 + H_2 \xrightarrow{Pd} H{-}\underset{H}{\overset{H}{C}}{-}\underset{H}{\overset{H}{C}}{-}H$$

ethylene → ethane (H_2 is added.)

SAMPLE PROBLEM 11.4

Draw the product of the following reaction.

$$CH_3CH_2CH_2CH{=}CH_2 + H_2 \xrightarrow{Pd}$$

Analysis

To draw the product of a hydrogenation reaction:

- Locate the C=C and mentally break one bond in the double bond.
- Mentally break the H—H bond of the reagent.
- Add one H atom to each C of the C=C, thereby forming two new C—H single bonds.

Solution

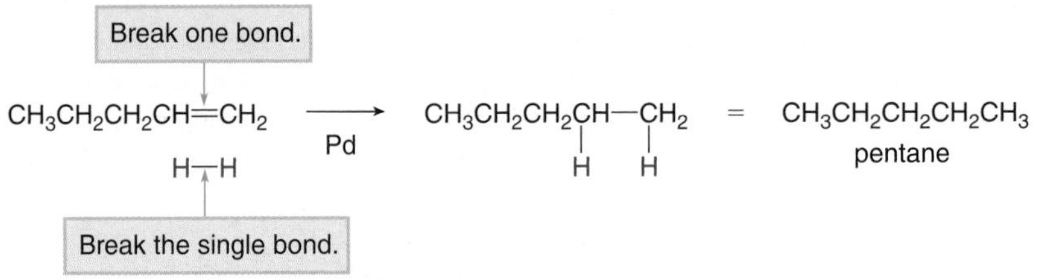

Break one bond.

$$CH_3CH_2CH_2CH{=}CH_2 + H{-}H \xrightarrow{Pd} CH_3CH_2CH_2\underset{H}{CH}{-}\underset{H}{CH_2} = CH_3CH_2CH_2CH_2CH_3$$

Break the single bond.

pentane

PROBLEM 11.12

What product is formed when each alkene is treated with H_2 and a Pd catalyst?

a. $CH_3CH_2CH{=}CHCH_2CH_3$

b. 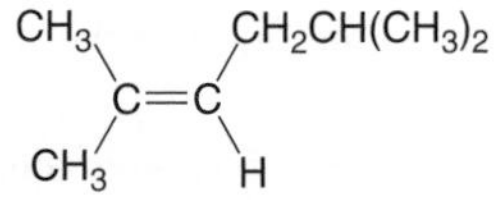 $(CH_3)_2C{=}CHCH_2CH(CH_3)_2$

c. 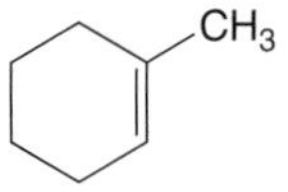 1-methylcyclohexene (cyclohexene ring with CH_3 on the double-bond carbon)

11.5B Addition of Water—Hydration

Hydration is the addition of water to an alkene. Two bonds are broken—one bond of the carbon–carbon double bond and the H—OH bond—and new C—H and C—OH bonds are formed.

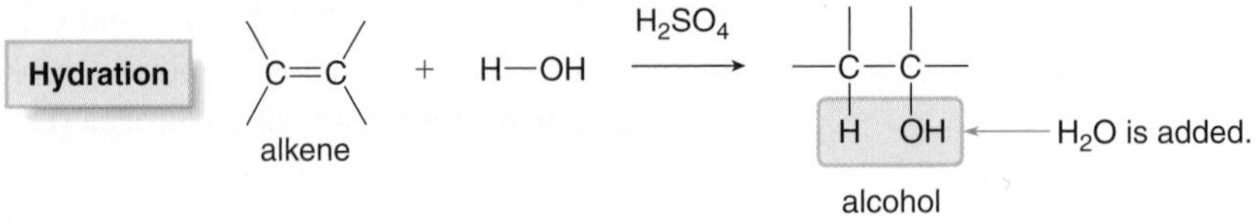

Hydration

$$>C{=}C< + H{-}OH \xrightarrow{H_2SO_4} {-}\underset{H}{\overset{|}{C}}{-}\underset{OH}{\overset{|}{C}}{-}$$

alkene → alcohol (H_2O is added.)

Hydration occurs only if a strong acid such as H_2SO_4 is added to the reaction mixture. The product of hydration is an **alcohol.** For example, hydration of ethylene forms ethanol.

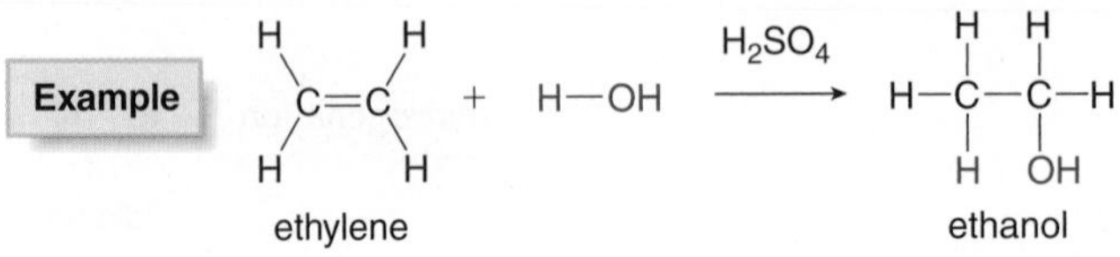

Example

$$H_2C{=}CH_2 + H{-}OH \xrightarrow{H_2SO_4} H{-}\underset{H}{\overset{H}{C}}{-}\underset{OH}{\overset{H}{C}}{-}H$$

ethylene → ethanol

Gasohol contains 10% ethanol.

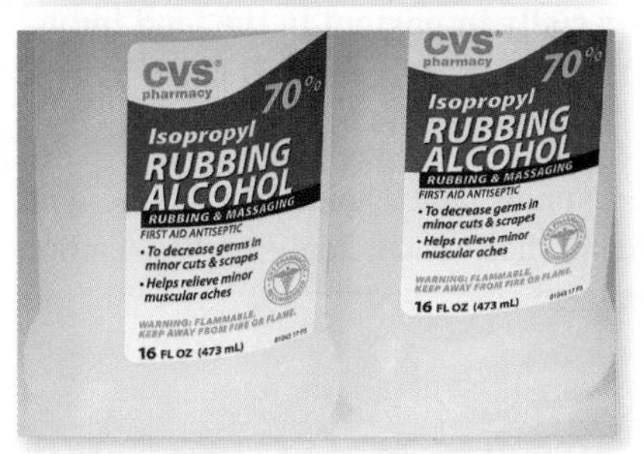

Addition of H_2O to propene forms 2-propanol [$(CH_3)_2CHOH$], the main component of rubbing alcohol.

Ethanol is used as a solvent in many reactions in the laboratory. **Ethanol is also used as a gasoline additive** because, like alkanes, it burns in the presence of oxygen to form CO_2 and H_2O with the release of a great deal of energy. Although ethanol can also be formed by the fermentation of carbohydrates in grains and potatoes, much of the ethanol currently used in gasoline and solvent comes from the hydration of ethylene.

There is one important difference in this addition reaction compared to the addition of H_2. In this case, addition puts different groups—H and OH—on the two carbons of the double bond. As a result, H_2O can add to the double bond to give two constitutional isomers when an unsymmetrical alkene is used as starting material.

$$CH_3CH{=}CH_2 \text{ (propene; C2 = CH, C1 = CH}_2) + H{-}OH \xrightarrow{H_2SO_4} CH_3{-}CH(OH){-}CH_2{-}H \text{ (2-propanol, C2)} \quad \text{or} \quad CH_3{-}CH_2{-}CH_2(OH) \text{ (1-propanol, C1)}$$

propene — 2-propanol (**only product**) — 1-propanol

For example, the addition of H_2O to propene could in theory form two products. If H adds to the end carbon (labeled C1) and OH adds to the middle carbon (C2), 2-propanol is formed. If OH adds to the end carbon (C1) and H adds to the middle carbon (C2), 1-propanol is formed. In fact, **addition forms *only* 2-propanol.** This is a specific example of a general trend called **Markovnikov's rule,** named for the Russian chemist who first observed the selectivity in alkene additions.

- **In the addition of H_2O to an unsymmetrical alkene, the H atom bonds to the less substituted carbon atom—that is, the carbon that has more H's to begin with.**

The end carbon (C1) of propene has two hydrogens while the middle carbon (C2) has just one hydrogen. Addition puts the H of H_2O on C1 since it had more hydrogens (two versus one) to begin with.

$$CH_3CH{=}CH_2 + H{-}OH \longrightarrow CH_3{-}CH(OH){-}CH_2{-}H$$

This C has only 1 H so the OH bonds here.

This C has 2 H's so the H bonds here.

SAMPLE PROBLEM 11.5

What product is formed when 2-methylpropene [$(CH_3)_2C{=}CH_2$] is treated with H_2O in the presence of H_2SO_4?

Analysis

Alkenes undergo addition reactions, so the elements of H and OH must be added to the double bond. Since the alkene is unsymmetrical, the H atom of H_2O bonds to the carbon that has more H's to begin with.

Solution

$$(CH_3)_2C{=}CH_2 + H{-}OH \longrightarrow CH_3{-}C(CH_3)(OH){-}CH_2{-}H$$

This C has no H's so the OH bonds here.

This C has 2 H's so the H bonds here.

PROBLEM 11.13

What product is formed when each alkene is treated with H_2O in the presence of H_2SO_4?

a. $CH_3CH{=}CHCH_3$

b. CH_3

c. $(CH_3)_2C{=}CHCH_3$

11.6 FOCUS ON HEALTH & MEDICINE Margarine or Butter?

One addition reaction of alkenes, hydrogenation, is especially important in the food industry. It lies at the heart of the debate over which product, butter or margarine, is better for the consumer.

As we learned in Section 11.3, butter is derived from saturated fatty acids like stearic acid [$CH_3(CH_2)_{16}COOH$], compounds with long carbon chains that contain only carbon–carbon single bonds. As a result, butter is a solid at room temperature.

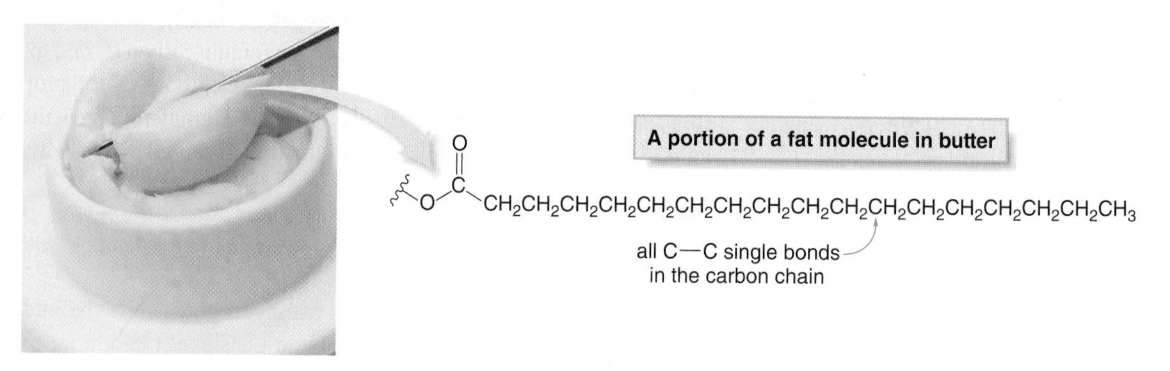

Margarine, on the other hand, is a synthetic product that mimics the taste and texture of butter. It is prepared from vegetable oils derived from unsaturated fatty acids like linoleic acid [$CH_3(CH_2)_4CH{=}CHCH_2CH{=}CH(CH_2)_7COOH$]. Margarine is composed mainly of *partially hydrogenated* vegetable oils formed by adding hydrogen to the double bonds in the carbon chain derived from unsaturated fatty acids.

EXTRA CRUNCHY
SKIPPY
SUPER CHUNK
INGREDIENTS: ROASTED PEANUTS, SUGAR, PARTIALLY HYDROGENATED VEGETABLE OILS (RAPESEED, COTTONSEED AND SOYBEAN) TO PREVENT SEPARATION, SALT.
MANUFACTURED AND UNCONDITIONALLY GUARANTEED BY ©UNILEVER BESTFOODS ENGLEWOOD CLIFFS, NJ 07632-9976
Comments and questions call 1-866-4SKIPPY.

Peanut butter is a common consumer product that contains partially hydrogenated vegetable oil.

When an unsaturated liquid vegetable oil is treated with hydrogen, some (or all) of the double bonds add H_2, as shown in Figure 11.4. This increases the melting point of the oil, thus giving it a semi-solid consistency that more closely resembles butter.

As we will learn in Section 15.4, unsaturated oils with carbon–carbon double bonds are healthier than saturated fats with no double bonds. Why, then, does the food industry partially hydrogenate oils, thus reducing the number of double bonds? The reasons relate to texture and shelf life. Consumers prefer the semi-solid consistency of margarine to a liquid oil. Imagine pouring vegetable oil on a piece of toast or pancakes!

Furthermore, unsaturated oils are more susceptible than saturated fats to oxidation, which makes the oil rancid and inedible. Hydrogenating the double bonds reduces the likelihood of oxidation, thereby increasing the shelf life

Figure 11.4 Partial Hydrogenation of the Double Bonds in a Vegetable Oil

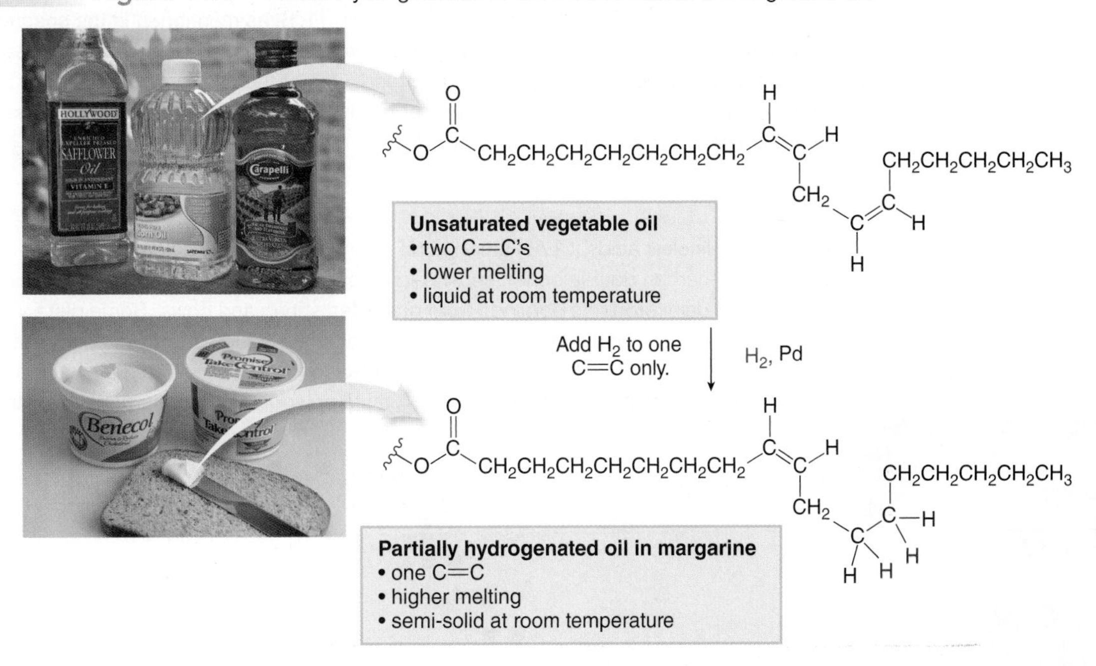

- When an oil is *partially* hydrogenated, some double bonds react with H_2, while some double bonds remain in the product. Since the product has fewer double bonds, it has a higher melting point. Thus, a liquid oil is converted to a semi-solid.

of the food product. This process reflects a delicate balance between providing consumers with healthier food products, while maximizing shelf life to prevent spoilage.

One other fact is worthy of note. During hydrogenation, some of the cis double bonds in vegetable oils are converted to trans double bonds, forming so-called "trans fats." The shape of the resulting fatty acid chain is very different, closely resembling the shape of a *saturated* fatty acid chain. As a result, trans fats are thought to have the same negative effects on blood cholesterol levels as saturated fats; that is, trans fats stimulate cholesterol synthesis in the liver, thus increasing blood cholesterol levels, a factor linked to increased risk of heart disease.

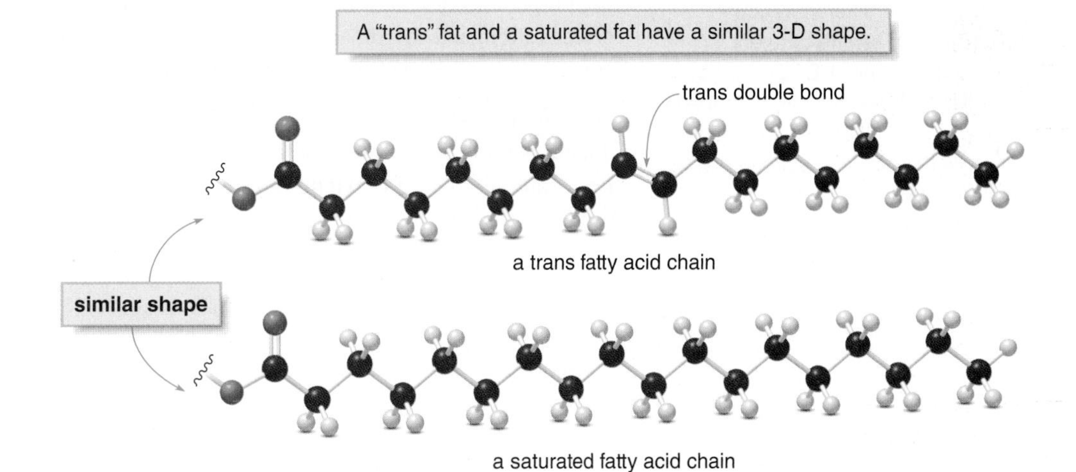

PROBLEM 11.14

Oil from the kukui nut contains linolenic acid (Table 11.1), as mentioned in the chapter opener. (a) When linolenic acid is partially hydrogenated with one equivalent of H_2 in the presence of a Pd catalyst, three constitutional isomers are formed. One of them is linoleic acid. Draw the structures of the other two possible products. (b) What product is formed when linolenic acid is completely hydrogenated with three equivalents of H_2?

PROBLEM 11.15

Conjugated linoleic acids (CLAs) are a group of unsaturated fatty acids that contain 18 carbons and two carbon–carbon double bonds separated by one single bond. Conjugated linoleic acids are found in meat and dairy products obtained from grass-fed sheep and cows. Some dietary supplements purported to reduce body fat contain CLAs. (a) Label the double bonds in conjugated linoleic acid **A** as cis or trans. (b) Draw the structure of a stereoisomer of **A.** (c) Would this stereoisomer have a higher or lower melting point than **A?** (d) Is **A** a stereoisomer or constitutional isomer of linoleic acid?

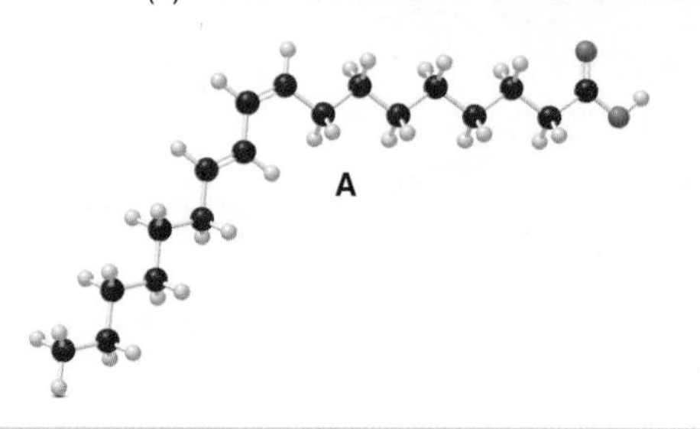

CONSUMER NOTE

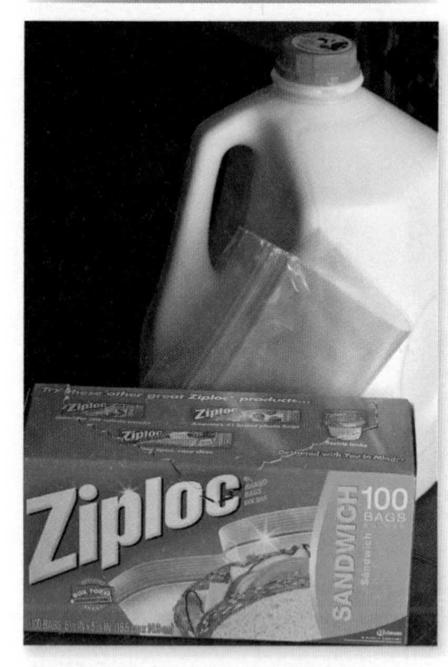

HDPE (high-density polyethylene) and **LDPE** (low-density polyethylene) are common types of polyethylene prepared under different reaction conditions and having different physical properties. HDPE is opaque and rigid, and used in milk containers and water jugs. LDPE is less opaque and more flexible, and used in plastic bags and electrical insulation.

11.7 Polymers—The Fabric of Modern Society

***Polymers* are large molecules made up of repeating units of smaller molecules—called *monomers*—covalently bonded together.** Polymers include the naturally occurring proteins that compose hair, tendons, and fingernails. They also include such industrially important plastics as polyethylene, poly(vinyl chloride) (PVC), and polystyrene. Since 1976, the U.S. production of synthetic polymers has exceeded its steel production.

11.7A Synthetic Polymers

Many synthetic polymers—that is, those synthesized in the lab—are among the most widely used organic compounds in modern society. Soft drink bottles, plastic bags, food wrap, compact discs, Teflon, and Styrofoam are all made of synthetic polymers. Figure 11.5 illustrates several consumer products and the polymers from which they are made.

To form a polymer from an alkene monomer, the weak bond that joins the two carbons of the double bond is broken and new strong carbon–carbon single bonds join the monomers together. For example, joining **ethylene monomers** together forms the polymer **polyethylene,** a plastic used in milk containers and sandwich bags.

Figure 11.5 Polymers in Some Common Products

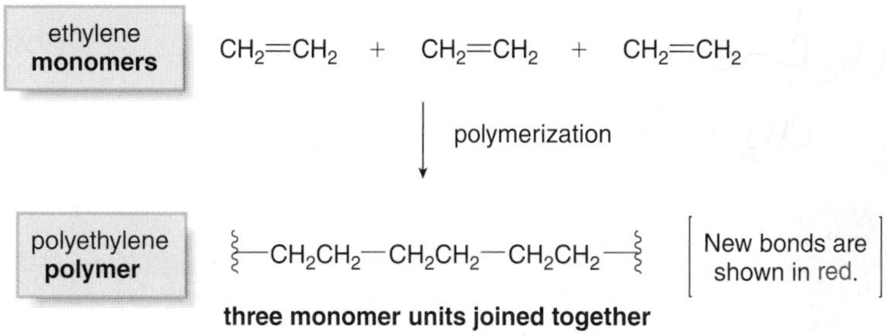

ethylene **monomers**	$CH_2{=}CH_2 + CH_2{=}CH_2 + CH_2{=}CH_2$

↓ polymerization

polyethylene **polymer**	$\sim CH_2CH_2{-}CH_2CH_2{-}CH_2CH_2 \sim$

three monomer units joined together

New bonds are shown in red.

- *Polymerization* is the joining together of monomers to make polymers.

Many ethylene derivatives having the general structure $\mathbf{CH_2{=}CHZ}$ are also used as monomers for polymerization. Polymerization of $CH_2{=}CHZ$ usually yields polymers with the Z groups on every other carbon atom in the chain.

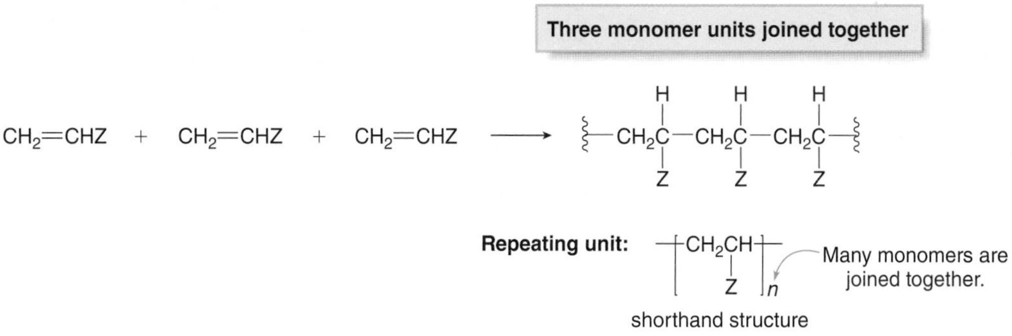

Three monomer units joined together

$$CH_2{=}CHZ + CH_2{=}CHZ + CH_2{=}CHZ \longrightarrow \sim CH_2CHZ{-}CH_2CHZ{-}CH_2CHZ \sim$$

Repeating unit: $\left[CH_2CH(Z) \right]_n$ — Many monomers are joined together.

shorthand structure

Polymer structures are often abbreviated by placing the atoms in the repeating unit in brackets, as shown. Table 11.2 lists some common monomers and polymers used in medicine and dentistry.

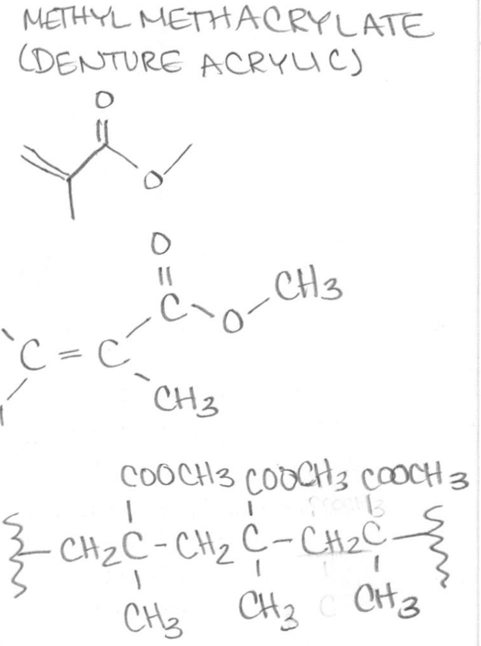

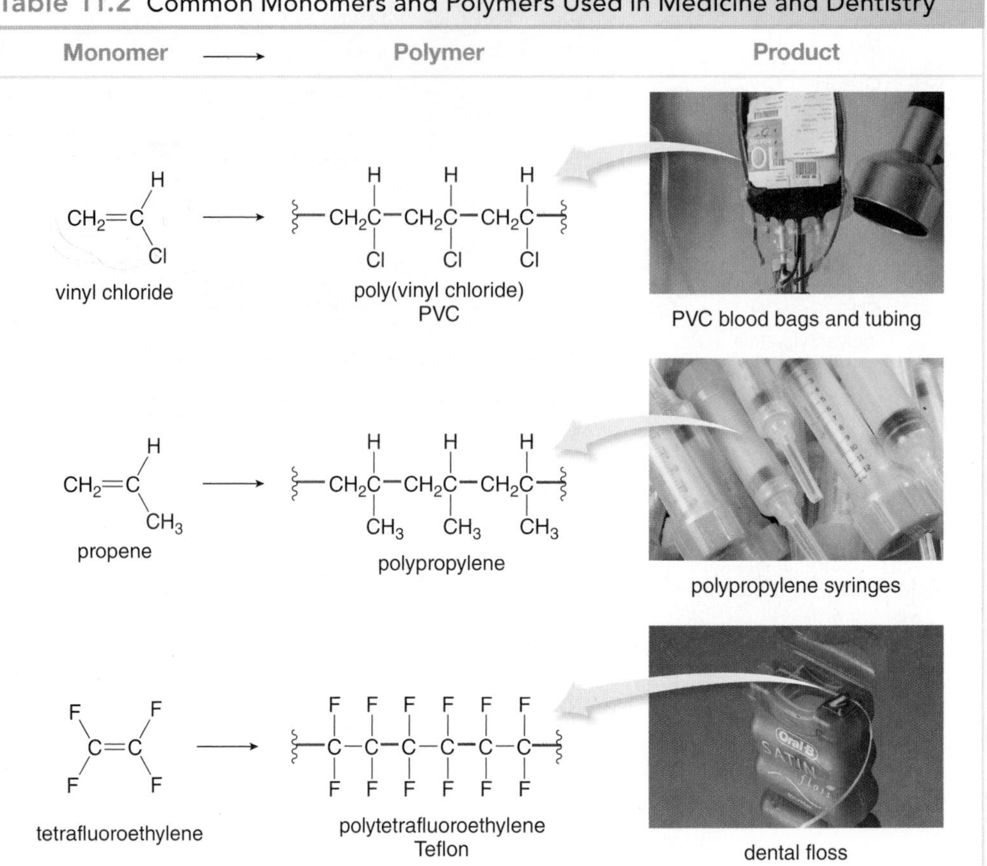

Table 11.2 Common Monomers and Polymers Used in Medicine and Dentistry

Monomer →	Polymer	Product
$CH_2{=}CHCl$ vinyl chloride	$\sim CH_2CHCl{-}CH_2CHCl{-}CH_2CHCl \sim$ poly(vinyl chloride) PVC	PVC blood bags and tubing
$CH_2{=}CHCH_3$ propene	$\sim CH_2CH(CH_3){-}CH_2CH(CH_3){-}CH_2CH(CH_3) \sim$ polypropylene	polypropylene syringes
$CF_2{=}CF_2$ tetrafluoroethylene	$\sim CF_2{-}CF_2{-}CF_2{-}CF_2{-}CF_2{-}CF_2 \sim$ polytetrafluoroethylene Teflon	dental floss

SAMPLE PROBLEM 11.6

What polymer is formed when $CH_2{=}CHCO_2CH_3$ (methyl acrylate) is polymerized?

Analysis

Draw three or more alkene molecules and arrange the carbons of the double bonds next to each other. Break one bond of each double bond, and join the alkenes together with single bonds. With unsymmetrical alkenes, substituents are bonded to every other carbon.

Solution

Join these 2 C's. Join these 2 C's.

$$CH_2{=}CH(CO_2CH_3) \quad CH_2{=}CH(CO_2CH_3) \quad CH_2{=}CH(CO_2CH_3) \longrightarrow \sim CH_2{-}CH(CO_2CH_3){-}CH_2{-}CH(CO_2CH_3){-}CH_2{-}CH(CO_2CH_3) \sim$$

Answer:

Break one bond that joins each C=C.

[New bonds are drawn in red.]

PROBLEM 11.16

What polymer is formed when each compound is polymerized?

a. $CH_2{=}C(CH_3)CH_2CH_3$

b. $CH_2{=}C(CH_3)CN$

c. $CH_2{=}CH$–(3-chlorophenyl) (benzene ring bearing Cl)

PROBLEM 11.17

What monomer is used to form poly(vinyl acetate), a polymer used in paints and adhesives?

$$\sim CH_2\underset{\underset{COCH_3}{|}\atop{O}}{\overset{H}{C}}-CH_2\underset{\underset{COCH_3}{|}\atop{O}}{\overset{H}{C}}-CH_2\underset{\underset{COCH_3}{|}\atop{O}}{\overset{H}{C}}\sim$$

poly(vinyl acetate)

11.7B FOCUS ON THE ENVIRONMENT Polymer Recycling

The same desirable characteristics that make polymers popular materials for consumer products—durability, strength, and lack of reactivity—also contribute to environmental problems. Polymers do not degrade readily, and as a result, billions of pounds of polymers end up in landfills every year. Recycling existing polymer types to make new materials is one solution to the waste problem created by polymers.

Although thousands of different synthetic polymers have now been prepared, six compounds, called the **"Big Six,"** account for 76% of the synthetic polymers produced in the United States each year. Each polymer is assigned a recycling code (1–6) that indicates its ease of recycling; **the lower the number, the easier it is to recycle.** Table 11.3 lists these six most common polymers, as well as the type of products made from each recycled polymer.

Table 11.3 Recyclable Polymers

Recycling Code	Polymer Name	Shorthand Structure	Recycled Product
1	PET polyethylene terephthalate	$\left[CH_2CH_2-O-C(=O)-C_6H_4-C(=O)-O\right]_n$	fleece jackets carpeting plastic bottles
2	HDPE high-density polyethylene	$\left[CH_2CH_2\right]_n$	Tyvek insulation outdoor furniture sports clothing
3	PVC poly(vinyl chloride)	$\left[CH_2\underset{Cl}{\underset{\vert}{CH}}\right]_n$	floor mats
4	LDPE low-density polyethylene	$\left[CH_2CH_2\right]_n$	trash bags
5	PP polypropylene	$\left[CH_2\underset{CH_3}{\underset{\vert}{CH}}\right]_n$	furniture
6	PS polystyrene	$\left[CH_2\underset{C_6H_5}{\underset{\vert}{CH}}\right]_n$	molded trays trash cans

Of the Big Six, only the polyethylene terephthalate (PET) in soft drink bottles and the high-density polyethylene (HDPE) in milk jugs and juice bottles are recycled to any great extent. Recycled HDPE is converted to Tyvek, an insulating wrap used in new housing construction, and outdoor furniture. Recycled PET is used to make fibers for fleece clothing and carpeting.

11.8 Aromatic Compounds

Aromatic compounds represent another example of unsaturated hydrocarbons. Aromatic compounds were originally named because many simple compounds in this family have characteristic odors. Today, the word **aromatic refers to compounds that contain a benzene ring,** or rings that react in a similar fashion to benzene.

HEALTH NOTE

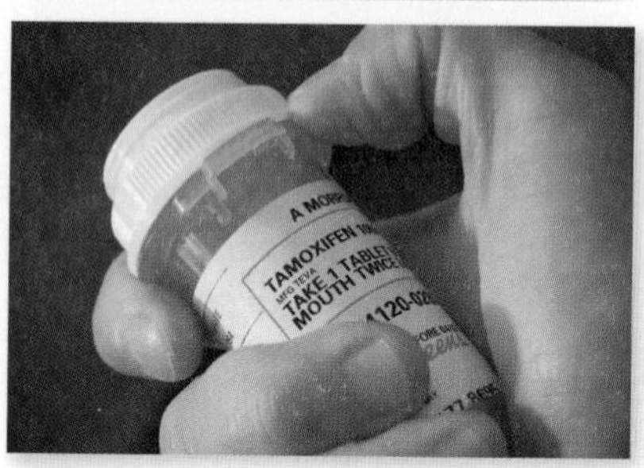

Tamoxifen, a potent anticancer drug sold under the trade name of Novaldex, contains three benzene rings.

$OCH_2CH_2N(CH_3)_2$

C=C

CH_3CH_2

tamoxifen

Benzene, the simplest and most widely known aromatic compound, contains a six-membered ring and three double bonds. Since each carbon of the ring is also bonded to a hydrogen atom, the molecular formula for benzene is C_6H_6. Each carbon is surrounded by three groups, making it trigonal planar. Thus, **benzene is a planar molecule,** and all bond angles are **120°.**

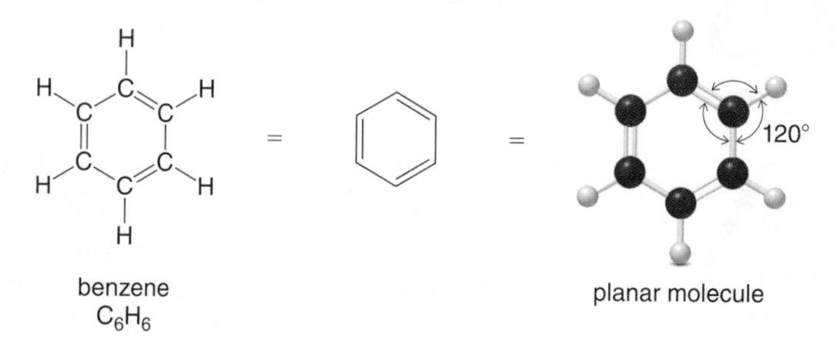

120°
benzene
C_6H_6
planar molecule

Although benzene is drawn with a six-membered ring and three double bonds, there are two different ways to arrange the double bonds so that they alternate with single bonds around the ring. Each of these representations is equivalent.

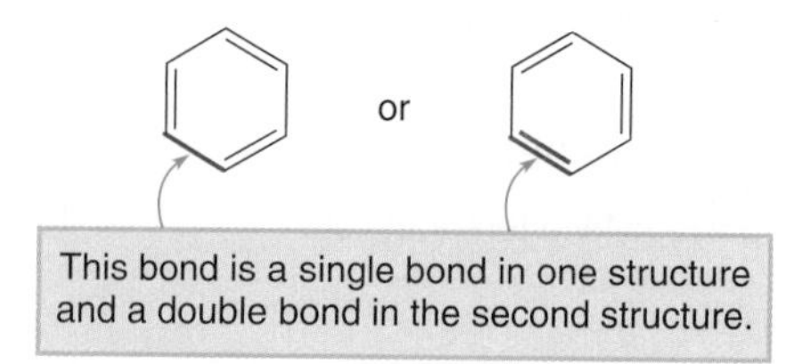

or
This bond is a single bond in one structure and a double bond in the second structure.

The physical properties of aromatic hydrocarbons are similar to other hydrocarbons—they have low melting points and boiling points and are water insoluble.

11.9 Nomenclature of Benzene Derivatives

Many organic molecules contain a benzene ring with one or more substituents, so we must learn how to name them.

11.9A Monosubstituted Benzenes

To name a benzene ring with one substituent, **name the substituent and add the word *benzene*.** Carbon substituents are named as alkyl groups. When a halogen is a substituent, name the halogen by changing the *-ine* ending of the name of the halogen to the suffix *-o*; for example, chlor***ine*** → chlor***o***.

CH_2CH_3 — ethyl group — ethylbenzene

$CH_2CH_2CH_2CH_3$ — butyl group — butylbenzene

Cl — chloro group — chlorobenzene

Many monosubstituted benzenes, such as those with methyl (CH_3–), hydroxyl (–OH), and amino (–NH_2) groups, have common names that you must learn, too.

toluene (methylbenzene) | phenol (hydroxybenzene) | aniline (aminobenzene)

The pain reliever acetaminophen (trade name Tylenol) contains a para-disubstituted benzene ring.

acetaminophen (Trade name: Tylenol)

11.9B Disubstituted Benzenes

There are three different ways that two groups can be attached to a benzene ring, so a prefix—**ortho, meta,** or **para**—is used to designate the relative position of the two substituents. Ortho, meta, and para are generally abbreviated as ***o*, *m*,** and ***p*,** respectively.

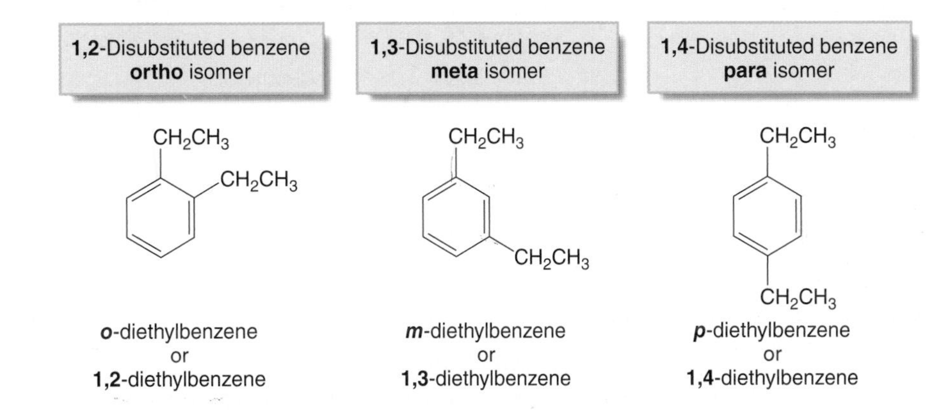

If the two groups on the benzene ring are different, **alphabetize the name of the substituents** preceding the word benzene. If one of the substituents is part of a **common root,** name the **molecule as a derivative of that monosubstituted benzene.**

Alphabetize two different substituent names:

o-**bromochloro**-benzene | *m*-**ethylfluoro**-benzene

Use a common root name:

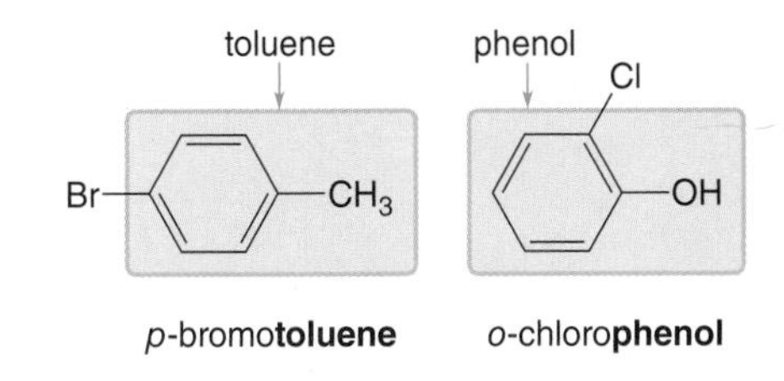

p-bromo**toluene** | *o*-chloro**phenol**

11.9C Polysubstituted Benzenes

For three or more substituents on a benzene ring:

1. **Number to give the lowest possible numbers around the ring.**
2. **Alphabetize the substituent names.**
3. **When substituents are part of common roots, name the molecule as a derivative of that monosubstituted benzene. The substituent that comprises the common root is located at C1, but the "1" is omitted from the name.**

Examples of naming polysubstituted benzenes

1, CH_2CH_3, Cl, $CH_2CH_2CH_3$, 4, 2

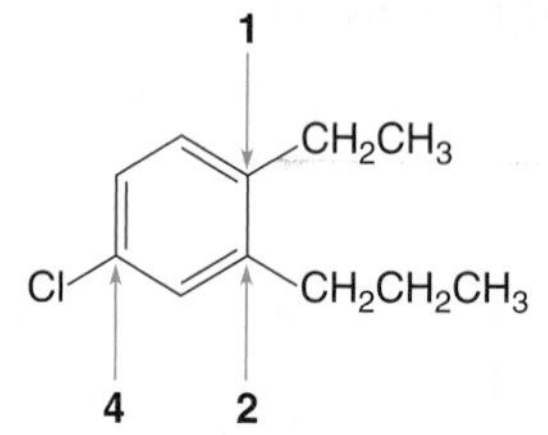

- Assign the lowest set of numbers.
- Alphabetize the names of all the substituents.

4-**c**hloro-**1**-**e**thyl-**2**-**p**ropylbenzene

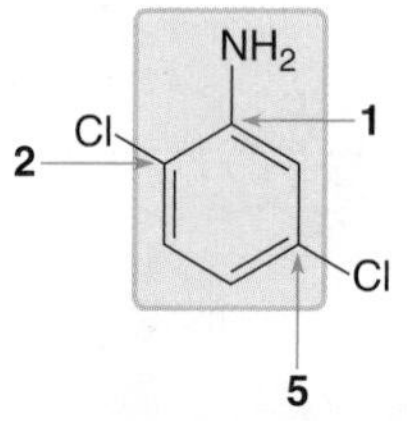

NH_2, Cl, 2, 1, Cl, 5

- Name the molecule as a derivative of the common root **aniline.**
- Designate the position of the NH_2 group as "1," and then assign the lowest possible set of numbers to the other substituents.

2,5-dichloro**aniline**

SAMPLE PROBLEM 11.7

Name each of the following aromatic compounds.

a. 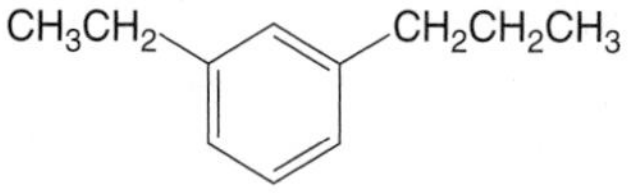 CH_3CH_2, $CH_2CH_2CH_3$

b. 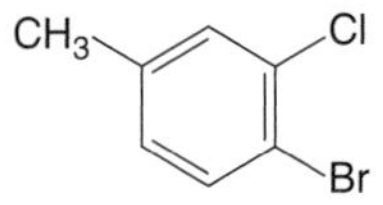 CH_3, Cl, Br

Analysis

Name the substituents on the benzene ring. With two groups, alphabetize the substituent names and use the prefix ortho, meta, or para to indicate their location. With three substituents, alphabetize the substituent names, and number to give the lowest set of numbers.

Solution

a. 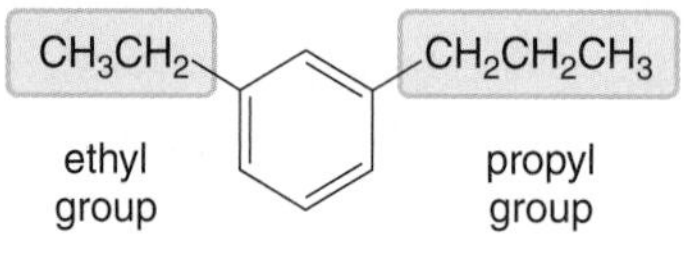 CH_3CH_2 (ethyl group), $CH_2CH_2CH_3$ (propyl group)

- The two substituents are located 1,3- or **meta** to each other.
- Alphabetize the *e* of **e**thyl before the *p* of **p**ropyl.

Answer: *m*-ethylpropylbenzene

b. 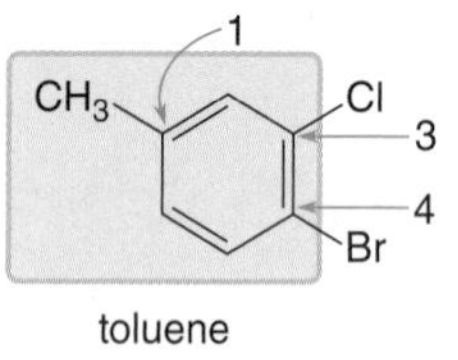 CH_3, 1, Cl, 3, 4, Br, toluene

- Since a CH_3– group is bonded to the ring, name the molecule as a derivative of toluene.
- Place the CH_3 group at the "1" position, and number to give the lowest set of numbers.

Answer: 4-bromo-3-chlorotoluene

PROBLEM 11.18

Give the IUPAC name of each compound.

a. 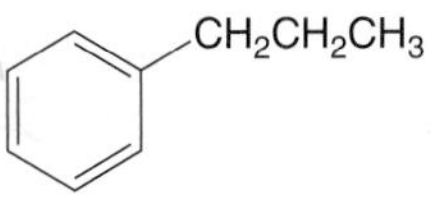 (benzene ring with $CH_2CH_2CH_3$)

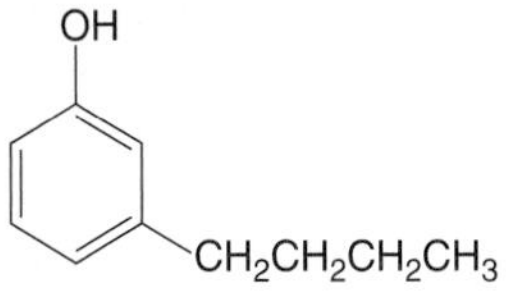

b. 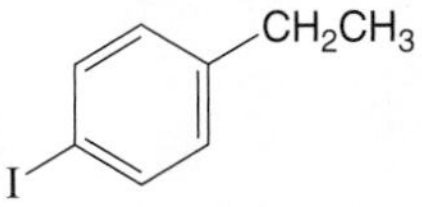 (benzene ring with CH_2CH_3 and I, para)

c. (benzene ring with OH and $CH_2CH_2CH_2CH_3$, meta)

d. 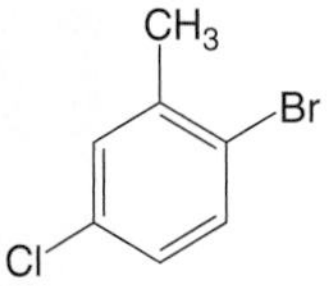 (benzene ring with CH_3, Br, and Cl)

PROBLEM 11.19

Draw the structure corresponding to each name.

a. pentylbenzene
b. *o*-dichlorobenzene
c. *m*-bromoaniline
d. 4-chloro-1,2-diethylbenzene

HEALTH NOTE

Commercial sunscreens are given an **SPF** rating (sun protection factor), according to the amount of sunscreen present. The higher the number, the greater the protection.

11.10 FOCUS ON HEALTH & MEDICINE Sunscreens and Antioxidants

11.10A Sunscreens

All commercially available sunscreens contain a benzene ring. A sunscreen absorbs ultraviolet radiation and thus shields the skin for a time from its harmful effects. Two sunscreens that have been used for this purpose are *p*-aminobenzoic acid (PABA) and Padimate O.

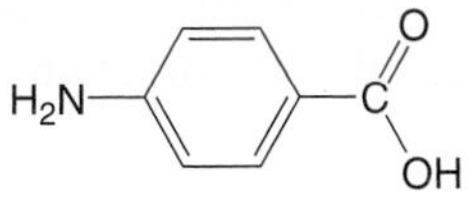

$H_2N-C_6H_4-C(=O)OH$
p-aminobenzoic acid (PABA)

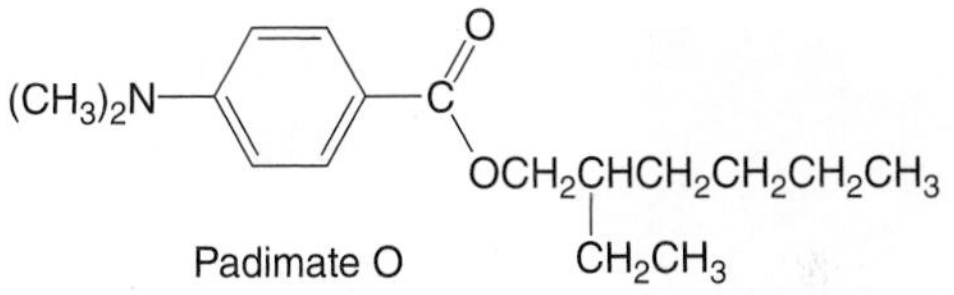

$(CH_3)_2N-C_6H_4-C(=O)OCH_2CH(CH_2CH_3)CH_2CH_2CH_2CH_3$
Padimate O

PROBLEM 11.20

Identify the functional groups in each sunscreen: (a) PABA; (b) Padimate O.

PROBLEM 11.21

Which of the following compounds might be an ingredient in a commercial sunscreen? Explain why or why not.

a. 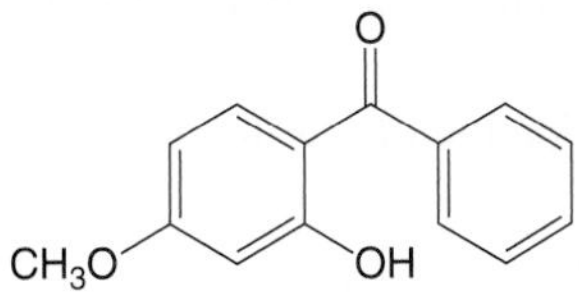 (benzophenone with CH_3O and OH substituents)

b. 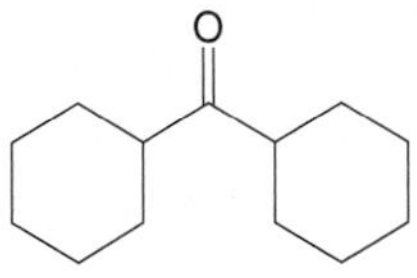 (dicyclohexyl ketone)

11.10B Phenols as Antioxidants

A wide variety of **phenols,** compounds that contain a hydroxyl group bonded to a benzene ring, occur in nature. **Vanillin** from the vanilla bean is a phenol, as is **curcumin,** a yellow pigment isolated from turmeric, a tropical perennial in the ginger family and a principal ingredient in curry powder. Curcumin has long been used as an anti-inflammatory agent in traditional eastern medicine.

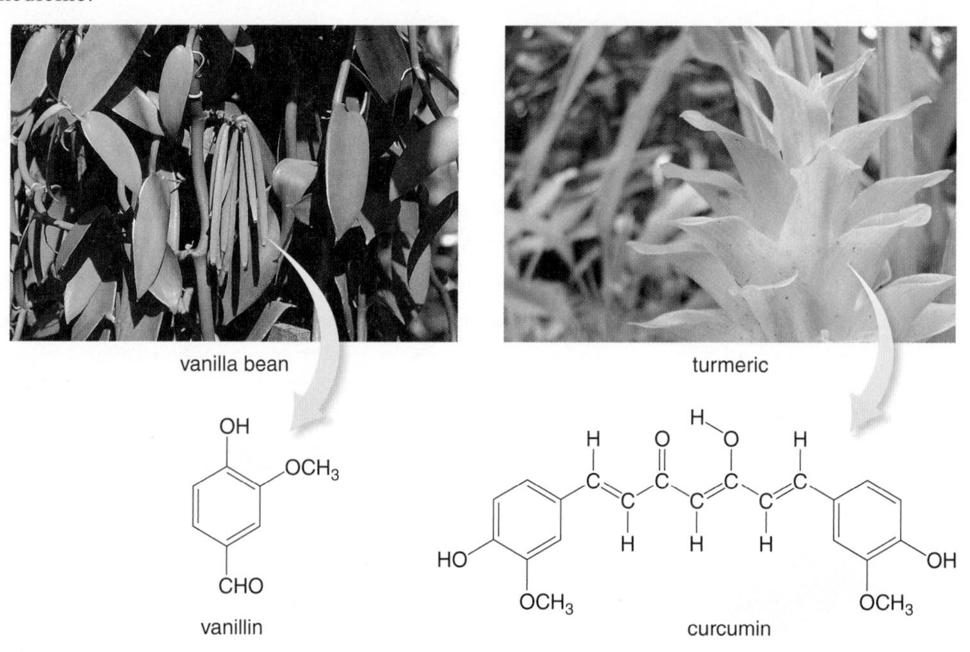

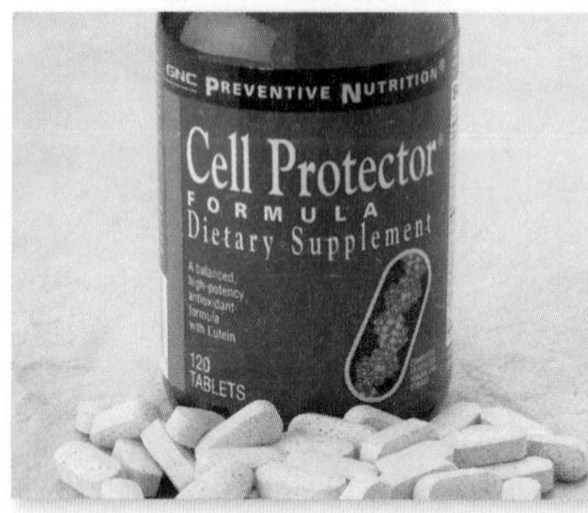

HEALTH NOTE

The purported health benefits of antioxidants have made them a popular component in anti-aging formulations.

Many **phenols are antioxidants,** compounds that prevent unwanted oxidation reactions from occurring. Two examples are naturally occurring vitamin E and synthetic BHT. **The OH group on the benzene ring is the key functional group that prevents oxidation reactions from taking place.**

vitamin E: HO, CH_3, CH_3, CH_3, O, $(CH_2)_3CH(CH_2)_3CH(CH_2)_3CH(CH_3)_2$ with CH_3 branches

BHT (**b**utylated **h**ydroxy **t**oluene): OH, $(CH_3)_3C$, $C(CH_3)_3$, CH_3

Vitamin E is a natural antioxidant found in fish oil, peanut oil, wheat germ, and leafy greens. Although the molecular details of its function remain obscure, it is thought that vitamin E prevents the unwanted oxidation of unsaturated fatty acid residues in cell membranes. In this way, vitamin E helps retard the aging process.

Synthetic antioxidants such as **BHT**—**b**utylated **h**ydroxy **t**oluene—are added to packaged and prepared foods to prevent oxidation and spoilage. BHT is a common additive in breakfast cereals.

HEALTH NOTE

Nuts are an excellent source of vitamin E.

PROBLEM 11.22

Which of the following compounds might be antioxidants?

a. CH_3–(cyclohexane ring, OH)–$CH(CH_3)_2$ menthol

b. curcumin

c. CH_3–(benzene ring)–CH_3 *p*-xylene (gasoline additive)

KEY TERMS

Addition reaction (11.5)
Alkene (11.1)
Alkyne (11.1)
Antioxidant (11.10)
Aromatic compound (11.8)
Cis isomer (11.3)
Fat (11.3)
Fatty acid (11.3)
Hydration (11.5)
Hydrogenation (11.5)
Markovnikov's rule (11.5)
Meta isomer (11.9)
Monomer (11.7)
Oil (11.3)
Ortho isomer (11.9)
Para isomer (11.9)
Partial hydrogenation (11.6)
Phenol (11.10)
Polymer (11.7)
Polymerization (11.7)
Stereoisomer (11.3)
Trans isomer (11.3)
Unsaturated hydrocarbon (11.1)

KEY CONCEPTS

1 What are the characteristics of alkenes, alkynes, and aromatic compounds?

- Alkenes are unsaturated hydrocarbons that contain a carbon–carbon double bond and have molecular formula C_nH_{2n}. Each carbon of the double bond is trigonal planar. (11.1)
- Alkynes are unsaturated hydrocarbons that contain a carbon–carbon triple bond and have molecular formula C_nH_{2n-2}. Each carbon of the triple bond is linear. (11.1)
- Benzene, molecular formula C_6H_6, is the most common aromatic hydrocarbon. Benzene contains a six-membered ring with three double bonds, and each carbon is trigonal planar. (11.8)

2 How are alkenes, alkynes, and substituted benzenes named?

- An alkene is identified by the suffix *-ene,* and the carbon chain is numbered to give the C═C the lower number. (11.2)
- An alkyne is identified by the suffix *-yne,* and the carbon chain is numbered to give the C≡C the lower number. (11.2)
- Substituted benzenes are named by naming the substituent and adding the word *benzene.* When two substituents are bonded to the ring, the prefixes ortho, meta, and para are used to show the relative positions of the two groups: 1,2-, 1,3- or 1,4-, respectively. With three substituents on a benzene ring, number to give the lowest possible numbers. (11.9)

3 What is the difference between constitutional isomers and stereoisomers? How are cis and trans isomers different? (11.3)

- Constitutional isomers differ in the way the atoms are bonded to each other.
- Stereoisomers differ only in the three-dimensional arrangement of atoms.
- Cis and trans isomers are one type of stereoisomer. A cis alkene has two alkyl groups on the same side of the double bond. A trans alkene has two alkyl groups on opposite sides of the double bond.

4 How do saturated and unsaturated fatty acids differ? (11.3B)

- Fatty acids are carboxylic acids (RCOOH) with long carbon chains. Saturated fatty acids have no double bonds in the carbon chain and unsaturated fatty acids have one or more double bonds in their long carbon chains.
- All double bonds in naturally occurring fatty acids are cis.
- As the number of double bonds in the fatty acid increases, the melting point decreases.

5 What types of reactions do alkenes undergo? (11.5)

- Alkenes undergo addition reactions with reagents X—Y. One bond of the double bond and the X—Y bond break and two new single bonds (C—X and C—Y) are formed.
- Alkenes react with H_2 (Pd catalyst) and H_2O (with H_2SO_4).

6 What products are formed when a vegetable oil is partially hydrogenated? (11.6)

- When an unsaturated oil is partially hydrogenated, some but not all of the cis C═C's add H_2, reducing the number of double bonds and increasing the melting point.
- Some of the cis double bonds are converted to trans double bonds, forming trans fats, whose shape and properties closely resemble those of saturated fats.

7 What are polymers, and how are they formed from alkene monomers? (11.7)

- Polymers are large molecules made up of repeating smaller molecules called monomers covalently bonded together. When alkenes are polymerized, one bond of the double bond breaks, and two new single bonds join the alkene monomers together in long carbon chains.

UNDERSTANDING KEY CONCEPTS

Selected in-chapter and odd-numbered end-of-chapter problems have brief answers at the end of each chapter. The *Student Study Guide and Solutions Manual* contains detailed solutions to all in-chapter and odd-numbered end-of-chapter problems, as well as additional worked examples and a chapter self-test.

11.23 Anethole, the major constituent of anise oil, is used in licorice-flavored sweets and flavored brandy. Answer the following questions using the ball-and-stick model of anethole.

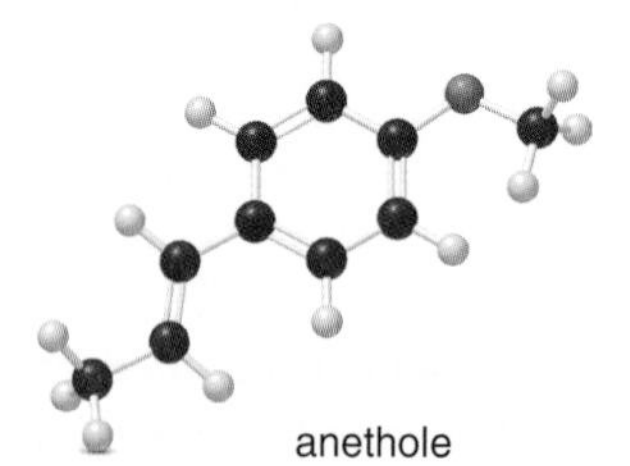

anethole

a. What is the molecular formula of anethole?
b. Identify the functional groups.
c. Label the carbon–carbon double bond as cis or trans.
d. Draw the structure of anethole and label each carbon as trigonal planar or tetrahedral.

11.24 Methyl cinnamate, isolated from red clover flowers, is used in perfumery and as a flavoring agent because of its strawberry-like taste and odor. Answer the following questions using the ball-and-stick model of methyl cinnamate.

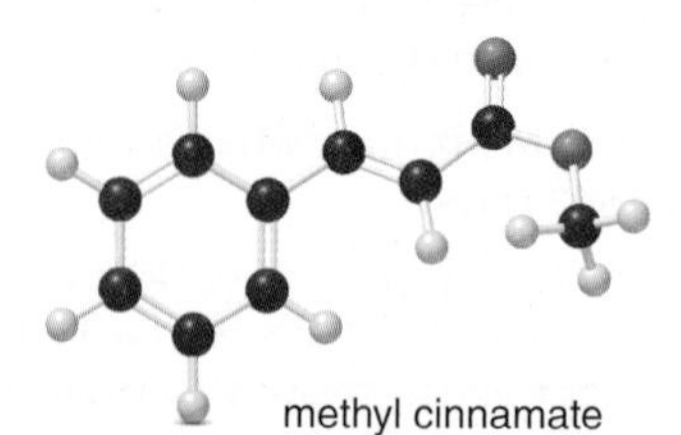

methyl cinnamate

a. What is the molecular formula of methyl cinnamate?
b. Identify the functional groups.
c. Label the carbon–carbon double bond as cis or trans.
d. Draw the structure of methyl cinnamate and label each carbon as trigonal planar or tetrahedral.

11.25 Give the IUPAC name for each compound.

a. $CH_2{=}CHCH_2CH_2C(CH_3)_3$ b. $CH_3C{\equiv}CCH_2C(CH_3)_3$

11.26 Give the IUPAC name for each compound.

a. $CH_2{=}CHCH_2CHCH_2CH_3$ (with CH_3 bonded to the fourth C)   b. $CH_3C{\equiv}CCHCH_3$ (with $CH_2CH_2CH_3$ bonded to the fourth C)

11.27 Answer the following questions about compound **A**, represented in the given ball-and-stick model.

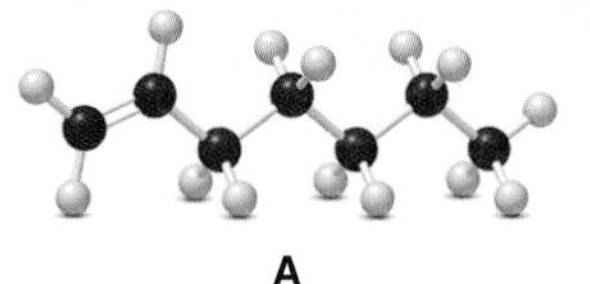

A

a. Convert **A** to a condensed structure and give its IUPAC name.
b. What product is formed when **A** is treated with H_2 in the presence of a metal catalyst?
c. What product is formed when **A** is treated with H_2O in the presence of H_2SO_4?

11.28 Answer the following questions about compound **B**, represented in the given ball-and-stick model.

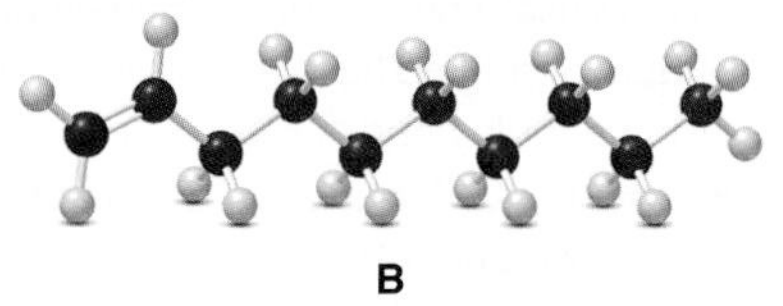

B

a. Convert **B** to a condensed structure and give its IUPAC name.
b. What product is formed when **B** is treated with H_2 in the presence of a metal catalyst?
c. What product is formed when **B** is treated with H_2O in the presence of H_2SO_4?

11.29 What polymer is formed when $CH_2{=}CCl_2$ is polymerized?

11.30 What polymer is formed when $CH_2{=}CHNHCOCH_3$ is polymerized?

11.31 Draw the structure of the three constitutional isomers that have a Cl atom and an NH_2 group bonded to a benzene ring. Name each compound using the IUPAC system.

11.32 Draw the structure of 2,4,6-trichlorotoluene.

ADDITIONAL PROBLEMS

Alkene, Alkyne, and Benzene Structure

11.33 What is the molecular formula for a hydrocarbon with 10 carbons that is (a) completely saturated; (b) an alkene; (c) an alkyne?

11.34 Draw the structure of a hydrocarbon with molecular formula C_6H_{10} that also contains: (a) a carbon–carbon triple bond; (b) two carbon–carbon double bonds; (c) one ring and one C=C.

11.35 Draw structures for the three alkynes having molecular formula C_5H_8.

11.36 Draw the structures of the five constitutional isomers of molecular formula C_5H_{10} that contain a double bond.

11.37 Label each carbon in the following molecules as tetrahedral, trigonal planar, or linear.

a. (bicyclic structure) b. C_6H_5—C≡CH

11.38 Falcarinol is a natural pesticide found in carrots that protects them from fungal diseases. Predict the indicated bond angles in falcarinol.

H—C(H)=C(H)—C(H)(OH)—C≡C—C≡C—CH_2—C(H)=C(H)—$(CH_2)_6CH_3$ (angles a, b, c, d, e indicated)

falcarinol

Nomenclature of Alkenes and Alkynes

11.39 Give the IUPAC name for each molecule depicted in the ball-and-stick models.

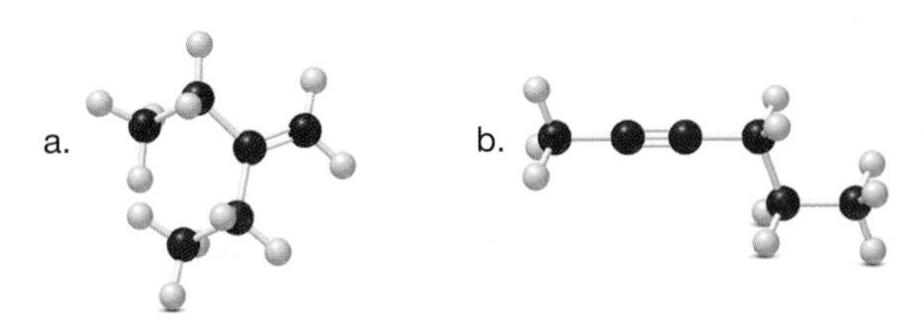

11.40 Give the IUPAC name for each molecule depicted in the ball-and-stick models.

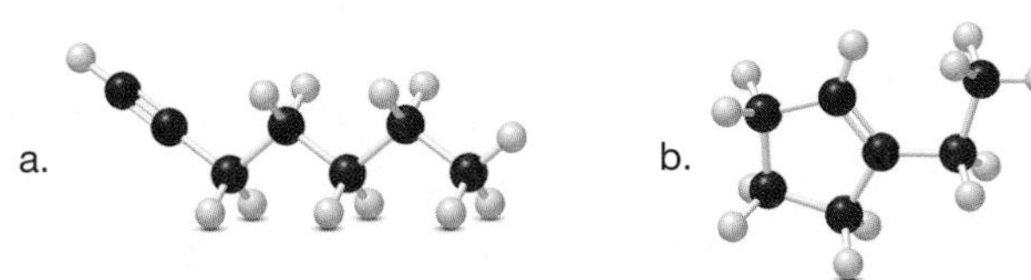

11.41 Give the IUPAC name for each compound.

a. $(CH_3CH_2)_2C{=}CHCH(CH_3)CH_2CH(CH_3)CH_3$

b. $CH_2{=}C(CH_2CH_2CH_2CH_2CH_3)CH_2CH_3$

c. $CH_3C{\equiv}C{-}CH_2{-}CH(CH_3){-}C(CH_3)(CH_2CH_3){-}CH_2CH_3$

11.42 Give the IUPAC name for each compound.

a. $(CH_3)_2C{=}CHCH_2CH(CH_2CH_2CH_3)CH_2CH_3$

b. $(CH_3)_3CC{\equiv}CC(CH_3)_3$

c. $(CH_3CH_2CH_2CH_2)_2C{=}CHCH_3$

11.43 Give the structure corresponding to each IUPAC name.

a. 3-methyl-1-octene
b. 2-methyl-3-hexyne
c. 3,5-diethyl-2-methyl-3-heptene
d. *cis*-7-methyl-2-octene

11.44 Give the structure corresponding to each IUPAC name.

a. 6-ethyl-2-octyne
b. *trans*-5-methyl-2-hexene
c. 5,6-dimethyl-2-heptyne
d. 3,4,5,6-tetramethyl-1-decyne

Isomers

11.45 Leukotriene C_4 is a key compound that causes the constriction of small airways, and in this way contributes to the asthmatic response. Label each double bond in leukotriene C_4 as cis or trans.

(structure: $CH_3CH_2CH_2CH_2CH_2$—CH=CH—CH_2—CH=CH—CH=CH—CH=CH—CH(OH)—$CH(S{-}CH_2CH(CONHCH_2COOH)NHCOCH_2CH_2CH(NH_2)COOH)$—$CH_2CH_2CH_2COOH$)

leukotriene C_4

11.46 Draw the complete structure of each naturally occurring compound using the proper cis or trans arrangement around the carbon–carbon double bond. Muscalure is the sex attractant of the housefly. Cinnamaldehyde is responsible for the odor of cinnamon.

a. $CH_3(CH_2)_8CH{=}CH(CH_2)_{12}CH_3$
muscalure
(cis double bond)

b. C_6H_5—CH=CHCHO
cinnamaldehyde
(trans double bond)

11.47 Label the carbon–carbon double bond as cis or trans, and give the IUPAC name for the alkene depicted in the ball-and-stick model.

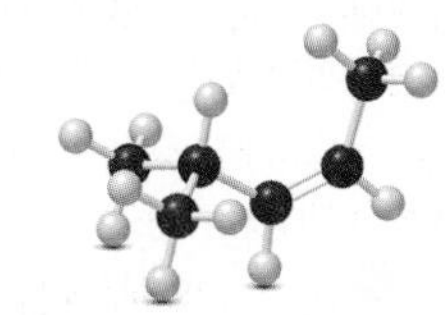

11.48 Label the carbon–carbon double bond as cis or trans, and give the IUPAC name for the alkene depicted in the ball-and-stick model.

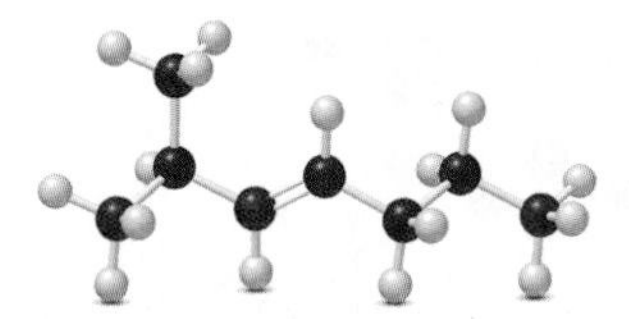

11.49 Draw the cis and trans isomers for each compound: (a) 2-nonene; (b) 2-methyl-3-heptene.

11.50 Draw the cis and trans isomers for each compound: (a) 3-heptene; (b) 4,4-dimethyl-2-hexene.

11.51 How are the compounds in each pair related? Choose from constitutional isomers, stereoisomers, or identical.

a. $CH_3CH_2C{\equiv}CCH_3$ and $HC{\equiv}CCH_2CH_2CH_3$

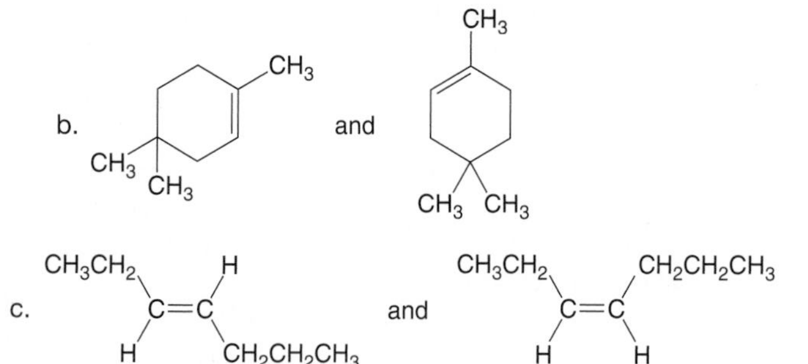

11.52 Consider alkenes **A**, **B**, and **C**. How are the compounds in each pair related? Choose from constitutional isomers, stereoisomers, or identical: (a) **A** and **B**; (b) **A** and **C**; (c) **B** and **C**.

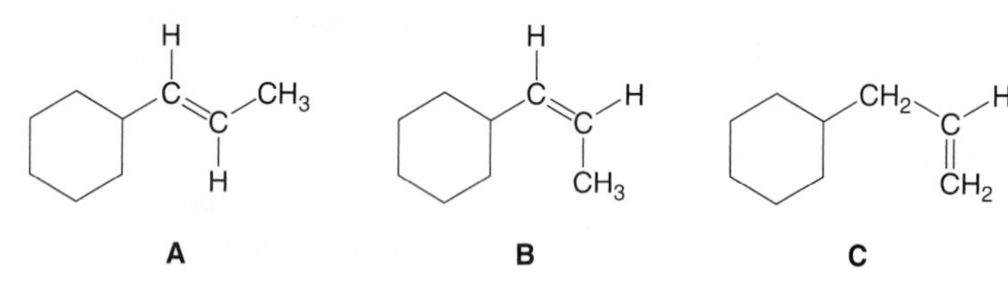

11.53 How are the compounds in each pair related? Choose from constitutional isomers, stereoisomers, or identical.

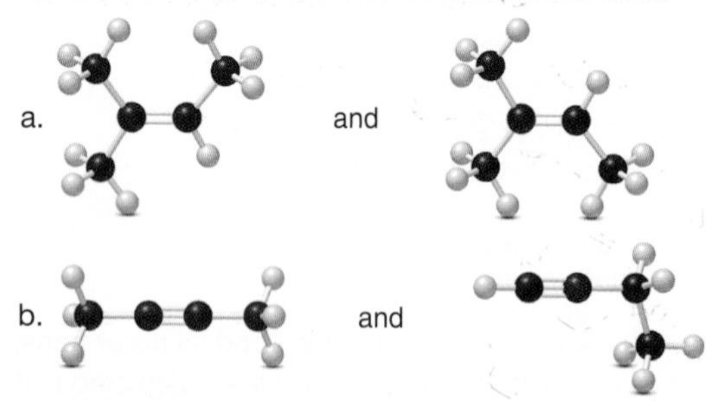

11.54 How are the compounds in each pair related? Choose from constitutional isomers, stereoisomers, or identical.

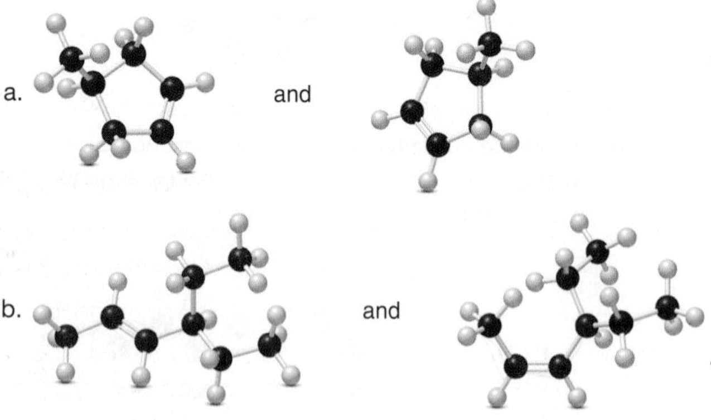

Reactions of Alkenes

11.55 What alkane is formed when each alkene is treated with H_2 in the presence of a Pd catalyst?

a. $CH_2{=}CHCH_2CH_2CH_2CH_3$

b. $(CH_3)_2C{=}CHCH_2CH_2CH_3$

c. [3-methylcyclohexene structure: CH_3 on ring]

d. [methylenecyclohexane structure: ring $=CH_2$]

11.56 What alcohol is formed when each alkene is treated with H_2O and H_2SO_4?

a. [cyclobutene structure]

b. $(CH_3)_2C{=}C(CH_3)_2$

c. $CH_2{=}CHCH_2CH(CH_3)_2$

d. [methylenecyclohexane structure: ring $=CH_2$]

11.57 Draw the product formed when 1-ethylcyclohexene is treated with each reagent: (a) H_2, Pd; (b) H_2O, H_2SO_4.

[structure: cyclohexene ring with CH_2CH_3]

1-ethylcyclohexene

11.58 Draw the products formed in each reaction.

a. [cyclopentene] + H_2 $\xrightarrow{Pd}$

b. [1,2-dimethylcyclohexene: ring with two CH_3] + H_2 $\xrightarrow{Pd}$

c. [cyclopentene ring with CH_3, CH_3, and CH_3 substituents] + H_2O $\xrightarrow{H_2SO_4}$

11.59 The hydration of 2-pentene ($CH_3CH{=}CHCH_2CH_3$) with H_2O and H_2SO_4 forms two alcohols. Draw the structure of both products and explain why more than one product is formed.

11.60 When myrcene is treated with three equivalents of H_2O in the presence of H_2SO_4, a single addition product of molecular formula $C_{10}H_{22}O_3$ is formed. Draw the structure of the product.

$$(CH_3)_2C{=}CHCH_2CH_2\underset{}{\overset{CH_2}{\overset{\|}{C}}}CH{=}CH_2$$

myrcene

11.61 What alkene is needed as a starting material to prepare each of the following alcohols?

a. CH_3CH_2OH

b. [cyclohexanol structure: ring with OH]

11.62 2-Butanol can be formed as the only product of the addition of H_2O to two different alkenes. In contrast, 2-pentanol can be formed as the *only* product of the addition of H_2O to just one alkene. Draw the structures of the alkene starting materials and explain the observed results.

$CH_3CH(OH)CH_2CH_3$ — 2-butanol

$CH_3CH(OH)CH_2CH_2CH_3$ — 2-pentanol

Polymers

11.63 Draw the structure of poly(acrylic acid), the polymer formed by polymerizing acrylic acid ($CH_2{=}CHCOOH$). Poly(acrylic acid) is used in disposable diapers because it can absorb 30 times its weight in water.

11.64 What polymer is formed when methyl α-methylacrylate [$CH_2{=}C(CH_3)CO_2CH_3$] is polymerized? This polymer is used in Lucite and Plexiglas, transparent materials that are lighter but more impact resistant than glass.

11.65 Draw the structure of each compound shown in the ball-and-stick models. Then draw the polymer that is formed when each compound is polymerized.

a. 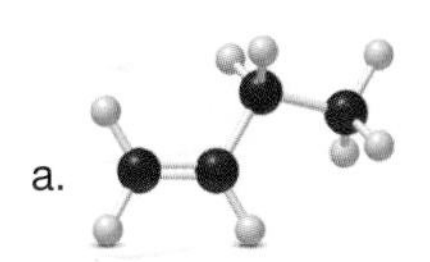

b. 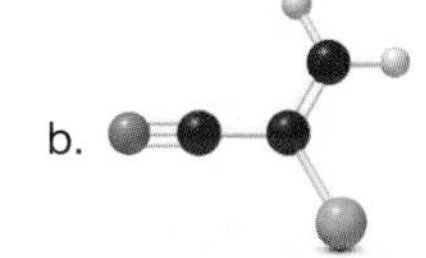

11.66 Draw the structure of each compound shown in the ball-and-stick models. Then draw the polymer that is formed when each compound is polymerized.

a. 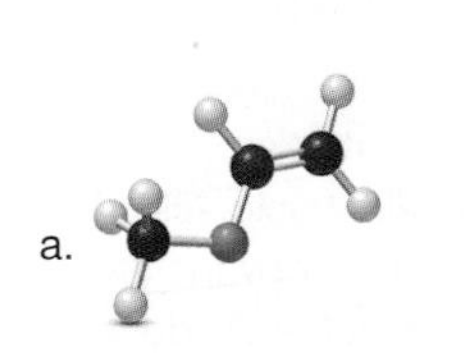

b. 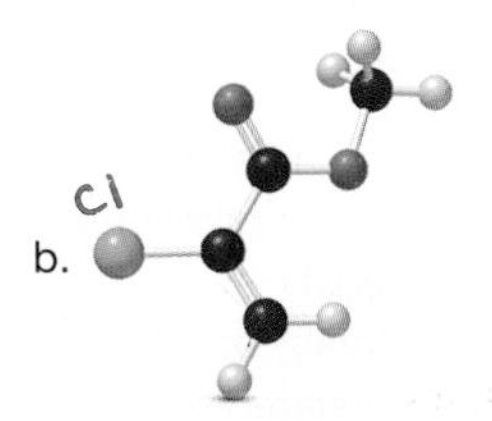

11.67 What monomer is used to form the following polymer?

$$\text{-}CH_2-C(Br)(Cl)-CH_2-C(Br)(Cl)-CH_2-C(Br)(Cl)\text{-}$$

11.68 What monomer is used to form the following polymer?

$$\text{-}CH_2-C(CH_3)(C{=}O\,OCH_2CH_3)-CH_2-C(CH_3)(C{=}O\,OCH_2CH_3)-CH_2-C(CH_3)(C{=}O\,OCH_2CH_3)\text{-}$$

Nomenclature of Benzene

11.69 Name each aromatic compound.

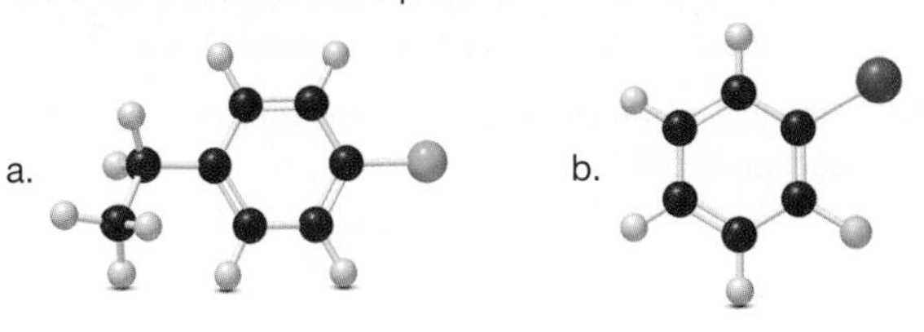

11.70 Name each aromatic compound.

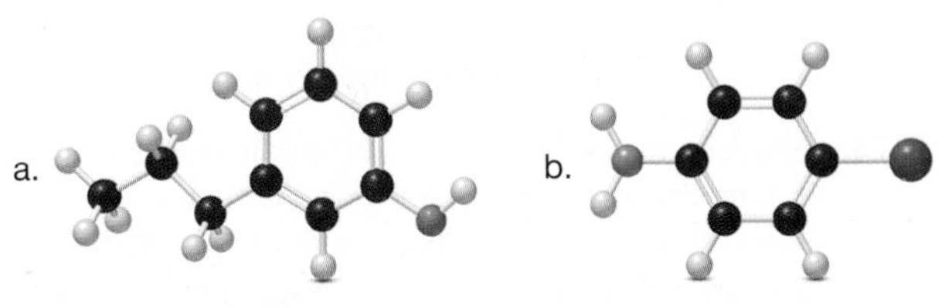

11.71 Give the IUPAC name for each substituted benzene.

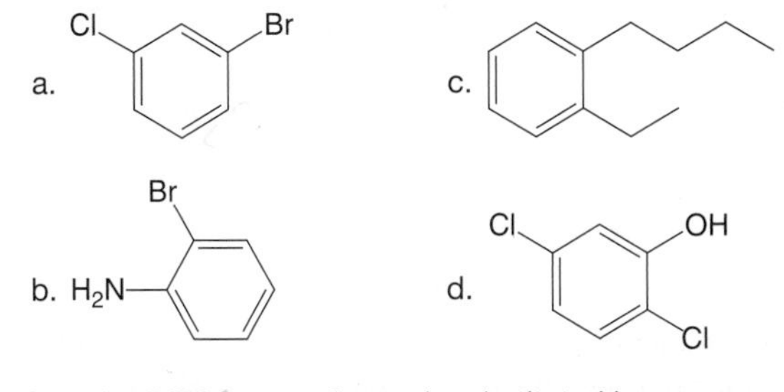

11.72 Give the IUPAC name for each substituted benzene.

a. $CH_3(CH_2)_3$–C_6H_4–$(CH_2)_3CH_3$ (para)

b. benzene ring with CH_2CH_3 and Br (ortho)

c. benzene ring with H_2N and CH_2CH_3 (meta)

d. benzene ring with Br, CH_3, and I

11.73 Give the structure corresponding to each IUPAC name.

a. *m*-dibutylbenzene
b. *o*-iodophenol
c. 2-bromo-4-chlorotoluene
d. 2-chloro-6-iodoaniline

11.74 Give the structure corresponding to each IUPAC name.

a. *o*-difluorobenzene
b. *p*-bromotoluene
c. 1,3,5-tributylbenzene
d. 2,4-dibromophenol

11.75 Each of the following IUPAC names is incorrect. Explain why it is incorrect and give the correct IUPAC name: (a) 5,6-dichlorophenol; (b) *m*-dibromoaniline.

11.76 Each of the following IUPAC names is incorrect. Explain why it is incorrect and give the correct IUPAC name: (a) 1,5-dichlorobenzene; (b) 1,3-dibromotoluene.

Applications

11.77 The breakfast cereal Cheerios lists vitamin E as one of its ingredients. What function does vitamin E serve?

11.78 Why is BHA an ingredient in some breakfast cereals and other packaged foods?

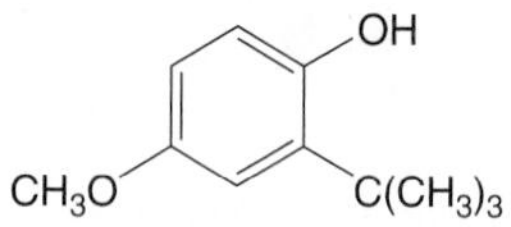

BHA
(**b**utylated **h**ydroxy **a**nisole)

11.79 Although nonpolar compounds tend to dissolve and remain in fatty tissues, polar substances are more water soluble, and more readily excreted into an environment where they may be degraded by other organisms. Explain why methoxychlor is more biodegradable than DDT (Problem 11.80).

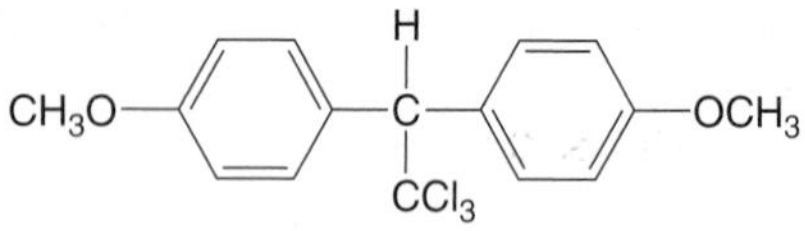

methoxychlor

11.80 Explain why the pesticide DDT is insoluble in water, but the herbicide 2,4-D is water soluble. 2,4-D is one component of the defoliant Agent Orange used extensively during the Vietnam War.

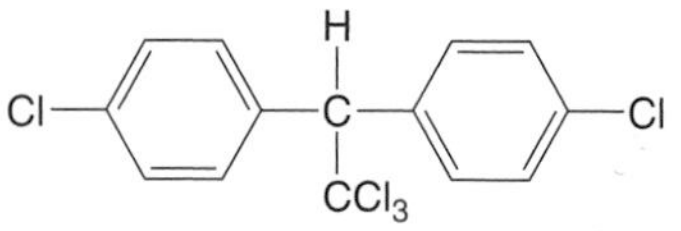

DDT 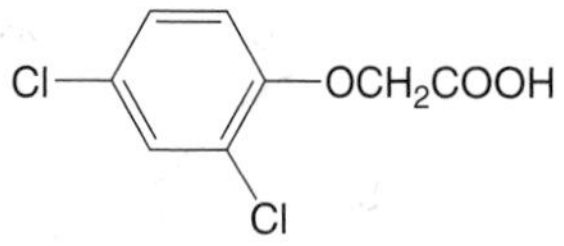 2,4-D

11.81 Kukui nuts contain oil that is high in linoleic acid content (Table 11.1). (a) What two constitutional isomers are formed when linoleic acid is partially hydrogenated with one equivalent of H_2? (b) What product is formed when linoleic acid is completely hydrogenated with H_2? (c) Draw a possible product that could be formed if, during hydrogenation, one equivalent of H_2 is added, *and* one of the cis double bonds is converted to a trans double bond.

11.82 Eleostearic acid is an unsaturated fatty acid found in tung oil, obtained from the seeds of the tung oil tree (*Aleurites fordii*), a deciduous tree native to China. Eleostearic acid is unusual in that the double bond at C9 is cis, but the other two double bonds are trans. (a) Draw the structure of eleostearic acid, showing the arrangement of groups around each double bond. (b) Draw a stereoisomer of eleostearic acid in which all of the double bonds are trans. (c) Which compound, eleostearic acid or its all-trans isomer, has the higher melting point? Explain your reasoning.

$CH_3(CH_2)_3CH{=}CHCH{=}CHCH{=}CH(CH_2)_7CO_2H$ (C9)

eleostearic acid

11.83 Which of the following compounds might be an antioxidant?

a. 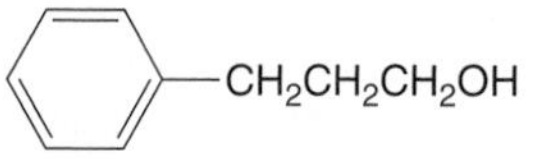 $C_6H_5CH_2CH_2CH_2OH$
b. 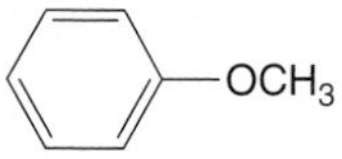 $C_6H_5OCH_3$
c. $HO{-}C_6H_4{-}OCH_3$

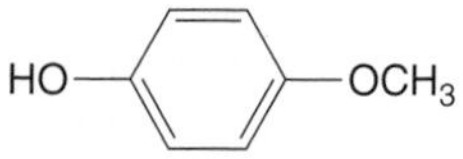

11.84 Which of the following compounds might be an ingredient in a commercial sunscreen?

a. 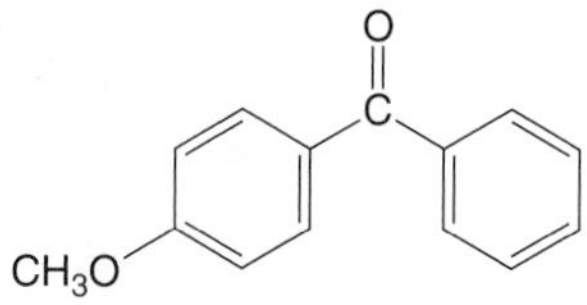 b.

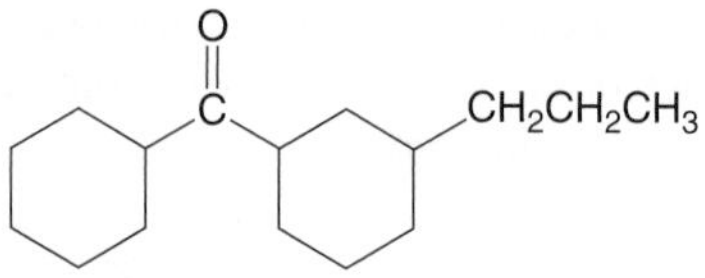

11.85 Macadamia nuts have a high concentration of unsaturated oils formed from palmitoleic acid [$CH_3(CH_2)_5CH{=}CH(CH_2)_7COOH$]. (a) Draw the structure of the naturally occurring fatty acid with a cis double bond. (b) Draw a stereoisomer of palmitoleic acid. (c) Draw a constitutional isomer of palmitoleic acid.

11.86 What products are formed when the hydrocarbon polyethylene is completely combusted?

CHALLENGE PROBLEMS

11.87 Are *cis*-2-hexene and *trans*-3-hexene constitutional isomers or stereoisomers? Explain.

11.88 Some polymers are copolymers, formed from two different alkene monomers joined together. An example is Saran, the polymer used in the well known plastic food wrap. What two alkene monomers combine to form Saran?

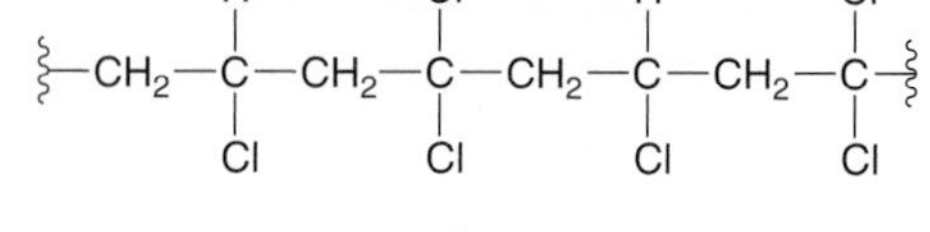

Saran

BEYOND THE CLASSROOM

11.89 Research the subject of using paper or plastic bags for supermarket purchases. What are the advantages and disadvantages of each? Consider the volume that the bags occupy and the raw materials used to make the bags. Also consider whether the bags decompose in a landfill, or whether they can be recycled. Draw the chemical structures for the principal organic compound in each bag type. In your opinion, which has the lesser negative environmental impact, using paper or plastic bags?

11.90 Pick a product that is composed of a polymer. Possibilities might include compact discs, nonstick pans, synthetic carpets, garden hoses, cling wrap, hot beverage cups, or water bottles. Draw the chemical structure of the polymer, as well as the monomer(s) from which the polymer is made. Suggest reasons why the polymer is used in that particular product. Does the polymer degrade easily in a landfill or can it be recycled?

11.91 Pick a spice used routinely in your household. Possibilities might include vanilla, basil, hot pepper, thyme, mint, caraway, or cloves. Determine the major organic compound in the spice, give its name, and draw its chemical structure. Identify all the functional groups. Has the spice been used in any therapeutic or medicinal preparations in any societies in the world, and if it has, what are its purported health benefits? Have there been any experimental studies to support these claims?

11.92 Bisphenol A (BPA) is an aromatic compound that has been used in resins found in baby bottles and children's cups. What is the structure of BPA and why have concerns been raised about its presence in baby products? What other products contain resins made with BPA? What is the current status of FDA regulations on the use of BPA in consumer products? What alternatives are now used in place of BPA in baby products? Be as specific as possible, and include chemical structures when they are known.

ANSWERS TO SELECTED PROBLEMS

11.1 a. H—C(H)=C(H)—C(H)(H)—Ö—H

b. H—C(H)(H)—C(CH₃)=C(H)—C(H)(H)—C(H)(H)—C(H)(H)—H

c. H—C(H)(H)—C(H)(CH₃)—C≡C—C(H)(H)—C(H)(CH₃)—C(H)(H)—H

11.3 a. C_4H_8 b. C_6H_{14} c. C_7H_{12} d. C_5H_{10}

11.4 a. 3-methyl-1-pentene
b. 3-ethyl-3-heptene

11.5 a. 2-methyl-3-nonyne b. 6,6-dimethyl-3-octyne

11.7 a. CH_3(H)C=C(H)$CH_2CH_2CH_2CH_2CH_3$ (cis)

b. CH_3CH_2(H)C=C(H)$CH_2CH_2CH_3$ (trans)

c. CH_3(H)C=C(H)$CHCH_3$ with CH_3 on CH (cis)

11.9 a. constitutional b. stereoisomers

11.11 palmitic acid, mp = 63 °C
palmitoleic acid, mp = 1 °C

11.12 a. $CH_3CH_2CH_2CH_2CH_2CH_3$

b. $CH_3CH(CH_3)CH_2CH_2CH(CH_3)_2$

c. methylcyclohexane (cyclohexane ring with CH_3)

11.13 a. $CH_3CH(OH)$—CH_2CH_3 (H H above; HO H below)

b. 1-methylcyclohexanol (cyclohexane ring with OH and CH_3 on the same carbon)

c. CH_3—$C(CH_3)(OH)$—CH_2CH_3 (CH₃ H above; HO H below)

11.15 a. H, H (C=C) H; $CH_3(CH_2)_5$–C=C(H)–CH=CH–$(CH_2)_7COOH$
trans, cis

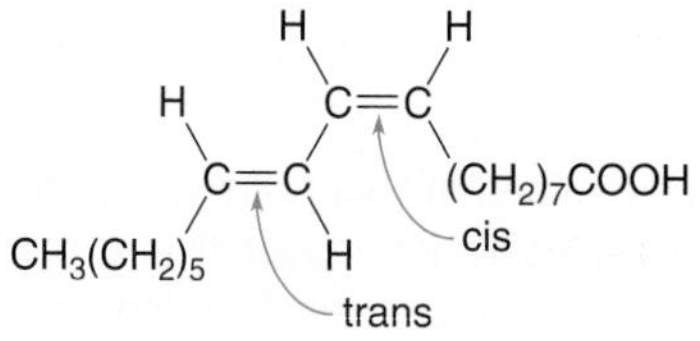

b. $CH_3(CH_2)_5$, H (C=C) H; H, H (C=C) $(CH_2)_7COOH$
cis, cis

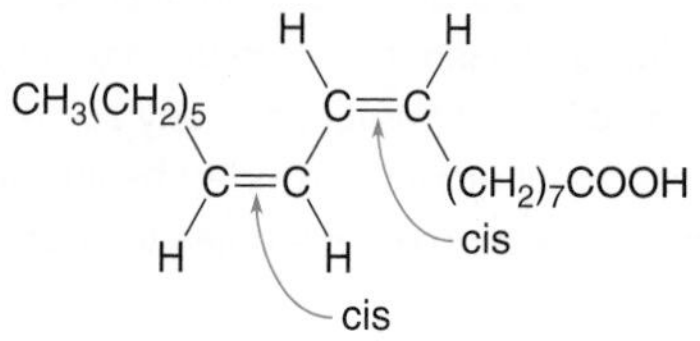

c. lower mp
d. constitutional isomer

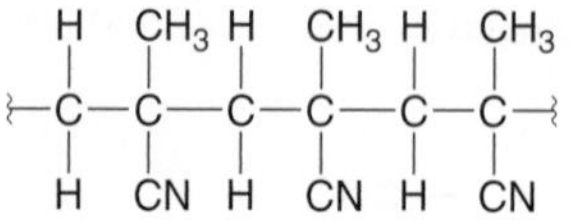

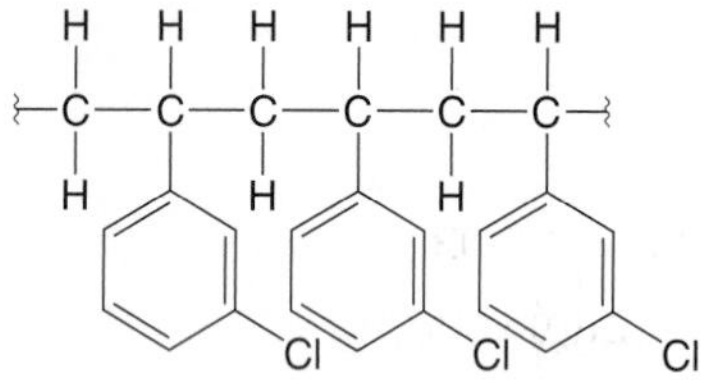

11.16 a. –[CH₂–C(CH₃)(CH₂CH₃)]₃–

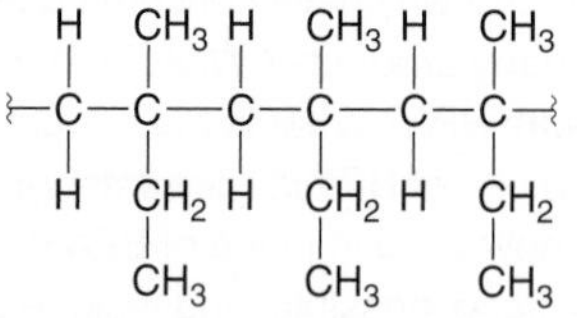

H	CH_3	H	CH_3	H	CH_3
–C–	–C–	–C–	–C–	–C–	–C–
H	CH_2	H	CH_2	H	CH_2
	CH_3		CH_3		CH_3

b.

H	CH_3	H	CH_3	H	CH_3
–C–	–C–	–C–	–C–	–C–	–C–
H	CN	H	CN	H	CN

c.

H	H	H	H	H	H
–C–	–C–	–C–	–C–	–C–	–C–
H	m-chlorophenyl (Cl)	H	m-chlorophenyl (Cl)	H	m-chlorophenyl (Cl)

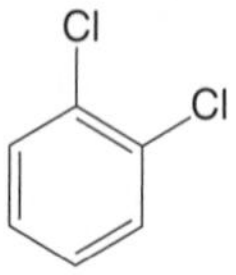

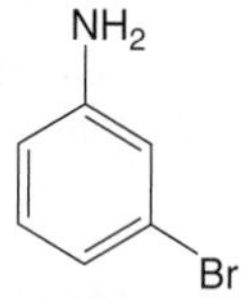

11.17 $CH_2{=}C(H)–O–COCH_3$

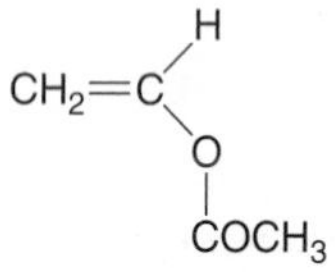

11.18 a. propylbenzene
b. *p*-ethyliodobenzene
c. *m*-butylphenol
d. 2-bromo-5-chlorotoluene

11.19 a. $C_6H_5CH_2CH_2CH_2CH_2CH_3$
b. Cl, Cl (1,2-dichlorobenzene)
c. NH_2, Br (3-bromoaniline)
d. CH_2CH_3, CH_2CH_3, Cl

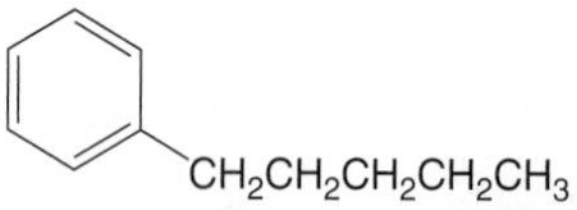

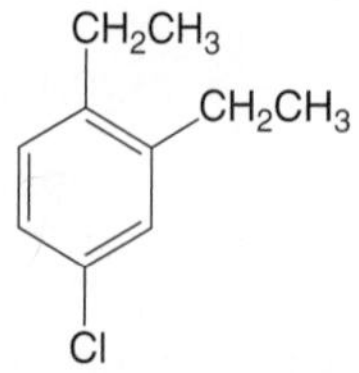

11.21 Compound (a) might be found in a sunscreen since it contains two aromatic rings.

11.23 a. $C_{10}H_{12}O$
b. aromatic ring, alkene, ether
c. trans
d. tetrahedral CH_3–C(H)=C(H)–C_6H_4–OCH_3 tetrahedral
All other C's are trigonal planar.

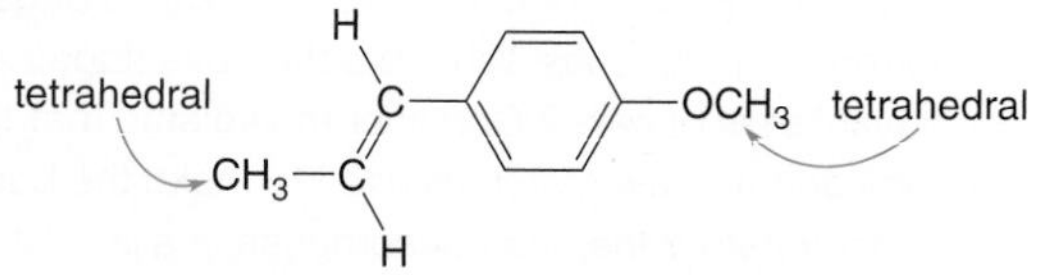

11.25 a. 5,5-dimethyl-1-hexene b. 5,5-dimethyl-2-hexyne

11.27 a. $CH_2{=}CHCH_2CH_2CH_2CH_2CH_3$
1-heptene
b. $CH_3(CH_2)_5CH_3$
c. $CH_3CH(OH)CH_2CH_2CH_2CH_2CH_3$

11.29 –CH_2CCl_2–CH_2CCl_2–CH_2CCl_2–

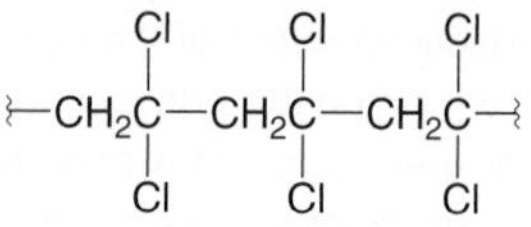

11.31 NH_2, Cl; NH_2, Cl; NH_2, Cl
o-chloroaniline *m*-chloroaniline *p*-chloroaniline

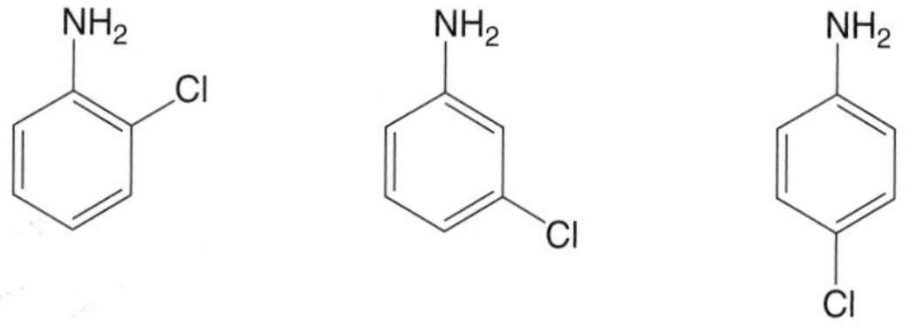

11.33 a. $C_{10}H_{22}$
b. $C_{10}H_{20}$
c. $C_{10}H_{18}$

11.35 $HC{\equiv}CCH_2CH_2CH_3$
$CH_3C{\equiv}CCH_2CH_3$
$HC{\equiv}C–C(CH_3)_2–H$

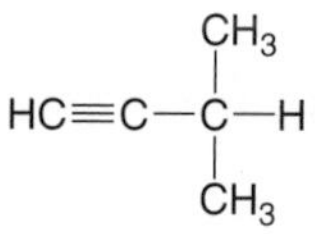

11.37 a. trigonal planar tetrahedral
trigonal planar

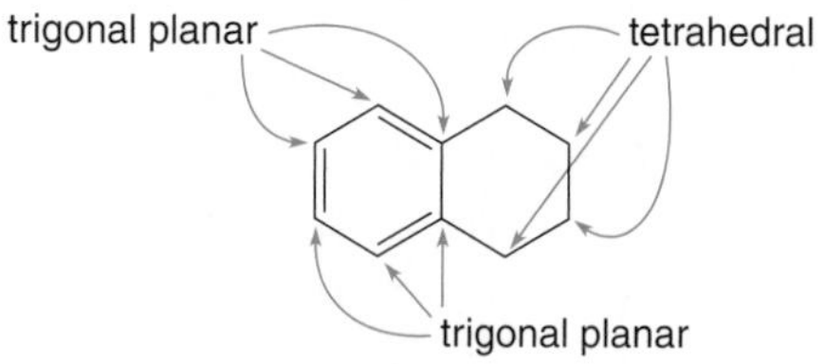

b. linear
C_6H_5–C≡CH
all trigonal planar

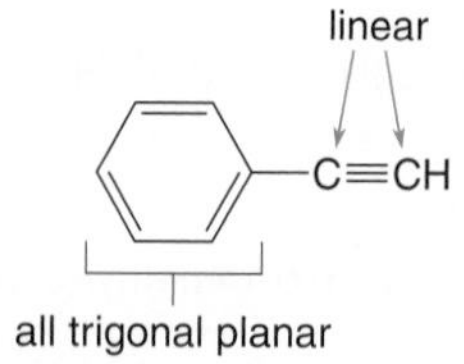

11.39 a. 2-ethyl-1-butene
b. 2-hexyne

11.41 a. 3-ethyl-5,7-dimethyl-3-octene
b. 2-ethyl-1-heptene
c. 6-ethyl-5,6-dimethyl-2-octyne

11.43 a. CH_3 on C3: $CH_2{=}CHCHCH_2CH_2CH_2CH_2CH_3$

b. $CH_3CH_2C{\equiv}CCCH_3$ (with CH_3 and H on the fifth C)

c. $CH_3CH_2CHC{=}CCHCH_3$ (with CH_2CH_3 and H on C3, CH_3 and CH_2CH_3 on C4)

d. CH_3 and H on one C, $CH_2CH_2CH_2CHCH_3$ (with CH_3 on CH) and H on the other: C=C

11.45 (structure with cis and trans double bonds labeled)

$CH_3CH_2CH_2CH_2CH_2$–CH=CH(cis)–CH_2–CH=CH(cis)–CH=CH(trans)–CH=CH(trans)–$CHCHCH_2CH_2CH_2COOH$, with OH on the first CH and S–CH_2 on the other CH; S–CH_2–$CHCONHCH_2COOH$; $NHCOCH_2CH_2CHCOOH$ with NH_2

11.47 *cis*-4-methyl-2-pentene

11.49 a. H and CH_3 on one C, H and $CH_2CH_2CH_2CH_2CH_2CH_3$ on the other (C=C)
cis-2-nonene

H and CH_3 on one C, $CH_2CH_2CH_2CH_2CH_2CH_3$ and H on the other (C=C)
trans-2-nonene

b. H and CH_3CH (with CH_3) on one C, H and $CH_2CH_2CH_3$ on the other (C=C)
cis-2-methyl-3-heptene

H and CH_3CH (with CH_3) on one C, $CH_2CH_2CH_3$ and H on the other (C=C)
trans-2-methyl-3-heptene

11.51 a. constitutional isomers
b. identical
c. stereoisomers

11.53 a. identical
b. constitutional isomers

11.55 a. $CH_3CH_2CH_2CH_2CH_2CH_3$
b. $(CH_3)_2CHCH_2CH_2CH_2CH_3$
c. CH_3–cyclohexane
d. cyclohexane–CH_3

11.57 a. cyclohexane–CH_2CH_3
b. cyclohexane with OH and CH_2CH_3 on the same C

11.59 CH_3–$C(H)(H)$–$C(OH)(H)$–CH_2CH_3 and CH_3–$C(OH)(H)$–$C(H)(H)$–CH_2CH_3

OH adds to either side of the C=C since both carbons have an equal number of H's.

11.61 a. $CH_2{=}CH_2$
b. cyclohexene

11.63 –$CH_2C(COOH)(H)$–$CH_2C(COOH)(H)$–$CH_2C(COOH)(H)$–

11.65 a. $CH_2{=}CHCH_2CH_3$ –$CH_2C(CH_2CH_3)(H)$–$CH_2C(CH_2CH_3)(H)$–$CH_2C(CH_2CH_3)(H)$–
b. $CH_2{=}C(Cl)(C{\equiv}N)$ –$CH_2C(Cl)(CN)$–$CH_2C(Cl)(CN)$–$CH_2C(Cl)(CN)$–

11.67 $CH_2{=}C(Cl)(Br)$

11.69 a. *p*-chloroethylbenzene
b. *o*-bromofluorobenzene

11.71 a. *m*-bromochlorobenzene
b. *o*-bromoaniline
c. *o*-butylethylbenzene
d. 2,5-dichlorophenol

11.73 a. 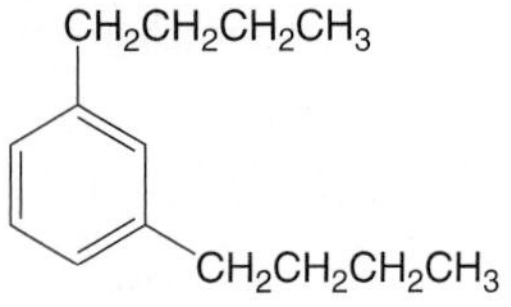 Benzene ring with $CH_2CH_2CH_2CH_3$ and $CH_2CH_2CH_2CH_3$ (meta)

c. 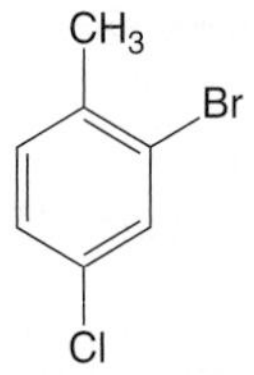 Benzene ring with CH_3, Br, Cl

b. 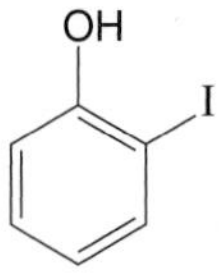 Benzene ring with OH, I

d. 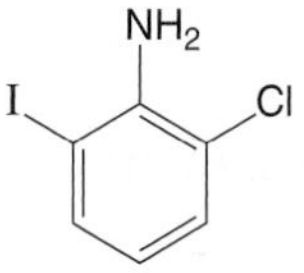 Benzene ring with NH_2, I, Cl

11.75 a. Assign the lowest numbers: 2,3-dichlorophenol.
b. You can only use ortho, meta, para for a disubstituted benzene: 3,5-dibromoaniline.

11.77 Vitamin E is an antioxidant.

11.79 Methoxychlor is more water soluble. The OCH_3 groups can hydrogen bond to water. This increase in water solubility makes methoxychlor more biodegradable.

11.81 a.

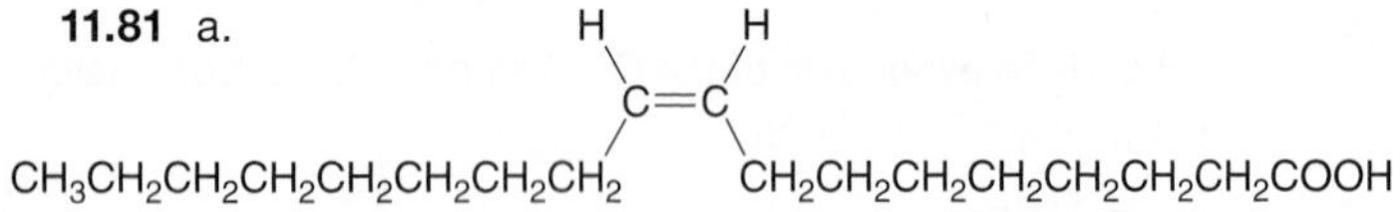

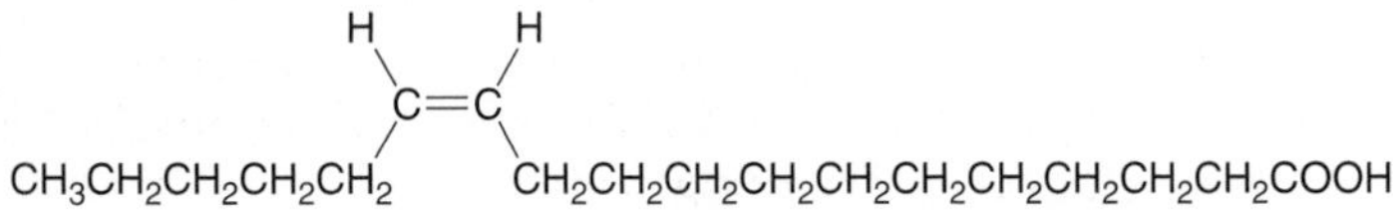

H H
C=C
$CH_3CH_2CH_2CH_2CH_2CH_2CH_2CH_2$ $CH_2CH_2CH_2CH_2CH_2CH_2CH_2COOH$

H H
C=C
$CH_3CH_2CH_2CH_2CH_2$ $CH_2CH_2CH_2CH_2CH_2CH_2CH_2CH_2CH_2CH_2COOH$

b.
$CH_3CH_2CH_2CH_2CH_2CH_2CH_2CH_2CH_2CH_2CH_2CH_2CH_2CH_2CH_2CH_2CH_2COOH$

c.
one possibility: H $CH_2CH_2CH_2CH_2CH_2CH_2CH_2COOH$
C=C
$CH_3CH_2CH_2CH_2CH_2CH_2CH_2CH_2$ H

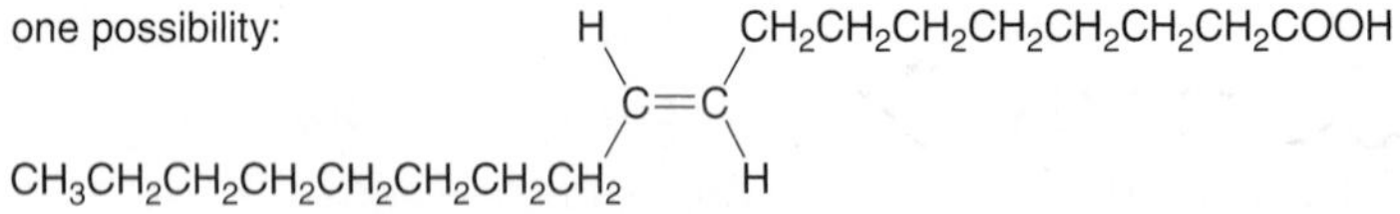

11.83 (c)

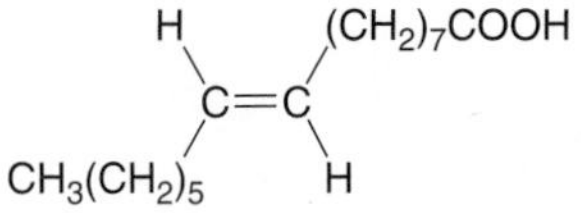

11.85 a. 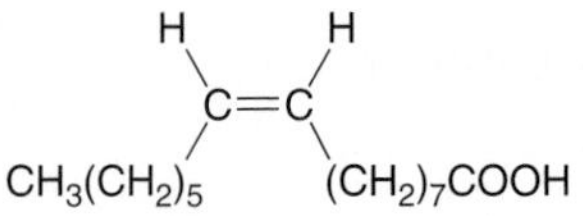
H H
C=C
$CH_3(CH_2)_5$ $(CH_2)_7COOH$

b.
H $(CH_2)_7COOH$
C=C
$CH_3(CH_2)_5$ H

c.
H H
C=C
$CH_3(CH_2)_4$ $(CH_2)_8COOH$

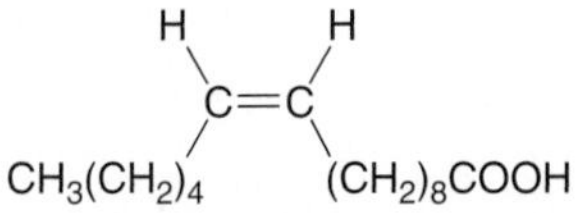

11.87 They are constitutional isomers because the double bond is located in a different place on the carbon chain.

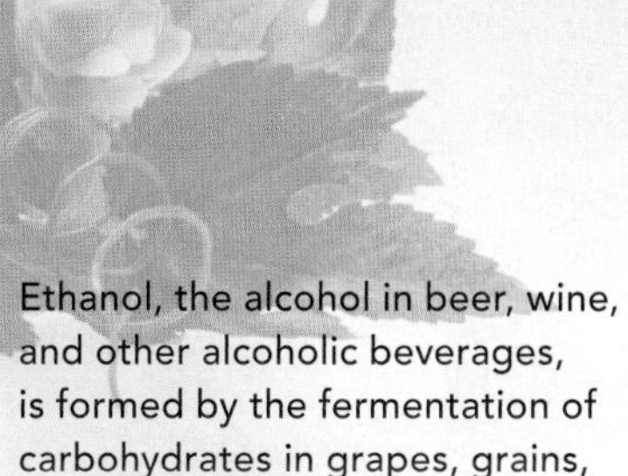

12

Ethanol, the alcohol in beer, wine, and other alcoholic beverages, is formed by the fermentation of carbohydrates in grapes, grains, and potatoes.

Organic Compounds That Contain Oxygen, Halogen, or Sulfur

CHAPTER OUTLINE

CHAPTER GOALS

In this chapter you will learn how to:

1. Identify alcohols, ethers, alkyl halides, thiols, aldehydes, and ketones
2. Determine the properties of each functional group
3. Name alcohols, ethers, alkyl halides, aldehydes, and ketones
4. Determine the products of alcohol dehydration and oxidation
5. Convert thiols to disulfides
6. Draw the products of oxidation reactions of aldehydes
7. Identify chirality centers and recognize when a molecule is chiral or achiral
8. Draw Fischer projection formulas

Chapter 12 concentrates on six families of compounds. **Alcohols** (ROH), **ethers** (ROR), **alkyl halides** (RX, X = F, Cl, Br, or I), and **thiols** (RSH) contain a carbon atom singly bonded to a heteroatom, while **aldehydes** (RCHO) and **ketones** (RCOR) contain a carbonyl group (C=O). Alcohols such as ethanol (CH_3CH_2OH) are widely occurring, and ethers are the most common anesthetics in use today. Alkyl halides are often used as solvents, and the —SH group of thiols plays an important role in protein chemistry. Many naturally occurring compounds are aldehydes and ketones, and these carbonyl-containing molecules serve as useful starting materials and solvents in industrial processes. In Chapter 12, we learn about the properties of these families of compounds.

12.1 Introduction

Chapter 12 focuses on families of organic compounds that contain the heteroatoms oxygen, halogen, or sulfur—alcohols, ethers, alkyl halides, thiols, aldehydes, and ketones.

Alcohols (ROH), **ethers** (ROR), **alkyl halides** (RX), and **thiols** (RSH) are four families of compounds that contain a carbon atom singly bonded to an oxygen, halogen, or sulfur.

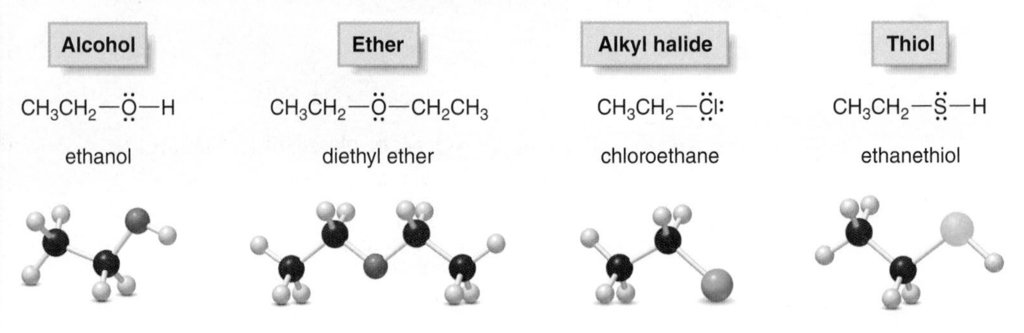

The characteristic odor of coffee is due to 2-mercaptomethylfuran, a compound that contains both an ether and a thiol (labeled in red).

CH_2SH

2-mercaptomethylfuran

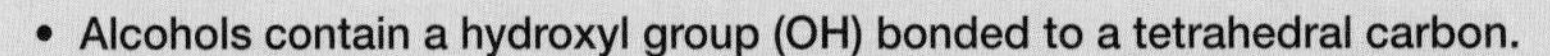

- Alcohols contain a hydroxyl group (OH) bonded to a tetrahedral carbon.
- Ethers contain two alkyl groups bonded to an oxygen atom.
- Alkyl halides contain a halogen atom (X = F, Cl, Br, or I) bonded to a tetrahedral carbon.
- Thiols contain a sulfhydryl group (SH) bonded to a tetrahedral carbon.

The oxygen atom in alcohols and ethers and the sulfur atom in thiols have two lone pairs of electrons, so each heteroatom is surrounded by eight electrons. The halogen atom in alkyl halides has three lone pairs of electrons.

Aldehydes (RCHO) and **ketones** (RCOR) contain a **carbonyl group (C=O)** with the carbonyl carbon bonded to carbon or hydrogen atoms.

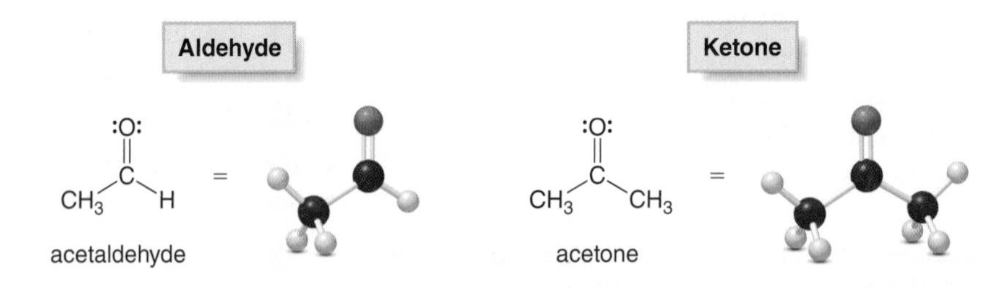

Octanal [$CH_3(CH_2)_6CHO$] and decanal [$CH_3(CH_2)_8CHO$] are two aldehydes that contribute to the flavor and odor of an orange.

- An aldehyde has at least one H atom bonded to the carbonyl carbon.
- A ketone has two alkyl groups bonded to the carbonyl carbon.

The double bond of a carbonyl group is usually omitted in shorthand structures. Acetaldehyde, for example, is written as CH_3CHO. Remember that the **H atom is bonded to the carbon atom,** not the oxygen. Likewise, acetone is written as CH_3COCH_3 or $(CH_3)_2CO$. Remember that **each compound contains a C=O.**

PROBLEM 12.1

a. Label the hydroxyl groups, thiols, halogens, and ether oxygens in each compound.
b. Which –OH group in salmeterol (**C**) is *not* part of an alcohol? Explain.

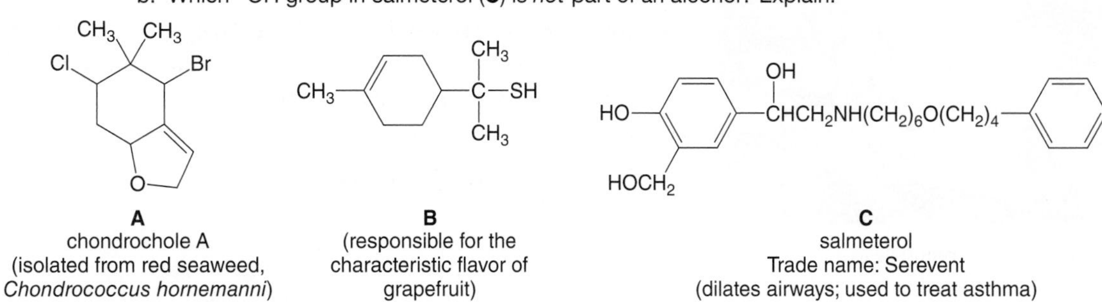

A
chondrochole A
(isolated from red seaweed, *Chondrococcus hornemanni*)

B
(responsible for the characteristic flavor of grapefruit)

C
salmeterol
Trade name: Serevent
(dilates airways; used to treat asthma)

PROBLEM 12.2

Draw out each compound to clearly show what groups are bonded to the carbonyl carbon. Label each compound as a ketone or aldehyde.

a. CH_3CH_2CHO b. $CH_3CH_2COCH_3$ c. $(CH_3)_3CCOCH_3$ d. $(CH_3CH_2)_2CHCHO$

12.2 Structure and Properties of Alcohols

Alcohols (ROH) are classified as **primary (1°), secondary (2°),** or **tertiary (3°)** based on the number of carbon atoms bonded to the carbon with the OH group.

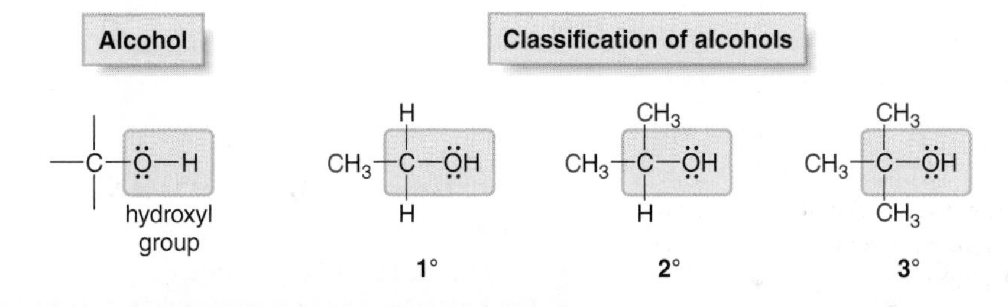

- **A primary (1°) alcohol has an OH group on a carbon bonded to one carbon.**
- **A secondary (2°) alcohol has an OH group on a carbon bonded to two carbons.**
- **A tertiary (3°) alcohol has an OH group on a carbon bonded to three carbons.**

SAMPLE PROBLEM 12.1

Classify each alcohol as 1°, 2°, or 3°.

a. (cyclohexane)–CH_2CH_2OH b. (cyclohexane)–OH

Analysis

To determine whether an alcohol is 1°, 2°, or 3°, locate the C with the OH group and count the number of C's bonded to it. A 1° alcohol has the OH group on a C bonded to one C, and so forth.

Solution

Draw out the structure or add H's to the skeletal structure to clearly see how many C's are bonded to the C bearing the OH group.

a. (cyclohexane)–C(H)(H)–C(H)(H)–OH
This C is bonded to 1 C.
1° alcohol

b. (cyclohexane with C bearing OH and H)
This C is bonded to 2 C's in the ring.
2° alcohol

CONSUMER NOTE

Sorbitol occurs naturally in some berries and fruits. It is used as a substitute sweetener in sugar-free—that is, sucrose-free—candy and gum.

PROBLEM 12.3

Classify each alcohol as 1°, 2°, or 3°.

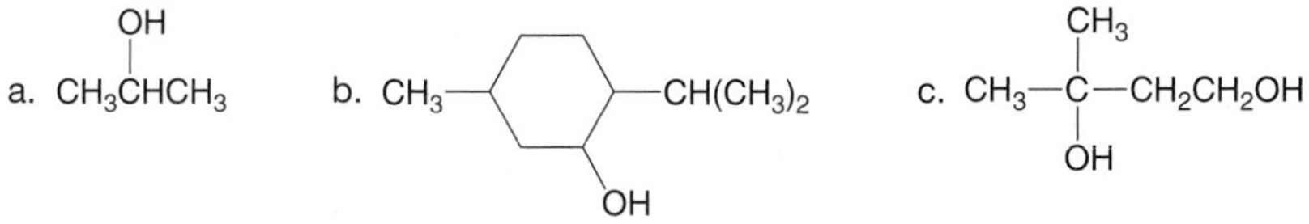

a. $CH_3CH(OH)CH_3$

b. CH_3–(cyclohexane ring)–$CH(CH_3)_2$, with OH on the ring

c. CH_3–$C(CH_3)(OH)$–CH_2CH_2OH

PROBLEM 12.4

Classify each hydroxyl group in sorbitol as 1°, 2°, or 3°.

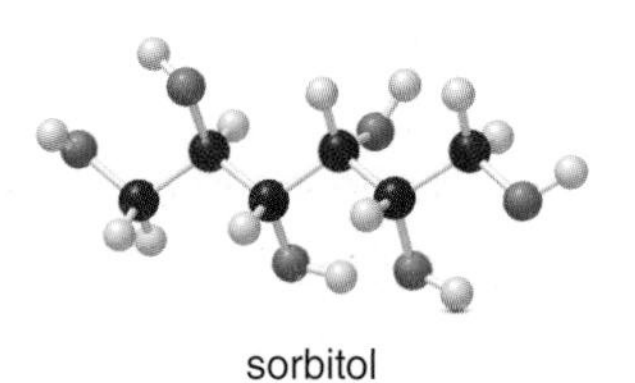

sorbitol

12.2A Physical Properties of Alcohols

An alcohol contains an oxygen atom with a **bent** shape like H_2O. The C—O—H bond angle is similar to the tetrahedral bond angle of 109.5°. **Alcohols are capable of intermolecular hydrogen bonding,** since they possess a hydrogen atom bonded to an oxygen. This gives alcohols much stronger intermolecular forces than the hydrocarbons of Chapters 10 and 11.

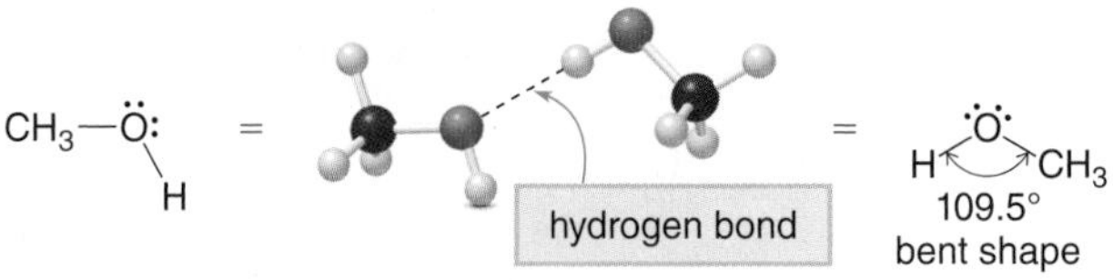

CH_3—Ö:—H = [ball-and-stick model] = H–Ö–CH_3

hydrogen bond

109.5°
bent shape

As a result:

- **Alcohols have higher boiling points and melting points than hydrocarbons of comparable molecular weight and shape.**

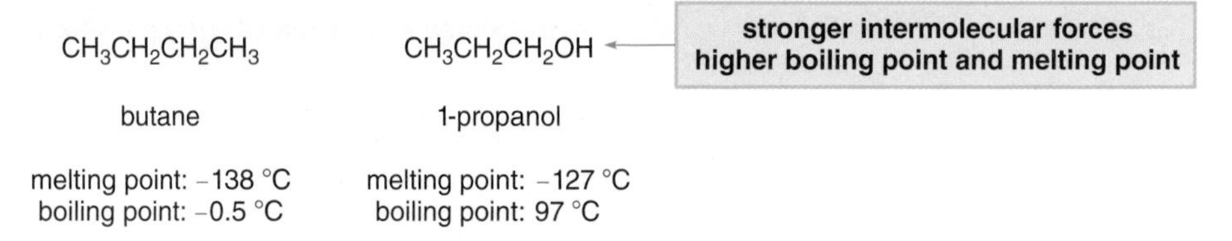

The general rule governing solubility—**"like dissolves like"**—explains the solubility properties of alcohols.

- **Alcohols are soluble in organic solvents.**
- **Low molecular weight alcohols (those having less than six carbons) are soluble in water.**
- **Higher molecular weight alcohols (those having six carbons or more) are not soluble in water.**

Thus, both ethanol (CH_3CH_2OH) and 1-octanol [$CH_3(CH_2)_7OH$] are soluble in organic solvents, but ethanol is water soluble and 1-octanol is not.

CH_3CH_2OH $CH_3CH_2CH_2CH_2CH_2CH_2CH_2CH_2OH$

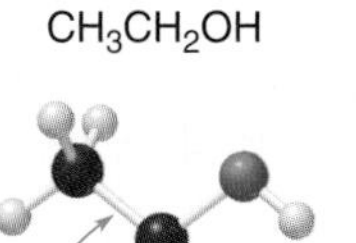
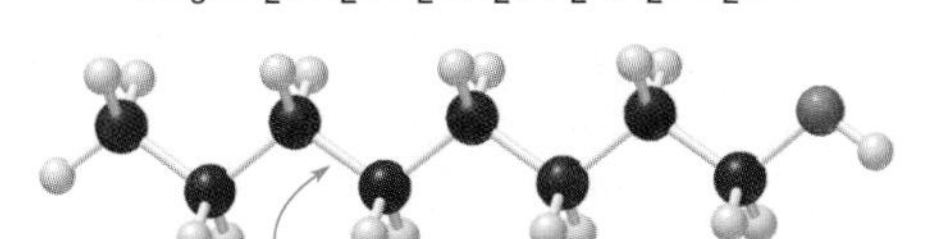

only 2 C's in the chain
water soluble

many nonpolar C–C and C–H bonds
water insoluble

PROBLEM 12.5

Which compound in each pair has the higher boiling point?

a. cyclohexanol (OH) or methylcyclohexane (CH_3)   b. $(CH_3)_3C—OH$ or $(CH_3)_4C$

PROBLEM 12.6

Label each compound as water soluble or water insoluble.

a. decalin   b. $(CH_3)_3C—OH$ c. $CH_2{=}CHCH_2CH_3$ d. decalin with OH

12.2B Nomenclature of Alcohols

In the IUPAC system, alcohols are identified by the suffix ***-ol.*** To name an alcohol:

- **Find the longest carbon chain containing the carbon bonded to the OH group.**
- **Number the carbon chain to give the OH group the lower number, and apply all other rules of nomenclature.**

When an OH group is bonded to a ring, the **ring is numbered beginning with the OH group,** and the "1" is usually omitted from the name. The ring is then numbered in a clockwise or counterclockwise fashion to give the next substituent the lower number.

CH_3 (3), OH (1)
3-methylcyclohexanol

[The OH group is at C1; the second substituent (CH_3) gets the lower number.]

OH (1), CH_2CH_3 (2)
2-ethylcyclopentanol

[The OH group is at C1; the second substituent (CH_3CH_2) gets the lower number.]

SAMPLE PROBLEM 12.2

Give the IUPAC name of the following alcohol.

$$CH_3CH_2C(CH_3)_2CH_2CH_2OH$$

Analysis and Solution

[1] Find the longest carbon chain that contains the carbon bonded to the OH group.

$CH_3CH_2C(CH_3)_2CH_2CH_2$—OH

5 C's in the longest chain ----> **pentanol**

- Change the **-*e*** ending of the parent alkane to the suffix ***-ol.***

[2] Number the carbon chain to give the OH group the lower number, and apply all other rules of nomenclature.

a. **Number** the chain.

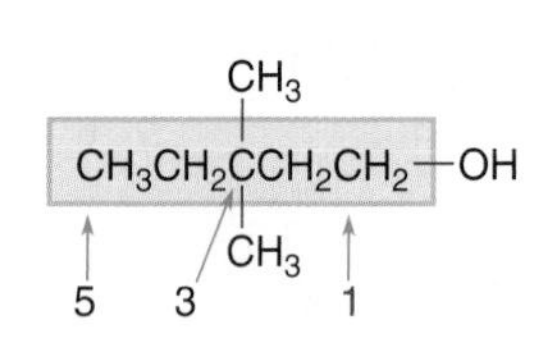

- Number the chain to put the OH group at C1, not C5.

1-pentanol

b. **Name** and **number** the substituents.

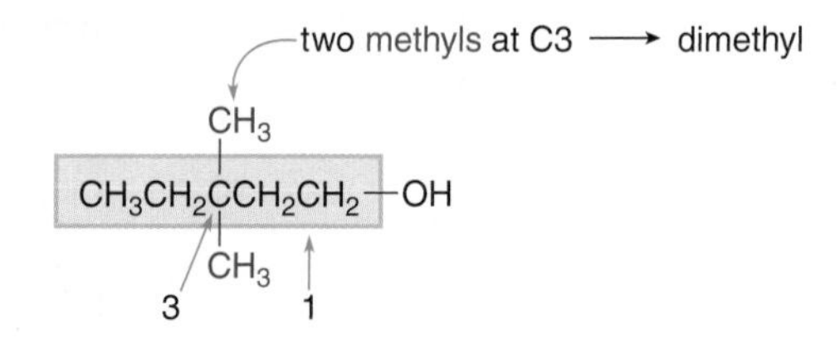

Answer: 3,3-dimethyl-1-pentanol

PROBLEM 12.7

Give the IUPAC name for each compound.

a. $CH_3CH(OH)(CH_2)_4CH_3$

b. $(CH_3CH_2)_2CHCH(OH)CH_2CH_3$

c. cyclohexane ring with CH_3 and OH on adjacent carbons

PROBLEM 12.8

Give the structure corresponding to each name.

a. 7,7-dimethyl-4-octanol

b. 5-methyl-4-propyl-3-heptanol

c. 2-ethyl-3-methylcyclohexanol

12.3 Structure and Properties of Ethers

***Ethers* (ROR) are organic compounds that have two alkyl groups bonded to an oxygen atom.** The oxygen atom of an ether is surrounded by two carbon atoms and two lone pairs of electrons, giving it a **bent** shape like the oxygen in H_2O. The C—O—C bond angle is similar to the tetrahedral bond angle of 109.5°.

CH_3—Ö—CH_3 = [ball-and-stick model, 109.5°]

dimethyl ether   bent shape

Simple ethers are usually assigned common names. To do so, **name both alkyl groups** bonded to the oxygen, arrange these names alphabetically, and add the word **ether.** For ethers with identical alkyl groups, name the alkyl group and add the prefix **di-.**

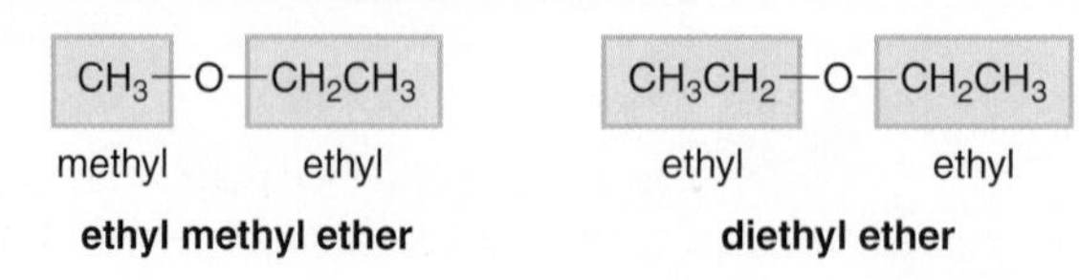

ethyl methyl ether   **diethyl ether**

PROBLEM 12.9

Name each ether.

a. $CH_3—O—CH_2CH_2CH_2CH_3$ b. $CH_3CH_2CH_2—O—CH_2CH_2CH_3$

PROBLEM 12.10

Draw the structure of the three constitutional isomers of molecular formula $C_4H_{10}O$ that contain an ether.

Because oxygen is more electronegative than carbon, the C—O bonds of an ether are both polar. Since an ether contains two polar bonds and a bent shape, it has a **net dipole.**

CH_3–Ö–CH_3 (net dipole; two polar bonds)

Ethers do not contain a hydrogen atom bonded to oxygen, so unlike alcohols, two ether molecules *cannot* intermolecularly hydrogen bond to each other. **This gives ethers stronger intermolecular forces than alkanes but weaker intermolecular forces than alcohols.** As a result:

- **Ethers have higher melting points and boiling points than hydrocarbons of comparable molecular weight and shape.**
- **Ethers have lower melting points and boiling points than alcohols of comparable molecular weight and shape.**

$CH_3CH_2CH_2CH_3$	$CH_3OCH_2CH_3$	$CH_3CH_2CH_2OH$
butane	ethyl methyl ether	1-propanol
boiling point −0.5 °C	boiling point 11 °C	boiling point 97 °C

Increasing boiling point →

All ethers are soluble in organic solvents. Like alcohols, **low molecular weight ethers are water soluble,** because the oxygen atom of the ether can hydrogen bond to one of the hydrogens of water. When the alkyl groups of the ether have more than a total of five carbons, the nonpolar portion of the molecule is too large, so the ether is water insoluble.

CH_3—Ö—CH_3 =

dimethyl ether

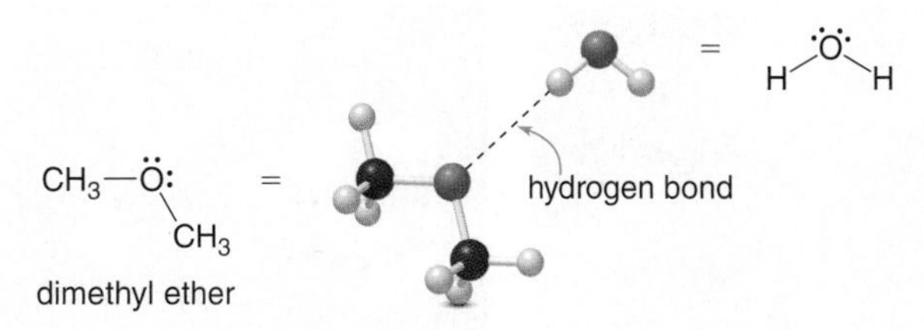

SAMPLE PROBLEM 12.3

Rank the following compounds in order of increasing boiling point:

cyclopentyl–OCH_3	cyclopentyl–CH_2CH_3	cyclopentyl–CH_2OH
A	**B**	**C**

Analysis

Look at the functional groups to determine the strength of the intermolecular forces—**the stronger the forces, the higher the boiling point.**

Solution

B is an alkane with nonpolar C—C and C—H bonds, so it has the weakest intermolecular forces and therefore the lowest boiling point. **C** is an alcohol capable of intermolecular hydrogen bonding, so it has the strongest intermolecular forces and the highest boiling point. **A** is an ether, so it contains a net dipole but is incapable of intermolecular hydrogen bonding. **A,** therefore, has intermolecular forces of intermediate strength and has a boiling point between the boiling points of **B** and **C.**

Cyclopentyl–CH_2CH_3 (**B**), cyclopentyl–OCH_3 (**A**), cyclopentyl–CH_2OH (**C**)

Increasing intermolecular forces
Increasing boiling point

PROBLEM 12.11

Which compound in each pair has the higher boiling point?

a. $CH_3(CH_2)_6CH_3$ or $CH_3(CH_2)_5OCH_3$

b. cyclohexane or tetrahydropyran (six-membered ring containing O)

c. $CH_3(CH_2)_6OH$ or $CH_3(CH_2)_5OCH_3$

d. $CH_3(CH_2)_5OCH_3$ or CH_3OCH_3

12.4 Interesting Alcohols and Ethers

12.4A Simple Alcohols

The most well known alcohol is ethanol, CH_3CH_2OH. **Ethanol** (Figure 12.1), formed by the fermentation of carbohydrates in grains and grapes, is the alcohol present in alcoholic beverages. Fermentation requires yeast, which provides the needed enzymes for the conversion. Ethanol is likely the first organic compound synthesized by humans, since alcohol has been produced for at least 4,000 years. Other simple alcohols are listed in Figure 12.2.

Figure 12.1 Ethanol—The Alcohol in Alcoholic Beverages

Ethanol is the alcohol in red wine, obtained by the fermentation of grapes. All alcoholic beverages are mixtures of ethanol and water in various proportions. Beer has 3–8% ethanol, wines have 10–14% ethanol, and other liquors have 35–90% ethanol.

Figure 12.2
Some Simple Alcohols

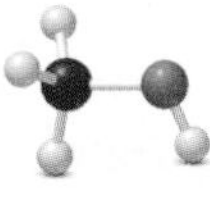

CH_3OH

- **Methanol (CH_3OH)** is a useful solvent and starting material for the synthesis of plastics. Methanol is extremely toxic. Ingestion of as little as 15 mL (about half an ounce) causes blindness and 100 mL causes death.

$(CH_3)_2CHOH$

- **2-Propanol [$(CH_3)_2CHOH$, isopropyl alcohol]** is the major component of rubbing alcohol. 2-Propanol is used to clean skin before medical procedures and to sterilize medical instruments.

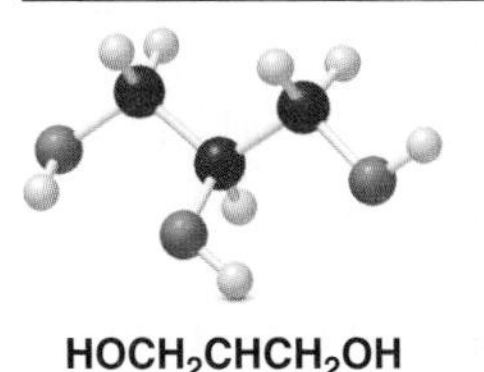

$HOCH_2CH(OH)CH_2OH$

- **Glycerol [$(HOCH_2)_2CHOH$]** is used in lotions, liquid soap, and shaving cream. Since it is sweet tasting and nontoxic, it is also an additive in candy and some prepared foods. Its three OH groups readily hydrogen bond to water, so it helps to retain moisture in these products.

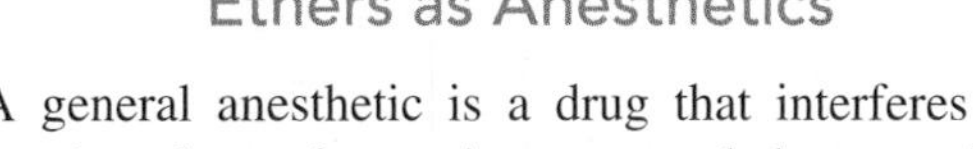

12.4B FOCUS ON HEALTH & MEDICINE Ethers as Anesthetics

This painting by Robert Hinckley depicts a public demonstration of the use of diethyl ether as an anesthetic at the Massachusetts General Hospital in Boston, MA in the 1840s.

A general anesthetic is a drug that interferes with nerve transmission in the brain, resulting in a loss of consciousness and the sensation of pain. The discovery that **diethyl ether** ($CH_3CH_2OCH_2CH_3$) is a general anesthetic dramatically changed surgery in the nineteenth century.

Diethyl ether is an imperfect anesthetic, but considering the alternatives at the time, it was considered revolutionary. It is safe and easy to administer with low patient mortality, but it is highly flammable, and it causes nausea in many patients.

For these reasons, alternatives to diethyl ether are now widely used. Many of these newer general anesthetics, which cause little patient discomfort, are also ethers. These include isoflurane, enflurane, and methoxyflurane. Replacing some of the hydrogen atoms in the ether by halogens results in compounds with similar anesthetic properties but decreased flammability.

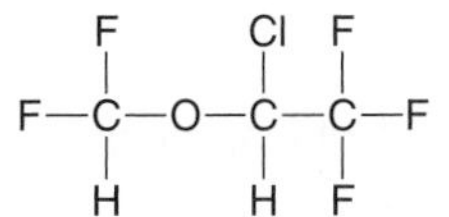

$CHF_2–O–CHCl–CF_3$
isoflurane
(Trade name: Forane)

$CHF_2–O–CF_2–CHClF$
enflurane
(Trade name: Ethrane)

$CH_3–O–CF_2–CHCl_2$
methoxyflurane
(Trade name: Penthrane)

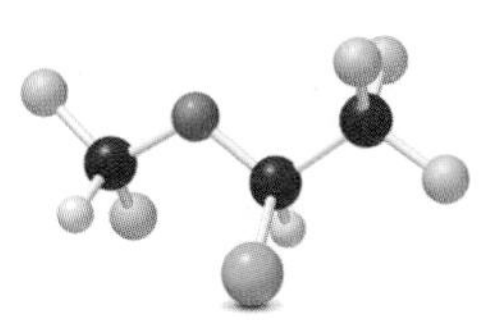

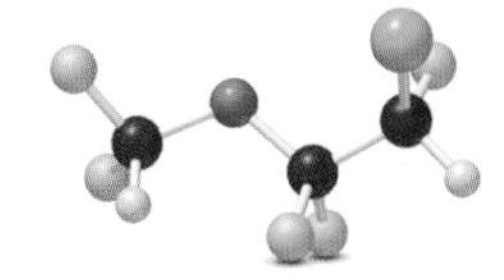

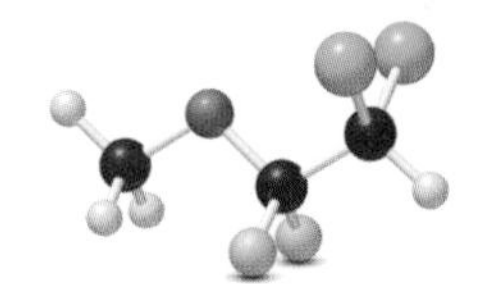

12.5 Reactions of Alcohols

Alcohols undergo two useful reactions—**dehydration** and **oxidation.**

12.5A Dehydration

When an alcohol is treated with a strong acid such as H_2SO_4, the elements of water are lost and an alkene is formed as product. **Loss of H_2O from a starting material is called *dehydration*.** Dehydration takes place by breaking bonds on two adjacent atoms—the C—OH bond and an adjacent C—H bond.

Dehydration

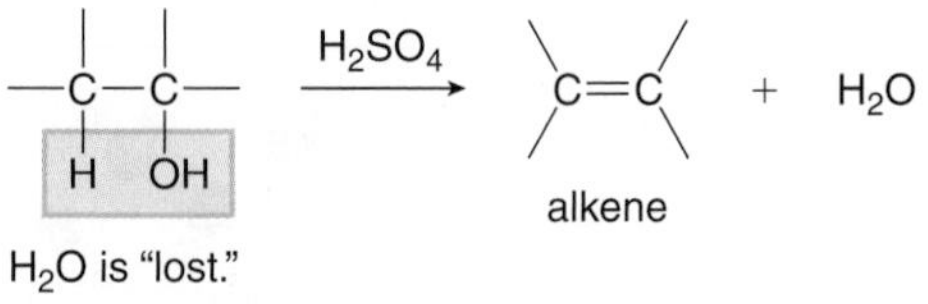

$$-\underset{H}{\overset{|}{C}}-\underset{OH}{\overset{|}{C}}- \xrightarrow{H_2SO_4} C=C + H_2O$$

H_2O is "lost."

alkene

Dehydration is an example of a general type of organic reaction called an **elimination reaction.**

- ***Elimination* is a reaction in which elements of the starting material are "lost" and a new multiple bond is formed.**

For example, dehydration of ethanol (CH_3CH_2OH) with H_2SO_4 forms ethylene ($CH_2{=}CH_2$), as shown. To draw the product of any dehydration, remove the elements of H and OH from two adjacent atoms and draw a carbon–carbon double bond in the product.

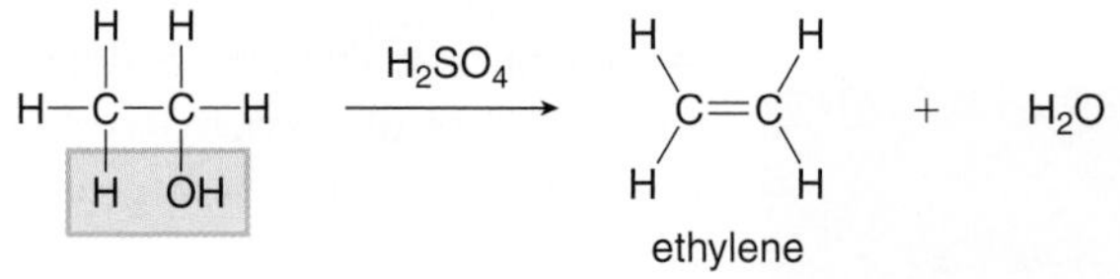

$$H-\underset{H}{\overset{H}{C}}-\underset{OH}{\overset{H}{C}}-H \xrightarrow{H_2SO_4} \underset{H}{\overset{H}{}}C=C\underset{H}{\overset{H}{}} + H_2O$$

ethylene

SAMPLE PROBLEM 12.4

Draw the product formed when cyclohexanol is dehydrated with H_2SO_4.

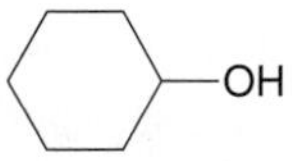

cyclohexanol

Analysis

To draw the products of dehydration, find the carbon bonded to the OH group. Remove the elements of H and OH from two adjacent C's, and draw a double bond between these C's in the product.

Solution

Elimination of H and OH from two adjacent atoms forms cyclohexene.

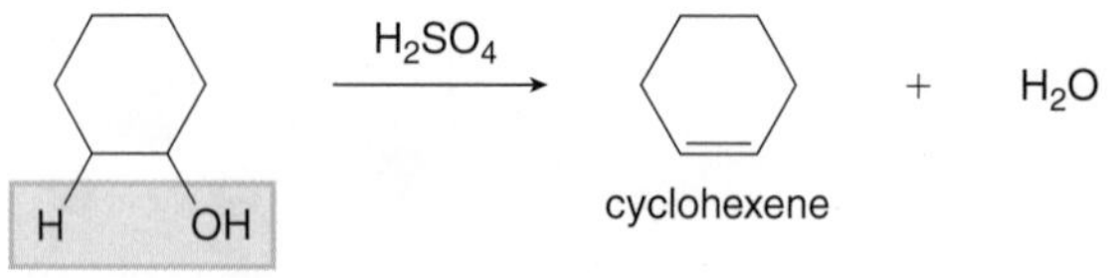

$\xrightarrow{H_2SO_4}$ cyclohexene $+ H_2O$

PROBLEM 12.12

Draw the product formed when each alcohol is dehydrated with H_2SO_4.

a. 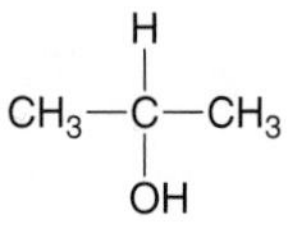 $CH_3-\underset{OH}{\overset{H}{C}}-CH_3$

b. $C_6H_5-CH_2CH_2OH$

c. cyclopentyl—OH

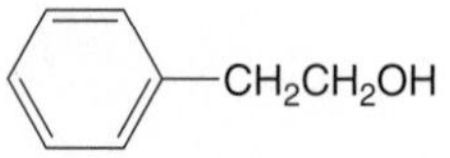

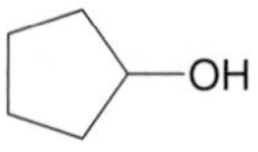

12.5B Oxidation

To determine if an organic compound has been oxidized, we compare the relative number of C—H and C—O bonds in the starting material and product.

- **Oxidation results in an increase in the number of C—O bonds or a decrease in the number of C—H bonds.**

Thus, an organic compound like CH_4 can be oxidized by replacing C—H bonds with C—O bonds, as shown in Figure 12.3. Organic chemists use the symbol [O] to indicate oxidation.

Figure 12.3 A General Scheme for Oxidizing an Organic Compound

CH_4 —[O]→ $H_3C{-}OH$ —[O]→ $H_2C{=}O$ —[O]→ $HO(H)C{=}O$ —[O]→ $O{=}C{=}O$

CH_4	CH_3OH	$H_2C{=}O$	$HCOOH$	$O{=}C{=}O$
4 C—H bonds **0** C—O bonds	**3** C—H bonds **1** C—O bond	**2** C—H bonds **2** C—O bonds	**1** C—H bond **3** C—O bonds	**0** C—H bonds **4** C—O bonds

Increasing number of C—O bonds →

Alcohols can be oxidized to a variety of compounds, depending on the type of alcohol and the reagent. **Oxidation occurs by replacing the C—H bonds *on the carbon bearing the OH group* by C—O bonds.** All oxidation products from alcohol starting materials contain a **C=O, a carbonyl group.**

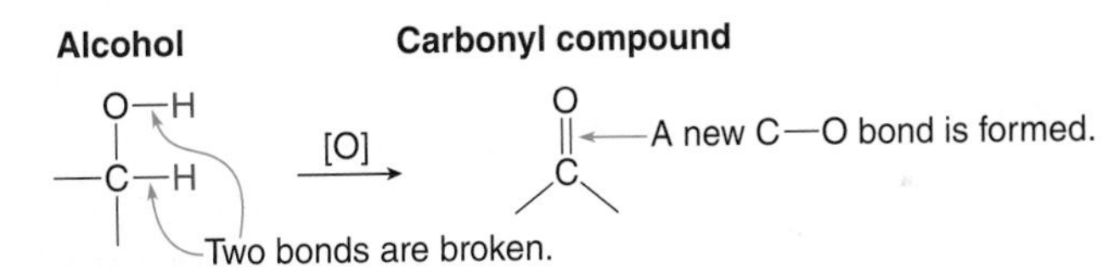

- **Primary (1°) alcohols are first oxidized to aldehydes (RCHO), which are further oxidized to carboxylic acids (RCOOH) by replacing one and then two C—H bonds by C—O bonds.**

CH_3CH_2OH —[O]→ CH_3CHO —[O]→ CH_3COOH

ethanol (1° alcohol) → acetaldehyde → acetic acid

Oxidation of one C—H bond of ethanol forms acetaldehyde. Since acetaldehyde still contains a hydrogen atom on the carbonyl carbon, converting this C—H bond to a C—O bond forms acetic acid, a carboxylic acid with three C—O bonds.

- **Secondary (2°) alcohols are oxidized to ketones (R_2CO), by replacing one C—H bond by one C—O bond.**

$CH_3CH(OH)CH_3$ —[O]→ CH_3COCH_3

2-propanol (2° alcohol) → acetone

Since 2° alcohols have only one hydrogen atom bonded to the carbon with the OH group, they can be oxidized to only one type of compound, a ketone. Thus, 2-propanol is oxidized to acetone.

- **Tertiary 3° alcohols have no H atoms on the carbon with the OH group, so they are not oxidized.**

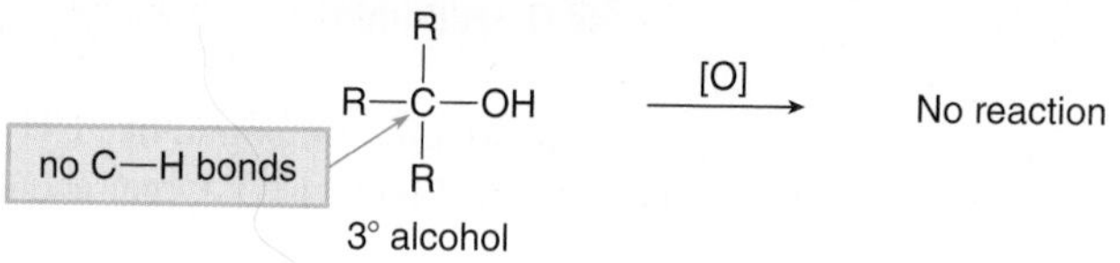

$$R_3C{-}OH \xrightarrow{[O]} \text{No reaction}$$

no C—H bonds

3° alcohol

SAMPLE PROBLEM 12.5

Draw the carbonyl products formed when each alcohol is oxidized.

a. 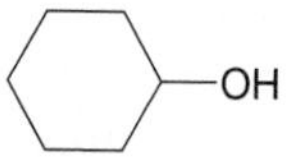 cyclohexanol (cyclohexane ring—OH)

b. $CH_3CH_2CH_2CH_2OH$

1-butanol

Analysis

Classify the alcohol as 1°, 2°, or 3° by drawing in all of the H atoms on the C with the OH. Then concentrate on the C with the OH group and replace H atoms by bonds to O. Keep in mind:

- RCH_2OH (1° alcohols) are oxidized to RCHO, which are then oxidized to RCOOH.
- R_2CHOH (2° alcohols) are oxidized to R_2CO.

Solution

a. Since cyclohexanol is a 2° alcohol with only one H atom on the C bonded to the OH group, it is oxidized to a ketone, cyclohexanone.

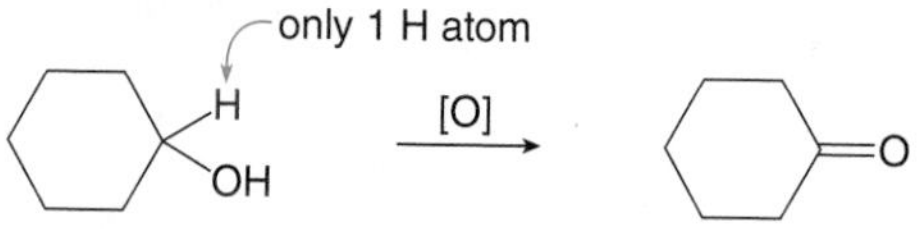

only 1 H atom

cyclohexanol (2° alcohol) $\xrightarrow{[O]}$ cyclohexanone (ketone)

b. 1-Butanol is a 1° alcohol with two H atoms on the C bonded to the OH group. Thus, it is first oxidized to an aldehyde and then to a carboxylic acid.

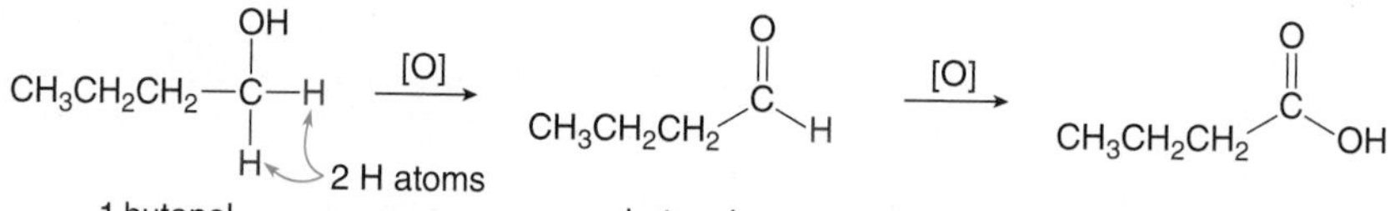

$$CH_3CH_2CH_2{-}CH_2OH \xrightarrow{[O]} CH_3CH_2CH_2CHO \xrightarrow{[O]} CH_3CH_2CH_2COOH$$

2 H atoms

1-butanol (1° alcohol) → butanal (aldehyde) → butanoic acid (carboxylic acid)

PROBLEM 12.13

What products are formed when each alcohol is oxidized? In some cases, no reaction occurs.

a. $CH_3CH(OH)CH_2CH_2CH_3$

b. $(CH_3)_3CCH_2CH_2OH$

c. 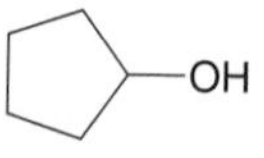 cyclopentanol (cyclopentane ring—OH)

d. 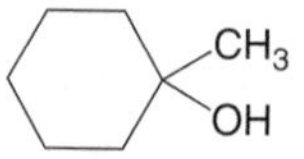 1-methylcyclohexanol (cyclohexane ring with CH_3 and OH on the same carbon)

12.5C FOCUS ON THE HUMAN BODY Oxidation and Blood Alcohol Screening

While alcohol use is socially acceptable, alcohol-related traffic fatalities are common with irresponsible alcohol consumption. In 2004, almost 40% of all fatalities in car crashes in the United States were alcohol-related.

A common reagent for alcohol oxidation is potassium dichromate, $K_2Cr_2O_7$, a red-orange solid. Oxidation with this chromium reagent is characterized by a color change, as the **red-orange reagent** is reduced to a **green Cr^{3+} product.** The first devices used to measure blood alcohol content in individuals suspected of "driving under the influence," made use of this color change. Oxidation of CH_3CH_2OH, the 1° alcohol in alcoholic beverages, with red-orange $K_2Cr_2O_7$ forms CH_3COOH and green Cr^{3+}.

$$\underset{\text{ethanol "alcohol"}}{CH_3CH_2{-}OH} + \underset{\text{red-orange}}{K_2Cr_2O_7} \longrightarrow \underset{\text{acetic acid}}{CH_3C(=O)OH} + \underset{\text{green}}{Cr^{3+}}$$

Blood alcohol level can be determined by having an individual blow into a tube containing $K_2Cr_2O_7$ and an inert solid. The alcohol in the exhaled breath is oxidized by the chromium reagent, which turns green in the tube (Figure 12.4). The higher the concentration of CH_3CH_2OH in the breath, the more chromium reagent is reduced, and the farther the green color extends down the length of the sample tube. This value is then correlated with blood alcohol content to determine if an individual has surpassed the legal blood alcohol limit.

Figure 12.4 Blood Alcohol Screening

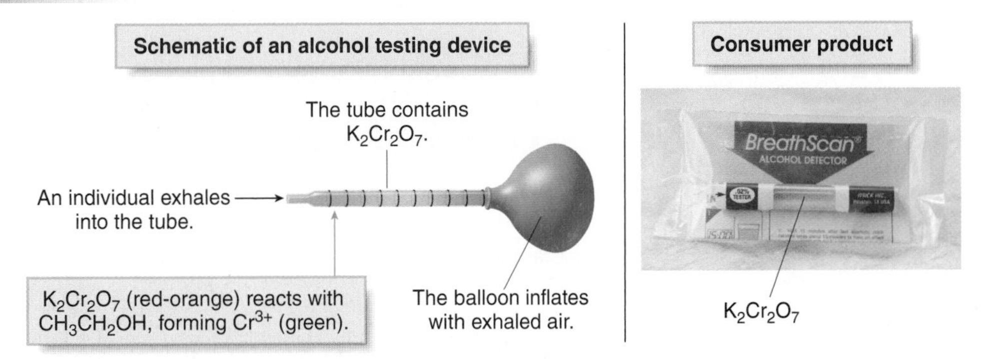

- The oxidation of CH_3CH_2OH with $K_2Cr_2O_7$ to form CH_3COOH and Cr^{3+} was the first available method for the routine testing of alcohol concentration in exhaled air. Some consumer products for alcohol screening are still based on this technology.
- A driver is considered "under the influence" in most states with a blood alcohol concentration of 0.08%.

12.5D FOCUS ON HEALTH & MEDICINE The Metabolism of Ethanol

Throughout history, humans have ingested alcoholic beverages for their pleasant taste and the feeling of euphoria they impart. When ethanol is consumed, it is quickly absorbed in the stomach and small intestines and then rapidly transported in the bloodstream to other organs. Ethanol is metabolized in the liver, by a two-step oxidation sequence. The body does not use chromium reagents as oxidants. Instead, high molecular weight enzymes, alcohol dehydrogenase and aldehyde dehydrogenase, and a small molecule called a **coenzyme** carry out these oxidations.

The products of the biological oxidation of ethanol are the same as the products formed in the laboratory. When ethanol (CH_3CH_2OH, a 1° alcohol) is ingested, it is oxidized in the liver first to CH_3CHO (acetaldehyde), and then to CH_3COOH (acetic acid).

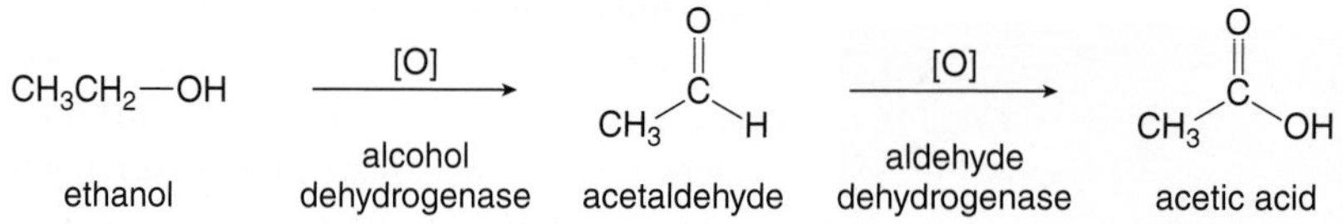

$$CH_3CH_2{-}OH \xrightarrow[\text{alcohol dehydrogenase}]{[O]} CH_3C(=O)H \xrightarrow[\text{aldehyde dehydrogenase}]{[O]} CH_3C(=O)OH$$

ethanol — acetaldehyde — acetic acid

If more ethanol is ingested than can be metabolized in a given time period, the concentration of acetaldehyde accumulates. This toxic compound is responsible for the feelings associated with a hangover.

12.6 Alkyl Halides

Alkyl halides have the general molecular formula $C_nH_{2n+1}X$, and are formally derived from an alkane by replacing a hydrogen atom with a halogen.

***Alkyl halides* are organic molecules containing a halogen atom X (X = F, Cl, Br, I) bonded to a tetrahedral carbon atom.** Alkyl halides are classified as **primary (1°)**, **secondary (2°)**, or **tertiary (3°)** depending on the number of carbons bonded to the carbon with the halogen.

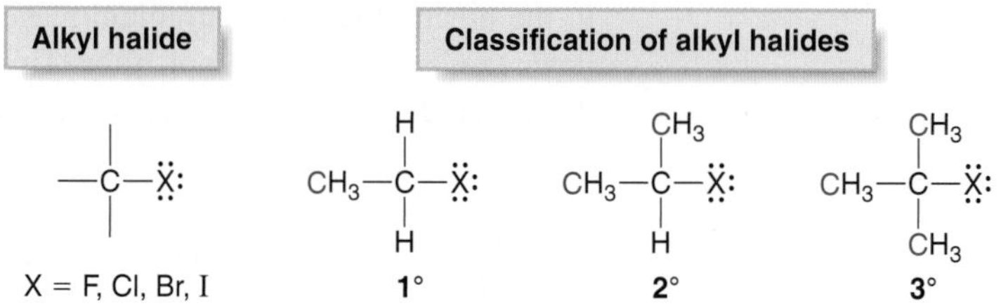

Alkyl halide

$-\overset{|}{\underset{|}{C}}-\ddot{X}:$

X = F, Cl, Br, I

Classification of alkyl halides

$CH_3-CH_2-\ddot{X}:$	$(CH_3)_2CH-\ddot{X}:$	$(CH_3)_3C-\ddot{X}:$
1°	**2°**	**3°**

- A primary (1°) alkyl halide has a halogen on a carbon bonded to one carbon.
- A secondary (2°) alkyl halide has a halogen on a carbon bonded to two carbons.
- A tertiary (3°) alkyl halide has a halogen on a carbon bonded to three carbons.

PROBLEM 12.14

Classify each alkyl halide as 1°, 2°, or 3°.

a. $CH_3CH_2CH_2CH_2CH_2{-}Br$

b. 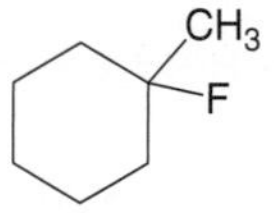 cyclohexane ring bearing CH_3 and F on the same carbon

c. 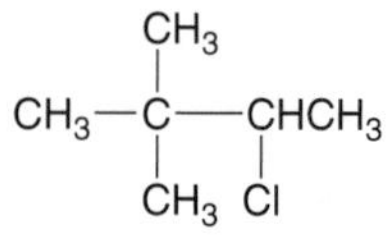 $CH_3-C(CH_3)_2-CHClCH_3$

12.6A Physical Properties

Alkyl halides contain a polar C—X bond, but since they have all of their hydrogens bonded to carbon, they are incapable of intermolecular hydrogen bonding. Alkyl halides with one halogen are polar molecules, therefore, because they contain a net dipole. As a result, they have higher melting points and boiling points than alkanes with the same number of carbons.

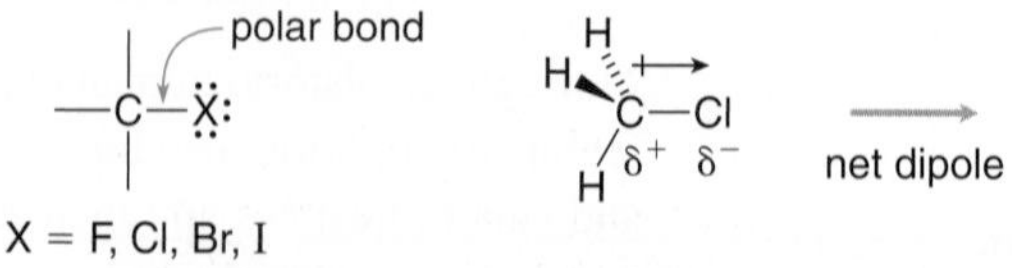

$-\overset{|}{\underset{|}{C}}-\ddot{X}:$ (polar bond)

X = F, Cl, Br, I

$H_3C^{\delta+}-Cl^{\delta-}$ → net dipole

The size of the alkyl groups and the halogen also affects the physical properties of an alkyl halide.

- The boiling point and melting point of an alkyl halide increase with the size of the alkyl group because of increased surface area.
- The boiling point and melting point of an alkyl halide increase with the size of the halogen.

Thus, $CH_3CH_2CH_2Cl$ has a higher boiling point than CH_3CH_2Cl because it has one more carbon, giving it a larger surface area. $CH_3CH_2CH_2Br$ has a higher boiling point than $CH_3CH_2CH_2Cl$ because Br is further down the column of the periodic table, making it larger than Cl.

CH_3CH_2Cl	$CH_3CH_2CH_2Cl$	$CH_3CH_2CH_2Br$
chloroethane	1-chloropropane	1-bromopropane
bp = 12 °C	bp = 47 °C	bp = 71 °C

Increasing boiling point
Increasing size of the alkyl group
Increasing size of the halogen

Since alkyl halides are incapable of hydrogen bonding, they are insoluble in water regardless of size.

PROBLEM 12.15

Rank the compounds in each group in order of increasing boiling point.

a. $CH_3CH_2CH_2I$, $CH_3CH_2CH_2Cl$, $CH_3CH_2CH_2F$
b. $CH_3(CH_2)_4CH_3$, $CH_3(CH_2)_5Br$, $CH_3(CH_2)_5Cl$

12.6B Nomenclature

In the IUPAC system, an alkyl halide is named as an alkane with a halogen substituent—that is, as a ***halo alkane.*** To name a halogen substituent, change the ***-ine*** ending of the name of the halogen to the suffix ***-o*** (e.g., chlor*ine* → chlor*o*).

How To Name an Alkyl Halide Using the IUPAC System

Example: Give the IUPAC name of the following alkyl halide.

$CH_3CH_2CH(CH_3)CH_2CH_2CH(Cl)CH_3$

Step [1] Find the parent carbon chain containing the carbon bonded to the halogen.

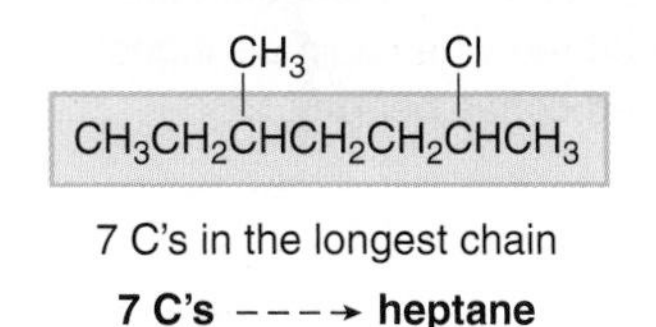

7 C's in the longest chain

7 C's ---→ heptane

- Name the parent chain as an ***alkane,*** with the halogen as a substituent bonded to the longest chain.

Step [2] Apply all other rules of nomenclature.

a. **Number** the chain.

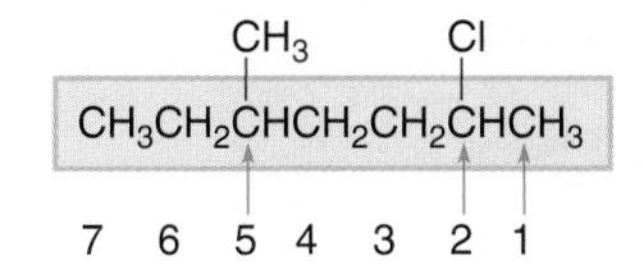

- Begin at the end nearest the first substituent, either alkyl or halogen.

b. **Name** and **number** the substituents.

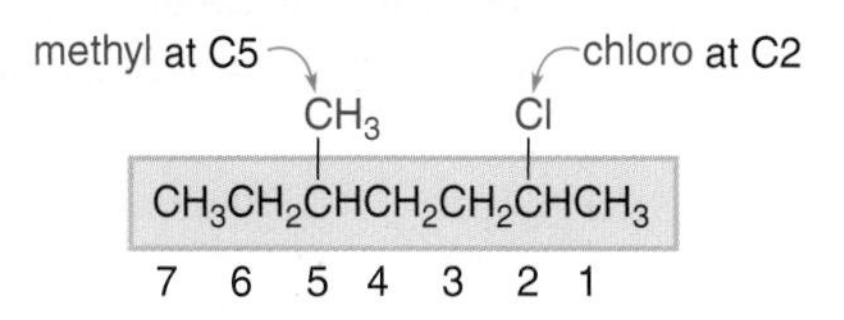

c. Alphabetize: *c* for **c**hloro, then *m* for **m**ethyl.

Answer: 2-chloro-5-methylheptane

Common names for alkyl halides are used only for simple alkyl halides. To assign a common name, name the carbon atoms as an alkyl group. Then name the halogen by changing the ***-ine*** ending of the halogen name to the suffix ***-ide;*** for example, **brom*ine* → brom*ide*.**

CH_3CH_2—Cl chlor*ine* ⟶ chlor*ide*

ethyl group

Common name: **ethyl chloride**

PROBLEM 12.16

Give the structure corresponding to each name.

a. 3-chloro-2-methylhexane

b. 4-ethyl-5-iodo-2,2-dimethyloctane

c. 1,1,3-tribromocyclohexane

d. propyl chloride

12.6C Interesting Alkyl Halides

Giant kelp forests in the ocean release chloromethane (CH_3Cl) into the atmosphere.

Many simple alkyl halides make excellent solvents because they are not flammable and they dissolve a wide variety of organic compounds. Two compounds in this category include $CHCl_3$ (chloroform or trichloromethane) and CCl_4 (carbon tetrachloride or tetrachloromethane). Large quantities of these solvents are produced industrially each year, but like many chlorinated organic compounds, both chloroform and carbon tetrachloride are toxic if inhaled or ingested. Two other simple alkyl halides are given in Figure 12.5.

Figure 12.5 Two Simple Alkyl Halides

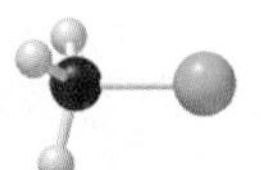

CH_3Cl

- **Chloromethane (CH_3Cl)** is produced by giant kelp and algae and is also found in emissions from volcanoes such as Hawaii's Kilauea. Almost all of the atmospheric chloromethane results from these natural sources.

CH_2Cl_2

- **Dichloromethane (or methylene chloride, CH_2Cl_2)** is an important solvent, once used to decaffeinate coffee. Coffee is now decaffeinated using liquid CO_2 due to concerns over the possible ill effects of trace amounts of residual CH_2Cl_2 in the coffee. Subsequent studies on rats have shown, however, that no cancers occurred when animals ingested the equivalent of over 100,000 cups of decaffeinated coffee per day.

12.7 Thiols

***Thiols* are organic compounds that contain a sulfhydryl group (SH group) bonded to a tetrahedral carbon.** Since sulfur is directly below oxygen in the periodic table, thiols can be considered sulfur analogues of alcohols.

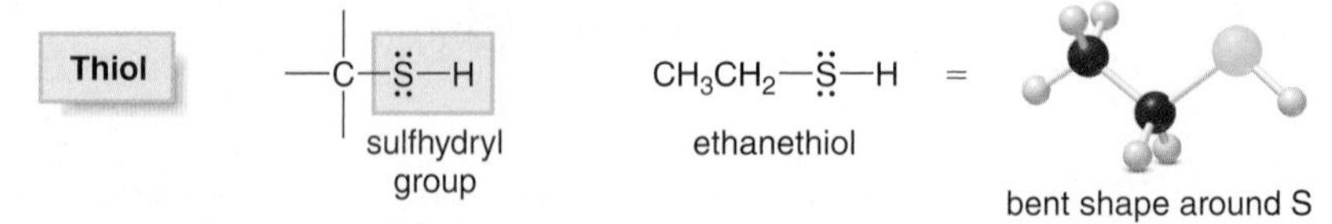

Thiols differ from alcohols in one important way. They contain no O—H bonds, so they are incapable of intermolecular hydrogen bonding. **This gives thiols lower boiling points and melting points compared to alcohols having the same size and shape.**

Hydrogen bonding can occur. stronger intermolecular forces **higher boiling point**	→ CH_3CH_2—OH ethanol bp 78 °C	CH_3CH_2—SH ← ethanethiol bp 35 °C	Hydrogen bonding is impossible. weaker intermolecular forces **lower boiling point**

1-Propanethiol ($CH_3CH_2CH_2SH$) is partly responsible for the characteristic odor of onions.

The most obvious physical property of thiols is their distinctive foul odor. 3-Methyl-1-butanethiol [$(CH_3)_2CHCH_2CH_2SH$] is one of the main components of the defensive spray of skunks.

Thiols undergo one important reaction: **thiols are oxidized to disulfides,** compounds that contain a sulfur–sulfur bond. This is an oxidation reaction because two hydrogen atoms are removed in forming the disulfide.

Oxidation

$$2\ CH_3CH_2\text{—S—H} \xrightarrow{[O]} CH_3CH_2\text{—S—S—}CH_2CH_3$$

thiol → disulfide (new S—S bond)

Disulfides can also be converted back to thiols with a reducing agent. The symbol for a general reducing agent is **[H],** since hydrogen atoms are often added to a molecule during reduction.

Reduction

$$CH_3CH_2\text{—S—S—}CH_2CH_3 \xrightarrow{[H]} 2\ CH_3CH_2\text{—S—H}$$

disulfide → thiol (new S—H bond)

The chemistry of thiols and disulfides plays an important role in determining the properties and shape of some proteins. For example, α-keratin, the protein in hair, contains many disulfide bonds. Straight hair can be made curly by cleaving the disulfide bonds in α-keratin, then rearranging and re-forming them, as shown schematically in Figure 12.6.

PROBLEM 12.17

(a) Draw the disulfide formed when $CH_3CH_2CH_2SH$ is oxidized. (b) Draw the products formed when the following disulfide is reduced: $CH_3CH_2CH_2CH_2SSCH_2CH_3$.

Figure 12.6 Focus on the Human Body: Making Straight Hair Curly

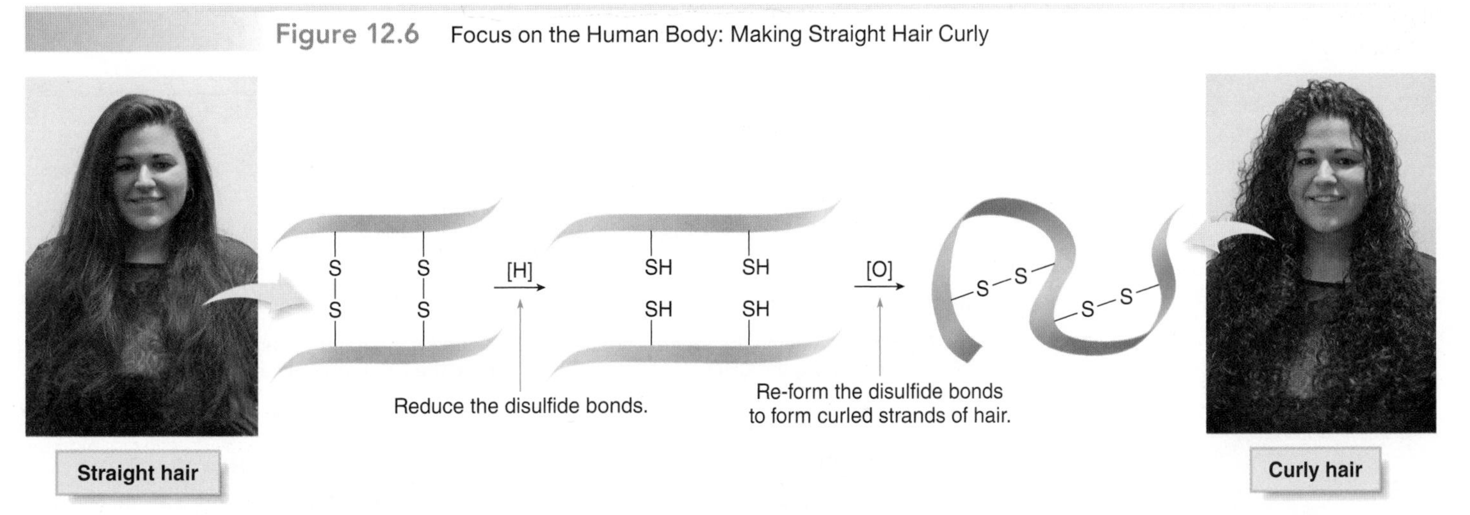

To make straight hair curly, the disulfide bonds holding the protein chains together are reduced. This forms free SH groups. The hair is turned around curlers and then an oxidizing agent is applied. This re-forms the disulfide bonds to the hair, now giving it a curly appearance.

12.8 Structure and Properties of Aldehydes and Ketones

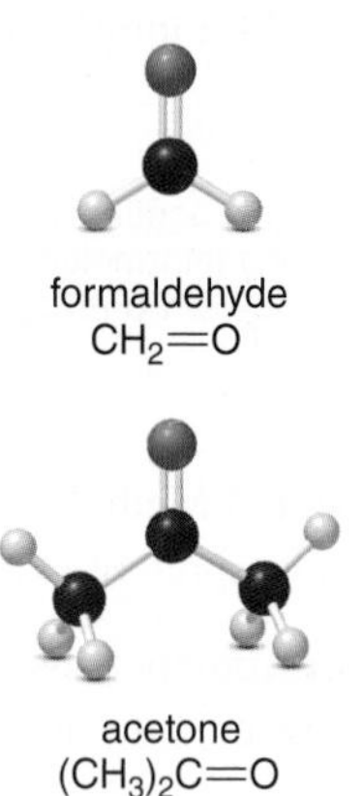

Aldehydes (RCHO) and ketones (RCOR or R_2CO) are two families of compounds that contain a carbonyl group. Two structural features dominate the properties and chemistry of the carbonyl group.

120° 120° trigonal planar = polar bond δ^- δ^+

- The carbonyl carbon atom is trigonal planar, and all bond angles are 120°.
- Since oxygen is more electronegative than carbon, a carbonyl group is polar. The carbonyl carbon is electron poor (δ^+) and the oxygen is electron rich (δ^-).

12.8A Naming Aldehydes

- In IUPAC nomenclature, aldehydes are identified by the suffix *-al.*

To name an aldehyde using the IUPAC system:

- Find the longest chain containing the CHO group, and change the **-e** ending of the parent alkane to the suffix ***-al.***
- Number the chain or ring to put the CHO group at C1, but omit this number from the name. Apply all of the other usual rules of nomenclature.

Simple aldehydes have common names that are virtually always used instead of their IUPAC names. Common names all contain the suffix ***-aldehyde.***

HCHO	CH_3CHO	C_6H_5CHO
formaldehyde (methanal)	acetaldehyde (ethanal)	benzaldehyde (benzenecarbaldehyde)

(IUPAC names are in parentheses.)

SAMPLE PROBLEM 12.6

Give the IUPAC name for the following aldehyde.

$CH_3CH(CH_3)CH(CH_3)-C(=O)-H$

Analysis and Solution

[1] Find and name the longest chain containing the CHO.

$CH_3CH(CH_3)CH(CH_3)-C(=O)-H$

butane ---→ butan*al*
(4 C's)

[2] Number and name substituents, making sure the CHO group is at C1.

$CH_3CH(CH_3)CH(CH_3)-C(=O)-H$ (C1 = CHO carbon, then 2, 3)

Answer: 2,3-dimethylbutanal

PROBLEM 12.18

Give the IUPAC name for each aldehyde.

a. $(CH_3)_2CHCH_2CH_2CH_2CHO$

b. $(CH_3)_3CC(CH_3)_2CH_2CHO$

c.
$$\begin{array}{c} \quad CH_2CH_3 \qquad\quad CHO \\ CH_3\overset{|}{C}H\overset{|}{C}HCH_2CH_2\overset{|}{C}HCH_3 \\ \;\; CH_3 \end{array}$$

PROBLEM 12.19

Give the structure corresponding to each IUPAC name.

a. 2-chloropropanal

b. 3,4,5-triethylheptanal

c. 3,6-diethylnonanal

d. *o*-ethylbenzaldehyde

PROBLEM 12.20

Give the IUPAC name for each aldehyde depicted in the ball-and-stick models. Both aldehydes are isolated from yuzu, a citrus fruit extensively cultivated in Japan.

a. 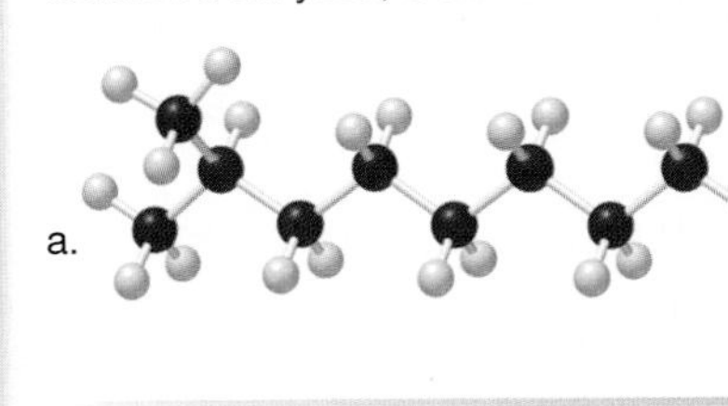

b. 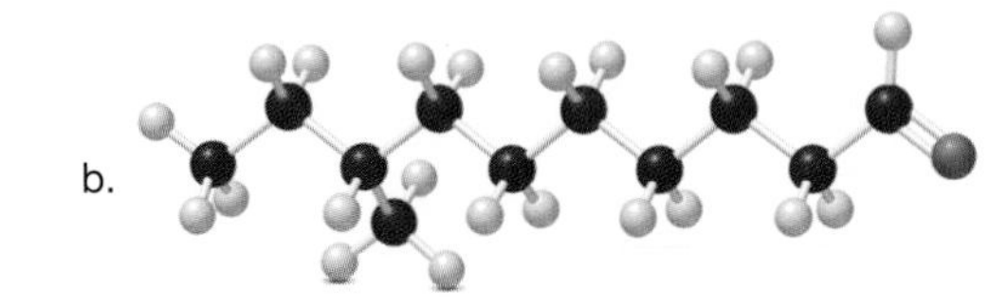

Yuzu is a small citrus fruit used widely in Japanese cooking. Its unusual aroma is partly due to the aldehydes depicted in Problem 12.20.

12.8B Naming Ketones

- **In the IUPAC system, ketones are identified by the suffix *-one*.**

To name an acyclic ketone using IUPAC rules:

- **Find the longest chain containing the carbonyl group, and change the *-e* ending of the parent alkane to the suffix *-one*.**
- **Number the carbon chain to give the carbonyl carbon the lower number. Apply all of the other usual rules of nomenclature.**

With cyclic ketones, numbering always begins at the carbonyl carbon, but the "1" is usually omitted from the name. The ring is then numbered clockwise or counterclockwise to give the first substituent the lower number.

Three simple ketones have widely used common names.

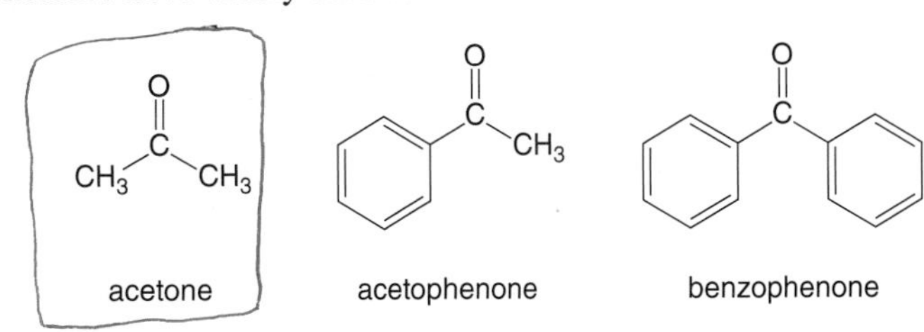

SAMPLE PROBLEM 12.7

Give the IUPAC name for the following ketone.

O

CH_2CH_3

CH_3

Analysis and Solution

[1] **Name the ring.**

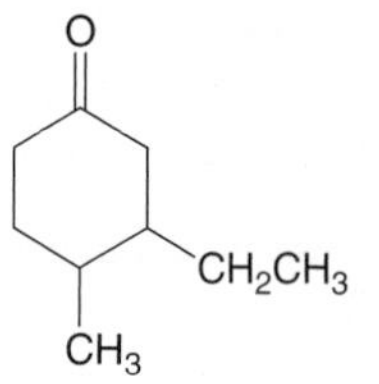

cyclohexane ---→ cyclohexan*one* (6 C's)

[2] **Number and name substituents, making sure the carbonyl carbon is at C1.**

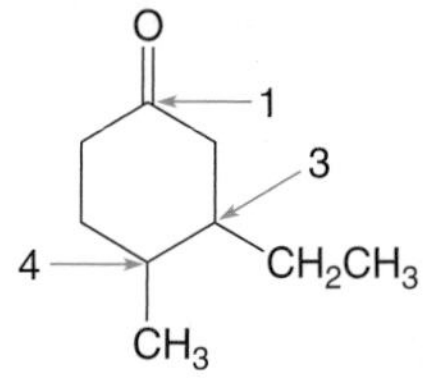

Answer: 3-ethyl-4-methylcyclohexanone

PROBLEM 12.21

Give the IUPAC name for each ketone.

a. 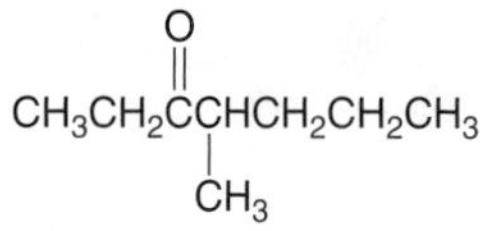 $CH_3CH_2COCH(CH_3)CH_2CH_2CH_3$

b. 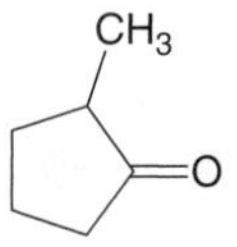 2-methylcyclopentanone structure (CH_3 on ring carbon adjacent to C=O)

c. $(CH_3)_3CCOCH_2CH_2CH_2CH_3$

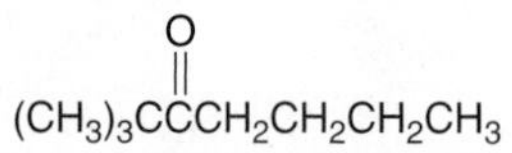

12.8C Physical Properties

Because aldehydes and ketones have a polar carbonyl group, they are **polar molecules** with stronger intermolecular forces than the hydrocarbons of Chapters 10 and 11. Since they have no O—H bond, two molecules of RCHO or RCOR are incapable of intermolecular hydrogen bonding, giving them weaker intermolecular forces than alcohols.

As a result:

- **Aldehydes and ketones have *higher* boiling points than hydrocarbons of comparable molecular weight.**
- **Aldehydes and ketones have *lower* boiling points than alcohols of comparable molecular weight.**

$CH_3CH_2CH_2CH_2CH_3$	$CH_3CH_2CH_2CHO$	$CH_3CH_2CH_2CH_2OH$
pentane	butanal	1-butanol
bp 36 °C	bp 76 °C	bp 118 °C
	$CH_3CH_2COCH_3$	
	2-butanone	
	bp 80 °C	

Increasing strength of intermolecular forces
Increasing boiling point

Based on the general rule governing solubility (i.e., **"like dissolves like"**), aldehydes and ketones are soluble in organic solvents. Moreover, because aldehydes and ketones contain an oxygen atom with an available lone pair, they can intermolecularly hydrogen bond to water.

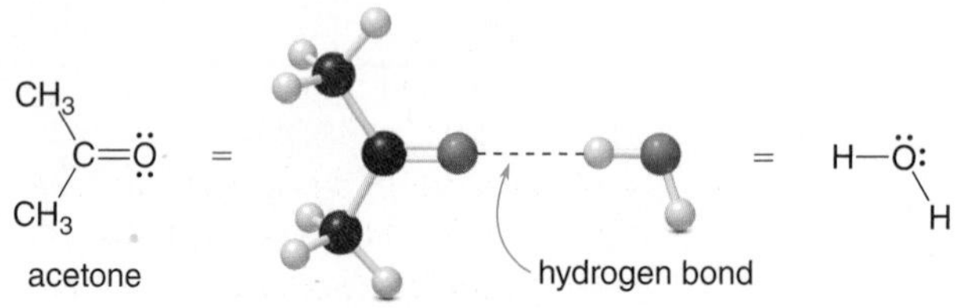

$(CH_3)_2C{=}\ddot{O}$: = [ball-and-stick model] = H—Ö:—H

acetone

hydrogen bond

As a result:

- **Low molecular weight aldehydes and ketones (those having less than six carbons) are soluble in both organic solvents and water.**
- **Higher molecular weight aldehydes and ketones (those having six carbons or more) are soluble in organic solvents, but insoluble in water.**

PROBLEM 12.22

Which compound in each pair has the higher boiling point?

a. =O or —CH_3 c. =O or —OH

b. $(CH_3CH_2)_2CO$ or $(CH_3CH_2)_2C{=}CH_2$ d. $CH_3(CH_2)_6CH_3$ or $CH_3(CH_2)_5CHO$

PROBLEM 12.23

Acetone and progesterone are two ketones that occur naturally in the human body. Discuss the solubility properties of both compounds in water and organic solvents.

O, C, CH_3, CH_3

acetone

O, C, CH_3, CH_3, CH_3, O

progesterone

12.9 FOCUS ON HEALTH & MEDICINE Interesting Aldehydes and Ketones

Many aldehydes with characteristic odors occur in nature, as shown in Figure 12.7.

Figure 12.7

Some Naturally Occurring Aldehydes with Characteristic Odors

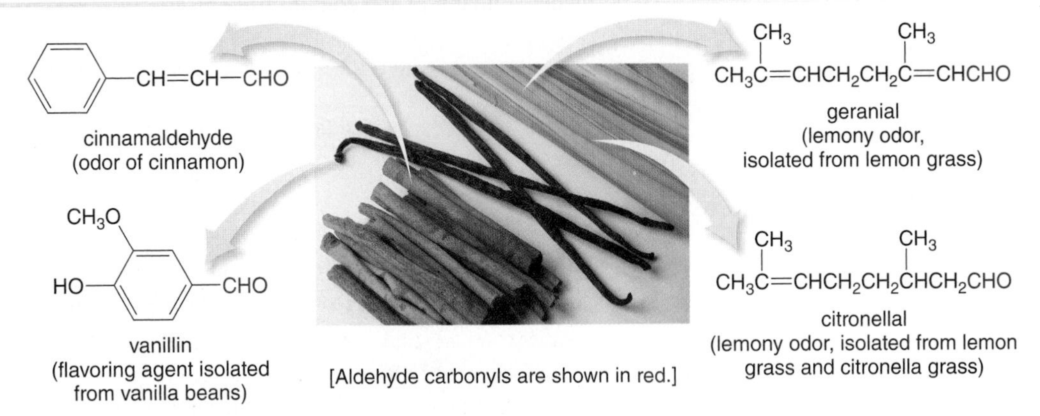

- **Cinnamaldehyde,** the major component of cinnamon bark, is a common flavoring agent.
- **Vanillin** is the primary component of the extract of the vanilla bean. Because natural sources cannot meet the high demand, most vanilla flavoring agents are made synthetically from starting materials derived from petroleum.
- **Geranial** has the lemony odor characteristic of lemon grass. Geranial is used in perfumery and as a starting material for synthesizing vitamin A.
- **Citronellal** gives the distinctive lemon odor to citronella candles, commonly used to repel mosquitoes.

Acetone [$(CH_3)_2C{=}O$, the simplest ketone] is produced naturally in cells during the breakdown of fatty acids. In diabetes, a disease where normal metabolic processes are altered because of the inadequate secretion of insulin, individuals often have unusually high levels of acetone in the bloodstream. The characteristic odor of acetone can be detected on the breath of diabetic patients when their disease is poorly controlled.

Ketones play an important role in the tanning industry. Dihydroxyacetone is the active ingredient in commercial tanning agents that produce sunless tans. Dihydroxyacetone reacts with proteins in the skin, producing a complex colored pigment that gives the skin a brown hue. In addition, many commercial sunscreens are ketones. Examples include avobenzone, oxybenzone, and dioxybenzone.

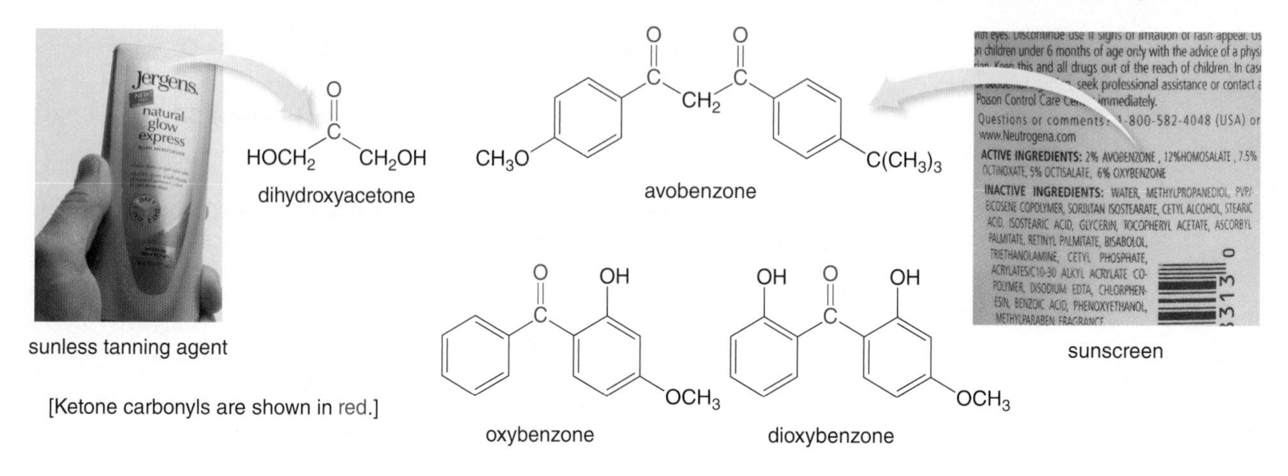

PROBLEM 12.24

Which sunscreen—avobenzone, oxybenzone, or dioxybenzone—is probably most soluble in water, and therefore most readily washed off when an individual goes swimming? Explain your choice.

12.10 Oxidation of Aldehydes

Since aldehydes contain a hydrogen atom bonded directly to the carbonyl carbon, they can be oxidized to carboxylic acids; that is, the aldehyde C—H bond can be converted to a C—OH bond. Since ketones have no hydrogen atom bonded to the carbonyl group, they are not oxidized under similar reaction conditions.

Oxidation

Replace 1 C—H bond by 1 C—O bond.

$$RCHO \xrightarrow{[O]} RCOOH$$

aldehyde → carboxylic acid

A common reagent for this oxidation is potassium dichromate, $K_2Cr_2O_7$, a red-orange solid that is converted to a green Cr^{3+} product during oxidation.

Examples

$$CH_3CH_2CH_2CHO \xrightarrow{K_2Cr_2O_7} CH_3CH_2CH_2COOH$$

butanal → butanoic acid

no C—H bond

$$CH_3CH_2COCH_3 \xrightarrow{K_2Cr_2O_7} \text{No reaction}$$

2-butanone

Aldehydes are said to give a *positive* Tollens test; that is, they react with Ag^+ to form RCOOH and Ag. When the reaction is carried out in a glass flask, a silver mirror is formed on its walls. Other functional groups give a *negative* Tollens test; that is, no silver mirror forms.

As we learned in Section 12.5, $K_2Cr_2O_7$ oxidizes 1° and 2° alcohols as well. Aldehydes can be oxidized *selectively* in the presence of other functional groups using **silver(I) oxide (Ag_2O) in aqueous ammonium hydroxide (NH_4OH).** This is called **Tollens reagent.** Only aldehydes react with Tollens reagent; all other functional groups are inert. Oxidation with Tollens reagent provides a distinct color change because the Ag^+ reagent is converted to silver metal (Ag), which precipitates out of the reaction mixture as a silver mirror.

HO–C₆H₁₀–CHO $\xrightarrow[NH_4OH]{Ag_2O}$ HO–C₆H₁₀–COOH + Ag (silver mirror)

Only the aldehyde is oxidized.

SAMPLE PROBLEM 12.8

What product is formed when each compound is treated with Tollens reagent (Ag_2O, NH_4OH)?

a. CH_3CHO b. $HOCH_2$–C₆H₁₀–CHO

Analysis

Only aldehydes (RCHO) react with Tollens reagent. Ketones and alcohols are inert to oxidation.

Solution

The aldehyde in both compounds is oxidized to RCO_2H, but the 1° alcohol in part (b) does not react with Tollens reagent.

a. $CH_3CHO \xrightarrow[NH_4OH]{Ag_2O} CH_3COOH$

Replace 1 C—H bond by 1 C—O bond.

b. $HOCH_2$–C₆H₁₀–CHO $\xrightarrow[NH_4OH]{Ag_2O}$ $HOCH_2$–C₆H₁₀–COOH

Only the aldehyde is oxidized.

PROBLEM 12.25

What product is formed when each compound is treated with Tollens reagent (Ag_2O, NH_4OH)? In some cases, no reaction occurs.

a. $CH_3(CH_2)_6CHO$ b. cyclopentanone (C₅H₈=O) c. cyclopentyl–CHO d. cyclohexyl–OH

12.11 Looking Glass Chemistry—Molecules and Their Mirror Images

The remainder of Chapter 12 concentrates on **stereochemistry**—the three-dimensional structure of compounds. The principles presented here apply to all of the families of molecules in this chapter, as well as the hydrocarbons we encountered in Chapters 10 and 11. To understand stereochemistry, recall the definition of stereoisomers learned in Section 11.3.

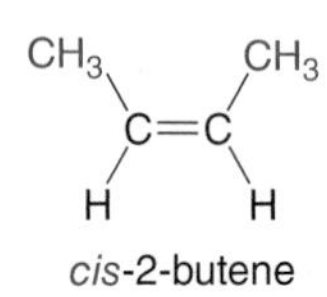

cis-2-butene

CH_3 H C=C H CH_3
trans-2-butene

- *Stereoisomers* are isomers that differ *only* in the three-dimensional arrangement of atoms.

The cis and trans isomers of 2-butene are one example of stereoisomers. The cis isomer has the two CH_3 groups on the same side of the double bond, while the trans isomer has the two CH_3 groups on opposite sides.

Another type of stereoisomer occurs at tetrahedral carbons. To learn more about this type of stereoisomer, we must turn our attention to molecules and their mirror images. **Everything, including molecules, has a mirror image.** What's important in chemistry is **whether a molecule is *identical* to or *different* from its mirror image.**

12.11A What It Means to Be Chiral or Achiral

The dominance of right-handedness over left-handedness occurs in all races and cultures. Despite this fact, even identical twins can exhibit differences in hand preference. Pictured are Zachary (left-handed) and Matthew (right-handed), identical twin sons of the author.

The adjective ***chiral*** (pronounced ky-rel) comes from the Greek *cheir,* meaning *hand.* Left and right hands are **chiral:** they are mirror images that do not superimpose on each other.

Some molecules are like hands. **Left and right hands are mirror images of each other, but they are *not* identical.** If you try to mentally place one hand inside the other hand you can never superimpose either all the fingers, or the tops and palms. To *superimpose* an object on its mirror image means to align *all* parts of the object with its mirror image. With molecules, this means aligning all atoms and all bonds.

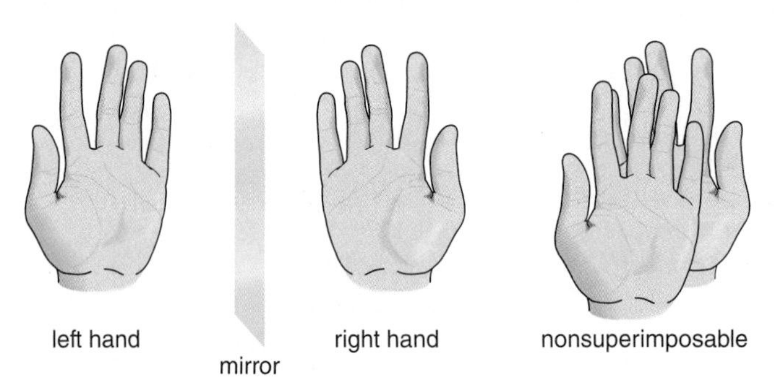

- A molecule (or object) that is *not* superimposable on its mirror image is said to be *chiral.*

Other molecules are like socks. **Two socks from a pair are mirror images that *are* superimposable.** One sock can fit inside another, aligning toes and heels, and tops and bottoms. A sock and its mirror image are *identical.*

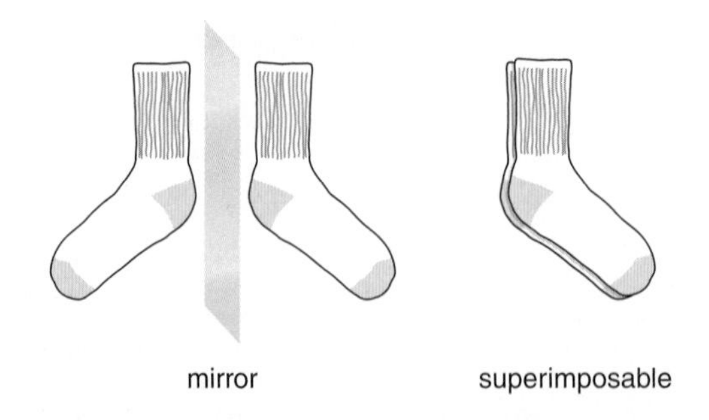

- A molecule (or object) that *is* superimposable on its mirror image is said to be *achiral.*

PROBLEM 12.26

Classify each object as chiral or achiral: (a) nail; (b) screw; (c) glove; (d) pencil.

When is a molecule chiral or achiral? To determine whether a molecule is chiral or achiral, we must examine what groups are bonded to each carbon atom.

- A chiral molecule has at least one carbon atom bonded to four *different* groups.
- An achiral molecule does *not* contain a carbon atom bonded to four different groups.

Using these definitions, we can classify CH_2BrCl and CHBrClF as chiral or achiral molecules.

CH_2BrCl is an achiral molecule. Its single carbon atom is bonded to two H atoms. Ball-and-stick models illustrate that this molecule is superimposable on its mirror image.

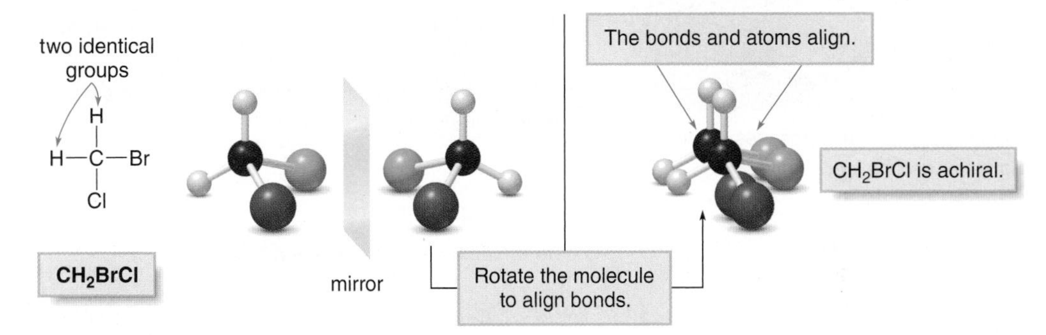

Naming a carbon atom with four different groups is a topic that currently has no firm agreement among organic chemists. IUPAC recommends the term *chirality center,* and so this is the term used in this text. Other terms in common use are chiral center, chiral carbon, asymmetric carbon, stereocenter, and stereogenic center.

With CHBrClF, the result is different. CHBrClF contains a carbon atom bonded to four different groups—H, Br, Cl, and F. **A carbon bonded to four different groups is called a *chirality center.*** The molecule (labeled **A**) and its mirror image (labeled **B**) are not superimposable. No matter how you rotate **A** and **B,** all of the atoms never align. **CHBrClF is thus a chiral molecule,** and **A** and **B** are different compounds.

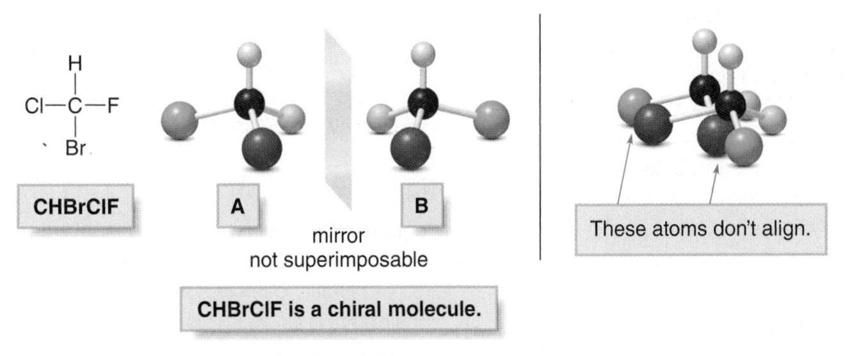

A and **B** are **stereoisomers** since they are isomers differing only in the three-dimensional arrangement of substituents. They represent a new type of stereoisomer that occurs at tetrahedral carbon atoms. These stereoisomers are called ***enantiomers.***

- Enantiomers are mirror images that are not superimposable.

Thus, CH_2BrCl is an achiral molecule with no chirality center, while CHBrClF is a chiral molecule with one chirality center.

12.11B Locating Chirality Centers

A necessary skill in studying the three-dimensional structure of molecules is the ability to locate chirality centers—carbon atoms bonded to four different groups. To locate a chirality center, examine each **tetrahedral** carbon atom in a molecule, and look at the four ***groups***—not the four

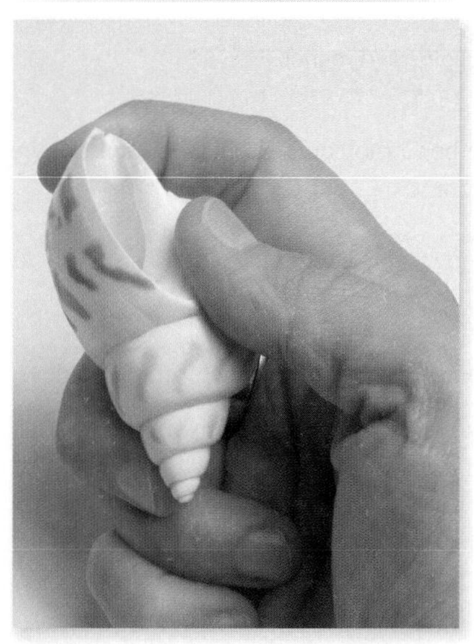

Many types of snails possess a chiral, right-handed helical shell.

atoms—bonded to it. CBrClFI has one chirality center since its carbon atom is bonded to four different elements—Br, Cl, F, and I. 3-Bromohexane also has one chirality center since one carbon is bonded to H, Br, CH_2CH_3, and $CH_2CH_2CH_3$. We consider all atoms in a group as a *whole unit,* not just the atom bonded directly to the carbon in question.

Br
Cl—C—I
F
chirality center

H
CH_3CH_2—C—$CH_2CH_2CH_3$
Br
chirality center
3-bromohexane

This C is bonded to: **H**
Br
CH_2CH_3
$CH_2CH_2CH_3$
two *different* alkyl groups

Keep in mind the following:

- **CH_2 and CH_3 groups have more than one H atom bonded to the same carbon, so each of these carbons is *never* a chirality center.**
- **A carbon that is part of a multiple bond does not have four groups around it, so it can *never* be a chirality center.**

SAMPLE PROBLEM 12.9

Locate the chirality center in each drug.

a. H_2N—C(COOH)(H)—C(CH_3)(CH_3)—SH
penicillamine

b. HO—(C_6H_3, $HOCH_2$)—C(OH)(H)—$CH_2NHC(CH_3)_3$
albuterol

Analysis

In compounds with many C's, look at each C individually and eliminate those C's that can't be chirality centers. Thus, omit all CH_2 and CH_3 groups and all multiply bonded C's. Check all remaining C's to see if they are bonded to four different groups.

Solution

a. **Penicillamine** is used to treat Wilson's disease, a genetic disorder that leads to a buildup of copper in the liver, kidneys, and the brain. The CH_3 groups and COOH group in penicillamine are not chirality centers. One C is bonded to two CH_3 groups, so it can be eliminated from consideration as well. This leaves one C with four different groups bonded to it.

b. **Albuterol** is a bronchodilator—that is, it widens air passages—so it is used to treat asthma. Omit all CH_2 and CH_3 groups and all C's that are part of a double bond. Also omit from consideration the one C bonded to three CH_3 groups. This leaves one C with four different groups bonded to it.

a. H_2N—C(COOH)(H)—C(CH_3)(CH_3)—SH — chirality center
penicillamine

b. HO—(C_6H_3, $HOCH_2$)—C(OH)(H)—$CH_2NHC(CH_3)_3$ — chirality center
albuterol

PROBLEM 12.27

Label the chirality center in the given molecules. The compounds contain zero or one chirality center.

a. 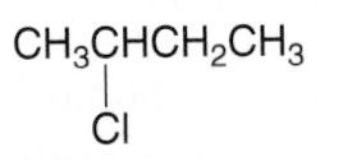 $CH_3CH(Cl)CH_2CH_3$ (Cl on the second carbon)

b. $(CH_3)_3CH$

c. 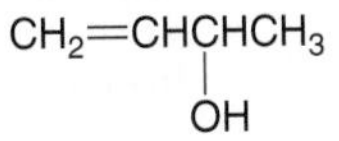 $CH_2{=}CHCH(OH)CH_3$ (OH on the third carbon)

d. 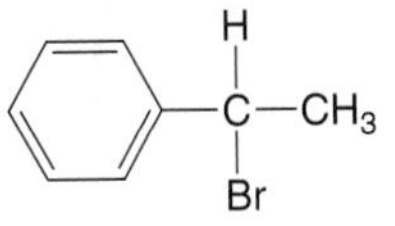 C_6H_5–C(H)(Br)–CH_3 (benzene ring bonded to a carbon bearing H, Br and CH_3)

PROBLEM 12.28

Label the chirality center in each drug.

a. 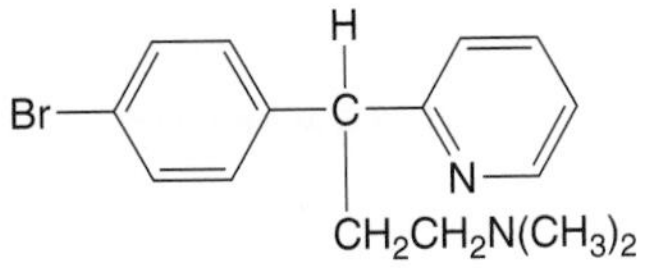 Br–C_6H_4–C(H)(pyridin-2-yl)–$CH_2CH_2N(CH_3)_2$

brompheniramine
(antihistamine)

b. 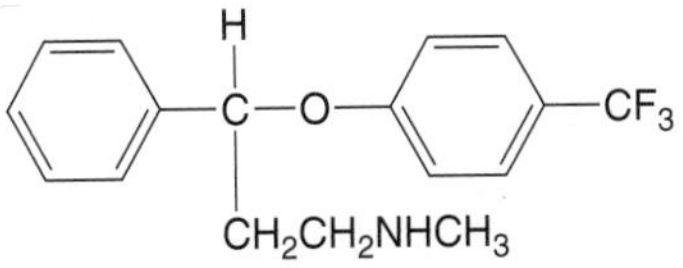 C_6H_5–C(H)(O–C_6H_4–CF_3)–$CH_2CH_2NHCH_3$

fluoxetine
Trade name: Prozac
(antidepressant)

PROBLEM 12.29

Locate the chirality center in the carbohydrate erythrulose, an ingredient in sunless tanning agents.

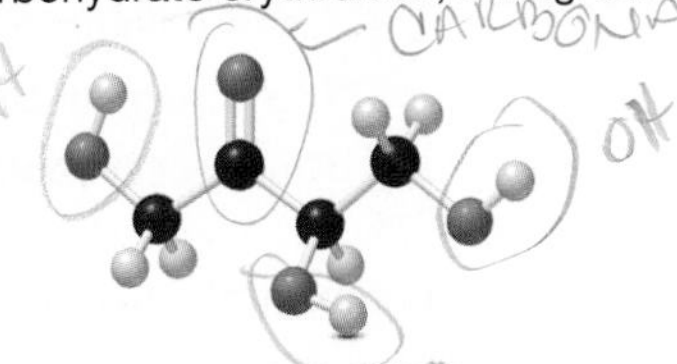

erythrulose

Larger organic molecules can have two, three, or even hundreds of chirality centers. Tartaric acid, isolated from grapes, and ephedrine, isolated from the herb ma huang (used to treat respiratory ailments in traditional Chinese medicine), each have two chirality centers. Once a popular drug to promote weight loss and enhance athletic performance, ephedrine use has now been linked to episodes of sudden death, heart attack, and stroke.

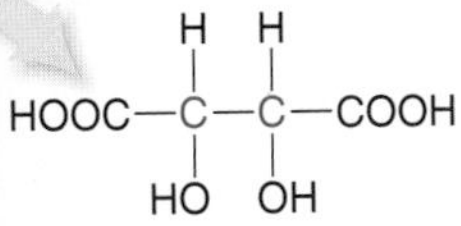

HOOC–C(H)(OH)–C(H)(OH)–COOH

tartaric acid

C_6H_5–C(H)(OH)–C(H)(CH_3)–$NHCH_3$

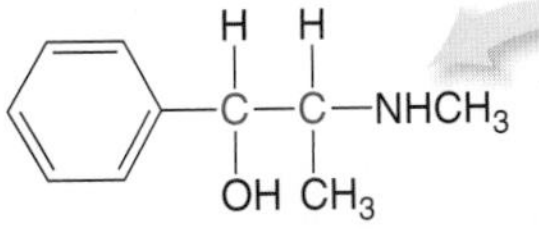

ephedrine
(bronchodilator, decongestant)

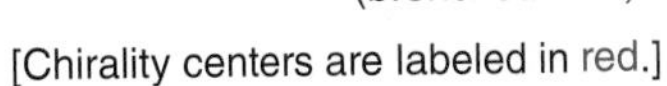

[Chirality centers are labeled in red.]

PROBLEM 12.30

Label the chirality centers in each molecule. Compounds may have one or two chirality centers.

a. 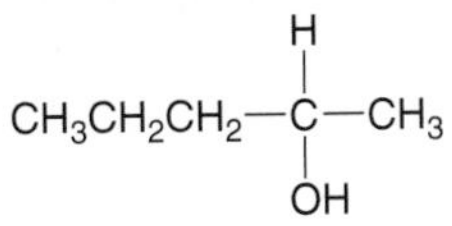 $CH_3CH_2CH_2$–C(H)(OH)–CH_3

b. H_2N–C(H)(COOH)–C(H)(CH_3)–CH_2CH_3

c. CH_3CH_2–C(Br)(H)–CH_2CH_2–C(Cl)(H)–CH_3

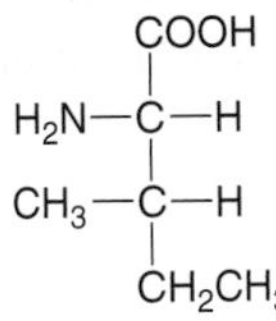

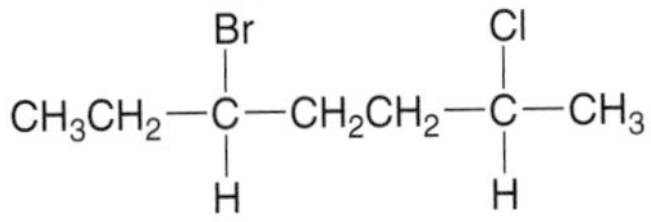

12.11C Fischer Projections

The chirality centers in some organic compounds, most notably carbohydrates (Chapter 14), are often drawn using the following convention.

Instead of drawing a tetrahedron with two bonds in the plane, one in front of the plane, and one behind it, the **tetrahedron is tipped so that both horizontal bonds come forward (drawn on wedges) and both vertical bonds go behind (on dashed lines).** This structure is then abbreviated by a **cross formula,** also called a **Fischer projection formula.** In a Fischer projection formula, therefore,

Draw a tetrahedron as: Abbreviate it as a cross formula:

W
Horizontal bonds come forward, on wedges. Z–C–X = Z–+–X chirality center
Y
Vertical bonds go back, on dashes. Fischer projection formula

- **A carbon atom is located at the intersection of the two lines of the cross.**
- **The horizontal bonds come forward, on wedges.**
- **The vertical bonds go back, on dashed lines.**

For example, to draw the chirality center of 2-butanol [$CH_3CH(OH)CH_2CH_3$] using this convention, **draw the tetrahedron with horizontal bonds on wedges and vertical bonds on dashed lines.** Then, **replace the chirality center with a cross** to draw the Fischer projection. The carbon skeleton is generally arranged vertically.

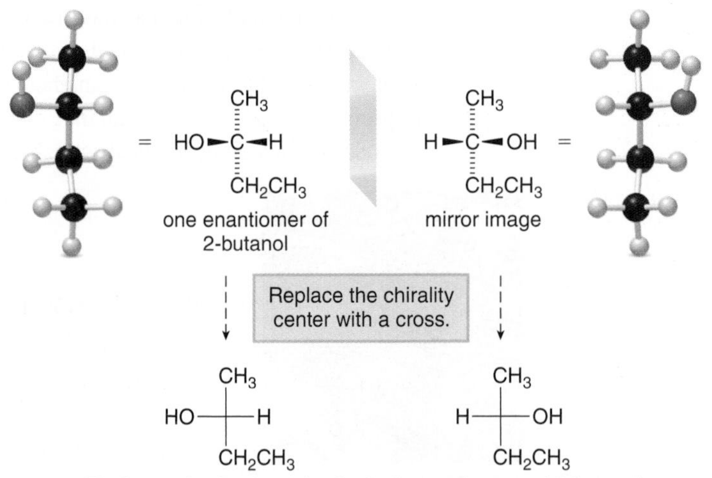

Fischer projection formulas for both enantiomers of 2-butanol

Sample Problem 12.10 illustrates another example of drawing two enantiomers of a carbohydrate using this convention. The carbonyl group of a carbohydrate is placed on the top vertical bond in the Fischer projection.

SAMPLE PROBLEM 12.10

Draw both enantiomers of glyceraldehyde, a simple carbohydrate, using Fischer projection formulas.

$HOCH_2CH(OH)CHO$

glyceraldehyde

Analysis

- Draw the tetrahedron of one enantiomer with the horizontal bonds on wedges, and the vertical bonds on dashed lines. Arrange the four groups on the chirality center—H, OH, CHO, and CH_2OH—with the CHO group at the top and the carbon skeleton drawn vertically.
- Draw the second enantiomer by arranging the substituents in the mirror image so they are a **reflection** of the groups in the first molecule.
- Replace the chirality center with a cross to draw the Fischer projections.

Solution

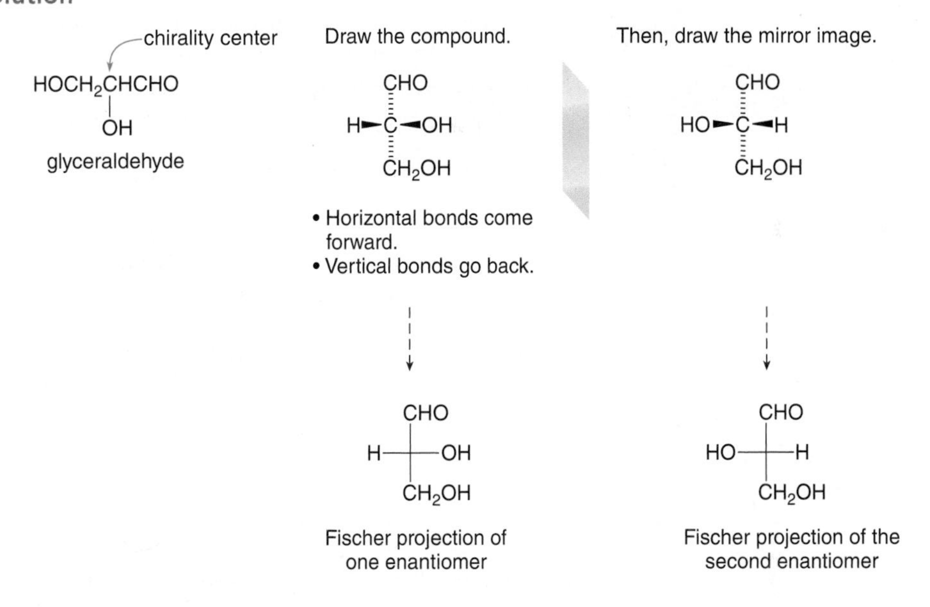

Fischer projection of one enantiomer

Fischer projection of the second enantiomer

PROBLEM 12.31

Convert each molecule into a Fischer projection formula.

a. CH_3 (top), H (wedge, left), C, Br (wedge, right), Cl (bottom)

b. COOH (top), H (wedge, left), C, OH (wedge, right), CH_2Cl (bottom)

PROBLEM 12.32

Draw Fischer projections of both enantiomers for each compound.

a. $CH_3CH(OH)CH_2CH_2CH_3$

b. $CH_3CH_2CH(Cl)CH_2Cl$

12.12 FOCUS ON HEALTH & MEDICINE Chiral Drugs

Ibuprofen is commonly used to relieve headaches and muscle and joint pain.

A living organism is a sea of chiral molecules. Many drugs are chiral, and often they must interact with a chiral receptor to be effective. One enantiomer of a drug may be effective in treating a disease whereas its mirror image may be ineffective.

Why should such a difference in biological activity be observed? One enantiomer may "fit" the chiral receptor and evoke a specific response. Its mirror image may not fit the same receptor, making it ineffective; or if it "fits" another receptor, it can evoke a totally different response. Figure 12.8 schematically illustrates this difference in binding between two enantiomers and a chiral receptor.

Ibuprofen and L-dopa are two drugs that illustrate how two enantiomers can have different biological activities.

Ibuprofen is the generic name for the pain relievers known as Motrin and Advil. Ibuprofen has one chirality center, and thus exists as a pair of enantiomers. Only one enantiomer, labeled **A,** is an active anti-inflammatory agent. Its enantiomer **B** is inactive. **B,** however, is slowly converted to **A** in the body. Ibuprofen is sold as a mixture of both enantiomers.

COOH, H, CH_3, $(CH_3)_2CHCH_2$ — **ibuprofen** active anti-inflammatory agent **A**

COOH, H, CH_3, $CH_2CH(CH_3)_2$ — inactive enantiomer **B**

Figure 12.8 The Interaction of Two Enantiomers with a Chiral Receptor

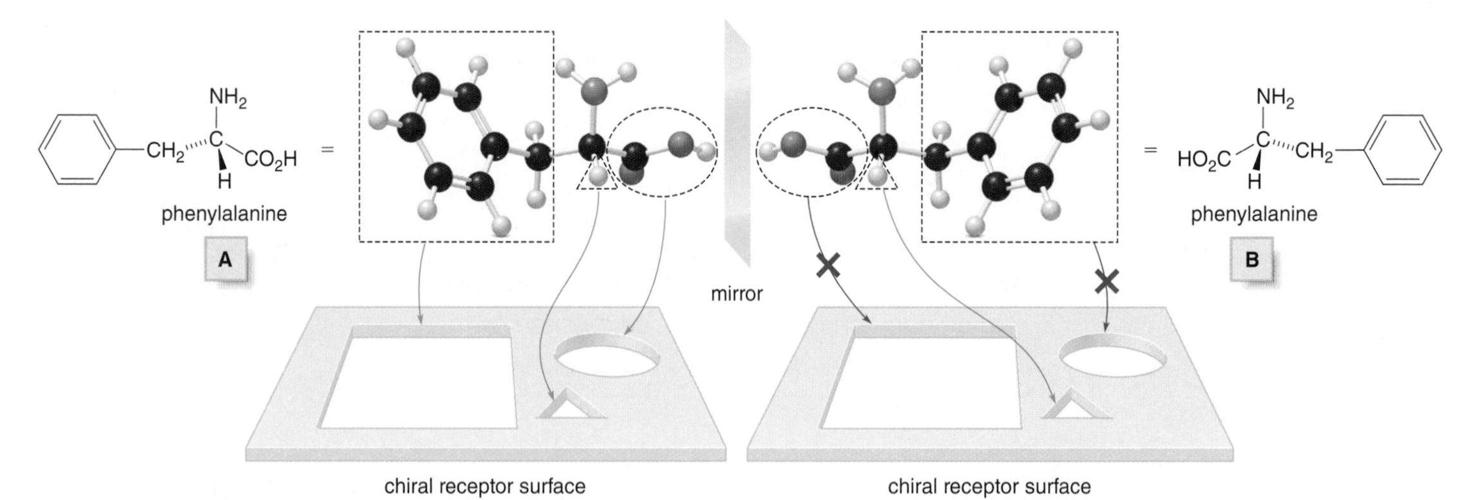

One enantiomer of the amino acid phenylalanine, labeled **A,** has three groups that can interact with the appropriate binding sites of the chiral receptor. The groups around the chirality center in the mirror image **B,** however, can never be arranged so that all three groups can bind to these same three binding sites. Thus, enantiomer **B** does not "fit" the same receptor, so it does not evoke the same response.

The broad bean *Vicia faba* produces the drug L-dopa. To meet the large demand for L-dopa today, the commercially available drug is synthesized in the laboratory.

L-Dopa was first isolated from the seeds of the broad bean *Vicia faba,* and since 1967 it has been the drug of choice for the treatment of Parkinson's disease. L-Dopa is a chiral molecule, which is sold as a single enantiomer. Like many chiral drugs, only one enantiomer is active against Parkinson's disease. The inactive enantiomer is also responsible for neutropenia, a decrease in certain white blood cells that help to fight infection.

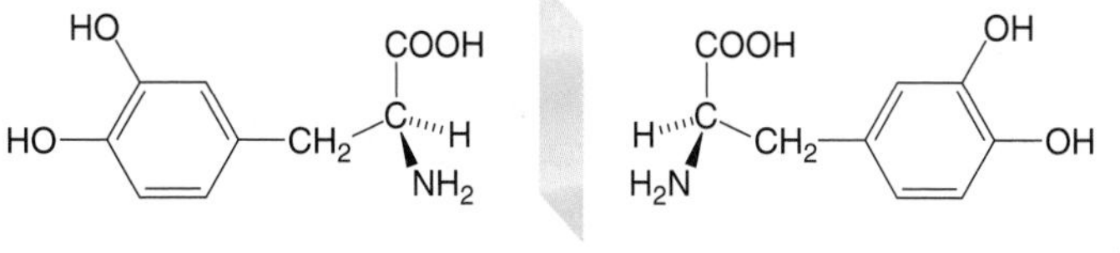

L-dopa
active against Parkinson's disease

D-dopa
enantiomer responsible for undesired side effects

Parkinson's disease, which afflicts 1.5 million individuals in the United States, results from the degeneration of neurons that produce the neurotransmitter dopamine in the brain. When the level of dopamine drops, the loss of motor control symptomatic of Parkinson's disease results.

L-Dopa is an oral medication that is transported to the brain by the bloodstream. In the brain it is converted to dopamine (Figure 12.9). Dopamine itself cannot be given as a medication because it cannot pass from the bloodstream into the brain; that is, it does not cross the blood–brain barrier.

pathways
Parkinson's disease is characterized by degeneration of dopamine nerve pathways in the basal ganglia.
Dopamine is released by one nerve cell, and then binds to a receptor site in a target cell.

Figure 12.9 L-Dopa, Dopamine, and Parkinson's Disease

COOH

CH2

NH2

CH2CH2NH2

basal ganglia
frontal lobe
nerve impulse

cell body

molecules

release

receptors
axon

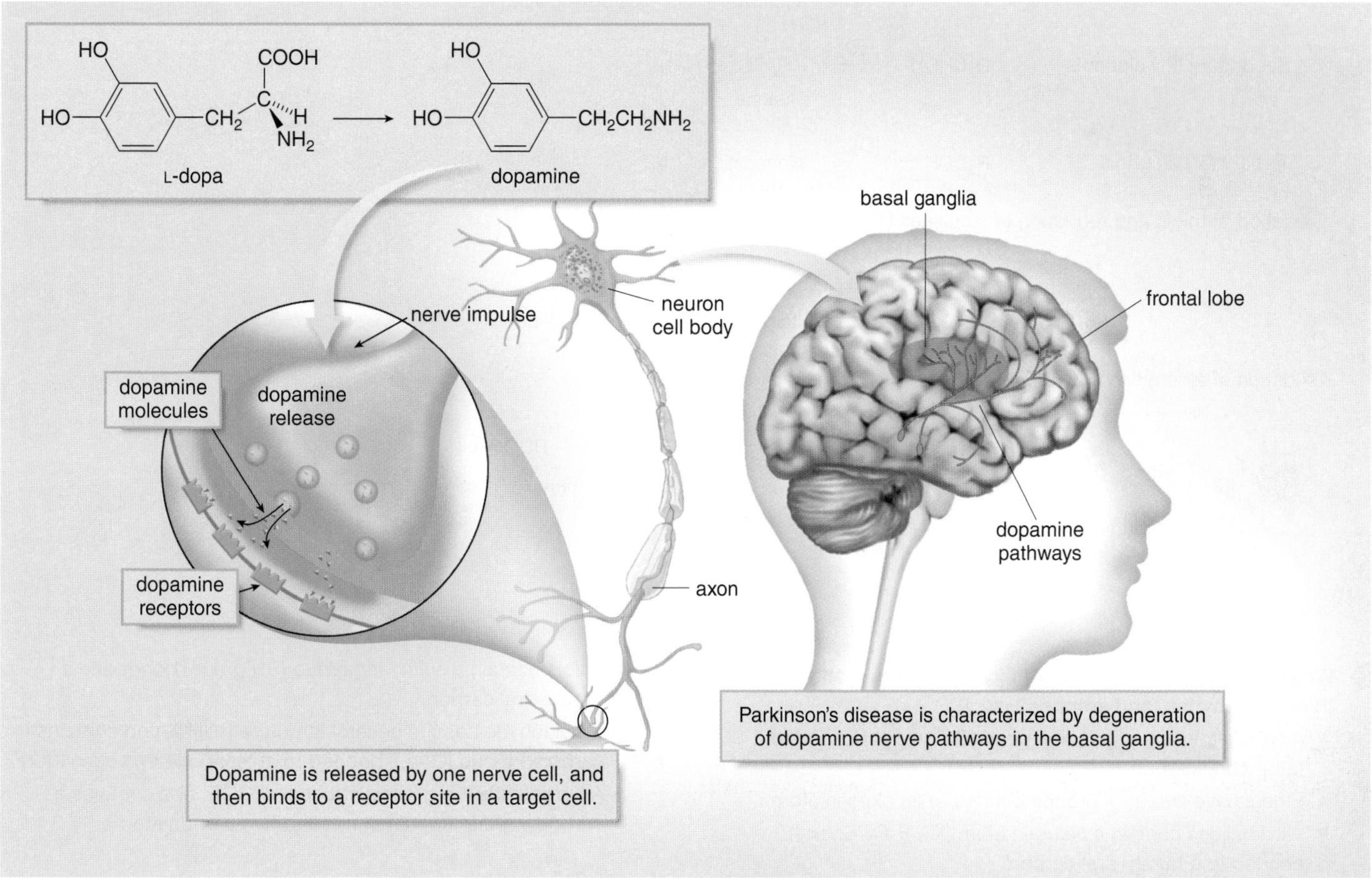

KEY TERMS

Achiral (12.11)
Alcohol (12.1)
Aldehyde (12.1)
Alkyl halide (12.1)
Carbonyl group (12.1)
Chiral (12.11)
Chirality center (12.11)
Cross formula (12.11)
Dehydration (12.5)
Disulfide (12.7)
Elimination (12.5)
Enantiomer (12.11)
Ether (12.1)
Fischer projection formula (12.11)
Hydroxyl group (12.1)
Ketone (12.1)
Oxidation (12.5)
Primary (1°) alcohol (12.2)
Primary (1°) alkyl halide (12.6)
Secondary (2°) alcohol (12.2)
Secondary (2°) alkyl halide (12.6)
Stereochemistry (12.11)
Stereoisomer (12.11)
Sulfhydryl group (12.1)
Tertiary (3°) alcohol (12.2)
Tertiary (3°) alkyl halide (12.6)
Thiol (12.1)
Tollens reagent (12.10)

KEY REACTIONS

[1] Dehydration of alcohols (12.5)

$$-\underset{H}{\overset{|}{\underset{|}{C}}}-\underset{OH}{\overset{|}{\underset{|}{C}}}- \xrightarrow[H_2SO_4]{} \;\rangle C{=}C\langle \; + \; H_2O$$

[2] Oxidation of alcohols (12.5)

a. Primary (1°) alcohols

$$RCH_2OH \xrightarrow{[O]} \underset{\text{aldehyde}}{RCHO} \xrightarrow{[O]} \underset{\text{carboxylic acid}}{RCOOH}$$

b. Secondary (2°) alcohols

$$R_2CHOH \xrightarrow{[O]} \underset{\text{ketone}}{R_2CO}$$

c. Tertiary (3°) alcohols

$$R_3COH \xrightarrow{[O]} \text{No reaction}$$

[3] Oxidation of thiols and reduction of disulfides (12.7)

$$2\ \underset{\text{thiol}}{R{-}S{-}H} \underset{[H]}{\overset{[O]}{\rightleftharpoons}} \underset{\text{disulfide}}{R{-}S{-}S{-}R}$$

[4] Oxidation of aldehydes (12.10)

$$RCHO \xrightarrow[\text{or Tollens reagent}]{K_2Cr_2O_7} \underset{\text{carboxylic acid}}{RCOOH}$$

KEY CONCEPTS

1 What are the characteristics of alcohols, ethers, alkyl halides, thiols, aldehydes, and ketones? (12.1)

- Alcohols contain a hydroxyl group (OH group) bonded to a tetrahedral carbon.
- Ethers have two alkyl groups bonded to an oxygen atom.
- Alkyl halides contain a halogen atom (X = F, Cl, Br, or I) bonded to a tetrahedral carbon.
- Thiols contain a sulfhydryl group (SH group) bonded to a tetrahedral carbon.
- An aldehyde has the general structure RCHO, and contains a carbonyl group (C=O) bonded to at least one hydrogen atom.
- A ketone has the general structure RCOR, and contains a carbonyl group (C=O) bonded to two carbon atoms.

2 What are the properties of each functional group?

- Alcohols have a bent shape and polar C—O and O—H bonds. Their OH bond allows for intermolecular hydrogen bonding between two alcohol molecules or between an alcohol molecule and water. As a result, alcohols have the strongest intermolecular forces of the families of molecules in this chapter. (12.2)
- Ethers have a bent shape and two polar C—O bonds, so they have a net dipole. (12.3)
- Alkyl halides with one halogen have one polar bond and a net dipole. (12.6)
- Thiols have a bent shape and lower boiling points than alcohols with the same number of carbons. (12.7)
- The carbonyl group is polar, giving an aldehyde or ketone stronger intermolecular forces than hydrocarbons. (12.8)
- Aldehydes and ketones have lower boiling points than alcohols, but higher boiling points than hydrocarbons of comparable molecular weight. (12.8)

3 How are alcohols, ethers, alkyl halides, aldehydes, and ketones named?

- Alcohols are identified by the suffix *-ol.* (12.2)
- Simple ethers are named by naming the alkyl groups bonded to the ether oxygen and adding the word *ether.* (12.3)
- Alkyl halides are named as *halo* alkanes. (12.6)
- Aldehydes are identified by the suffix *-al,* and the carbon chain is numbered to put the carbonyl group at C1. (12.8)
- Ketones are identified by the suffix *-one,* and the carbon chain is numbered to give the carbonyl group the lower number. (12.8)

4 What products are formed when an alcohol undergoes dehydration or oxidation? (12.5)

- Alcohols form alkenes on treatment with strong acid. The elements of H and OH are lost from two adjacent atoms and a new double bond is formed.
- Primary alcohols (RCH_2OH) are oxidized to aldehydes (RCHO), which are further oxidized to carboxylic acids (RCO_2H).
- Secondary alcohols (R_2CHOH) are oxidized to ketones.
- Tertiary alcohols have no C—H bond on the carbon with the OH group, so they are not oxidized.

5 What product is formed when a thiol is oxidized? (12.7)

- Thiols (RSH) are oxidized to disulfides (RSSR).
- Disulfides are reduced to thiols.

6 What products are formed when aldehydes are oxidized? (12.10)

- Aldehydes are oxidized to carboxylic acids (RCOOH) with $K_2Cr_2O_7$ or Tollens reagent.
- Ketones are not oxidized since they contain no H atom on the carbonyl carbon.

7 What is a chirality center and when is a molecule chiral or achiral? (12.11)

- A chirality center is a carbon with four different groups around it.
- A chiral molecule is not superimposable on its mirror image.
- An achiral molecule is superimposable on its mirror image.

8 What is a Fischer projection? (12.11)

- A Fischer projection is a specific way of depicting a chirality center. The chirality center is located at the intersection of a cross. The horizontal lines represent bonds that come out of the plane on wedges, and the vertical lines represent bonds that go back on dashed lines.

UNDERSTANDING KEY CONCEPTS

Selected in-chapter and odd-numbered end-of-chapter problems have brief answers at the end of each chapter. The *Student Study Guide and Solutions Manual* contains detailed solutions to all in-chapter and odd-numbered end-of-chapter problems, as well as additional worked examples and a chapter self-test.

12.33 Locate the chirality center in the neurotransmitter norepinephrine. Classify the hydroxyl group of the alcohol (not the phenols) as 1°, 2°, or 3°.

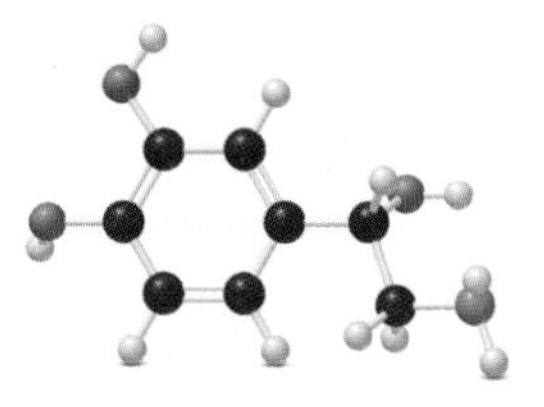

norepinephrine

12.34 Locate the chirality center in the general anesthetic isoflurane. Why does isoflurane have a bent arrangement around its O atom?

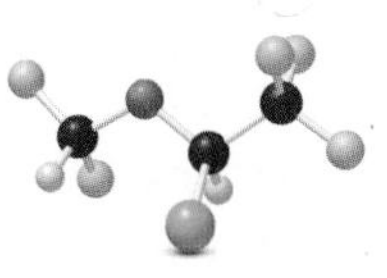

isoflurane

12.35 Consider the following ball-and-stick model of an alcohol.

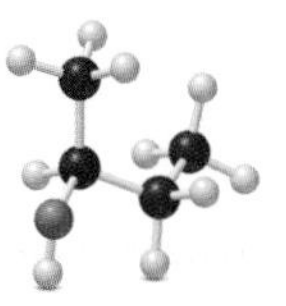

a. Locate the chirality center.

b. Classify the alcohol as 1°, 2°, or 3°.

c. Name the alcohol.

d. Draw the structure of the product formed when the alcohol is oxidized.

12.36 Consider the following ball-and-stick model.

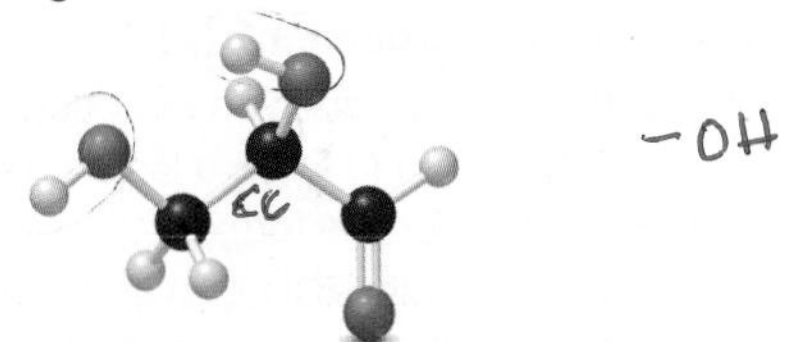

a. Locate the chirality center.
b. Is the carbonyl group part of an aldehyde or a ketone?
c. Classify each hydroxyl group as 1°, 2°, or 3°.
d. What product is formed when this compound is treated with Tollens reagent?

12.37 Name each compound.

a. F (on C2)
$CH_3CHCH_2CH_2CH_3$

b. $CH_3CH_2OCH_2CH_2CH_2CH_3$

c. $CH_3CH_2CH_2CHCH_2CHO$
CH_3 (on C4 of the chain as drawn)

d. $(CH_3)_2CHCH_2-C(=O)-CH_3$

12.38 Name each compound.

a. $CH_3CH_2CHCH_2CH_2CH(CH_3)_2$
Cl (on the third carbon)

b. $CH_3OCH_2CH_2CH_3$

c. $(CH_3)_3CCH_2CHO$

d. $CH_3CH_2CH_2CH_2CHCH_2CH_3$
$C(=O)CH_3$ (on the CH)

12.39 Answer the following questions about alcohol **A**.

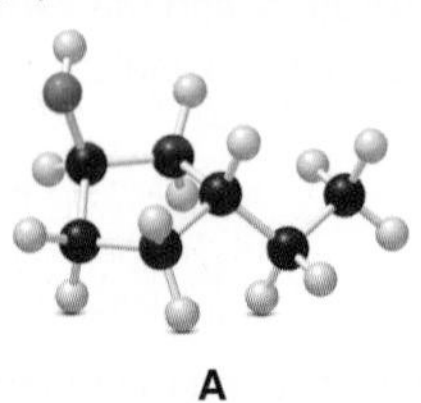

A

a. Give the IUPAC name.
b. Classify the alcohol as 1°, 2°, or 3°.
c. Draw the products formed when **A** is dehydrated with H_2SO_4.
d. What product is formed when **A** is oxidized with $K_2Cr_2O_7$?
e. Draw a constitutional isomer of **A** that contains an OH group.
f. Draw a constitutional isomer of **A** that contains an ether.

12.40 Answer the following questions about alcohol **B**.

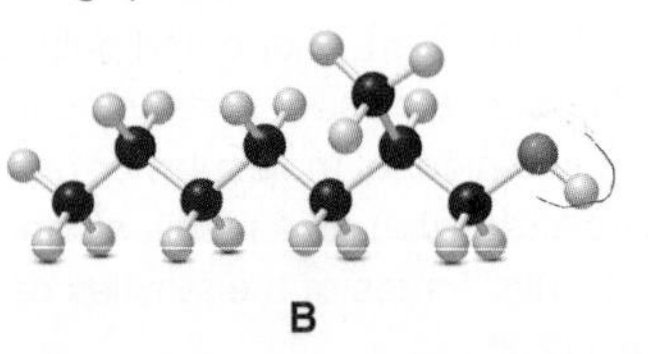

B

a. Give the IUPAC name.
b. Classify the alcohol as 1°, 2°, or 3°.
c. Draw the products formed when **B** is dehydrated with H_2SO_4.
d. What product is formed when **B** is oxidized with $K_2Cr_2O_7$?
e. Draw a constitutional isomer of **B** that contains an OH group.
f. Draw a constitutional isomer of **B** that contains an ether.

12.41 Rank the following compounds in order of increasing melting point.

cyclopentane=O   cyclopentane–OH   cyclopentane–CH_3

12.42 Menthone and menthol are both isolated from mint. Explain why menthol is a solid at room temperature but menthone is a liquid.

CH_3, =O, $CH(CH_3)_2$ on cyclohexane — menthone

CH_3, OH, $CH(CH_3)_2$ on cyclohexane — menthol

12.43 Hydroxydihydrocitronellal is an example of a compound that has two enantiomers that smell differently. One enantiomer smells minty, and the other smells like lily of the valley.

$CH_3C(OH)(CH_3)CH_2CH_2CH_2CH(CH_3)CH_2-CHO$

hydroxydihydrocitronellal

a. Locate the chirality center in this compound.
b. Draw Fischer projection formulas for both enantiomers.

12.44 Two enantiomers can sometimes taste very different. L-Leucine, a naturally occurring amino acid used in protein synthesis, tastes bitter, but its enantiomer, D-leucine, tastes sweet.

COOH
H_2N-C-H
$CH_2CH(CH_3)_2$

leucine

a. Locate the chirality center in leucine.
b. Draw Fischer projection formulas for both enantiomers.

ADDITIONAL PROBLEMS

Alcohols, Ethers, Alkyl Halides, and Thiols

12.45 Classify each alcohol as 1°, 2°, or 3°.

a. $CH_3CH_2CH_2OH$

b. (cyclohexane ring bearing OH, and CH_3, CH_3 on the adjacent carbon)

12.46 Classify each alkyl halide as 1°, 2°, or 3°.

a. (cyclohexane ring bearing Cl and $CH_2CH_2CH_3$ on the same carbon)

b. $CH_3CH_2CH_2CHBr$ with CH_3 on the CHBr carbon

12.47 Draw the structure of a molecule that fits each description:

a. a 2° alcohol of molecular formula $C_6H_{14}O$
b. an ether with molecular formula $C_6H_{14}O$ that has a methyl group bonded to oxygen
c. a thiol that contains four carbons
d. a 2° alkyl bromide that contains five carbons

12.48 Draw the structure of a molecule that fits each description:

a. a 2° alcohol of molecular formula $C_6H_{12}O$
b. a cyclic ether with molecular formula $C_5H_{10}O$
c. a thiol of molecular formula C_3H_8S
d. a 1° alkyl iodide with molecular formula $C_6H_{13}I$

Nomenclature of Alcohols, Ethers, and Alkyl Halides

12.49 Name each molecule depicted in the ball-and-stick models.

a. 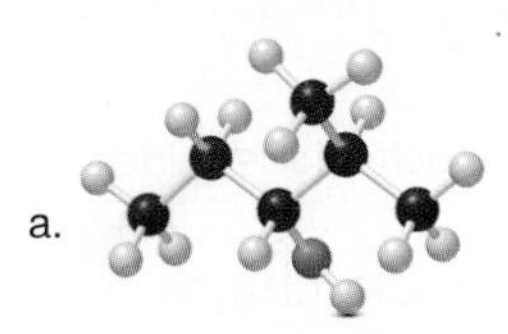

b. 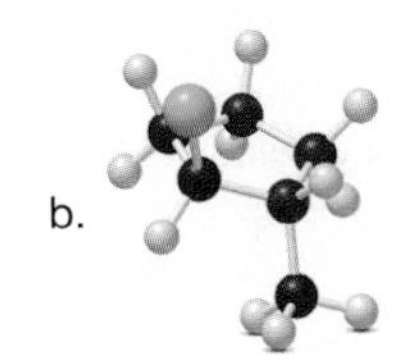

12.50 Name each molecule depicted in the ball-and-stick models.

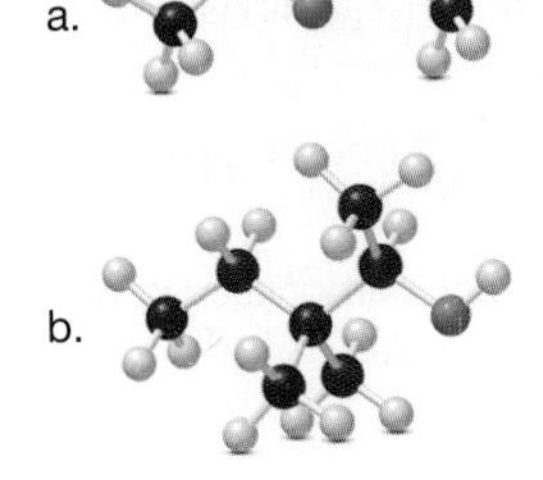

a.

b.

12.51 Give the IUPAC name for each alcohol.

a. $CH_3CH(OH)C(CH_3)_2CH_2CH_2CH_3$ (drawn as $CH_3\overset{HO}{CH}\overset{CH_3}{C}CH_2CH_2CH_3$ with a second CH_3 below)

b. $(CH_3)_2CHCH_2CHCH_2CH_3$ with $CH_2CH_2CH_2OH$ on the second CH

c. (cyclohexane ring bearing HO and, at the para position, $CH_2CH_2CH_2CH_3$)

12.52 Give the IUPAC name for each alcohol.

a. $CH_3CH_2CH_2CHCH_2CH_3$ with CH_2OH on the CH

b. $CH_3(CH_2)_3CHCH_3$ with CH_2CH_2OH on the CH

c. (cyclopentane ring bearing OH and CH_2CH_3 on one carbon, CH_3 on the adjacent carbon)

12.53 Give the structure corresponding to each name.

a. 3-hexanol
b. dibutyl ether
c. 2-methylcyclopropanol
d. 1-chloro-3,5-dimethylheptane

12.54 Give the structure corresponding to each name.

a. 3-methyl-3-pentanol
b. ethyl propyl ether
c. 2-bromo-2,4-dimethylhexane
d. 3,5-dimethylcyclohexanol

12.55 Draw the structures and give the IUPAC names for all constitutional isomers of molecular formula $C_7H_{16}O$ that contain an OH group and have seven carbons in the longest chain.

12.56 Draw structures for the four constitutional isomers of molecular formula C_4H_9Br. Give the IUPAC name for each alkyl halide.

Physical Properties

12.57 Explain the following observation. Dimethyl ether [$(CH_3)_2O$] and ethanol (CH_3CH_2OH) are both water-soluble compounds. The boiling point of ethanol (78 °C), however, is much higher than the boiling point of dimethyl ether (–24 °C).

12.58 Rank the following compounds in order of increasing boiling point: $CH_3CH_2OCH_2CH_3$, $CH_3(CH_2)_3CH_3$, $CH_3CH_2CH_2CH_2OH$.

12.59 Explain why two four-carbon organic molecules have very different solubility properties: 1-butanol ($CH_3CH_2CH_2CH_2OH$) is water soluble but 1-chlorobutane ($CH_3CH_2CH_2CH_2Cl$) is water insoluble.

12.60 Explain why the boiling point of $CH_3CH_2CH_2CH_2OH$ (118 °C) is higher than the boiling point of $CH_3CH_2CH_2CH_2SH$ (98 °C), even though $CH_3CH_2CH_2CH_2SH$ has a higher molecular weight.

12.61 Which compound in each pair has the higher boiling point?

a. $(CH_3)_3CCH_2CH_2CH_3$ or $(CH_3)_3CCH_2CHO$

b. cyclopentyl—$COCH_3$ or cyclopentyl—CH_2CH_2OH

12.62 Which compound in each pair is more water soluble?

a. $CH_3(CH_2)_6CHO$ or $CH_3(CH_2)_7OH$

b. benzene ring—C(=O)CH_3 or benzene ring—$CH(CH_3)_2$

Reactions of Alcohols and Thiols

12.63 Draw the product formed when each alcohol is dehydrated with H_2SO_4.

a. cyclobutane ring with OH

b. benzene ring—$CH(OH)CH_2CH_3$

12.64 Draw the product formed when each alcohol is dehydrated with H_2SO_4.

a. $(CH_3)_2CHCH_2CH_2CH_2OH$ b. $(CH_3CH_2)_3COH$

12.65 Draw the product formed when each alcohol is oxidized with $K_2Cr_2O_7$. In some cases, no reaction occurs.

a. $CH_3(CH_2)_6CH_2OH$

b. $(CH_3CH_2)_2CHOH$

c. $CH_3CH_2C(CH_3)_2OH$

12.66 Draw the product formed when each alcohol is oxidized with $K_2Cr_2O_7$. In some cases, no reaction occurs.

a. cyclopentane ring with OH and adjacent CH_3

b. $CH_3(CH_2)_8CH_2OH$

c. $(CH_3CH_2)_3COH$

12.67 Cortisol is an anti-inflammatory agent that also regulates carbohydrate metabolism. What oxidation product is formed when cortisol is treated with $K_2Cr_2O_7$?

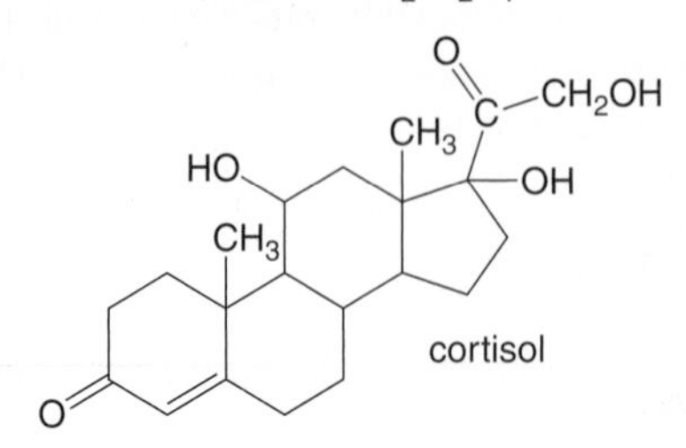

cortisol

12.68 Xylitol is a nontoxic compound as sweet as table sugar but with only one-third the calories, so it is often used as a sweetening agent in gum and hard candy.

CH_2OH
H—C—OH
HO—C—H
H—C—OH
CH_2OH

xylitol

a. Classify the OH groups as 1°, 2°, or 3°.

b. What product is formed when only the 1° OH groups are oxidized?

c. What product is formed when all of the OH groups are oxidized?

12.69 What products are formed when **X** is treated with each reagent: (a) H_2SO_4; (b) $K_2Cr_2O_7$?

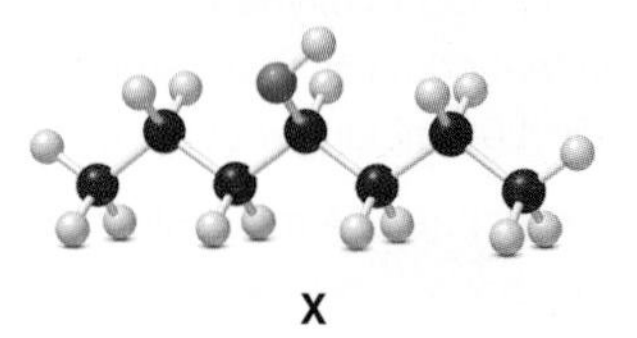

X

12.70 What products are formed when **Y** is treated with each reagent: (a) H_2SO_4; (b) $K_2Cr_2O_7$?

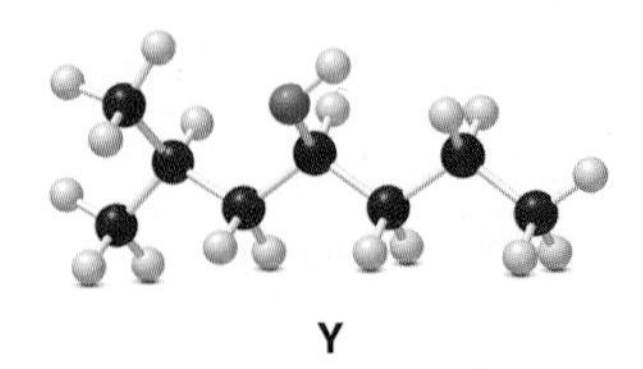

Y

12.71 What disulfide is formed when each thiol is oxidized?

a. cyclohexane ring—SH b. $CH_3(CH_2)_4SH$

12.72 The smell of fried onions is in part determined by the two disulfides below. What thiols are formed when each disulfide is reduced?

a. $CH_3S—SCH_2CH_2CH_3$ b. $CH_2{=}CHCH_2S—SCH_2CH_2CH_3$

Structure and Nomenclature of Aldehydes and Ketones

12.73 Give an acceptable name for each carbonyl compound depicted in the ball-and-stick models.

a. 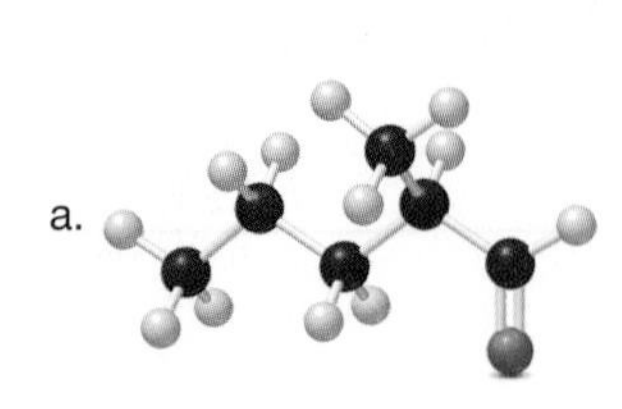

b. 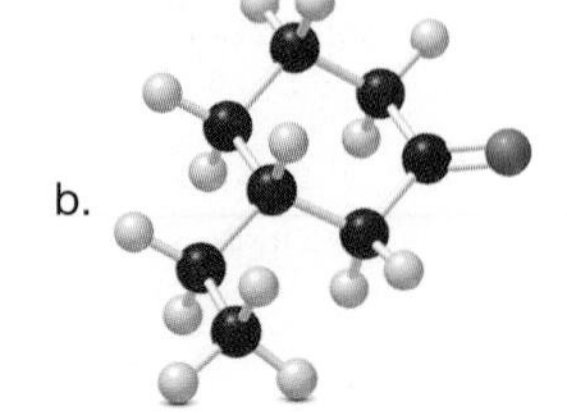

12.74 Give an acceptable name for each carbonyl compound depicted in the ball-and-stick models.

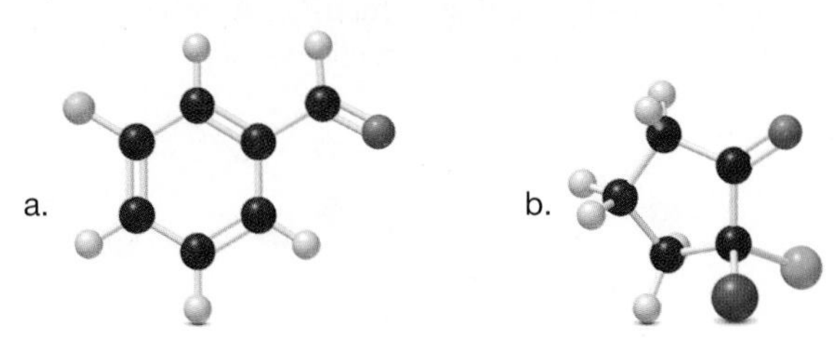

12.75 Give an acceptable name for each aldehyde or ketone.

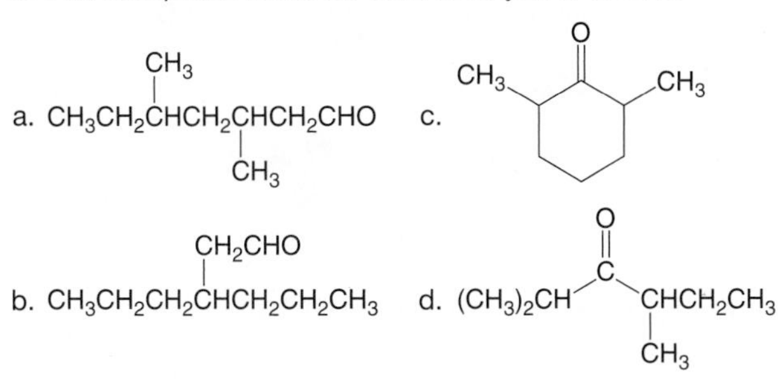

12.76 Give an acceptable name for each aldehyde or ketone.

a. $CH_3CH_2CH_2CH_2CHCHCH_3$ (with CH_3 on C5 and CH_2CHO on C6)

b. $CH_3CH_2CH_2$—C(CH_3)$_2$—$CH_2CH_2CHCH_2CHO$ (with CH_2CH_3 on the CH)

c. CH_3CH_2–C(=O)–$(CH_2)_5CH(CH_3)_2$

d. cyclopentanone with CH_2CH_3 and CH_3 substituents

12.77 Draw the structure corresponding to each name.

a. 3,4-dimethylhexanal
b. 4-hydroxyheptanal
c. 3,3-dimethyl-2-hexanone
d. 2,4,5-triethylcyclohexanone

12.78 Draw the structure corresponding to each name.

a. 2-propylheptanal
b. 3,4-dihydroxynonanal
c. 1-chloro-3-pentanone
d. 3-hydroxycyclopentanone

12.79 Draw the structure of the four constitutional isomers of molecular formula $C_6H_{12}O$ that contain an aldehyde and four carbons in the longest chain. Give the IUPAC name for each aldehyde.

12.80 Draw the structure of the three isomeric ketones of molecular formula $C_5H_{10}O$. Give the IUPAC name for each ketone.

Oxidation Reactions

12.81 What product is formed when each compound is treated with $K_2Cr_2O_7$?

a. $CH_3(CH_2)_4CHO$
b. cyclopentyl–CH_2CHO
c. $CH_3(CH_2)_4CH_2OH$

12.82 What product is formed when each compound is treated with $K_2Cr_2O_7$?

a. benzene ring with Cl and CHO (ortho)
b. $CH_3(CH_2)_8CHO$
c. $CH_3CHCH_2CH_2CH_3$ (OH on C2)

12.83 What product is formed when each compound in Problem 12.81 is treated with Tollens reagent (Ag_2O, NH_4OH)? With some compounds, no reaction occurs.

12.84 What product is formed when each compound in Problem 12.82 is treated with Tollens reagent (Ag_2O, NH_4OH)? With some compounds, no reaction occurs.

12.85 Benzaldehyde is the compound principally responsible for the odor of almonds. What products are formed when benzaldehyde is treated with each reagent?

C_6H_5–C(=O)H

benzaldehyde

a. $K_2Cr_2O_7$
b. Ag_2O, NH_4OH

12.86 Anisaldehyde is a component of anise, a spice with a licorice odor used in cooking and aromatherapy. What products are formed when anisaldehyde is treated with each reagent?

CH_3O–C_6H_4–C(=O)H

anisaldehyde

a. $K_2Cr_2O_7$
b. Ag_2O, NH_4OH

Chiral Compounds, Chirality Centers, and Enantiomers

12.87 Label each of the following objects as chiral or achiral: (a) chalk; (b) shoe; (c) baseball glove; (d) soccer ball.

12.88 Label each of the following objects as chiral or achiral: (a) boot; (b) index card; (c) scissors; (d) drinking glass.

12.89 Label each compound as chiral or achiral.

a. C with CF_3, Br, Cl, H

halothane (general anesthetic)

b. C with Cl, H, H, Cl

dichloromethane (common solvent)

12.90 Label each compound as chiral or achiral.

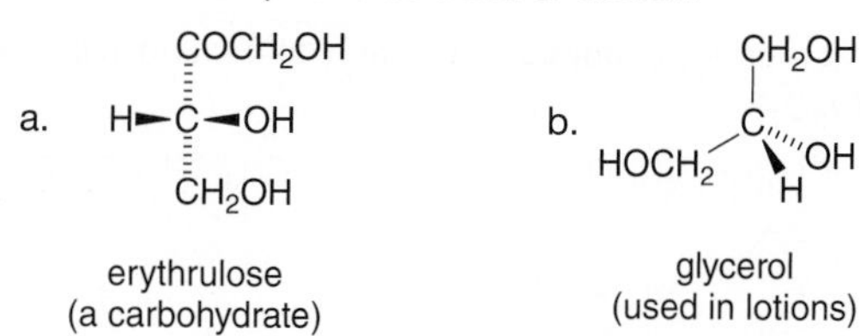

a. erythrulose (a carbohydrate)

b. glycerol (used in lotions)

12.91 Label the chirality center (if one exists) in each compound. Compounds contain zero or one chirality center.

a. CH_3CH_2CHCl (with Cl on the CH)

b. $CH_3CHCHCl_2$ (with Cl on the second C)

c. CH_3CHCH_3 (with Cl on the central C)

d. $CH_3CCH_2CHCH_3$ (with CH_3 and H on the second C; Cl on the fourth C)

12.92 Label the chirality center (if one exists) in each compound. Compounds contain zero or one chirality center.

a. $CH_3C(=O)CH_2CH_3$

b. $CH_3C(=O)CHCH_2CH_3$ (with CH_3 on the CH)

c. CH_3CHCHO (with CH_3 on the CH)

d. CH_3CHCHO (with Cl on the CH)

12.93 Locate the chirality center(s) in each biologically active compound.

a. $HO_2CCH_2CHCNHCHCH_2$—C_6H_5 (with NH_2 on the first CH, C=O, and CO_2CH_3 on the second CH)

aspartame
Trade name: Equal
(synthetic sweetener)

b. C_6H_5—C(=O)—C_6H_4—$CHCO_2H$ (with CH_3 on the CH)

ketoprofen
(anti-inflammatory agent)

12.94 Locate the chirality center in each compound.

a. C_6H_5—CH_2CHNCH_3 (with CH_3 on the CH and H on the N)

methamphetamine
(addictive stimulant)

b. HO—C_6H_3(HO)—$CH_2CH_2NCHCH_2CH_2$—C_6H_4—OH (with H on the N and CH_3 on the CH)

dobutamine
(heart stimulant)

Isomers and Fischer Projections

12.95 Consider the ball-and-stick model of **A,** and label **B** and **C** as either identical to **A** or an enantiomer of **A.**

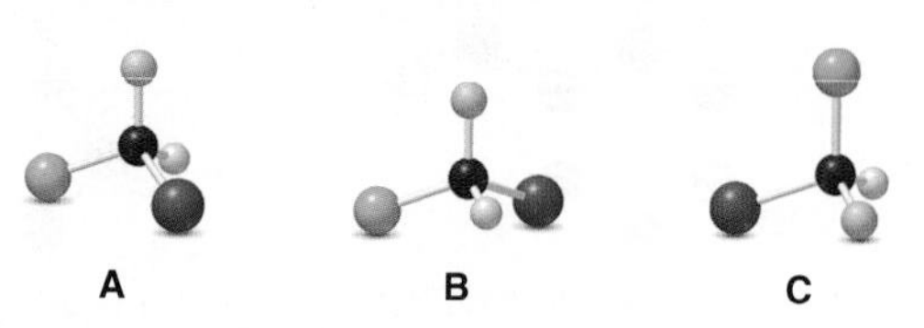

12.96 Consider the ball-and-stick model of **D,** and label **E** and **F** as either identical to **D** or an enantiomer of **D.**

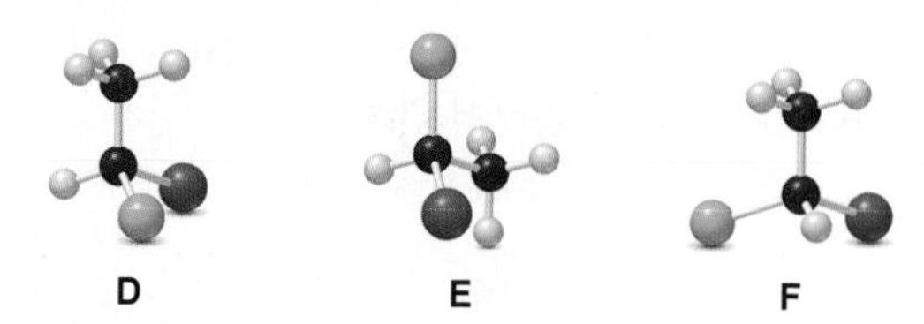

12.97 How are the compounds in each pair related? Are they identical molecules or enantiomers?

a. H—C(—CH_3) with CH_3 above and OH below **and** CH_3—C—H with CH_3 above and OH below

b. Br above C; CH_3, Cl (hashed), H (wedged) **and** Br above C; Cl (hashed), H (wedged), CH_3

12.98 How are the compounds in each pair related? Are they identical molecules or enantiomers?

a. $HOCH_2$—C—OH with COOH above and CH_3 below **and** HO—C—CH_2OH with COOH above and CH_3 below

b. OCH_3 above C; CH_3O, CHO (hashed), CH_3 (wedged) **and** OCH_3 above C; OHC (hashed), CH_3 (wedged), OCH_3

12.99 Convert each three-dimensional representation into a Fischer projection.

a. H—C—OH with CHO above and CH_3 below

b. H—C—NH_2 with $COCH_3$ above and CH_2CH_3 below

12.100 Convert each three-dimensional representation into a Fischer projection.

a. CH_3—C—H with OCH_2CH_3 above and $N(CH_3)_2$ below

b. HO—C—CH_3 with CHO above and $CH(CH_3)_2$ below

12.101 Convert the given ball-and-stick model to a Fischer projection.

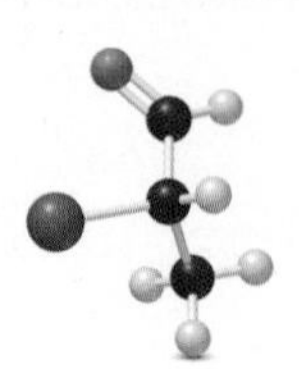

12.102 Convert the given ball-and-stick model to a Fischer projection.

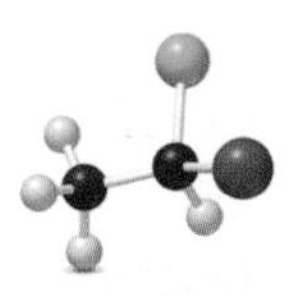

Applications

12.103 In contrast to ethylene glycol ($HOCH_2CH_2OH$), which is extremely toxic, propylene glycol [$CH_3CH(OH)CH_2OH$] is nontoxic because it is oxidized in the body to a product produced during the metabolism of carbohydrates. What product is formed when propylene glycol is oxidized?

12.104 Lactic acid [$CH_3CH(OH)CO_2H$] gives sour milk its distinctive taste. What product is formed when lactic acid is oxidized by $K_2Cr_2O_7$?

12.105 Lactic acid is a product of glucose metabolism. During periods of strenuous exercise, lactic acid forms faster than it can be oxidized, resulting in the aching feeling of tired muscles.

CH_3CHCO_2H (with OH on the central C)
lactic acid

a. Locate the chirality center in lactic acid.
b. Draw Fischer projection formulas for both enantiomers of lactic acid.

12.106 Plavix, the trade name for the generic drug clopidogrel, is used in the treatment of coronary artery disease. Like many newer drugs, Plavix is sold as a single enantiomer.

CO_2CH_3
S
N—C—
H
Cl

clopidogrel
Trade name: Plavix

a. Locate the chirality center.
b. Draw Fischer projection formulas of both enantiomers.

CHALLENGE PROBLEMS

12.107 Identify the nine chirality centers in sucrose, the sweet-tasting carbohydrate we use as table sugar.

$HOCH_2$ OH
O $HOCH_2$ OH
HO O
O CH_2OH
HO OH

sucrose

12.108 Dehydration of alcohol **C** forms two products of molecular formula $C_{14}H_{12}$ that are isomers, but they are not constitutional isomers. Draw the structures of these two products.

OH H
—C—C—
H H

C

BEYOND THE CLASSROOM

12.109 Nitrous oxide, chloroform, diethyl ether, halothane, and sevoflurane are compounds that have been used in general anesthesia. Pick two or more of these compounds and look up their structures. What are the advantages and disadvantages of each anesthetic? Is the anesthetic still widely used, and if not, why has its use been curtailed or discontinued?

12.110 Go to your local drug store or supermarket and examine the ingredients of several mouthwashes and liquid cold remedies. List any components whose names end in the suffix *-ol,* and research the chemical structure. Does the compound contain a hydroxyl group? What purpose does the ingredient serve?

12.111 Thalidomide and naproxen are two drugs that contain a chirality center. Research the structure of each drug and label the chirality center. Each drug has two enantiomers, one of which has beneficial therapeutic effects, while the other is harmful. How does the biological activity of the enantiomers of each drug differ? Can the drug be used effectively now that the difference between the enantiomers is understood? See if you can find other examples of drugs whose enantiomers have very different properties.

12.112 Many alcohols and ketones derived from plant sources have pleasant odors. Fragrant alcohols include linalool, sclareol, patchoulol, and nerolidol, while fragrant ketones include camphor, carvone, beta-ionone, and *cis*-jasmone. Pick one alcohol and one ketone from these lists or use other compounds of your choosing that contain these functional groups. Draw the chemical structure. Where does each compound occur in nature? Is the molecule used as a flavor or fragrance in commercial products? Does the compound have any useful medicinal activity? Is the nature-made compound used in commercial products, or is a synthetic compound that has been prepared in the laboratory used?

ANSWERS TO SELECTED PROBLEMS

12.1 a.

halide, CH_3, CH_3, Cl, Br, halide, O, ether

CH_3, CH_3, thiol, C—SH, CH_3

hydroxyl, hydroxyl, OH, HO—, $CHCH_2NH(CH_2)_6O(CH_2)_4$—, ether, $HOCH_2$, hydroxyl

b. The OH on the benzene ring of salmeterol is part of a phenol.

12.3 a. 2° b. 2° c. 3° and 1°

12.5 a. (cyclohexanol, OH) b. $(CH_3)_3C—OH$

12.7 a. 2-heptanol c. 2-methylcyclohexanol
b. 4-ethyl-3-hexanol

12.9 a. butyl methyl ether b. dipropyl ether

12.11 a. $CH_3(CH_2)_5OCH_3$ c. $CH_3(CH_2)_6OH$
b. (ring with O) d. $CH_3(CH_2)_5OCH_3$

12.12 a. $CH_3—C{=}CH_2$ (with H on C) b. (benzene)—$CH{=}CH_2$ c. (cyclopentene)

12.13 a. $CH_3CCH_2CH_2CH_3$ (C=O) c. (cyclopentanone, =O)
b. $(CH_3)_3CCH_2C—OH$ (C=O) d. no reaction

12.15 a. $CH_3CH_2CH_2F < CH_3CH_2CH_2Cl < CH_3CH_2CH_2I$
b. $CH_3(CH_2)_4CH_3 < CH_3(CH_2)_5Cl < CH_3(CH_2)_5Br$

12.17 a. $CH_3CH_2CH_2—S—S—CH_2CH_2CH_3$
b. $CH_3CH_2CH_2CH_2SH$ and CH_3CH_2SH

12.18 a. 5-methylhexanal c. 2,5,6-trimethyloctanal
b. 3,3,4,4-tetramethylpentanal

12.19 a. CH_3CHCHO (Cl on C2)
b. $CH_3CH_2CHCHCHCH_2CHO$ (substituents CH_3CH_2, CH_2CH_3, CH_2CH_3)
c. $CH_3CH_2CH_2CHCH_2CH_2CHCH_2CHO$ (substituents CH_2CH_3, CH_2CH_3)
d. (benzene)—CHO with CH_2CH_3

12.21 a. 4-methyl-3-heptanone c. 2,2-dimethyl-3-heptanone
b. 2-methylcyclopentanone

12.23 Acetone will be soluble in water and organic solvents. Progesterone will be soluble only in organic solvents.

12.25 a. $CH_3(CH_2)_6C—OH$ (C=O) c. (cyclopentane)—C(=O)—OH
b. and d. no reaction

12.27 a. $CH_3CHCH_2CH_3$ (Cl) c. $CH_2{=}CHCHCH_3$ (OH)
b. none d. (benzene)—C(H)(Br)—CH_3

12.29 $HOCH_2CCHCH_2OH$ (C=O, OH) — chirality center

12.31 a. H—C(CH_3)(Cl)—Br b. H—C(COOH)(CH_2Cl)—OH

12.33 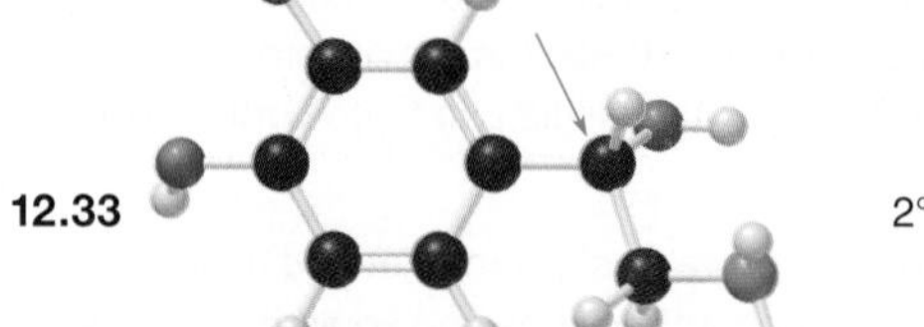 2° alcohol

12.35 a. c. 2-butanol
b. 2° alcohol d. $CH_3—C—CH_2CH_3$ (C=O)

12.37 a. 2-fluoropentane
b. butyl ethyl ether
c. 3-methylhexanal
d. 4-methyl-2-pentanone

12.39 a. 3-ethylcyclopentanol
b. 2°
c.  (cyclopentene)—CH_2CH_3 and (cyclopentene)—CH_2CH_3
d. 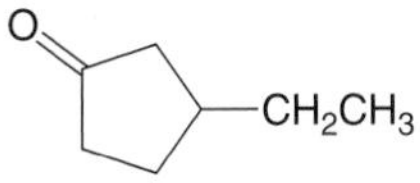 (cyclopentanone)—CH_2CH_3
e. 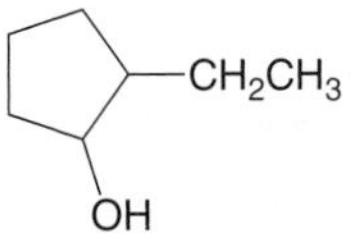 (cyclopentanol)—CH_2CH_3
f. 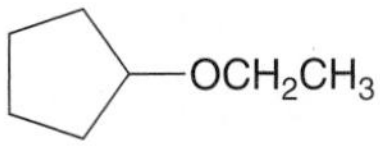 (cyclopentyl)—OCH_2CH_3

12.41 (cyclopentyl)—CH_3 (cyclopentyl)=O (cyclopentyl)—OH

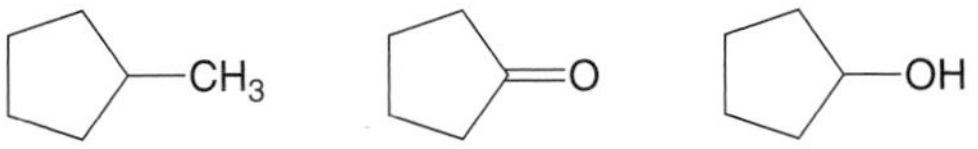

12.43 a. 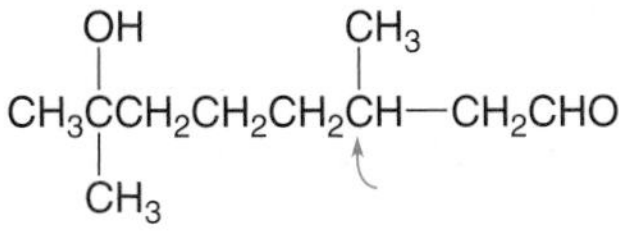 $CH_3C(OH)(CH_3)CH_2CH_2CH_2CH(CH_3){-}CH_2CHO$
b. 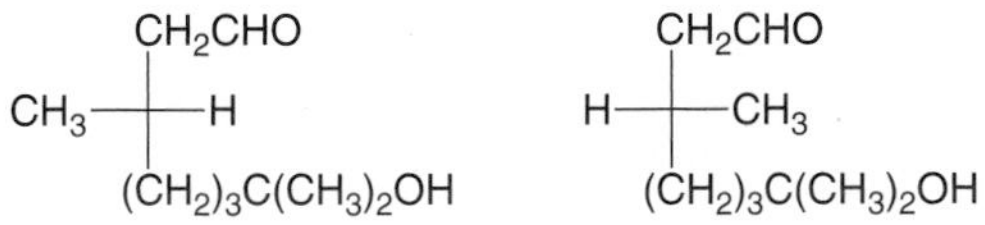 CH_3—C(H)(CH_2CHO)($(CH_2)_3C(CH_3)_2OH$) and H—C(CH_3)(CH_2CHO)($(CH_2)_3C(CH_3)_2OH$)

12.45 a. 1° b. 2°

12.47 a. 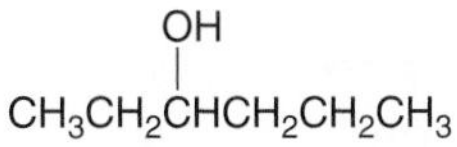 $CH_3CH_2CH(OH)CH_2CH_2CH_3$
b. $CH_3{-}O{-}CH_2CH_2CH_2CH_2CH_3$
c. $CH_3CH_2CH_2CH_2SH$
d. 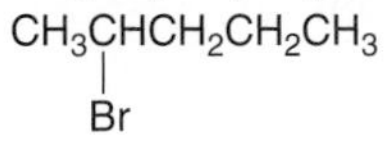 $CH_3CH(Br)CH_2CH_2CH_3$

12.49 a. 2-methyl-3-pentanol
b. 1-chloro-2-methylcyclopentane

12.51 a. 3,3-dimethyl-2-hexanol
b. 4-ethyl-6-methyl-1-heptanol
c. 4-butylcyclohexanol

12.53 a. 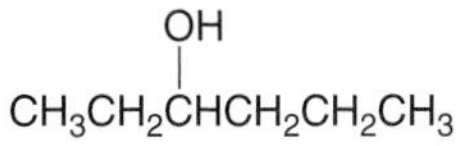 $CH_3CH_2CH(OH)CH_2CH_2CH_3$
b. $CH_3CH_2CH_2CH_2OCH_2CH_2CH_2CH_3$
c. 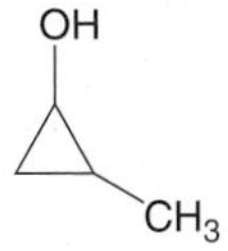 (cyclopropane with OH and CH_3)
d. 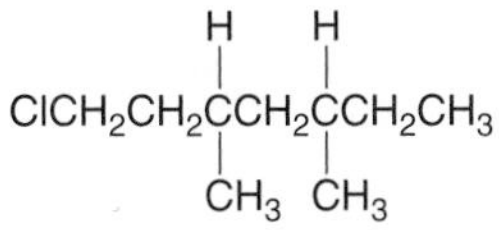 $ClCH_2CH_2CH(CH_3)CH_2CH(CH_3)CH_2CH_3$

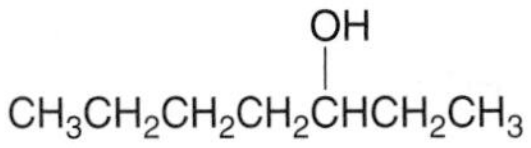

12.55 $CH_3CH_2CH_2CH_2CH_2CH_2CH_2OH$ 1-heptanol
$CH_3CH_2CH_2CH_2CH_2CH(OH)CH_3$ 2-heptanol
$CH_3CH_2CH_2CH_2CH(OH)CH_2CH_3$ 3-heptanol
$CH_3CH_2CH_2CH(OH)CH_2CH_2CH_3$ 4-heptanol

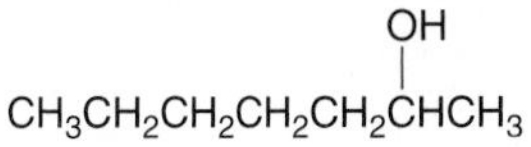

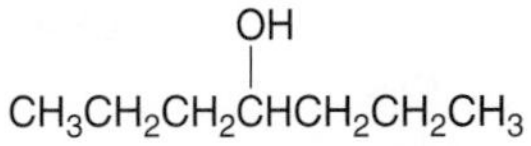

12.57 Ethanol is a polar molecule capable of hydrogen bonding with itself and water. The stronger intermolecular forces make its boiling point higher. Both ethanol and dimethyl ether have only two carbons and can hydrogen bond to H_2O, so both are water soluble.

12.59 1-Butanol is capable of hydrogen bonding with H_2O and is therefore water soluble. 1-Chlorobutane cannot hydrogen bond with H_2O, so it is not water soluble.

12.61 a. $(CH_3)_3CCH_2CHO$ b. 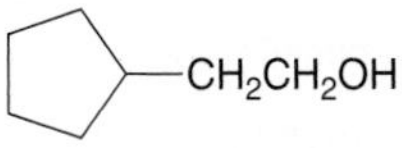 (cyclopentyl)—CH_2CH_2OH

12.63 a.  (cyclobutene) b. 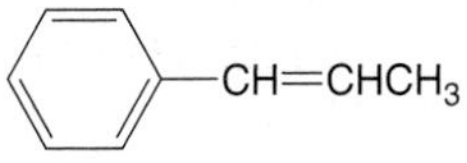 (benzene)—$CH{=}CHCH_3$

12.65 a. 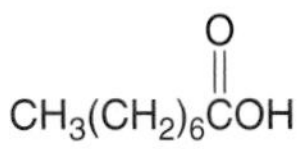 $CH_3(CH_2)_6COOH$ (shown as $CH_3(CH_2)_6\overset{O}{\overset{\|}{C}}OH$)
b. $CH_3CH_2\overset{O}{\overset{\|}{C}}CH_2CH_3$
c. no reaction

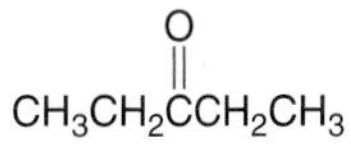

12.67 (steroid structure with C=O groups, CH_3 groups, OH, and $C(=O)COH$)

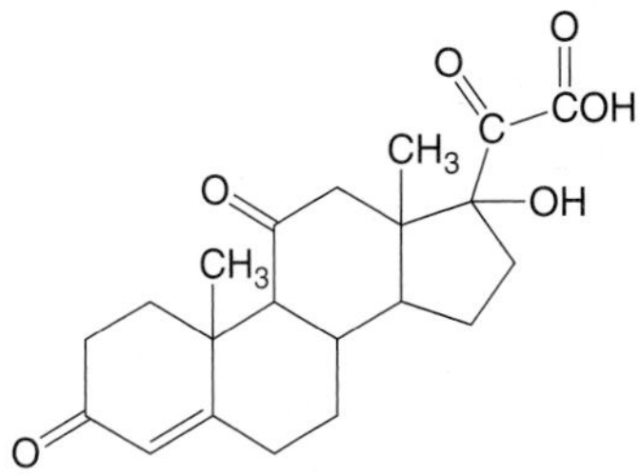

12.69 a. $CH_3CH_2CH_2CH{=}CHCH_2CH_3$
b. 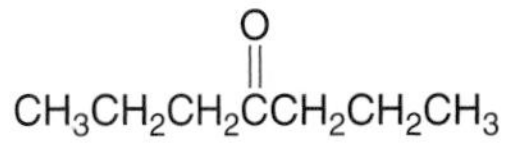 $CH_3CH_2CH_2\overset{O}{\overset{\|}{C}}CH_2CH_2CH_3$

12.71 a. 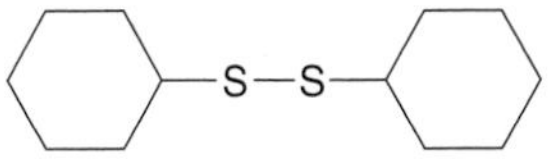 (cyclohexyl)—S—S—(cyclohexyl)
b. $CH_3(CH_2)_4S{-}S(CH_2)_4CH_3$

12.73 a. 2-methylpentanal
b. 3-ethylcyclohexanone

12.75 a. 3,5-dimethylheptanal
b. 3-propylhexanal
c. 2,6-dimethylcyclohexanone
d. 2,4-dimethyl-3-hexanone

12.77 a. 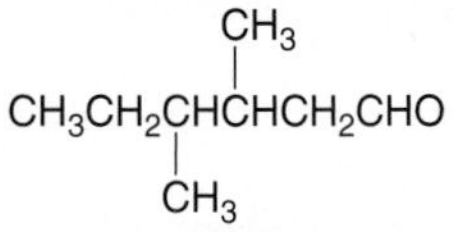 $CH_3CH_2CH(CH_3)CH(CH_3)CH_2CHO$

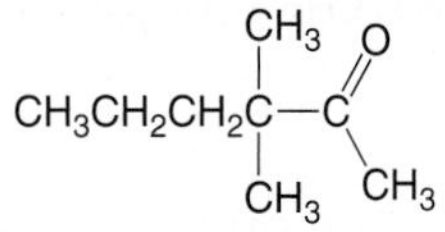

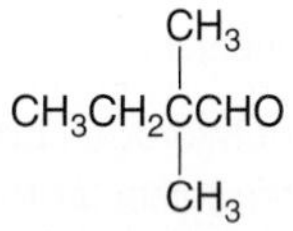

b. 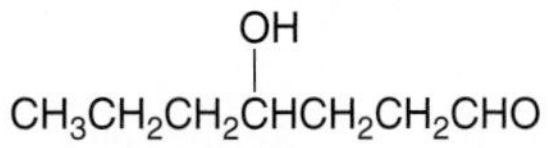 $CH_3CH_2CH_2CH(OH)CH_2CH_2CHO$

c. $CH_3CH_2CH_2C(CH_3)_2C(=O)CH_3$

d. 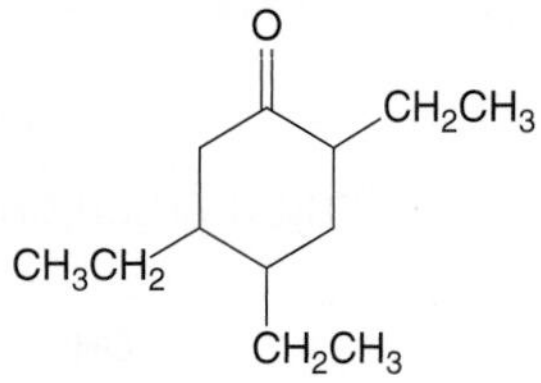 O CH_2CH_3 CH_3CH_2 CH_2CH_3

12.79 $CH_3CH_2C(CH_3)_2CHO$ 2,2-dimethylbutanal

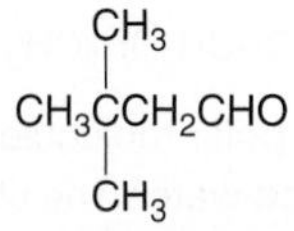

$CH_3C(CH_3)_2CH_2CHO$ 3,3-dimethylbutanal

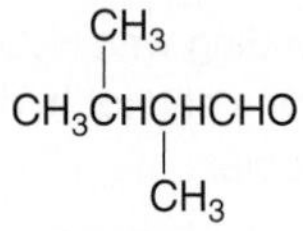

$CH_3CH(CH_3)CH(CH_3)CHO$ 2,3-dimethylbutanal

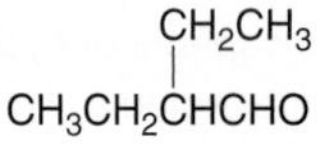

$CH_3CH_2CH(CH_2CH_3)CHO$ 2-ethylbutanal

12.81 a. $CH_3(CH_2)_4COOH$

b. 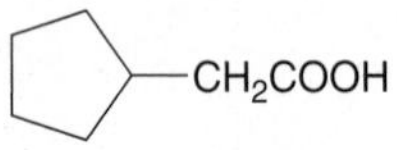 $-CH_2COOH$

c. $CH_3(CH_2)_4COOH$

12.83 a. $CH_3(CH_2)_4COOH$

b. 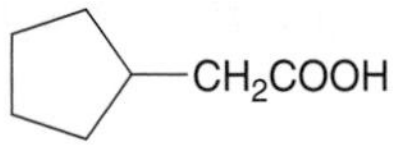 $-CH_2COOH$

c. no reaction

12.85 a. and b. 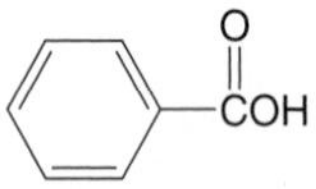 $-COOH$

12.87 a,d: achiral b,c: chiral

12.89 a. chiral b. achiral

12.91 a. none

b. 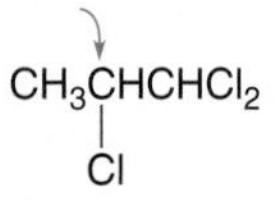 $CH_3CH(Cl)CHCl_2$

c. none

d. 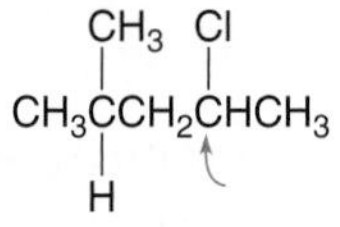 $CH_3CH(CH_3)CH_2CH(Cl)CH_3$

12.93 a. 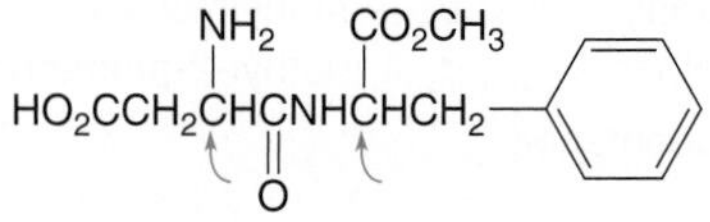 $HO_2CCH_2CH(NH_2)C(=O)NHCH(CO_2CH_3)CH_2-$

b. 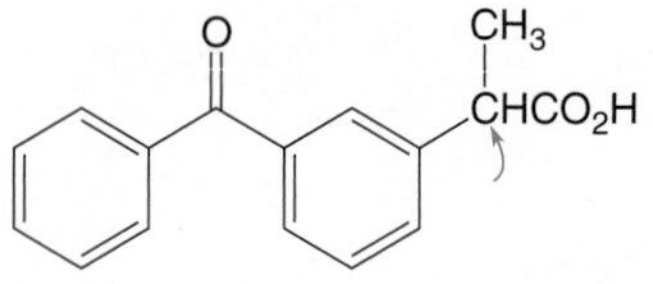 O CH_3 $CHCO_2H$

12.95 **A** and **B** are enantiomers.
A and **C** are identical.

12.97 a. identical b. enantiomers

12.99 a. 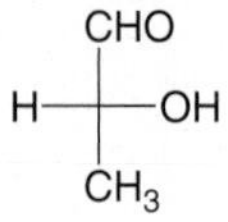 CHO H OH CH_3

b. 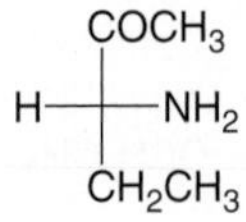 $COCH_3$ H NH_2 CH_2CH_3

12.101 CHO Br H CH_3

12.103 $CH_3C(=O)COH$ (with second $C=O$)

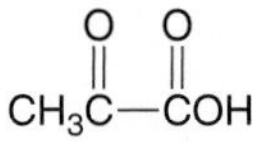

12.105 a. 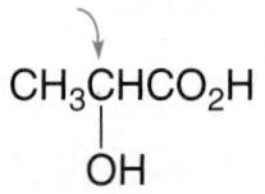 $CH_3CH(OH)CO_2H$

b. 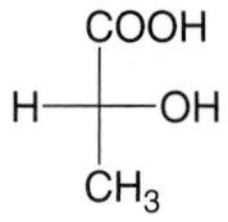 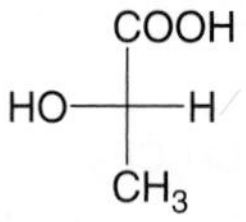 $COOH$ H OH CH_3 $COOH$ HO H CH_3

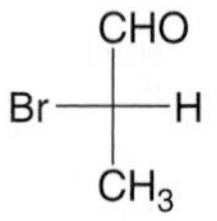

12.107 $HOCH_2$ O HO OH HO $HOCH_2$ O OH O CH_2OH OH

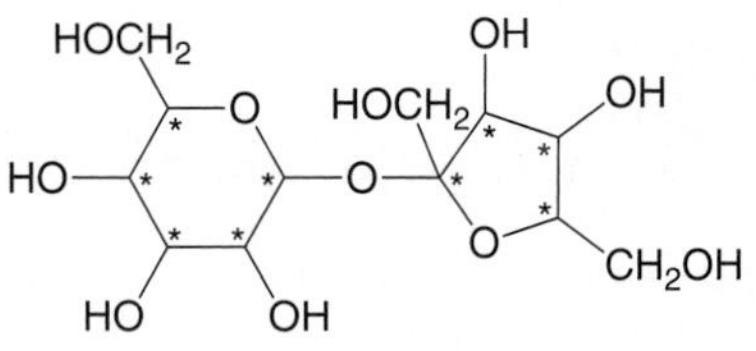

* = chirality center

13

Epinephrine, an amine synthesized by the adrenal glands, increases blood pressure and heart rate, and prepares an individual for "fight or flight" when faced with a particularly challenging or strenuous activity.

Carboxylic Acids, Esters, Amines, and Amides

CHAPTER OUTLINE

CHAPTER GOALS

In this chapter you will learn how to:

1. Identify the characteristics of carboxylic acids, esters, amines, and amides
2. Name carboxylic acids, esters, amines, and amides
3. Draw the products of acid–base reactions of carboxylic acids
4. Explain how soap cleans away dirt
5. Draw the products of reactions involving carboxylic acids, esters, and amides
6. Draw the products of acid–base reactions of amines
7. Identify and name ammonium salts
8. Explain how penicillin works

Chapter 13 discusses the chemistry and properties of **carboxylic acids (RCOOH), esters (RCOOR), amines (RNH_2, R_2NH, or R_3N), and amides ($RCONR_2$).** All four families of organic compounds occur widely in nature, and many useful synthetic compounds have been prepared as well. Simple carboxylic acids like acetic acid (CH_3COOH) have a sour taste and a biting odor, while simple esters have easily recognized fragrances. Common amines include the caffeine in coffee and soft drinks and the nicotine in tobacco products. Several useful drugs are amides, and all proteins, such as those that form hair, muscle, and connective tissue, contain many amide units.

13.1 Introduction

Carboxylic acids and **esters** are two families of organic molecules that contain a carbonyl group (C=O) singly bonded to an oxygen atom.

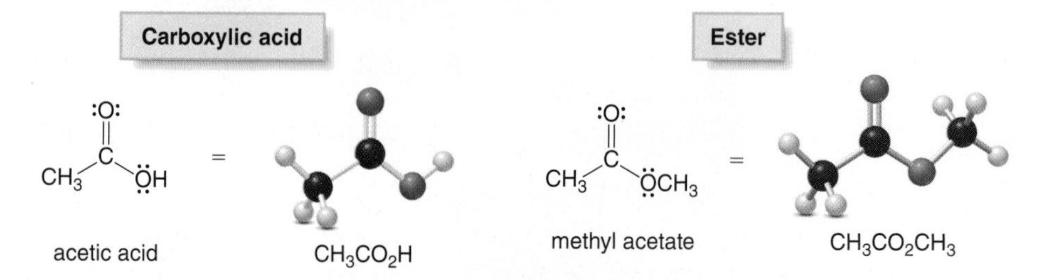

- **A carboxylic acid contains a carboxyl group (COOH).**
- **An ester contains an alkoxy group (OR) bonded to the carbonyl carbon.**

The structure of a carboxylic acid is often abbreviated as RCOOH or RCO_2H, while the structure of an ester is abbreviated as RCOOR or RCO_2R. Keep in mind that the central carbon in both functional groups has a double bond to one oxygen and a single bond to another.

Amines and **amides** are nitrogen-containing organic molecules.

Amine | Amide

methylamine = CH_3NH_2 | acetamide = CH_3CONH_2

- **An amine is an organic nitrogen compound formed by replacing one or more hydrogen atoms of ammonia (NH_3) with alkyl groups.**
- **An amide is a carbonyl compound that contains a nitrogen atom bonded to the carbonyl carbon.**

Each oxygen has two lone pairs and each nitrogen has one lone pair so each atom is surrounded by eight electrons.

PROBLEM 13.1

Draw out each compound to clearly show what groups are bonded to the carbonyl carbon. Label each compound as a carboxylic acid, ester, or amide.

a. $CH_3CH_2CO_2CH_2CH_3$ b. $CH_3CONHCH_3$ c. $(CH_3)_3CCO_2H$ d. $(CH_3)_2CHCON(CH_3)_2$

PROBLEM 13.2

Draw out each compound to clearly show what groups are bonded to the nitrogen atom. Label each compound as an amine or amide.

a. $CH_3CH_2CONHCH_2CH_3$

b. $(CH_3)_3N$

c. $CH_3COCH_2CH_2NH_2$

d. $(CH_3)_2NCH_2CH_2OCH_3$

13.2 Nomenclature of Carboxylic Acids and Esters

To name carboxylic acids and esters, we must learn the suffix that identifies each functional group in the IUPAC system.

HEALTH NOTE

Hexanoic acid $[CH_3(CH_2)_4CO_2H]$ contributes to the unpleasant odor of ginkgo seeds. Ginkgo trees have existed for over 280 million years. Extracts of the ginkgo tree have long been used in China, Japan, and India in medicine and cooking.

13.2A Naming a Carboxylic Acid—RCOOH

- **In the IUPAC system, carboxylic acids are identified by the suffix *-oic acid.***

To name a carboxylic acid using the IUPAC system:

1. **Find the longest chain containing the COOH group, and change the *-e* ending of the parent alkane to the suffix *-oic acid.***
2. **Number the carbon chain to put the COOH group at C1, but omit this number from the name. Apply all of the other usual rules of nomenclature.**

Many simple carboxylic acids are often referred to by their common names. A common name uses the suffix ***-ic acid.*** Three common names are virtually always used.

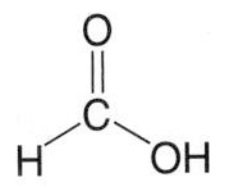

formic acid
(methanoic acid)

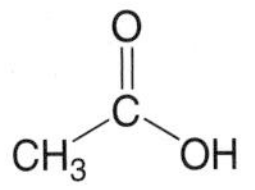

acetic acid
(ethanoic acid)

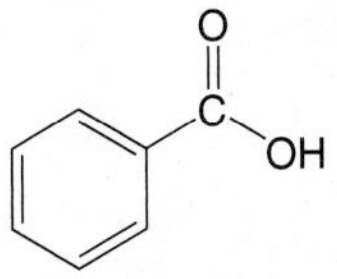

benzoic acid
(benzenecarboxylic acid)

[IUPAC names are given in parentheses, and are rarely used.]

SAMPLE PROBLEM 13.1

Give the IUPAC name for each carboxylic acid.

a. 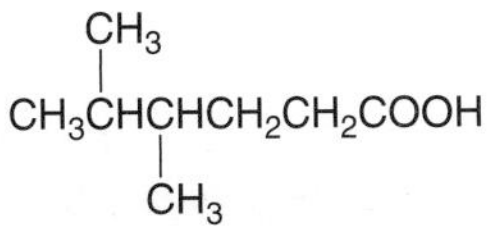

b. 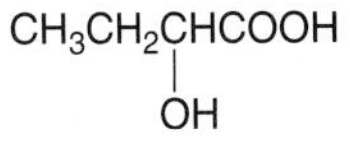

Analysis and Solution

a. [1] **Find and name the longest chain containing COOH.**

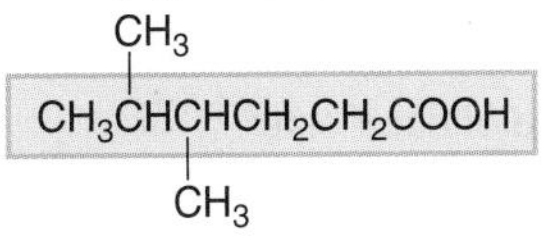

hexane ---→ hexan*oic acid*
(6 C's)

The COOH contributes one C to the longest chain.

[2] **Number and name the substituents, making sure the COOH group is at C1.**

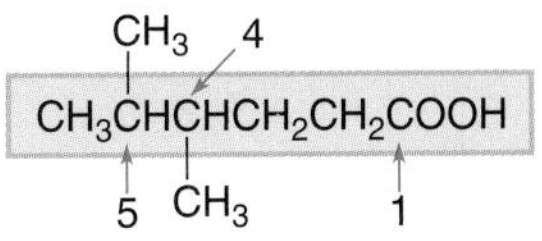

two methyl substituents on C4 and C5

Answer: 4,5-dimethylhexanoic acid

b. [1] **Find and name the longest chain containing COOH.**

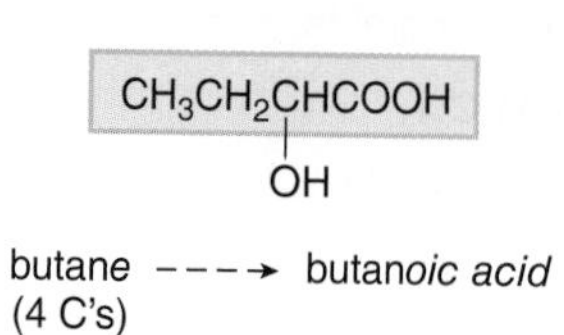

butane (4 C's) ---→ butan*oic acid*

[2] **Number and name the substituent, making sure the COOH group is at C1.**

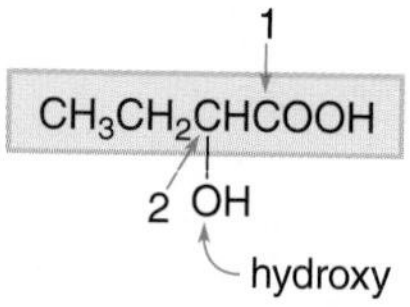

$CH_3CH_2CHCOOH$ (C1 at the COOH carbon; 2 OH → hydroxy)

Answer: 2-hydroxybutanoic acid

PROBLEM 13.3

Give the IUPAC name for each compound.

a. 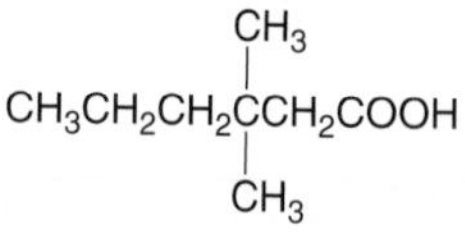 $CH_3CH_2CH_2C(CH_3)_2CH_2COOH$

b. 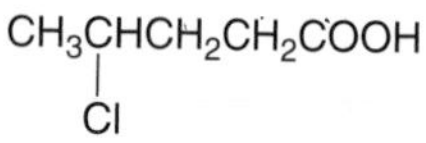 $CH_3CHClCH_2CH_2COOH$

c. 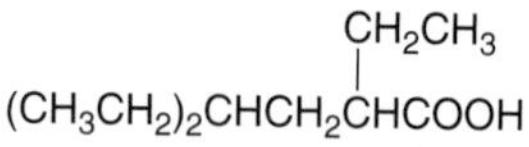 $(CH_3CH_2)_2CHCH_2CH(CH_2CH_3)COOH$

PROBLEM 13.4

Give the structure corresponding to each IUPAC name.

a. 2-bromobutanoic acid

b. 2,3-dimethylpentanoic acid

c. 2-ethyl-5,5-dimethyloctanoic acid

d. 3,4,5,6-tetraethyldecanoic acid

Ethyl butanoate [$CH_3(CH_2)_2CO_2CH_2CH_3$] is an ester isolated from mangoes.

13.2B Naming an Ester—RCOOR'

The names of esters are derived from the names of the parent carboxylic acids. Keep in mind that the common names **formic acid, acetic acid,** and **benzoic acid** are used for the parent acid, so these common parent names are used for their derivatives as well.

An ester has two parts to its structure, each of which must be named separately: the **RCO– group** that contains the carbonyl group and is derived from a carboxylic acid, and an **alkyl group** (designated as **R'**) bonded to the oxygen atom.

R–C(=O)–O–R'

from RCO_2H (R–C(=O) part); an alkyl group (R')

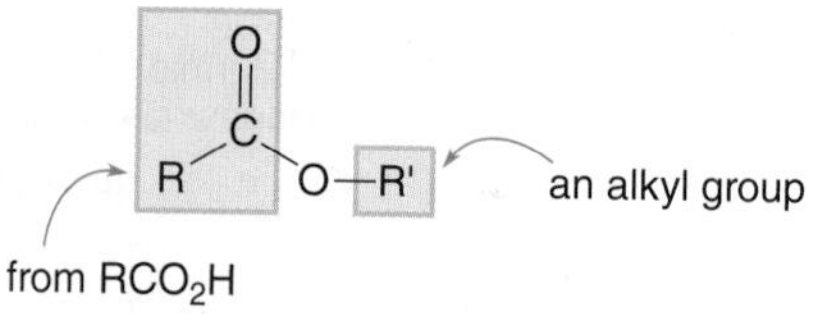

- **In the IUPAC system, esters are identified by the suffix** ***-ate.***

To name an ester:

Esters are often written as RCOOR', where the alkyl group (R') is written *last*. When an ester is named, however, the R' group appears *first* in the name.

- **Name the R' group bonded to the oxygen atom as an alkyl group.**
- **Name the RCO– group by changing the** ***-ic acid*** **ending of the parent carboxylic acid to the suffix** ***-ate.***

SAMPLE PROBLEM 13.2

Give the IUPAC name of each ester.

a. 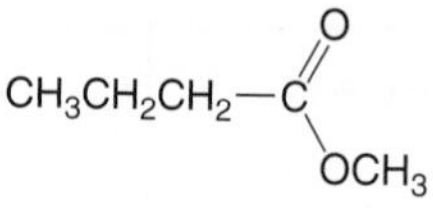 $CH_3CH_2CH_2–C(=O)OCH_3$

b. 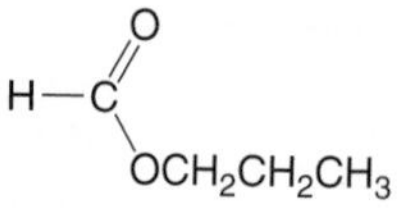 $H–C(=O)OCH_2CH_2CH_3$

Analysis and Solution

a. [1] **Name the alkyl group on the O atom.**

$CH_3CH_2CH_2-C(=O)OCH_3$ — methyl group

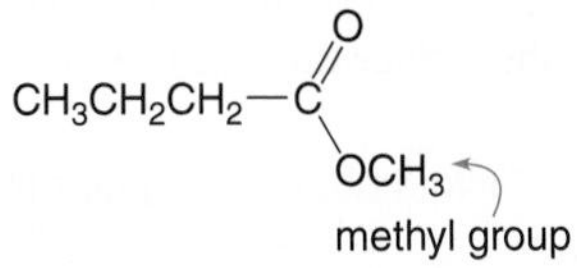

The word *methyl* becomes the first part of the name.

[2] **To name the RCO– group, find and name the longest chain containing the carbonyl group, placing the C=O at C1.**

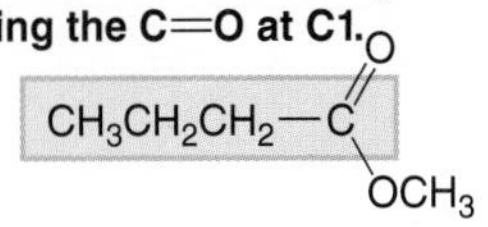

$CH_3CH_2CH_2-C(=O)OCH_3$

butano*ic acid* ---→ butano*ate*
(4 C's)

Answer: methyl butanoate

b. [1] **Name the alkyl group on the O atom.**

$H-C(=O)OCH_2CH_2CH_3$ — propyl group

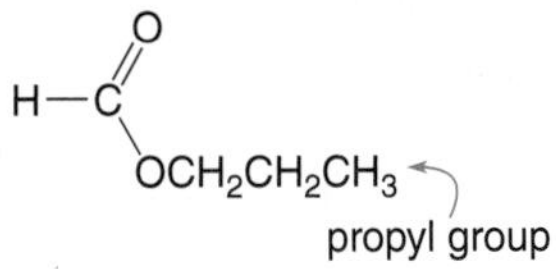

The word *propyl* becomes the first part of the name.

[2] **Name the acyl group (RCO–).**

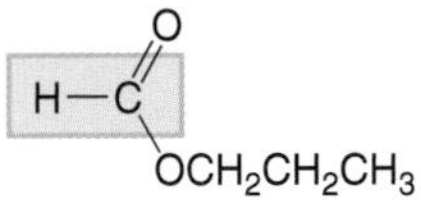

$H-C(=O)OCH_2CH_2CH_3$

form*ic acid* ---→ form*ate*

Answer: propyl formate

PROBLEM 13.5

Give an acceptable name for each ester.

a. $CH_3(CH_2)_4CO_2CH_3$ b. 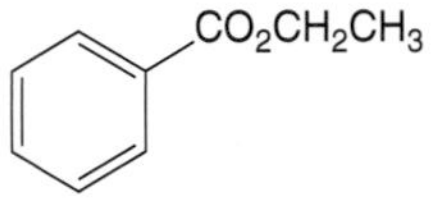 (benzene ring)–$CO_2CH_2CH_3$ c. $CH_3CH_2CH_2CH_2COOCH_2CH_2CH_3$

PROBLEM 13.6

Give the structure corresponding to each name.

a. propyl propanoate b. butyl acetate c. ethyl hexanoate d. methyl benzoate

13.3 Physical Properties of Carboxylic Acids and Esters

Carboxylic acids and esters are polar compounds because they possess a polar carbonyl group. Carboxylic acids also exhibit intermolecular **hydrogen bonding** since they possess a hydrogen atom bonded to an electronegative oxygen atom. Carboxylic acids often exist as **dimers,** held together by *two* intermolecular hydrogen bonds, as shown in Figure 13.1. The carbonyl oxygen atom of one molecule hydrogen bonds to the hydrogen atom of another molecule. As a result:

- **Carboxylic acids have stronger intermolecular forces than esters, giving them higher boiling points and melting points, when comparing compounds of similar molecular weight.**

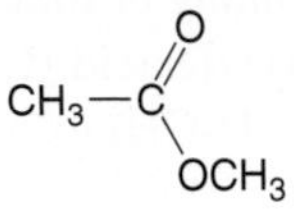

$CH_3-C(=O)OCH_3$
bp 58 °C
no hydrogen bonding possible

$CH_3CH_2-C(=O)OH$ ← hydrogen bonding possible 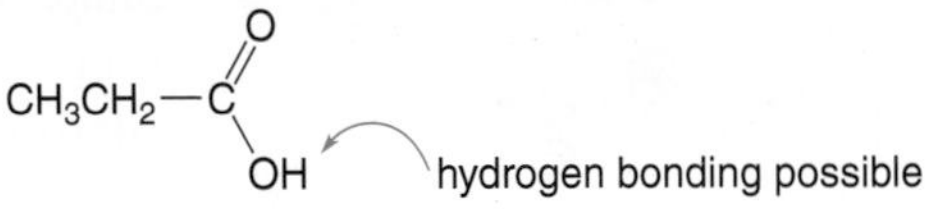
bp 141 °C
stronger intermolecular forces
higher boiling point

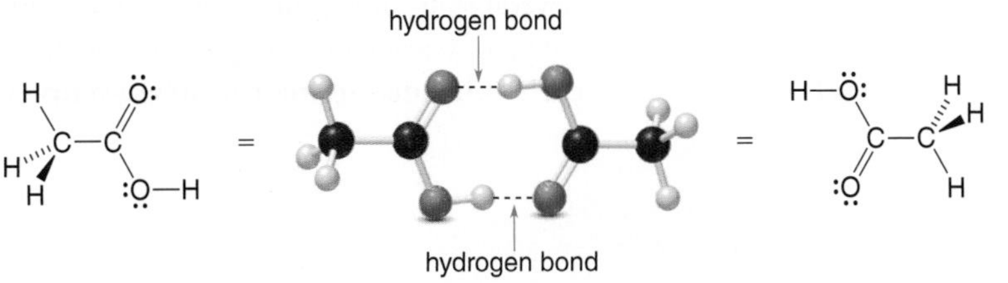

Figure 13.1
Two Molecules of Acetic Acid (CH_3COOH) Held Together by Two Hydrogen Bonds

hydrogen bond
hydrogen bond

Carboxylic acids also have stronger intermolecular forces than alcohols, even though both functional groups exhibit hydrogen bonding; two molecules of a carboxylic acid are held together by *two* intermolecular hydrogen bonds, whereas only *one* hydrogen bond is possible between two molecules of an alcohol (Section 12.2). As a result:

- **Carboxylic acids have higher boiling points and melting points than alcohols of comparable molecular weight.**

$CH_3CH_2CH_2OH$
1-propanol
bp 97 °C

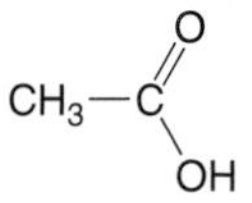

$CH_3-C(=O)OH$
acetic acid
bp 118 °C

more hydrogen-bonding interactions possible
higher boiling point

Like other oxygen-containing compounds, carboxylic acids and esters having fewer than six carbons are soluble in water. Higher molecular weight compounds are insoluble in water because the nonpolar portion of the molecule, the C—C and C—H bonds, gets larger than the polar carbonyl group.

PROBLEM 13.7

Which compound in each pair has the higher boiling point?

a. CH_3COOH or CH_3CH_2CHO b. C_6H_5–COOH or C_6H_5–CH(OH)CH_3

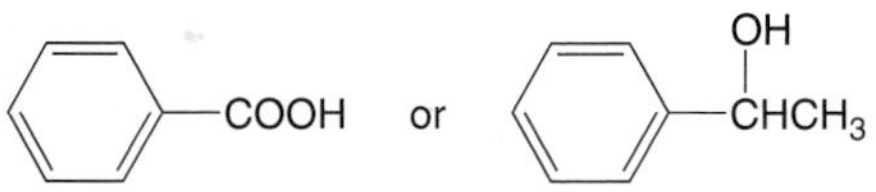

PROBLEM 13.8

Rank the following compounds in order of increasing boiling point. Which compound is the most water soluble? Which compound is the least water soluble?

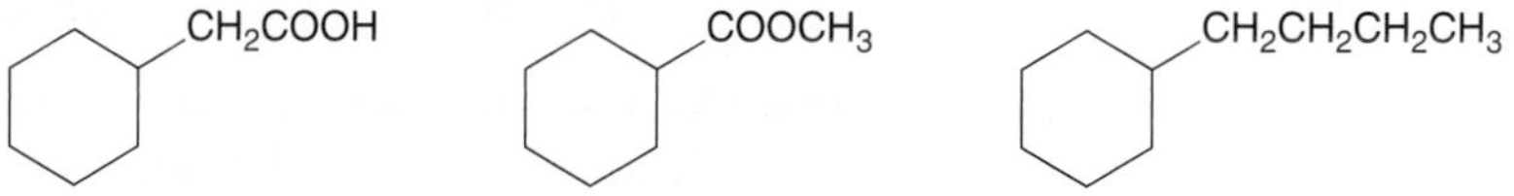

cyclohexyl–CH_2COOH cyclohexyl–$COOCH_3$ cyclohexyl–$CH_2CH_2CH_2CH_3$

13.4 Interesting Carboxylic Acids in Consumer Products and Medicines

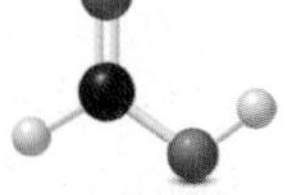

HCO_2H

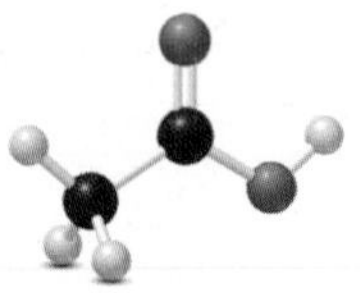

CH_3CO_2H

Simple carboxylic acids have biting or foul odors. **Formic acid** (HCO_2H) is responsible for the sting of some types of ants. **Acetic acid** (CH_3CO_2H) is the sour-tasting component of vinegar. Air oxidation of ethanol (CH_3CH_2OH) to acetic acid makes "bad" wine taste sour.

13.4A FOCUS ON HEALTH & MEDICINE Skin Care Products

Several skin care products that purportedly smooth fine lines and improve skin texture contain α-hydroxy acids (alpha-hydroxy acids). **α-Hydroxy acids contain a hydroxyl group on the carbon bonded to the carboxyl group.** Two common α-hydroxy acids are glycolic acid and

CONSUMER NOTE

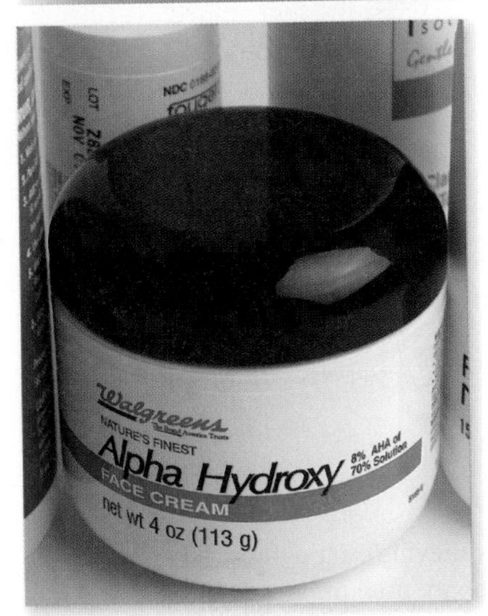

α-Hydroxy acids are used in skin care products that are promoted to remove wrinkles and age spots.

lactic acid. Glycolic acid occurs naturally in sugarcane, and lactic acid gives sour milk its distinctive taste.

General structure

α carbon

α-hydroxy acid glycolic acid lactic acid

Do products that contain α-hydroxy acids make the skin look younger? α-Hydroxy acids react with the outer layer of skin cells, causing them to loosen and flake off. Underneath is a layer of healthier looking skin that has not been exposed to the sun. In this way, these skin products do not actually reverse the aging process; rather, they remove a layer of old skin that may be less pliant or contain small age spots. They can, however, give the appearance of younger skin for a time.

PROBLEM 13.9

Which compounds are α-hydroxy acids?

a. tartaric acid (isolated from grapes)

b. $CH_3CH(OH)CH_2COOH$ — 3-hydroxybutanoic acid (starting material for polymer synthesis)

c. salicylic acid (isolated from meadowsweet)

13.4B FOCUS ON HEALTH & MEDICINE Aspirin and Anti-Inflammatory Agents

Aspirin and ibuprofen are common pain relievers and anti-inflammatory agents that contain a carboxyl group.

aspirin ibuprofen

How does aspirin relieve pain and reduce inflammation? In the 1970s, it was shown that aspirin blocks the synthesis of **prostaglandins,** carboxylic acids containing 20 carbons that are responsible for mediating pain, inflammation, and a wide variety of other biological functions.

Prostaglandins are not stored in cells. Rather, they are synthesized on an as-needed basis from arachidonic acid, an unsaturated fatty acid with four cis double bonds. **Aspirin relieves pain and decreases inflammation because it prevents the synthesis of prostaglandins, the compounds responsible for both of these physiological responses, from arachidonic acid.**

HEALTH NOTE

The modern history of aspirin dates back to 1763 when Reverend Edmund Stone reported on the analgesic effects of chewing on the bark of the willow tree. Willow bark is now known to contain salicin, which is structurally similar to aspirin.

$CH_3CH_2CH_2CH_2CH_2$–CH=CH–CH_2–CH=CH–CH_2–CH=CH–CH_2–CH=CH–$CH_2CH_2CH_2CO_2H$

arachidonic acid

Aspirin blocks this process. ⟹ ↓

HO, HO (cyclopentane ring) with $CH_2CH{=}CHCH_2CH_2CH_2COOH$ and $CH{=}CHCH(CH_2)_4CH_3$ (OH)

$PGF_{2\alpha}$
a prostaglandin

$PGF_{2\alpha}$ and compounds like it cause pain and inflammation.

Aspirin was first used in medicine for its analgesic (pain-relieving), antipyretic (fever-reducing), and anti-inflammatory properties. Today it is also commonly used to prevent blood clots from forming in arteries. In this way, it is used to treat and prevent heart attacks and strokes.

PROBLEM 13.10

Which compound, $PGF_{2\alpha}$ (a prostaglandin) or arachidonic acid, is more water soluble? Explain your choice.

13.5 The Acidity of Carboxylic Acids

As their name implies, **carboxylic acids are acids;** that is, **they are *proton donors.*** When a carboxylic acid is dissolved in water, an acid–base reaction occurs: the carboxylic acid donates a proton to H_2O, forming its conjugate base, a **carboxylate anion,** and water gains a proton, forming its conjugate acid, H_3O^+.

$$RCOOH + H_2O \rightleftharpoons RCOO^- + H_3O^+$$

carboxylic acid — This proton is removed. — carboxylate anion

While carboxylic acids are more acidic than other families of organic compounds, they are weak acids compared to inorganic acids like HCl or H_2SO_4. Thus, only a small percentage of a carboxylic acid is ionized in aqueous solution.

13.5A Reaction with Bases

Carboxylic acids react with bases such as NaOH to form water-soluble salts. In this reaction, essentially all of the carboxylic acid is converted to its carboxylate anion.

$$CH_3COOH + Na^+OH^- \longrightarrow CH_3COO^-\ Na^+ + H{-}O{-}H$$

acetic acid — base — sodium acetate

This proton is transferred from the acid to the base.

- **A proton is removed from acetic acid (CH_3COOH) to form its conjugate base, the acetate anion (CH_3COO^-), which is present in solution as its sodium salt, sodium acetate.**
- **Hydroxide (OH^-) gains a proton to form neutral H_2O.**

Similar acid–base reactions occur with other hydroxide bases (KOH), sodium bicarbonate ($NaHCO_3$), and sodium carbonate (Na_2CO_3). **In each reaction, a proton is transferred from the acid (RCOOH) to the base.**

SAMPLE PROBLEM 13.3

What products are formed when propanoic acid (CH_3CH_2COOH) reacts with potassium hydroxide (KOH)?

Analysis

In any acid–base reaction with a carboxylic acid:

- Remove a proton from the carboxyl group (COOH) and form the carboxylate anion ($RCOO^-$).
- Add a proton to the base.
- Balance the charge of the carboxylate anion by drawing it as a salt with a metal cation.

Solution

$$CH_3CH_2C(=O)O{-}H + K^+OH^- \longrightarrow CH_3CH_2C(=O)O^-\ K^+ + H{-}O{-}H$$

propanoic acid; base; The carboxylate anion is formed as a potassium salt.

This proton is transferred from the acid to the base.

Thus, CH_3CH_2COOH loses a proton to form $CH_3CH_2COO^-$, which is present in solution as its potassium salt, $CH_3CH_2COO^-\ K^+$. Hydroxide (OH^-) gains a proton to form H_2O.

PROBLEM 13.11

Draw the products of each acid–base reaction.

a. $C_6H_{11}{-}C(=O)OH + NaOH \longrightarrow$ (cyclohexanecarboxylic acid)

b. $CH_3CH_2CH_2C(=O)OH + Na_2CO_3 \longrightarrow$

To name the metal salts of carboxylate anions, three parts must be put together: the name of the metal cation, the parent name indicating the number of carbons in the parent carboxylic acid, and the suffix indicating that the species is a salt. The suffix is added by changing the ***-ic acid*** ending of the parent carboxylic acid to the suffix ***-ate.***

name of the metal cation + **parent** + ***-ate* (suffix)**

SAMPLE PROBLEM 13.4

Give an acceptable name for each salt.

a. $CH_3C(=O)O^-\ Na^+$

b. $CH_3CH_2C(=O)O^-\ K^+$

Analysis

Name the carboxylate salt by putting three parts together:

- the name of the metal cation
- the parent name that indicates the number of carbons in the parent chain
- the suffix, *-ate*

Solution

a. $CH_3C(=O)O^-\ Na^+$ — sodium cation

parent + suffix
acet- -ate

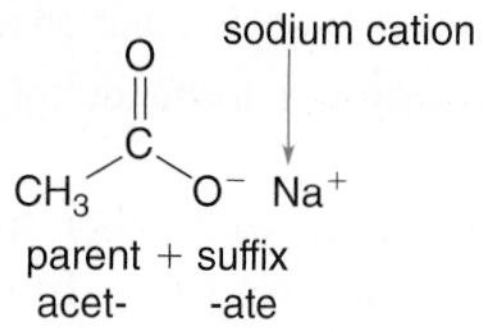

- The first part of the name is the metal cation, sodium.
- The parent name is derived from the common name, acetic acid. Change the ***-ic acid*** ending to ***-ate;*** acetic acid → acetate.

Answer: sodium acetate

b. $CH_3CH_2C(=O)O^-\ K^+$ — potassium cation

parent + suffix
propano- -ate

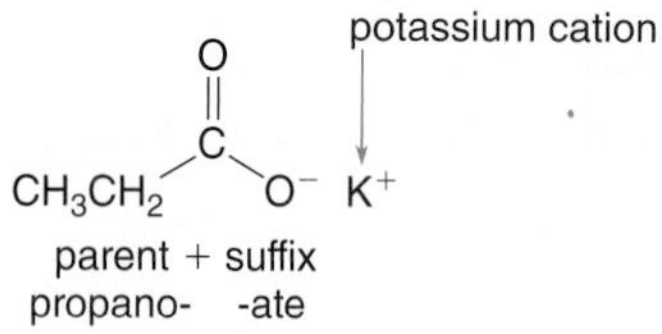

- The first part of the name is the metal cation, potassium.
- The parent name is derived from the IUPAC name, propanoic acid. Change the ***-ic acid*** ending to ***-ate;*** propanoic acid → propanoate.

Answer: potassium propanoate

PROBLEM 13.12

Name each salt of a carboxylic acid.

a. $CH_3CH_2CH_2CO_2^-Na^+$

b. 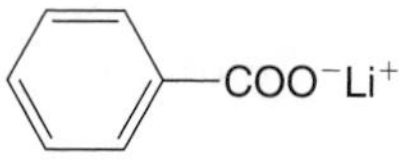 C_6H_5—COO^-Li^+

The salts of carboxylic acids that are formed by acid–base reactions are water-soluble ionic solids. Thus, a water-*insoluble* carboxylic acid like octanoic acid can be converted to its water-*soluble* sodium salt by reaction with NaOH.

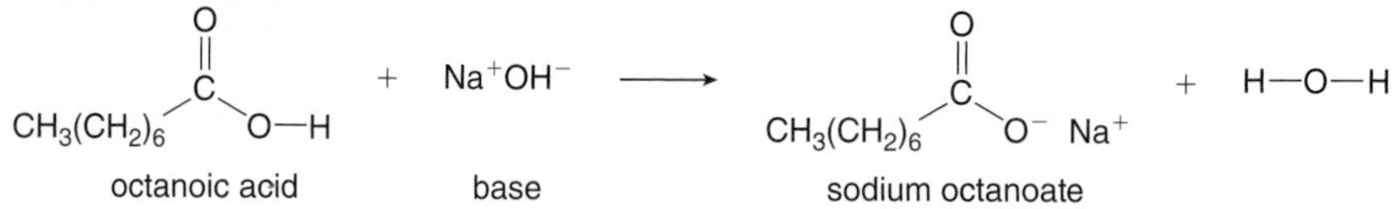

$$CH_3(CH_2)_6C(=O)O{-}H + Na^+OH^- \longrightarrow CH_3(CH_2)_6C(=O)O^-\ Na^+ + H{-}O{-}H$$

octanoic acid; base; sodium octanoate

The acid has more than 5 C's, so it is **insoluble** in water.

The ionic salt is **soluble** in water.

Salts of carboxylic acids are commonly used as preservatives. **Sodium benzoate,** which inhibits the growth of fungus, is a preservative used in soft drinks, and potassium sorbate is an additive that prolongs the shelf-life of baked goods and other foods. These salts do not kill bacteria or fungus. They increase the pH of the product, thus preventing further growth of microorganisms.

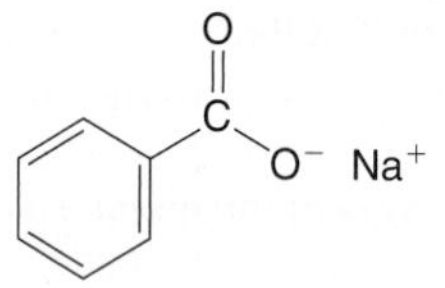

$C_6H_5C(=O)O^-\ Na^+$ — sodium benzoate

$CH_3CH{=}CHCH{=}CHC(=O)O^-\ K^+$ — potassium sorbate

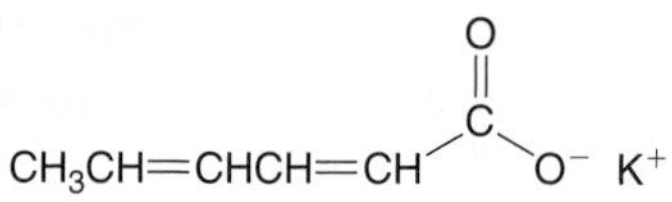

13.5B How Does Soap Clean Away Dirt?

Soap has been used by humankind for some 2,000 years. **Soaps are salts of carboxylic acids that have many carbon atoms in a long hydrocarbon chain.** A soap molecule has two parts.

- The ionic end is called the *polar head.*
- The carbon chain of nonpolar C—C and C—H bonds is called the *nonpolar tail.*

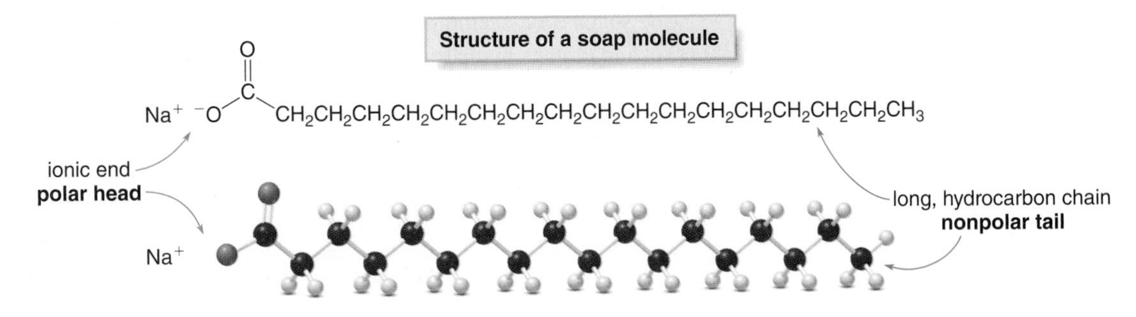

Dissolving soap in water forms **spherical droplets having the ionic heads on the surface and the nonpolar tails packed together in the interior.** These spherical droplets are called ***micelles*** and are illustrated in Figure 13.2. In this arrangement, the ionic heads can interact with the polar solvent water, and this brings the nonpolar, "greasy" hydrocarbon portion of the soap into solution.

Figure 13.2 Dissolving Soap in Water

soap molecule

$Na^+ \ ^-O-C(=O)-CH_2CH_2CH_2CH_2CH_2CH_2CH_2CH_2CH_2CH_2CH_2CH_2CH_2CH_2CH_2CH_2CH_3$

polar head

nonpolar tail

nonpolar interior

polar exterior solvated by H_2O

soap micelle

soap

+

H_2O

soap micelles in H_2O

When soap is dissolved in H_2O, the molecules form spherical droplets with the nonpolar tails in the interior and the polar heads on the surface.

CONSUMER NOTE

All soaps are metal salts of carboxylate anions. The main difference between brands is the addition of other ingredients that do not alter their cleaning properties: dyes for color, scents for a pleasing odor, and oils for lubrication. Soaps that float have been aerated so that they are less dense than water.

How does soap dissolve grease and oil? The polar solvent water alone cannot dissolve dirt, which is composed largely of nonpolar hydrocarbons. When soap is mixed with water, however, the nonpolar hydrocarbon tails dissolve the dirt in the interior of the micelle. The polar head of the soap remains on the surface to interact with water. The nonpolar tails of the soap molecules are so well sealed off from the water by the polar head groups that the micelles are water soluble, so they can separate from the fibers of our clothes and be washed down the drain with water. In this way, soaps do a seemingly impossible task: they remove nonpolar hydrocarbon material from skin and clothes by dissolving it in the polar solvent water.

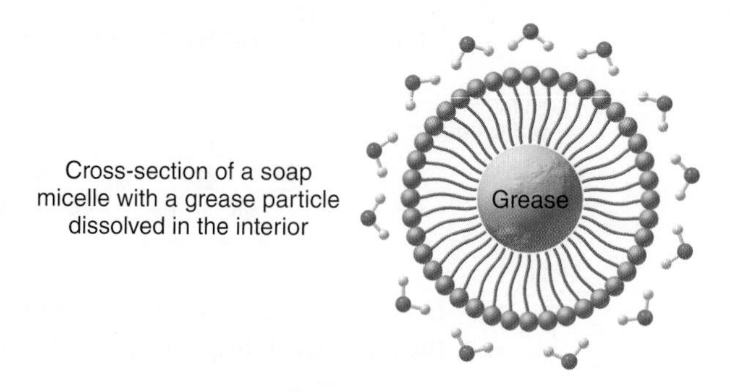

Cross-section of a soap micelle with a grease particle dissolved in the interior

PROBLEM 13.13

Draw the structure of a soap molecule that has a potassium cation and a carboxylate anion containing 16 carbons.

13.6 Reactions Involving Carboxylic Acids and Esters

Carboxylic acids and esters undergo a common type of reaction—*substitution.* When a carboxylic acid or ester (RCOZ) undergoes substitution, the group Z (Z = OH or OR') bonded to the carbonyl carbon is *replaced* by another group of atoms (Y = OR' or OH).

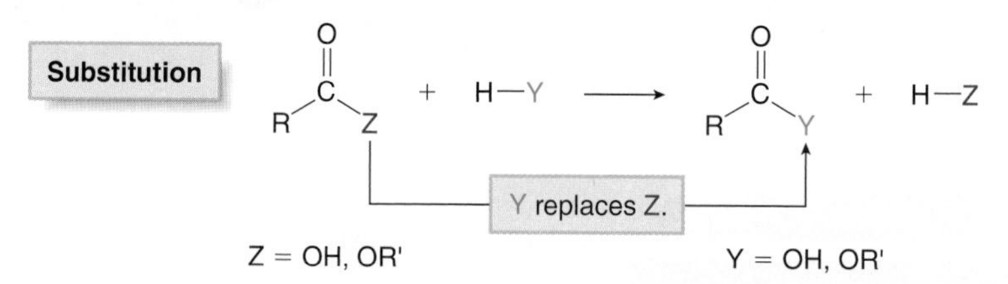

CONSUMER NOTE

Ethyl acetate, a common organic solvent with a very characteristic odor, is used in nail polish remover.

13.6A Ester Formation

Treatment of a carboxylic acid (RCOOH) with an alcohol (R'OH) in the presence of an acid catalyst forms an ester (RCOOR'). This reaction is called a **Fischer esterification.** Esterification is a substitution because the OR' group of an alcohol replaces the OH group of the starting carboxylic acid.

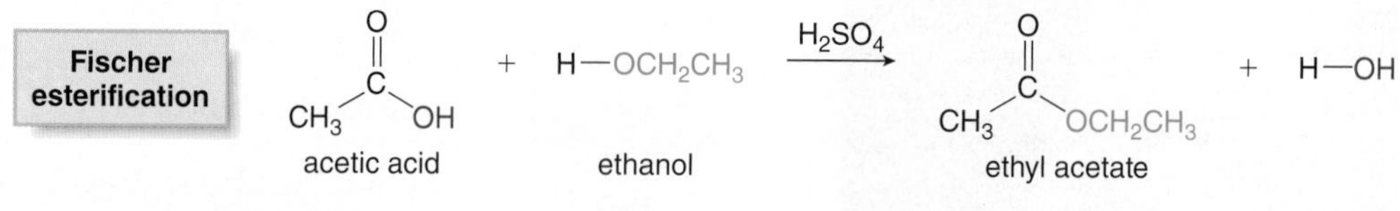

SAMPLE PROBLEM 13.5

What ester is formed when propanoic acid (CH_3CH_2COOH) is treated with methanol (CH_3OH) in the presence of H_2SO_4?

Analysis

To draw the products of esterification, replace the OH group of the carboxylic acid by the OR' group of the alcohol, forming a new C—O bond at the carbonyl carbon.

Remove OH and H to form H_2O.

new C—O bond

carboxylic acid + H—OR' ⟶ ester + H—OH

Solution

Replace the OH group of propanoic acid by the OCH_3 group of methanol to form the ester.

$$CH_3CH_2COOH + H{-}OCH_3 \xrightarrow{H_2SO_4} CH_3CH_2COOCH_3 + H{-}OH$$

new C—O bond

OCH_3 replaces OH.

PROBLEM 13.14

What ester is formed when each carboxylic acid is treated with ethanol (CH_3CH_2OH) in the presence of H_2SO_4?

a. $(CH_3)_2CHCOOH$

b. HCO_2H

c. $CH_3(CH_2)_6COOH$

d. cyclohexane–CO_2H

PROBLEM 13.15

What ester **A** is formed in the following reaction? **A** was converted in one step to blattellaquinone, the sex pheromone of the female German cockroach, *Blattella germanica*.

$$\text{2,5-dimethoxybenzyl alcohol } (OCH_3, CH_2OH, OCH_3) + HOCOCH_2CH(CH_3)_2 \xrightarrow{H_2SO_4} \mathbf{A} \longrightarrow \text{quinone–}CH_2O{-}COCH_2CH(CH_3)_2$$

blattellaquinone

ENVIRONMENTAL NOTE

A short laboratory synthesis of the ester blattellaquinone, the sex pheromone of the female German cockroach (Problem 13.15), opens new possibilities for cockroach population control using pheromone-baited traps.

13.6B Ester Hydrolysis

Esters are hydrolyzed with water in the presence of acid or base. Treatment of an ester (RCOOR') with water in the presence of an acid catalyst forms a carboxylic acid (RCOOH) and a molecule of alcohol (R'OH). This reaction is a **hydrolysis, since bonds are cleaved by reaction with water.**

$$\underset{\text{ester}}{CH_3COOCH_2CH_3} + H{-}OH \xrightarrow{H_2SO_4} \underset{\text{carboxylic acid}}{CH_3COOH} + \underset{\text{alcohol}}{H{-}OCH_2CH_3}$$

SAMPLE PROBLEM 13.6

What products are formed when ethyl propanoate ($CH_3CH_2CO_2CH_2CH_3$) is hydrolyzed with water in the presence of H_2SO_4?

Analysis

To draw the products of hydrolysis in acid, replace the OR' group of the ester by an OH group from water, forming a new C—O bond at the carbonyl carbon. A molecule of alcohol (R'OH) is also formed from the alkoxy group (OR') of the ester.

$$\underset{\text{ester}}{RCOOR'} + H{-}OH \longrightarrow RCOOH + H{-}OR'$$

new C—O bond

OH replaces OR'.

Solution

Replace the OCH_2CH_3 group of ethyl propanoate by the OH group of water to form propanoic acid ($CH_3CH_2CO_2H$) and ethanol (CH_3CH_2OH).

$$CH_3CH_2C(=O)OCH_2CH_3 + H{-}OH \xrightarrow{H_2SO_4} CH_3CH_2C(=O)OH + H{-}OCH_2CH_3$$

ethyl propanoate

OH replaces OCH_2CH_3.

PROBLEM 13.16

What products are formed when each ester is treated with H_2O and H_2SO_4?

a. $CH_3(CH_2)_8C(=O)OCH_3$

b. $CH_3CH(CH_3)CH_2C(=O)OCH_2CH_3$

c. $C_6H_{11}{-}CO_2CH_2CH_2CH_3$ (cyclohexane ring)

PROBLEM 13.17

Aspirin cannot be sold as a liquid solution for children because it slowly undergoes hydrolysis in water. What products are formed when aspirin is hydrolyzed?

(benzene ring bearing COOH and $O{-}C(=O){-}CH_3$ on adjacent carbons)

aspirin

Esters are hydrolyzed in aqueous base to form carboxylate anions and a molecule of alcohol. Basic hydrolysis of an ester is called **saponification.**

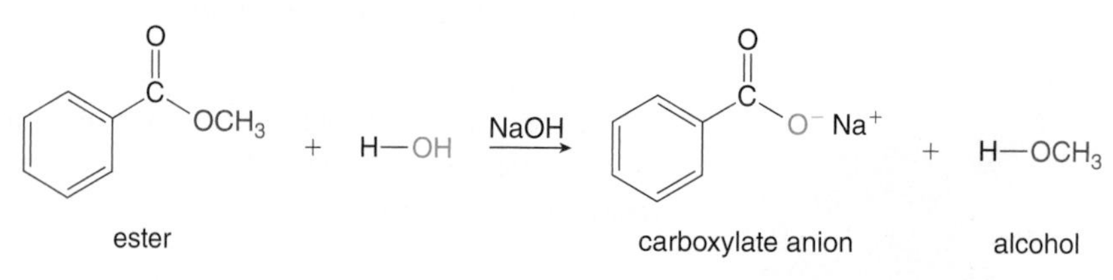

PROBLEM 13.18

What products are formed when each ester in Problem 13.16 is treated with H_2O and NaOH?

13.6C FOCUS ON HEALTH & MEDICINE Olestra, a Synthetic Fat

The most prevalent naturally occurring esters are the **triacylglycerols.** ***Triacylglycerols* contain three ester groups, each having a long carbon chain (abbreviated as R, R', and R'') bonded to the carbonyl group.** Triacylglycerols are **lipids;** that is, they are water-insoluble organic compounds found in biological systems. Animal fats and vegetable oils are composed of triacylglycerols.

$$\begin{array}{l} CH_2{-}O{-}C(=O){-}R \\ |\\ CH{-}O{-}C(=O){-}R' \\ | \\ CH_2{-}O{-}C(=O){-}R'' \end{array}$$

R groups have 11–19 C's.

[Three ester groups are labeled in red.]

triacylglycerol

CONSUMER NOTE

This product contains the "fake fat" olestra, giving it fewer calories for the calorie-conscious consumer.

Animals store energy in the form of triacylglycerols kept in a layer of fat cells below the surface of the skin. This fat serves to insulate the organism, as well as provide energy for its metabolic needs for long periods of time. The first step in the metabolism of a triacylglycerol is hydrolysis of the ester bonds to form glycerol and three fatty acids—long-chain carboxylic acids. **This reaction is simply ester hydrolysis.** In cells, this reaction is carried out with enzymes called **lipases.**

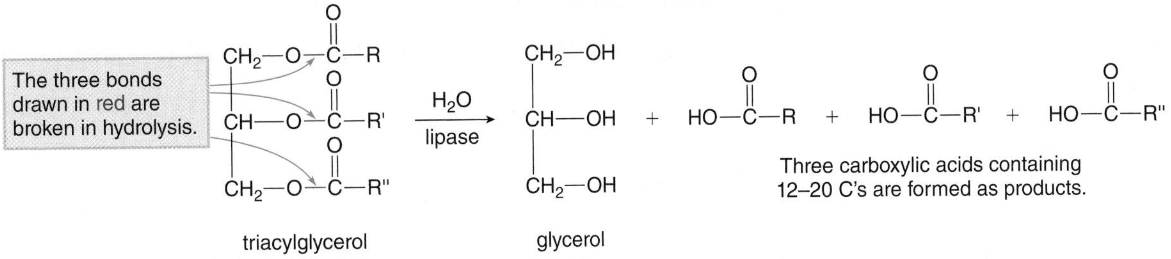

The three bonds drawn in red are broken in hydrolysis.

$$\begin{array}{l} CH_2-O-C(=O)-R \\ CH-O-C(=O)-R' \\ CH_2-O-C(=O)-R'' \end{array} \xrightarrow[\text{lipase}]{H_2O} \begin{array}{l} CH_2-OH \\ CH-OH \\ CH_2-OH \end{array} + HO-C(=O)-R + HO-C(=O)-R' + HO-C(=O)-R''$$

triacylglycerol — glycerol

Three carboxylic acids containing 12–20 C's are formed as products.

The fatty acids produced on hydrolysis are then oxidized, yielding CO_2 and H_2O, as well as a great deal of energy. Diets high in fat content can lead to a large amount of stored fat, ultimately causing an individual to be overweight. One recent attempt to reduce calories in common snack foods has been to substitute "fake fats" such as **olestra** (trade name: **Olean**) for triacylglycerols.

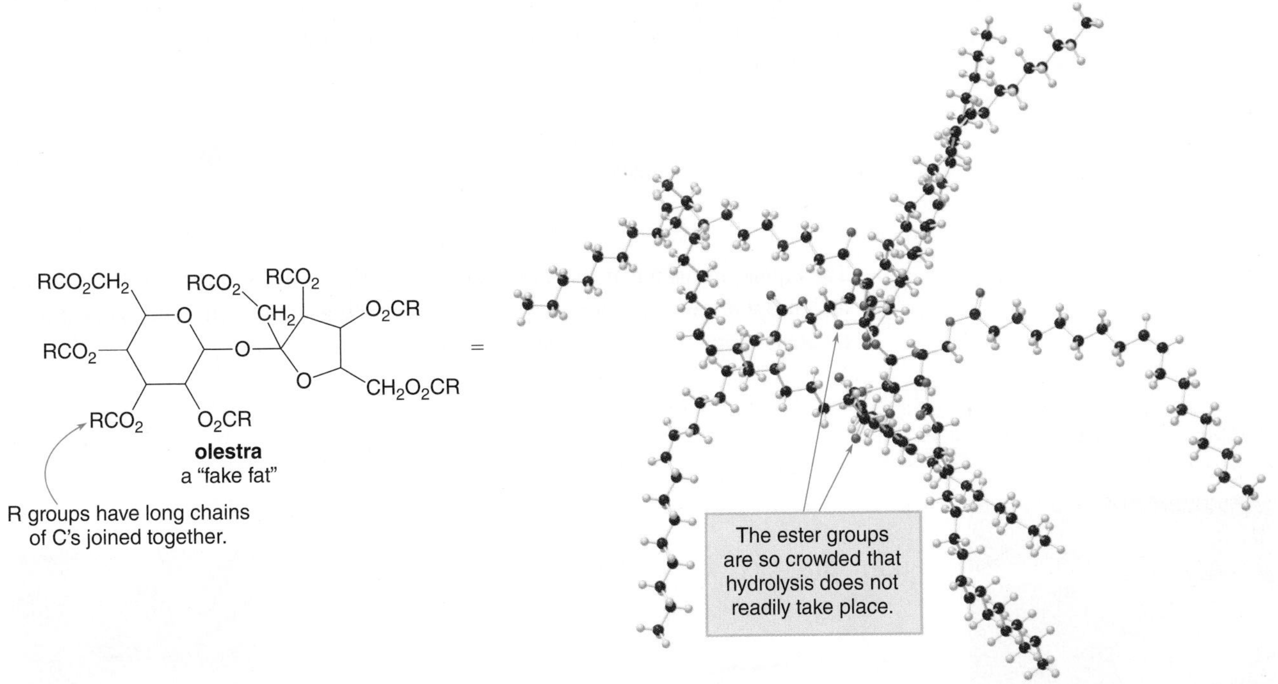

RCO_2CH_2, RCO_2, RCO_2, RCO_2, O_2CR, CH_2, O_2CR, CH_2O_2CR, O, O, O

olestra
a "fake fat"

R groups have long chains of C's joined together.

The ester groups are so crowded that hydrolysis does not readily take place.

Olestra has many ester groups formed from long-chain carboxylic acids and sucrose, the sweet-tasting carbohydrate in table sugar. Olestra has many properties similar to the triacylglycerols in fats and oils. In one way, however, olestra is different. Olestra has so many ester units clustered together that they are too crowded to be hydrolyzed. As a result, olestra is *not* metabolized nor is it absorbed. Instead, it passes through the body unchanged, *providing no calories to the consumer.*

PROBLEM 13.19

What products are formed when the following triacylglycerol is hydrolyzed with water and H_2SO_4?

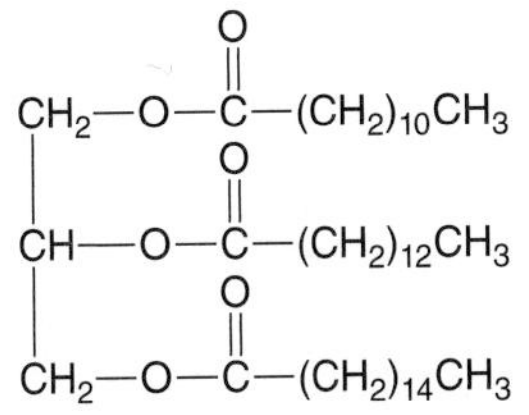

$$\begin{array}{l} CH_2-O-C(=O)-(CH_2)_{10}CH_3 \\ CH-O-C(=O)-(CH_2)_{12}CH_3 \\ CH_2-O-C(=O)-(CH_2)_{14}CH_3 \end{array}$$

13.7 Amines

***Amines* are organic nitrogen compounds,** formed by replacing one or more hydrogen atoms of ammonia (NH_3) with alkyl groups.

13.7A Structure and Classification

Amines are classified as 1°, 2°, or 3° by the number of alkyl groups bonded to the *nitrogen* atom.

R—N̈—H (with H on N)	R—N̈—H (with R on N)	R—N̈—R (with R on N)
1° amine	2° amine	3° amine

- **A primary (1°) amine has one C—N bond and the general structure RNH_2.**
- **A secondary (2°) amine has two C—N bonds and the general structure R_2NH.**
- **A tertiary (3°) amine has three C—N bonds and the general structure R_3N.**

Like ammonia, **the amine nitrogen atom has a lone pair of electrons,** which is generally omitted in condensed structures. An amine nitrogen atom is surrounded by three atoms and one nonbonded electron pair, making it trigonal pyramidal in shape, with bond angles of approximately 109.5°.

CH_3NH_2 = CH_3—N̈(H)(H) = [ball-and-stick model]

methylamine; 109.5°; trigonal pyramidal

The amine nitrogen can also be part of a ring. Figure 13.3 illustrates the structures of morphine and atropine, two **alkaloids—naturally occurring amines derived from plant sources.** Each alkaloid contains a nitrogen atom in a ring.

Figure 13.3 Biologically Active Alkaloids

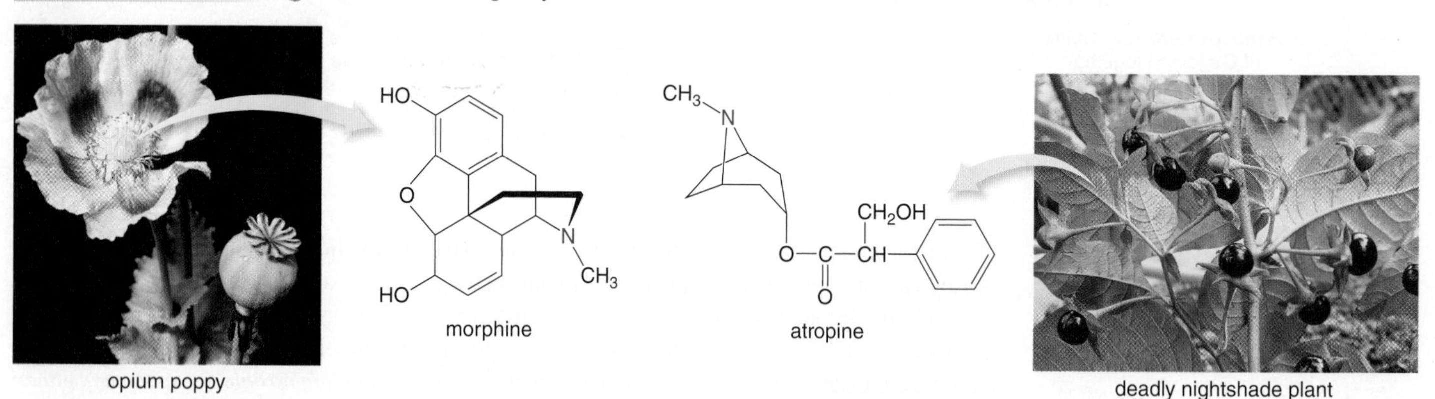

- The analgesic and narcotic effects of opium are due largely to the alkaloid morphine.
- Atropine is isolated from *Atropa belladonna,* the deadly nightshade plant. Atropine dilates pupils, increases heart rate, and relaxes smooth muscles.

SAMPLE PROBLEM 13.7

Classify each amine in the following compounds as 1°, 2°, or 3°. Putrescine is partly responsible for the foul odor of decaying fish. MDMA is the illegal stimulant commonly called "Ecstasy."

a. $H_2N(CH_2)_4NH_2$

putrescine

b. $CH_2CHNHCH_3$ / CH_3

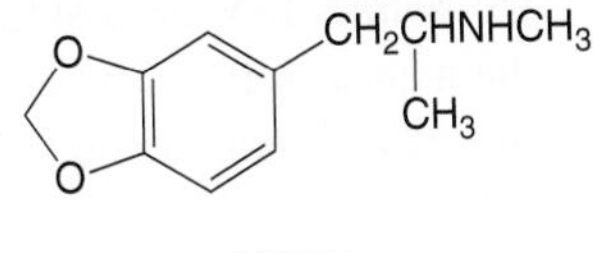

MDMA
"Ecstasy"

Analysis

To determine whether an amine is 1°, 2°, or 3°, count the number of carbons bonded to the nitrogen atom. A 1° amine has one C—N bond, and so forth.

Solution

Draw out the structure or add H's to the skeletal structure to clearly see how many C—N bonds the amine contains.

a. H—N(H)—$CH_2CH_2CH_2CH_2$—N(H)—H

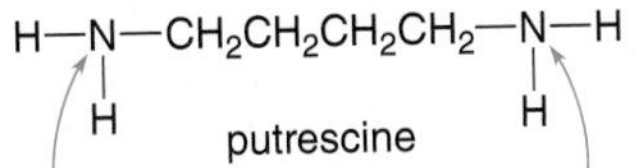

putrescine

Each N is bonded to only 1 C.
Both amines are **1°**.

b. $CH_2C(H)(CH_3)$—N(H)—CH_3

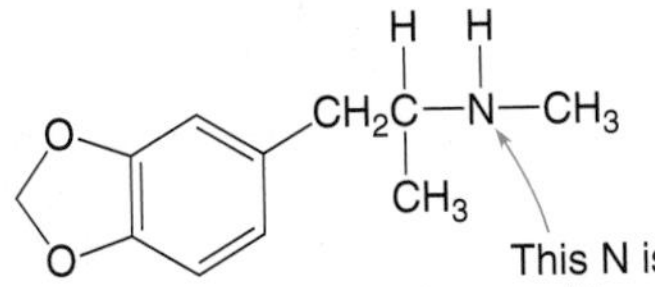

MDMA
"Ecstasy"

This N is bonded to 2 C's, making it a **2° amine.**

PROBLEM 13.20

Classify each amine in the following compounds as 1°, 2°, or 3°.

a. $H_2N(CH_2)_3NH(CH_2)_4NH(CH_2)_3NH_2$

spermine
(isolated from semen)

b. CH_3CH_2O—C(=O)—, C_6H_5, N—CH_3

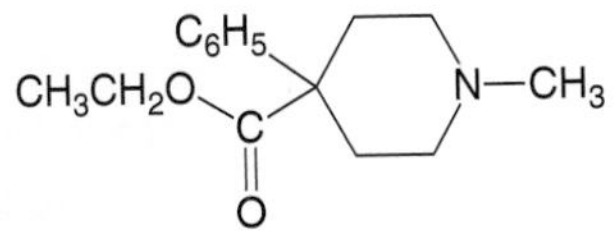

meperidine
(Trade name: Demerol)

PROBLEM 13.21

Label the amine and hydroxyl group in scopolamine, a drug used to treat motion sickness, as 1°, 2°, or 3°. Scopolamine, shown in a ball-and-stick model on the cover of this text, is obtained from angel's trumpets, ornamental plants native to South America.

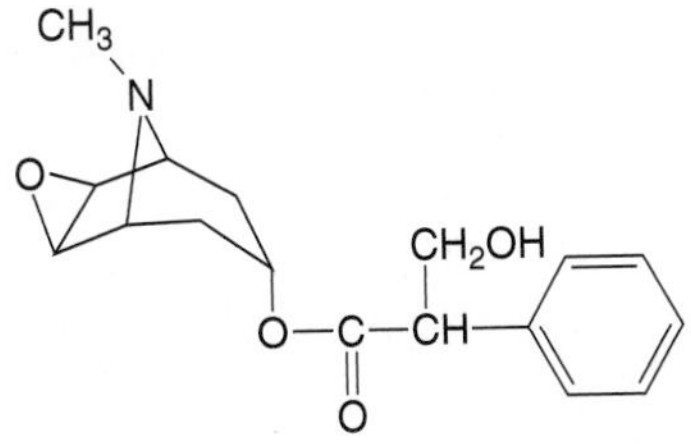

CH_3, N, O, CH_2OH, O—C(=O)—CH—

scopolamine

13.7B Nomenclature

To name a primary (1°) amine, name the alkyl group bonded to the nitrogen atom and add the suffix ***-amine,*** forming a single word. For 2° and 3° amines with different alkyl groups, alphabetize the names of the alkyl groups. Secondary (2°) and 3° amines having identical alkyl groups are named by using the prefix **di-** or **tri-** with the name of the primary amine.

CH_3NH_2 — methylamine

$CH_3CH_2-N(CH_2CH_3)-CH_2CH_3$ — triethylamine

$CH_3CH_2CH_2-N(H)-CH_2CH_3$ — ethylpropylamine

SAMPLE PROBLEM 13.8

Name each amine.

a. $CH_3CH_2CH_2CH_2CH_2NH_2$ b. $CH_3-N(H)-CH_2CH_3$

Analysis and Solution

a. For a 1° amine, name the alkyl group and add the suffix *-amine.*

$CH_3CH_2CH_2CH_2CH_2NH_2$ (5 C's)

Answer: pentylamine

b. For a 2° amine, name each alkyl group, alphabetize the names, and add the suffix *-amine.*

$CH_3-N(H)-CH_2CH_3$ (a methyl and ethyl group on N)

Answer: ethylmethylamine

PROBLEM 13.22

Name each amine.

a. $CH_3CH_2CH_2NH_2$ b. $CH_3CH_2CH_2NHCH_3$ c. $(CH_3CH_2CH_2CH_2)_2NH$

Aromatic amines, amines having a nitrogen atom bonded directly to a benzene ring, are named as derivatives of aniline. Use the prefix *N-* before any alkyl group bonded to the amine nitrogen.

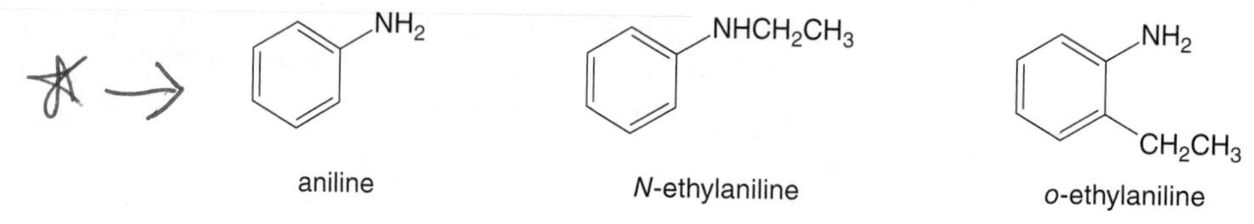

PROBLEM 13.23

Draw a structure corresponding to each name: (a) *N*-methylaniline; (b) *m*-ethylaniline; (c) 3,5-diethylaniline; (d) *N,N*-diethylaniline.

13.7C Physical Properties

Many low molecular weight amines have *very* foul odors. **Trimethylamine** [$(CH_3)_3N$], formed when enzymes break down certain fish proteins, has the characteristic odor of rotting fish. **Cadaverine** ($NH_2CH_2CH_2CH_2CH_2CH_2NH_2$) is a poisonous diamine with a putrid odor also present in rotting fish, and partly responsible for the odor of semen, urine, and bad breath.

Because nitrogen is much more electronegative than carbon or hydrogen, amines contain polar C—N and N—H bonds. **Primary (1°) and 2° amines are also capable of intermolecular hydrogen bonding,** because they contain N—H bonds.

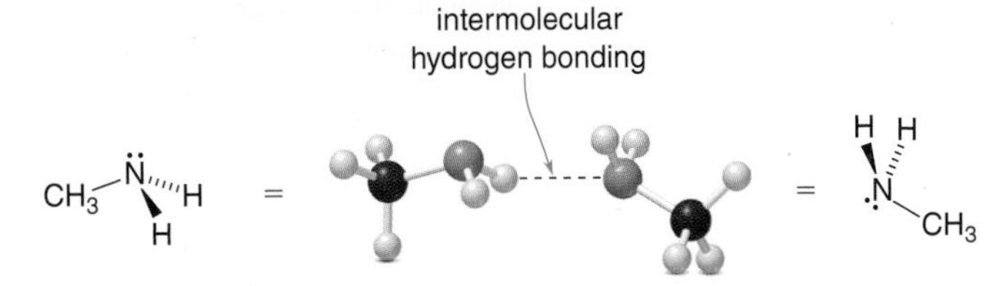

Since nitrogen is less electronegative than oxygen, however, intermolecular hydrogen bonds between N and H are *weaker* than those between O and H. As a result:

- **In comparing compounds of similar molecular weight, 1° and 2° amines have higher boiling points than compounds incapable of hydrogen bonding, but lower boiling points than alcohols that have stronger intermolecular hydrogen bonds.**

$CH_3CH_2OCH_2CH_3$	$CH_3CH_2CH_2CH_2NH_2$	$CH_3CH_2CH_2CH_2OH$
diethyl ether	butylamine	1-butanol
bp = 38 °C	bp = 78 °C	bp = 118 °C

Increasing intermolecular forces
Increasing boiling point

- **Tertiary (3°) amines have lower boiling points than 1° and 2° amines of comparable molecular weight, because they have no N—H bonds.**

3° amine	$CH_3CH_2N(CH_3)_2$	$CH_3CH_2NHCH_2CH_3$	2° amine
no N—H bond	bp = 38 °C	bp = 56 °C	N—H bond
		higher boiling point	

Amines are soluble in organic solvents regardless of size. Amines with fewer than six carbons are water soluble since they can hydrogen bond with water. Larger amines are water insoluble since the nonpolar alkyl portion is too large to dissolve in the polar water solvent.

SAMPLE PROBLEM 13.9

Which compound in each pair has the higher boiling point: (a) $CH_3CH_2NHCH_3$ or $CH_3CH_2OCH_3$; (b) $(CH_3)_3N$ or $CH_3CH_2CH_2NH_2$?

Analysis

Keep in mind the general rule: For compounds of comparable molecular weight, **the stronger the intermolecular forces, the higher the boiling point.** Compounds that can hydrogen bond have higher boiling points than compounds that are polar but cannot hydrogen bond. Polar compounds have higher boiling points than nonpolar compounds.

Solution

a. The 2° amine ($CH_3CH_2NHCH_3$) has an N—H bond, so intermolecular hydrogen bonding is possible. The ether ($CH_3CH_2OCH_3$) has only C—H bonds, so there is no possibility of intermolecular hydrogen bonding. $CH_3CH_2NHCH_3$ has a higher boiling point because it has stronger intermolecular forces.

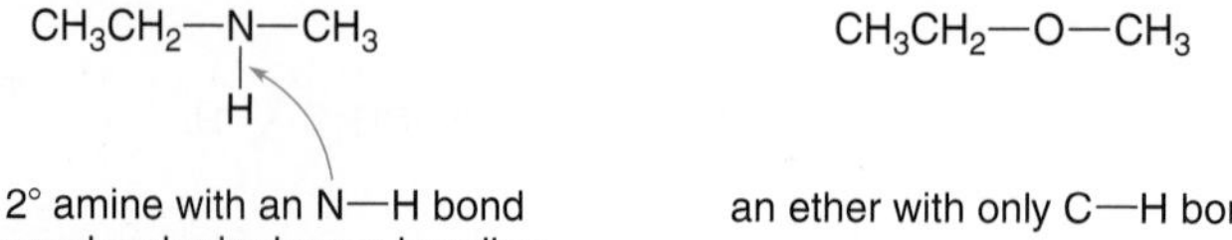

$CH_3CH_2—N(H)—CH_3$ — a 2° amine with an N—H bond, intermolecular hydrogen bonding, higher boiling point

$CH_3CH_2—O—CH_3$ — an ether with only C—H bonds

b. The 1° amine ($CH_3CH_2CH_2NH_2$) has N—H bonds, so intermolecular hydrogen bonding is possible. The 3° amine [$(CH_3)_3N$] has only C—H bonds, so there is no possibility of intermolecular hydrogen bonding. $CH_3CH_2CH_2NH_2$ has a higher boiling point because it has stronger intermolecular forces.

$CH_3CH_2CH_2NH_2$ — a 1° amine with N—H bonds, intermolecular hydrogen bonding, higher boiling point

$CH_3—N(CH_3)—CH_3$ — a 3° amine with only C—H bonds

PROBLEM 13.24

Which compound in each pair has the higher boiling point?

a. 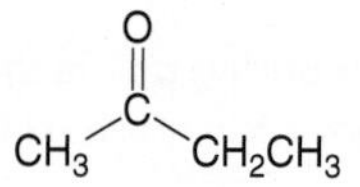 $CH_3C(=O)CH_2CH_3$ or $(CH_3)_2CHCH_2NH_2$

b. $(CH_3)_2CHCH_2NH_2$ or $(CH_3)_2CHCH_2OH$

c. 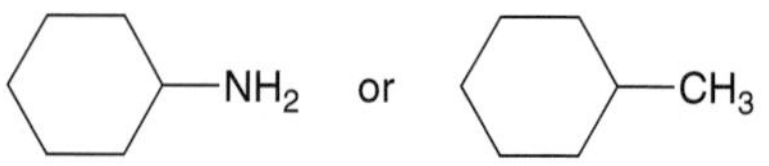 cyclohexyl—NH_2 or cyclohexyl—CH_3

13.8 Amines as Bases

Like ammonia (NH_3), **amines are bases;** that is, **they are *proton acceptors.*** When an amine is dissolved in water, an acid–base reaction occurs: the amine accepts a proton from H_2O, forming its conjugate acid, an **ammonium ion,** and water loses a proton, forming hydroxide, OH^-.

This electron pair forms a new bond to a proton.

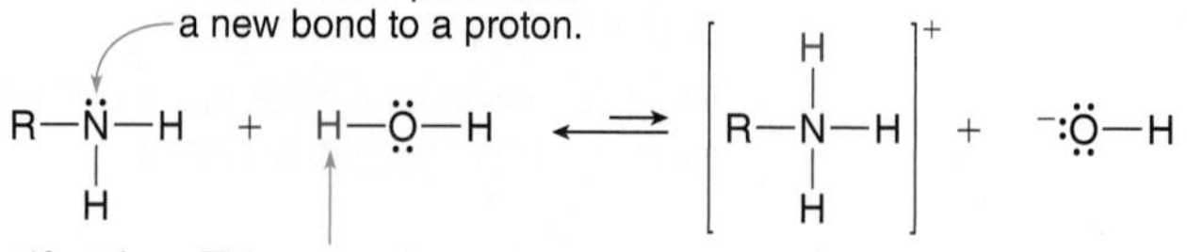

$$R—\ddot{N}H_2 + H—\ddot{O}—H \rightleftharpoons [R—NH_3]^+ + {}^-:\ddot{O}—H$$

1° amine; This proton is removed.; ammonium ion

This acid–base reaction occurs with 1°, 2°, and 3° amines. While amines are more basic than other families of organic compounds, they are weak bases compared to inorganic bases like NaOH.

13.8A Reaction of Amines with Acids

Amines also react with acids such as HCl to form water-soluble salts. The lone pair of electrons from the amine nitrogen atom is always used to form a new bond to a proton from the acid.

Example

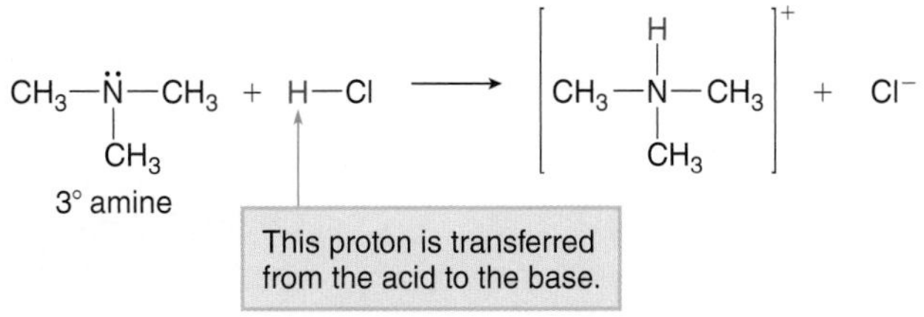

$$CH_3-\ddot{N}(CH_3)-CH_3 + H-Cl \longrightarrow [CH_3-N(H)(CH_3)-CH_3]^+ + Cl^-$$

3° amine

This proton is transferred from the acid to the base.

- **The amine $[(CH_3)_3N]$ gains a proton to form its conjugate acid, an ammonium cation $[(CH_3)_3NH^+]$.**
- **A proton is removed from the acid (HCl) to form its conjugate base, the chloride anion (Cl^-).**

Similar acid–base reactions occur with other inorganic acids (H_2SO_4), and with organic acids like CH_3COOH, as well.

- **In an acid–base reaction of an amine, the amine nitrogen always forms a new bond to a proton, forming an ammonium ion.**

SAMPLE PROBLEM 13.10

What products are formed when methamphetamine reacts with HCl?

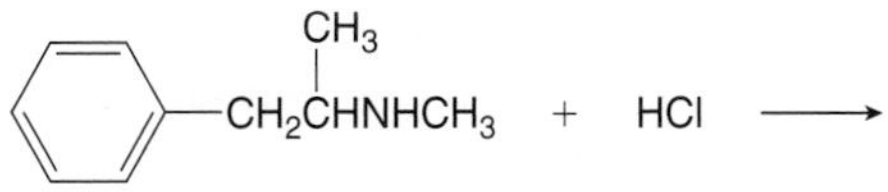

$$C_6H_5-CH_2CH(CH_3)NHCH_3 + HCl \longrightarrow$$

methamphetamine

Analysis

In any acid–base reaction with an amine:

- Locate the N atom of the amine and add a proton to it.
- Remove a proton from the acid (HCl) and form its conjugate base (Cl^-).

Solution

Transfer a proton from the acid to the base. Use the lone pair on the N atom to form the new bond to the proton of the acid.

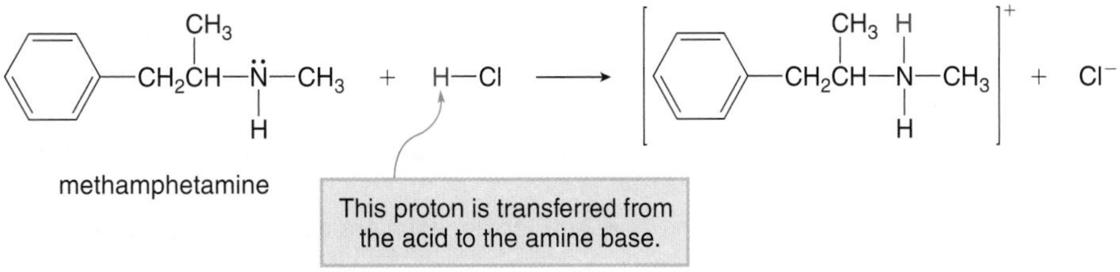

$$C_6H_5-CH_2CH(CH_3)-\ddot{N}(H)-CH_3 + H-Cl \longrightarrow [C_6H_5-CH_2CH(CH_3)-N(H)_2-CH_3]^+ + Cl^-$$

methamphetamine

This proton is transferred from the acid to the amine base.

Thus, HCl loses a proton to form Cl^-, and the N atom of methamphetamine gains a proton to form an ammonium cation.

PROBLEM 13.25

What products are formed when each of the following amines is treated with HCl: (a) $CH_3CH_2NH_2$; (b) $(CH_3CH_2)_2NH$; (c) $(CH_3CH_2)_3N$?

13.8B Ammonium Salts

When an amine reacts with an acid, the product is an ***ammonium salt:*** **the amine forms its conjugate acid, a positively charged ammonium ion, and the acid forms its conjugate base, an anion.**

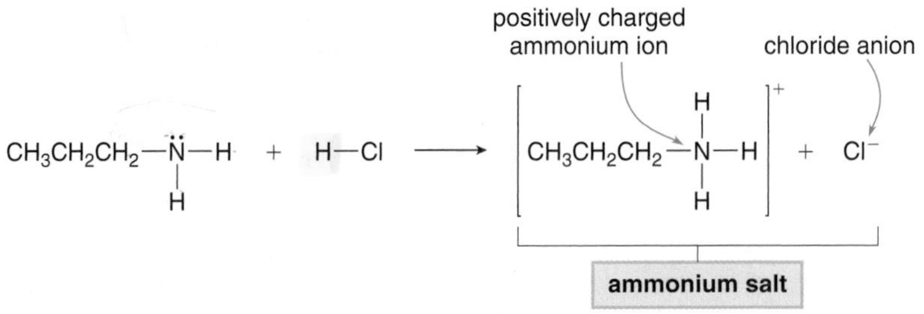

$$CH_3CH_2CH_2{-}\ddot{N}H_2 + H{-}Cl \longrightarrow [CH_3CH_2CH_2{-}NH_3]^+ + Cl^-$$

positively charged ammonium ion; chloride anion; **ammonium salt**

To name an ammonium salt, change the suffix *-amine* of the parent amine from which the salt is formed to the suffix *-ammonium.* Then add the name of the anion as a separate word.

SAMPLE PROBLEM 13.11

Name the ammonium salt: $[(CH_3)_3NH]^+\ CH_3COO^-$.

Analysis

To name an ammonium salt, identify the four groups bonded to the N atom. Remove one hydrogen from the N atom to draw the structure of the parent amine. Then put the two parts of the name together.

- Name the ammonium ion by changing the suffix *-amine* of the parent amine to the suffix *-ammonium.*
- Add the name of the anion.

Solution

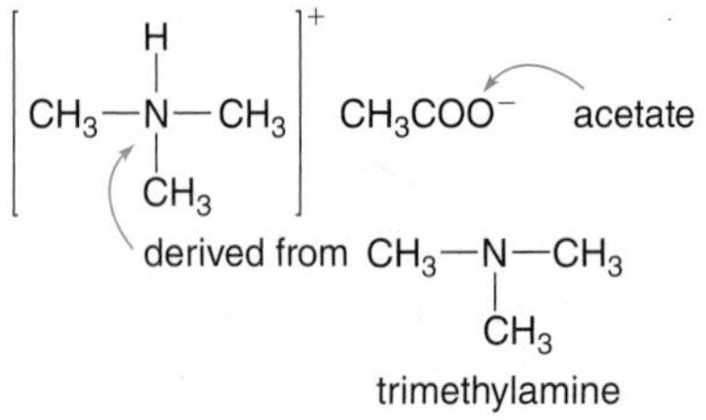

$$[(CH_3)_3NH]^+\ CH_3COO^- \text{ acetate}$$

derived from $CH_3{-}N(CH_3){-}CH_3$ trimethylamine

- Change the name trimethyl*amine* to trimethyl*ammonium.*
- Add the name of the anion, acetate.

Answer: trimethylammonium acetate

PROBLEM 13.26

Name each ammonium salt.

a. $(CH_3NH_3)^+\ Cl^-$ b. $[(CH_3CH_2CH_2)_2NH_2]^+\ Br^-$ c. $[(CH_3)_2NHCH_2CH_3]^+\ CH_3COO^-$

Ammonium salts are ionic compounds, and as a result:

- **Ammonium salts are water-soluble solids.**

In this way, the solubility properties of an amine can be changed by treatment with acid. For example, octylamine has eight carbons, making it water insoluble. Reaction with HCl forms octylammonium chloride. This ionic solid is now soluble in water.

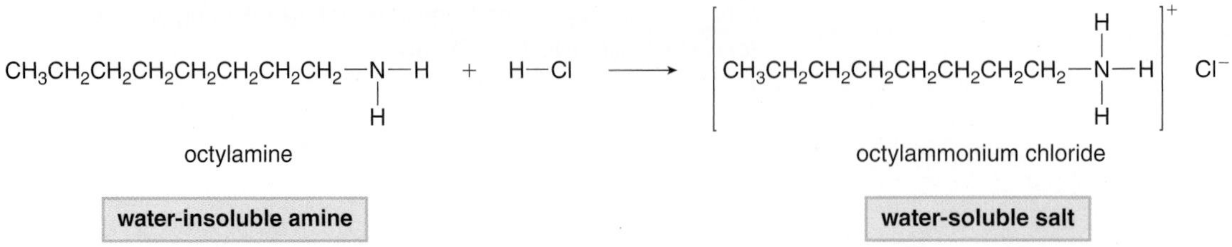

$$CH_3CH_2CH_2CH_2CH_2CH_2CH_2CH_2{-}NH_2 + H{-}Cl \longrightarrow [CH_3CH_2CH_2CH_2CH_2CH_2CH_2CH_2{-}NH_3]^+\ Cl^-$$

octylamine — **water-insoluble amine**; octylammonium chloride — **water-soluble salt**

PROBLEM 13.27

Label each compound as water soluble or water insoluble.

a. $(CH_3CH_2)_3N$ b. $[(CH_3CH_2)_3NH]^+ Br^-$ c. $CH_3CH_2NH_2$ d. [piperidinium ring with N bearing two H]$^+$ Cl^-

Ammonium salts can be re-converted to amines by treatment with base. Base removes a proton from the nitrogen atom of the ammonium ion, regenerating the neutral amine.

This proton is removed by the base.

$$[CH_3CH_2CH_2CH_2CH_2CH_2CH_2CH_2{-}NH_3]^+ Cl^- + Na^+OH^- \longrightarrow CH_3CH_2CH_2CH_2CH_2CH_2CH_2CH_2{-}NH_2 + H{-}OH + NaCl$$

octylammonium chloride **water-soluble salt** → octylamine **water-insoluble amine**

PROBLEM 13.28

Draw the product formed when each ammonium salt is treated with NaOH.

a. $[(CH_3CH_2)_3NH]^+ Br^-$ b. $(CH_3CH_2NH_3)^+ HSO_4^-$ c. [piperidinium ring with N bearing two H]$^+$ Cl^-

CONSUMER NOTE

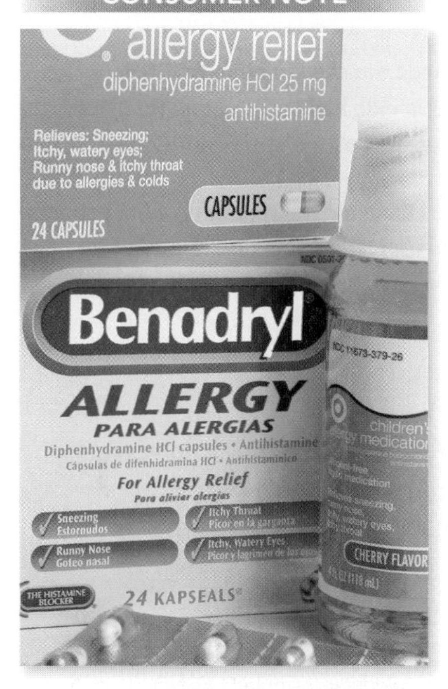

Many antihistamines and decongestants are sold as their hydrochloride salts.

13.8C FOCUS ON HEALTH & MEDICINE Ammonium Salts as Useful Drugs

Many amines with useful medicinal properties are sold as their ammonium salts. Since the ammonium salts are more water soluble than the parent amine, they are easily transported through the body in the aqueous medium of the blood.

For example, diphenhydramine is a 3° amine that is sold as its ammonium salt under the name of Benadryl. Benadryl, formed by treating diphenhydramine with HCl, is an over-the-counter antihistamine that is used to relieve the itch and irritation of skin rashes and hives.

$$(C_6H_5)_2CH{-}OCH_2CH_2{-}N(CH_3){-}CH_3 + HCl \longrightarrow [(C_6H_5)_2CH{-}OCH_2CH_2{-}NH(CH_3){-}CH_3]^+ Cl^-$$

diphenhydramine → ammonium salt Benadryl

The names of medicines sold as ammonium salts are often derived from the names of the amine and the acid used to form them. Since Benadryl is formed from diphenhydramine and HCl, it is called **diphenhydramine hydrochloride.**

HEALTH NOTE

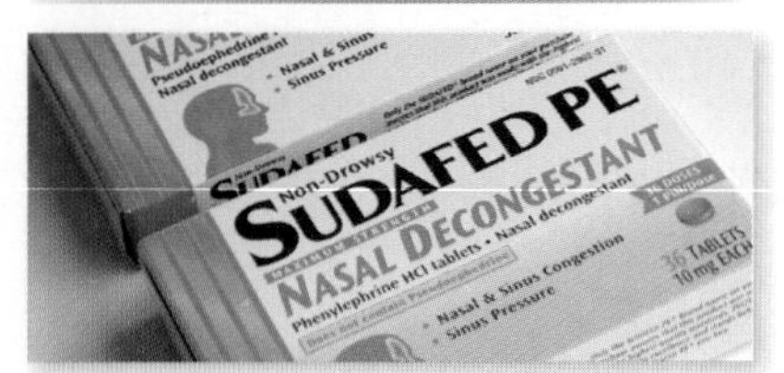

Sudafed PE is an over-the-counter decongestant that contains phenylephrine hydrochloride (Problem 13.29).

PROBLEM 13.29

Phenylephrine hydrochloride is the decongestant in Sudafed PE. Draw the amine and acid used to form phenylephrine hydrochloride.

phenylephrine hydrochloride

13.9 Amides

Amides contain a carbonyl group bonded to a nitrogen atom.

General structure

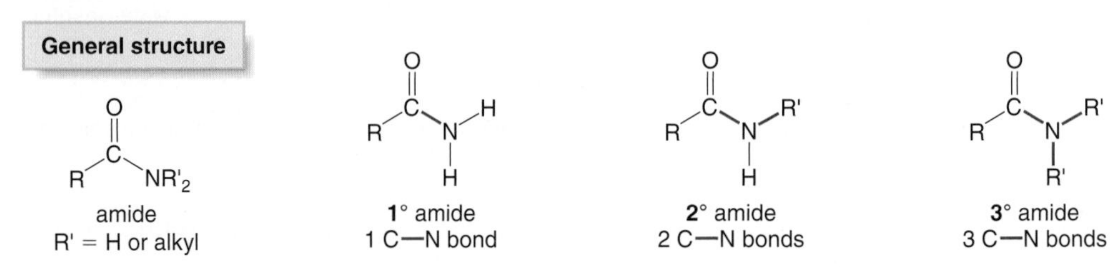

The N atom of an amide may be bonded to other hydrogen atoms or alkyl groups. Amides are classified as **1°, 2°,** or **3°** depending on the number of carbon atoms bonded directly to the *nitrogen* atom.

- A primary (1°) amide contains one C—N bond. A 1° amide has the structure $RCONH_2$.
- A secondary (2°) amide contains two C—N bonds. A 2° amide has the structure RCONHR'.
- A tertiary (3°) amide contains three C—N bonds. A 3° amide has the structure $RCONR'_2$.

For example, **capsaicin,** the compound responsible for the characteristic spiciness of hot peppers, is a 2° amide. Capsaicin is the active ingredient in several topical creams for pain relief.

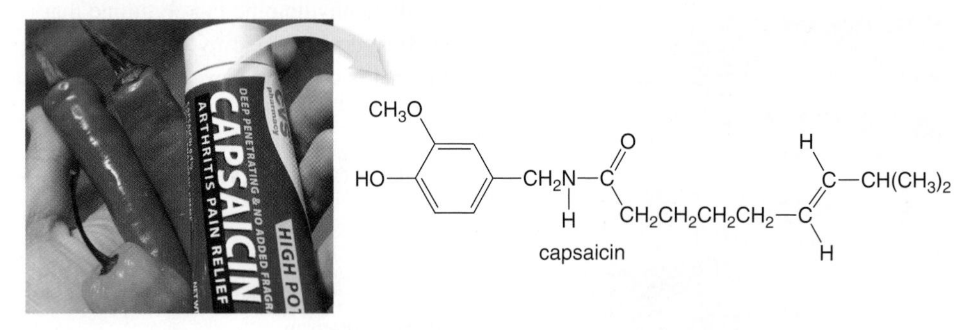

PROBLEM 13.30

Draw the structure of a 1°, 2°, and 3° amide, each having the molecular formula C_4H_9NO.

13.9A Naming an Amide

- **In the IUPAC system, amides are identified by the suffix *-amide*.**

All 1° amides are named by replacing the ***-oic acid*** ending (or *-ic acid* ending of a common name) with the suffix ***-amide.***

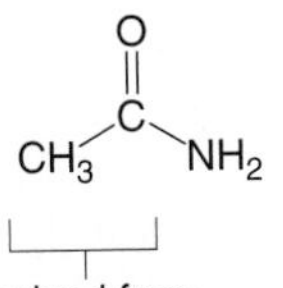

CH_3CONH_2: derived from acet*ic acid* ---→ **acetamide**

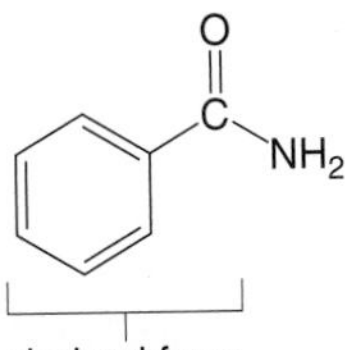

$C_6H_5CONH_2$: derived from benzo*ic acid* ---→ **benzamide**

A 2° or 3° amide has two parts to its structure: the **RCO– group** that contains the carbonyl and one or two **alkyl groups** bonded to the nitrogen atom. To name a 2° or 3° amide:

- **Name the alkyl group (or groups) bonded to the N atom of the amide. Use the prefix "*N*-" preceding the name of each alkyl group.**
- **Name the RCO– group with the suffix *-amide.***

SAMPLE PROBLEM 13.12

Name the following amide: $HCONHCH_2CH_3$.

Analysis and Solution

[1] **Name the alkyl group on the N atom, and precede its name with *N*-.**

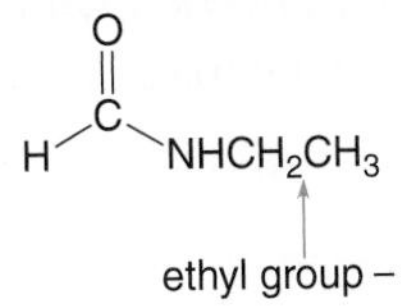

$HCONHCH_2CH_3$: ethyl group ---→ *N*-ethyl

[2] **Name the RCO– group. Change the *-ic acid* ending of the parent carboxylic acid to *-amide*, and combine the parts together.**

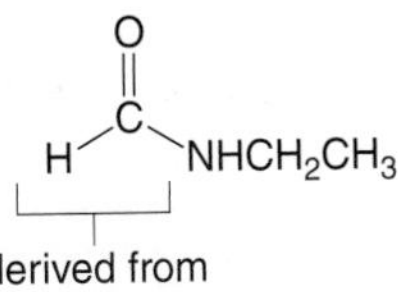

$HCONHCH_2CH_3$: derived from form*ic acid* ---→ **formamide**

Answer: *N*-ethylformamide

PROBLEM 13.31

Name each amide.

a. 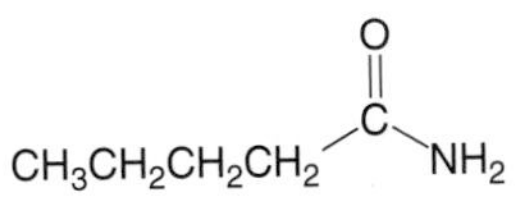 $CH_3CH_2CH_2CH_2CONH_2$

b. $C_6H_5CONHCH_3$

c. $HCON(CH_2CH_2CH_3)_2$

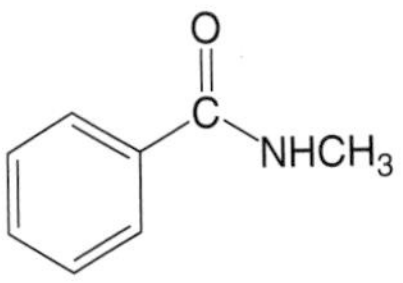

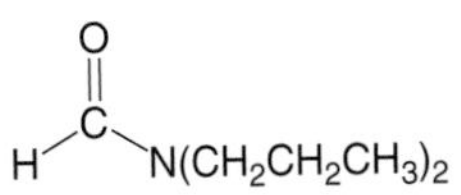

PROBLEM 13.32

Draw the structure of each amide.

a. propanamide
b. *N*-ethylhexanamide
c. *N,N*-dimethylacetamide

13.9B Physical Properties

Primary (1°) and 2° amides contain N—H bonds, so hydrogen bonding is possible between two molecules of the amide. This gives 1° and 2° amides stronger intermolecular forces than 3° amides and esters, which can't intermolecularly hydrogen bond. As a result:

- **Primary (1°) and 2° amides have higher boiling points and melting points than esters and 3° amides of comparable molecular weight.**

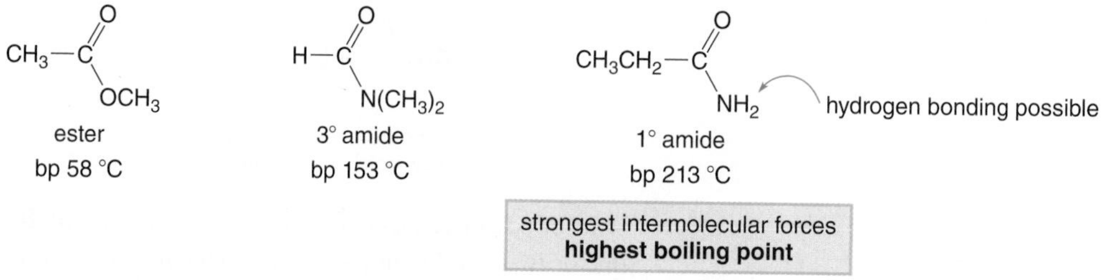

$CH_3-C(=O)-OCH_3$ — ester, bp 58 °C

$H-C(=O)-N(CH_3)_2$ — 3° amide, bp 153 °C

$CH_3CH_2-C(=O)-NH_2$ — 1° amide, bp 213 °C (hydrogen bonding possible)

strongest intermolecular forces
highest boiling point

Amides having fewer than six carbons are soluble in water. Higher molecular weight amides are insoluble in water because the nonpolar portion of the molecule, the C—C and C—H bonds, gets larger than the polar carbonyl group.

PROBLEM 13.33

Why is the boiling point of CH_3CONH_2 (221 °C) higher than the boiling point of $CH_3CON(CH_3)_2$ (166 °C), even though the latter compound has a higher molecular weight and more surface area?

13.9C Amide Hydrolysis

Like esters, amides undergo hydrolysis, but amides are much less reactive than esters. Nonetheless, under forcing conditions amides can be hydrolyzed with water in the presence of acid or base. **Treatment of an amide (RCONHR') with water in the presence of an acid catalyst (HCl) forms a carboxylic acid (RCOOH) and an ammonium salt ($R'NH_3^+$ Cl^-).**

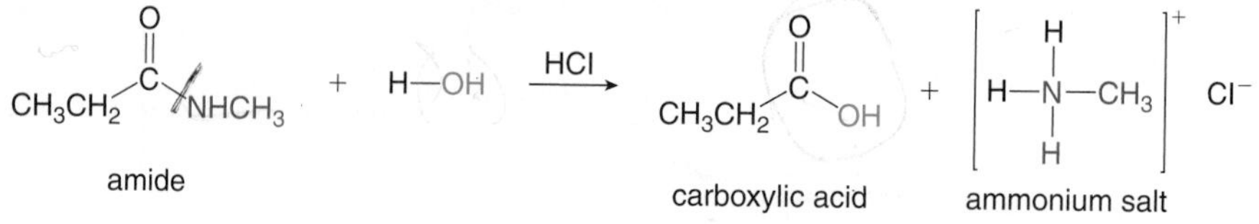

$$CH_3CH_2C(=O)NHCH_3 + H{-}OH \xrightarrow{HCl} CH_3CH_2C(=O)OH + [H{-}NH_2{-}CH_3]^+ \; Cl^-$$

amide → carboxylic acid + ammonium salt

Amides are also hydrolyzed in aqueous base to form carboxylate anions and a molecule of ammonia (NH_3) or amine.

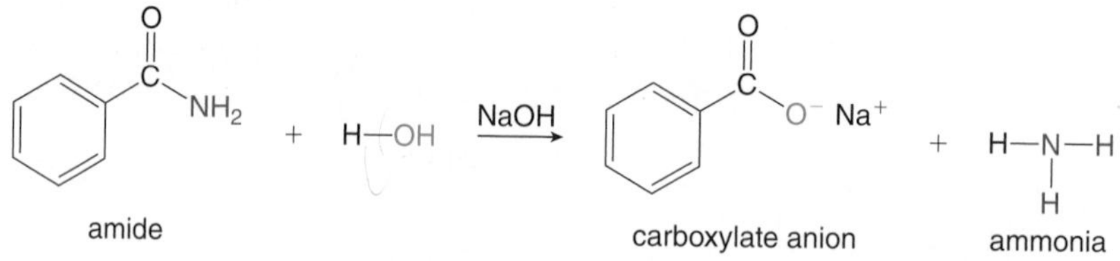

$$C_6H_5C(=O)NH_2 + H{-}OH \xrightarrow{NaOH} C_6H_5C(=O)O^- \; Na^+ + H{-}NH{-}H$$

amide → carboxylate anion + ammonia

The relative lack of reactivity of the amide bond is important in the proteins in the body. Proteins are polymers connected by amide linkages, as we will learn in Chapter 16. Proteins are stable in water in the absence of acid or base, so they can perform their various functions in the cell without breaking down. The hydrolysis of the amide bonds in proteins requires a variety of specific enzymes.

SAMPLE PROBLEM 13.13

What products are formed when *N*-methylacetamide ($CH_3CONHCH_3$) is hydrolyzed with water in the presence of NaOH?

Analysis

To draw the products of amide hydrolysis in base, replace the NHR' group of the amide by an oxygen anion (O^-), forming a new C—O bond at the carbonyl carbon. A molecule of amine ($R'NH_2$) is also formed from the nitrogen group (NHR') of the amide.

$$\underset{\text{amide}}{\text{—C(=O)—NHR'}} + \text{H—OH} \xrightarrow{\text{NaOH}} \text{—C(=O)—O}^- \ \text{Na}^+ + \text{H—N(H)—R'}$$

new C—O bond

O^- replaces NHR'.

Solution

Replace the $NHCH_3$ group of *N*-methylacetamide by a negatively charged oxygen atom (O^-) to form sodium acetate ($CH_3CO_2^-\ Na^+$) and methylamine (CH_3NH_2).

$$\underset{N\text{-methylacetamide}}{CH_3\text{—C(=O)—}NHCH_3} + \text{H—OH} \xrightarrow{\text{NaOH}} CH_3\text{—C(=O)—O}^-\ \text{Na}^+ + \text{H—N(H)—}CH_3$$

O^- replaces $NHCH_3$.

PROBLEM 13.34

What products are formed when each amide is treated with H_2O and H_2SO_4?

a. $CH_3(CH_2)_8\text{—C(=O)—}NH_2$

b. $CH_3CH(CH_3)CH_2\text{—C(=O)—}NHCH_3$

c. cyclohexyl—$CON(CH_2CH_2CH_3)_2$

PROBLEM 13.35

What products are formed when each amide in Problem 13.34 is treated with H_2O and NaOH?

13.10 Interesting Amines and Amides

Caffeine and **nicotine** are widely used stimulants of the central nervous system that contain nitrogen atoms in rings. Caffeine and nicotine, like the amines in Figure 13.3, are ***alkaloids*, naturally occurring amines derived from plant sources.**

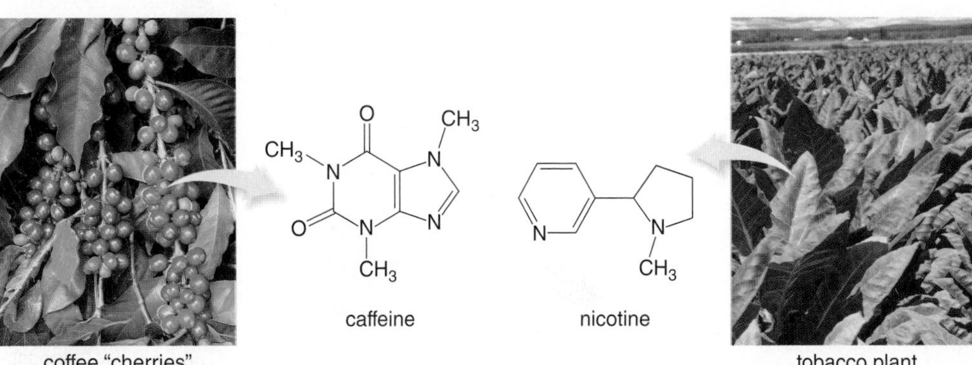

coffee "cherries" tobacco plant

13.10A FOCUS ON THE HUMAN BODY
Epinephrine and Related Compounds

Epinephrine, or **adrenaline** as it is commonly called, is an amine synthesized in the adrenal glands from norepinephrine (noradrenaline).

HO, HO–(benzene ring)–$CH(OH)CH_2NH_2$ ⟶ HO, HO–(benzene ring)–$CH(OH)CH_2NHCH_3$

norepinephrine (noradrenaline) → epinephrine (adrenaline)

When an individual senses danger or is confronted by stress, the hypothalamus region of the brain signals the adrenal glands to synthesize and release epinephrine, which enters the bloodstream and then stimulates a response in many organs (Figure 13.4). Stored carbohydrates are metabolized in the liver to form glucose, which is further metabolized to provide an energy boost. Heart rate and blood pressure increase, and lung passages are dilated. These physiological changes are commonly referred to as a "rush of adrenaline," and they prepare an individual for "fight or flight."

The search for drugs that were structurally related to epinephrine but exhibited only some components of its wide range of biological activities led to the discovery of some useful medications. Both **albuterol** and **salmeterol** dilate lung passages; that is, they are ***bronchodilators.*** They do not, however, stimulate the heart. This makes both compounds useful for the treatment of asthma. Albuterol is a short-acting drug used to relieve the wheezing associated with asthma. Salmeterol is much longer acting, and thus it is often used before bedtime to keep an individual symptom free overnight.

Figure 13.4

Epinephrine—The "Fight or Flight" Hormone

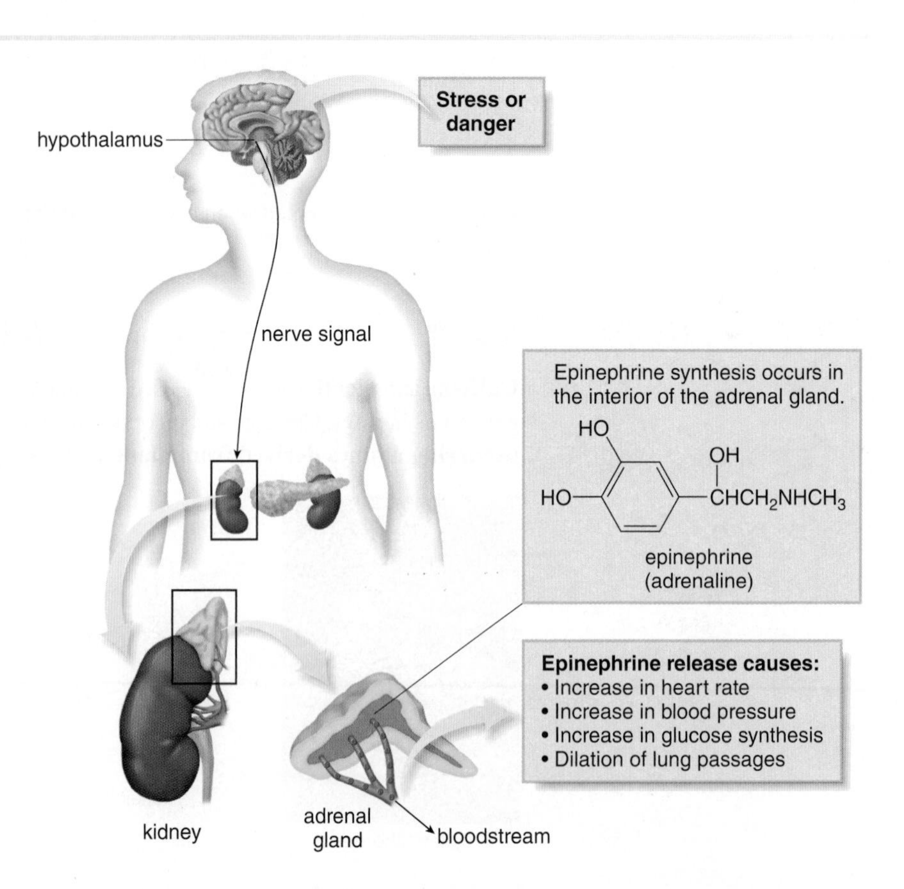

albuterol
(Trade names: Ventolin, Proventil)

salmeterol
(Trade name: Serevent)

PROBLEM 13.36

Classify the amines and the alcohols (not the phenols) in albuterol and salmeterol as 1°, 2°, or 3°.

A RACE AGAINST DEATH!
The Faster this building is completed...the quicker our wounded men get
Penicillin
THE NEW LIFE-SAVING DRUG
Give this job EVERYTHING You've got!

Penicillin was used to treat injured soldiers in World War II before its structure had been conclusively determined.

13.10B FOCUS ON HEALTH & MEDICINE Penicillin

In the twenty-first century it is hard to imagine that an infected cut or scrape could be life-threatening. Before antibiotics were discovered in the early twentieth century, however, that was indeed the case.

The antibiotic properties of **penicillin** were first discovered in 1928 by Sir Alexander Fleming, who noticed that a mold of the genus *Penicillium* inhibited the growth of certain bacteria. After years of experimentation, penicillin was first used to treat a female patient who had developed a streptococcal infection in 1942. By 1944, penicillin production was given high priority by the United States government, because it was needed to treat the many injured soldiers in World War II.

The penicillins are a group of related antibiotics. All penicillins contain two amide units. One amide is part of a four-membered ring called a **β-lactam.** The second amide is bonded to the four-membered ring. Particular penicillins differ in the identity of the R group in the amide side chain. The first penicillin to be discovered was penicillin G. Amoxicillin is another example in common use today.

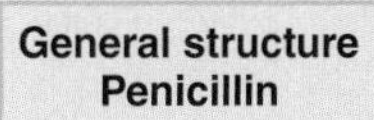

β-lactam

penicillin G

amoxicillin

Unlike mammalian cells, bacterial cells are surrounded by a fairly rigid cell wall, which allows the bacterium to live in many different environments. **Penicillin interferes with the synthesis of the bacterial cell wall.** The β-lactam ring of the penicillin molecule reacts with an enzyme needed to synthesize the cell wall, and this deactivates the enzyme. Cell wall construction is halted, killing the bacterium.

KEY TERMS

Alkaloid (13.7)
Amide (13.1)
Amine (13.1)
Ammonium salt (13.8)
Carboxyl group (13.1)
Carboxylate anion (13.5)
Carboxylic acid (13.1)
Ester (13.1)
Fischer esterification (13.6)
Hydrolysis (13.6)
α-Hydroxy acid (13.4)
β-Lactam (13.10)
Lipid (13.6)
Micelle (13.5)
Penicillin (13.10)
Primary (1°) amide (13.9)
Primary (1°) amine (13.7)
Prostaglandin (13.4)
Saponification (13.6)
Secondary (2°) amide (13.9)
Secondary (2°) amine (13.7)
Soap (13.5)
Tertiary (3°) amide (13.9)
Tertiary (3°) amine (13.7)
Triacylglycerol (13.6)

KEY REACTIONS

[1] Acid–base reaction of carboxylic acids (13.5)

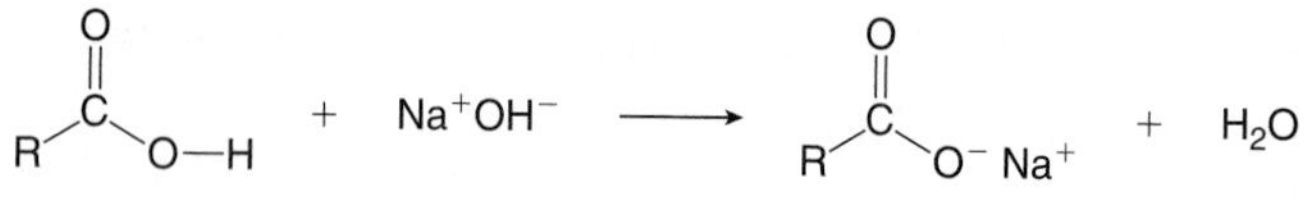

$$RCOOH + Na^+OH^- \longrightarrow RCOO^- Na^+ + H_2O$$

[2] Reactions involving carboxylic acids, esters, and amides

a. Formation of esters (13.6)

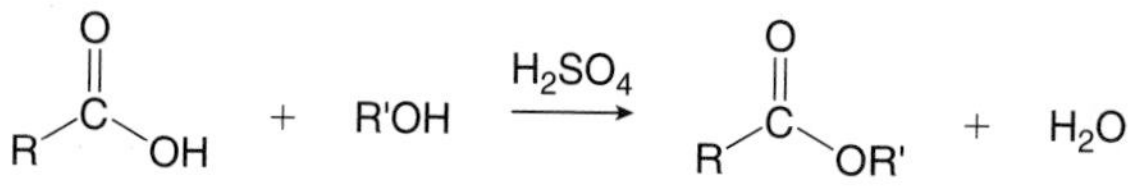

$$RCOOH + R'OH \xrightarrow{H_2SO_4} RCOOR' + H_2O$$

b. Ester hydrolysis (13.6)

$$RCOOR' + H_2O \xrightarrow{H^+} RCOOH + R'OH$$

$$RCOOR' + H_2O \xrightarrow{OH^-} RCOO^- + R'OH$$

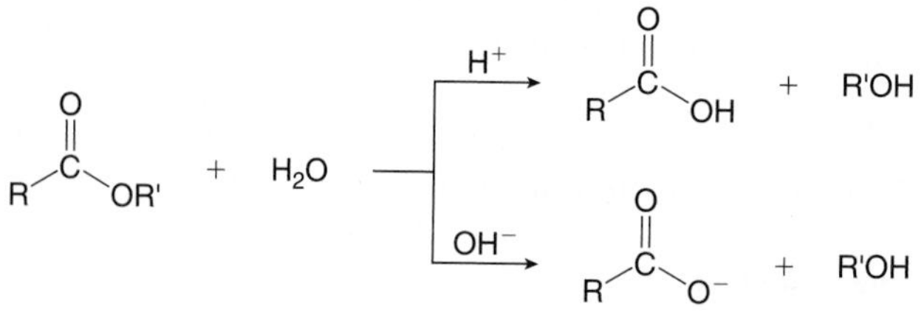

c. Amide hydrolysis (13.6)

$$RCONR'_2 + H_2O \xrightarrow{H^+} RCOOH + (R'_2NH_2)^+$$

$$RCONR'_2 + H_2O \xrightarrow{OH^-} RCOO^- + R'_2NH$$

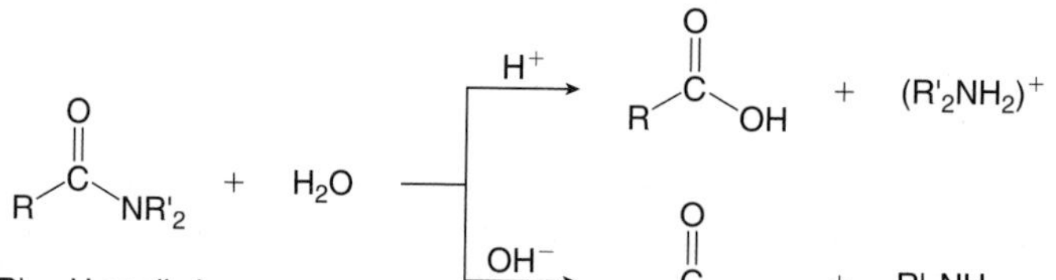

R' = H or alkyl

[3] Acid–base reaction of amines (13.8)

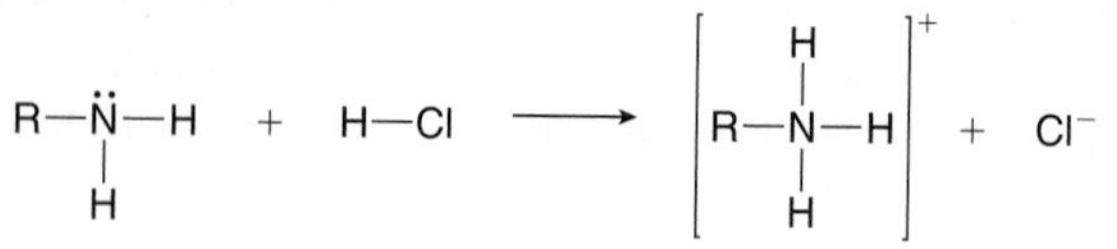

$$R\ddot{N}H_2 + H{-}Cl \longrightarrow [RNH_3]^+ + Cl^-$$

KEY CONCEPTS

1 What are the characteristics of carboxylic acids, esters, amines, and amides?

- Carboxylic acids have the general structure RCOOH; esters have the general structure RCOOR'; amides have the general structure $RCONR'_2$, where R' = H or alkyl. (13.1)
- All carbonyl compounds have a polar C═O. RCO_2H, $RCONH_2$, and RCONHR' are capable of intermolecular hydrogen bonding. (13.3, 13.9)
- Amines are formed by replacing one or more H atoms of NH_3 by alkyl groups. The N atom has a lone pair of electrons. Primary (1°) and 2° amines (RNH_2 and R_2NH) can hydrogen bond. (13.7)

2 What suffixes are used to identify a carboxylic acid, ester, amine, and amide?

- Carboxylic acids are identified by the suffix *-oic acid.* (13.2)
- Esters are identified by the suffix *-ate.* (13.2)
- Amines are identified by the suffix *-amine.* (13.7)
- Amides are identified by the suffix *-amide.* (13.9)

3 What products are formed when carboxylic acids are treated with base? (13.5)

- Carboxylic acids react with bases to form water-soluble salts of carboxylate anions ($RCOO^-$).
- Carboxylate anions are identified by the suffix *-ate.*

4 How does soap clean away dirt? (13.5B)

- Soaps are salts of carboxylic acids that have many carbon atoms in a long hydrocarbon chain. A soap molecule has an ionic head and a nonpolar hydrocarbon tail.
- Soap forms micelles in water with the polar heads on the surface and the hydrocarbon tails in the interior. Grease and dirt dissolve in the nonpolar tails, so they can be washed away with water.

5 What reactions involve carboxylic acids, esters, and amides? (13.6, 13.9)

- Carboxylic acids are converted to esters by reaction with alcohols (R'OH) and acid (H_2SO_4).
- Esters are hydrolyzed to carboxylic acids (RCOOH) in the presence of an acid catalyst (H_2SO_4). Esters are converted to carboxylate anions ($RCOO^-$) with aqueous base (NaOH in H_2O).
- Amides are hydrolyzed to carboxylic acids (RCOOH) and ammonium ions ($R'_2NH_2^+$) in the presence of an acid catalyst (HCl). Amides are converted to carboxylate anions ($RCOO^-$) and amines (R'_2NH) with aqueous base (NaOH in H_2O).

6 What products are formed when an amine is treated with acid? (13.8)

- Amines act as proton acceptors in water and acid. For example, the reaction of RNH_2 with HCl forms the water-soluble ammonium salt $RNH_3^+\ Cl^-$.

7 What are the characteristics of ammonium salts and how are they named? (13.8)

- An ammonium salt consists of a positively charged ammonium ion and an anion.
- An ammonium salt is named by changing the suffix *-amine* of the parent amine to the suffix *-ammonium* followed by the name of the anion.
- Ammonium salts are water-soluble solids.
- Water-insoluble amine drugs are sold as their ammonium salts to increase their solubility in the aqueous environment of the blood.

8 How does penicillin act as an antibiotic? (13.10)

- The β-lactam of penicillin reacts with an enzyme needed to synthesize the cell wall of a bacterium. Without a cell wall, the bacterium dies.

UNDERSTANDING KEY CONCEPTS

Selected in-chapter and odd-numbered end-of-chapter problems have brief answers at the end of each chapter. The *Student Study Guide and Solutions Manual* contains detailed solutions to all in-chapter and odd-numbered end-of-chapter problems, as well as additional worked examples and a chapter self-test.

13.37 Label each nitrogen-containing functional group in lidocaine, a local anesthetic, as an amine or amide, and classify it as 1°, 2°, or 3°.

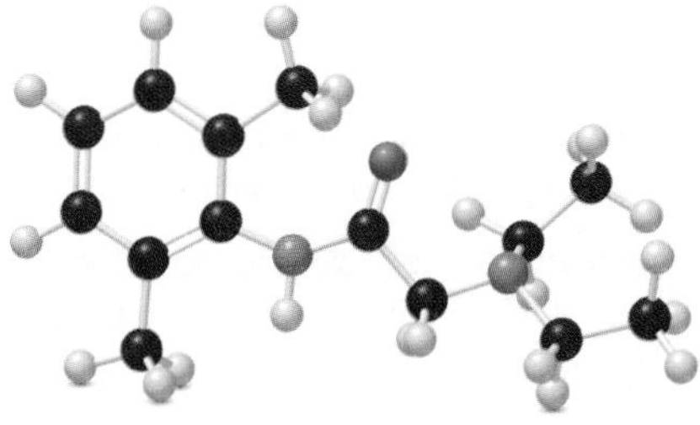

lidocaine

13.38 Label each nitrogen-containing functional group in fentanyl, a narcotic pain reliever, as an amine or amide, and classify it as 1°, 2°, or 3°.

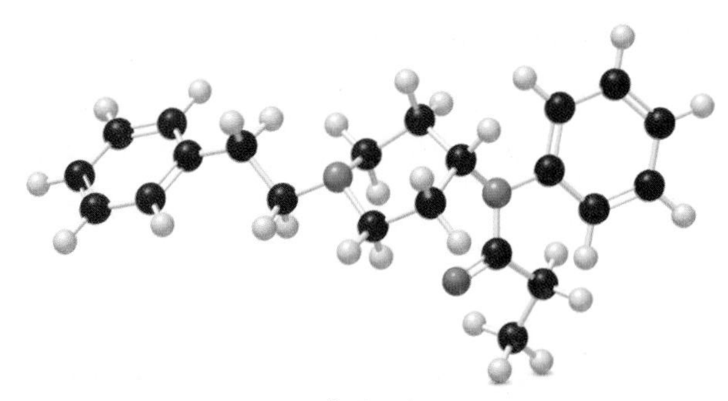

fentanyl

13.39 Which compound(s) can hydrogen bond to another molecule like itself? Which compound(s) can hydrogen bond to water? (a) HCO_2CH_3; (b) $CH_3CH_2CO_2H$.

13.40 Which compound(s) can hydrogen bond to another molecule like itself? Which compound(s) can hydrogen bond to water? (a) $CH_3CONHCH_3$; (b) $HCON(CH_3)_2$.

13.41 Answer the following questions about **A**, depicted in the ball-and-stick model.

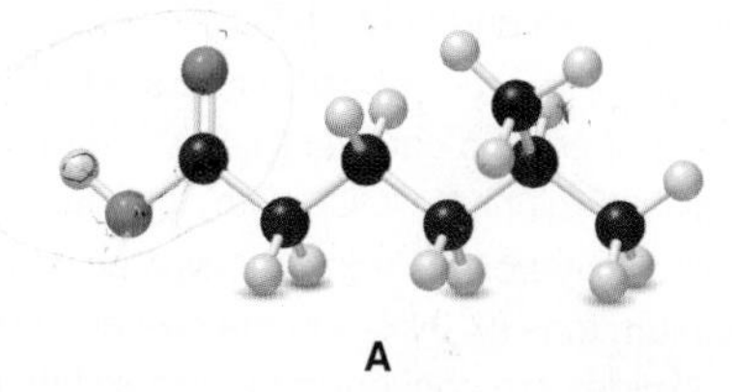

A

a. What is the IUPAC name for **A?**
b. Draw an isomer of **A** that has the same functional group.
c. Draw an isomer of **A** that has a different functional group.
d. What products are formed when **A** is treated with NaOH?
e. What product is formed when **A** is treated with CH_3CH_2OH and H_2SO_4?

13.42 Answer the questions in Problem 13.41 for **B,** depicted in the ball-and-stick model.

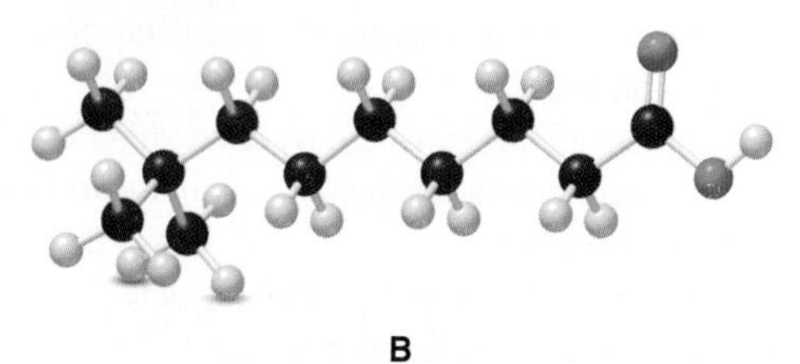

B

13.43 Sunscreens that contain an ester can undergo hydrolysis. What products are formed when octyl salicylate, a commercial sunscreen, is hydrolyzed with water?

O
C
$OCH_2CH_2CH_2CH_2CH_2CH_2CH_2CH_3$
OH

octyl salicylate

13.44 What products are formed when octinoxate, a commercial sunscreen, is hydrolyzed with water?

H O
C C
C O—$CH_2CHCH_2CH_2CH_2CH_3$
H CH_2CH_3
CH_3O

octinoxate

13.45 What products are formed when the pain reliever phenacetin is hydrolyzed with water and H_2SO_4?

CH_3CH_2O— —N—C
H O
CH_3

phenacetin

13.46 What products are formed when phenacetin (Problem 13.45) is hydrolyzed with water and NaOH?

13.47 Benzphetamine (trade name: Didrex) is a habit-forming diet pill.

CH_3
$-CH_2CHNCH_2-$
CH_3

benzphetamine

a. Label the amine as 1°, 2°, or 3°.
b. Draw a constitutional isomer that contains a 1° amine.
c. Draw a constitutional isomer that contains a 3° amine.
d. What products are formed when benzphetamine is treated with acetic acid, CH_3COOH?

13.48 Phentermine is one component of the banned diet drug fen–phen.

CH_3
$-CH_2CNH_2$
CH_3

phentermine

a. Label the amine as 1°, 2°, or 3°.
b. Draw a constitutional isomer that contains a 1° amine.
c. Draw a constitutional isomer that contains a 2° amine.
d. What products are formed when phentermine is treated with benzoic acid, C_6H_5COOH?

13.49 Explain why a 1° amine and a 3° amine having the same number of carbons are soluble in water to a similar extent, but the 1° amine has a higher boiling point.

13.50 Which compound has the higher water solubility: $CH_3(CH_2)_5NH_2$ or $CH_3(CH_2)_5NH_3^+ Cl^-$? Explain your choice.

ADDITIONAL PROBLEMS

Structure and Bonding

13.51 Draw the structure of a compound that fits each description:

a. a carboxylic acid of molecular formula $C_8H_{16}O_2$ that has six carbons in its longest chain
b. an ester of molecular formula $C_6H_{12}O_2$ that contains a methoxy group (OCH_3) bonded to the carbonyl group
c. an ester of molecular formula $C_6H_{10}O_2$ that contains a ring

13.52 Draw the structure of four constitutional isomers of molecular formula $C_4H_8O_2$ that contain an ester.

13.53 Draw the structure of a compound of molecular formula $C_5H_{11}NO$ that contains: (a) a 1° amide; (b) a 2° amide; (c) a 3° amide.

13.54 Draw the structure of a compound of molecular formula $C_9H_{11}NO$ that contains a benzene ring and: (a) a 1° amide; (b) a 2° amide; (c) a 3° amide.

13.55 Classify each amine as 1°, 2°, or 3°.

a. $CH_3CHCH_2CH_2NHCH_3$ (with CH_3 on C2)

b. (six-membered ring)N—CH_3

13.56 Draw the structure of a compound of molecular formula $C_4H_{11}NO$ that fits each description:

a. a compound that contains a 1° amine and a 1° alcohol
b. a compound that contains a 2° amine and a 2° alcohol
c. a compound that contains a 1° amine and a 3° alcohol

Nomenclature

13.57 Give an acceptable name for each compound.

a.

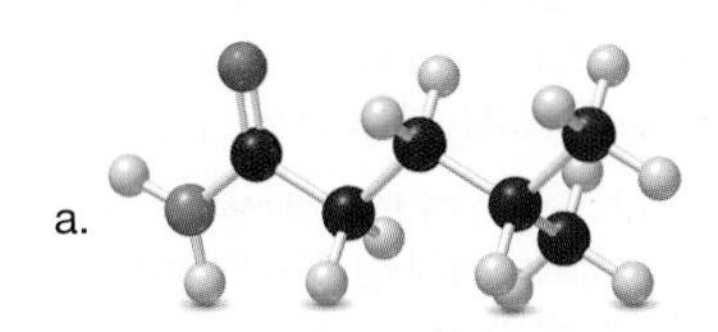

b.

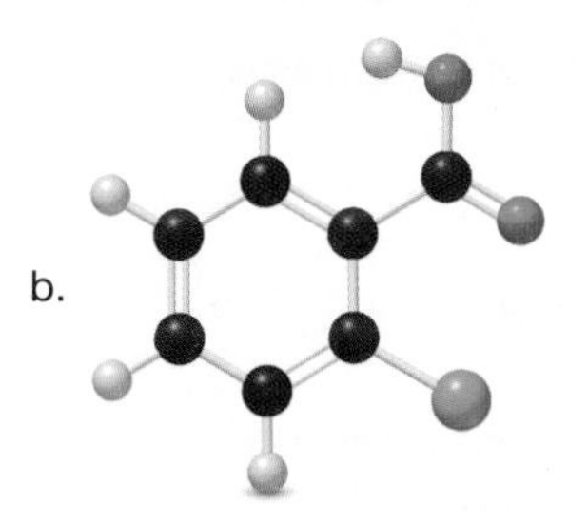

13.58 Give an acceptable name for each compound.

a.

b.

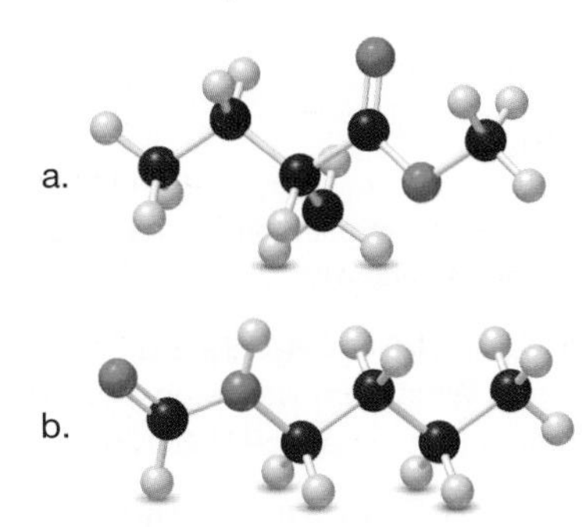

13.59 Give an acceptable name for each compound.

a. $(CH_3)_2CHCH_2CH_2COOH$

b. $CH_3CH_2CH_2CHCHCH_2CH_2COOH$ (with CH_2CH_3 on C4 and C5)

c. $HC(=O)O^-\ Li^+$

d. C_6H_5—$CO_2(CH_2)_3CH_3$

13.60 Give an acceptable name for each compound.

a. $(CH_3)_2CHCH_2CH_2CH_2CH_2CH_2COOH$

b. $CH_3CHCHCHCH_2COOH$ (with CH_3 on C2, C3, and C4)

c. $CH_3CO_2(CH_2)_4CH_3$

d. $CH_3(CH_2)_5COO^-\ K^+$

13.61 Give an acceptable name for each amine.

a.

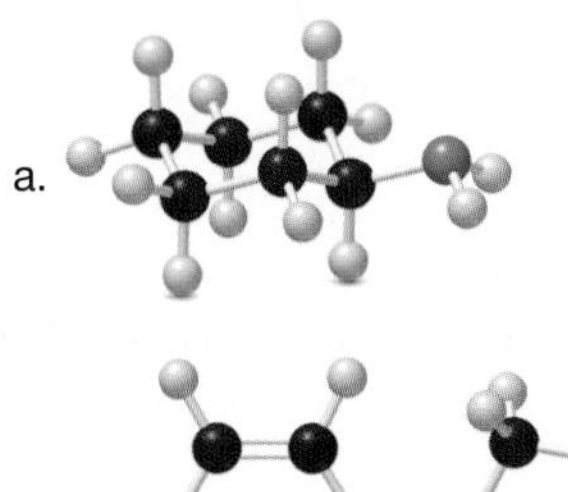

b.

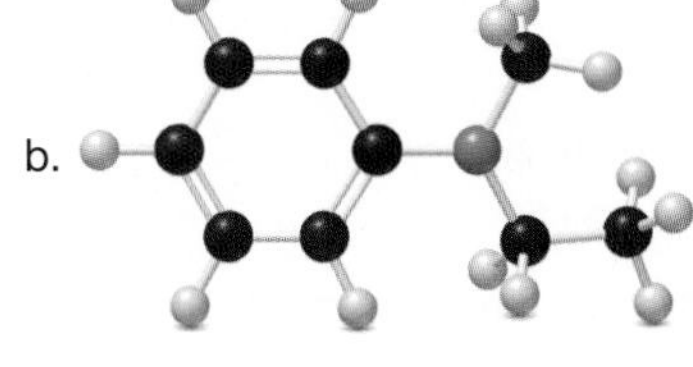

13.62 Give an acceptable name for each amine.

a.

b.

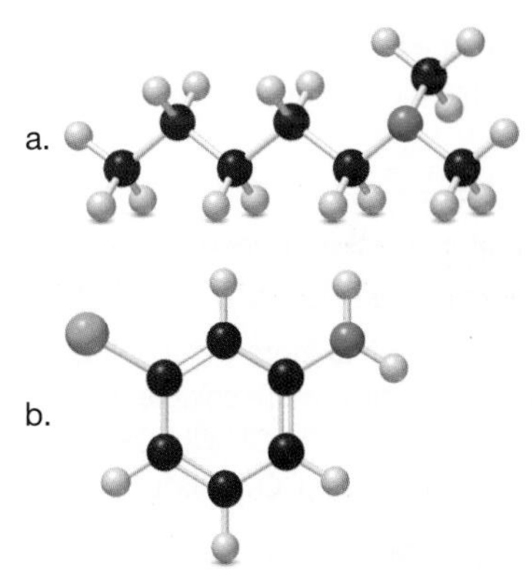

13.63 Give an acceptable name for each amine or amide.

a. CH_3CH_2—N(H)—CH_2CH_3

b. $CH_3(CH_2)_4CONH_2$

c. $HCONHCH_2CH_2CH_2CH_3$

13.64 Give an acceptable name for each amine or amide.

a. $CH_3(CH_2)_6NH_2$

b. $CH_3(CH_2)_6CONH_2$

c. C_6H_5—C(=O)—$N(CH_2CH_3)_2$

13.65 Draw the structure corresponding to each name.
- a. 2-hydroxyheptanoic acid
- b. 4-chlorononanoic acid
- c. butyl butanoate
- d. heptyl benzoate
- e. *N*-ethylhexanamide
- f. *N*-ethyl-*N*-methylheptanamide

13.66 Draw the structure corresponding to each name.
- a. 3-methylhexanoic acid
- b. 3-hydroxy-4-methylheptanoic acid
- c. hexyl pentanoate
- d. propyl hexanoate
- e. *N*-butylbenzamide
- f. *N,N*-dimethyloctanamide

13.67 Draw the structure of each amine or ammonium salt.
- a. *p*-bromoaniline
- b. ethylhexylamine
- c. dipropylammonium chloride
- d. butylammonium bromide

13.68 Draw the structure of each amine or ammonium salt.
- a. butylamine
- b. *N*-pentylaniline
- c. triethylammonium iodide
- d. ethylmethylammonium chloride

Physical Properties and Intermolecular Forces

13.69 Rank the following compounds in order of increasing boiling point: $CH_3CH_2CH(CH_3)_2$, $CH_3CH_2CO_2H$, and $CH_3CH_2COCH_3$.

13.70 Which compound in each pair is more water soluble? Which compound in each pair is more soluble in an organic solvent? (a) $CH_3CH_2CH_2CH_3$ or $CH_3CH_2CH_2CH_2CO_2H$; (b) $CH_3(CH_2)_4COOH$ or $CH_3(CH_2)_4COO^- Na^+$.

13.71 Which compound in each pair has the higher boiling point?

a. $CH_3(CH_2)_6OH$ or $CH_3(CH_2)_6NH_2$

b. (cyclohexyl)–$CH_2N(CH_3)_2$ or (cyclohexyl)–$CH_2CH_2CH_2NH_2$

13.72 Which compound in each pair has the higher boiling point?

a. (cyclohexane) or (piperidine, ring with NH)

b. $CH_3CH(Cl)CH_2CH(NH_2)CH_3$ or $[CH_3CH_2CH_2CH(NH_3)CH_3]^+ Cl^-$

Acidity

13.73 Draw the products of each acid–base reaction.

a. $CH_3(CH_2)_3C(=O)OH$ + KOH ⟶

b. $(CH_3)_2CHCH_2CH_2COOH$ + Na_2CO_3 ⟶

13.74 Draw the products of each acid–base reaction.

a. CH_3–C_6H_4–$C(=O)OH$ (para) + NaOH ⟶

b. Cl_3CCOOH + $NaHCO_3$ ⟶

Esterification and Hydrolysis

13.75 What ester is formed when butanoic acid ($CH_3CH_2CH_2CO_2H$) is treated with each of the following alcohols in the presence of H_2SO_4: (a) CH_3OH; (b) $CH_3CH_2CH_2OH$?

13.76 What ester is formed when each of the following carboxylic acids is treated with 2-propanol [$(CH_3)_2CHOH$] in the presence of H_2SO_4: (a) CH_3CH_2COOH; (b) HCO_2H?

13.77 What products are formed when each ester is hydrolyzed with water and H_2SO_4?

a. $CH_3CH_2CH_2C(=O)OCH(CH_3)_2$

b. (cyclopentyl)–O–$C(=O)CH_2CH_2CH_3$

13.78 What products are formed when each ester in Problem 13.77 is hydrolyzed with water and NaOH?

13.79 What products are formed when each amide is hydrolyzed with water and HCl?

a. $(CH_3)_3CCON(CH_3)_2$

b. $HO(CH_2)_4CONHCH_3$

13.80 What products are formed when each amide in Problem 13.79 is hydrolyzed with water and NaOH?

13.81 Ethyl phenylacetate ($C_6H_5CH_2CO_2CH_2CH_3$) is a naturally occurring ester in honey. What hydrolysis products are formed when this ester is treated with water and H_2SO_4?

13.82 Benzyl acetate ($CH_3CO_2CH_2C_6H_5$) is a naturally occurring ester in peaches. What hydrolysis products are formed when this ester is treated with water and NaOH?

13.83 Draw the products formed in each reaction.

a. $CH_3C(=O)OH$ + CH_3OH $\xrightarrow{H_2SO_4}$

b. $(CH_3)_2CHOC(=O)CH_3$ + H_2O $\xrightarrow{H_2SO_4}$

13.84 Draw the products formed in each reaction.

a. $(CH_3)_2CHCH_2NH{-}C({=}O){-}CH_3 + H_2O \xrightarrow{HCl}$

b. $(CH_3)_2CHNH{-}C({=}O){-}CH_2CH_3 + H_2O \xrightarrow{NaOH}$

Acid–Base Reactions of Amines

13.85 Draw the acid–base reaction that occurs when each amine dissolves in water: (a) $CH_3CH_2NH_2$; (b) $(CH_3CH_2)_2NH$; (c) $(CH_3CH_2)_3N$.

13.86 What ammonium salt is formed when each amine is treated with HCl?

a. $C_6H_5{-}NH_2$ b. $C_6H_5{-}CH_2NHCH_3$

13.87 Draw the products of each acid–base reaction.

a. $CH_3CH_2CH_2N(CH_3)_2 + HCl \longrightarrow$

b. $CH_3CH_2CH(NH_2)CH_2CH_3 + H_2SO_4 \longrightarrow$

13.88 Draw the products of each acid–base reaction.

a. $CH_3CH_2CH(NHCH_3)CH_2CH_3 + H_2SO_4 \longrightarrow$

b. $CH_3NH_2 + CH_3COOH \longrightarrow$

13.89 What ammonium salt is formed when coniine, an alkaloid from hemlock, is treated with HCl?

(piperidine ring with N–H and $CH_2CH_2CH_3$ substituent)

coniine

13.90 What ammonium salt is formed when amphetamine is treated with H_2SO_4?

$C_6H_5{-}CH_2CH(CH_3)NH_2$

amphetamine

Applications

13.91 Which of the following structures represent soaps? Explain your answers.

a. $CH_3CO_2^- \ Na^+$
b. $CH_3(CH_2)_{14}CO_2^- \ Na^+$
c. $CH_3(CH_2)_{12}COOH$

13.92 Explain how soap is able to dissolve nonpolar hydrocarbons in a polar solvent like H_2O.

13.93 Ritalin is the trade name for methylphenidate, a drug used to treat attention deficit hyperactivity disorder (ADHD).

$C_6H_5{-}CH(COOCH_3){-}$(2-piperidinyl, HN)

methylphenidate
(Trade name: Ritalin)

a. Identify the functional groups.
b. Label the amine as 1°, 2°, or 3°.
c. Draw the structure of methylphenidate hydrochloride.

13.94 Seldane is the trade name for terfenadine, an antihistamine once used in the United States but withdrawn from the market because of cardiac side effects observed in some patients.

$(CH_3)_3C{-}C_6H_4{-}CH(OH)CH_2CH_2CH_2{-}N$(piperidine)${-}C(OH)(C_6H_5)_2$

terfenadine
(Trade name: Seldane)

a. Identify the functional groups.
b. Label the amine as 1°, 2°, or 3°.
c. Draw the structure of terfenadine hydrochloride.

CHALLENGE PROBLEMS

13.95 Draw the products formed by acidic hydrolysis of **aspartame,** the artificial sweetener used in Equal and many diet beverages. One of the hydrolysis products of this reaction is the naturally occurring amino acid phenylalanine. Infants afflicted with phenylketonuria cannot metabolize this amino acid, so it accumulates, causing mental retardation. When identified early, a diet limiting the consumption of phenylalanine (and compounds like aspartame that are converted to it) can allow for a normal life.

$HO{-}C({=}O){-}CH_2CH(NH_2){-}C({=}O){-}NH{-}CH(CH_2C_6H_5){-}C({=}O){-}OCH_3$

aspartame

13.96 Today, synthetic detergents like the compound drawn below are used to clean clothes, not soaps. Synthetic detergents are similar to soaps in that they contain an ionic head bonded to a large hydrocarbon group (the nonpolar tail). Explain how this detergent cleans away dirt.

$$CH_3(CH_2)_9\text{—}C_6H_4\text{—}SO_3^-\ Na^+$$

a detergent

13.97 Polyhydroxybutyrate (PHB) is a biodegradable polyester; that is, the polymer is degraded by microorganisms that naturally occur in the environment. Such biodegradable polymers are attractive alternatives to the commonly used polymers that persist in landfills for years. What is the single product formed when PHB is hydrolyzed with acid and water?

$$\text{—O—CH(CH}_3\text{)—CH}_2\text{—C(=O)—O—CH(CH}_3\text{)—CH}_2\text{—C(=O)—}$$

PHB

BEYOND THE CLASSROOM

13.98 Examine the labels of several anti-aging creams and skin care products that remove fine lines and wrinkles. Research the chemical structures of the active ingredients. Do any of these products contain carboxylic acids, esters, amines, or amides? Are any of the products significantly more expensive than others, and if so, is cost related to the amount of the active ingredients?

13.99 Norepinephrine, dopamine, and serotonin are three neurotransmitters, molecules that transmit nerve impulses from one nerve cell to another. All three compounds contain an amine, in addition to other functional groups. Determine the structure of one or more of these neurotransmitters and research why proper levels are needed for an individual's mental health. What conditions result if the level is too high or too low? What medications are available to treat these conditions?

13.100 Pick a polymer that is composed of ester or amide units. Possibilities include PET, PTT, Kevlar, or nylon. Draw the structure of the polymer and the monomers from which it is made. What products are made from the polymer? Can the polymer be recycled?

13.101 Several different β-lactam antibiotics are sold as prescription medicines to treat a variety of bacterial infections. All β-lactam antibiotics contain an amide in a four-membered ring, but differ in the presence of functionality elsewhere in the molecule. Examples include penicillin V, ampicillin, dicloxacillin, and cefalexin. Determine the structure of two of these antibiotics, or choose other drugs that contain a β-lactam. What are the structural similarities and differences between the two drugs you chose? What type of bacteria are the drugs active against? How do bacteria develop resistance to β-lactam-containing drugs? Augmentin is a combination drug that contains both amoxicillin and clavulanic acid. What is the structure of clavulanic acid and what purpose does it serve?

13.102 Research the nicotine content (in mg) in different cigarettes. Are there significant variations among different types of cigarettes: regular cigarettes (70 mm in length), king size varieties (100 mm in length), those with filters, "light" brands, or menthols? How does the nicotine content of cigarettes compare with smokeless tobacco or cigars? Ask family members or classmates who smoke to estimate how many cigarettes they smoke in a day. Calculate how many grams of nicotine the smoker would inhale in a year with various cigarette types. Research how nicotine can be both a stimulant and a relaxant, two apparently conflicting effects.

ANSWERS TO SELECTED PROBLEMS

13.1 a. ester c. carboxylic acid
b. amide d. amide

13.3 a. 3,3-dimethylhexanoic acid
b. 4-chloropentanoic acid
c. 2,4-diethylhexanoic acid

13.5 a. methyl hexanoate
b. ethyl benzoate
c. propyl pentanoate

13.7 a. CH_3COOH b. $C_6H_5\text{—}COOH$

13.9 a.

13.11 a. $C_6H_{11}\text{—}C(=O)O^-\ Na^+ \; + \; H_2O$

b. $CH_3CH_2CH_2\text{—}C(=O)O^-\ Na^+ \; + \; Na^+HCO_3^-$

13.12 a. sodium butanoate
b. lithium benzoate

13.13 $CH_3(CH_2)_{14}CO_2^-\ K^+$

13.14 a. $(CH_3)_2CHCOOCH_2CH_3$ (ester: $(CH_3)_2CH-C(=O)-OCH_2CH_3$)
b. $H-C(=O)-OCH_2CH_3$
c. $CH_3(CH_2)_6-C(=O)-OCH_2CH_3$
d. $C_6H_{11}-C(=O)-OCH_2CH_3$ (cyclohexyl)

13.15 A: benzene ring bearing OCH_3, $CH_2OCCH_2CH(CH_3)_2$ (with C=O), and OCH_3

13.16 a. $CH_3(CH_2)_8-C(=O)-OH$ + CH_3OH
b. $CH_3CH(CH_3)CH_2-C(=O)-OH$ + CH_3CH_2OH
c. cyclohexyl$-COH$ (C=O) + $CH_3CH_2CH_2OH$

13.17 benzene ring with CO_2H and OH (ortho) + CH_3CO_2H

13.19 $(HOCH_2)_2CHOH + CH_3(CH_2)_{10}COOH + CH_3(CH_2)_{12}COOH + CH_3(CH_2)_{14}COOH$

13.20 a. $H_2N(CH_2)_3NH(CH_2)_4NH(CH_2)_3NH_2$ (1°, 2°, 2°, 1°)
b. $CH_3CH_2O-C(=O)-$ on a piperidine ring carbon that also bears C_6H_5; ring $N-CH_3$ (3°)

13.21 Scopolamine contains a 3° amine and a 1° alcohol.

13.22 a. propylamine
b. methylpropylamine
c. dibutylamine

13.23 a. benzene ring$-NHCH_3$
b. benzene ring with H_2N and CH_2CH_3 (meta)
c. benzene ring with CH_3CH_2, CH_2CH_3, and NH_2 (1,3,5)
d. benzene ring$-N(CH_2CH_3)_2$

13.24 a. $(CH_3)_2CHCH_2NH_2$
b. $(CH_3)_2CHCH_2OH$
c. cyclohexyl$-NH_2$

13.25 a. $(CH_3CH_2NH_3)^+$ + Cl^-
b. $[(CH_3CH_2)_2NH_2]^+$ + Cl^-
c. $[(CH_3CH_2)_3NH]^+$ + Cl^-

13.26 a. methylammonium chloride
b. dipropylammonium bromide
c. ethyldimethylammonium acetate

13.27 a. insoluble b,c,d: soluble

13.29 benzene ring with HO (meta) and $-CH(OH)CH_2NHCH_3$ and HCl

13.31 a. pentanamide
b. *N*-methylbenzamide
c. *N,N*-dipropylformamide

13.33 CH_3CONH_2 has 2 H's available for hydrogen bonding, giving it stronger intermolecular forces and a higher boiling point than $CH_3CON(CH_3)_2$.

13.34 a. $CH_3(CH_2)_8-C(=O)-OH$ + $NH_4^+\ HSO_4^-$
b. $CH_3CH(CH_3)CH_2-C(=O)-OH$ + $(CH_3NH_3)^+\ HSO_4^-$
c. cyclohexyl$-COH$ (C=O) + $[(CH_3CH_2CH_2)_2NH_2]^+\ HSO_4^-$

13.35 a. $CH_3(CH_2)_8COO^-\ Na^+ + NH_3$
b. $(CH_3)_2CHCH_2COO^-\ Na^+ + CH_3NH_2$
c. cyclohexyl$-CO_2^-\ Na^+$ + $(CH_3CH_2CH_2)_2NH$

13.37 Lidocaine contains a 2° amide and a 3° amine.

13.39 (a) HCO_2CH_3 can hydrogen bond to water.
(b) CH_3CH_2COOH can hydrogen bond to itself and to water.

13.41 a. 5-methylhexanoic acid
b. $CH_3CH_2CH(CH_3)CH_2CH_2-C(=O)-OH$
c. $CH_3CH_2CH_2CH_2CH_2-C(=O)-OCH_3$
d. $CH_3CH(CH_3)CH_2CH_2CH_2-C(=O)-O^-Na^+$ + H_2O
e. $CH_3CH(CH_3)CH_2CH_2CH_2-C(=O)-OCH_2CH_3$

13.43 [2-hydroxybenzoic acid: benzene ring bearing C(=O)OH and OH] + $HOCH_2CH_2CH_2CH_2CH_2CH_2CH_2CH_3$

13.45 $[CH_3CH_2O\text{–}C_6H_4\text{–}NH_3]^+$ + $CH_3C(=O)OH$ (HO–C(=O)–CH_3)

13.47 a. 3° amine

b. [benzene ring bearing CH_3, $CH_2CH(CH_3)NH_2$, and $CH_2C_6H_5$]

c. [benzene ring bearing $CH_2CH_2N(CH_3)\text{–}CH_3$ and $CH_2C_6H_5$]

d. $[C_6H_5\text{–}CH_2CH(CH_3)NH(CH_3)CH_2\text{–}C_6H_5]^+$ + CH_3COO^-

13.49 Primary amines can hydrogen bond to each other, whereas 3° amines cannot. Therefore, 1° amines will have a higher boiling point than 3° amines of similar size. Any amine can hydrogen bond to water, so both 1° and 3° amines have similar solubility properties.

13.51 a. $CH_3C(CH_3)_2CH_2CH_2CH_2\text{–}C(=O)\text{–}OH$

b. $CH_3CH_2CH_2CH_2\text{–}C(=O)\text{–}OCH_3$

c. [cyclobutane ring]–$C(=O)\text{–}OCH_3$

13.53 a. $CH_3CH_2CH_2CH_2\text{–}C(=O)\text{–}NH_2$

b. $CH_3CH_2\text{–}C(=O)\text{–}NH\text{–}CH_2CH_3$

c. $CH_3CH_2\text{–}C(=O)\text{–}N(CH_3)\text{–}CH_3$

13.55 a. 2° b. 3°

13.57 a. 4-methylpentanamide
b. *o*-chlorobenzoic acid

13.59 a. 4-methylpentanoic acid
b. 4,5-diethyloctanoic acid
c. lithium formate
d. butyl benzoate

13.61 a. cyclohexylamine
b. *N*-ethyl-*N*-methylaniline

13.63 a. diethylamine
b. hexanamide
c. *N*-butylformamide

13.65 a. $CH_3CH_2CH_2CH_2CH_2CH(OH)\text{–}C(=O)\text{–}OH$

b. $CH_3CH_2CH_2CH_2CH_2CH(Cl)CH_2CH_2\text{–}C(=O)\text{–}OH$

c. $CH_3CH_2CH_2\text{–}C(=O)\text{–}OCH_2CH_2CH_2CH_3$

d. $C_6H_5\text{–}C(=O)\text{–}OCH_2CH_2CH_2CH_2CH_2CH_2CH_3$

e. $CH_3CH_2CH_2CH_2CH_2\text{–}C(=O)\text{–}NHCH_2CH_3$

f. $CH_3CH_2CH_2CH_2CH_2CH_2\text{–}C(=O)\text{–}N(CH_3)\text{–}CH_2CH_3$

13.67 a. [benzene ring bearing NH_2 and Br (para)]

b. $CH_3CH_2NHCH_2CH_2CH_2CH_2CH_2CH_3$

c. $[(CH_3CH_2CH_2)_2NH_2]^+\ Cl^-$

d. $(CH_3CH_2CH_2CH_2NH_3)^+\ Br^-$

13.69 $CH_3CH_2CH(CH_3)_2 < CH_3CH_2COCH_3 < CH_3CH_2CO_2H$

13.71 a. $CH_3(CH_2)_6OH$
b. [cyclohexane ring]–$CH_2CH_2CH_2NH_2$

13.73 a. $CH_3(CH_2)_3\text{–}C(=O)\text{–}O^-K^+$ + H_2O

b. $(CH_3)_2CHCH_2CH_2C(=O)O^-Na^+$ + $NaHCO_3$

13.75 a. $CH_3CH_2CH_2\text{–}C(=O)\text{–}OCH_3$

b. $CH_3CH_2CH_2\text{–}C(=O)\text{–}OCH_2CH_2CH_3$

13.77 a. 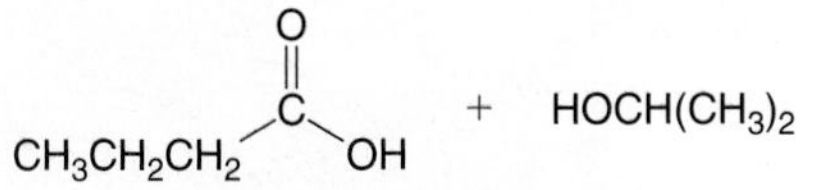 $CH_3CH_2CH_2C(=O)OH$ + $HOCH(CH_3)_2$

b. 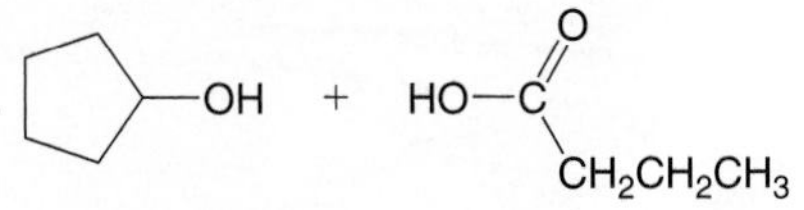 cyclopentyl–OH + $HO-C(=O)CH_2CH_2CH_3$

13.79 a. 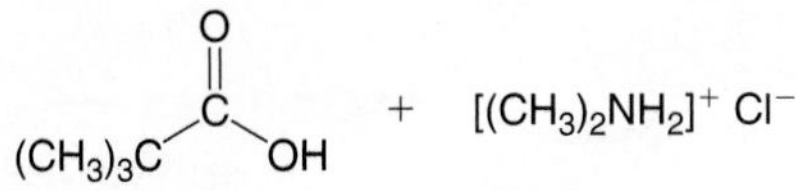 $(CH_3)_3C-C(=O)OH$ + $[(CH_3)_2NH_2]^+\ Cl^-$

b.  $HO(CH_2)_4COOH$ + $(CH_3NH_3)^+\ Cl^-$

13.81 $C_6H_5-CH_2-C(=O)OH$ + CH_3CH_2OH

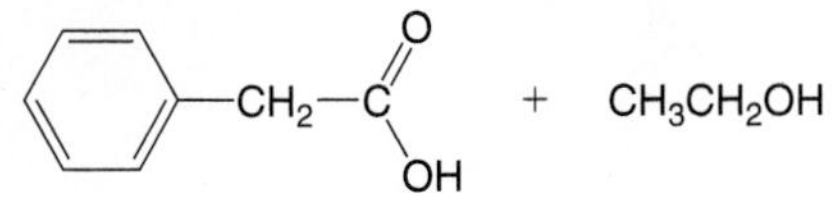

13.83 a. 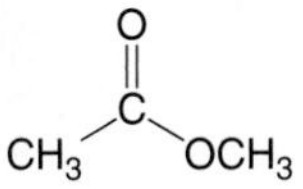 $CH_3C(=O)OCH_3$

b. 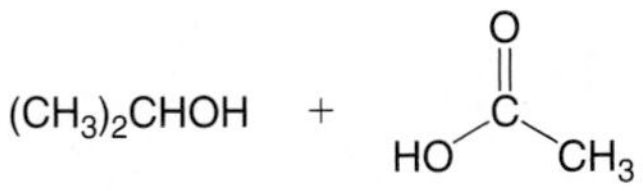 $(CH_3)_2CHOH$ + $HOC(=O)CH_3$

13.85 a. $CH_3CH_2NH_2 + H_2O \longrightarrow (CH_3CH_2NH_3)^+ + OH^-$

b. $(CH_3CH_2)_2NH + H_2O \longrightarrow [(CH_3CH_2)_2NH_2]^+ + OH^-$

c. $(CH_3CH_2)_3N + H_2O \longrightarrow [(CH_3CH_2)_3NH]^+ + OH^-$

13.87 a. $[CH_3CH_2CH_2NH(CH_3)_2]^+ + Cl^-$

b. 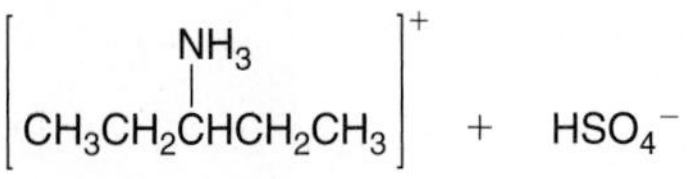 $[CH_3CH_2CH(NH_3)CH_2CH_3]^+ + HSO_4^-$

13.89 [2-propylpiperidinium (NH_2^+ ring with $CH_2CH_2CH_3$ substituent)]$^+$ Cl^-

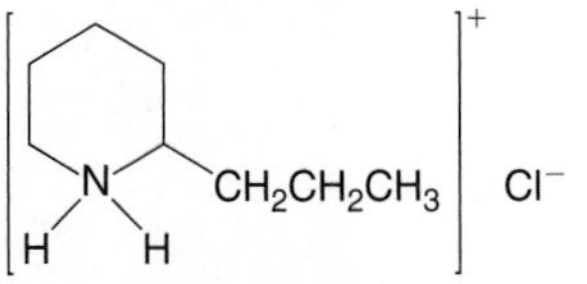

13.91 b. This is a sodium salt of a long-chain carboxylic acid.

13.93 a,b: aromatic ring, ester, 2° amine

c. 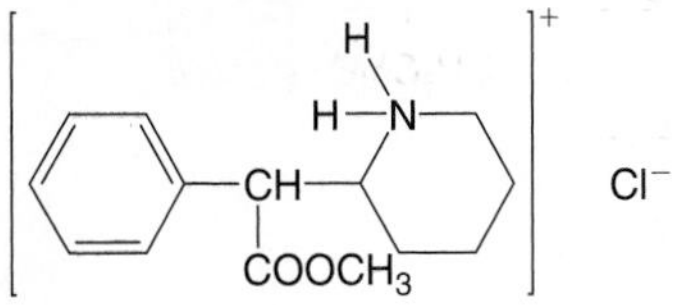 [$C_6H_5-CH(COOCH_3)-$(2-piperidinyl, N bearing two H)]$^+$ Cl^-

13.95

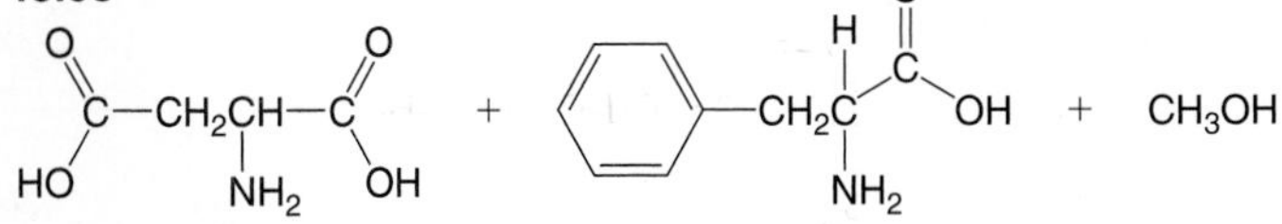

$HOC(=O)-CH_2CH(NH_2)-C(=O)OH$ + $C_6H_5-CH_2C(H)(NH_2)C(=O)OH$ + CH_3OH

13.97 $HO-C(H)(CH_3)-CH_2-C(=O)-OH$

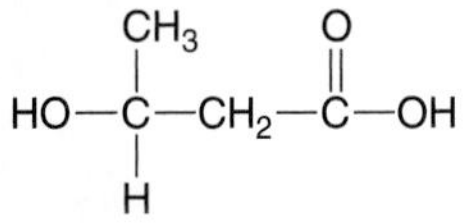

Milk contains **lactose,** a carbohydrate formed from two simple sugars, glucose and galactose.

Carbohydrates

CHAPTER OUTLINE

CHAPTER GOALS

In this chapter you will learn how to:

1. Identify the three major types of carbohydrates
2. Recognize the major structural features of monosaccharides
3. Draw the cyclic forms of monosaccharides and classify them as α or β isomers
4. Draw reduction and oxidation products of monosaccharides
5. Recognize the major structural features of disaccharides
6. Describe the characteristics of cellulose, starch, and glycogen
7. Describe the role that carbohydrates play in determining blood type

Chapter 14 is the first of four chapters that deal with the chemistry of ***biomolecules*, organic molecules found in biological systems.** Chapter 14 discusses carbohydrates, the largest group of biomolecules in nature, while Chapter 15 focuses on lipids, biomolecules that contain many carbon–carbon and carbon–hydrogen bonds, making them soluble in organic solvents and insoluble in water. Chapter 16 focuses on proteins and the amino acids that compose them. Finally, the properties of DNA, the polymer responsible for the storage of genetic information in the chromosomes of cells, is presented in Chapter 17. These compounds are all organic molecules, so many of the principles and chemical reactions that you have already learned will be examined once again. But, as you will see, each class of compound has its own unique features that we will discuss as well.

14.1 Introduction

Carbohydrates, commonly referred to as sugars and starches, are polyhydroxy aldehydes and ketones, or compounds that can be hydrolyzed to them. Carbohydrates can be simple or complex, having as few as three or as many as thousands of carbon atoms. They are the largest group of organic molecules in nature, comprising approximately 50% of the earth's biomass.

Carbohydrates are classified into three groups:

- Monosaccharides (Sections 14.1–14.4)
- Disaccharides (Section 14.5)
- Polysaccharides (Sections 14.6–14.7)

Monosaccharides or simple sugars are the simplest carbohydrates. Glucose and fructose, the two major constituents of honey, are monosaccharides. Glucose contains an aldehyde at one end of a six-carbon chain, and fructose contains a ketone. Every other carbon atom has a hydroxyl group bonded to it. Monosaccharides cannot be converted to simpler compounds by hydrolysis.

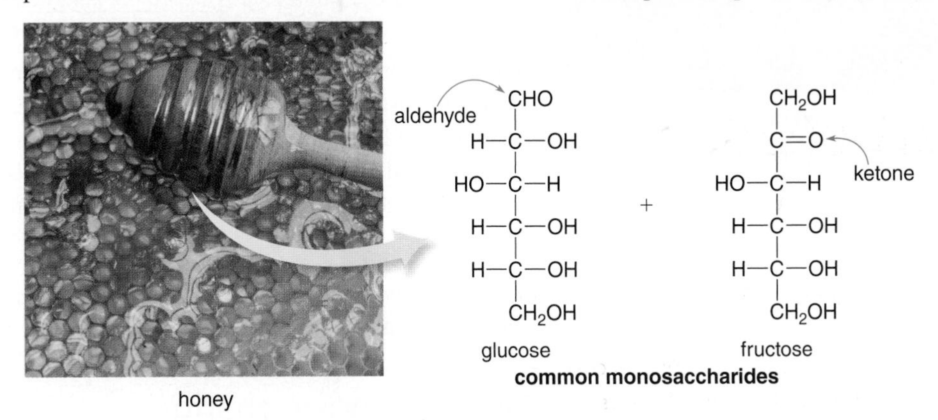

honey

glucose fructose
common monosaccharides

Disaccharides are composed of two monosaccharides joined together. Lactose, the principal carbohydrate in milk, is a disaccharide. Although disaccharides contain no carbonyl groups, they are hydrolyzed to simple monosaccharides that contain an aldehyde or ketone, as we will learn in Section 14.5.

lactose
a common disaccharide

Polysaccharides have three or more monosaccharides joined together. Starch, the main carbohydrate found in the seeds and roots of plants, is a polysaccharide composed of hundreds of glucose molecules joined together. Like disaccharides, polysaccharides are hydrolyzed to simple monosaccharides that contain carbonyl groups. Pasta, bread, rice, and potatoes are foods that contain a great deal of starch.

one form of starch
a polysaccharide

Carbohydrates are storehouses of chemical energy. Carbohydrates are synthesized in green plants and algae through **photosynthesis,** a process that uses the energy from the sun to convert carbon dioxide and water into glucose and oxygen. Plants store glucose in the form of polysaccharides like starch and cellulose (Section 14.6).

Chlorophyll in green leaves converts CO_2 and H_2O to glucose and O_2 during photosynthesis.

Energy is stored in **photosynthesis.**

$$6\ CO_2 + 6\ H_2O \xrightarrow[\text{chlorophyll}]{\text{sunlight}} \underset{\text{glucose}}{C_6H_{12}O_6} + 6\ O_2$$

Energy is released in **metabolism.**

The energy stored in glucose bonds is released when glucose is metabolized. The oxidation of glucose is a multistep process that forms carbon dioxide, water, and a great deal of energy. Although the metabolism of lipids provides more energy per gram than the metabolism of carbohydrates, glucose is the preferred source when a burst of energy is needed during exercise. Glucose is water soluble, so it can be quickly and easily transported through the bloodstream to tissues.

PROBLEM 14.1

Draw a Lewis structure for glucose that clearly shows the aldehyde carbonyl group and all lone pairs on the oxygen atoms.

14.2 Monosaccharides

Monosaccharides, the simplest carbohydrates, generally have three to six carbon atoms in a chain, with a **carbonyl group** at either the terminal carbon, numbered C1, or the carbon adjacent to it, numbered C2. In most carbohydrates, each of the remaining carbon atoms has a **hydroxyl group.** Monosaccharides are drawn vertically, with the carbonyl group at (or near) the top.

3–6 C's: C1 or C2 → C=O
carbonyl at C1 ---→ aldehyde ---→ **aldose**
carbonyl at C2 ---→ ketone ---→ **ketose**
C3, C4: OH on all (or most) other C's

monosaccharide

- Monosaccharides with a carbonyl group at C1 are aldehydes called *aldoses.*
- Monosaccharides with a carbonyl group at C2 are ketones called *ketoses.*

CONSUMER NOTE

Dihydroxyacetone (DHA) is the active ingredient in many artificial tanning agents.

Glyceraldehyde is the simplest aldose and dihydroxyacetone is the simplest ketose. Glyceraldehyde and dihydroxyacetone both have molecular formula $C_3H_6O_3$, so they are **constitutional isomers;** that is, they have the same molecular formula but a different arrangement of atoms. Glucose is the most prevalent aldose.

```
aldehyde ---> aldose        ketone ---> ketose        aldehyde ---> aldose

    CHO                          CH2OH                     CHO
    |                            |                         |
H—C—OH                           C=O                   H—C—OH
    |                            |                         |
    CH2OH                        CH2OH                HO—C—H
                                                           |
glyceraldehyde              dihydroxyacetone           H—C—OH
                                                           |
                                                       H—C—OH
                                                           |
                                                           CH2OH
                                                       glucose
```

A monosaccharide is characterized by the number of carbons in its chain.

- **A triose has three carbons.**
- **A tetrose has four carbons.**
- **A pentose has five carbons.**
- **A hexose has six carbons.**

These terms are then combined with the words *aldose* and *ketose* to indicate both the number of carbon atoms in the monosaccharide and whether it contains an aldehyde or ketone. Thus, glyceraldehyde is an aldotriose (three carbons and an aldehyde), dihydroxyacetone is a ketotriose (three carbons and a ketone), and glucose is an aldohexose (six carbons and an aldehyde).

SAMPLE PROBLEM 14.1

Classify each monosaccharide by the type of carbonyl group and the number of carbons in the chain.

```
                             CH2OH
                             |
a.     CHO          b.       C=O
       |                     |
   H—C—OH                HO—C—H
       |                     |
   H—C—OH                 H—C—OH
       |                     |
   H—C—OH                 H—C—OH
       |                     |
       CH2OH                 CH2OH
     ribose                fructose
```

Analysis

Identify the type of carbonyl group to label the monosaccharide as an aldose or ketose. An aldose has the C=O at C1 so that a hydrogen atom is bonded to the carbonyl carbon. A ketose has two carbons bonded to the carbonyl carbon. Count the number of carbons in the chain to determine the suffix—namely, *-triose, -tetrose,* and so forth.

Solution

a.

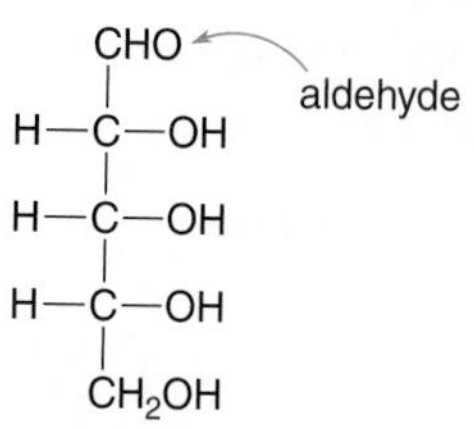

CHO ← aldehyde
H—C—OH
H—C—OH
H—C—OH
CH_2OH

ribose
5 C's in the chain
Answer: aldopentose

b.

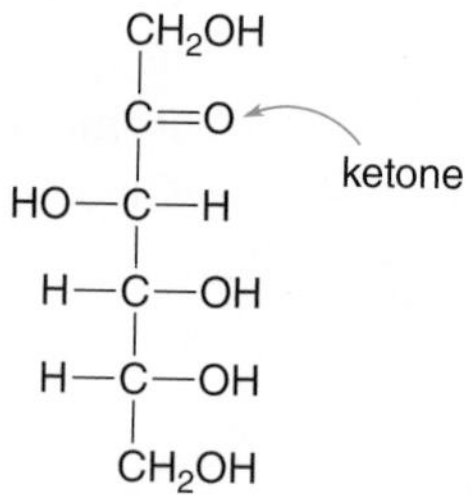

CH_2OH
C=O ← ketone
HO—C—H
H—C—OH
H—C—OH
CH_2OH

fructose
6 C's in the chain
Answer: ketohexose

PROBLEM 14.2

Classify each monosaccharide by the type of carbonyl group and the number of carbons in the chain.

a.

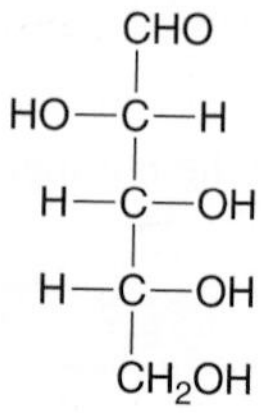

CHO
HO—C—H
H—C—OH
H—C—OH
CH_2OH

arabinose

b.

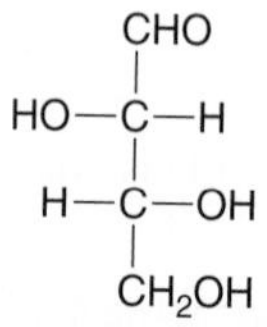

CHO
HO—C—H
H—C—OH
CH_2OH

threose

c.

CH_2OH
C=O
H—C—OH
CH_2OH

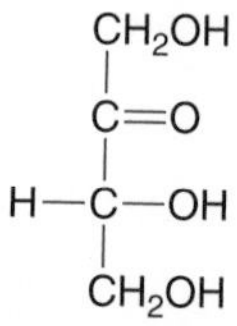

erythrulose

PROBLEM 14.3

Classify the monosaccharide by the type of carbonyl group and the number of carbons in the chain.

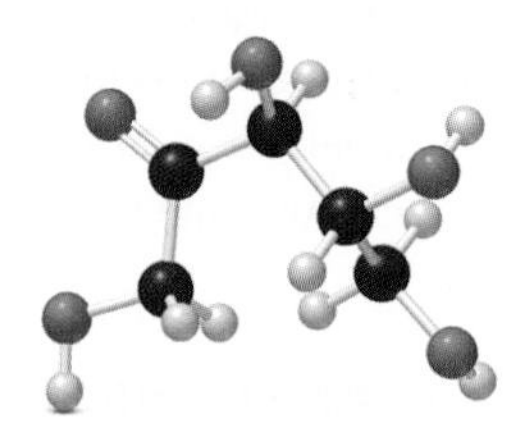

PROBLEM 14.4

Draw the structure of (a) an aldotetrose; (b) a ketopentose; (c) an aldohexose.

Monosaccharides are all sweet tasting, but their relative sweetness varies a great deal. Monosaccharides are polar compounds with high melting points. The presence of so many polar functional groups capable of hydrogen bonding makes them very water soluble.

PROBLEM 14.5

Rank the following compounds in order of increasing water solubility: glucose, hexane [$CH_3(CH_2)_4CH_3$], and 1-decanol [$CH_3(CH_2)_9OH$]. Explain your choice.

14.2A Fischer Projection Formulas

A striking feature of carbohydrate structure is the presence of chirality centers. **All carbohydrates except for dihydroxyacetone contain one or more chirality centers.** The simplest aldose, glyceraldehyde, has one chirality center—one carbon atom bonded to four different groups. Thus, there are two possible enantiomers—mirror images that are not superimposable.

Chirality centers and enantiomers were first encountered in Section 12.10.

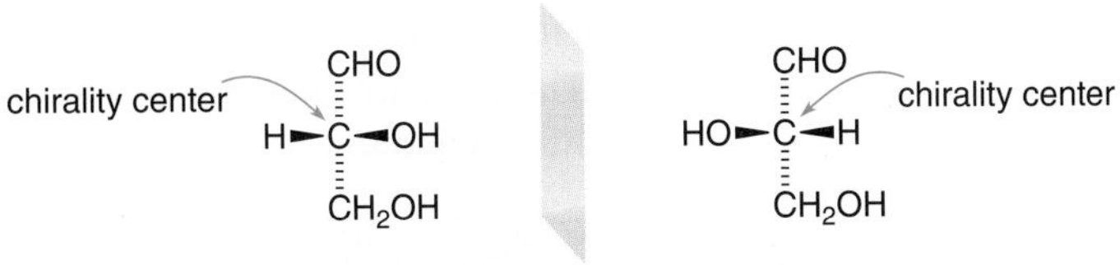

chirality center
CHO
H—C—OH
CH_2OH

naturally occurring isomer
D-glyceraldehyde

CHO
HO—C—H
CH_2OH
chirality center

unnatural isomer
L-glyceraldehyde

Only one enantiomer of glyceraldehyde occurs in nature. When the carbon chain is drawn vertically with the aldehyde at the top, the naturally occurring enantiomer has the OH group drawn on the right side of the carbon chain. **To distinguish the two enantiomers, the prefixes D and L precede the name.** Thus, the naturally occurring enantiomer is labeled D-glyceraldehyde, while the unnatural isomer is L-glyceraldehyde.

Fischer projection formulas are commonly used to depict the chirality centers in monosaccharides. Recall from Section 12.10 that a Fischer projection formula uses a cross to represent a tetrahedral carbon. In a Fischer projection formula:

- **A carbon atom is located at the intersection of the two lines of the cross.**
- **The horizontal bonds come forward, on wedges.**
- **The vertical bonds go back, on dashed lines.**

In a carbohydrate, the carbon skeleton is drawn vertically, with the carbonyl group at the top. Using a Fischer projection formula, D-glyceraldehyde becomes:

CHO
H—C—OH
CH_2OH

D-glyceraldehyde

CHO
H—+—OH
CH_2OH

Fischer projection formula

HEALTH NOTE

A 5% intravenous glucose (dextrose) solution provides a patient with calories and hydration.

PROBLEM 14.6

Draw L-glyceraldehyde using a Fischer projection formula.

14.2B Monosaccharides with More Than One Chirality Center

Fischer projection formulas are also used for compounds like aldohexoses that contain several chirality centers. Glucose, for example, contains four chirality centers labeled in the structure below. To convert the molecule to a Fischer projection, the molecule is drawn with a vertical carbon skeleton with the aldehyde at the top, and the horizontal bonds are assumed to come forward (on wedges). In the Fischer projection, each chirality center is replaced by a cross.

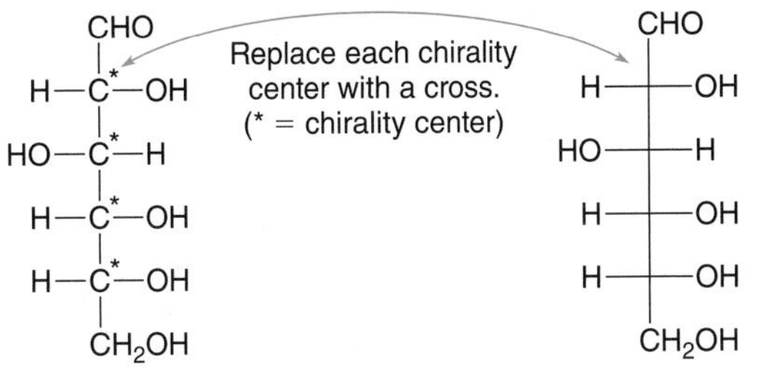

CHO
H—C*—OH
HO—C*—H
H—C*—OH
H—C*—OH
CH_2OH

glucose

Replace each chirality center with a cross.
(* = chirality center)

CHO
H—+—OH
HO—+—H
H—+—OH
H—+—OH
CH_2OH

Fischer projection formula

The letters D and L are used to label all monosaccharides, even those with many chirality centers. **The configuration of the chirality center *farthest* from the carbonyl group determines whether a monosaccharide is D or L.**

- **A D monosaccharide has the OH group on the chirality center farthest from the carbonyl on the right (like D-glyceraldehyde).**
- **An L monosaccharide has the OH group on the chirality center farthest from the carbonyl on the left (like L-glyceraldehyde).**

chirality center farthest from the C=O

CHO	CHO
H—C—OH	HO—C—H
CH_2OH	CH_2OH
OH on the right D sugar	OH on the left L sugar

Glucose and all other naturally occurring sugars are D sugars. L-Glucose, a compound that does not occur in nature, is the enantiomer of D-glucose. L-Glucose has the opposite configuration at *every* chirality center.

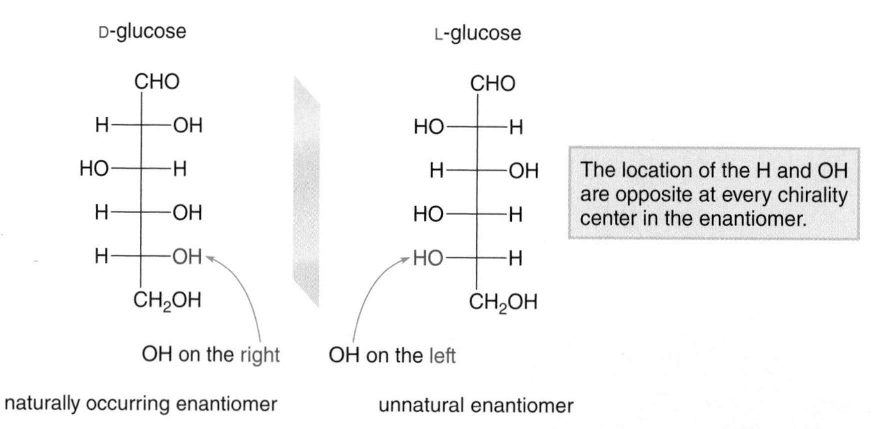

SAMPLE PROBLEM 14.2

Consider the aldopentose ribose. (a) Label all chirality centers. (b) Classify ribose as a D or L monosaccharide. (c) Draw the enantiomer.

CHO
H—C—OH
H—C—OH
H—C—OH
CH_2OH

ribose

Analysis

- A chirality center has four different groups around a carbon atom.
- The labels D and L are determined by the position of the OH group on the chirality center farthest from the carbonyl group: a D sugar has the OH group on the right and an L sugar has the OH group on the left.
- To draw an enantiomer, draw the mirror image so that each group is a reflection of the group in the original compound.

Solution

a. The three carbons that contain both H and OH groups in ribose are chirality centers.

b. Ribose is a D sugar since the OH group on the chirality center farthest from the carbonyl is on the right.

c. The enantiomer of D-ribose, L-ribose, has all three OH groups on the left side of the carbon chain.

```
      CHO                 CHO                  CHO
 H ---*--- OH        H ------- OH        HO ------- H
 H ---*--- OH        H ------- OH        HO ------- H
 H ---*--- OH        H ------- OH        HO ------- H
      CH2OH               CH2OH                CH2OH
* = chirality center     D sugar            enantiomer
```

PROBLEM 14.7

For each monosaccharide: [1] label all chirality centers; [2] classify the monosaccharide as D or L; [3] draw the enantiomer.

```
a.     CHO          b.     CHO          c.     CHO
  H ------- OH        HO ------- H        HO ------- H
  H ------- OH         H ------- OH       HO ------- H
       CH2OH          HO ------- H         H ------- OH
                           CH2OH               CH2OH
```

PROBLEM 14.8

For each type of monosaccharide, [1] give an example of a D sugar; [2] label each chirality center: (a) an aldopentose; (b) a ketohexose; (c) a ketotetrose.

14.2C Common Monosaccharides

The most common monosaccharides in nature are the aldohexoses D-glucose and D-galactose, and the ketohexose D-fructose.

```
       CHO                  CHO                 CH2OH
        |                    |                    |
   H—C—OH               H—C—OH                 C=O
        |                    |                    |
  HO—C—H               HO—C—H               HO—C—H
        |                    |                    |
   H—C—OH              HO—C—H                H—C—OH
        |                    |                    |
   H—C—OH               H—C—OH               H—C—OH
        |                    |                    |
      CH2OH                CH2OH                CH2OH

   D-glucose            D-galactose           D-fructose
```

HEALTH NOTE

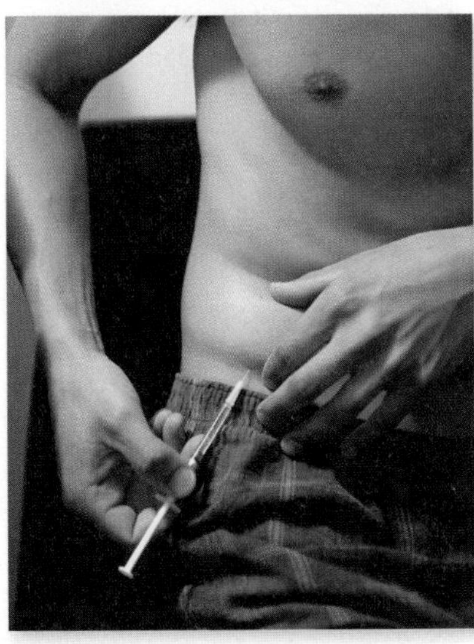

Insulin injections taken by diabetic patients help to maintain a proper blood glucose level.

Glucose, also called dextrose, is the sugar referred to when blood sugar is measured. It is the most abundant monosaccharide. Glucose is the building block for the polysaccharides starch and cellulose. Glucose, the carbohydrate that is transported in the bloodstream, provides energy for cells when it is metabolized. Normal blood glucose levels are in the range of 70–110 mg/dL.

Insulin, a protein produced in the pancreas, regulates blood glucose levels. When glucose concentration increases after eating, insulin stimulates the uptake of glucose in tissues and its conversion to glycogen. Patients with diabetes produce insufficient insulin to adequately regulate blood glucose levels, and the concentration of glucose rises. With close attention to diet and daily insulin injections or other medications, a normal level of glucose can be maintained in most diabetic patients. Individuals with poorly controlled diabetes can develop many other significant complications, including cardiovascular disease, chronic renal failure, and blindness.

HEALTH NOTE

An individual with galactosemia must avoid cow's milk and all products derived from cow's milk (Section 14.2C).

CONSUMER NOTE

Some "lite" food products have fewer calories because they use only half as much fructose as sucrose for the same level of sweetness (Section 14.2C).

Galactose is one of the two monosaccharides that form the disaccharide lactose (Section 14.5). Individuals with galactosemia, a rare inherited disease, lack an enzyme needed to metabolize galactose. Galactose accumulates, causing a variety of physical problems, including cataracts, cirrhosis, and mental retardation. Galactosemia can be detected in newborn screening, and affected infants must be given soy-based formula to avoid all milk products with lactose.

Fructose is one of two monosaccharides that form the disaccharide sucrose (Section 14.5). Fructose is a ketohexose found in honey and is almost twice as sweet as normal table sugar with about the same number of calories per gram.

PROBLEM 14.9

(a) Draw a Fischer projection formula for D-galactose. (b) Draw a Fischer projection formula for the enantiomer of D-galactose.

14.3 The Cyclic Forms of Monosaccharides

Although the monosaccharides in Section 14.2 were drawn as acyclic carbonyl compounds, the hydroxyl and carbonyl groups can react together to form a ring. Let's illustrate the process with D-glucose, and then learn a general method for drawing the cyclic forms of any aldohexose.

Which of the five OH groups reacts with the aldehyde carbonyl? In glucose, the OH group on C5 reacts with the carbonyl carbon to form a six-membered ring.

1 CHO
H—2C—OH
HO—3C—H
H—4C—OH
H—5C—OH
6 CH_2OH
D-glucose

This OH group reacts to form a six-membered ring.

To convert this acyclic form (labeled **A**) into a cyclic monosaccharide, first rotate the carbon skeleton clockwise 90° to form **B.** Note that groups that were drawn on the right side of the carbon skeleton in **A** end up *below* the carbon chain in **B.** Then twist the chain to put the OH group on C5 close to the aldehyde carbonyl, forming **C.** In this process, the CH_2OH group at the end of the chain ends up *above* the carbon skeleton.

D-glucose

1 CHO
H—2C—OH
HO—3C—H
H—4C—OH
H—5C—OH
6 CH_2OH

A

[1] Rotate

H H OH H
6 $HOCH_2$—5C—4C—3C—2C—1CHO
OH OH H OH

B

[2] Twist

Draw the CH_2OH up.
6 CH_2OH
5 H OH
4 H HO
H OH 3 H OH
H 2 OH 1 CHO

C

To draw the cyclic form, the OH group on C5 reacts with the aldehyde carbonyl to form a six-membered ring with a new chirality center. Cyclization yields two isomers, since the OH group on the new chirality center can be located above or below the six-membered ring.

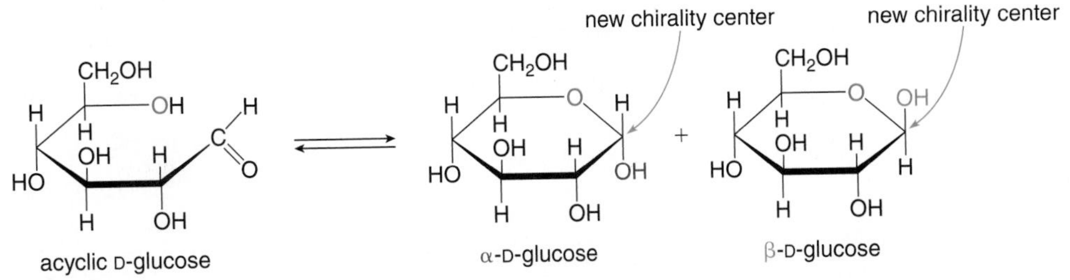

acyclic D-glucose ⇌ α-D-glucose + β-D-glucose

- **The α isomer, called α-D-glucose, has the OH group on the new chirality center drawn down (shown in red).**
- **The β isomer, called β-D-glucose, has the OH group on the new chirality center drawn up (shown in blue).**

These flat, six-membered rings used to represent the cyclic forms of glucose and other sugars are called **Haworth projections.**

Thus, D-glucose really exists in three different forms—an acyclic aldehyde and two cyclic compounds. The mixture has 37% of the α isomer, 63% of the β isomer, and only a trace amount of the acyclic aldehyde. Three-dimensional models for the three forms of D-glucose are shown in Figure 14.1.

Figure 14.1 The Three Forms of D-Glucose

α-D-glucose ⇌ acyclic D-glucose ⇌ β-D-glucose

α isomer 37% — aldehyde C=O trace — β isomer 63%

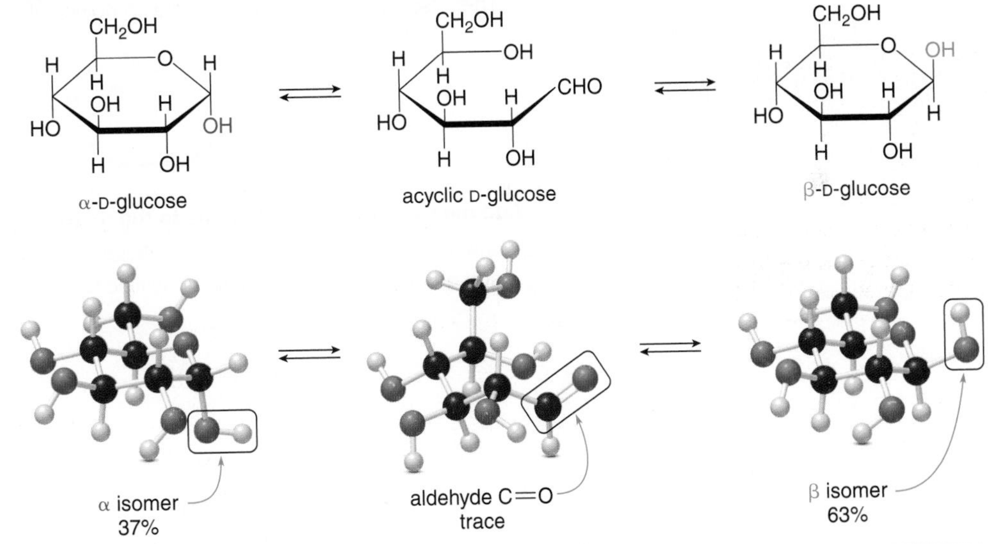

14.3A Haworth Projections

All aldohexoses exist primarily as cyclic compounds typically drawn in Haworth projections. Sample Problem 14.3 shows how to convert an acyclic monosaccharide to a Haworth projection.

SAMPLE PROBLEM 14.3

Draw the α isomer of the cyclic form of D-galactose.

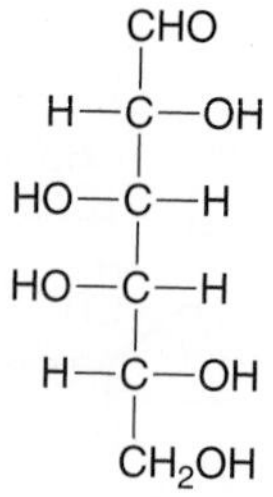

D-galactose

Analysis and Solution

[1] Draw a hexagon with an O atom in the upper right corner. Add the CH_2OH above the ring on the first carbon to the left of the O atom.

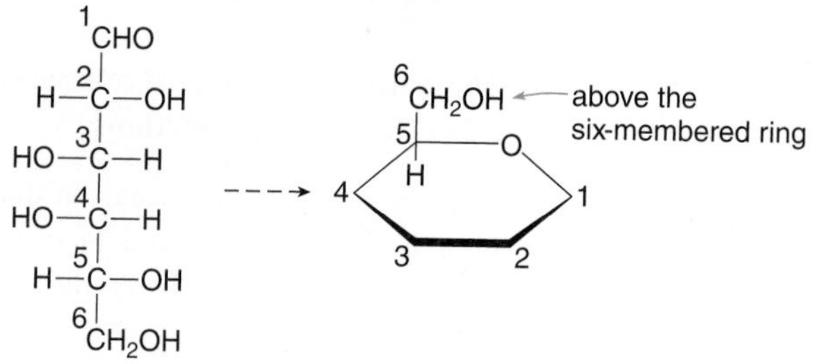

D-galactose

above the six-membered ring

[2] Draw the new chirality center on the first carbon clockwise from the O atom.

- The α isomer has the OH group drawn down, while a β isomer has the OH group drawn up.

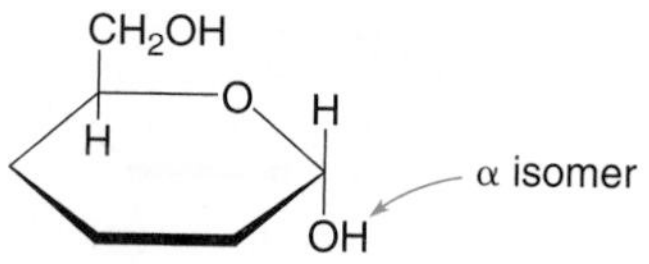

α isomer

[3] Add the OH groups and H atoms to the three remaining carbons (C2–C4).

- Groups on the *right* side in the acyclic form are drawn *down,* below the six-membered ring, and groups on the *left* side in the acyclic form are drawn *up,* above the six-membered ring.

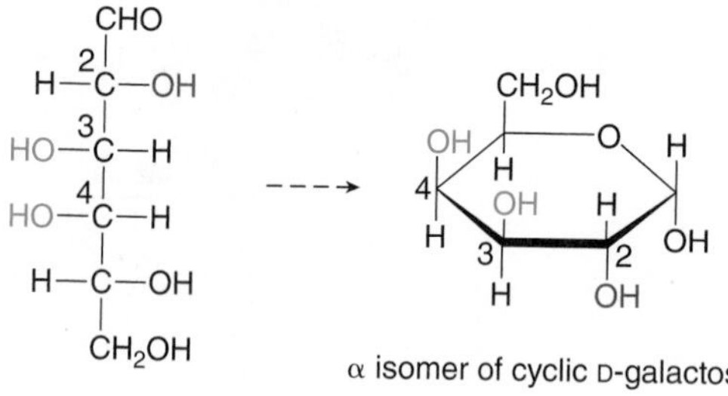

α isomer of cyclic D-galactose

PROBLEM 14.10

Convert each aldohexose to the indicated isomer using a Haworth projection.

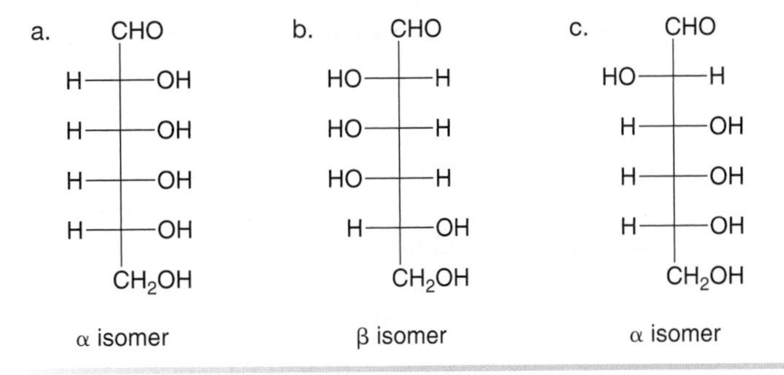

α isomer β isomer α isomer

14.3B The Cyclic Forms of Fructose, a Ketohexose

Certain monosaccharides—notably aldopentoses and ketohexoses—form five-membered rings, *not* six-membered rings, in solution. The same principles apply to drawing these structures as for drawing six-membered rings, except the ring size is one atom smaller.

- **Cyclization forms two isomers. For a D sugar, the OH group is drawn down on the new chirality center in the α isomer and up in the β isomer.**

For example, D-fructose forms a five-membered ring when it cyclizes because the carbonyl group is a ketone at C2, instead of an aldehyde at C1.

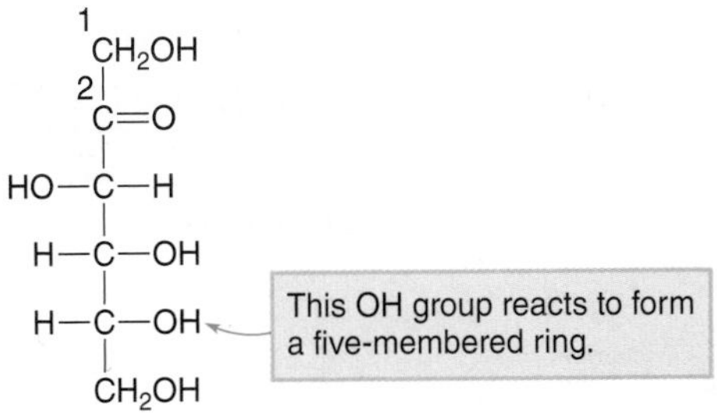

This OH group reacts to form a five-membered ring.

acyclic D-fructose

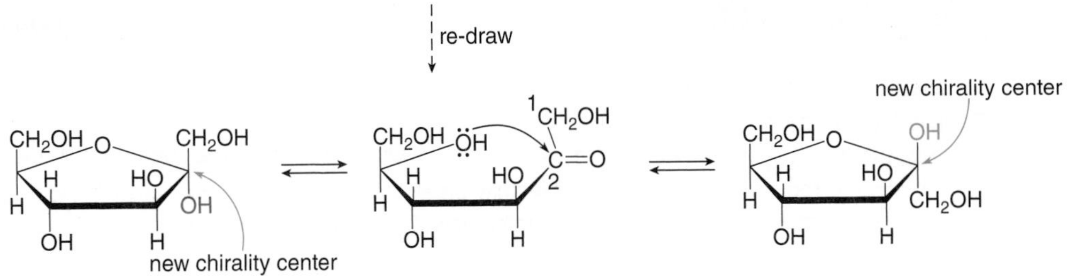

re-draw

new chirality center

α isomer

α-D-fructose

β isomer

β-D-fructose

PROBLEM 14.11

Label each compound as an α or β isomer.

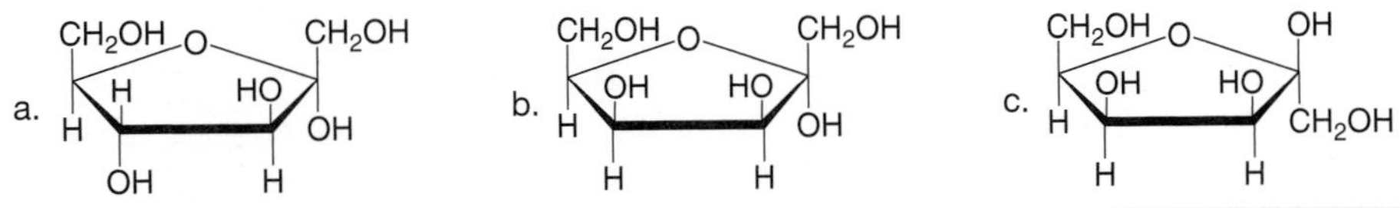

14.4 Reactions of Monosaccharides

The aldehyde carbonyl group of a monosaccharide undergoes two common reactions—**reduction to an alcohol** and **oxidation to a carboxylic acid.**

14.4A Reduction of the Aldehyde Carbonyl Group

Like the double bond in an alkene, the double bond in the carbonyl of an aldose reacts with hydrogen (H_2) in the presence of a palladium (Pd) metal catalyst (Section 11.5). The product, an alcohol called an **alditol,** is sometimes referred to as a "sugar alcohol." For example, reduction of D-glucose with H_2 and Pd yields glucitol, commonly called sorbitol. This reaction is a **reduction,** since the number of carbon–oxygen bonds decreases.

H_2 adds to the C=O.

```
  H—C=O                               H   H
    |                                 |   |
  H—C—OH                            H—C—O              CH2OH
    |                                 |                  |
 HO—C—H     + H2   ——Pd——>          H—C—OH           H—C—OH
    |                                 |                  |
  H—C—OH                           HO—C—H      =    HO—C—H
    |                                 |                  |
  H—C—OH                            H—C—OH           H—C—OH
    |                                 |                  |
   CH2OH                            H—C—OH           H—C—OH
                                      |                  |
                                     CH2OH              CH2OH
```

D-glucose (D-sorbose)   glucitol (sorbitol)

The alditol contains an OH group on every C.

Sorbitol is 60% as sweet as table sugar (sucrose) and contains two-thirds the calories per gram. Sorbitol is used as a sweetening agent in a variety of sugar-free candies and gum.

SAMPLE PROBLEM 14.4

Draw the structure of xylitol, the product formed when D-xylose is treated with H_2 in the presence of a Pd catalyst.

```
     CHO
      |
    H—C—OH
      |
   HO—C—H
      |
    H—C—OH
      |
     CH2OH
```

D-xylose

Analysis

- Locate the C=O and mentally break one bond in the double bond.
- Mentally break the H—H bond of the reagent and add one H atom to each atom of the C=O, forming new C—H and O—H bonds.

Solution

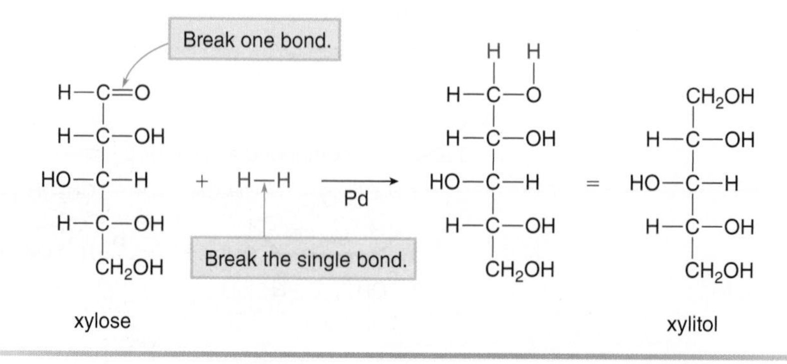

CONSUMER NOTE

Xylitol (Sample Problem 14.4) is used as a substitute sweetener in sugar-free—that is, sucrose-free—products.

PROBLEM 14.12

What compound is formed when each aldose is treated with H_2 in the presence of a Pd catalyst?

a. 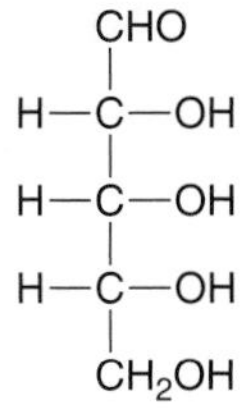
CHO
H—C—OH
H—C—OH
H—C—OH
CH_2OH

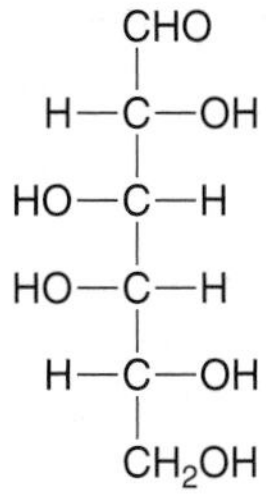

b.
CHO
H—C—OH
HO—C—H
HO—C—H
H—C—OH
CH_2OH

c.
CHO
H—C—OH
H—C—OH
CH_2OH

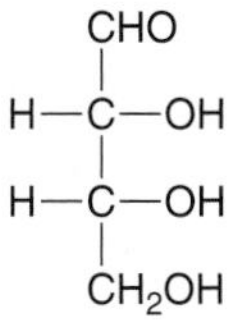

14.4B Oxidation of the Aldehyde Carbonyl Group

The aldehyde carbonyl of an aldose is easily oxidized with a variety of reagents to form a carboxyl group, yielding an **aldonic acid.** For example, D-glucose is oxidized with a Cu^{2+} reagent called **Benedict's reagent** to form gluconic acid. In the process, a characteristic color change occurs as the blue Cu^{2+} is reduced to Cu^{+}, forming brick-red Cu_2O.

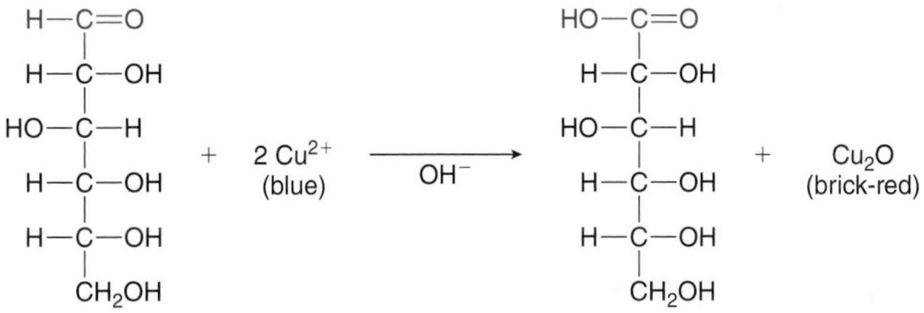

H—C=O
H—C—OH
HO—C—H
H—C—OH
H—C—OH
CH_2OH
D-glucose

\+ $2\ Cu^{2+}$ (blue) —OH^-→

HO—C=O
H—C—OH
HO—C—H
H—C—OH
H—C—OH
CH_2OH
gluconic acid

\+ Cu_2O (brick-red)

This reaction is an oxidation since the product carboxylic acid has one more C—O bond than the starting aldehyde. Carbohydrates that are oxidized with Benedict's reagent are called **reducing sugars,** because the Cu^{2+} in Benedict's reagent is reduced to Cu^{+} during the reaction. Those that do not react with Benedict's reagent are called **nonreducing sugars. All aldoses are reducing sugars.**

SAMPLE PROBLEM 14.5

Draw the product formed when L-threose is oxidized with Benedict's reagent.

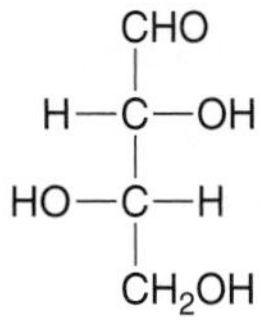

CHO
H—C—OH
HO—C—H
CH_2OH
L-threose

Analysis

To draw the oxidation product of an aldose, convert the CHO group to COOH.

Solution

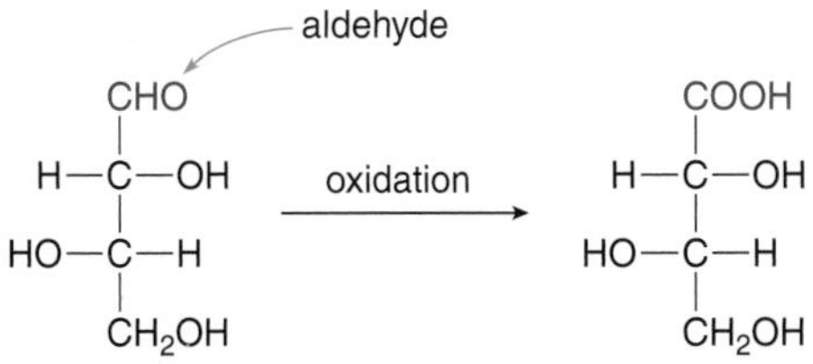

aldehyde

CHO
H—C—OH
HO—C—H
CH_2OH
L-threose

—oxidation→

COOH
H—C—OH
HO—C—H
CH_2OH

PROBLEM 14.13

What aldonic acid is formed by oxidation of each monosaccharide?

a.
$$\begin{array}{c} \text{CHO} \\ | \\ \text{H—C—OH} \\ | \\ \text{H—C—OH} \\ | \\ \text{H—C—OH} \\ | \\ \text{CH}_2\text{OH} \end{array}$$

b.
$$\begin{array}{c} \text{CHO} \\ | \\ \text{H—C—OH} \\ | \\ \text{H—C—OH} \\ | \\ \text{CH}_2\text{OH} \end{array}$$

PROBLEM 14.14

Label each compound as an aldonic acid or alditol.

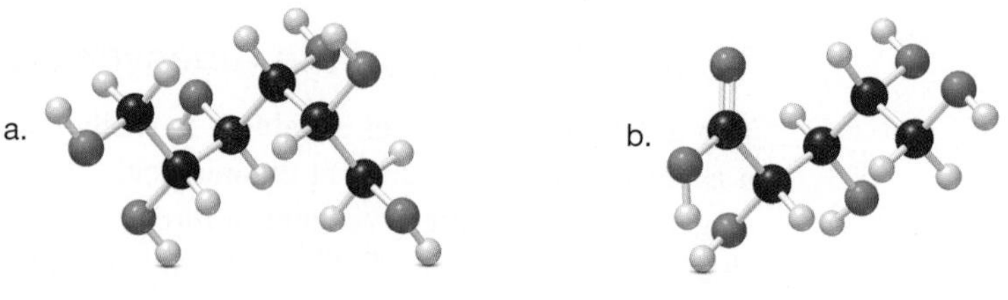

PROBLEM 14.15

What product is formed when D-arabinose is treated with each reagent: (a) H_2, Pd; (b) Benedict's reagent?

$$\begin{array}{c} \text{CHO} \\ | \\ \text{HO—C—H} \\ | \\ \text{H—C—OH} \\ | \\ \text{H—C—OH} \\ | \\ \text{CH}_2\text{OH} \end{array}$$

D-arabinose

14.4C FOCUS ON HEALTH & MEDICINE Monitoring Glucose Levels

In order to make sure that their blood glucose levels are in the proper range, individuals with diabetes frequently measure the concentration of glucose in their blood. A common method for carrying out this procedure today involves the oxidation of glucose to gluconic acid using the enzyme glucose oxidase.

$$\begin{array}{c} \text{CHO} \\ | \\ \text{H—C—OH} \\ | \\ \text{HO—C—H} \\ | \\ \text{H—C—OH} \\ | \\ \text{H—C—OH} \\ | \\ \text{CH}_2\text{OH} \\ \text{D-glucose} \end{array} + O_2 \xrightarrow[\text{oxidase}]{\text{glucose}} \begin{array}{c} \text{COOH} \\ | \\ \text{H—C—OH} \\ | \\ \text{HO—C—H} \\ | \\ \text{H—C—OH} \\ | \\ \text{H—C—OH} \\ | \\ \text{CH}_2\text{OH} \\ \text{gluconic acid} \end{array} + H_2O_2$$

HEALTH NOTE

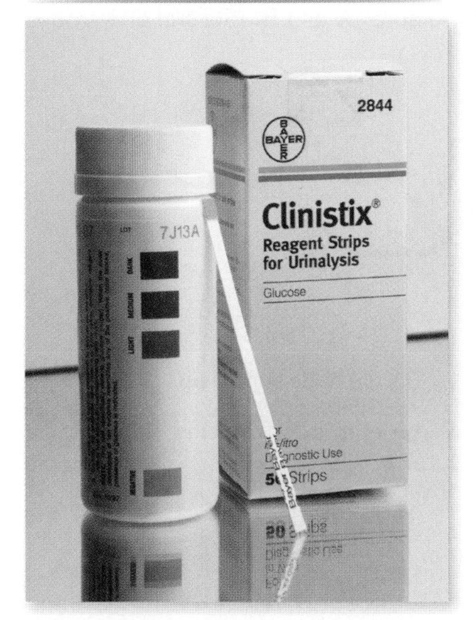

Test strips that contain glucose oxidase are used to measure glucose concentration in urine.

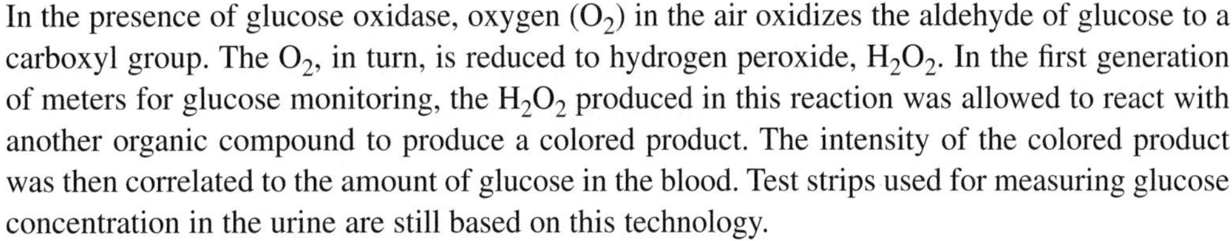

In the presence of glucose oxidase, oxygen (O_2) in the air oxidizes the aldehyde of glucose to a carboxyl group. The O_2, in turn, is reduced to hydrogen peroxide, H_2O_2. In the first generation of meters for glucose monitoring, the H_2O_2 produced in this reaction was allowed to react with another organic compound to produce a colored product. The intensity of the colored product was then correlated to the amount of glucose in the blood. Test strips used for measuring glucose concentration in the urine are still based on this technology.

Modern glucose meters are electronic devices that measure the amount of oxidizing agent that reacts with a known amount of blood (Figure 14.2). This value is correlated with blood glucose concentration and the result is displayed digitally. A high blood glucose level may mean that an individual needs more insulin, while a low level may mean that it is time to ingest some calories.

Figure 14.2 Monitoring Blood Glucose Levels

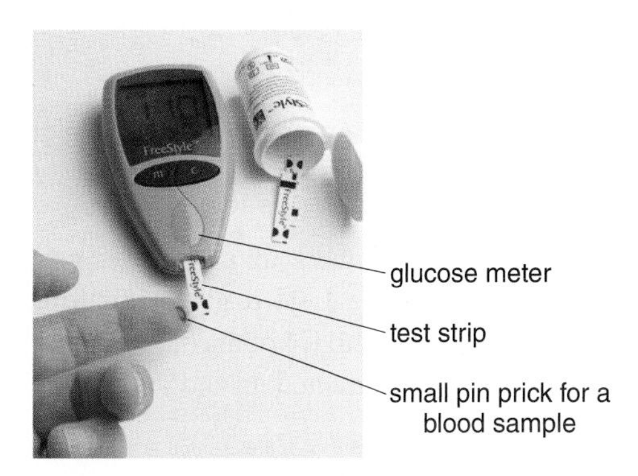

A small drop of blood is placed on a disposable test strip that is inserted in an electronic blood glucose meter, and the glucose concentration is read on a digital display.

Maltose gets its name from malt, the liquid obtained from barley used in the brewing of beer.

14.5 Disaccharides

Disaccharides are carbohydrates composed of two monosaccharides. A disaccharide is formed when a hydroxyl group of one monosaccharide reacts with a hydroxyl group of a second monosaccharide. The new C—O bond that joins the two rings together is called a **glycosidic linkage.** The carbon in a glycosidic linkage is bonded to two O atoms—one O atom is part of a ring, and the other O atom joins the two rings together.

For example, maltose is a disaccharide formed from two molecules of glucose. Maltose, which is found in grains such as barley, is a product of the hydrolysis of starch. Each ring in maltose is numbered beginning at the carbon bonded to two oxygen atoms. In maltose, the glycosidic linkage joins C1 of one ring to C4 of the other ring.

glucose
glycosidic linkage
glucose
maltose

The glycosidic linkage that joins the two monosaccharides in a disaccharide can be oriented in two different ways, shown with Haworth projections in structures **A** and **B.** (Several OH groups are omitted for clarity.)

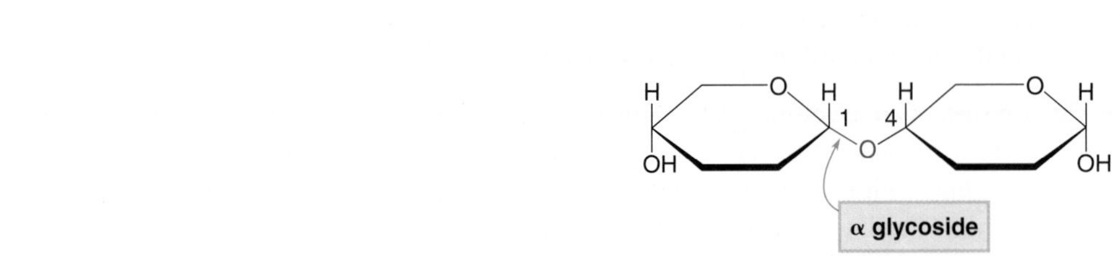

α glycoside

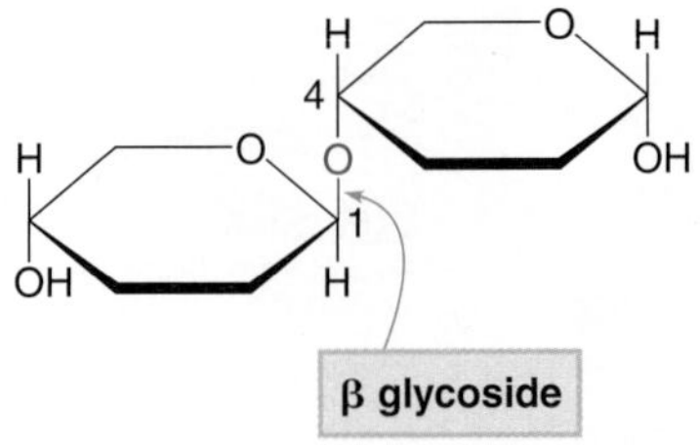

The glycoside bond is **down.**

1→4-α-glycosidic linkage

A

β glycoside

The glycoside bond is **up.**

1→4-β-glycosidic linkage

B

- **An α glycoside has the glycosidic linkage oriented down.**
- **A β glycoside has the glycosidic linkage oriented up.**

Numbers are used to designate which ring atoms are joined in the disaccharide. Disaccharide **A** has a **1→4-α-glycosidic linkage** since the glycoside bond is oriented down and joins C1 of one ring to C4 of the other. Disaccharide **B** has a **1→4-β-glycosidic linkage** since the glycoside bond is oriented up and joins C1 of one ring to C4 of the other.

SAMPLE PROBLEM 14.6

(a) Locate the glycosidic linkage in the following disaccharide. (b) Number the carbon atoms in both rings. (c) Classify the glycosidic linkage as α or β, and use numbers to designate its location.

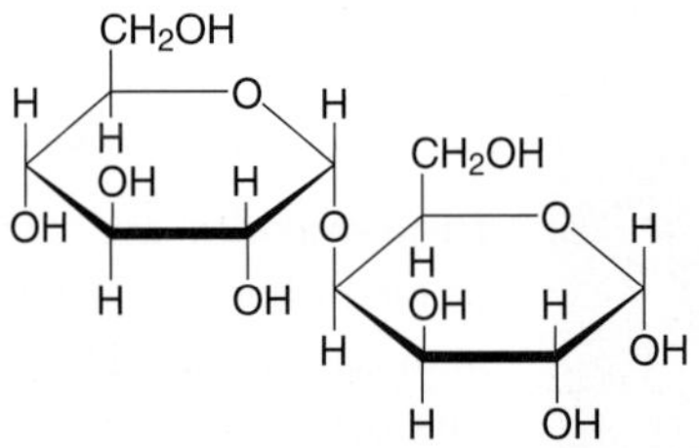

Analysis and Solution

a. and b. The glycosidic linkage labeled in red contains a carbon bonded to two oxygens—one ring oxygen as well as the oxygen that joins the two rings together. Each ring is numbered beginning at the carbon bonded to two oxygen atoms.

glycosidic linkage

c. The disaccharide has an α glycosidic linkage since the C—O bond is drawn down. The glycosidic linkage joins C1 of one ring to C4 of the other ring, so the disaccharide contains a 1→4-α-glycosidic linkage.

PROBLEM 14.16

(a) Locate the glycosidic linkage in cellobiose. (b) Number the carbon atoms in both rings. (c) Classify the glycosidic linkage as α or β, and use numbers to designate its location.

cellobiose

The hydrolysis of a disaccharide cleaves the C—O glycosidic linkage and forms two monosaccharides. For example, hydrolysis of maltose yields two molecules of glucose.

Cleave the glycosidic linkage.

maltose + H—OH ⟶ glucose + glucose

Water is added in hydrolysis.

PROBLEM 14.17

What monosaccharides are formed when cellobiose (Problem 14.16) is hydrolyzed with water?

Individuals who are lactose intolerant can drink lactose-free milk. Tablets that contain the lactase enzyme can also be taken when ice cream or other milk products are ingested.

14.5A FOCUS ON HEALTH & MEDICINE Lactose Intolerance

Lactose is the principal disaccharide found in milk from both humans and cows. Unlike many mono- and disaccharides, lactose is not appreciably sweet. Lactose consists of one galactose ring and one glucose ring, joined by a 1→4-β-glycoside linkage.

galactose — glucose — β glycoside bond

lactose

Lactose is digested in the body by first cleaving the 1→4-β-glycoside bond using the enzyme *lactase*. **Individuals who are lactose intolerant no longer produce this enzyme,** and so lactose cannot be properly digested, causing abdominal cramps and diarrhea. Lactose intolerance is especially prevalent in Asian and African populations whose diets have not traditionally included milk beyond infancy.

PROBLEM 14.18

What products are formed when lactose is hydrolyzed with water?

14.5B FOCUS ON HEALTH & MEDICINE
Sucrose and Artificial Sweeteners

Sucrose, the disaccharide found in sugarcane and the compound generally referred to as "sugar," is the most common disaccharide in nature. It contains one glucose ring and one fructose ring. Unlike maltose and lactose, which contain only six-membered rings, sucrose contains one six-membered and one five-membered ring.

sugarcane

two varieties of refined sugar

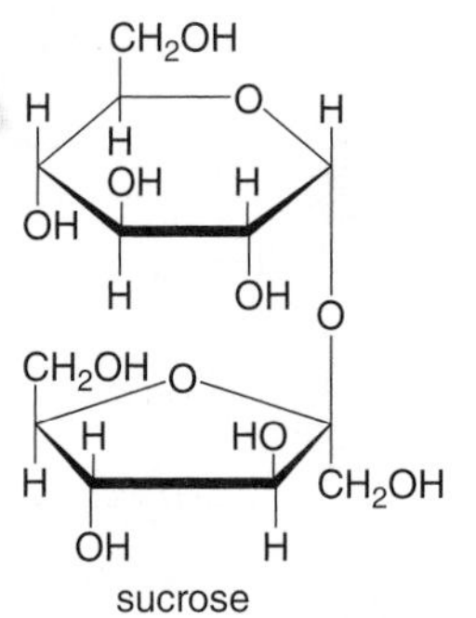

sucrose

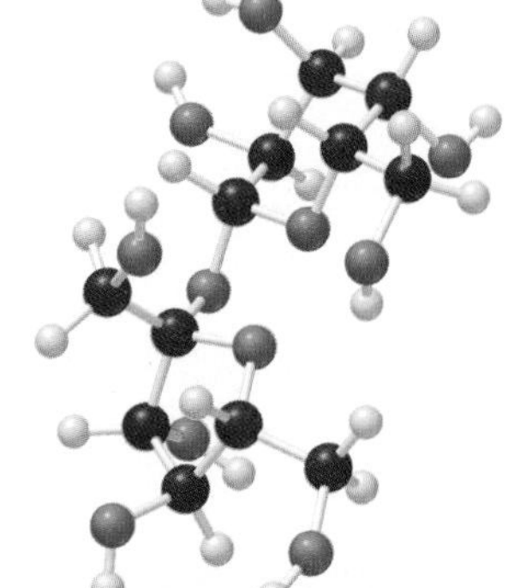

Sucrose's pleasant sweetness has made it a widely used ingredient in baked goods, cereals, bread, and many other products. It is estimated that the average American ingests 100 lb of sucrose annually. Like other carbohydrates, however, sucrose contains many calories. To reduce caloric intake while maintaining sweetness, a variety of artificial sweeteners have been developed. These include aspartame, saccharin, and sucralose (Figure 14.3). These compounds are much sweeter than sucrose, so only a small amount of each compound is needed to achieve the same level of perceived sweetness. A relative sweetness scale ranks the sweetness of carbohydrates and synthetic sweeteners, as shown in Table 14.1.

PROBLEM 14.19

Identify the functional groups in aspartame.

Figure 14.3 Artificial Sweeteners

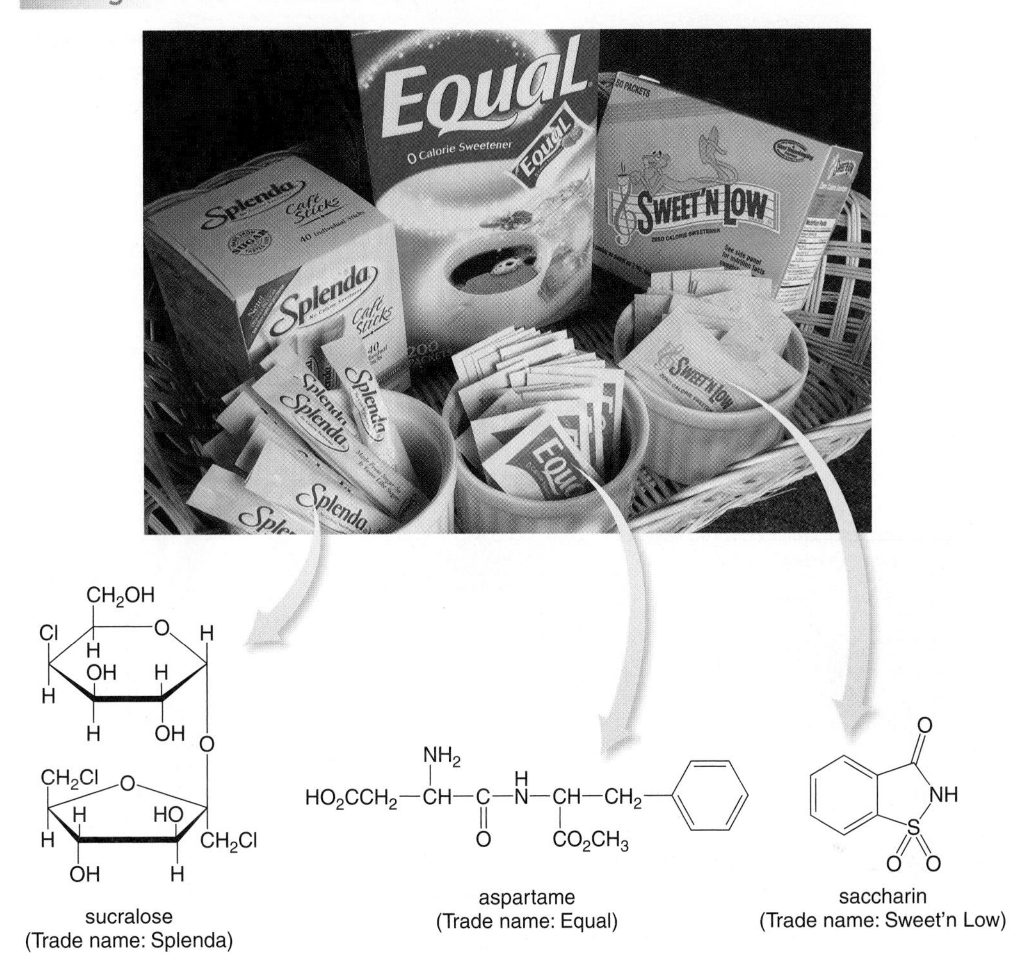

The sweetness of these artificial sweeteners was discovered accidentally. The sweetness of sucralose was discovered in 1976 when a chemist misunderstood his superior, and so he *tasted* rather than *tested* his compound. Aspartame was discovered in 1965 when a chemist licked his dirty fingers in the lab and tasted its sweetness. Saccharin, the oldest known artificial sweetener, was discovered in 1879 by a chemist who failed to wash his hands after working in the lab. Saccharin was not used extensively until sugar shortages occurred during World War I. Although there were concerns in the 1970s that saccharin causes cancer, there is no proven link between cancer occurrence and saccharin intake at normal levels.

Table 14.1 Relative Sweetness of Some Carbohydrates and Artificial Sweeteners

Compound	Relative Sweetness
Sorbitol	0.60
Glucose	0.75
Sucrose	1.00
Fructose	1.75
Aspartame	150
Saccharin	350
Sucralose	600

14.6 Polysaccharides

Polysaccharides contain three or more monosaccharides joined together. Three prevalent polysaccharides in nature are **cellulose, starch,** and **glycogen,** each of which consists of repeating glucose units joined by glycosidic bonds.

14.6A Cellulose

Cellulose is found in the cell walls of nearly all plants, where it gives support and rigidity to wood, plant stems, and grass (Figure 14.4). Wood, cotton, and flax are composed largely of cellulose.

Figure 14.4 Cellulose

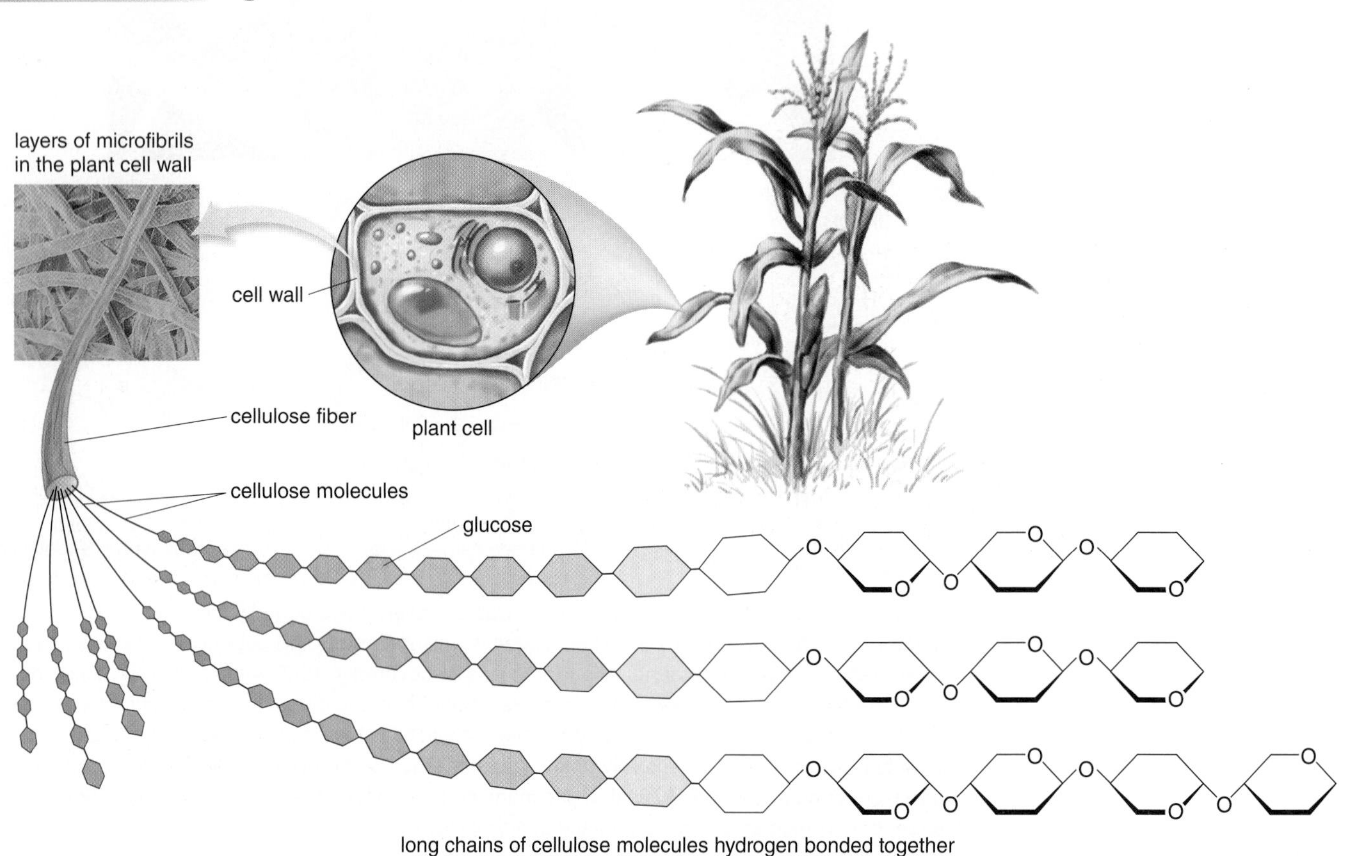

Cellulose is an unbranched polymer composed of repeating glucose units joined in a 1→4-β-glycosidic linkage. The β glycosidic linkages create long linear chains of cellulose molecules that stack in sheets, making an extensive three-dimensional array.

cellulose

β glycosidic linkage (shown in blue)

In some cells, cellulose is hydrolyzed by an enzyme that cleaves all of the β glycoside bonds, forming glucose. **Humans do not possess this enzyme, and therefore *cannot* digest cellulose.** Ruminant animals, on the other hand, such as cattle, deer, and camels, have bacteria containing this enzyme in their digestive systems, so they can derive nutritional benefit from eating grass and leaves.

Much of the insoluble fiber in our diet is cellulose, which passes through the digestive system without being metabolized. Foods rich in cellulose include whole wheat bread, brown rice, and bran cereals. Fiber is an important component of the diet even though it gives us no nutrition; fiber adds bulk to solid waste, so that it is eliminated more readily.

14.6B Starch

Starch is the main carbohydrate found in the seeds and roots of plants. Corn, rice, wheat, and potatoes are common foods that contain a great deal of starch. **Starch is a polymer composed of repeating glucose units joined in α glycosidic linkages.** The two common forms of starch are **amylose** and **amylopectin.**

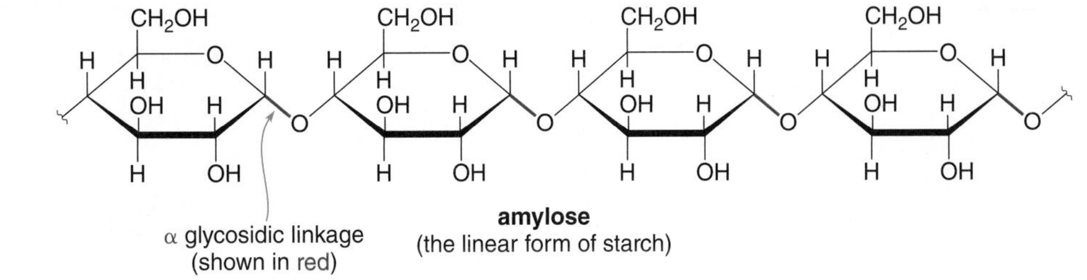

α glycosidic linkage (shown in red)

amylose
(the linear form of starch)

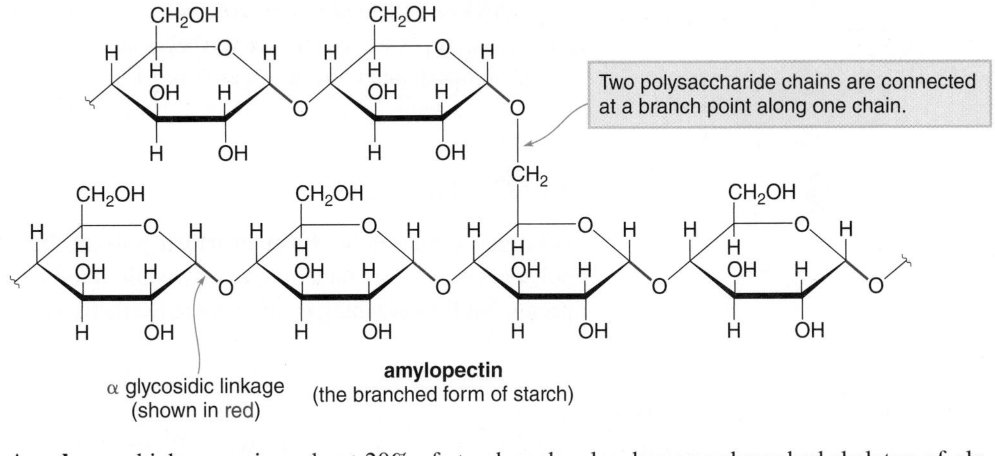

Two polysaccharide chains are connected at a branch point along one chain.

α glycosidic linkage (shown in red)

amylopectin
(the branched form of starch)

Amylose, which comprises about 20% of starch molecules, has an unbranched skeleton of glucose molecules with **1→4-α-glycoside bonds.** Because of this linkage, an amylose chain adopts a helical arrangement, giving it a very different three-dimensional shape from the linear chains of cellulose (Figure 14.5).

Amylopectin, which comprises about 80% of starch molecules, consists of a backbone of glucose units joined in **α glycosidic bonds,** but it also contains considerable branching along the chain. The linear linkages of amylopectin are formed by **1→4-α-glycoside bonds,** similar to amylose.

Both forms of starch are water soluble. Since the OH groups in these starch molecules are not buried in a three-dimensional network, they are available for hydrogen bonding with water molecules, leading to greater water solubility than cellulose.

Figure 14.5
Starch—Amylose and Amylopectin

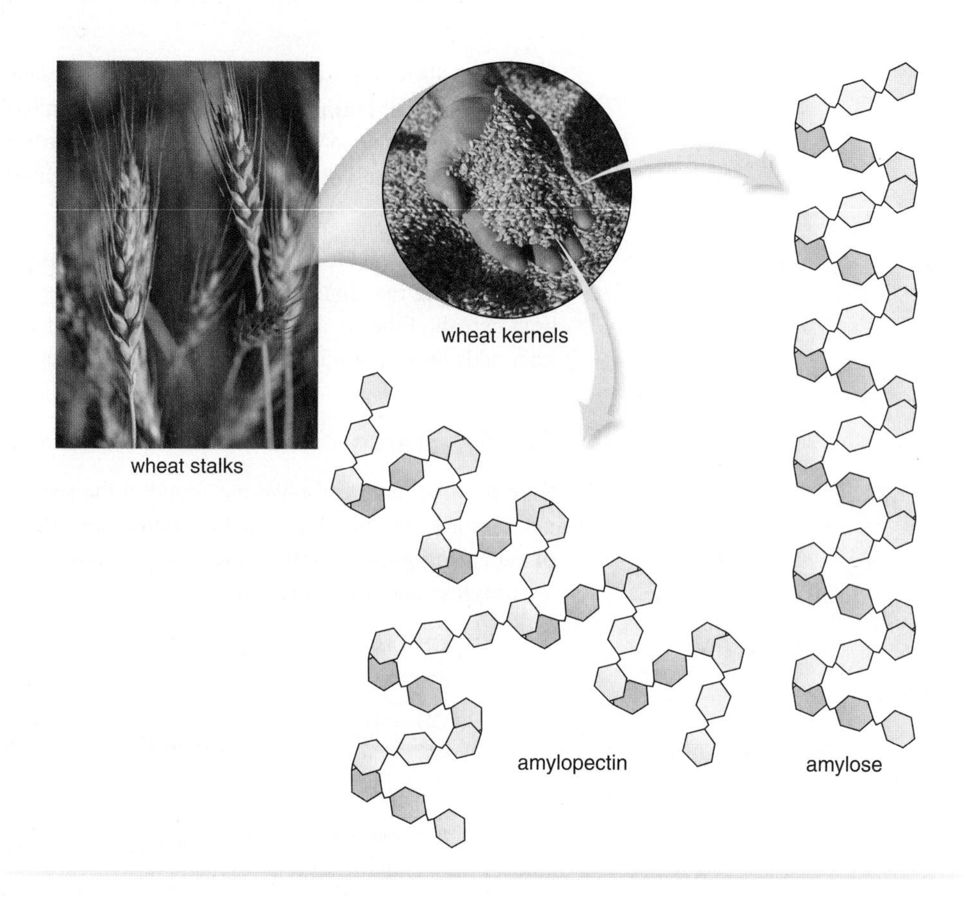

Both amylose and amylopectin are hydrolyzed to glucose with cleavage of the glycosidic bonds. The human digestive system has the necessary amylase enzymes needed to catalyze this process. Bread and pasta made from wheat flour, rice, and corn tortillas are all sources of starch that are readily digested.

14.6C Glycogen

Glycogen is the major form in which polysaccharides are stored in animals. Glycogen, a polymer of glucose containing **α glycosidic bonds,** has a branched structure similar to amylopectin, but the branching is much more extensive (Figure 14.6).

Figure 14.6
Glycogen

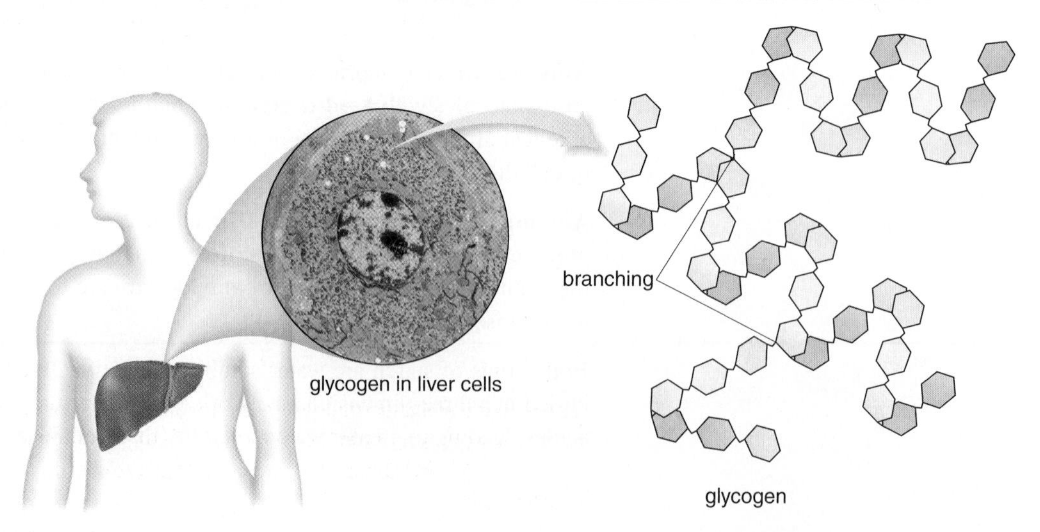

Glycogen is stored principally in the liver and muscle. When glucose is needed for energy in the cell, glucose units are hydrolyzed from the ends of the glycogen polymer, and then further metabolized with the release of energy. Because glycogen has a highly branched structure, there are many glucose units at the ends of the branches that can be cleaved whenever the body needs them.

PROBLEM 14.20

Cellulose is water *in*soluble, despite its many OH groups. Based on its three-dimensional structure, why do you think this is so?

HEALTH NOTE

The blood type of a blood donor and recipient must be compatible, so donated blood is clearly labeled with the donor's blood type.

14.7 FOCUS ON THE HUMAN BODY Blood Type

Human blood is classified into one of four types—A, B, AB, and O. An individual's blood type is determined by three or four monosaccharides attached to a membrane protein of red blood cells. These monosaccharides include:

D-galactose — L-fucose — *N*-acetyl-D-glucosamine — *N*-acetyl-D-galactosamine

Each blood type is associated with a different carbohydrate structure, as shown in Figure 14.7. The short polysaccharide chains distinguish one type of red blood cell from another, and signal the cells about foreign viruses, bacteria, and other agents. When a foreign substance enters the blood, the body's immune system uses antibodies to attack and destroy the invading substance so that it does the host organism no harm.

Figure 14.7
Carbohydrates and Blood Types

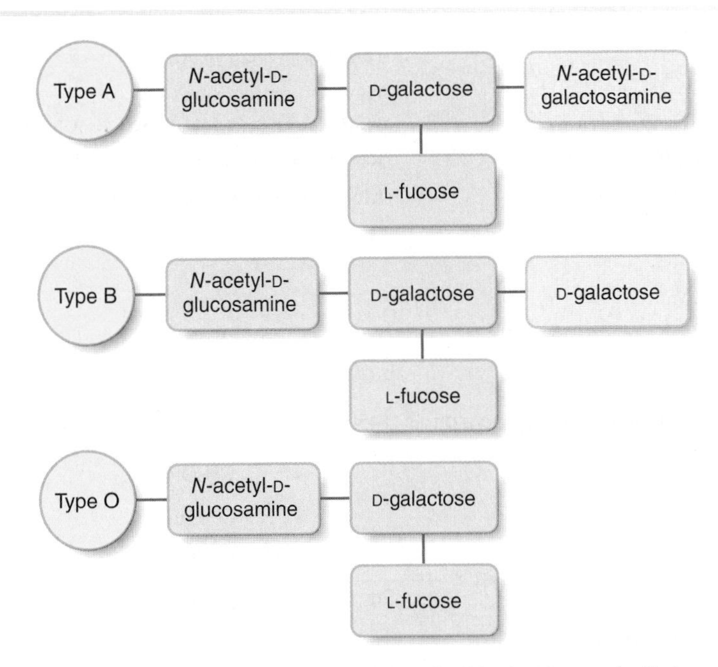

Each blood type is characterized by a different polysaccharide that is covalently bonded to a membrane protein of the red blood cell. There are three different carbohydrate sequences, one each for A, B, and O blood types. Blood type AB contains the sequences for both blood type A and blood type B.

Table 14.2 Compatibility Chart of Blood Types

Blood Type	Can Receive Blood Type:	Can Donate to Blood Type:
A	A, O	A, AB
B	B, O	B, AB
AB	A, B, AB, O	AB
O	O	A, B, AB, O

Knowing an individual's blood type is necessary before receiving a blood transfusion. Because the blood of an individual may contain antibodies to another blood type, the types of blood that can be given to a patient are often limited. An individual with blood type A produces antibodies to type B blood, and an individual with blood type B produces antibodies to type A blood. Type AB blood contains no antibodies to other blood types, while type O blood contains antibodies to both types A and B. As a result:

- **Individuals with type O blood are called *universal donors* because type O blood can be given to individuals of any blood type.**
- **Individuals with type AB blood are called *universal recipients* because individuals with type AB blood can receive blood of any type.**

Table 14.2 lists what blood types can be safely given to an individual. Blood must be carefully screened to make sure that the blood types of the donor and recipient are compatible.

STUDY SKILLS PART III: BIOMOLECULES

Chapters 14–18 focus on biomolecules, the organic compounds found in biological systems. The first four chapters each concentrate on a specific type of biomolecule—carbohydrates, lipids, proteins, and nucleic acids—while the final chapter focuses on the metabolism of these molecules as well as energy production in the cell.

Most of these molecules are more complex than the organic compounds in Chapters 10–13. For this reason we will spend less time on learning specific reactions and nomenclature, and more time on learning general structural features and properties. To master this material, *concentrate on the basic structure* of each class of biomolecule, and *learn how that structure translates into its role in the cell.*

Understanding biomolecules requires remembering concepts in both general and organic chemistry. Review the following topics before tackling problems on biomolecules: bond polarity (Section 3.11), intermolecular forces (Section 4.3), solubility (Section 7.3), functional groups (Section 10.4), and chirality centers (Section 12.10).

KEY TERMS

Alditol (14.4)
Aldonic acid (14.4)
Aldose (14.2)
Benedict's reagent (14.4)
Carbohydrate (14.1)
Disaccharide (14.1)
Glycosidic linkage (14.5)
Haworth projection (14.3)
Hexose (14.2)
Ketose (14.2)
Monosaccharide (14.1)
Pentose (14.2)
Polysaccharide (14.1)
Reducing sugar (14.4)
Tetrose (14.2)
Triose (14.2)

KEY REACTIONS

[1] Reduction of monosaccharides to alditols (14.4A)

CHO
H—C—OH
HO—C—H
H—C—OH
H—C—OH
CH_2OH

$+ H_2 \xrightarrow{Pd}$

CH_2OH
H—C—OH
HO—C—H
H—C—OH
H—C—OH
CH_2OH

alditol

[2] Oxidation of monosaccharides to aldonic acids (14.4B)

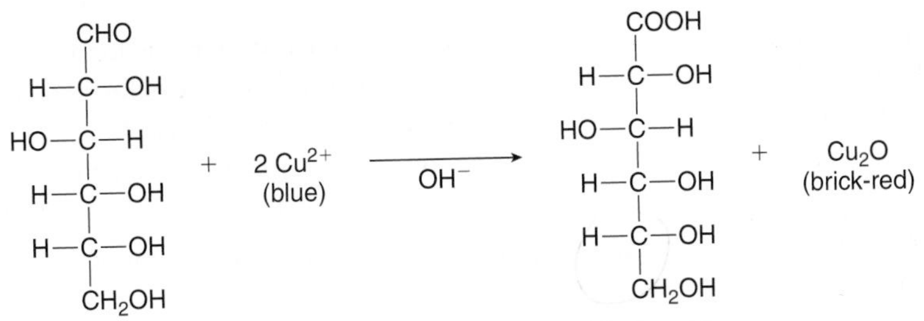

$$\text{aldose} + 2\,Cu^{2+} \text{ (blue)} \xrightarrow{OH^-} \text{aldonic acid} + Cu_2O \text{ (brick-red)}$$

aldonic acid

[3] Hydrolysis of disaccharides (14.5)

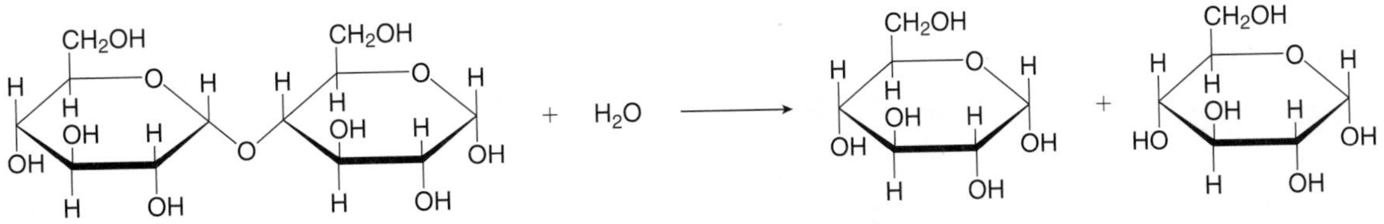

$$\text{disaccharide} + H_2O \longrightarrow \text{monosaccharide} + \text{monosaccharide}$$

KEY CONCEPTS

1 What are the three major types of carbohydrates? (14.1)

- Monosaccharides have three to six carbons with a carbonyl group at either the terminal carbon or the carbon adjacent to it. Generally, all other carbons have OH groups bonded to them.
- Disaccharides are composed of two monosaccharides.
- Polysaccharides are composed of three or more monosaccharides.

2 What are the major structural features of monosaccharides? (14.2)

- Monosaccharides with a carbonyl group at C1 are called aldoses and those with a carbonyl at C2 are called ketoses. Generally, all other carbons have OH groups bonded to them. The terms triose, tetrose, and so forth are used to indicate the number of carbons in the chain.
- The acyclic form of monosaccharides is drawn with Fischer projection formulas. A D sugar has the OH group of the chirality center farthest from the carbonyl on the right side. An L sugar has the OH group of the chirality center farthest from the carbonyl on the left side.

3 How are the cyclic forms of monosaccharides drawn? (14.3)

- In aldohexoses the OH group on C5 reacts with the aldehyde carbonyl to give two isomers. The α isomer has the OH group on the new chirality center drawn down for a D sugar and the β isomer has the OH group drawn up.

4 What reduction and oxidation products are formed from monosaccharides? (14.4)

- Monosaccharides are reduced to alditols with H_2 and Pd.
- Monosaccharides are oxidized to aldonic acids with Benedict's reagent.

5 What are the major structural features of disaccharides? (14.5)

- Disaccharides contain two monosaccharide units joined by a glycosidic linkage. An α glycoside has the glycosidic linkage oriented down and a β glycoside has the glycosidic linkage oriented up.
- Disaccharides are hydrolyzed to two monosaccharides by the cleavage of the glycosidic C—O bond with water.

6 What are the differences in the polysaccharides cellulose, starch, and glycogen? (14.6)

- Cellulose is an unbranched polymer composed of repeating glucose units joined in $1\rightarrow4$-β-glycosidic linkages.
- There are two forms of starch—amylose, which is an unbranched polymer, and amylopectin, which is a branched polysaccharide polymer. Both forms contain $1\rightarrow4$-α-glycosidic linkages.
- Glycogen resembles amylopectin but is more extensively branched.

7 What role do carbohydrates play in determining blood type? (14.7)

- Human blood type—A, B, AB, or O—is determined by three or four monosaccharides attached to a membrane protein on the surface of red blood cells. Since the blood of an individual may contain antibodies to another blood type, blood type must be known before receiving a transfusion.

UNDERSTANDING KEY CONCEPTS

Selected in-chapter and odd-numbered end-of-chapter problems have brief answers at the end of each chapter. The *Student Study Guide and Solutions Manual* contains detailed solutions to all in-chapter and odd-numbered end-of-chapter problems, as well as additional worked examples and a chapter self-test.

14.21 (a) Classify each monosaccharide by the type of carbonyl group and number of carbons in the chain. (b) Locate the chirality centers in each compound.

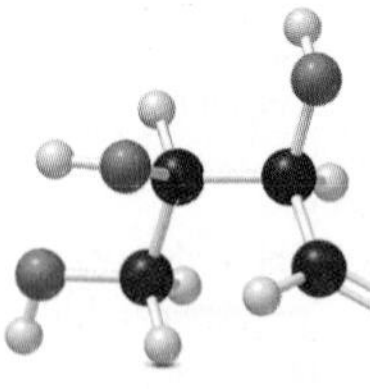

A

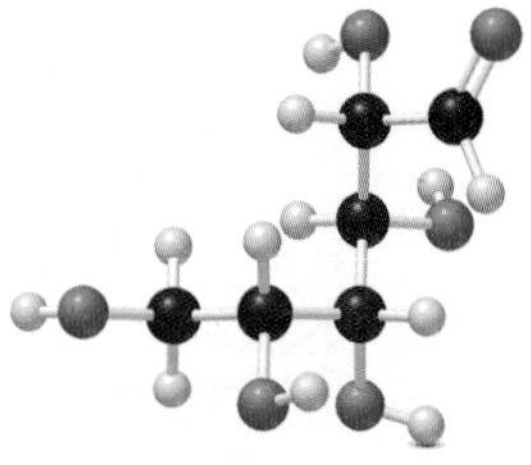

B

14.22 (a) Classify each monosaccharide by the type of carbonyl group and number of carbons in the chain. (b) Locate the chirality centers in each compound.

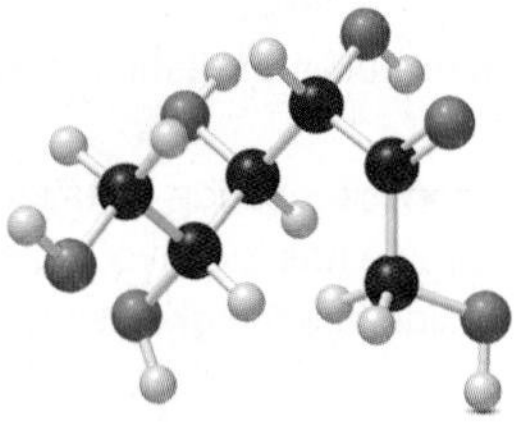

C

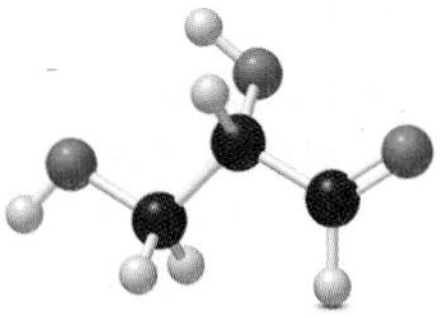

D

14.23 Draw the structure of each type of compound.

a. an L-aldopentose

b. a D-aldotetrose

c. a five-carbon alditol

14.24 Draw the structure of each type of compound.

a. a D-aldotriose

b. an L-ketohexose

c. a four-carbon aldonic acid

14.25 Consider the following monosaccharide.

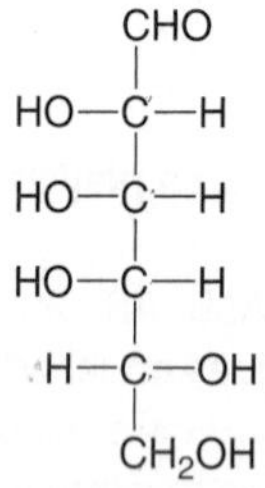

a. Is the monosaccharide a D or L sugar?

b. Classify the monosaccharide by the type of carbonyl and the number of atoms in the chain.

c. Label the chirality centers.

d. Draw the enantiomer.

e. Draw the α isomer of the cyclic form.

f. What product is formed when the monosaccharide is treated with Benedict's reagent?

g. What product is formed when the monosaccharide is treated with H_2 and Pd?

14.26 Answer Problem 14.25 using the following monosaccharide.

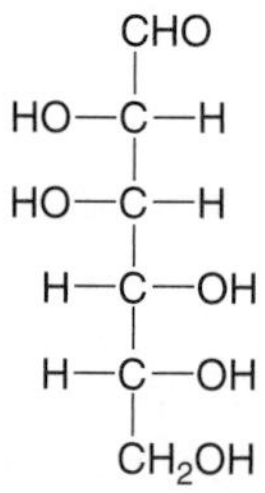

14.27 If a monosaccharide is represented by a single blue sphere, which of the following representations corresponds to (a) lactose; (b) cellulose; (c) amylopectin?

A B C

14.28 If a monosaccharide is represented by a single blue sphere, which of the following representations corresponds to (a) a disaccharide; (b) glycogen; (c) amylose?

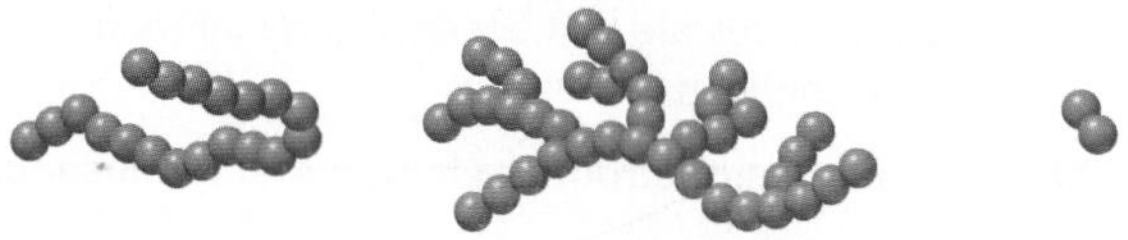

A B C

14.29 (a) Locate the glycosidic linkage in the following disaccharide. (b) Number the carbon atoms in both rings. (c) Classify the glycosidic linkage as α or β, and use numbers to designate its location.

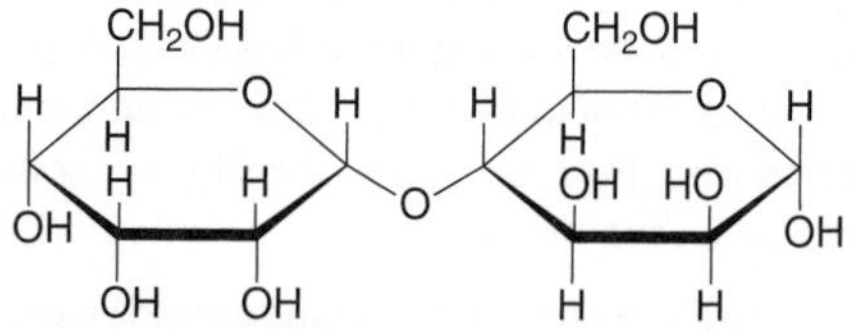

14.30 Answer Problem 14.29 using the following disaccharide.

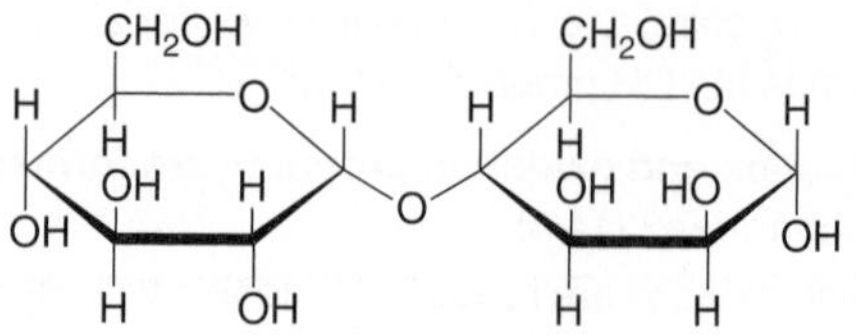

ADDITIONAL PROBLEMS

Monosaccharides

14.31 Classify each monosaccharide by the type of carbonyl group and the number of carbons in the chain.

a.
CHO
HO—C—H
HO—C—H
CH_2OH

b.
CHO
HO—C—H
HO—C—H
HO—C—H
H—C—OH
CH_2OH

c.
CH_2OH
C=O
H—C—OH
HO—C—H
CH_2OH

14.32 Classify each monosaccharide by the type of carbonyl group and the number of carbons in the chain.

a.
CHO
HO—C—H
CH_2OH

b.
CHO
H—C—OH
HO—C—H
HO—C—H
CH_2OH

c.
CHO
H—C—OH
H—C—OH
HO—C—H
HO—C—H
CH_2OH

14.33 For each compound in Problem 14.31: [1] label all the chirality centers; [2] classify the compound as a D or L monosaccharide; [3] draw the enantiomer; [4] draw a Fischer projection.

14.34 For each compound in Problem 14.32: [1] label all the chirality centers; [2] classify the compound as a D or L monosaccharide; [3] draw the enantiomer; [4] draw a Fischer projection.

14.35 (a) Draw a Fischer projection of the monosaccharide depicted in the ball-and-stick model and label it as a D or L sugar. (b) Classify the monosaccharide by the type of carbonyl group and number of carbons it contains.

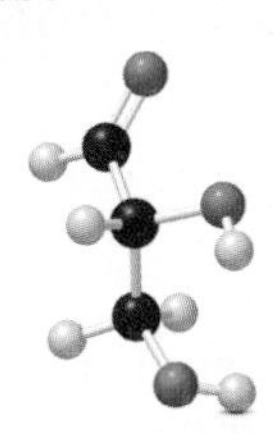

14.36 (a) Draw a Fischer projection of the monosaccharide depicted in the ball-and-stick model and label it as a D or L sugar. (b) Classify the monosaccharide by the type of carbonyl group and number of carbons it contains.

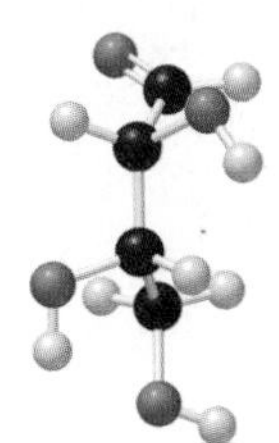

14.37 Are α-D-glucose and β-D-glucose enantiomers? Explain your choice.

14.38 Are D-fructose and L-fructose enantiomers? Explain your choice.

14.39 Consider monosaccharides **A**, **B**, and **C**.

A
CHO
H—C—OH
H—C—OH
CH_2OH

B
CHO
H—C—OH
HO—C—H
CH_2OH

C
CH_2OH
C=O
H—C—OH
CH_2OH

a. Which two monosaccharides are stereoisomers?
b. Identify two compounds that are constitutional isomers.
c. Draw the enantiomer of **B**.
d. Draw a Fischer projection for **A**.

14.40 Consider monosaccharides **D**, **E**, and **F**.

D
CHO
H—C—OH
H—C—OH
H—C—OH
CH_2OH

E
CH_2OH
C=O
HO—C—H
HO—C—H
CH_2OH

F
CHO
HO—C—H
HO—C—H
H—C—OH
CH_2OH

a. Which two monosaccharides are stereoisomers?
b. Identify two compounds that are constitutional isomers.
c. Draw the enantiomer of **F**.
d. Draw a Fischer projection for **D**.

14.41 Using Haworth projections, draw the α and β isomers of the cyclic form of the following D monosaccharide.

CHO
HO—C—H
H—C—OH
HO—C—H
H—C—OH
CH_2OH

14.42 Using Haworth projections, draw the α and β isomers of the cyclic form of the following D monosaccharide.

CHO
H—C—OH
H—C—OH
HO—C—H
H—C—OH
CH_2OH

14.43 Consider the following cyclic monosaccharide.

a. Label the monosaccharide as an α or β isomer.

b. Draw the other cyclic form.

14.44 Consider the following cyclic monosaccharide.

a. Label the monosaccharide as an α or β isomer.

b. Draw the other cyclic form.

14.45 Label each monosaccharide as an α or β isomer.

14.46 Label each monosaccharide as an α or β isomer.

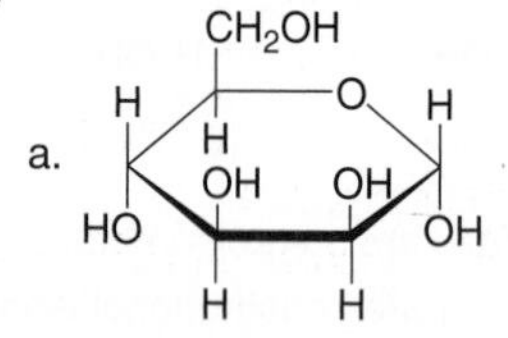

14.47 Why are no D, L labels used for a ketotriose?

14.48 Explain how the labels D and L are assigned in a monosaccharide with three chirality centers.

Reactions of Carbohydrates

14.49 Label each compound as a monosaccharide, an aldonic acid, or an alditol.

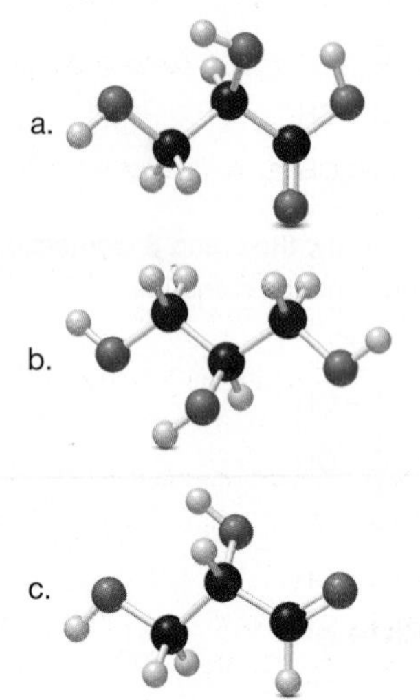

14.50 Label each compound as a monosaccharide, an aldonic acid, or an alditol.

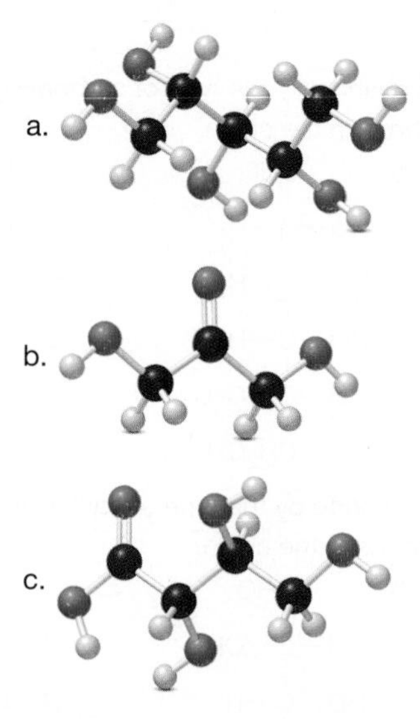

14.51 Draw the organic products formed when each monosaccharide is treated with each of the following reagents: [1] H_2, Pd; [2] Cu^{2+}, OH^-.

a. CHO / HO—C—H / H—C—OH / CH_2OH

b. CHO / H—C—OH / HO—C—H / HO—C—H / CH_2OH

c. CHO / H—C—OH / H—C—OH / HO—C—H / HO—C—H / CH_2OH

14.52 Draw the organic products formed when each monosaccharide is treated with each of the following reagents: [1] H_2, Pd; [2] Cu^{2+}, OH^-.

a. CHO / HO—C—H / HO—C—H / CH_2OH

b. CHO / HO—C—H / H—C—OH / H—C—OH / CH_2OH

c. CHO / H—C—OH / H—C—OH / H—C—OH / H—C—OH / CH_2OH

14.53 What product is formed when the given aldopentose is treated with Benedict's reagent?

CHO / H—C—OH / H—C—OH / HO—C—H / CH_2OH

14.54 What product is formed when the given aldopentose is treated with Benedict's reagent?

14.55 What monosaccharides are formed when the given disaccharide is hydrolyzed?

14.56 What monosaccharides are formed when the given disaccharide is hydrolyzed?

Disaccharides and Polysaccharides

14.57 (a) Locate the glycosidic linkage in the following disaccharide, and label it as α or β. (b) Draw the structure of the monosaccharides formed on hydrolysis.

14.58 Answer Problem 14.57 with the following disaccharide.

14.59 Draw the structure of a disaccharide that contains two six-membered rings and an α glycosidic linkage.

14.60 Draw the structure of a disaccharide that contains one six-membered ring and one five-membered ring, as well as a β glycosidic linkage.

14.61 Describe the similarities and differences in the structures of maltose and lactose.

14.62 Describe the similarities and differences in the structures of lactose and sucrose.

14.63 Consider the disaccharide isomaltose.

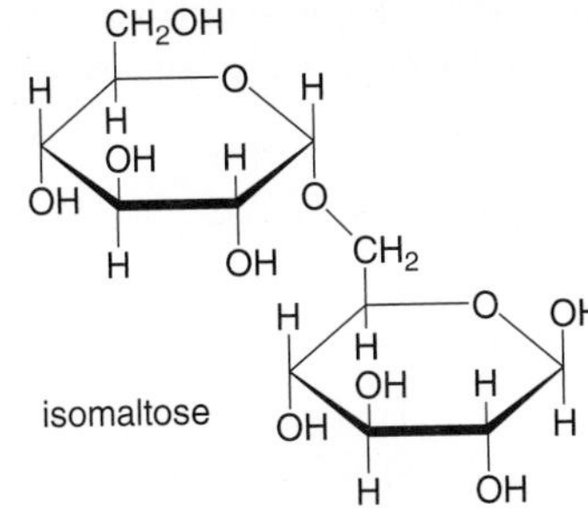

a. Classify the glycosidic linkage as α or β.

b. What monosaccharides are formed when isomaltose is hydrolyzed?

14.64 Consider the disaccharide sophorose.

sophorose

a. Classify the glycosidic linkage as α or β.

b. What monosaccharides are formed when sophorose is hydrolyzed?

14.65 Draw the structure of a disaccharide formed from two galactose units joined by a 1→4-β-glycosidic linkage. (The cyclic structure of galactose appears in Sample Problem 14.3.)

14.66 In what ways are cellulose and amylose similar? How do the structures of cellulose and amylose differ?

Applications

14.67 Describe the difference between lactose intolerance and galactosemia.

14.68 Explain why cellulose is a necessary component of our diet even though we don't digest it.

14.69 Explain why fructose is called a reduced calorie sweetener while sucralose is called an artificial sweetener.

14.70 How do oxidation reactions help an individual with diabetes monitor blood glucose levels?

14.71 Why can an individual with type A blood receive only blood types A and O, but he or she can donate to individuals with either type A or AB blood?

14.72 Why can an individual with type B blood receive only blood types B and O, but he or she can donate to individuals with either type B or AB blood?

CHALLENGE PROBLEMS

14.73 Draw a short segment of a polysaccharide that contains three galactose units joined together in 1→4-α-glycosidic linkages. (The cyclic structure of galactose appears in Sample Problem 14.3.)

14.74 Convert each cyclic form to an acyclic monosaccharide.

a. 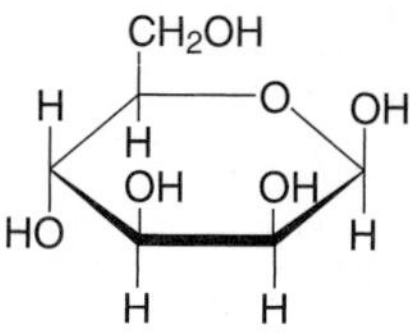 CH_2OH; O; H; OH; H; OH; OH; HO; H; H; H

b. 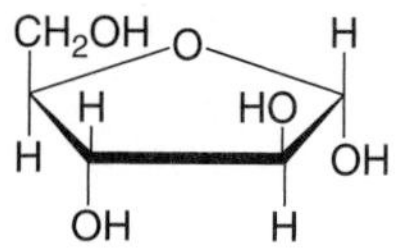 CH_2OH; O; H; H; HO; H; OH; OH; H

BEYOND THE CLASSROOM

14.75 In addition to the compounds shown in Figure 14.3, cyclamate, Truvia, acesulfame K, and alitame are artificial or low-calorie sweeteners. Research one or more of these compounds and draw its chemical structure. Is the compound isolated from a natural source or synthesized in the laboratory? How sweet is the compound compared to sucrose? Is the compound used in any commercial products? Does this sweetener offer any advantages or disadvantages over current widely used artificial sweeteners?

14.76 As mentioned in Section 14.6, cotton fiber is composed almost entirely of the polysaccharide cellulose. How does the structure of cotton compare with other polymers such as wool or silk that are derived from animal sources? How does the structure of cotton compare to synthetic polymers such as nylon or polyester? Pick five articles of clothing and list what polymers they are composed of. Are there advantages or disadvantages of using one polymer over another in particular types of clothing (underwear, t-shirts, athletic clothing, etc.)?

14.77 Pick a recipe for a favorite cookie, cake, or other baked good. Determine the number of grams of sugar used for one batch of the recipe, as well as the number of Calories that this amount contains. How much sucralose or saccharin would be used instead of sugar for an equivalent level of sweetness? Conversion factors you will need: density of granulated sugar = 0.70 g/cc; Calories from carbohydrates = 4 Cal/g.

14.78 What is the difference between high-fructose corn syrup and sucrose? Why has high-fructose corn syrup replaced sucrose in many processed food products? Where does it come from? Why have health concerns been raised over its presence in consumer products? Can you find evidence that the replacement of sucrose by high-fructose corn syrup is directly related to the rise in obesity in the United States?

ANSWERS TO SELECTED PROBLEMS

14.1

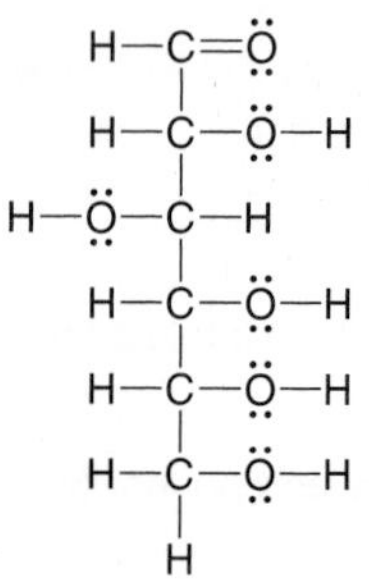

```
      H—C=Ö:
        |
      H—C—Ö—H
        |
  H—Ö—C—H
        |
      H—C—Ö—H
        |
      H—C—Ö—H
        |
      H—C—Ö—H
        |
        H
```

14.2 a. aldopentose b. aldotetrose c. ketotetrose

14.3 ketopentose

14.5 hexane < 1-decanol < glucose

14.7 a.

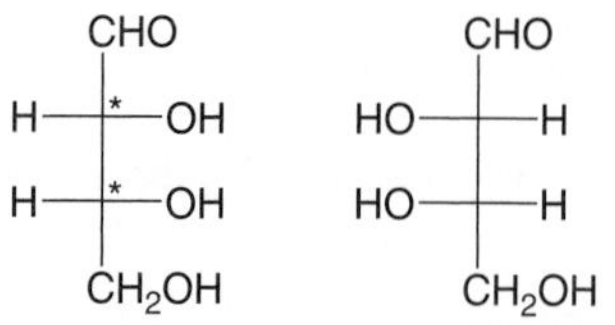

```
      CHO              CHO
  H—*—OH          HO——H
  H—*—OH          HO——H
     CH2OH            CH2OH
       D            enantiomer
```

b.

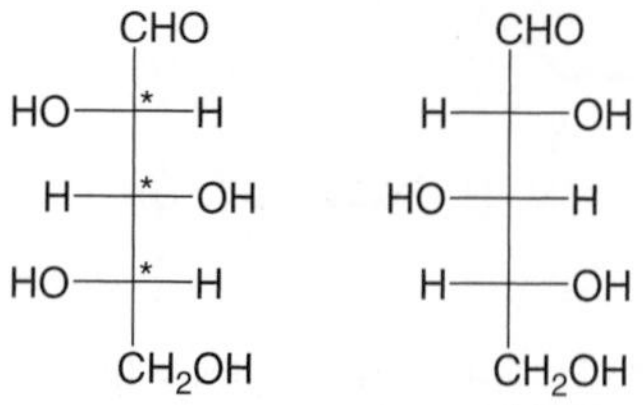

```
      CHO              CHO
 HO—*—H            H——OH
  H—*—OH          HO——H
 HO—*—H            H——OH
     CH2OH            CH2OH
       L            enantiomer
```

c.

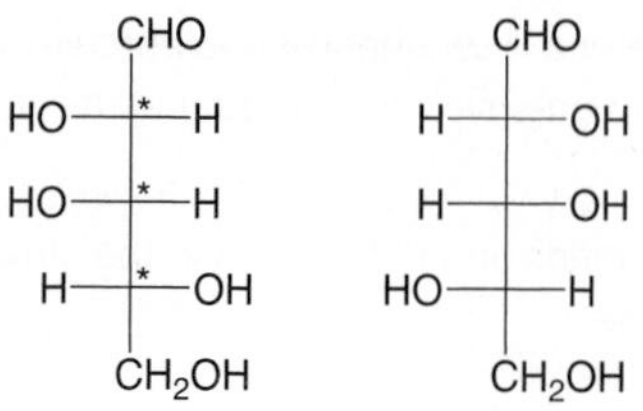

```
      CHO              CHO
 HO—*—H            H——OH
 HO—*—H            H——OH
  H—*—OH          HO——H
     CH2OH            CH2OH
       D            enantiomer
```

* chirality center

14.9 a.

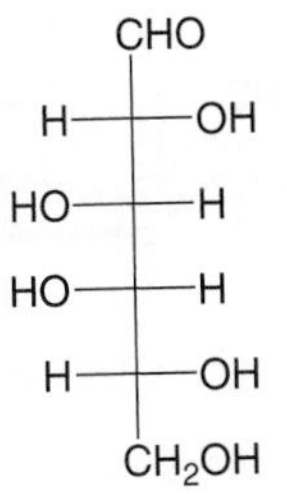

CHO
H—OH
HO—H
HO—H
H—OH
CH_2OH
D-galactose

b.

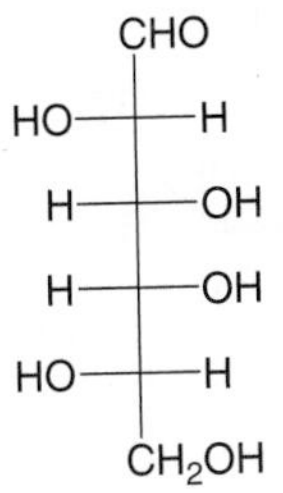

CHO
HO—H
H—OH
H—OH
HO—H
CH_2OH
enantiomer

14.10 a.

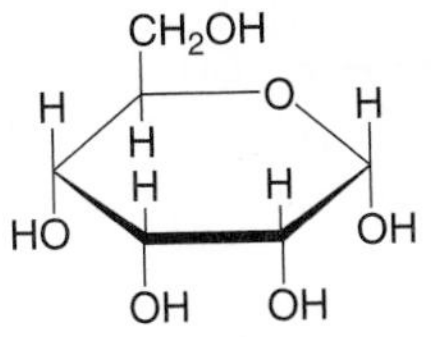

CH_2OH
H O H
H
H H
HO OH
OH OH

b.

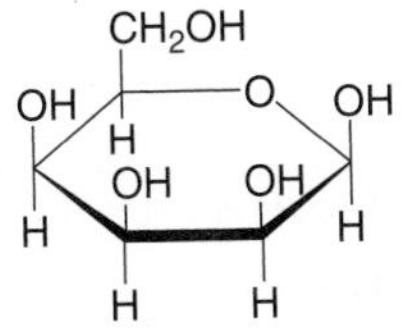

CH_2OH
OH O OH
H
OH OH
H H
H H

c.

CH_2OH
H O H
H
H OH
HO OH
OH H

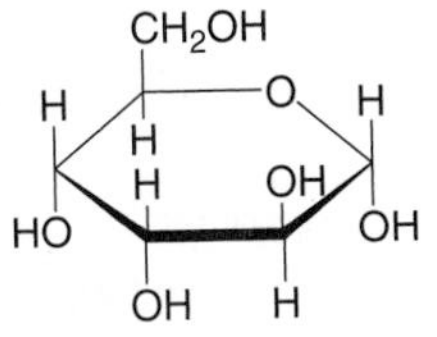

14.11 a. α isomer b. α isomer c. β isomer

14.12 a.

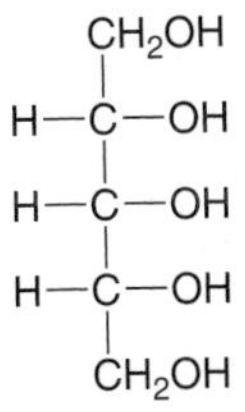

CH_2OH
H—C—OH
H—C—OH
H—C—OH
CH_2OH

b.

CH_2OH
H—C—OH
HO—C—H
HO—C—H
H—C—OH
CH_2OH

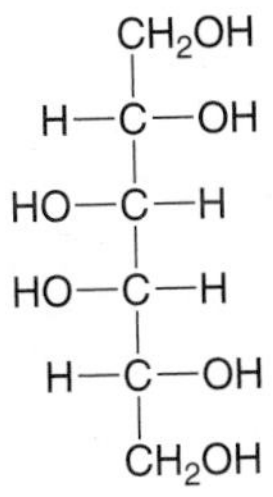

c.

CH_2OH
H—C—OH
H—C—OH
CH_2OH

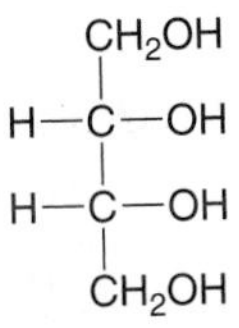

14.13 a.

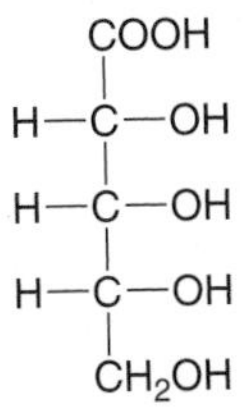

COOH
H—C—OH
H—C—OH
H—C—OH
CH_2OH

b.

COOH
H—C—OH
H—C—OH
CH_2OH

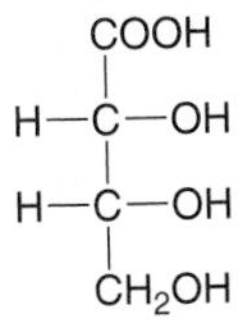

14.14 a. alditol b. aldonic acid

14.15 a.

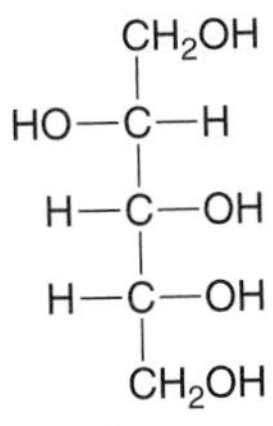

CH_2OH
HO—C—H
H—C—OH
H—C—OH
CH_2OH

b.

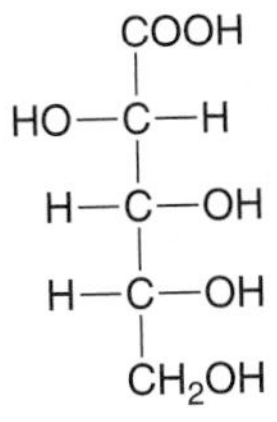

COOH
HO—C—H
H—C—OH
H—C—OH
CH_2OH

14.16 1 ⟶ 4-β-disaccharide

CH_2OH
H 5 O H
4 H 1
OH H OH
3 2
O H OH
CH_2OH
H 5 O
4 H 1
OH H
OH 3 2 H
H OH
glycosidic linkage

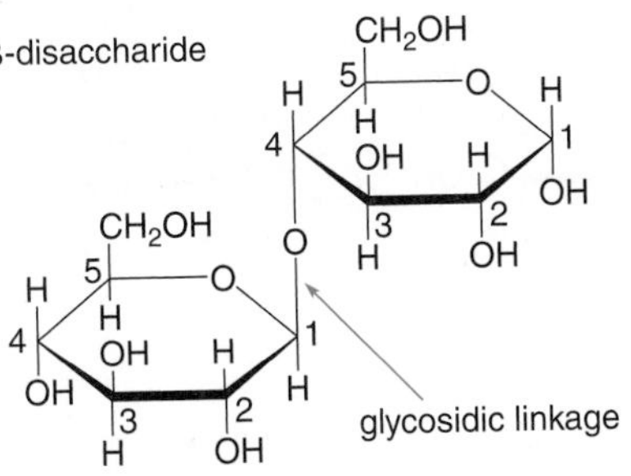

14.17

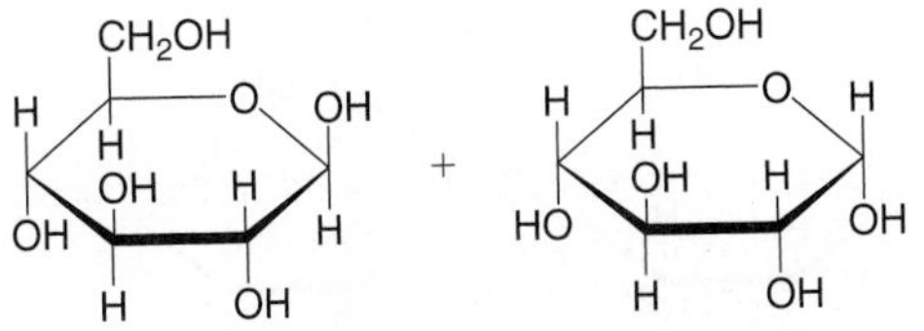

CH_2OH
H O OH
H
OH H
OH H
H OH

\+

CH_2OH
H O H
H
OH H
HO OH
H OH

14.19 carboxylic acid, amine, amide, ester, aromatic ring

14.21 $HOCH_2\overset{*}{C}H\overset{*}{C}HCHO$ (with OH OH below)
A
aldotetrose
* chirality center

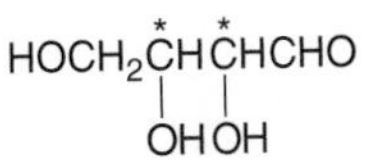

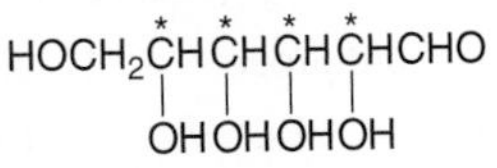

$HOCH_2\overset{*}{C}H\overset{*}{C}H\overset{*}{C}H\overset{*}{C}HCHO$ (with OH OH OH OH below)
B
aldohexose

14.23 a.

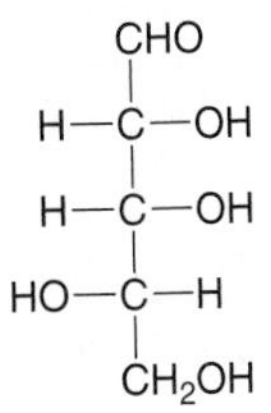

CHO
H—C—OH
H—C—OH
HO—C—H
CH_2OH

b.

CHO
H—C—OH
H—C—OH
CH_2OH

c.

CH_2OH
H—C—OH
H—C—OH
HO—C—H
CH_2OH

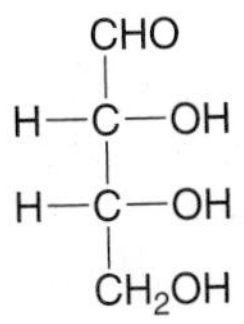

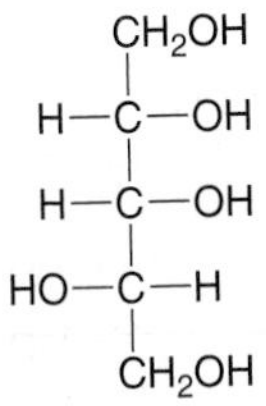

14.25 a. D sugar

b. aldohexose

c.

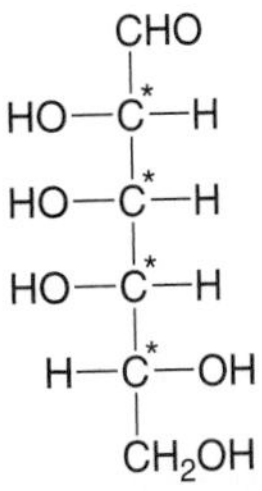

CHO
HO—C*—H
HO—C*—H
HO—C*—H
H—C*—OH
CH_2OH
* chirality center

d.

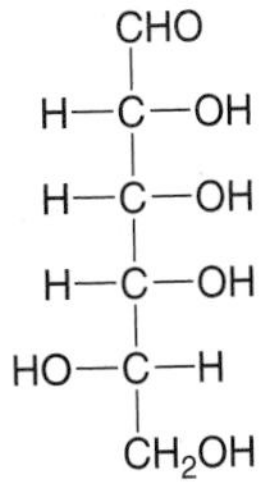

CHO
H—C—OH
H—C—OH
H—C—OH
HO—C—H
CH_2OH
enantiomer

e.

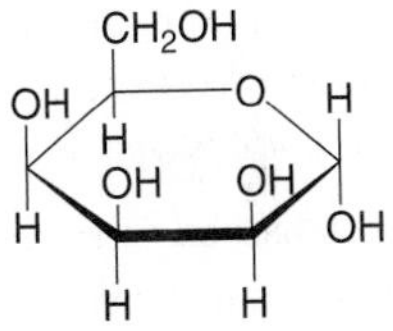

CH_2OH
OH O H
H
OH OH
H OH
H H

f.

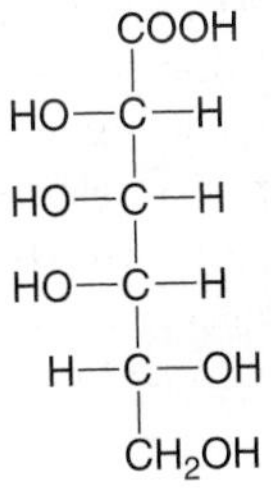

COOH
HO—C—H
HO—C—H
HO—C—H
H—C—OH
CH_2OH

g.

CH_2OH
HO—C—H
HO—C—H
HO—C—H
H—C—OH
CH_2OH

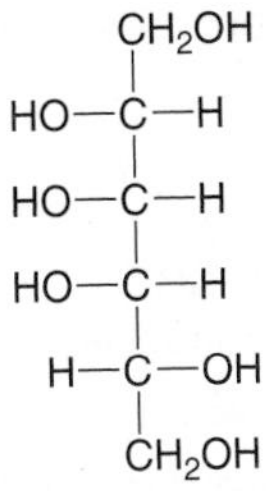

14.27 a. **A** b. **C** c. **B**

14.29 a,b: glycosidic linkage

c. 1→4-α-glycosidic linkage

14.31 a. aldotetrose b. aldohexose c. ketopentose

14.33 a.

CHO / HO—C*—H / HO—C*—H / CH_2OH — L-tetrose

CHO / H—C—OH / H—C—OH / CH_2OH — enantiomer

CHO / HO—H / HO—H / CH_2OH — Fischer projection

b.

CHO / HO—C*—H / HO—C*—H / HO—C*—H / H—C*—OH / CH_2OH — D-hexose

CHO / H—C—OH / H—C—OH / H—C—OH / HO—C—H / CH_2OH — enantiomer

CHO / HO—H / HO—H / HO—H / H—OH / CH_2OH — Fischer projection

c.

CH_2OH / C=O / H—C*—OH / HO—C*—H / CH_2OH — L-pentose

CH_2OH / C=O / HO—C—H / H—C—OH / CH_2OH — enantiomer

CH_2OH / C=O / H—OH / HO—H / CH_2OH — Fischer projection

* chirality center

14.35 a. CHO / H—OH / CH_2OH — D-monosaccharide

b. aldotriose

14.37 α-D-Glucose and β-D-glucose are not enantiomers because they differ in the orientation of only one OH at C1.

14.39 a. **A** and **B**

b. **B** and **C** (or **A** and **C**)

c. CHO / HO—C—H / H—C—OH / CH_2OH

d. CHO / H—OH / H—OH / CH_2OH

14.41 β isomer α isomer

14.43 a. β isomer b.

14.45 a. β isomer b. α isomer

14.47 A ketotriose has no stereoisomers.

14.49 a. aldonic acid

b. alditol

c. monosaccharide

14.51 a. [1] CH_2OH / HO—C—H / H—C—OH / CH_2OH [2] COOH / HO—C—H / H—C—OH / CH_2OH

b. [1] CH_2OH / H—C—OH / HO—C—H / HO—C—H / CH_2OH [2] COOH / H—C—OH / HO—C—H / HO—C—H / CH_2OH

c. [1] CH_2OH / H—C—OH / H—C—OH / HO—C—H / HO—C—H / CH_2OH [2] COOH / H—C—OH / H—C—OH / HO—C—H / HO—C—H / CH_2OH

14.53 COOH / H—C—OH / H—C—OH / HO—C—H / CH_2OH

14.55 +

14.57 a.

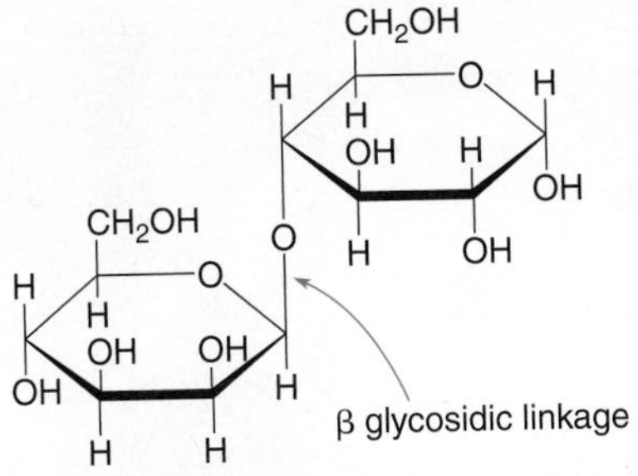

β glycosidic linkage

b.

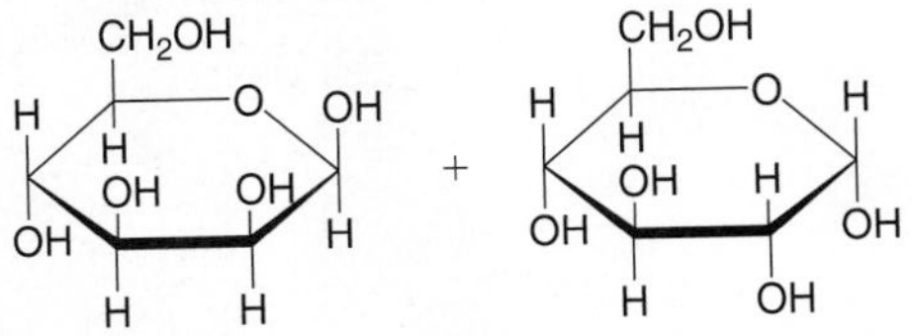

14.59

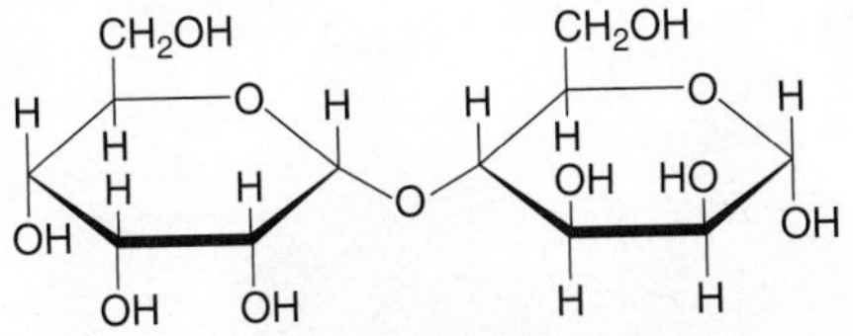

14.61 Maltose and lactose are both disaccharides composed of two hexoses. Maltose has an α glycosidic linkage and lactose has a β glycosidic linkage. Maltose contains two linked glucose molecules and lactose has one glucose molecule linked with galactose.

14.63 a. α glycoside

b.

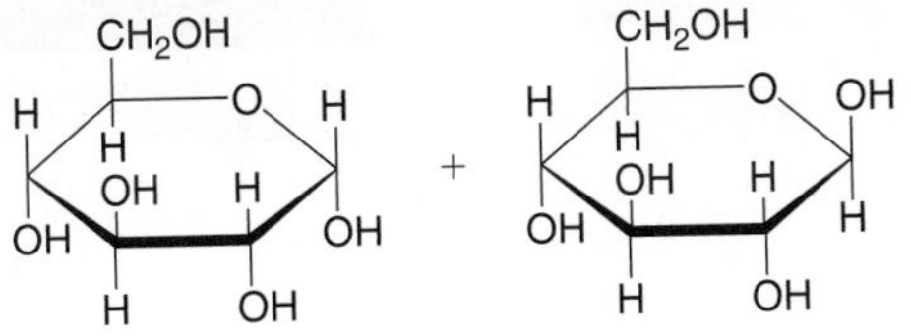

14.65

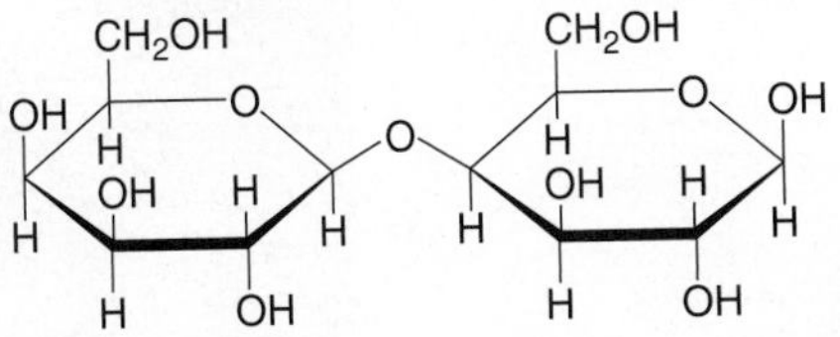

14.67 Lactose intolerance results from a lack of the enzyme lactase. It causes abdominal cramping and diarrhea. Galactosemia results from the inability to metabolize galactose. Therefore, it accumulates in the liver, causing cirrhosis, and in the brain, leading to mental retardation.

14.69 Fructose is a naturally occurring sugar with more perceived sweetness per gram than sucrose. Sucralose is a synthetic sweetener; that is, it is not naturally occurring.

14.71 An individual with type A blood can receive only blood types A and O, because he or she will produce antibodies and an immune response to B or AB blood. He or she can donate to individuals with either type A or AB blood as the type A polysaccharides are common to both and no immune response will be generated.

14.73

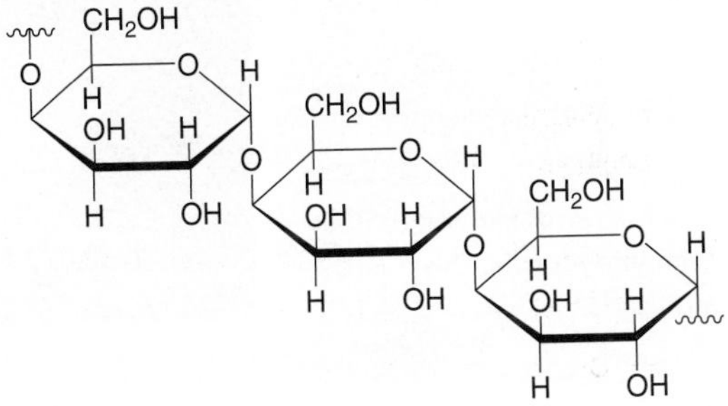

The cocoa butter in chocolate is rich in **triacylglycerols**, the most abundant lipids.

Lipids

CHAPTER OUTLINE

CHAPTER GOALS

In this chapter you will learn how to:

1. Describe the general characteristics of lipids
2. Classify fatty acids and describe the relationship between melting point and the number of double bonds
3. Draw the structure of a wax and identify the carboxylic acid and alcohol components
4. Draw the structure of triacylglycerols and describe the difference between a fat and an oil
5. Draw the hydrolysis products of triacylglycerols
6. Identify the structural features of phospholipids
7. Describe the structure of a cell membrane, as well as different mechanisms of transport across the membrane
8. Recognize the main structural features of steroids like cholesterol
9. Define what a hormone is and list several examples of steroid hormones
10. Identify fat-soluble vitamins

HEALTH NOTE

Common lipids include triacylglycerols in vegetable oils, cholesterol in egg yolk, and vitamin E in leafy greens.

In Chapter 15, we turn our attention to lipids, biomolecules that are soluble in organic solvents and insoluble in water. Unlike the carbohydrates in Chapter 14, lipids contain few functional groups and are composed mainly of carbon–carbon and carbon–hydrogen bonds. Since lipids are the biomolecules that most closely resemble the hydrocarbons discussed in Chapters 10 and 11, we have already learned many facts that directly explain their properties. In Chapter 15 we learn more details about the chemistry of lipids.

15.1 Introduction to Lipids

- *Lipids* are biomolecules that are soluble in organic solvents and insoluble in water.

Lipids are unique among organic molecules because their identity is defined on the basis of a *physical property* and not by the presence of a particular functional group. Because of this, lipids come in a wide variety of structures and they have many different functions.

Lipids contain a large number of nonpolar carbon–carbon and carbon–hydrogen bonds. In addition, most lipids have a few polar bonds that may be found in a variety of functional groups. As a result, lipids are nonpolar or weakly polar molecules that are very soluble in organic solvents like hexane (C_6H_{14}) and carbon tetrachloride (CCl_4), and insoluble in a polar medium like water.

Because lipids share many properties with hydrocarbons, several features of lipid structure and properties have been discussed in previous chapters, as summarized in Table 15.1.

Table 15.1 Summary of Lipid Chemistry Prior to Chapter 15

Topic	Section
Fatty acids	11.3B
Oral contraceptives	11.4
Margarine and butter	11.6
Aspirin and prostaglandins	13.4B
Soap	13.5B
Olestra, a synthetic fat	13.6C

Lipids can be categorized as hydrolyzable or nonhydrolyzable.

1. *Hydrolyzable lipids* can be converted into smaller molecules by hydrolysis with water. We will examine three subgroups: waxes, triacylglycerols, and phospholipids.

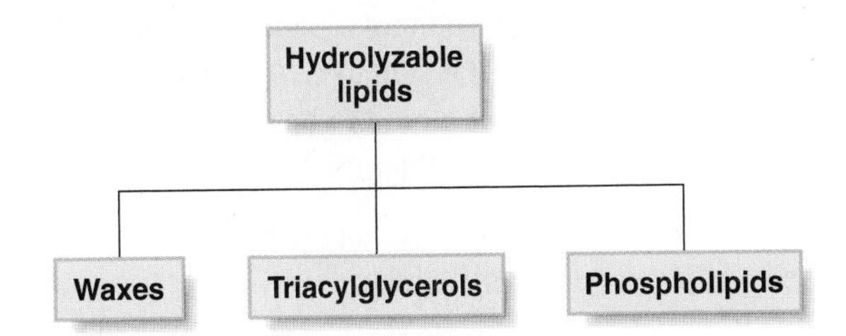

Most hydrolyzable lipids contain an ester.

2. *Nonhydrolyzable lipids* cannot be cleaved into smaller units by aqueous hydrolysis. Nonhydrolyzable lipids tend to be more varied in structure. We will examine two different types: steroids and fat-soluble vitamins.

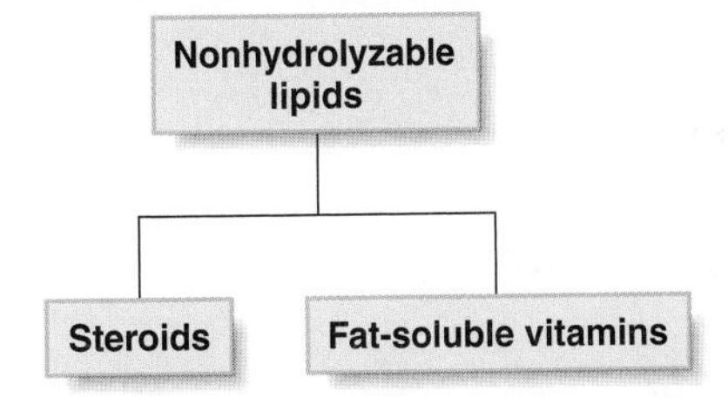

Lipids have many important roles in biological systems. Since lipids release over twice the amount of energy per gram than carbohydrates or proteins (9 kcal/g for lipids compared to 4 kcal/g for carbohydrates and proteins, Section 4.1), lipids are an excellent source of energy. Moreover, lipids are key components of the cell membrane, and they serve as chemical messengers in the body.

PROBLEM 15.1

In which solvents or solutions will a lipid be soluble: (a) CH_2Cl_2; (b) 5% aqueous NaCl solution; (c) $CH_3CH_2CH_2CH_2CH_3$?

15.2 Fatty Acids

Hydrolyzable lipids are derived from **fatty acids,** carboxylic acids that were first discussed in Section 11.3B.

- ***Fatty acids*** **are carboxylic acids (RCOOH) with long carbon chains of 12–20 carbon atoms.**

Palmitic acid is a common 16-carbon fatty acid whose structure is given in condensed, skeletal, and three-dimensional representations.

$CH_3CH_2CH_2CH_2CH_2CH_2CH_2CH_2CH_2CH_2CH_2CH_2CH_2CH_2CH_2COOH$

palmitic acid

$C_{16}H_{32}O_2$

skeletal structure

polar C—O and O—H bonds

hydrophilic portion

nonpolar C—C and C—H bonds

hydrophobic portion

The nonpolar part of the molecule (comprised of C—C and C—H bonds) is not attracted to water, so it is said to be ***hydrophobic*** (water fearing). The polar part of the molecule is attracted to water, so it is said to be ***hydrophilic*** (water loving). In a lipid, the hydrophobic portion is always much larger than the hydrophilic portion.

Naturally occurring fatty acids have an even number of carbon atoms. There are two types of fatty acids.

- ***Saturated*** **fatty acids have no double bonds in their long hydrocarbon chains.**
- ***Unsaturated*** **fatty acids have one or more double bonds in their long hydrocarbon chains. Generally, double bonds in naturally occurring fatty acids are cis.**

Recall from Section 11.3 that a **cis** alkene has two alkyl groups on the *same* side of the double bond, while a **trans** alkene has two alkyl groups on *opposite* sides of the double bond.

Three-dimensional models for stearic acid, a saturated fatty acid, and oleic acid, an unsaturated fatty acid with one cis double bond, are shown. Table 15.2 lists the structures of the most common saturated and unsaturated fatty acids.

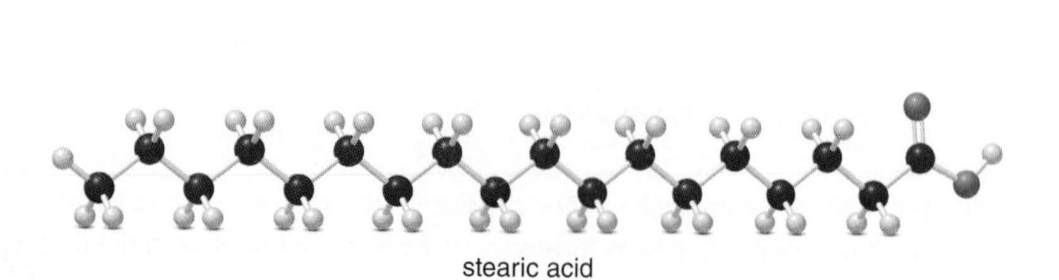

stearic acid

a saturated fatty acid
no double bonds in the long chain

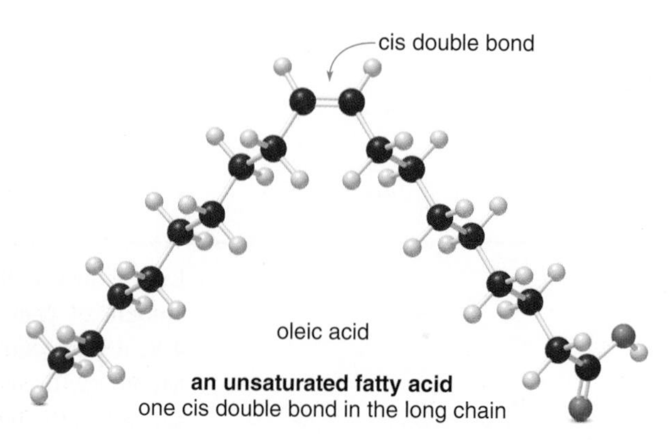

oleic acid

an unsaturated fatty acid
one cis double bond in the long chain

Table 15.2 Common Fatty Acids

Number of C's	Number of C=C's	Structure	Name	Mp (°C)
		Saturated Fatty Acids		
12	0	$CH_3(CH_2)_{10}COOH$	Lauric acid	44
14	0	$CH_3(CH_2)_{12}COOH$	Myristic acid	58
16	0	$CH_3(CH_2)_{14}COOH$	Palmitic acid	63
18	0	$CH_3(CH_2)_{16}COOH$	Stearic acid	71
20	0	$CH_3(CH_2)_{18}COOH$	Arachidic acid	77
		Unsaturated Fatty Acids		
16	1	$CH_3(CH_2)_5CH{=}CH(CH_2)_7COOH$	Palmitoleic acid	1
18	1	$CH_3(CH_2)_7CH{=}CH(CH_2)_7COOH$	Oleic acid	16
18	2	$CH_3(CH_2)_4CH{=}CHCH_2CH{=}CH(CH_2)_7COOH$	Linoleic acid	–5
18	3	$CH_3CH_2CH{=}CHCH_2CH{=}CHCH_2CH{=}CH(CH_2)_7COOH$	Linolenic acid	–11
20	4	$CH_3(CH_2)_4(CH{=}CHCH_2)_4(CH_2)_2COOH$	Arachidonic acid	–49

The most common saturated fatty acids are palmitic and stearic acid. The most common unsaturated fatty acid is oleic acid. Linoleic and linolenic acids are called **essential fatty acids** because humans cannot synthesize them and must acquire them in our diets.

Oils formed from omega-3 fatty acids may provide health benefits to individuals with cardiovascular disease, as discussed in Section 15.4.

Unsaturated fatty acids are sometimes classified as **omega-*n* acids,** where *n* is the carbon at which the first double bond occurs in the carbon chain, beginning at the end of the chain that contains the CH_3 group. Thus, linoleic acid is an omega-6 acid and linolenic acid is an omega-3 acid.

first C=C at C6 ⟶ an **omega-6 acid**

$$\underset{1}{CH_3}\underset{2}{CH_2}\underset{3}{CH_2}\underset{4}{CH_2}\underset{5}{CH_2}\underset{6}{CH}{=}CHCH_2CH{=}CHCH_2CH_2CH_2CH_2CH_2CH_2CH_2COOH$$

linoleic acid

first C=C at C3 ⟶ an **omega-3 acid**

$$\underset{1}{CH_3}\underset{2}{CH_2}\underset{3}{CH}{=}CHCH_2CH{=}CHCH_2CH{=}CHCH_2CH_2CH_2CH_2CH_2CH_2CH_2COOH$$

linolenic acid

As we learned in Section 11.3B, the presence of cis double bonds affects the melting point of these fatty acids greatly.

- **As the number of double bonds in the fatty acid *increases*, the melting point *decreases*.**

SAMPLE PROBLEM 15.1

(a) Draw a skeletal structure of gadoleic acid, $CH_3(CH_2)_9CH{=}CH(CH_2)_7COOH$, a 20-carbon fatty acid obtained from fish oils. (b) Label the hydrophobic and hydrophilic portions. (c) Predict how its melting point compares to the melting points of arachidic acid and arachidonic acid (Table 15.2).

Analysis

- Skeletal structures have a carbon at the intersection of two lines and at the end of every line. The double bond must have the cis arrangement in an unsaturated fatty acid.
- The nonpolar C—C and C—H bonds comprise the hydrophobic portion of a molecule and the polar bonds comprise the hydrophilic portion.
- For the same number of carbons, increasing the number of double bonds decreases the melting point of a fatty acid.

Gadoleic acid (Sample Problem 15.1) and DHA (Problem 15.3) are two fatty acids derived from tuna fish oil.

Solution

a. and b. The skeletal structure for gadoleic acid is drawn below. The hydrophilic portion is the COOH group and the hydrophobic portion is the rest of the molecule.

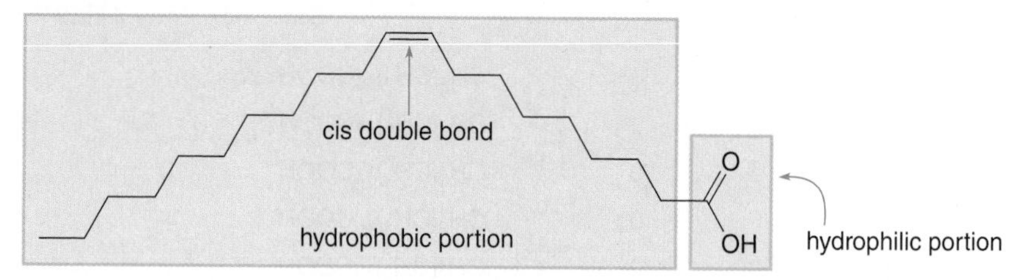

c. Gadoleic acid has one cis double bond, giving it a lower melting point than arachidic acid (no double bonds), but a higher melting point than arachidonic acid (four double bonds).

PROBLEM 15.2

(a) Draw a skeletal structure for each fatty acid. (b) Label the hydrophobic and hydrophilic portions of each molecule. (c) Without referring to Table 15.2, which fatty acid has the higher melting point, **A** or **B?** Explain your choice.

$CH_3(CH_2)_{16}COOH$ — **A**

$CH_3(CH_2)_4CH{=}CHCH_2CH{=}CH(CH_2)_7COOH$ — **B**

PROBLEM 15.3

DHA (4,7,10,13,16,19-docosahexaenoic acid) is a common fatty acid present in tuna fish oil. (a) Draw a skeletal structure for DHA. (b) Give the omega-*n* designation for DHA.

$CH_3CH_2CH{=}CHCH_2CH{=}CHCH_2CH{=}CHCH_2CH{=}CHCH_2CH{=}CHCH_2CH{=}CHCH_2CH_2COOH$

DHA

CONSUMER NOTE

When commercial whaling was commonplace, spermaceti wax obtained from sperm whales was used extensively in cosmetics and candles.

PROBLEM 15.4

Give the omega-*n* designation for (a) oleic acid; (b) arachidonic acid (Table 15.2).

15.3 Waxes

Waxes are the simplest hydrolyzable lipids.

- ***Waxes* are esters (RCOOR') formed from a fatty acid (RCOOH) and a high molecular weight alcohol (R'OH).**

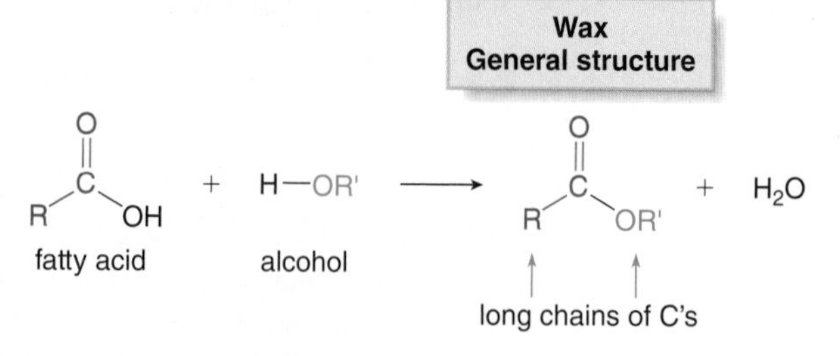

Water beads up on the surface of a leaf because of its waxy coating.

For example, spermaceti wax, isolated from the heads of sperm whales, is largely cetyl palmitate, an ester with the structure $CH_3(CH_2)_{14}COO(CH_2)_{15}CH_3$. Cetyl palmitate is formed from a 16-carbon fatty acid [$CH_3(CH_2)_{14}COOH$] and a 16-carbon alcohol [$CH_3(CH_2)_{15}OH$].

$$CH_3(CH_2)_{14}C(=O)OH + H{-}O(CH_2)_{15}CH_3 \longrightarrow CH_3(CH_2)_{14}C(=O)O(CH_2)_{15}CH_3 + H_2O$$

palmitic acid → cetyl palmitate (major component of spermaceti wax); the C(=O)O linkage is labeled "ester".

Waxes form a protective coating on the feathers of birds to make them water repellent, and on leaves to prevent water evaporation. Beeswax, a complex mixture of over 200 different compounds, contains the wax myricyl palmitate as its major component.

SAMPLE PROBLEM 15.2

Draw the structure of cetyl myristate, a minor component of spermaceti wax, formed from a 14-carbon fatty acid and a 16-carbon straight chain alcohol.

Analysis

To draw the wax, arrange the carboxyl group of the fatty acid (RCOOH) next to the OH group of the alcohol (R'OH) with which it reacts. Then, replace the OH group of the fatty acid with the OR' group of the alcohol, forming an ester RCOOR' with a new C—O bond at the carbonyl carbon.

Solution

Draw the structures of the fatty acid and the alcohol, and replace the OH group of the 14-carbon acid with the $O(CH_2)_{15}CH_3$ group of the alcohol.

$$\underset{\text{14-carbon fatty acid}}{CH_3(CH_2)_{12}C(=O)OH} + \underset{\text{16-carbon alcohol}}{H{-}O(CH_2)_{15}CH_3} \longrightarrow \underset{\text{cetyl myristate}}{CH_3(CH_2)_{12}C(=O){-}O(CH_2)_{15}CH_3} + H_2O$$

(new C—O bond)

PROBLEM 15.5

Draw the structure of a wax formed from stearic acid [$CH_3(CH_2)_{16}COOH$] and each alcohol.

a. $CH_3(CH_2)_9OH$ b. $CH_3(CH_2)_{11}OH$ c. $CH_3(CH_2)_{29}OH$

PROBLEM 15.6

One component of jojoba oil is a wax formed from eicosenoic acid [$CH_3(CH_2)_7CH{=}CH(CH_2)_9CO_2H$] and $CH_3(CH_2)_7CH{=}CH(CH_2)_8OH$. Draw the structure of the wax, including the cis geometry of both carbon–carbon double bonds.

The seeds of the jojoba plant grown in the southwestern United States are rich in waxes used in cosmetics and personal care products.

Like other esters, **waxes (RCOOR') are hydrolyzed with water in the presence of acid or base to re-form the carboxylic acid (RCOOH) and alcohol (R'OH) from which they are prepared** (Section 13.6). Thus, hydrolysis of cetyl palmitate in the presence of H_2SO_4 forms a fatty acid and a long chain alcohol by cleaving the carbon–oxygen single bond of the ester.

$$\underset{\text{cetyl palmitate}}{CH_3(CH_2)_{14}C(=O){-}O(CH_2)_{15}CH_3} + H_2O \xrightarrow{H_2SO_4} \underset{\text{fatty acid}}{CH_3(CH_2)_{14}C(=O)OH} + \underset{\text{alcohol}}{H{-}O(CH_2)_{15}CH_3}$$

(The C—O bond is broken.)

PROBLEM 15.7

What hydrolysis products are formed when cetyl laurate (shown in the ball-and-stick model) is treated with aqueous acid?

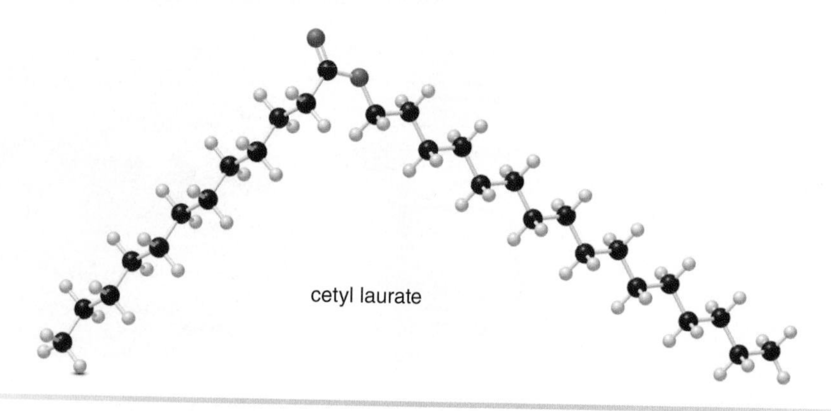

15.4 Triacylglycerols—Fats and Oils

Animal fats and vegetable oils, the most abundant lipids, are composed of **triacylglycerols.**

- ***Triacylglycerols,* or triglycerides, are triesters formed from glycerol and three molecules of fatty acids.**

CH_2—OH, CH—OH, CH_2—OH (glycerol) + HO—C(=O)—R, HO—C(=O)—R', HO—C(=O)—R'' (fatty acids) ⟶ CH_2—O—C(=O)—R, CH—O—C(=O)—R', CH_2—O—C(=O)—R'' (triacylglycerol)

glycerol

fatty acids
R groups have 11–19 C's.

triacylglycerol
[Three ester groups are labeled in red.]

glycerol — fatty acid, fatty acid, fatty acid

block diagram

15.4A General Features

Triacylglycerols may be composed of three identical fatty acid side chains, or they may be derived from two or three different fatty acids. The fatty acids may be saturated or unsaturated. **Fats** and **oils** are triacylglycerols with different physical properties.

- **Fats have higher melting points—they are *solids* at room temperature.**
- **Oils have lower melting points—they are *liquids* at room temperature.**

The identity of the three fatty acids in the triacylglycerol determines whether it is a fat or an oil. ***Increasing* the number of double bonds in the fatty acid side chains *decreases* the melting point of the triacylglycerol.**

- **Fats are derived from fatty acids having few double bonds.**
- **Oils are derived from fatty acids having a larger number of double bonds.**

Solid fats have a relatively high percentage of saturated fatty acids and are generally animal in origin. Thus, lard (hog fat), butter, and whale blubber contain a high percentage of saturated fats. With no double bonds, the three side chains of the saturated lipid lie parallel with each other, leading to a high melting point.

no double bonds in the long carbon chains

animal products
source of saturated fats

a saturated triacylglycerol

HEALTH NOTE

Oil in the coconuts from plantations like this one is high in saturated fats, which are believed to contribute to a greater risk of heart disease.

Liquid oils have a higher percentage of unsaturated fatty acids and are generally vegetable in origin. Thus, oils derived from corn, soybeans, and olives contain more unsaturated lipids. In the unsaturated lipid, a cis double bond places a kink in the side chain, making it more difficult to pack efficiently in the solid state, thus leading to a lower melting point.

vegetable oil
source of unsaturated oils

an unsaturated triacylglycerol

HEALTH NOTE

Fish oils are high in unsaturated triacylglycerols derived from omega-3 fatty acids.

Unlike other vegetable oils, oils from palm and coconut trees are very high in saturated fats. Considerable evidence currently suggests that diets high in saturated fats lead to a greater risk of heart disease (Sections 15.4B and 15.8). For this reason, the demand for coconut and palm oil has decreased considerably in recent years, and many coconut plantations previously farmed in the South Pacific are no longer in commercial operation.

Oils derived from fish such as salmon, herring, mackerel, and sardines are very rich in polyunsaturated triacylglycerols. These triacylglycerols pack so poorly that they have very low melting points, and they remain liquids even in very cold water. Fish oils derived from omega-3 fatty acids are thought to be especially beneficial for individuals at risk for developing coronary artery disease.

SAMPLE PROBLEM 15.3

Draw the structure of a triacylglycerol formed from glycerol, one molecule of stearic acid, and two molecules of oleic acid. Bond the stearic acid to the 2° OH group (OH on the middle carbon atom) of glycerol.

Analysis

To draw the triacylglycerol, arrange each OH group of glycerol next to the carboxyl group of a fatty acid. Then join each O atom of glycerol to a carbonyl carbon of a fatty acid, to form three new C—O bonds.

Solution

Form three new ester bonds (RCOOR') from OH groups of glycerol and the three fatty acids (RCOOH).

$$\begin{array}{l} CH_2-OH \\ CH-OH \\ CH_2-OH \end{array} + \begin{array}{l} HO-C(=O)-(CH_2)_7CH=CH(CH_2)_7CH_3 \\ HO-C(=O)-(CH_2)_{16}CH_3 \\ HO-C(=O)-(CH_2)_7CH=CH(CH_2)_7CH_3 \end{array} \longrightarrow \begin{array}{l} CH_2-O-C(=O)-(CH_2)_7CH=CH(CH_2)_7CH_3 \\ CH-O-C(=O)-(CH_2)_{16}CH_3 \\ CH_2-O-C(=O)-(CH_2)_7CH=CH(CH_2)_7CH_3 \end{array} + 3\,H_2O$$

glycerol; 2° OH

oleic acid (R group in blue)
stearic acid (R group in red)

triacylglycerol

PROBLEM 15.8

Draw the structure of a triacylglycerol that contains (a) three molecules of stearic acid; (b) three molecules of oleic acid.

PROBLEM 15.9

Unlike many fats and oils, the cocoa butter used to make chocolate is remarkably uniform in composition. All triacylglycerols contain oleic acid esterified to the 2° OH group of glycerol, and either palmitic acid or stearic acid esterified to the 1° OH groups. Draw the structure of a possible triacylglycerol that composes cocoa butter.

CONSUMER NOTE

In addition to triacylglycerols, the cocoa used to prepare chocolate contains antioxidants and caffeine. One ounce of unsweetened chocolate has as much caffeine as two ounces of brewed coffee.

15.4B FOCUS ON HEALTH & MEDICINE Fats and Oils in the Diet

Fats and oils in our diet come from a variety of sources—meat, dairy products, seeds and nuts, salad dressing, fried foods, and any baked good or packaged food made with oil. Some fat is required in the diet. Fats are the building blocks of cell membranes, and stored body fat insulates an organism and serves as an energy source that can be used at a later time.

Currently, the United States Food and Drug Administration recommends that no more than 20–35% of an individual's calorie intake come from lipids. Moreover, a high intake of *saturated* triacylglycerols is linked to an increased incidence of heart disease. Saturated fats stimulate cholesterol synthesis in the liver and transport to the tissues, resulting in an increase in cholesterol concentration in the blood. An elevated cholesterol level in the blood can lead to cholesterol deposits or plaques on arteries, causing a narrowing of blood vessels, heart attack, and stroke (Section 15.8).

In contrast, *unsaturated* triacylglycerols lower the risk of heart disease by decreasing the amount of cholesterol in the blood. Unsaturated triacylglycerols from omega-3 fatty acids appear to reduce the risk of heart attack, especially in individuals who already have heart disease.

Wheat Thins
BIG
Nutrition Facts
Serving Size 11 Crackers (31g)
Servings Per Container About 7
Amount Per Serving
Calories 150 Calories from Fat 50
% Daily Value*
Total Fat 6g 9%
Saturated Fat 1g 5%
Trans Fat 0g
Polyunsaturated Fat 3g
Monounsaturated Fat 1g
Cholesterol 0mg 0%
Sodium 270mg 11%
Total Carbohydrate 21g 7%
Dietary Fiber 1g 4%
Sugars 4g

Dietary recommendations on fat intake must also take into account trans triacylglycerols, so-called *trans fats*. As we learned in Section 11.6, trans fats are formed when liquid oils are partially hydrogenated to form semi-solid triacylglycerols. The three-dimensional structure of a trans triacylglycerol shows its similarity to saturated triacylglycerols. Like saturated fats, trans fats also *increase* the amount of cholesterol in the bloodstream, thus increasing an individual's risk of developing coronary artery disease.

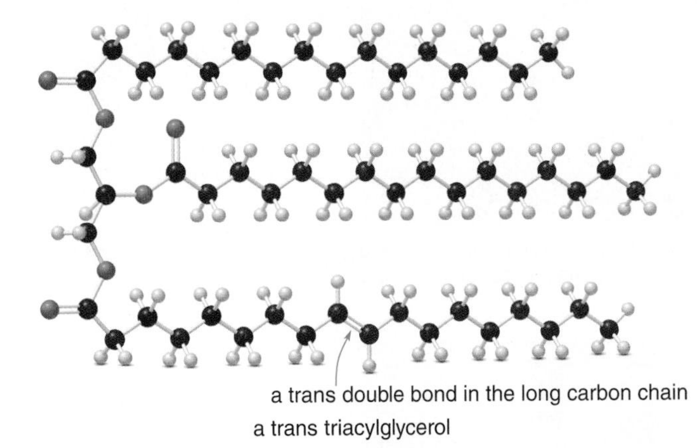

a trans triacylglycerol

Thus, while lipids are a necessary part of anyone's diet, the amount of saturated fat and trans fat should be limited. Calories from lipids should instead be obtained from unsaturated oils. Figure 15.1 illustrates some foods high in saturated, unsaturated, and trans triacylglycerols.

PROBLEM 15.10

Draw the structure of a triacylglycerol that fits each description:

a. a saturated triacylglycerol formed from three 12-carbon fatty acids
b. an unsaturated triacylglycerol that contains three cis double bonds
c. a trans triacylglycerol that contains a trans double bond in each hydrocarbon chain

15.5 Hydrolysis of Triacylglycerols

Like other esters, **triacylglycerols are hydrolyzed with water in the presence of acid, base, or enzymes (in biological systems) to form glycerol and three molecules of fatty acids.** Thus, hydrolysis of tristearin with aqueous sulfuric acid forms glycerol and three molecules of stearic acid.

The three bonds drawn in red are broken in hydrolysis.

$$\begin{array}{l} CH_2-O-C(=O)-(CH_2)_{16}CH_3 \\ CH-O-C(=O)-(CH_2)_{16}CH_3 \\ CH_2-O-C(=O)-(CH_2)_{16}CH_3 \end{array} + 3\ H_2O \xrightarrow{H_2SO_4} \begin{array}{l} CH_2-OH \\ CH-OH \\ CH_2-OH \end{array} + 3\ HO-C(=O)-(CH_2)_{16}CH_3$$

tristearin — glycerol — stearic acid

Hydrolysis cleaves the three single bonds between the carbonyl carbons and the oxygen atoms of the esters (Section 13.6). Since tristearin contains three identical R groups on the carbonyl carbons, three molecules of a single fatty acid, stearic acid, are formed. Triacylglycerols that contain different R groups bonded to the carbonyl carbons form mixtures of fatty acids, as shown in Sample Problem 15.4.

Figure 15.1 Saturated, Unsaturated, and Trans Triacylglycerols in the Diet

Food source | **Triacylglycerol**

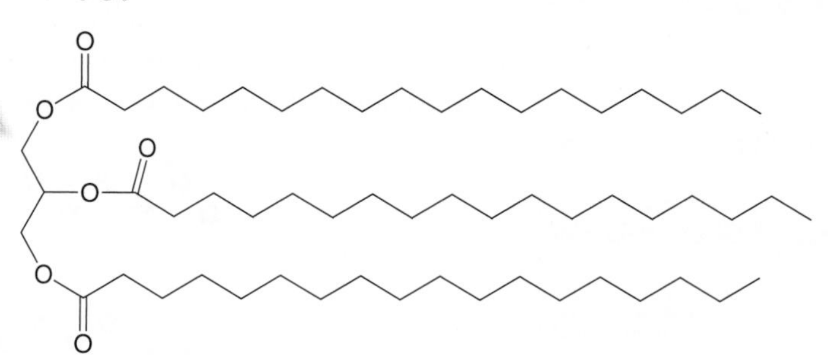

Foods rich in saturated triacylglycerols: fatty red meat, cheese, butter, fried foods, ice cream

- no double bonds in the carbon chains
- solid at room temperature
- increases blood cholesterol level

Foods rich in unsaturated triacylglycerols: plant oils, nuts, soybeans, fish (salmon, herring, mackerel)

- one (or more) double bonds in the carbon chains
- liquid at room temperature
- decreases blood cholesterol level

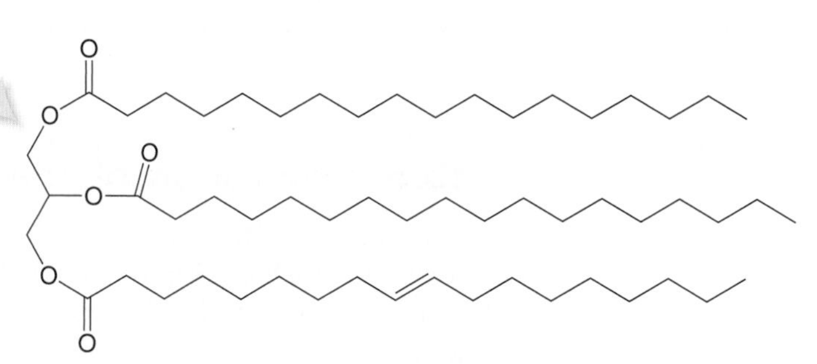

Foods rich in trans triacylglycerols: margarine, processed foods, fried foods, some baked goods

- one (or more) trans double bonds in the carbon chains
- semi-solid at room temperature
- increases blood cholesterol level

SAMPLE PROBLEM 15.4

Draw the products formed when the given triacylglycerol is hydrolyzed with water in the presence of sulfuric acid.

$$\begin{array}{l} CH_2-O-\overset{O}{\overset{\|}{C}}-(CH_2)_{16}CH_3 \\ | \\ CH-O-\overset{O}{\overset{\|}{C}}-(CH_2)_{14}CH_3 \\ | \\ CH_2-O-\overset{O}{\overset{\|}{C}}-(CH_2)_7CH{=}CH(CH_2)_5CH_3 \end{array}$$

Analysis

To draw the products of ester hydrolysis, cleave the three C—O single bonds at the carbonyl carbons to form glycerol and three fatty acids (RCOOH).

Solution

Hydrolysis forms glycerol and stearic, palmitic, and palmitoleic acids.

Bonds broken during hydrolysis

$$\begin{array}{l} CH_2-O-C(=O)-(CH_2)_{16}CH_3 \\ CH-O-C(=O)-(CH_2)_{14}CH_3 \\ CH_2-O-C(=O)-(CH_2)_7CH=CH(CH_2)_5CH_3 \end{array} \xrightarrow[H_2SO_4]{3\ H-OH} \underset{\text{glycerol}}{\begin{array}{l} CH_2-OH \\ CH-OH \\ CH_2-OH \end{array}} + \begin{array}{ll} HO-C(=O)-(CH_2)_{16}CH_3 & \text{stearic acid} \\ HO-C(=O)-(CH_2)_{14}CH_3 & \text{palmitic acid} \\ HO-C(=O)-(CH_2)_7CH=CH(CH_2)_5CH_3 & \text{palmitoleic acid} \end{array}$$

PROBLEM 15.11

Draw the products formed from hydrolysis of each triacylglycerol.

a. $$\begin{array}{l} CH_2-O-C(=O)-(CH_2)_{12}CH_3 \\ CH-O-C(=O)-(CH_2)_{12}CH_3 \\ CH_2-O-C(=O)-(CH_2)_{12}CH_3 \end{array}$$

b. $$\begin{array}{l} CH_2-O-C(=O)-(CH_2)_{12}CH_3 \\ CH-O-C(=O)-(CH_2)_7CH=CH(CH_2)_5CH_3 \\ CH_2-O-C(=O)-(CH_2)_7CH=CH(CH_2)_7CH_3 \end{array}$$

15.5A FOCUS ON THE HUMAN BODY Metabolism of Triacylglycerols

Humans store energy in the form of triacylglycerols, kept in a layer of fat cells, called **adipose cells,** below the surface of the skin, in bone marrow, in the breast area (of women), around the kidneys, and in the pelvis (Figure 15.2). In adulthood, the number of adipose cells is constant. When weight is lost or gained the amount of stored lipid in each cell changes, but the number of adipose cells does not change.

Figure 15.2

The Storage and Metabolism of Triacylglycerols

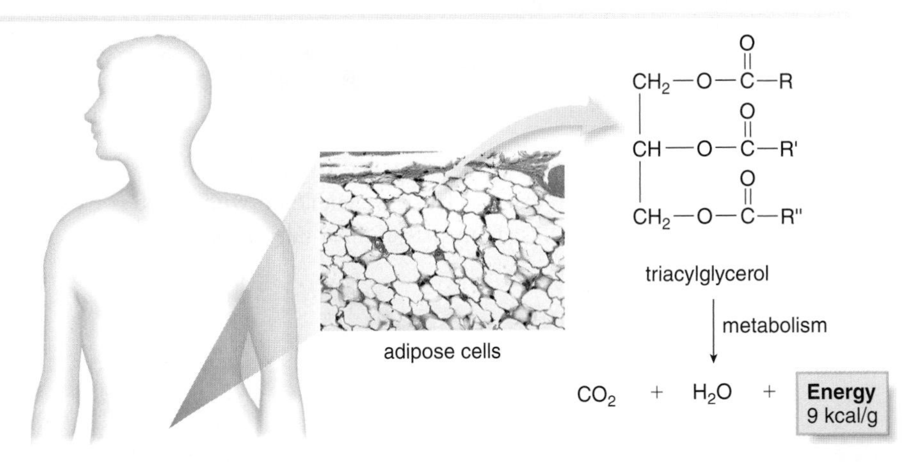

Triacylglycerols are stored in adipose cells below the skin and concentrated in some regions of the body. The average fat content of men and women is ~20% and ~25%, respectively. This stored fat provides two to three months of the body's energy needs.

A grizzly bear uses its stored body fat as its sole energy source during its many months of hibernation.

Adipose tissue serves to insulate the organism, as well as provide energy for its metabolic needs for long periods of time. The first step in the metabolism of a triacylglycerol is hydrolysis of the ester bonds to form glycerol and three fatty acids. **This reaction is simply ester hydrolysis.** In cells, this reaction is carried out with enzymes called **lipases.**

$$\underset{\text{triacylglycerol}}{\begin{array}{l} CH_2-O-C(=O)-R \\ CH-O-C(=O)-R' \\ CH_2-O-C(=O)-R'' \end{array}} \xrightarrow[\text{lipase}]{3\,H_2O} \underset{\text{glycerol}}{\begin{array}{l} CH_2-OH \\ CH-OH \\ CH_2-OH \end{array}} + \underbrace{HO-C(=O)-R + HO-C(=O)-R' + HO-C(=O)-R''}_{\text{three fatty acids}}$$

Complete metabolism of a triacylglycerol yields CO_2 and H_2O, and a great deal of energy. This overall reaction is reminiscent of the combustion of alkanes in fossil fuels, a process that also yields CO_2 and H_2O and provides energy to heat homes and power automobiles (Section 10.10). Fundamentally, both processes convert C—C and C—H bonds to C—O and O—H bonds, a highly exothermic reaction. Carbohydrates provide an energy boost, but only for the short term, such as during strenuous exercise. Our long-term energy needs are provided by triacylglycerols, because they store 9 kcal/g, whereas carbohydrates and proteins store only 4 kcal/g.

Because triacylglycerols release heat on combustion, they can in principle be used as fuels for vehicles. In fact, coconut oil was used as a fuel during both World Wars I and II, when gasoline and diesel supplies ran short. Since coconut oil is more viscous than petroleum products and freezes at 24 °C, engines must be modified to use it and it can't be used in cold climates. Nonetheless, a limited number of trucks and boats can now use vegetable oils, sometimes blended with diesel—*biodiesel*—as a fuel source.

ENVIRONMENTAL NOTE

When the price of crude oil is high, the use of **biofuels** such as biodiesel becomes economically attractive. Biofuels are prepared from renewable resources such as vegetable oils and animal fats.

15.5B Soap Synthesis

As we learned in Section 13.5, soaps are metal salts of carboxylic acids that contain many carbon atoms in a long hydrocarbon chain; that is, **soaps are metal salts of fatty acids.** For example, sodium stearate is the sodium salt of stearic acid, an 18-carbon saturated fatty acid.

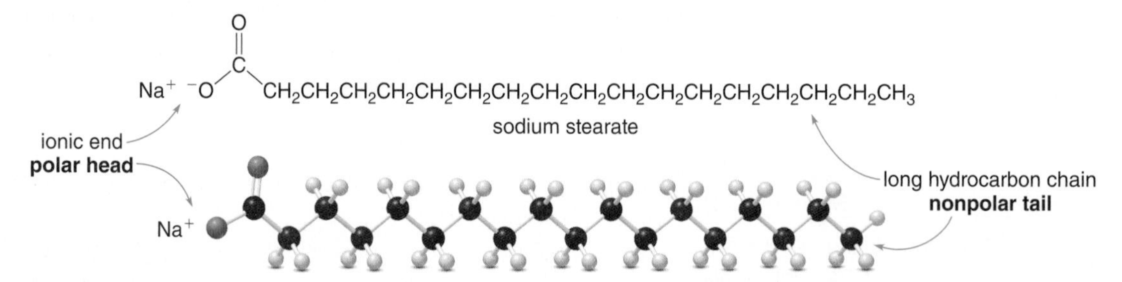

Soap is prepared by the basic hydrolysis (saponification) of a triacylglycerol. Heating an animal fat or vegetable oil with aqueous base hydrolyzes the three esters to form glycerol and sodium salts of three fatty acids.

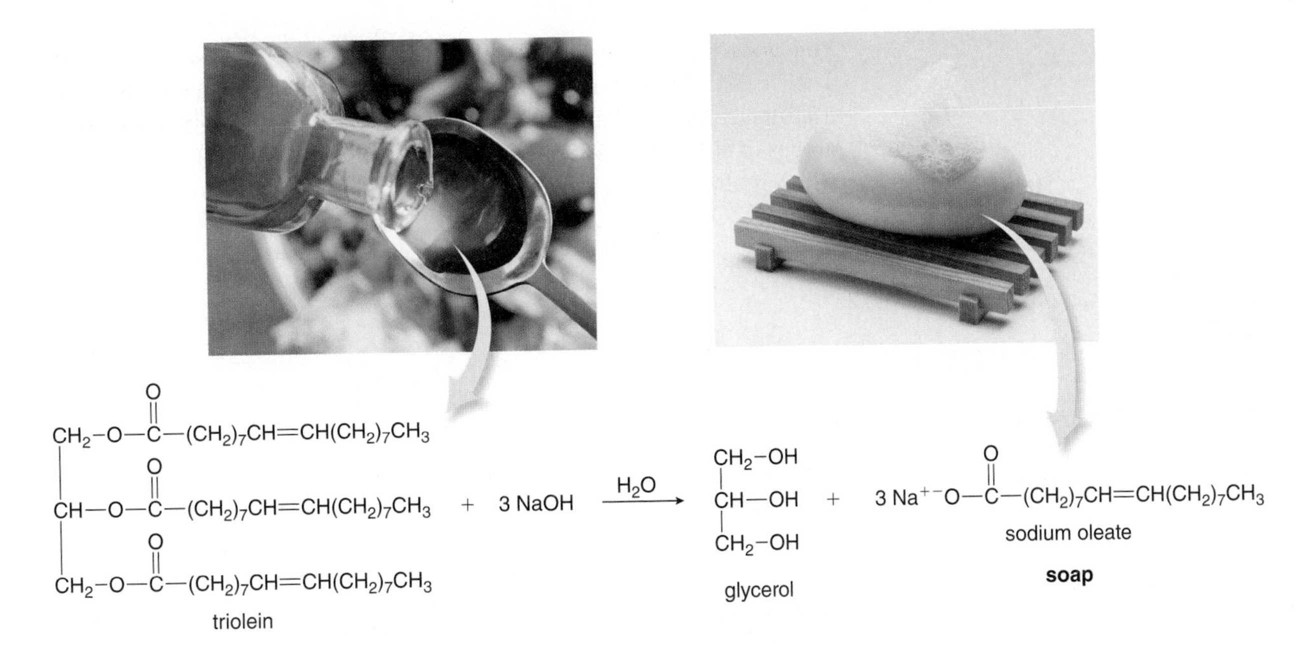

Saponification comes from the Latin *sapo* meaning *soap.*

These carboxylate salts are **soaps,** which clean away dirt, as shown in Figure 13.2. The nonpolar tail dissolves grease and oil and the polar head makes it soluble in water. Most triacylglycerols have two or three different R groups in their hydrocarbon chains, so soaps are usually mixtures of two or three different carboxylate salts.

PROBLEM 15.12

What is the composition of the soap formed by basic hydrolysis of each triacylglycerol?

a.
$$\begin{array}{l} CH_2-O-C(=O)-(CH_2)_{12}CH_3 \\ CH-O-C(=O)-(CH_2)_{12}CH_3 \\ CH_2-O-C(=O)-(CH_2)_{12}CH_3 \end{array}$$

b.
$$\begin{array}{l} CH_2-O-C(=O)-(CH_2)_{12}CH_3 \\ CH-O-C(=O)-(CH_2)_7CH=CH(CH_2)_5CH_3 \\ CH_2-O-C(=O)-(CH_2)_7CH=CH(CH_2)_7CH_3 \end{array}$$

15.6 Phospholipids

Phospholipids are lipids that contain a phosphorus atom. Phospholipids can be considered organic derivatives of phosphoric acid (H_3PO_4), formed by replacing two of the H atoms by R groups. This type of functional group is called a **phosphodiester.** In cells, the remaining OH group on phosphorus loses its proton, giving the phosphodiester a net negative charge.

$$H-O-P(=O)(OH)-O-H$$
phosphoric acid
H_3PO_4

$$R-O-P(=O)(OH)-O-R'$$
phosphodiester

$$R-O-P(=O)(O^-)-O-R'$$
phosphodiester in cells

Phosphoacylglycerols (or **phosphoglycerides**) are the most common type of phospholipid. They form the principal lipid component of most cell membranes. Their structure resembles the triacylglycerols of the preceding section with one important difference. Only *two* of the hydroxyl groups of glycerol are esterified with fatty acids. The third OH group is part of a phosphodiester, which is also bonded to an alkyl group (R") derived from a low molecular weight alcohol.

General structure of a phosphoacylglycerol

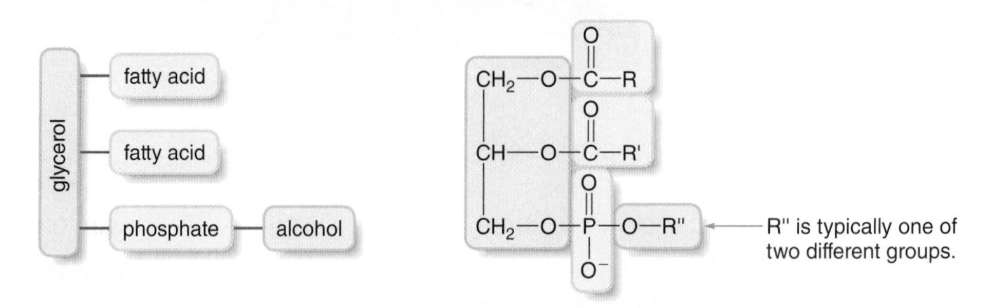

There are two prominent types of phosphoacylglycerols. They differ in the identity of the R" group in the phosphodiester.

- **When R" = $CH_2CH_2NH_3^+$, the compound is called a *cephalin.***
- **When R" = $CH_2CH_2N(CH_3)_3^+$, the compound is called a *lecithin.***

cephalin

Derived from the alcohol: $H{-}OCH_2CH_2\overset{+}{N}H_3$

ethanolamine

lecithin

Derived from the alcohol: $H{-}OCH_2CH_2\overset{+}{N}(CH_3)_3$

choline

The phosphorus side chain of a phosphoacylglycerol makes it different from a triacylglycerol. **The two fatty acid side chains form two nonpolar "tails" that lie parallel to each other, while the phosphodiester end of the molecule is a charged or polar "head."** A three-dimensional structure of a phosphoacylglycerol is shown in Figure 15.3.

SAMPLE PROBLEM 15.5

Draw the structure of a cephalin formed from two molecules of stearic acid.

Analysis

Substitute the 18-carbon saturated fatty acid stearic acid for the R and R' groups in the general structure of a cephalin molecule. In a cephalin, $-CH_2CH_2NH_3^+$ forms part of the phosphodiester.

Solution

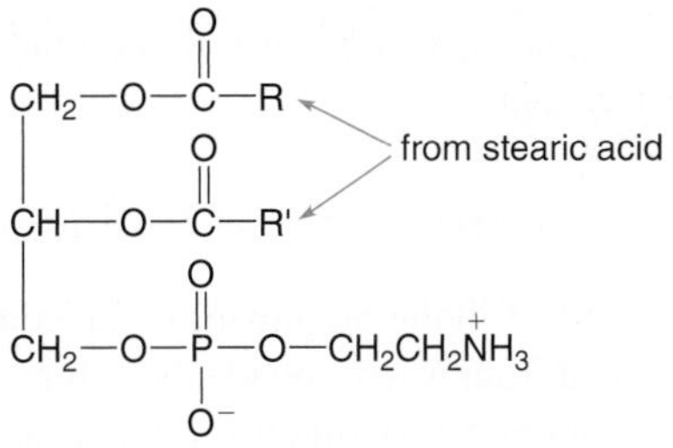

general structure:

$$\begin{array}{l} CH_2{-}O{-}\overset{O}{\overset{\|}{C}}{-}R \\ | \\ CH{-}O{-}\overset{O}{\overset{\|}{C}}{-}R' \\ | \\ CH_2{-}O{-}\overset{O}{\overset{\|}{P}}{-}O{-}CH_2CH_2\overset{+}{N}H_3 \\ \qquad\qquad | \\ \qquad\qquad O^- \end{array}$$

R and R' from stearic acid

general structure

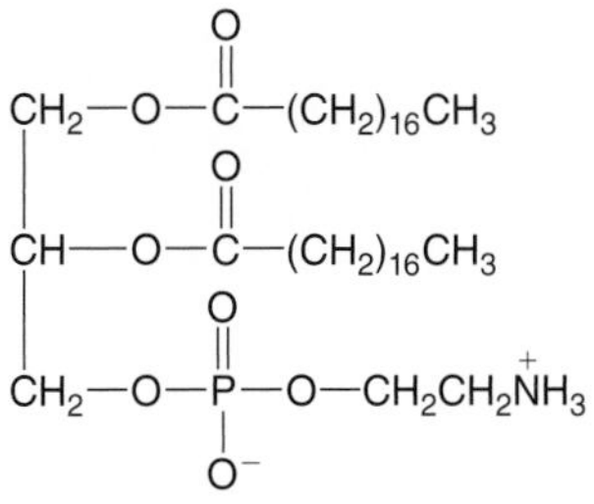

$$\begin{array}{l} CH_2{-}O{-}\overset{O}{\overset{\|}{C}}{-}(CH_2)_{16}CH_3 \\ | \\ CH{-}O{-}\overset{O}{\overset{\|}{C}}{-}(CH_2)_{16}CH_3 \\ | \\ CH_2{-}O{-}\overset{O}{\overset{\|}{P}}{-}O{-}CH_2CH_2\overset{+}{N}H_3 \\ \qquad\qquad | \\ \qquad\qquad O^- \end{array}$$

Answer

PROBLEM 15.13

Draw the structure of two different cephalins containing oleic acid and palmitic acid as fatty acid side chains.

PROBLEM 15.14

Classify each lipid as a triacylglycerol, cephalin, or lecithin.

a.

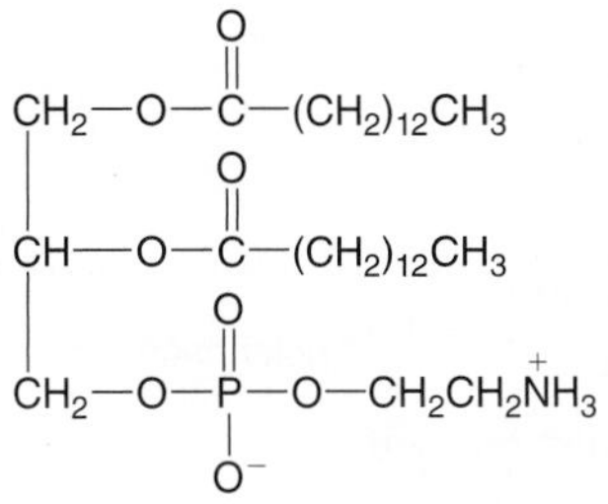

$$\begin{array}{l} CH_2{-}O{-}\overset{O}{\overset{\|}{C}}{-}(CH_2)_{12}CH_3 \\ | \\ CH{-}O{-}\overset{O}{\overset{\|}{C}}{-}(CH_2)_{12}CH_3 \\ | \\ CH_2{-}O{-}\overset{O}{\overset{\|}{P}}{-}O{-}CH_2CH_2\overset{+}{N}H_3 \\ \qquad\qquad | \\ \qquad\qquad O^- \end{array}$$

b.

$$\begin{array}{l} CH_2{-}O{-}\overset{O}{\overset{\|}{C}}{-}(CH_2)_{12}CH_3 \\ | \\ CH{-}O{-}\overset{O}{\overset{\|}{C}}{-}(CH_2)_7CH{=}CH(CH_2)_5CH_3 \\ | \\ CH_2{-}O{-}\overset{O}{\overset{\|}{C}}{-}(CH_2)_{12}CH_3 \end{array}$$

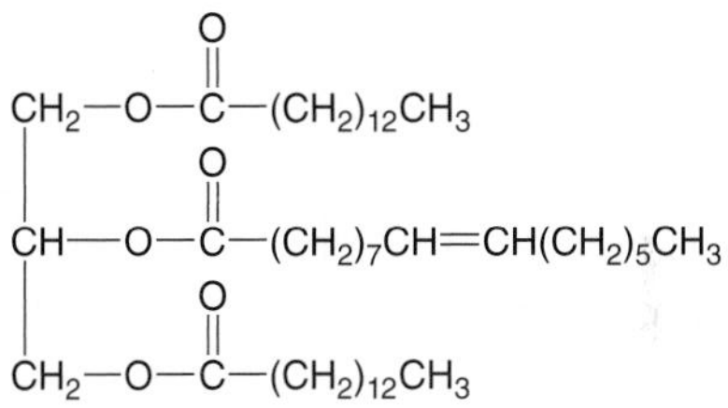

Figure 15.3 Three-Dimensional Structure of a Phosphoacylglycerol

$$\begin{array}{l} \qquad\qquad\qquad\qquad\quad CH_2{-}O{-}\overset{O}{\overset{\|}{C}}{-}CH_2CH_2CH_2CH_2CH_2CH_2CH_2CH_2CH_2CH_2CH_2CH_2CH_2CH_2CH_2CH_2CH_3 \\ \qquad\qquad\qquad\qquad\quad | \\ \qquad\qquad\qquad\qquad\quad CH{-}O{-}\overset{O}{\overset{\|}{C}}{-}CH_2CH_2CH_2CH_2CH_2CH_2CH_2CH_2CH_2CH_2CH_2CH_2CH_2CH_2CH_2CH_2CH_3 \\ \qquad\qquad\qquad\qquad\quad | \\ (CH_3)_3\overset{+}{N}CH_2CH_2{-}O{-}\overset{O}{\overset{\|}{P}}{-}O{-}CH_2 \\ \qquad\qquad\qquad\quad | \\ \qquad\qquad\qquad\quad O^- \end{array}$$

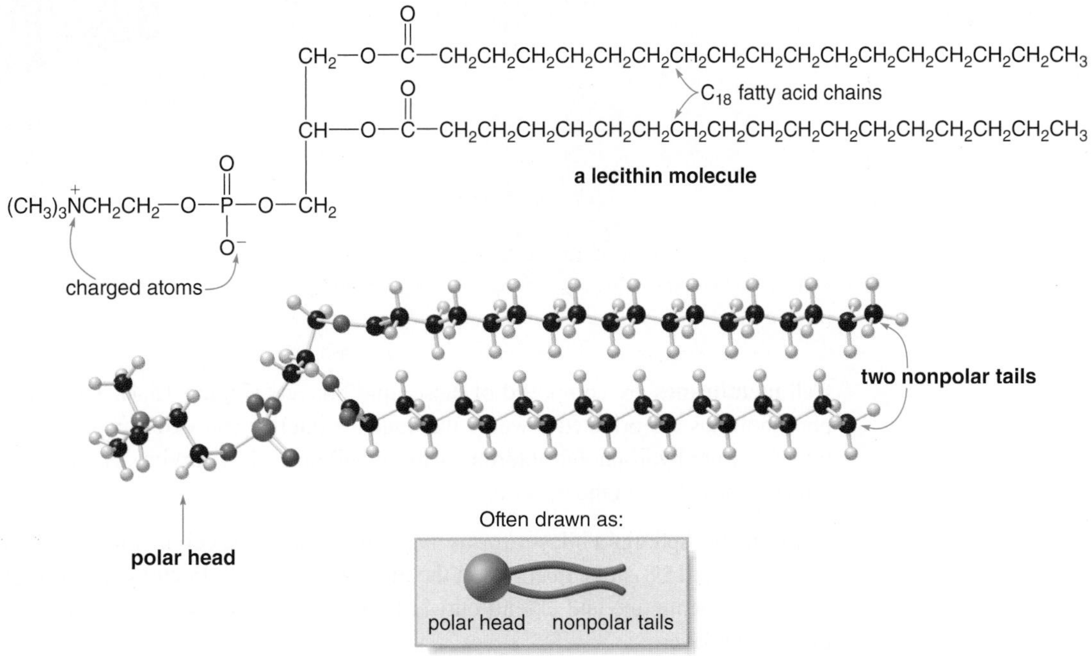

C18 fatty acid chains

a lecithin molecule

charged atoms

two nonpolar tails

polar head

Often drawn as:

polar head nonpolar tails

A phosphoacylglycerol has two distinct regions: two nonpolar tails due to the long-chain fatty acids, and a very polar head from the charged phosphodiester.

15.7 Cell Membranes

The cell membrane is a beautifully complex example of how chemistry comes into play in a biological system.

15.7A Structure of the Cell Membrane

The basic unit of living organisms is the **cell.** The cytoplasm is the aqueous medium inside the cell, separated from water outside the cell by the **cell membrane.** The cell membrane serves two apparently contradictory functions. It acts as a barrier to the passage of ions and molecules into and out of the cell, but it is also selectively permeable, allowing nutrients in and waste out.

Phospholipids are the major component of the cell membrane. **Phospholipids contain a hydrophilic polar head and two nonpolar tails composed of C—C and C—H bonds.** When phospholipids are mixed with water, they assemble in an arrangement called a **lipid bilayer,** with the ionic heads oriented on the outside and the nonpolar tails on the inside. The polar heads electrostatically interact with the polar solvent H_2O, while the nonpolar tails are held in close proximity by numerous London dispersion forces.

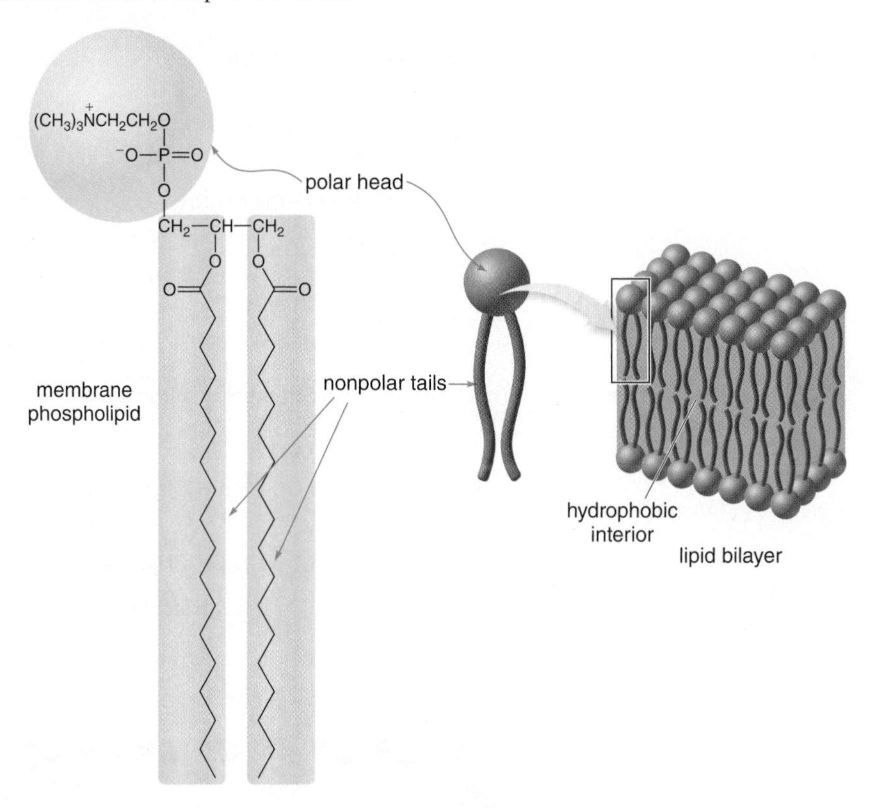

Cell membranes are composed of these lipid bilayers (Figure 15.4). The charged heads of the phospholipids are oriented towards the aqueous interior and exterior of the cell. The nonpolar tails form the hydrophobic interior of the membrane, thus serving as an insoluble barrier that protects the cell from the outside.

While the phospholipid bilayer forms the main fabric of the cell membrane, proteins and cholesterol (Section 15.8) are embedded in the membrane as well. **Peripheral proteins** are embedded within the membrane and extend outward on one side only. **Integral proteins** extend through the entire bilayer.

PROBLEM 15.15

Why are phospholipids rather than triacylglycerols present in cell membranes?

Figure 15.4

Composition of the Cell Membrane

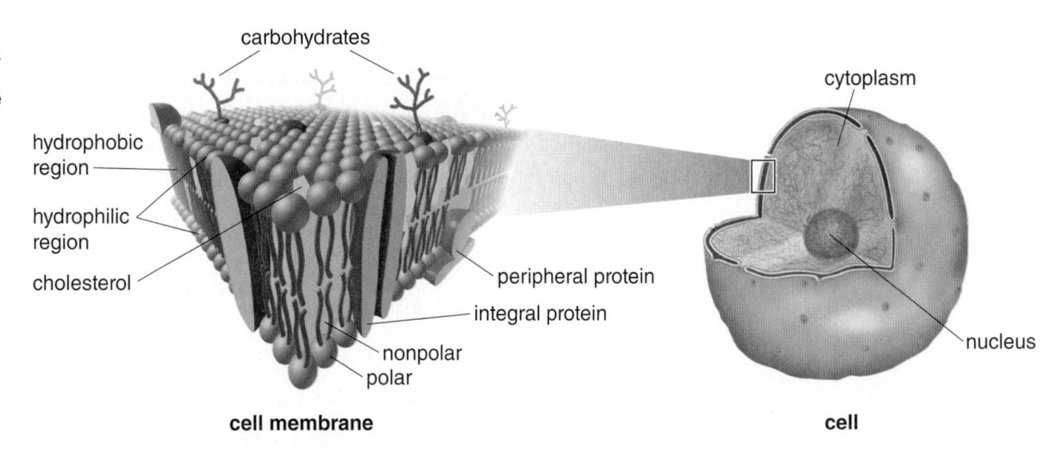

Cell membranes are composed of a lipid bilayer having the hydrophilic polar heads of phospholipids arranged on the exterior of the bilayer, where they can interact with the polar aqueous environment inside and outside the cell. The hydrophobic tails of the phospholipid are arranged in the interior of the bilayer, forming a "greasy" layer that is only selectively permeable to the passage of species from one side to the other.

15.7B Transport Across a Cell Membrane

How does a molecule or ion in the water on one side of a cell membrane pass through the non-polar interior of the cell membrane to the other side? A variety of transport mechanisms occur (Figure 15.5).

Small molecules like O_2 and CO_2 can simply diffuse through the cell membrane, traveling from the side of higher concentration to the side of lower concentration. With larger polar molecules and some ions, simple diffusion is too slow or not possible, so a process of **facilitated transport** occurs. Ions such as Cl^- or HCO_3^- and glucose molecules travel through the channels created by integral proteins.

Figure 15.5

How Substances Cross a Cell Membrane

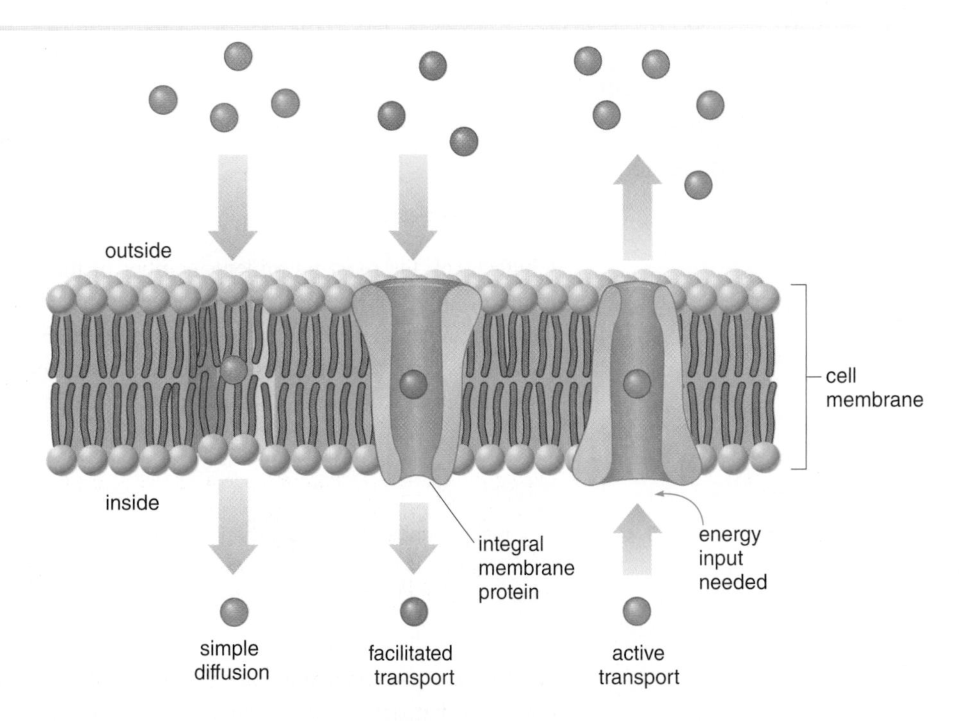

Some ions, notably Na^+, K^+, and Ca^{2+}, must move across the cell membrane against the concentration gradient—that is, from a region of lower concentration to a region of higher concentration. To move an ion across the membrane in this fashion requires energy input, and the process is called **active transport.** Active transport occurs whenever a nerve impulse causes a muscle to contract. In this process, energy is supplied to move K^+ ions from outside to inside a cell, against a concentration gradient.

PROBLEM 15.16

Why don't ions readily diffuse through the interior of the cell membrane?

15.8 FOCUS ON HEALTH & MEDICINE Cholesterol, the Most Prominent Steroid

The steroids are a group of lipids whose carbon skeletons contain three six-membered rings and one five-membered ring. This tetracyclic carbon skeleton is drawn below.

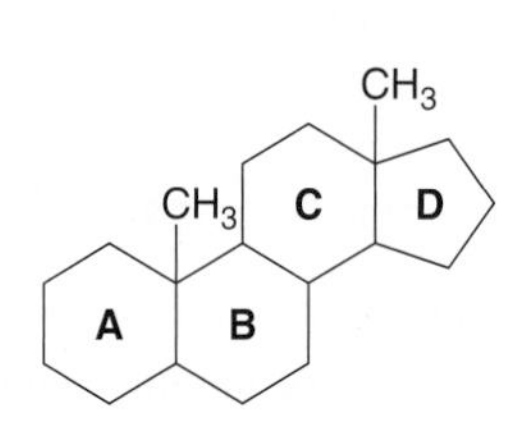

steroid skeleton

numbering the steroid skeleton

Many steroids also contain two methyl groups that are bonded to the rings. The steroid rings are lettered **A, B, C,** and **D,** and the 17 ring carbons are numbered as shown. The two methyl groups are numbered C18 and C19. Steroids differ in the identity and location of the substituents attached to the skeleton.

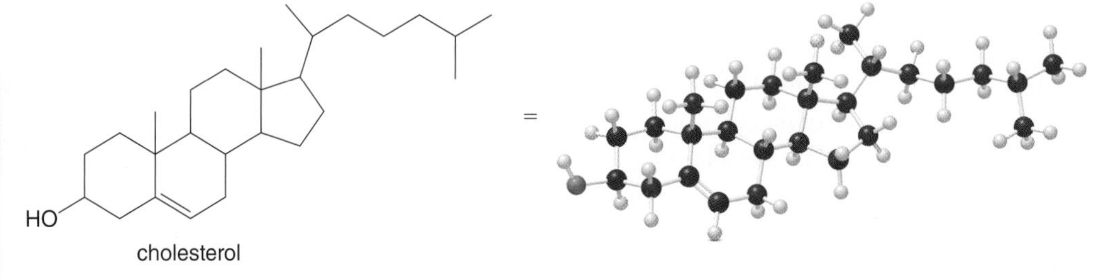

cholesterol

Cholesterol, the most prominent member of the steroid family, is synthesized in the liver and found in almost all body tissues. Cholesterol is obtained in the diet from a variety of sources, including meat, cheese, butter, and eggs. Table 15.3 lists the cholesterol content in some foods. While the American Heart Association currently recommends that the daily intake of cholesterol should be less than 300 mg, the average American diet includes 400–500 mg of cholesterol each day.

PROBLEM 15.17

(a) Label the rings of the steroid nucleus in cholesterol. (b) Give the number of the carbon to which the OH group is bonded. (c) Between which two carbons is the double bond located? (d) Label the polar bonds in cholesterol and explain why it is insoluble in water.

While health experts agree that the amount of cholesterol in the diet should be limited, it is also now clear that elevated *blood* cholesterol (serum cholesterol) can lead to coronary artery disease. High blood cholesterol levels are associated with an increased risk of developing coronary artery

HEALTH NOTE

Plants do not synthesize cholesterol, so fresh fruits and vegetables, nuts, and whole grains are cholesterol free.

Table 15.3 Cholesterol Content in Some Foods

Food	Serving Size	Cholesterol (mg)
Boiled egg	1	225
Cream cheese	1 oz	27
Cheddar cheese	1 oz	19
Butter	3.5 oz	250
Beefsteak	3.5 oz	70
Chicken	3.5 oz	60
Ice cream	3.5 oz	45
Sponge cake	3.5 oz	260

disease, heart attack, and stroke. To understand the relationship between cholesterol and heart disease, we must learn about how cholesterol is transported through the bloodstream.

Like other lipids, cholesterol is insoluble in the aqueous medium of the blood, since it has only one polar OH group and many nonpolar C—C and C—H bonds. In order for it to be transported from the liver where it is synthesized, to the tissues, cholesterol combines with phospholipids and proteins to form small water-soluble spherical particles called **lipoproteins.**

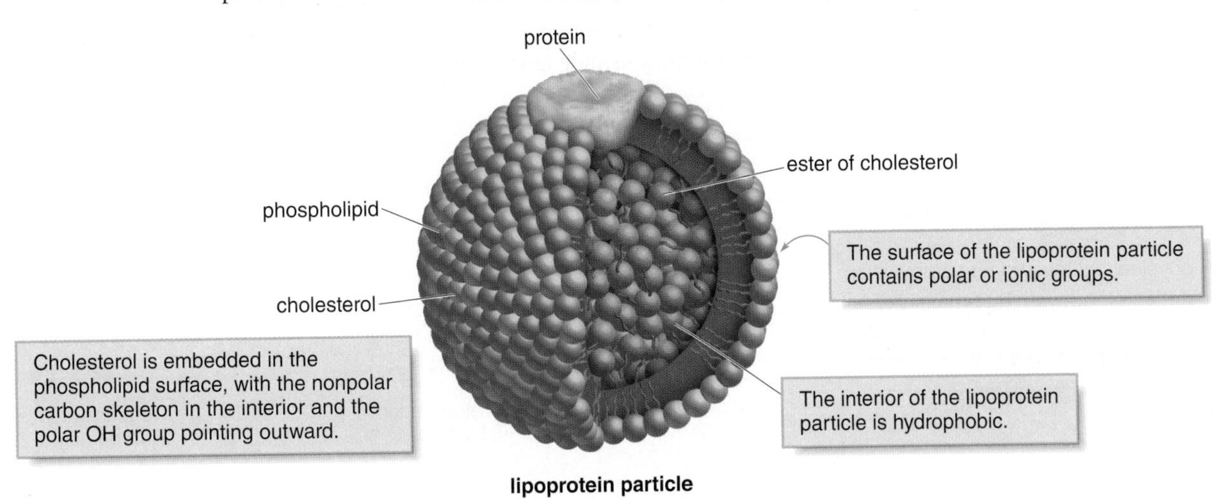
protein
ester of cholesterol
phospholipid
cholesterol
The surface of the lipoprotein particle contains polar or ionic groups.
Cholesterol is embedded in the phospholipid surface, with the nonpolar carbon skeleton in the interior and the polar OH group pointing outward.
The interior of the lipoprotein particle is hydrophobic.
lipoprotein particle

In a lipoprotein, the polar heads of phospholipids and the polar portions of protein molecules are arranged on the surface. The nonpolar molecules are buried in the interior of the particle. In this way, the nonpolar material is "dissolved" in an aqueous environment.

Lipoproteins are classified on the basis of their density, with two types being especially important in determining serum cholesterol levels.

- **Low-density lipoproteins (LDLs) transport cholesterol from the liver to the tissues.**
- **High-density lipoproteins (HDLs) transport cholesterol from the tissues back to the liver.**

LDL particles transport cholesterol to tissues where it is incorporated in cell membranes. When LDLs supply more cholesterol than is needed, LDLs deposit cholesterol on the wall of arteries, forming plaque (Figure 15.6). Atherosclerosis is a disease that results from the buildup of these

Figure 15.6 Plaque Formation in an Artery

a. Open artery

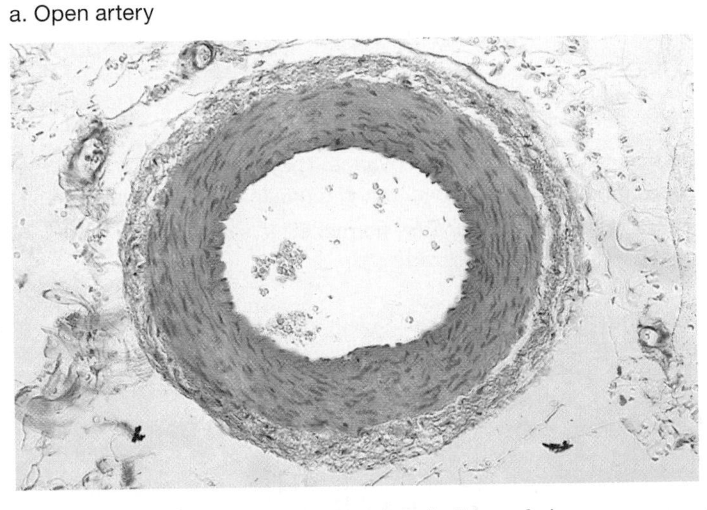

b. Blocked artery

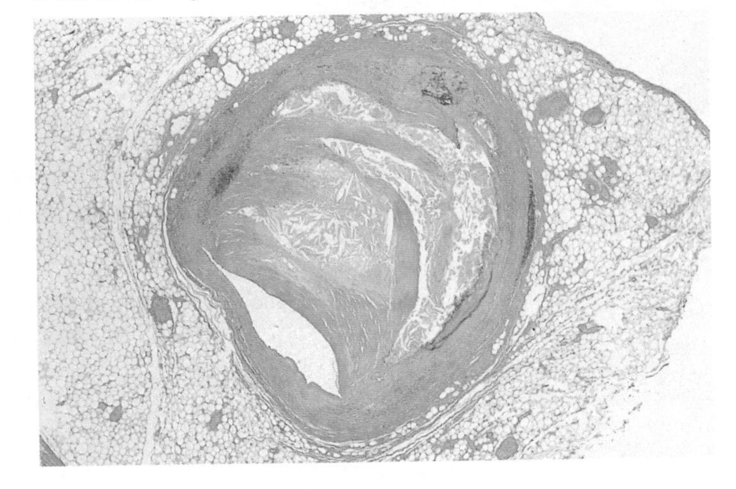

a. Cross-section of a clear artery with no buildup of plaque
b. Artery almost completely blocked by the buildup of plaque

HEALTH NOTE

A physical examination by a physician includes blood work that measures three quantities: total serum cholesterol, HDL cholesterol, and LDL cholesterol.

fatty deposits, restricting the flow of blood, increasing blood pressure, and increasing the likelihood of a heart attack or stroke. As a result, LDL cholesterol is often called "bad" cholesterol.

HDL particles transport excess cholesterol from the tissues back to the liver, where it is converted to other substances or eliminated. Thus, HDLs reduce the level of serum cholesterol, so HDL cholesterol is often called "good" cholesterol.

Several drugs called **statins** are now available to reduce the level of cholesterol in the bloodstream. These compounds act by blocking the synthesis of cholesterol at its very early stages. Two examples include atorvastatin (Lipitor) and simvastatin (Zocor).

Generic name: atorvastatin
Trade name: Lipitor

Generic name: simvastatin
Trade name: Zocor

PROBLEM 15.18

Would you expect triacylglycerols to be contained in the interior of a lipoprotein particle or on the surface with the phospholipids? Explain your choice.

15.9 Steroid Hormones

HEALTH NOTE

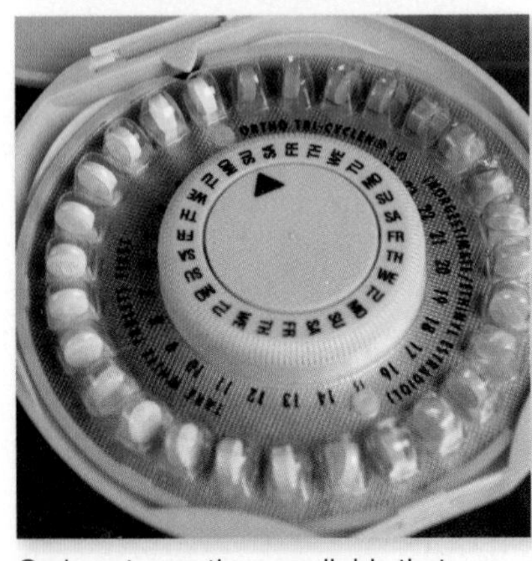

Oral contraceptives are lipids that artificially elevate hormone levels in a woman, thereby preventing pregnancy (Section 11.4).

Many biologically active steroids are hormones secreted by the endocrine glands. **A *hormone* is a molecule that is synthesized in one part of an organism, which then elicits a response at a different site.** Two important classes of steroid hormones are the **sex hormones** and the **adrenal cortical steroids.**

There are two types of female sex hormones, **estrogens** and **progestins.**

estradiol

estrone

progesterone

- Estradiol and estrone are estrogens synthesized in the ovaries. They control the development of secondary sex characteristics in females and regulate the menstrual cycle.
- Progesterone is a progestin often called the "pregnancy hormone." It is responsible for the preparation of the uterus for implantation of a fertilized egg.

The male sex hormones are called **androgens.**

testosterone

androsterone

HEALTH NOTE

Some body builders use anabolic steroids to increase muscle mass. Long-term or excessive use can cause many health problems, including high blood pressure, liver damage, and cardiovascular disease.

- **Testosterone and androsterone are androgens synthesized in the testes. They control the development of secondary sex characteristics in males—growth of facial hair, increase in muscle mass, and deepening of the voice.**

Synthetic androgen analogues, called **anabolic steroids,** promote muscle growth. They were first developed to help individuals whose muscles had atrophied from lack of use following surgery. They have since come to be used by athletes and body builders, although their use is not permitted in competitive sports. Many physical and psychological problems result from their prolonged use.

Anabolic steroids, such as stanozolol, nandrolone, and tetrahydrogestrinone have the same effect on the body as testosterone, but they are more stable, so they are not metabolized as quickly. Tetrahydrogestrinone (also called THG or The Clear), the performance-enhancing drug used by track star Marion Jones during the 2000 Sydney Olympics, was considered a "designer steroid" because it was initially undetected in urine tests for doping. After its chemical structure and properties were determined, it was added to the list of banned anabolic steroids in 2004.

stanozolol nandrolone tetrahydrogestrinone

A second group of steroid hormones includes the **adrenal cortical steroids.** Three examples of these hormones are aldosterone, cortisone, and cortisol. All of these compounds are synthesized in the outer layer of the adrenal gland. Aldosterone regulates blood pressure and volume by controlling the concentration of Na^+ and K^+ in body fluids. Cortisone and cortisol serve as anti-inflammatory agents and they regulate carbohydrate metabolism.

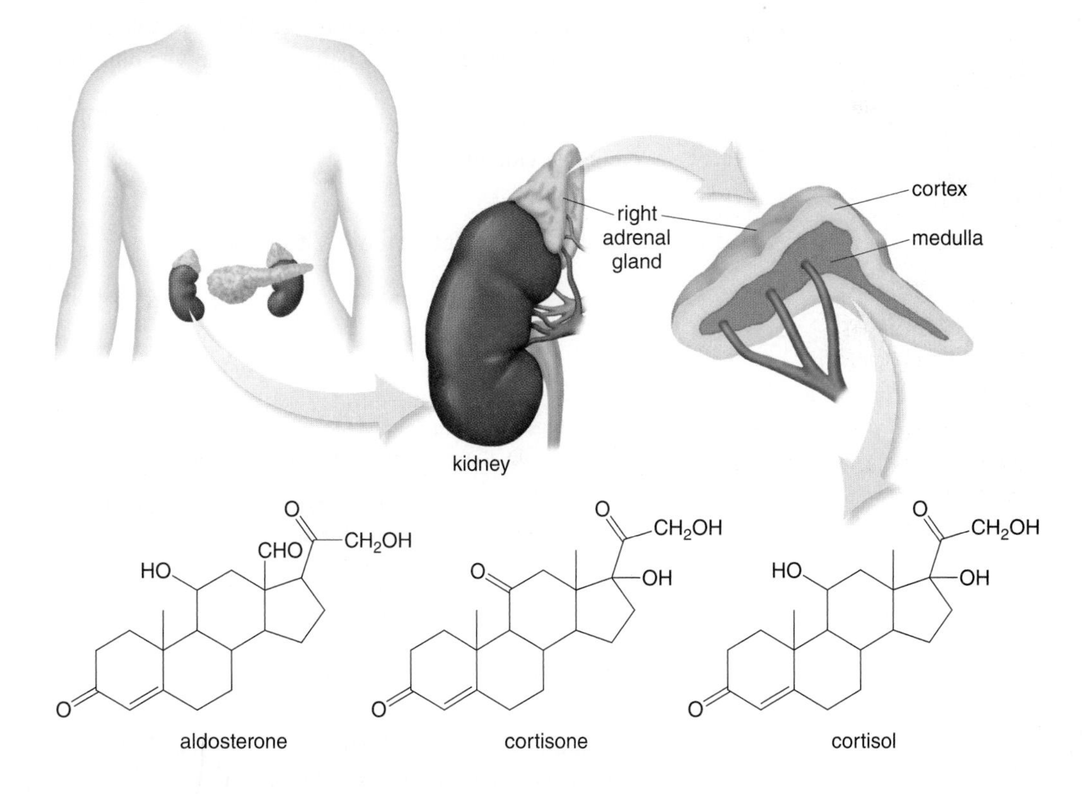

Cortisone and related compounds are used to suppress organ rejection after transplant surgery and to treat many allergic and autoimmune disorders. Prolonged use of these steroids can have undesired side effects, including bone loss and high blood pressure. Prednisone, a widely used synthetic alternative, has similar anti-inflammatory properties but can be taken orally.

prednisone
(synthetic steroid)

PROBLEM 15.19

Compare the structures of estrone and progesterone. (a) Identify the differences in the A ring of these hormones. (b) How do these hormones differ in functionality at C17?

PROBLEM 15.20

Identify the functional groups in aldosterone. Classify each alcohol as 1°, 2°, or 3°.

15.10 FOCUS ON HEALTH & MEDICINE Fat-Soluble Vitamins

Vitamins are organic compounds required in small quantities for normal metabolism. Since our cells cannot synthesize these compounds, they must be obtained in the diet. Vitamins can be categorized as fat soluble or water soluble. **The fat-soluble vitamins are lipids.**

The four fat-soluble vitamins—**A, D, E,** and **K**—are found in fruits and vegetables, fish, liver, and dairy products. Although fat-soluble vitamins must be obtained from the diet, they do not have to be ingested every day. Excess vitamins are stored in adipose cells, and then used when needed. Table 15.4 summarizes the dietary sources and recommended daily intake of the fat-soluble vitamins.

Table 15.4 Fat-Soluble Vitamins

Vitamin	Food Source	Recommended Daily Intake
A	Liver, kidney, oily fish, dairy products, eggs, fortified breakfast cereals	900 μg (men) 700 μg (women)
D	Fortified milk and breakfast cereals	5 μg
E	Sunflower and safflower oils, nuts, beans, whole grains, leafy greens	15 mg
K	Cauliflower, soybeans, broccoli, leafy greens, green tea	120 μg (men) 90 μg (women)

Source: Data from Harvard School of Public Health.

Vitamin A is obtained from liver, oily fish, and dairy products, and is synthesized from β-carotene, the orange pigment in carrots. In the body, vitamin A is converted to 11-*cis*-retinal, the light-sensitive compound responsible for vision in all vertebrates. It is also needed for healthy mucous membranes. A deficiency of vitamin A causes night blindness, as well as dry eyes and skin.

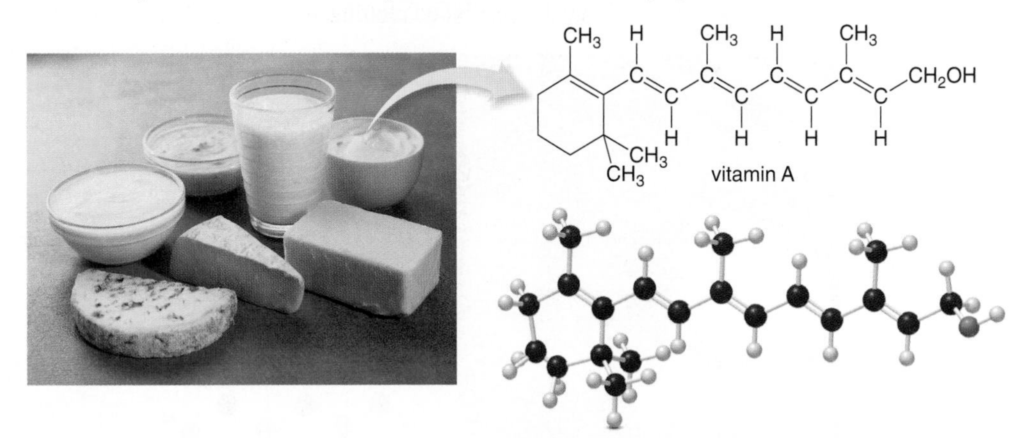

Vitamin D, strictly speaking, is not a vitamin because it can be synthesized in the body from cholesterol. Nevertheless, it is classified as such, and many foods (particularly milk) are fortified with vitamin D so that we get enough of this vital nutrient. Vitamin D helps regulate both calcium and phosphorus metabolism. A deficiency of vitamin D causes rickets, a bone disease characterized by knock-knees, spinal curvature, and other skeletal deformities.

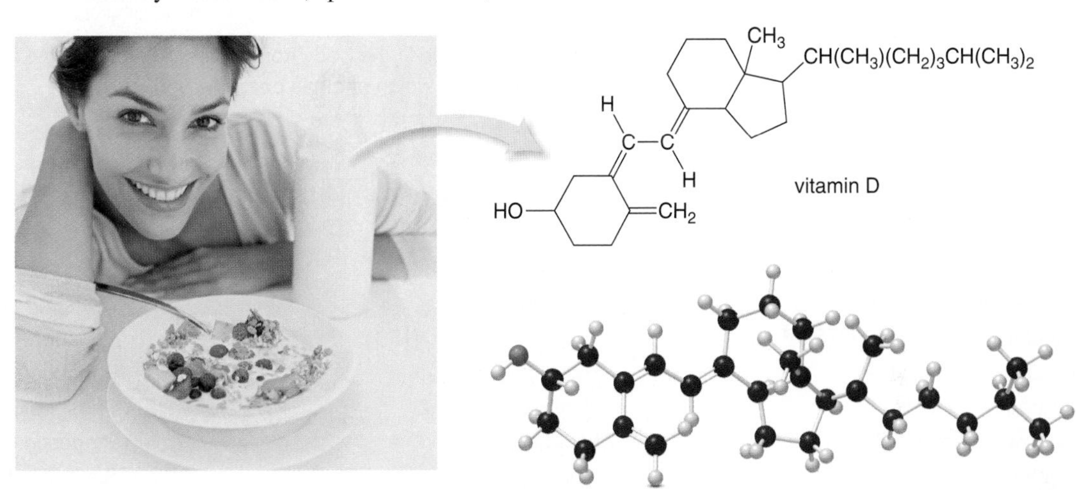

Vitamin E is an antioxidant, and in this way it protects unsaturated side chains in fatty acids from unwanted oxidation. A deficiency of vitamin E causes numerous neurological problems, although it is rare for vitamin E deficiency to occur.

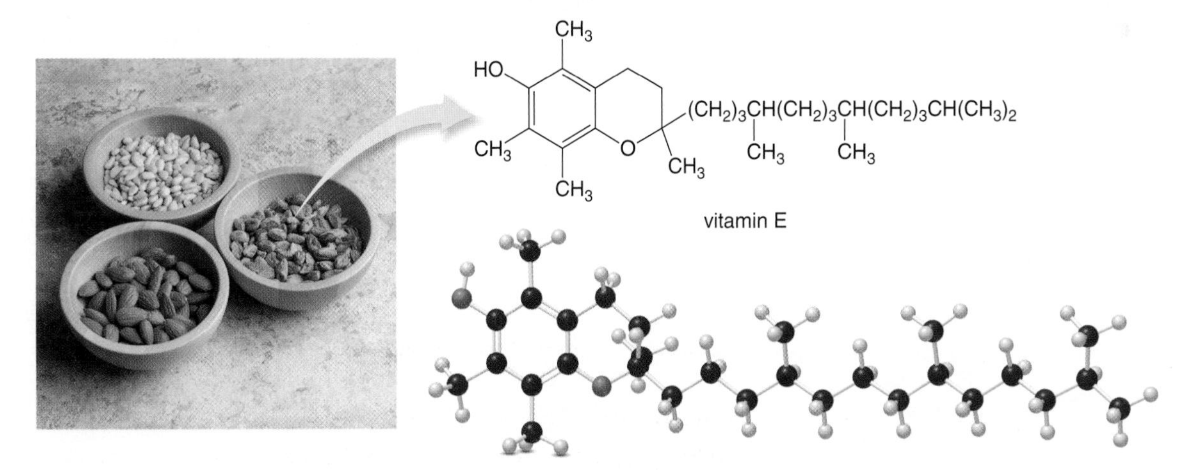

Vitamin K regulates the synthesis of prothrombin and other proteins needed for blood to clot. A severe deficiency of vitamin K leads to excessive and sometimes fatal bleeding because of inadequate blood clotting.

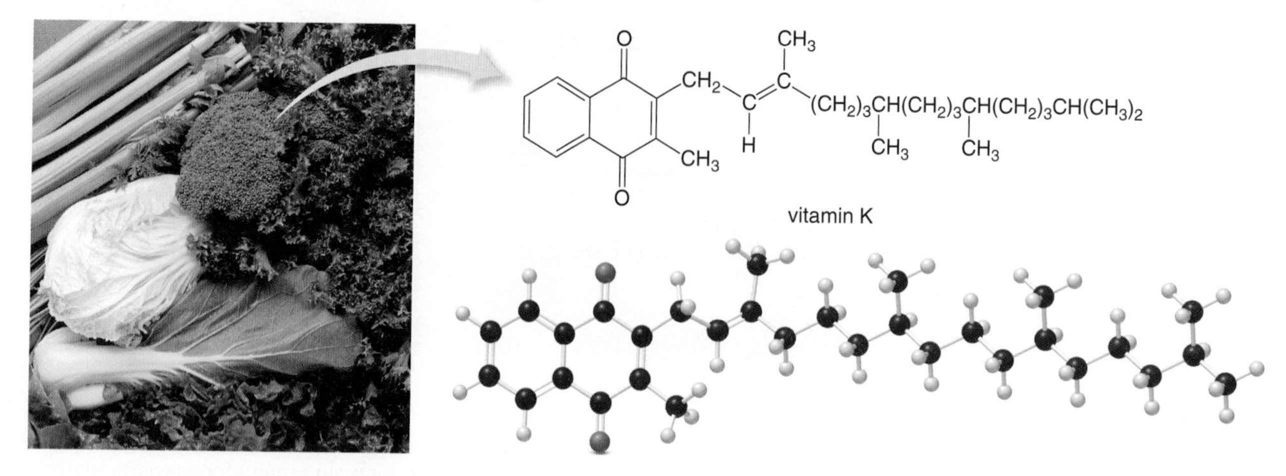

PROBLEM 15.21

Why is it much easier to overdose on a fat-soluble vitamin than a water-soluble vitamin?

PROBLEM 15.22

Vitamin D is synthesized in the body from a steroid. Which of the steroid rings—A, B, C, or D—are intact in vitamin D and which ring has been cleaved?

KEY TERMS

Active transport (15.7)
Adrenal cortical steroid (15.9)
Anabolic steroid (15.9)
Androgen (15.9)
Cell membrane (15.7)
Cephalin (15.6)
Estrogen (15.9)
Facilitated transport (15.7)
Fat (15.4)
Fat-soluble vitamin (15.10)
Fatty acid (15.2)
High-density lipoprotein (15.8)
Hormone (15.9)
Hydrolyzable lipid (15.1)
Hydrophilic (15.2)
Hydrophobic (15.2)
Lecithin (15.6)
Lipase (15.5)
Lipid (15.1)
Lipid bilayer (15.7)
Lipoprotein (15.8)
Low-density lipoprotein (15.8)
Nonhydrolyzable lipid (15.1)
Oil (15.4)
Omega-*n* acid (15.2)
Phosphoacylglycerol (15.6)
Phosphodiester (15.6)
Phospholipid (15.6)
Progestin (15.9)
Saponification (15.5)
Saturated fatty acid (15.2)
Soap (15.5)
Steroid (15.8)
Triacylglycerol (15.4)
Unsaturated fatty acid (15.2)
Wax (15.3)

KEY REACTIONS

[1] Hydrolysis of waxes (15.3)

$$\underset{\text{wax}}{RCOOR'} + H_2O \xrightarrow{H_2SO_4} \underset{\text{fatty acid}}{RCOOH} + \underset{\text{alcohol}}{H{-}OR'}$$

[2] Hydrolysis of triacylglycerols in the presence of acid or enzymes (15.5)

$$\underset{\text{triacylglycerol}}{\begin{array}{l} CH_2-O-\overset{O}{\overset{\|}{C}}-R \\ CH-O-\overset{O}{\overset{\|}{C}}-R' \\ CH_2-O-\overset{O}{\overset{\|}{C}}-R'' \end{array}} + 3\,H_2O \xrightarrow[\text{lipase}]{H_2SO_4 \text{ or}} \underset{\text{glycerol}}{\begin{array}{l} CH_2-OH \\ CH-OH \\ CH_2-OH \end{array}} + \underbrace{HO-\overset{O}{\overset{\|}{C}}-R + HO-\overset{O}{\overset{\|}{C}}-R' + HO-\overset{O}{\overset{\|}{C}}-R''}_{\text{three fatty acids}}$$

[3] Hydrolysis of triacylglycerols in the presence of base—Saponification (15.5)

$$\underset{\text{triacylglycerol}}{\begin{array}{l} CH_2-O-\overset{O}{\overset{\|}{C}}-R \\ CH-O-\overset{O}{\overset{\|}{C}}-R' \\ CH_2-O-\overset{O}{\overset{\|}{C}}-R'' \end{array}} + 3\,NaOH \xrightarrow{H_2O} \underset{\text{glycerol}}{\begin{array}{l} CH_2-OH \\ CH-OH \\ CH_2-OH \end{array}} + \underbrace{Na^{+\,-}O-\overset{O}{\overset{\|}{C}}-R + Na^{+\,-}O-\overset{O}{\overset{\|}{C}}-R' + Na^{+\,-}O-\overset{O}{\overset{\|}{C}}-R''}_{\text{soaps}}$$

KEY CONCEPTS

1 What are the general characteristics of lipids? (15.1)

- Lipids are biomolecules that contain many nonpolar C—C and C—H bonds, making them soluble in organic solvents and insoluble in water.
- Hydrolyzable lipids, including waxes, triacylglycerols, and phospholipids, can be converted to smaller molecules on reaction with water.
- Nonhydrolyzable lipids, including steroids and fat-soluble vitamins, cannot be cleaved into smaller units by hydrolysis.

2 How are fatty acids classified and what is the relationship between their melting points and the number of double bonds they contain? (15.2)

- Fatty acids are saturated if they contain no carbon–carbon double bonds and unsaturated if they contain one or more double bonds. Naturally occurring unsaturated fatty acids generally contain cis double bonds.
- As the number of double bonds in the fatty acid increases, its melting point decreases.

3 What are waxes? (15.3)

- A wax is an ester (RCOOR') formed from a fatty acid (RCOOH) and a high molecular weight alcohol (R'OH).
- Waxes (RCOOR') are hydrolyzed to fatty acids (RCOOH) and alcohols (R'OH).

4 What are triacylglycerols, and how do the triacylglycerols in a fat and oil differ? (15.4)

- Triacylglycerols, or triglycerides, are triesters formed from glycerol and three molecules of fatty acids.
- Fats are triacylglycerols derived from fatty acids having few double bonds, making them solids at room temperature.
- Oils are triacylglycerols derived from fatty acids having a larger number of double bonds, making them liquid at room temperature.

5 What hydrolysis products are formed from a triacylglycerol? (15.5)

- Triacylglycerols are hydrolyzed in acid or with enzymes (in biological systems) to form glycerol and three molecules of fatty acids. Base hydrolysis of a triacylglycerol forms glycerol and sodium salts of fatty acids—soaps.

6 What are the major structural features of phospholipids? (15.6)

- All phospholipids contain a phosphorus atom, and have a polar (ionic) head and two nonpolar tails. Phosphoacylglycerols are derived from glycerol, two molecules of fatty acids, phosphate, and an alcohol (either ethanolamine or choline).

7 Describe the structure of the cell membrane. How do molecules and ions cross the cell membrane? (15.7)

- The main component of the cell membrane is phospholipids, arranged in a lipid bilayer with the ionic heads oriented towards the outside of the bilayer, and the nonpolar tails on the interior.
- Small molecules like O_2 and CO_2 diffuse through the membrane from the side of higher concentration to the side of lower concentration. Larger polar molecules and some ions (Cl^-, HCO_3^-, and glucose) travel through channels created by integral membrane proteins (facilitated diffusion). Some cations (Na^+, K^+, and Ca^{2+}) must travel against the concentration gradient, a process called active transport, which requires energy input.

8 What are the main structural features of steroids? (15.8)

- Steroids like cholesterol are tetracyclic lipids that contain three six-membered rings and one five-membered ring. Because cholesterol is insoluble in the aqueous medium of the blood, it is transported through the bloodstream in water-soluble particles called lipoproteins.

9 What is a hormone? Give examples of steroid hormones. (15.9)

- A hormone is a molecule that is synthesized in one part of an organism, and elicits a response at a different site. Steroid hormones include estrogens and progestins (female sex hormones), androgens (male sex hormones), and adrenal cortical steroids such as cortisone, which are synthesized in the adrenal gland.

10 Which vitamins are fat soluble? (15.10)

- Fat-soluble vitamins are lipids required in small quantities for normal cell function, and which cannot be synthesized in the body. Vitamins A, D, E, and K are fat soluble.

UNDERSTANDING KEY CONCEPTS

Selected in-chapter and odd-numbered end-of-chapter problems have brief answers at the end of each chapter. The *Student Study Guide and Solutions Manual* contains detailed solutions to all in-chapter and odd-numbered end-of-chapter problems, as well as additional worked examples and a chapter self-test.

15.23 (a) Which food is high in unsaturated triacylglycerols? (b) Which food is high in saturated fat? (c) Which food likely contains trans fats?

A

B

C

15.24 Match the nutrition label to the type of food: (a) corn oil; (b) butter; (c) margarine.

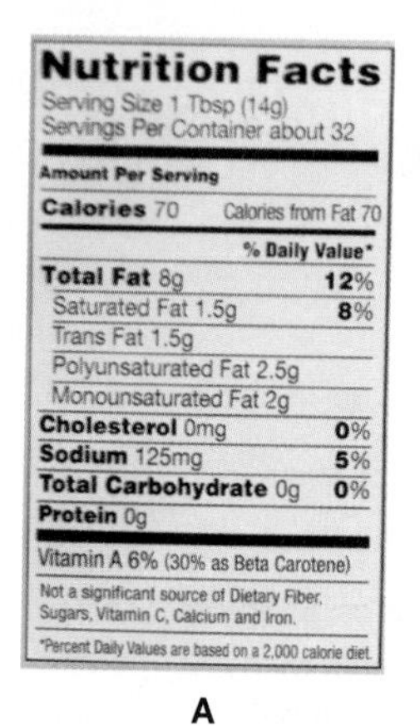

A

Nutrition Facts
Serv. Size 1 Tbsp (14g)
Servings about 64
Calories 120
Fat Cal. 120
*Percent Daily Values (DV) are based on a 2,000 calorie diet.

Amount Per Serving	% DV*	Amount Per Serving	% DV*
Total Fat 14g	22%	**Cholest.** 0mg	0%
Sat. Fat 2g	10%	**Sodium** 0mg	0%
Trans Fat 0g		**Total Carb.** 0g	0%
Polyunsat. Fat 8g		**Protein** 0g	
Monounsat. Fat 4g			

Not a significant source of dietary fiber, sugars, vitamin A, vitamin C, calcium and iron.

B

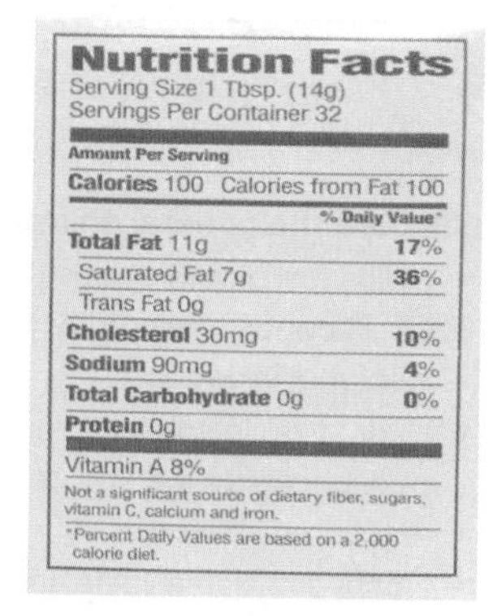

C

15.25 Draw the structure of a wax formed from palmitic acid [$CH_3(CH_2)_{14}COOH$] and $CH_3(CH_2)_{21}OH$.

15.26 What hydrolysis products are formed when the wax $CH_3(CH_2)_{12}COO(CH_2)_{15}CH_3$ is treated with aqueous sulfuric acid?

15.27 Consider the following four types of compounds: [1] fatty acids; [2] soaps; [3] waxes; [4] triacylglycerols. For each type of compound: (a) give the general structure; (b) draw the structure of a specific example; (c) label the compound as water soluble or water insoluble; (d) label the compound as soluble or insoluble in the organic solvent hexane [$CH_3(CH_2)_4CH_3$].

15.28 How do fats and oils compare with respect to each of the following features?

a. identity and number of functional groups present
b. number of carbon–carbon double bonds present
c. melting point
d. natural source

15.29 Draw the products formed when the given triacylglycerol is hydrolyzed under each of the following conditions: (a) water and H_2SO_4; (b) water and NaOH.

$$CH_2-O-\overset{\overset{O}{\|}}{C}-(CH_2)_{14}CH_3$$
$$|$$
$$CH-O-\overset{\overset{O}{\|}}{C}-(CH_2)_{14}CH_3$$
$$|$$
$$CH_2-O-\overset{\overset{O}{\|}}{C}-(CH_2)_{16}CH_3$$

15.30 Draw the products formed when the given triacylglycerol is hydrolyzed under each of the following conditions: (a) water and H_2SO_4; (b) water and NaOH.

$$CH_2-O-\overset{\overset{O}{\|}}{C}-(CH_2)_{10}CH_3$$
$$|$$
$$CH-O-\overset{\overset{O}{\|}}{C}-(CH_2)_{14}CH_3$$
$$|$$
$$CH_2-O-\overset{\overset{O}{\|}}{C}-(CH_2)_{16}CH_3$$

15.31 Block diagrams representing the general structures of two types of lipids are drawn. Which terms describe each diagram: (a) phospholipid; (b) triacylglycerol; (c) hydrolyzable lipid; (d) phosphoacylglycerol? More than one term may apply to a diagram.

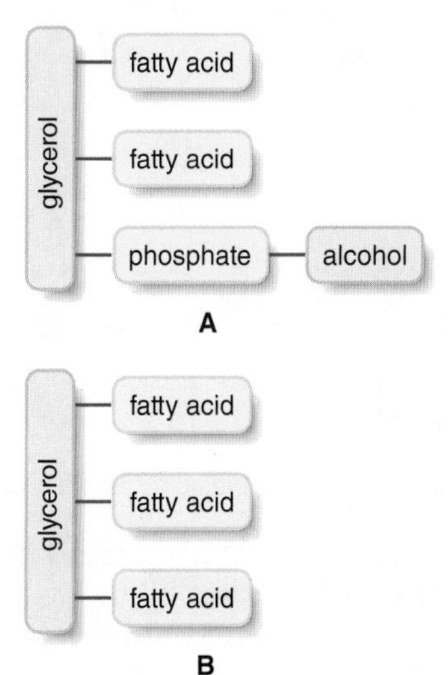

15.32 Which "cartoon" represents a soap and which represents a phosphoacylglycerol? What structural features are present in the polar head and nonpolar tails of each compound?

ADDITIONAL PROBLEMS

General Characteristics of Lipids

15.33 Label each compound as a hydrolyzable or nonhydrolyzable lipid.

a. triacylglycerol
b. vitamin A
c. lecithin
d. cholesterol

15.34 Label each compound as a hydrolyzable or nonhydrolyzable lipid.

a. phospholipid
b. cephalin
c. wax
d. estrogen

15.35 In which solvents might a wax be soluble: (a) H_2O; (b) CH_2Cl_2; (c) $CH_3CH_2OCH_2CH_3$?

15.36 In which solvents or solutions might a steroid be soluble: (a) blood plasma; (b) CCl_4; (c) 5% NaCl solution?

Fatty Acids, Waxes, and Triacylglycerols

15.37 Rank the fatty acids in order of increasing melting point: $CH_3(CH_2)_{16}COOH$, $CH_3(CH_2)_7CH{=}CH(CH_2)_7COOH$, and $CH_3(CH_2)_5CH{=}CH(CH_2)_7COOH$.

15.38 How does each of the following affect the melting point of a fatty acid: (a) increasing the number of carbon atoms; (b) increasing the number of double bonds?

15.39 Rank the fatty acids in order of increasing melting point.

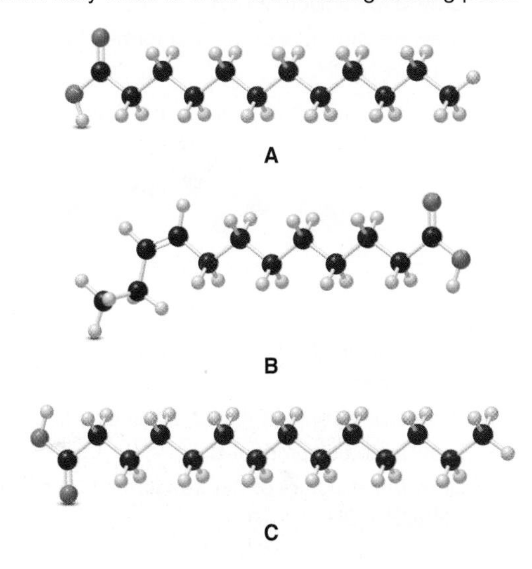

15.40 How would you expect the melting points of the following unsaturated fatty acids to compare? Explain your choice.

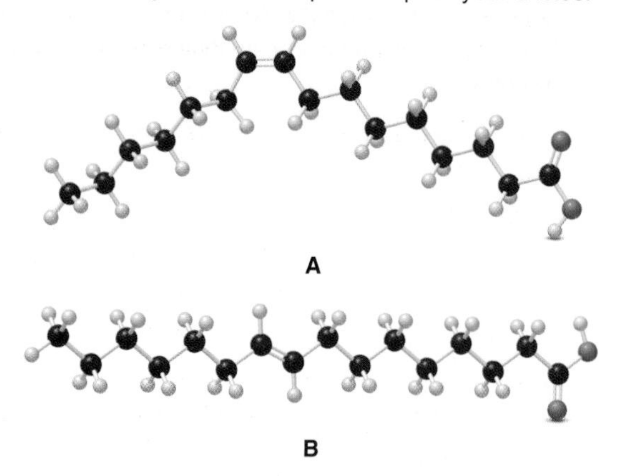

A

B

15.41 Is a fatty acid a hydrolyzable lipid? Explain your choice.

15.42 Why are soaps water soluble, but the fatty acids from which they are derived, water insoluble?

15.43 Draw the structure of a wax formed from palmitic acid [$CH_3(CH_2)_{14}COOH$] and each alcohol.

a. $CH_3(CH_2)_{11}OH$ b. $CH_3(CH_2)_9OH$

15.44 Draw the structure of a wax formed from a 30-carbon straight chain alcohol and each carboxylic acid.

a. lauric acid b. myristic acid

15.45 Consider fatty acid **A** depicted in the ball-and-stick model.

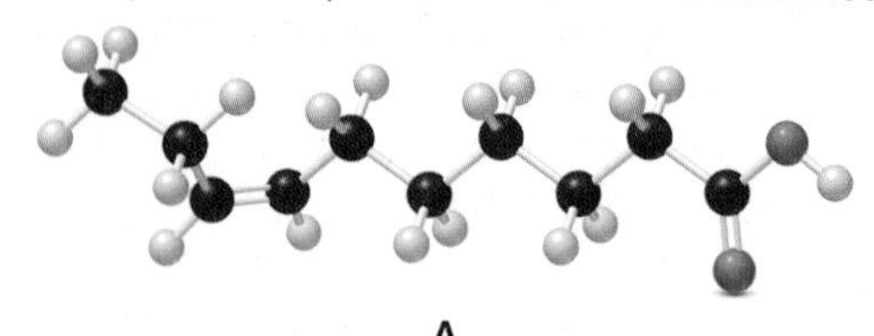

A

a. Draw a skeletal structure for **A** and clearly show the cis double bond.

b. What type of omega-*n* acid is this fatty acid?

c. Draw a stereoisomer of **A.**

d. Draw the structure of the wax formed from **A** and $CH_3(CH_2)_{10}OH$.

15.46 Consider fatty acid **B** depicted in the ball-and-stick model.

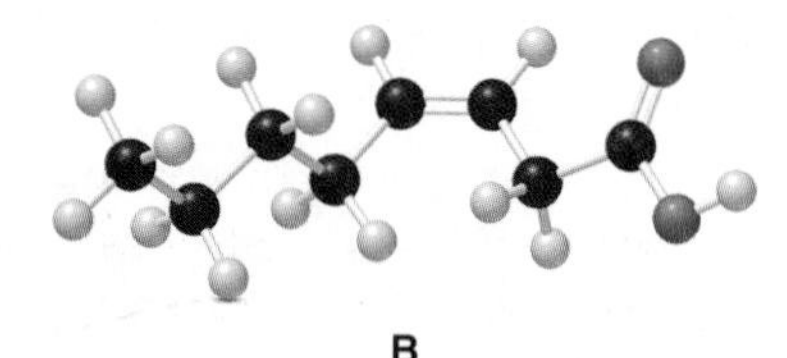

B

a. Draw a skeletal structure for **B** and clearly show the cis double bond.

b. What type of omega-*n* acid is this fatty acid?

c. Draw a stereoisomer of **B.**

d. Draw the structure of the wax formed from **B** and $CH_3(CH_2)_8OH$.

15.47 What hydrolysis products are formed when each wax is treated with aqueous sulfuric acid?

a. $CH_3(CH_2)_{16}COO(CH_2)_{17}CH_3$

b. $CH_3(CH_2)_{12}COO(CH_2)_{25}CH_3$

15.48 What hydrolysis products are formed when each wax is treated with aqueous sulfuric acid?

a. $CH_3(CH_2)_{18}COO(CH_2)_{29}CH_3$

b. $CH_3(CH_2)_{24}COO(CH_2)_{23}CH_3$

15.49 Draw a triacylglycerol that fits each description:

a. a triacylglycerol formed from lauric, myristic, and linoleic acids

b. an unsaturated triacylglycerol that contains two cis double bonds in one fatty acid side chain

c. a saturated triacylglycerol formed from three 14-carbon fatty acids

15.50 Draw a triacylglycerol that fits each description:

a. a triacylglycerol formed from two molecules of lauric acid and one molecule of palmitic acid

b. a polyunsaturated triacylglycerol formed from three molecules of linoleic acid

c. a trans triacylglycerol that contains two trans double bonds

15.51 Answer the following questions about the given triacylglycerol.

$$\begin{array}{l} CH_2-O-C(=O)-(CH_2)_{18}CH_3 \\ | \\ CH-O-C(=O)-(CH_2)_{16}CH_3 \\ | \\ CH_2-O-C(=O)-(CH_2)_{10}CH_3 \end{array}$$

a. What fatty acids are used to form this triacylglycerol?

b. Would you expect this triacylglycerol to be a solid or a liquid at room temperature?

c. What regions are hydrophobic?

d. What regions are hydrophilic?

e. What hydrolysis products are formed when the triacylglycerol is treated with aqueous sulfuric acid?

15.52 Answer the following questions about the given triacylglycerol.

$$\begin{array}{l} CH_2-O-C(=O)-(CH_2)_{14}CH_3 \\ | \\ CH-O-C(=O)-(CH_2)_7(CH=CHCH_2)_2(CH_2)_3CH_3 \\ | \\ CH_2-O-C(=O)-(CH_2)_7CH=CH(CH_2)_5CH_3 \end{array}$$

a. What fatty acids are used to form this triacylglycerol?

b. Would you expect this triacylglycerol to be a solid or a liquid at room temperature?

c. What regions are hydrophobic?

d. What regions are hydrophilic?

e. What hydrolysis products are formed when the triacylglycerol is treated with aqueous sulfuric acid?

15.53 Draw the structure of a triacylglycerol that yields only the sodium carboxylate depicted in the ball-and-stick model when it is hydrolyzed in base.

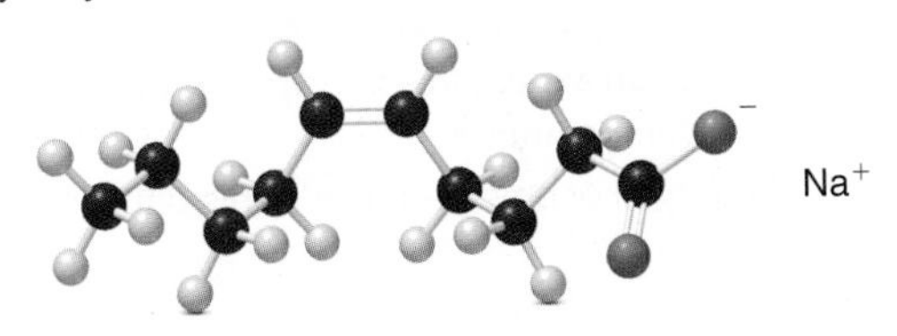

15.54 Draw the structure of a triacylglycerol that yields only the sodium carboxylate depicted in the ball-and-stick model when it is hydrolyzed in base.

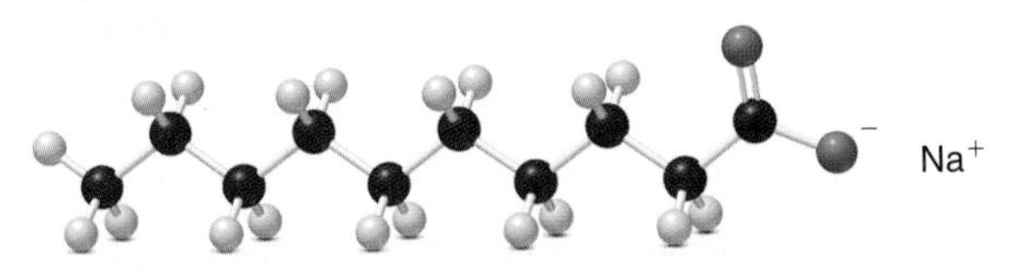

15.55 Draw the products formed when the given triacylglycerol is hydrolyzed under each of the following conditions: (a) water and H_2SO_4; (b) water and NaOH.

$$\begin{array}{l} CH_2-O-\overset{\displaystyle O}{\overset{\|}{C}}-(CH_2)_{14}CH_3 \\ | \\ CH-O-\overset{\displaystyle O}{\overset{\|}{C}}-(CH_2)_7CH{=}CH(CH_2)_7CH_3 \\ | \\ CH_2-O-\overset{\displaystyle O}{\overset{\|}{C}}-(CH_2)_7CH{=}CH(CH_2)_5CH_3 \end{array}$$

15.56 Draw the products formed when the given triacylglycerol is hydrolyzed under each of the following conditions: (a) water and H_2SO_4; (b) water and NaOH.

$$\begin{array}{l} CH_2-O-\overset{\displaystyle O}{\overset{\|}{C}}-(CH_2)_7CH{=}CH(CH_2)_7CH_3 \\ | \\ CH-O-\overset{\displaystyle O}{\overset{\|}{C}}-(CH_2)_{16}CH_3 \\ | \\ CH_2-O-\overset{\displaystyle O}{\overset{\|}{C}}-(CH_2)_7CH{=}CH(CH_2)_5CH_3 \end{array}$$

Phospholipids and Cell Membranes

15.57 Draw a phospholipid that fits each description.
- a. a cephalin formed from two molecules of palmitoleic acid
- b. a lecithin formed from two molecules of lauric acid

15.58 Draw a phospholipid that fits each description.
- a. a lecithin formed from two molecules of oleic acid
- b. a cephalin formed from two molecules of myristic acid

Steroids

15.59 Draw the structure of the anabolic steroid 4-androstene-3,17-dione, also called "andro," from the following description. Andro contains the tetracyclic steroid skeleton with carbonyl groups at C3 and C17, a double bond between C4 and C5, and methyl groups bonded to C10 and C13.

15.60 Draw the structure of the anabolic steroid methenolone from the following description. Methenolone contains the tetracyclic steroid skeleton with a carbonyl group at C3, a hydroxyl at C17, a double bond between C1 and C2, and methyl groups bonded to C1, C10, and C13.

15.61 Why must cholesterol be transported through the bloodstream in lipoprotein particles?

15.62 Why are LDLs soluble in the blood?

15.63 Describe the role of HDLs and LDLs in cholesterol transport in the blood. What is the relationship of HDL and LDL levels to cardiovascular disease?

15.64 What are anabolic steroids? Give an example. What adverse effects arise from using anabolic steroids?

15.65 (a) Draw the structure of an estrogen and an androgen. (b) What structural features are similar in the two steroids? (c) What structural features are different? (d) Describe the biological activity of each steroid.

15.66 (a) Draw the structure of an androgen and a progestin. (b) What structural features are similar in the two steroids? (c) What structural features are different? (d) Describe the biological activity of each steroid.

Vitamins

15.67 Answer each question with regards to vitamins A and D.
- a. How many tetrahedral carbons does the vitamin contain?
- b. How many trigonal planar carbons does the vitamin contain?
- c. Identify the functional groups.
- d. Label all polar bonds.
- e. What function does the vitamin serve in the body?
- f. What problems result when there is a deficiency of the vitamin?
- g. Give a dietary source.

15.68 Answer each question in Problem 15.67 for vitamins E and K.

General Questions

15.69 Give an example of each type of lipid.
- a. an unsaturated fatty acid with one C=C
- b. a wax that contains a total of 30 carbons
- c. a saturated triacylglycerol

15.70 Give an example of each type of lipid.
- a. an unsaturated fatty acid with more than one C=C
- b. a wax derived from a 12-carbon fatty acid
- c. a cephalin

15.71 Why are phosphoacylglycerols more water soluble than triacylglycerols?

15.72 How are soaps and phosphoacylglycerols similar in structure? How do they differ?

15.73 Some fish oils contain triacylglycerols formed from the polyunsaturated fatty acid, 7,10,13,16,19-docosapentaenoic acid.

$$CH_3CH_2CH{=}CHCH_2CH{=}CHCH_2CH{=}CHCH_2CH{=}CHCH_2CH{=}CH(CH_2)_5COOH$$

7,10,13,16,19-docosapentaenoic acid

a. Draw a skeletal structure showing the cis arrangement at each double bond.

b. Label the hydrophobic and hydrophilic portions of the fatty acid.

c. How does the melting point of this fatty acid compare to its all trans isomer?

d. Would you expect this fatty acid to be a solid or a liquid at room temperature?

e. What type of omega-*n* acid is this fatty acid?

15.74 Some marine plankton contain triacylglycerols formed from the polyunsaturated fatty acid, 3,6,9,12,15-octadecapentaenoic acid.

$$CH_3CH_2CH{=}CHCH_2CH{=}CHCH_2CH{=}CHCH_2CH{=}CHCH_2CH{=}CHCH_2COOH$$

3,6,9,12,15-octadecapentaenoic acid

a. Draw a skeletal structure showing the cis arrangement at each double bond.

b. Label the hydrophobic and hydrophilic portions of the fatty acid.

c. How does the melting point of this fatty acid compare to the melting point of oleic acid?

d. Would you expect this fatty acid to be a solid or a liquid at room temperature?

e. What type of omega-*n* acid is this fatty acid?

Applications

15.75 The main fatty acid component of the triacylglycerols in coconut oil is lauric acid, $CH_3(CH_2)_{10}COOH$. Explain why coconut oil is a liquid at room temperature despite the fact that it contains a large fraction of this saturated fatty acid.

15.76 Can an individual survive on a cholesterol-free diet?

15.77 Can an individual survive on a completely fat-free diet?

15.78 Why should saturated fats in the diet be avoided?

CHALLENGE PROBLEMS

15.79 If the serum cholesterol level in an adult is 167 mg/dL, how many grams of cholesterol are contained in 5.0 L of blood?

15.80 How many triacylglycerols can be prepared from three different fatty acids? Draw all possible structures, excluding stereoisomers.

BEYOND THE CLASSROOM

15.81 Flaxseed, chia seed, fish oil, and macadamia nuts are all advertised as "healthy" sources of triacylglycerols. Research one or more of these products and determine what types of triacylglycerols are present. What is beneficial about adding the product to your diet? From what you have read, do you agree or disagree with the claim that the oil, nut, or seed provides health benefits when regularly consumed?

15.82 Pick a favorite snack food such as potato chips, corn chips, crackers, or peanut butter. Using the nutrition information on the package, determine how much "fat" (i.e., triacylglycerol) is contained in a serving. Calculate what percentage of saturated, unsaturated, and trans fat the product contains per serving. Calculate the number of grams of fat you would consume if you ate this snack food every day for a month.

15.83 Compare the fat content of butter, Crisco, and one or two soft butter substitutes available at your local food store. Record the total fat content per tablespoon, as well as the percentage of saturated, unsaturated, and trans fat. What are the major differences among the products? Do any of the butter substitutes offer a significant health benefit over the others? Are healthy products more or less expensive?

15.84 Locate a biodiesel facility in your state. Where do the oils processed in the facility come from, and how are the oils converted to usable fuels? In what ways is the biodiesel used? What are the advantages and disadvantages of using biodiesel as an energy source? What fraction of the energy needs in your area is supplied by biodiesel?

15.85 A by-product in the synthesis of biodiesel from vegetable oil is glycerol [$HOCH(CH_2OH)_2$]. Research what consumer products contain glycerol, and why glycerol is added to these products. A worldwide glut of glycerol in the last decade prompted companies to search for new processes that convert glycerol to other compounds that are also used in consumer products. What is the current status of these research efforts?

ANSWERS TO SELECTED PROBLEMS

15.1 a. and c.

15.2 a,b:

A = (skeletal structure of a saturated fatty acid ending in C(=O)OH)

hydrophilic portion: the COOH group; hydrophobic portion: the hydrocarbon chain

B = (skeletal structure of an unsaturated fatty acid with two cis C=C, ending in C(=O)OH)

hydrophilic portion: the COOH group; hydrophobic portion: the hydrocarbon chain

c. **A** will have the higher melting point because **A** is saturated and the molecules can pack together better.

15.3 a. (skeletal structure of a polyunsaturated fatty acid with four cis C=C, ending in COOH)

b. omega-3

15.5 a. $CH_3(CH_2)_{16}CO(CH_2)_9CH_3$ (C=O)

b. $CH_3(CH_2)_{16}CO(CH_2)_{11}CH_3$ (C=O)

c. $CH_3(CH_2)_{16}CO(CH_2)_{29}CH_3$ (C=O)

15.7 $CH_3(CH_2)_{10}COOH$ + $HO(CH_2)_{15}CH_3$

15.8 a.
$CH_2{-}O{-}C(=O){-}(CH_2)_{16}CH_3$
$CH{-}O{-}C(=O){-}(CH_2)_{16}CH_3$
$CH_2{-}O{-}C(=O){-}(CH_2)_{16}CH_3$

b.
$CH_2{-}O{-}C(=O){-}(CH_2)_7CH{=}CH(CH_2)_7CH_3$ (cis, H H)
$CH{-}O{-}C(=O){-}(CH_2)_7CH{=}CH(CH_2)_7CH_3$ (cis, H H)
$CH_2{-}O{-}C(=O){-}(CH_2)_7CH{=}CH(CH_2)_7CH_3$ (cis, H H)

15.9
$CH_2{-}O{-}C(=O){-}(CH_2)_{16}CH_3$
$CH{-}O{-}C(=O){-}(CH_2)_7CH{=}CH(CH_2)_7CH_3$
$CH_2{-}O{-}C(=O){-}(CH_2)_{16}CH_3$

15.11 a. $CH_2{-}OH$, $CH{-}OH$, $CH_2{-}OH$ (glycerol) + 3 $HO{-}C(=O){-}(CH_2)_{12}CH_3$

b. $CH_2{-}OH$, $CH{-}OH$, $CH_2{-}OH$ (glycerol) + $HO{-}C(=O){-}(CH_2)_{12}CH_3$, $HO{-}C(=O){-}(CH_2)_7CH{=}CH(CH_2)_5CH_3$, $HO{-}C(=O){-}(CH_2)_7CH{=}CH(CH_2)_7CH_3$

15.13
$CH_2{-}O{-}C(=O){-}(CH_2)_{14}CH_3$
$CH{-}O{-}C(=O){-}(CH_2)_7CH{=}CH(CH_2)_7CH_3$ (H H)
$CH_2{-}O{-}P(=O)(O^-){-}O{-}CH_2CH_2\overset{+}{N}H_3$

$CH_2{-}O{-}C(=O){-}(CH_2)_7CH{=}CH(CH_2)_7CH_3$ (H H)
$CH{-}O{-}C(=O){-}(CH_2)_{14}CH_3$
$CH_2{-}O{-}P(=O)(O^-){-}O{-}CH_2CH_2\overset{+}{N}H_3$

15.15 Because they have a polar head (from the ammonium ion) and two nonpolar tails (from the fatty acids), phospholipids can form a lipid bilayer needed for cell membrane function. Triacylglycerols are basically nonpolar compounds so they have no polar head to attract water on the outside of a membrane.

15.17 a,b,c: (structure of cholesterol with rings A, B, C, D labeled)

HO — OH at C3

double bond between C5 and C6

d. The C–O and O–H bonds are polar. The large number of C–C and C–H bonds makes cholesterol water insoluble.

15.19 a. Estrone has a phenol (a benzene ring with a hydroxyl group) and progesterone has a ketone and C=C in ring A. Progesterone also has a methyl group bonded to C10.

b. Estrone has a ketone at C17 and progesterone has a C–C bond, which is attached to a ketone.

15.21 Water-soluble vitamins are excreted in the urine, whereas fat-soluble vitamins are stored in the body.

15.23 a. **B** b. **C** c. **A**

15.25 $CH_3(CH_2)_{14}C(=O)O(CH_2)_{21}CH_3$

15.27

Compound	a. General Structure	b. Example	c. Water Soluble (Y/N)	d. Hexane Soluble (Y/N)
[1] Fatty acid	RCOOH	$CH_3(CH_2)_{10}COOH$	N	Y
[2] Soap	$RCOO^- Na^+$	$CH_3(CH_2)_{10}COO^- Na^+$	Y	N
[3] Wax	RCOOR'	$CH_3(CH_2)_6CO_2(CH_2)_7CH_3$	N	Y
[4] Triacylglycerol	$CH_2-O-C(=O)-R$ $CH-O-C(=O)-R'$ $CH_2-O-C(=O)-R''$	$CH_2-O-C(=O)-(CH_2)_{12}CH_3$ $CH-O-C(=O)-(CH_2)_{12}CH_3$ $CH_2-O-C(=O)-(CH_2)_{12}CH_3$	N	Y

15.29 a. CH_2-OH, $CH-OH$, CH_2-OH + $2\ HOC(=O)-(CH_2)_{14}CH_3$, $HOC(=O)-(CH_2)_{16}CH_3$

b. CH_2-OH, $CH-OH$, CH_2-OH + $2\ Na^{+}\ {}^{-}OC(=O)-(CH_2)_{14}CH_3$, $Na^{+}\ {}^{-}OC(=O)-(CH_2)_{16}CH_3$

15.31 **A:** a,c,d **B:** b,c

15.33 a,c: hydrolyzable b,d: nonhydrolyzable

15.35 b. and c.

15.37 $CH_3(CH_2)_5CH{=}CH(CH_2)_7COOH$, $CH_3(CH_2)_7CH{=}CH(CH_2)_7COOH$, $CH_3(CH_2)_{16}COOH$

15.39 **B < A < C**

15.41 Fatty acids are not hydrolyzable because they contain a very long hydrocarbon chain attached to a carboxylic acid group.

15.43 a. $CH_3(CH_2)_{14}C(=O)O(CH_2)_{11}CH_3$ b. $CH_3(CH_2)_{14}C(=O)O(CH_2)_9CH_3$

15.45 a. (skeletal structure; labels: cis, CO_2H)

b. omega-3 acid

c. (skeletal structure; labels: trans, CO_2H)

d. (skeletal structure of an ester with a cis double bond; labels: O, O)

15.47 a. $CH_3(CH_2)_{16}COOH + HO(CH_2)_{17}CH_3$

b. $CH_3(CH_2)_{12}COOH + HO(CH_2)_{25}CH_3$

15.49 a. $CH_2-O-C(=O)-(CH_2)_{10}CH_3$
$CH-O-C(=O)-(CH_2)_{12}CH_3$
$CH_2-O-C(=O)-(CH_2)_7CH{=}CHCH_2CH{=}CH(CH_2)_4CH_3$

b. $CH_2-O-C(=O)-(CH_2)_{10}CH_3$
$CH-O-C(=O)-(CH_2)_8CH_3$
$CH_2-O-C(=O)-(CH_2)_7C(H){=}C(H)CH_2C(H){=}C(H)(CH_2)_4CH_3$

c. $CH_2-O-C(=O)-(CH_2)_{12}CH_3$
$CH-O-C(=O)-(CH_2)_{12}CH_3$
$CH_2-O-C(=O)-(CH_2)_{12}CH_3$

15.51 a. arachidic acid, stearic acid, and lauric acid

b. solid

c. The long hydrocarbon chains are hydrophobic.

d. The ester linkages are hydrophilic.

e. CH_2-OH, $CH-OH$, CH_2-OH + $HOOC(CH_2)_{10}CH_3$, $HOOC(CH_2)_{16}CH_3$, $HOOC(CH_2)_{18}CH_3$

15.53 $CH_2-O-C(=O)-(CH_2)_3CH{=}CH(CH_2)_3CH_3$
$CH-O-C(=O)-(CH_2)_3CH{=}CH(CH_2)_3CH_3$
$CH_2-O-C(=O)-(CH_2)_3CH{=}CH(CH_2)_3CH_3$

15.55 a.
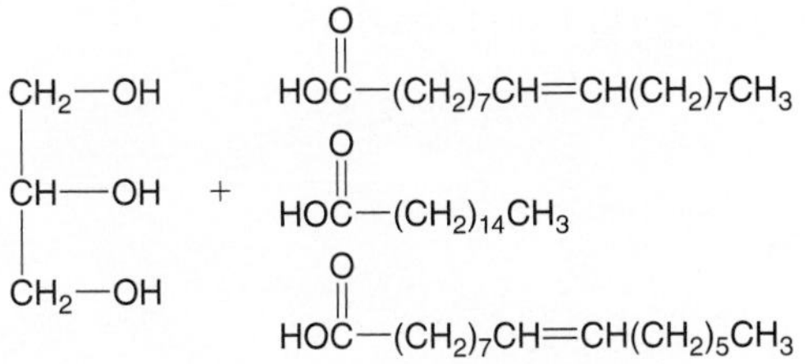

b.
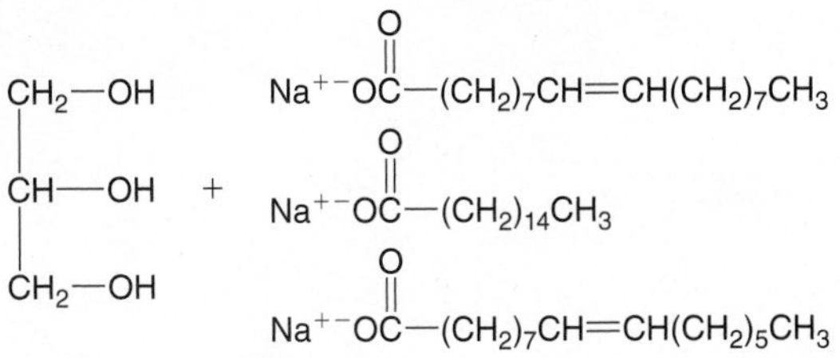

15.57 a.
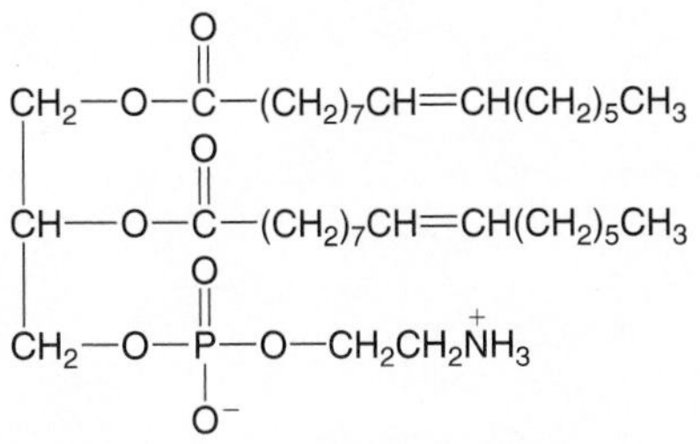

b.
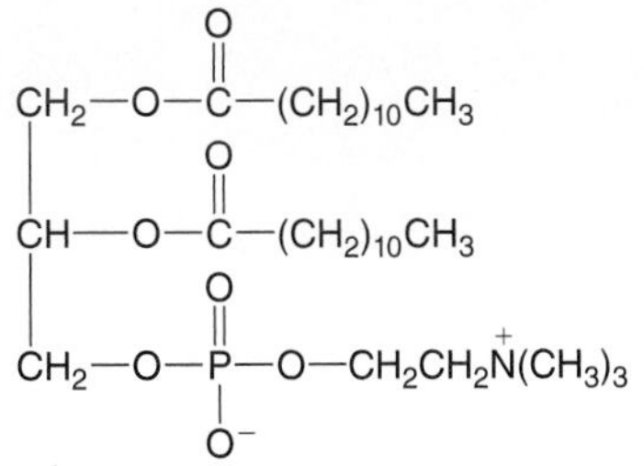

15.59
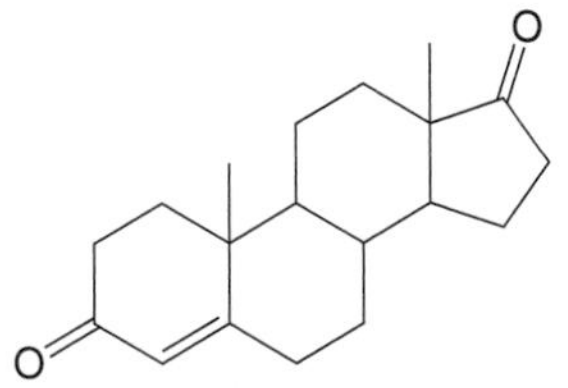

15.61 Cholesterol is insoluble in the aqueous medium of the bloodstream. By being bound to a lipoprotein particle, it can be transported in the aqueous solution.

15.63 Low-density lipoproteins (LDLs) transport cholesterol from the liver to the tissues where it is incorporated in cell membranes. High-density lipoproteins (HDLs) transport cholesterol from the tissues back to the liver. When LDLs supply more cholesterol than is needed, LDLs deposit cholesterol on the wall of arteries, forming plaque. Atherosclerosis is a disease that results from the buildup of these fatty deposits, restricting the flow of blood, increasing blood pressure, and increasing the likelihood of a heart attack or stroke. As a result, LDL cholesterol is often called "bad" cholesterol.

15.65 a.
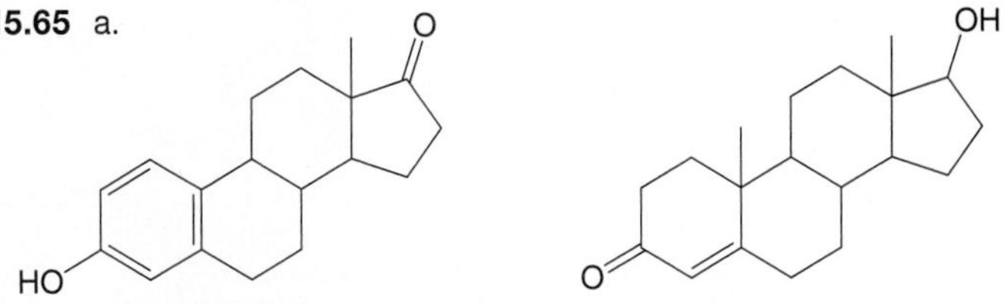

estrone testosterone

b. The estrogen (left) and androgen (right) both contain the four rings of the steroid skeleton. Both contain a methyl group bonded to C13.

c. The estrogen has an aromatic A ring and a hydroxyl group on this ring. The androgen has a carbonyl on the A ring but does not contain an aromatic ring. The androgen also contains a C=C in the A ring and an additional CH_3 group at C10. The D rings are also different. The estrogen contains a carbonyl at C17 and the androgen has an OH group.

d. Estrogens, synthesized in the ovaries, control the menstrual cycle and secondary sexual characteristics of females. Androgens, synthesized in the testes, control the development of male secondary sexual characteristics.

15.67

	Vitamin A	Vitamin D
a.	10	21
b.	10	6
c.	Five alkenes, one hydroxyl group	Three alkenes, one hydroxyl group
d.	Polar C–O and O–H bonds	Polar C–O and O–H bonds
e.	Required for normal vision	Regulates calcium and phosphorus metabolism
f.	Night blindness	Rickets and skeletal deformities
g.	Liver, kidney, oily fish, dairy	Milk and breakfast cereals

15.69 a. $CH_3(CH_2)_5CH{=}CH(CH_2)_7COOH$

b. $CH_3(CH_2)_{12}COO(CH_2)_{15}CH_3$

c.
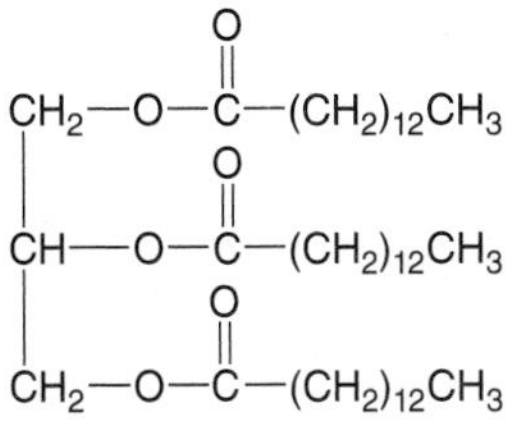

15.71 They contain an ionic head, making them more polar than triacylglycerols.

15.73 a,b:
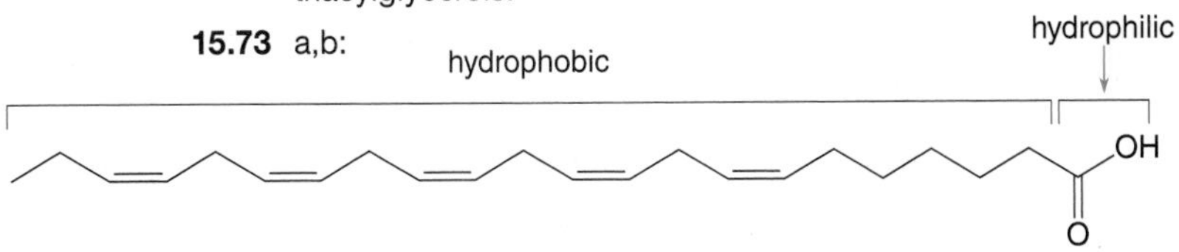

c. The melting point would be lower than the melting point of the trans isomer.

d. liquid

e. omega-3 fatty acid

15.75 The hydrocarbon chains have only 12 carbons in them, making them short enough so that the triacylglycerol remains a liquid at room temperature.

15.77 No, certain fatty acids and fat-soluble vitamins are required in the diet.

15.79 8.4 g

Whales have a high concentration of the protein myoglobin in their muscles. Myoglobin binds oxygen, so it serves as an oxygen reservoir while the whale is submerged for long periods of time.

Amino Acids, Proteins, and Enzymes

CHAPTER OUTLINE

CHAPTER GOALS

In this chapter you will learn how to:

1. Identify the general structural features of amino acids
2. Describe the acid–base properties of amino acids
3. Label the N- and C-terminal amino acids of simple peptides
4. Describe the characteristics of the primary, secondary, tertiary, and quaternary structure of proteins
5. Describe the features of fibrous proteins like α-keratin and collagen
6. Describe the features of globular proteins like hemoglobin and myoglobin
7. Draw the products of protein hydrolysis
8. Describe protein denaturation
9. Describe the main features of enzymes
10. Describe the use of enzymes to diagnose and treat disease

Of the four major groups of biomolecules—lipids, carbohydrates, proteins, and nucleic acids—proteins have the widest array of functions. **Keratin** and **collagen,** for example, form long insoluble fibers, giving strength and support to tissues. Hair, horns, hooves, and fingernails are all made up of keratin. **Collagen** is found in bone, connective tissue, tendons, and cartilage. **Membrane proteins** transport small organic molecules and ions across cell membranes. **Insulin,** the hormone that regulates blood glucose levels, and **hemoglobin,** which transports oxygen from the lungs to tissues, are proteins. **Enzymes** are proteins that catalyze and regulate all aspects of cellular function. In Chapter 16 we discuss proteins and their primary components, the amino acids.

HEALTH NOTE

Meat, fish, beans, and nuts are all high-protein foods.

16.1 Introduction

***Proteins* are biomolecules that contain many amide bonds, formed by joining amino acids together.**

$$H_2N-\underset{R}{\overset{H}{C}}-\overset{O}{\overset{\|}{C}}-OH \longrightarrow -\underset{R_1}{CH}-\overset{O}{\overset{\|}{C}}-\underset{H}{N}-\underset{R_2}{CH}-\overset{O}{\overset{\|}{C}}-\underset{H}{N}-\underset{R_3}{CH}-\overset{O}{\overset{\|}{C}}-\underset{H}{N}-\underset{R_4}{CH}-\overset{O}{\overset{\|}{C}}-\underset{H}{N}-$$

amino acid

protein
[Amide bonds are shown in red.]

Proteins occur widely in the human body, accounting for approximately 50% of its dry weight (Figure 16.1). Fibrous proteins, like keratin in hair, skin, and nails and collagen in connective tissue, give support and structure to tissues and cells. Protein hormones and enzymes regulate the body's metabolism. Transport proteins carry substances through the blood, and storage proteins store elements and ions in organs. Contractile proteins control muscle movements, and immunoglobulins are proteins that defend the body against foreign substances.

Figure 16.1
Proteins in the Human Body

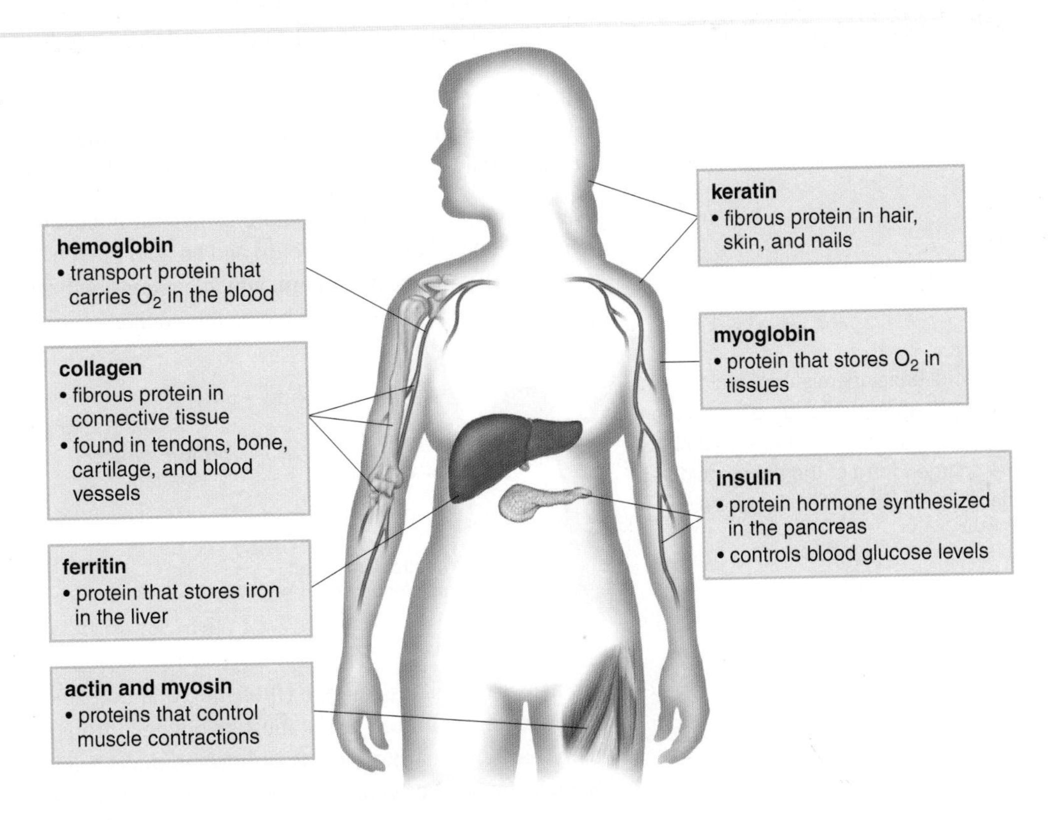

keratin
• fibrous protein in hair, skin, and nails
hemoglobin
• transport protein that carries O_2 in the blood
myoglobin
• protein that stores O_2 in tissues
collagen
• fibrous protein in connective tissue
• found in tendons, bone, cartilage, and blood vessels
insulin
• protein hormone synthesized in the pancreas
• controls blood glucose levels
ferritin
• protein that stores iron in the liver
actin and myosin
• proteins that control muscle contractions

Table 16.1 Recommended Daily Protein Intake

Group	Daily Protein Intake (g protein/kg body weight)
Children (1–3 years)	1.1
Children (4–13 years)	0.95
Children (14–18 years)	0.85
Adult	0.8

Source: Data from U.S. Food and Drug Administration.

Unlike lipids and carbohydrates, which the body stores for use when needed, protein is not stored so it must be consumed on a daily basis. The current recommended daily intake for adults is 0.8 grams of protein per kilogram of body weight. Since children need protein for both growth and maintenance, the recommended daily intake is higher, as shown in Table 16.1.

16.2 Amino Acids

To understand protein properties and structure, we must first learn about the amino acids that compose them.

16.2A General Features of Amino Acids

Amino acids contain two functional groups—an amino group (NH_2) and a carboxyl group (COOH). In most naturally occurring amino acids, the amino group is bonded to the α carbon, the carbon adjacent to the carbonyl group, making them **α-amino acids.**

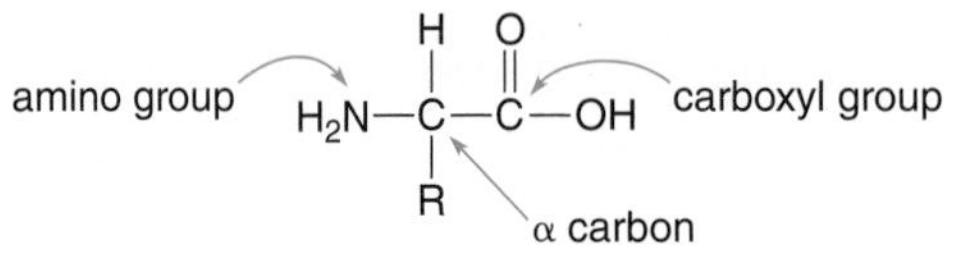

amino group — $H_2N-C(H)(R)-C(=O)-OH$ — carboxyl group; α carbon

α-amino acid

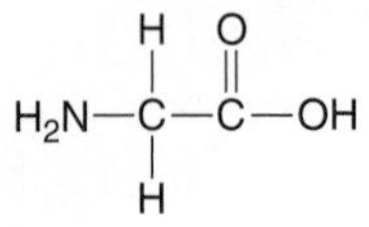

$H_2N-CH_2-C(=O)-OH$

glycine
the simplest amino acid

The 20 amino acids that occur naturally in proteins differ in the identity of the R group bonded to the α carbon. The R group is called the **side chain** of the amino acid. The simplest amino acid, called **glycine,** has R = H. Other side chains may be simple alkyl groups, or have additional functional groups such as OH, SH, COOH, or NH_2. Table 16.2 lists the structures of the 20 common amino acids that occur in proteins.

- Amino acids with an additional COOH group in the side chain are called acidic amino acids.
- Those with an additional basic N atom in the side chain are called basic amino acids.
- All others are neutral amino acids.

All amino acids have common names, which are abbreviated by a three-letter or one-letter designation. For example, glycine is often written as the three-letter abbreviation **Gly,** or the one-letter abbreviation **G.** These abbreviations are also given in Table 16.2.

Amino acids never exist in nature as neutral molecules with all uncharged atoms. Since amino acids contain a base (NH_2 group) and an acid (COOH), proton transfer from the acid to the base forms a salt called a **zwitterion,** which contains both a positive and a negative charge. These salts have high melting points and are water soluble.

The acid–base chemistry of amino acids is discussed in greater detail in Section 16.3. The structures in Table 16.2 show the charged form of the amino acids at the physiological pH of the blood.

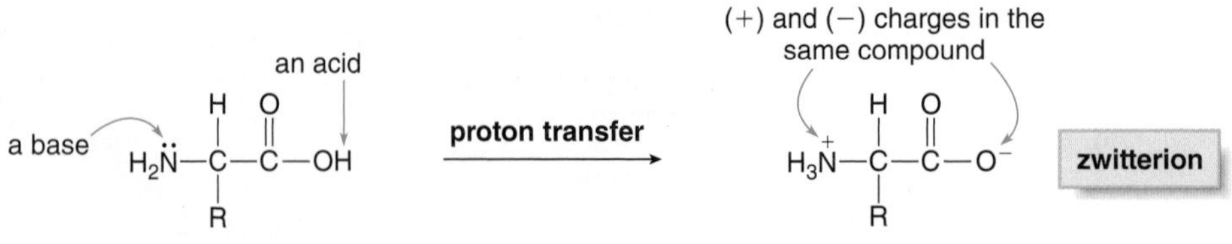

a base — $H_2\ddot{N}-C(H)(R)-C(=O)-OH$ — an acid

This neutral form of an amino acid does ***not*** exist.

proton transfer →

(+) and (−) charges in the same compound

$H_3\overset{+}{N}-C(H)(R)-C(=O)-O^-$ **zwitterion**

a **salt**

This salt is the neutral form of an amino acid.

Humans can synthesize only 10 of the 20 amino acids needed for proteins. The remaining 10, called **essential amino acids,** must be obtained from the diet and consumed on a regular, almost daily basis. Diets that include animal products readily supply all of the needed amino acids. Since plant sources generally do not have sufficient amounts of all of the essential amino acids, vegetarian diets must be carefully balanced. Grains—wheat, rice, and corn—are low in lysine, and legumes—beans, peas, and peanuts—are low in methionine, but a combination of these foods provides all the needed amino acids.

HEALTH NOTE

A diet of rice and tofu provides all essential amino acids. A peanut butter sandwich on wheat bread does the same.

PROBLEM 16.1

In addition to the amino and carboxyl groups, what other functional groups are present in each amino acid: (a) asparagine; (b) serine; (c) cysteine?

PROBLEM 16.2

How do the OH groups in Ser, Thr, and Tyr differ?

16.2B Stereochemistry of Amino Acids

Except for the simplest amino acid, glycine, **all other amino acids have a chirality center—a carbon bonded to four different groups—on the α carbon.** Thus, an amino acid like alanine ($R = CH_3$) has two possible enantiomers, drawn below in both three-dimensional representations with wedges and dashed bonds, and Fischer projections.

L isomer: COO^- (top), $H_3\overset{+}{N}$ (left), H (right), CH_3 (bottom) = L-alanine: $H_3\overset{+}{N}$–C–H with COO^- and CH_3

naturally occurring enantiomer

D-alanine: H–C–$\overset{+}{N}H_3$ with COO^- and CH_3 = D isomer: COO^- (top), H (left), $\overset{+}{N}H_3$ (right), CH_3 (bottom)

Like monosaccharides, the prefixes D and L are used to designate the arrangement of groups on the chirality center of amino acids. When drawn with a vertical carbon chain having the $-COO^-$ group at the top and the R group at the bottom,

HEALTH NOTE

The essential amino acid leucine is sold as a dietary supplement that is used by body builders to help prevent muscle loss and heal muscle tissue after injury.

- **L Amino acids have the $-NH_3^+$ group on the *left* side in the Fischer projection. Common naturally occurring amino acids are L isomers.**
- **D Amino acids have the $-NH_3^+$ group on the *right* side in the Fischer projection. D Amino acids occur infrequently in nature.**

SAMPLE PROBLEM 16.1

Draw the Fischer projection for each amino acid: (a) L-leucine; (b) D-cysteine.

Analysis

To draw an amino acid in a Fischer projection, place the $-COO^-$ group at the top and the R group at the bottom. The L isomer has the $-NH_3^+$ on the left side and the D isomer has the $-NH_3^+$ on the right side.

Solution

a. For leucine, $R = CH_2CH(CH_3)_2$

COO^- (top), $H_3\overset{+}{N}$ (left), H (right), $CH_2CH(CH_3)_2$ (bottom)

$\overset{+}{N}H_3$ on left
L isomer

b. For cysteine, $R = CH_2SH$

COO^- (top), H (left), $\overset{+}{N}H_3$ (right), CH_2SH (bottom)

$\overset{+}{N}H_3$ on right
D isomer

Table 16.2 The 20 Common Naturally Occurring Amino Acids

Neutral Amino Acids					
Name	**Structure**	**Abbreviations**	**Name**	**Structure**	**Abbreviations**
Alanine	$H_3\overset{+}{N}-\underset{CH_3}{\overset{H}{C}}-COO^-$	Ala A	Phenylalanine*	$H_3\overset{+}{N}-\underset{CH_2-C_6H_5}{\overset{H}{C}}-COO^-$	Phe F
Asparagine	$H_3\overset{+}{N}-\underset{CH_2CONH_2}{\overset{H}{C}}-COO^-$	Asn N	Proline	(pyrrolidine ring) $-COO^-$; $\overset{+}{N}$ with H, H	Pro P
Cysteine	$H_3\overset{+}{N}-\underset{CH_2SH}{\overset{H}{C}}-COO^-$	Cys C	Serine	$H_3\overset{+}{N}-\underset{CH_2OH}{\overset{H}{C}}-COO^-$	Ser S
Glutamine	$H_3\overset{+}{N}-\underset{CH_2CH_2CONH_2}{\overset{H}{C}}-COO^-$	Gln Q	Threonine*	$H_3\overset{+}{N}-\underset{CH(OH)CH_3}{\overset{H}{C}}-COO^-$	Thr T
Glycine	$H_3\overset{+}{N}-\underset{H}{\overset{H}{C}}-COO^-$	Gly G	Tryptophan*	$H_3\overset{+}{N}-\underset{CH_2-\text{(indole, N–H)}}{\overset{H}{C}}-COO^-$	Trp W
Isoleucine*	$H_3\overset{+}{N}-\underset{CH(CH_3)CH_2CH_3}{\overset{H}{C}}-COO^-$	Ile I	Tyrosine	$H_3\overset{+}{N}-\underset{CH_2-C_6H_4-OH}{\overset{H}{C}}-COO^-$	Tyr Y
Leucine*	$H_3\overset{+}{N}-\underset{CH_2CH(CH_3)_2}{\overset{H}{C}}-COO^-$	Leu L	Valine*	$H_3\overset{+}{N}-\underset{CH(CH_3)_2}{\overset{H}{C}}-COO^-$	Val V
Methionine*	$H_3\overset{+}{N}-\underset{CH_2CH_2SCH_3}{\overset{H}{C}}-COO^-$	Met M			

Essential amino acids are labeled with an asterisk (*).

(continued on next page)

Table 16.2 *(continued)*

Acidic Amino Acids			Basic Amino Acids		
Name	**Structure**	**Abbreviations**	**Name**	**Structure**	**Abbreviations**
Aspartic acid	$H_3\overset{+}{N}-\underset{CH_2COO^-}{\overset{H}{C}}-COO^-$	Asp D	Arginine*	$H_3\overset{+}{N}-\underset{(CH_2)_3-\overset{H}{N}-\overset{\overset{+}{N}H_2}{\overset{\Vert}{C}}-NH_2}{\overset{H}{C}}-COO^-$	Arg R
Glutamic acid	$H_3\overset{+}{N}-\underset{CH_2CH_2COO^-}{\overset{H}{C}}-COO^-$	Glu E	Histidine*	$H_3\overset{+}{N}-\underset{CH_2-(\text{imidazole ring, N–H})}{\overset{H}{C}}-COO^-$	His H
			Lysine*	$H_3\overset{+}{N}-\underset{(CH_2)_4\overset{+}{N}H_3}{\overset{H}{C}}-COO^-$	Lys K

Essential amino acids are labeled with an asterisk (*).

CONSUMER NOTE

MSG (monosodium glutamate), the sodium salt of glutamic acid (Table 16.2), is a common food additive used as a flavor enhancer in canned soups and other processed products.

PROBLEM 16.3

Draw both enantiomers of each amino acid in Fischer projections and label them as D or L: (a) phenylalanine; (b) methionine.

PROBLEM 16.4

Which of the following amino acids is naturally occurring? By referring to the structures in Table 16.2, name each amino acid and include its D or L designation in the name.

a. Fischer projection: top COO^-; left $H_3\overset{+}{N}$; right H; bottom CH_2OH

b. Fischer projection: top COO^-; left H; right $\overset{+}{N}H_3$; bottom $CH_2CH_2COO^-$

c. Fischer projection: top COO^-; left $H_3\overset{+}{N}$; right H; bottom CH_2COO^-

16.3 Acid–Base Behavior of Amino Acids

As mentioned in Section 16.2, an amino acid contains both a basic amino group (NH_2) and an acidic carboxyl group (COOH). As a result, proton transfer from the acid to the base forms a **zwitterion,** a salt that contains both a positive and a negative charge. The zwitterion is *neutral;* that is, the net charge on the salt is zero.

$$H_2\ddot{N}-\underset{R}{\overset{H}{C}}-\overset{O}{\overset{\Vert}{C}}-OH \xrightarrow{\text{proton transfer}} H_3\overset{+}{N}-\underset{R}{\overset{H}{C}}-\overset{O}{\overset{\Vert}{C}}-O^-$$

The lone pair can bond to a proton.

The carboxyl group can donate a proton.

zwitterion

In actuality, **an amino acid can exist in different forms, depending on the pH of the aqueous solution in which it is dissolved.** When the pH of a solution is around 6, alanine (R = CH_3) and other neutral amino acids exist in their zwitterionic form (**A**), having no net charge. In this form, the carboxyl group bears a net negative charge—it is a **carboxylate anion**—and the amino group bears a net positive charge (an **ammonium cation**).

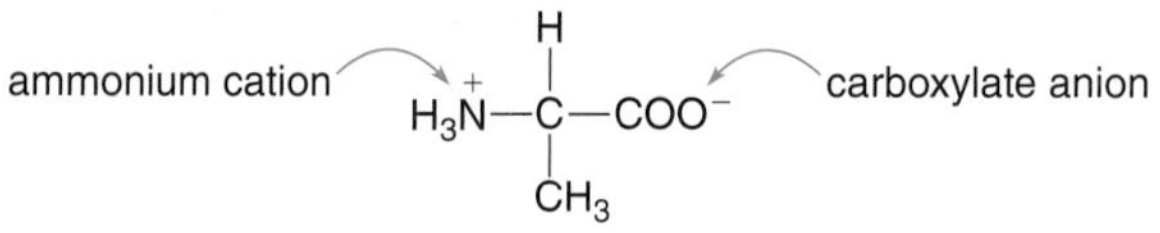

ammonium cation → $H_3\overset{+}{N}-C(H)(CH_3)-COO^-$ ← carboxylate anion

alanine
no net charge
A
pH ≈ 6

When strong acid is added to lower the pH to 2 or less, the carboxylate anion gains a proton and the **amino acid has a net positive charge** (form **B**).

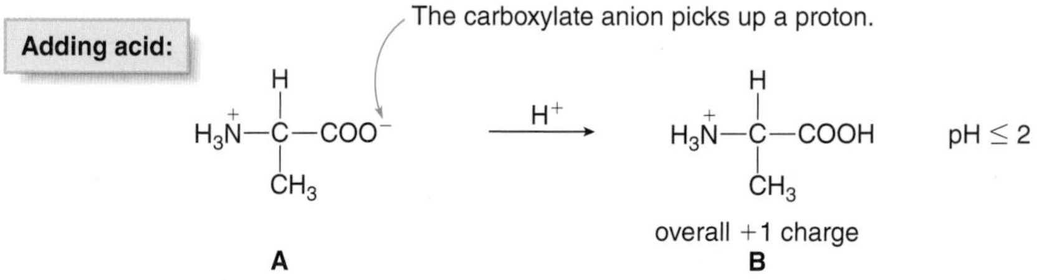

Adding acid:

The carboxylate anion picks up a proton.

$$H_3\overset{+}{N}-C(H)(CH_3)-COO^- \xrightarrow{H^+} H_3\overset{+}{N}-C(H)(CH_3)-COOH \quad pH \leq 2$$

A → overall +1 charge **B**

When strong base is added to **A** to raise the pH to 10 or higher, the ammonium cation loses a proton and the **amino acid has a net negative charge** (form **C**).

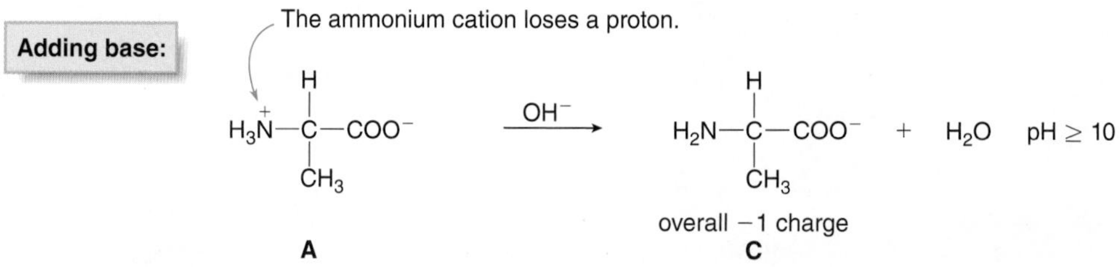

Adding base:

The ammonium cation loses a proton.

$$H_3\overset{+}{N}-C(H)(CH_3)-COO^- \xrightarrow{OH^-} H_2N-C(H)(CH_3)-COO^- + H_2O \quad pH \geq 10$$

A → overall −1 charge **C**

Thus, **alanine exists in one of three different forms depending on the pH of the solution in which it is dissolved.** At the physiological pH of 7.4, neutral amino acids are primarily in their zwitterionic forms.

- **The pH at which the amino acid exists primarily in its neutral form is called its *isoelectric point*, abbreviated as p*I*.**

The isoelectric points of neutral amino acids are generally around 6. Acidic amino acids (Table 16.2), which have an additional carboxyl group that can lose a proton, have lower p*I* values (around 3). The three basic amino acids, which have an additional basic nitrogen atom that can accept a proton, have higher p*I* values (7.6–10.8).

SAMPLE PROBLEM 16.2

Draw the structure of the amino acid glycine at each pH: (a) 6; (b) 2; (c) 11.

Analysis

A neutral amino acid exists in its zwitterionic form (no net charge) at its isoelectric point, which is pH ≈ 6. The zwitterionic forms of neutral amino acids appear in Table 16.2. At low pH (≤ 2), the carboxylate anion is protonated and the amino acid has a net positive (+1) charge. At high pH (≥ 10), the ammonium cation loses a proton and the amino acid has a net negative (–1) charge.

Solution

a. At pH = 6, the neutral, zwitterionic form of glycine predominates.

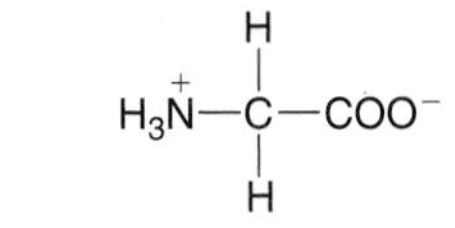

$$H_3\overset{+}{N}-CH_2-COO^-$$

neutral
pH = 6

b. At pH = 2, glycine has a net +1 charge.

$$H_3\overset{+}{N}-CH_2-COOH$$

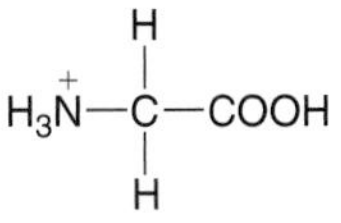

+1 charge
pH = 2

c. At pH = 11, glycine has a net −1 charge.

$$H_2N-CH_2-COO^-$$

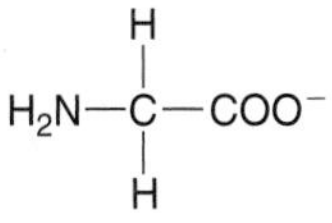

−1 charge
pH = 11

PROBLEM 16.5

Draw the structure of the amino acid valine at each pH: (a) 6; (b) 2; (c) 11. Which form predominates at valine's isoelectric point?

PROBLEM 16.6

Identify the amino acid shown with all uncharged atoms in the ball-and-stick model, and draw the neutral, positively charged, and negatively charged forms of the amino acid.

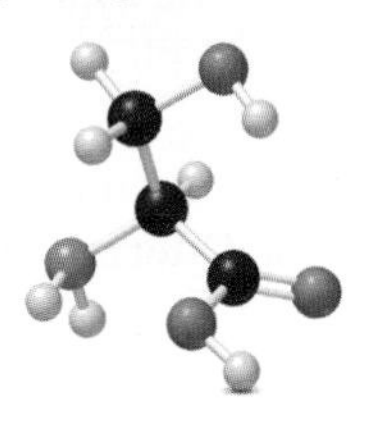

PROBLEM 16.7

Draw the positively charged, neutral, and negatively charged forms for the amino acid phenylalanine. Which species predominates at pH 11? Which species predominates at pH 1?

16.4 Peptides

When amino acids are joined together by amide bonds, they form larger molecules called **peptides** and **proteins.**

- A *dipeptide* has two amino acids joined together by *one* amide bond.
- A *tripeptide* has three amino acids joined together by *two* amide bonds.

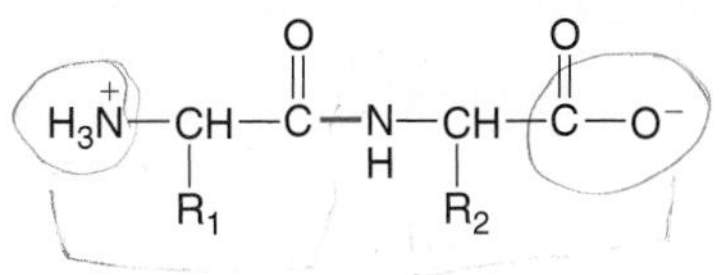

$$H_3\overset{+}{N}-CH(R_1)-C(=O)-NH-CH(R_2)-C(=O)-O^-$$

dipeptide

$$H_3\overset{+}{N}-CH(R_1)-C(=O)-NH-CH(R_2)-C(=O)-NH-CH(R_3)-C(=O)-O^-$$

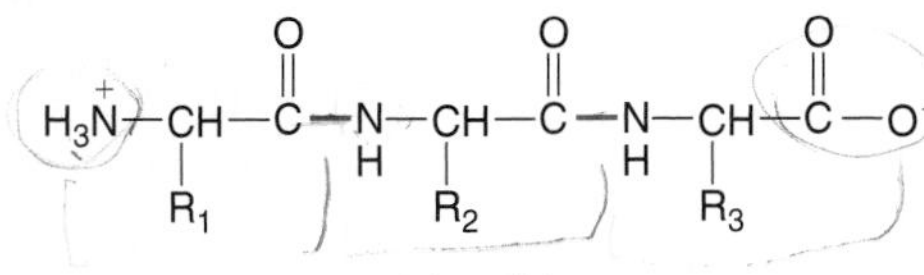

tripeptide

[Amide bonds are shown in red.]

Polypeptides and **proteins** both have many amino acids joined together in long linear chains, but the term **protein** is usually reserved for polymers of more than 40 amino acids.

- The amide bonds in peptides and proteins are called *peptide bonds.*
- The individual amino acids are called *amino acid residues.*

To form a dipeptide, **the $-NH_3^+$ group of one amino acid forms an amide bond with the carboxylate ($-COO^-$) of another amino acid, and the elements of H_2O are removed.** For example, reaction of the $-COO^-$ group of alanine with the $-NH_3^+$ group of serine forms a dipeptide with one new amide bond, as shown. The dipeptide has an ammonium cation ($-NH_3^+$) at one end of its chain and a carboxylate anion ($-COO^-$) at the other.

$$H_3\overset{+}{N}-CH(CH_3)-CO-O^- + H-\overset{+}{N}H_2-CH(CH_2OH)-CO-O^- \longrightarrow H_3\overset{+}{N}-CH(CH_3)-CO-NH-CH(CH_2OH)-CO-O^- + H_2O$$

Ala Ser

reacting functional groups

new amide bond

- **The amino acid with the free $-NH_3^+$ group on the α carbon is called the N-terminal amino acid.**
- **The amino acid with the free $-COO^-$ group on the α carbon is called the C-terminal amino acid.**

By convention, **the N-terminal amino acid is always written at the *left* end of the chain and the C-terminal amino acid at the *right*.**

$$H_3\overset{+}{N}-CH(CH_3)-CO-NH-CH(CH_2OH)-CO-O^-$$

N-terminal amino acid

C-terminal amino acid

alanylserine
Ala–Ser

Peptides are named as derivatives of the C-terminal amino acid. To name a peptide:

- **Name the C-terminal amino acid using the names in Table 16.2.**
- **Name all other amino acids from left to right as substituents of the C-terminal amino acid. Change the *-ine* or *-ic acid* ending of the amino acid name to the suffix *-yl*.**

Thus, the dipeptide, which has serine as its C-terminal amino acid, is named as *alanylserine.*

The peptide can be abbreviated by writing the one- or three-letter symbols for the amino acids in the chain from the N-terminal to the C-terminal end. Thus, Ala–Ser has alanine at the N-terminal end and serine at the C-terminal end.

SAMPLE PROBLEM 16.3

Label the N-terminal and C-terminal amino acids in the following tripeptide. Identify the individual amino acids. What is the name of the tripeptide?

$$H_3\overset{+}{N}-CH(CH_2CH(CH_3)_2)-CO-NH-CH(CH_3)-CO-NH-CH(CH_2-C_6H_4-OH)-CO-O^-$$

Analysis

- The N-terminal amino acid has an $-NH_3^+$ group on the α carbon, and the C-terminal amino acid has a $-COO^-$ group on the α carbon.

- To identify the individual amino acids, locate the amide bonds, and compare the side chain of each amino acid with the structures in Table 16.2.
- To name the peptide: [1] Name the C-terminal amino acid. [2] Name the other amino acids as substituents by changing the *-ine* (or *-ic acid*) ending to the suffix *-yl.* Place the names of the substituent amino acids in order from left to right.

Solution

The amide bonds that join the amino acids together are shown in red. The tripeptide contains leucine (N-terminal), alanine, and tyrosine (C-terminal).

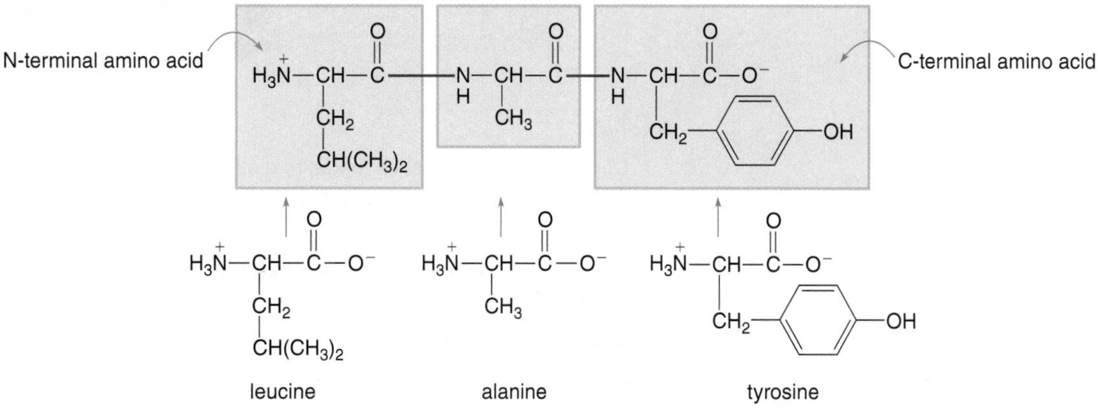

N-terminal amino acid → $H_3\overset{+}{N}-CH(CH_2CH(CH_3)_2)-CO-NH-CH(CH_3)-CO-NH-CH(CH_2-C_6H_4-OH)-COO^-$ ← C-terminal amino acid

$H_3\overset{+}{N}-CH(CH_2CH(CH_3)_2)-COO^-$ leucine

$H_3\overset{+}{N}-CH(CH_3)-COO^-$ alanine

$H_3\overset{+}{N}-CH(CH_2-C_6H_4-OH)-COO^-$ tyrosine

The tripeptide is named as a derivative of the C-terminal amino acid, tyrosine, with leucine and alanine as substituents; thus, the tripeptide is named: **leucylalanyltyrosine.**

PROBLEM 16.8

Identify the N-terminal and C-terminal amino acid in each peptide.

a. 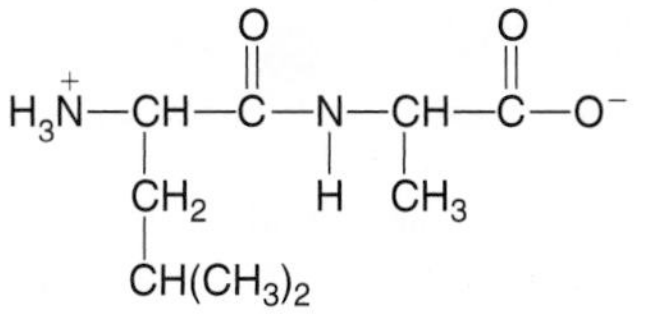 $H_3\overset{+}{N}-CH(CH_2CH(CH_3)_2)-CO-NH-CH(CH_3)-COO^-$

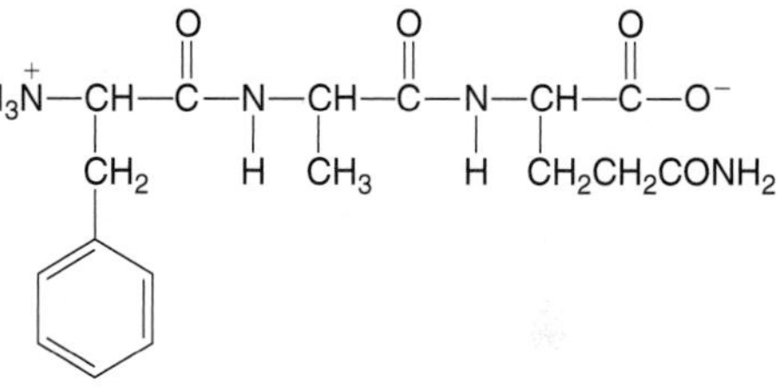

b. Arg–His–Asn–Tyr

c. $H_3\overset{+}{N}-CH(CH_2C_6H_5)-CO-NH-CH(CH_3)-CO-NH-CH(CH_2CH_2CONH_2)-COO^-$

d. Val–Thr–Pro–Phe

PROBLEM 16.9

(a) Identify the N-terminal amino acid in the tetrapeptide alanylglycylleucylmethionine. (b) What is the C-terminal amino acid? (c) Write the peptide using three-letter symbols for the amino acids.

PROBLEM 16.10

Identify the individual amino acids in each dipeptide, and then name the dipeptide using three-letter abbreviations.

a. 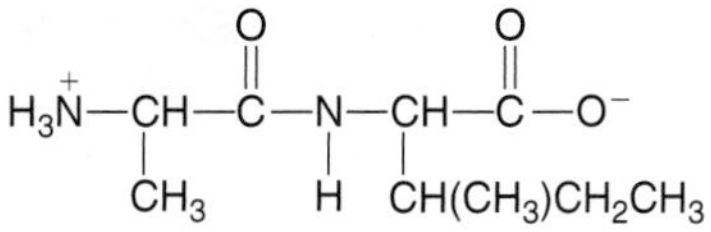 $H_3\overset{+}{N}-CH(CH_3)-CO-NH-CH(CH(CH_3)CH_2CH_3)-COO^-$

b. 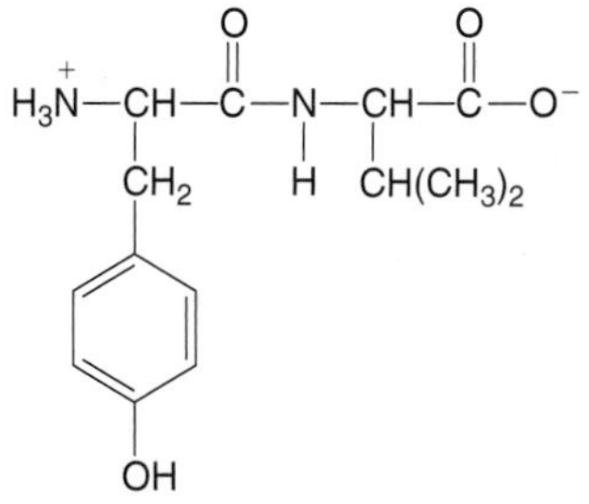 $H_3\overset{+}{N}-CH(CH_2-C_6H_4-OH)-CO-NH-CH(CH(CH_3)_2)-COO^-$

SAMPLE PROBLEM 16.4

Locate the peptide bond in the dipeptide shown in the ball-and-stick model, and identify the amino acids that form the dipeptide. Name the dipeptide using the three-letter abbreviations for the amino acids.

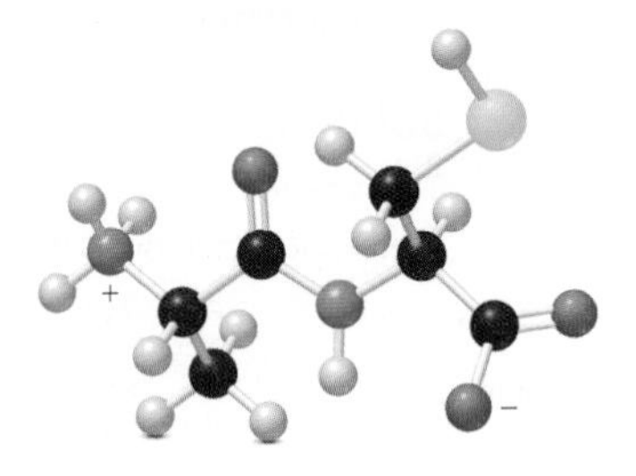

Analysis

- Convert the ball-and-stick model to a structural formula. The peptide bond is the amide bond that joins the two amino acids together.
- Identify the side chains of the amino acids and use Table 16.2 to determine the names.
- Name the dipeptide by writing the three-letter abbreviations for the amino acids with the N-terminal amino acid first, followed by the C-terminal amino acid.

Solution

The peptide bond is the amide bond that contains a carbon atom doubly bonded to oxygen and singly bonded to nitrogen.

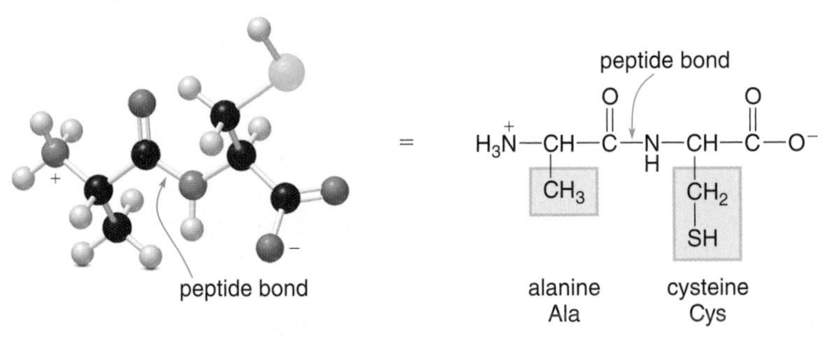

The side chains of the two amino acids, the R groups that extend from the peptide backbone, identify them as alanine and cysteine (Table 16.2). The dipeptide is named with the N-terminal amino acid (alanine) first, followed by the C-terminal amino acid (cysteine), using the three-letter abbreviations: **Ala–Cys.**

PROBLEM 16.11

Locate the peptide bond in the dipeptide shown in the ball-and-stick model, and identify the amino acids that form the dipeptide. Name the dipeptide using the three-letter abbreviations for the amino acids.

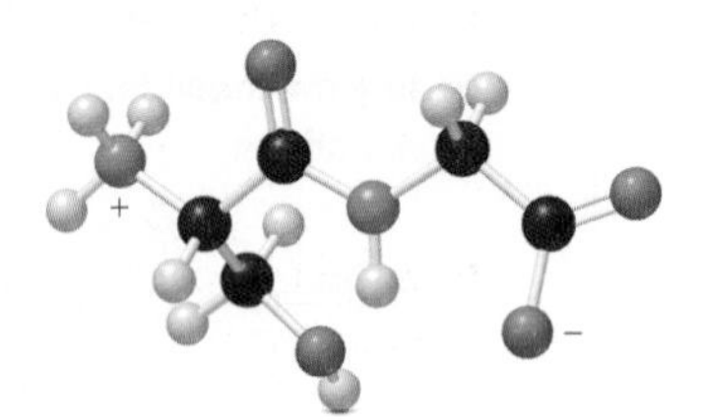

16.5 FOCUS ON THE HUMAN BODY Biologically Active Peptides

Many relatively simple peptides have important biological functions.

16.5A Neuropeptides—Enkephalins and Pain Relief

Enkephalins, peptides synthesized in the brain, act as pain killers and sedatives by binding to pain receptors. Two enkephalins that differ in the identity of only one amino acid are known. Met-enkephalin contains a C-terminal methionine residue, while leu-enkephalin contains a C-terminal leucine.

Tyr–Gly–Gly–Phe–Met
met-enkephalin

Tyr–Gly–Gly–Phe–Leu
leu-enkephalin

The addictive narcotic analgesics morphine and heroin bind to the same receptors as the enkephalins, and thus produce a similar physiological response. Enkephalins are related to a group of larger polypeptides called **endorphins** that contain 16–31 amino acids. Endorphins also block pain and are thought to produce the feeling of well-being experienced by an athlete after excessive or strenuous exercise.

PROBLEM 16.12

(a) Label the four amide bonds in met-enkephalin. (b) What N-terminal amino acid is present in both enkephalins?

16.5B Peptide Hormones—Oxytocin and Vasopressin

Oxytocin and **vasopressin** are cyclic peptide hormones secreted by the pituitary gland. Their sequences are identical except for two amino acids.

N-terminal amino acid

oxytocin: Cys—Tyr—Ile—Gln—Asn—Cys—Pro—Leu—GlyNH$_2$ (Cys 1 and Cys 6 joined by S—S, disulfide bond)

vasopressin: Cys—Tyr—Phe—Gln—Asn—Cys—Pro—Arg—GlyNH$_2$ (Cys 1 and Cys 6 joined by S—S, disulfide bond)

Oxytocin stimulates the contraction of uterine muscles, and it initiates the flow of milk in nursing mothers (Figure 16.2). Oxytocin, sold under the trade names Pitocin and Syntocinon, is used to induce labor.

Vasopressin, also called antidiuretic hormone (ADH), targets the kidneys and helps to keep the electrolytes in body fluids in the normal range. Vasopressin is secreted when the body is dehydrated and causes the kidneys to retain fluid, thus decreasing the volume of the urine (Figure 16.3).

The N-terminal amino acid in both hormones is a cysteine residue, and the C-terminal residue is glycine. Instead of a free carboxylate (–COO$^-$), both peptides have an amide (–CONH$_2$) at the C-terminal end, so this is indicated with the additional NH$_2$ group drawn at the end of the chain. The structure of both peptides includes a **disulfide bond,** a form of covalent bonding in which the –SH groups from two cysteine residues are oxidized to form a sulfur–sulfur bond (Section 12.7). In oxytocin and vasopressin, the disulfide bonds make the peptides cyclic.

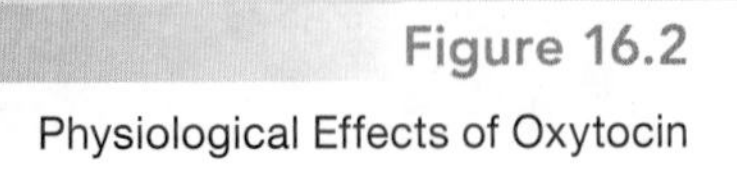

Figure 16.2
Physiological Effects of Oxytocin

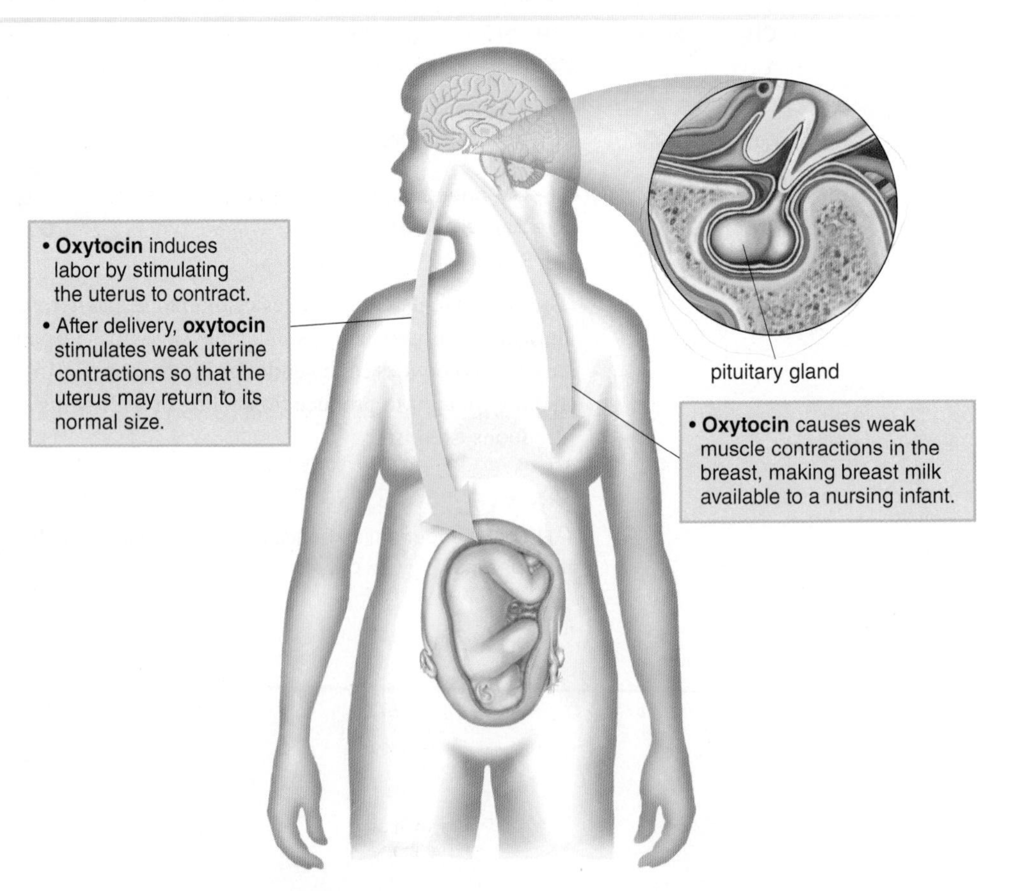
• Oxytocin induces labor by stimulating the uterus to contract.
• After delivery, oxytocin stimulates weak uterine contractions so that the uterus may return to its normal size.
pituitary gland
• Oxytocin causes weak muscle contractions in the breast, making breast milk available to a nursing infant.

Figure 16.3

Vasopressin—An Antidiuretic Hormone

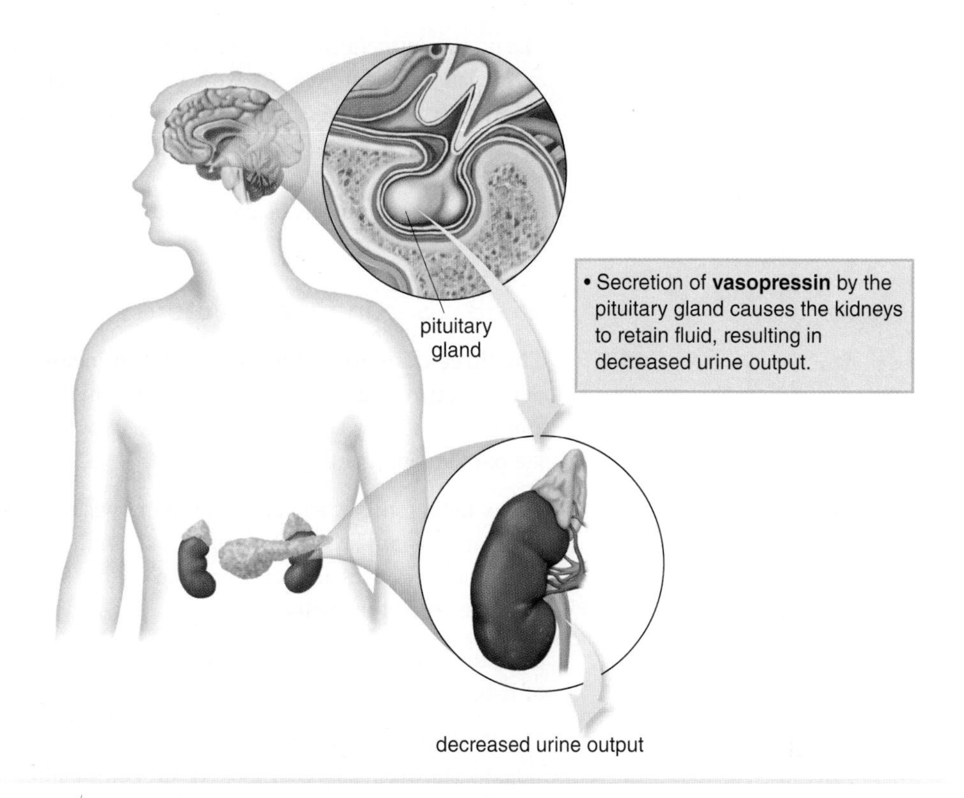
pituitary gland
• Secretion of **vasopressin** by the pituitary gland causes the kidneys to retain fluid, resulting in decreased urine output.
decreased urine output

16.6 Proteins

To understand proteins, the large polymers of amino acids that are responsible for so much of the structure and function of all living cells, we must learn about four levels of structure, called the **primary, secondary, tertiary, and quaternary structure** of proteins.

16.6A Primary Structure

The *primary structure* of a protein is the particular sequence of amino acids that is joined together by peptide bonds. The most important element of this primary structure is the amide bond that joins the amino acids.

The carbonyl carbon of the amide has **trigonal planar** geometry. All six atoms involved in the peptide bond lie in the same plane. All bond angles are 120° and the C═O and N—H bonds are oriented 180° from each other. As a result, the backbone of the protein adopts a zigzag arrangement as shown in the three-dimensional structure of a portion of a protein molecule.

C—C(═O)—N(H)—C (amide bond)

peptide bond

—N(H)—CH(R_1)—C(═O)—N(H)—CH(R_2)—C(═O)—N(H)—CH(R_3)—C(═O)—N(H)—CH(R_4)—C(═O)—

[Amide bonds are shown in red.]

120°

120°

These six atoms lie in a plane.

R_1 R_2 R_3 R_4

The primary structure of a protein—the exact sequence of amino acids—determines all properties and function of a protein. As we will see in Section 16.7, substitution of a single amino acid by a different amino acid can result in very different properties.

PROBLEM 16.13

Can two *different* proteins be composed of the same number and type of amino acids?

16.6B Secondary Structure

The three-dimensional arrangement of localized regions of a protein is called its secondary structure. These regions arise due to hydrogen bonding between the N—H proton of one amide and the C═O oxygen of another. Two arrangements that are particularly stable are called the **α-helix** and the **β-pleated sheet.**

The **α-helix** forms when a peptide chain twists into a right-handed or clockwise spiral, as shown in Figure 16.4a. The C═O group of one amino acid is hydrogen bonded to an N—H group four amino acid residues farther along the chain. The R groups of the amino acids extend outward from the core of the helix.

Both the myosin in muscle and the α-keratin in hair are proteins composed almost entirely of α-helices.

The **β-pleated sheet** forms when two or more peptide chains, called **strands,** line up side-by-side, as shown in Figure 16.4b. Hydrogen bonding often occurs between the N—H and C═O groups of nearby amino acid residues. The R groups are oriented above and below the plane of the sheet, and alternate from one side to the other along a given strand.

Figure 16.4 Secondary Structure of Proteins

a. The right-handed α-helix

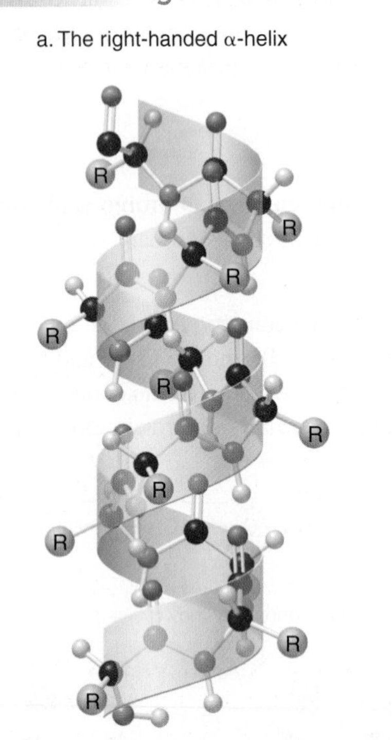

In the α-helix, all C═O bonds are pointing up and all N—H bonds are pointing down.

b. The β-pleated sheet

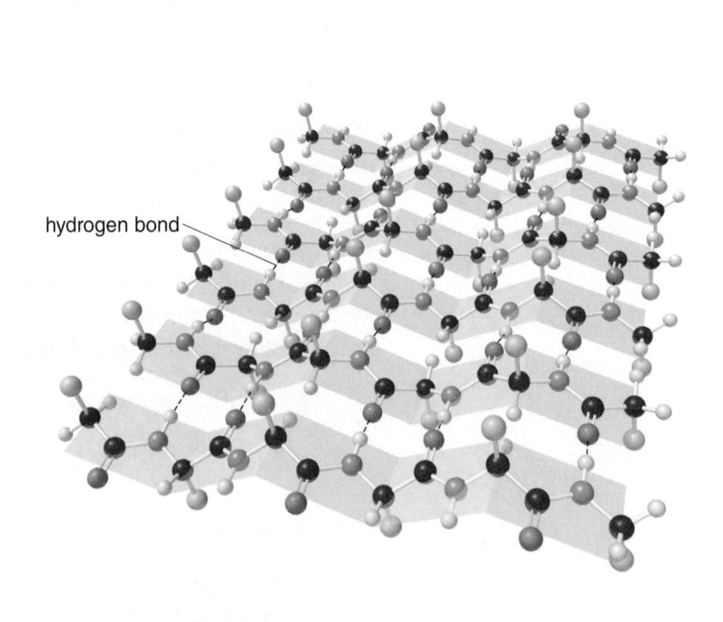

The β-pleated sheet consists of extended strands of the peptide chains held together by hydrogen bonding. The C═O and N—H bonds lie in the plane of the sheet, while the R groups (shown as yellow balls) alternate above and below the plane.

Figure 16.5 Lysozyme

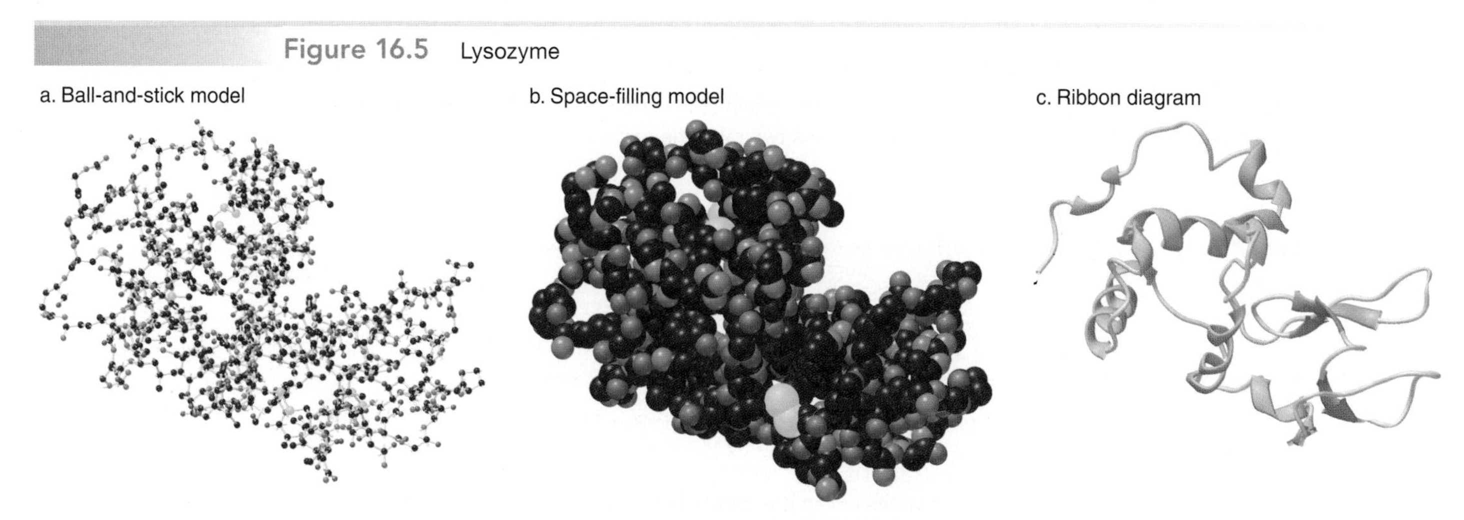

(a) The ball-and-stick model of lysozyme shows the protein backbone with color-coded C, N, O, and S atoms. Individual amino acids are most clearly located using this representation. (b) The space-filling model uses color-coded balls for each atom in the backbone of the enzyme and illustrates how the atoms fill the space they occupy. (c) The ribbon diagram shows regions of α-helix and β-sheet that are not clearly in evidence in the other two representations.

The β-pleated sheet arrangement is favored by amino acids with small R groups, like alanine and glycine. With larger R groups, steric interactions prevent the chains from getting close together, so the sheet cannot be stabilized by hydrogen bonding.

Most proteins have regions of α-helix and β-pleated sheet, in addition to other regions that cannot be characterized by either of these arrangements. Shorthand symbols are often used to indicate these regions of secondary structure. In particular, a flat helical ribbon is used for the α-helix, while a flat wide arrow is used for the β-pleated sheet. These representations are often used in **ribbon diagrams** to illustrate protein structure.

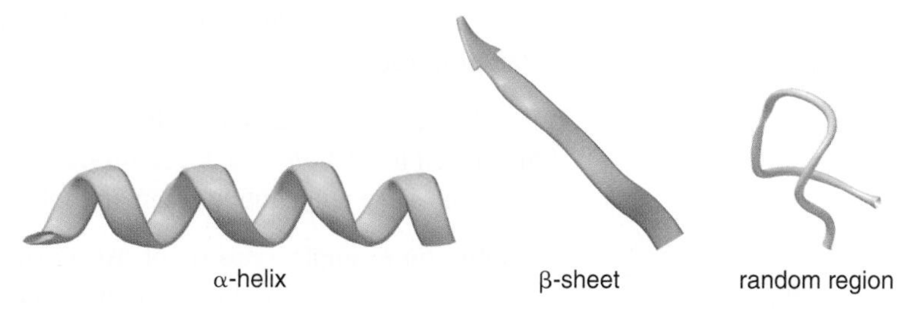

Proteins are drawn in a variety of ways to show different aspects of their structure. Figure 16.5 illustrates three different representations of the protein lysozyme, an enzyme found in both plants and animals. Lysozyme catalyzes the hydrolysis of bonds in bacterial cell walls, weakening them, often causing the bacteria to burst.

16.6C Tertiary and Quaternary Structure

The three-dimensional shape adopted by the entire peptide chain is called its tertiary structure. A peptide generally folds into a shape that maximizes its stability. In the aqueous environment of the cell, proteins often fold in such a way as to place a large number of polar and charged groups on their outer surface, to maximize the dipole–dipole and hydrogen bonding interactions with water. This generally places most of the nonpolar side chains in the interior of the protein, where **London dispersion forces** between these hydrophobic groups help stabilize the molecule, too.

Figure 16.6 Insulin

Insulin is a small protein consisting of two polypeptide chains (designated as the **A** and **B** chains), held together by two disulfide bonds. An additional disulfide bond joins two cysteine residues within the **A** chain.

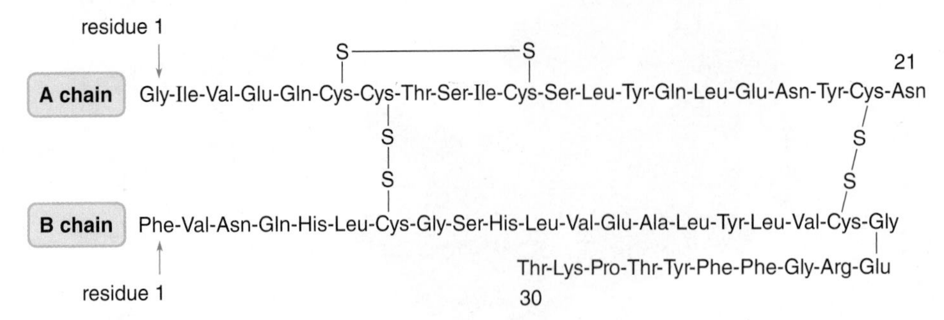

Synthesized by groups of cells in the pancreas called the islets of Langerhans, insulin is the protein that regulates blood glucose levels. A relative or complete lack of insulin results in diabetes. Many of the abnormalities associated with this disease can be controlled by the injection of insulin.

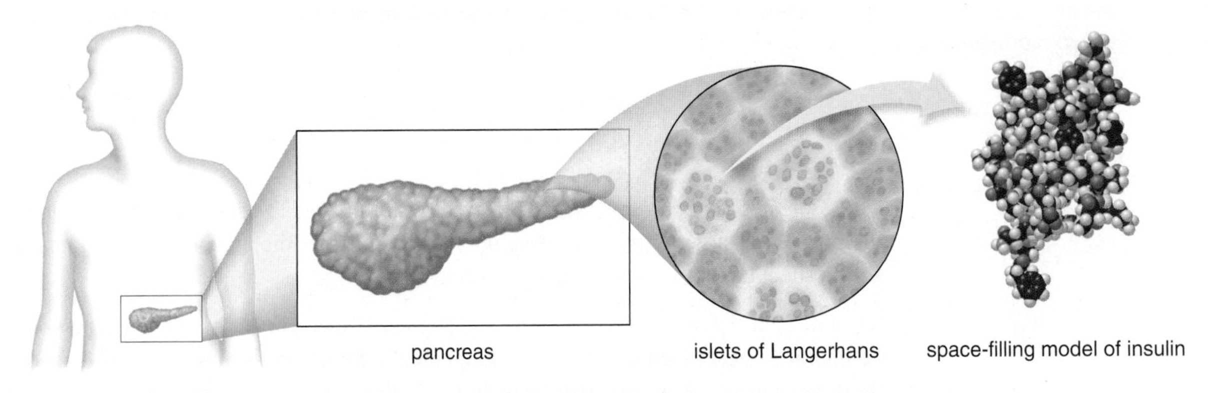

In addition, polar functional groups hydrogen bond with each other (not just water), and amino acids with charged side chains like $-COO^-$ and $-NH_3^+$ can stabilize tertiary structure by **electrostatic interactions.**

Finally, **disulfide bonds are the only covalent bonds that stabilize tertiary structure.** As mentioned in Section 16.5, these strong bonds form by the oxidation of two cysteine residues on either the same polypeptide chain or another polypeptide chain of the same protein.

Insulin, for example, consists of two separate polypeptide chains (labeled the **A** and **B** chains) that are covalently linked by two intermolecular disulfide bonds, as shown in Figure 16.6. The **A** chain, which also has an intramolecular disulfide bond, has 21 amino acid residues, whereas the **B** chain has 30.

Figure 16.7 schematically illustrates the many different kinds of intramolecular forces that stabilize the secondary and tertiary structures of polypeptide chains. Nearby amino acid residues that have only nonpolar carbon–carbon and carbon–hydrogen bonds are stabilized by London dispersion forces. Amino acids that contain hydroxyl (OH) and amino groups (NH_2) in their side chains can intermolecularly hydrogen bond to each other.

The shape adopted when two or more folded polypeptide chains come together into one protein complex is called the **quaternary structure** of the protein. Each individual polypeptide chain is called a **subunit** of the overall protein. **Hemoglobin,** for example, consists of two α and two β subunits held together by intermolecular forces in a compact three-dimensional shape. The unique function of hemoglobin is possible only when all four subunits are together.

Figure 16.7

The Stabilizing Interactions in Secondary and Tertiary Protein Structure

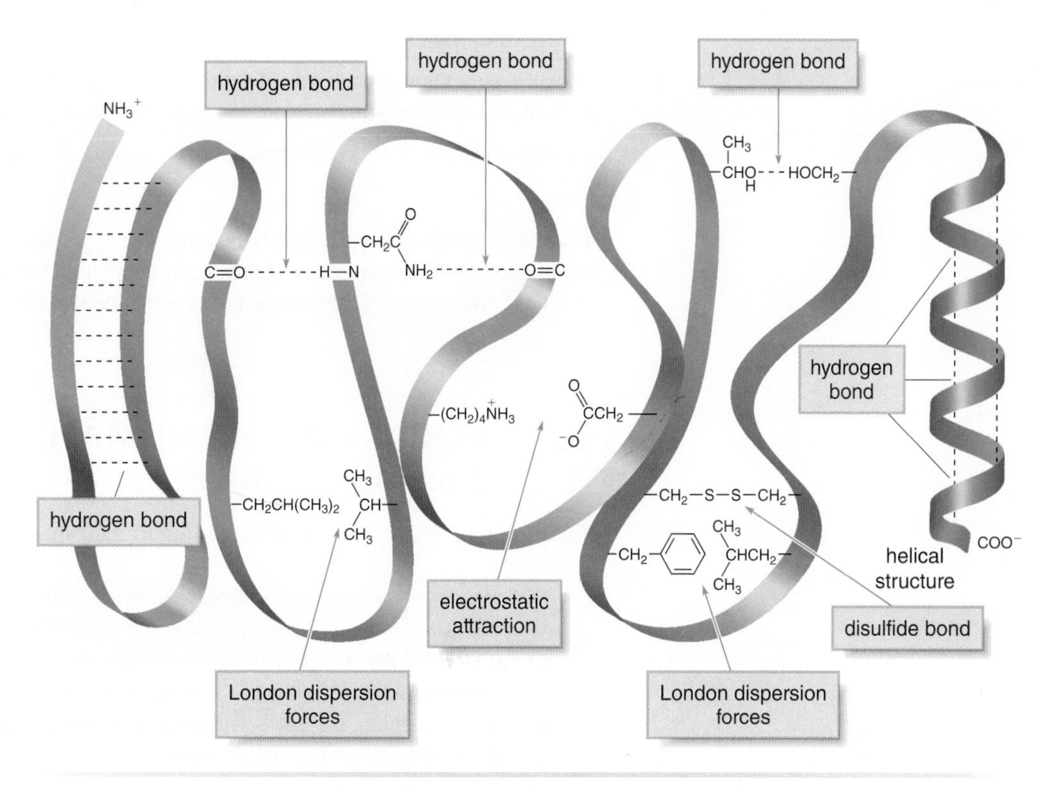

The four levels of protein structure are summarized in Figure 16.8.

Figure 16.8 The Primary, Secondary, Tertiary, and Quaternary Structure of Proteins

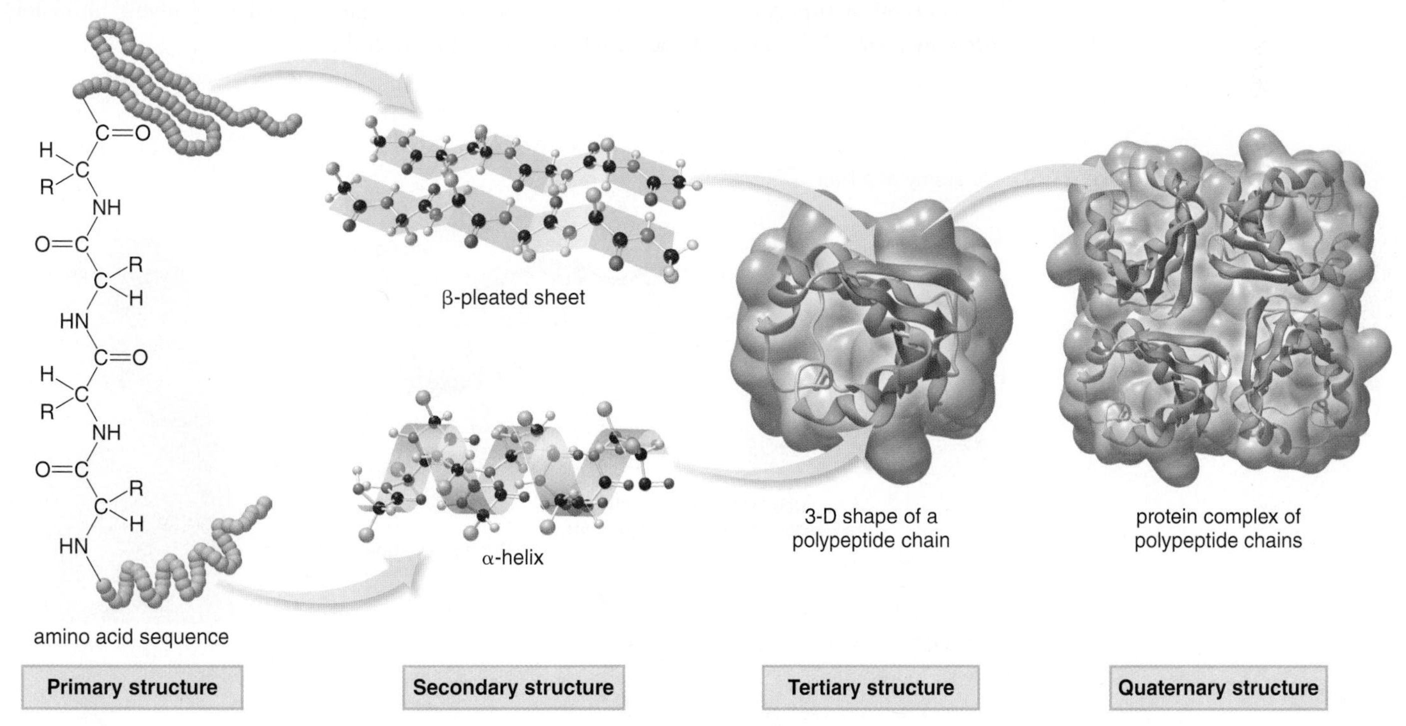

PROBLEM 16.14

Draw the structures of each pair of amino acids and indicate what types of intermolecular forces are present between the side chains. Use Figure 16.7 as a guide.

a. Ser and Tyr b. Val and Leu c. 2 Phe residues

PROBLEM 16.15

The fibroin proteins found in silk fibers consist of large regions of β-pleated sheets stacked one on top of another. The polypeptide sequence in these regions has the amino acid glycine at every other residue. Explain how this allows the β-pleated sheets to stack on top of each other.

16.7 FOCUS ON THE HUMAN BODY Common Proteins

Proteins are generally classified according to their three-dimensional shapes.

- ***Fibrous proteins*** **are composed of long linear polypeptide chains that are bundled together to form rods or sheets. These proteins are insoluble in water and serve structural roles, giving strength and protection to tissues and cells.**
- ***Globular proteins*** **are coiled into compact shapes with hydrophilic outer surfaces that make them water soluble. Enzymes and transport proteins are globular to make them soluble in blood and other aqueous environments.**

16.7A α-Keratins

α-Keratins are the proteins found in hair, hooves, nails, skin, and wool. They are composed almost exclusively of long sections of α-helix units, having large numbers of alanine and leucine residues. Since these nonpolar amino acids extend outward from the α-helix, these proteins are very insoluble in water. Two α-keratin helices coil around each other, forming a structure called a **supercoil** or **superhelix.** These, in turn, form larger and larger bundles of fibers, ultimately forming a strand of hair, as shown schematically in Figure 16.9.

Figure 16.9 Anatomy of a Hair

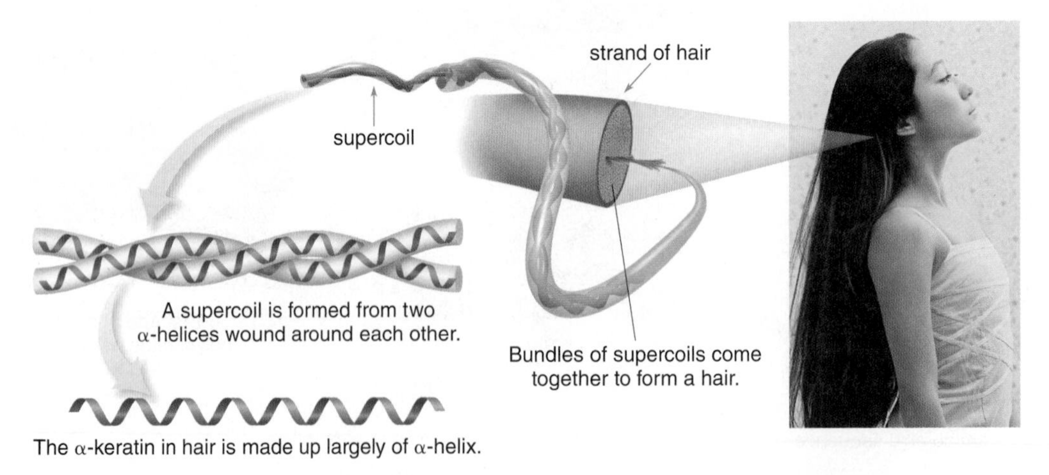
strand of hair
supercoil
A supercoil is formed from two α-helices wound around each other.
Bundles of supercoils come together to form a hair.
The α-keratin in hair is made up largely of α-helix.

α-Keratins also have a number of cysteine residues, and because of this, disulfide bonds are formed between adjacent helices. The number of disulfide bridges determines the strength of the material. Claws, horns, and fingernails have extensive networks of disulfide bonds, making them extremely hard.

16.7B Collagen

Collagen, the most abundant protein in vertebrates, is found in connective tissues such as bone, cartilage, tendons, teeth, and blood vessels. Glycine and proline account for a large fraction of its amino acid residues. Collagen forms an elongated left-handed helix, and then three of these helices wind around each other to form a right-handed **superhelix** or **triple helix.** The side chain of glycine is only a hydrogen atom, so the high glycine content allows the collagen superhelices to lie compactly next to each other, thus stabilizing the superhelices via hydrogen bonding (Figure 16.10).

Figure 16.10 The Triple Helix of Collagen

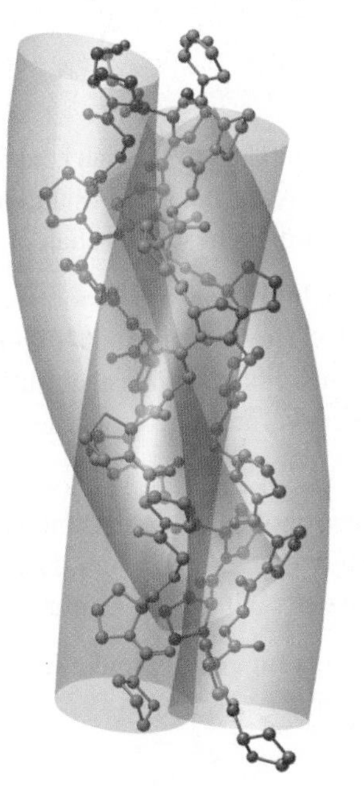

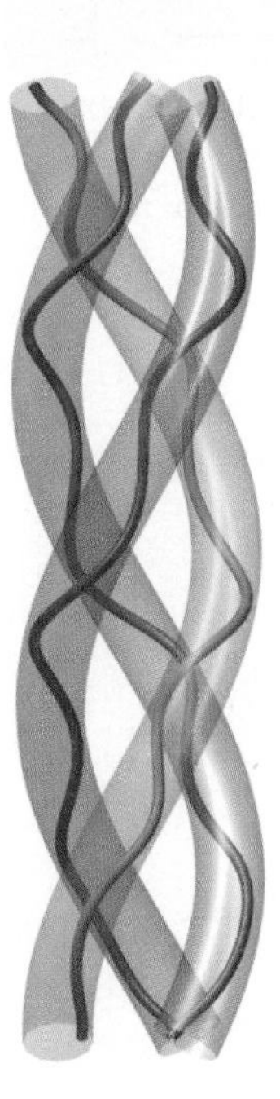

In collagen, three polypeptide chains having an unusual left-handed helix wind around each other in a right-handed triple helix.

Vitamin C is required for a reaction that modifies the original amino acids incorporated into the collagen chain so that strong hydrogen bonds form between the helices. When there is a deficiency of vitamin C in the diet, the collagen fibers do not form properly and scurvy results. Weakened blood vessels and poorly formed cartilage lead to spongy and bloody gums and dark purple skin lesions.

16.7C Hemoglobin and Myoglobin

Hemoglobin and **myoglobin,** two globular proteins, are called **conjugated proteins** because they are composed of a protein unit and a nonprotein molecule. In hemoglobin and myoglobin, the nonprotein unit is called **heme,** a complex organic compound containing the Fe^{2+} ion complexed with a large nitrogen-containing ring system. The Fe^{2+} ion of hemoglobin and myoglobin binds oxygen. Hemoglobin, which is present in red blood cells, transports oxygen to wherever it is needed in the body, whereas myoglobin stores oxygen in tissues.

The high concentration of myoglobin in a whale's muscles allows it to remain underwater for long periods of time.

heme

Myoglobin has 153 amino acid residues in a single polypeptide chain (Figure 16.11a). It has eight separate α-helical sections and a heme group held in a cavity inside the polypeptide. Myoglobin gives cardiac muscle its characteristic red color.

Hemoglobin consists of four polypeptide chains (two α subunits and two β subunits), each of which carries a heme unit (Figure 16.11b). Hemoglobin has more nonpolar amino acid residues than myoglobin. When each subunit is folded, some of these remain on the surface. The London dispersion forces between these hydrophobic groups are what stabilize the quaternary structure of the four subunits.

Carbon monoxide is poisonous because it binds to the Fe^{2+} of hemoglobin 200 times more strongly than does oxygen. Hemoglobin complexed with CO cannot carry O_2 from the lungs to the tissues. Without O_2 available to the tissues for metabolism, cells cannot function, and they die.

The properties of all proteins depend on their three-dimensional shape, and their shape depends on their primary structure—that is, their amino acid sequence. This is particularly well exemplified by

Figure 16.11

Protein Ribbon Diagrams for Myoglobin and Hemoglobin

a. Myoglobin

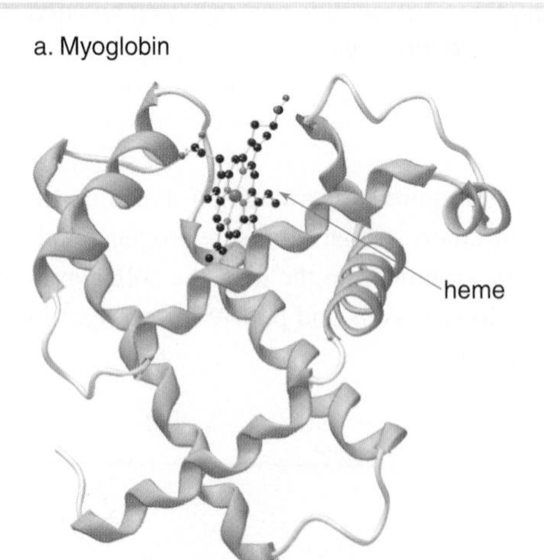

Myoglobin consists of a single polypeptide chain with a heme unit shown in a ball-and-stick model.

b. Hemoglobin

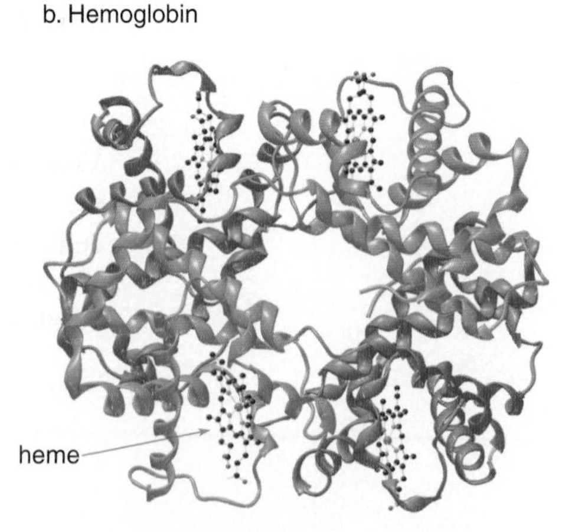

Hemoglobin consists of two α and two β chains shown in red and blue, respectively, and four heme units shown in ball-and-stick models.

HEALTH NOTE

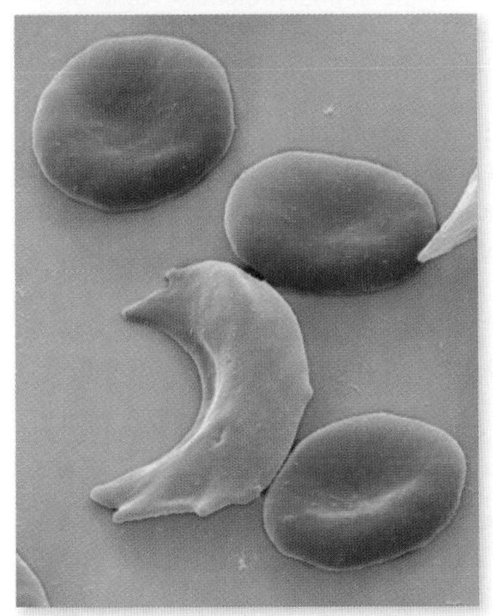

When red blood cells take on a "sickled" shape in persons with sickle cell disease, they block capillaries causing organ injury and they break easily leading to profound anemia. This devastating illness results from the change of a single amino acid in hemoglobin. Note the single sickled cell surrounded by three red cells with normal shape.

comparing normal hemoglobin with **sickle cell hemoglobin,** a mutant variation in which a single amino acid of both β subunits is changed from glutamic acid to valine. The replacement of one acidic amino acid (Glu) with one nonpolar amino acid (Val) changes the shape of hemoglobin, which has profound effects on its function. Red blood cells with sickle cell hemoglobin become elongated and crescent shaped, and they are unusually fragile. As a result, they rupture and occlude capillaries, causing pain and inflammation, leading to severe anemia and organ damage. The end result is often a painful and premature death.

This disease, called **sickle cell anemia,** is found almost exclusively among people originating from central and western Africa, where malaria is an enormous health problem. Sickle cell hemoglobin results from a genetic mutation in the DNA sequence that is responsible for the synthesis of hemoglobin. Individuals who inherit this mutation from both parents develop sickle cell anemia, whereas those who inherit it from only one parent are said to have the sickle cell trait. They do not develop sickle cell anemia and they are more resistant to malaria than individuals without the mutation. The relative benefit of this mutation apparently accounts for this detrimental gene being passed on from generation to generation.

PROBLEM 16.16

Why is hemoglobin more water soluble than α-keratin?

PROBLEM 16.17

Why is it possible to discuss the quaternary structure of hemoglobin but not the quaternary structure of myoglobin?

16.8 Protein Hydrolysis and Denaturation

The properties of a protein are greatly altered and often entirely destroyed when any level of protein structure is disturbed.

16.8A Protein Hydrolysis

Like other amide bonds, the peptide bonds in proteins are hydrolyzed by treatment with aqueous acid, base, or certain enzymes.

- **The hydrolysis of the amide bonds in a protein forms the individual amino acids that comprise the primary structure.**

For example, hydrolysis of the amide bonds in the tripeptide Ile–Gly–Phe forms the amino acids isoleucine, glycine, and phenylalanine. When each amide bond is broken, the elements of H_2O are added, forming a carboxylate anion ($-COO^-$) in one amino acid and an ammonium cation ($-NH_3^+$) in the other.

Break the peptide bonds.

Ile–Gly–Phe: $H_3\overset{+}{N}-CH(CH(CH_3)CH_2CH_3)-CO-NH-CH(H)-CO-NH-CH(CH_2C_6H_5)-COO^-$

H_2O H_2O H_2O

$H_3\overset{+}{N}-CH(CH(CH_3)CH_2CH_3)-COO^-$ Ile

$H_3\overset{+}{N}-CH(H)-COO^-$ Gly

$H_3\overset{+}{N}-CH(CH_2C_6H_5)-COO^-$ Phe

The first step in the digestion of dietary protein is hydrolysis of the amide bonds of the protein backbone. The enzyme pepsin in the acidic gastric juices of the stomach cleaves some of the amide bonds to form smaller peptides, which pass into the small intestines and are further broken down into individual amino acids by the enzymes trypsin and chymotrypsin.

Proteins in the diet serve a variety of nutritional needs. Like carbohydrates and lipids, proteins can be metabolized for energy. Moreover, the individual amino acids formed by hydrolysis are used as starting materials to make new proteins that the body needs. Likewise the N atoms in the amino acids are incorporated into other biomolecules that contain nitrogen.

SAMPLE PROBLEM 16.5

Draw the structures of the amino acids formed by hydrolysis of the neuropeptide leu-enkephalin (Section 16.5).

Analysis

Locate each amide bond in the protein or peptide backbone. To draw the hydrolysis products, break each amide bond by adding the elements of H_2O to form a carboxylate anion ($-COO^-$) in one amino acid and an ammonium cation ($-NH_3^+$) in the other.

Solution

[1] Locate the amide bonds in the peptide backbone.
[2] Break each bond by adding H_2O.

$H_3\overset{+}{N}-CH(CH_2C_6H_4OH)-CO-NH-CH_2-CO-NH-CH_2-CO-NH-CH(CH_2C_6H_5)-CO-NH-CH(CH_2CH(CH_3)_2)-COO^-$

leu-enkephalin

$H_3\overset{+}{N}-CH(CH_2C_6H_4OH)-COO^-$ Tyr

$H_3\overset{+}{N}-CH(H)-COO^-$ Gly

$H_3\overset{+}{N}-CH(H)-COO^-$ Gly

$H_3\overset{+}{N}-CH(CH_2C_6H_5)-COO^-$ Phe

$H_3\overset{+}{N}-CH(CH_2CH(CH_3)_2)-COO^-$ Leu

PROBLEM 16.18

Draw the structure of the products formed by hydrolysis of each tripeptide: (a) Ala–Leu–Gly; (b) Ser–Thr–Phe; (c) Leu–Tyr–Asn.

PROBLEM 16.19

What hydrolysis products are formed from the neuropeptide met-enkephalin (Section 16.5)?

16.8B Protein Denaturation

When the secondary, tertiary, or quaternary structure of a protein is disturbed, the properties of a protein are also altered and the biological activity is often lost.

- ***Denaturation* is the process of altering the shape of a protein without breaking the amide bonds that form the primary structure.**

High temperature, acid, base, and even agitation can disrupt the noncovalent interactions that hold a protein in a specific shape. Heat breaks up weak London forces between nonpolar amino acids. Heat, acid, and base disrupt hydrogen bonding interactions between polar amino acids, which account for much of the secondary and tertiary structure. As a result, denaturation causes a globular protein to uncoil into an undefined randomly looped structure.

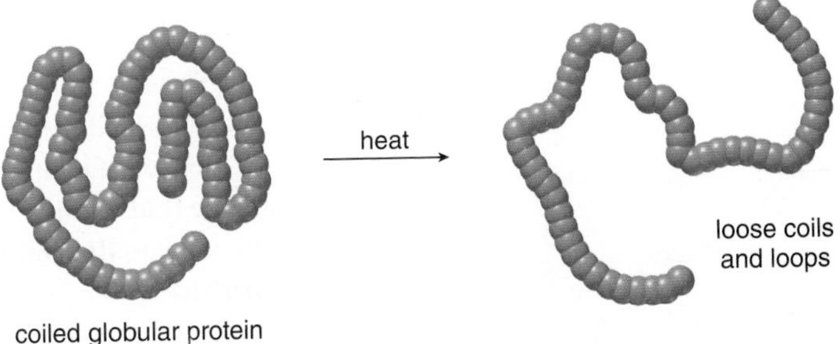

Cooking or whipping egg whites denatures the globular proteins they contain, forming insoluble protein.

Denaturation often makes globular proteins less water soluble. Globular proteins are typically folded with hydrophobic regions in the interior to maximize the interaction of polar residues on the outside surface with water. This makes them water soluble. When the protein is denatured, more hydrophobic regions are exposed and the protein often loses water solubility.

We witness many examples of protein denaturation in the kitchen. As milk ages it becomes sour from enzymes that produce lactic acid. The acid also denatures milk proteins, which precipitate as an insoluble curd. Ovalbumin, the major protein in egg white, is denatured when an egg is boiled or fried, forming a solid. Even vigorously whipping egg whites denatures its protein, forming the stiff meringue used to top a lemon meringue pie.

PROBLEM 16.20

Heating collagen, a water-insoluble fibrous protein, forms the jelly-like substance called gelatin. Explain how this process may occur.

16.9 Enzymes

We conclude the discussion of proteins with ***enzymes*, proteins that serve as biological catalysts for reactions in all living organisms.** Like all catalysts (Section 5.9B), enzymes increase the rate of reactions, but they themselves are not permanently changed in the process. Enzymes are crucial to the biological reactions that occur in the body, which would otherwise often proceed too slowly to be of any use. In humans, enzymes must catalyze reactions under very specific physiological conditions, usually a pH around 7.4 and a temperature of 37 °C.

16.9A Characteristics of Enzymes

Enzymes are generally water-soluble, globular proteins that exhibit two characteristic features.

- **Enzymes greatly enhance reaction rates.**
- **Enzymes are very specific.**

The specificity of an enzyme varies. Some enzymes, such as catalase, catalyze a single reaction. Catalase catalyzes the conversion of hydrogen peroxide (H_2O_2) to O_2 and H_2O (Section 5.1).

$$2\,H_2O_2(aq) \xrightarrow{\text{catalase}} 2\,H_2O(l) + O_2(g)$$

Other enzymes, such as carboxypeptidase A, catalyze a particular type of reaction with a variety of substrates. Carboxypeptidase A, a digestive enzyme that breaks down proteins, catalyzes the hydrolysis of a specific type of peptide bond—the amide bond closest to the C-terminal end of the protein.

$$\text{—NH—CH(R}_1\text{)—CO—NH—CH(R}_2\text{)—CO—NH—CH(R}_3\text{)—COO}^- \xrightarrow{\text{carboxypeptidase A}} \text{—NH—CH(R}_1\text{)—CO—NH—CH(R}_2\text{)—COO}^- + H_3\overset{+}{N}\text{—CH(R}_3\text{)—COO}^-$$

Only this amide bond is broken.

As is the case with catal*ase* and carboxypeptid*ase* A, **the names of most enzymes end in the suffix *-ase*.** Enzymes are classified into groups depending on the type of reaction they catalyze. For example, **an enzyme that catalyzes a hydrolysis reaction is called a *hydrolase*.** Hydrolases can be further subdivided into **lipases,** which catalyze the hydrolysis of the ester bonds in lipids (Section 15.5), or **proteases,** which catalyze the hydrolysis of proteins. Carboxypeptidase A is a protease.

HEALTH NOTE

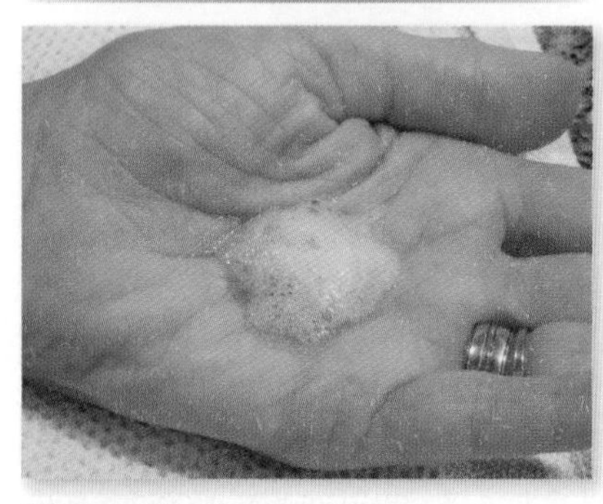

When a bloody wound is cleaned with hydrogen peroxide, the enzyme catalase in the blood converts the H_2O_2 to H_2O and O_2, forming a white foam of oxygen bubbles.

The conversion of lactate to pyruvate illustrates other important features of enzyme-catalyzed reactions.

$$\underset{\text{lactate}}{CH_3\text{—CH(OH)—COO}^-} + \underset{\text{coenzyme}}{NAD^+} \xrightarrow{\text{lactate dehydrogenase}} \underset{\text{pyruvate}}{CH_3\text{—CO—COO}^-} + NADH + H^+$$

The 2 H's in red are removed.

NADH: formed from the coenzyme

The enzyme for this process is *lactate dehydrogenase,* a name derived from the substrate (lactate) and the type of reaction, a **dehydrogenation**—that is, the removal of two hydrogen atoms. To carry out this reaction a **cofactor** is needed.

A **dehydrogenase** catalyzes the removal of two hydrogen atoms from a substrate to form a double bond.

- A *cofactor* is a metal ion or a nonprotein organic molecule needed for an enzyme-catalyzed reaction to occur.

NAD^+ (nicotinamide adenine dinucleotide) is the cofactor that oxidizes lactate to pyruvate. **An organic compound that serves as an enzyme cofactor is called a *coenzyme.*** While lactate dehydrogenase greatly speeds up this oxidation reaction, NAD^+ is the coenzyme that actually oxidizes the substrate, lactate. NAD^+ is a common biological oxidizing agent used as a coenzyme (Section 18.3).

The need for metal ions as enzyme cofactors explains why trace amounts of certain metals must be present in our diet. For example, certain **oxidases,** enzymes that catalyze oxidation reactions, require either Fe^{2+} or Cu^{2+} as a cofactor needed for transferring electrons.

PROBLEM 16.21

From the name alone, decide which of the following might be enzymes: (a) sucrose; (b) sucrase; (c) lactose; (d) lactase; (e) phosphofructokinase.

16.9B How Enzymes Work

An enzyme contains a region called the **active site** that binds the substrate, forming an **enzyme–substrate complex.** The active site is often a small cavity that contains amino acids that are attracted to the substrate with various types of intermolecular forces. Sometimes polar amino acids of the enzyme hydrogen bond to the substrate or nonpolar amino acids have stabilizing hydrophobic interactions.

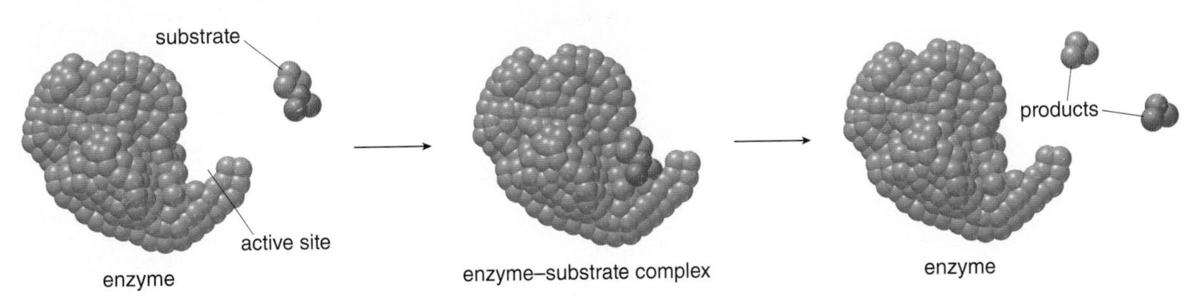

Two models have been proposed to explain the specificity of a substrate for an enzyme's active site: the **lock-and-key model** and the **induced-fit model.**

In the lock-and-key model, the shape of the active site is rigid. The three-dimensional geometry of the substrate must exactly match the shape of the active site for catalysis to occur. The lock-and-key model explains the high specificity observed in many enzyme reactions.

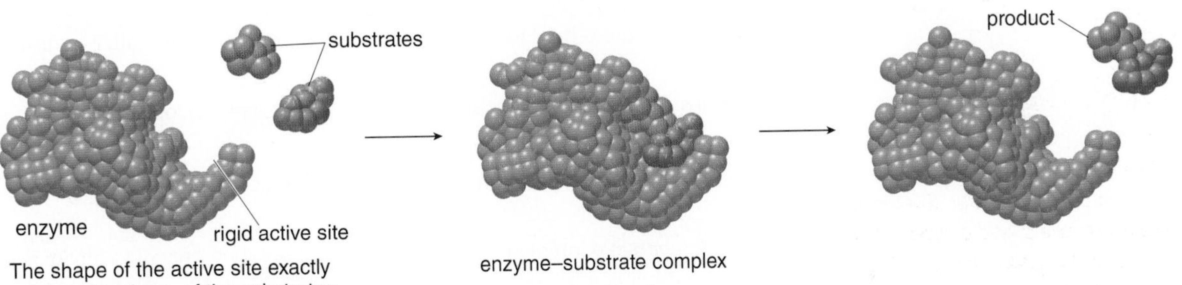

In the induced-fit model, the shape of the active site is more flexible. It is thought that when the substrate and the enzyme interact, the shape of the active site can adjust to fit the shape of the substrate.

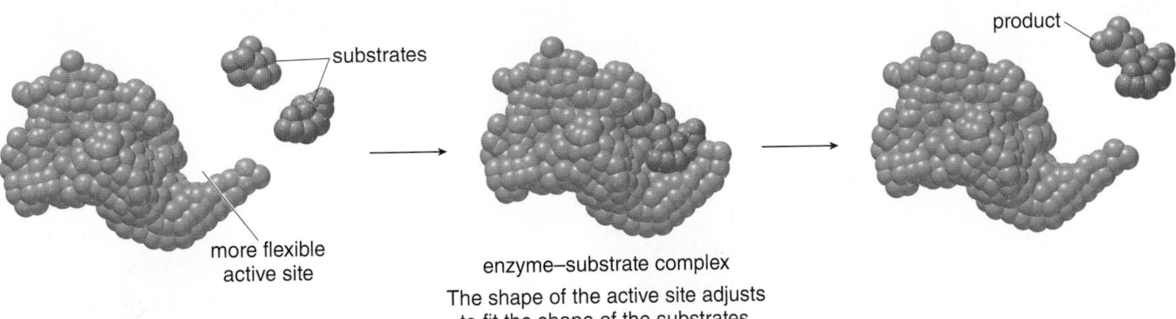

This model is often used to explain why some enzymes catalyze reactions with a wider variety of substrates. The shape of the active site and the substrate must still be reasonably similar, but once bound, the shape of the active site is different from the active site in the unbound enzyme. A well-characterized example of the induced-fit model is seen in the binding of glucose to the enzyme hexokinase, shown in Figure 16.12.

Figure 16.12 Hexokinase—An Example of the Induced-Fit Model

a. Shape of the active site before binding

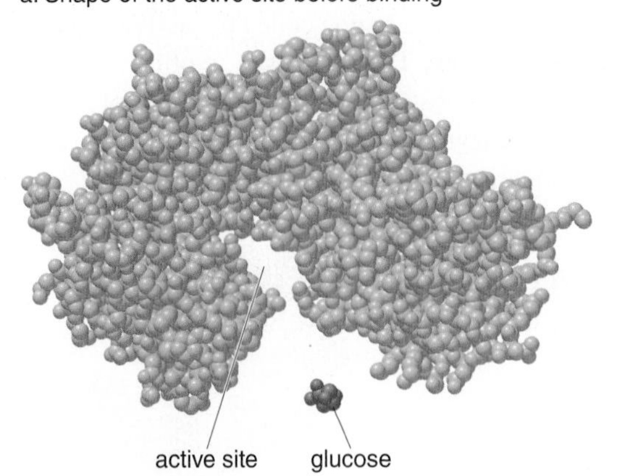

b. Shape of the active site after binding

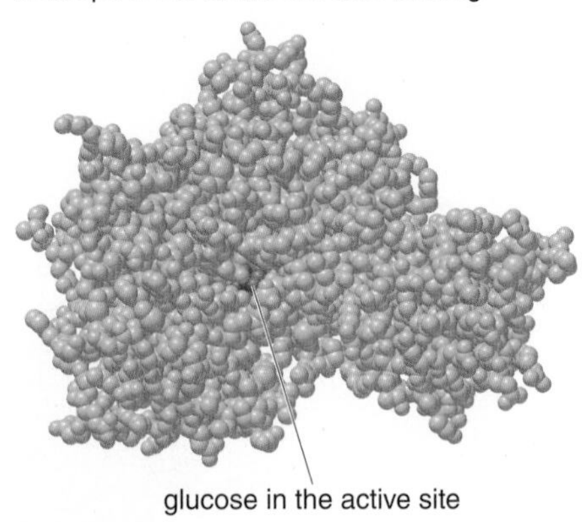

The open cavity of the active site of the enzyme closes around the substrate for a tighter fit.

(a) The free enzyme has a small open cavity that fits a glucose molecule. (b) Once bound, the cavity encloses the glucose molecule more tightly in the enzyme–substrate complex.

16.9C Enzyme Inhibitors

Some substances bind to enzymes and in the process, greatly alter or destroy the enzyme's activity.

- **An *inhibitor* is a molecule that causes an enzyme to lose activity.**

An inhibitor can bind to an enzyme reversibly or irreversibly.

- **A *reversible* inhibitor binds to an enzyme but then enzyme activity is restored when the inhibitor is released.**
- **An *irreversible* inhibitor covalently binds to an enzyme, permanently destroying its activity.**

Penicillin is an antibiotic that kills bacteria because it irreversibly binds to glycopeptide transpeptidase, an enzyme required for the synthesis of a bacterial cell wall (Section 13.10B). Penicillin binds to a hydroxyl group (OH) of the enzyme, thus inactivating it, halting cell wall construction, and killing the bacterium.

penicillin G

enzyme

inactive enzyme

Penicillin is covalently bound to the enzyme, inactivating it.

Reversible inhibition can be **competitive** or **noncompetitive.**

A *noncompetitive* inhibitor binds to the enzyme but does not bind at the active site. The inhibitor causes the enzyme to change shape so that the active site can no longer bind the substrate. When the noncompetitive inhibitor is no longer bound to the enzyme, normal enzyme activity resumes.

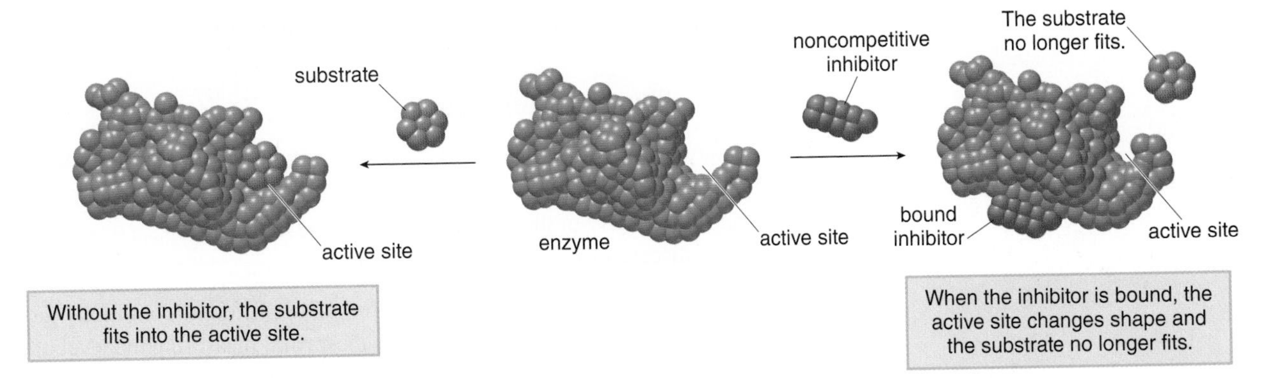

Without the inhibitor, the substrate fits into the active site.

When the inhibitor is bound, the active site changes shape and the substrate no longer fits.

A *competitive* inhibitor has a shape and structure similar to the substrate, so it competes with the substrate for binding to the active site. The antibiotic **sulfanilamide** is a competitive inhibitor of the enzyme needed to synthesize the vitamin folic acid from *p*-aminobenzoic acid in bacteria. When sulfanilamide binds to the active site, *p*-aminobenzoic acid cannot be converted to folic acid, so a bacterium cannot grow and reproduce. Sulfanilamide does not affect human cells because humans do not synthesize folic acid, and must obtain it in the diet.

sulfanilamide
a competitive inhibitor

p-aminobenzoic acid
PABA

→

folic acid

Both compounds can bind to the active site.

PROBLEM 16.22

Use the given representations for an enzyme, substrate, and inhibitor to illustrate the process of competitive inhibition.

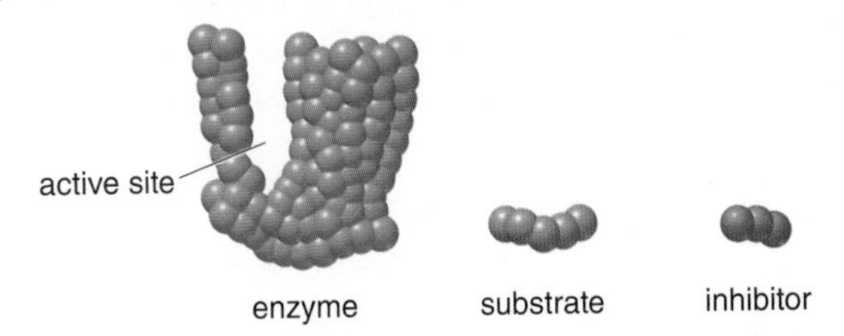

PROBLEM 16.23

The nerve gas sarin acts as a poison by covalently bonding to a hydroxyl group in the active site of the enzyme acetylcholinesterase. This binding results in a higher-than-normal amount of acetylcholine, resulting in muscle spasms. From this description, would you expect sarin to be a competitive, noncompetitive, or irreversible inhibitor?

16.10 FOCUS ON HEALTH & MEDICINE Using Enzymes to Diagnose and Treat Diseases

Measuring enzyme levels in the blood and understanding the key role of enzymes in biological reactions have aided greatly in both diagnosing and treating diseases.

16.10A Enzyme Levels as Diagnostic Tools

Certain enzymes are present in a higher concentration in particular cells. When the cells are damaged by disease or injury, the cells rupture and die, releasing the enzymes into the bloodstream. Measuring the activity of the enzymes in the blood then becomes a powerful tool to diagnose the presence of disease or injury in some organs.

Measuring enzyme levels is now routine for many different conditions. For example, when a patient comes into an emergency room with chest pain, measuring the level of creatine phosphokinase (CPK) and other enzymes indicates whether a heart attack, which results in damage to some portion of the heart, has occurred. Common enzymes used for diagnosis are listed in Table 16.3.

Table 16.3 Common Enzymes Used for Diagnosis

Enzyme	Condition
Creatine phosphokinase	Heart attack
Alkaline phosphatase	Liver or bone disease
Acid phosphatase	Prostate cancer
Amylase, lipase	Diseases of the pancreas

16.10B Treating Disease with Drugs That Interact with Enzymes

Molecules that inhibit an enzyme can be useful drugs. The antibiotics penicillin and sulfanilamide are two examples discussed in Section 16.9. Drugs used to treat high blood pressure and acquired immune deficiency syndrome (AIDS) are also enzyme inhibitors.

HEALTH NOTE

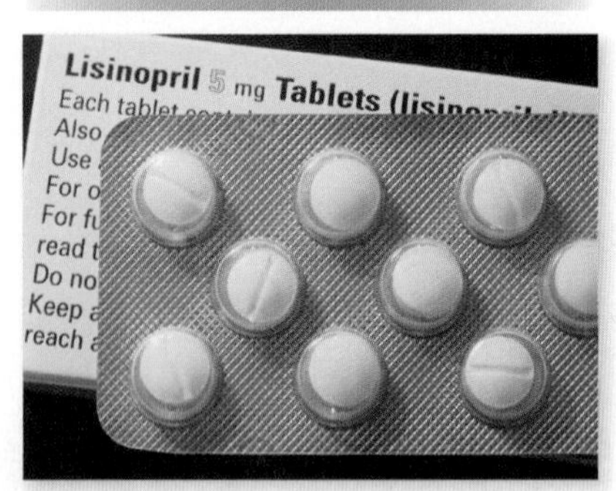

ACE inhibitors such as lisinopril (trade name: Zestril) decrease the concentration of angiotensin, thus decreasing blood pressure.

ACE inhibitors are a group of drugs used to treat individuals with high blood pressure. Angiotensin is a peptide that narrows blood vessels, thus increasing blood pressure. Angiotensin is formed from angiotensinogen by action of **ACE, the angiotensin-converting enzyme,** which cleaves two amino acids from the inactive peptide.

ACE cleaves here.

Arg–Arg–Val–Tyr–Ile–His–Pro–Phe–His–Leu ⟶ Arg–Arg–Val–Tyr–Ile–His–Pro–Phe + His–Leu

angiotensinogen → angiotensin

increases blood pressure

By blocking the conversion of angiotensinogen to angiotensin, blood pressure is decreased. Several effective ACE inhibitors are currently available, including captopril and enalapril.

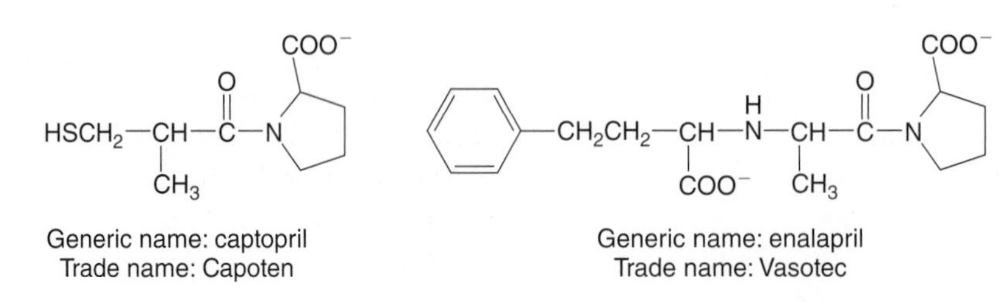

Generic name: captopril
Trade name: Capoten

Generic name: enalapril
Trade name: Vasotec

Several enzyme inhibitors are also available to treat human immunodeficiency virus (HIV), the virus that causes AIDS. The most effective treatments are **HIV protease inhibitors.** These drugs inhibit the action of the HIV protease enzyme, an essential enzyme needed by HIV to make copies of itself that go on to infect other cells. Amprenavir (trade name: Agenerase) is a protease inhibitor taken twice daily by individuals who are HIV positive. The three-dimensional structure of the HIV-1 protease enzyme is shown in Figure 16.13.

Generic name: amprenavir
Trade name: Agenerase

PROBLEM 16.24

How are the structures of the ACE inhibitors captopril and enalapril similar? How are they different?

Figure 16.13 The HIV Protease Enzyme

a. Ball-and-stick model of the HIV protease enzyme

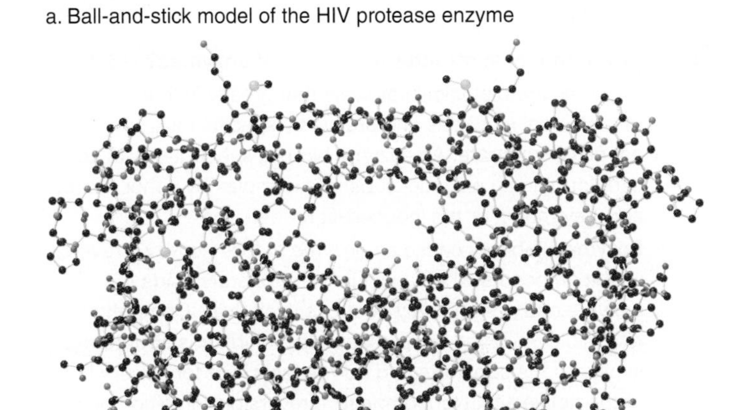

b. Ribbon diagram with the protease inhibitor amprenavir in the active site

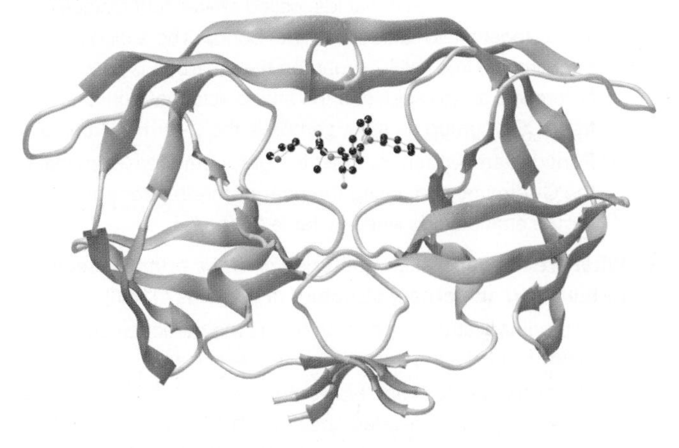

KEY TERMS

Active site (16.9)
Amino acid (16.2)
Coenzyme (16.9)
Cofactor (16.9)
Competitive inhibitor (16.9)
Conjugated protein (16.7)
C-Terminal amino acid (16.4)
Denaturation (16.8)
Dipeptide (16.4)
Enzyme (16.9)
Enzyme–substrate complex (16.9)
Fibrous protein (16.7)
Globular protein (16.7)
α-Helix (16.6)
Heme (16.7)
Induced-fit model (16.9)
Inhibitor (16.9)
Irreversible inhibitor (16.9)
Isoelectric point (16.3)
Lock-and-key model (16.9)
Noncompetitive inhibitor (16.9)
N-Terminal amino acid (16.4)
Peptide (16.4)
Peptide bond (16.4)
β-Pleated sheet (16.6)
Primary structure (16.6)
Protein (16.1)
Quaternary structure (16.6)
Reversible inhibitor (16.9)
Secondary structure (16.6)
Tertiary structure (16.6)
Tripeptide (16.4)
Zwitterion (16.2)

KEY CONCEPTS

1 What are the main structural features of an amino acid? (16.2)

- Amino acids contain an amino group (NH_2) on the α carbon to the carboxyl group (COOH). Amino acids exist in their neutral form as zwitterions having the general structure $^+H_3NCH(R)COO^-$. Because they are salts, amino acids are water soluble and have high melting points.
- All amino acids except glycine (R = H) have a chirality center on the α carbon. L Amino acids are naturally occurring.
- Amino acids are subclassified as neutral, acidic, or basic by the functional groups present in the R group, as shown in Table 16.2.

2 Describe the acid–base properties of amino acids. (16.3)

- Neutral, uncharged amino acids exist as zwitterions containing an ammonium cation ($-NH_3^+$) and a carboxylate anion ($-COO^-$).
- When strong acid is added, the carboxylate anion gains a proton and the amino acid has a net +1 charge. When strong base is added, the ammonium cation loses a proton and the amino acid has a net –1 charge.

3 What are the main structural features of peptides? (16.4)

- Peptides contain amino acids, called amino acid residues, joined together by amide (peptide) bonds. The amino acid that contains the free $-NH_3^+$ group on the α carbon is called the N-terminal amino acid, and the amino acid that contains the free $-COO^-$ group on the α carbon is the C-terminal amino acid.
- Peptides are written from left to right, from the N-terminal to the C-terminal end, using the one- or three-letter abbreviations for the amino acids listed in Table 16.2.

4 What are the general characteristics of the primary, secondary, tertiary, and quaternary structure of proteins? (16.6)

- The primary structure of a protein is the particular sequence of amino acids joined together by amide bonds.
- The two most common types of secondary structure are the α-helix and the β-pleated sheet. Both structures are stabilized by hydrogen bonds between the N—H and C=O groups.
- The tertiary structure is the three-dimensional shape adopted by the entire peptide chain.
- When a protein contains more than one polypeptide chain, the quaternary structure describes the shape of the protein complex formed by two or more chains.

5 What are the basic features of fibrous proteins like α-keratin and collagen? (16.7)

- Fibrous proteins are composed of long linear polypeptide chains that serve structural roles and are water insoluble.
- α-Keratin in hair is a fibrous protein composed almost exclusively of α-helix units that wind together to form a superhelix. Disulfide bonds between chains make the resulting bundles of protein chains strong.
- Collagen, found in connective tissue, is composed of a superhelix formed from three elongated left-handed helices.

6 What are the basic features of globular proteins like hemoglobin and myoglobin? (16.7)

- Globular proteins have compact shapes and are folded to place polar amino acids on the outside to make them water soluble. Hemoglobin and myoglobin are both conjugated proteins composed of a protein unit and a heme molecule. The Fe^{2+} ion of the heme binds oxygen.

7 What products are formed when a protein is hydrolyzed? (16.8)

- Hydrolysis breaks up the primary structure of a protein to form the amino acids that compose it. All of the amide bonds are broken by the addition of water, forming a carboxylate anion ($-COO^-$) in one amino acid and an ammonium cation ($-NH_3^+$) in the other.

8 What is denaturation? (16.8)

- Denaturation is a process that alters the shape of a protein by disrupting the secondary, tertiary, or quaternary structure. High temperature, acid, base, and agitation can denature a protein. Compact water-soluble proteins uncoil and become less water soluble.

9 What are the main structural features of enzymes? (16.9)

- Enzymes are biological catalysts that greatly increase the rate of biological reactions and are highly specific for a substrate or a type of substrate. An enzyme binds a substrate at its active site, forming an enzyme–substrate complex by either the lock-and-key model or the induced-fit model.
- Enzyme inhibitors cause an enzyme to lose activity. Irreversible inhibition occurs when an inhibitor covalently binds the enzyme and permanently destroys its activity. Competitive reversible inhibition occurs when the inhibitor is structurally similar to the substrate and competes with it for occupation of the active site. Noncompetitive reversible inhibition occurs when an inhibitor binds to a location other than the active site, altering the shape of the active site.

10 How are enzymes used in medicine? (16.10)

- Measuring blood enzyme levels is used to diagnose heart attacks and diseases that cause higher-than-normal concentrations of certain enzymes to enter the blood.
- Drugs that inhibit the action of an enzyme can be used to kill bacteria. ACE inhibitors are used to treat high blood pressure. HIV protease inhibitors are used to treat HIV by binding to an enzyme needed by the virus to replicate itself.

UNDERSTANDING KEY CONCEPTS

Selected in-chapter and odd-numbered end-of-chapter problems have brief answers at the end of each chapter. The *Student Study Guide and Solutions Manual* contains detailed solutions to all in-chapter and odd-numbered end-of-chapter problems, as well as additional worked examples and a chapter self-test.

16.25 (a) Identify the amino acid shown with all uncharged atoms in the ball-and-stick model. (b) Give the three-letter and one-letter abbreviations for the amino acid. (c) Draw the form present at the isoelectric point of the amino acid.

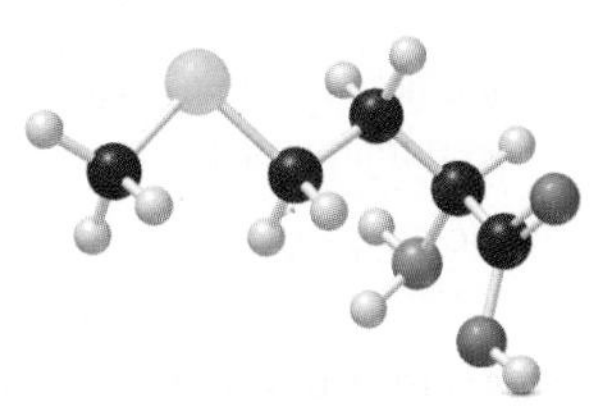

16.26 (a) Identify the amino acid shown with all uncharged atoms in the ball-and-stick model. (b) Give the three-letter and one-letter abbreviations for the amino acid. (c) Draw the form of the amino acid present at pH 11.

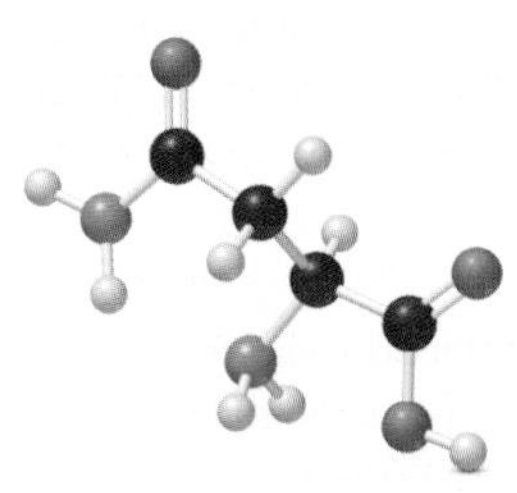

16.27 For each amino acid: [1] draw the L enantiomer in a Fischer projection; [2] classify the amino acid as neutral, acidic, or basic; [3] give the three-letter symbol; [4] give the one-letter symbol.

a. leucine b. tryptophan

16.28 For each amino acid: [1] give the name; [2] give the three-letter abbreviation; [3] give the one-letter abbreviation; [4] classify the amino acid as neutral, acidic, or basic.

a. COO^- ; $H_3\overset{+}{N}$—C—H ; $CH_2CH_2SCH_3$

b. COO^- ; $H_3\overset{+}{N}$—C—H ; CH_2OH

16.29 For the given tripeptide: (a) identify the amino acids that form the peptide; (b) label the N- and C-terminal amino acids; (c) name the tripeptide using three-letter symbols.

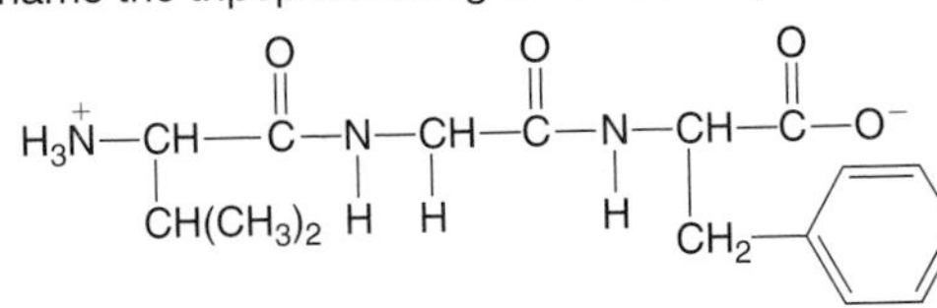

16.30 For the given tripeptide: (a) identify the amino acids that form the peptide; (b) label the N- and C-terminal amino acids; (c) name the tripeptide using three-letter symbols.

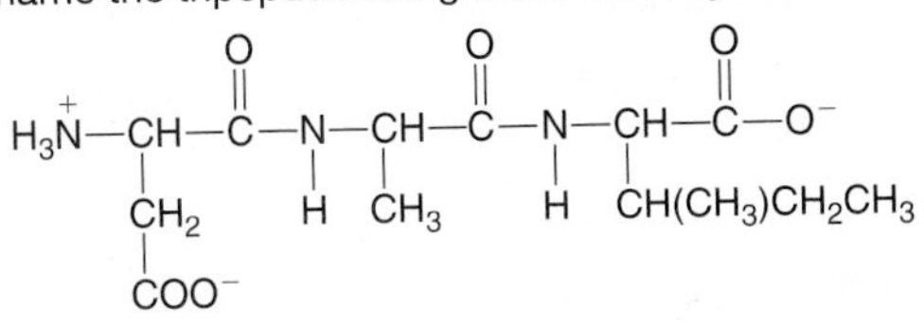

16.31 Label the regions of secondary structure in the following protein ribbon diagram.

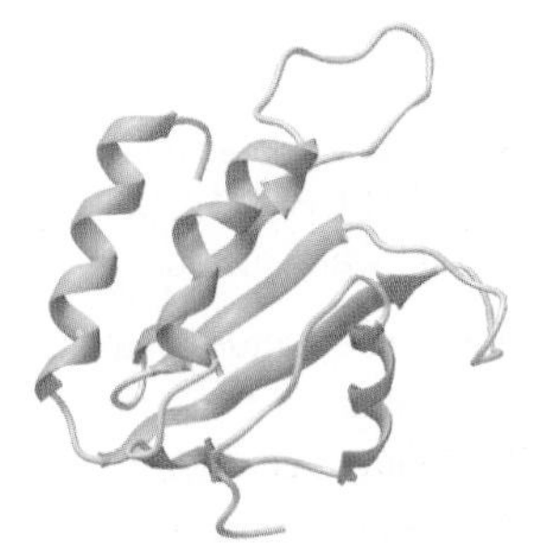

16.32 Label the regions of secondary structure in the following protein ribbon diagram.

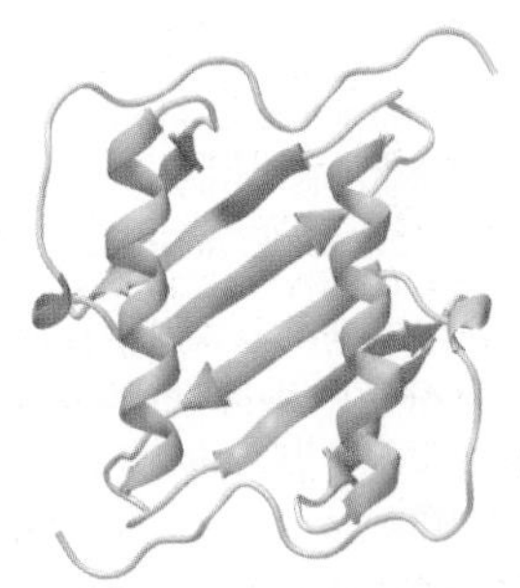

16.33 What type of interactions occur at each of the labeled sites in the portion of the protein ribbon shown?

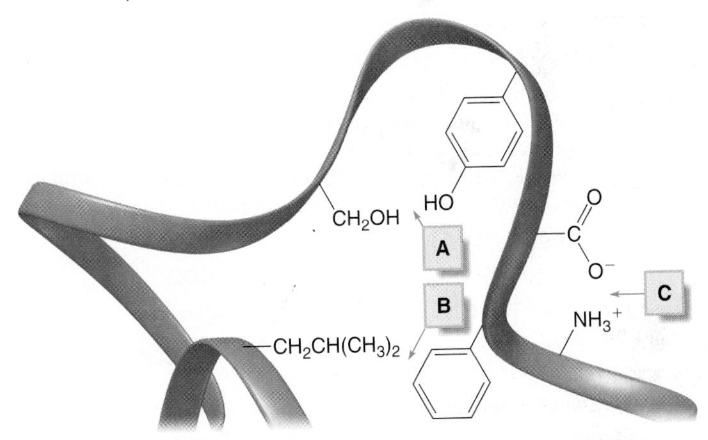

16.34 Using the given representations for an enzyme and substrate, which diagram represents (a) the enzyme–substrate complex; (b) competitive inhibition; (c) noncompetitive inhibition?

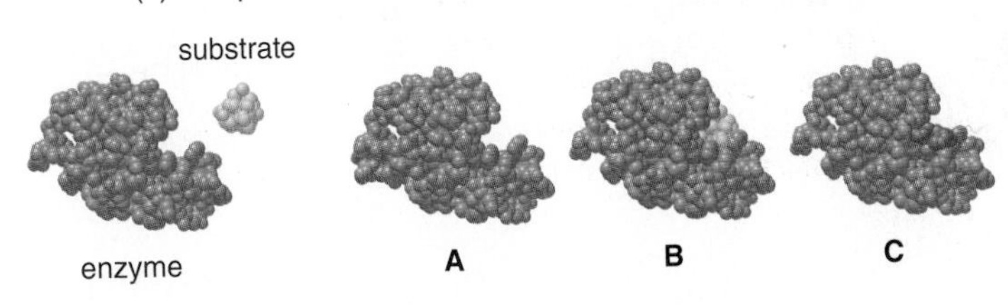

ADDITIONAL PROBLEMS

Amino Acids

16.35 Naturally occurring amino acids are L-α-amino acids. What do the L and α designations represent?

16.36 Why do neutral amino acids exist as zwitterions with no net charge?

16.37 The amino acid alanine is a solid at room temperature and has a melting point of 315 °C, while pyruvic acid (CH_3COCO_2H) has a similar molecular weight but is a liquid at room temperature with a boiling point of 165 °C. Account for the difference.

16.38 Why is phenylalanine water soluble but 4-phenylbutanoic acid ($C_6H_5CH_2CH_2CH_2COOH$), a compound of similar molecular weight, water insoluble?

16.39 Draw the structure of a naturally occurring amino acid that:

a. contains a 1° alcohol

b. contains an amide

c. is an essential amino acid with an aromatic ring

16.40 Draw the structure of a naturally occurring amino acid that:

a. contains a 2° alcohol

b. contains a thiol

c. is an acidic amino acid

16.41 For each amino acid: [1] draw the L enantiomer in a Fischer projection; [2] classify the amino acid as neutral, acidic, or basic; [3] give the three-letter symbol; [4] give the one-letter symbol.

a. lysine b. aspartic acid

16.42 For each amino acid: [1] draw the L enantiomer in a Fischer projection; [2] classify the amino acid as neutral, acidic, or basic; [3] give the three-letter symbol; [4] give the one-letter symbol.

a. arginine b. tyrosine

16.43 Draw both enantiomers of each amino acid and label them as D or L: (a) methionine; (b) asparagine.

16.44 Which of the following Fischer projections represent naturally occurring amino acids? Name each amino acid and designate it as a D or L isomer.

a. COO^- on top; $H_3\overset{+}{N}$ left, H right; $CH_2CH(CH_3)_2$ on bottom

b. COO^- on top; H left, $\overset{+}{N}H_3$ right; CH_2COO^- on bottom

16.45 For each amino acid: [1] give the name; [2] give the three-letter abbreviation; [3] give the one-letter abbreviation; [4] classify the amino acid as neutral, acidic, or basic.

a. COO^- on top; $H_3\overset{+}{N}$ left, H right; $CH_2CH_2CONH_2$ on bottom

b. COO^- on top; $H_3\overset{+}{N}$ left, H right; $CH(OH)CH_3$ on bottom

16.46 For each amino acid: [1] give the name; [2] give the three-letter abbreviation; [3] give the one-letter abbreviation; [4] classify the amino acid as neutral, acidic, or basic.

a. COO^- on top; $H_3\overset{+}{N}$ left, H right; CH_2–(benzene ring)–OH on bottom

b. COO^- on top; $H_3\overset{+}{N}$ left, H right; CH_2–(imidazole ring: N, HN) on bottom

Acid–Base Properties of Amino Acids

16.47 Draw the amino acid leucine at each pH: (a) 6; (b) 10; (c) 2. Which form predominates at leucine's isoelectric point?

16.48 Draw the amino acid isoleucine at each pH: (a) 6; (b) 10; (c) 2. Which form predominates at isoleucine's isoelectric point?

16.49 Draw the structure of the neutral, positively charged, and negatively charged forms of the amino acid tyrosine. Which form predominates at pH 1? Which form predominates at pH 11? Which form predominates at the isoelectric point?

16.50 Draw the structure of the neutral, positively charged, and negatively charged forms of the amino acid proline. Which form predominates at pH 1? Which form predominates at pH 11? Which form predominates at the isoelectric point?

Peptides

16.51 For each tripeptide: [1] identify the N-terminal and C-terminal amino acids; [2] name the peptide using three-letter symbols for the amino acids.

a. leucylvalyltryptophan b. alanylglycylvaline

16.52 For each tripeptide: [1] identify the N-terminal and C-terminal amino acids; [2] name the peptide using three-letter symbols for the amino acids.

a. tyrosylleucylisoleucine b. methionylisoleucylcysteine

16.53 For the given tripeptide: (a) identify the amino acids that form the peptide; (b) label the N- and C-terminal amino acids; (c) name the tripeptide using three-letter symbols.

$H_3\overset{+}{N}$—CH($CH_2CH(CH_3)_2$)—C(=O)—N(H)—CH(CH_2–(benzene ring)–OH)—C(=O)—N(H)—CH($CH_2CH_2SCH_3$)—C(=O)—O^-

16.54 For the given tripeptide: (a) identify the amino acids that form the peptide; (b) label the N- and C-terminal amino acids; (c) name the tripeptide using three-letter symbols.

```
              O              O              O
   +          ||             ||             ||
H3N—CH—C—N—CH—C—N—CH—C—O⁻
       |         |   |            |   |
    CH2OH   H  CH2           H  CH2CH2CONH2
                       |
                    (benzene ring)
```

16.55 Locate the peptide bond in the dipeptide shown in the ball-and-stick model, and identify the amino acids that form the dipeptide. Name the dipeptide using the three-letter abbreviations for the amino acids.

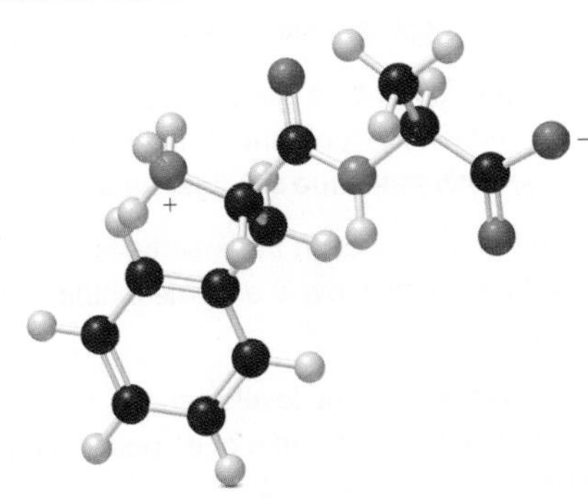

16.56 Label the N-terminal and C-terminal amino acids in the dipeptide shown in the ball-and-stick model. Name the dipeptide using the one-letter abbreviations for the amino acids.

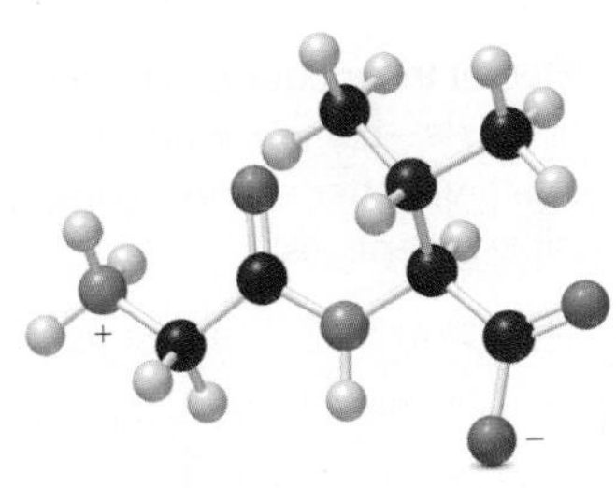

16.57 Draw the structures of the amino acids formed when the tripeptide in Problem 16.53 is hydrolyzed.

16.58 Draw the structures of the amino acids formed when the tripeptide in Problem 16.54 is hydrolyzed.

16.59 What amino acids are formed by hydrolysis of the tripeptide depicted in the ball-and-stick model?

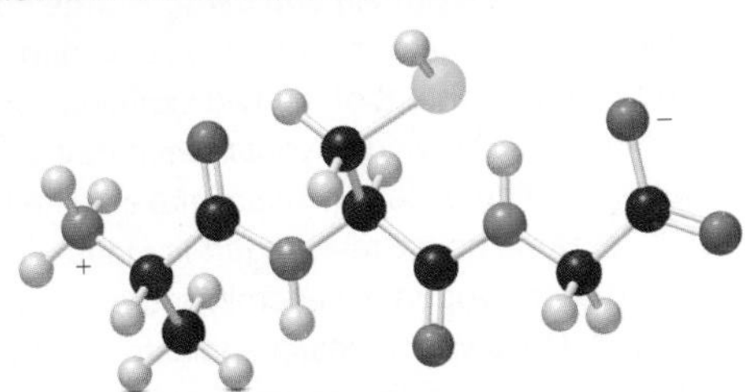

16.60 Give the three-letter abbreviations for the amino acids formed by hydrolysis of the tripeptide depicted in the ball-and-stick model.

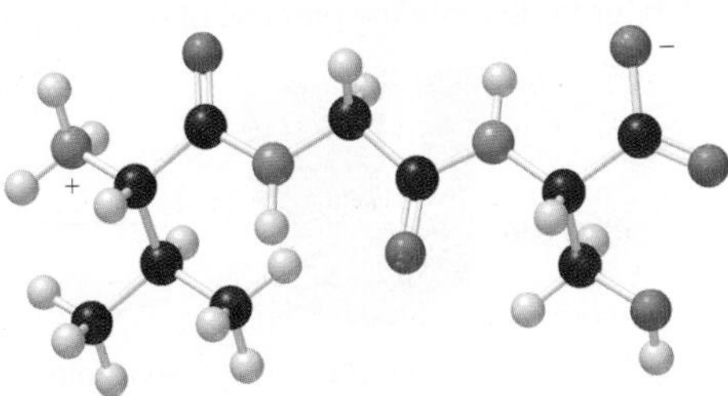

Proteins

16.61 What is the difference between the primary and secondary structure of a protein?

16.62 What is the difference between the tertiary and quaternary structure of a protein?

16.63 What type of intermolecular forces exist between the side chains of each of the following pairs of amino acids?

a. isoleucine and valine   b. Lys and Glu

16.64 Which of the following pairs of amino acids can have intermolecular hydrogen bonding between the functional groups in their side chains?

a. two tyrosine residues   b. alanine and threonine

16.65 List two amino acids that would probably be located in the interior of a globular protein.

16.66 List two amino acids that would probably be located on the exterior of a globular protein.

16.67 Compare α-keratin and hemoglobin with regards to each of the following: (a) secondary structure; (b) water solubility; (c) function; (d) location in the body.

16.68 Compare collagen and myoglobin with regards to each of the following: (a) secondary structure; (b) water solubility; (c) function; (d) location in the body.

16.69 When a protein is denatured, how is its primary, secondary, tertiary, and quaternary structure affected?

16.70 Hydrogen bonding stabilizes both the secondary and tertiary structures of a protein. (a) What functional groups hydrogen bond to stabilize secondary structure? (b) What functional groups hydrogen bond to stabilize tertiary structure?

16.71 Describe the function or biological activity of each protein or peptide: (a) insulin; (b) myoglobin; (c) α-keratin; (d) chymotrypsin; (e) oxytocin.

16.72 Describe the function or biological activity of each protein or peptide: (a) collagen; (b) hemoglobin; (c) vasopressin; (d) pepsin; (e) met-enkephalin.

Enzymes

16.73 Use the given representations for an enzyme, substrate, and inhibitor to illustrate the process of noncompetitive inhibition.

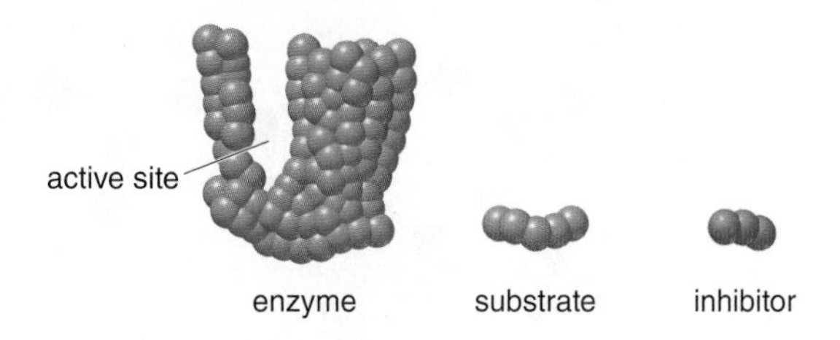

16.74 Use the given representations for an enzyme and substrate to illustrate the difference between the lock-and-key model and the induced-fit model of enzyme specificity.

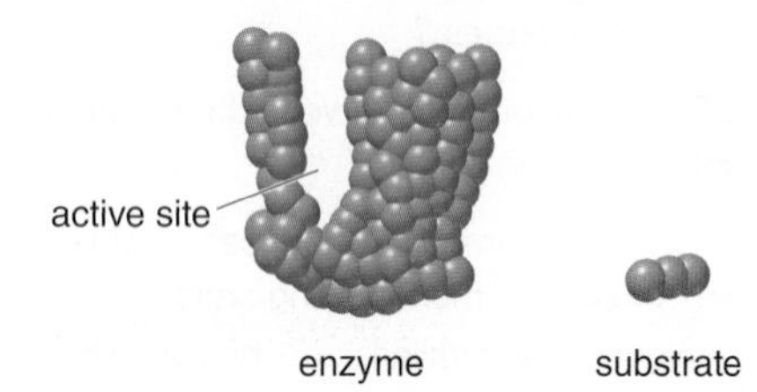

16.75 How are enzyme inhibitors used to treat high blood pressure? Give a specific example of a drug used and an enzyme inhibited.

16.76 How are enzyme inhibitors used to treat HIV? Give a specific example of a drug used and an enzyme inhibited.

Applications

16.77 What structural feature in α-keratin makes fingernails harder than skin?

16.78 Why does the α-keratin in hair contain many cysteine residues?

16.79 Why must vegetarian diets be carefully balanced?

16.80 Why does cooking meat make it easier to digest?

16.81 Sometimes an incision is cauterized (burned) to close the wound and prevent bleeding. What does cauterization do to protein structure?

16.82 Why is insulin administered by injection instead of taken in tablet form?

16.83 How is sickle cell disease related to hemoglobin structure?

16.84 The silk produced by a silkworm is a protein with a high glycine and alanine content. With reference to the structure, how does this make the silk fiber strong?

16.85 Explain the difference in the mechanism of action of penicillin and sulfanilamide. How is enzyme inhibition involved in both mechanisms?

16.86 How are blood enzyme levels used to diagnose certain diseases? Give an example of a specific condition and enzyme used for diagnosis.

CHALLENGE PROBLEMS

16.87 Explain why two amino acids—aspartic acid and glutamic acid—have a +1 net charge at low pH, but a –2 net charge at high pH.

16.88 How many different tripeptides can be formed from three different amino acids—namely, methionine, histidine, and arginine? Using three-letter abbreviations, give the names for all of the possible tripeptides.

BEYOND THE CLASSROOM

16.89 Compare the ingredients in two different types of shampoos. What is the active cleaning agent in each product? Are the ingredients in any of the following shampoos—baby shampoo, anti-dandruff shampoo, "all-natural" shampoo—different from those marketed for the general consumer? What is the optimum pH for a shampoo? Is there evidence that added vitamins or amino acids in a shampoo produce healthier hair protein? What other common ingredients are used in shampoos, and what are their roles?

16.90 Pick one or more of the following enzymes or types of enzymes, which are used in a commercial process: cellulases in the biofuels industry, rennin in dairy products, glucose isomerase in the food industry, proteases in eye care products, or another enzyme of your choice. What is known about the structure of the enzyme? What process does the enzyme catalyze? Be as specific as possible. How is the enzyme used commercially or in consumer products?

16.91 Pick an ethnic diet and give two or three examples of vegetarian dishes that would provide all of the essential amino acids.

16.92 Greek yogurt is considered especially nutritious, because it has more protein and fewer carbohydrates than most regular yogurts. Compare the protein and carbohydrate content of a popular Greek yogurt with that of a regular yogurt. How does the fat and sodium content of the two varieties compare? If you ingest one 8-oz serving of yogurt each day for a month, then what is the difference in protein intake in the two varieties? Research how Greek yogurt is made, and suggest reasons why Greek yogurt is generally more expensive than regular yogurt. Does your research indicate that there are considerable health benefits derived from regularly eating Greek yogurt?

ANSWERS TO SELECTED PROBLEMS

16.1 a. amide b. alcohol c. thiol

16.3 a. COO^-; $H_3\overset{+}{N}$—C—H; CH_2 (L) COO^-; H—C—$\overset{+}{N}H_3$; CH_2 (D)

b. COO^-; $H_3\overset{+}{N}$—C—H; $CH_2CH_2SCH_3$ (L) COO^-; H—C—$\overset{+}{N}H_3$; $CH_2CH_2SCH_3$ (D)

16.5 a. $H_3\overset{+}{N}-C(H)(CH(CH_3)_2)-COO^-$ predominant at pI

b. $H_3\overset{+}{N}-C(H)(CH(CH_3)_2)-COOH$

c. $H_2N-C(H)(CH(CH_3)_2)-COO^-$

16.7

$H_3\overset{+}{N}-C(H)(CH_2-C_6H_5)-COOH$ predominant at pH 1

$H_3\overset{+}{N}-C(H)(CH_2-C_6H_5)-COO^-$ neutral

$H_2N-C(H)(CH_2-C_6H_5)-COO^-$ predominant at pH 11

16.8 a. N-terminal: leucine C-terminal: alanine
b. N-terminal: arginine C-terminal: tyrosine
c. N-terminal: phenylalanine C-terminal: glutamine
d. N-terminal: valine C-terminal: phenylalanine

16.9 a. alanine b. methionine c. Ala–Gly–Leu–Met

16.11 The peptide bond joins the C=O of one amino acid to the N atom of the second amino acid. The dipeptide (Ser–Gly) is composed of serine and glycine.

16.13 Yes, the amino acids may be ordered differently.

16.15 Glycine has no large side chain and this allows for the β-pleated sheets to stack well together.

16.17 Hemoglobin has four chains that combine in a quaternary structure. Myoglobin has a single protein chain, so it has no quaternary structure.

16.18 a. $H_3\overset{+}{N}-C(H)(CH_3)-COO^-$ $H_3\overset{+}{N}-C(H)(CH_2CH(CH_3)_2)-COO^-$ $H_3\overset{+}{N}-C(H)(H)-COO^-$

b. $H_3\overset{+}{N}-C(H)(CH_2OH)-COO^-$ $H_3\overset{+}{N}-C(H)(CH(OH)CH_3)-COO^-$ $H_3\overset{+}{N}-C(H)(CH_2-C_6H_5)-COO^-$

c. $H_3\overset{+}{N}-C(H)(CH_2CH(CH_3)_2)-COO^-$ $H_3\overset{+}{N}-C(H)(CH_2-C_6H_4-OH)-COO^-$ $H_3\overset{+}{N}-C(H)(CH_2CONH_2)-COO^-$

16.19 tyrosine, glycine (two equivalents), phenylalanine, methionine

16.21 b,d,e

16.23 Sarin is an irreversible inhibitor since it forms a covalent bond at the enzyme's active site.

16.25 a,b,c: COO^-; $H_3\overset{+}{N}$—C—H; $CH_2CH_2SCH_3$
methionine
Met, M

16.27

	[1]	[2]	[3]	[4]
a.	COO^-; $H_3\overset{+}{N}$—C—H; $CH_2CH(CH_3)_2$	neutral	Leu	L
b.	COO^-; $H_3\overset{+}{N}$—C—H; CH_2–(indole ring, N–H)	neutral	Trp	W

16.29 a,b: valine: N-terminal, glycine, phenylalanine: C-terminal
c. Val–Gly–Phe

16.31

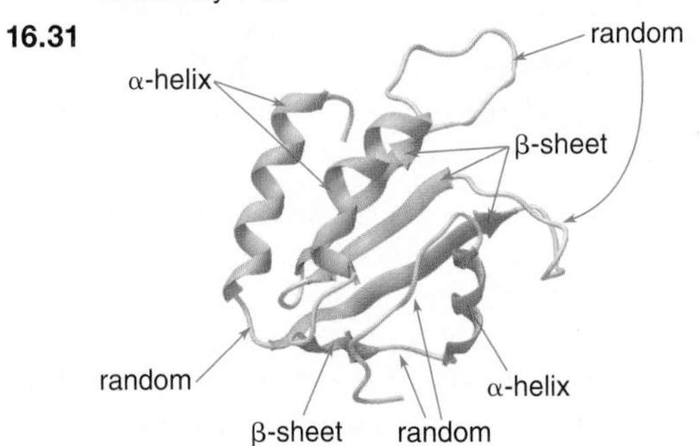

16.33 **A** hydrogen bonding, **B** London dispersion forces, **C** electrostatic attraction

16.35 The L designation refers to the configuration at the chirality center. With a vertical carbon chain in the Fischer projection, the L isomer has the $-NH_3^+$ drawn on the left side. The α amino acid designation indicates that the amino group is bonded to the carbon adjacent to the carbonyl group.

16.37 Alanine is an ionic salt with strong electrostatic forces, leading to its high melting point, and making it a solid at room temperature. Pyruvic acid has hydrogen bonding but this is a much weaker intermolecular force, so it is a liquid at room temperature.

16.39 a. $H_3\overset{+}{N}-C(H)(CH_2OH)-COO^-$

b. $H_3\overset{+}{N}-C(H)(CH_2CONH_2)-COO^-$

c. $H_3\overset{+}{N}-C(H)(CH_2C_6H_5)-COO^-$

16.41

	[1]	[2]	[3]	[4]
a.	COO^-; $H_3\overset{+}{N}$—C—H; $(CH_2)_4\overset{+}{N}H_3$	basic	Lys	K
b.	COO^-; $H_3\overset{+}{N}$—C—H; CH_2COO^-	acidic	Asp	D

16.43 a. L: COO^-; $H_3\overset{+}{N}$—C—H; $CH_2CH_2SCH_3$ D: COO^-; H—C—$\overset{+}{N}H_3$; $CH_2CH_2SCH_3$

b. L: COO^-; $H_3\overset{+}{N}$—C—H; CH_2CONH_2 D: COO^-; H—C—$\overset{+}{N}H_3$; CH_2CONH_2

16.45

	[1]	[2]	[3]	[4]
a.	glutamine	Gln	Q	neutral
b.	threonine	Thr	T	neutral

16.47 a. $H_3\overset{+}{N}-C(H)(CH_2CH(CH_3)_2)-COO^-$
predominant form at pI

b. $H_2N-C(H)(CH_2CH(CH_3)_2)-COO^-$

c. $H_3\overset{+}{N}-C(H)(CH_2CH(CH_3)_2)-COOH$

16.49 $H_3\overset{+}{N}-C(H)(CH_2-C_6H_4-OH)-COO^-$
neutral
predominant form at pI

$H_3\overset{+}{N}-C(H)(CH_2-C_6H_4-OH)-COOH$
positive charge
pH = 1

$H_2N-C(H)(CH_2-C_6H_4-OH)-COO^-$
negative charge
pH = 11

16.51 a. [1] N-terminal: leucine
C-terminal: tryptophan

[2] Leu–Val–Trp

b. [1] N-terminal: alanine
C-terminal: valine

[2] Ala–Gly–Val

16.53 a,b: leucine: N-terminal, tyrosine, methionine: C-terminal
c. Leu–Tyr–Met

16.55

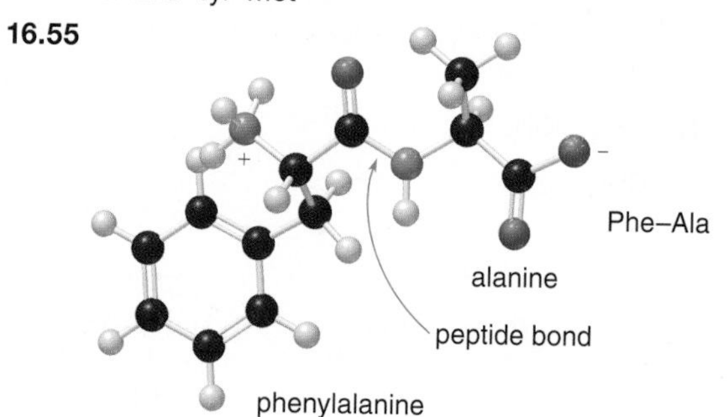

16.57 $H_3\overset{+}{N}-C(H)(CH_2CH(CH_3)_2)-COO^-$ $H_3\overset{+}{N}-C(H)(CH_2-C_6H_4-OH)-COO^-$ $H_3\overset{+}{N}-C(H)(CH_2CH_2SCH_3)-COO^-$

16.59 alanine, cysteine, glycine

16.61 The primary structure of a protein is the order of its amino acids. The secondary structure refers to the three-dimensional arrangements of regions within the protein.

16.63 a. London dispersion forces   b. electrostatic

16.65 Any two of the following amino acids: valine, alanine, phenylalanine, leucine, glycine, and isoleucine

16.67

	a. Secondary Structure	b. H_2O Solubility	c. Function	d. Location
Hemo-globin	Globular with much α-helix	Soluble	Carries oxygen to tissues	Blood
Keratin	α-Helix	Insoluble	Firm tissues	Nail, hair

16.69 When heated, a protein's primary structure is unaffected. The 2°, 3°, and 4° structures may be altered.

16.71 a. Insulin is a hormone that controls glucose levels.
b. Myoglobin stores oxygen in muscle.
c. α-Keratin forms hard tissues such as hair and nails.
d. Chymotrypsin is a protease that hydrolyzes peptide bonds.
e. Oxytocin is a hormone that stimulates uterine contractions and induces the release of breast milk.

16.73

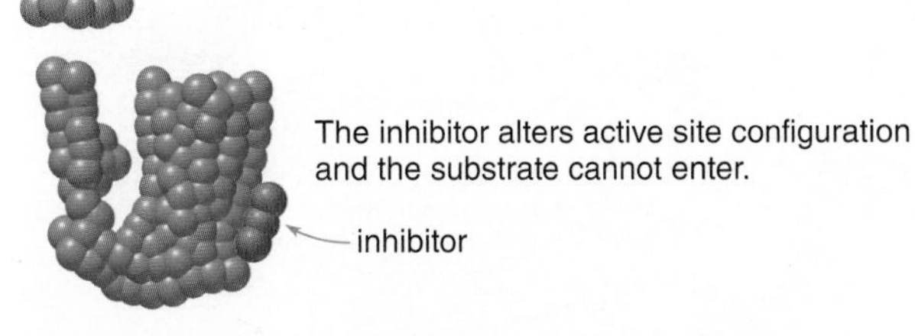
substrate
The inhibitor alters active site configuration and the substrate cannot enter.
inhibitor

16.75 Captopril inhibits the angiotensin-converting enzyme, blocking the conversion of angiotensinogen to angiotensin. This reduces the concentration of angiotensin, which in turn lowers blood pressure.

16.77 The α-keratin in nails has more cysteine residues to form disulfide bonds. The larger the number of disulfide bonds, the harder the substance.

16.79 Humans cannot synthesize the amino acids methionine and lysine. Diets that include animal products readily supply all the needed amino acids, but plant sources generally do not have sufficient amounts of all the essential amino acids. Grains—wheat, rice, and corn—are low in lysine, and legumes—beans, peas, and peanuts—are low in methionine, but a combination of these foods provides all the needed amino acids.

16.81 Cauterization denatures the proteins in a wound.

16.83 In sickle hemoglobin there is a substitution of a single amino acid—namely, valine for glutamic acid.

16.85 Penicillin inhibits the formation of the bacterial cell wall by irreversibly binding to an enzyme needed for its construction. Sulfanilamide inhibits the production of folic acid and therefore reproduction in bacteria.

16.87 Both aspartic acid and glutamic acid have two carboxylic acid groups. At low pH they have a +1 charge with both acid groups protonated, but at a high pH both acid groups are ionized, leading to a net charge of –2.

$H_3\overset{+}{N}—CH(CH_2COOH)—COOH$

Form of Asp at low pH

$H_2N—CH(CH_2COO^-)—COO^-$

Form of Asp at high pH

Techniques to analyze DNA samples are commonly utilized in criminal investigations and to screen for genetic diseases.

Nucleic Acids and Protein Synthesis

CHAPTER OUTLINE

CHAPTER GOALS

In this chapter you will learn how to:

1. Draw the structure of nucleosides and nucleotides
2. Draw short segments of the nucleic acids DNA and RNA
3. Describe the basic features of the DNA double helix
4. Outline the main steps of replication
5. List the three types and functions of RNA molecules
6. Explain the process of transcription
7. Describe the basic elements of the genetic code
8. Explain the process of translation
9. Define the terms "mutation" and "genetic disease"
10. Describe the basic features of DNA fingerprinting
11. Describe the main characteristics of viruses

Whether you are tall or short, fair-skinned or dark-complexioned, blue-eyed or brown-eyed, your unique characteristics are determined by the nucleic acid polymers that reside in the chromosomes of your cells. The nucleic acid **DNA** stores the genetic information of a particular organism, while the nucleic acid **RNA** translates this genetic information into the synthesis of proteins needed by cells for proper function and development. Even minor alterations in the nucleic acid sequence can have significant effects on an organism, sometimes resulting in devastating diseases like sickle cell anemia and cystic fibrosis. In Chapter 17, we study nucleic acids and learn how the genetic information stored in DNA is translated into protein synthesis.

17.1 Nucleosides and Nucleotides

Nucleic acids are unbranched polymers composed of repeating monomers called *nucleotides*. There are two types of nucleic acids.

- **DNA, deoxyribonucleic acid, stores the genetic information of an organism and transmits that information from one generation to another.**
- **RNA, ribonucleic acid, translates the genetic information contained in DNA into proteins needed for all cellular functions.**

The nucleotide monomers that compose DNA and RNA consist of three components—a **monosaccharide, a nitrogen-containing base, and a phosphate group.**

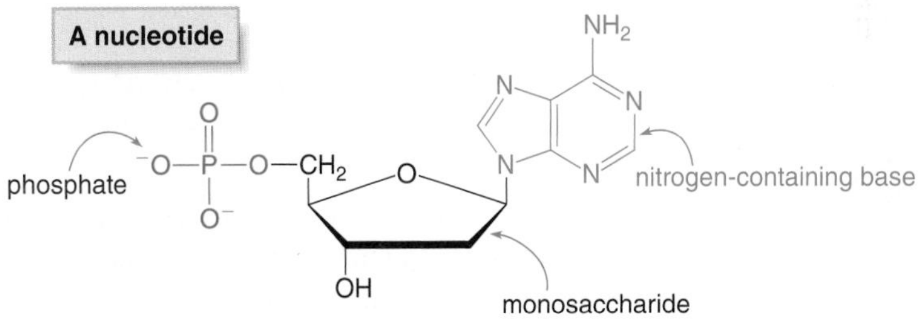

A nucleotide

phosphate

nitrogen-containing base

monosaccharide

DNA molecules contain several million nucleotides while RNA molecules are much smaller, containing perhaps a few thousand nucleotides. DNA is contained in the chromosomes of the nucleus, each chromosome having a different type of DNA. The number of chromosomes differs from species to species. Humans have 46 chromosomes (23 pairs). An individual chromosome is composed of many genes. **A *gene* is a portion of the DNA molecule responsible for the synthesis of a single protein.**

We begin our study of nucleic acids with a look at the structure and formation of the nucleotide monomers.

17.1A Nucleosides—Joining a Monosaccharide and a Base

The nucleotides of both DNA and RNA contain a five-membered ring monosaccharide, often called simply the *sugar* component.

- **In RNA, the monosaccharide is the aldopentose D-ribose.**
- **In DNA the monosaccharide is D-2-deoxyribose, an aldopentose that lacks a hydroxyl group at C2.**

The prefix *deoxy* means *without oxygen.*

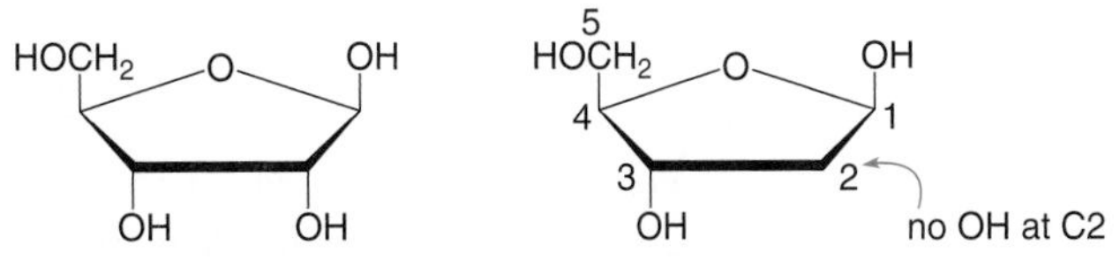

no OH at C2

D-ribose
(present in RNA)

D-2-deoxyribose
(present in DNA)

Only five common nitrogen-containing bases are present in nucleic acids. Three bases with one ring (**cytosine, uracil,** and **thymine**) are derived from the parent compound **pyrimidine.** Two bases with two rings (**adenine** and **guanine**) are derived from the parent compound **purine.** Each base is designated by a one-letter abbreviation as shown.

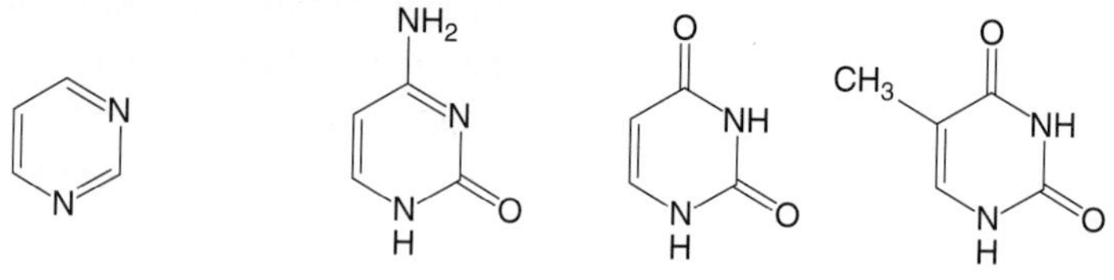

pyrimidine (parent compound) | cytosine **C** | uracil **U** | thymine **T**

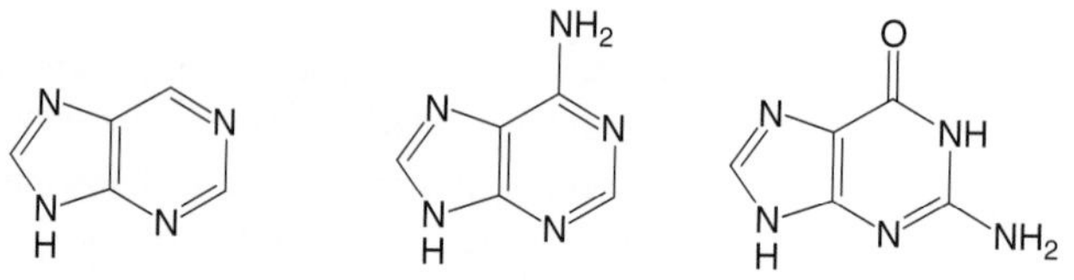

purine (parent compound) | adenine **A** | guanine **G**

Uracil (U) occurs only in RNA, while thymine (T) occurs only in DNA. As a result:

- DNA contains the bases A, G, C, and T.
- RNA contains the bases A, G, C, and U.

A *nucleoside* is formed by joining a carbon of the monosaccharide with a nitrogen atom of the base. A nucleoside is called an *N*-glycoside. Primes (') are used to number the carbons of the monosaccharide in a nucleoside.

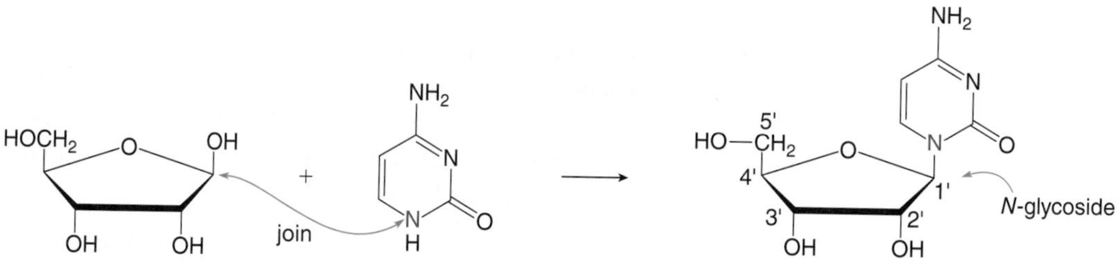

D-ribose + cytosine **C** (join) → cytidine, a ribonucleoside (*N*-glycoside at 1')

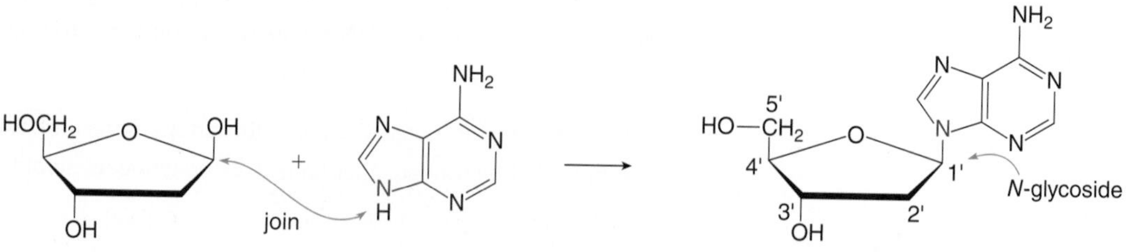

D-2-deoxyribose + adenine **A** (join) → deoxyadenosine, a deoxyribonucleoside (*N*-glycoside at 1')

For example, joining cytosine with ribose forms the ribonucleoside **cytidine.** Joining adenine with 2-deoxyribose forms the deoxyribonucleoside **deoxyadenosine.**

Nucleosides are named as derivatives of the bases from which they are formed.

- **To name a nucleoside derived from a pyrimidine base, use the suffix *-idine* (cytosine → cyt*idine*).**
- **To name a nucleoside derived from a purine base, use the suffix *-osine* (adenine → aden*osine*).**
- **For deoxyribonucleosides, add the prefix *deoxy-*, as in *deoxy*adenosine.**

SAMPLE PROBLEM 17.1

Identify the base and monosaccharide used to form the following nucleoside, and then name it.

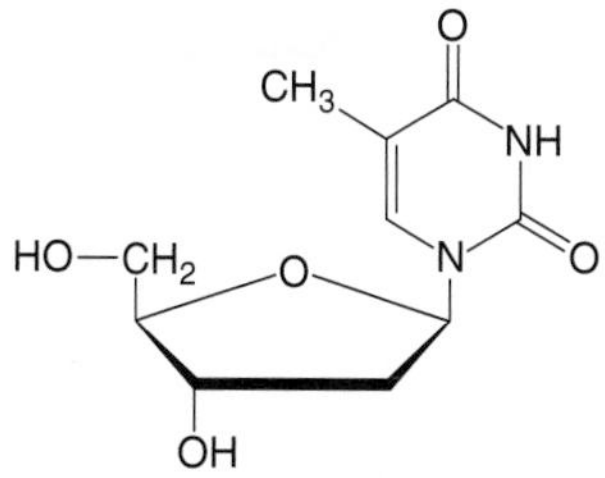

HO—CH$_2$ ... OH (nucleoside structure with CH$_3$, O, NH, N, O)

Analysis

- The sugar portion of a nucleoside contains the five-membered ring. If there is an OH group at C2', the sugar is ribose, and if there is no OH group at C2', the sugar is deoxyribose.
- The base is bonded to the five-membered ring. A pyrimidine base has one ring, and is derived from either cytosine, uracil, or thymine. A purine base has two rings, and is derived from either adenine or guanine.
- Nucleosides derived from pyrimidines end in the suffix *-idine.* Nucleosides derived from purines end in the suffix *-osine.*

Solution

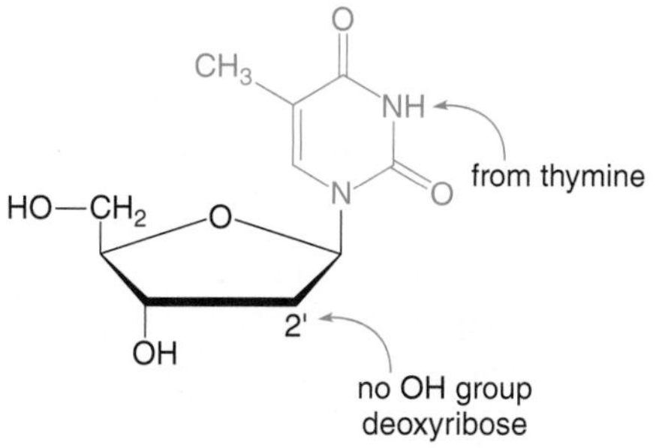

from thymine

2'

no OH group
deoxyribose

The sugar contains no OH at C2', so it is derived from deoxyribose. The base is thymine. To name the deoxyribonucleoside, change the suffix of the base to *-idine* and add the prefix *deoxy;* thus, thymine → deoxythymidine.

PROBLEM 17.1

Identify the base and monosaccharide used to form the following nucleosides, and then assign names.

a. 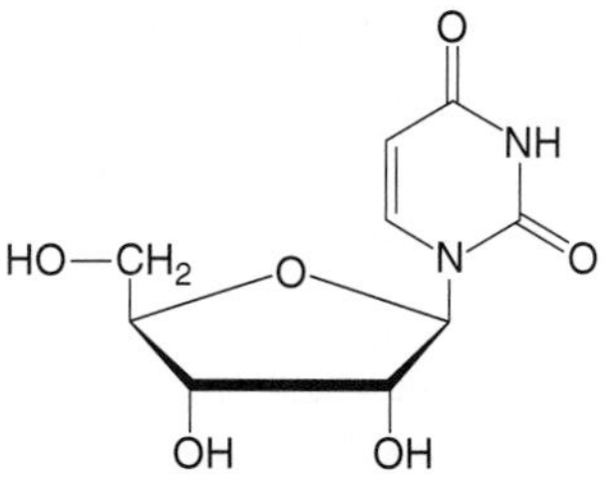 HO—CH$_2$ (structure with OH, OH; base with O, NH, N, O)

b. 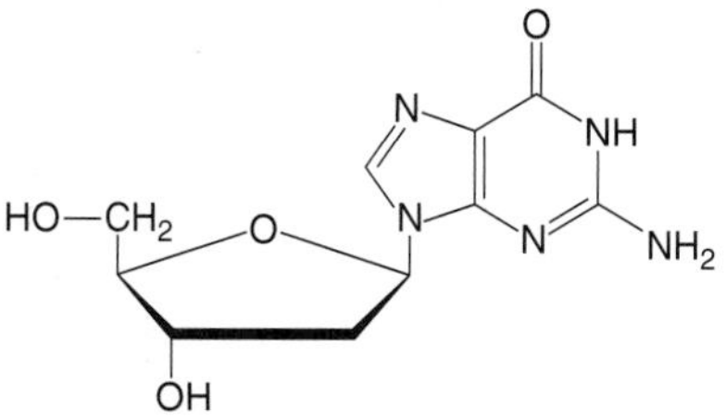 HO—CH$_2$ (structure with OH; base with N, N, O, NH, N, NH$_2$)

PROBLEM 17.2

Draw the structure of guanosine. Classify the compound as a ribonucleoside or a deoxyribonucleoside.

17.1B Nucleotides—Joining a Nucleoside with a Phosphate

Nucleotides are formed by adding a phosphate group to the 5'-OH of a nucleoside. Nucleotides are named by adding the term *5'-monophosphate* to the name of the nucleoside from which they are derived. **Ribonucleotides** are derived from ribose, while **deoxyribonucleotides** are derived from 2-deoxyribose.

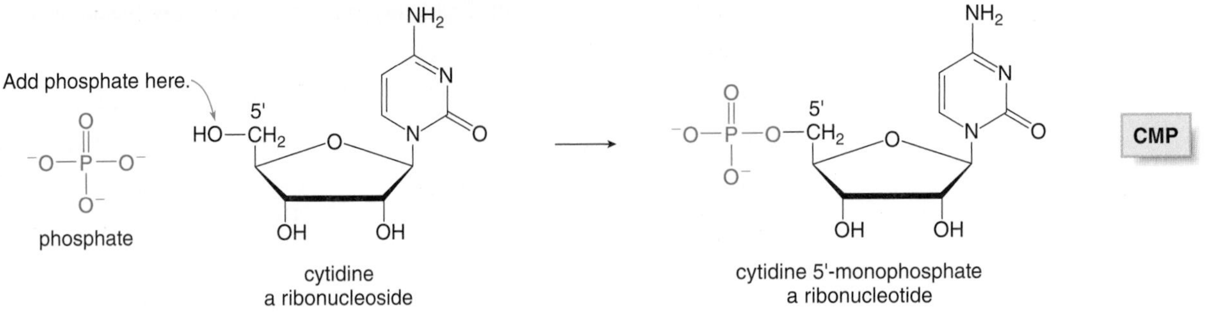

Add phosphate here.

phosphate

cytidine
a ribonucleoside

cytidine 5'-monophosphate
a ribonucleotide

CMP

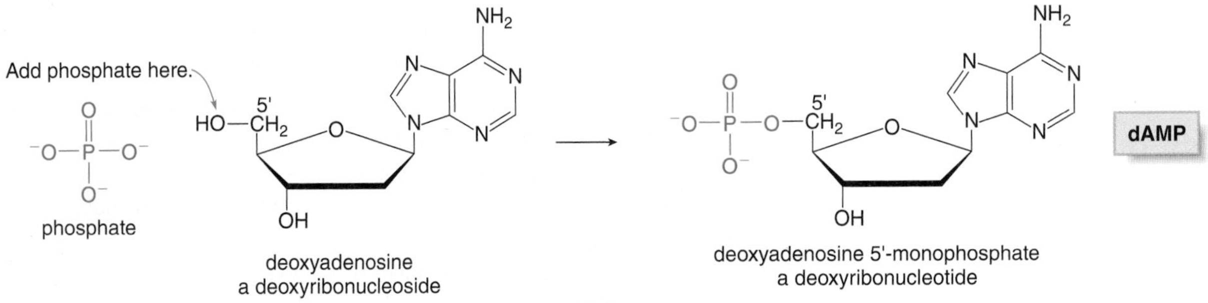

Add phosphate here.

phosphate

deoxyadenosine
a deoxyribonucleoside

deoxyadenosine 5'-monophosphate
a deoxyribonucleotide

dAMP

Because of the lengthy names of nucleotides, three- or four-letter abbreviations are commonly used instead. Thus, **c**ytidine 5'-**mono**phosphate is **CMP** and **d**eoxy**a**denosine 5'-**mono**phosphate is **dAMP.**

Figure 17.1 summarizes the information about nucleic acids and their components learned thus far. Table 17.1 summarizes the names and abbreviations used for the bases, nucleosides, and nucleotides needed in nucleic acid chemistry.

Di- and triphosphates can also be prepared from nucleosides by adding two and three phosphate groups, respectively, to the 5'-OH. For example, adenosine can be converted to adenosine 5'-diphosphate and adenosine 5'-triphosphate, abbreviated as ADP and ATP, respectively. We

Figure 17.1
Summary of the Components of Nucleosides, Nucleotides, and Nucleic Acids

Type of Compound	Components
Nucleoside	A monosaccharide + a base A ribonucleoside contains the monosaccharide ribose. A deoxyribonucleoside contains the monosaccharide 2-deoxyribose.
Nucleotide	A nucleoside + phosphate = a monosaccharide + a base + phosphate A ribonucleotide contains the monosaccharide ribose. A deoxyribonucleotide contains the monosaccharide 2-deoxyribose.
DNA	A polymer of deoxyribonucleotides The monosaccharide is 2-deoxyribose. The bases are A, G, C, and T.
RNA	A polymer of ribonucleotides The monosaccharide is ribose. The bases are A, G, C, and U.

Table 17.1 Names of Bases, Nucleosides, and Nucleotides in Nucleic Acids

Base	Abbreviation	Nucleoside	Nucleotide	Abbreviation
DNA				
Adenine	A	Deoxyadenosine	Deoxyadenosine 5'-monophosphate	dAMP
Guanine	G	Deoxyguanosine	Deoxyguanosine 5'-monophosphate	dGMP
Cytosine	C	Deoxycytidine	Deoxycytidine 5'-monophosphate	dCMP
Thymine	T	Deoxythymidine	Deoxythymidine 5'-monophosphate	dTMP
RNA				
Adenine	A	Adenosine	Adenosine 5'-monophosphate	AMP
Guanine	G	Guanosine	Guanosine 5'-monophosphate	GMP
Cytosine	C	Cytidine	Cytidine 5'-monophosphate	CMP
Uracil	U	Uridine	Uridine 5'-monophosphate	UMP

will learn about the central role of these phosphates, especially ATP, in energy production in Chapter 18.

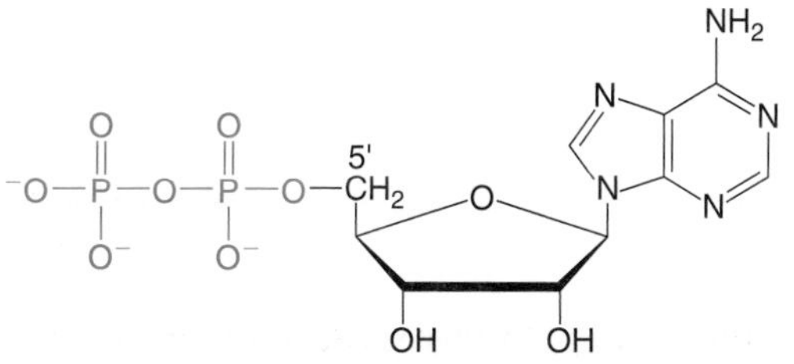

adenosine 5'-diphosphate
ADP

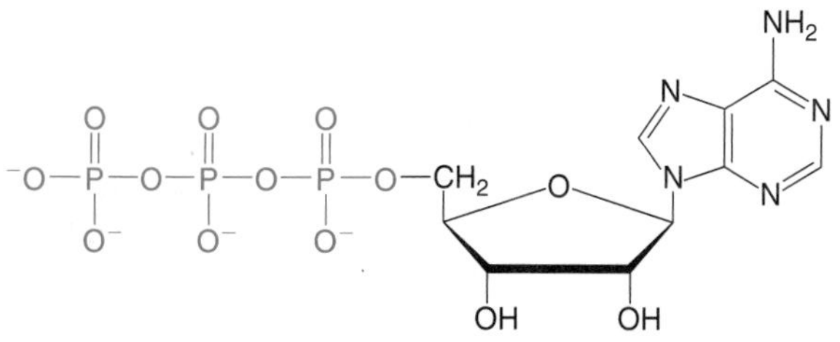

adenosine 5'-triphosphate
ATP

SAMPLE PROBLEM 17.2

Identify the base and monosaccharide in the following nucleotide. Name the nucleotide and give its three- or four-letter abbreviation.

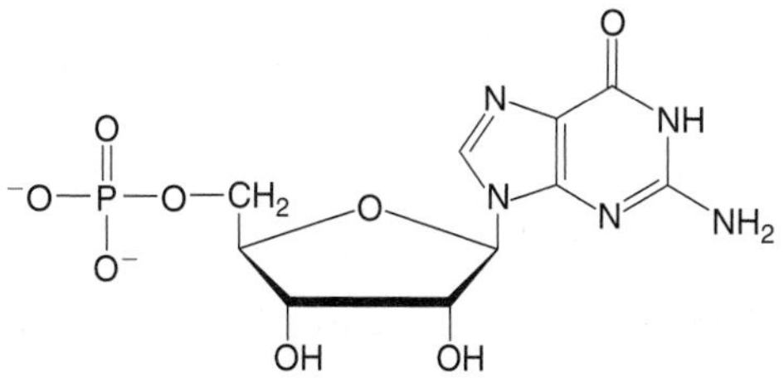

Analysis

- Identify the sugar and base as in Sample Problem 17.1.
- Name the nucleotide by naming the nucleoside and adding the term *5'-monophosphate,* since the molecule contains only one phosphorus bonded to the 5'-OH.
- Use the abbreviations from Table 17.1. When the sugar portion has an OH group at C2', a three-letter abbreviation is used. When the sugar portion has no OH group at C2', the four-letter abbreviation that begins with "d" is used.

Solution

The sugar portion is derived from ribose and the base is guanine.

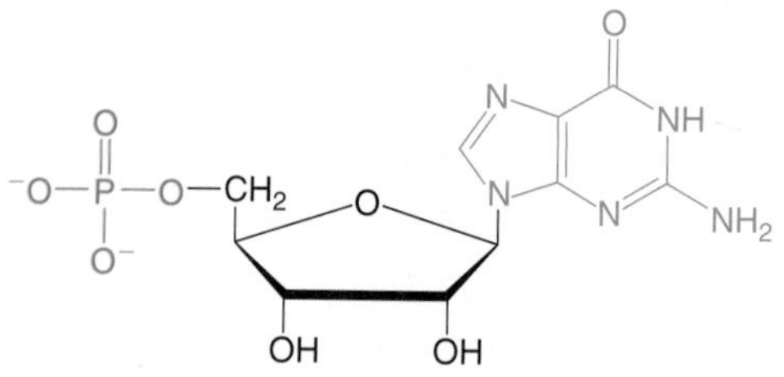

guanosine 5'-monophosphate
GMP

PROBLEM 17.3

Identify the base and monosaccharide in each nucleotide. Name the nucleotide and give its three- or four-letter abbreviation.

a.

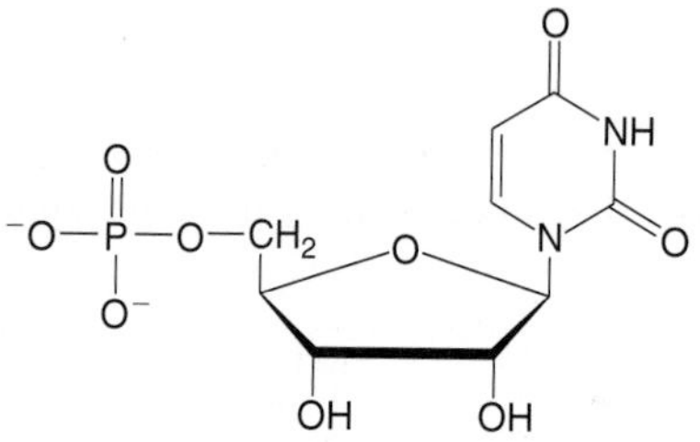

b.

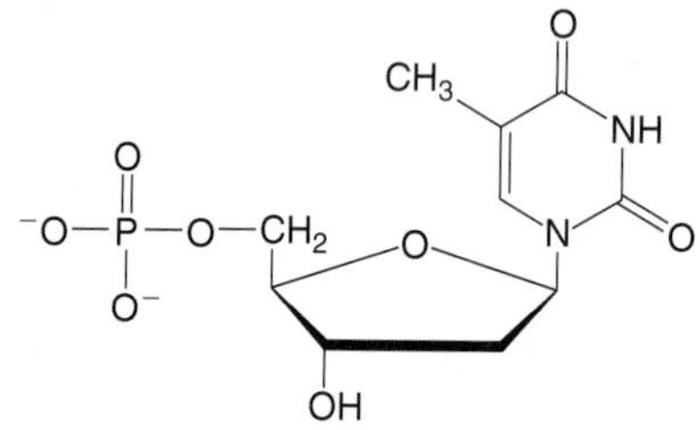

PROBLEM 17.4

Give the name that corresponds to each abbreviation: (a) GTP; (b) dCDP; (c) dTTP; (d) UDP.

PROBLEM 17.5

Which nucleic acid (DNA or RNA) contains each of the following components?

a. the sugar ribose
b. the sugar deoxyribose
c. the base T
d. the base U
e. the nucleotide GMP
f. the nucleotide dCMP

PROBLEM 17.6

Label each statement about the compound deoxycytidine as true or false.

a. Deoxycytidine is a nucleotide.
b. Deoxycytidine is a nucleoside.
c. Deoxycytidine contains a phosphate at its 5'-OH group.
d. Deoxycytidine contains a pyrimidine base.

PROBLEM 17.7

Identify each component as a base, nucleoside, or nucleotide.

a. adenine
b. cytidine
c. uridine 5'-monophosphate
d. deoxythymidine

17.2 Nucleic Acids

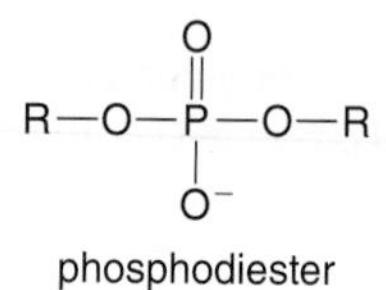

phosphodiester

Nucleic acids—both DNA and RNA—are polymers of nucleotides, formed by joining the 3'-OH group of one nucleotide with the 5'-phosphate of a second nucleotide in a **phosphodiester** linkage (Section 15.6).

For example, joining the 3'-OH group of dCMP (deoxycytidine 5'-monophosphate) and the 5'-phosphate of dAMP (deoxyadenosine 5'-monophosphate) forms a dinucleotide that contains a 5'-phosphate on one end (called the **5' end**) and a 3'-OH group on the other end (called the **3' end**).

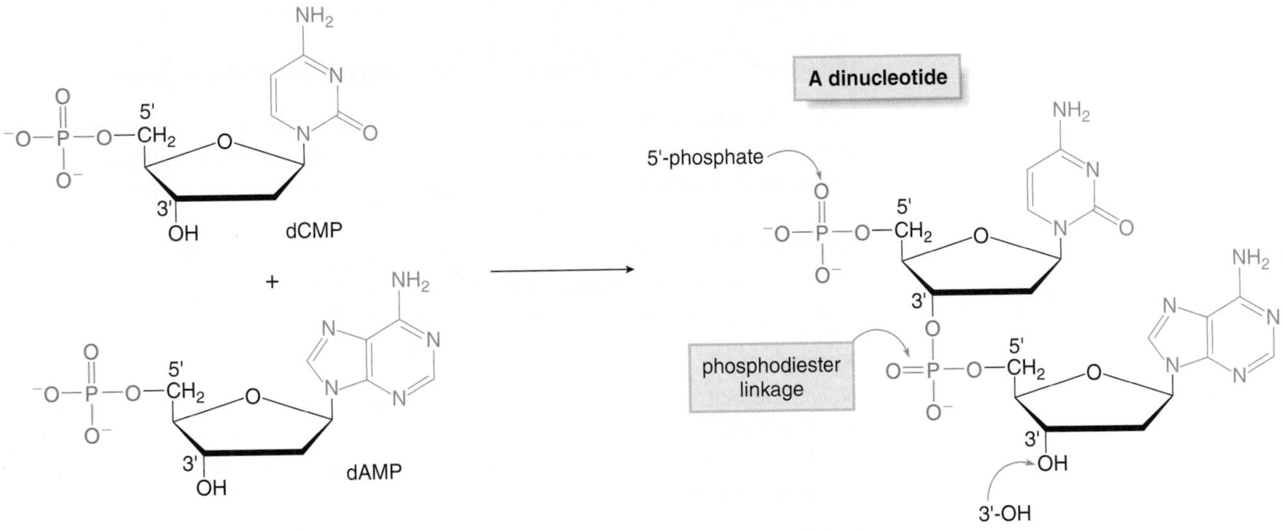

As additional nucleotides are added, the nucleic acid grows, each time forming a new phosphodiester linkage that holds the nucleotides together. Figure 17.2 illustrates the structure of a polynucleotide formed from four different nucleotides. Several features are noteworthy.

- **A polynucleotide contains a backbone consisting of alternating sugar and phosphate groups. All polynucleotides contain the same sugar–phosphate backbone.**
- **A polynucleotide has one free phosphate group at the 5' end.**
- **A polynucleotide has a free OH group at the 3' end.**

Figure 17.2 Primary Structure of a Polynucleotide

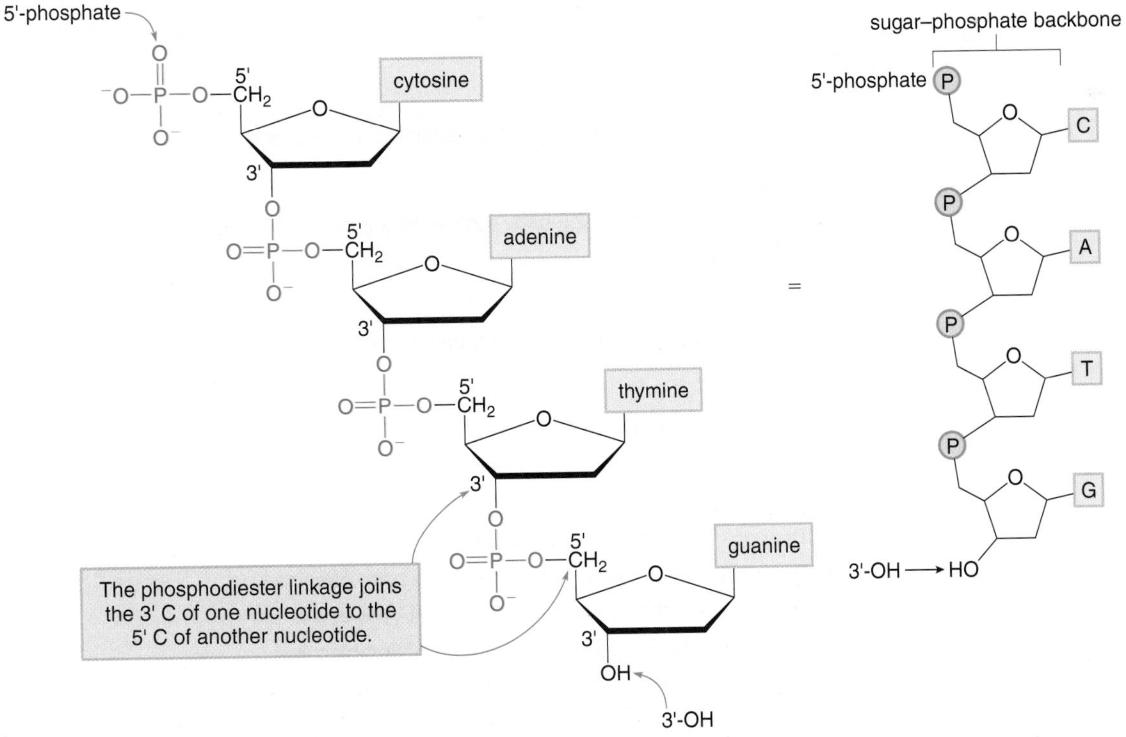

The name of a polynucleotide is read from the 5' end to the 3' end, using the one-letter abbreviations for the bases it contains. Drawn is the structure of the polynucleotide CATG.

The **primary structure** of a polynucleotide is the sequence of nucleotides that it contains. This sequence, which is determined by the identity of the bases, is unique to a nucleic acid. **In DNA, the sequence of bases carries the genetic information of the organism.**

Polynucleotides are named by the sequence of the bases they contain, beginning at the 5' end and using the one-letter abbreviation for the bases. Thus, the polynucleotide in Figure 17.2 contains the bases cytosine, adenine, thymine, and guanine, in order from the 5' end; thus, it is named CATG.

SAMPLE PROBLEM 17.3

(a) Draw the structure of a dinucleotide formed by joining the 3'-OH group of AMP to the 5'-phosphate in GMP. (b) Label the 5' and 3' ends. (c) Name the dinucleotide.

Analysis

Draw the structure of each nucleotide. In this case the sugar is ribose since the names of the mononucleotides do not contain the prefix *deoxy.* Bond the 3'-OH group to the 5'-phosphate to form the phosphodiester bond. The name of the dinucleotide begins with the nucleotide that contains the free phosphate at the 5' end.

Solution

a. and b.

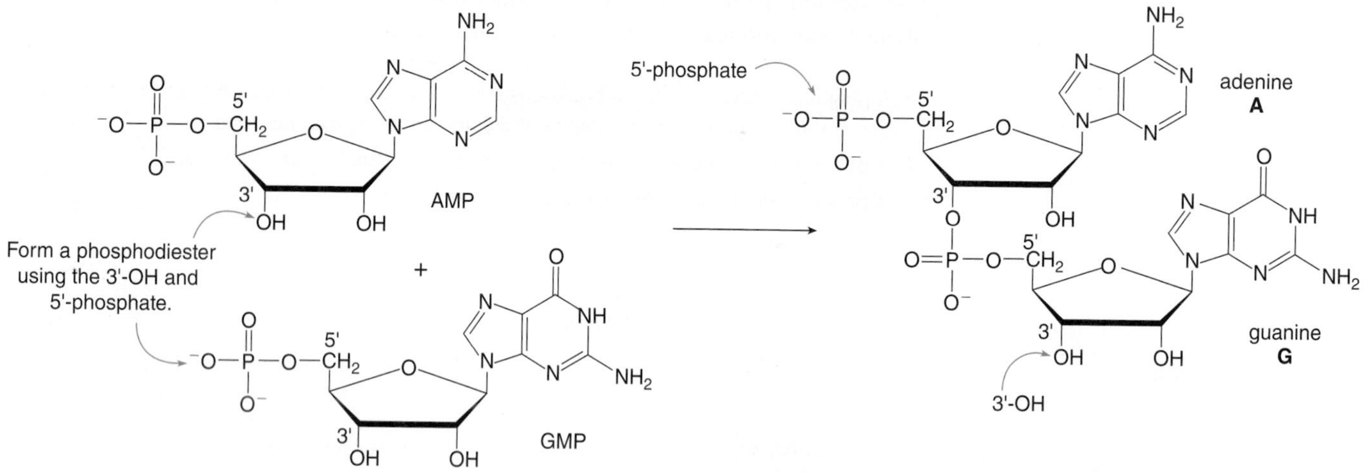

Form a phosphodiester using the 3'-OH and 5'-phosphate.

AMP + GMP → 5'-phosphate ... adenine A ... guanine G ... 3'-OH

c. Since polynucleotides are named beginning at the 5' end, this dinucleotide is named AG.

PROBLEM 17.8

Draw the structure of a dinucleotide formed by joining the 3'-OH group of dTMP to the 5'-phosphate in dGMP.

PROBLEM 17.9

Label the 5' end and the 3' end in each polynucleotide: (a) ATTTG; (b) CGCGUU; (c) GGACTT.

PROBLEM 17.10

Label each statement about the polynucleotide ATGGCG as true or false.

a. The polynucleotide has six nucleotides.
b. The polynucleotide contains six phosphodiester linkages.
c. The nucleotide at the 5' end contains the base guanine.
d. The nucleotide at the 3' end contains the base guanine.
e. The polynucleotide could be part of a DNA molecule.
f. The polynucleotide could be part of an RNA molecule.

17.3 The DNA Double Helix

Our current understanding of the structure of DNA is based on the model proposed initially by James Watson and Francis Crick in 1953 (Figure 17.3).

- **DNA consists of two polynucleotide strands that wind into a right-handed double helix.**

The sugar–phosphate backbone lies on the outside of the helix and the bases lie on the inside, perpendicular to the axis of the helix. The two strands of DNA run in *opposite* directions; that is, one strand runs from the 5' end to the 3' end, while the other runs from the 3' end to the 5' end.

Figure 17.3

The Three-Dimensional Structure of DNA—A Double Helix

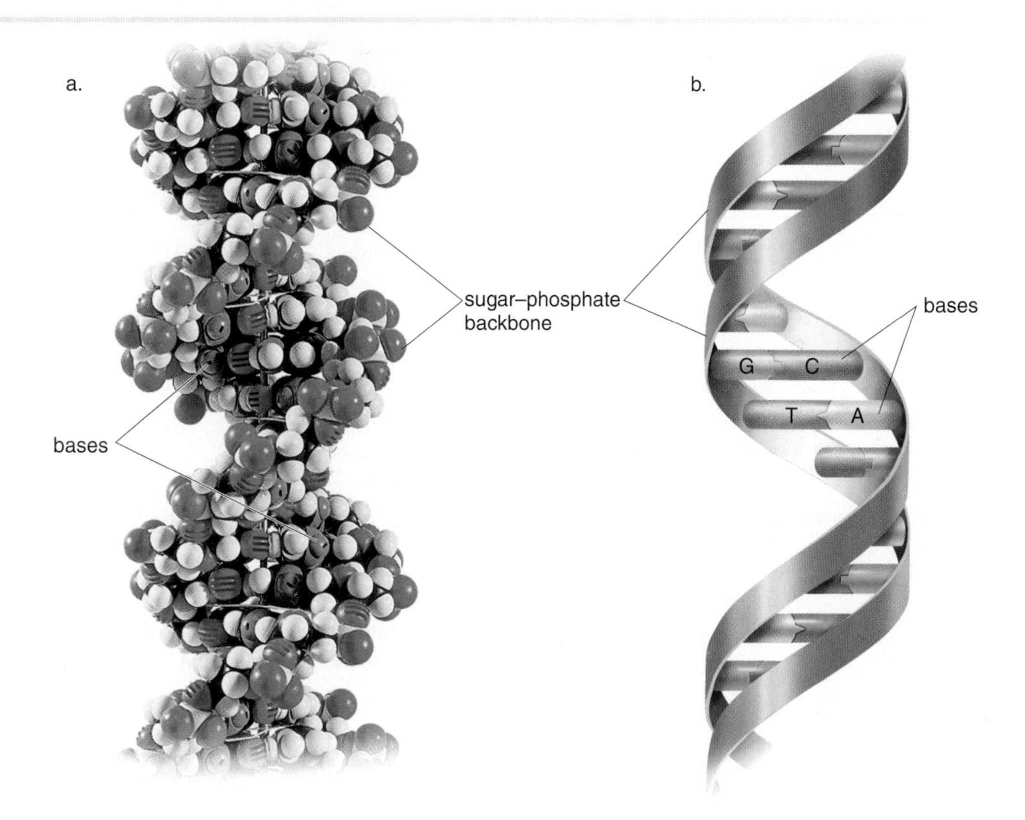

DNA consists of a double helix of polynucleotide chains. In view (a), the three-dimensional molecular model shows the sugar–phosphate backbone with the red (O), black (C), and white (H) atoms visible on the outside of the helix. In view (b), the bases on the interior of the helix are labeled.

The double helix is stabilized by hydrogen bonding between the bases of the two DNA strands as shown in Figure 17.4. A purine base on one strand always hydrogen bonds with a pyrimidine base on the other strand. Two bases hydrogen bond together in a predictable manner, forming **complementary base pairs.**

- **Adenine pairs with thymine using two hydrogen bonds, forming an A–T base pair.**
- **Cytosine pairs with guanine using three hydrogen bonds, forming a C–G base pair.**

Figure 17.4
Hydrogen Bonding in the DNA Double Helix

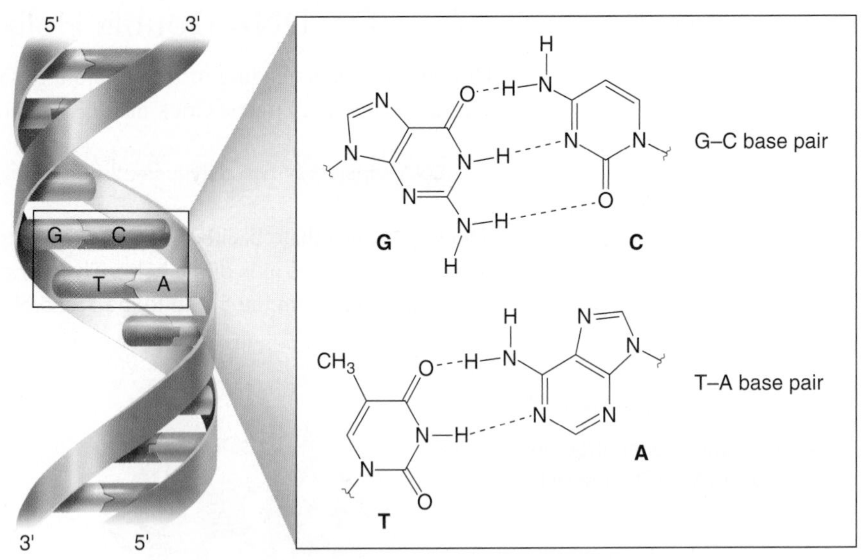

Hydrogen bonding of base pairs (A–T and C–G) holds the two strands of DNA together.

Because of this consistent pairing of bases, knowing the sequence of one strand of DNA allows us to write the sequence of the other strand, as shown in Sample Problem 17.4.

SAMPLE PROBLEM 17.4

Write the sequence of the complementary strand of the following portion of a DNA molecule: 5'–TAGGCTA–3'.

Analysis

The complementary strand runs in the opposite direction, from the 3' to the 5' end. Use base pairing to determine the corresponding sequence on the complementary strand: A pairs with T and C pairs with G.

Solution

Original strand: 5'–T A G G C T A–3'

Complementary strand: 3'–A T C C G A T–5'

PROBLEM 17.11

Write the complementary strand for each of the following strands of DNA.

a. 5'–AAACGTCC–3'
b. 5'–TATACGCC–3'
c. 5'–ATTGCACCCGC–3'
d. 5'–CACTTGATCGG–3'

Identical twins have the same genetic makeup, so that characteristics determined by DNA—such as hair color, eye color, or complexion—are also identical.

The enormously large DNA molecules that compose the **human genome**—the total DNA content of an individual—pack tightly into the nucleus of the cell. The double-stranded DNA helices wind around a core of protein molecules called histones to form a chain of nucleosomes, as shown in Figure 17.5. The chain of nucleosomes winds into a supercoiled fiber called chromatin, which composes each of the 23 pairs of chromosomes in humans.

The **genetic information of an organism is stored in the sequence of bases of its DNA molecules.** How is this information transferred from one generation to another? How, too, is the information stored in DNA molecules used to direct the synthesis of proteins?

Figure 17.5 The Structure of a Chromosome

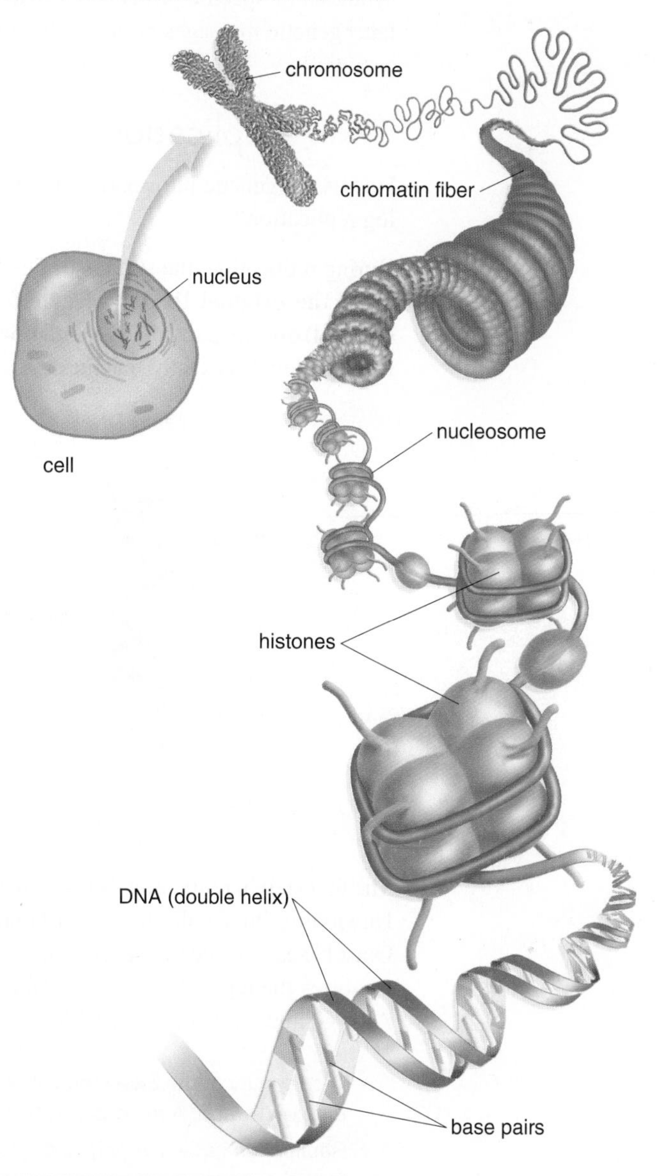

To answer these questions we must understand three key processes.

- *Replication* is the process by which DNA makes a copy of itself when a cell divides.
- *Transcription* is the ordered synthesis of RNA from DNA. In this process, the genetic information stored in DNA is passed onto RNA.
- *Translation* is the synthesis of proteins from RNA. In this process, the genetic message contained in RNA determines the specific amino acid sequence of a protein.

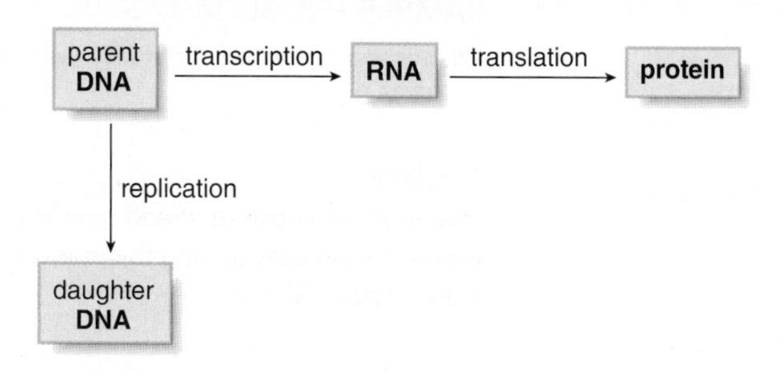

Each chromosome contains many **genes,** those portions of the DNA molecules that result in the synthesis of specific proteins. Only a small fraction (1–2%) of the DNA in a chromosome contains genetic messages or genes that result in protein synthesis.

17.4 Replication

How is the genetic information in the DNA of a parent cell passed onto new daughter cells during replication?

During replication, the strands of DNA separate and each serves as a template for a new strand. Thus, **the original DNA molecule forms two DNA molecules, each of which contains one strand from the parent DNA and one new strand.** The sequence of both strands of the daughter DNA molecules exactly matches the sequence in the parent DNA.

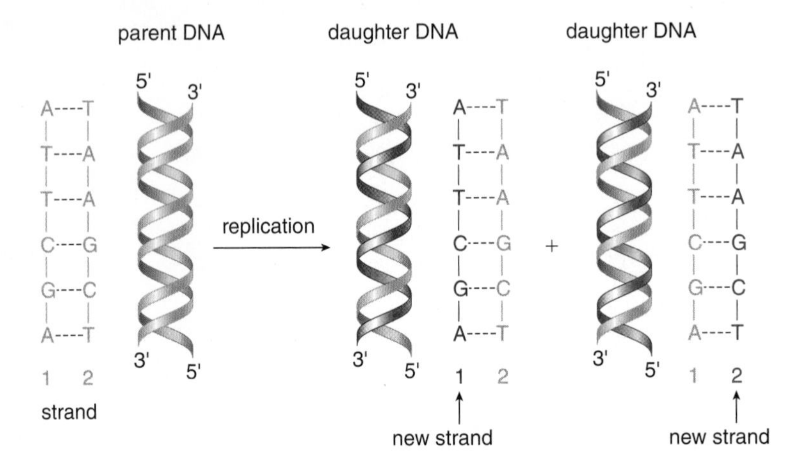

The first step in replication is the unwinding of the DNA helix to expose the bases on each strand. Unwinding breaks the hydrogen bonds that hold the two strands of the double helix together. Once bases have been exposed on the unwound strands of DNA, the enzyme DNA polymerase catalyzes the replication process using the four nucleoside triphosphates (derived from the bases A, T, G, and C) that are available in the nucleus (Figure 17.6).

- **The identity of the bases on the template strand determines the order of the bases on the new strand: A must pair with T, and G must pair with C.**
- **Replication occurs in only one direction on the template strand, from the 3' end to the 5' end.**

Since replication proceeds in only one direction—that is, from the 3' end to the 5' end of the template—the two new strands of DNA must be synthesized by somewhat different techniques. One strand, called the **leading strand,** grows continuously. The other strand, called the **lagging strand,** is synthesized in small fragments, which are then joined together by an enzyme. The end result is two new strands of DNA, one in each of the daughter DNA molecules, both with complementary base pairs joining the two DNA strands together.

SAMPLE PROBLEM 17.5

What is the sequence of a newly synthesized DNA segment if the template strand has the sequence 3'–TGCACC–5'?

Analysis

The newly synthesized strand runs in the opposite direction, from the 5' end to the 3' end in this example. Use base pairing to determine the corresponding sequence on the new strand: A pairs with T and C pairs with G.

Figure 17.6
DNA Replication

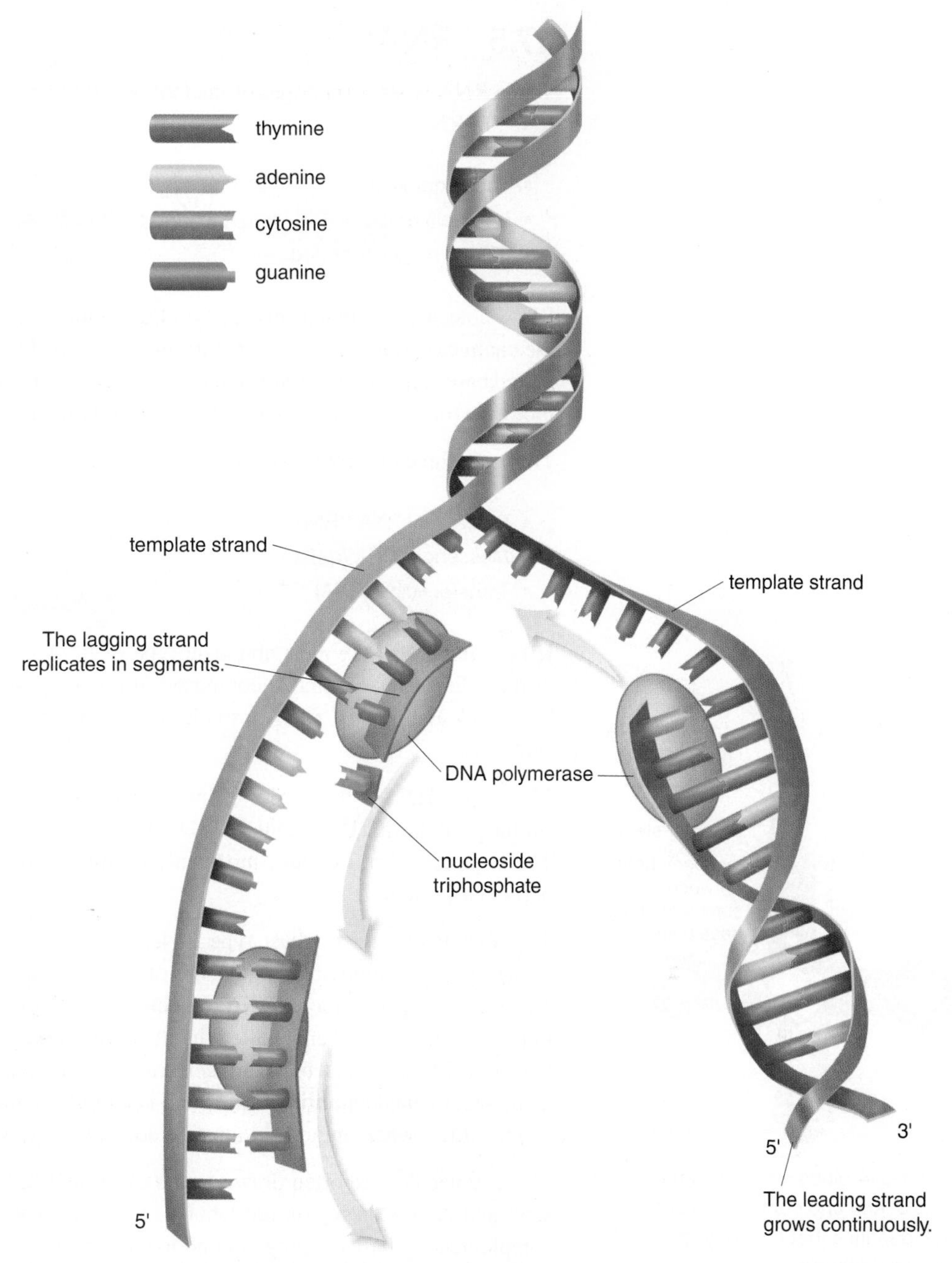

Replication proceeds along both strands of unwound DNA. The leading strand grows continuously, while the lagging strand must be synthesized in fragments that are joined together by a DNA ligase enzyme.

Solution

Template strand: 3'–T G C A C C–5'

New strand: 5'–A C G T G G–3'

PROBLEM 17.12

What is the sequence of a newly synthesized DNA segment if the template strand has each of the following sequences?

a. 3'–AGAGTCTC–5' b. 5'–ATTGCTC–3' c. 3'–ATCCTGTAC–5' d. 5'–GGCCATACTC–3'

17.5 RNA

While RNA is also composed of nucleotides, there are important differences between DNA and RNA. In RNA,

- The sugar is ribose.
- U (uracil) replaces T (thymine) as one of the bases.
- RNA is single stranded.

RNA molecules are much smaller than DNA molecules. Although RNA contains a single strand, the chain can fold back on itself, forming loops, and intermolecular hydrogen bonding between paired bases on a single strand can form helical regions. When base pairing occurs within an RNA molecule (or between RNA and DNA), **C and G form base pairs,** and **A and U form base pairs.**

There are three different types of RNA molecules.

- Ribosomal RNA (rRNA)
- Messenger RNA (mRNA)
- Transfer RNA (tRNA)

Ribosomal RNA, the most abundant type of RNA, is found in the ribosomes in the cytoplasm of the cell. Each ribosome is composed of one large subunit and one small subunit that contain both RNA and protein. rRNA provides the site where polypeptides are assembled during protein synthesis.

Figure 17.7 Transfer RNA

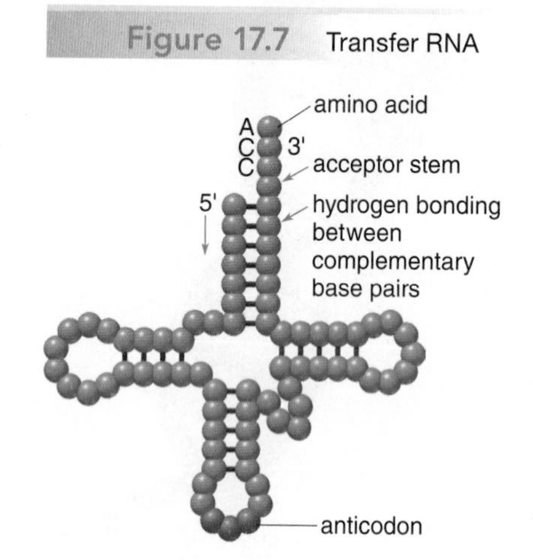

Each tRNA binds a specific amino acid to its 3' end and contains an anticodon that identifies that amino acid for protein synthesis.

Messenger RNA is the carrier of information from DNA (in the cell nucleus) to the ribosomes (in the cytoplasm). Each gene of a DNA molecule corresponds to a specific mRNA molecule. The sequence of nucleotides in the mRNA molecule determines the amino acid sequence in a particular protein.

Transfer RNA, the smallest type of RNA, interprets the genetic information in mRNA and brings specific amino acids to the site of protein synthesis in the ribosome. Each amino acid is recognized by one or more tRNA molecules, which contain 70–90 nucleotides. tRNAs have two important sites. The 3' end, called the **acceptor stem,** always contains the nucleotides ACC and has a free OH group that binds a specific amino acid. Each tRNA also contains a sequence of three nucleotides called an **anticodon,** which is complementary to three bases in an mRNA molecule, and identifies what amino acid must be added to a growing polypeptide chain.

tRNA molecules are often drawn in the cloverleaf fashion shown in Figure 17.7. The acceptor stem and anticodon region are labeled. Folding creates regions of the tRNA in which nearby complementary bases hydrogen bond to each other.

Table 17.2 summarizes the characteristics of the three types of RNAs.

Table 17.2 Three Types of RNA Molecules

Type of RNA	Abbreviation	Function
Ribosomal RNA	rRNA	The site of protein synthesis, found in the ribosomes
Messenger RNA	mRNA	Carries the information from DNA to the ribosomes
Transfer RNA	tRNA	Brings specific amino acids to the ribosomes for protein synthesis

17.6 Transcription

The conversion of the information in DNA to the synthesis of proteins begins with ***transcription*—that is, the synthesis of messenger RNA from DNA.**

RNA synthesis begins in the same manner as DNA replication: the double helix of DNA unwinds (Figure 17.8). Since RNA is single stranded, however, only one strand of DNA is needed for RNA synthesis. The **template** strand is the strand of DNA used for RNA synthesis.

Each mRNA molecule corresponds to a small segment of a DNA molecule. Transcription proceeds from the 3' end to the 5' end of the template strand. Complementary base pairing determines what RNA nucleotides are added to the growing RNA chain: C pairs with G, T pairs with A, and A pairs with U. Thus, the RNA chain grows from the 5' to 3' direction. When transcription is completed, the new mRNA molecule is released and the double helix of the DNA molecule re-forms.

- **Transcription forms a messenger RNA molecule with a sequence that is *complementary* to the DNA template from which it is prepared.**

SAMPLE PROBLEM 17.6

If a portion of the template strand of a DNA molecule has the sequence 3'–CTAGGATAC–5', what is the sequence of the mRNA molecule produced from this template?

Analysis

mRNA has a base sequence that is complementary to the template from which it is prepared.

Figure 17.8
Transcription

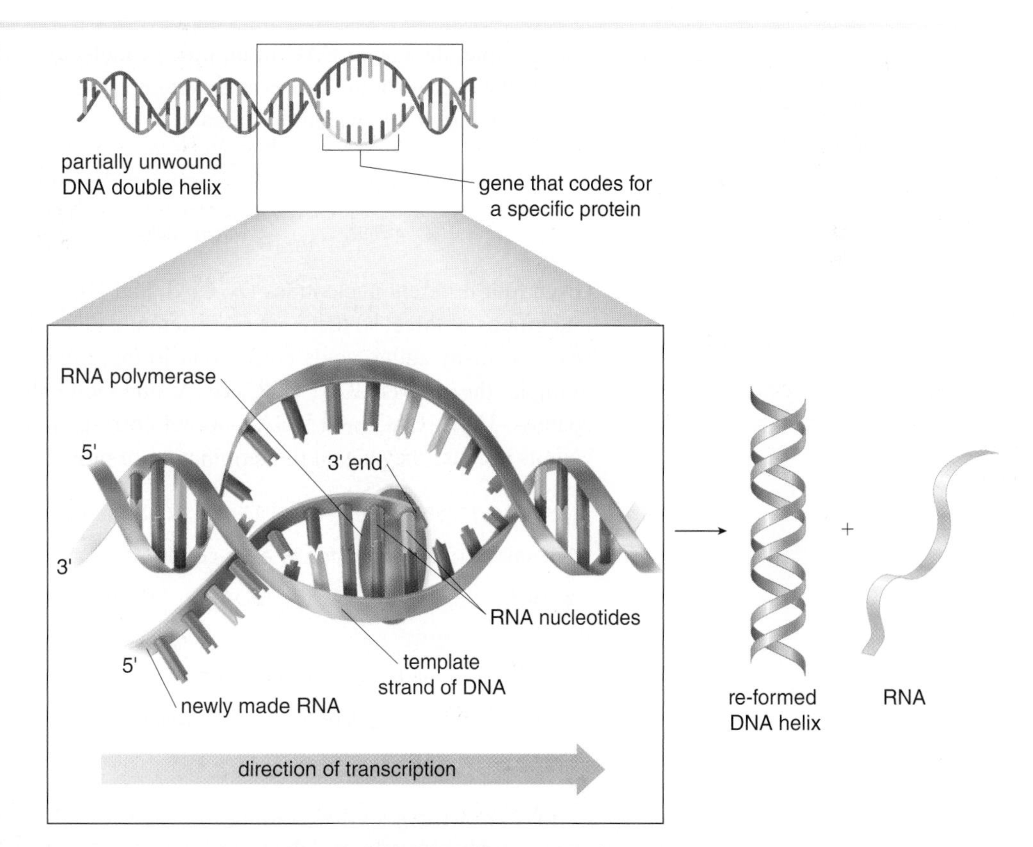

In transcription, the DNA helix unwinds and RNA polymerase catalyzes the formation of mRNA along the DNA template strand. Transcription forms an mRNA molecule with base pairs complementary to the DNA template strand.

Solution

Template strand of DNA: 3'–C T A G G A T A C–5'
mRNA sequence: 5'–G A U C C U A U G–3'
complementary

PROBLEM 17.13

What is the sequence of the mRNA molecule synthesized from each DNA template?

a. 3'–TGCCTAACG–5' b. 3'–GACTCC–5' c. 3'–TTAACGCGA–5' d. 3'–CAGTGACCGTAC–5'

PROBLEM 17.14

What is the sequence of the DNA template strand from which each of the following mRNA strands was synthesized?

a. 5'–UGGGGCAUU–3' b. 5'–GUACCU–3' c. 5'–CCGACGAUG–3' d. 5'–GUAGUCACG–3'

17.7 The Genetic Code

Once the genetic information of DNA has been transcribed in a messenger RNA molecule, RNA can direct the synthesis of an individual protein. How can RNA, which is composed of only four different nucleotides, direct the synthesis of polypeptides that are formed from 20 different amino acids? The answer lies in the **genetic code.**

- **A sequence of three nucleotides (a triplet) codes for a specific amino acid. Each triplet is called a *codon.***

For example, the codon UAC in an mRNA molecule codes for the amino acid serine, and the codon UGC codes for the amino acid cysteine. The same genetic code occurs in almost all organisms, from bacteria to whales to humans.

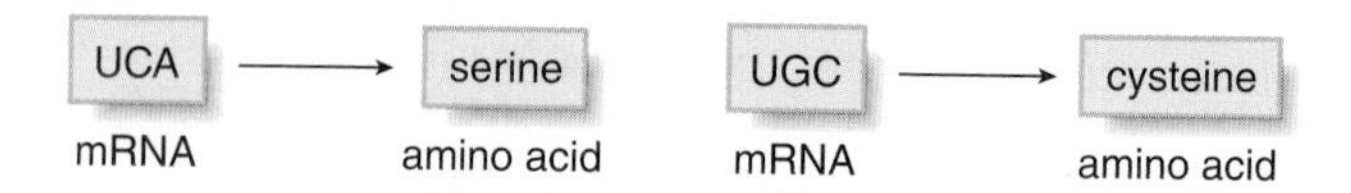

Given four different nucleotides (A, C, G, and U), there are 64 different ways to combine them into groups of three, so there are 64 different codons. Sixty-one codons code for specific amino acids, so many amino acids correspond to more than one codon, as shown in Table 17.3. For example, the codons GGU, GGC, GGA, and GGG all code for the amino acid glycine. Three codons—UAA, UAG, and UGA—do not correspond to any amino acids; they are called **stop codons** because they signal the termination of protein synthesis.

PROBLEM 17.15

What amino acid is coded for by each codon?

a. GCC b. AAU c. CUA d. AGC e. CAA f. AAA

PROBLEM 17.16

What codons code for each amino acid?

a. glycine b. isoleucine c. lysine d. glutamic acid

Codons are written so that reading from left to right, the first triplet codes for the N-terminal amino acid in a protein, and the last triplet codes for the C-terminal amino acid. Sample Problem 17.7 illustrates the conversion of a sequence of bases in mRNA to a sequence of amino acids in a peptide.

Table 17.3 The Genetic Code—Triplets in Messenger RNA

First Base (5' end)	Second Base								Third Base (3' end)
	U		C		A		G		
U	UUU	Phe	UCU	Ser	UAU	Tyr	UGU	Cys	U
	UUC	Phe	UCC	Ser	UAC	Tyr	UGC	Cys	C
	UUA	Leu	UCA	Ser	UAA	Stop	UGA	Stop	A
	UUG	Leu	UCG	Ser	UAG	Stop	UGG	Trp	G
C	CUU	Leu	CCU	Pro	CAU	His	CGU	Arg	U
	CUC	Leu	CCC	Pro	CAC	His	CGC	Arg	C
	CUA	Leu	CCA	Pro	CAA	Gln	CGA	Arg	A
	CUG	Leu	CCG	Pro	CAG	Gln	CGG	Arg	G
A	AUU	Ile	ACU	Thr	AAU	Asn	AGU	Ser	U
	AUC	Ile	ACC	Thr	AAC	Asn	AGC	Ser	C
	AUA	Ile	ACA	Thr	AAA	Lys	AGA	Arg	A
	AUG	Met	ACG	Thr	AAG	Lys	AGG	Arg	G
G	GUU	Val	GCU	Ala	GAU	Asp	GGU	Gly	U
	GUC	Val	GCC	Ala	GAC	Asp	GGC	Gly	C
	GUA	Val	GCA	Ala	GAA	Glu	GGA	Gly	A
	GUG	Val	GCG	Ala	GAG	Glu	GGG	Gly	G

SAMPLE PROBLEM 17.7

Derive the amino acid sequence that is coded for by the following mRNA sequence.

CAU AAA ACG GUG UUA AUA

Analysis

Use Table 17.3 to identify the codons that correspond to each amino acid. Codons correspond to a peptide written from the N-terminal to C-terminal end.

Solution

CAU AAA ACG GUG UUA AUA

↓ ↓ ↓ ↓ ↓ ↓

His — Lys — Thr — Val — Leu — Ile

N-terminal amino acid C-terminal amino acid

PROBLEM 17.17

Derive the amino acid sequence that is coded for by each mRNA sequence.

a. CAA GAG GUA UCC UAC AGA
b. GUC AUC UGG AGG GGC AUU
c. CUA UGC AGU AGG ACA CCC

PROBLEM 17.18

Write a possible mRNA sequence that codes for each of the following peptides.

a. Met–Arg–His–Phe b. Gly–Ala–Glu–Gln c. Gln–Asn–Gly–Ile–Val d. Thr–His–Asp–Cys–Trp

PROBLEM 17.19

Considering the given sequence of nucleotides in an mRNA molecule, (a) what is the sequence of the DNA template strand from which the RNA was synthesized? (b) What peptide is synthesized by this mRNA sequence?

GAG CCC GUA UAC GCC ACG

17.8 Translation and Protein Synthesis

The translation of the information in messenger RNA to protein synthesis occurs in the ribosomes. Each type of RNA plays a role in protein synthesis.

- **mRNA contains the sequence of codons that determines the order of amino acids in the protein.**
- **Individual tRNAs bring specific amino acids to add to the peptide chain.**
- **rRNA contains binding sites that provide the platform on which protein synthesis occurs.**

Each individual tRNA contains an **anticodon** of three nucleotides that is complementary to the codon in mRNA and identifies individual amino acids (Section 17.5). For example, a codon of UCA in mRNA corresponds to an anticodon of AGU in a tRNA molecule, which identifies serine as the amino acid. Other examples are shown in Table 17.4.

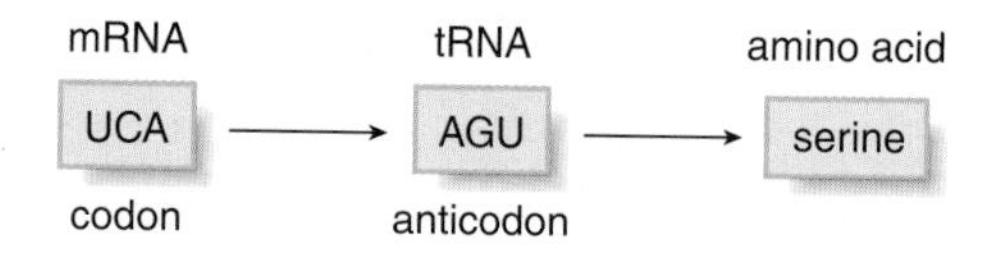

PROBLEM 17.20

For each codon: [1] Write the anticodon. [2] What amino acid does each codon represent?

a. CGG b. GGG c. UCC d. AUA e. CCU f. GCC

Table 17.4 Relating Codons, Anticodons, and Amino Acids

mRNA Codon		tRNA Anticodon		Amino Acid
ACA	→	UGU	→	threonine
GCG	→	CGC	→	alanine
AGA	→	UCU	→	arginine
UCC	→	AGG	→	serine

There are three stages in translation: initiation, elongation, and termination. Figure 17.9 depicts the main features of translation.

[1] Initiation

Translation begins when an mRNA molecule binds to the smaller subunit of the ribosome and a tRNA molecule carries the first amino acid of the peptide chain to the binding site. Translation always begins at the codon AUG, which codes for the amino acid methionine. The arriving tRNA contains an anticodon with the complementary base sequence UAC.

Figure 17.9

Translation—The Synthesis of Proteins from RNA

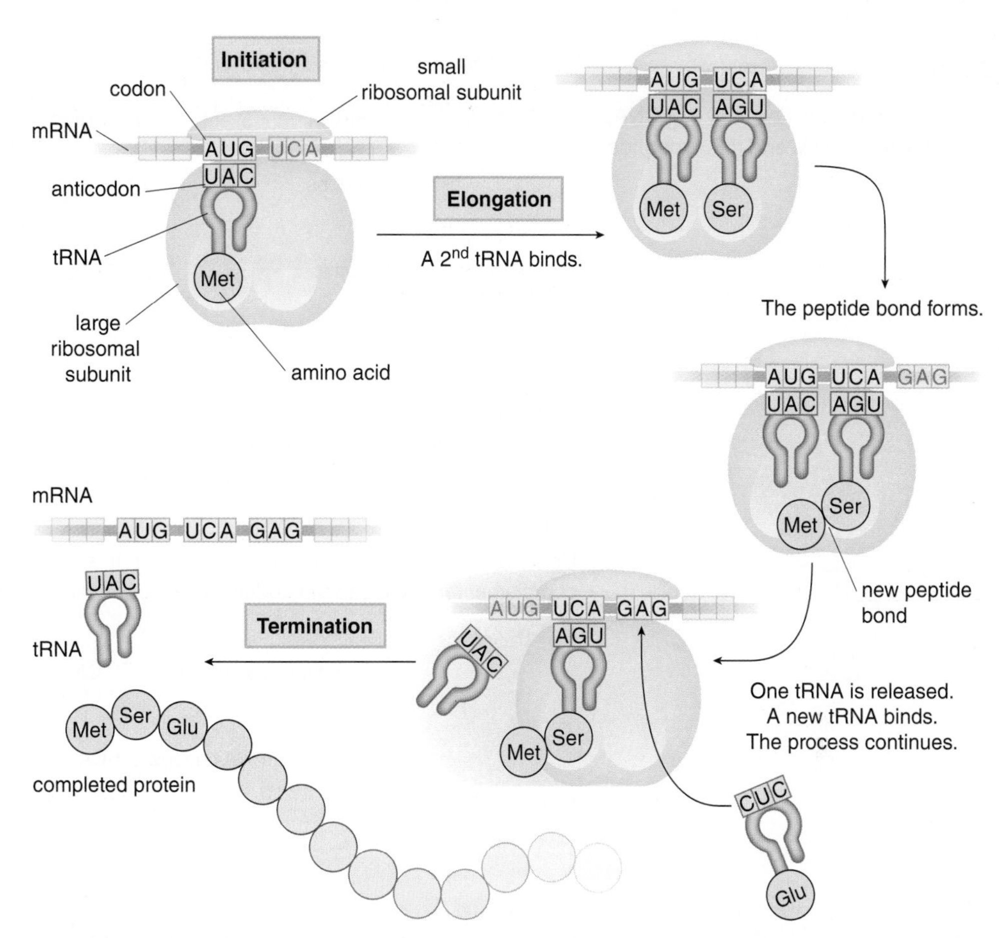

- **Initiation** consists of the binding of the ribosomal subunits to mRNA and the arrival of the first tRNA carrying its amino acid.
- The protein is synthesized during **elongation.** One by one a tRNA with its designated amino acid binds to a site on the ribosome adjacent to the first tRNA. A peptide bond forms and a tRNA is released. The ribosome shifts to the next codon and the process continues.
- **Termination** occurs when a stop codon is reached. The synthesis is complete and the protein is released from the complex.

[2] Elongation

The next tRNA molecule containing an anticodon for the second codon binds to mRNA, delivering its amino acid, and a peptide bond forms between the two amino acids. The first tRNA molecule, which has delivered its amino acid and is no longer needed, dissociates from the complex. The ribosome shifts to the next codon along the mRNA strand and the process continues when a new tRNA molecule binds to the mRNA.

[3] Termination

Translation continues until a stop codon is reached. There is no tRNA that contains an anticodon complementary to any of the three stop codons (UAA, UAG, and UGA), so protein synthesis ends and the protein is released from the ribosome. Often the first amino acid in the chain, methionine, is not needed in the final protein and so it is removed after protein synthesis is complete.

Figure 17.10 shows a representative segment of DNA, and the mRNA, tRNA, and amino acid sequences that correspond to it.

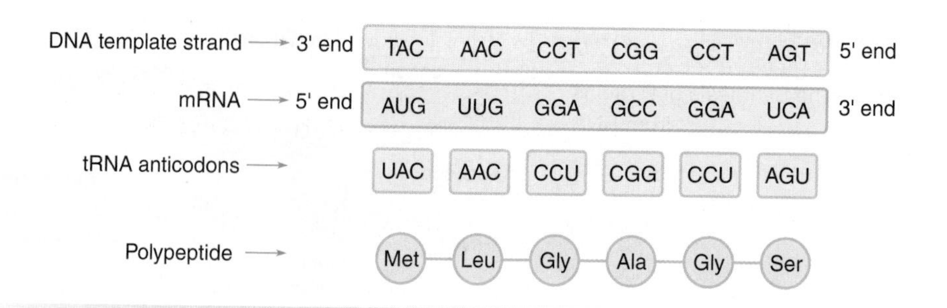

Figure 17.10
Comparing the Sequence of DNA, mRNA, tRNA, and a Polypeptide

SAMPLE PROBLEM 17.8

What sequence of amino acids would be formed from the following mRNA sequence:
CAA AAG ACG UAC CGA? List the anticodons contained in each of the needed tRNA molecules.

Analysis

Use Table 17.3 to determine the amino acid that is coded for by each codon. The anticodons contain complementary bases to the codons: A pairs with U, and C pairs with G.

Solution

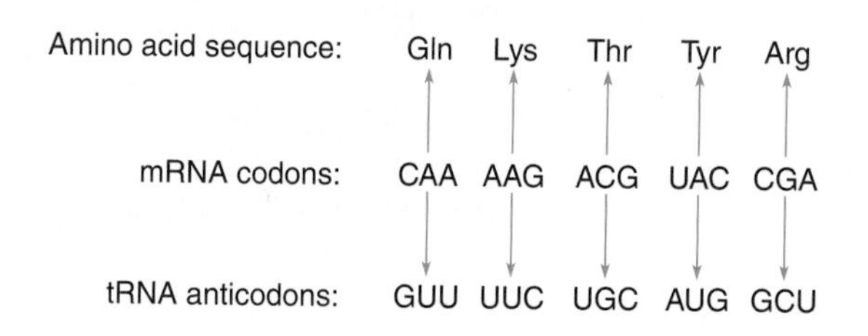

PROBLEM 17.21

What sequence of amino acids would be formed from each mRNA sequence? List the anticodons contained in each of the needed tRNA molecules.

a. CCA CCG GCA AAC GAA GCA
b. GCA CCA CUA AGA GAC

SAMPLE PROBLEM 17.9

What polypeptide would be synthesized from the following template strand of DNA:
CGG TGT CTT TTA?

Analysis

To determine what polypeptide is synthesized from a DNA template, two steps are needed. First use the DNA sequence to determine the transcribed mRNA sequence: C pairs with G, T pairs with A, and A (on DNA) pairs with U (on mRNA). Then use the codons in Table 17.3 to determine what amino acids are coded for by a given codon in mRNA.

Solution

DNA template strand: ⟶ CGG TGT CTT TTA

mRNA: ⟶ GCC ACA GAA AAU

Polypeptide: ⟶ Ala — Thr — Glu — Asn

PROBLEM 17.22

What polypeptide would be synthesized from each of the following template strands of DNA?

a. TCT CAT CGT AAT GAT TCG
b. GCT CCT AAA TAA CAC TTA

17.9 Mutations and Genetic Diseases

Although replication provides a highly reliable mechanism for making an exact copy of DNA, occasionally an error occurs, thus producing a DNA molecule with a slightly different nucleotide sequence.

- A *mutation* is a change in the nucleotide sequence in a molecule of DNA.

If the mutation occurs in a nonreproductive cell, the mutation is passed on to daughter cells within the organism, but is not transmitted to the next generation. If the mutation occurs in an egg or sperm cell, it is passed on to the next generation of an organism. Some mutations are random events, while others are caused by **mutagens,** chemical substances that alter the structure of DNA. Exposure to high-energy radiation such as X-rays or ultraviolet light can also produce mutations.

Mutations can be classified according to the change that results in a DNA molecule.

- A *point* mutation is the *substitution* of one nucleotide for another.

Original DNA: G A G T T C

replacement of G by C

Point mutation: G A C T T C

- A *deletion* mutation occurs when one or more nucleotides is *lost* from a DNA molecule.

Original DNA: G A G T T C

loss of G

Deletion mutation: G A T T C

- An *insertion* mutation occurs when one or more nucleotides is *added* to a DNA molecule.

Original DNA: G A G T T C

addition of C

Insertion mutation: G A G C T T C

A mutation can have a negligible, minimal, or catastrophic effect on an organism. To understand the effect of a mutation, we must determine the mRNA sequence that is transcribed from the DNA sequence as well as the resulting amino acid for which it codes. A point mutation in the three-base sequence CTT in a gene that codes for a particular protein illustrates some possible outcomes of a mutation.

The sequence CTT in DNA is transcribed to the codon GAA in mRNA, and using Table 17.3, this triplet codes for the amino acid glutamic acid. If a point mutation replaces CTT by CTC in DNA, CTC is transcribed to the codon GAG in mRNA. Since GAG codes for the *same* amino acid—glutamic acid—this mutation does not affect the protein synthesized by this segment of DNA. Such a mutation is said to be **silent.**

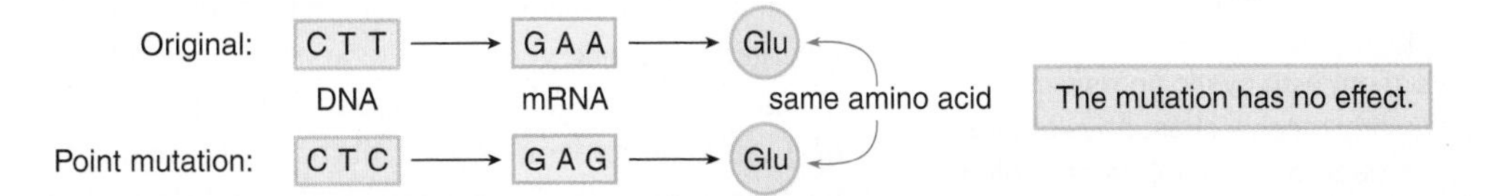

Alternatively, suppose a point mutation replaces CTT by CAT in DNA. CAT is transcribed to the codon GUA in mRNA, and GUA codes for the amino acid valine. Now, the mutation produces a protein with one *different* amino acid—namely, valine instead of glutamic acid. In some proteins this alteration of the primary sequence may have little effect on the protein's secondary and tertiary structure. In other proteins, such as hemoglobin, the substitution of valine for glutamic acid produces a protein with vastly different properties, resulting in the fatal disease sickle cell anemia (Section 16.7).

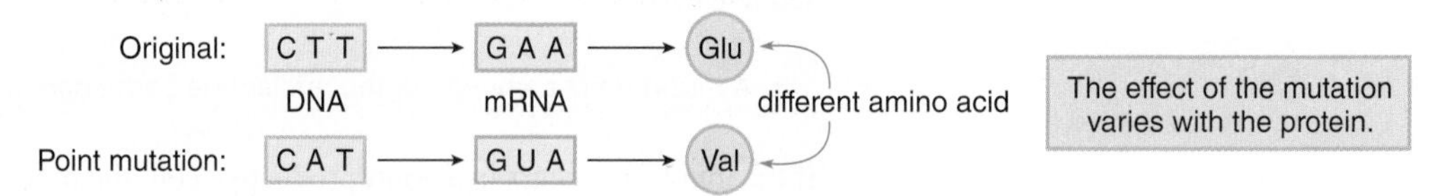

Finally, suppose a point mutation replaces CTT by ATT in DNA. ATT is transcribed to the codon UAA in mRNA, and UAA is a stop codon. This terminates protein synthesis and no more amino acids are added to the protein chain. In this case, a needed protein is not synthesized and depending on the protein's role, the organism may die.

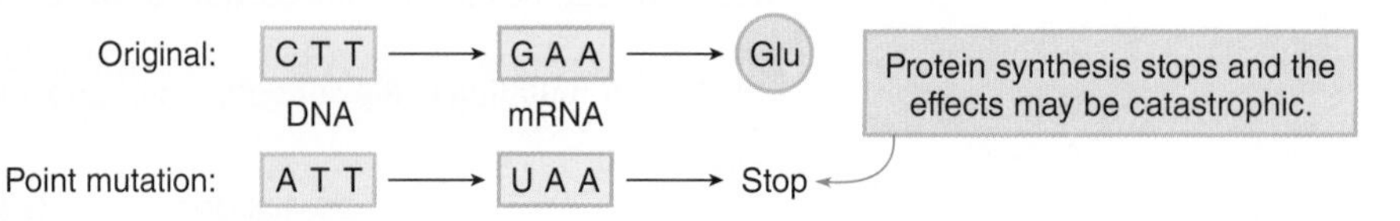

SAMPLE PROBLEM 17.10

(a) What dipeptide is produced from the following segment of DNA: AGAGAT? (b) What happens to the dipeptide when a point mutation occurs and the DNA segment contains the sequence ATAGAT instead?

Analysis

Transcribe the DNA sequence to an mRNA sequence with complementary base pairs. Then use Table 17.3 to determine what amino acids are coded for by each codon.

Solution

a. Since UCU codes for serine and CUA codes for leucine, the dipeptide Ser–Leu results.

b. Since UAU codes for tyrosine, the point mutation results in the synthesis of the dipeptide Tyr–Leu.

PROBLEM 17.23

Consider the following sequence of DNA: AACTGA. (a) What dipeptide is formed from this DNA after transcription and translation? (b) How is the amino acid sequence affected when point mutations produce each of the following DNA sequences: [1] AACGGA; [2] ATCTGA; [3] AATTGA?

HEALTH NOTE

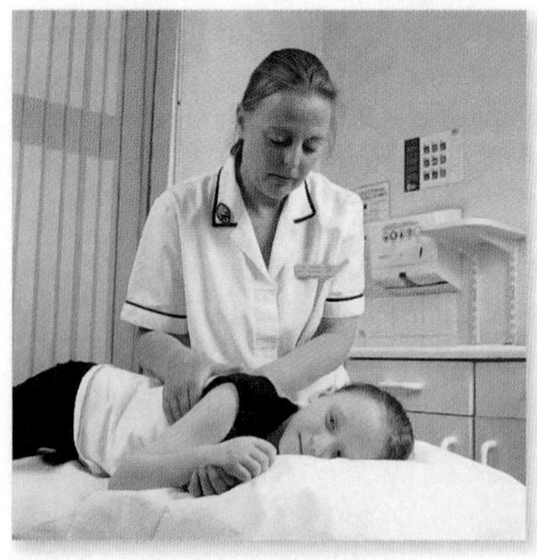

A patient with cystic fibrosis must regularly receive chest physiotherapy, a procedure that involves pounding on the chest to dislodge thick mucus clogging the lungs.

When a mutation causes a protein deficiency or results in a defective protein and the condition is inherited from one generation to another, a **genetic disease** results. For example, cystic fibrosis, the most common genetic disease in Caucasians, is caused by a mutation resulting in the synthesis of a defective protein—cystic fibrosis transmembrane conductance regulator (CFTR)—needed for proper passage of ions across cell membranes. Individuals with cystic fibrosis have decreased secretions of pancreatic enzymes, resulting in a failure to thrive (poor growth), and they produce thick sticky mucus in the lungs that leads to devastating lung infections and a shortened life span.

17.10 FOCUS ON THE HUMAN BODY DNA Fingerprinting

Because the DNA of each individual is unique, **DNA fingerprinting** (Figure 17.11) is now routinely used as a method of identification.

Almost any type of cell—skin, saliva, semen, blood, and so forth—can be used to obtain a DNA fingerprint. The DNA is cut into fragments with various enzymes and the fragments are separated by size using a technique called gel electrophoresis. DNA fragments can be visualized on X-ray film after they react with a radioactive probe. The result is an image consisting

Figure 17.11

DNA Fingerprinting in Forensic Analysis

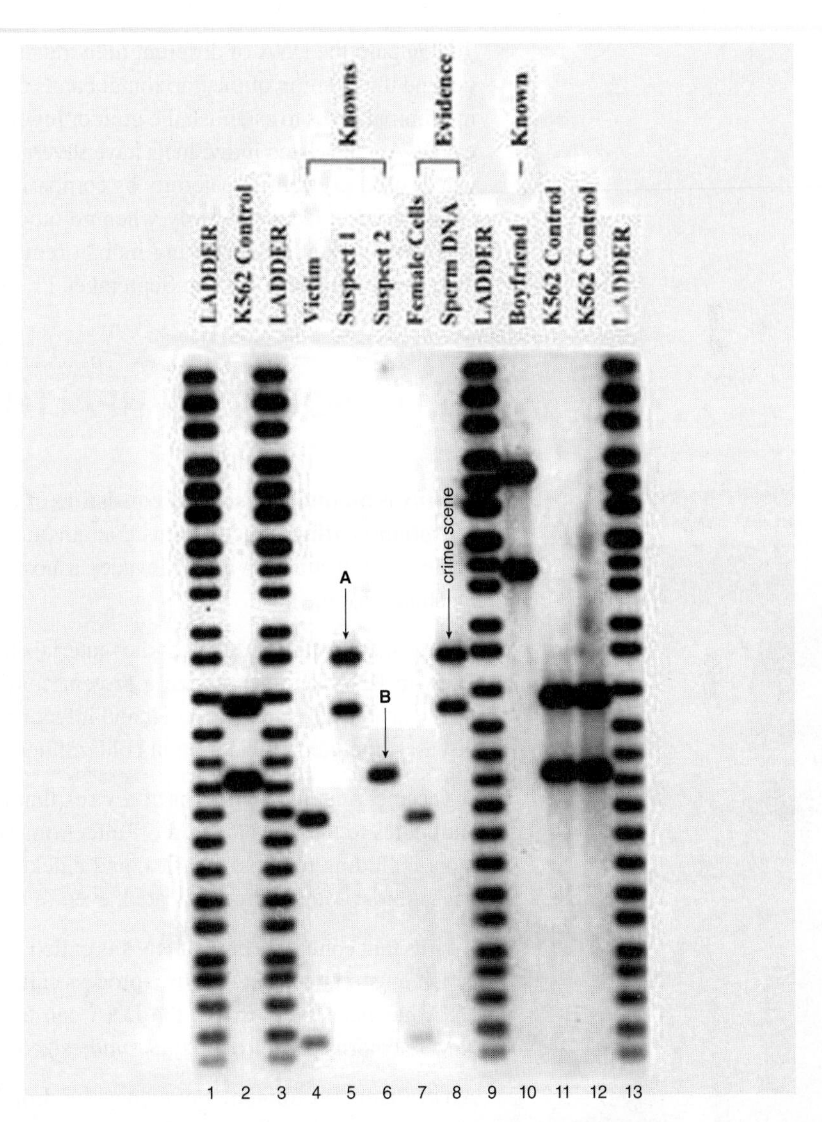

Each vertical lane (numbered 1–13) corresponds to a DNA sample.

- Lanes 1, 3, 9, and 13 are called DNA ladders. They correspond to DNA fragments of known size and are used to show the approximate size of the DNA fragments of the fingerprints.
- Lane 4 is the DNA obtained from a female assault victim and lane 7 is the female DNA from the crime scene. The two lanes match, indicating that they are from the same individual.
- Two individuals (**A** and **B**), whose DNA appears in lanes 5 and 6, were considered suspects.
- The male DNA obtained at the crime scene is shown in lane 8. The horizontal bands correspond to those of suspect **A,** incriminating him and eliminating individual **B** as a suspect.

of a set of horizontal bands, each band corresponding to a segment of DNA, sorted from low to high molecular weight.

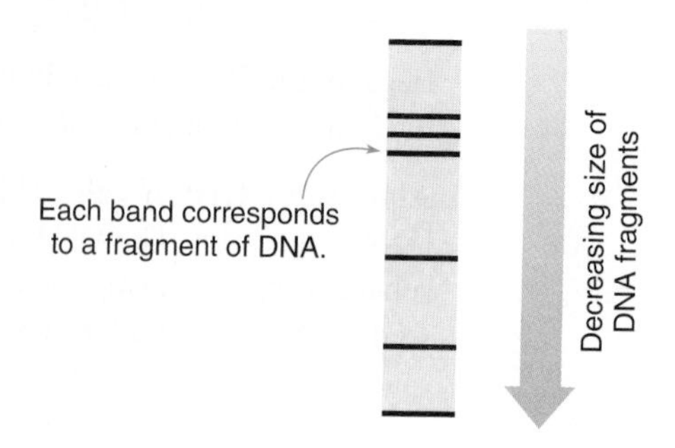

To compare the DNA of different individuals, samples are placed next to each other on the same gel and the position of the horizontal bands compared. DNA fingerprinting is now routinely used in criminal cases to establish the guilt or innocence of a suspect. Only identical twins have identical DNA, but related individuals have several similar DNA fragments. Thus, DNA fingerprinting can be used to establish paternity by comparing the DNA of a child with that of each parent. DNA is also used to identify a body when no other means of identification is possible. DNA analysis was instrumental in identifying human remains found in the rubble of the World Trade Center after the towers collapsed on September 11, 2001.

17.11 FOCUS ON HEALTH & MEDICINE Viruses

HEALTH NOTE

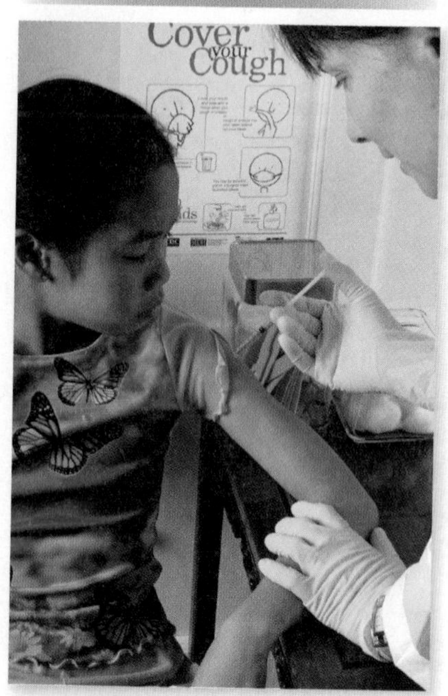

Childhood vaccinations have significantly decreased the incidence of once common diseases such as chickenpox, measles, and mumps.

A *virus* is an infectious agent consisting of a DNA or RNA molecule that is contained within a protein coating. Since a virus contains no enzymes or free nucleotides of its own, it is incapable of replicating. When it invades a host organism, however, it takes over the biochemical machinery of the host.

A virus that contains DNA uses the materials in the host organism to replicate its DNA, transcribe DNA to RNA, and synthesize a protein coating, thus forming new virus particles. These new virus particles leave the host cell and infect new cells and the process continues. Many prevalent diseases, including the common cold, influenza, and herpes are viral in origin.

A vaccine is an inactive form of a virus that causes an individual's immune system to produce antibodies to the virus to ward off infection. Many childhood diseases that were once very common, including mumps, measles, and chickenpox, are now prevented by vaccination. Polio has been almost completely eradicated, even in remote areas worldwide, by vaccination.

A virus that contains a core of RNA is called a **retrovirus.** Once a retrovirus invades a host organism, it must first make DNA by a process called **reverse transcription** (Figure 17.12). Once viral DNA has been synthesized, the DNA can transcribe RNA, which can direct protein synthesis. New retrovirus particles are thus synthesized and released to infect other cells.

AIDS (acquired immune deficiency syndrome) is caused by HIV (human immunodeficiency virus), a retrovirus that attacks lymphocytes central to the body's immune response against invading organisms. As a result, an individual infected with HIV becomes susceptible to life-threatening bacterial infections. HIV is spread by direct contact with the blood or other body fluids of an infected individual.

Major progress in battling the AIDS epidemic has occurred in recent years. HIV is currently best treated with a "cocktail" of drugs designed to destroy the virus at different stages of its reproductive cycle. One group of drugs, the protease inhibitors such as amprenavir (Section 16.10), act as enzyme inhibitors that prevent viral RNA from synthesizing needed proteins.

Figure 17.12

How a Retrovirus Infects an Organism

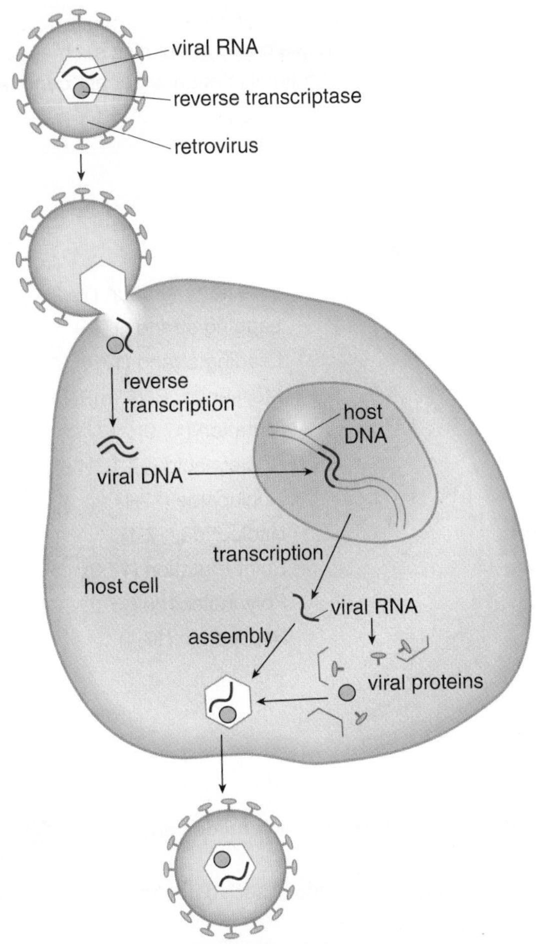

A retrovirus (containing viral RNA and a reverse transcriptase) binds and infects a host cell. Reverse transcription forms viral DNA from RNA. One strand of viral DNA becomes a template to transcribe viral RNA, which is then translated into viral proteins. A new virus is assembled and the virus leaves the host cell to infect other cells.

HEALTH NOTE

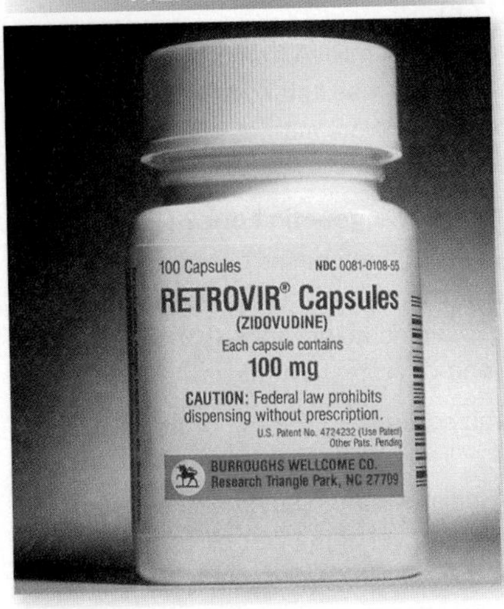

AZT, also called zidovudine and originally sold under the trade name Retrovir, has been available since the 1990s for the treatment of HIV.

Other drugs are designed to interfere with reverse transcription, an essential biochemical process unique to the virus. Two drugs in this category are **AZT** (azidodeoxythymidine) and **ddI** (dideoxyinosine). The structure of each drug closely resembles the nucleotides that must be incorporated in viral DNA, and thus they are inserted into a growing DNA strand. Each drug lacks a hydroxyl group at the 3' position, however, so no additional nucleotide can be added to the DNA chain, thus halting DNA synthesis.

no OH group → 3' (N_3)

azidodeoxythymidine
AZT

no OH group → 3'

dideoxyinosine
ddI

PROBLEM 17.24

Explain why antibiotics such as sulfanilamide and penicillin (Section 16.9) are effective in treating bacterial infections but are completely ineffective in treating viral infections.

KEY TERMS

Anticodon (17.5, 17.8)
Codon (17.7)
Complementary base pairs (17.3)
Deletion mutation (17.9)
Deoxyribonucleic acid (DNA, 17.1, 17.3)
Deoxyribonucleoside (17.1)
Deoxyribonucleotide (17.1)
DNA fingerprinting (17.10)
Gene (17.1)
Genetic code (17.7)
Genetic disease (17.9)
Insertion mutation (17.9)
Lagging strand (17.4)
Leading strand (17.4)
Messenger RNA (mRNA, 17.5)
Mutation (17.9)
Nucleic acid (17.1, 17.2)
Nucleoside (17.1)
Nucleotide (17.1)
Point mutation (17.9)
Polynucleotide (17.2)
Replication (17.4)
Retrovirus (17.11)
Reverse transcription (17.11)
Ribonucleic acid (RNA, 17.1, 17.5)
Ribonucleoside (17.1)
Ribonucleotide (17.1)
Ribosomal RNA (rRNA, 17.5)
Transcription (17.6)
Transfer RNA (tRNA, 17.5)
Translation (17.8)
Virus (17.11)

KEY CONCEPTS

1 What are the main structural features of nucleosides and nucleotides? (17.1)

- A nucleoside contains a monosaccharide joined to a nitrogen-containing base.
- A nucleotide contains a monosaccharide joined to a nitrogen-containing base, and a phosphate bonded to the 5'-OH group of the monosaccharide.
- The monosaccharide is either ribose or 2-deoxyribose, and the bases are abbreviated as A, G, C, T, and U.

2 How do the nucleic acids DNA and RNA differ in structure? (17.2)

- DNA is a polymer of deoxyribonucleotides, where the sugar is 2-deoxyribose and the bases are A, G, C, and T. DNA is double stranded.
- RNA is a polymer of ribonucleotides, where the sugar is ribose and the bases are A, G, C, and U. RNA is single stranded.

3 Describe the basic features of the DNA double helix. (17.3)

- DNA consists of two polynucleotide strands that wind into a right-handed double helix. The sugar–phosphate backbone lies on the outside of the helix and the bases lie on the inside. The double helix is stabilized by hydrogen bonding between complementary base pairs; A pairs with T and C pairs with G.

4 Outline the main steps in the replication of DNA. (17.4)

- An original DNA molecule forms two DNA molecules, each of which has one strand from the parent DNA and one new strand.
- In replication, DNA unwinds and the enzyme DNA polymerase catalyzes replication on both strands. The identity of the bases on the template strand determines the order of the bases on the new strand, with A pairing with T and C pairing with G.

5 List the three types of RNA molecules and describe their functions. (17.5)

- Ribosomal RNA (rRNA) provides the site where proteins are assembled.
- Messenger RNA (mRNA) contains the sequence of nucleotides that determines the amino acid sequence in a protein.
- Transfer RNA (tRNA) contains an anticodon that identifies the amino acid that it carries on its acceptor stem and delivers that amino acid to a growing polypeptide.

6 What is transcription? (17.6)

- Transcription is the synthesis of mRNA from DNA. The DNA helix unwinds and RNA polymerase catalyzes RNA synthesis from the 3' to 5' end of the template strand, forming mRNA with complementary bases.

7 What are the main features of the genetic code? (17.7)

- mRNAs contain sequences of three bases called codons that code for individual amino acids. There are 61 codons that correspond to the 20 amino acids, as well as three stop codons that signal the end of protein synthesis.

8 How are proteins synthesized by the process of translation? (17.8)

- Translation begins with initiation, the binding of the ribosomal subunits to mRNA and the arrival of the first tRNA with an amino acid. During elongation, tRNAs bring individual amino acids to the ribosome one after another, and new peptide bonds are formed. Termination occurs when a stop codon is reached.

9 What is a mutation and how are mutations related to genetic diseases? (17.9)

- Mutations are changes in the nucleotide sequence in a DNA molecule. A point mutation results in the substitution of one nucleotide for another. Deletion and insertion mutations result in the loss or addition of nucleotides, respectively. A mutation that causes an inherited condition may result in a genetic disease.

10 What are the principal features of DNA fingerprinting? (17.10)

- DNA fingerprinting is used to identify an individual by cutting DNA with enzymes to give a unique set of fragments.

11 What are the main characteristics of viruses? (17.11)

- A virus is an infectious agent that contains either DNA or RNA within a protein coat. When the virus invades a host cell, it uses the biochemical machinery of the host to replicate. A retrovirus contains RNA and a reverse transcriptase that allow the RNA to synthesize viral DNA, which then transcribes RNA that directs protein synthesis.

UNDERSTANDING KEY CONCEPTS

Selected in-chapter and odd-numbered end-of-chapter problems have brief answers at the end of each chapter. The *Student Study Guide and Solutions Manual* contains detailed solutions to all in-chapter and odd-numbered end-of-chapter problems, as well as additional worked examples and a chapter self-test.

17.25 Label each statement as pertaining to DNA, RNA, or both.

a. The polynucleotide is double stranded.
b. The polynucleotide may contain adenine.
c. The polynucleotide may contain dGMP.
d. The polynucleotide is a polymer of ribonucleotides.

17.26 Label each statement as pertaining to DNA, RNA, or both.

a. The polynucleotide is single stranded.
b. The polynucleotide may contain guanine.
c. The polynucleotide may contain UMP.
d. The polynucleotide is a polymer of deoxyribonucleotides.

17.27 (a) Identify the base and monosaccharide in the following nucleotide. (b) Give the name and three-letter abbreviation for the compound.

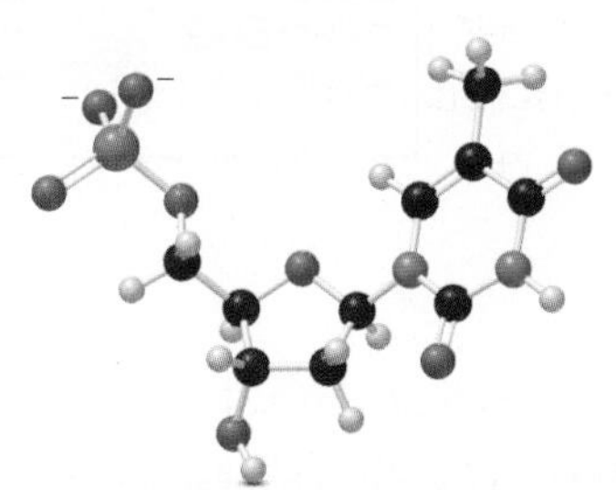

17.28 (a) Identify the base and monosaccharide in the following nucleotide. (b) Give the name and three-letter abbreviation for the compound.

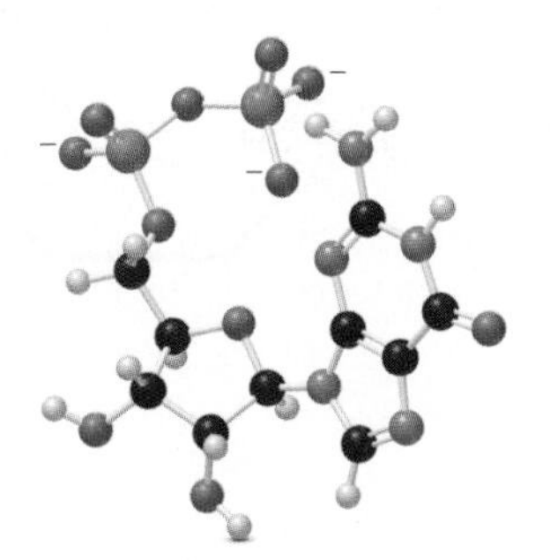

17.29 Consider the given dinucleotide.

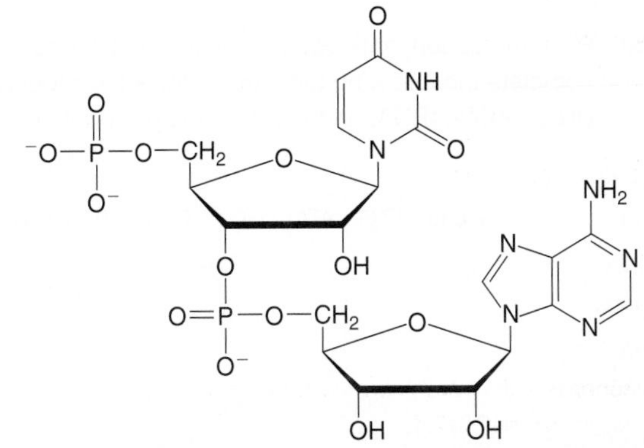

a. Identify the bases present in the dinucleotide.
b. Label the 5' and 3' ends.
c. Give the three- or four-letter abbreviations for the two nucleotides.
d. Is this dinucleotide a ribonucleotide or a deoxyribonucleotide? Explain your choice.
e. Name the dinucleotide.

17.30 Answer Problem 17.29 for the following dinucleotide.

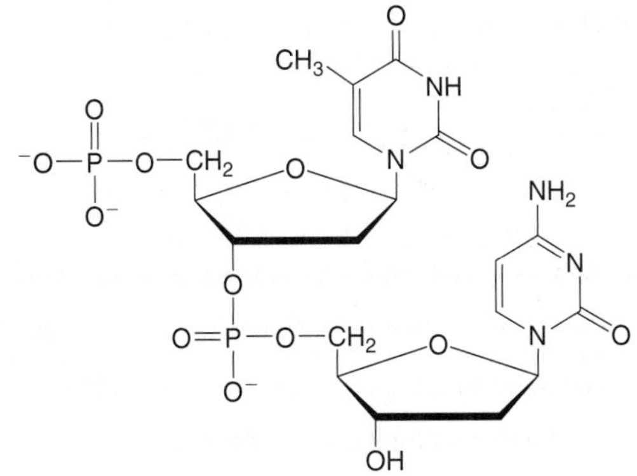

17.31 Fill in the missing information in the schematic of a tRNA during the elongation phase of translation.

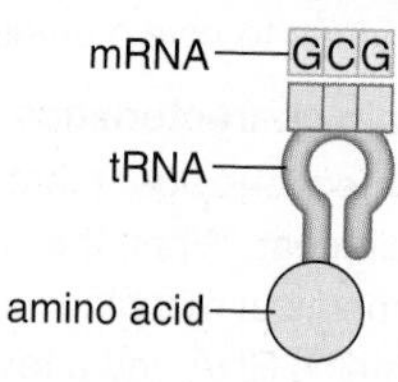

17.32 Fill in the missing information in the schematic of a tRNA during the elongation phase of translation.

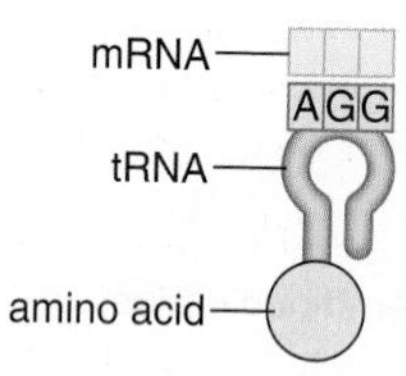

17.33 Fill in the codon, anticodon, or amino acid needed to complete the following table that relates the sequences of DNA, mRNA, tRNA, and the resulting polypeptide.

DNA template strand:	3' end	TTG	ATA	GGT	TGC	TTC	TAC	5' end
mRNA codons:	5' end							3' end
tRNA anticodons:								
Polypeptide:								

17.34 Fill in the codon, anticodon, or amino acid needed to complete the following table that relates the sequences of DNA, mRNA, tRNA, and the resulting polypeptide.

DNA template strand:	3' end	TCC	GAC	TTG	TGC	CAT	CAC	5' end
mRNA codons:	5' end							3' end
tRNA anticodons:								
Polypeptide:								

17.35 The lanes in the given gel show the DNA of a father and three children. Which lane represents the DNA of the father?

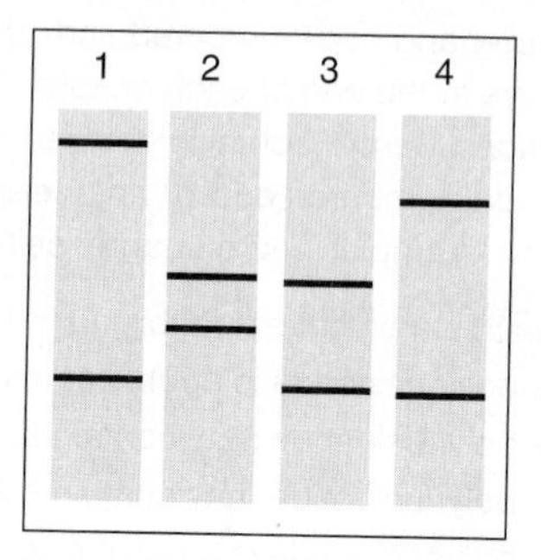

17.36 Which lane in the given gel represents an identical twin of the individual whose DNA is shown in Lane 1?

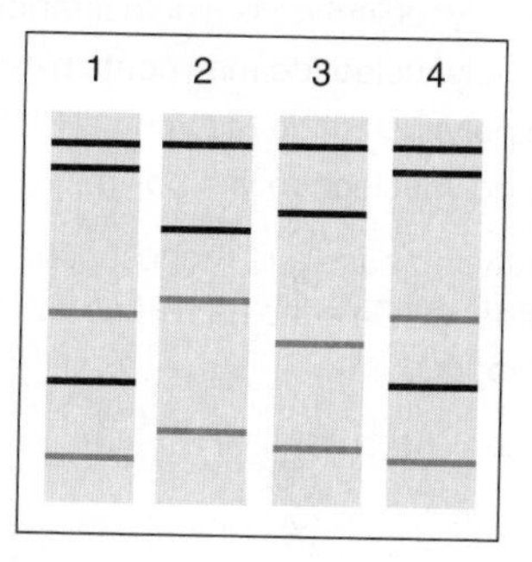

ADDITIONAL PROBLEMS

Nucleosides, Nucleotides, and Nucleic Acids

17.37 What is the difference between a gene and a chromosome?

17.38 What is the difference between uracil and uridine?

17.39 List three structural differences between DNA and RNA.

17.40 List three structural similarities in DNA and RNA.

17.41 Identify the base and monosaccharide in the following compound and then name it.

NH_2

N

HO—CH_2 O N O

OH

17.42 Identify the base and monosaccharide in the following compound and then name it.

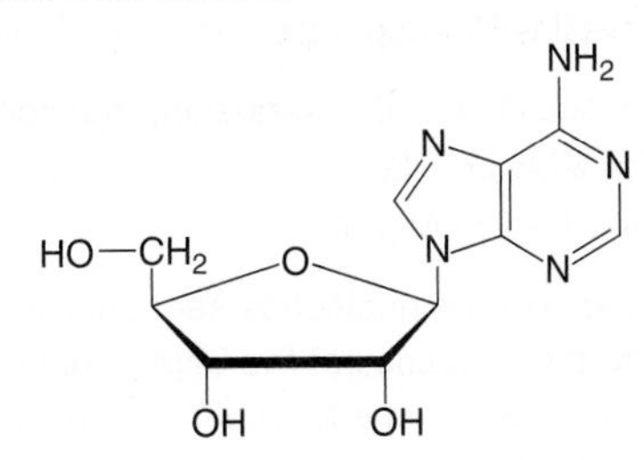

17.43 (a) Give the name of each compound shown as a ball-and-stick model. (b) Classify the compound as a base, nucleoside, or nucleotide. (c) Would the compound be a component of DNA, RNA, or both?

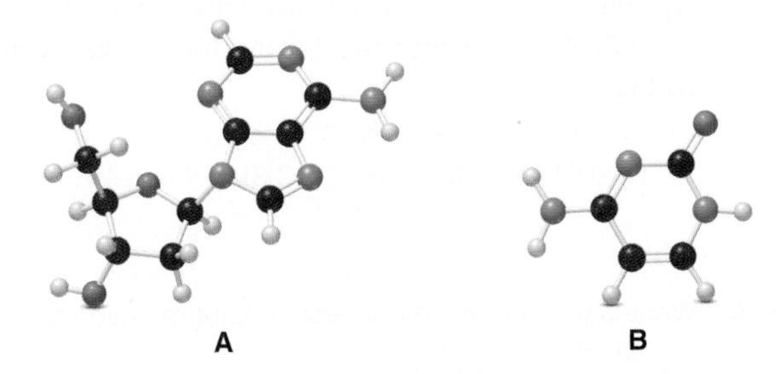

17.44 (a) Give the name of each compound shown as a ball-and-stick model. (b) Classify the compound as a base, nucleoside, or nucleotide. (c) Would the compound be a component of DNA, RNA, or both?

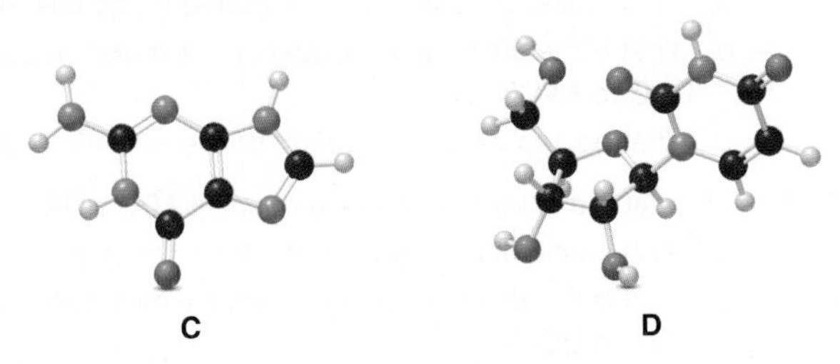

17.45 Give the name, abbreviation, or structure of each of the following:

a. a purine base
b. a nucleoside that contains 2-deoxyribose and a pyrimidine base
c. a nucleotide that contains ribose and a purine base

17.46 Give the name, abbreviation, or structure of each of the following:

a. a pyrimidine base
b. a nucleoside that contains 2-deoxyribose and a purine base
c. a nucleotide that contains ribose and a pyrimidine base

17.47 Classify each molecule as a nucleoside or nucleotide.

a. adenosine
b. deoxyguanosine
c. GDP
d. dTDP

17.48 Classify each molecule as a nucleoside or nucleotide.

a. uridine
b. deoxycytidine
c. dGMP
d. UTP

17.49 Draw the structure of the deoxyribonucleotide formed by joining the 3'-OH group of dTMP with the 5'-phosphate of dAMP.

17.50 Draw the structure of the ribonucleotide formed by joining the 5'-phosphate of UMP with the 3'-OH of AMP.

17.51 Describe in detail the DNA double helix with reference to each of the following features: (a) the sugar–phosphate backbone; (b) the functional groups at the end of each strand; (c) the hydrogen bonding between strands.

17.52 Describe in detail the DNA double helix with reference to each of the following features: (a) the location of the bases; (b) the complementary base pairing; (c) the phosphodiester linkages.

17.53 Write the sequence of the complementary strand of each segment of a DNA molecule.

a. 5'–AAATAAC–3'
b. 5'–ACTGGACT–3'
c. 5'–CGATATCCCG–3'
d. 5'–TTCCCGGGATA–3'

17.54 Write the sequence of the complementary strand of each segment of a DNA molecule.

a. 5'–TTGCGA–3'
b. 5'–CGCGTAAT–3'
c. 5'–ACTTCAGGT–3'
d. 5'–CCGGTTAATACGGC–3'

17.55 If 27% of the nucleotides in a sample of DNA contain the base adenine (A), what are the percentages of bases T, G, and C?

17.56 If 19% of the nucleotides in a sample of DNA contain the base cytosine (C), what are the percentages of bases G, A, and T?

Replication, Transcription, Translation, and Protein Synthesis

17.57 Consider the following simple graphic that illustrates the main features of DNA replication. Label each of the letters (**A–G**) with one of the following terms: leading strand, lagging strand, template strand, 5' end, or 3' end. A term may be used more than once.

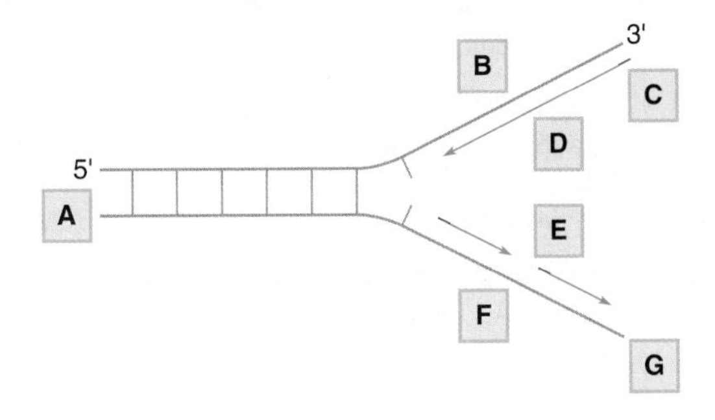

17.58 Consider the following graphic that illustrates the transcription of DNA to RNA. Label each of the letters (**A–F**) with one of the following terms: nucleotide, RNA, template strand, RNA polymerase, 5' end, or 3' end.

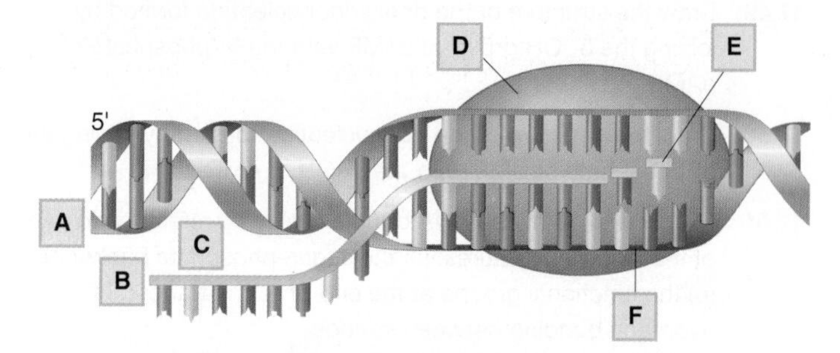

17.59 What is the sequence of a newly synthesized DNA segment if the template strand has the sequence 3'–ATGGCCTATGCGAT–5'?

17.60 What is the sequence of a newly synthesized DNA segment if the template strand has the sequence 3'–CGCGATTAGATATTGCCGC–5'?

17.61 Explain the roles of messenger RNA and transfer RNA in converting the genetic information coded in DNA into protein synthesis.

17.62 What are the two main structural features of transfer RNA molecules?

17.63 What mRNA is transcribed from each DNA sequence in Problem 17.53?

17.64 What mRNA is transcribed from each DNA sequence in Problem 17.54?

17.65 What is the sequence of the mRNA molecule synthesized from each DNA template?

a. 3'–ATGGCTTA–5' b. 3'–CGGCGCTTA–5'

17.66 What is the sequence of the mRNA molecule synthesized from each DNA template?

a. 3'–GGCCTATA–5' b. 3'–GCCGAT–5'

17.67 For each codon, give its anticodon and the amino acid for which it codes.

a. CUG b. UUU c. AAG d. GCA

17.68 For each codon, give its anticodon and the amino acid for which it codes.

a. GUU b. AUA c. CCC d. GCG

17.69 Derive the amino acid sequence that is coded for by each mRNA sequence.

a. CCA ACC UGG GUA GAA
b. AUG UUU UUA UGG UGG
c. GUC GAC GAA CCG CAA

17.70 Derive the amino acid sequence that is coded for by each mRNA sequence.

a. AAA CCC UUU UGU
b. CCU UUG GAA GUA CUU
c. GGG UGU AUG CAC CGA UUG

17.71 Write a possible mRNA sequence that codes for each peptide.

a. Ile–Met–Lys–Ser–Tyr
b. Pro–Gln–Glu–Asp–Phe

17.72 Write a possible mRNA sequence that codes for each peptide.

a. Phe–Phe–Leu–Lys
b. Val–Gly–Gln–Asp–Asn

17.73 Considering each nucleotide sequence in an mRNA molecule: [1] write the sequence of the DNA template strand from which the mRNA was synthesized; [2] give the peptide synthesized by the mRNA.

a. 5' UAU UCA AUA AAA AAC 3'
b. 5' GAU GUA AAC AAG CCG 3'

17.74 Considering each nucleotide sequence in an mRNA molecule: [1] write the sequence of the DNA template strand from which the mRNA was synthesized; [2] give the peptide synthesized by the mRNA.

a. 5' UUG CUC AAC CAA 3'
b. 5' AUU GUA CCA CAA CCC 3'

Mutations

17.75 What is the difference between a point mutation and a silent mutation?

17.76 Explain why some mutations cause no effect on a cell whereas others have a major effect.

17.77 Consider the following mRNA sequence: CUU CAG CAC.

a. What amino acid sequence is coded for by this mRNA?
b. What is the amino acid sequence if a mutation converts CAC to AAC?
c. What occurs when a mutation converts CAG to UAG?

17.78 Consider the following mRNA sequence: ACC UUA CGA.

a. What amino acid sequence is coded for by this mRNA?
b. What is the amino acid sequence if a mutation converts UUA to UCA?
c. What is the amino acid sequence if a mutation converts CGA to AGA?

17.79 Consider the following sequence of DNA: 3'–TTA CGG–5'.

a. What dipeptide is formed from this DNA after transcription and translation?
b. If a mutation converts CGG to AGG in DNA, what dipeptide is formed?

17.80 Consider the following sequence of DNA: 3'–ATA GGG–5'.

a. What dipeptide is formed from this DNA after transcription and translation?
b. What occurs when a mutation converts ATA to ATT in DNA?

DNA Fingerprinting

17.81 The given gel contains the DNA fingerprint of a mother (lane 1), father (lane 2), and four children (lanes 3–6). One child is adopted, two children are the offspring of both parents, and one child is the offspring of one parent only.

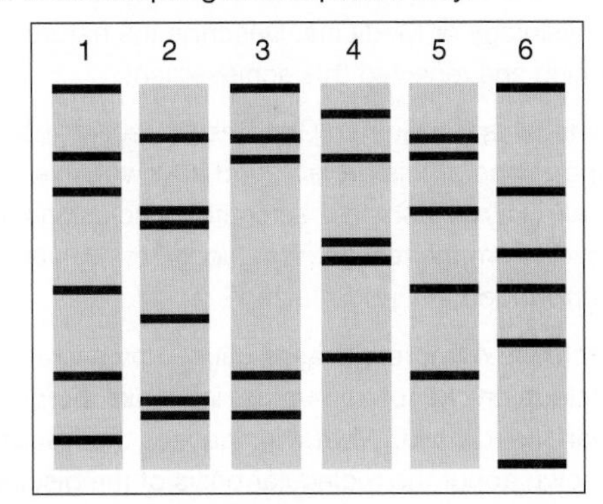

a. Which lanes, if any, represent a biological child of both parents? Explain your choice.

b. Which lanes, if any, represent an adopted child? Explain your choice.

17.82 With reference to the gel in Problem 17.81: (a) Which lanes (if any) represent a biological child of the mother only? (b) Which lanes (if any) represent a biological child of the father only? (c) Which lanes (if any) represent twins? Explain each choice.

Viruses

17.83 What is a retrovirus?

17.84 What is reverse transcription?

17.85 Lamivudine, also called 3TC, is a nucleoside analogue used in the treatment of HIV and the virus that causes hepatitis B. (a) What nucleoside does lamivudine resemble? (b) How does lamivudine inhibit reverse transcription?

lamivudine
3TC

17.86 How does a vaccine protect an individual against certain viral infections?

General Questions

17.87 Fill in the bases, codon, anticodon, or amino acid needed to complete the following table that relates the sequences of DNA, mRNA, tRNA, and the resulting polypeptide.

DNA template strand:	3' end	CAT					5' end
mRNA codons:	5' end		UCA			AUG	3' end
tRNA anticodons:					GUG		
Polypeptide:				Thr			

17.88 Fill in the bases, codon, anticodon, or amino acid needed to complete the following table that relates the sequences of DNA, mRNA, tRNA, and the resulting polypeptide.

DNA template strand:	3' end	CAA				GTC	5' end
mRNA codons:	5' end		UAC				3' end
tRNA anticodons:				ACA			
Polypeptide:					Lys		

17.89 If there are 325 amino acids in a polypeptide, how many bases are present in a single strand of the gene that codes for it, assuming that every base is transcribed and then translated to polypeptide?

17.90 If a single strand of a gene contains 678 bases, how many amino acids result in the polypeptide prepared from it, assuming every base of the gene is transcribed and then translated?

17.91 Met-enkephalin (Tyr–Gly–Gly–Phe–Met) is a pain killer and sedative. What is a possible nucleotide sequence in the template strand of the gene that codes for met-enkephalin, assuming that every base of the gene is transcribed and then translated?

17.92 Leu-enkephalin (Tyr–Gly–Gly–Phe–Leu) is a pain killer and sedative. What is a possible nucleotide sequence in the template strand of the gene that codes for leu-enkephalin, assuming that every base of the gene is transcribed and then translated?

CHALLENGE PROBLEMS

17.93 Give a possible nucleotide sequence in the template strand of the gene that codes for the following tripeptide.

17.94 Give a possible nucleotide sequence in the template strand of the gene that codes for the following tripeptide.

BEYOND THE CLASSROOM

17.95 Genetic disorders can result from a mutation, gene deletion, or an extra or missing chromosome. Pick one or more of the following conditions: Down syndrome, Duchene muscular dystrophy, hemophilia, Tay–Sach's disease, phenylketonuria, or galactosemia. What is the underlying genetic defect? What biochemical abnormality results? How many individuals have this disorder in the United States? What are the symptoms of the disease and what is the status of current treatment options?

17.96 Although James Watson and Francis Crick receive much of the credit for unraveling the structure of the double helix of DNA, much early research in this area was carried out by Rosalind Franklin, a woman who was ineligible to receive the Nobel Prize because she had passed away before it was awarded for this work in 1962. Few women have earned the Nobel Prize since it was first awarded in 1901. Locate a list of Nobel Prize winners on the web or in the library, and pick one woman who has received this honor in Chemistry or in Physiology or Medicine. Describe the nature of the work for which she received this achievement.

17.97 What was the Human Genome Project? What were the goals and scope of this project, and how was this project carried out? List at least three scientific conclusions that were made from the results. What legal, ethical, and social issues surrounded this undertaking?

17.98 Pick one or more diseases caused by a virus. Examples include chickenpox, mumps, influenza, herpes, hepatitis, warts, and polio. What virus causes the disease? What is known about the molecular basis of the disease and how is the disease treated or cured?

ANSWERS TO SELECTED PROBLEMS

17.1 a. base = uracil, sugar = ribose, nucleoside = uridine
b. base = guanine, sugar = deoxyribose, nucleoside = deoxyguanosine

17.3 a. base = uracil, sugar = ribose, uridine 5'-monophosphate, UMP
b. base = thymine, sugar = deoxyribose, deoxythymidine 5'-monophosphate, dTMP

17.5 a,d,e: RNA b,c,f: DNA

17.7 a. base b. nucleoside c. nucleotide d. nucleoside

17.8

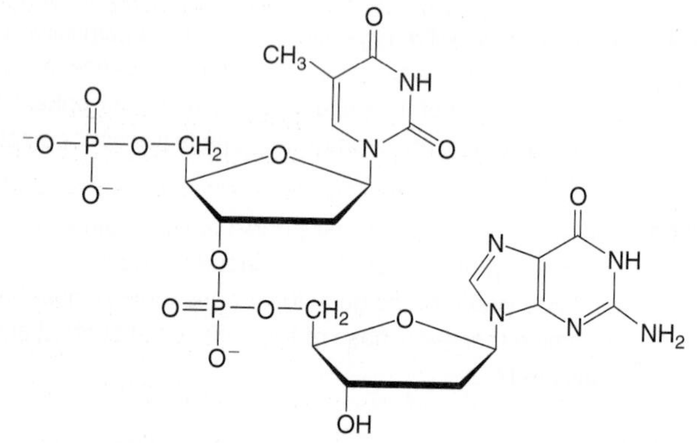

17.9 a. 5'–ATTTG–3' b. 5'–CGCGUU–3' c. 5'–GGACTT–3'

17.11 a. 3'–TTTGCAGG–5' b. 3'–ATATGCGG–5' c. 3'–TAACGTGGGCG–5' d. 3'–GTGAACTAGCC–5'

17.12 a. 5'–TCTCAGAG–3' b. 3'–TAACGAG–5' c. 5'–TAGGACATG–3' d. 3'–CCGGTATGAG–5'

17.13 a. 5'–ACGGAUUGC–3' b. 5'–CUGAGG–3' c. 5'–AAUUGCGCU–3' d. 5'–GUCACUGGCAUG–3'

17.15 a. Ala b. Asn c. Leu d. Ser e. Gln f. Lys

17.17 a. Gln–Glu–Val–Ser–Tyr–Arg
b. Val–Ile–Trp–Arg–Gly–Ile
c. Leu–Cys–Ser–Arg–Thr–Pro

17.19 a. CTC GGG CAT ATG CGG TGC
b. Glu–Pro–Val–Tyr–Ala–Thr

17.21 a. Pro–Pro–Ala–Asn–Glu–Ala
GGU GGC CGU UUG CUU CGU
b. Ala–Pro–Leu–Arg–Asp
CGU GGU GAU UCU CUG

17.22 a. Arg–Val–Ala–Leu–Leu–Ser
b. Arg–Gly–Phe–Ile–Val–Asn

17.23 a. Leu–Thr
b. [1] Leu–Pro
[2] Stop codon, so no dipeptide is formed.
[3] Leu–Thr

17.25 a,c: DNA b. both d. RNA

17.27 a. thymine, deoxyribose
b. deoxythymidine 5'-monophosphate, dTMP

17.29 a. uracil, adenine
b. The 5' end has the free phosphate and the 3' end has two OH groups on the five-membered sugar ring.
c. UMP, AMP
d. This dinucleotide is a ribonucleotide because the sugar rings contain an OH group on C2'.
e. UA

17.31 anticodon: CGC; amino acid: alanine

17.33

DNA template strand:	3' end	TTG	ATA	GGT	TGC	TTC	TAC	5' end
mRNA codons:	5' end	AAC	UAU	CCA	ACG	AAG	AUG	3' end
tRNA anticodons:		UUG	AUA	GGU	UGC	UUC	UAC	
Polypeptide:		Asn	Tyr	Pro	Thr	Lys	Met	

17.35 Lane 3

17.37 A gene is a portion of the DNA molecule responsible for the synthesis of a single protein. Many genes form each chromosome.

17.39 [1] In RNA, the monosaccharide is the aldopentose ribose. In DNA, the monosaccharide is deoxyribose. [2] Uracil (U) occurs only in RNA, while thymine (T) occurs only in DNA. As a result: DNA contains the bases A, G, C, and T. RNA contains the bases A, G, C, and U. [3] DNA forms a double helix with complementary base pairs. RNA is a single chain composed of nucleotides.

17.41 base: cytosine
monosaccharide: deoxyribose
deoxycytidine

17.43 **A:** a. deoxyadenosine b. nucleoside c. DNA
B: a. cytosine b. base c. both DNA and RNA

17.45 a. A or G c. AMP or GMP
b. deoxycytidine or deoxythymidine

17.47 a,b: nucleoside c,d: nucleotide

17.49 a.

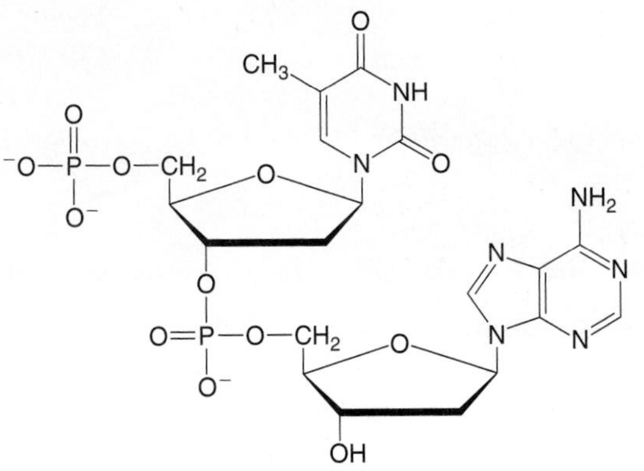

17.51 a. The DNA double helix has deoxyribose as the only sugar. The sugar–phosphate groups are on the outside of the helix.
b. The 5' end has a phosphate and the 3' end has an OH group.
c. Hydrogen bonding occurs in the interior of the helix between base pairs: A pairs with T and G pairs with C.

17.53 a. 3'–TTTATTG–5' c. 3'–GCTATAGGGC–5'
b. 3'–TGACCTGA–5' d. 3'–AAGGGCCCTAT–5'

17.55 27% T, 23% G, 23% C

17.57 **A:** 3' end **E:** lagging strand
B: template strand **F:** template strand
C: 5' end **G:** 5' end
D: leading strand

17.59 5'–TACCGGATACGCTA–3'

17.61 Messenger RNA carries the specific sequence of the DNA code from the cell nucleus to the ribosomes in the cytoplasm to make a protein. Each transfer RNA brings a specific amino acid to the growing protein chain on the ribosome according to the sequence specified by the mRNA.

17.63 a. 3'–UUUAUUG–5' c. 3'–GCUAUAGGGC–5'
b. 3'–UGACCUGA–5' d. 3'–AAGGGCCCUAU–5'

17.65 a. 5'–UACCGAAU–3' b. 5'–GCCGCGAAU–3'

17.67 a. GAC:Leu b. AAA:Phe c. UUC:Lys d. CGU:Ala

17.69 a. Pro–Thr–Trp–Val–Glu
b. Met–Phe–Leu–Trp–Trp
c. Val–Asp–Glu–Pro–Gln

17.71 a. AUU AUG AAA AGU UAU b. CCU CAA GAA GAU UUU

17.73 a. [1] 3' ATA AGT TAT TTT TTG 5'
[2] Tyr–Ser–Ile–Lys–Asn
b. [1] 3' CTA CAT TTG TTC GGC 5'
[2] Asp–Val–Asn–Lys–Pro

17.75 A point mutation results in the substitution of one nucleotide for another in a DNA molecule. A silent mutation is a point mutation in DNA that results in no change in an amino acid sequence.

17.77 a. Leu–Gln–His
b. Leu–Gln–Asn
c. This is a stop codon so the chain is terminated.

17.79 a. Asn–Ala b. Asn–Ser

17.81 a. Lanes 3 and 5 represent DNA of children that share both parents because they both have DNA fragments common to both parents.
b. Lane 4 represents DNA from an adopted child because the DNA fragments have little relationship to the parental DNA fragments.

17.83 A retrovirus is a virus that has RNA rather than DNA in its core. Once it invades a cell it uses reverse transcriptase to synthesize DNA for replication.

17.85 a. cytidine
b. Lamivudine is a nucleoside analogue that gets incorporated into a DNA chain, but since it does not contain a 3' hydroxyl group, synthesis is terminated.

17.87

DNA template strand:	3' end	CAT	AGT	TGA	GTG	TAC	5' end
mRNA codons:	5' end	GUA	UCA	ACU	CAC	AUG	3' end
tRNA anticodons:		CAU	AGU	UGA	GUG	UAC	
Polypeptide:		Val	Ser	Thr	His	Met	

17.89 975

17.91 ATA CCA CCA AAA TAC

17.93 TAA CCT AAA

A complex set of biochemical pathways converts ingested carbohydrates, lipids, and proteins to usable materials and energy to meet the body's needs.

Energy and Metabolism

CHAPTER OUTLINE

CHAPTER GOALS

In this chapter you will learn how to:

1. Define metabolism and describe the four stages of catabolism
2. Explain the role of ATP in energy production
3. Describe the roles of the main coenzymes used in metabolism
4. Describe the main aspects of glycolysis
5. List the pathways for pyruvate metabolism
6. List the main features of the citric acid cycle
7. Describe the main components of the electron transport chain and oxidative phosphorylation
8. Calculate the energy yield from glucose metabolism
9. Summarize the process of the β-oxidation of fatty acids
10. Identify the structures of ketone bodies and describe their role in metabolism
11. Describe the main components of amino acid catabolism

Despite the wide diversity among life forms, virtually all organisms contain the same types of biomolecules—lipids, carbohydrates, proteins, and nucleic acids—and use the same biochemical reactions. These reactions provide both the raw materials and the energy for growth and maintenance. The metabolism of ingested food begins with the hydrolysis of large biomolecules into small compounds that can be absorbed through the intestinal wall. Then specific pathways that often involve many steps convert these compounds into lower molecular weight molecules and generate energy for movement, thought, and a myriad of other processes. In Chapter 18, we learn about the metabolism of carbohydrates, lipids, and proteins.

18.1 An Overview of Metabolism

Each moment thousands of reactions occur in a living cell: large molecules are broken down into smaller components, small molecules are converted into larger molecules, and energy changes occur.

- *Metabolism* is the sum of all of the chemical reactions that take place in an organism.

There are two types of metabolic processes called **catabolism** and **anabolism.**

- *Catabolism* is the breakdown of large molecules into smaller ones. Energy is generally released during catabolism.
- *Anabolism* is the synthesis of large molecules from smaller ones. Energy is generally absorbed during anabolism.

The oxidation of glucose ($C_6H_{12}O_6$) to carbon dioxide (CO_2) and water (H_2O) is an example of catabolism, while the synthesis of a protein from component amino acids is an example of anabolism. An organized series of consecutive reactions that converts a starting material to a final product is called a **metabolic pathway.**

18.1A Energy Production in the Cell

Catabolism breaks down the carbohydrates, lipids, and proteins in food into smaller molecules, releasing energy to supply the body's needs. Where does energy production occur in cells?

A typical animal cell, such as the one shown in Figure 18.1, is surrounded by a cell membrane (Section 15.7) and has a nucleus that contains DNA in chromosomes (Section 17.3). The **cytoplasm,** the region

Figure 18.1 The Mitochondrion in a Typical Animal Cell

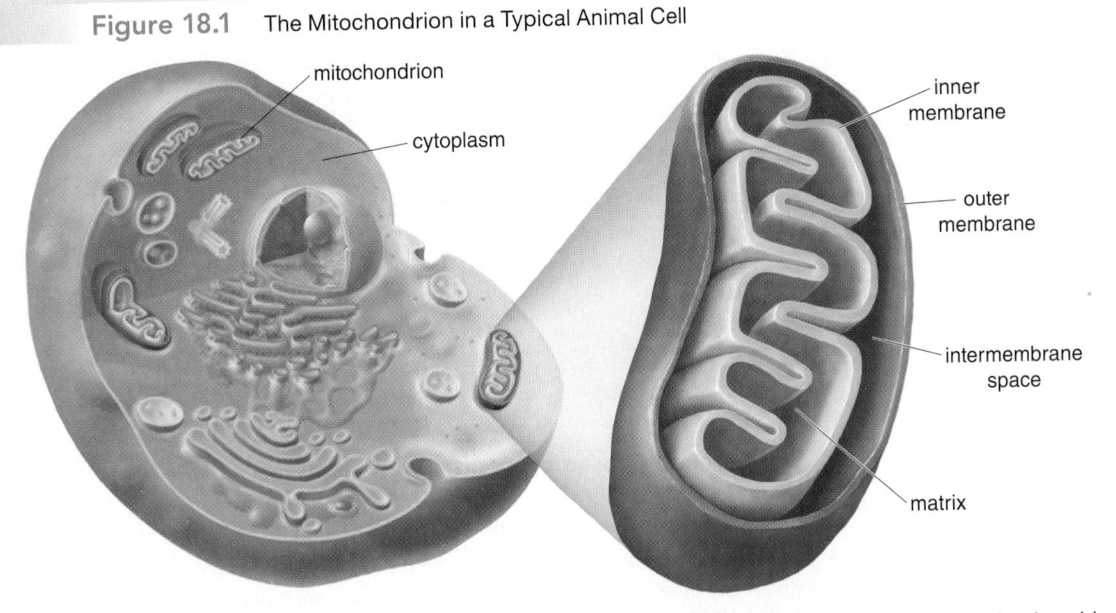

A mitochondrion is a small organelle located in the cytoplasm of the cell. Cellular energy production occurs in the mitochondria.

of the cell between the cell membrane and the nucleus, contains various specialized structures called **organelles,** each of which has a specific function. **Mitochondria** are small sausage-shaped organelles in which energy production takes place. Mitochondria contain an outer membrane and an inner membrane with many folds. The area between these two membranes is called the **intermembrane space.** Energy production occurs within the **matrix,** the area surrounded by the inner membrane of the mitochondrion. The number of mitochondria in a cell varies depending on its energy needs. Cells in the heart, brain, and muscles of the human body typically contain many mitochondria.

Figure 18.2

The Four Stages in Biochemical Catabolism and Energy Production

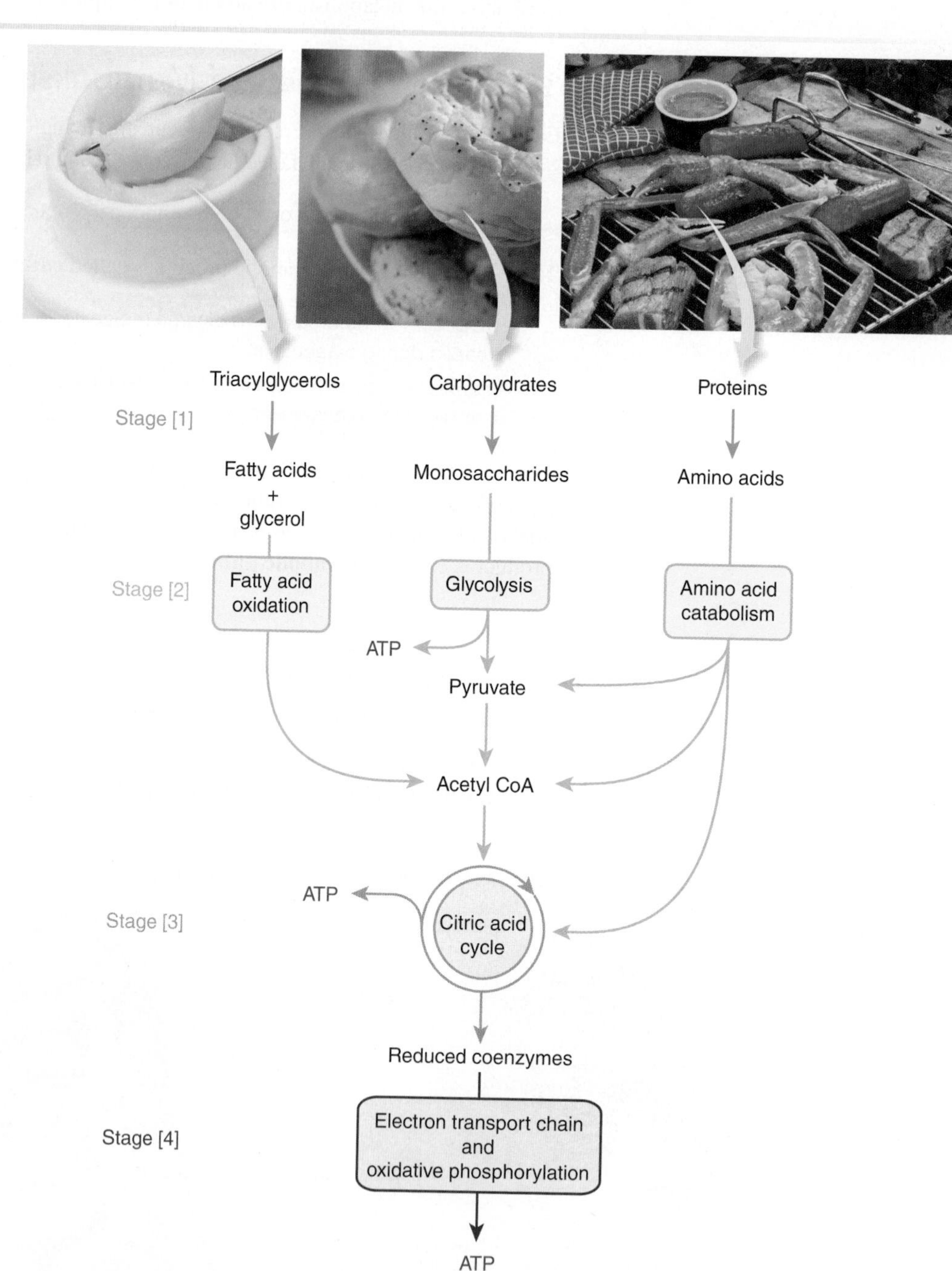

- **Carbohydrates: Glycolysis** converts glucose to pyruvate, which is then metabolized to acetyl CoA (Sections 18.4–18.5).
- **Lipids:** Fatty acids are oxidized by a stepwise procedure to form acetyl CoA (Section 18.9).
- **Amino acids:** The amino acids formed from protein hydrolysis are often assembled into new proteins without any other modification. Excess amino acids are catabolized for energy as discussed in Section 18.11. The amino groups (NH_2) are converted to urea [$(NH_2)_2C{=}O$], which is excreted in urine.

18.1B The Four Stages of Catabolism

The catabolic pathways can be organized into four stages, as shown in Figure 18.2.

Stage [1]—Digestion

The catabolism of food begins with **digestion,** which is catalyzed by enzymes in the saliva, stomach, and small intestines (Figure 18.3). The hydrolysis of carbohydrates to monosaccharides begins with amylase enzymes in the saliva, and continues in the small intestines. Protein digestion begins in the stomach, where acid denatures the protein, and the protease pepsin begins to cleave the protein backbone into smaller polypeptides and amino acids. Digestion continues in the small intestines, where trypsin and chymotrypsin further cleave the protein backbone to form amino acids. Triacylglycerols, the most common lipids, are first emulsified by bile secreted by the liver, and then hydrolyzed to glycerol and fatty acids by lipases in the small intestines.

These small molecules are each absorbed through the intestinal wall into the bloodstream and transported to other cells in the body. Some substances such as cellulose are not metabolized as they travel through the digestive tract, so they pass into the large intestines and are excreted.

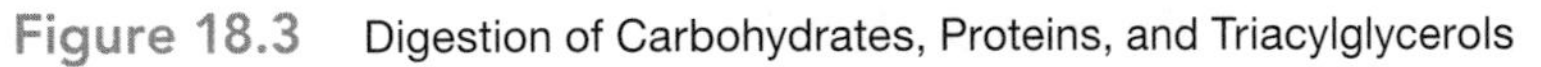

Figure 18.3 Digestion of Carbohydrates, Proteins, and Triacylglycerols

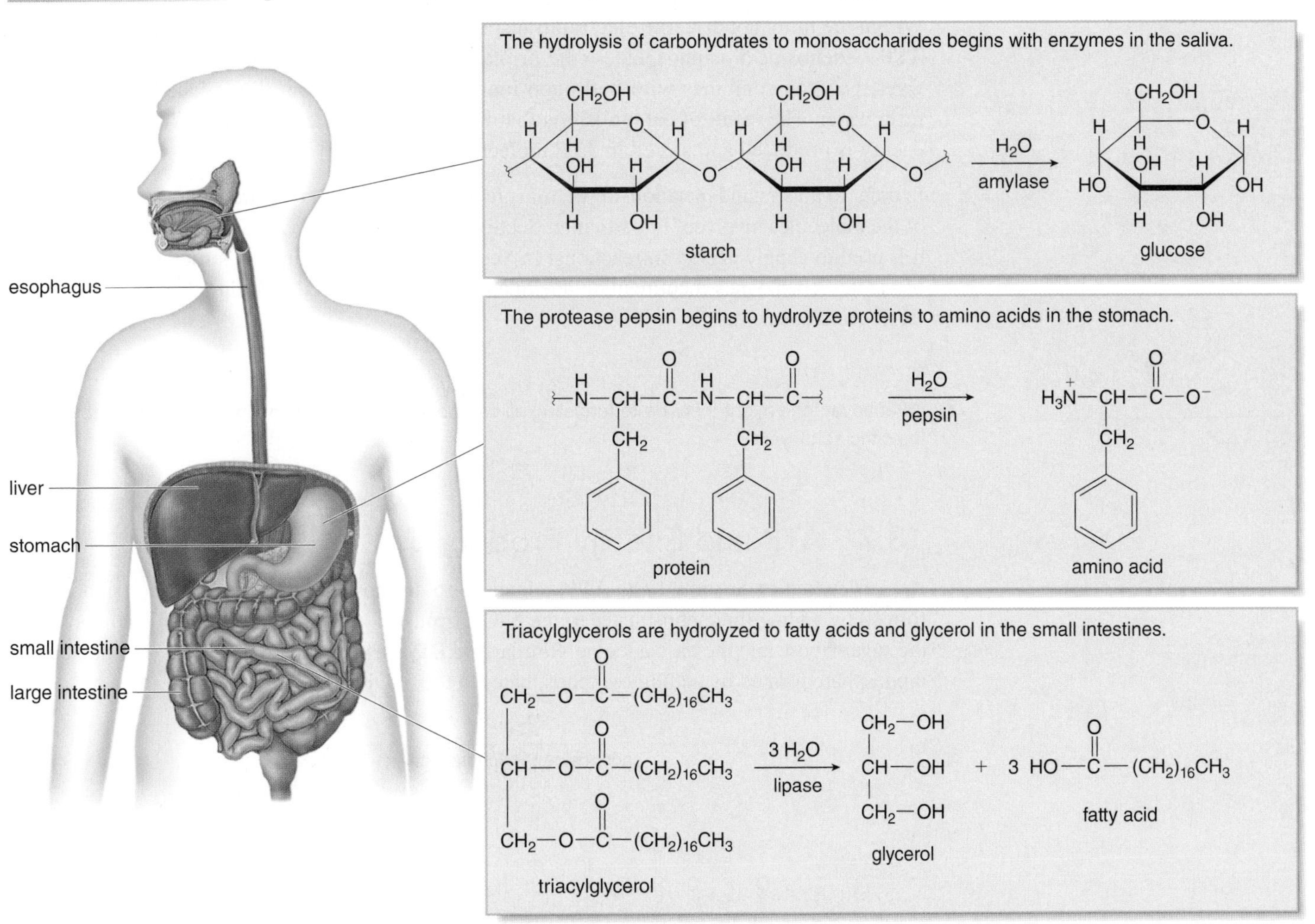

The first stage of catabolism, **digestion,** is the hydrolysis of large molecules to small molecules: polysaccharides such as starch are hydrolyzed to monosaccharides (Section 14.6), proteins are hydrolyzed to their component amino acids (Section 16.8), and triacylglycerols are hydrolyzed to glycerol and fatty acids (Section 15.5). Each of these molecules enters its own metabolic pathway to be further broken down into smaller components, releasing energy.

Once small molecules are formed, catabolism continues to break down each type of molecule to smaller units releasing energy in the process.

Stage [2]—Formation of Acetyl CoA

Monosaccharides, amino acids, and fatty acids are degraded into **acetyl groups (CH_3CO–),** two-carbon units that are bonded to coenzyme A (a coenzyme), forming **acetyl CoA.** Details of the structure of acetyl CoA are given in Section 18.3C.

The product of the catabolic pathways is the ***same*** for all three types of molecules. As a result, a common catabolic pathway, the **citric acid cycle,** continues the processing of all types of biomolecules to generate energy.

Stage [3]—The Citric Acid Cycle

The citric acid cycle is based in the mitochondria. In this biochemical cycle, the acetyl groups of acetyl CoA are oxidized to carbon dioxide. Some energy produced by this process is stored in the bonds of a nucleoside triphosphate (Section 17.1) and reduced coenzymes, whose structures are shown in Section 18.3.

Stage [4]—The Electron Transport Chain and Oxidative Phosphorylation

Within the mitochondria, the electron transport chain and oxidative phosphorylation produce **ATP**—adenosine 5'-triphosphate—the primary energy-carrying molecule in metabolic pathways. Oxygen combines with hydrogen ions and electrons from the reduced coenzymes to form water. The result of catabolism is that biomolecules are converted to CO_2 and H_2O and energy is produced and stored in ATP molecules.

In order to understand metabolism, we must first learn about the structure and properties of some of the molecules involved. In particular, Section 18.2 is devoted to a discussion of ATP and how it is used to supply energy in reactions. In Section 18.3 we examine the structure and reactions of the key coenzymes nicotinamide adenine dinucleotide (NAD^+), flavin adenine dinucleotide (FAD), and coenzyme A.

PROBLEM 18.1

What advantage might there be to funneling all catabolic pathways into a single common pathway, the citric acid cycle?

18.2 ATP and Energy Production

As we learned in Section 17.1, **ATP,** adenosine 5'-triphosphate, is a nucleoside triphosphate formed by adding three phosphates to the 5'-OH group of adenosine, a nucleoside composed of the sugar ribose and the base adenine. Similarly, **ADP,** adenosine 5'-diphosphate, is a nucleoside diphosphate formed by adding two phosphates to the 5'-OH group of adenosine.

ATP

adenosine 5'-**tri**phosphate

$$^{-}O-\overset{O}{\overset{\|}{\underset{O^-}{\underset{|}{P}}}}-O-\overset{O}{\overset{\|}{\underset{O^-}{\underset{|}{P}}}}-O-\overset{O}{\overset{\|}{\underset{O^-}{\underset{|}{P}}}}-O-\overset{5'}{C}H_2-\text{ribose}-\text{adenine}$$

NH2 N N N N 5' O OH OH adenine ribose

ADP

adenosine 5'-**di**phosphate

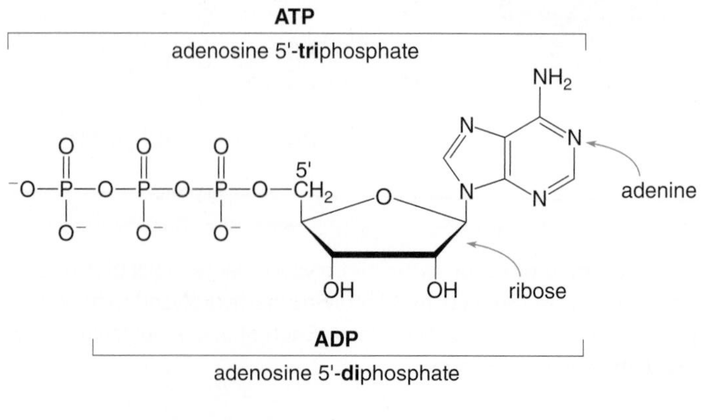

In metabolic pathways, the interconversion of ATP and ADP is the most important process for the storage and release of energy.

- **Hydrolysis of ATP cleaves one phosphate group, forming ADP and hydrogen phosphate, HPO_4^{2-}, often abbreviated as P_i (inorganic phosphate). This reaction *releases* 7.3 kcal/mol of energy.**

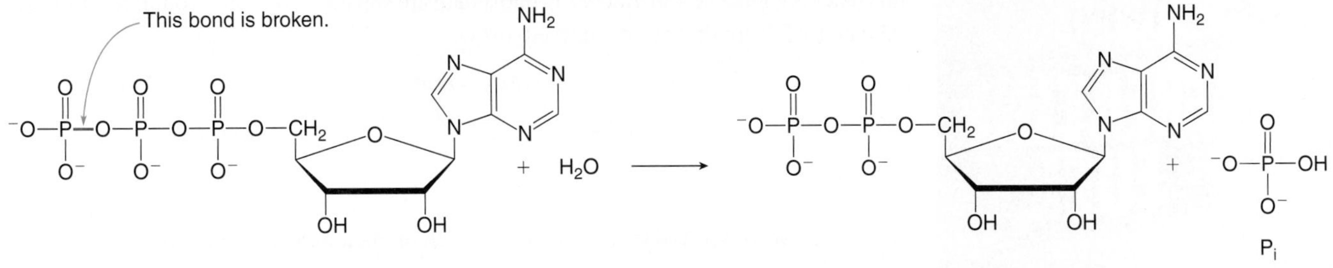

This bond is broken.

$$\text{ATP} + H_2O \longrightarrow \text{ADP} + {}^{-}O-P(=O)(O^{-})-OH \; (P_i)$$

Any process, such as walking, running, swallowing, or breathing, is fueled by the release of energy from the hydrolysis of ATP to ADP. ATP is the most prominent member of a group of "high-energy" molecules, reactive molecules that release energy by cleaving a bond.

- **The reverse reaction, phosphorylation, adds a phosphate group to ADP, forming ATP. Phosphorylation *requires* 7.3 kcal/mol of energy.**

$$\text{ADP} + P_i \longrightarrow \text{ATP} + H_2O$$

Figure 18.4 summarizes the reactions and energy changes that occur when ATP is synthesized and hydrolyzed. As we learned in Section 5.8, when energy is *released* in a reaction, the energy change is reported as a *negative* (–) value, so the energy change for ATP hydrolysis is –7.3 kcal/mol. When energy is *absorbed* in a reaction, the energy change is reported as a *positive* (+) value, so the energy change for the phosphorylation of ADP is +7.3 kcal/mol. The synthesis of ATP with P_i is the reverse of ATP hydrolysis, and the energy changes for such reactions are *equal* in value but *opposite* in sign.

ATP is constantly synthesized and hydrolyzed. It is estimated that each ATP molecule exists for about a minute before it is hydrolyzed and its energy released. Even though the body contains only about one gram of ATP at a given time, the energy needs of the body are such that an average individual synthesizes 40 kg of ATP daily!

PROBLEM 18.2

GTP, guanosine 5'-triphosphate, is another high-energy molecule that releases 7.3 kcal of energy when it is hydrolyzed to GDP. Write the equation for the hydrolysis of GTP to GDP. Is the energy change reported as a positive (+) or negative (–) value?

The hydrolysis of ATP to ADP is a favorable reaction because energy is released and lower energy products are formed. In metabolism, this reaction provides the energy to drive reactions that require energy.

Figure 18.4
ATP Hydrolysis and Synthesis

		Energy change
ATP Hydrolysis	ATP + H_2O ⟶ ADP + P_i	–7.3 kcal/mol
ATP Synthesis	ADP + P_i ⟶ ATP + H_2O	+7.3 kcal/mol

Energy is released. Energy is absorbed.

CONSUMER NOTE

Since the 1990s, creatine supplements have been used by some athletes who wish to increase muscle mass and strength. Increased creatine intake increases the amount of creatine phosphate (Problem 18.4) in the muscle.

Reactions that use ATP or coenzymes (Section 18.3) are often drawn using a combination of horizontal and curved arrows. The principal organic reactants and products are drawn from left to right with a reaction arrow as usual, but additional compounds like ATP and ADP are drawn on a curved arrow. This technique is meant to emphasize the organic substrates of the reaction, while making it clear that other materials are needed for the reaction to occur.

For example, the phosphorylation of glucose with ATP forms glucose 6-phosphate and ADP. In this reaction, glucose and glucose 6-phosphate are separated by a horizontal reaction arrow, while ATP and ADP are drawn on a curved arrow.

$$\text{glucose} \xrightarrow{\text{ATP} \quad \text{ADP}} \text{glucose 6-phosphate}$$

PROBLEM 18.3

Use curved arrow symbolism to write the reaction of fructose with ATP to form fructose 6-phosphate and ADP.

PROBLEM 18.4

Creatine phosphate is stored in muscles. When existing supplies of ATP are depleted during strenuous exercise, creatine phosphate reacts with ADP to form creatine and a new supply of ATP as a source of more energy. (a) Write this reaction using curved arrow symbolism. (b) If the energy change in this reaction is –3.0 kcal/mol, is energy absorbed or released?

18.3 Coenzymes in Metabolism

Many reactions in metabolic pathways involve coenzymes. As we learned in Section 16.9, a ***coenzyme*** **is an organic compound needed for an enzyme-catalyzed reaction to occur.** Some coenzymes serve as important oxidizing and reducing agents (Sections 18.3A and 18.3B), whereas coenzyme A activates acetyl groups (CH_3CO–), resulting in the transfer of a two-carbon unit to other substrates (Section 18.3C).

18.3A Coenzymes NAD^+ and NADH

Many coenzymes are involved in oxidation and reduction reactions. A coenzyme may serve as an oxidizing agent or a reducing agent in a biochemical pathway.

- **An oxidizing agent causes an oxidation reaction to occur, so the oxidizing agent is reduced.**
- **A reducing agent causes a reduction reaction to occur, so the reducing agent is oxidized.**

In examining what happens to a coenzyme during oxidation and reduction, it is convenient to think in terms of hydrogen atoms being composed of protons (H^+) and electrons (e^-).

- **When a coenzyme gains hydrogen atoms—that is, H^+ and e^-—the coenzyme is reduced; thus, the coenzyme is an *oxidizing* agent.**
- **When a coenzyme loses hydrogen atoms—that is, H^+ and e^-—the coenzyme is oxidized; thus, the coenzyme is a *reducing* agent.**

The coenzyme **nicotinamide adenine dinucleotide, NAD^+,** is a common biological oxidizing agent. When NAD^+ reacts with two hydrogen atoms, it gains one proton and two electrons and one proton is left over. Thus, NAD^+ is reduced and a new C—H bond is formed in the product, written as **NADH,** and referred to as the *reduced form* of nicotinamide adenine dinucleotide.

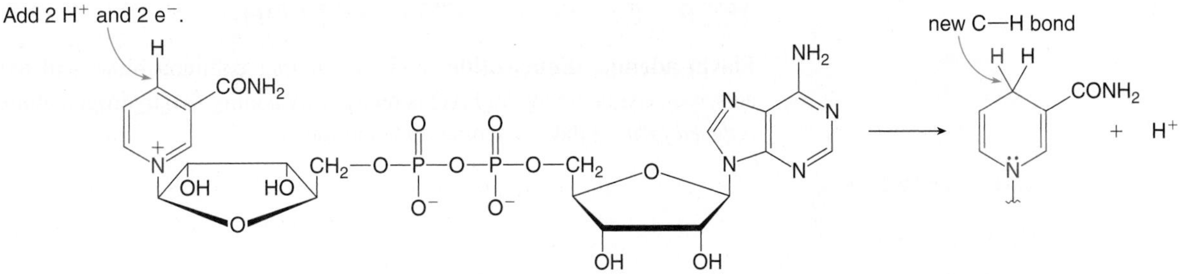

Add 2 H^+ and 2 e^-.

new C—H bond

NAD$^+$
nicotinamide adenine dinucleotide

NADH
(reduced form of NAD$^+$)

Curved arrow symbolism is often used to depict reactions with coenzymes.

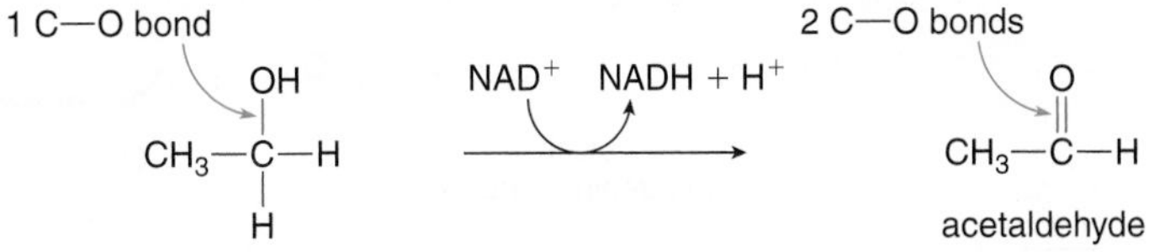

1 C—O bond — 2 C—O bonds

$$CH_3CH_2OH \xrightarrow{NAD^+ \rightarrow NADH + H^+} CH_3CHO$$

ethanol — acetaldehyde

The conversion of ethanol to acetaldehyde is an **oxidation** since the number of C—O bonds in the substrate *increases*. **NAD$^+$ serves as the oxidizing agent, and in the process, is reduced to NADH.** The reduced form of the coenzyme, NADH, is a biological reducing agent. When NADH reacts, it forms NAD$^+$ as a product. Thus, **NAD$^+$ and NADH are interconverted by oxidation and reduction reactions.**

SAMPLE PROBLEM 18.1

Label the reaction as an oxidation or reduction, and give the reagent, NAD$^+$ or NADH, that would be used to carry it out.

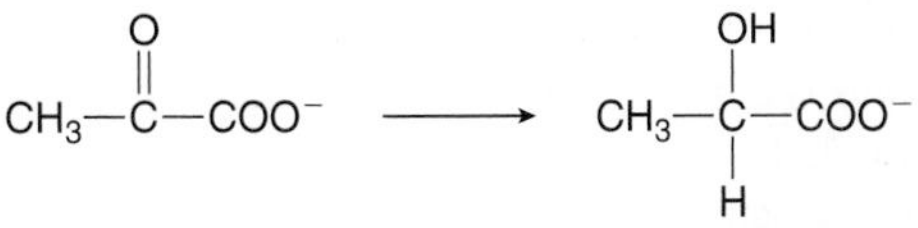

$$CH_3COCOO^- \longrightarrow CH_3CH(OH)COO^-$$

pyruvate — lactate

Analysis

Count the number of C—O bonds in the starting material and product. Oxidation increases the number of C—O bonds and reduction decreases the number of C—O bonds. NAD$^+$ is the coenzyme needed for an oxidation, and NADH is the coenzyme needed for a reduction.

Solution

The conversion of pyruvate to lactate is a reduction, since the product has one fewer C—O bond than the reactant. To carry out the reduction, the reducing agent NADH could be used.

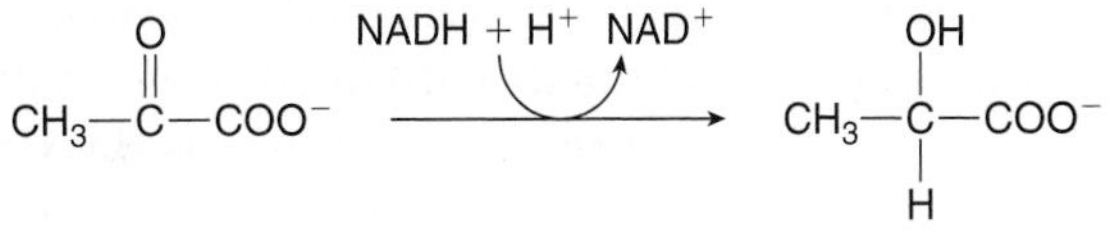

$$CH_3COCOO^- \xrightarrow{NADH + H^+ \rightarrow NAD^+} CH_3CH(OH)COO^-$$

pyruvate — lactate

PROBLEM 18.5

Label each reaction as an oxidation or reduction, and give the reagent, NAD$^+$ or NADH, that would be used to carry out the reaction.

a. $H_2C{=}O \longrightarrow CH_3OH$

b. 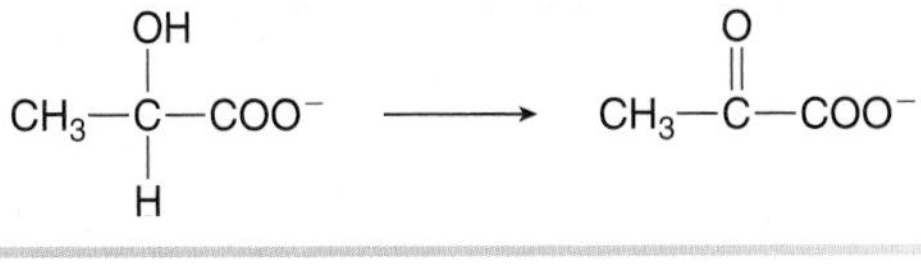 $CH_3CH(OH)COO^- \longrightarrow CH_3COCOO^-$

18.3B Coenzymes FAD and $FADH_2$

Flavin adenine dinucleotide, FAD, is another common biological oxidizing agent. When it acts as an oxidizing agent, FAD is reduced by adding two hydrogen atoms, forming **$FADH_2$,** the *reduced form* of flavin adenine dinucleotide.

Add 2 H^+ and 2 e^-.

FAD
flavin adenine dinucleotide

$FADH_2$
(reduced form of FAD)

Table 18.1 summarizes the common coenzymes used in oxidation and reduction reactions.

FAD is synthesized in cells from vitamin B_2, **riboflavin.** Riboflavin is a yellow, water-soluble vitamin obtained in the diet from leafy green vegetables, soybeans, almonds, and liver. When large quantities of riboflavin are ingested, excess amounts are excreted in the urine, giving it a bright yellow appearance.

riboflavin
vitamin B_2

HEALTH NOTE

Leafy green vegetables, soybeans, and almonds are good sources of riboflavin, vitamin B_2. Since this vitamin is light sensitive, riboflavin-fortified milk contained in glass or clear plastic bottles should be stored in the dark.

Table 18.1 Coenzymes in Oxidation and Reduction

Coenzyme	Role
NAD^+	Oxidizing agent
NADH	Reducing agent
FAD	Oxidizing agent
$FADH_2$	Reducing agent

PROBLEM 18.6

What makes riboflavin a water-soluble vitamin?

18.3C Coenzyme A

Coenzyme A differs from other coenzymes in this section because it is not an oxidizing or a reducing agent. In addition to many other functional groups, coenzyme A contains a **sulfhydryl group (SH group),** making it a **thiol (RSH).** To emphasize this functional group, we sometimes abbreviate the structure as **HS–CoA.**

thiol

coenzyme A = HS—CoA

HEALTH NOTE

Avocados are an excellent dietary source of pantothenic acid, vitamin B_5.

The sulfhydryl group of coenzyme A reacts with acetyl groups ($CH_3CO–$) or other acyl groups ($RCO—$) to form thioesters, RCOSR'. When an acetyl group is bonded to coenzyme A, the product is called **acetyl coenzyme A,** or simply **acetyl CoA.**

$$CH_3{-}\overset{O}{\overset{\|}{C}}{-} \; + \; HS{-}CoA \longrightarrow CH_3{-}\overset{O}{\overset{\|}{C}}{-}S{-}CoA$$

acetyl CoA

General structure of a thioester

$$R{-}\overset{O}{\overset{\|}{C}}{-}S{-}R'$$

Thioesters such as acetyl CoA are another group of high-energy compounds that release energy on reaction with water. In addition, acetyl CoA reacts with other substrates in metabolic pathways to deliver its two-carbon acetyl group, as in the citric acid cycle in Section 18.6. Coenzyme A is synthesized in cells from **pantothenic acid, vitamin B_5.**

$$HO{-}\overset{O}{\overset{\|}{C}}{-}CH_2CH_2{-}\overset{H}{N}{-}\overset{O}{\overset{\|}{C}}{-}\overset{OH}{\overset{|}{C}H}{-}\underset{CH_3}{\overset{CH_3}{\overset{|}{\underset{|}{C}}}}{-}CH_2{-}OH$$

pantothenic acid
vitamin B_5

PROBLEM 18.7

Predict the water solubility of vitamin B_5.

18.4 Glycolysis

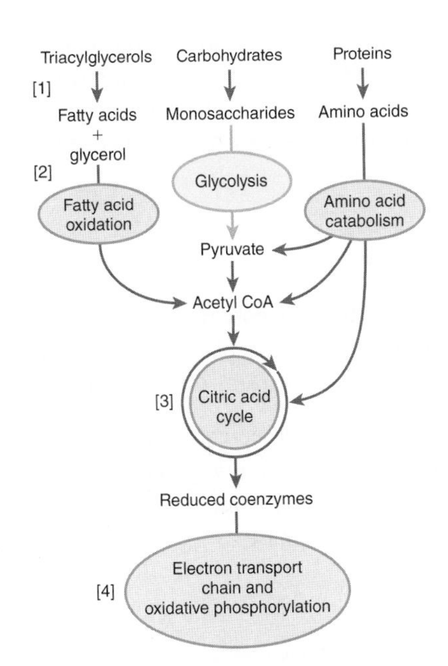

The metabolism of monosaccharides centers around glucose. Whether it is obtained by the hydrolysis of ingested polysaccharides or stored glycogen (Section 14.6), glucose is the principal monosaccharide used for energy in the human body.

- *Glycolysis* is a linear, 10-step pathway that converts glucose, a six-carbon monosaccharide, to two molecules of pyruvate ($CH_3COCO_2^-$).

Glycolysis is an anaerobic (without air) pathway that takes place in the cytoplasm and can be conceptually divided into two parts (Figure 18.5).

- Steps [1]–[5] comprise the *energy-investment phase.* The addition of two phosphate groups requires the energy stored in two ATP molecules. Cleavage of a carbon–carbon bond forms two three-carbon products.
- Steps [6]–[10] comprise the *energy-generating phase.* Each of the three-carbon products is ultimately oxidized, forming NADH, and two high-energy phosphate bonds are broken to form two ATP molecules.

18.4A The Steps in Glycolysis

The specific steps and all needed enzymes in glycolysis are shown in Figures 18.6 and 18.7.

Figure 18.5

An Overview of Glycolysis

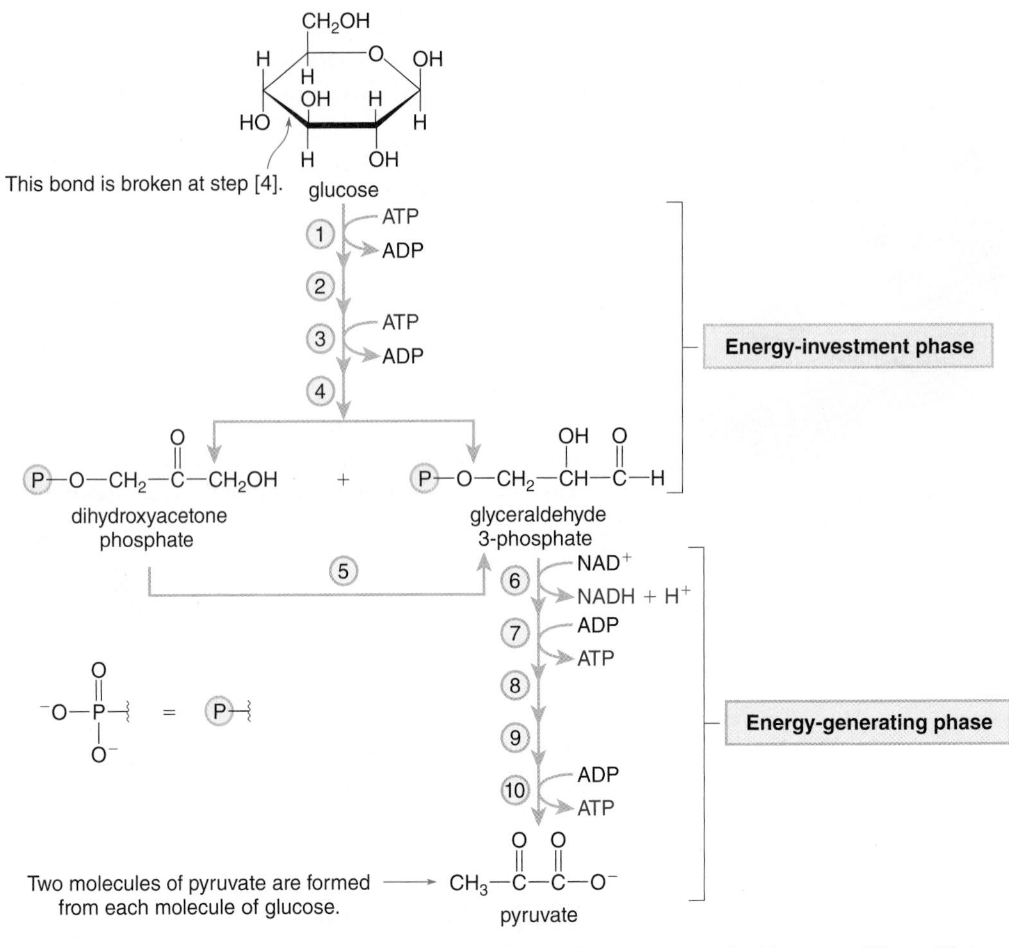

In the energy-investment phase of glycolysis, ATP supplies energy needed for steps [1] and [3]. In the energy-generating phase, each three-carbon product from step [5] forms one NADH and two ATP molecules. Since two glyceraldehyde 3-phosphate molecules are formed from each glucose molecule, a total of two NADH and four ATP molecules are formed in the energy-generating phase.

Glycolysis: Steps [1]–[5]

Glycolysis begins with the phosphorylation of glucose with ATP to form glucose 6-phosphate (Figure 18.6). Isomerization of glucose 6-phosphate to fructose 6-phosphate takes place with an isomerase enzyme in step [2]. Phosphorylation with ATP in step [3] yields fructose 1,6-bisphosphate.

Cleavage of the six-carbon chain of fructose 1,6-bisphosphate forms two three-carbon products—dihydroxyacetone phosphate and glyceraldehyde 3-phosphate. Since only glyceraldehyde 3-phosphate continues on in glycolysis, dihydroxyacetone phosphate is isomerized to glyceraldehyde 3-phosphate in step [5], completing the energy-investment phase of glycolysis.

In summary:

- **The first phase of glycolysis converts glucose to *two* molecules of glyceraldehyde 3-phosphate.**
- **The energy from two ATP molecules is utilized.**

PROBLEM 18.8

Identify the type of carbonyl groups present in dihydroxyacetone phosphate and glyceraldehyde 3-phosphate. Classify the OH groups in each compound as 1°, 2°, or 3°.

Figure 18.6 Glycolysis: Steps [1]–[5]

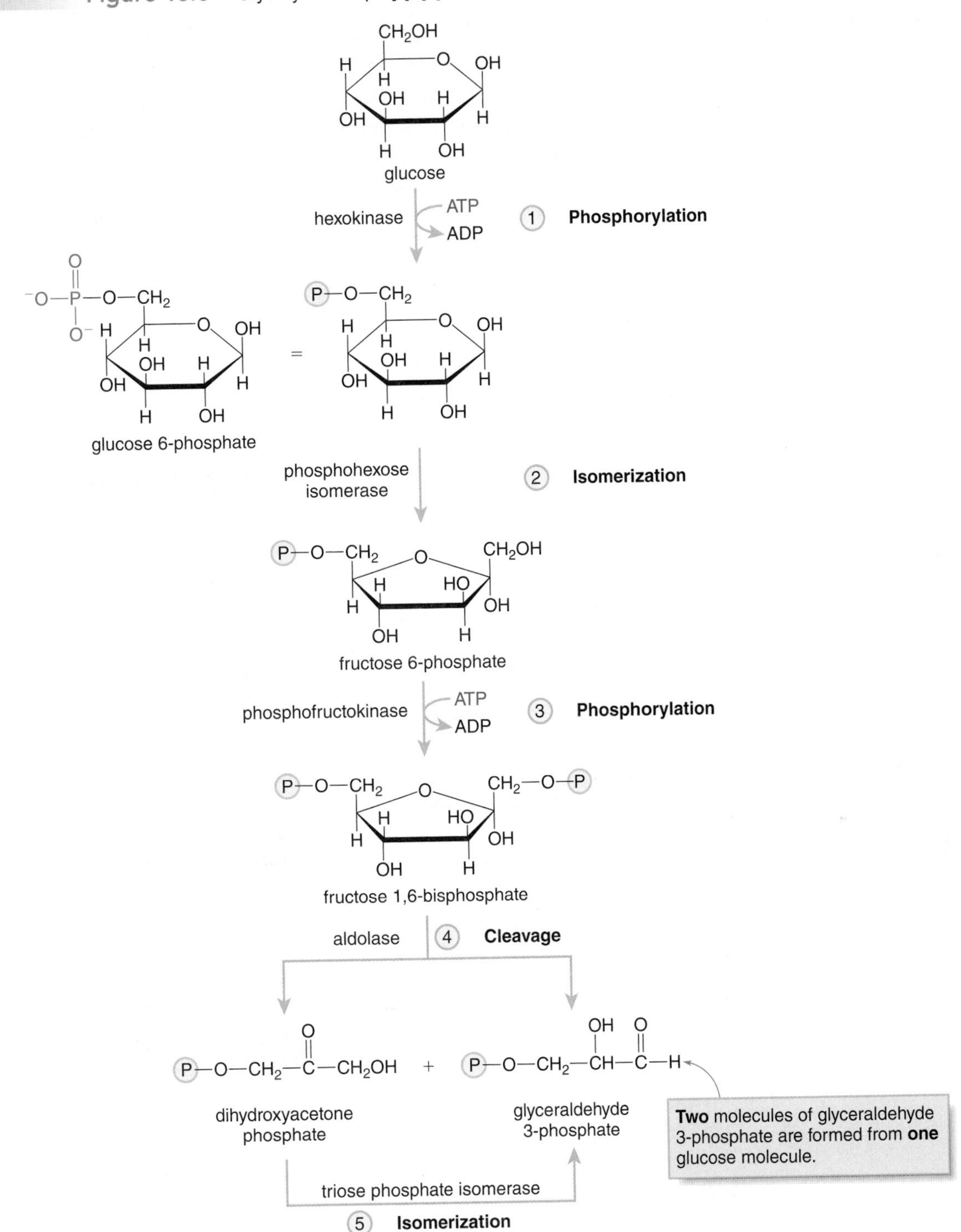

- All $-PO_3^{2-}$ groups in glycolysis are abbreviated as —Ⓟ.
- The energy from two ATP molecules is used for phosphorylation in steps [1] and [3].
- Cleavage of a carbon–carbon bond and isomerization form two molecules of glyceraldehyde 3-phosphate from glucose, completing the energy-investment phase of glycolysis.

Glycolysis: Steps [6]–[10]

Each three-carbon aldehyde (glyceraldehyde 3-phosphate) produced in step [5] of glycolysis is carried through a series of five reactions that ultimately form pyruvate (Figure 18.7).

In step [6], oxidation of the –CHO group of glyceraldehyde 3-phosphate and phosphorylation form 1,3-bisphosphoglycerate. In this process, the oxidizing agent NAD^+ is reduced to NADH. Transfer of a phosphate group from 1,3-bisphosphoglycerate to ADP forms 3-phosphoglycerate and generates ATP in step [7]. Isomerization of the phosphate group in step [8] and loss of water in step [9] form phosphoenolpyruvate. Finally, transfer of a phosphate to ADP forms ATP and pyruvate

Figure 18.7 Glycolysis: Steps [6]–[10]

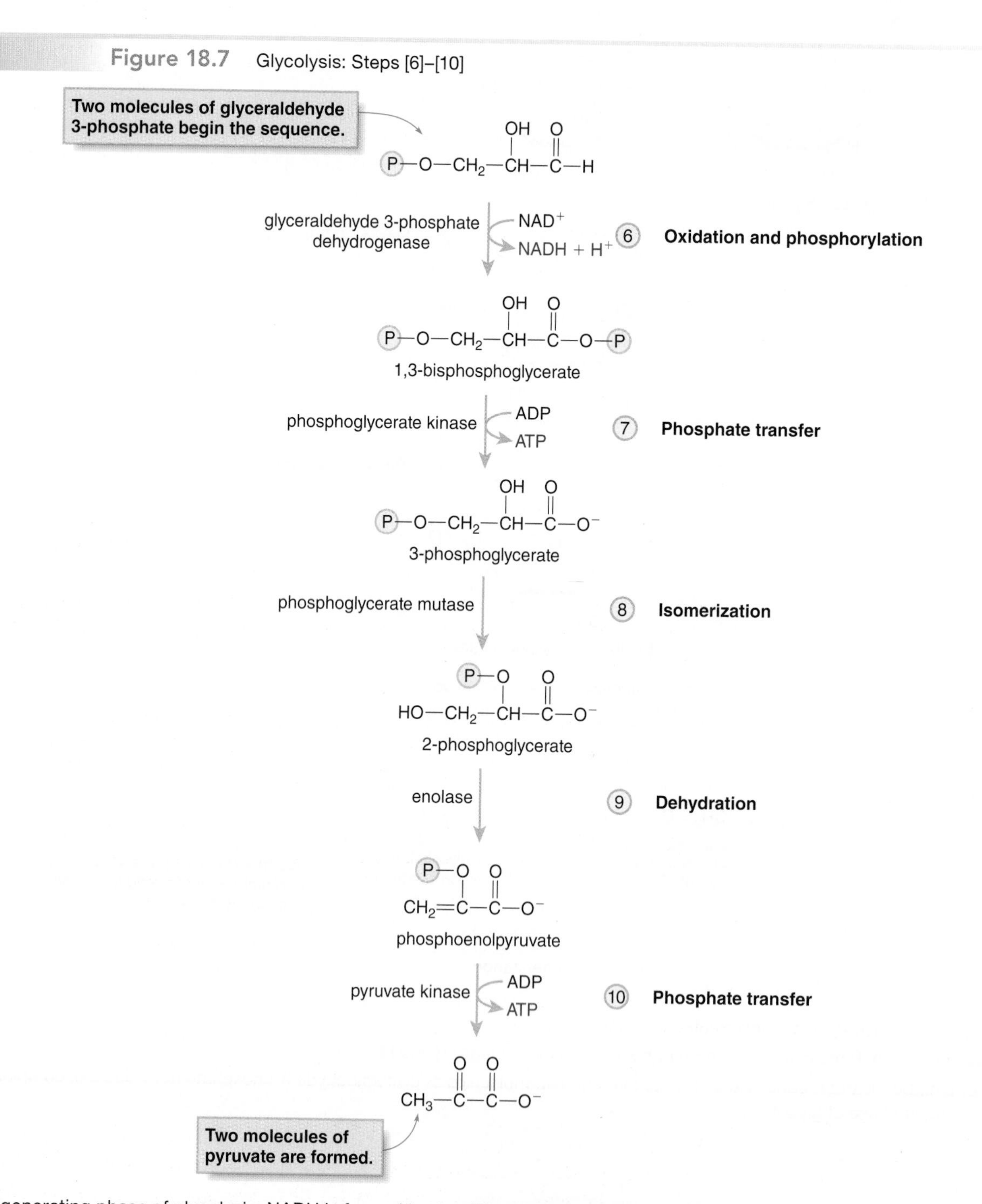

In the energy-generating phase of glycolysis, NADH is formed in step [6] and ATP is formed in steps [7] and [10]. Since one glucose molecule yields two molecules of glyceraldehyde that begin the sequence, 2 $CH_3COCO_2^-$, 2 NADH, and 4 ATP are ultimately formed in steps [6]–[10].

in step [10]. Thus, **one NADH molecule is produced in step [6] and two ATPs are formed in steps [7] and [10] for each glyceraldehyde 3-phosphate.**

- Since each glucose molecule yielded *two* glyceraldehyde 3-phosphate molecules in step [5], overall *two* NADH molecules and *four* ATP molecules are formed in the energy-generating phase of glycolysis.

PROBLEM 18.9

A kinase is an enzyme that catalyzes the transfer of a phosphate group from one substrate to another. Identify all of the reactions of glycolysis that utilize kinases. What species is phosphorylated in each reaction?

PROBLEM 18.10

(a) Convert the ball-and-stick model of **A,** one of the intermediates in glycolysis, to a structural formula, and name the compound. (b) What is the immediate precursor of **A** in glycolysis; that is, what compound forms **A** as a reaction product? (c) What compound is formed from **A** in glycolysis?

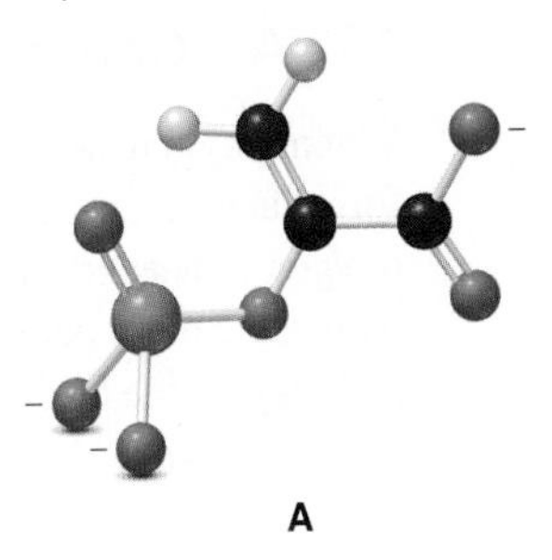

A

18.4B The Net Result of Glycolysis

Three major products are formed in glycolysis—**ATP, NADH,** and **pyruvate.**

- Two ATP molecules are used in the energy-investment phase (steps [1] and [3]), and four molecules of ATP are formed in the energy-generating phase (steps [7] and [10]). The net result is the *synthesis of two molecules of ATP from glycolysis.*

- Two molecules of NADH are formed from two glyceraldehyde 3-phosphate molecules in step [6]. The NADH formed in glycolysis must be transported from the cytoplasm to the mitochondria for use in the electron transport chain to generate more ATP.

- Two three-carbon molecules of pyruvate ($CH_3COCO_2^-$) are formed from the six carbon atoms of glucose. The fate of pyruvate depends on oxygen availability, as discussed in Section 18.5.

The overall process of glycolysis can be summarized in the following equation.

$$\underset{\text{glucose}}{C_6H_{12}O_6} + 2\ NAD^+ \xrightarrow{2\ ADP \quad 2\ ATP} \underset{\text{pyruvate}}{2\ CH_3-\overset{\displaystyle O}{\overset{\|}{C}}-\overset{\displaystyle O}{\overset{\|}{C}}-O^-} + 2\ NADH + 2\ H^+$$

Although glycolysis is an ongoing pathway in cells, the rate of glycolysis depends on the body's need for the products it forms—that is, pyruvate, ATP, and NADH. When ATP levels are high, glycolysis is inhibited at various stages. When ATP levels are depleted, such as during strenuous exercise, glycolysis is activated so that more ATP is synthesized.

18.5 The Fate of Pyruvate

While pyruvate is the end product of glycolysis, it is not the final product of glucose metabolism. What happens to pyruvate depends on the existing conditions and the organism. In particular, there are three possible products:

$$\underset{\text{pyruvate}}{CH_3-\overset{O}{\overset{\|}{C}}-\overset{O}{\overset{\|}{C}}-O^-} \longrightarrow \underset{\text{acetyl CoA}}{CH_3-\overset{O}{\overset{\|}{C}}-SCoA} \quad \text{or} \quad \underset{\text{lactate}}{CH_3-\overset{OH}{\overset{|}{\underset{H}{\underset{|}{C}}}}-\overset{O}{\overset{\|}{C}}-O^-} \quad \text{or} \quad \underset{\text{ethanol}}{CH_3CH_2OH}$$

- **Acetyl CoA,** $CH_3COSCoA$, **is formed under aerobic conditions.**
- **Lactate,** $CH_3CH(OH)CO_2^-$, **is formed under anaerobic conditions.**
- **Ethanol,** CH_3CH_2OH, **is formed in fermentation.**

18.5A Conversion to Acetyl CoA

When oxygen is plentiful, oxidation of pyruvate by NAD^+ in the presence of coenzyme A forms acetyl CoA and carbon dioxide. Although oxygen is not needed for this specific reaction, oxygen is needed to oxidize NADH back to NAD^+. Without an adequate supply of NAD^+, this pathway cannot occur. The acetyl CoA formed in this process then enters the common metabolic pathways—the citric acid cycle, the electron transport chain, and oxidative phosphorylation—to generate a great deal of ATP.

$$\underset{\text{pyruvate}}{CH_3-\overset{O}{\overset{\|}{C}}-\overset{O}{\overset{\|}{C}}-O^-} + HS-CoA \xrightarrow{NAD^+ \quad NADH + H^+} \underset{\text{acetyl CoA}}{CH_3-\overset{O}{\overset{\|}{C}}-SCoA} + CO_2$$

18.5B FOCUS ON HEALTH & MEDICINE Conversion to Lactate

When oxygen levels are low, the metabolism of pyruvate must follow a different course. Oxygen is needed to oxidize the NADH formed in step [6] of glycolysis back to NAD^+. If there is not enough O_2 to re-oxidize NADH, cells must get NAD^+ in a different way. The conversion of pyruvate to lactate provides the solution.

Reduction of pyruvate with NADH forms lactate and NAD^+, which can now re-enter glycolysis and oxidize glyceraldehyde 3-phosphate at step [6].

$$\underset{\text{pyruvate}}{CH_3-\overset{O}{\overset{\|}{C}}-\overset{O}{\overset{\|}{C}}-O^-} \xrightarrow{NADH + H^+ \quad NAD^+} \underset{\text{lactate}}{CH_3-\overset{OH}{\overset{|}{\underset{H}{\underset{|}{C}}}}-\overset{O}{\overset{\|}{C}}-O^-}$$

NAD^+ can re-enter glycolysis at step [6].

During periods of strenuous exercise, when ATP needs are high, there is an inadequate level of oxygen to re-oxidize NADH. At these times, pyruvate is reduced to lactate for the sole purpose of re-oxidizing NADH to NAD^+ to maintain glycolysis. Under anaerobic conditions, therefore,

HEALTH NOTE

Muscle fatigue is caused by lactate buildup due to the anaerobic metabolism of glucose. Deep breaths after exercise replenish O_2 in oxygen-depleted tissues and allow the re-oxidation of lactate to pyruvate.

lactate becomes the main product of glucose metabolism and only two ATP molecules are formed per glucose.

Anaerobic conditions

$$\underset{\text{glucose}}{C_6H_{12}O_6} \xrightarrow{\text{2 ADP} \;\to\; \text{2 ATP}} \underset{\text{lactate}}{2\ CH_3\text{—}CH(OH)\text{—}CO_2^-}$$

Anaerobic metabolism leads to an increase in lactate levels in muscles, which in turn is associated with soreness and cramping. During these periods an "oxygen debt" is created. When vigorous activity ceases, an individual inhales deep breaths of air to repay the oxygen debt caused by heavy exercise. Lactate is then gradually re-oxidized to pyruvate, which can once again be converted to acetyl CoA, and muscle soreness, fatigue, and shortness of breath resolve.

In any tissue deprived of oxygen, pyruvate is converted to lactate rather than acetyl CoA. The pain produced during a heart attack, for example, is caused by an increase in lactate concentration that results when the blood supply to part of the heart muscle is blocked (Figure 18.8). The lack of oxygen delivery to heart tissue results in the anaerobic metabolism of glucose to lactate rather than acetyl CoA.

Measuring lactate levels in the blood is a common diagnostic tool used by physicians to assess how severely ill an individual is. A higher-than-normal lactate concentration generally indicates inadequate oxygen delivery to some tissues. Lactate levels increase transiently during exercise, but can remain elevated because of lung disease, congestive heart failure, or the presence of a serious infection.

Figure 18.8 Lactate Production During a Heart Attack

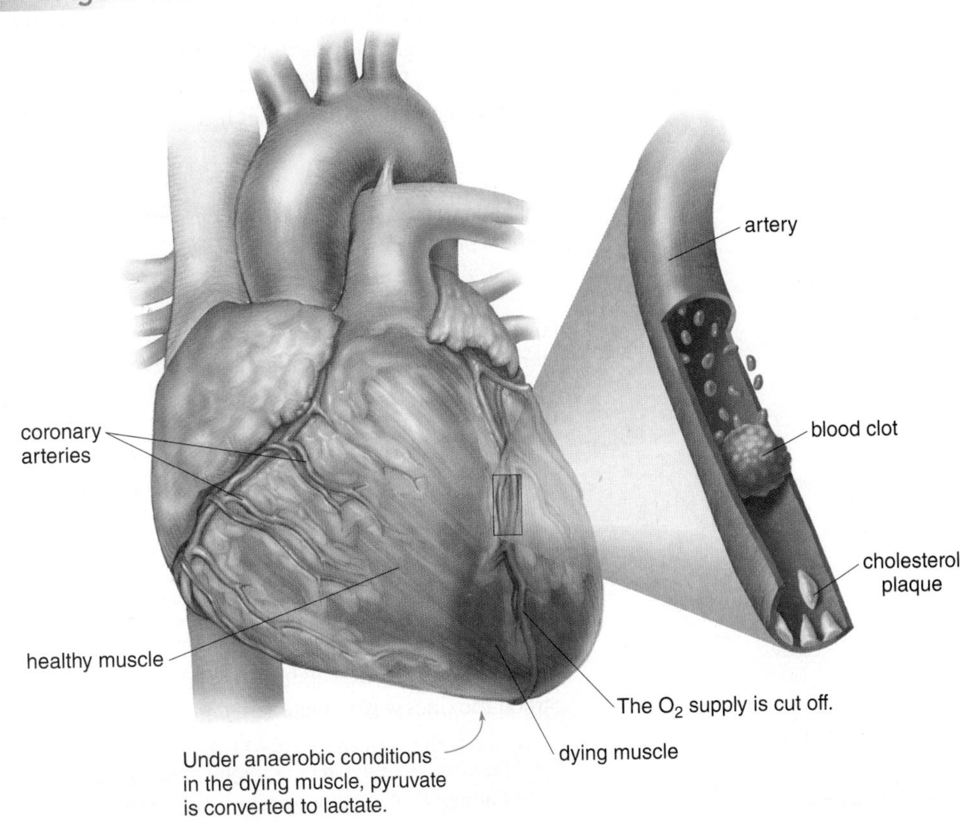
artery
coronary arteries
blood clot
cholesterol plaque
healthy muscle
The O_2 supply is cut off.
Under anaerobic conditions in the dying muscle, pyruvate is converted to lactate.
dying muscle

Fermentation plays a key role in the production of bread, beer, and cheese.

18.5C Conversion to Ethanol

In yeast and other microorganisms, pyruvate is converted to ethanol (CH_3CH_2OH) and carbon dioxide (CO_2) by a two-step process: **decarboxylation** (loss of CO_2) to acetaldehyde followed by reduction to ethanol.

$$\underset{\text{pyruvate}}{CH_3-\overset{O}{\overset{\|}{C}}-\overset{O}{\overset{\|}{C}}-O^-} \xrightarrow[\text{pyruvate decarboxylase}]{} CH_3-\overset{O}{\overset{\|}{C}}-H + CO_2 \xrightarrow[\text{alcohol dehydrogenase}]{NADH + H^+ \quad NAD^+} \underset{\text{ethanol}}{CH_3CH_2OH}$$

The NAD^+ generated during reduction can enter glycolysis as an oxidizing agent in step [6]. As a result, glucose can be metabolized by yeast under anaerobic conditions: glycolysis forms pyruvate, which is further metabolized to ethanol and carbon dioxide. Two molecules of ATP are generated during glycolysis.

- ***Fermentation* is the anaerobic conversion of glucose to ethanol and CO_2.**

Fermentation

$$\underset{\text{glucose}}{C_6H_{12}O_6} \xrightarrow{2\,ADP \quad 2\,ATP} \underset{\text{ethanol}}{2\,CH_3CH_2OH} + 2\,CO_2$$

The ethanol in beer, wine, and other alcoholic beverages is obtained by the fermentation of sugar, quite possibly the oldest example of chemical synthesis. The carbohydrate source determines the type of alcoholic beverage formed.

Fermentation plays a role in forming other food products. Cheese is produced by fermenting curdled milk, while yogurt is prepared by fermenting fresh milk. When yeast is mixed with flour, water, and sugar, the enzymes in yeast carry out fermentation to produce the CO_2 that causes bread to rise. Some of the characteristic and "intoxicating" odor associated with freshly baked bread is due to the ethanol that has evaporated during the baking process.

PROBLEM 18.11

What role does NADH play in the conversion of pyruvate to lactate? What role does NADH play in the conversion of pyruvate to ethanol?

PROBLEM 18.12

(a) In what way(s) is the conversion of pyruvate to acetyl CoA similar to the conversion of pyruvate to ethanol? (b) In what way(s) are the two processes different?

PROBLEM 18.13

(a) In what way(s) is the conversion of pyruvate to lactate similar to the conversion of pyruvate to ethanol? (b) In what way(s) are the two processes different?

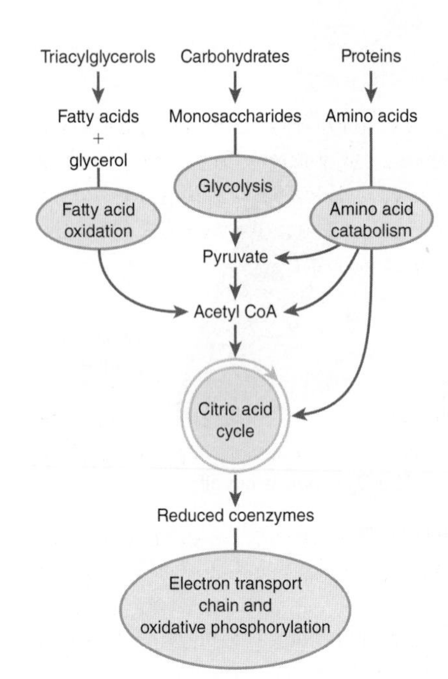

18.6 The Citric Acid Cycle

The **citric acid cycle,** a series of enzyme-catalyzed reactions that occur in mitochondria, comprises the third stage of the catabolism of biomolecules—carbohydrates, lipids, and amino acids—to carbon dioxide, water, and energy.

- **The citric acid cycle is an eight-step cyclic metabolic pathway that begins with the addition of acetyl CoA to a four-carbon substrate.**
- **The citric acid cycle produces high-energy compounds for ATP synthesis in stage [4] of catabolism.**

18.6A Overview of the Citric Acid Cycle

The citric acid cycle is also called the tricarboxylic acid cycle (TCA cycle) or the Krebs cycle. The general scheme (Figure 18.9) illustrates the key features.

- The citric acid cycle begins when two carbons of acetyl CoA ($CH_3COSCoA$) react with a four-carbon organic substrate to form a six-carbon product (step [1]).
- Two carbon atoms are removed to form two molecules of CO_2 (steps [3] and [4]).
- Four molecules of reduced coenzymes (NADH and $FADH_2$) are formed in steps [3], [4], [6], and [8]. These molecules serve as carriers of electrons to the electron transport chain in stage [4] of catabolism, which ultimately results in the synthesis of a great deal of ATP.
- One mole of GTP is synthesized in step [5]. GTP is a high-energy nucleoside triphosphate similar to ATP.

Figure 18.9 General Features of the Citric Acid Cycle

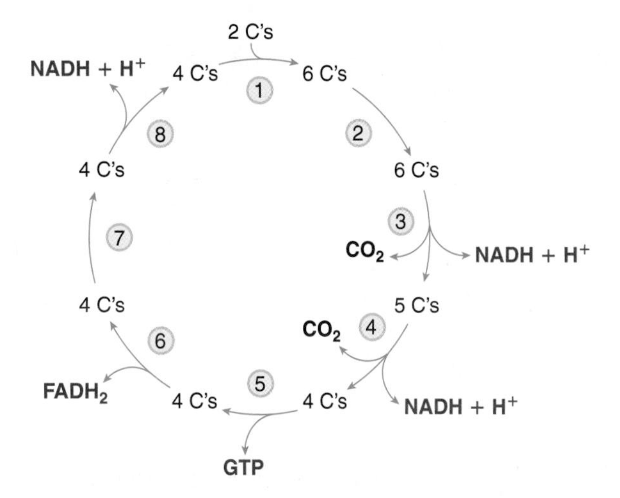

The citric acid cycle begins with the addition of two carbons from acetyl CoA in step [1] to a four-carbon organic substrate, drawn at the top of the cyclic pathway. Each turn of the citric acid cycle forms two molecules of CO_2, four molecules of reduced coenzymes (3 NADH and 1 $FADH_2$), and one high-energy GTP molecule.

18.6B Specific Steps of the Citric Acid Cycle

The eight reactions of the citric acid cycle, which can be conceptually divided into two parts, are shown in Figure 18.10. The first part of the cycle includes the addition of acetyl CoA to oxaloacetate to form the six-carbon product citrate, which undergoes two separate **decarboxylations**—reactions that give off CO_2. In part [2], functional groups are added and oxidized to re-form oxaloacetate, the substrate needed to begin the cycle again. Each step is enzyme catalyzed.

Part [1] of the Citric Acid Cycle

As shown in Figure 18.10, acetyl CoA enters the cycle by reaction with oxaloacetate at step [1] of the pathway. This reaction adds two carbons to oxaloacetate, forming citrate. In step [2], the 3° alcohol of citrate is isomerized to the 2° alcohol isocitrate. These first two steps add carbon atoms and rearrange functional groups.

Loss of two carbon atoms begins in step [3], by decarboxylation of isocitrate. The oxidizing agent NAD^+ also converts the 2° alcohol to a ketone to form α-ketoglutarate, which now contains one fewer carbon atom. This reaction forms NADH and H^+, which will carry electrons and protons gained in this reaction to the electron transport chain. In step [4], decarboxylation releases a

Figure 18.10
Steps in the Citric Acid Cycle

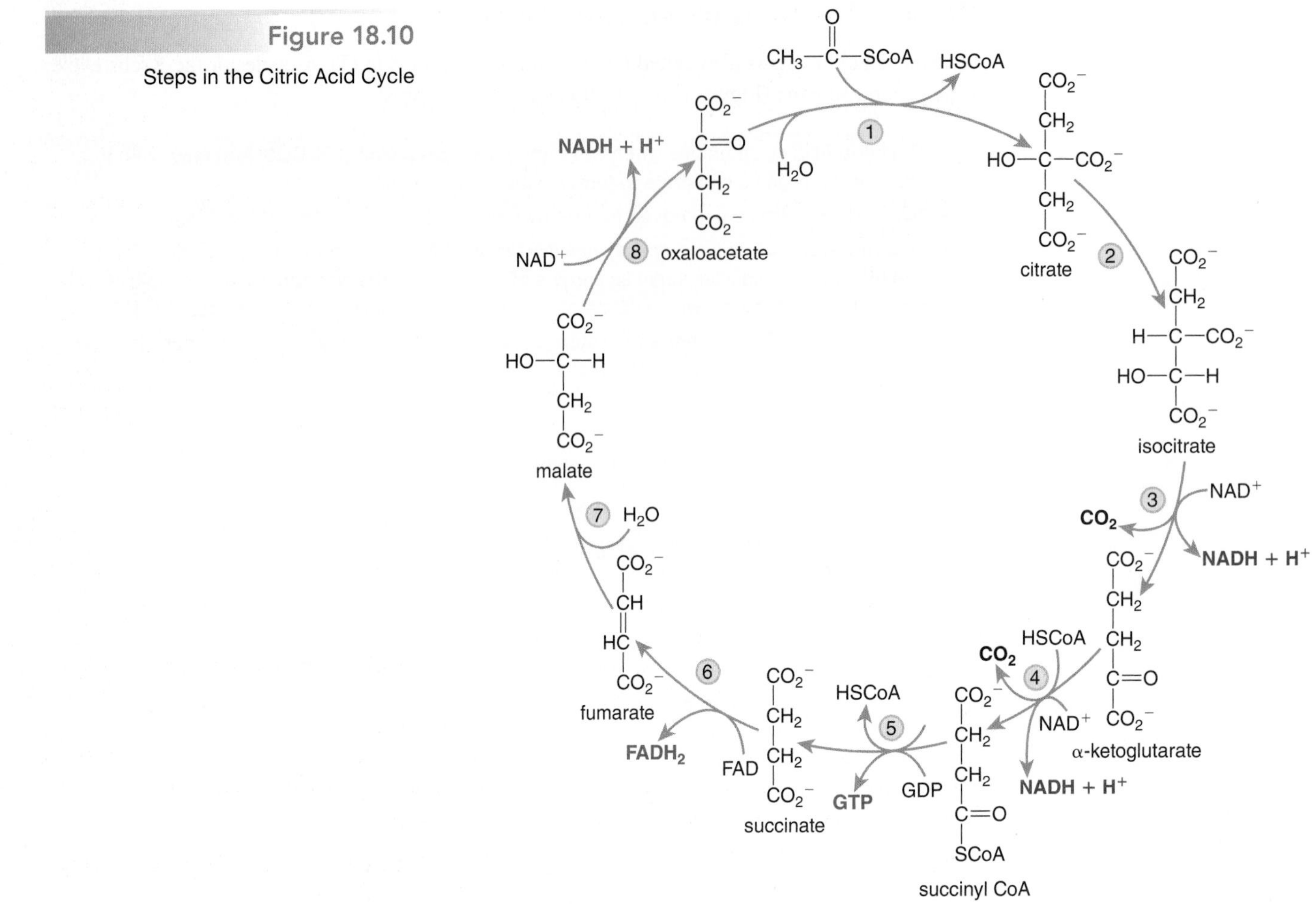

Each step of the citric acid cycle is enzyme catalyzed. The net result of the eight-step cycle is the conversion of the two carbons added to oxaloacetate to two molecules of CO_2. Reduced coenzymes (NADH and $FADH_2$) are also formed, which carry electrons to the electron transport chain to synthesize ATP. One molecule of high-energy GTP is synthesized in step [5].

second molecule of CO_2. Also, oxidation with NAD^+ in the presence of coenzyme A forms the thioester succinyl CoA. **By the end of step [4], two carbons are lost as CO_2 and two molecules of NADH are formed.**

Part [2] of the Citric Acid Cycle

Part [2] consists of four reactions that manipulate the functional groups of succinyl CoA to re-form oxaloacetate. In step [5], succinyl CoA is hydrolyzed to form succinate, releasing energy to convert GDP to GTP. GTP, guanosine 5'-triphosphate, is similar to ATP: GTP is a high-energy molecule that releases energy during hydrolysis. This is the only step of the citric acid cycle that directly generates a triphosphate.

In step [6], succinate is converted to fumarate with FAD. This reaction forms the reduced coenzyme $FADH_2$, which will carry electrons and protons to the electron transport chain. Addition of water in step [7] forms malate and oxidation of the 2° alcohol in malate with NAD^+ forms oxaloacetate in step [8]. Another molecule of NADH is also formed in step [8]. By the end of step [8], two more molecules of reduced coenzymes ($FADH_2$ and NADH) are formed. **Since the product of step [8] is the starting material of step [1], the cycle can continue as long as additional acetyl CoA is available for step [1].**

Overall, the citric acid cycle results in formation of

- two molecules of CO_2
- four molecules of reduced coenzymes (3 NADH and 1 $FADH_2$)
- one molecule of GTP

The net equation for the citric acid cycle can be written as shown. The ultimate fate of each product is also indicated.

$$CH_3-\overset{\overset{O}{||}}{C}-SCoA \; + \; 2\,H_2O \; + \; 3\,NAD^+ \; + \; FAD \; + \; GDP \; + \; P_i$$

overall reaction

$$2\,CO_2 \; + \; HSCoA \; + \; 3\,NADH \; + \; 3\,H^+ \; + \; FADH_2 \; + \; GTP$$

- $2\,CO_2$: exhaled gas
- HSCoA: The coenzyme re-enters the cycle.
- $3\,NADH + 3\,H^+ + FADH_2$: The reduced coenzymes enter the electron transport chain.
- GTP: energy source

- **The main function of the citric acid cycle is to produce reduced coenzymes that enter the electron transport chain and ultimately produce ATP.**

The rate of the citric acid cycle depends on the body's need for energy. When energy demands are high and the amount of available ATP is low, the cycle is activated. When energy demands are low and NADH concentration is high, the cycle is inhibited.

Although the citric acid cycle is complex, many individual reactions can be understood by applying the basic principles of organic chemistry learned in previous chapters.

SAMPLE PROBLEM 18.2

(a) Write out the reaction that converts succinate to fumarate with FAD using curved arrow symbolism. (b) Classify the reaction as an oxidation, reduction, or decarboxylation.

Analysis

Use Figure 18.10 to draw the structures for succinate and fumarate. Draw the organic reactant and product on the horizontal arrow and the oxidizing reagent FAD, which is converted to $FADH_2$, on the curved arrow. Oxidation reactions result in a loss of electrons, a loss of hydrogen, or a gain of oxygen. Reduction reactions result in a gain of electrons, a gain of hydrogen, or a loss of oxygen. A decarboxylation results in the loss of CO_2.

Solution

a. Equation:

$$\underset{\text{succinate}}{{}^-O_2C-CH_2-CH_2-CO_2^-} \xrightarrow{FAD \;\curvearrowright\; FADH_2} \underset{\text{fumarate}}{{}^-O_2C-CH=CH-CO_2^-}$$

b. Since succinate contains four C—H bonds and fumarate contains only two C—H bonds, hydrogen atoms have been lost, making this reaction an oxidation. In the process FAD is reduced to $FADH_2$.

PROBLEM 18.14

(a) Draw out the structures for the reaction that converts malate to oxaloacetate with NAD^+ using curved arrow symbolism. (b) Classify the reaction as an oxidation, reduction, or decarboxylation.

PROBLEM 18.15

(a) Convert the ball-and-stick model of **A**, one of the eight synthetic intermediates in the citric acid cycle, to a structural formula, and name the compound. (b) What is the immediate precursor of **A** in the citric acid cycle; that is, what compound forms **A** as a reaction product? (c) What compound is formed from **A** in the citric acid cycle?

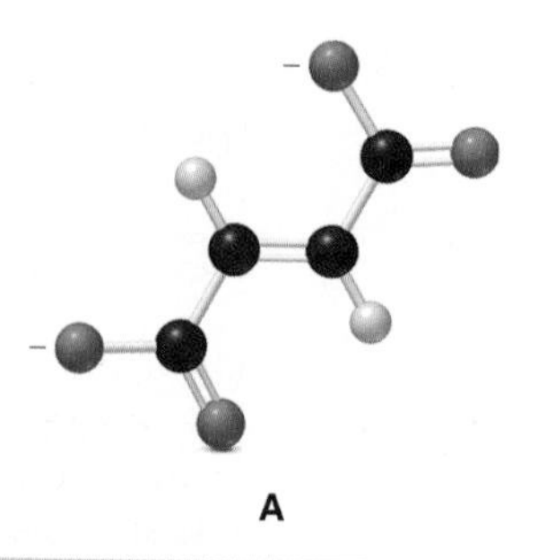

18.7 The Electron Transport Chain and Oxidative Phosphorylation

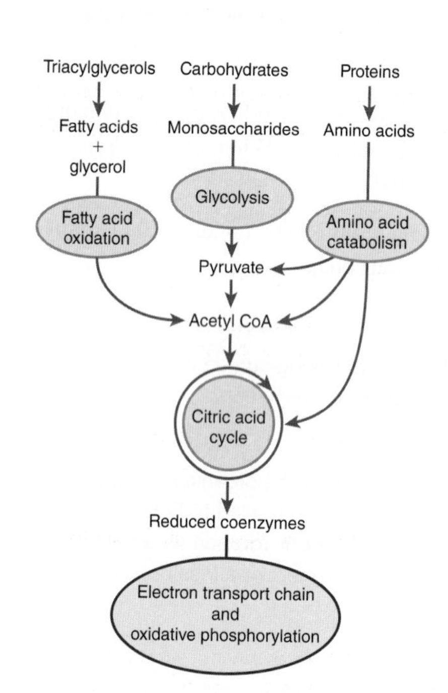

Most of the energy generated during the breakdown of biomolecules is formed during stage [4] of catabolism. Because oxygen is required, this process is called **aerobic respiration.** There are two facets to this stage:

- the electron transport chain, or the respiratory chain
- oxidative phosphorylation

The reduced coenzymes formed in the citric acid cycle enter the **electron transport chain** and the electrons they carry are transferred from one molecule to another by a series of oxidation–reduction reactions. Each reaction releases energy until electrons and protons react with oxygen to form water. Electron transfer also causes H^+ ions to be pumped across the inner mitochondrial cell membrane, creating an energy reservoir that is used to synthesize ATP by the **phosphorylation** of ADP.

18.7A The Electron Transport Chain

The electron transport chain is a multistep process that relies on four enzyme systems, called **complexes I, II, III, and IV,** as well as mobile electron carriers. Each complex is composed of enzymes, additional protein molecules, and metal ions that can gain and lose electrons in oxidation and reduction reactions. The complexes are situated in the inner membrane of the mitochondria, arranged so that electrons can be passed to progressively stronger oxidizing agents (Figure 18.11).

The electron transport chain begins with the reduced coenzymes—3 NADH and 1 $FADH_2$—formed during the citric acid cycle. These reduced coenzymes are electron rich and as such, they are capable of donating electrons to other species. Thus, **NADH and $FADH_2$ are *reducing agents* and when they donate electrons, they are oxidized.** When NADH donates two electrons, it is oxidized to NAD^+, which can re-enter the citric acid cycle. Likewise, when $FADH_2$ donates two electrons, it is oxidized to FAD, which can be used as an oxidant in step [6] of the citric acid cycle once again.

Once in the electron transport chain, the electrons are passed down from complex to complex in a series of redox reactions, and small packets of energy are released along the way. At the end of

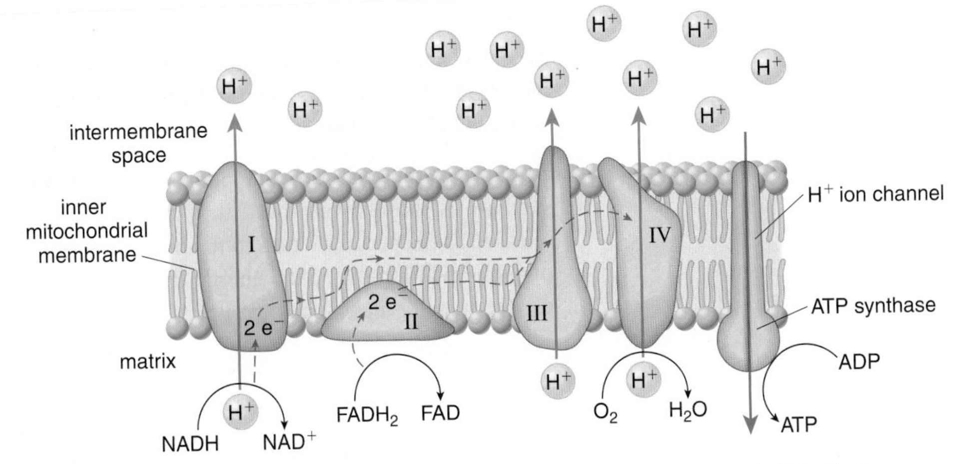

Figure 18.11

The Electron Transport Chain and ATP Synthesis in a Mitochondrion

The four enzyme complexes (I–IV) of the electron transport chain are located within the inner membrane of a mitochondrion, between the matrix and the intermembrane space. Electrons enter the chain when NADH and $FADH_2$ are oxidized and then transported through a series of complexes along the pathway shown in red. The electrons ultimately combine with O_2 to form H_2O. Protons (H^+) are pumped across the inner membrane into the intermembrane space at three locations shown by blue arrows. The energy released when protons return to the matrix by traveling through a channel (in green) in the ATP synthase enzyme is used to convert ADP to ATP.

HEALTH NOTE

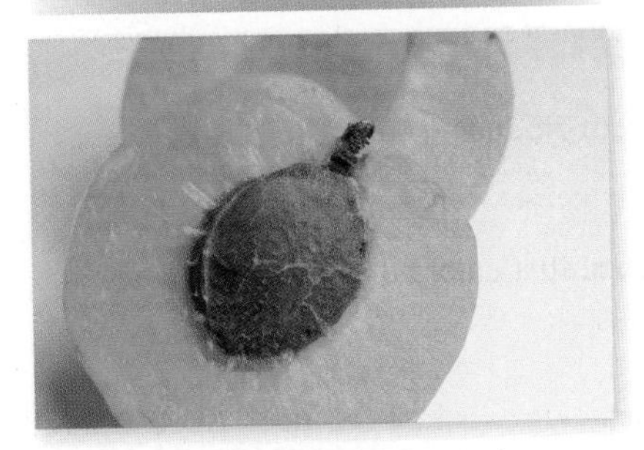

Hydrogen cyanide (HCN) is a poison because it disrupts the electron transport chain. The pits of apricots and peaches contain amygdalin, which forms HCN in the presence of certain enzymes.

the chain, **the electrons and protons (obtained from the reduced coenzymes or the matrix of the mitochondrion) react with inhaled oxygen to form water** and this facet of the process is complete.

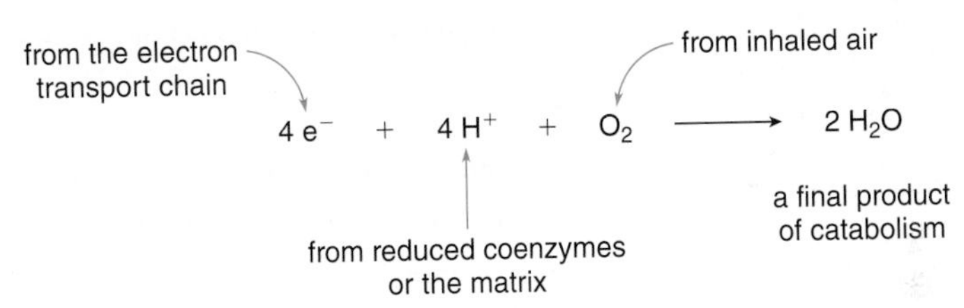

Because oxygen is needed for the final stage of electron transport, this process is **aerobic.**

PROBLEM 18.16

If NADH and $FADH_2$ were not oxidized in the electron transport chain, what would happen to the citric acid cycle?

PROBLEM 18.17

At several points in the electron transport chain, iron cations gain or lose electrons by reactions that interconvert Fe^{2+} and Fe^{3+} cations. (a) When Fe^{3+} is converted to Fe^{2+}, is the reaction an oxidation or a reduction? (b) Is Fe^{3+} an oxidizing agent or a reducing agent?

18.7B ATP Synthesis by Oxidative Phosphorylation

Although the electron transport chain illustrates how electrons carried by reduced coenzymes ultimately react with oxygen to form water, we have still not learned how ATP is synthesized. The answer lies in what happens to the H^+ ions in the mitochondrion.

H^+ ions generated by reactions in the electron transport chain, as well as H^+ ions present in the matrix of the mitochondria, are pumped across the inner mitochondrial membrane into the intermembrane space at three different sites (Figure 18.11). This process requires energy, since it moves protons against the concentration gradient. The energy comes from redox reactions in the electron transport chain. Since the concentration of H^+ ions is then higher on one side of the membrane, this creates a potential energy gradient, much like the potential energy of water that is stored behind a dam.

To return to the matrix, the H^+ ions travel through a channel in the ATP synthase enzyme. ATP synthase is the enzyme that catalyzes the phosphorylation of ADP to form ATP. The energy released as the protons return to the matrix converts ADP to ATP. This process is called **oxidative phosphorylation,** since the energy that results from the oxidation of the reduced coenzymes is used to transfer a phosphate group.

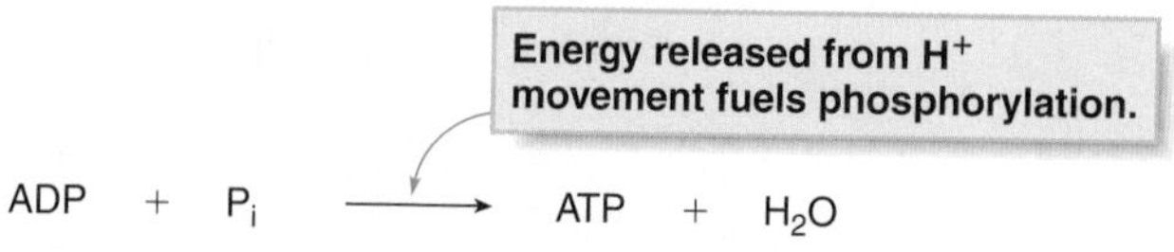

Energy released from H^+ movement fuels phosphorylation.

$$\text{ADP} + \text{P}_i \longrightarrow \text{ATP} + H_2O$$

PROBLEM 18.18

In which region of the mitochondrion—the matrix or the intermembrane space—would the pH be lower? Explain your choice.

18.7C ATP Yield from Oxidative Phosphorylation

How much ATP is generated during stage [4] of catabolism?

- **Each NADH enters the electron transport chain at complex I in the inner mitochondrial membrane and the resulting cascade of reactions produces enough energy to synthesize 2.5 ATPs.**
- **$FADH_2$ enters the electron transport chain at complex II, producing energy for the synthesis of 1.5 ATPs.**

How much ATP is generated for each acetyl CoA fragment that enters the entire common catabolic pathway—that is, stages [3] and [4] of catabolism?

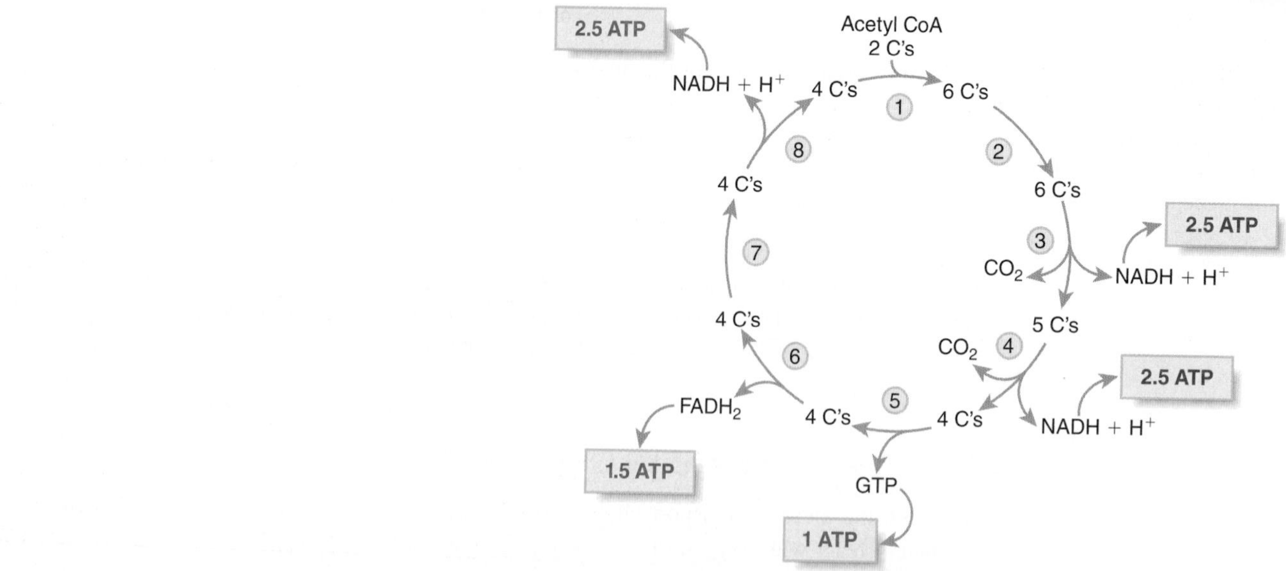

Acetyl CoA
2 C's
[1] 4 C's → 6 C's
[2] 6 C's → 6 C's
[3] 6 C's → 5 C's + CO_2 + NADH + H^+ → 2.5 ATP
[4] 5 C's → 4 C's + CO_2 + NADH + H^+ → 2.5 ATP
[5] 4 C's → 4 C's + GTP → 1 ATP
[6] 4 C's → 4 C's + $FADH_2$ → 1.5 ATP
[7] 4 C's → 4 C's
[8] 4 C's → 4 C's + NADH + H^+ → 2.5 ATP

For each turn of the citric acid cycle, three NADH molecules and one $FADH_2$ molecule are formed. In addition, one GTP molecule is produced directly during the citric acid cycle (step [5]); one GTP molecule is equivalent in energy to one ATP molecule. These facts allow us to calculate the total number of ATP molecules formed for each acetyl CoA.

$$3\ NADH \times 2.5\ ATP/NADH = 7.5\ ATP$$
$$1\ FADH_2 \times 1.5\ ATP/FADH_2 = 1.5\ ATP$$
$$1\ GTP = 1\ ATP$$
$$10\ ATP$$

- **Complete catabolism of each acetyl CoA molecule that enters the citric acid cycle forms 10 ATP molecules.**

Each ATP molecule can now provide energy for other reactions.

18.8 The ATP Yield from Glucose

How much ATP is generated from the complete catabolism of glucose ($C_6H_{12}O_6$) to carbon dioxide (CO_2)? To answer this question we must take into account the number of ATP molecules formed in the following sequential pathways:

- the glycolysis of glucose to two pyruvate molecules
- the oxidation of two pyruvate molecules to two molecules of acetyl CoA
- the citric acid cycle
- the electron transport chain and oxidative phosphorylation

To calculate how much ATP is generated, we must consider both the ATP formed directly in reactions, as well as the ATP produced from reduced coenzymes (NADH and $FADH_2$) after oxidative phosphorylation. As we learned in Section 18.7, each NADH provides the energy to yield 2.5 ATPs, while each $FADH_2$ yields 1.5 ATPs.

This calculation must also take into account that glucose is split into *two* three-carbon molecules at step [4] of glycolysis, so the **ATP yield in each reaction must be *doubled* after this step.** With this information and Figure 18.12 in hand, we can now determine the total yield of ATP from the complete catabolism of glucose.

Figure 18.12

The ATP Yield from the Aerobic Metabolism of Glucose to CO_2

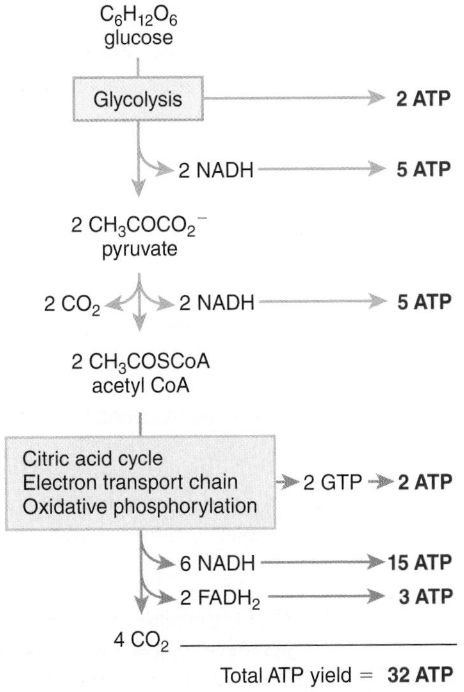

$C_6H_{12}O_6$ glucose

Glycolysis → **2 ATP**

→ 2 NADH → **5 ATP**

2 $CH_3COCO_2^-$ pyruvate

2 CO_2 ← → 2 NADH → **5 ATP**

2 $CH_3COSCoA$ acetyl CoA

Citric acid cycle
Electron transport chain
Oxidative phosphorylation → 2 GTP → **2 ATP**

→ 6 NADH → **15 ATP**

→ 2 $FADH_2$ → **3 ATP**

4 CO_2

Total ATP yield = **32 ATP**

The complete catabolism of glucose forms six CO_2 molecules and 32 ATP molecules.

- **Glycolysis yields a net of two ATP molecules. The two molecules of NADH formed during step [6] of glycolysis yield five additional ATPs.**

$$\underset{\text{glucose}}{C_6H_{12}O_6} \longrightarrow \underset{\text{pyruvate}}{2\ CH_3COCO_2^-} + 2\ ATP + 2\ NADH$$

2 NADH → $2 \times (2.5\ ATP/NADH) = 5\ ATP$

- **Oxidation of two molecules of pyruvate to acetyl CoA in the mitochondria forms two NADH molecules that yield five ATP molecules.**

$$\underset{\text{pyruvate}}{2\ CH_3COCO_2^-} \longrightarrow \underset{\text{acetyl CoA}}{2\ CH_3COSCoA} + 2\ CO_2 + 2\ NADH$$

2 NADH → $2 \times (2.5\ ATP/NADH) = 5\ ATP$

- **Beginning with two acetyl CoA molecules, the citric acid cycle (Figure 18.10) forms two GTP molecules, the energy equivalent of two ATPs, in step [5]. The six NADH molecules and two $FADH_2$ molecules also formed yield an additional 18 ATPs from the electron transport chain and oxidative phosphorylation. Thus, 20 ATPs are formed from two acetyl CoA molecules.**

$$\underset{\text{acetyl CoA}}{2\ CH_3COSCoA} \longrightarrow 4\ CO_2 + 2\ GTP + 6\ NADH + 2\ FADH_2$$

2 GTP → 2 ATP

6 NADH → $6 \times (2.5\ ATP/NADH) = 15\ ATP$

2 $FADH_2$ → $2 \times (1.5\ ATP/FADH_2) = 3\ ATP$

Adding up the ATP formed in each pathway gives a **total of 32 molecules of ATP for the complete catabolism of each glucose molecule.** Most of the ATP generated from glucose metabolism results from the citric acid cycle, electron transport chain, and oxidative phosphorylation.

Glucose is the main source of energy for cells and the only source of energy used by the brain. When energy demands are low, glucose is stored as the polymer glycogen in the liver and muscles. When blood levels of glucose are low, glycogen is hydrolyzed to keep adequate blood glucose levels to satisfy the body's energy needs.

Blood glucose levels are carefully regulated by two hormones. When blood glucose concentration rises after a meal, **insulin** stimulates the passage of glucose into cells for metabolism. When blood glucose levels are low, the hormone **glucagon** stimulates the conversion of stored glycogen to glucose.

PROBLEM 18.19

How much ATP results from each transformation?

a. glucose → 2 pyruvate
b. pyruvate → acetyl CoA
c. glucose → 2 acetyl CoA
d. 2 acetyl CoA → 4 CO_2

PROBLEM 18.20

What three reactions form CO_2 when glucose is completely catabolized?

PROBLEM 18.21

What three reactions form a nucleoside triphosphate (GTP or ATP) directly when glucose is completely catabolized?

PROBLEM 18.22

What is the difference in ATP generation between the aerobic oxidation of glucose to CO_2 and the anaerobic conversion of glucose to lactate?

Triacylglycerols Carbohydrates Proteins
Fatty acids + glycerol Monosaccharides Amino acids
Fatty acid oxidation Glycolysis Amino acid catabolism
Pyruvate
Acetyl CoA
Citric acid cycle
Reduced coenzymes
Electron transport chain and oxidative phosphorylation

18.9 The Catabolism of Triacylglycerols

The first step in the catabolism of triacylglycerols, the most common lipids, is the hydrolysis of the three ester bonds to form glycerol and fatty acids (Figure 18.3).

18.9A Fatty Acid Catabolism by β-Oxidation

Fatty acids are catabolized by **β-oxidation,** a process in which two-carbon acetyl CoA units are sequentially cleaved from the fatty acid. Key to this process is the oxidation of the β carbon to the carbonyl group of a thioester (RCOSR'), which then undergoes cleavage between the α and β carbons.

$$-CH_2-CH_2-CH_2-\underset{\beta}{CH_2}-\underset{\alpha}{CH_2}-\overset{O}{\overset{\|}{C}}-$$

This carbon is oxidized. These C's come off as acetyl CoA. This C—C bond is cleaved.

Fatty acid oxidation begins with conversion of the fatty acid to a thioester with coenzyme A. This process requires energy, which comes from the hydrolysis of two phosphate bonds in ATP to form AMP, adenosine *mono*phosphate. Much like the beginning of glycolysis requires an energy investment, so, too, the initial step of fatty acid oxidation requires energy input.

$$\underset{\text{stearic acid}}{CH_3(CH_2)_{14}-CH_2-CH_2-\overset{O}{\overset{\|}{C}}-OH} + HS-CoA \xrightarrow[\text{acyl CoA synthetase}]{ATP \quad AMP + 2\ P_i} \underset{\text{an acyl CoA}}{CH_3(CH_2)_{14}-CH_2-CH_2-\overset{O}{\overset{\|}{C}}-SCoA}$$

Once the product, an **acyl CoA,** is inside the mitochondrion, the process of β-oxidation is set to begin. β-Oxidation requires four steps to cleave a two-carbon acetyl CoA unit from the acyl CoA, as shown with the 18-carbon fatty acid stearic acid in Figure 18.13.

In step [1], FAD removes two hydrogen atoms to form $FADH_2$ and a double bond between the α and β carbons of the thioester. Water is added to the double bond in step [2] to place an OH group on the β carbon to the carbonyl group, which is then oxidized in step [3] to form a carbonyl group. The NAD^+ oxidizing agent is reduced to NADH in step [3] as well. Finally, cleavage of the bond between the α and β carbons forms acetyl CoA and a 16-carbon acyl CoA in step [4].

- **As a result, a new acyl CoA having two carbons fewer than the original acyl CoA is formed.**

The following equation summarizes the important components of β-oxidation for a general acyl CoA, $RCH_2CH_2COSCoA$. **Each four-step sequence forms one molecule each of acetyl CoA, NADH, and $FADH_2$.**

$$R-CH_2-CH_2-\overset{O}{\overset{\|}{C}}-SCoA + NAD^+ + FAD$$

$$\downarrow HSCoA + H_2O$$

$$R-\overset{O}{\overset{\|}{C}}-SCoA + CH_3-\overset{O}{\overset{\|}{C}}-SCoA + NADH + H^+ + FADH_2$$

Figure 18.13

β-Oxidation of a Fatty Acid

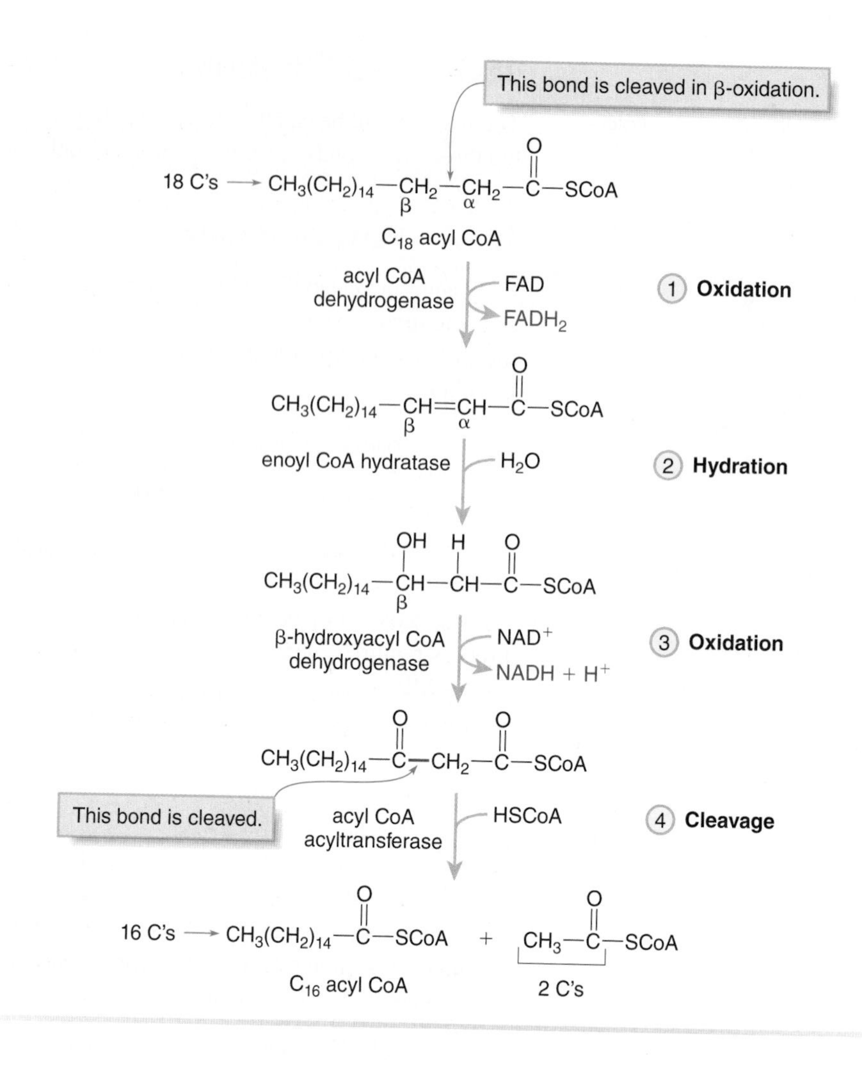

Once the 16-carbon acyl CoA is formed in step [4], it becomes the substrate for a new four-step β-oxidation sequence. The process continues until a four-carbon acyl CoA is cleaved to form *two* acetyl CoA molecules. As a result:

- **An 18-carbon acyl CoA is cleaved to *nine* two-carbon acetyl CoA molecules.**
- **A total of *eight* cycles of β-oxidation are needed to cleave the eight carbon–carbon bonds.**

$$CH_3CH_2-CH_2CH_2-CH_2CH_2-CH_2CH_2-CH_2CH_2-CH_2CH_2-CH_2CH_2-CH_2CH_2-CH_2-\overset{O}{\overset{\|}{C}}-SCoA$$

C_{18} acyl CoA

↓ → acetyl CoA

C_{16} acyl CoA

↓ → acetyl CoA

C_{14} acyl CoA

5 cycles ↓ → 5 acetyl CoA

C_4 acyl CoA

↓

2 acetyl CoA

All bonds in red are cleaved, one at a time, beginning near the carbonyl end of the acyl CoA.

Thus, complete β-oxidation of the acyl CoA derived from stearic acid forms:

- **9 $CH_3COSCoA$ molecules** (from the 18-carbon fatty acid)
- **8 NADH** (from eight cycles of β-oxidation)
- **8 $FADH_2$** (from eight cycles of β-oxidation)

β-Oxidation of unsaturated fatty acids proceeds in a similar fashion, although an additional step is required. **Ultimately every carbon in the original fatty acid ends up as a carbon atom of acetyl CoA.**

SAMPLE PROBLEM 18.3

Consider lauric acid, $CH_3(CH_2)_{10}CO_2H$. (a) How many molecules of acetyl CoA are formed from complete β-oxidation? (b) How many cycles of β-oxidation are needed for complete catabolism?

Analysis

The number of carbons in the fatty acid determines the number of molecules of acetyl CoA formed and the number of times β-oxidation occurs.

- The number of molecules of acetyl CoA equals one-half the number of carbons in the original fatty acid.
- Because the final turn of the cycle forms *two* molecules of acetyl CoA, the number of cycles is one fewer than the number of acetyl CoA molecules formed.

Solution

Since lauric acid has 12 carbons, it forms six molecules of acetyl CoA from five cycles of β-oxidation.

PROBLEM 18.23

For each fatty acid: [1] How many molecules of acetyl CoA are formed from complete catabolism? [2] How many cycles of β-oxidation are needed for complete oxidation?

a. arachidic acid ($C_{20}H_{40}O_2$) b. palmitoleic acid ($C_{16}H_{30}O_2$)

18.9B The Energy Yield from Fatty Acid Oxidation

How much energy—in terms of the number of molecules of ATP formed—results from the complete catabolism of a fatty acid? To determine this quantity, we must take into account the ATP cost for the conversion of the fatty acid to the acyl CoA, as well as the ATP production from the coenzymes (NADH and $FADH_2$) and acetyl CoA formed during β-oxidation. The steps are shown in the accompanying *How To* procedure.

How To Determine the Number of Molecules of ATP Formed from a Fatty Acid

Example How much ATP is formed by the complete catabolism of stearic acid, $C_{18}H_{36}O_2$?

Step [1] Determine the amount of ATP required to synthesize the acyl CoA from the fatty acid.

- Since the conversion of stearic acid ($C_{17}H_{35}COOH$) to an acyl CoA ($C_{17}H_{35}COSCoA$) requires the hydrolysis of two phosphate bonds, this is equivalent to the energy released when 2 ATPs are converted to 2 ADPs.
- Thus, the **first step in catabolism costs the equivalent of 2 ATPs—that is, −2 ATPs.**

Step [2] Add up the ATP generated from the coenzymes produced during β-oxidation.

- As we learned in Section 18.9A, each cycle of β-oxidation produces one molecule each of NADH and $FADH_2$. To cleave eight carbon–carbon bonds in stearic acid requires eight cycles of β-oxidation, so 8 NADH and 8 $FADH_2$ are produced.

$$8 \text{ NADH} \times 2.5 \text{ ATP/NADH} = 20 \text{ ATP}$$

$$8 \text{ FADH}_2 \times 1.5 \text{ ATP/FADH}_2 = 12 \text{ ATP}$$

$$\text{From reduced coenzymes: } 32 \text{ ATP}$$

- Thus, 32 ATPs would be produced from oxidative phosphorylation after the reduced coenzymes enter the electron transport chain.

—Continued

How To, continued . . .

Step [3] Determine the amount of ATP that results from each acetyl CoA, and add the results for steps [1]–[3].

- From Section 18.9A, stearic acid generates nine molecules of acetyl CoA, which then enter the citric acid cycle and go on to produce ATP by the electron transport chain and oxidative phosphorylation. As we learned in Section 18.7C, each acetyl CoA results in 10 ATPs.

$$9 \text{ acetyl CoA} \times 10 \text{ ATP/acetyl CoA} = 90 \text{ ATP}$$

- Totaling the values obtained in steps [1]–[3]:

$$(-2) + 32 + 90 = 120 \text{ ATP molecules from stearic acid}$$

Answer

PROBLEM 18.24

Calculate the number of molecules of ATP formed by the complete catabolism of palmitic acid, $C_{16}H_{32}O_2$.

PROBLEM 18.25

Calculate the number of molecules of ATP formed by the complete catabolism of arachidic acid, $C_{20}H_{40}O_2$.

18.10 Ketone Bodies

When carbohydrates do not meet energy needs, the body turns to catabolizing stored triacylglycerols, which generate acetyl CoA by β-oxidation of fatty acids. Normally the acetyl CoA is metabolized in the citric acid cycle. When acetyl CoA levels exceed the capacity of the citric acid cycle, however, acetyl CoA is converted to three compounds that are collectively called **ketone bodies**—acetoacetate, β-hydroxybutyrate, and acetone.

$$2\ CH_3-\overset{O}{\overset{\|}{C}}-SCoA \xrightarrow{\text{several steps}} \underset{\text{acetoacetate}}{CH_3-\overset{O}{\overset{\|}{C}}-CH_2-\overset{O}{\overset{\|}{C}}-O^-} \xrightarrow{NADH + H^+ \rightarrow NAD^+} \underset{\beta\text{-hydroxybutyrate}}{CH_3-\overset{OH}{\overset{|}{\underset{H}{\underset{|}{C}}}}-CH_2-\overset{O}{\overset{\|}{C}}-O^-}$$

$$\text{acetoacetate} \xrightarrow{-CO_2} \underset{\text{acetone}}{CH_3-\overset{O}{\overset{\|}{C}}-CH_3}$$

- *Ketogenesis* is the synthesis of ketone bodies from acetyl CoA.

HEALTH NOTE

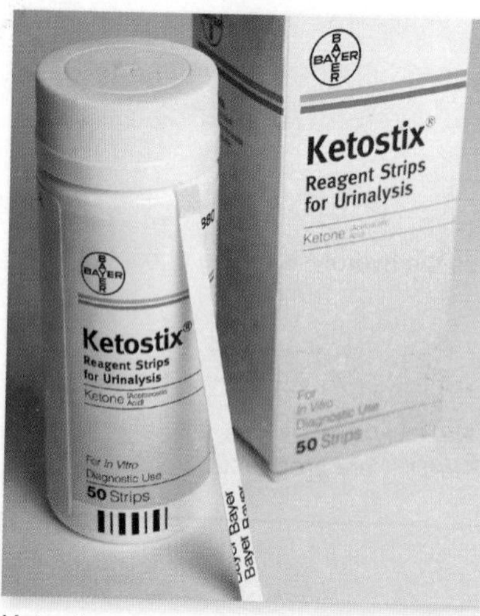

Ketostix are a brand of test strips that detect ketone bodies in urine.

Ketone bodies are produced in the liver, and since they are small molecules that can hydrogen bond with water, they are readily soluble in blood and urine. Once they reach tissues, β-hydroxybutyrate and acetoacetate can be re-converted to acetyl CoA and metabolized for energy.

Under some circumstances—notably starvation, vigorous dieting, and uncontrolled diabetes—when glucose is unavailable or cannot pass into a cell for use as fuel, ketone bodies accumulate, a condition called **ketosis.** As a result, ketone bodies are eliminated in urine and the sweet odor of acetone can be detected in exhaled breath. Sometimes the first indication of diabetes in a patient is the detection of excess ketone bodies in a urine test.

HEALTH NOTE

Low carbohydrate diets such as the Atkins plan induce the use of stored fat for energy production to assist weight loss.

An abnormally high concentration of ketone bodies can lead to **ketoacidosis**—that is, a lowering of the blood pH caused by the increased level of β-hydroxybutyrate and acetoacetate.

Low carbohydrate diets, popularized by Dr. Robert Atkins in a series of diet books published in the 1990s, restrict carbohydrate intake to induce the utilization of the body's stored fat as its main energy source. This increased level of fatty acid metabolism leads to an increased concentration of ketone bodies in the blood and urine.

PROBLEM 18.26

Are any structural features common to the three ketone bodies?

PROBLEM 18.27

Why is the term "ketone body" a misleading name for β-hydroxybutyrate?

18.11 Amino Acid Metabolism

After proteins are hydrolyzed in the stomach and intestines, the individual amino acids are reassembled to new proteins or converted to intermediates in other metabolic pathways. The catabolism of amino acids provides energy when the supply of carbohydrates and lipids is exhausted.

The catabolism of amino acids can be conceptually divided into two parts: the fate of the amino group and the fate of the carbon atoms. As shown in Figure 18.14, amino acid carbon skeletons are converted to pyruvate, acetyl CoA, or various carbon compounds that are part of the citric acid cycle.

Figure 18.14 An Overview of the Catabolism of Amino Acids

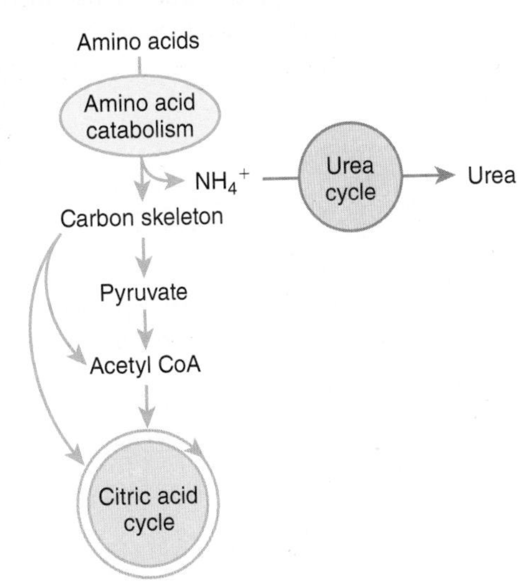

The breakdown of amino acids forms NH_4^+, which enters the urea cycle to form urea, and a carbon skeleton that is metabolized to either pyruvate, acetyl CoA, or an intermediate in the citric acid cycle.

18.11A Degradation of Amino Acids—The Fate of the Amino Group

The catabolism of carbohydrates and triacylglycerols deals with the oxidation of carbon atoms only. With amino acids, an **amino group ($-NH_2$)** must be metabolized, as well. The catabolism of amino acids begins with the removal of the amino group from the carbon skeleton to form an α-keto acid and an ammonium ion (NH_4^+) by a stepwise procedure.

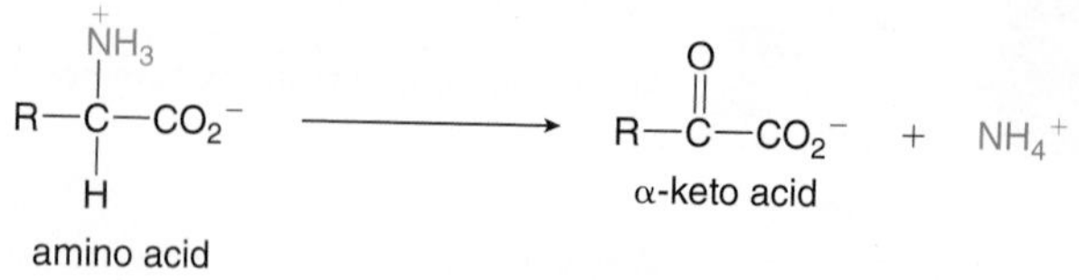

$$R-\underset{H}{\overset{\overset{+}{N}H_3}{C}}-CO_2^- \longrightarrow R-\overset{O}{\overset{\|}{C}}-CO_2^- + NH_4^+$$

amino acid → α-keto acid

For example, removal of the amino group from alanine forms pyruvate and NH_4^+.

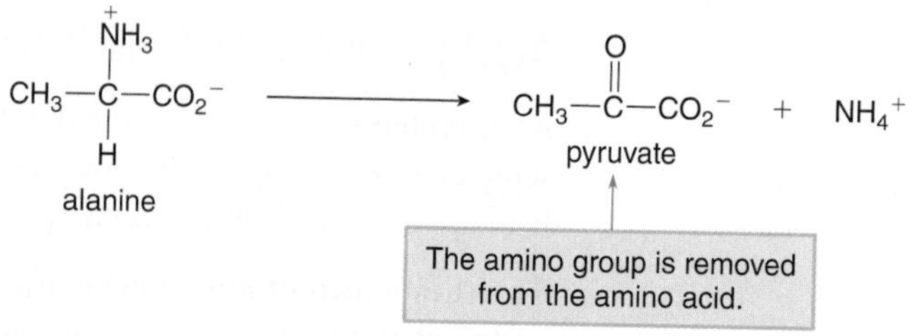

$$CH_3-\underset{H}{\overset{\overset{+}{N}H_3}{C}}-CO_2^- \longrightarrow CH_3-\overset{O}{\overset{\|}{C}}-CO_2^- + NH_4^+$$

alanine → pyruvate

The amino group is removed from the amino acid.

α-Keto acids like pyruvate are then degraded along the catabolic pathways described in Section 18.11B. The ammonium ion (NH_4^+) enters the **urea cycle,** where it is converted to urea, $(NH_2)_2C{=}O$, in the liver. Urea is then transported to the kidneys and eliminated in urine.

SAMPLE PROBLEM 18.4

What products are formed when the amino group of leucine is removed during the early stages of amino acid catabolism?

$$(CH_3)_2CHCH_2-\underset{H}{\overset{\overset{+}{N}H_3}{C}}-CO_2^-$$

leucine

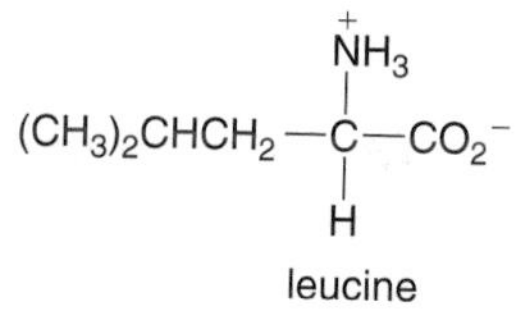

Analysis

To draw the organic product, replace the C—H and C—NH_3^+ on the α carbon of the amino acid by C=O. NH_4^+ is formed from the amino group.

Solution

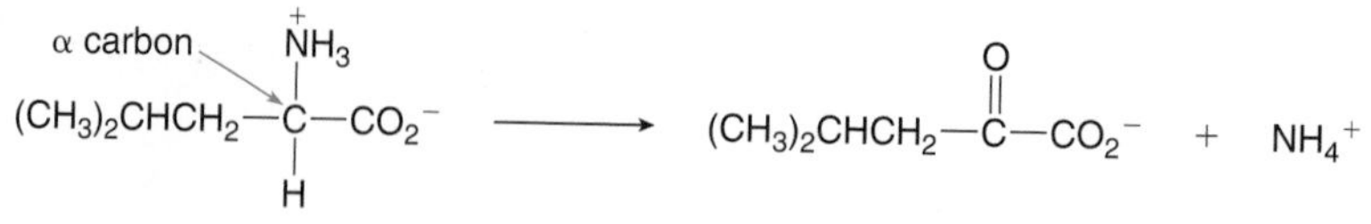

α carbon

$$(CH_3)_2CHCH_2-\underset{H}{\overset{\overset{+}{N}H_3}{C}}-CO_2^- \longrightarrow (CH_3)_2CHCH_2-\overset{O}{\overset{\|}{C}}-CO_2^- + NH_4^+$$

PROBLEM 18.28

What products are formed when the amino group of each amino acid is removed during the early stages of amino acid catabolism: (a) threonine; (b) glycine; (c) isoleucine? Use the structures in Table 16.2.

18.11B Degradation of Amino Acids—The Fate of the Carbon Skeleton

Once the nitrogen has been removed from an amino acid, the carbon skeletons of individual amino acids are catabolized in a variety of ways. There are three common fates of the carbon skeletons of amino acids, shown in Figure 18.15:

- **conversion to pyruvate, $CH_3COCO_2^-$**
- **conversion to acetyl CoA, $CH_3COSCoA$**
- **conversion to an intermediate in the citric acid cycle**

Some amino acids such as alanine (Section 18.11A) are catabolized to pyruvate. Pyruvate can be broken down for energy or used to synthesize glucose. In considering catabolism, amino acids are often divided into two groups.

- ***Glucogenic* amino acids are catabolized to pyruvate or an intermediate in the citric acid cycle. Glucogenic amino acids can be used to synthesize glucose.**

- ***Ketogenic* amino acids are converted to acetyl CoA, or the related thioester acetoacetyl CoA, $CH_3COCH_2COSCoA$. These catabolic products cannot be used to synthesize glucose, but they can be converted to ketone bodies and yield energy by this path.**

We will not examine the specific pathways that convert the carbon skeletons of individual amino acids into other products. Figure 18.15 illustrates where each amino acid feeds into the metabolic pathways we have already discussed.

PROBLEM 18.29

What metabolic intermediate is produced from the carbon atoms of each amino acid?

a. cysteine b. aspartic acid c. valine d. threonine

PROBLEM 18.30

Which amino acid(s) in Problem 18.29 are ketogenic?

Figure 18.15

Amino Acid Catabolism

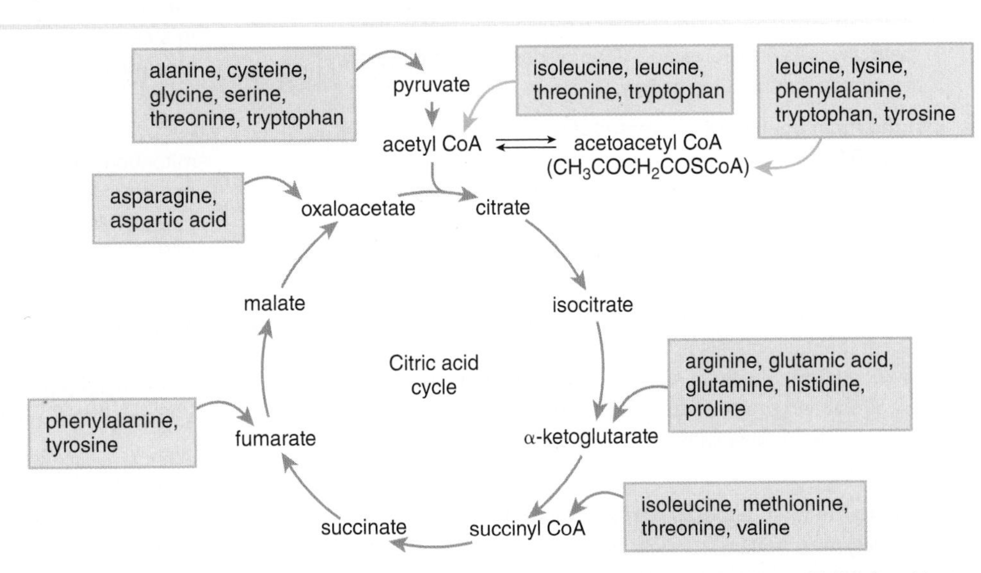

Glucogenic amino acids are highlighted in blue, while ketogenic amino acids are highlighted in tan. Amino acids that appear more than once in the scheme can be degraded by multiple routes.

KEY TERMS

Acetyl CoA (18.3)
Acyl CoA (18.9)
Adenosine 5'-diphosphate (ADP, 18.2)
Adenosine 5'-triphosphate (ATP, 18.2)
Anabolism (18.1)
Catabolism (18.1)
Citric acid cycle (18.6)
Coenzyme A (18.3)
Electron transport chain (18.7)
Fermentation (18.5)
Flavin adenine dinucleotide (FAD, 18.3)
Glucogenic amino acid (18.11)
Glycolysis (18.4)
Ketogenesis (18.10)
Ketogenic amino acid (18.11)
Ketone bodies (18.10)
Ketosis (18.10)
Metabolism (18.1)
Mitochondrion (18.1)
NADH (18.3)
Nicotinamide adenine dinucleotide (NAD^+, 18.3)
β-Oxidation (18.9)
Oxidative phosphorylation (18.7)
Phosphorylation (18.2)
Urea cycle (18.11)

KEY CONCEPTS

1 What is metabolism and what are the four stages of metabolism? (18.1)

- Metabolism is the sum of all of the chemical reactions that take place in an organism. Catabolic reactions break down large molecules and release energy, while anabolic reactions synthesize larger molecules and require energy.
- Metabolism begins with digestion in stage [1], in which large molecules are hydrolyzed to smaller molecules.
- In stage [2], biomolecules are degraded into two-carbon acetyl units.
- The citric acid cycle (stage [3]) converts two carbon atoms to two molecules of CO_2, and forms reduced coenzymes, NADH and $FADH_2$.
- In stage [4], the electron transport chain and oxidative phosphorylation produce ATP, and oxygen is converted to water.

2 What is ATP and how do reactions utilize ATP to drive energetically unfavorable reactions? (18.2)

- ATP is the primary energy-carrying molecule in metabolic pathways. The hydrolysis of ATP provides the energy to drive a reaction that requires energy.

3 List the main coenzymes in metabolism and describe their roles. (18.3)

- Nicotinamide adenine dinucleotide (NAD^+) is a biological oxidizing agent that accepts electrons and protons, thus generating its reduced form NADH. NADH is a reducing agent that donates electrons and protons, re-forming NAD^+.
- Flavin adenine dinucleotide (FAD) is a biological oxidizing agent that accepts electrons and protons, thus yielding its reduced form, $FADH_2$. $FADH_2$ is a reducing agent that donates electrons and protons, re-forming FAD.
- Coenzyme A reacts with acetyl groups (CH_3CO–) to form acetyl CoA.

4 What are the main aspects of glycolysis? (18.4)

- Glycolysis is a linear, 10-step pathway that converts glucose to two three-carbon pyruvate molecules.
- The net result of glycolysis is 2 $CH_3COCO_2^-$, 2 NADHs, and 2 ATPs.

5 What are the major pathways for pyruvate metabolism? (18.5)

- When oxygen is plentiful, pyruvate is converted to acetyl CoA, which can enter the citric acid cycle.
- When the oxygen level is low, the anaerobic metabolism of pyruvate forms lactate and NAD^+.
- In yeast and other microorganisms, pyruvate is converted to ethanol and CO_2 by fermentation.

6 What are the main features of the citric acid cycle? (18.6)

- The citric acid cycle is an eight-step cyclic pathway that begins with the addition of acetyl CoA to a four-carbon substrate. In the citric acid cycle, two carbons are converted to CO_2 and four molecules of reduced coenzymes are formed. One molecule of a high-energy nucleoside triphosphate is also formed.

7 What are the main components of the electron transport chain and oxidative phosphorylation? (18.7)

- Electrons from reduced coenzymes enter the electron transport chain and are passed from one molecule to another in a series of redox reactions. At the end of the chain, electrons and protons react with inhaled oxygen to form water.
- H^+ ions are pumped across the inner membrane of the mitochondrion, forming a high concentration of H^+ ions in the intermembrane space, thus creating a potential energy gradient. When the H^+ ions travel through the channel in the ATP synthase enzyme, this energy is used to convert ADP to ATP—a process called oxidative phosphorylation.

8 How much ATP is formed by the complete catabolism of glucose? (18.8)

- To calculate the amount of ATP formed in the catabolism of glucose, we must take into account the ATP yield from glycolysis, the oxidation of two molecules of pyruvate to two molecules of acetyl CoA, the citric acid cycle, and oxidative phosphorylation.
- As shown in Figure 18.12, the complete catabolism of glucose forms six CO_2 molecules and 32 molecules of ATP.

9 Describe the main features of the β-oxidation of fatty acids. (18.9)

- β-Oxidation cleaves two-carbon acetyl CoA units from an acyl CoA derived from a fatty acid. Each cycle of β-oxidation consists of a four-step sequence that forms one molecule each of acetyl CoA, NADH, and $FADH_2$.
- To determine the ATP yield from the complete catabolism of a fatty acid, we must consider the ATP used up in the synthesis of the acyl CoA, the ATP generated from coenzymes produced during β-oxidation, and the ATP that results from the catabolism of each acetyl CoA.

10 What are ketone bodies and how do they play a role in metabolism? (18.10)

- Ketone bodies—acetoacetate, β-hydroxybutyrate, and acetone—are formed when acetyl CoA levels exceed the capacity of the citric acid cycle. Ketone bodies can be re-converted to acetyl CoA and metabolized for energy.

11 What are the main features of amino acid catabolism? (18.11)

- The catabolism of amino acids forms an α-keto acid and NH_4^+. The NH_4^+ ion enters the urea cycle where it is converted to urea and eliminated in urine. The carbon skeletons of the amino acids are catabolized by a variety of pathways to yield pyruvate, acetyl CoA, or an intermediate in the citric acid cycle.

UNDERSTANDING KEY CONCEPTS

Selected in-chapter and odd-numbered end-of-chapter problems have brief answers at the end of each chapter. The *Student Study Guide and Solutions Manual* contains detailed solutions to all in-chapter and odd-numbered end-of-chapter problems, as well as additional worked examples and a chapter self-test.

18.31 In what stage of catabolism does each of the following processes occur?
- a. cleavage of a protein with chymotrypsin
- b. oxidation of a fatty acid to acetyl CoA
- c. oxidation of malate to oxaloacetate with NAD^+
- d. conversion of ADP to ATP with ATP synthase
- e. hydrolysis of starch to glucose with amylase

18.32 In what stage of catabolism does each of the following processes occur?
- a. conversion of a monosaccharide to acetyl CoA
- b. hydrolysis of a triacylglycerol with lipase
- c. reaction of oxygen with protons and electrons to form water
- d. conversion of succinate to fumarate with FAD
- e. degradation of a fatty acid to acetyl CoA

18.33 When a substrate is oxidized, is NAD^+ oxidized or reduced? Is NAD^+ an oxidizing agent or a reducing agent?

18.34 When a substrate is reduced, is $FADH_2$ oxidized or reduced? Is $FADH_2$ an oxidizing agent or a reducing agent?

18.35 What steps in the citric acid cycle have each of the following characteristics?
- a. The reaction generates NADH.
- b. CO_2 is removed.
- c. The reaction utilizes FAD.
- d. The reaction forms a new carbon–carbon single bond.

18.36 What steps in the citric acid cycle have each of the following characteristics?
- a. The reaction generates $FADH_2$.
- b. The organic substrate is oxidized.
- c. The reaction utilizes NAD^+.
- d. The reaction breaks a carbon–carbon bond.

18.37 Compare the energy-investment phase and the energy-generating phase of glycolysis with regards to each of the following: (a) the reactant that begins the phase and the final product formed; (b) the amount of ATP used or formed; (c) the number of reduced coenzymes used or formed.

18.38 Write the overall equation with key coenzymes for each process.
- a. glucose → pyruvate
- b. glucose → ethanol
- c. pyruvate → lactate

18.39 (a) How many molecules of acetyl CoA are formed from complete β-oxidation of the fatty acid depicted in the ball-and-stick model? (b) How many cycles of β-oxidation are needed for complete oxidation?

18.40 (a) How many molecules of acetyl CoA are formed from complete β-oxidation of the fatty acid depicted in the ball-and-stick model? (b) How many cycles of β-oxidation are needed for complete oxidation?

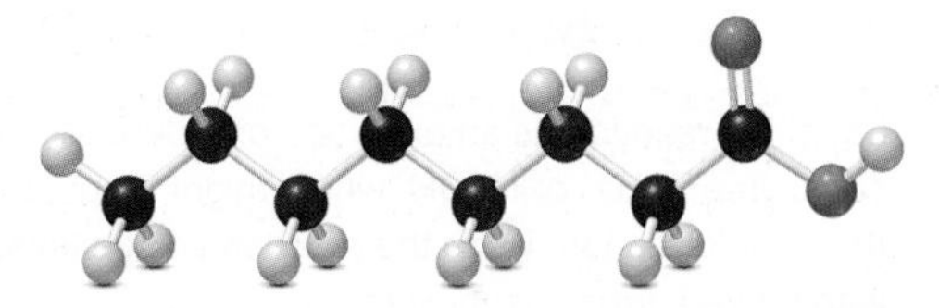

18.41 What products are formed when the amino group is removed from methionine in the early stages of amino acid catabolism?

18.42 What products are formed when the amino group is removed from tyrosine in the early stages of amino acid catabolism?

18.43 How might pyruvate be metabolized in the cornea, tissue that has little blood supply?

18.44 How is pyruvate metabolized in red blood cells, which contain no mitochondria?

ADDITIONAL PROBLEMS

Metabolism

18.45 What is the difference between catabolism and anabolism?

18.46 What is the difference between metabolism and digestion?

18.47 Describe the main features of a mitochondrion. Where does energy production occur in a mitochondrion?

18.48 Explain why mitochondria are called the powerhouses of the cell.

18.49 Place the following steps in the catabolism of carbohydrates in order: the electron transport chain, the conversion of glucose to acetyl CoA, the hydrolysis of starch, oxidative phosphorylation, and the citric acid cycle.

18.50 Describe the main features of the four stages of catabolism.

ATP and Coenzymes

18.51 (a) Using curved arrow symbolism, write the equation for the reaction of fructose 1,6-bisphosphate with ADP to form fructose 6-phosphate and ATP. (b) If the energy change in this reaction is +3.4 kcal/mol, does this reaction absorb or release energy? (c) Is this reaction energetically favorable?

18.52 (a) Using curved arrow symbolism, write the equation for the reaction of glucose with ATP to form glucose 1-phosphate and ADP. (b) If this reaction releases 2.3 kcal/mol of energy, is the energy change reported as a positive (+) or negative (–) value?

18.53 Classify each substance as an oxidizing agent, a reducing agent, or neither: (a) $FADH_2$; (b) ATP; (c) NAD^+.

18.54 Classify each substance as an oxidizing agent, a reducing agent, or neither: (a) NADH; (b) ADP; (c) FAD.

18.55 Label the reaction as an oxidation or reduction and give the coenzyme, NAD^+ or NADH, which might be used to carry out the transformation. Write the reaction using skeletal structures with curved arrow symbolism.

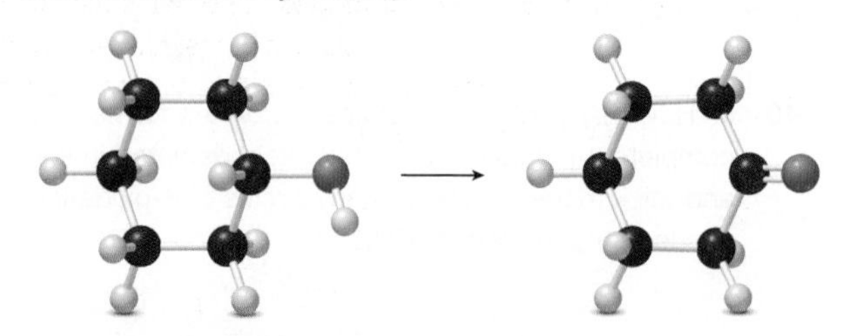

18.56 Label the reaction as an oxidation or reduction and give the coenzyme, NAD^+ or NADH, which might be used to carry out the transformation. Write the reaction using skeletal structures with curved arrow symbolism.

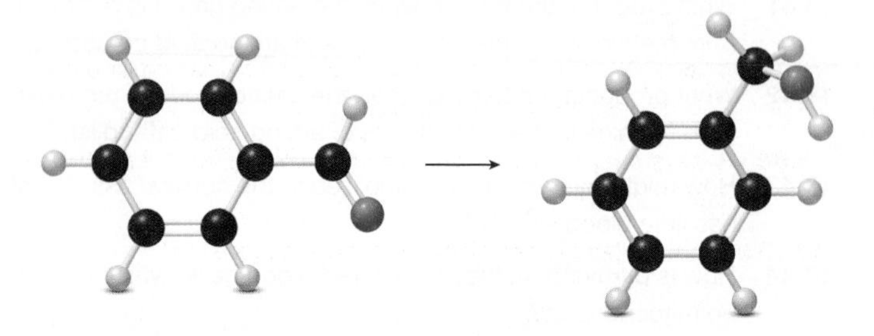

Glucose Metabolism

18.57 Considering the individual steps in glycolysis:
a. Which steps form ATP?
b. Which steps use ATP?
c. Which steps form a reduced coenzyme?
d. Which step breaks a C—C bond?

18.58 Explain the role of the coenzymes NAD^+ and NADH in each reaction.
a. pyruvate → acetyl CoA
b. pyruvate → lactate
c. pyruvate → ethanol

18.59 Glucose is completely metabolized to six molecules of CO_2. What specific reactions generate each molecule of CO_2?

18.60 Why is glycolysis described as an anaerobic process?

18.61 Consider the aerobic and anaerobic avenues of pyruvate metabolism in the human body.
a. Where do the carbon atoms of pyruvate end up in each pathway?
b. What coenzymes are used and formed?

18.62 What is the main purpose for the conversion of pyruvate to lactate under anaerobic conditions?

18.63 What metabolic products are formed from pyruvate in each case: (a) anaerobic conditions in the body; (b) anaerobic conditions in yeast; (c) aerobic conditions?

18.64 Why must the NADH produced in glycolysis be re-oxidized to NAD^+? How is this accomplished aerobically? How is this accomplished anaerobically?

18.65 Explain in detail how 32 ATP molecules are generated during the complete catabolism of glucose to CO_2.

18.66 In fermentation, where do the six carbon atoms of glucose end up?

Citric Acid Cycle, Electron Transport Chain, and Oxidative Phosphorylation

18.67 (a) Which intermediate(s) in the citric acid cycle contain two chirality centers? (b) Which intermediate(s) contain a 2° alcohol?

18.68 What products of the citric acid cycle are funneled into the electron transport chain?

18.69 Why are the reactions that occur in stage [4] of catabolism sometimes called aerobic respiration?

18.70 What is the role of each of the following in the electron transport chain: (a) NADH; (b) O_2; (c) complexes I–IV; (d) H^+ ion channel?

18.71 What is the role of each of the following in the electron transport chain: (a) $FADH_2$; (b) ADP; (c) ATP synthase; (d) the inner mitochondrial membrane?

18.72 Explain the importance of the movement of H^+ ions across the inner mitochondrial membrane and then their return passage through the H^+ ion channel in ATP synthase.

18.73 What are the final products of the electron transport chain?

18.74 What product is formed from each of the following compounds during the electron transport chain: (a) NADH; (b) $FADH_2$; (c) ADP; (d) O_2?

18.75 Why does one NADH that enters the electron transport chain ultimately produce 2.5 ATPs, while one $FADH_2$ produces 1.5 ATPs?

18.76 How does the energy from the proton gradient result in ATP synthesis?

Triacylglycerol Metabolism

18.77 How much ATP is used or formed when a fatty acid is converted to an acyl CoA? Explain your reasoning.

18.78 How much ATP is ultimately generated from each cycle of β-oxidation of a fatty acid?

18.79 How many molecules of acetyl CoA are formed from complete β-oxidation of myristic acid, $C_{13}H_{27}CO_2H$? How many cycles of β-oxidation are needed for complete oxidation?

18.80 How many molecules of acetyl CoA are formed from complete β-oxidation of oleic acid, $C_{17}H_{33}CO_2H$? How many cycles of β-oxidation are needed for complete oxidation?

18.81 How much ATP is generated by the complete catabolism of myristic acid in Problem 18.79?

18.82 How much ATP is generated by the complete catabolism of oleic acid in Problem 18.80?

18.83 Consider decanoic acid, $C_9H_{19}CO_2H$.

a. Label the α and β carbons.
b. Draw the acyl CoA derived from this fatty acid.
c. How many acetyl CoA molecules are formed by complete β-oxidation?
d. How many cycles of β-oxidation are needed for complete oxidation?
e. How many molecules of ATP are formed from the complete catabolism of this fatty acid?

18.84 Answer Problem 18.83 for docosanoic acid, $C_{21}H_{43}CO_2H$.

Ketone Bodies

18.85 What is the difference between ketosis and ketogenesis?

18.86 How is the production of ketone bodies related to ketoacidosis?

18.87 Why are more ketone bodies produced in an individual whose diabetes is poorly managed?

18.88 Why do some individuals use test strips to measure the presence of ketone bodies in their urine?

Amino Acid Metabolism

18.89 Draw the structure of the α-keto acid formed by removal of the amino group during the catabolism of each amino acid: (a) glycine; (b) phenylalanine.

18.90 Draw the structure of the α-keto acid formed by removal of the amino group during the catabolism of each amino acid: (a) tyrosine; (b) asparagine.

18.91 What metabolic intermediate is formed from the carbon skeleton of each amino acid?

a. phenylalanine
b. glutamic acid
c. asparagine
d. glycine

18.92 What metabolic intermediate is formed from the carbon skeleton of each amino acid?

a. lysine
b. tryptophan
c. methionine
d. serine

18.93 What is the difference between ketogenic and glucogenic amino acids?

18.94 Can an amino acid be considered both glucogenic and ketogenic? Explain your choice.

General Questions and Applications

18.95 (a) Convert the ball-and-stick model of **A,** one of the intermediates in glycolysis, to a structural formula, and name the compound. (b) What is the immediate precursor of **A** in glycolysis; that is, what compound forms **A** as a reaction product? (c) What compound is formed from **A** in glycolysis?

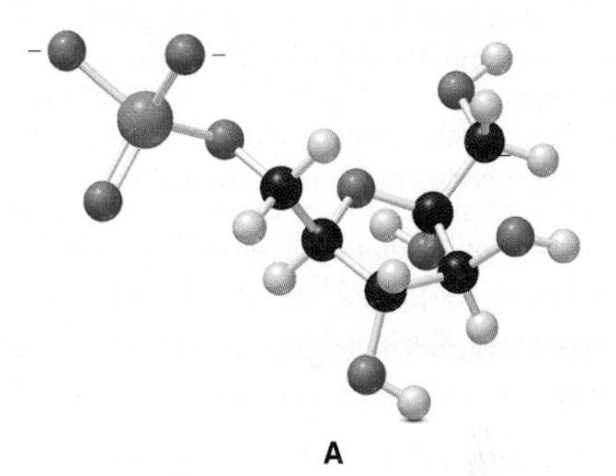

A

18.96 (a) Convert the ball-and-stick model of **B,** one of the eight synthetic intermediates in the citric acid cycle, to a structural formula, and name the compound. (b) What is the immediate precursor of **B** in the citric acid cycle; that is, what compound forms **B** as a reaction product? (c) What compound is formed from **B** in the citric acid cycle?

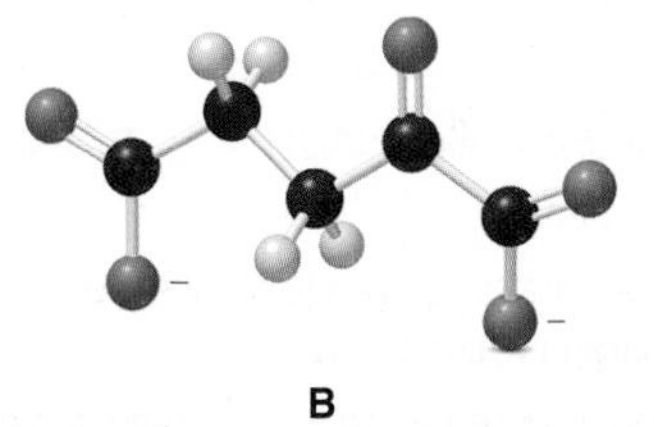

B

18.97 What is the cause of the pain and cramping in a runner's muscles?

18.98 Explain the reaction that occurs during the baking of bread that causes the bread to rise.

18.99 Why is the Atkins low carbohydrate diet called a ketogenic diet?

18.100 What metabolic conditions induce ketogenesis?

CHALLENGE PROBLEMS

18.101 How many moles of ATP can be synthesized from ADP using the 500. Calories ingested during a fast-food lunch? How many molecules of ATP does this correspond to?

18.102 Determine the amount of ATP generated per gram of glucose (molar mass 180.2 g/mol) compared to the amount of ATP generated per gram of stearic acid (284 g/mol) during catabolism. Does your result support or refute the fact that lipids are more effective energy-storing molecules than carbohydrates?

BEYOND THE CLASSROOM

18.103 Examine the nutrition label of a favorite breakfast cereal. Determine the number of grams of digestible carbohydrates in a serving by subtracting the number of grams of dietary fiber (due to indigestible cellulose) from the total carbohydrate content. Assume this value gives the number of grams of glucose per serving. How many moles of ATP are formed from the glucose in one serving of cereal? (Molar mass of glucose = 180.2 g/mol.)

18.104 We have now discussed diabetes in Chapters 16 (Section 16.6) and 18 (Section 18.8). What is the difference between type I and type II diabetes? What causes each type and what treatment options are typically used? What dietary recommendations are given to individuals with each type of diabetes?

18.105 Many diet plans (Weight Watchers, Nutrisystem, South Beach Diet) are advertised to help individuals lose weight. Pick one program and discuss how the regimen proposes to foster weight loss. Besides restricting Calorie intake, is there any biochemical basis for the weight loss plan? Many fat-burning dietary supplements are now available. Is there clear research data that any of these supplements aid in weight loss, and if so, how does the supplement work?

18.106 Many examples of interactions between foods, herbal supplements, and various medications are now well known. Because chemicals in food or herbal remedies can alter the activity of liver enzymes or drug delivery systems in the body, patients are sometimes advised to avoid a certain item in their diet while taking a medication. As an example, individuals who take the cholesterol-lowering drug atorvastatin (trade name Lipitor) are told to avoid grapefruit juice. Research what is known on the molecular level about the grapefruit juice effect. Is a similar effect observed with other cholesterol-lowering medications such as simvastatin (Zocor), lovastatin (Mevacor), and ezetimibe (Zetia)? Can you find other examples of how foods and herbal supplements affect the properties of a prescription medication?

ANSWERS TO SELECTED PROBLEMS

18.1 This makes it possible to use the same reactions and the same enzyme systems to metabolize all types of biomolecules.

18.3 fructose —(ATP → ADP)→ fructose 6-phosphate

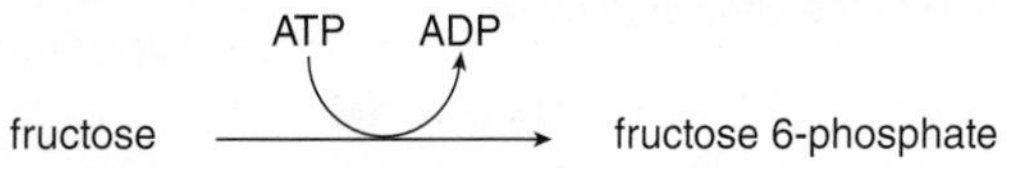

18.5 a. reduction using NADH b. oxidation using NAD^+

18.7 water soluble

18.9 step [1]: glucose; step [3]: fructose 6-phosphate; step [7]: ADP; step [10]: ADP

18.11 NADH is the reducing agent that converts pyruvate to both lactate and ethanol.

18.13 a. Both conversions use NADH.
b. One process involves decarboxylation and one process involves reduction.

18.14 a. $^-O_2C-CH_2-CH(OH)-CO_2^-$ —(NAD^+ → $NADH + H^+$)→ $^-O_2C-CH_2-C(=O)-CO_2^-$
b. oxidation

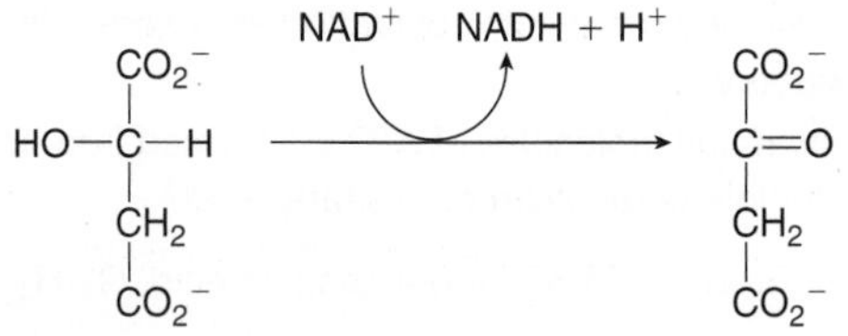

18.15 a. A: (H)(CO_2^-)C=C(H)(^-O_2C) — fumarate b. succinate c. malate

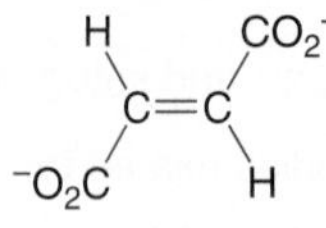

18.17 a. reduction b. oxidizing agent

18.19 a. 7 b. 2.5 for each pyruvate c. 12 d. 20

18.21 Steps [7] and [10] of the glycolysis pathway and step [5] of the citric acid cycle form ATP or GTP directly.

18.23 a. [1] 10; [2] 9 cycles b. [1] 8; [2] 7 cycles

18.25 134

18.27 β-Hydroxybutyrate contains an alcohol and a carboxylate anion, but no ketone.

18.28 a. $CH_3CH(OH)-C(=O)-CO_2^- \; + \; NH_4^+$

b. $H-C(=O)-CO_2^- \; + \; NH_4^+$

c. $CH_3CH_2CH(CH_3)-C(=O)-CO_2^- \; + \; NH_4^+$

18.29 a. pyruvate b. oxaloacetate c. succinyl CoA d. acetyl CoA, pyruvate, succinyl CoA

18.31 a. stage [1] b. stage [2] c. stage [3] d. stage [4] e. stage [1]

18.33 When a substrate is oxidized, NAD^+ is reduced. NAD^+ is an oxidizing agent.

18.35 a. steps [3], [4], and [8] b. steps [3] and [4] c. step [6] d. step [1]

18.37 Energy-investment phase:
a. glucose, glyceraldehyde 3-phosphate
b. 2 ATP utilized
c. no coenzymes used or formed
Energy-generating phase:
a. glyceraldehyde 3-phosphate, pyruvate
b. 4 ATP/glucose produced
c. 2 NAD^+ used and 2 NADH produced per glucose

18.39 3 molecules of acetyl CoA 2 cycles

18.41 $CH_3SCH_2CH_2-C(=O)-CO_2^- \; + \; NH_4^+$

18.43 Pyruvate is metabolized anaerobically to lactate.

18.45 Catabolism is the breakdown of large molecules into smaller ones and anabolism is the synthesis of large molecules from smaller ones.

18.47 Mitochondria contain an outer membrane and an inner membrane with many folds. The area between these two membranes is called the intermembrane space. Energy production occurs within the matrix, the area surrounded by the inner membrane.

18.49 hydrolysis of starch, the conversion of glucose to acetyl CoA, the citric acid cycle, electron transport chain, oxidative phosphorylation

18.51 a. fructose 1,6-bisphosphate $\xrightarrow{\text{ADP} \rightarrow \text{ATP}}$ fructose 6-phosphate
b. The reaction absorbs energy.
c. This reaction is energetically unfavorable.

18.53 a. reducing agent b. neither c. oxidizing agent

18.55 cyclohexanol (cyclohexane ring–OH) $\xrightarrow[\text{oxidation}]{NAD^+ \rightarrow NADH + H^+}$ cyclohexanone (cyclohexane ring=O)

18.57 a. 7,10 b. 1,3 c. 6 d. 4

18.59 The conversion of pyruvate to acetyl CoA forms one CO_2 for each pyruvate. The conversion of isocitrate to α-ketoglutarate and α-ketoglutarate to succinyl CoA in the citric acid cycle forms one CO_2 for each acetyl CoA.

18.61 a. aerobic: CO_2
anaerobic: lactate
b. aerobic: NAD^+ and FAD are used and NADH and $FADH_2$ are formed.
anaerobic: NADH is used to convert pyruvate to lactate and NAD^+ is formed.

18.63 a. lactate b. ethanol c. CO_2

18.65 During glycolysis: 2 ATP and 2 NADH generated.
Each NADH in turn leads to 2.5 ATP,
so glycolysis leads to: total 7 ATP
Conversion of 2 pyruvate to 2 acetyl CoA yields 2 NADH, which leads to: 5 ATP
In the citric acid cycle, 20 additional ATP are formed for 2 $CH_3COSCoA$:
2 GTP (ATP), 6 NADH $\longrightarrow$ 15 ATP,
2 $FADH_2$ $\longrightarrow$ 3 ATP: total 20 ATP
= 32 ATP

18.67 a. isocitrate b. malate, isocitrate

18.69 They require oxygen.

18.71 a. $FADH_2$ donates electrons to the electron transport chain.
b. ADP is a substrate for the formation of ATP.
c. ATP synthase catalyzes the formation of ATP from ADP.
d. The inner mitochondrial membrane contains the four complexes for the electron transport chain. ATP synthase is also embedded in the membrane and contains the H^+ ion channel that allows H^+ to return to the matrix.

18.73 NAD^+, FAD, and H_2O

18.75 $FADH_2$ enters the electron transport chain at complex II, whereas NADH enters at complex I.

18.77 Two phosphate bonds of ATP are hydrolyzed, forming AMP; therefore, 2 ATP equivalents are used.

18.79 7 molecules of acetyl CoA 6 cycles

18.81 92

18.83 a. $CH_3(CH_2)_6-\underset{\beta}{CH_2}-\underset{\alpha}{CH_2}-C(=O)-OH$

b. $CH_3(CH_2)_6-\underset{\beta}{CH_2}-\underset{\alpha}{CH_2}-C(=O)-SCoA$

c. 5 d. 4 e. 64

18.85 Ketosis is the condition under which ketone bodies accumulate. Ketogenesis is the synthesis of ketone bodies from acetyl CoA.

18.87 In poorly controlled diabetes, glucose cannot be metabolized. Fatty acids are used for metabolism and ketone bodies are formed to a greater extent.

18.89 a. 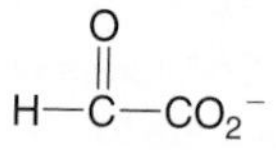 $H{-}C({=}O){-}CO_2^-$ b. 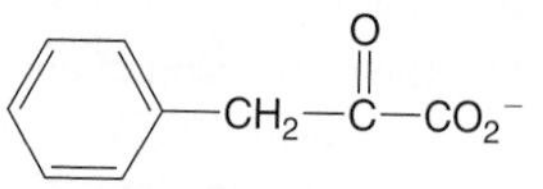 $C_6H_5{-}CH_2{-}C({=}O){-}CO_2^-$

18.91 a. acetoacetyl CoA, fumarate
b. α-ketoglutarate
c. oxaloacetate
d. pyruvate

18.93 Ketogenic amino acids are converted to acetyl CoA or a related thioester and can be converted to ketone bodies. Glucogenic amino acids are catabolized to pyruvate or another intermediate in the citric acid cycle.

18.95 a. 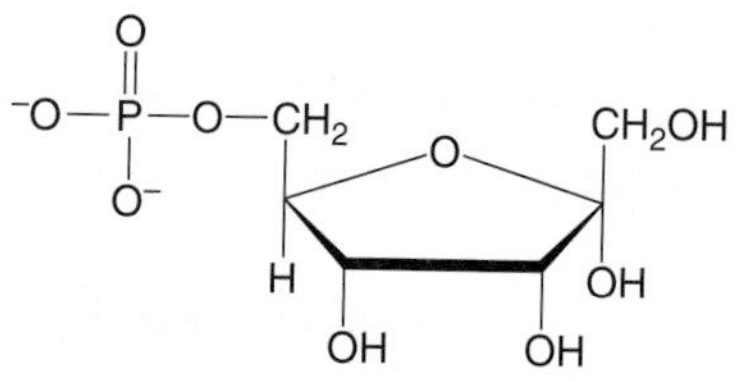 $^-O{-}P({=}O)(O^-){-}O{-}CH_2$ (furanose ring with O, H, OH, OH, CH_2OH, OH)

A
fructose 6-phosphate

b. glucose 6-phosphate
c. fructose 1,6-bisphosphate

18.97 Cramping is due to lactic acid buildup due to anaerobic metabolism of glucose.

18.99 The diet calls for ingestion of protein and fat, so in the absence of carbohydrates to be metabolized, ketone bodies are formed.

18.101 68 moles, 4.1×10^{25} molecules

APPENDIX

Useful Mathematical Concepts

Three common mathematical concepts are needed to solve many problems in chemistry:

- Using scientific notation
- Determining the number of significant figures
- Using a scientific calculator

Scientific Notation

To write numbers that contain many leading zeros (at the beginning) or trailing zeros (at the end), scientists use **scientific notation.**

- **In scientific notation, a number is written as $y \times 10^x$, where y (the coefficient) is a number between 1 and 10, and x is an exponent, which can be any positive or negative whole number.**

To convert a standard number to scientific notation:

1. Move the decimal point to give a number between 1 and 10.
2. Multiply the result by 10^x, where x is the number of places the decimal point was moved.
   - If the decimal point is moved to the **left,** x is **positive.**
   - If the decimal point is moved to the **right,** x is **negative.**

$2822. = 2.822 \times 10^3$ — the number of places the decimal point was moved to the left

Move the decimal point three places to the left.

$0.000\,004\,5 = 4.5 \times 10^{-6}$ — the number of places the decimal point was moved to the right

Move the decimal point six places to the right.

To convert a number in scientific notation to a standard number, use the value of x in 10^x to indicate the number of places to move the decimal point in the coefficient.

- **Move the decimal point to the right when x is positive.**
- **Move the decimal point to the left when x is negative.**

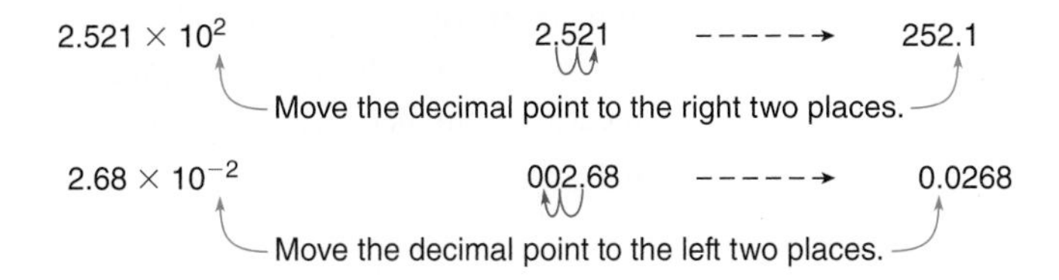
2.521 × 10^2
2.521
252.1
Move the decimal point to the right two places.
2.68 × 10^-2
002.68
0.0268
Move the decimal point to the left two places.

Table A.1 shows how several numbers are written in scientific notation.

Table A.1 Numbers in Standard Form and Scientific Notation

Number	Scientific Notation
26,200	2.62×10^{4}
0.006 40	6.40×10^{-3}
3,000,000	3×10^{6}
0.000 000 139	1.39×10^{-7}
2,000.20	2.00020×10^{3}

Often, numbers written in scientific notation must be multiplied or divided.

- To multiply two numbers in scientific notation, *multiply* the coefficients together and *add* the exponents in the powers of 10.

Add exponents. (8 + 3)

$$(3.0 \times 10^{8}) \times (2.0 \times 10^{3}) = 6.0 \times 10^{11}$$

Multiply coefficients. (3.0 × 2.0)

- To divide two numbers in scientific notation, *divide* the coefficients and *subtract* the exponents in the powers of 10.

Divide coefficients. (6.0 ÷ 2.0)

$$\frac{6.0 \times 10^{6}}{2.0 \times 10^{10}} = 3.0 \times 10^{-4}$$

Subtract exponents. (6 − 10)

Table A.2 shows the result of multiplying or dividing several numbers written in scientific notation.

Table A.2 Calculations Using Numbers Written in Scientific Notation

Calculation	Answer
$(3.5 \times 10^{3}) \times (2.2 \times 10^{22}) =$	7.7×10^{25}
$(3.5 \times 10^{3}) \div (2.2 \times 10^{22}) =$	1.6×10^{-19}
$(3.5 \times 10^{3}) \times (2.2 \times 10^{-10}) =$	7.7×10^{-7}
$(3.5 \times 10^{3}) \div (2.2 \times 10^{-10}) =$	1.6×10^{13}

Significant Figures

Whenever we measure a number, there is a degree of uncertainty associated with the result. The last number (furthest to the right) is an estimate. **Significant figures** are all of the digits in a measured number including one estimated digit. How many significant figures are contained in a number?

- All nonzero digits are always significant.
- A zero *counts* as a significant figure when it occurs between two nonzero digits, or at the end of a number with a decimal point.
- A zero does *not* count as a significant figure when it occurs at the beginning of a number, or at the end of a number that does not have a decimal point.

Table A.3 lists the number of significant figures in several quantities.

Table A.3 Examples Illustrating Significant Figures

Quantity	Number of Significant Figures	Quantity	Number of Significant Figures
1,267 g	Four	203 L	Three
24,345 km	Five	6.10 atm	Three
1.200 mg	Four	0.3040 g	Four
0.000 001 mL	One	1,200 m	Two

The number of significant figures must also be taken into account in calculations. To avoid reporting a value with too many digits, we must often **round off the number** to give the correct number of significant figures. Two rules are used in rounding off numbers.

- If the first number that must be dropped is 4 or fewer, drop it and all remaining numbers.
- If the first number that must be dropped is 5 or greater, *round the number up* by adding one to the last digit that will be retained.

To round 63.854 to two significant figures:

These digits must be retained.

63.854 — first digit to be dropped

These digits must be dropped.

- Since the first digit to be dropped is 8 (5 or greater), add 1 to the first digit to its left.
- The number 63.854 rounded to two digits is **64.**

Table A.4 gives other examples of rounding off numbers.

Table A.4 Rounding Off Numbers

Original Number	Rounded to	Rounded Number
15.2538	Two places	15
15.2538	Three places	15.3
15.2538	Four places	15.25
15.2538	Five places	15.254

The first number to be dropped is indicated in red in each original number.

The number of significant figures in the answer of a problem depends on the type of mathematical calculation—multiplication (and division) or addition (and subtraction).

- **In multiplication and division, the answer has the same number of significant figures as the original number with the *fewest* significant figures.**

five significant figures → 5.5067 × 2.6 = 14.31742 rounded to **14**

two significant figures → 2.6; first digit to be dropped → 3 (in 14.31742)

The answer must contain only **two** significant figures.

- **In addition and subtraction, the answer has the same number of decimal places as the original number with the *fewest* decimal places.**

two digits after the decimal point → 10.17 + 3.5 = 13.67 rounded to **13.7**

one digit after the decimal point → 3.5; last significant digit → 7 (in 13.7)

The answer can have only **one** digit after the decimal point.

Table A.5 lists other examples of calculations that take into account the number of significant figures.

Table A.5 Calculations Using Significant Figures

Calculation	Answer
3.2 × 699 =	2,236.8 rounded to **2,200**
4.66892 ÷ 2.13 =	2.191981221 rounded to **2.19**
25.3 + 3.668 + 29.1004 =	58.0684 rounded to **58.1**
95.1 – 26.335 =	68.765 rounded to **68.8**

Using a Scientific Calculator

A scientific calculator is capable of carrying out more complicated mathematical functions than simple addition, subtraction, multiplication, and division. For example, these calculators allow the user to convert a standard number to scientific notation, as well as readily determine the logarithm (log) or antilogarithm (antilog) of a value. Carrying out these operations is especially important in determining pH or hydronium ion concentration in Chapter 8.

Described in this section are the steps that can be followed in calculations with some types of calculators. Consult your manual if these steps do not produce the stated result.

Converting a Number to Scientific Notation

To convert a number, such as 1,200, from its standard form to scientific notation:

- Enter 1200.
- Press 2nd and then SCI.
- The number will appear as 1.2^{03}, indicating that $1,200 = 1.2 \times 10^3$.

Entering a Number Written in Scientific Notation

To enter a number written in scientific notation with a positive exponent, such as 1.5×10^8:

- Enter 1.5.
- Press EE.
- Enter 8.
- The number will appear as 1.5^{08}, indicating that it is equal to 1.5×10^8.

To enter a number written in scientific notation with a negative exponent, such as 3.5×10^{-4}:

- Enter 3.5.
- Press EE.
- Enter 4.
- Press CHANGE SIGN (+ → –).
- The number will appear as 3.5^{-04}, indicating that it is equal to 3.5×10^{-4}.

Taking the Logarithm of a Number: Calculating pH from a Known $[H_3O^+]$

Since pH = –log $[H_3O^+]$, we must learn how to calculate logarithms on a calculator in order to determine pH values. To determine the pH from a known hydronium ion concentration, say $[H_3O^+] = 1.8 \times 10^{-5}$, carry out the following steps:

- Enter 1.8×10^{-5} (Enter 1.8; press EE; enter 5; press CHANGE SIGN). The number 1.8^{-05} will appear.
- Press LOG.
- Press CHANGE SIGN (+ → –).
- The number 4.744 727 495 will appear. Since the coefficient, 1.8, contains two significant figures, round the logarithm to 4.74, which has two digits to the right of the decimal point. Thus, the pH of the solution is 4.74.

Taking the Antilogarithm of a Number: Calculating $[H_3O^+]$ from a Known pH

Since $[H_3O^+]$ = antilog(–pH), we must learn how to calculate an antilogarithm—that is, the number that has a given logarithm value—using a calculator. To determine the hydronium ion concentration from a given pH, say 3.91, carry out the following steps:

- Enter 3.91.
- Press CHANGE SIGN (+ → –).
- Press 2nd and then LOG.
- The number 0.000 123 027 will appear. To convert this number to scientific notation, press 2nd and SCI.
- The number $1.230\ 268\ 771^{-04}$ will appear, indicating that $[H_3O^+] = 1.230\ 268\ 771 \times 10^{-4}$. Since the original pH (a logarithm) had two digits to the right of the decimal point, the answer must have two significant figures in the coefficient in scientific notation. As a result, $[H_3O^+] = 1.2 \times 10^{-4}$.

Table A.6 lists pH values that correspond to given $[H_3O^+]$ values. You can practice using a calculator to determine pH or $[H_3O^+]$ by entering a value in one column, following the listed steps, and then checking to see if you obtain the corresponding value in the other column.

Table A.6 The pH of a Solution from a Given Hydronium Ion Concentration $[H_3O^+]$

$[H_3O^+]$	pH	$[H_3O^+]$	pH
1.8×10^{-10}	9.74	4.0×10^{-13}	12.40
3.8×10^{-2}	1.42	6.6×10^{-4}	3.18
5.0×10^{-12}	11.30	2.6×10^{-9}	8.59
4.2×10^{-7}	6.38	7.3×10^{-8}	7.14

Glossary

A

Acetyl CoA (18.3) A compound formed when an acetyl group ($CH_3CO–$) is bonded to coenzyme A (HS–CoA); $CH_3COSCoA$.

Achiral (12.11) Being superimposable on a mirror image.

Acid (8.1) In the Arrhenius definition, a substance that contains a hydrogen atom and dissolves in water to form a hydrogen ion, H^+.

Acidic solution (8.4) A solution in which $[H_3O^+] > [OH^-]$; thus, $[H_3O^+] > 10^{-7}$ M.

Actinides (2.4) A group of elements in the periodic table beginning with thorium (Z = 90) and immediately following the element actinium (Ac).

Active site (16.9) The region in an enzyme that binds a substrate, which then undergoes a very specific reaction with an enhanced rate.

Active transport (15.7) The process of moving an ion across a cell membrane that requires energy input.

Acyclic alkane (10.5) An alkane with molecular formula C_nH_{2n+2}, which contains a chain of carbon atoms but no rings. An acyclic alkane is also called a saturated hydrocarbon.

Acyl CoA (18.9) The thioester formed from a fatty acid and coenzyme A that undergoes β-oxidation in mitochondria; general structure RCOSCoA.

Addition reaction (11.5) A reaction in which elements are added to a compound.

Adenosine 5'-diphosphate (ADP, 18.2) A nucleoside diphosphate formed by adding two phosphates to the 5'-OH group of adenosine.

Adenosine 5'-triphosphate (ATP, 18.2) A nucleoside triphosphate formed by adding three phosphates to the 5'-OH group of adenosine. ATP is the most prominent member of a group of "high-energy" molecules that release energy during hydrolysis.

Adrenal cortical steroid (15.9) A steroid hormone synthesized in the outer layer of the adrenal gland.

Alcohol (10.4, 12.1) A compound containing a hydroxyl group (OH) bonded to a tetrahedral carbon atom; general formula ROH.

Aldehyde (10.4, 12.1) A compound that has a hydrogen atom bonded directly to a carbonyl carbon; general formula RCHO.

Alditol (14.4) A compound produced when the carbonyl group of an aldose is reduced to a 1° alcohol.

Aldonic acid (14.4) A compound produced when the aldehyde carbonyl of an aldose is oxidized to a carboxyl group.

Aldose (14.2) A monosaccharide with an aldehyde carbonyl group at C1.

Alkali metal (2.4) An element located in group 1A (group 1) of the periodic table. Alkali metals include lithium (Li), sodium (Na), potassium (K), rubidium (Rb), cesium (Cs), and francium (Fr).

Alkaline earth element (2.4) An element located in group 2A (group 2) of the periodic table. Alkaline earth elements include beryllium (Be), magnesium (Mg), calcium (Ca), strontium (Sr), barium (Ba), and radium (Ra).

Alkaloid (13.7) A naturally occurring amine derived from a plant source.

Alkane (10.4) A compound having only C—C and C—H single bonds.

Alkene (10.4, 11.1) A compound having a carbon–carbon double bond.

Alkyl group (10.6) A group formed by removing one hydrogen from an alkane.

Alkyl halide (10.4, 12.6) A compound with the general structure R–X that contains a halogen atom (X = F, Cl, Br, or I) bonded to a tetrahedral carbon.

Alkyne (10.4, 11.1) A compound with a carbon–carbon triple bond.

Alpha (α) particle (9.1) A high-energy particle that is emitted from a radioactive nucleus and contains two protons and two neutrons.

Amide (10.4, 13.1) A compound that contains a nitrogen atom bonded directly to a carbonyl carbon; general structure $RCONR'_2$, where R' = H or alkyl.

Amine (10.4, 13.1) An organic compound that contains a nitrogen atom bonded to one, two, or three alkyl groups; general structure RNH_2, R_2NH, or R_3N.

Amino acid (16.2) A compound that contains two functional groups—an amino group (NH_2) and a carboxyl group (COOH) bonded to the same carbon.

Amino group (10.4) An $–NH_2$ group.

Ammonium ion (3.6) An NH_4^+ ion.

Ammonium salt (13.8) An ionic compound that contains a positively charged ammonium ion and an anion. $(CH_3CH_2CH_2NH_3)^+Cl^-$ is an ammonium salt.

Amphoteric compound (8.2) A compound that contains both a hydrogen atom and a lone pair of electrons, so that it can be either an acid or a base.

Anabolic steroid (15.9) A synthetic androgen analogue that promotes muscle growth.

Anabolism (18.1) The synthesis of large molecules from smaller ones in a metabolic pathway.

Androgen (15.9) A hormone that controls the development of secondary sex characteristics in males.

Anion (3.2) A negatively charged ion with more electrons than protons.

Anticodon (17.8) Three nucleotides in a tRNA molecule that are complementary to the codon in mRNA and identify an individual amino acid.

Antioxidant (11.10) A compound that prevents an unwanted oxidation reaction from occurring.

Aqueous solution (7.1) A solution with water as the solvent.

Aromatic compound (10.4, 11.8) A compound that contains a benzene ring, a six-membered ring with three double bonds.

Atmosphere (6.1) A unit used to measure pressure; 1 atm = 760 mm Hg.

Atom (2.2) The basic building block of matter composed of a nucleus and an electron cloud.

Atomic mass unit (2.2) A unit abbreviated as amu, which equals one-twelfth the mass of a carbon atom that has six protons and six neutrons; 1 amu = 1.661×10^{-24} g.

Atomic number (2.2) The number of protons in the nucleus of an atom; symbolized as Z.

Atomic weight (2.3) The weighted average of the mass of all naturally occurring isotopes of a particular element, reported in atomic mass units.

Avogadro's law (6.6) A gas law that states that the volume of a gas is proportional to the number of moles present when the pressure and temperature are held constant.

Avogadro's number (5.3) A quantity that contains 6.02×10^{23} items—usually atoms, molecules, or ions.

B

Balanced chemical equation (5.2) An equation written so that an equal number of atoms of each element is present on both sides.

Barometer (6.1) A device for measuring atmospheric pressure.

Base (8.1) In the Arrhenius definition, a substance that contains hydroxide and dissolves in water to form OH^-.

Basic solution (8.4) A solution in which $[OH^-] > [H_3O^+]$; thus, $[OH^-] > 10^{-7}$ M.

Becquerel (9.4) An SI unit used to measure radioactivity, abbreviated as Bq; 1 Bq = 1 disintegration/s.

Benedict's reagent (14.4) A Cu^{2+} reagent that oxidizes the aldehyde carbonyl of an aldose to a carboxyl group, yielding an aldonic acid.

Beta (β) particle (9.1) A high-energy electron emitted from a radioactive nucleus.

Boiling point (4.4) The temperature at which a liquid is converted to the gas phase.

Bonding (3.1) The joining of two atoms in a stable arrangement.

Boyle's law (6.2) A gas law that relates pressure and volume. Boyle's law states that for a fixed amount of gas at constant temperature, the pressure and volume of the gas are inversely related.

Branched-chain alkane (10.5) An alkane that contains one or more carbon branches bonded to a carbon chain.

Brønsted–Lowry acid (8.1) A proton donor.

Brønsted–Lowry base (8.1) A proton acceptor.

Buffer (8.8) A solution whose pH changes very little when acid or base is added. Most buffers are solutions composed of approximately equal amounts of a weak acid and the salt of its conjugate base.

Building-block element (2.1) One of the four nonmetals—oxygen, carbon, hydrogen, and nitrogen—that comprise 96% of the mass of the human body.

C

Calorie (4.1) A unit of energy that equals the amount of energy needed to raise the temperature of 1 g of water by 1 °C; abbreviated as cal, where 1 cal = 4.184 J.

Carbohydrate (14.1) A polyhydroxy aldehyde or ketone, or a compound that can be hydrolyzed to a polyhydroxy aldehyde or ketone.

Carbonate (3.6) A polyatomic anion with the structure CO_3^{2-}.

Carbonyl group (10.4, 12.1) A carbon–oxygen double bond (C=O).

Carboxylate anion (13.5) The conjugate base of a carboxylic acid; general structure $RCOO^-$.

Carboxyl group (10.4, 13.1) A COOH group.

Carboxylic acid (10.4, 13.1) A compound that contains an OH group bonded directly to the carbonyl carbon; general structure RCOOH or RCO_2H.

Catabolism (18.1) The breakdown of large molecules into smaller ones during metabolism.

Catalyst (5.9) A substance that increases the rate of a reaction but is recovered unchanged at the end of the reaction.

Cation (2.8) A positively charged particle with fewer electrons than protons.

Cell membrane (15.7) The semipermeable membrane that surrounds the cell, composed of a lipid bilayer.

Celsius scale (1.9) One of three temperature scales in which water freezes at 0 °C and boils at 100 °C.

Cephalin (15.6) A phosphoacylglycerol in which the identity of the R group esterified to the phosphodiester is $-CH_2CH_2NH_3^+$.

Chain reaction (9.6) The process by which each neutron produced during fission can go on to bombard three other nuclei to produce more nuclei and more neutrons.

Charles's law (6.3) A gas law that states that for a fixed amount of gas at constant pressure, the volume of the gas is proportional to its Kelvin temperature.

Chemical equation (5.1) An expression that uses chemical formulas and other symbols to illustrate what reactants constitute the starting materials in a reaction and what products are formed.

Chemical formula (2.1) A representation that uses element symbols to show the identity of elements in a compound, and subscripts to show the number of atoms of each element contained in the compound.

Chemical properties (1.2) Those properties that determine how a substance can be converted to another substance by a chemical reaction.

Chemistry (1.1) The study of matter—its composition, properties, and transformations.

Chiral (12.11) Not superimposable on a mirror image.

Chirality center (12.11) A carbon atom bonded to four different groups.

Chlorofluorocarbons (CFCs, 2.4, 6.9) Compounds containing the elements of carbon, chlorine, and fluorine that are implicated in the destruction of ozone in the upper atmosphere.

Cis isomer (11.3) An alkene with two R groups on the same side of the double bond.

Citric acid cycle (18.6) A cyclic metabolic pathway that begins with the addition of acetyl CoA to a four-carbon substrate and ends when the same four-carbon compound is formed as a product eight steps later.

Codon (17.7) A sequence of three nucleotides (triplet) in mRNA that codes for a specific amino acid.

Coenzyme (16.9) An organic molecule needed for an enzyme-catalyzed reaction to occur.

Coenzyme A (18.3) A coenzyme that contains a sulfhydryl group (SH group) making it a thiol (RSH), and abbreviated as HS–CoA.

Cofactor (16.9) A metal ion or a nonprotein organic molecule needed for an enzyme-catalyzed reaction to occur.

Colloid (7.1) A homogeneous mixture with large particles, often having an opaque appearance.

Combined gas law (6.5) A gas law that relates pressure, volume, and temperature. For a constant number of moles, the product of pressure and volume divided by temperature is a constant.

Combustion (10.10) An oxidation reaction in which carbon-containing compounds react with oxygen to form carbon dioxide (CO_2) and water.

Competitive inhibitor (16.9) An enzyme inhibitor that has a shape and structure similar to the substrate, and competes with the substrate for binding to the active site.

Complementary base pairs (17.3) The predictable pairing of bases between two strands of DNA. Adenine pairs with thymine using two hydrogen bonds, forming an A–T base pair, and cytosine pairs with guanine using three hydrogen bonds, forming a C–G base pair.

Compound (1.3) A pure substance formed by chemically combining two or more elements.

Concentration (7.5) The amount of solute dissolved in a given amount of solution.

Condensation (4.6) The conversion of a gas to a liquid.

Condensed structure (10.3) A representation used for a compound having a chain of atoms bonded together. The atoms are drawn in, but the two-electron bond lines and lone pairs on heteroatoms are generally omitted.

Conjugate acid (8.2) The product formed by the gain of a proton by a base.

Conjugate acid–base pair (8.2) Two species that differ by the presence of a proton.

Conjugate base (8.2) The product formed by loss of a proton from an acid.

Conjugated protein (16.7) A compound composed of a protein unit and a nonprotein molecule.

Constitutional isomers (10.5) Isomers that differ in the way the atoms are connected to each other.

Conversion factor (1.7) A term that converts a quantity in one unit to a quantity in another unit.

Cooling curve (4.7) A graph that shows how the temperature of a substance changes as heat is removed.

Covalent bond (3.7) A chemical bond that results from the sharing of electrons between two atoms.

Critical mass (9.6) The amount of a radioactive element required to sustain a chain reaction.

Cross formula (12.11) A Fischer projection formula that replaces a chirality center with a cross. The horizontal lines represent wedged bonds and the vertical lines represent dashed bonds.

C-Terminal amino acid (16.4) In a peptide, the amino acid with the free $-COO^-$ group on the α carbon.

Cubic centimeter (1.4) A unit of volume equal to one milliliter; one cubic centimeter = 1 cm^3 = 1 cc.

Curie (9.4) A unit used to measure radioactivity and equal to 3.7×10^{10} disintegrations/s. A curie corresponds to the decay rate of one gram of the element radium.

Cycloalkane (10.5) A compound with the general formula C_nH_{2n} that contains carbons joined in one or more rings.

D

Dalton's law (6.8) A law that states that the total pressure (P_{total}) of a gas mixture is equal to the sum of the partial pressures of its component gases.

Dehydration (12.5) The loss of water (H_2O) from a starting material.

Deletion mutation (17.9) The loss of one or more nucleotides from a DNA molecule.

Denaturation (16.8) The process of altering the shape of a protein without breaking the amide bonds that form the primary structure.

Density (1.10) A physical property that relates the mass of a substance to its volume; density = g/(mL or cc).

Deoxyribonucleic acid (DNA, 17.1, 17.3) A polymer of deoxyribonucleotides that stores the genetic information of an organism and transmits that information from one generation to another.

Deoxyribonucleoside (17.1) A compound that contains the monosaccharide 2-deoxyribose and a purine or pyrimidine base.

Deoxyribonucleotide (17.1) A compound that contains the monosaccharide 2-deoxyribose bonded to a purine or pyrimidine base, as well as a phosphate at the 5'-OH group.

Deposition (4.6) The conversion of a gas directly to a solid.

Deuterium (2.3) A hydrogen atom having one proton and one neutron, giving it a mass number of two.

Dialysis (7.8) A process that involves the selective passage of substances across a semipermeable membrane, called a dialyzing membrane.

Diatomic molecule (3.7) A molecule that contains two atoms. Hydrogen (H_2) is a diatomic molecule.

Dilution (7.7) The addition of solvent to a solution to decrease the concentration of solute.

Dipeptide (16.4) A peptide formed from two amino acids joined together by one amide bond.

Dipole (3.11) The partial separation of charge in a bond or molecule.

Dipole–dipole interactions (4.3) The attractive intermolecular forces between the permanent dipoles of two polar molecules.

Disaccharide (14.1) A carbohydrate composed of two monosaccharides joined together.

Dissociation (8.3) The process that occurs when an acid or base dissolves in water to form ions.

Disulfide (12.7) A compound that contains a sulfur–sulfur bond.

DNA fingerprinting (17.10) A technique in which DNA is cut into fragments that are separated by size using gel electrophoresis. This forms a set of horizontal bands, each band corresponding to a segment of DNA, sorted from low to high molecular weight.

Double bond (3.8) A multiple bond that contains four electrons—that is, two two-electron bonds.

E

Electrolyte (7.2) A substance that conducts an electric current in water.

Electron (2.2) A negatively charged subatomic particle.

Electron cloud (2.2) The space surrounding the nucleus of an atom, which contains electrons and comprises most of the volume of an atom.

Electron-dot symbol (2.7) A symbol that shows the number of valence electrons around an atom.

Electronegativity (3.11) A measure of an atom's attraction for electrons in a bond.

Electronic configuration (2.6) The arrangement of electrons in an atom's orbitals.

Electron transport chain (18.7) A series of reactions that transfers electrons from reduced coenzymes to progressively stronger oxidizing agents, ultimately converting oxygen to water.

Element (1.3) A pure substance that cannot be broken down into simpler substances by a chemical reaction.

Elimination (12.5) A reaction in which elements of the starting material are "lost" and a new multiple bond is formed.

Enantiomers (12.11) Mirror images that are not superimposable.

Endothermic (4.6, 5.8) Absorbing energy. In an endothermic chemical reaction, ΔH is positive (+) and energy is absorbed.

Energy (4.1) The capacity to do work.

Energy diagram (5.8) A schematic representation of the energy changes in a reaction, which plots energy on the vertical axis and the progress of the reaction—the reaction coordinate—on the horizontal axis.

Energy of activation (5.8) The difference in energy between the reactants and the transition state; symbolized by E_a.

English system of measurement (1.4) A system of measurement used primarily in the United States in which units are not systematically related to each other and require memorization.

Enthalpy change (5.8) The energy absorbed or released in any reaction—also called the heat of reaction and symbolized by ΔH.

Enzyme (16.9) A biological catalyst composed of one or more chains of amino acids in a very specific three-dimensional shape.

Enzyme–substrate complex (16.9) A structure composed of a substrate bonded to the active site of an enzyme.

Equilibrium (5.10) A reaction that consists of forward and reverse reactions that have equal reaction rates, so the concentration of each species does not change.

Equivalent (7.2) The number of moles of charge that a mole of ions contributes to a solution.

Ester (10.4, 13.1) A compound that contains an OR group bonded directly to the carbonyl carbon; general structure RCOOR.

Estrogen (15.9) A hormone that controls the development of secondary sex characteristics in females and regulates the menstrual cycle.

Ether (10.4, 12.1) A compound that has two alkyl groups bonded to an oxygen atom; general structure ROR.

Exact number (1.5) A number that results from counting objects or is part of a definition.

Exothermic (4.6, 5.8) Releasing energy. In an exothermic reaction, energy is released and ΔH is negative (–).

F

Facilitated transport (15.7) The process by which some ions and molecules travel through the channels in a cell membrane created by integral proteins.

FAD (flavin adenine dinucleotide, 18.3) A biological oxidizing agent synthesized in cells from vitamin B_2, riboflavin. FAD is reduced by adding two hydrogen atoms, forming $FADH_2$.

Fahrenheit scale (1.9) One of three temperature scales in which water freezes at 32 °F and boils at 212 °F.

Fat (11.3, 15.4) A triacylglycerol with few double bonds, making it a solid at room temperature.

Fat-soluble vitamin (15.10) A vitamin that dissolves in an organic solvent but is insoluble in water. Vitamins A, D, E, and K are fat soluble.

Fatty acid (11.3, 15.2) A carboxylic acid (RCOOH) with a long carbon chain, usually containing 12–20 carbon atoms.

Fermentation (18.5) The anaerobic conversion of glucose to ethanol and CO_2.

Fibrous protein (16.7) A water-insoluble protein composed of long linear polypeptide chains that are bundled together to form rods or sheets.

Fischer esterification (13.6) Treatment of a carboxylic acid (RCOOH) with an alcohol (R'OH) and an acid catalyst to form an ester (RCOOR').

Fischer projection formula (12.11) A method of drawing chiral compounds with the chirality center at the intersection of a cross. The horizontal bonds are assumed to be wedges and the vertical bonds are assumed to be dashed lines.

Formula weight (5.4) The sum of the atomic weights of all the atoms in a compound, reported in atomic mass units (amu).

Forward reaction (5.10) In equilibrium, a reaction that proceeds from left to right as drawn.

Freezing (4.6) The conversion of a liquid to a solid.

Functional group (10.4) An atom or a group of atoms with characteristic chemical and physical properties.

G

Gamma (γ) ray (9.1) High-energy radiation released from a radioactive nucleus.

Gas (1.2) A state of matter that has no definite shape or volume. The particles of a gas move randomly and are separated by a distance much larger than their size.

Gas laws (6.2) A series of laws that relate the pressure, volume, and temperature of a gas.

Gay–Lussac's law (6.4) A gas law that states for a fixed amount of gas at constant volume, the pressure of the gas is proportional to its Kelvin temperature.

Geiger counter (9.4) A small portable device used for measuring radioactivity.

Gene (17.1) A portion of a DNA molecule responsible for the synthesis of a single protein.

Genetic code (17.7) The sequence of nucleotides in mRNA (coded in triplets) that specifies the amino acid sequence of a protein. Each triplet is called a codon.

Genetic disease (17.9) A disease resulting from a mutation that causes a condition to be inherited from one generation to another.

Globular protein (16.7) A protein that is coiled into a compact shape with a hydrophilic outer surface to make it water soluble.

Glucogenic amino acid (18.11) An amino acid that can be used to synthesize glucose.

Glycolysis (18.4) A linear, 10-step pathway that converts glucose, a six-carbon monosaccharide, to two three-carbon pyruvate molecules.

Glycosidic linkage (14.5) The C—O bond that joins two monosaccharides together. The carbon in a glycosidic linkage is bonded to two oxygens.

Gram (1.4) The basic unit of mass in the metric system; abbreviated as g.

Gray (9.4) A unit that measures absorbed radiation; abbreviated as Gy.

Ground state (2.6) The lowest energy arrangement of electrons.

Group (2.4) A column in the periodic table.

Group number (2.4) A number that identifies a particular column in the periodic table.

H

Half-life (9.3) The time it takes for one-half of a sample to decay.

Half reaction (5.7) An equation written for an individual oxidation or reduction that shows how many electrons are gained or lost.

Halogen (2.4) An element located in group 7A (group 17) of the periodic table. Halogens include fluorine (F), chlorine (Cl), bromine (Br), iodine (I), and astatine (At).

Haworth projection (14.3) A planar, six-membered ring used to represent the cyclic form of glucose and other sugars.

Heating curve (4.7) A graph that shows how the temperature of a substance changes as heat is added.

Heat of fusion (4.6) The amount of energy needed to melt one gram of a substance.

Heat of reaction (5.8) The energy absorbed or released in any reaction and symbolized by ΔH—also called the enthalpy change.

Heat of vaporization (4.6) The amount of energy needed to vaporize one gram of a substance.

α-Helix (16.6) A secondary structure of a protein formed when a peptide chain twists into a right-handed or clockwise spiral.

Heme (16.7) A complex organic compound containing an Fe^{2+} ion complexed with a large nitrogen-containing ring system.

Henry's law (7.4) A law that states that the solubility of a gas in a liquid is proportional to the partial pressure of the gas above the liquid.

Heteroatom (10.2) Any atom in an organic compound that is not carbon or hydrogen.

Heterogeneous mixture (7.1) A mixture that does not have a uniform composition throughout a sample.

Hexose (14.2) A monosaccharide with six carbons.

High-density lipoprotein (15.8) A spherical particle that transports cholesterol from the tissues to the liver.

Homogeneous mixture (7.1) A mixture that has a uniform composition throughout a sample.

Hormone (15.9) A compound synthesized in one part of an organism, which then travels through the bloodstream to elicit a response at a target tissue or organ.

Hydration (11.5) The addition of water to a molecule.

Hydrocarbon (10.4) A compound that contains only the elements of carbon and hydrogen.

Hydrogenation (11.5) The addition of hydrogen (H_2) to an alkene.

Hydrogen bonding (4.3) An attractive intermolecular force that occurs when a hydrogen atom bonded to O, N, or F is electrostatically attracted to an O, N, or F atom in another molecule.

Hydrolysis (13.6) A cleavage reaction that uses water.

Hydrolyzable lipid (15.1) A lipid that can be converted to smaller molecules by hydrolysis with water.

Hydronium ion (3.6) The H_3O^+ ion.

Hydrophilic (15.2) The polar part of a molecule that is attracted to water.

Hydrophobic (15.2) The nonpolar part of a molecule (C—C and C—H bonds) that is not attracted to water.

Hydroxide (3.6) The OH^- ion.

α-Hydroxy acid (13.4) A compound that contains a hydroxyl group on the α carbon to a carboxyl group.

Hydroxyl group (10.4, 12.1) An OH group.

Hypertonic solution (7.8) A solution that has a higher osmotic pressure than body fluids.

Hypotonic solution (7.8) A solution that has a lower osmotic pressure than body fluids.

I

Ideal gas law (6.7) A gas law that relates the pressure (P), volume (V), temperature (T), and number of moles (n) of a gas in a single equation; $PV = nRT$, where R is a constant.

Incomplete combustion (10.10) An oxidation reaction that forms carbon monoxide (CO) instead of carbon dioxide (CO_2) because insufficient oxygen is available.

Induced-fit model (16.9) The binding of a substrate to an enzyme such that the shape of the active site adjusts to fit the shape of the substrate.

Inexact number (1.5) A number that results from a measurement or observation and contains some uncertainty.

Inhibitor (16.9) A molecule that causes an enzyme to lose activity.

Inner transition metal elements (2.4) A group of elements consisting of the lanthanides and actinides.

Insertion mutation (17.9) The addition of one or more nucleotides to a DNA molecule.

Intermolecular forces (4.3) The attractive forces that exist between molecules.

Ion (3.1) A charged species in which the number of protons and electrons in an atom is not equal.

Ion–dipole interaction (7.3) The attraction of an ion to a dipole in another molecule.

Ionic bond (3.1) A bond that results from the transfer of electrons from one element to another.

Ionization energy (2.8) The energy needed to remove an electron from a neutral atom.

Ion–product constant (8.4) The product of the concentrations of H_3O^+ and OH^- in water or an aqueous solution—symbolized by K_w and equal to 1×10^{-14}.

Irreversible inhibitor (16.9) An inhibitor that covalently binds to an enzyme, permanently destroying its activity.

Isoelectric point (16.3) The pH at which an amino acid exists primarily in its neutral form; abbreviated as p*I*.

Isomers (10.5) Two different compounds with the same molecular formula.

Isotonic solution (7.8) Two solutions with the same osmotic pressure.

Isotopes (2.3) Atoms of the same element having a different number of neutrons.

IUPAC nomenclature (10.6) A systematic method of naming compounds developed by the International Union of Pure and Applied Chemistry.

J

Joule (4.1) A unit of measurement for energy; abbreviated as J, where 1 cal = 4.184 J.

K

Kelvin scale (1.9) A temperature scale commonly used by scientists. The Kelvin scale is divided into kelvins (K); $T_K = T_C + 273$.

Ketogenesis (18.10) The synthesis of ketone bodies from acetyl CoA.

Ketogenic amino acid (18.11) An amino acid that cannot be used to synthesize glucose, but can be converted to ketone bodies.

Ketone (10.4, 12.1) A compound that has two alkyl groups bonded to the carbonyl group; general structure RCOR.

Ketone bodies (18.10) Three compounds—acetoacetate, β-hydroxybutyrate, and acetone—formed when acetyl CoA levels exceed the capacity of the citric acid cycle.

Ketose (14.2) A monosaccharide with a carbonyl group at C2.

Ketosis (18.10) The accumulation of ketone bodies during starvation and uncontrolled diabetes.

Kinetic energy (4.1) The energy of motion.

Kinetic-molecular theory (6.1) A theory that describes the fundamental characteristics of gas particles.

L

β-Lactam (13.10) An amide contained in a four-membered ring.

Lagging strand (17.4) The strand of DNA synthesized in small fragments during replication, which are then joined together by an enzyme.

Lanthanides (2.4) A group of 14 elements in the periodic table beginning with the element cerium ($Z = 58$) and immediately following the element lanthanum (La).

Law of conservation of energy (4.1) A law that states that the total energy in a system does not change. Energy cannot be created or destroyed.

Law of conservation of matter (5.1) A law that states that atoms cannot be created or destroyed in a chemical reaction.

LD_{50} (9.4) The lethal dose of radiation (or a poison) that kills 50% of a population.

Leading strand (17.4) The strand of DNA that grows continuously during replication.

Le Châtelier's principle (5.10) A principle that states that if a chemical system at equilibrium is disturbed or stressed, the system will react in the direction that counteracts the disturbance or relieves the stress.

Lecithin (15.6) A phosphoacylglycerol in which the identity of the R group esterified to the phosphodiester is $-CH_2CH_2N(CH_3)_3^+$.

Lewis structure (3.7) An electron-dot structure for a molecule that shows the location of all valence electrons in the molecule, both the shared electrons in bonds and the nonbonded electron pairs.

Lipase (15.5) An enzyme that catalyzes the hydrolysis of the esters in a triacylglycerol.

Lipid (13.6, 15.1) A biomolecule that is soluble in organic solvents and insoluble in water.

Lipid bilayer (15.7) The basic structure of the cell membrane formed from two layers of phospholipids having their ionic heads oriented on the outside and their nonpolar tails on the inside.

Lipoprotein (15.8) A small water-soluble spherical particle composed of proteins and lipids.

Liquid (1.2) A state of matter that has a definite volume, but takes on the shape of the container it occupies. The particles of a liquid are close together but they can randomly move past each other.

Liter (1.4) The basic unit of volume in the metric system; abbreviated as L.

Lock-and-key model (16.9) The binding of a substrate to a rigid active site, such that the three-dimensional geometry of the substrate exactly matches the shape of the active site.

London dispersion forces (4.3) Very weak intermolecular interactions due to the momentary changes in electron density in a molecule.

Lone pair (3.7) An unshared electron pair.

Low-density lipoprotein (15.8) A spherical particle containing proteins and lipids, which transports cholesterol from the liver to the tissues.

M

Main group element (2.4) An element in groups 1A–8A of the periodic table.

Major mineral (macronutrient, 2.1) One of the seven elements present in the body in small amounts (0.1–2% by mass) and needed in the daily diet.

Markovnikov's rule (11.5) The rule that states that in the addition of H_2O to an unsymmetrical alkene, the H atom bonds to the less substituted carbon atom.

Mass (1.4) A measure of the amount of matter in an object.

Mass number (2.2) The total number of protons and neutrons in a nucleus; symbolized as *A*.

Matter (1.1) Anything that has mass and takes up volume.

Melting (4.6) The conversion of a solid to a liquid.

Melting point (4.4) The temperature at which a solid is converted to the liquid phase.

Messenger RNA (mRNA, 17.5) The carrier of information from DNA (in the cell nucleus) to the ribosomes (in the cell cytoplasm). Each gene of a DNA molecule corresponds to a specific mRNA molecule.

Metabolism (18.1) The sum of all of the chemical reactions that take place in an organism.

Meta isomer (11.9) A 1,3-disubstituted benzene.

Metal (2.1) A shiny element that is a good conductor of heat and electricity.

Metalloid (2.1) An element with properties intermediate between a metal and a nonmetal. Metalloids include boron (B), silicon (Si), germanium (Ge), arsenic (As), antimony (Sb), tellurium (Te), and astatine (At).

Meter (1.4) A unit used to measure length; abbreviated as m.

Metric system (1.4) A measurement system in which each type of measurement has a base unit and all other units are related to the base unit by a prefix that indicates if the unit is larger or smaller than the base unit.

Micelle (13.5) A spherical droplet formed when soap is dissolved in water. The ionic heads of the soap molecules are oriented on the surface and the nonpolar tails are packed in the interior.

Millimeters mercury (6.1) A unit used to measure pressure; abbreviated as mm Hg and also called "torr."

Mitochondrion (18.1) A small sausage-shaped organelle within a cell in which energy production takes place.

Mixture (1.3) Matter composed of more than one component.

Molarity (7.6) The number of moles of solute per liter of solution; abbreviated as M.

Molar mass (5.4) The mass of one mole of any substance, reported in grams per mole.

Mole (5.3) A quantity that contains 6.02×10^{23} items—usually atoms, molecules, or ions.

Molecular formula (3.8) A formula that shows the number and identity of all of the atoms in a compound, but it does not indicate what atoms are bonded to each other.

Molecular weight (5.4) The formula weight of a covalent compound.

Molecule (3.1) A compound or element containing two or more atoms joined together with covalent bonds.

Monomers (11.7) Small molecules that covalently bond together to form polymers.

Monosaccharide (14.1) A carbohydrate that cannot be hydrolyzed to simpler compounds.

Mutation (17.9) A change in the nucleotide sequence in a molecule of DNA.

N

NAD$^+$ (nicotinamide adenine dinucleotide, 18.3) A biological oxidizing agent and coenzyme synthesized from the vitamin niacin. NAD$^+$ and NADH are interconverted by oxidation and reduction reactions.

NADH (18.3) A biological reducing agent and coenzyme formed when NAD$^+$ is reduced.

Net ionic equation (8.6) An equation that contains only the species involved in a reaction.

Neutralization reaction (8.6) An acid–base reaction that produces a salt and water as products.

Neutral solution (8.4) Any solution that has an equal concentration of H_3O^+ and OH^- ions and a pH = 7.

Neutron (2.2) A neutral subatomic particle in the nucleus.

Noble gases (2.4) Elements located in group 8A (group 18) of the periodic table. The noble gases are helium (He), neon (Ne), argon (Ar), krypton (Kr), xenon (Xe), and radon (Rn).

Nomenclature (3.4) The system of assigning an unambiguous name to a compound.

Nonbonded electron pair (3.7) An unshared electron pair or lone pair.

Noncompetitive inhibitor (16.9) An inhibitor that binds to an enzyme but does not bind at the active site.

Nonelectrolyte (7.2) A substance that does not conduct an electric current when dissolved in water.

Nonhydrolyzable lipid (15.1) A lipid that cannot be cleaved into smaller units by aqueous hydrolysis.

Nonmetal (2.1) An element that does not have a shiny appearance and poorly conducts heat and electricity.

Nonpolar bond (3.11) A bond in which electrons are equally shared.

N-Terminal amino acid (16.4) In a peptide, the amino acid with the free $-NH_3^+$ group on the α carbon.

Nuclear fission (9.6) The splitting apart of a nucleus into lighter nuclei and neutrons.

Nuclear fusion (9.6) The joining together of two nuclei to form a larger nucleus.

Nuclear reaction (9.2) A reaction that involves the subatomic particles of the nucleus.

Nucleic acid (17.1, 17.2) An unbranched polymer composed of nucleotides. DNA and RNA are nucleic acids.

Nucleoside (17.1) A compound formed by joining a monosaccharide with a nitrogen atom of a purine or pyrimidine base.

Nucleotide (17.1) A compound formed by adding a phosphate group to the 5'-OH of a nucleoside.

Nucleus (2.2) The dense core of the atom that contains protons and neutrons.

O

Octet rule (3.2) The rule in bonding that states that main group elements are especially stable when they possess eight electrons (an octet) in the outer shell.

Oil (11.3, 15.4) A triacylglycerol that is liquid at room temperature.

Omega-*n* acid (15.2) An unsaturated fatty acid where *n* is the carbon at which the first double bond occurs in the carbon chain. The numbering begins at the end of the chain with the CH_3 group.

Orbital (2.5) A region of space where the probability of finding an electron is high.

Organic chemistry (10.1) The study of compounds that contain the element carbon.

Ortho isomer (11.9) A 1,2-disubstituted benzene.

Osmosis (7.8) The selective diffusion of solvent (usually water) across a semipermeable membrane from a less concentrated solution to a more concentrated solution.

Osmotic pressure (7.8) The pressure that prevents the flow of additional solvent into a solution on one side of a semipermeable membrane.

Oxidation (5.7, 12.5) The loss of electrons from an atom. Oxidation may result in a gain of oxygen atoms or a loss of hydrogen atoms.

β-Oxidation (18.9) A process in which two-carbon acetyl CoA units are sequentially cleaved from a fatty acid.

Oxidative phosphorylation (18.7) The process by which the energy released from the oxidation of reduced coenzymes is used to convert ADP to ATP using the enzyme ATP synthase.

Oxidizing agent (5.7) A compound that gains electrons (i.e., is reduced), causing another compound to be oxidized.

P

Para isomer (11.9) A 1,4-disubstituted benzene.

Parent name (10.6) The root that indicates the number of carbons in the longest continuous carbon chain in a molecule.

Partial hydrogenation (11.6) The hydrogenation of some, but not all, of the double bonds in a molecule.

Partial pressure (6.8) The pressure exerted by one component of a mixture of gases.

Parts per million (7.5) A concentration term (abbreviated ppm)—the number of "parts" in 1,000,000 parts of solution.

Penicillin (13.10) An antibiotic that contains a β-lactam and interferes with the synthesis of the bacterial cell wall.

Pentose (14.2) A monosaccharide with five carbons.

Peptide (16.4) A compound that contains many amino acids joined together by amide bonds.

Peptide bond (16.4) An amide bond in peptides and proteins.

Period (2.4) A row in the periodic table.

Periodic table (2.1) A schematic arrangement of all known elements that groups elements with similar properties.

Phenol (11.10) A compound that contains an OH group bonded to a benzene ring.

Pheromone (10.5) A chemical substance used for communication in a specific animal species, most commonly an insect population.

Phosphate (3.6) A PO_4^{3-} anion.

Phosphoacylglycerol (15.6) A lipid with a glycerol backbone that contains two of the hydroxyls esterified with fatty acids and the third hydroxyl as part of a phosphodiester.

Phosphodiester (15.6) A derivative of phosphoric acid (H_3PO_4) that is formed by replacing two of the H atoms by R groups.

Phospholipid (15.6) A lipid that contains a phosphorus atom.

Phosphorylation (18.2) A reaction that adds a phosphate group to a molecule.
pH scale (8.5) The scale used to report the H_3O^+ concentration; pH = –log [H_3O^+].
Physical properties (1.2) Those properties of a substance that can be observed or measured without changing the composition of the material.
β-Pleated sheet (16.6) A secondary structure formed when two or more peptide chains, called strands, line up side-by-side.
Point mutation (17.9) The substitution of one nucleotide for another.
Polar bond (3.11) A bond in which electrons are unequally shared and pulled towards the more electronegative element.
Polyatomic ion (3.6) A cation or anion that contains more than one atom.
Polymer (11.7) A large molecule made up of repeating units of smaller molecules—called monomers—covalently bonded together.
Polymerization (11.7) The joining together of monomers to make polymers.
Polynucleotide (17.2) A polymer of nucleotides that contains a sugar–phosphate backbone.
Polysaccharide (14.1) Three or more monosaccharides joined together.
***p* Orbital** (2.5) A dumbbell-shaped orbital higher in energy than an *s* orbital in the same shell.
Positron (9.1) A radioactive particle that has a negligible mass and a +1 charge.
Potential energy (4.1) Energy that is stored.
Pressure (6.1) The force (*F*) exerted per unit area (*A*); symbolized by *P.*
Primary (1°) alcohol (12.2) An alcohol having the general structure RCH_2OH.
Primary (1°) alkyl halide (12.6) An alkyl halide having the general structure RCH_2X.
Primary (1°) amide (13.9) A compound having the general structure $RCONH_2$.
Primary (1°) amine (13.7) A compound having the general structure RNH_2.
Primary structure (16.6) The particular sequence of amino acids that is joined together by peptide bonds in a protein.
Product (5.1) A substance formed in a chemical reaction.
Progestin (15.9) A hormone responsible for the preparation of the uterus for implantation of a fertilized egg.
Prostaglandins (13.4) A group of carboxylic acids that contain a five-membered ring, are synthesized from arachidonic acid, and have a wide range of biological activities.
Proteins (16.1) Biomolecules that contain many amide bonds, formed by joining amino acids together.
Proton (2.2) A positively (+) charged subatomic particle that resides in the nucleus of the atom.
Proton transfer reaction (8.2) A Brønsted–Lowry acid–base reaction in which a proton is transferred from an acid to a base.
Pure substance (1.3) A substance that contains a single component, and has a constant composition regardless of the sample size.

Q

Quaternary structure (16.6) The shape adopted when two or more folded polypeptide chains come together into one protein complex.

R

Rad (9.4) The radiation absorbed dose; the amount of radiation absorbed by one gram of a substance.
Radioactive decay (9.2) The process by which an unstable radioactive nucleus emits radiation, forming a nucleus of new composition.
Radioactive isotope (9.1) An isotope that is unstable and spontaneously emits energy to form a more stable nucleus.
Radioactivity (9.1) The energy emitted by a radioactive isotope.
Radiocarbon dating (9.3) A method to date artifacts that is based on the ratio of the radioactive carbon-14 isotope to the stable carbon-12 isotope.
Reactant (5.1) The starting material in a reaction.
Reaction rate (5.9) A measure of how fast a chemical reaction occurs.
Redox reaction (5.7) A reaction that involves the transfer of electrons from one element to another.
Reducing agent (5.7) A compound that loses electrons (i.e., is oxidized), causing another compound to be reduced.
Reducing sugar (14.4) A carbohydrate that is oxidized with Benedict's reagent.
Reduction (5.7) The gain of electrons by an atom. Reduction may result in the loss of oxygen atoms or the gain of hydrogen atoms.
Rem (9.4) The radiation equivalent for man; the amount of radiation absorbed by a substance that also factors in its energy and potential to damage tissue.
Replication (17.4) The process by which DNA makes a copy of itself when a cell divides.
Retrovirus (17.11) A virus that contains a core of RNA.
Reverse reaction (5.10) In equilibrium, a reaction that proceeds from right to left as drawn.
Reverse transcription (17.11) A process by which a retrovirus produces DNA from RNA.
Reversible inhibitor (16.9) An inhibitor that binds to an enzyme, but enzyme activity is restored when the inhibitor is released.
Reversible reaction (5.10) A reaction that can occur in either direction, from reactants to products or from products to reactants.
Ribonucleic acid (RNA, 17.1, 17.5) A polymer of ribonucleotides that translates genetic information to protein synthesis.
Ribonucleoside (17.1) A compound that contains the monosaccharide ribose and a purine or pyrimidine base.
Ribonucleotide (17.1) A compound that contains the monosaccharide ribose bonded to either a purine or pyrimidine base as well as a phosphate at the 5'-OH group.
Ribosomal RNA (rRNA, 17.5) The most abundant type of RNA. rRNA is found in the ribosomes of the cell and provides the site where polypeptides are assembled during protein synthesis.

S

Saponification (13.6, 15.5) The basic hydrolysis of an ester.
Saturated fatty acids (15.2) Fatty acids that have no double bonds in their long hydrocarbon chains.
Saturated hydrocarbon (10.5) An alkane with molecular formula C_nH_{2n+2} that contains a chain of carbon atoms but no rings.
Saturated solution (7.3) A solution that has the maximum number of grams of solute that can be dissolved.
Scientific notation (1.6) A system in which numbers are written as $y \times 10^x$, where y is a number between 1 and 10 and x can be either positive or negative.
Secondary (2°) alcohol (12.2) An alcohol having the general structure R_2CHOH.
Secondary (2°) alkyl halide (12.6) An alkyl halide having the general structure R_2CHX.
Secondary (2°) amide (13.9) A compound that has the general structure RCONHR'.
Secondary (2°) amine (13.7) A compound that has the general structure R_2NH.
Secondary structure (16.6) The three-dimensional arrangement of localized regions of a protein. The α-helix and β-pleated sheet are two kinds of secondary structure.
Semipermeable membrane (7.8) A membrane that allows only certain molecules to pass through.
Shell (2.5) A region where an electron that surrounds a nucleus is confined. A shell is also called a principal energy level.
Sievert (9.4) A unit that measures absorbed radiation; abbreviated as Sv.
Significant figures (1.5) All of the digits in a measured number, including one estimated digit.

SI units (1.4) The International System of Units formally adopted as the uniform system of units for the sciences.

Skeletal structure (10.3) A shorthand method used to draw organic compounds in which carbon atoms are assumed to be at the junction of any two lines or at the end of a line, and all H's on C's are omitted.

Soap (13.5, 15.5) A salt of a long-chain carboxylic acid.

Solid (1.2) A state of matter that has a definite shape and volume. The particles of a solid lie close together, and are arranged in a regular, three-dimensional array.

Solubility (7.3) The amount of solute that dissolves in a given amount of solvent.

Solute (7.1) The substance present in the lesser amount in a solution.

Solution (7.1) A homogeneous mixture that contains small particles. Liquid solutions are transparent.

Solvation (7.3) The process of surrounding particles of a solute with solvent molecules.

Solvent (7.1) The substance present in the larger amount in a solution.

***s* Orbital** (2.5) A spherical orbital that is lower in energy than other orbitals in the same shell.

Specific gravity (1.10) A unitless quantity that compares the density of a substance with the density of water at 4 °C.

Specific heat (4.5) The amount of heat energy needed to raise the temperature of 1 g of a substance 1 °C.

Spectator ion (8.6) An ion that appears on both sides of an equation but undergoes no change in a reaction.

Standard molar volume (6.6) The volume of one mole of any gas at STP—22.4 L.

States of matter (1.2) The forms in which most matter exists—that is, gas, liquid, and solid.

Stereochemistry (12.11) The three-dimensional structure of molecules.

Stereoisomers (11.3, 12.11) Isomers that differ only in their three-dimensional arrangement of atoms.

Steroid (15.8) A lipid whose carbon skeleton contains three six-membered rings and one five-membered ring.

STP (6.6) Standard conditions of temperature and pressure—1 atm (760 mm Hg) for pressure and 273 K (0 °C) for temperature.

Straight-chain alkane (10.5) An alkane that has all of its carbons in one continuous chain.

Sublimation (4.6) A phase change in which the solid phase enters the gas phase without passing through the liquid state.

Sulfate (3.6) An SO_4^{2-} ion.

Sulfhydryl group (12.1) An SH group.

Supersaturated solution (7.4) A solution that contains more than the predicted maximum amount of solute at a given temperature.

Suspension (7.1) A heterogeneous mixture that contains large particles (> 1 μm in diameter) suspended in a liquid.

T

Temperature (1.9) A measure of how hot or cold an object is.

Tertiary (3°) alcohol (12.2) An alcohol that has the general structure R_3COH.

Tertiary (3°) alkyl halide (12.6) An alkyl halide having the general structure R_3CX.

Tertiary (3°) amide (13.9) A compound that has the general structure $RCONR'_2$.

Tertiary (3°) amine (13.7) A compound that has the general structure R_3N.

Tertiary structure (16.6) The three-dimensional shape adopted by an entire peptide chain.

Tetrose (14.2) A monosaccharide with four carbons.

Thiol (12.1) A compound that contains a sulfhydryl group (SH group) bonded to a tetrahedral carbon atom; general structure RSH.

Titration (8.7) A technique for determining an unknown molarity of an acid by adding a base of known molarity to a known volume of acid.

Tollens reagent (12.10) An oxidizing agent that contains silver(I) oxide (Ag_2O) in aqueous ammonium hydroxide (NH_4OH).

Trace element (micronutrient, 2.1) An element required in the daily diet in small quantities—usually less than 15 mg.

Transcription (17.6) The process that synthesizes RNA from DNA.

Transfer RNA (tRNA, 17.5) The smallest type of RNA, which brings a specific amino acid to the site of protein synthesis on a ribosome.

Trans isomer (11.3) An alkene with two R groups on opposite sides of a double bond.

Transition metal element (2.4) An element contained in one of the 10 columns in the periodic table numbered 1B–8B.

Transition state (5.8) The unstable energy maximum located at the top of the energy hill in an energy diagram.

Translation (17.8) The synthesis of proteins from RNA.

Triacylglycerol (13.6, 15.4) A triester formed from glycerol and three molecules of fatty acids.

Triose (14.2) A monosaccharide with three carbons.

Tripeptide (16.4) A peptide that contains three amino acids joined together by two amide bonds.

Triple bond (3.8) A multiple bond that contains six electrons—that is, three two-electron bonds.

Tritium (2.3) A hydrogen atom that has one proton and two neutrons, giving it a mass number of three.

U

Universal gas constant (6.7) The constant, symbolized by R, that equals the product of the pressure and volume of a gas, divided by the product of the number of moles and Kelvin temperature; $R = PV/nT$.

Unsaturated fatty acid (15.2) A fatty acid that has one or more double bonds in its long hydrocarbon chain.

Unsaturated hydrocarbon (11.1) A compound that contains fewer than the maximum number of hydrogen atoms per carbon.

Unsaturated solution (7.3) A solution that has less than the maximum number of grams of solute.

Urea cycle (18.11) The process by which an ammonium ion is converted to urea, $(NH_2)_2C{=}O$.

V

Valence electron (2.7) An electron in the outermost shell that takes part in bonding and chemical reactions.

Valence shell electron pair repulsion (VSEPR) theory (3.10) A theory that predicts molecular geometry based on the fact that electron pairs repel each other; thus, the most stable arrangement keeps these groups as far away from each other as possible.

Vaporization (4.6) The conversion of a liquid to a gas.

Virus (17.11) An infectious agent consisting of a DNA or RNA molecule that is contained within a protein coating.

Volume/volume percent concentration (7.5) The number of milliliters of solute dissolved in 100 mL of solution.

W

Wax (15.3) An ester (RCOOR') formed from a fatty acid (RCOOH) and a high molecular weight alcohol (R'OH).

Weight (1.4) The force that matter feels due to gravity.

Weight/volume percent concentration (7.5) The number of grams of solute dissolved in 100 mL of solution.

X

X-ray (9.7) A high-energy form of radiation.

Z

Zwitterion (16.2) A neutral compound that contains both a positive and a negative charge.

Credits

Image Research by Mary Reeg

Chapter 1

Opener: ©Alexander S. Berk; Fig 1.1a,b: ©McGraw-Hill Education/Jill Braaten; 1.2a: ©Bob Krist/Corbis; 1.2b: ©Corbis RF; 1.3a: ©Douglas Peebles/Alamy; 1.3b: ©Daniel C. Smith; 1.3c: ©Dr. Parvinder Sethi RF; p. 6(left): ©Daniel C. Smith; (middle, right): ©McGraw-Hill Education/Jill Braaten; 1.4a: ©Daniel C. Smith; 1.4b: ©Richard T. Nowitz/Science Source; 1.5a: ©Daniel C. Smith; 1.5b: ©Keith Eng, 2008; 1.5c: ©McGraw-Hill Education/Jill Braaten; 1.5d: ©Daniel C. Smith; p. 9(top): ©Doug Wilson/Corbis; (bottom): ©McGraw-Hill Education/Jill Braaten; p. 10: ©Chuck Savage/Corbis; p. 10(top): ©PhotoDisc/Getty Images RF; (bottom): ©PhotoDisc/Getty Images RF; p. 12: ©Zachary D.-K. Smith; p. 13: ©Daniel C. Smith; p. 17: ©Stockbyte/Getty Images RF; 1.7: ©Creative Studios/Alamy RF; p. 20: ©McGraw-Hill Education/Elite Images; p. 22: ©PhotoDisc/Getty Images RF; p. 23: ©McGraw-Hill Education/Jill Braaten; p. 24: ©McGraw-Hill Education/Mark Dierker; p. 25: ©McGraw-Hill Education/Jill Braaten; p. 27: ©McGraw-Hill Education/Jill Braaten; p. 28: ©Photodisc/Getty Images RF.

Chapter 2

Opener: ©Eyewire/Photodisc/PunchStock RF; p. 37: ©Michael S. Yamashita/Getty Images; p. 43(top): ©Jill Braaten; (bottom): ©McGraw-Hill Education/Charles D. Winters/Timeframe Photography, Inc.; 2.4: ©Ingram Publishing/SuperStock RF; p. 49: ©Steven Snodgrass; 2.5b,c: ©Dr. A. Leger/Phototake; p. 50(top): ©Anna Kari - Photojournalist; (bottom): ©Peter Dench/Peter Dench/Corbis; p. 52(lithium): ©McGraw-Hill Education/Jill Braaten and Anthony Arena, Chemistry Consultant; (sodium): ©McGraw-Hill Education/Stephen Frisch; (potasium): ©McGraw-Hill Education/Jill Braaten; p. 53(chlorine): ©McGraw-Hill Education/Jill Braaten and Anthony Arena, Chemistry Consultant; (bromine): ©McGraw-Hill Education/Joe Franek; (Iodine): ©McGraw-Hill Education/Jill Braaten and Anthony Arena, Chemistry Consultant; p. 53(CFCs): ©McGraw-Hill Education/Jill Braaten; (radon detector): ©Keith Eng, 2008; 2.7a: ©Charles O'Rear/Corbis; 2.7b: ©Lester V. Bergman/Corbis; 2.7c: ©McGraw-Hill Education/Jill Braaten and Anthony Arena, Chemistry Consultant; p. 58: ©Diamond Images/Getty Images; p. 59: ©Dr. Parvinder Sethi RF.

Chapter 3

Opener: ©Stockbyte/PunchStock RF; 3.1(Sodium metal): ©McGraw-Hill Education/Stephen Frisch; (Chlorine gas): ©Dane S. Johnson/Visuals Unlimited; (Sodium chloride crystals): ©Dane S. Johnson/Visuals Unlimited; p. 75(Hydrogen peroxide), p. 81: ©McGraw-Hill Education/Jill Braaten; p. 83: ©James L. Amos/Science Source; p. 84(exercise): ©Digital Vision Ltd./SuperStock RF; (labels (both)): ©McGraw-Hill Education/Jill Braaten; p. 87: ©BananaStock/PunchStock RF; 3.5: ©McGraw-Hill Education/Jill Braaten; p. 90: ©McGraw-Hill Education/Elite Images; p. 91: ©CNRI/Science Source; p. 92(oysters): ©Image Source/Corbis; (Tums, Maalox, Iron & supplements): ©McGraw-Hill Education/Jill Braaten; p. 93(top): ©Photodisc/Alamy; (bottom): Photo by Tim McCabe, USDA Natural Resources Conservation Service; p. 95: U.S. Dept. of Commerce photoset Hawaii Volcanism: Lava Forms; p. 97: ©Ted Nelson/Dembinsky Photo Associates; p. 99(top): ©Daniel C. Smith; (bottom): ©Image Source Black/Alamy RF; p. 105: ©John A. Rizzo/Getty Images RF.

Chapter 4

Opener: ©John A. Rizzo/Getty Images RF; p. 115: ©Daniel C. Smith; p. 117(top): ©Jack Star/PhotoLink/Getty Images RF; (bottom): ©Royalty-Free/Corbis; 4.2: ©Doug Menuez/Getty Images RF; p. 120(top): ©Martin Harvey/Corbis; (bottom): ©Andrew Syred/Science Source; p. 124: ©Royalty-Free/Corbis; p. 129(ice, liquid water): ©McGraw-Hill Education/Jill Braaten; (ice water): ©McGraw-Hill Education/Suzie Ross; p. 130(Chloroethane): Courtesy of Gebauer Company, Cleveland, Ohio; p. 130(liquid water, steam): ©McGraw-Hill Education/Jill Braaten; p. 131(all): ©McGraw-Hill Education/Jill Braaten and Anthony Arena, Chemistry Consultant.

Chapter 5

Opener: ©Jill Braaten; 5.1(left): ©McGraw-Hill Education/Jill Braaten and Anthony Arena, Chemistry Consultant; (right): ©Daniel C. Smith; p. 148: ©BananaStock/Punchstock RF; p. 149: ©Jim Erickson/Corbis; 5.2: ©Jack Sullivan/Alamy; p. 153: ©McGraw-Hill Education/Jill Braaten and Anthony Arena, Chemistry Consultant; p. 155: ©McGraw-Hill Education/Jill Braaten; p. 156: ©McGraw-Hill Education/Mark Dierker; p. 158(top): ©Hisham F. Ibrahim/Getty Images RF; (bottom): ©Keith Eng, 2008; p. 159: ©Comstock Images/Punchstock RF; p. 161(top): ©Royalty-Free/Corbis; (bottom): ©Scott Olson/Getty Images; 5.3: ©McGraw-Hill Education/Jill Braaten and Anthony Arena, Chemistry Consultant; p. 166: ©Jill Braaten; p. 167(left): ©Tony Cordoza/Alamy RF; (right): ©Clair Dunn/Alamy; p. 170(top): ©Digital Vision/PunchStock RF; (bottom): ©Comstock/PunchStock RF; p. 173: ©McGraw-Hill Education/Jill Braaten; p. 174: ©Ingram Publishing/SuperStock RF.

Chapter 6

Opener: ©Daniel C. Smith; p. 190 & 194: ©Open Door/Alamy RF; p. 198: ©Bill Aaron/PhotoEdit; p. 200: ©Jill Braaten; p. 205: ©National Geographic/Getty Images; p. 206: ©Spencer Grant/PhotoEdit; p. 208: NASA.

Chapter 7

Opener: ©Zachary D.-K. Smith; p. 217: ©Jill Braaten; 7.1a: ©Brian Evans/Science Source; 7.1b: ©Digital Vision/Alamy RF; 7.1c: ©Keith Eng, 2008 RF; p. 218: ©Don Farrall/Getty Images RF; p. 219(both) & 220: ©McGraw-Hill Education/Jill Braaten and Anthony Arena, chemistry consultant; 7.2(both): ©McGraw-Hill Education/Jill Braaten and Anthony Arena, chemistry consultant; p. 223: ©Royalty-Free/Corbis; 7.3: ©Tom Pantages; p. 225: ©Keith Eng, 2008 RF; p. 226: ©Oliver Anlauf/Getty Images; 7.4(both): ©McGraw-Hill Education/Jill Braaten; p. 227: ©McGraw-Hill Education/Suzi Ross; 7.5(both): ©McGraw-Hill Education/Jill Braaten and Anthony Arena, chemistry consultant; p. 228: ©BananaStock/PunchStock RF; p. 229: ©David Hoffman Photo Library/Alamy; p. 230(top): ©McGraw-Hill Education/Mark Dierker; (bottom): ©Comstock/PunchStock RF; p. 234: ©Jill Braaten; 7.6a-c: ©Dennis Kunkel Microscopy, Inc.

Chapter 8

Opener: ©Photolink/Getty Images RF; p. 251(both): ©McGraw-Hill Education/Jill Braaten and Anthony Arena, chemistry consultant; 8.1a: ©Colin Dutton/Grand Tour/Corbis; 8.1b: ©Photolink/Getty Images RF; 8.1c: ©BananaStock/PunchStock RF; 8.2a-c: ©McGraw-Hill Education/Jill Braaten; p. 257: ©McGraw-Hill Education/Jill Braaten; 8.4 & 8.5(all): ©McGraw-Hill Education/Jill Braaten and Anthony Arena, chemistry consultant; p. 262: ©Digital Vision Ltd./SuperStock RF; p. 264: ©Comstock Images/PicturesQuest RF; p. 265: ©Royalty-Free/Corbis; 8.6a: ©McGraw-Hill Education/Stephen Frisch; 8.6b: ©sciencephoto/Alamy; 8.6c: ©Leslie Garland Picture Library/Alamy; 8.7(lemons): ©Humberto Olarte Cupas/Alamy RF; (strawberries): ©Photolink/Getty Images RF; (tomatoes): ©Burke/Triolo/Getty Images RF; (milk): ©Mitch Hrdlicka/Getty Images RF; (Phillips): ©McGraw-Hill Education/Jill Braaten; (Clorox): ©McGraw-Hill Education/Terry Wild Studio; p. 267: ©Bruce Heinemann/Getty Images RF; p. 270 & 271: ©McGraw-Hill Education/Jill Braaten and Anthony Arena, chemistry consultant; 8.9a-c: ©McGraw-Hill Education/Jill Braaten and Anthony Arena, chemistry consultant; p. 277: ©JupiterImages/Brand X/Alamy RF.

Chapter 9

Opener: ©Martin Dohrn/Science Source; p. 289: ©S. Wanke/PhotoLink/Getty Images RF; 9.2a: ©Owen Franken/Corbis; 9.2b,c: ©Custom Medical Stock Photo; 9.3: ©Science VU/Visuals Unlimited; p. 297(top): ©Hank Morgan/Science Source; (bottom): Landauer Corporation; p. 298: ©AFP/Getty Images; p. 299: ©Cordelia Molloy/Science Source; 9.4b: ©Living Art Enterprises, LLC/Science Source; 9.5a-9.7c: ©Daniel C. Smith; 9.8a: ©Martin Bond/Science Source; 9.9a: ©Royalty-Free/Corbis; 9.9b: ©Scott Camazine/Science Source; 9.9c: With permission, Daniel C. Smith.

Chapter 10

Opener: Photo courtesy of Poly-Wood, Inc.; 10.1a: ©Comstock/PunchStock RF; 10.1b: ©Corbis Premium/Alamy RF; 10.1c: ©Steven May/Alamy; 10.1d: ©Layne Kennedy/Corbis; Table 10.1(lighter): ©PIXTAL/PunchStock RF; (salt): ©Royalty-Free/Corbis; p. 317(top): ©Ingram Publishing/SuperStock RF; (middle): ©PhotoDisc/Getty Images RF; (bottom): ©Vol. 129 PhotoDisc/Getty Images RF; p. 323: ©Goodshoot/PunchStock RF; p. 324(top): ©McGraw-Hill Education/Jill Braaten; (bottom): ©Ted Kinsman/Science Source; p. 326, 327: ©Jill Braaten; p. 328(top): ©Graham Titchmarsh/Alamy RF; (bottom): Fresenius Kabi USA, LLC; p. 329(left): God of Insects; (right): ©Douglas Peebles/Corbis; p. 330: ©Bloomberg/Getty Images; p. 334: ©PhotoDisc Website RF; p. 339: ©Jill Braaten; 10.2: ©PhotoDisc Website RF; p. 342: ©McGraw-Hill Education/Jill Braaten; p. 343(top): ©Vince Bevan/Alamy; (bottom): ©Jill Braaten; p. 344: ©Thinkstock/Getty Images RF.

Chapter 11

Opener: ©Inga Spence/Science Source; p. 356: ©McGraw-Hill Education/Jill Braaten; p. 361: ©Pascal Goetgheluck/Science Source; p. 362: ©Jill Braaten; p. 364: ©Comstock/Punchstock RF; p. 367(top): Ethanol Promotion and Information Council (EPIC); (bottom): ©McGraw-Hill Education/Jill Braaten; p. 368(butter): ©Royalty-Free/Corbis; (peanut butter/both): ©McGraw-Hill Education/Elite Images; 11.4(both): ©McGraw-Hill Education/Jill Braaten; p. 370: ©Jill Braaten; 11.5(pvc): ©McGraw-Hill Education/John Thoeming; (balls): ©Dynamicgraphics/Jupiterimages RF; (blankets): ©Fernando Bengoe/Corbis; (Styrofoam): ©McGraw-Hill Education/John Thoeming; Table 11.2(blood bags): ©Image Source/Getty Images RF; (syringes): ©Royalty-Free/Corbis; (floss): ©Jill Braaten; p. 374: ©Jill Braaten; p. 375: ©McGraw-Hill Education/Jill Braaten; p. 377: ©McGraw-Hill Education/John Thoeming; p. 378(vanilla bean): ©CD16/Author's Image RF; (tumeric): Mr. Napat Kitipanangkul, Science School, Walailak University with financial support by Biodiversity; Research and training program (BRT); (cell protector): ©McGraw-Hill Education/Elite Images; (nuts): ©Royalty-Free/Corbis.

Chapter 12

Opener: ©John A. Rizzo/Getty Images RF; p. 390(top): ©Creatas/PunchStock RF; (bottom): ©Iconotec/Alamer and Cali, photographers RF; p. 392: ©McGraw-Hill Education/Jill Braaten; 12.1(grapes): ©Dynamic Graphics Group/PunchStock RF; (barrels): ©Goodshoot/PunchStock RF; (wine): ©BananaStock/PunchStock RF; p. 397: The Boston Medical Library in the Francis A. Countway Library of Medicine; p. 401(top): Gwinnet County Police Department/Centers for Disease Control and Prevention; 12.4(BreathScan): ©McGraw-Hill Education/Elite Images; p. 404: Claire Fackler, CINMS, NOAA; p. 405: ©Mark Downey/Getty Images RF; 12.6(both): ©McGraw-Hill Education/Suzie Ross; p. 407: ©TTH/a.collection/Getty Images RF; 12.7 & p. 410(both): ©McGraw-Hill Education/Jill Braaten.

Chapter 13

Opener: ©Blend Images/Pete Saloutos/Getty Images RF; p. 433: Photo by Forest and Kim Starr; p. 434: ©CD16/Author's Image RF; p. 437: ©McGraw-Hill Education/Jill Braaten; p. 438: ©National Geographic/Getty Images; p. 441, 442: ©Jill Braaten; p. 443: ©Coby Schal/NC State University; p: 444: ©McGraw-Hill Education/Jill Braaten; 13.3(left): ©Rene Dulhoste/Science Source; (right): Werner Arnold; p. 453, 454(both): ©McGraw-Hill Education/Jill Braaten; p. 457(left): ©Mediacolor's/Alamy; (right): ©Flora Torrance/Life File RF; p. 459: Library of Congress/LC-USZC4-1986.

Chapter 14

Opener: ©Isabelle Rozenbaum & Frederic Cirou/Photo Alto/PhotoStock RF; p. 471(top): ©Michelle Garrett/Corbis; (bottom): ©ImageSource/Corbis RF; p. 472(top): ©Alan Rosenberg/Cole Group/Getty Images RF; (bottom): ©Daniel C. Smith; p. 473: ©McGraw-Hill Education/Elite Images; p. 475: ©Alamy; p. 477: ©Keith Brofsky/Getty Images RF; p. 478(top): ©Comstock/PunchStock RF; (bottom): ©Jill Braaten; p. 482: ©McGraw-Hill Education/Mark Dierker; p. 485(top): ©McGraw-Hill Education/Jill Braaten; (bottom): ©Vol. 3/PhotoDisc/Getty Images RF; 14.2: ©McGraw-Hill Education/Jill Braaten; p. 487: ©McGraw-Hill Education/Jill Braaten; p. 488(left): ©Pixtal/AGE Fotostock RF; (right): ©McGraw-Hill Education/Images; 14.3: ©McGraw-Hill Education/Jill Braaten; 14.5: ©Vol. 1 PhotoDisc/Getty Images RF; (inset): ©Borland/PhotoLink/Getty Images RF; p. 493: ©Royalty-Free/Corbis.

Chapter 15

Opener: ©Deborah Van Kirk/Stone/Getty Images RF; p. 505: ©McGraw-Hill Education/Jill Braaten; p. 508(top): ©Brand X Pictures/PunchStock RF; (middle): ©Dennis Scott/Corbis; (bottom): ©Daniel C. Smith; p. 509(top,left): ©McGraw-Hill Education/Jill Braaten; (top,right): ©Photodisc/Getty Images RF; (bottom): USDA, ARS, National Genetic Resources Program; p. 511(coconut trees): ©Ingo Jezierski/Corbis RF; (animal products): ©McGraw-Hill Education/Jill Braaten; (vegetable oil): ©Jupiterimages/Image Source RF; (fish): ©Daniel C. Smith; p. 512: ©Johner Images/Getty Images RF; p. 513, 15.1(all): ©McGraw-Hill Education/Jill Braaten; 15.2: ©McGraw-Hill Education/Dennis Strete; p. 516(top): ©Daniel C. Smith; (bottom): ©AP Images; p. 517(left): ©Jupiterimages/Image Source RF; (right): ©David Buffington/Getty Images RF; p. 522: ©Getty Images RF; 15.6a,b: ©Ed Reschke; p. 524(top): ©Andrew Brookes/Corbis; (bottom): ©McGraw-Hill Education/Chris Kerrigan; p. 525: ©Comstock/JupiterImages RF; p. 526: ©A.T. Willett/Alamy; p. 527(top): ©BananaStock RF; (middle): ©Stockbyte/PunchStock RF; (bottom): ©IT Stock/PunchStock RF; p. 528: ©AGStockUSA, Inc./Alamy; p 530(all): ©McGraw-Hill Education/Mark Dierker.

Chapter 16

Opener: ©Royalty-Free/Corbis; p. 539: ©Jill Braaten; p. 541(top): Copyright ©FoodCollection RF; (bottom): ©Jill Braaten; p. 543: ©Comstock/PunchStock RF; 16.9: ©Image Source/PunchStock RF; p. 558: ©Daniel C. Smith; p. 559: ©Eye of Science/Science Source; p. 561: ©McGraw-Hill Education/Jill Braaten; p. 562: ©Daniel C. Smith; p. 566: ©Science Source.

Chapter 17

Opener: ©Lawrence Lawry/Getty Images RF; p. 586: ©Daniel C. Smith; p. 598: CF Trust; 17.11: Courtesy Genelex Corp., www.HealthandDNA.com; p. 600: ©McGraw-Hill Education/Jill Braaten; p. 601: ©Time & Life Pictures/Getty Images.

Chapter 18

Opener: ©Dennis Gray/Cole Group Getty Images RF; Fig 18.2(butter): ©Dennis Gray/Cole Group Getty Images RF; (bagels): ©Nancy R. Cohen/Getty Images RF; (lobster): ©PhotoLink/Getty Images RF; p. 616: ©Jill Braaten; p. 618: ©McGraw-Hill Education/Jill Braaten; p. 619: ©Digital Vision/Getty Images RF; p. 625: ©Warren Morgan/Corbis; p. 626: ©John A. Rizzo/Getty Images RF; p. 631-639: ©McGraw-Hill Education/Jill Braaten.

Index

Page numbers followed by *f*, indicate figures; *mn*, marginal notes; and *t*, tables. Page numbers preceded by "A-" indicate pages in the *Appendix*.

A

B

C

D

E

F

G

H

M

N

O

P